Challenges and Opportunities in Industrial and Mechanical Engineering: A Progressive Research Outlook

Proceedings of the International Conference on Progressive Research in Industrial & Mechanical Engineering (Prime 2021), August 05–07, 2021, Patna, India

Edited by
S M Pandey
Ambrish Maurya
Chetan Kumar Hirwani
Om Ji Shukla

CRC Press
Taylor & Francis Group
Boca Raton London New York

CRC Press is an imprint of the
Taylor & Francis Group, an **informa** business

First edition published 2024
by CRC Press
4 Park Square, Milton Park, Abingdon, Oxon, OX14 4RN

and by CRC Press
2385 NW Executive Center Drive, Suite 320, Boca Raton FL 33431

British Library Cataloguing-in-Publication Data
A catalogue record for this book is available from the British Library

ISBN: 9781032713212 (pbk)
ISBN: 9781032713229 (ebk)

DOI: 10.1201/9781032713229

Typeset in Sabon LT Std
by HBK Digital

Contents

List of figures

List of tables

Preface

Present time Industry 4.0 is the need of all industries because it connects industries to AI, high productivity, safety, and flexibility, ensures the 100% utilization of resources across diverse manufacturing systems, and could accelerate normal manufacturing systems to advanced manufacturing systems by using robotics, additive manufacturing, and many more.

In this book, the collection of selected papers is constituted from the International Conference on Progressive Research in Industrial & Mechanical Engineering (PRIME 2021), which was at the National Institute of Technology (NIT), Patna, India from August 5 to 7, 2021. This conference brings together all academic people, industry experts, and researchers from India as well as abroad for involving thoughts on the needs, challenges, new technology, opportunities threats in the current transformational field of aspire. This book deliberates on several elements and their relevance to hard-core areas of industrial and mechanical engineering including design engineering, production engineering, industrial engineering, automobile engineering, thermal and fluid engineering, mechatronics control robotics, interdisciplinary, and many new emerging topics that keep potential in several areas of applications. This book focuses on providing versatile knowledge of cutting-edge practices to all readers, helping to develop a clear vision toward Industry 4.0, robotics automation, and additive manufacturing in this demanding and evolving time. The book will be a treasured reference for students, researchers, and professionals interested in mechanical engineering and allied fields.

1   A time series model for forecasting milk production in North India: a case study of Bihar dairy sector

Abhishek Kashyap[1,a] and Om Ji Shukla[1,b]

[1]Department of Mechanical Engineering, National Institute of Technology Patna, India

Abstract: Milk is a product that has been an essential component for a large segment of people, particularly in a country like India. In order to prevent any unnecessary wastage, it is important that the value of its demand and supply match up as closely as possible. This work is an attempt to predict milk production in north India with the help of the secondary data gathered from a company known as the Bihar State Milk Co-Operative Federation Ltd. (COMFED). The study has incorporated a time series model named the autoregressive integrated moving average (ARIMA) model. This study with the help of a statistical software named Statistical Package for Social Sciences (SPSS) forecasts the milk production of COMFED. Moreover, the paper has also discussed the role of the nature of autocorrelation function (ACF) and partial autocorrelation function (PACF) components in the selection of the ARIMA model. Based on the values of ACF and PACF, two models namely ARIMA (2, 1, 0) and ARIMA (3, 1, 0) were nominated for the discussion. However, based on the error terms like R-squared, stationary R-square, and normalized BIC value, model ARIMA (2, 1, 0) is the best-suited model for the given data. Eventually, the paper validates the model by comparing the forecasted value with the actual value.

Keywords: ACF, ARIMA, milk production forecasting, PACF, SPSS, time series analysis.

Introduction

Dairy products are an integral part of everyday life as dairy products provide more than 10% protein requirement of the body. Over the course of history, they have developed from products of luxury to the product of necessity. One of the dairy products is milk. Milk is one of the natural foods that supplies all the vital nutrients needed for the growth and development of the body [1]. The dairy sector plays an essential part in the country's social and economic growth, it also comprises a major rural segment within [2]. With a population of 125.34 million milch animals, India ranks first in milk production and consumption worldwide. India's milk output has climbed from 55.6 million tons in 1991–1992 to 187.7 million tons in 2018–2019, an increase of 237.58%. During the last decade, milk production in India has increased by around 4.8% each year [2]. Similarly, milk availability per capita has grown from 130 grams per day in 1950 to 1951 to 374 grams per day from 2017 to 2018, which is more than the predicted global average consumption for 2017.

[a]abhishekk.phd20.me@nitp.ac.in, [b]om.mechanical@gmail.com

India claimed a 21.29 percent share of the world's total milk output in 2017. Dairy is the greatest contributor to Agriculture's gross domestic product (GDP); thus, it is vital to anticipate milk production in order to determine the availability and demand for milk so that the appropriate policy intervention may be implemented to fill the resulting gap [10,16].

In this study milk production in north India, specifically, Bihar has been considered. The dairy industry in Bihar was distinguished by having an immense quantity of number of animals that were low-productive. The government of Bihar allotted Rs100 crore to milk processing plants at Hajipur, which is located in the Vaishali district of Bihar [3]. These plants have the potential to produce 4 lakh liters of milk, 30 tons of powdered milk, and tetra packs of milk. These actions are quite significant in terms of the development of this industry. This study uses the secondary data of a dairy company known as COMFED which is located in Northern India (Bihar). It began in the year 1983 with 1030 co-operatives in it and barely handled 23.62 kiloliters of milk every day, by the year 2012, it had more than 13000 cooperatives with a milk handling capacity of up to 11 lakh liters every day; such growth is very appreciable [3]. In light of this context, this paper makes an effort to examine and predict milk production using the established time series modeling approach ARIMA.

This paper seeks to address the following research objectives (ROs):

a. investigation of milk production in north India from the year 2005–2018 using COMFED data
b. formulation of the ARIMA models to forecast the milk production for the year 2019 with the help of AR coefficient (p) and MA coefficient (q)
c. selection of the best-suited model considering the error and the actual milk production for the year 2019.

The following is the structure of the paper: Section 2 contains the method and materials followed by the model formation in Section 3. Section 4 contains the results and discussions. Eventually, the paper concludes in Section 5.

Materials and Methods

The process of forecasting is a method that makes use of historical data as inputs to generate educated estimates that may be used to make accurate predictions about the path that future trends will take. This kind of demand forecasting is essentially a technique that makes use of local sales history and also incorporates a search for the foreseeable future [4]. Forecasting, in its most basic form, refers to making predictions, which can be done with or without the assistance of historical information [15]. This study has incorporated the following methodology as shown in Figure 1.1.

Initially, an extensive literature review based on forecasting using the ARIMA model is carried out followed by the collection of the data. Data has been collected from COMFED, Patna. Data from thirteen years (2005–2018) has been considered to formulate the ARIMA model to predict milk production for the coming year. Collected data is then checked for stationarity to the execution of the ARIMA model. ARIMA, a time series model, is an extension of ARMA (Autoregressive and moving average) models. In an ARIMA model, it is assumed that the analyzed time series is a linear function of previous values and random shocks [5–8]. An ARIMA model, denoted ARIMA (p,d,q) consists of three components: p, the order of auto-regression (AR); d, the order of integration (or differencing) to establish

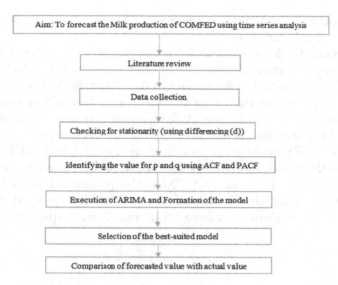

Figure 1.1 Detailed methodology.

stationarity; and q, the order of Moving Average (MA). The mathematical form of the model is shown in equations (1), (2), and (3) [11, 12].

$$y_t = c + \mu_1 y_{t-1} + \mu_2 y_{t-2} + \ldots\ldots + \mu_p y_{t-p} + \varepsilon \ (AR) \tag{1}$$
$$y_t = c - \phi_1 \varepsilon_{t-1} - \phi_2 \varepsilon_{t-2} - \ldots\ldots - \phi_q \varepsilon_{t-q} + \varepsilon \ (MA) \tag{2}$$
$$y_t = c + \mu_1 y_{t-1} + \mu_2 y_{t-2} + \ldots\ldots + \mu_p y_{t-p} + \varepsilon - \phi_1 \varepsilon_{t-1} - \phi_2 \varepsilon_{t-2} - \ldots\ldots - \phi_q \varepsilon_{t-q} \ (ARMA) \tag{3}$$

where, y_t is milk production in time t (year or month or week), c is a constant, μ is AR coefficients, ϕ is MA coefficients, ε is the error term.

The first and most significant prerequisite for ARIMA modeling is ensuring that the series under study is stationary since the estimation technique is only viable for stationary series [13]. We say that a series is stationary if its meaning and autocorrelation does not vary over time [14]. A time series' stationarity may be determined using a time plot or unit root tests. A non-stationary series may be converted or identified into a stationary series by differencing. The number of times a series is differentiated to get stationarity is known as the order of integration/difference, denoted by d. After getting the values of p, d, and, q the ARIMA model is deployed for the prediction of milk production. In this paper, a statistical software named Statistical Package for Social Sciences (SPSS) is used for the ARIMA modeling. SPSS is a statistical software that may be used for data transformation, regression analysis, analysis of variance, multivariate analysis of variance, analysis of covariance, t-tests, non-parametric tests, time series forecasting, and a variety of other statistical applications [17–19].

Model Formation

The ARIMA model contains multiple processes to arrive at a final forecasting value [17]. The forecasting process uses a variety of tools, including ACF and PACF, both of which play an important role. The data obtained from COMFED is plotted in the graph as shown in Figure 1.2. The graph reveals that there is an increasing trend in the data and in order to make the data applicable for ARIMA, we give it a difference of one, denoted by the symbol d = 1, as can be seen in Figure 1.3.

After the difference d is determined, which in this case is 1, the data will begin to display variation along a constant mean as shown in Figure 1.4. The ACF and the PACF are used to determine the value of p and q [9]. As shown in Table 1.1, if the ACF is geometric in nature and the PACF is significant to lag p then the MA component will be taken zero i.e., the value of q becomes zero. Similarly, if the PACF is geometric in nature and ACF is significant to lag q then the AR component will be taken zero i.e., p becomes 0.

This study uses the SPSS software for the formation of ACF and PACF followed by the final formation of the model. Initially, the data is put into the software followed by integrating the difference (d = 1) to deploy ARIMA. This process is followed by the identification of the p and q values with the help of ACF and PACF given by SPSS. The ACF and PACF is given by SPSS are shown in Figure 1.4(a) and Figure 1.4(b) respectively.

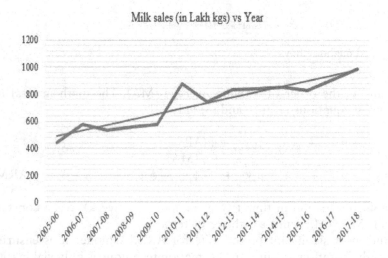

Figure 1.2 Annual milk production.

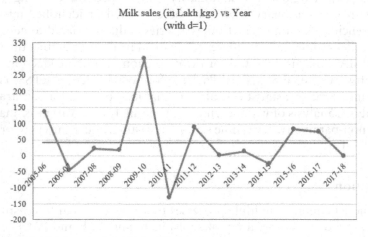

Figure 1.3 Annual production data with d = 1.

The formulation of the model is completely dependent on the values of p, d, and q. As shown in Figure 1.3, d = 1 provides stationarity to the data. Moreover, the value of p and q are determined with the help of Figure 1.4(a) and 1.4(b). ACF shows geometric nature whereas PACF is significant till lag 2 and lag 3. With the help of Table 1.1, ARIMA (2, 1, 0) and ARIMA (3, 1, 0) are the identified model. After the integration of the model into the SPSS software, the software provides the forecasting values for both models as shown in Figures 1.5 and 1.6.

Results and Discussions

The first stage in time series analysis is to visually evaluate the behavior of the investigated data by plotting it. Figure 1.2 depicts the yearly output of milk production by COMFED from 2005 to 2018 as a function of time which shows an upward trend. The first differentiation revealed that the series is stationary. Figure 1.3 displays the time plots of the differenced series, revealing that the first-order differenced series is stagnant. After achieving the stagnancy in the data, ACF and PACF are used to identify the value of p and q. Nevertheless, the methodology shown in Figure 1.1 briefly discusses the processes involved in the incorporation of ARIMA into this study. As discussed in section 3 the SPSS shows two ARIMA models for the forecasting of milk production. ARIMA (2, 1, 0) contains values for p, d, and q as 2, 1, and 0 respectively. Similarly, the model ARIMA (3, 1, 0) contains the value of p, d, and q as 3, 1, and 0 respectively. In against the actual value of milk

Table 1.1: Model selection using ACF and PACF.

	ACF	*PACF*	*Model*
AR(p)	Geometric	Significant till p lags	(p, d, 0)
MA(q)	Significant till q lags	Geometric	(0, d, q)

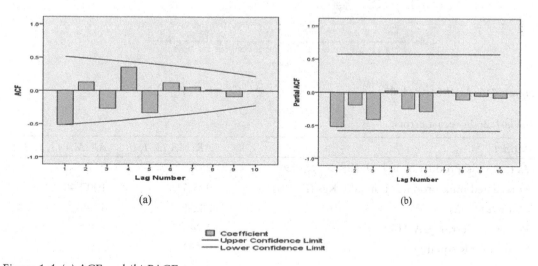

Figure 1.4 (a) ACF and (b) PACF.

production in the year 2019 which is 967 lakh kgs, ARIMA (2, 1, 0) and ARIMA (3, 1, 0) give the respective forecasted value for milk production for the year 2019 as 995.51 lakh kgs and 1007.40 lakh kgs as shown in Figures 1.5 and 1.6. Moreover, the software also identifies different errors involved in both the models as shown in Table 1.2.

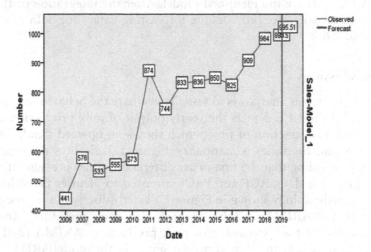

Figure 1.5 ARIMA (2, 1, 0).

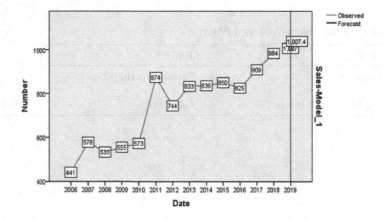

Figure 1.6 ARIMA (3, 1, 0).

Table 1.2: Forecast errors.

Model	ARIMA (2, 1, 0)	ARIMA (3, 1, 0)
Actual milk production in lakh Kgs (A) (2019)	967	967
Forecasted milk production in lakh Kgs (F) (2019)	995.51	1007.40
Error (e) (F-A)	28.51	40.40
Percentage error (e/A)*100	2.94	4.17
Stationary R-squared	0.295	0.401
R-squared	0.672	0.721
Normalized BIC	9.818	9.981

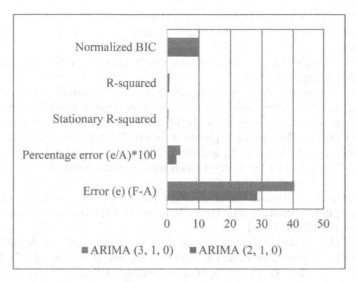

Figure 1.7 Error graph.

The percentage error between the actual value and the forecasted value for both the models i.e., model (2, 1, 0) and (3, 1, 0) as mentioned in Table 1.2 are 2.94 and 4.17 respectively. Moreover, the stationary R-squared, R-squared, and normalized BIC are shown in Figure 1.7. The model with the least normalized Bayesian Information Criterion (BIC) value is selected as the most appropriate model for the data under consideration [16]. In all the aspects model 1, i.e., ARIMA (2,1,0) is the best-suited model for the considered data.

Conclusion, Limitations, and Future Work

Agricultural and associated areas, such as the animal sciences, make substantial use of time series analysis tools to examine data. These strategies have been beneficial for anticipating potential implications and preparing for the situation accordingly. Based on stationary R-squared, R-squared, and normalized BIC, the ARIMA (2,1,0) model was determined to be the most effective for predicting milk production between the two models. India is the highest producer and consumer of dairy products and needs very close supervision of the production and consumption of milk to avoid wastage. Forecasting of milk production and other dairy products may work as a driver to eliminate unnecessary losses or wastage of milk. The ARIMA model, which is often used in such research, is only capable of collecting linear data components. Moreover, the data considered were available on an annual basis, the monthly and weekly data may provide a more precise result. However, future research may also be led to the use of machine learning techniques like linear regression techniques and random forest techniques to validate the result of this study. Unlike this work, to broaden the implication of this research, future research may consider the data for the entire country and forecast the values of milk production for the coming years.

References

1. Deshmukh, S. S., and Paramasivam, R. (2016). Forecasting of milk production in India with ARIMA and VAR time series models. *Asian Journal of Dairy and Food Research*, 35(1), 17–22.

2. Mishra, P., Fatih, C., Niranjan, H. K., Tiwari, S., Devi, M., and Dubey, A. (2020). Modelling and forecasting of milk production in Chhattisgarh and India. *Indian Journal of Animal Research*, 54(7), 912–917.
3. Dhir, S., and Dhir, S. (2017). COMFED: the new challenges of diversification. *Emerald Emerging Markets Case Studies*, 7(2), 1–26.
4. Fradinata, E., Suthummanon, S., Sirivongpaisal, N., and Suntiamorntuthq, W. (2014). ANN, ARIMA and MA timeseries model for forecasting in cement manufacturing industry: Case study at lafarge cement Indonesia—Aceh. In 2014 International Conference of Advanced Informatics: Concept, Theory and Application (ICAICTA), (pp. 39–44). IEEE.
5. Taye, B. A., Alene, A. A., Nega, A. K., and Yirsaw, B. G. (2021). Time series analysis of cow milk production at Andassa dairy farm, west Gojam zone, Amhara region, Ethiopia. *Modeling Earth Systems and Environment*, 7, 181–189.
6. Yadav, A. K., Das, K. K., Das, P., Raman, R. K., Kumar, J., and Das, B. K. (2020). Growth trends and forecasting of fish production in Assam, India using ARIMA model. *Journal of Applied and Natural Science*, 12(3), 415–421.
7. Pal, S., Ramasubramanian, V., and Mehta, S. C. (2007). Statistical models for forecasting milk production in India. *Journal of Indian Society of Agricultural Statistics*, 61(2), 80–83.
8. Ljung, G. M., and Box, G. E. (1978). On a measure of lack of fit in time series models. *Biometrika*, 65(2), 297–303.
9. Rahardja, D. (2020). Statistical methodological review for time-series data. *Journal of Statistics and Management Systems*, 23(8), 1445–1461.
10. Minhas, K. S., Sidhu, J. S., Mudahar, G. S., and Singh, A. K. (2002). Flow behavior characteristics of ice cream mix made with buffalo milk and various stabilizers. *Plant Foods for Human Nutrition*, 57, 25–40.
11. Almasarweh, M., and Alwadi, S. (2018). ARIMA model in predicting banking stock market data. *Modern Applied Science*, 12(11), 309.
12. Siami-Namini, S., Tavakoli, N., and Namin, A. S. (2018). A comparison of ARIMA and LSTM in forecasting time series. In 2018 17th IEEE International Conference on Machine Learning and Applications (ICMLA), (pp. 1394–1401). IEEE.
13. Schaffer, A. L., Dobbins, T. A., and Pearson, S. A. (2021). Interrupted time series analysis using autoregressive integrated moving average (ARIMA) models: a guide for evaluating large-scale health interventions. *BMC Medical Research Methodology*, 21(1), 1–12.
14. Ogasawara, E., Martinez, L. C., De Oliveira, D., Zimbrão, G., Pappa, G. L., and Mattoso, M. (2010). Adaptive normalization: a novel data normalization approach for non-stationary time series. In The 2010 International Joint Conference on Neural Networks (IJCNN), (pp. 1–8). IEEE.
15. Subbanna, Y. B., Kumar, S., and Puttaraju, S. K. M. (2021). Forecasting buffalo milk production in India: Time series approach. *Buffalo Bulletin*, 40(2), 335–343.
16. Ravichandran, M., Naresh, R., and Kandasamy, J. (2020). Supply chain routing in a diary industry using heterogeneous fleet system: simulation-based approach. *Journal of The Institution of Engineers (India): Series C*, 101, 891–911.
17. Argyrous, G. (2011). Statistics for Research: With a Guide to SPSS. Sage Publications.
18. Bryman, A., and Cramer, D. (2012). Quantitative Data Analysis with IBM SPSS 17, 18 & 19: A Guide for Social Scientists. Routledge.
19. Levesque, R. (2007). SPSS Programming and Data Management. A Guide for SPSS and SAS Users, Fourth Edition, SPSS Inc., Chicago, 3.

2 Nonlinear characteristics of hollow steel piles under axial harmonic loading

Amit Kumar[1,a], Shiva Shankar Choudhary[2,b], and Avijit Burman[2,c]

[1]Research Scholar, Department of Civil Engineering, National Institute of Technology Patna, India

[2]Assistant Professor, Department of Civil Engineering, National Institute of Technology Patna, India

Abstract: The primary objective of this study is to investigate the dynamic axial response characteristics of pile groups subjected to machinery-generated axial harmonic loading. To achieve this objective, piles with a length of 3 meters and a diameter of 0.114 meters are used to conduct forced vibration field experiments. In this experimental investigation, dynamic frequency-amplitude responses are measured with four distinct eccentric moments. Field experiments conducted on the soil-pile system disclosed typical nonlinear behaviour as the resonant frequencies declined and the resonant amplitudes disproportionally rise with excitation forces. A theoretical study is also done by incorporating the inverse methodology of Novak [6]. In this method, the field results based on frequency versus amplitude response are back-calculated to quantify the changes in stiffness, damping, and effective mass of the piles. A comparative study is performed to analyze the soil-pile responses obtained through the field and theoretical approaches, and it can be seen from the comparison that the theoretically predicted responses are almost close to the field results. Furthermore, it can also be observed that the evaluated damping values of the three-pile group system increased and the effective mass and stiffness decreased with excitation forces.

Keywords: Axial harmonic loading, frequency versus amplitude responses, pile group.

Introduction

Now a day, machine-based foundations are frequently used by many heavy industries like the power or petroleum sectors, and due to the generation of dynamic forces, piles are used to counter these forces applied to machine-based foundations. The effects of dynamic behaviour and characteristics of pile response subjected to machine foundations are still being investigated and studied by many researchers. The axial loading tests and theoretical analysis of piles to predict the characteristics of the soil-pile systems are still a very complicated research area, and more field and laboratory tests are required to enhance the prediction accuracy of frequency and amplitude responses with combined soil-pile stiffness and damping of the system. Field or laboratory investigations are always necessary to test the relevance and efficiency of various theoretical approaches. The variations of nonlinearity on the pile responses in frequencies, amplitudes, and impedance characteristics have been studied by Elkasabgy and El Naggar [3], Biswas and Manna, [1] and Choudhary et al. [2]

[a]amitk.pg18.ce@nitp.ac.in, [b]shiva@nitp.ac.in, [c]avijit@nitp.ac.in

using theoretical analysis applying the continuum approach method. In order to quantify the dynamic resistance the soil exerts against the lateral surface of the pile during longitudinal vibration, Liu et al. [5] analyzed the stiffness and damping coefficients of a 3D Voigt model. Wen-Jie et al. [7] used a tapered hypothetical soil-pile model and the sophisticated stiffness transfer model to simulate compaction effects in their study of torsional vibration. On the other hand, Khalil et al. [4] investigated the dynamic behaviour of the soil-pile system and discovered that the loading frequency greatly affects the dynamic impedance characteristics and their corresponding amplitudes. A comprehensive investigation is required to gain a complete understanding of the non-linear characteristics exhibited by pile foundations. Predicting the dynamic frequency-amplitude response and its variations under harmonic loading is of utmost importance, as it is influenced by the stiffness and damping characteristics due to pile-soil-pile interaction. Moreover, there has been a notable absence of acknowledgment regarding the importance of carrying out empirical validation for diverse concepts. The aim of this study is to examine the interactions among a cluster of three piles and the nonlinear theoretical approach introduced by Novak [6] for addressing the inverse problem. The subject of inquiry pertains to the behaviour of piles when subjected to mechanically induced axial harmonic loading. The method under consideration entails the use of a mechanical oscillator to administer harmonic forces to the soil-pile structures. The magnitude of the force exhibits direct proportionality to the square of the frequency. The evaluation of the dynamic characteristics of the soil-pile system involves a comparison between the empirical field data and the theoretical nonlinear response obtained through the application of Novak's [6] inverse problem methodology. The utilization of the dynamic response curve facilitates the computation of the effective mass, damping, and restoring force parameters for the 3-pile group.

Testing Site and Soil Properties

A series of axial vibration field tests are carried out at the campus of IIT Delhi. in order to examine the soil characteristics at the location. Both in-situ and laboratory soil tests are conducted on soil samples obtained from the borehole, including both disturbed and undisturbed samples. Two distinct soil layers are observed, with a shear modulus of 1.3×10^4 kN/m^2 for the first layer (0–3.5 m) and 2.3×10^4 kN/m^2 for the second (3.5–6.5 m). The first layer comprises 39% sand, 43% silt, and 18% clay, whereas the second layer comprises 3% gravel, 36% sand, 42% silt, and 19% clay. Results from various soil tests indicate that the soil consists primarily of clayey silt at various depths.

Dynamic Axial Loading Test

In this study, to conduct a axial loading field tests on three pile group, hollow steel pipes are used with the follows dimensions: pile length (l) = 3.0 m, inner diameter (d) = 11.1 cm, and thickness (t) = 0.3 cm. The group pile setup is lodged in undersized boreholes to maximize pile-to-soil interaction. Using a tripod and an SPT hammer, the piles are pushed into the ground with hammer blows. Steel plates are used at the pile's bottom, ensuring end bearing. While driving the piles for a group installation, the piles are kept at a $3d$ distance. A mechanical oscillator is used to create the dynamic forces required for the pile foundation. The two revolving masses of a mechanical oscillator are aligned in a horizontal plane, and a axial excitation force is generated through the oscillator. The force can be adjusted by changing the eccentricity (θ) of the masses in rotation. Eccentric moment (*m.e*) value can be expressed as:

$$me = (W/g) \cdot e = \left[0.9 \sin (\theta/2)\right]/g \quad \text{Nsec}^2 \tag{1}$$

where, where *W, m, e,* and *g* denote the eccentric rotating parts weight, mass, eccentric distance of rotating masses, and gravitational acceleration respectively.

The dynamic responses of a pile group under four different eccentric moments are examined using dynamic loading testing. The placement of the oscillator on the pile-cap loading system was oriented horizontally in order to establish a configuration that would produce harmonic forces in an axial direction. In this work, a data acquisition system is used to quantify the soil-pile responses of the combined pile-soil setup at various frequencies. In order to examine the dynamic properties of the soil-pile system, an accelerometer is attached to the upper and central portions of the experimental arrangement to record time-acceleration responses. A frequency sensor is hooked to the DC (direct current) motor in order to quantify its frequency, and all these generated dynamic field responses are recorded through a laptop and used for plotting the frequency-amplitude curves. Figure 2.1 shows the experimental setup utilized for the axial loading field test.

Theoretical Investigation

Different approximation or asymptotic techniques can be employed to examine nonlinear axial loading and estimate the characteristics of soil-pile. The efficiency of these techniques is contingent upon their distinct dynamic properties, including frequency, amplitude, and loading forces. The concept of the inverse problem as an analytical methodology was introduced by Novak in [6]. The process entails the utilization of experimental datasets to assess the soil-pile system's nonlinear dynamic response and impedance parameters. Therefore, in this study, this methodology is used to analyze the frequency-amplitude response curves of pile groups. The procedure entails the generation of a consistent oscillation state through the application of a dynamic force. The observed phenomenon in this process involves a proportional increase in the axial amplitude of the soil-pile system, which is directly related to the square of the frequency, as in most applications. The correlation between

Figure 2.1 Axial loading test setup of pile group.

the dynamic nonlinear response and backbone curves is essential to this methodology. The analysis of the backbone curve provides valuable insights into the relationship between undamped natural frequencies $\Omega(A)$ and amplitude variations.

$$\Omega = \sqrt{\omega_1 \omega_2} \qquad (2)$$

The values of ω_1 and ω_2 are determined by identifying the frequencies at which the frequency-amplitude response curve intersects with a line that passes through the origin of the coordinates. In the context of elastic nonlinearity, the restoring force demonstrates linear characteristics when subjected to harmonic excitation, but only within certain amplitude ranges. The linearity of the harmonic excitation is preserved when varying spring constants are employed, resulting in a consistent amplitude function. When examining the function $F(A)$ in the context of steady excitation, it is possible to represent the squared values of the natural frequencies in the following manner:

$$\Omega^2(A) = \frac{F(A)}{A m_{eff}} \qquad (3)$$

where the notation of $\Omega(A)$ is for the undamped natural frequencies and the notation of m_{eff} for the effective mass of the soil-pile loading system. Several techniques exist for determining the damping and effective mass, all of which rely on an analysis of the geometric properties of the nonlinear response curve. To calculate the damping, ω_1 and ω_2 from the response curves are used, corresponding to the resonant amplitude A_T and the resonant frequency (ω_T), along with their corresponding points on the Ω-curve. These points are used in the formula for calculating damping.

$$D = \frac{1}{2} \frac{\omega_2 - \omega_1}{\omega_0} \left[2 \left(\frac{\Omega}{\omega_T} \right)^2 - 1 \right]^{-\frac{1}{2}} \qquad (4)$$

When the information regarding the mass is known, examining the characteristics of the pseudo-nonlinear restoring forces becomes possible by utilizing the backbone curve Ω. To achieve the desired Ω-curve, it is required to select the restoring force $F(A)$.

$$F(A) = A m_{eff} \Omega^2 \qquad (5)$$

The frequency-amplitude response curves can calculate the $F(A)$ and associated displacement.

Theory Versus Experiment

The theory of nonlinear vibration proposed by Novak in [6] is employed in this study to retrospectively ascertain the frequency-amplitude response curves. These curves are determined based on the dynamic field response data obtained for the pile group.

Both the field and back-calculated response curves are compared for all four excitation forces. Figure 2.2 shows the experimental response curves used to plot the backbone curve by intersecting lines representing various eccentric moments or excitation forces. The comparison of estimated and measured responses from theoretical and experimental data revealed that the resonant frequencies declined and the amplitudes raised with eccentric

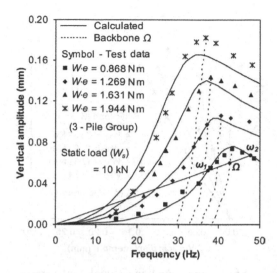

Figure 2.2 Test and theoretical responses of piles under axial loading.

Table 2.1: Measured dynamic parameters.

Eccentric moment (Nm)	Effective mass		Damping	Stiffness (kN/mm)
	Mass m_{eff} (kg)	Mass coeff. (ξ)		
0.868	11958	10.95	0.17	443.35
1.269	10098	9.09	0.18	301.42
1.631	8934	7.93	0.19	241.49
1.944	7176	6.18	0.20	177.06

moments. The pattern of variation with respect to eccentric forces indicates the nonlinearity of the system. From the results, it is noted that the stiffness characteristics of the pile group change with the change in these eccentric forces. The behaviour of soil-pile systems in terms of damping, stiffness, and effective mass is determined and presented in Table 2.1.

The tabular results indicate that the evaluated damping values are raised and effective mass and avg. stiffness are decline with the increase of dynamic forces. This decreasing behavior in soil stiffness indicates the decrease of the soil resistance around the pile.

The higher values of additional effective mass at lower eccentric moments show the partial bonding break/separation of pile-soil under higher eccentric moments. The experimental peak displacements, or resonant amplitudes, are higher as compared to predicted or theoretical amplitude values under higher excitation forces. These types of variations or differences between response curves may occur due to the loosening of soil around the pile during higher excitation or vibration. From the plotting of $F(A)$ versus axial displacements (Figure 2.3), it is observed that the overall stiffness of the pile group reduced under an increase in axial dynamic loading, which is based on eccentric moments/forces. The theoretical response curves and the measured field responses in terms of frequency and amplitude with axial loading almost match each other. Therefore, a feasible and approximate theoretical estimation can be done through this method, with the system having a dynamic/nonlinear restoring force and linear damping.

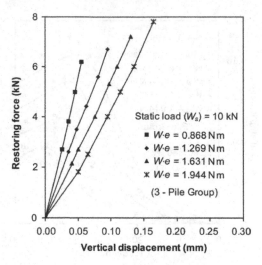

Figure 2.3 Measured restoring force corresponding to axial displacement.

Conclusions

The primary goal of this study is to experimentally as well as theoretically investigate the dynamic behaviour of three pile group with axial loading condition. The output results of the machine-based foundation system show the nonlinear behaviour of the combined system, with a reduction of resonant frequencies and a disproportional increase in amplitudes as eccentric moments continue to increase. In this present work, the changes in stiffness, damping, and effective mass produced by various eccentric moments using field response curves and Novak's inverse back-calculation technique [6] are also determined. The results indicate a rise in damping values and a decline in effective mass and average stiffness with dynamic forces. This decreasing characteristic of soil stiffness indicates the decrease in soil resistance around the piles. Theoretical prediction revealed that when eccentric moments increased, soil resistance around the piles decreased, leading to a drop in pile group stiffness. A close comparison of the soil-pile response curves generated via back calculation and the test results is found. The theoretical method, based on the inverse technique, provides feasible and adaptable results for understanding pile dynamics. The proposed inverse approach for dynamic pile analysis is a robust and efficient method that can be utilized to analyze the performance of pile-supported structures throughout the design phase under machine foundation.

References

1. Biswas, S., and Manna, B. (2018). Experimental and theoretical studies on the nonlinear characteristics of soil-pile systems under coupled vibrations. *Journal of Geotechnical and Geoenvironmental Engineering ASCE*, 144(3), 1–14. doi:10.1061/(ASCE)GT.1943 5606.0001850.
2. Choudhary, S. S., Biswas, S., and Manna, B. (2021). Effect of pile arrangements on the dynamic coupled response of pile groups. *Geotechnical and Geological Engineering*, 39, 1963–1978. https://doi.org/10.1007/s10706-020-01599-6.
3. Elkasabgy, M., and El Naggar, M. H. (2013). Dynamic response of vertically loaded helical and driven steel piles. *Canadian Geotechnical Journal*, 50, 521–535.

4.   Khalil, M. M., Hassan, A. M., and Elmamlouk, H. H. (2020). Dynamic behavior of pile foundations under vertical and lateral vibrations: review of existing codes and manuals. *HBRC Journal*, 16(1), 39–58.

5.   Liu, X., Wang, K., and El Naggar, M. H. (2020). Dynamic pile-side soil resistance during longitudinal vibration. *Soil Dynamics and Earthquake Engineering*, 134, 1–10.

6.   Novak, M. (1971). Data reduction from nonlinear response curves. *Journal of Engineering Mechanics ASCE*, 97(4), 1187–1204.

7.   Wen-jie, G., Wen-bing, W. U., Guo-sheng, J., Leo, C. J., and Guo-dong, D. (2020). Torsional dynamic response of tapered pile considering compaction effect and stress diffusion effect. *Journal of Central South University*, 27, 3839–3851.

3 Machine induced field response of steel piles under rocking vibrations

Sumit Kumar[1,a], Shiva Shankar Choudhary[2,b], and Avijit Burman[2,c]

[1]Research Scholar, Department of Civil Engineering, National Institute of Technology Patna, Patna, India

[2]Assistant Professor, Department of Civil Engineering, National Institute of Technology Patna, Patna, India

Abstract: The vibration of piles subjected to rocking mode with lateral translation is a typical dynamic characteristic of pile foundations supporting rotating machines and piles under seismic conditions. In many cases, it has been observed that the horizontal and rocking motions are critical and often govern design. Therefore, in this investigation, field-based forced vibration tests on hollow steel piles were conducted to investigate the frequency-amplitude responses of piles subjected to varying eccentric moments caused by machine-induced rocking vibration. The frequency-amplitude responses discovered in the dynamic field tests show two resonant peaks between 0 and 50 Hz. As the excitation forces are increased, the resonant frequencies are dropped, and the resonant amplitudes are increased, but not in a linear way, and it is found that the dynamic response curves exhibit nonlinear properties. The fundamental properties of soil-pile responses remain the same for any given group pile configuration. However, due to the dynamic complex stiffness of the pile-soil-pile interaction, resonant frequency indicate greater and resonant amplitude indicate lower values as the number of piles increases.

Keywords: Dynamic characteristic, field test, frequency-amplitude responses, machine vibration.

Introduction

In recent decades, there has been a notable increase in the prevalence of machine-based socioeconomic sectors such as power generation, petrochemical production, oil refining, food processing, pharmaceutical manufacturing, and cement and steel industries. It is necessary to employ pile foundations for machine foundations in industries where mechanical devices like turbines and rotary compressors are installed because of the large magnitude of the vibratory loads they generate. There is an increasing demand for precise vibration prediction of machine-based pile foundations, which has led to a sustained interest among engineers in the dynamic modeling of machine foundations. Pile foundations support machinery and other vibrating equipment, and their primary function is to dampen vibrations to a tolerance limit. In order to design foundations subjected to vertical, horizontal, and rocking loading, it is necessary to assess the frequency versus amplitude responses of

[a]sumitk.phd19.ce@nitp.ac.in, [b]shiva@nitp.ac.in, [c]avijit@nitp.ac.in

the soil-pile systems. The dynamic response in any given situation can now be determined using any one of a variety of theoretical formulations and computer-based programs.

Experimental investigations are always necessary to test the relevance and efficiency of various theoretical approaches. The accuracy of the dynamic linear and nonlinear formulations was verified by conducting a few experiments. Novak and Grigg [7] conducted dynamic tests on small-scale piles to validate the suitability of a linear-elastic theoretical approach. The testing results showed that the dynamic response could be quantified using the static pile group interaction parameters proposed by Poulos [8]. El Marsafawi et al. [6] performed vibration experiments under dynamic load on two groups of piles with varied soil strata. In order to analyze the effects of soil-pile spacing and diameter ratio, Burr et al. [2] conducted field vibration testing on piles' nonlinear response on 13 model pile groups in various soil layers. In addition, the variations of nonlinearity on the pile responses in frequencies, amplitudes, and impedance characteristics have been studied by Elkasabgy and El Naggar [4] and Sinha et al. [9] using theoretical analysis applying the continuum approach method. Other researchers have conducted field tests on piles Elkasabgy and Naggar [5], Biswas and Manna [1], Choudhary et al. [3]. The results were compared to analytical curves generated using a continuum technique. The analytical investigation was compared to field data collected while machines were vibrating, and it was found that the analytical analysis produced a realistic estimation of the horizontal and rocking response curves. Such condition was successfully attained by incorporating a boundary zone surrounding the piles and establishing separation lengths between the pile and the surrounding soil.

The pile-soil-pile behaviour and its dynamic properties, as measured by frequency and amplitude responses under machine-based vibrations, have been the subject of very few investigations. These studies regarding field vibration tests under rocking mode are also getting very narrow. Furthermore, very limited research works are available on dynamic characteristics of soil-pile systems through rocking vibration tests under various soil-pile setups. Hence, this study carried out four distinct patterns of soil-pile arrangements (single pile, 2 × 2, 2 × 3, and 3 × 3 group of piles) to acquire the dynamic nonlinear response of piles. A number of dynamic field testing outcomes for piles under the rocking mode of vibrations are presented in the present study, which may be useful to the academic community and practising engineers. Such experimental datasets could be useful in understanding the complex behaviour of individual piles and groups of piles when subjected to machine vibration.

Site Characterization and Location

At IIT Delhi in India, rocking vibration field tests have been carried out between blocks II and III. Borehole soil samples are tested in situ and in the laboratory on soil at the testing site. Two distinct soil layers are observed, with a shear modulus of 1.3×10^4 kN/m^2 for the first layer (0–3.5 m) and 2.3×10^4 kN/m^2 for the second (3.5–6.5 m). The first layer comprises 39% sand, 43% silt, and 18% clay, whereas the second layer comprises 3% gravel, 36% sand, 42% silt, and 19% clay. Results from various soil tests indicate that the soil consists primarily of clayey silt at various depths.

Description of Pile and Installation

The dimensions of the hollow steel pipes used in the studies are as follows: pile length (l) = 3.0 m, inner diameter (d) = 11.1 cm, and thickness (t) = 0.3 cm. The experiments are performed in a rocking mode with a dynamically forced vibration condition. Individual and group piles are lodged in undersized boreholes to maximize pile-to-soil interaction. Using a tripod and an SPT hammer, the piles are pushed into the ground with hammer blows. Steel plates are used at the pile's bottom, ensuring end bearing. While driving the piles for a group installation, the piles are kept at a $3d$ distance.

Rocking Vibration Test

A mechanical oscillator is used to create the dynamic forces required for the pile foundation. The two revolving masses of a rocking mechanical oscillator are aligned in a vertical plane, and a horizontal excitation force is generated through the oscillator. Figure 3.1 illustrates an oscillator that includes rotating masses and force direction. The force can be adjusted by changing the eccentricity (θ) of the masses in rotation. Eccentric moment ($m.e$) value can be expressed as:

$$me = (W/g) \cdot e = [0.9 \sin(\theta/2)]/g \quad \text{Nsec}^2 \tag{1}$$

The eccentric degree (θ)'s dynamic vibration force (P) can be described as follows at any frequency:

$$P = me\omega^2 = \frac{[0.9 \sin(\theta/2)]}{g} \times \omega^2 \quad \text{N} \tag{2}$$

where W, m, e, g, and ω denote the eccentric rotating parts weight, mass, eccentric distance of rotating masses, gravitational acceleration, and circular frequency, respectively.

A steel plate of 90 cm × 90 cm × 3.7 cm is positioned at the top of the pile arrangements to use as a pile cap. Number of plates and an oscillator are affixed on the pile cap's upper surface after the pile cap has been installed. The plates and oscillator are bolted to the pile cap to ensure the system for loading the pile cap functions as a single unit. The tests are conducted for all soil-pile setups under 12 kN static loads. The measurement of the rocking component is conducted using an accelerometer, which is positioned vertically along the axis of the pile cap. This accelerometer is set up on the pile-cap loading system's top at a predetermined distance from the known pile-cap center. During vibration testing,

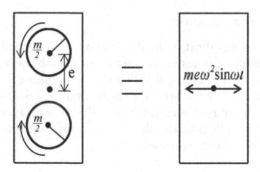

Figure 3.1 Forcing direction of oscillator.

a frequency sensor is glue into 5 hp DC motor to determine the operational frequency. At the time of experimental tests, the time versus acceleration and time versus frequency responses are assessed, and subsequently, the frequency-amplitude curves are obtained from these response curves. The testing setup is shown in Figure 3.2.

Field Response of Soil-Pile Systems

Figure 3.3a to 3d shows the frequency-amplitude response curves for single pile, 2 × 2 group pile, 2 × 3 group piles and 3 × 3 group piles respectively. Under the rocking vibration mode, the dynamic response of these curves shows two resonant peaks at different frequencies for all soil-pile arrangements.

The second peak exhibits more dominance in comparison to the initial peak. Well-spaced and separated first and second resonant peaks are observed under these excitation tests. The second peak's resonant frequencies and amplitudes are not observed for the 3 × 3-pile group because of the mechanical oscillator's limitation of operating frequency range (which is up to 0–50 Hz). Typical behaviour of the soil-pile system is observed in the measured dynamic responses as the eccentric moments are varied; resonant frequencies are found to decrease, and resonant amplitudes are found to increase asymmetrically. It is found from the figures that there is an increase in the resonant rocking frequency values, accompanied by a decrease in the resonant rocking amplitude values, as the number of piles within a pile group increases. This observation suggests that adding more piles increases the pile groups' complex soil-pile stiffness. The resonant frequency values of the first peak for 2 × 2, 2 × 3 and 3 × 3 pile groups are observed approximately 85%, 160%, and 224% more than the initial resonant frequency of an single pile. The resonant amplitude for the

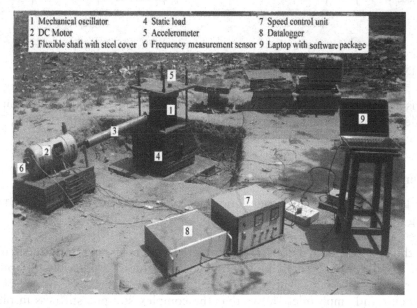

1 Mechanical oscillator	4 Static load	7 Speed control unit
2 DC Motor	5 Accelerometer	8 Datalogger
3 Flexible shaft with steel cover	6 Frequency measurement sensor	9 Laptop with software package

Figure 3.2 Complete rocking vibration test setup.

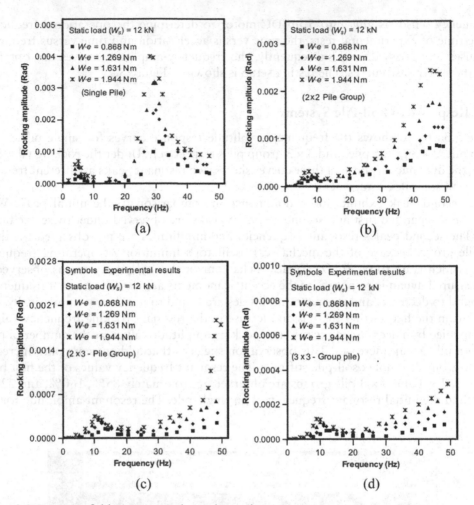

Figure 3.3 Dynamic field response under rocking vibration.

first peak is significantly reduced in the 2 × 2, 2 × 3, and 3 × 3-pile groups compared to a single pile under the same static load, with approximate reductions of 59%, 79%, and 86% respectively.

Conclusions

In this work, the impact on single piles and pile groups (2 × 2, 2 × 3, and 3 × 3) based on the rocking mode of machine-based dynamic vibrations is studied.

- From the test results, it is observed that the frequencies and amplitudes under resonance conditions measured higher and lower values, respectively, for pile groups compared to single piles. This pattern of higher and lower observations of resonant frequencies and amplitudes shows that the complex soil-pile stiffness in pile groups increases with pile numbers.
- It is observed from the experiment with various pile arrangements that the values of rocking resonant frequency and amplitude decline and rise with eccentric moments.

This behavior highlights the nonlinearity of the combined systems in rocking vibration mode.

- Two separate resonant peaks under rocking excitation are observed where second peak indicates higher dominated nature as compared to the initial peak.
- The first peak values of resonant frequencies for 2 × 2, 2 × 3, and 3 × 3-pile groups are observed approximately 85%, 160%, and 224% higher, respectively, and for resonant amplitudes, 59%, 79%, and 86% lower as compared to the single pile.

References

1. Biswas, S., and Manna, B. (2018). Experimental and theoretical studies on the nonlinear characteristics of soil-pile systems under coupled vibrations. *Journal of Geotechnical and Geoenvironmental Engineering ASCE*, 144(3), 1–14. doi:10.1061/(ASCE)GT.1943 5606.0001850.
2. Burr, J. P., Pender, M. J., and Larkin, T. J. (1997). Dynamic response of laterally excited pile groups. *Journal of Geotechnical and Geoenvironmental Engineering ASCE*, 123(1), 1–8.
3. Choudhary, S. S., Biswas, S., and Manna, B. (2021). Effect of pile arrangements on the dynamic coupled response of pile groups. *Geotechnical and Geological Engineering*, 39, 1963–1978. DOI: 10.1007/s10706-020-01599.
4. Elkasabgy, M., and El Naggar, M. H. (2013). Dynamic response of vertically loaded helical and driven steel piles. *Canadian Geotechnical Journal*, 50, 521–535.
5. Elkasabgy, M., and El Naggar, M. H. (2018). Lateral vibration of helical and driven steel piles installed in cohesive soils. *Journal of Geotechnical and Geoenvironmental Engineering ASCE*, 144(9), 1–8.
6. El Marsafawi, H., Han, Y. C., and Novak, M. (1992). Dynamic experiments on two pile groups. *Journal of Geotechnical Engineering ASCE*, 118(4), 576–592.
7. Novak, M., and Grigg, R. F. (1976). Dynamic experiments with small pile foundations. *Canadian Geotechnical Journal*, 13, 372–385.
8. Poulos, H. G. (1971). Behaviour of laterally loaded piles II: pile groups. *Journal of Soil Mechanics and Foundation Engineering ASCE*, 97, 733–751.
9. Sinha, S. K., Biswas, S., and Manna, B. (2015). Nonlinear characteristics of floating piles under rotating machine induced vertical vibration. *Geotechnical and Geological Engineering Springer*, 33, 1031–1046.

4 Performance evaluation of a silica gel–methanol adsorption chiller running by diesel truck engine exhaust

Ramesh P. Sah[1,a], Anirban Sur[2,b], and Palash Soni[3,c]

[1]Asansol Engineering College Asansol, West Bengal, India

[2]Symbiosis Institute of technology, Symbiosis International (Deemed University), Pune, Maharashtra, India

[3]National Institute of technology Raipur, Chhattisgarh, India

Abstract: In this paper how the performance of an adsorption chiller (silica gel–methanol working pair) varied with different input parameters has been discussed with the help of a mathematical model. The proposed adsorption chiller used to cool diesel truck cabins and run by engine exhaust. Adsorption heat of the adsorber bed has been carried out by water and released in the radiator. The desorber bed has been heated by waste heat from engine exhaust. Here hot water has been used as a heat carrier from the engine exhaust gas. Analysis has been done based on the adsorption kinetic equation at the adsorption equilibrium condition. Mass and energy balances of the basic parts of the adsorption chiller have been analyzed here. Influence of inlet temperature (hot water for desorber bed, cooling water for adsorber bed and chilled water for evaporator), flow rate on cycle time, switching time on the performances of adsorption cooler have been observed. Simulated results predicted that the maximum system COP value of the chiller is 0.214, which proves that the proposed adsorption chiller can be used to maintain the low temperature at the driver cabin of the truck during summer.

Keywords: Adsorption chiller, cooling capacity, cycle performance, engine exhaust, lumped parameter model, operating parameters.

Introduction

Literature reviews say popular conventional cooling (vapor compression refrigeration systems) are running by electricity (fossil fuels) and responsible for global warming and ozone layer duplication (due to using of CFC/HCFC and electrical energy). The adsorption chiller can be an alternative to a conventional cooling system for commercial uses as it uses zero GWP and ODP refrigerants (natural like water/methanol etc), run by renewable energy (Waste heat/solar heat), less movable components and less noise [1, 2]. Researchers have developed various mathematical models and computer codes and analyzed different types of adsorption refrigeration at different working conditions [3–24]. In this paper, a silica gel–methanol adsorption refrigeration system powered by diesel engine waste heat has been analyzed. Literature shows that the efficiency of the diesel engine of a truck is about 40% and in total fuel use in transportation sectors, 20% of that has been consumed by heavy trucks [25–28]. In India, a truck driver cabin should be air-conditioned. So the consumption of fuel to run the truck in India is more. So this adsorption chiller will reduce

---

[a]sramesh2031@gmail.com, [b]anirbansur26@gmail.com, [c]sonipalash752@gmail.com

fuel cost as it is run by engine exhaust. In this paper, the performance of an adsorption cooling system (adsorption chiller) running by a truck diesel engine exhaust heat has been analyzed. Analysis has been done based on adsorption kinetic equation at adsorption equilibrium condition, mass and energy balance of the basics adsorption chiller has been analyzed here. Influence of inlet temperature (hot water for desorber bed, cooling water for adsorber bed and chilled water for evaporator), flow rate on cycle time, switching time of the adsorber to desorber, on the performance of the adsorption chiller has been observed.

System Details

Mathematical Model of BLDCM

Figure 4.1 shows an adsorption system installed in the diesel truck engine. It has been run by exhaust heat from the diesel engine. The desorber bed has been heated by the thermite fluid flows inside the engine exhaust manifold through the heat exchanger (shell and tube type). Cold coolant has been circulated from the radiator to the condenser and adsorption bed to maintain them cool. Diesel engine containing 3000cc, six cylinders and exhaust temperatures varied between 400–300°C. Figure 4.2 shows a ray diagram of the adsorption refrigeration system. At the heat exchanger, normal DS water is heated by engine

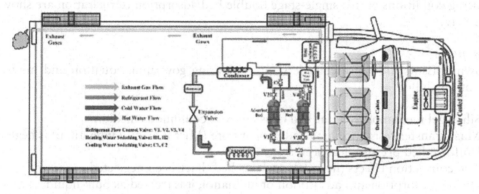

Figure 4.1 Adsorption chiller fitted inside the truck.

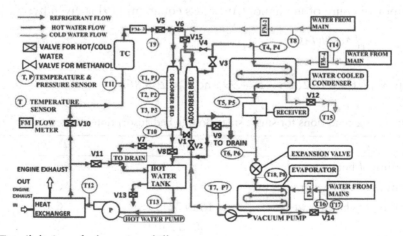

Figure 4.2 Detail design of adsorption chiller.

exhaust gas. After achieving the desired temperature, hot water sends to the storage tank (TC) through the flow valve V10. The adsorption refrigeration system has four basic components: an adsorber/desorber bed, a condenser, an expansion valve and an evaporator. Cooling water coming from the radiator, cool the adsorber when the valve V15 is opened and V6 is closed. Cooling water cools the adsorber which results in the adsorption of the refrigerant vapor in the adsorber bed.

At the same time, hot water flows through the desorber bed, when valve V5 is opened. Due to the heating of the desorber bed refrigerants starts vaporized. After achieving condenser pressure, V4 open and desorbs refrigerant flows from the desorber bed to the condenser. The refrigerant vapor coming from the bed through V4 is condensed by cold water circulation at the condenser. From the condenser, it liquid refrigerant stores in the receiver and returns to the evaporator through an expansion valve. After this process when the adsorber bed reaches its maximum concentration level, and the desorber bed reaches the lowest concentration level, the adsorber bed is switched to a desorber bed and the desorber bed is switched to an adsorber bed. The adsorber bed is pre-cooled as after adsorption of refrigerants it losses its kinetic energy and comes to static and release heat to the adsorber bed. To stop unwanted vapor migration and for achieving pre-decided vapor refrigerant pressure, all the gas valves are closed properly during this process. The whole system initially maintains sub-atmospheric pressure using a vacuum pump shown in Figure 4.2. The operating conditions of this single-stage double bed adsorption refrigeration are shown in Table 4.1.

Assumption
Following assumptions are considers for generating governing equation and model the adsorption chiller

- Silica gel methanol granular bed consider as a continuous medium
- Mass transfer resistance in the adsorbent pore (Micro/ macro/ interstitial) is neglected.
- Uniform bed pressure has been considered
- The convection effects (inside the porous bed) have been neglected.
- During desorption and adsorption of methanol, it is treated as bulk liquid

Governing Equations Used for modelling:
Saturated vapor pressure of methanol (Antoine's equation) [19] is given by;

$$P_s = 133.3 \times 10^{\left(A1 - \left(B/(T+C)\right)\right)} \tag{1}$$

where T is in °C, A1 = 8.072, B = 1574.99 and C = 238.86.

Table 4.1: Operating conditions for single stage double bed adsorption refrigeration system.

Parameter	Range	Unit
Inlet temperature of hot water	60 to 120	°C
Inlet temperature of cold water	20 to 32	°C
Inlet temperature of chilled water	10 to 22	°C
Cycle time	400 to 2400	Sec
Switching time	10 to 100	Sec

Adsorber/desorber energy balance

$$\left[m_{sg}\left(C_{p,sg}+C_p^m x\right)+m_{hex}C_{p,hex}\right]\frac{dT_{bed}^m}{dt}=m_{sg}Q_{st}\frac{dx_{bed}^m}{dt}-km_{sg}C_{p,v}^m\left(T_{bed}-T_{eva}\right)\frac{dx_{bed}^m}{dt} \quad (2)$$
$$+\dot{m}_f C_{p,f}\left(T_{bed,in}-T_{bed,out}\right)$$

The cooling/heating fluid coming from engine exhaust outlet temperature of bed is calculated by following equation

$$\frac{T_{bed,out}-T_{bed}}{T_{bed,in}-T_{bed}}=exp\left[-\frac{(UA)_{bed}^m}{\dot{m}_f C_{p,f}}\right] \quad (3)$$

Evaporator energy balance

$$\left(m_{eva}C_{p,eva}+m_{eva,hex}C_{p,hex}\right)\frac{dT_{eva}^m}{dt}=-h_{fg}m_{sg}\frac{dx_{ads}}{dt}+m_{sg}C_{p,l}^m\frac{dx_{des}}{dt}\left(T_{eva}-T_{con}\right) \quad (4)$$
$$+\dot{m}_{f,chill}C_{p,f}\left(T_{chill,in}-T_{chill,out}\right)$$

Chilled water outlet temperature is calculated by

$$\frac{T_{chill,out}-T_{eva}}{T_{chill,in}-T_{eva}}=exp\left[\frac{(UA)_{eva}^m}{\dot{m}_{f,chill}C_{p,f}}\right] \quad (5)$$

Condenser energy balance

$$\left(m_{con}C_{p,con}+m_{con,hex}C_{p,hex}\right)\frac{dT_{con}^m}{dt}=-h_{fg}m_{sg}\frac{dx_{des}}{dt}+m_{sg}C_{p,v}^m\frac{dx_{des}}{dt}\left(T_{con}-T_{bed}\right) \quad (6)$$
$$+\dot{m}_{f,con}C_{p,f}\left(T_{con,in}-T_{con,out}\right)$$

Temperature of water coming out from condenser is calculated by

$$\frac{T_{con,out}-T_{con}}{T_{con,in}-T_{con}}=exp\left[\frac{(UA)_{con}^m}{\dot{m}_{f,con}C_f}\right]$$

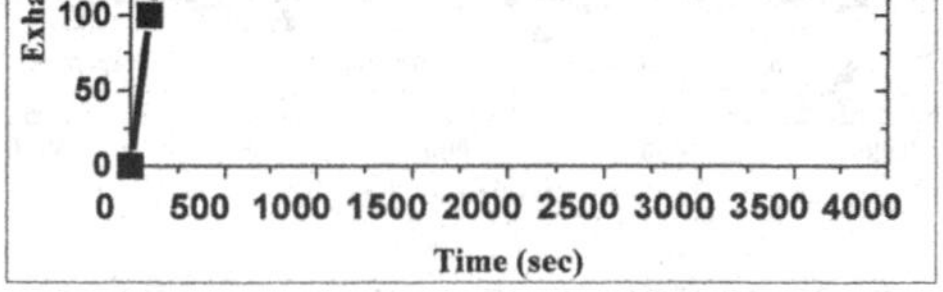

Figure 4.3 Exhaust temperature variation with respect to engine RPM.

Result

With the help of the above described governing equations, a simulation model has been developed and solved using MATLAB. Operating parameters values provide in Figure 4.3 shows the variation of exhaust gas temperature for different engine speeds. This temperature has been calculated using simple diesel engine exhaust gas temperature formula. Figure 4.4 shows cooling capacity and COP both increase with increases in hot inlet temperature. Cooling capacity decrease after 80°C of hot water, because at that time desorber bed mass concentration ratio goes to a minimum. Figure 4.5 shows the temperature profile of the adsorber and desorber bed for a cycle time of 600 s, the desorber bed nearly reaches the hot water temperature but the adsorbent bed temperature becomes 5–6°C higher than the temperature of cooling water. This figure also says that heating of desorber bed is much easy compare to remove of adsorbing heat. The temperature profile of the condenser represents that at the beginning of the cycle the temperature becomes maximum because the cooling water temperature is high initially as the adsorber bed switches from adsorption to desorption mode. Figure 4.5 also represents that all the four components of the adsorption refrigeration system reach their steady-state condition after around 34 cycles.

Figure 4.6 shows that the cooling capacity increases with a decrease in cooling water temperatures from 32 to 20°C. This is because the lower adsorption temperatures result

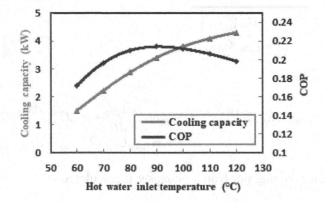

Figure 4.4 Influence of hot water inlet temperature in desorber bed on COP and cooling capacity.

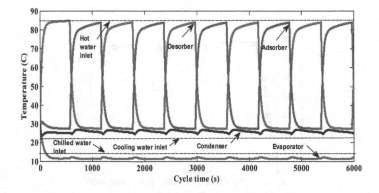

Figure 4.5 Variation of all four components temperatures with cycle time.

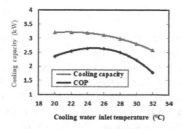

Figure 4.6 Influence of cooling water inlet temperature on COP and Cooling capacity.

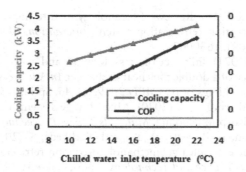

Figure 4.7 Variation of cooling capacity and COP with variation chilling water entering evaporator.

in a large amount of refrigerant being adsorbed and desorbed during each cycle. The COP also increases with lower values of cooling water inlet temperature as adsorption of the refrigerant increases with lower temperatures and it has an optimum COP value of 0.2127 at an inlet temperature of cooling water 24°C. Below this temperature, the adsorbent adsorbs more refrigerant with lower cooling water inlet temperature but at the same time, the adsorbent needs more amount of heat to reach the desorption temperature. The overall effect of both processes during the cycle decreases the value of COP.

Figure 4.7 represents COP and the cooling capacity both increase with increases in chilled water temperature. The larger chilled water temperature increases the evaporation rate of refrigerants, which increases results refrigeration effect also. But practically this advantage is not possible to take, as higher chiller water increases the system's cooling load also.

Concluding

A simulation study of a single-stage double bed silica gel-methanol adsorption refrigeration system for a diesel truck has been discussed. The total time requires for the adsorber and desorber cycle is 600sec each. Results show how the system performance (COP and cooling effect) varies with the variation of operating conditions. Maximum COP (0.214) can be achieved at 90°C hot water inlet temperature and 24°C cold water temperature, for that maximum Cooling capacity of 4kW observed. In this study, thermodynamic analysis is also done to find the optimum operating values of the proposed adsorption chiller. To achieve the best performances, the adsorption chiller must be operated at its optimum operating values.

References

1. Lorentzen, G. (1995). The use of natural refrigerants: a complete solution to the CFC/HCFC predicament. *International Journal of Refrigeration*, 18, 190–197.
2. Calm, J. M. (2002). Emissions and environmental impacts from air-conditioning and refrigeration systems. *International Journal of Refrigeration*, 25, 293–305.
3. Sah, R. P., Choudhury, B., Das, R. K., and Sur, A. (2017). An overview of modelling techniques employed for performance simulation of low–grade heat operated adsorption cooling systems. *Renewable and Sustainable Energy Reviews*, 74, 364–376.
4. Sur, A., and Das, R. K. (2015). Numerical modeling and thermal analysis of an adsorption refrigeration system. *International Journal of Air-Conditioning and Refrigeration*, 23(04), 1550033.
5. Sur, A., and Das, R. K. (2016). Review of technology used to improve heat and mass transfer characteristics of adsorption refrigeration system. *International Journal of Air-Conditioning and Refrigeration*, 24(02), 1630003.
6. Gulia, V., and Sur, A. (2021). Influence of mass flow rate and concentration of Al_2O_3 nanofluid on thermal performance of a double pipe heat exchanger. *In Advances in Mechanical Processing and Design: Select Proceedings of ICAMPD 2019*, 33–45. Springer Singapore, 2021.
7. Sur, A., Sah, R. P., and Pandya, S. (2020). Milk storage system for remote areas using solar thermal energy and adsorption cooling. *Materials Today: Proceedings*, 28, 1764–1770.
8. Sur, A., Pandya, S., Sah, R. P., Kotecha, K., and Narkhede, S. (2021). Influence of bed temperature on performance of silica gel/methanol adsorption refrigeration system at adsorption equilibrium. *Particulate Science and Technology*, 39(5), 624–631.
9. Sur, A., Das, R. K., and Sah, R. P. (2018). Influence of initial bed temperature on bed performance of an adsorption refrigeration system. *Thermal Science*, 22(6PartA), 2583–2595.
10. Lemmini, F., and Errougani, A. (2007). Experimentation of a solar adsorption refrigerator in Morocco. *Renew Energy*, 32, 2629–2641.
11. Headley, O. S., Kothdiwala, A. F., and McDoom, I. A. (1994). Charcoal–methanol adsorption refrigerator powered by a compound parabolic concentrating solar collector. *Solar Energy*, 53(No. 2), 191–197.
12. El-Sharkawy, I. I., Saha, B. B., Koyama, S., He, J., Ng, K. C., and Yap, C. (2008). Experimental investigation on activated carbon–ethanol pair for solar powered adsorption cooling applications. *International Journal of Refrigeration*, 31, 1407–1413.
13. EI-Sharkawy, I. I., Kuwahara, K., Saha, B. B., Koyama, S., and Ng, K. C. (2006). Experimental investigation of activated carbon fibres/ethanol pairs for adsorption cooling system application. *Applied Thermal Engineering*, 26, 859–865.
14. Tamainot, Z., and Critoph, R. E. (1997). Adsorption refrigerator using monolithic carbon–ammonia pair. *International Journal of Refrigeration*, 20(2), 146–155.
15. Critoph, R. E. (2002). Multiple bed regenerative adsorption cycle using the monolithic carbon–ammonia pair. *Applied Thermal Engineering*, 22, 667–677.
16. Wang, X., and Chua, H. T. (2007). Two bed silica gel–water adsorption chillers: an effectual lumped parameter model. *International Journal of Refrigeration*, 30, 1417–1426.
17. Sah, R. P., Choudhury, B., and Das, R. K. (2015). A review on adsorption cooling systems with silica gel and carbon as adsorbents. *Renewable and Sustainable Energy Reviews*, 45, 123–134.
18. Meunier, F. (1985). Second law analysis of a solid adsorption heat pump operating on reversible cascade cycles: application to the zeolite–water pair. *Journal of Heat Recovery Systems*, 5, 133–141.
19. Meunier, F., and Douss, N. (1990). Performance of adsorption heat pumps. Active carbon–methanol and zeolite–water pairs. *ASHRAE Transactions*, 2, 267–274.
20. Sakoda, A., and Suzuki, M. (1984). Fundamental study on solar powered adsorption cooling system. *Journal of Chemical Engineering of Japan*, 17(1), 52–57.

21. Saha, B. B., Boelman, E. C., and Kashiwagi, T. (1995). Computer simulation of a silica gel-water adsorption refrigeration cycle—the influence of operating conditions on cooling output and COP. *ASHRAE Transaction: Research*, 101(2), 348–355.
22. Chua, H. T., Ng, K. C., Malek, A., Kashiwagi, T., Akisawa, A., and Saha, B. B. (1999). Modeling the performance of two-bed, silica gel–water adsorption chillers. *International Journal of Refrigeration*, 22, 194–204.
23. Rezk, A. R. M., and Al-Dadah, R. K. (2012). Physical and operating conditions effects on silica gel/water adsorption chiller performance. *Applied Energy*, 89, 142–149.
24. Zheng, W., Worek, W. M., and Nowakowski, G. (1995). Effect of design and operating parameters on the performance of two-bed sorption heat pump systems. *ASME Journal of Energy Resources Technology*, 117, 67–74.
25. Davis, S. C., Diegel, S. W., and Boundy, R. G. (2008). Transportation Energy Data Book. 28th ed. Oak Ridge National Laboratory, ORNL-6984.
26. Sur, A., and Das, R. K. (2017). Experimental investigation on waste heat driven activated carbon-methanol adsorption cooling system. *Journal of the Brazilian Society of Mechanical Sciences and Engineering*, 39(7), 2735–2746.
27. Hamamoto, Y., Alam, K. C. A., Akisawa, A., and Kashiwagi, T. (2003). Experimental study of heat and mass transfer in activated carbon fiber with adsorbing refrigerant. In Proceedings of 21st IIR International Congress of Refrigeration, CD-ROM; Washington, DC, USA, 21.
28. Chemical Society of Japan (1975). 2nd ed. Chemical Handbook Fundamentals. Tokyo: Maruzen, (pp. 892–902).

5 Increasing heat dissipation of a Li-ion battery electrode using lattice structures in cells

Anirban Sur[a], Swapnil Narkhede[b], Gaurav Kothari[c], and Mandar Sapre[d]

Symbiosis Institute of Technology (SIT), Symbiosis International (Deemed University) (SIU), Pune, Maharashtra India

Abstract: Lattices have the ability to organize materials into designs that form static or dynamic structures, which include very minuscule framework of an array of conjunction of beams struts and nodes. This architecture results in reduction of weight and simultaneously retains the integrity of the structure by providing a better degree of freedom and enhanced freedom of control over certain characteristics. The nodes and points present in the lattice portions can ameliorate interlinked strength between the nodes and points, which help in unprecedented performance while using very less material while retaining its strength and integrity. The battery capacity of Lithium-ion and heat dissipation can be refined considering electrodes contain micro-scale pores and channels. In this work, first we have analyzed the internal heat generation inside the lithium-ion battery using the multi-scale multi-domain model available in commercial solver ANSYS-fluent. The contours of internal heat generation and temperature depicted that the majority of the heat is being generated near the electrodes. In order to reduce the heat accumulation near the electrodes and the effective dissipation of heat, here we propose porous electrodes made up of lattice structures (1 mm ligament diameter). The porous nature of lattice structures provides a higher surface area for ion deposition. The thermal and structural analysis of lattice structures is performed and the relevant conclusions are drawn and presented.

Keywords: Battery thermal management system, electrode, electrode modification, Li-ion cell.

Introduction

Lattices can be seen everywhere around us, from metal crystallography to, the internal structure of bones [1]. Their application includes heat exchangers or batteries, where energy transmission is efficient and optimal [2]. The structure of Lattice is with respect to its porosity. This part takes into higher strength to weight ratio and allows building of complex parts using less material. In Li-ion batteries, the charge capacities could be increased by introduction of porous electrodes [3–9]. This is because the porosity will increase the surface area of the electrode and the deposition sites for lithium deposition which will increase the usage of electrodes by a significant amount [10]. Generally, in normal batteries a large portion (30–50%) of the electrode volume is not utilized. With the use

[a]anirbansur26@gmail.com, [b]swapnil.narkhede.phd2018@sitpune.edu.in,
[c]gaurav.kothari.btech2017@sitpune.edu.in, [d]mandar.sapre@sitpune.edu.in

of lattice structures this can be overcome and the electrolyte is able to penetrate the entire electrode efficiently, that will further increase the charging rate of the battery [11–13]. Most of the experiments conducted in literature till day, deal with the change in the electrolyte and its properties in order to drastically increase the capacity of the electrochemical reactions. But if the surface area of the electrode is increased in a significant way the chemical reactions taking place in the electrode due to its porosity increases and due to this increased exposure, this can result to the increase in capacity of said reactions. On the basis of this hypothesis, we have further devised in analysis of a 3-D structure for the electrode which will have significantly higher surface area [15–19]. However, after the initial analysis we aim to find a way to increase the heat dissipation through the electrodes in the battery. Materials use in lithium-ion batteries are, **Cathode materials**: Cathode materials such as rechargeable lithium oxides and lithium-metal oxides ($LiMn_2O_4$, Olivines, Vanadium oxides, $LiCoO_2$, **Anode materials**: Anode materials are silicon graphite and lithium alloys **Electrolytes**: Solid-state electrolytes, liquid and **Separator**: Polyethylene, polypropylene, polyolefin based materials with semi-crystalline structure.

In the Li-ion battery the thermal runaway causes a chain reaction. The energy stored in the battery is suddenly released in milliseconds when the temperature rises rapidly. Therefore, it is also an important factor to be considered, all these factors combined would be considered, from a basic design of lithium ion battery. As the lattice structure would be introduced as a substitute for a solid structure for the electrodes therefore it is crucial to consider the basic design of the battery and all the changes that would arise in case of this trend. Octahedral structures are basically advantageous than other equally complex structures because of their structural design and composition. With solid struts, it is a single length scale micro-structure that can be manufactured relatively and easily and the lattice core can act as heat transfer material while carrying loads.

For the present analysis, a prismatic cell is chosen with specifications mentioned, (capacity—75Ah, voltage (nominal)—3.65V, maximum cont. discharge current @ 25–152A, Upper voltage limit—4.3V, Lower voltage limit—3V, Charge temperature

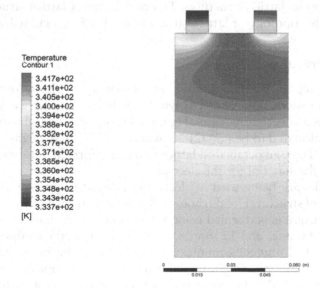

Figure 5.1 Temperature contour on the prismatic Li-ion battery.

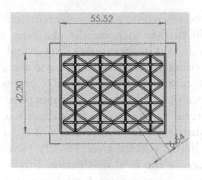

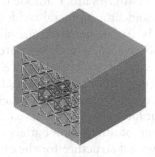

Figure 5.2 Lattice structured porous electrode design 2.

range: 10–50°C, Discharge temperature range: –30–60°C. Mechanical specifications: Dimensions (60 × 24 × 116 mm) and is simulated for constant discharge current corresponding to 1C rate. Figure 5.1 shows the temperature contour on the battery surface at the end of discharge cycle (3600S). The maximum temperature is observed near the electrodes and the tab region as a result of higher heat generation. In order to effectively dissipate this heat to the electrolyte and reduce the heat accumulation, electrodes are modified with the porous lattice structures. Figure 5.2 shows lattice structured Porous electrode design 2. The topology of lattice structure used here is octahedral.

Design of Lattice Structure

The reason for choosing an octahedral structure is that it has the second largest void, which will help with ease of flow of electrolyte while being more space efficient than a cubic void, since it has a lower radius ratio. Figure 5.3 shows octahedral unit cell. The edge of the octahedral is calculated to be ($\sqrt{3}/2$) *L, where L is the total height of the structure or length of diagonal. Thus, an octahedral lattice unit cell is formed. Lithium Cobalt Oxide $LiCoO_2$ is selected as the material for the electrodes.

Since Li-ion batteries are being used in electric vehicles, the battery should be able to absorb some amount of structural load (shocks). So, the structural analysis of a unit cell of octahedral lattice structure is performed using forces occurred in daily usage of a car and light crashes, being 2.4 tonnes and 12 tonnes, respectively. Also, the analysis of unit cell is extended to an array of 12 unit cells, mimicking the electrode of a Li-ion battery.

From the above analysis, we saw elastic deformation and but not any point of critical failure has been observed. So, its proves that the proposed design is structurally stable for such kind of application.

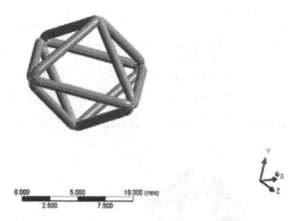

Figure 5.3 Octahedral unit cell.

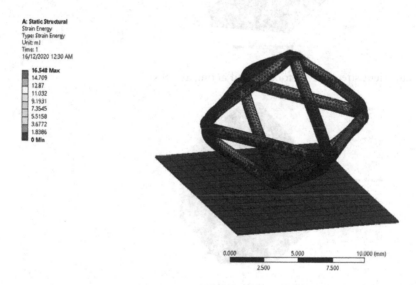

Figure 5.4 Strain energy for 2.4T.

Thermal Analysis of Lattice Structure as an Electrode

After the structural analysis of above-mentioned combinations of lattice members, the CFD analysis is performed for finding convective heat dissipation capability of lattice electrodes. The energy equation is solved, and the flow of electrolyte is assumed to be laminar. The materials for the analysis were chosen as mentioned above, as $LiCoO_2$ and Lithium Hexa-fluorophosphate ($LiPF_6$). Table 5.1 shows properties of materials used for analysis.

Boundary conditions for the present analysis are as follows: The internal heat generation is specified for the lattice which is calculated with the volume of the lattice and the heat generation during the chemical reaction-taking place between the electrode and the electrolyte (obtained from MSMD model). The value determined is 8.025×10^{-5} W/m^3. At the inlet, velocity and the temperature of electrolyte are specified. Figure 5.4 shows strain energy for 2.4T. Figure 5.5 shows equivalent stress for 12 tonnes load acting at apex. Figure 5.6 shows Strain energy for 12 tonnes acting at apex.

Table 5.1: Properties of materials used for analysis.

Properties	$LiCoO_2$	$LiPF_6$
Viscosity	0.0028 kg/m-s	0.0093 kg/m-s
Thermal conductivity	5.4 W/m-K	0.38 W/m-K
Density	4790 kg/m³	1500 kg/m³
Specific heat	750 J/kg-K	g-K

Figure 5.5 Equivalent stress for 12 tonnes load acting at apex.

Figure 5.6 Strain energy for 12 tonnes acting at apex.

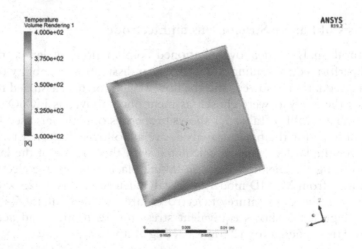

Figure 5.7 Result for fluent analysis temperature volume rendering near the electrode.

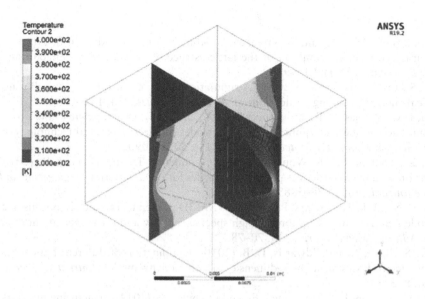

Figure 5.8 Result for fluent analysis temperature rendering using contour planes near electrode.

Results and Discussion

The result of the analysis performed on the unit cell of lattice structure depicts its multi-functionality in a sense that it can simultaneously be used as a structural element as well as a heat dissipating media. Also, the porous nature of the lattice structured electrode provides additional sites for absorption of lithium-ions as compared to a solid electrode, thereby increasing its energy density. Utilization of Multi-Scale Multi-Domain Model to predict the heat generation inside the Li-ion cell provides accurate estimation of heat generation and temperature distribution inside the Li-ion cell. The temperature contours in Figures 5.7 and 8 shows the temperature distribution inside the unit cell of lattice structure. The red zone corresponding to the higher temperature occurring at the ligament of lattice structure and the temperature gradient in the surrounding (electrolyte) is seen depicting the heat transfer.

Conclusion

The thermo-structural analysis of an octahedral lattice structure for possible application as an electrode of a Li-ion battery is presented here. The result of the study depicts that the lattice structure can be effectively utilized as an electrode of a Li-ion battery for efficient thermal management. Also, the structural analysis presented here shows the energy absorption capability and the stability of lattice structure. The porous nature of the lattice structure proves to be beneficial in a sense that it provides increased surface area when utilized as an electrode, which helps increase the energy density of a Li-ion battery within the same dimensional constraints. Further studies will investigate the effects of hollow lattice structured electrodes in which the externally cooled electrolyte can be used as a coolant for effective thermal management of the electrodes.

References

1. Nagesha, B. K., Dhinakaran, V., Shree, M. V., Kumar, K. P. M., and Sathish, T. (2020). Review on characterization and impacts of the lattice structure in additive manufacturing. *Material Today Proceeding*, 21(1), 916–919.

2. Wang, S., Lu, L., and Liu, X. (2013). A simulation on safety of LiFePO4/C cell using electrochemical-thermal coupling model. *Journal of Power Sources*, 244, 101–108.

3. Walsh, F. C., Arenas, L. F., and Ponce de León, C. (2018). Developments in electrode design: structure, decoration and applications of electrodes for electrochemical technology. *Journal of Chemical Technology and Biotechnology*, 93(11), 3073–3090.

4. Chen, X., Gallant, B. M., Wunderlich, P. U., Lohmann, T., and Greer, J. R. (2015). Three-dimensional au micro lattices as positive electrodes for LiO2 batteries. *Engineering and Applied Science Journal*, 9(6), 5876–5883.

5. Saleh, M. S., Li, J., Park, J., and Panat, R. (2018). 3D printed hierarchically-porous micro lattice electrode materials for exceptionally high specific capacity and areal capacity lithium ion batteries. *Additive Manufacturing*, 23, 70–78.

6. Zhang, S., Bean, T., and Edwards, D. B. (2010). Examination of different lattice structures in porous electrodes using a three-dimensional conductivity model. *Journal of Power Sources*, 195(3), 883–889.

7. Osiak, M., Geaney, H., Armstronga, E., and O'Dwyer, C. (2014). Structuring materials for lithium-ion batteries: advancements in nanomaterial structure, composition, and defined assembly on cell performance. *Journal of Materials Chemistry*, 2, 9433–9460.

8. Arunkumar, R., Anbumalar, S., Priya, F. A., and Poongothai, K. (2015). Computational fluid analysis of lithium-ion battery using ANSYS fluent. *International Journal of Scientific and Engineering Research*, 6(4), 74.

9. Maloney, K. J., Fink, K. D., Schaedler, T. A., Kolodziejska, J. A., Jacobsen, A. J., and Roper, C. S. (2011). Multifunctional heat exchangers derived from three-dimensional micro-lattice structures, HRL Laboratories LLC, 3011 Malibu Canyon Road, Malibu, CA 90265, United States. *International Journal of Heat and Mass Transfer*, 32(1), 239–248.

10. Xu, J., Chao, J., Li, T., Yan, T., Wu, S., Wu, M., Zhao, B., and Wang, R. (2020). Near-zero-energy smart battery thermal management enabled by sorption energy harvesting from air. *ACS Central Science*, 6(9), 1542-1554.

11. Grzegorz M. Kardasz (Richmond Hill), Barbara Anna Kardasz (Richmond Hill), Seungwoo Chu (Hwaseong-si Gyeonggi-do), David Mark Pascoe (Aurora), Jean-Yves St.Gelais (Stouffville), Soeren Striepe (Mt. Albert), Battery having internal electrolyte flow path and/or integral, heat sink, United States Patent Application Publication Publication number: 20120088140 Type: Application Filed: Oct 7, 2011 Publication Date: Apr 12, 2012 Inventors: v Application Number: 13/269, 029.

12. Fu, L. J., Liu, H., Li, C., Wu, Y. P., Rahm, E., Holze, R., and Wu, H. Q. (2006). Surface modifications of electrode materials for lithium ion batteries. *Solid State Sciences*, 8(2), 113–128. Liu, H., Wei, Z., He, W., and Zhao, J. (2017). Thermal issues about Li-ion batteries and recent progress in battery thermal management systems: A review. *Energy Conversion and Management*, 150, 304-330.

13. Lim, C., Yan, B., Yin, L., and Zhu, L. (2012). Simulation of diffusion-induced stress using reconstructed electrodes particle structures generated by micro/nano-CT. *Electrochimica Acta*, 75, 279–287.

14. Singh, M., Kaiser, J., and Hahn, H. (2015). Thick electrodes for high energy lithium ion batteries. *Journal of The Electrochemical Society*, 162(7), A1196.

15. Richard, M. N., and Dahn, J. R. (1997). Thermal stability of lithium ion battery electrode materials in organic electrolytes. *MRS Online Proceedings Library (OPL)*, 496, 445–456.

16. Zhang, M., Mei, H., Chang, P., and Cheng, L. (2020). 3D printing of structured electrodes for rechargeable batteries. *Journal of Materials Chemistry A*, 8(21), 10670–10694.

17. Zhao, R., Zhang, S., Liu, J., and Gu, J. (2015). A review of thermal performance improving methods of lithium ion battery: Electrode modification and thermal management system. *Journal of Power Sources*, 299, 557–577.

18. Kim, U. S., Shin, C. B., and Kim, C. S. (2008). Effect of electrode configuration on the thermal behavior of a lithium-polymer battery. *Journal of Power Sources*, 180(2), 909–916.

19. Mao-Sung Wua, K. H. Liub, Yung-Yun Wangb, Chi-Chao Wanb. (2002). Heat dissipation design for lithium-ion batteries. *Journal of Power Sources*, 109(1), 160–166.

6 Fractional derivative model of 0-3 viscoelastic composite for geometrically nonlinear vibration control of sandwich plate

Abhay Gupta, Rajidi Shashidhar Reddy, and Satyajit Panda[a]

Department of Mechanical Engineering, Indian Institute of Technology Guwahati, Guwahati, India

Abstract: In this article, passive damping performance of the sandwich plate with 0-3 viscoelastic composite (VEC) core is analyzed. The 0–3 VEC is composed by incorporating uniformly oriented graphite wafers in viscoelastic layer. This 0-3 VEC is modelled using fractional order derivative (FOD) viscoelastic constitutive relation in the time-domain. The overall sandwich plate is modelled in the finite element framework using layer wise first-order shear deformation theory considering von Kármán type geometric nonlinearity. Initially, the influence of geometric parameters such as the number of graphite wafers and viscoelastic thickness ratio of 0-3 VEC on nonlinear transient responses are analyzed. The demonstrated results refer that the enhancement in passive damping is highly dependent on these geometric parameters. Thus, the optimal geometric configuration of 0-3 VEC is decided based on the maximum performance index, and its effectiveness is compared with the conventional viscoelastic material (VEM) layer. The results reveal that the passive damping performance of the sandwich plate with 0-3 VEC core is higher compared to the conventional VEM core.

Keywords: Finite element method, fractional order derivative, nonlinear vibration control, viscoelastic composite.

Introduction

Viscoelastic materials (VEMs) consist of viscus and elastic properties and have magnificent energy absorption/dissipation capability under harmonic load. So VEMs are commonly used in the suppression of structural vibration. Besides, sandwich structures which consist of VEM, constrained between two stiff face layers, have widely used in many engineering applications such as aircraft, spacecraft, train, car, wind turbine blades, ship, etc. [1]. Due to its enormous properties like high strength to weight ratio and vibration suppression capability. The significant research work has been done to analyze the influence of damping on different sandwich structures with conventional VEM core [2–5]. Besides, viscoelastic composites (VECs) have also been utilized in structural applications [6–13].

However, these viscoelastic damping materials can be modelled in the frequency domain and time domain. In this queue, various mathematical methods such as Maxwell, Voigt, augmenting thermodynamic fields (ATF), anelastic displacement field (ADF),

[a]spanda@iitg.ac.in

Golla-Hughes-McTavish (GHM), fractional order derivative (FOD) method [14–18] have been introduced for the modelling of viscoelastic damping materials. Out of these models, FOD model is better for modelling VEMs in the time domain with less computational cost. Since, FOD model does not require additional dissipative coordinates, so the additional generalized degree of freedom of the structural system may not increase [17–20].

However, in the present work, the nonlinear transient analysis of the sandwich plate with 0-3 VEC core is done using FOD model. The 0-3 VEC layer consists of incorporating the graphite wafers within a conventional Butyl rubber VEM matrix in the form of a rectangular array [10]. The damping performance of the overall plate is dependent on different geometric parameters, such as the inclusion of graphite wafers and thickness of the VEM phase within 0-3 VEC [10]. So, presently the influence of these geometric parameters on the transient responses of the sandwich plate is analyzed under geometrically nonlinear vibrations. Further, at different geometric parameters of 0-3 VEC, passive damping of the sandwich plate is estimated through the performance index. And the optimal geometric configuration is decided based on the maximal performance index. Then its passive damping performance is compared with conventional VEM for controlling the nonlinear vibrations of the sandwich plate.

Finite Element Formulation of the Sandwich Plate

The diagrammatic representation of a sandwich plate with 0-3 VEC core is shown in Figure 6.1. The thickness of face layers and core of sandwich plate are designated as h_f and h_c, while the length, width, thickness of the overall plate are denoted by a, b and h (Figure 6.1). The core is made of 0-3 VEC and it consist by incorporating the graphite wafers within conventional VEM Butyl rubber matrix in a form of rectangular array [10]. Since the number of graphite wafers (n_g), the gap (Δ) between two consequent graphite wafers in x and y direction are considered as same, and top/bottom thickness (h_v) of VEM phase is also same within 0-3 VEC (Figure 6.1). The overall plate is modelled in finite element framework using layer wise approach. Since sandwich plate consist of five thin layers so the plane stress assumption is considered in the along the z-axis and the first order shear deformation theory is used in all the layers. Accordingly, for the kth layer, the displacements coordinates (u^k, v^k, w^k) at any point along x, y and z directions, can be exhibited as,

$$u^k = u_0 + z^k \phi_x,\ v^k = v_0 + z^k \phi_y,\ w^k = w_0 \tag{1}$$

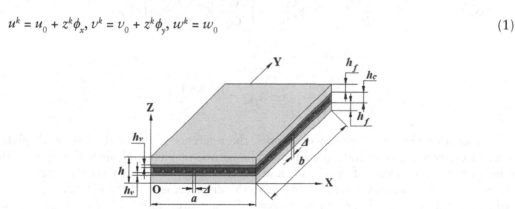

■ VEM phase ■ Graphite □ Face layer

Figure 6.1 Diagrammatic representation of a sandwich plate with 0-3 VEC core.

Since the von Kármán type geometric nonlinearity is considered in the present model. So based on plane stress assumption, for the kth layer, the state of strain/stress (ε/σ) at any point can be exhibited as,

$$\varepsilon_b^k = \{\varepsilon_x^k \quad \varepsilon_y^k \quad \varepsilon_{xy}^k\}^{\mathrm{T}}, \quad \varepsilon_s^k = \{\varepsilon_{xz}^k \quad \varepsilon_{yz}^k\}^{\mathrm{T}},$$

$$\sigma_b^k = \{\sigma_x^k \quad \sigma_y^k \quad \sigma_{xy}^k\}^{\mathrm{T}}, \tag{2}$$

$$\sigma_s^k = \{\sigma_{xz}^k \quad \sigma_{yz}^k\}^{\mathrm{T}}; \text{ or,}$$

$$\varepsilon_b^k = \left(\varepsilon_{bL} + \varepsilon_{bNL} + Z_b^k \kappa_b\right), \quad \varepsilon_s^k = \left(\varepsilon_{sL} + Z_s^k \kappa_s\right)$$

Since sandwich plate with 0-3 VEC core consists of five layers. The face layers $(k = 1, 5)$ are made of isotropic material while core $(k = 2, 3, 4)$ is made of 0-3 VEC. So the constitutive relation for the isotropic material is expressed as in Eq. (3). However, the VEM phase within 0-3 VEC is modelled using FOD viscoelastic constitutive relation. The detailed procedure for the modelling VEM phase using FOD constitutive relation are explained in [17–20]. Although, the final discretized forms of the constitutive relations of the VEM phase in time domain using FOD parameters $(E_0, E_\infty, t, \alpha)$ at any $(n + 1)^{th}$ time step is exhibited in Eq. (4).

$$\sigma_b^k = C_b^k \varepsilon_b^k, \quad \sigma_s^k = C_s^k \varepsilon_s^k,$$

$$C_b^k = \frac{E^k}{1 - \left(v^k\right)^2} \begin{bmatrix} 1 & v^k & 0 \\ v^k & 1 & 0 \\ 0 & 0 & \left(1 - v^k\right)/2 \end{bmatrix}, \tag{3}$$

$$C_s^k = G^k \begin{bmatrix} 1 & 0 \\ 0 & 1 \end{bmatrix}$$

$$\left(\sigma_b^k\right)_{n+1} = \left\langle 1 - c\left(\frac{E_\infty - E_0}{E_0}\right)\right\rangle C_b^k \left(\varepsilon_b^k\right)_{n+1} + \left(\frac{cE_\infty}{E_0}\right) C_b^k \sum_{j=1}^{N_t} A_{j+1}\left(\bar{\varepsilon}_b^k\right)_{n+1-j}$$

$$\left(\sigma_s^k\right)_{n+1} = \left\langle 1 - c\left(\frac{G_\infty - G_0}{G_0}\right)\right\rangle C_s^k \left(\varepsilon_s^k\right)_{n+1} + \left(\frac{cG_\infty}{G_0}\right) C_s^k \sum_{j=1}^{N_t} A_{j+1}\left(\bar{\varepsilon}_s^k\right)_{n+1-j} \tag{4}$$

$$\text{where,} \quad c = \frac{\tau^\alpha}{\tau^\alpha + \left(\Delta t\right)^\alpha}$$

$$A_{j+1} = \frac{\Gamma(j - \alpha)}{\Gamma(-\alpha)\Gamma(j+1)} \text{ or } A_{j+1} = \frac{j - \alpha - 1}{j} A_j$$

Further, in order to derive the FE model, the reference plane of the sandwich plate is discretized by using nine-node quadrilateral isoperimetric elements. Since the layers within sandwich plate are made of face layers and VEM or graphite so the typical element is considered with different stacking sequences. So the displacement vector (d) can expressed as Nd^e, where designates shape function vector. By substituting Eq. (2)–(4) in the potential and kinetic energy variation $(\delta T_P, \delta T_K)$ (Equations. (5) and (6)). Further implementing it

(Eq. (5) and (6)) in extended Hamilton's principle $\left(\int\limits_{t1}^{t2} \left(\delta T_K - \delta T_P \right) dt = 0 \right)$ and by assembling the elemental equations, the following global equation of motion of the overall plate at $(n + 1)^{th}$ time step can be obtained as in Eq. (7).

$$\delta T_p = \int\limits_0^a \int\limits_0^b \left[\sum_{k=1}^5 \int\limits_{h_k}^{h_{k+1}} \left\langle (\delta \varepsilon_b^k)^{\mathrm{T}} \sigma_b^k + (\delta \varepsilon_s^k)^{\mathrm{T}} \sigma_s^k \right\rangle dz - \left\langle \delta w_0 \, p(t) \right\rangle \right] dy dx \tag{5}$$

$$\delta T_K = \int\limits_0^a \int\limits_0^b \left[\sum_{k=1}^5 \int\limits_{h_k}^{h_{k+1}} \left\langle \{\delta(\dot{u}^k) \quad \delta(\dot{v}^k) \quad \delta(\dot{w}^k)\} \rho^k \{(\dot{u}^k) \quad (\dot{v}^k) \quad (\dot{w}^k)\}^{\mathrm{T}} \right\rangle dz \right] dy dx \tag{6}$$

$$M \ddot{d}_{n+1} + \left(K_L + K_{bNL}(d_{n+1}) \right) d_{n+1} = P_M p(t) - \left(P_{vL} + P_{vbN}(d_{n+1}) \right) \tag{7}$$

Since Eq. (7) is a nonlinear transient equation, it is solved using the Newmark-beta method and direct iteration procedures [18,21]. However, in the present analysis, FE code is written, and its solution is estimated using the MATLAB software package.

Results and Discussion

The nonlinear transient analysis of the sandwich plate with 0-3 VEC core is done in the present section. The dimensions of the overall plate are taken as $a = 0.4$ m, $b = 0.4$ m, $h_f = h_c = 2$ mm (Figure 6.1). The face layers are made of Aluminium ($E = 69$ GPa, $v = 0.3$, $\rho = 2740$ kg/m^3), while the material properties of graphite wafers within 0-3 VEC are taken as $E = 250$ GPa, $v = 0.3$, $\rho = 1400$ kg/m^3 [10,11]. For conventional Butyl rubber VEM phase, Poisson's ratio (v) and mass density (ρ) are taken as 0.49 and 920 kg/m^3, respectively, while, the FOD parameters are evaluated as $E_0 = 9.0483$ MPa, $E_\infty = 194.1$ MPa, $\tau = 5.55$ μs and $\alpha = 0.84$, using genetic algorithm (GA) at constant room temperature (32° C) and $500 \leq \omega \leq 4500$ rad/s frequency range [14]. However, the edges of sandwich plate are clamped, and uniformly distributed step load is applied in transverse direction at the bottom surface of the plate ($z = 0$) with intensity $p_0 = 200$ kN/m^2.

Since 0-3 VEC composed with conventional Butyl rubber VEM phase and the graphite wafers. The gap between two consequent graphite wafers (Δ) in any (x/y) direction is considered as 100 μm. However, VEM thickness ratio ($r_v = h_v/h_c$) and number of graphite wafers (n_g) in any (x/y) direction within 0-3 VEC layer are dependent on passive damping performance [10]. So, presently, the influence of these geometric parameters (n_g, r_v) on nonlinear transient responses of the sandwich plate are analysed. Moreover, the effect of 0-3 VEC core for the improved passive damping of the sandwich plate is examined and compared with the conventional VEM without incorporating graphite wafers within VEM phase. And this damping performance can be evaluated from performance index

$$I_d = \left[\left\langle (W_{max}/h)_{t=0} - (W_{max}/h)_{t=0.04 \, s} \right\rangle / (W_{max}/h)_{t=0} \right],$$

where, W_{max}/h denotes the maximum transverse displacement-amplitude of the nonlinear transient vibration of the sandwich plate.

Initially, the influence of the inclusion of graphite wafers on the maximum vibration amplitude (W_{max}/h) of the plate are studied. Figures 6.2 and 6.3 illustrate the nonlinear transient responses of the sandwich plate with 0-3 VEC core for different n_g at $r_v = 0.08$ and different r_v at $n_g = 4$, respectively, within 0-3 VEC core. It may be observed from Figure 6.2 and 6.3 that the attenuation in the maximum vibration amplitude (W_{max}/h) of sandwich plate varies with n_g and r_v of 0-3 VEC.

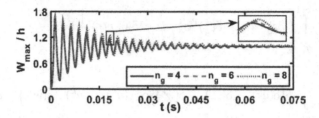

Figure 6.2 Transient response of sandwich plate with different n_g and at $r_v = 0.08$ within 0-3 VEC core ($p_0 = 200$ kN/m²).

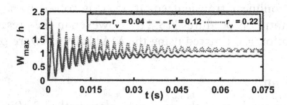

Figure 6.3 Transient response of sandwich plate with different r_v and n_g at $= 4$ within 0-3 VEC core ($p_0 = 200$ kN/m²).

It may be noticed from the results present in Figures 6.2 and 6.3 that the attenuation in the maximum vibration amplitude (W_{max}/h) is dependent on both and So in order to study the combined effect of these parameters (n_g, r_v) on damping performance of the sandwich plate explicitly, the performance index (I_d) is evaluated corresponding to different values of configured parameters (n_g, r_v) of 0-3 VEC within limits, $1 \leq n_g \leq 10$ and $0.02 \leq r_v \leq 0.22$. A 2-D grid is prepared within these limits and correspondingly performance index (I_d) are computed at every grid point. These results are presented through the contour as in Figure 6.4.

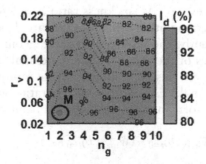

Figure 6.4 Variation of performance index (I_d) within two-dimensional domain of n_g and r_v in any x/y direction within 0-3 VEC core of sandwich plate ($p_0 = 200$ kN/m²).

Figure 6.4 illustrates that the performance index (I_d) of sandwich plate increases by incorporating less number of graphite wafers (n_g) and VEM thickness ratio (r_v). However, maximum performance index (I_d) of the sandwich plate with 0-3 VEC is obtained at n_g = 2, r_v = 0.04 (point M, Figure 6.4).

Further, the damping performance of the sandwich plate with 0-3 VEC core is compared with the conventional VEM without incorporating graphite wafers within the VEM phase. Figure 6.5 illustrate the nonlinear transient response of the sandwich plate with 0-3 VEC core corresponding to optimal geometric parameters (n_g = 2, r_v = 0.04). Similar results for the conventional VEM core are also evaluated and illustrated in the same figure (Figure 6.5). Figure 6.5 illustrates that the attenuation in maximum vibration amplitude (W_{max}/h) of the sandwich plate is significantly more for 0-3 VEC core compared to conventional VEM core. However, the performance index (I_d) of the sandwich plate with 0-3 VEC core is estimated as 97.38 while 60.65 for conventional VEM core. These results indicate that the 0-3 VEC core may be beneficial compared to the conventional VEM core to control the geometrically nonlinear vibrations of the sandwich plate.

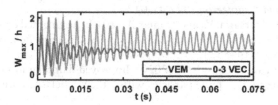

Figure 6.5 Transient responses of the sandwich plate with 0-3 VEC/VEM core (p_0 = 200 kN/m^2).

Conclusions

In this work, the passive control of nonlinear transient responses of sandwich plate with 0-3 VEC core is done. The sandwich plate is modelled using layer wise FSDT in the FE framework. However, the VEM phase within 0-3 VEC is modelled using FOD viscoelastic constitutive relation. Initially, the influence of the number of graphite wafers and VEM thickness ratio on nonlinear transient responses of the sandwich plate is evaluated. It is observed that the attenuation in the vibration amplitude of the sandwich plate depends on these geometric parameters of 0-3 VEC. So, for maximum improvement in damping, the optimal geometric parameters of 0-3 VEC are obtained based on the maximal performance index of the overall plate. It is found that the performance index and the attenuation in vibration amplitude of the sandwich plate is more for 0-3 VEC core at optimal geometric parameters compared to conventional VEM core (without incorporation of graphite wafers within VEM phase). And these results reveal that the 0-3 VEC may be beneficial compared to conventional VEM for controlling the geometrically nonlinear vibrations of the sandwich plate.

References

1. Birman, V., and Kardomateas, G. A. (2018). Review of current trends in research and applications of sandwich structures. *Composites Part B: Engineering, 142*, 221–240.
2. Yang, C., Jin, G., Liu, Z., Wang, X., and Miao, X. (2015). Vibration and damping analysis of thick sandwich cylindrical shells with a viscoelastic core under arbitrary boundary conditions. *International Journal of Mechanical Sciences, 92*, 162–177.

3. Gupta, A., Panda, S., and Reddy, R. S. (2021). Passive control of parametric instability of layered beams using graphite particle-filled viscoelastic damping layers. *Mechanics of Advanced Materials and Structures,* 13, 1–6.

4. Madeira, J. F. A., Araújo, A. L., Soares, C. M., and Ferreira, A. J. M. (2015). Multiobjective design of viscoelastic laminated composite sandwich panels. *Composites Part B: Engineering,* 77, 391–401.

5. Khalfi, B., and Ross, A. (2016). Transient and harmonic response of a sandwich with partial constrained layer damping: A parametric study. *Composites Part B: Engineering,* 91, 44–55.

6. Huang, C. Y., and Tsai, J. L. (2015). Characterizing vibration damping response of composite laminates containing silica nanoparticles and rubber particles. *Journal of Composite Materials,* 49(5), 545–557.

7. Rajoria, H., and Jalili, N. (2005). Passive vibration damping enhancement using carbon nanotube-epoxy reinforced composites. *Composites Science and Technology,* 65(14), 2079–2093.

8. Gupta, A., Reddy, R. S., Panda, S., and Kumar, N. (2020). Damping treatment of beam with unconstrained/constrained 1–3 smart viscoelastic composite layer. *Materials Today: Proceedings,* 26, 956–962.

9. Gupta, A., Rajidi, S. R., and Panda, S. (2020). Design of a 1–3 smart viscoelastic composite layer for augmented constrained layer damping treatment of plates. *IOP Conference Series: Materials Science and Engineering,* 872(1), 012067.

10. Kumar, A., Panda, S., Narsaria, V., and Kumar, A. (2018). Augmented constrained layer damping in plates through the optimal design of a 0-3 viscoelastic composite layer. *Journal of Vibration and Control,* 24(23), 5514–5524.

11. Gupta, A., Panda, S., and Reddy, R. S. (2020). Improved damping in sandwich beams through the inclusion of dispersed graphite particles within the viscoelastic core. *Composite Structures,* 247, 112–424.

12. Gupta, A., Panda, S., and Reddy, R. S. (2020). An actively constrained viscoelastic layer with the inclusion of dispersed graphite particles for control of plate vibration. *Journal of Vibration and Control,* 1077546320956533.

13. Gupta, A., and Panda, S. (2021). Hybrid damping treatment of a layered beam using a particle-filled viscoelastic composite layer. *Composite Structures,* 113–623.

14. Jones, D. I. G. (2001). Handbook of Viscoelastic Vibration Damping. John Wiley and Sons.

15. Lesieutre, G. A., and Bianchini, E. (1995). Time domain modeling of linear viscoelasticity using anelastic displacement fields. *Journal of Vibration and Acoustics,* 117(4), 424–430.

16. Golla, D. F., and Hughes, P. C. (1985). Dynamics of viscoelastic structures—a time-domain, finite element formulation. *Journal of Applied Mechanics,* 52(4), 897–906.

17. Bagley, R. L., and Torvik, J. (1983). Fractional calculus-a different approach to the analysis of viscoelastically damped structures. *AIAA Journal,* 21(5), 741–748.

18. Datta, P., and Ray, M. C. (2015). Fractional order derivative model of viscoelastic layer for active damping of geometrically nonlinear vibrations of smart composite plates. *Computers, Materials and Continua,* 49(1), 47–80.

19. Schmidt, A., and Gaul, L. (2002). Finite element formulation of viscoelastic constitutive equations using fractional time derivatives. *Nonlinear Dynamics,* 29(1–4), 37–55.

20. Galucio, A. C., Deü, J. F., and Ohayon, R. (2004). Finite element formulation of viscoelastic sandwich beams using fractional derivative operators. *Computational Mechanics,* 33(4), 282–291.

21. Reddy, J. N. (2004). Nonlinear Finite Element Analysis. Oxford University Press.

7 Experimental analysis of Open Well Monoblock centrifugal pump with impeller sizing

Maitrik Shah[a], Beena Baloni[b], and Salim Channiwala

Sardar Vallabhbhai National Institute of Technology, Surat, India

Abstract: The centrifugal pump is a widely used turbomachine for industrial, agricultural, and domestic applications. The energy efficient pump consumes less electricity and delivers efficient performance at duty point operations. The operating points of the pump may vary based on field application. The impeller sizing technique is used for appropriate selection of pump with desired operational points. The Monoblock centrifugal pump is preferred for low power transmission losses due to the absence of couple joint. The Open Well Monoblock centrifugal pump is used in fire prevention, pumping, and processing of fluids in agriculture, hospitality, processing, HVAC applications, and many more. The present study focuses on experimental investigation for enhancement in performance of an Open Well Monoblock centrifugal pump with impeller sizing technique. The testing is carried out as per IS 9137. Repeatability as well as the uncertainty of experimental data is checked and found in an acceptable range. The analysis is carried out with different impeller outer diameter i.e., 174 mm (design) to 135 mm in the interval of 5 mm. The results are compared based on obtained efficiency and optimum impeller outer diameter is suggested for the present pump.

Keywords: Energy efficient, experimental analysis, impeller sizing, IS 9137, Open Well Monoblock centrifugal pump, performance test.

Introduction

A turbomachine called 'centrifugal pump' converts mechanical energy into pressure energy. As per statistical data, energy consumption of total pumps are around 20% of total power consumption in the world [1]. The installed and operated number of centrifugal pumps account for 70–80% of the total number of pumps. The flow field of a centrifugal pump, which is influenced by system revolution and curvature, is exceedingly turbulent and unstable, as well as quite complicated due to separation and recirculation. Further selection of improved centrifugal pump behavior, at design and off-design working conditions, is going to be extremely difficult with traditional empirical law of impeller sizing and numerical analysis. It depends on assembly of parts, vibration, noise, testing condition and instrument used for performance measurement. It is very inconvenient, time consuming and extravagant to design and develop the site-specific centrifugal pumps corresponding to local site

---

[a]maitrikshah2006@gmail.com, [b]pbr@med.svnit.ac.in

conditions. Researchers performed experiment analysis of original impeller to trimmed impeller up to 30% reduction in size in gradual increment [2–4].

The pump industries adopt the practice of trimmed impeller with a specific volute of pump assembly to enlarge the operational range, which helps to prevent additional development cost. While the trimming of the impeller increases the annular cavity space in the centrifugal pump assembly. This affects the modification of fluid flow velocity, pressure distribution, radial forces across the volute, and efficiency for different diameters of the impeller [2]. In the centrifugal pump, the perturbations are minimum for flow rate around the best efficiency point and increases for both low and high flow rates [5, 6]. The fluid dynamic excitation, at the impeller blade passing frequency, mainly depends on the radial cavity space between the impeller and the volute tongue [7].

Arrio et al. [2], studied total fluid dynamic load induced in a centrifugal pump with a single stage specific volute with four different impellers outer diameter by progressive trimming of preliminary geometry of impeller. The researcher suggested 5–10% of the radial gap with respect to the impeller radius for a volute pump. Rikke et al. [8] claimed that, despite the decrement in flow rate to 25% of design load, the significance difference was reveled between the two adjacent impeller passages. One passage was dominated by high velocities along the blade pressure side, and the other passage exhibits a high separated flow field.

The fluid leaving to the impeller and entering the volute is very turbulent and resulting in high hydraulic losses while passing through rotating curvature flow passage to stationary. The sizing of the impeller provides annular space between the impeller and the volute, facilitate further space to the fluid for, streamline and smoothen the flow. Hence, the efficiency of the pump with enlarge radial gap between the impeller to volute may find to be better than the pump efficiency. In view of this, within the present study the performance of the centrifugal pump is studied with the original impeller (Ø 174 mm) to trimmed impeller (Ø 135 mm) size with reduction in outer diameter in the interval of 5 mm.

Investigated Pump

The investigated pump in this study is an Open Well Monoblock centrifugal pump. The less mechanical losses due to the absence of bearing, sealing, and coupling parts in Open Well Monoblock centrifugal pump is the main advantage of Open Well Monoblock centrifugal pump. These pumps are easy to mount into a sump, water tank in the underground or overhead. These pumps can be used for daily fluid pumping application as well as emergency application like fire prevention too. It is economically acceptable pump in pumping industries. The parameters of pump are as per Table 7.1. The 3D CAD model of Open Well Monoblock centrifugal pump used in fire prevention application and original impeller with outer diameter of 174 mm and smallest impeller after trimming, i.e., 135 mm are shown in Figure 7.1.

The operational running of the centrifugal pump at duty points is desired for efficient performance. These machines are installed for various applications in agriculture, hospitality, HVAC, processing industries and many more places having different operating heads and flow rate. The lesser head or discharge requirement can be fulfilled by impeller trimming of replacing the prevailing impeller with a smaller one, this might be the cost-effective solution and help to cover the required operating range of centrifugal pump for a specific site [9].

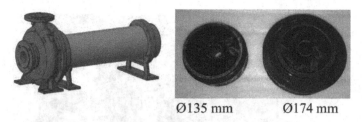

Ø135 mm   Ø174 mm

Figure 7.1 3D CAD geometry of openwell monoblock centrifugal pump, Trimmed OD (Ø135 mm) and full OD (Ø174 mm) of impeller.

Table 7.1: Specification of centrifugal pump.

Parameter	Value
Pump inlet diameter (D_i), mm	100
Pump outlet diameter (D_o), mm	90
Impeller outer diameter (D_2), mm	174
Number of blades	6
Impeller outlet width (b_2), mm	19
Rated speed (n_d), rpm	2980
Rated flow (Q_n), m³/hr	110
Rated head (H_n), meter	32
Specific speed (n_s)	39

Experimental Setup

An open loop test setup used to evaluate the performance of developed centrifugal pump with total loss measurements with modified and non-modified design stages and measured efficiency. Accurate instruments and closed loop control systems are used to capture minor performance improvements resulting from design modifications. Friction losses due to surface roughness, bend in piping and head losses by obstacle due to measuring and flow control instruments in systems are measured. The pictorial view of the test set up is shown in Figure 7.2. The main components in the centrifugal pump test setup comprise a test pump mounted on a submersible electrical motor and discharge piping system with a flow control ball valve. In addition, sophisticated measuring instruments like pressure transducer, electromagnetic flow meter are installed in the discharge pipe. The test setup is equipped with an energy meter for measuring motor power inputs, current and supply voltage.

A submersible electric motor of 20 hp of capacity with 2980 rpm is used for pump shaft prime mover. The three-phase, 415 V connection is used to run the motor. The seamless pipe with mild steel material is installed for the discharge line of the pumping system. The 150# class of flange connection is used for joints at the inline of the piping system. Pumping fluid water is contained in a sump having a larger capacity. Ball valve type inline flow control valve with flange ended is used in the discharge line for flow control to determine H v/s Q performance of the test pump.

The developed total head across the centrifugal pump is measured by pressure transducers at discharge conditions of the pump according to IS 9137 standards. Elevation

Figure 7.2 Experiment test setup.

difference between pump inlet centerline and discharge pressure transducers is added in the total calculated head. Fluid flow measurement in test setup discharge pipeline is performed by an electromagnetic flow meter and remote transmitter type display used for visually measurement of readings. The slip coil based principal use for rpm measurement. The output feedback signals are 4–20 mA for all measuring instruments and sent to PLC as an analog input signal. The hydro test is performed for the centrifugal pump assembly and confirmed leakage free joints.

Experimental Procedure

Before starting the pump testing, care should be taken for leakage less flange connection and proper fittings of electric connection. Initially, the discharge line control valve should fully close to operate the pump at shut-off condition, that will help to forestall power overloads [10]. The speed of the impeller gradually reached to rated speed and maintained it during pump start. The highest radial force was observed during the pump starting condition with zero flow rate. The radial force may lead to fatigue failure of the rotating shaft or wear of the sealing ring [11, 12].

The power supply voltage should be set as per electrical standards. Desired speed of centrifugal pump achieved by VFD system. Loading on the pump is done by opening of the discharge line flow control valve gradually at the desired head level and by maintaining the speed of the pump. Once the steady state conditions are reached, the performance parameters measured and stored into the computer system. These practices continuing for shut off head to minimum head and maximum flow. The best efficiency point would be a claim based on measured highest pump power output at the lowest pump power input. The test repeated with trimmed sizes of impeller.

Experimental Results

The permissible systematic error and uncertainty of used instruments are under permissible limits as per IS 9137 [13]. The relative uncertainty in measured head, discharge, motor input power and overall error in pump efficiency were 0.95%, 0.89%, 0.92% and 2.1% respectively. The repeatability of results is checked and found in an acceptable range of 95% confidence limit.

The performance of the centrifugal pump was calculated by measuring the total head, discharge flow rate and shaft input power. The impeller trimming effect at pump performance was analyzed with nine different impellers with outer diameter of 174 mm (original size), 170 mm to 135 mm with an interval of 5 mm. The performance curve of dimensional parameters H-Q, η-Q, P_{in}-Q measured under a steady state condition are shown in Figures 7.3–7.5. These diagrams are plot for PL (part load), BEP (best efficiency point) and OL (overload) working condition to represent the effect of impeller trimming on total head, shaft input power and pump efficiency at different flow rate for original and trimmed impeller sizes. The result claims that the best efficiency point is shifted from the design duty point in the experiment. The efficiency of the submersible motor is almost constant to 76% and the input electric power is measured by an energy meter.

Figures 7.3–7.5 indicate that trimming of the impeller result in a decrease in the value of discharge, head, and shaft input power. The rate of decrement is increased with increase in impeller trimming. The performance parameters of trimmed impeller are matched with theoretical relation $\emptyset D \propto Q$, $\emptyset D \propto H^{0.5}$, $\emptyset D \propto P^{0.33}$. The highest efficiency is achieved at nominal impeller size of $\emptyset160$ mm with optimum BEP (best efficiency point) discharge due to flow stabilization, reduce eddies and flow separation at increased radial space by impeller trimming. The raise in efficiency, fall in total head and reduce in power inputs observed 2.7%, 29%, and 24% at BEP (best efficiency point) for 9% trimmed OD ($\emptyset160$ mm) to the full OD ($\emptyset174$ mm) of the impeller. On the other hand, more trimming of the impeller lowers the flow rate, backshift the BEP (best efficiency point) towards lower discharge, decrease in pump efficiency due to increase of secondary flow losses at increased annular space developed by trimmed impeller.

The change in impeller diameter leads to change in impeller geometrical parameters, outlet width, blade outlet angle, and blade wrap angle. These parameters and increased annular space, due to trimming of the impeller, impact more on the performance of the centrifugal pump.

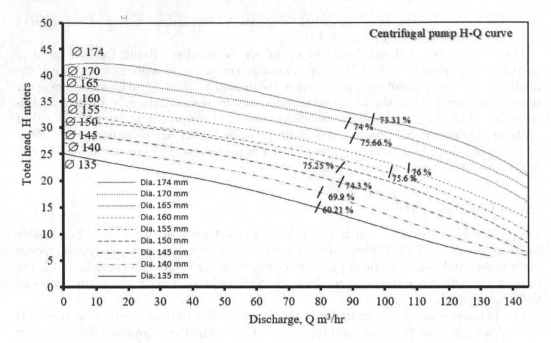

Figure 7.3 Total head v/s discharge curve for different OD of impeller.

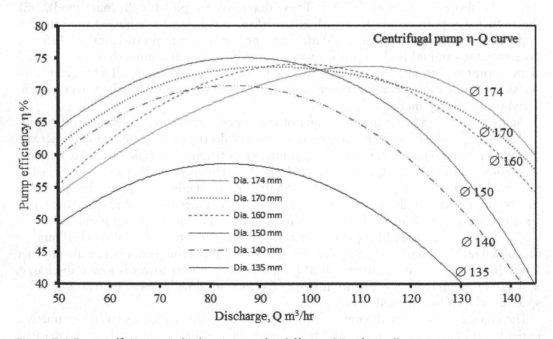

Figure 7.4 Pump efficiency v/s discharge curve for different OD of impeller.

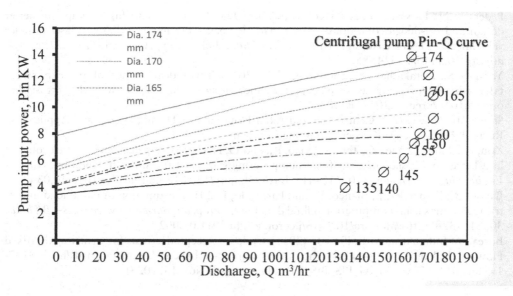

Figure 7.5 Pump input power v/s discharge curve for different OD of impeller.

Conclusion

Effect of impeller sizing on performance of Open Well Monoblock centrifugal pump is analyzed. The experimental results suggest that for present pump the highest efficiency of 76% is achieved with 9% reduce trimmed size of impeller i.e. 160 mm than original one.

References

1. Shankar V. K. A., Umashankar, S., Paramasivam, S., and Hanigovszki, N. (2016). A comprehensive review on energy efficiency enhancement initiatives in centrifugal pumping system. *Applied Energy*, 181, 495–513. https://doi.org/10.1016/j.apenergy.2016.08.070.
2. Blanco, E., and Parrondo, J. (2008). The effect of impeller cutback on the fluid-dynamic pulsations and load at the blade-passing. *Journal of Fluids Engineering*, 130, 1–11. https://doi.org/10.1115/1.2969273.
3. Tan, X., and Engeda, A. (2016). Performance of centrifugal pumps running in reverse as turbine: part II—systematic specific speed and specific diameter based performance prediction. *Renew Energy*, 99, 188–197. https://doi.org/10.1016/j.renene.2016.06.052.
4. Jain, S. V., Swarnkar, A., Motwani, K. H., and Patel, R. N. (2015). Effects of impeller diameter and rotational speed on performance of pump running in turbine mode. *Energy Conversion and Management*, 89, 808–824. https://doi.org/10.1016/j.enconman.2014.10.036.
5. Kaupert, K. A., and Staubli, T. (1999). The unsteady pressure field in a high specific speed centrifugal pump impeller—part I: influence of the volute. *ASME Journal of Fluids Engineering*, 121, 621–626.
6. Parrondo-gayo, J. L., Gonzalez-Perez, J., and Fernandez-Francos, J. (2002). The effect of the operating point on the pressure fluctuations at the blade passage frequency in the volute of a centrifugal pump. *Transactions ASME*, 124, 784–790. https://doi.org/10.1115/1.1493814.
7. Spence, R., and Amaral-Teixeira, J. (2008). Investigation into pressure pulsations in a centrifugal pump using numerical methods supported by industrial tests. *Computers and Fluids*, 37, 690–704. https://doi.org/10.1016/j.compfluid.2007.10.001.

8. Pedersen, N., Larsen, P. S., and Jacobsen, C. B. (2003). Flow in a centrifugal pump impeller at design and off-design conditions—part I: particle image velocimetry (PIV) and laser doppler velocimetry (LDV) measurements. *Journal of Fluids Engineering ASME*, 125, 61–72. https://doi.org/10.1115/1.1524585.

9. Yang, S. S., Derakhshan, S., and Kong, F. Y. (2012). Theoretical, numerical and experimental prediction of pump as turbine performance. *Renew Energy*, 48, 507–513. https://doi.org/10.1016/j.renene.2012.06.002.

10. Stepanoff, A. J. (1993). Centrifugal and Axial Flow Pumps: Theory, Design and Application, Revised. Florida, USA: Krieger Publishing Company.

11. Zou, Z., Wang, F., Yao, Z., Tao, R., Xiao, R., and Li, H. (2016). Impeller radial force evolution in a large double-suction centrifugal pump during startup at the shut-off condition. *Nuclear Engineering and Design*, 310, 410–417. https://doi.org/10.1016/j.nucengdes.2016.10.034.

12. Barrio, R., Fernández, J., Blanco, E., and Parrondo, J. (2011). Estimation of radial load in centrifugal pumps using computational fluid dynamics. *European Journal of Mechanics—B/Fluids*, 30, 316–324. https://doi.org/10.1016/j.euromechflu.2011.01.002

13. Bureau of Indian Standards, IS 9137 (2011). Code for Acceptance Tests for Centrifugal, Mixed Flow and Axial Pumps—Class. https://infostore.saiglobal.com/en-us/standards/bis-is-9137-1978-r2011-179827_SAIG_BIS_BIS_434900/ (accessed Nov. 13, 2023).

8   Finite element analysis of mechanical and thermal properties of graphene/polyvinylidene fluoride (PVDF) composite

S. K. Pradhan[1,a], Amit Kumar[2,b], Vedant Kedia[1,c], and Paramjit Kour[3,d]

[1]Department of Mechanical Engineering, Birla Institute of Technology, Mesra, Patna Campus, India

[2]Department of Mechanical Engineering., National Institute of Technology Patna, India

[3]Department of Physics, Birla Institute of Technology, Mesra, Patna Campus, India

Abstract: Defence-related equipment requires certain components which are sensitive to mechanical and thermal parameters. Polyvinylidene fluoride (PVDF) is a semi-crystalline thermoplastic polymer of fluorine. It is known for its piezoelectric properties as well as mechanical strength and has significant applications in various industries like electronics, automobiles, petrochemical, military aircraft, etc. Similarly, graphene has its own fascinating electrical and mechanical characteristics. A composite film made of these two materials in different proportions can be geometrized and simulated for analysis purpose. Recent developments have shown quite significant results; however, considering the two most extreme boundary conditions of high and low temperature; the behavior of mechanical properties in that condition becomes extremely vulnerable. Low-temperature loading finds its use in piezoelectric sensors for aircraft whereas high-temperature loading is the need of the hour in extreme weather conditions for equipment. This paper focuses on finite element analysis (FEA) of this composite material along with its simulation and comparison with its individual elemental properties and brings in significant conclusive applications for its use in the defense sector.

Keywords: FEA, graphene, piezoelectric, polyvinylidene fluoride (PVDF), simulation.

Introduction

Defense-related industries have grown over the years on a wide scale across the globe for security reasons in all prime aspects i.e. land, air and water. Different types of equipment are being designed, manufactured and tested to prove its efficacy and utilization holistically. The prime focus is to reduce the weight and size of the equipment, to increase the output efficiency of the instruments, the multidimensional approach of its use in a unidirectional line of attack. Polymer composites have played a crucial role in fulfilling the main objectives. However, inducing pure polymer composites doesn't fetch promising results. Hence, nanoparticles have come into effect since their discovery. It has not only increased the strength of the material in every aspect but also contributed to reducing the size of the equipment to micro-levels. Polymer nanocomposites have been designed for various defense applications like high-performance fiber/fabrics, ballistic protection, microwave absorbers, refractive index tuning, solid lubricants, porous nanocomposites in improving

[a]sudiptabitpatna@gmail.com, [b]amit@nitp.ac.in, [c]vedant25019@gmail.com, [d]paramjit.kour@bitmesra.ac.in

foaming properties, electrostatic charge dissipation in a space environment, ultraviolet irradiation resistance, fire retardation, corrosion protection, diffusion barriers and most importantly sensors and actuators [1].

Being a semi-crystalline thermoplastic polymer, PVDF (polyvinyl diene fluoride) is known for its mechanical elasticity, low acoustic resistance, low dielectric constant, high piezoelectric, pyroelectric coefficients, high thermal stability, high resistance to halogen and acids [2]. It is one of the most utilized polymers in the industry and finds its quantum applications in manufacturing piezoelectric sensors, ultrasound transducers, etc. This copolymer has four crystalline forms i.e. I and II, III and IIp or β, α, γ and α p respectively among which the most common form is I and II [3]. Similarly, Graphene is one of the most exceptional researched Nano fillers known for its unparalleled electrical conductivity. This nanocomposite produces multilayer platelets rather than single-layer sheets [4]. This single layer graphite is also known for its excellent physical, optical, mechanical, thermal characteristics [5]. Graphene is highly transparent and has high degree of flexibility. It also has various forms like GO (graphene oxide), GNPs(graphene Nanoplatelets), GNRs (graphene nanoribbons), rGO (reduced graphene oxide), GQDs (graphene quantum dots), etc.

By reviewing various articles, it has been observed that certain experiments that were conducted using PVDF and graphene as composite were mainly done to extrapolate the piezoelectric properties, enhancement of thermal stability, etc. The mechanical properties were also observed but that was specific to pure PVDF materials at low temperatures [6]. Various experiments and modelling were being prepared in various articles for calculation, however, designing of a composite layer using Ansys was not performed to have a theoretical check-in error percentage between the experimental data and the former one. Most of the articles derived the application of this composite in the field of developing sensors and electronic devices [7]. However, their data shows a possibility of using such composite in the defense sector as the results for thermal properties were quite promising. The mechanical properties obtained at low temperatures for pure PVDF have shown significant enhancement apart from a few shortcomings like cavitation during deformation, high strain rate etc [8].

Focusing on enhancement of mechanical properties of the material and their behavior in adverse condition, this paper aims to explore the finite element analysis of the fabricated composite layer of a polymer and nanoparticle and driving in simulation to get a brief knowledge about the theoretical improvement of the properties under ideal conditions and derive its application in the defense sector for its adverse use for a long period. The comparison would also be brought in with the pure elemental state to showcase the enhanced properties of the composite.

Material and Methods

The prime focus is to exercise the finite element analysis model using the ANSYS software where simulation work will be carried out. The materials used were PVDF and graphene. The sandwich layer composite was created. The parameter input was given to carry out the operations. Table 8.1 shows property of materials parameters like density, melting point, tensile yield and ultimate strength, Elastic modulus, young's modulus are the crucial ones mentioned for stimulating results and understanding the change in mechanical properties. The following result obtained from the simulation was validated with few research-oriented articles for justification. The geometry was being kept in adverse environmental conditions and based on that forces were applied. The complete methodology to perform the process is mentioned below:-

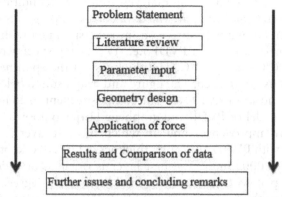

The table, shown below, states the property of the material being used for the analysis of stress condition on a thin surface with a composite layer.

Table 8.1: Property of materials [9–13].

Parameters	PVDF	Graphene
Density (g/cc)	1.78	2.267
Melting Temp(°C)	167.5	3652
Young's modulus (MPa)	2450	10E+6
Poisson's ratio	0.34	0.456
Yield tensile strength (Mpa)	0	97.5E+3
Ultimate Tensile Strength (MPa)	53.5	130E+3
Isotropic thermal Conductivity(W/mK)	0.2	4000

The dimension of the specimen surface was taken to be 50 × 50 mm. The meshing element size was 1 mm with a target quality of 0.95. The core material was assigned as PVDF. The skewness was nearly zero for 90% of the elements. The number of nodes was 2601 and elements standing at 2500. Figure 8.1 shows specimen meshing, graph showing skewness of elements (Composite 1).

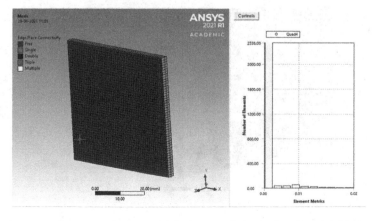

Figure 8.1 Specimen meshing, graph showing skewness of elements (Composite 1).

Results and Discussion

The analysis was done to explore the material in adverse condition when it is in a pure state or a composite state and compare the data corresponding to the results obtained. The force was applied in a downward direction into the specimen with a magnitude of 1000 N. Because of the melting point of PVDF i.e., 177°C, the specimen was kept within two temperatures i.e.,100°C and –100°C. The thickness of the specimen for pure PVDF and graphene was taken as 2 mm each. The figure and graph stated below address the issue of deformation, stress and strain at 100°C. The composite mentioned in the figure demarcates the sandwich layer model of PVDF and graphene [14]. Composite 1 depicts a three-layer fabric in a top-down approach manner in which the first layer is of PVDF followed by graphene and lastly with PVDF. The fabric thickness of PVDF is kept at 0.005 mm whereas graphene is kept at 0.00025mm. The first layer is oriented at 0 degree followed by the second layer piled up at 45 degree and the final layer at 90 degree. Composite 2 is a five-layer fabric, similar to the former, starting with PVDF. The orientation of material for the respective layer from top to bottom is aligned at 0-45-90-30-60 degrees. The fixed support was assigned to the four edges of the surface.

The red color Figures 8.2 and 8.3 depicts the highest value whereas the blue color depicts the least value. Figure 8.4 shows force applied in (composite 1) in downward direction. Figure 8.5 shows Thermal condition applied to the specimen (composite 1 form at –100°C).

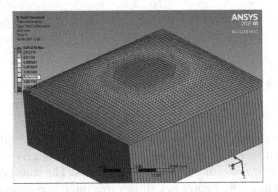

Figure 8.2 Deformation concentration (Composite 1).

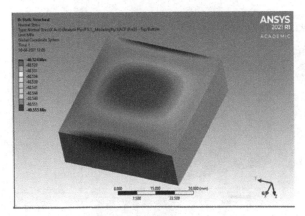

Figure 8.3 Normal Stress Concentration (Composite 1).

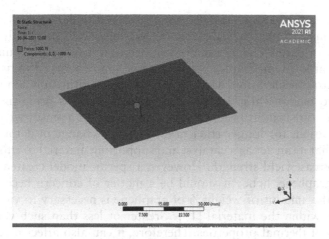

Figure 8.4 Force applied in (composite 1) in downward direction.

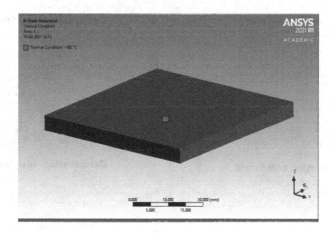

Figure 8.5 Thermal condition applied to the specimen (composite 1 form at –100 deg. Cel).

Deformation and Strain

Deformation and strain are some of the most important parameters when it comes to analyze the mechanical properties of the materials at adverse conditions. Figure 8.6 shows the graph showing deformation, normal and shear strain at two different temperatures i.e., 100 and –100°C with three different specimens one with pure PVDF and graphene and others with composite fabric in sandwich form. Figure 8.7 shows the graph showing normal and shear stress at two different temperatures i.e., 100 and –100°C with three different specimens one with pure PVDF and graphene and others with composite fabric in sandwich form The graph shown below is plotted based on data received from Ansys after simulation. It is quite evident from the fact that deformation keeps on decreasing with an increase in layers. The change in deformation is negligible to the change in temperature for composite material. The graph for pure materials shows a change in value. The value for deformation is higher at low temperature than at high temperature. However, the value of deformation for graphene material shows a minimal change. Various articles also suggest a similar trend of decreasing deformation with an increment of layer and become asymptomatic thereafter [15]. It has also been observed under various experiments that during deformation, cavitation is visible, but this was observed specifically with pure PVDF [16]. The porosity of pure PVDF is a major concern as it was observed under experiments from articles that at low-temperature porosity is maximum across a section of area for the specimen taken into consideration, whereas at room temperature it is converged only at the center. Lower deformation values at higher temperature seek a promising outcome of its use in high-temperature conditions in various equipment like a sensor which can withstand long adverse conditions for longer efficiency. Negligible change in deformation values of graphene shows that it can act as a catalyst and supporting material to enhance mechanical toughness and increase yield strength. However, a specific model created doesn't mention the weight% of graphene, hence it should be a matter of concern because some articles show that a specific amount of weight% of graphene is necessary to yield better thermal related properties within the material [17]. Greater or less than such values doesn't give significant results of thermal properties. Therefore, it can also affect mechanical strength in some way or the other.

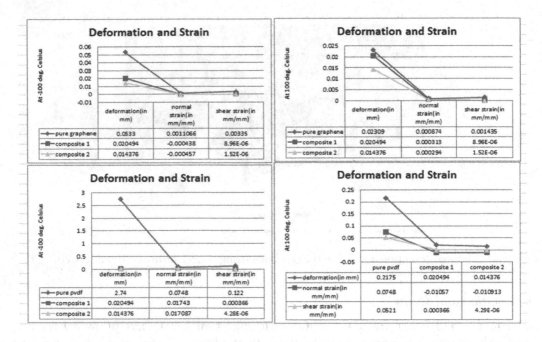

Figure 8.6 The graph showing deformation, normal and shear strain at two different temperatures i.e., 100 and −100 deg. Celsius with three different specimens one with pure PVDF and graphene and others with composite fabric in sandwich form.

Normal and Shear Strain
Normal and Shear strains are also an important parameter for analyzing mechanical properties. Normal strain is generally defined as a change in deformation in one of the sides to change in length to that side whereas a shear strain is generally defined as deformation in the direction parallel to the shear force over the actual length in the direction perpendicular to it. The graph shown above also depicts the values for these two parameters for different materials. Applying force vertically downwards, it is, however, predicted that shear values would be negligible and hence it is quite evident from the values obtained after simulation.

Normal strain is, however, more than shear strain irrespective of materials. Negative values are observed at −100°C for composite material showing compression within the material after application of force and a similar observation was seen at 100°C indicating that materials start compressing at adverse temperature condition. However, the values are so minimal compared to pure materials that it gives an assurance of high mechanical strength at both high and low temperatures. Various articles working on PVDF to characterize its mechanical strength have also shown similar trends. Due to high porosity in the material, it was observed that fracture occurred at low temperature before necking [18]. It was also observed from an article that a parameter named brittle-ductile area ratio was plotted against inverse strain rate [16]. The plot depicts the lowering of values as the temperature increases and with an increased inverse strain rate. However, the quantity of graphene is a cause of concern as it has clearly shown uncertainty in mechanical properties at different weight% [19].

Normal and Shear Stress

Graphene is a very strong nanoparticle. Its yield strength is very high. It is also evident from the graph that pure graphene exhibits normal stress at multiple times higher than pure PVDF. However, keeping the temperature constant, the value of stress decreases as the composite layer was stimulated. At low temperature, pure graphene is much higher than the value obtained at high temperature but for composite, the value is negative hence showing that another material in the composite is showing compressional phenomenon. Pure PVDF stress values are comprehensively low which shows that the material doesn't sustain at adverse temperatures compared to pure graphene. For validation, certain articles have also stated that PVDF with graphene coating decreases stress concentration and

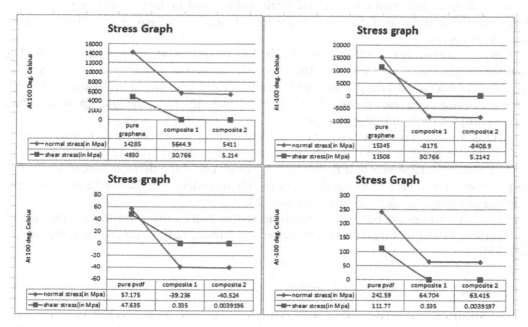

Figure 8.7 The graph showing normal and shear stress at two different temperatures i.e., 100 and −100°C with three different specimens one with pure PVDF and graphene and others with composite fabric in sandwich form.

hence mechanical strength is improved [15]. In certain cases, negative stress is also visible leading to distress in the material, however, deformation is low and the material is not subjected to distortion.

This study helps in analyzing how composite can be a key to various applications due to stress-strain concentration and deformation characteristics [20]. The results obtained are quite promising. This analysis simply focuses on the mechanical strength of the composite under extreme thermal conditions. However, various research articles also extrapolate data that this composite has also shown quite significant electrical properties making its inroads for use in nano generators [21]. The optical properties of graphene are also quite interesting parameters hence we can look for more scope into this field for its use [22]. Certain issues, which need to be addressed, is the quantity of graphene added in the PVDF matrix, adding multiple forces for analysis on the similar composite to understand the further change in mechanical characteristics.

Conclusion

A series of composite layers of PVDF and graphene was prepared and simulated on ANSYS software using the ACP Pre feature. The input parameters were density, melting point, tensile and yield strength, Poisson ratio etc.

After applying a single perpendicular force on the surface, keeping lower edges fixed and meshing at low values, the following data for deformation, stress, and strain were obtained. The data were plotted and analyzed.

Results were quite promising to see that under adverse conditions the materials were not losing their mechanical property. PVDF fibers gained strength after the addition of the graphene layer in between.

Low values of deformation, minimal strain values, and high-stress values have shown that material can be used for manufacturing of application which can survive under extreme pressure.

However, certain issues which were required to be dealt with on a practical basis. The quantity of graphene was a concern, the composite needed to be addressed under extreme pressure to understand its application at high altitude or higher depth underwater. Multiple force techniques can also be carried out to extrapolate its use in defense equipment with highly complex data and safer method.

Various articles have shown concerns of cavitation at lower temperature and porosity with PVDF at room temperature which affects volume change and strain rate. However, ANSYS results have shown that there is a hope of its use in the defense sector. Heat transfer parameters also need to be checked thoroughly to understand the conduciveness of the polymer composite under pressure situation [23].

Hence, this composite has quite promising results and can be used for defense-related applications because of its lightweight, better mechanical [24], electrical, thermal properties [25].

References

1. Kurahatti, R. V., Surendranathan, A. O., Kori, S. A., Singh, N., Kumar, A. V. R., and Srivastava, S. (2010). Defence application of polymer nanocomposites. *Defence Science Journal*, 60(5), 551–563.

2. Liangke, W., Jing, M., Liu, Y., Ning, H., Liu, X., Liu, S., Lin, L., Hu, N., and Liu, L. (2019). Power generation by PVDF-TrFE/graphene nanocomposite films. *Composites Part B*, 164, 703–799.

3. Hoimes-Siedle, A. G., Wilson, P. D., and Verrall, A. P. (1984). PVDF: An electronically-active polymer for industry. *Material and Design*, 4, 910–918.

4. Rahman, M. A., and Chung, G. S. (2013). Synthesis of PVDF-graphene nanocomposites and their properties. *Journal of Alloys and Compounds*, 581, 724–730.

5. Yu, C., Li, D., Wu, W., Luo, C., Zhang, Y., and Pan, C. (2014). Mechanical property enhancement of PVDF/graphene composite based on a high-quality graphene. *Journal of Materials Science*, 49, 8311–8316.

6. Gonazalez, G. D., Hernandez, G. S., Rusinek, A., Bernier, R., and Arias, A. (2020). Low temperature mechanical behaviour of PVDF: cryogenic pre-treatment, quasi-static, cyclic and dynamic experimental testing and modelling. *Mechanics of Materials*, 147, 103436.

7. Elashmawi, I. S., Alatawia, N. S., and Elsayed, N. H. (2017). Preparation and characterization of polymer nanocomposites based on 4 PVDF/PVC doped with graphene nanoparticles. *Results Physics*, 7, 636–640.

8. Wu, Q., Li, L., Zhang, D. Y., and Shui, J. W. (2017). Absorption and mechanical properties of SiCp/PVDF composites. *Composites Part B: Engineering*, 131, 1–7.

9. Huang, W., Feng, P., Gao, C., Shuai, X., Xiao, T., Shuai, C., and Peng, S. (2015). Microstructure, mechanical and biological properties of porous poly(vinylidene fluoride) scaffolds Fabricated by selecting laser sintering. *International Journal of Polymer Science*, 2015, 132–965.

10. Clausi, M., Grasselli, S., Malchiodi, A., and Bayer, I. S. (2020). Thermally conductive PVDF-graphene nanoplatelet (GnP) coatings. *Applied Surface Science*, 529, 147070.

11. Yang, B., Shi, Y., Miao, J. B., Xia, R., Su, L. F., Qian, J. S., Chen, P., Zhang, Q. L., and Liu, J. W. (2018). Evaluation of rheological and thermal properties of polyvinylidene fluoride(PVDF)/graphene nanoplatelets (GNP) composites. *Polymer Testing*, 67, 122–135.

12. Rashad, M., Pan, F., Tang, A., and Asif, M. (2014). Effect of graphene nanoplatelets addition on mechanical properties of pure aluminium using semi-powder method. *Progress In Natural Science: Materials International*, 24(2), 101–108.

13. Wu, Q., Xie, D. J., Zhang, Y. D., Jia, Z. M., and Zhang, H. Z. (2019). Mechanical properties and simulation of nanographene/ Polyvinylidene fluoride composite films. *Composite Part B: Engineering*, 156, 148–155.

14. Silibin, M. V., Bystrov, V. S., Karpinsky, D. V., Nasani, N., Goncalves, G., Gavrilin, I. M., Solnyshkin, A. V., Marques, P. A. A. P., Sing, B., and Bdikin, I. K. (2017). Local mechanical and electromechanical properties of the PVDF TrFE graphene oxide thin films. *Applied Surface Science*, 421, 42–51.

15. Arefi, M., Tabatabaeian, A., and Mohammadi, M. (2021). Bending and Stress analysis of polymeric composite plates reinforced with functionally graded graphene platelets based on sinusoidal shear-deformation plate theory. *Defence Technology*, 17(1), 64–74.

16. Laiarinandrasana, L., Besson, J., Lafarge, M., and Hochstetter, G. (2009). Temperature dependent mechanical behaviour of PVDF: experiments and numerical modelling. *International Journal of Plasticity*, 25(7), 1301–1324.

17. Deshmukh, K., and Joshi, M. G. (2014). Thermo-mechanical properties of poly (vinyl chloride)/graphene oxide as high performance nanocomposites. *Polymer Testing*, 34, 211–219.

18. Zhao, Y., Zhang, Y., Junhao, X., Zhang, M., Yu, P., and Zhao, Q. (2020). Frequency domain analysis of mechanical properties and failure modes of PVDF at high strain rate. *Construction and Building Materials*, 235, 117506.

19. Zheng, G. P., Jiang, Z. Y., Han, Z., and Yang, J. H. (2016). Mechanical and electro-mechanical properties of three-dimensional nanoporous graphene-poly(vinylidene fluoride)composites. *Express Polymer Letters*, 10(9), 730–741.

20. Yu, C., Li, D., Wu, W., Luo, C., Zhang, Y., and Pan, C. (2014). Mechanical property enhancement of PVDF/graphene composite based on a high-quality graphene. *Journal of Materials Science,* 49, 8311–8316.
21. Abolhasani, M. M., Shirvanimoghaddam, K., and Naebe, M. (2016). PVDF/Graphene Composite Nanofibers with enhanced piezoelectric performance for development of robust nano generators. *Composites Science and Technology,* 138, 49–56.
22. Ismail, A. M., Mohammed, M. I., and Fouad, S. S. (2018). Optical and Structural properties of Polyvinylidene fluoride(PVDF)/reduced graphene oxide(RGO) nanocomposites. *Journal of Molecular Structure,* 1170, 51–59.
23. Guo, H., Li, X., Li, B., Wang, J., and Wang, S. (2017). Thermal conductivity of graphene/poly (vinylidene fluoride) nanocomposite membrane. *Materials and Design,* 114, 355–363.
24. Lee, J., Eom, Y., Shin, E. Y., Hwang, H. S., Ko, H., and Chae, G. H. (2019). The effect of interfacial interaction on the conformational variation of poly(vinylidene fluoride) (PVDF) chains in PVDF/graphene oxide (GO) nano composite fibers and corresponding mechanical properties. *ACS Applied Materials and Interfaces,* 11, 14.
25. Kumar, P., Yadav, M. K., Panwar, N., Kumar, A., and Singhal, R. (2019). Temperature dependent thermal conductivity of free-standing reduced graphene oxide/poly(vinylidene fluoride-co-hexafluoropropylene) composite thin film. *Material Research Express,* 6, 115604.

9 Electromagnetic radiation of aluminum under tension and compression

S. K. Pradhan[1,a], Amit Kumar[2,b], and P. Kour[3,c]

[1]Department of Mechanical Engineering, Birla Institute of Technology, Mesra, Patna Campus, India

[2]Department of Mechanical Engineering., National Institute of Technology Patna, India

[3]Departmentof Physics, Birla Institute of Technology, Mesra, Patna Campus, India

Abstract: Changes in the electromagnetic radiation (EMR) were observed during plastic deformation under tension and compression of metals. Variations in EMR observed by changing the diameter of the aluminum specimen were also investigated. There are some basic differences between tensile and compressive tests of metallic materials. Macroscopically, the resulting stress–strain behavior for the plastic zone under compression should be similar to the tensile counterpart (yielding and the associated curvature). There is no maximum, in compression since necking does not occur. The mode of fracture is different in tension and compression. Increase of EMR amplitude due to an increase in diameter was observed as more charge is accumulated in larger diameter specimen resulting in the accelerated electric line dipoles which give EMR emission. Since, aluminum is a ductile material, the flow of charge and dipole formation is conveniently distributed within the volume. This results in more EMR amplitude values for tension than compression. The EMR release rate is again higher for tension as compared to compression; as charges are located on the skin-depth of the metal surface. EMR emission rate varies as FCC>HCP>BCC due to packaging efficiency; thus a larger EMR emission is observed in the case of aluminum than mild steel, as aluminum is FCC and mild steel is BCC.

Keywords: Dipole formation, EMR, Plastic deformation.

Introduction

The physics forms an important aspect of materials research for plastic deformation [1,2]. Electromagnetic radiation, magnetic field generation and thermal emission takes place during plastic deformation [3,4]. These emissions reflect upon the various internal processes which occur within the materials at an atomic level, many of which are not yet fully understood [5]. Dislocation structure for transient condition was observed [6]. Molotskii observed the accelerated dislocations [7]. EMR emission gives the dislocation phenomenon [8]. Yield point measurement can be possible by EMR amplitude phenomenon [9]. EMR emissions are very prominent in different mechanism such as in different metal-forming processes, chip formation process, for detection of plastic deformation, crack propagation, underground mine stability, prediction of earthquakes, in bridge, building maintenance and understanding their life, machine life and their failure [10, 11]. For plastic and crack development a model has been developed for electromagnetic radiation. Also, crystal structure

[a]sudiptabitpatna@gmail.com, [b]amit@nitp.ac.in, [c]paramjit.kour@bitmesra.ac.in

of metal under compression and tension is observed. For calculation of toughness and dislocation velocity, is used EMR technique. Charge oscillation takes place during crack development. Stress, intensity factor, elastic strain energy release rate has been analyzed. Some shortcomings are addressed. Variation of electromagnetic radiation in compression for diameter 10mm and 12 mm aluminum material. Comparison of electromagnetic radiation of metal during compression and tension. The main properties of aluminum are good thermal and electrical properties, reflectivity, nontoxic feature, high strength to weight ratio, recyclability, ease of fabrication. For CFRP generation of EMR during fracture was observed [12,13]. For ferroelectric material electromagnetic radiation was analyzed by mathematical model [14]. Soft PZT loading at low temperature EMR is observed [15]. The soft PZT shows higher EMR signals as compared to the hard PZT [16]. Also, microcrack development can be analyzed by advanced instruments [17,18].

Experimental

First of all, we performed. The tensile and compression test were performed with aluminum rods of diameter 10mm, 12mm to validate our experimental setup with the previous works as presented by Ranjana [19]. For the tensile test the work piece was cut of 45cm length which consists of 17 cm gauge length and 14cm for each side in upper and lower crosshead of UTM and for the compression test the work piece is in the standard L/D ratio of 1.5 as mentioned before due to dead metal zone i.e., for 10mm diameter specimen 15mm length is used and for 12 mm specimen 18mm length is used. The circular shaped notch has been created of radius 0.5mm at the middle of work piece so that the fracture occurs at that point due to stress concentration. Two copper chips having cylindrical electrode having dimensions 8 mm X 6 mm X 0.3 mm, wire was soldered on one longitudinal surface of the copper chip also called as "Antenna" as it transmits the movements and the variation in the structure of work piece. Then, this antenna was pasted on insulation paper using Feviquick and then fixed on diametrically opposite longitudinal surfaces of each specimen again using Feviquick. Direct contact between antenna and specimen was prevented by insulation paper. Two copper chips were then joined electrically as both the wires from the copper chip is short-circuited. The wire from the antenna is connected to the Red port of the crocodile wire and the earthed wire from the body of UTM is connected to the black port of crocodile wire, the crocodile wire is connected to the CH1 of the oscilloscope. Further, in some cases the filter of range 40 kHz to 100 kHz were employed to reduce the noise.

During the experiment peaks observed on the oscilloscope. Also the variation of voltage in the oscilloscope was observed with respect to time. After the complete fracture of the material the oscilloscope is stopped, the CSV file is generated from the software.

Result and Discussion

Various waveforms, using a graph of volt versus time and result have been shown which is generated using Ultrascope and origin 2016 software. As mentioned in the Research methodology, we have to validate the experimental setup as described by Ranjana [12]. The output from the Storage oscilloscope Rigol DS1102CD Digital Oscilloscope 100 MHz, 400MSa/s was stored after completion of the experiment. For signal Ultrascope DS1000 series software was used for analysis by the CSV data file. Few signals obtained were very noisy due to surrounding activities. The noise has been reduced using signal processing option under analysis in origin software.

Figure 9.1 graph has been plotted from the CSV file generated by the ultrascope software, in the figure peaks are clearly visible up to 26 mv. Again the obtained values are noisy and filtered using the "Savitzky-Golay". The obtained graph shows better peaks and the maximum voltage comes out to be 15.04 mv. These results are obtained from the 1st tensile specimen. After getting high noisy output again filter has been added to 2nd specimen to reduce it. Hence the better result obtained from later case. Figure 9.2 shows smooth graph between voltage and time.

(Figure 9.3) Graph has been plotted from the CSV file generated by the ultrascope software, in the figure peaks are clearly visible up to 14.7 mv. The obtained values are again noisy and was filtered using the "Savitzky-Golay". The obtained graph shows better peaks, and the maximum voltage comes out to be 14.7 mv. The voltage output was almost same because of less noise and filter. Figure 9.4 represents smoothened plot.

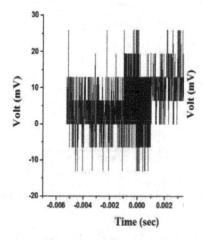

Figure 9.1 Actual voltage versus time graph for 10 mm Al specimen under Tension First specimen.

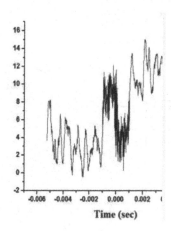

Figure 9.2 Smoothed graph between the voltage and time graph for 10 mm Al for tension first specimen.

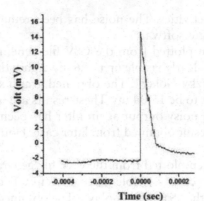

Figure 9.3 Actual voltage versus time graph for 10mm Al for tension Second specimen.

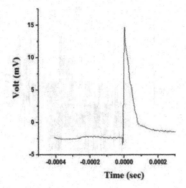

Figure 9.4 Smoothed voltage versus time graph for 10mm Al for tension Second specimen.

The peak values is 4.7 mv(Figure 9.5), for the 1st specimen of compression and the peak value of Figure 9.6 is 5.4 mv i.e. for the 2nd specimen of compression.

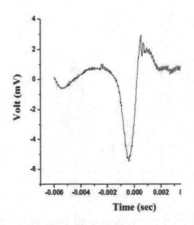

Figure 9.5 Voltage versus time graph for 10mm Al for compression First specimen.

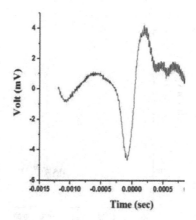

Figure 9.6 Voltage versus time graph for 10mm Al for compression Second specimen.

The (Figure 9.7) graph has been plotted from the CSV file generated by the ultrascope software and in the figure peaks are clearly visible up to 20 mv. Again the obtained values are noisy and filtered using the "Savitzky-Golay". The obtained graph shows better peaks and the maximum voltage comes out to be 20.49mV. These results are obtained from the 1st tensile specimen. After getting high noisy output again filter has been added to the 2nd specimen to reduce it. Hence, the better result obtained for a later. Figure 9.8 shows a smooth graph between voltage and time.

A graph (Figure 9.9) has been plotted from the CSV file generated by the ultrascope software, in the figure peaks are clearly visible up to 31.96 mv. Again the obtained values are noisy and filtered using the "Savitzky-Golay". The obtained graph shows better peaks and the maximum voltage comes out to be 17.71 mv. Figure 9.10 shows a smooth figure. The peak values are shown in Figure 9.11 is 7.9 mv i.e., for the 1st specimen of compression. The peak value for the 2nd specimen of compression is 8.4 mv (Figure 9.12).

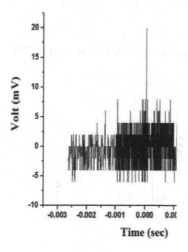

Figure 9.7 Shows actual Voltage versus time graph for 12 mm Al specimen under tension first specimen.

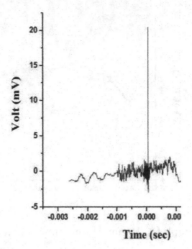

Figure 9.8 Shows the smoothed graph between the voltage and time for 12mm Al specimen under tension first specimen.

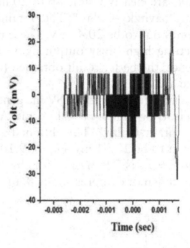

Figure 9.9 Actual Voltage versus time graph for 12 mm Al for Tension Second specimen.

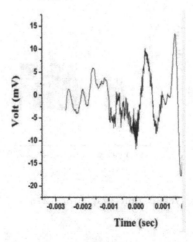

Figure 9.10 Smoothed voltage versus time graph for 12 mm Al for Tension Second specimen.

Smooth curve can be obtained by following procedure through Origin software.

1. Column curve is selected on the analysis menu. By click on signal processing, smooth, open dialog.
2. In smooth dialog box, Savizky-Golay method is selected. Set points of window is set to 100,bouandary condition is set for periodic and polynomial order is set for 3.

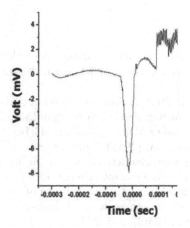

Figure 9.11 Voltage versus time graph for 12 mm Al for compression First specimen.

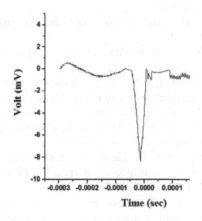

Figure 9.12 Voltage versus time graph for 12 mm Al for compression Second specimen.

Conclusion

1. In this paper, an experimental setup for the measurement of the variation of electromagnetic radiation in tension and compression have been discussed.
2. Experiments have been performed on different diameters, i.e. 10 mm and 12 mm Aluminium rod. EMR amplitude increases with increase in diameter.
3. As more charge is accumulated in 12 mm dia rod than 10 mm dia rod, the accelerated electric line dipoles which give EMR emission is more in 12mm.Since, aluminum is a ductile material, the flow of charge and dipole formation is easily distributed within the volume, this results in more EMR amplitude values for tension than compression.

4. The EMR release rate is again higher for tension as compared to compression as charges are located on the skin-depth on the metal surface. As according to structure EMR emission rate FCC>HCP>BCC due to packaging efficiency.
5. Since aluminum is FCC and mild steel is BCC, hence a large EMR emission is observed in the case of aluminum than mild steel. This concludes that the EMR emission with an increase in diameter and also with the material. Still, there is a lot of work to be done in this field.

References

1. Kumar, R., and Misra, A. (2007). Some basic aspects of electromagnetic radiation emission during plastic deformation and crack propagation in Cu–Zn alloys. *Materials Science and Engineering: A,* 454, 203–210.
2. Chauhan, V. S., and Misra, A. (2008). Effects of strain rate and elevated temperature on electromagnetic radiation emission during plastic deformation and crack propagation in ASTM B 265 grade 2 titanium sheets. *Journal of Materials Science,* 43(16), 5634–5643.
3. Singh, R., Lal, S. P., and Misra, A. (2015). Correlation of plastic deformation induced intermittent electromagnetic radiation characteristics with mechanical properties of Cu–Ni alloys. *International Journal of Materials Research,* 106(2), 137–150.
4. Kumar, M. S., Sharma, V., and Misra, A. (2014). Effect of rate of deformation on electromagnetic radiation during quasi-static compression of sintered aluminium preforms. *International Journal of Materials Research,* 105(3), 265–271.
5. Srilakshmi, B., and Misra, A. (2005). Electromagnetic radiation during opening and shearing modes of fracture in commercially pure aluminium at elevated temperature. *Materials Science and Engineering: A,* 404(1), 99–107.
6. Misra, A. (1978). A physical model for the stress-induced electromagnetic effect in metals. *Applied Physics,* 16(2), 195–199.
7. Molotskii, M. I. (1980). Dislocation mechanism for the Misra effect. *Soviet Technical Physics Letters,* 6(1), 22–23.
8. Misra, A., Prasad, R. C., Chauhan, V. S., and Srilakshmi, B. (2007). A theoretical model for the electromagnetic radiation emission during plastic deformation and crack propagation in metallic materials. *International Journal of Fracture,* 145(2), 99–121.
9. Misra, A., et al. (2010). Effect of Peierls' stress on the electromagnetic radiation during yielding of metals. *Mechanics of Materials,* 42(5), 505–521.
10. Chauhan, V. S., and Misra, A. (2010). Electromagnetic radiation during plastic deformation under unrestricted quasi-static compression in metals and alloys. *International Journal of Materials Research,* 101(7), 857–864.
11. Srilakshmi, B., and Misra, A. (2005). Secondary electromagnetic radiation during plastic deformation and crack propagation in uncoated and tin coated plain-carbon steel. *Journal of Materials Science,* 40(23), 6079–6086.
12. Gade, S., and Sause, M. (2017). Measurement and study of electromagnetic emission generated by tensile fracture of polymers and carbon fibres. *Journal of Nondestructive Evaluation,* 36(1), 9.
13. Gade, S., Alaca, B., and Sause, M. (2017). Determination of crack surface orientation in carbon fibre reinforced polymers by measuring electromagnetic emission. *Journal of Nondestructive Evaluation,* 36(2), 21.
14. Sharma, S. K., Chauhan, V. S., and Yadav, C. (2018). A theoretical model for the electromagnetic radiation emission from ferroelectric ceramics. *Materials Today Communications,* 14, 180–187.

15. Sharma, S. K., Sivarathri, A. K., Chauhan, V. S., and Sinapius, M. (2018). Effect of low temperature on electromagnetic radiation from soft PZT SP-5A under impact loading. *Journal of Electronics Material*, 47, 5930–5938.
16. Sharma, S. K., Chauhan, V. S., and Jain, S. C. (2019). Experimental and theoretical investigations of electromagnetic radiation emission from soft and hard PZT ceramics. *Journal Electron Mater*, 48(11), 7441–7451.
17. Singh, R., Lal, S. P., and Misra, A. (2019). Effect of notch-depth ratio on intermittent electromagnetic radiation from Cu–Ni alloy under tension. *Mater Test*, 61(9), 885–893.
18. Sharma, S. K., Kumar, A., Chauhan, V. S., Kiran, R., and Kumar, R. (2020). Electromagnetic radiation detection from cubical mortar sample and its theoretical model. *Materials Science and Engineering B*, 260, 114–638.
19. Singh, R., Lal, S. P., and Misra, A. (2014). Variation in electromagnetic radiation during plastic deformation under tension and compression of metals. *Applied Physics A*, 117(3), 1203–1215.

10 Estimation of heat transfer from a tube surface under varying reynolds number

Apoorva Deep Roy[a] and S. K. Dhiman[b]

Department of Mechanical Engineering, Birla Institute of Technology, Mesra, Ranchi, India

Abstract: Nusselt, Reynolds and Prandtl numbers are correlated by Power law relationship as Nu = C Re^m Pr^n around a circular tube in the cross flow of air. In present study, method of Matrix least squares have been used for the estimation of values of constant (C) and power indices (m) utilizing the experimental data of Local Nu through a MATLAB program. Correlations (19) through (22) have been developed for estimation of average Nu on overall tube surface and for different flow regimes. An empirical logarithmic expression (23) has also been proposed for estimating the overall Nu_{av} by curve fitting in MATLAB with minimum residual. Comparisons of correlations were made with the reported literatures. Estimates obtained from correlations (19, 20 and 21) have shown a good agreement with reported literature [20]. Deviations of (19) and (20) falls within 3.80% while that for (21) falls under limits of 2.80%. The overall Nu_{av} from the present study developed correlations (22 and 23) shows a good agreement with reported literature [2], having minor deviations within 2.75% for correlation (22) and 1.1% for correlation (23) respectively.

Keywords: Coefficients and indices, correlations, Nusselt number, Prandtl numbers, Reynolds number.

Introduction

Since many decades researchers have attempted many experimental, numerical and computational approaches by varying the fluid type and thermal conditions of the tube as well as of fluid to examine flow field and heat transfer. In the field of heat exchangers, such as recuperators, radiators, regenerators, etc., economy and compactness are being given the prime concern in recent days.

Many researchers have attempted to provide the correlations for average heat transfer such as Sanitjai and Goldstein [1, 2], Churchill et al. [3], Fand [4], Perkins et al. [5, 6], Whitaker [7], Zukauskas and Ziugzda [8] so that the phenomenon of heat transfer over the tube surface may be established, however the attempts are still in progress for providing segment-wise heat transfer over the tube surfaces. Lists of reported correlations are mentioned in Table 10.1. Considering Knudsen and Katz [9] analysis, Holman [10] explained that Nu_{av} is given by power law: $Nu = CRe^m Pr^{1/3}$ for flow past around a circular tube in air

[a]apoorvadeep12@gmail.com, [b]skdhiman@bitmesra.ac.in

where *Nu* is Nusselt number, *Re* is Reynolds number, *Pr* is Prandtl numbers, C and m are Coefficients and Indices, respectively.

Due to change in flow regime, there is a variation of Nu across the circumference of a circular tube (Eckert and Soehngen [11], Krall and Eckert [12]). Mohammed et al. [13] have experimentally carried out the investigation of the heat transfer around a tube and correlated the average Nu with Re. Sahu et al. [14] have numerically investigated the effects on the rate of heat transfer by Re and Pr over a tube and developed the correlations. Kumar et al. [15] have made investigation on heat transfer with effects of Re and Pr around a semi-circular cylinder. Giordano et al. [16] have conducted a study on heat transfer coefficient upon cylindrical wall surface based on of cylinder aspect ratio and Re. Haeri and Shrimpton [17] and Mortean and Mantelli [18] have proposed a correlation to calculate Nu over a circular tube at low to moderate Re.

From numerous literature reviews mentioned above, it can be observed that so far not enough work has been done to determine C and m of Re in different flow regimes. A matrix least square method programmed in MATLAB software has been used and inputs of local Nu data have been given to deter mine constant C and m around the tube surface. Development of correlations for estimation of overall average Nu by using calculated values of C and m around the tube surface is a novel work carried out under this study.

Table 10.1: A list of reported correlations.

Researchers	Correlations	Eqⁿ No.
Perkins & Leppert	$Nu_{\delta\theta overall}\left(\dfrac{\mu_w}{\mu_b}\right)^{0.25} = \left[0.31(Re)^{0.5} + 0.11(Re)^{0.67}\right](Pr)^{0.4}$	(1)
R M Fand	$Nu_{\delta\theta overall} = \left[0.35 + 0.34(Re)^{0.5} + 0.15(Re)^{0.58}\right](Pr)^{0.3}$	(2)
Sanitjai & Goldstein	$Nu_{\delta\theta overall} = \begin{aligned}&0.446(Re)^{0.5}(Pr)^{0.35} + 0.528\Big(6.5\big(e^{(Re/5000)}\big)\Big)^{-5} + \\ &\left(\big(0.031(Re)^{0.8}\big)^{-5}\right)^{-0.2}(Pr)^{0.42}\end{aligned}$	(3)
Zukauskas & Ziugzda	$Nu_{\delta\theta overall}\left(\dfrac{Pr_w}{Pr_b}\right)^{0.25} = 0.26(Re)^{0.6}(Pr)^{0.37}$	(4)
Churchill & Burnstein	$Nu_{\delta\theta overall} = \begin{aligned}&0.3 + \left(\big(0.62(Re)^{0.5}(Pr)^{0.33}\big) / \big(1 + (0.4/Pr)^{0.66}\big)^{0.25}\right) \\ &x\left[1 + (Re/282000)^{0.625}\right]^{0.8}\end{aligned}$	(5)
Whitaker	$Nu_{\delta\theta overall} = \left[0.4(Re)^{0.5} + 0.06(Re)^{0.67}\right](Pr)^{0.4}\left(\dfrac{\mu_f}{\mu_w}\right)^{0.25}$	(6)

Numerical Methodology

Power law relationship typically correlates Nu distribution as:

$$Nu = C\, Re^m\, Pr^n. \tag{7}$$

On applying log for Eq. (7) both sides,

$$Log\left[\frac{Nu}{Pr^n}\right] = LogC + mLog(Re) \tag{8}$$

Matrix least square method calculates the Ordinary least square (OLS). OLS is the maximum like hood estimator. It is commonly used to analyse the experimental and observational data by minimizing the SSR leading to a closed form of expression for the evaluation of the value of an unknown parametric vector $\hat{\beta}$.

Parametric equation is:

$$y = f(X) + \epsilon \tag{9}$$

where, y= Expected output variable, ϵ = Irreducible error and $f(X) = \beta_0 + \beta_1 X_1 + \beta_2 X_2 + \cdots\cdots + \beta_p X_p$ Eq. (9) in matrix form is represented as:

$$\begin{bmatrix} y_1 \\ y_2 \\ \vdots \\ \vdots \\ y_N \end{bmatrix} = \begin{bmatrix} \beta_0 + & \beta_1 X_{1,1} + & \beta_2 X_{1,2} + & \cdots + & \beta_p X_{1,p} + & \epsilon_1 \\ \beta_0 + & \beta_1 X_{2,1} + & \beta_2 X_{2,2} + & \cdots + & \beta_p X_{2,p+} & \epsilon_2 \\ \vdots & \vdots & \vdots & \cdots & \vdots & \vdots \\ \vdots & \vdots & \vdots & \cdots & \vdots & \vdots \\ \beta_0 + & \beta_1 X_{N,1} + & \beta_2 X_{N,2} + & \cdots + & \beta_p X_{N,p} + & \epsilon_N \end{bmatrix} \tag{10}$$

where, regression estimates are:

$$\hat{y} = \widehat{f(X)} = \widehat{\beta_0} + \widehat{\beta_1} X_1 + \widehat{\beta_2} X_2 + \cdots + \widehat{\beta_p} X_p = X\hat{\beta} \tag{11}$$

Matrix form of Eq. (11) can be written as:

$$\begin{bmatrix} \widehat{y_1} \\ \widehat{y_2} \\ \vdots \\ \vdots \\ \widehat{y_n} \end{bmatrix} = \begin{bmatrix} \widehat{\beta_0} + & \widehat{\beta_1} X_{1,1} + & \widehat{\beta_2} X_{1,2} + & \cdots + & \widehat{\beta_p} X_{1,p} \\ \widehat{\beta_0} + & \widehat{\beta_1} X_{2,1} + & \widehat{\beta_2} X_{2,2} + & \cdots + & \widehat{\beta_p} X_{2,p} \\ \vdots & \vdots & \vdots & \cdots & \vdots \\ \vdots & \vdots & \vdots & \cdots & \vdots \\ \widehat{\beta_0} + & \widehat{\beta_1} X_{N,1} + & \widehat{\beta_2} X_{N,2} + & \cdots + & \widehat{\beta_p} X_{N,p} \end{bmatrix} \tag{12}$$

From Eq. (9) & Eq. (11) residual error is:

$$e = \begin{bmatrix} e_1 \\ e_2 \\ \vdots \\ e_N \end{bmatrix} = \begin{bmatrix} y_1 - \widehat{y_1} \\ y_2 - \widehat{y_2} \\ \vdots \\ y_N - \widehat{y_N} \end{bmatrix} = \begin{bmatrix} y_1 \\ y_2 \\ \vdots \\ y_N \end{bmatrix} - \begin{bmatrix} \widehat{y_1} \\ \widehat{y_2} \\ \vdots \\ \widehat{y_N} \end{bmatrix} = (y - \hat{y}) \tag{13}$$

Now, the objective function is sum of squared residuals (SSR):

$$SSR = \sum_{i=1}^{n}(e_i)^2 = \sum_{i=1}^{n}(y - \hat{y})^2 \tag{14}$$

Eq. (14) in Matrix form is represented as:

$$SSR = e^T e = (y - \hat{y})^T (y - \hat{y})$$
$$SSR = y^T y - y^T X \hat{\beta} - \widehat{\beta^T} X^T y + \widehat{\beta^T} X^T X \hat{\beta} \tag{15}$$

To minimize SSR, Eq. (15) will be differentiated w.r.t $\hat{\beta}$ and equated to zero:

$$\widehat{\beta^T} X^T X = y^T X$$

Applying $(X^T X)^{-1}$ and taking transpose to both sides we will get (OLS) of β:

$$\hat{\beta} = (X^T X)^{-1} X^T y \tag{16}$$

Comparing each term of Eq. (9)) by Eq. (8) we get a Matrix as:

$$y = \begin{bmatrix} Log\left[\dfrac{Nu_1}{Pr^n}\right] \\ Log\left[\dfrac{Nu_2}{Pr^n}\right] \\ Log\left[\dfrac{Nu_3}{Pr^n}\right] \\ \vdots \\ \vdots \\ Log\left[\dfrac{Nu_n}{Pr^n}\right] \end{bmatrix}; \quad X = \begin{bmatrix} 1 & Log(Re_1) \\ 1 & Log(Re_2) \\ 1 & Log(Re_3) \\ \vdots & \vdots \\ \vdots & \vdots \\ 1 & Log(Re_N) \end{bmatrix}; \tag{17}$$

Combining Equations. (16) and (17) for Nu_i ($i = 1$ to N) C and m are represented as:

$$\hat{\beta} = \begin{bmatrix} \beta_0 \\ \beta_1 \end{bmatrix} = \begin{bmatrix} LogC \\ m \end{bmatrix}, \tag{18}$$

Development of Correlations

Nusselt number can be defined as ratio of the heat transfer by convection to that of conduction. Nu of value one represents heat transfer by pure conduction. A smaller value of Nu represents laminar flow whereas a larger Nu value corresponds to more active convection with turbulent flow. Experimentally determined Local Nu values around a tube surface of Dhiman et al. [19] has been used to calculate values of C and m. Figure 10.1 shows the variations of the estimates of C and m with angular position of tube. A linear declination of C has been observed from front stagnation point (fsp) to angular location 60° followed by a slight bump in the region around 75° and then it further declines till 90°. Thereafter the values of C are almost constant showing a minor bump between 90°–135°

followed by a slight increment till rsp. The estimates of m show a reverse trend to that of estimates of C. Different flow regions have been identified based on the variations in slope.

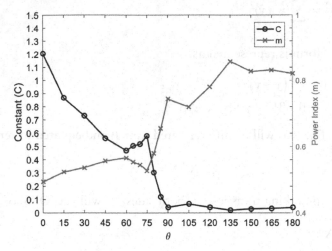

Figure 10.1 Variation of C *and* m *for* $11000 \leq Re \leq 62000$.

The estimated values of C and m region wise (as per slope variations) is shown in Table 10.2.

Table 10.2: The estimated values of C and m region wise.

Angular position (j = 1, 2, ..., 8)	C_j	m_j
0°	$C_1 = 1.20$	$m_1 = 0.490$
0° to 60°	0.769	0.533
60° to 75°	0.520	0.548
75° to 90°	0.261	0.626
90° to 105°	0.055	0.732
105° to 135°	0.043	0.787
135° to 180°	0.03	0.830
180°	0.039	0.820

Present study developed Correlations are tabulated in Table 10.3.

Table 10.3: Correlations developed in the present study.

θ	Region	Correlation	Correlation no.
$0° \leq \theta \leq 75°$	$\delta\theta_1$	$Nu_{\delta\theta1} = 0.5 Pr^{0.33} \sum_{j=2}^{3} C_j Re^{mj}$	(19)
$75° \leq \theta \leq 135°$	$\delta\theta_2$	$Nu_{\delta\theta2} = 0.33 Pr^{0.33} \sum_{=4}^{6} C\,Re$	(20)

θ	Region	Correlation	Correlation no.
$135° \leq \theta \leq 180°$	$\delta\theta_3$	$$\text{Nu}_{\delta\theta3} = C_7 \left(Re\right)^{m}_{7} \left(Pr\right)^{0.33}$$	(21)
$0° \leq \theta \leq 180°$	$\delta\theta_{\text{overall}}$	$$\text{Nu}_{\delta\theta\text{overall}} = 0.16Pr^{0.33} \sum_{j=2}^{7} C_j Re^{m_j}$$	(22)
$0° \leq \theta \leq 180°$	$\delta\theta_{\text{overall}}$ (Equation)	$$log_{10}\text{Nu}_{\delta\theta\text{overall}} = 0.66\ log_{10}Re - 0.87$$	(23)

Results and Discussion

A comparison of estimated Nu distribution over the entire tube surface with the reported literature of Igarashi and Hirata [20] has been presented at $10000 \leq Re \leq 62000$ in Figure 10.2. Local Nu values have been determined from the power law relation by using the values of C and m as shown in Figure 10.1.

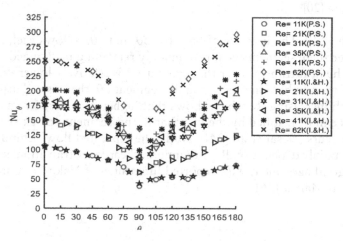

Figure 10.2 $Nu_\theta\, v/s\ \theta$ for $11000 \leq Re \leq 62000$.
(P.S: Present study, I&H: Igarashi & Hirata [20])

    Comparison of estimated local Nu shown in Figure 10.2 depicts a good agreement with the reported literature of Igarashi and Hirata [20] and hence it depicts the usual phenomenon of heat transfer around the tubular surface i.e., there is a decline in the heat transfer from fsp to the separating point and further reduction in heat transfer continues till 90°. Afterwards there is still an increment in the heat transfer to the downstream side of tube with a bump at Re of 40000. With increase in Re there is a rise in the heat transfer which agrees with the reported results of [21].

    Figure 10.3 shows the comparison of Nu_{av} normalized with Pr calculated from developed correlations (19, 20, and 21) for various segments $\delta\theta_1$, $\delta\theta_2$, and $\delta\theta_3$ respectively onto the circumference of tube with that of experimental data as stated by Igarashi & Hirata [20]. This figure also depicts that with increasing Re, Nu_{av} also increases for all segments, whereas their slopes reveal that different heat transfer occur which is influenced by the variation in flow structures of fluid onto these segments (Sanitjai and Goldstein [1, 2],

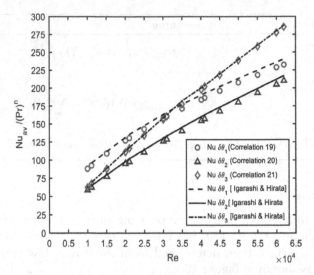

Figure 10.3 $Nu_{av}/(Pr)^n$ obtained through present study developed correlations (19, 20, 21) with that of Igarashi & Hirata [20].

Bloor [22], Giedt [23]). The increase of Nu_{av} for $\delta\theta_1$ and $\delta\theta_2$ follows each other indicating that the flow over these segments is approximately remaining unchanged with the change in Re, however the heat transfer over the segment $\delta\theta_2$ is relatively lower due to the forma-tion of the separation bubble in this region. In segment $\delta\theta_3$ there is a sharp increase in the heat transfer with increasing Re which shows that higher heat transfer can be obtained at a larger Re. This can be justified by the fact that in this segment vortex shedding of a very high frequency occurs (Desai et al. [24]). Nu in the separated flow behind a circular tube in the crossflow mainly depends on Re for laminar, shear-layer transition and wake regions which shows a good agreement with reported literatures of Nakamura and Igarashi [25] and Lebouche and Martin [26].

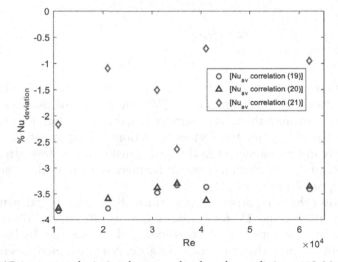

Figure 10.4 $Nu_{av}/(Pr)^n$ percent deviation between developed correlations (19,20,21) with reported results under [20].

The estimates of all the Nu_{av} from developed correlations (19, 20 and 21) have shown a good agreement to data as stated by Igarashi & Hirata [20] which can be depicted by the deviation plots of Figure 10.4. For $\delta\theta_1$ and $\delta\theta_2$ their deviations fall inside the upper limits of 3.80% whereas for $\delta\theta_3$ it lies within the upper limits of 2.80%.

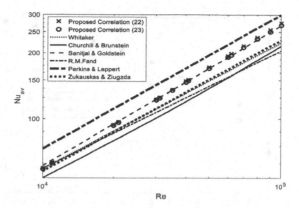

Figure 10.5 Reported correlations comparisons with the correlations (22 & 23) for overall Nu_{av} estimates within *10000≤Re≤100000*.

Figure 10.5 shows the plots of estimates of overall Nu_{av} from correlations (22) and (23), and their comparison with that of the equations (1) through (6), as shown in Table 10.1, proposed by benchmark of various authors. Correlation (23) has been obtained by linear fitting of correlation (22) with respect to Re within the residuals 0.0023, as can be observed in Figure 10.6.

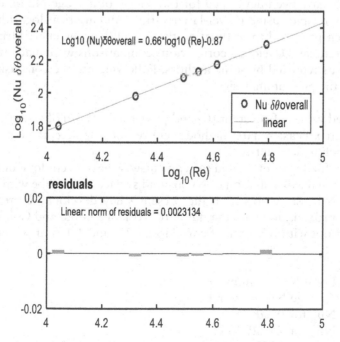

Figure 10.6 $Log_{10}(Nu_{\delta\theta overall})$ *v/s* $Log_{10}(Re)$.

The overall Nu_{av} estimate from the present study developed correlations shows a good agreement with benchmark work of Sanitjai and Goldstein [2], having minor deviations within 2.75% of correlation (22) and 1.1% of correlation (23) as represented in Figure 10.7.

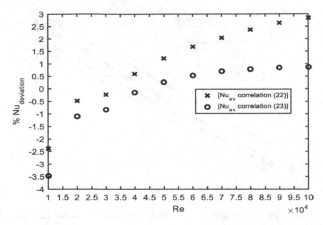

Figure 10.7 Percentage deviation of estimates of overall Nu_{av} obtained from correlations (22, 23) with that of reported correlation estimate of [2].

Conclusions

Under present study, method of Matrix least squares have been used for the estimation of values of constant (C) and power indices (m) of power law relationship in the cross flow of air around a circular tube utilizing the experimental data of Local Nu. Correlations (19) through (22) have been developed for estimation of average Nu on overall and also for different flow regimes using power-law relation. An empirical logarithmic expression (23) has also been proposed for estimating the overall Nu_{av} by curve fitting in MATLAB with minimum residual. Developed correlations comparisons were made with some of the benchmark data as reported by some authors. Following are the conclusions that can be drawn based on the present analysis:

1. The estimated values of constant (C) and power indices (m) of power law relationship by using Matrix Least square method were reasonably acceptable as it validates the reported benchmark results.
2. Correlations development related to segment-wise have been done due to the various discrepancies in the distribution of Nu around surface of the tube with variation in Re.
3. The overall Nu_{av} estimation from the present study developed correlations shows a very close approximation with the standard work of Sanitjai and Goldstein [2], having minor deviations within 2.75% of correlation (22) and 1.1% of correlation (23).

Nomenclature:

Nu_θ	Local Nusselt number
Nu_{av}	Average Nusselt number
$Nu_{\delta\theta1}$	Nu_{av} for $0° \leq \theta \leq 75°$
$Nu_{\delta\theta2}$	Nu_{av} for $75° \leq \theta \leq 135°$
$Nu_{\delta\theta3}$	Nu_{av} for $135° \leq \theta \leq 180°$

$\text{Nu}_{\delta\theta\text{overall}}$	Nu_{av} for $0°\leq\theta\leq180°$
θ	Angular position on circular cylinder

References

1. Sanitjai, S., and Goldstein, R. J. (2004). Heat transfer from a circular cylinder to mixtures of water and ethylene glycol. *International Journal of Heat and Mass Transfer*, 47, 4785–4794.
2. Sanitjai, S., and Goldstein, R. J. (2004). Forced convection heat transfer from a circular cylinder in cross flow to air and liquids. *International Journal of Heat and Mass Transfer*, 47, 4795–4805.
3. Churchill, S. W., and Bernstein, M. (1977). A correlating equation for forced convection from gases and liquids to a circular cylinder in cross flow. *Journal of Heat Transfer*, 99, 300–306.
4. Fand, R. M. (1965). Heat transfer by forced convection from a cylinder to water in cross-flow, Int. *Journal of Heat and Mass Transfer*, 8, 995–1010.
5. Perkins, H. C., and Leppert, G. (1962). Forced convection heat transfer from a uniformly heated cylinder. *ASME Journal of Heat Transfer*, 84(3), 257–261.
6. Perkins, H. C., and Leppert, G. (1964). Local heat-transfer coefficients on a uniformly heated cylinder. *International Journal of Heat and Mass Transfer*, 7, 143–158.
7. Whitaker, S. (1972). Forced convection heat transfer calculations for flow in pipes, past flat plate, single cylinder, and for flow in packed beds and tube bundles. *AIChE Journal*, 18, 361–371.
8. Zukauskas, A., and Ziugzda, J. (1985). Heat Transfer of a Cylinder in Cross Flow. Washington, New York: Hemisphere Pub.
9. Knudsen, J. D., and Katz, D. L. (1958). Fluid Dynamics and Heat Transfer. New York: McGraw-Hill.
10. Holman, J. P. (2010). Heat Transfer. 10th ed. New York: McGraw-Hill Higher Education.
11. Eckert, E., and Soehngen, E. (1952). Distribution of heat-transfer coefficients around circular cylinders in cross flow at reynolds numbers from 20 to 500. *Transactions, ASME*, 74, 343–347.
12. Krall, K. M., and Eckert, E. R. G. (1973). Local heat transfer around a cylinder at low reynolds number. *Transactions ASME Journal of Heat Transfer*, 95, 273–275.
13. Mohammed, H. A., and Salman, Y. K. (2007). Experimental investigation of mixed convection heat transfer for thermally developing flow in a horizontal circular cylinder. *Applied Thermal Engineering*, 27(8–9), 1522–1533.
14. Sahu, A. K., Chhabra, R. P., and Eswaran, V. (2009). Effects of Reynolds and Prandtl numbers on heat transfer from a square cylinder in the unsteady flow regime. *International Journal of Heat and Mass Transfer*, 52, 839–850.
15. Kumar, A., Dhiman, A., and Baranyi, L. (2016). Fluid flow and heat transfer around a confined semi-circular cylinder: onset of vortex shedding and effects of reynolds and prandt numbers. *International Journal of Heat and Mass Transfer*, 102, 417–425.
16. Giordano, R., Ianiro, A., Astarita, T., and Carlomagno, G. M. (2012). Flow field and heat transfer on the base surface of a finite circular cylinder in cross flow. *Applied Thermal Engineering*, 49, 79–88.
17. Haeri, S., and Shrimpton, J. S. (2013). A correlation for the calculation of the local Nusselt number around circular cylinders in the range $10 \leq \text{Re} \leq 250$ and $0.1 \leq \text{Pr} \leq 40$. *International Journal of Heat and Mass Transfer*, 59, 19–229.
18. Mortean, M. V. V., and Mantelli, M. B. H. (2019). Nusselt number correlation for compact heat exchangers in transition regimes. *Applied Thermal Engineering*, 151, 514–522.
19. Dhiman, S. K., and Prasad, J. K. (2017). Inverse estimation of heat flux from a hollow cylinder in cross-flow of air. *Applied Thermal Engineering*, 113, 952–961.
20. Igarashi, T., and Hirata, M. (1977). Heat transfer in separated flows part 2: theoretical analysis. *Heat Transfer-Japanese Research*, 6, 60–78.

21. Sarkar, S., Dalal, A., and Biswas, G. (2011). Unsteady wake dynamics and heat transfer in forced and mixed convection past a circular cylinder in cross flow for high Prandtl numbers. *International Journal of Heat and Mass Transfer*, 54(15–16), 3536–3551.

22. Bloor, M. S. (1964). The transition to turbulence in the wake of a circular cylinder. *Journal of Fluid Mechanics,* 19, 290–304.

23. Giedt, W. H. (1949). Investigation of variation of point unit-heat-transfer coefficient around a cylinder normal to an air stream. *Transactions ASME*, 71, 375–381.

24. Desai, A., Mittal, S., and Mittal, S. (2020). Experimental investigation of vortex shedding past a circular cylinder in the high subcritical regime. *Physics of Fluids,* 32, 14–105.

25. Nakamura, H., and Igarashi, T. (2004). Variation of Nusselt number with flow regimes behind a circular cylinder for Reynolds numbers from 70 to 30000. *International Journal of Heat and Mass Transfer*, 47, 5169–5173.

26. Lebouche, M., and Martin, M. (1975). Convection forcee autour du cylindre; sensibilite aux pulsations de l'ecoulement externe. *International Journal of Heat and Mass Transfer*, 18, 1161–1175.

11 Metamaterial for seismic design

Aman Thakur[1,a], Arpan Gupta[2,b], and Sandip Kumar Saha[1,c]

[1]School of Engineering, Indian Institute of Technology Mandi, India

[2]Department of Mechanical Engineering, Indian Institute of Technology Delhi, India

Abstract: The current work discusses the foundation design based on metamaterial to prohibit seismic wave propagation. Metamaterials–as the name implies, are artificial materials having properties superior to natural materials. These materials or structures are artificially designed and possess the potential to exhibit properties that can be counterintuitive and sporadically available in nature. One such property that could be used for protection from seismic waves–P-waves and S-waves–is frequency-dependent wave attenuation. In this work, seismic metamaterials are designed, and the frequency zone (bandgap) for high wave attenuation is obtained and compared with the literature for a repeating unit cell. A new layout of structural foundation, stimulated by the principles of conventional seismic base-isolation systems, can be developed that could help in minimizing the transmission of seismic energy from the base to the super-structure. The basic theory of 3-D periodic foundation, incorporating solid-state physics, is postulated, finite element models were developed, and different studies were conducted on these periodic foundations/panels to investigate their efficiency in attenuating incoming waves. These periodic structures fundamentally consist of repeating unit cells spanning finitely in vertical and horizontal directions. Separate models of these periodic foundations were formulated and subjected to excitations in transverse and longitudinal directions. The major response reduction was observed in the pre-determined attenuation (frequency bandgap) zones. Periodic panels with more layers in vertical directions show significant pruning within the bandgap zones. The method devised can be used to design and develop metamaterials for futuristic applications in seismic design.

Keywords: Band-Gap, bragg's Scattering, periodic foundation, seismic design.

Introduction

In the event of an earthquake, energy is released in the form of seismic waves, which causes ground shaking. Compression or primary wave (P-wave) and secondary wave or shear wave (S-wave) are the body waves contributing to the tremor. P-wave emerges first, followed by the S-wave, and travels to the earth's surface, which causes vibration in buildings. The latter cause more structural damage as it has more amplitude of shaking in the horizontal direction. Typical buildings are more susceptible to damage from horizontal motion than vertical motion. Earthquake-resistant design of structure aims to design a

[a]amanthakurbbps@gmail.com, [b]arpan.gupta@mech.iitd.ac.in, [c]sandip_saha@iitmandi.ac.in

building or any structure in such a way that it can withstand these severe and vigorous ground shakings.

Several systems for earthquake-resistant design have been developed for over a century now. Various studies on the seismic behavior of buildings have been carried out, and new technologies have been exhaustively manifested to minimize earthquake damage [1]. One such innovation is seismic base isolation systems. As the name says, these systems isolate the foundation of the building from its super-structure by introducing structural elements with lower lateral stiffness amidst them, that in turn reduces the fundamental frequency of vibration of the building [2]. Hence, it decouples the isolated super-structure from the horizontal component of ground motion. Some applications have also tried to isolate the building against vertical excitation. These systems prevent destructive seismic wave propagation to the building and minimize the damage that could occur to the super-structure [3]. Over the past half-century, base isolation systems have been used, and many new types are being developed. The traditional and most commonly used techniques are elastomeric rubber and sliding systems. These systems, however, provide effective reduction to horizontal ground motions and are not very efficient in reducing vertical ground accelerations. Further rocking motions and permanent deformations were challenges faced by newly developed base isolation systems such as rolling seal type and cable-reinforced air spring systems [4]. However, these systems cater to the above-mentioned limitations faced by former systems. Its practical deployment and efficiency are still in research. Looking at the inherent challenges the traditional base isolation systems face in protecting structures from strong ground shaking, a new approach to isolating the building from ground shaking is being studied. Herein a new form of periodic foundation is proposed to be designed incorporating the concept of solid-state physics and the use of novel phononic crystals. These crystals are formed from the unit cell, the most fundamental part of a periodic structure. These infinitely periodic structure made by the repetition of unit cell has the inherent property to attenuate applied mechanical and seismic waves [5, 6]. Periodic materials have attenuation in a specific frequency zone, known as the frequency bandgap zone. When the frequency of applied excitation falls in this bandgap zone, the wave's energy cannot propagate through these periodic structures. It is envisioned that the foundation of a structure can be designed and constructed as a periodic structure. This way, a major component of the seismic waves can be restricted from passing through the periodic foundation, and the seismic energy imparted on the super-structure can be minimized. Hence, the periodic foundation will isolate the structure at its base to protect it from earthquake shaking, but with a new approach.

There are two types of periodic structures, locally resonant structure, and bragg scattering structures. Locally resonant structures are generally composed of more than two materials or two distinct scatterers inside a matrix. Bragg scattering structures have only one type of scatterer inside a matrix and consist of two materials, one in the matrix and the other in reinforcement. The materials used are commonly available construction materials such as rubber, concrete, and steel. The idea here is to design these periodic structures, pre-determine their frequency bandgap zone, and then study its effectiveness in response reduction to applied excitation.

Periodic materials are further classified based on the direction of periodicity as one-dimensional (1-D), two-dimensional (2-D), and three-dimensional (3-D) periodic materials. Bao et al. [7], Xiang et al. [8], and Shi et al. [9], Gaofeng and Zhifei [10] conducted several studies pertaining to one-dimensional and two-dimensional periodic foundations as novel seismic isolation systems. Further, Thakur and Gupta [11] performed a comparative

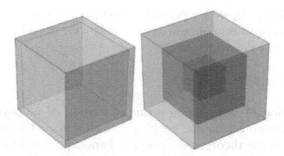

Figure 11.1 (a) Bragg scattering unit cell (b) Locally resonant unit cell.

study of the response of 3-D periodic foundations subjected to dynamic excitation of body waves with reinforced concrete foundations. In the present work, we have analyzed a three-dimensional (3-D) periodic foundation of both Bragg scattering and locally resonant materials. The unit cell of both these periodic materials is shown in Figure 11.1. Its frequency bandgap zone is then computed. These unit cells are duplicated in the vertical and horizontal direction to produce a 3-D periodic foundation or a finite periodic structure. Further, base excitation is applied as P-wave and S-wave on Bragg scattering periodic foundation in the transverse and longitudinal direction, and finite element analysis is employed to numerically compute the response in the form of the frequency response function, also known as transmission loss.

Theory of 3-D Periodic Material/Foundation/Meta-material

The frequency bandgap zone of a 3-D periodic material can be obtained using the following set of equations reducing to the eigenvalue problem, as shown below. Using a set of assumptions for the 3-D periodic material to be perfectly elastic, isotropic, continuous, with infinitesimal deformation, and ignoring the damping of the material, the governing equation of motion [12] is given by Equation (1).

$$\rho(r)\frac{\partial^2 u}{\partial t^2} = \nabla\left\{\left[\lambda(r)+2\mu(r)\right](\nabla \cdot u)\right\} - \nabla\times\left[\mu(r)\nabla\times u\right] \qquad (1)$$

Here, u is the displacement vector in coordinate vector r, $\mu(r)$, $\lambda(r)$, are Lame constants and $\rho(r)$ is the density. Displacement, $u(r)$ in Equation (1), satisfies the Bloch Fouquet theorem [13], shown by Equation (2).

$$u(r,t) = e^{i(K\cdot r-\omega t)}u_a(r) \qquad (2)$$

Where t is the time, ω is the angular frequency, K is the wave vector, $u_a(r)$ is the wave amplitude. Since the material is periodic in nature, the amplitude of the wave $u_a(r)$ will be periodic from $r \in (0, a)$, as shown in the form of function in Equation (3), where a is the unit cell's thickness.

$$u_a(r+a) = u_a(r) \qquad (3)$$

From Equation (2) and (3), we get periodic boundary conditions, given in Equation (4). The periodic boundary condition obtained is applied on the opposite faces of the cubic unit cell while solving in finite element software.

$$u(r+a,t) = e^{iK \cdot a} u(r,t) \tag{4}$$

Substituting Equation (4) into the governing equation of motion, it is observed that Equation (1) reduces to an eigenvalue problem, which is given by Equation (5). The solution of which produces the theoretical frequency bandgap. Here Ω is the stiffness matrix, and M represents the unit cell's mass matrix. K is the wave vector, varied along the edges of the first Brillouin zone [5], corresponding to a series of angular frequencies ω.

$$\left[\Omega(K) - \omega^2 M \right] \cdot u = 0 \tag{5}$$

Finite Element Modeling of 3-D Periodic Foundation

Salient details of the modeling done in COMSOL software are discussed here.

Frequency Bandgap Computation

For the Bragg scattering cubic unit cell, Figure 11.1(a), the stiffer core is composed of concrete of a length of 0.9 m, surrounded by a rubber matrix with a thickness of 0.05 m. For locally resonant material, Figure 11.1(b), the cubic core comprises cast iron, having a length of 0.102 m. The intermediate material is polyurethane, having an edge length of 0.203 m, and the concrete matrix with a side of 0.305 m. The properties of the material are given in Table 11.1.

Table 11.1: Material properties.

S. No	Material	Young's Modulus (GPa)	Poisson Ratio	Density (kg/m³)
1	Rubber	0.0001586	0.463	1277
2	Concrete	40	0.2	2300
3	Cast Iron	1650	0.275	7184

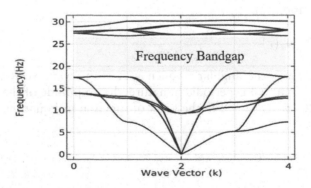

Figure 11.2 Frequency bandgap of bragg scattering panel.

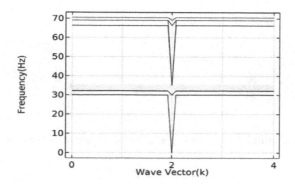

Figure 11.3 Frequency bandgap of locally resonant material.

The yellow shaded area in Figures 11.2 and 11.3 represents the frequency bandgap zone which ranges from 18.5–27.2 Hz for Bragg scattering panel and 32.8–35.7 Hz for locally resonant material. This implies that waves in this shaded frequency zone are unable to propagate in any direction. The obtained results are in good correlation with the literature [14, 15].

Frequency Response Function/Transmission Loss

We have adopted to conduct further study on Bragg scattering periodic materials, as a broader range of frequency bandgap is evident in these materials. The response of periodic foundations based on P-wave and S-wave excitation can be obtained by plotting the frequency response function. It is defined as per Equation (6):

$$FRF = 20\log\left(\frac{u_{out}}{u_{in}}\right) \qquad (6)$$

u_{in} is the input unit amplitude displacement which is generally provided at the base of the periodic panel and is the displacement at the top surface, corresponding to the excitation direction. For S-wave, the unit amplitude is applied in the horizontal direction, either in the x or y-axis. Rotation and displacement in other directions are restricted. Correspondingly, for P-wave, the unit amplitude is applied in the vertical direction i.e. z-axis, and rotation and displacement in other directions are constricted. The corresponding response output (u_{out}) is measured for both cases. Unit cells are repeated in the vertical and horizontal direction for producing a finite 3-D periodic foundation. Figure 11.4 shows a layout of the 3-D periodic foundation of Bragg scattering periodic material, where unit cells are repeated four times in the horizontal plane, and thereafter, the arrangement is repeated in two and three vertical layers. The negative value of FRF in the curve represents response reduction, while the positive represents response amplification.

FRF subjected to S-wave for bragg scattering 3-D periodic foundation is shown in Figure 11.5. The maximum response reduction can be seen in the yellow highlighted region, the theoretical frequency bandgap computed in the previous section. The same can be seen for the FRF subjected to P-wave, plotted in Figure 11.6. It is also observed that the 3-D periodic foundation has attenuation zones in different frequency ranges, other than frequency bandgap zones. Periodic panels with more layers in vertical directions show significant pruning within the bandgap zones.

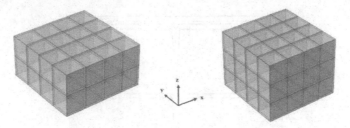

Figure 11.4 3-D periodic foundation with 4 × 4 unit cells repeated in two and three vertical layers.

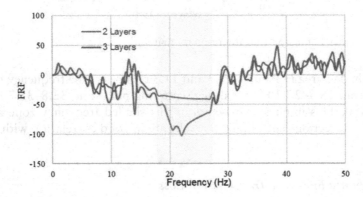

Figure 11.5 FRF of Bragg scattering 3-D periodic foundation subjected to S-wave.

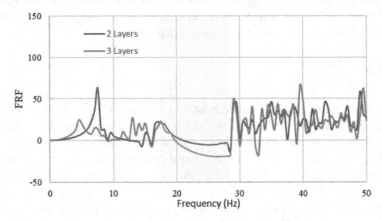

Figure 11.6 FRF of Bragg scattering 3-D periodic foundation subjected to P-wave.

Conclusions

The concept of 3-D periodic foundation in attenuating seismic excitations and deployment as an ingenious and futuristic earthquake-resistant design has been studied. Theoretical frequency bandgap for two types of periodic material was computed, and then the response towards applied excitation was analyzed in the form of FRF using the frequency-domain study. Conclusively the analysis yielded the following:

- The major response attenuation of the periodic foundations lies within the pre-determined, calculated frequency bandgap zones.

- With increase in periodicity or layers in vertical directions, higher attenuation peaks are observed within the frequency bandgap zones.

Calculating these frequency bandgaps and having prior information on these attenuation zones can efficiently help engineers design earthquake-resistant structures. Further, there exists a great level of uncertainty in selecting the bandgap when exhaustive seismic hazard data is not available for the location of the structure. Therefore, more thorough research and exhaustive experimental validation are needed for future practical purposes.

Acknowledgments

This work is financially supported by the Department of Science and Technology under Project No. DST/INT/CAN/P-04/2020. The authors acknowledge the use of AVL (Acoustics and Vibration Laboratory), IIT Mandi for the computational works, and other resources carried out in this work.

References

1. Elsesser, E. (2004). Seismically resistant design—past, present, future. In 13th World Conference Earthquake Engineering Vancouver, Canada, no. 1235, (pp. 1–7).
2. Naeim, F., and Kelly, J. M. (1999). Design of Seismic Isolated Structures: From Theory to Practice. NJ, USA: JW&S.
3. Inoue, K., Fushimi, M., Moro, S., Morishita, M., Kitamura, S., and Fujita. (2004). Development of three-dimensional seismic isolation system for next generation nuclear power plant. In Procedure 13th World Conference Earthquake Engineering, no. 3445.
4. Zhou, Z., Wong, J., and Mahin, S. (2016). Potentiality of using vertical and three-dimensional isolation systems in nuclear structures. *Nuclear Engineering and Technology*, 48(5), 1237–1251. DOI: 10.1016/j.net.
5. Kittel, C. (2004). Introduction to Solid State Physics. NY, USA: JW&S.
6. Shi, Z., Cheng, Z., and Xiang, H. (2014). Seismic isolation foundations with effective attenuation zones. *Soil Dynamics and Earthquake Engineering*, 57, 143–151. DOI: 10.1016/j.soildyn.2013.11.009.
7. Bao, J., Shi, Z., and Xiang, H. (2012). Dynamic responses of a structure with periodic foundations. *Journal Engineering Mechanics*, 138(7), 761–769. DOI: 10.1061/(ASCE) EM.19437889.0000383.
8. Xiang, H. J., Shi, Z. F., Wang, S. J., and Mo, Y. L. (2012). Periodic materials-based vibration attenuation in layered foundations: experimental validation. *Smart Materials and Structures*, 21(11), 112003. DOI: 10.1088/0964-1726/21/11/112003.
9. Shi, Z. F., Cheng, Z. B., and Xiong, C. (2010). A new seismic isolation method by using a periodic foundation. *Earth and Space 2010: Engineering, Science, Construction, and Operations in Challenging Environments*, 2586–2594. DOI: 10.1061/41096(366)242.
10. Jia, G., and Shi, Z. (2010). A new seismic isolation system and its feasibility study. *Earthquake Engineering and Engineering Vibration*, 9(1), 75–82. DOI: 10.1007/s11803-010-8159-8.
11. Thakur, A., and Gupta, A. (2021). Computational study of seismic wave propagation through metamaterial foundation. *International Journal for Computational Methods in Engineering Science and Mechanics*, 22(3), 200–207. DOI: 10.1080/15502287.2021.1916217.
12. Cheng, Z., and Shi, Z. (2013). Novel composite periodic structures with attenuation zones. *Engineering Structures*, 56, 1271–1282. DOI: 10.1016/j.engstruct.2013.07.003.

13. Kushwaha, M. S., Halevi, P., Martínez, G., Dobrzynski, L., and Djafari-Rouhani, B. (1994). Theory of acoustic band structure of periodic elastic composites. *Physical Review B*, 49(4), 2313–2322. DOI: 10.1103/PhysRevB.49.2313.
14. Witarto, W., Wang, S. J., Yang, C. Y., Wang, J., Mo, Y. L., Chang, K. C., and Tang, Y. (2019). Three-dimensional periodic materials as seismic base isolator for nuclear infrastructure. *AIP Advances*, 9(4), 045–014. DOI: 10.1063/1.5088609.
15. Yan, Y., Cheng, Z., Menq, F., Mo, Y. L., Tang, Y., and Shi Z. (2015). Three dimensional periodic foundations for base seismic isolation. *Smart Materials and Structures*, 24(7), 075006. DOI: 10.1088/0964-1726/24/7/075006.

12 A comparative analysis of musculoskeletal models to predict musculoskeletal forces

Abhijeet Singh[1,a], Shibendu Shekhar Roy[1], and Bidyut Pal[2,b]

[1]Mechanical Engineering Department, National Institute of Technology Durgapur, WB, India

[2]Mechanical Engineering Department, Indian Institute of Engineering Science and Technology, Shibpur, WB, India

Abstract: Musculoskeletal models are used to study human gait and predict realistic musculoskeletal forces. This paper presents a comparative analysis of two lower-extremity musculoskeletal models, Gait2354 and Gait2392, for walking and running activities. The objective of the study was to determine the variations in the predicted joint reaction forces between the said models. The method involved scaling, inverse kinematics, inverse dynamics, static optimization, and analysis of joint reaction using OpenSim. The experimental marker data in the static and dynamic trial and experimental ground reaction forces and moments were used as the input. For walking activity, the peak hip, knee, and ankle joint reaction forces were 421, 402, and 586% of the body-weight (BW), respectively, for the Gait 2354 model. Whereas for the Gait 2392 model, the respective forces were 421, 359, and 561% BW. For running activity, the peak hip, knee, and ankle joint reaction forces were 848, 1217, and 1162% BW, respectively, for the Gait 2354 model. Whereas for the Gait 2392 model, the respective forces were 972, 1239, and 1242% BW. The estimated values of hip and knee joint reaction forces were then compared with the literature on in-vivo measurements obtained using the instrumented prostheses. The root-mean-square-errors and the Correlation Coefficients were found to corroborate well with the published measured data. With reference to the published measured data, the Gait 2392 predicted the maximum knee reaction force better than the Gait 2354 model and, the Gait 2354 and Gait 2392 model predicted the same maximum hip reaction force.

Keywords: Gait 2354, Gait 2392, joint reaction force, musculoskeletal model, walking, running.

Introduction

Musculoskeletal models are extensively used to study normal and pathological gait [1], aiming at predicting musculoskeletal forces. The predicted musculoskeletal forces can be used to investigate the mechanical behavior of orthopedic implants and to study the muscular coordination of daily activities [2]. The joint reaction forces have a considerable influence on the mechanical stimuli generated in periprosthetic bone, influencing bone ingrowth and adaptive bone remodeling [3]. The accurate estimation of muscle forces can help to diagnose musculoskeletal disorders and can also help to develop improved treatments [4].

---

[a]jeetubit96@gmail.com, [b]bidyutpal@mech.iiests.ac.in

Direct measurement of joint reaction forces is only possible in patients having instrumented implants [5]. However, it is hard to carry out owing to ethical reasons. The musculoskeletal simulation may be used as an alternative way to predict musculoskeletal forces [6]. A musculoskeletal model incorporates bone geometry, joint kinematics, muscles, ligaments, and wrapping objects [7]. Markers placed on the musculoskeletal models are used to scale the segments and track the kinematics of the subject in dynamic trials in order to do musculoskeletal simulations [8]. Muscles are represented by musculotendinous actuators [9]. The intrinsic muscle properties influence the magnitude of muscle forces [10].

There are different software systems developed for musculoskeletal simulation. OpenSim is a musculoskeletal modeling and simulation freeware that is widely used to study dynamic simulations of human movement and musculoskeletal forces [8]. Delp et al. [11] developed Gait 2354 model featuring 54 Hill-type musculotendon actuators and Gait 2392 model featuring 92 Hill-type musculotendon actuators in the lower extremities and torso. The muscle architecture data were derived from five cadaver datasets [7]. These models have 23 degrees of freedom (dof) and contain 12 body segments such as the femur, talus, calcaneus, and toes in the two legs, including the pelvis and torso [10]. Weinhandl and Bennett [12] and Mathai and Gupta [13] compared different musculoskeletal models (Gait 2392 model [11], Arnold Lower Limb Model [14], Hip 2372 model [15] and London Lower Limb Model [16]). They observed that the London Lower Limb Model most accurately predicted joint contact forces during walking at various paces in comparison to the other musculoskeletal models. As the number of OpenSim musculoskeletal models in recent years has increased, so the selection of musculoskeletal models for the prediction of musculoskeletal forces is becoming more critical [17].

The objective of this study was to predict joint angles and musculoskeletal forces in the lower limb for walking and running activities using Gait 2392 and Gait 2354 models. The specific objectives are to compare the predicted joint reaction forces between the said models and to verify the accuracy of the predicted joint forces by comparing them with the experimentally measured (using instrumented implants) loading data published in the literature.

Materials and Methods

Two freely available musculoskeletal models, i.e., Gait 2354 and Gait 2392 models [11] were used. The kinematic data for static and dynamic trials and the ground reaction forces and moments data were obtained from the literature that studied a healthy subject (65.9 kg, walking speeds of 1.36 m/s [10] and 3.96 m/s [18]. The kinematic data included one trial for walking and running activity. The step-by-step methods followed to estimate musculoskeletal forces are presented in Figure 12.1.

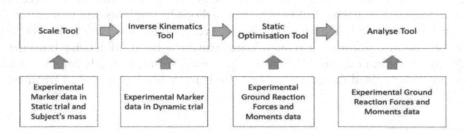

Figure 12.1 Schematic overview of musculoskeletal simulation.

Scaling

The dimensions of each segment of the musculoskeletal models were scaled using the Scale tool of OpenSim so that it accurately fits the particular subject, matching the distances between the virtual markers and those between experimental markers [8]. The inputs to the Scale Tool were experimental marker data in static trial and the mass of the subject (65.9 kg). The mass of the subject was attuned so that the total mass of the body equals the mass of the specified subject. The mass distribution was preserved during scaling. The virtual markers were positioned on these musculoskeletal models to track the segment movement. The markers were assigned a weight. The marker weights were adjusted such that the maximum marker error and root mean square error (RMSE) were kept within 1 cm [13].

Inverse Kinematics

The joint angles were determined using Inverse Kinematics tool. The tool reduces the error between experimental markers and virtual markers during the dynamic trial at each time step by solving a weighted least square optimization problem. The input to the Inverse Kinematics Tool was experimental marker data in dynamic trial and the subject's mass. The marker weights were adjusted such that the maximum marker error and RMSE were within 2 cm [13].

Weighed least square optimization objective function:

$$
Min \sum_{i=1}^{\substack{\text{markers}}} w_i \left(\overline{x}_i^{\text{subject}} - \overline{x}_i^{\text{model}} \right)^2 + \sum_{j=1}^{\substack{\text{joint angles}}} \omega_j \left(\theta_j^{\text{subject}} - \theta_j^{\text{model}} \right)^2 \tag{1}
$$

where $\overline{x}_i^{\text{subject}}$ and $\overline{x}_i^{\text{model}}$ are the three-dimensional positions of the ith marker or joint center for the subject and model, $\theta_j^{\text{subject}}$ and θ_j^{model} are the values of the jth joint angle for the subject and model, and w_i and ω_j are factors that allow markers and joint angles to be weighted differently [8].

Static Optimization

The muscle forces were computed using the Static Optimization tool of OpenSim from inverse kinematics results, experimental ground reaction forces, and moments data based on distributing the intersegmental joint moments to individual muscles. Static Optimization Tool minimizes a time-independent performance criterion (sum of some powers of muscle activations) subjected to joint moment equilibrium as a constraint such that the redundancy of muscular load sharing can be solved [19]. The infinite combination of muscle forces can be produced for the same net joint moment [16]. This muscle-moment redundancy problem is indeterminate in musculoskeletal models as musculotendon actuators attached to the segment of musculoskeletal models exceed the degree of freedom of the joints [20]. The inputs to the Static Optimization Tool were experimental marker data in dynamic trial and, ground reaction forces and moments data. Muscle force-length-velocity relation, which is the intrinsic property of muscle, was selected as the additional constraint. It helps to stabilize human walking and running by producing physiological realistic muscle forces for the given movement [10]. The power of muscle activation was taken as 2, which is the user-defined power term for walking and running activity [21]. The external ground reaction forces and moments were applied to the right and left calcaneus bodies of musculoskeletal models.

Static optimisation problem is given by

$$\text{Minimise} \quad \phi(Fi) = \sum_{i=1}^{n} \left(\frac{F_i}{F_{i,\ max}} \right)^p \tag{1}$$

$$\text{Subject to} \quad \sum_{i=1}^{n} \overline{r}_{ij} X \ \overline{F}_i = \overline{M}_j \quad i = 1,....,n; \quad j = 1,....,d \tag{2}$$

$$0 \le F_i \le F_{i,max} \ i = 1, ..., n \tag{3}$$

where F_i is the magnitude of i-th muscle force, $F_{i,max}$ is the value of maximal force the i-th muscle can exert, p is the power of the objective function, n is the total number of actuators, $\overline{r}_{ij}$ is the moment arm of i-th muscle with respect to the j-th joint axis, d is the total number of joint axes in the model, $\overline{M}_j$ is the moment acting around the j-th joint axis [16].

Joint Reaction Analysis

The resultant joint reaction forces and moments were determined using the Analyze tool using inverse kinematics results, static optimization results, and the experimental ground reaction forces and moment data. The inputs to the Analyze Tool were experimental marker data in the dynamic trial, ground reaction forces and moments data and, muscle forces obtained from Static Optimization Tool of OpenSim. A low-pass filter (6 Hz) was used to reduce noise [13].

Validation

The musculoskeletal forces were predicted for walking and running activity. The predicted joint reaction forces were verified against in vivo measured joint contact forces obtained from an instrumented prosthesis for walking activity. The RMSE and Pearson correlation coefficients were enumerated to quantify the goodness of fit between the simulated joint reaction forces and in vivo contact forces. These values were used for comparison of maximum joint reaction forces and the shape of joint reaction force profiles.

Results

Joint angles, joint moments, muscle forces, and joint reaction forces were obtained for the joints in the lower limb for walking and running activities. However, the results corresponding to the joint reaction forces for both activities have been presented here.

Hip Joint Reaction Forces for Walking and Running Activity

The maximum hip joint reaction forces for walking activity were 421% of the body-weight for both the Gait 2354 model and Gait 2392 model (Figure 12.2). The maximum hip joint reaction force was observed at 53.03% and 12.12% of the gait cycle for the Gait 2354 model and Gait 2392 model, respectively (Figure 12.2).

The maximum hip joint reaction forces for running activity were 848% and 972% of body weight for the Gait 2354 model and Gait 2392 model, respectively (Figure 12.3). The maximum hip joint reaction forces were observed at 21.43% and 19.05% of the gait cycle for the Gait 2354 model and Gait 2392 model, respectively (Figure 12.3).

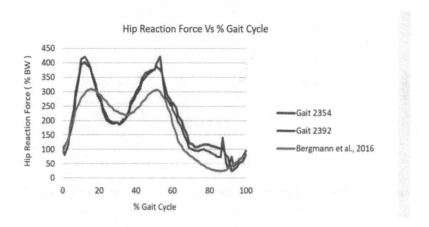

Figure 12.2 Variation of hip joint reaction forces for walking activity over a gait cycle.

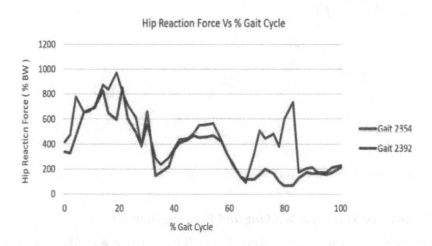

Figure 12.3 Variation of hip joint reaction forces for running activity over a gait cycle.

Knee Join Reaction Forces for Walking and Running Activity

The maximum knee joint reaction for walking activity was 402% and 359% of body weight for the Gait 2354 model and Gait 2392 model, respectively (Figure 12.4). The maximum knee joint reaction forces for walking activity were observed at 12.12% and 13.64% of the gait cycle for the Gait 2354 model and Gait 2392 model, respectively (Figure 12.4).

The maximum knee joint reaction for running activity was 1217% and 1239% of body weight for the Gait 2354 model and Gait 2392 model, respectively (Figure 12.5). The maximum knee joint reaction forces were observed at 16.67% of the gait cycle for both the Gait 2354 model and Gait 2392 model (Figure 12.5).

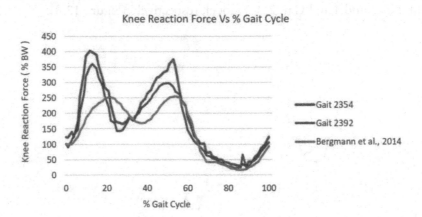

Figure 12.4 Variation of knee joint reaction forces for walking activity over a gait cycle.

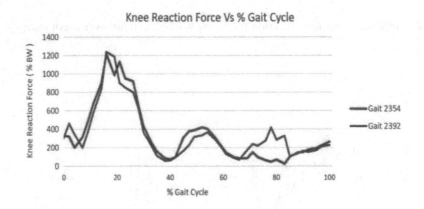

Figure 12.5 Variation of knee joint reaction force for running activity over a gait cycle.

Ankle Joint Reaction Forces for Walking and Running Activity

The maximum ankle joint reaction force for walking activity was 586% and 561% of body weight for the Gait 2354 model and Gait 2392 model, respectively (Figure 12.6). The maximum ankle joint reaction forces were observed at 51.51% of the gait cycle for both the Gait 2354 model and Gait 2392 model (Figure 12.6).

The maximum ankle joint reaction force for running activity was 1163% and 1242% of body weight for the Gait 2354 model and Gait 2392 model, respectively (Figure 12.7). The maximum ankle joint reaction forces for running activity were observed at 19.05% and 21.43% of the gait cycle for the Gait 2354 model and Gait 2392 model, respectively (Figure 12.7).

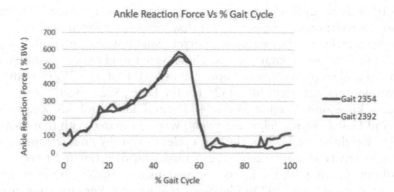

Figure 12.6 Variation of ankle joint reaction force for walking activity over a gait cycle.

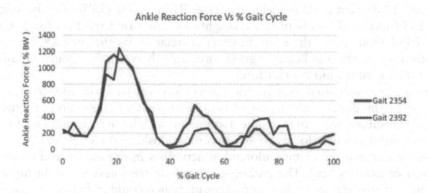

Figure 12.7 Variation of ankle joint reaction force for running activity over a gait cycle.

Discussion

This study compared predictions of two musculoskeletal models (Gait 2392 and Gait 2354) against in vivo measured joint contact forces reported in literature. The estimated hip, knee, and ankle joint reaction forces for walking activity for the models were verified with the published literature. The RMSE and Pearson correlation coefficients were calculated for the two musculoskeletal models for walking activity in order to compare the simulated joint reaction forces, and in-vivo measured joint contact forces. The Gait 2354 and Gait 2392 models are found to predict maximum hip, knee, and ankle joint reaction forces reasonably accurately.

The predicted hip joint reaction forces (about 421% of the body-weight for both the Gait 2354 model and Gait 2392 model) were compared with that measured (peak 308% of body weight at 15.15% of the gait cycle) in vivo by an instrumented prosthesis [4] for walking activity (Figure 12.1). The comparison of estimated hip joint reaction force for walking activity with the in vivo measurements by instrumented prosthesis indicated that Gait 2392 predicted the maximum hip joint reaction force only slightly better than the Gait 2354 model as the relative error for maximum hip reaction forces were 37% and 36% of body weight for the Gait 2354 model and Gait 2392 model, respectively (Figure 12.1). The Gait 2392 model (56% of body weight) had less RMSE than the Gait 2354 model (62% of body weight) for the hip joint reaction force. The Gait 2392 model had a higher Pearson correlation coefficient (0.93) than the Gait 2354 model (0.9) for the hip joint reaction force, which indicated a good correlation between the simulated hip reaction forces and the measured hip contact forces. These were found to corroborate well with the study by Mathai and Gupta [13], who reported an RMSE of 58.77% of body weight and a Pearson Correlation Coefficient of 0.8529 for the Gait 2392 model.

The predicted knee joint reaction forces (402% and 359% of body weight for the Gait 2354 model and Gait 2392 model, respectively) were compared with that measured (peak 255% of body weight at 54.55% of the gait cycle) in vivo by an instrumented prosthesis [2] for walking activity (Figure 12.3). The comparison of estimated knee joint reaction force for walking activity with the in vivo measurements obtained by instrumented prosthesis suggested that the Gait 2392 model predicted the peak knee joint reaction force better than Gait 2354 model. The relative errors for maximum knee joint reaction force were 57.77% and 41% of body weight for the Gait 2354 model and Gait 2392 model, respectively (Figure 12.3). Gait 2392 model had a lesser RMSE value (53% of body weight) than the Gait 2354 model (72.43% of body weight) for knee joint reaction force. Gait 2392 and Gait 2354 models had the same Pearson correlation coefficient (0.87) for the knee joint reaction force, which indicated a good correlation between the simulated hip reaction forces and the measured hip contact forces.

The difference in simulated joint reaction forces as compared with in vivo measurement may be due to the direct comparison of simulated joint reaction forces of the healthy subject and the subject with an instrumented prosthesis. The overestimation of maximum hip and knee joint reaction forces by the two musculoskeletal models may be attributed to the insufficient number of musculotendon actuators in the model and compensatory coactivation of muscles [20]. The pathological muscle weakness found in hip and knee replacement patients was not taken into account, which could influence the joint contact forces during walking and running activities [21]. The differences in the magnitude of simulated joint reaction forces between the two models were due to the difference in the number of musculotendon actuators present on the lower extremities and torso with 54 musculotendon actuators in the Gait 2354 model and 92 musculotendon actuators in the Gait 2392s model [19].

The predicted maximum hip and knee joint reaction forces for walking activity were observed at 53.03 % and 12.12% of the gait cycle, respectively, for the Gait 2354 model. For the Gait 2392 model, the maximum hip and knee joint reaction forces were observed at 12.12% and 13.64% of the Gait cycle, respectively. On the other hand, the in-vivo measured maximum hip and knee contact forces for walking activity were reported at 15.15% and 54.55% of the gait cycle, respectively. As expected, the maximum hip, knee, and ankle joint reaction forces were found to be higher for running activity (Figures 12.2, 12.4, and 12.6) (Gait 2354 model: 848%, 1217% and 1163% of body weight, respectively; and Gait 2392 model: 971.77%, 1239% and 1242% of body weight, respectively), than those for walking activity (Figures 12.1, 12.3, and 12.5) (Gait 2354 model: 421%, 402% and 586% of body weight, respectively; Gait 2392 model: 421%, 359% and 561% of body weight, respectively). The maximum hip joint reaction force for Gait 2354 model was found to be at the second peak, whereas the maximum hip joint reaction force for Gait 2392 model was found to be at the first peak for walking activity. The maximum knee joint reaction force was observed at the first peak for both the musculoskeletal models for walking activity. The difference in timing of maximum joint reaction forces may be due to the different activation profiles of the muscles which are mainly responsible for joint reaction forces.

There are some sources of inaccuracies in the prediction of the musculoskeletal models [17]. The models are based on measurements of limited number of cadavers. This could potentially affect the input related to the muscle parameters such as maximum isometric force, optimum fiber length, pennation angle, and tendon slack length. Another important source may be due to the consideration of insufficient number of muscles, ligaments, and soft tissues in the human lower limb model. Moreover, the path of the muscles may have been simplified due to inadequate via points and wrapping surfaces available. In this study, two lower limb models were selected, excluding the upper arms. Virtual markers in the arm's segment can lead to differences in scaled models and inverse kinematic results [17]. Gait 2354 and Gait 2392 models use a planar model of the knee to represent motion in the sagittal plane, and ankle and subtalar joints are idealized as one dof revolute joints [11].

Conclusion

Simulations were carried out for two musculoskeletal models for walking and running activity. The predicted values of the joint reaction forces were found to collaborate well with the published literature. The maximum joint reaction forces were lesser for Gait 2392 than the Gait 2354 model for walking activity. The maximum joint reaction forces were higher for Gait 2392 than the Gait 2354 model for running activity. The Gait 2392 model predicted maximum knee joint reaction forces better than the Gait 2354 model and, the Gait 2354 and Gait 2392 model predicted the same hip reaction force as compared to the measured values reported in the literature for walking.

References

1. Rajagopal, A., Dembia, C., DeMers, M., Delp, D., Hicks, J., and Delp, S. (2016). Full-body musculoskeletal model for muscle-driven simulation of human gait. *IEEE Transactions on Biomedical Engineering*, 63(10), 2068–2079.
2. Bergmann, G., Bender, A., Graichen, F., Dymke, J., Rohlmann, A., Trepczynski, A., Heller, M., and Kutzner, I. (2014). Standardized loads acting in knee implants. *PLoS ONE*, 9(1), e86035.
3. Mathai, B., and Gupta, S. (2020). The influence of loading configurations on numerical evaluation of failure mechanisms in an uncemented femoral prosthesis. *International Journal for Numerical Methods in Biomedical Engineering*, 36(8), e3353.
4. Bergmann, G., Bender, A., Dymke, J., Duda, G., and Damm, P. (2016). Standardized Loads Acting in Hip Implants. *PLOS One*, 11(5), e0155612.
5. Bergmann, G., Deuretzbacher, G., Heller, M., Graichen, F., Rohlmann, A., Strauss, J., and Duda, G. (2001). Hip contact forces and gait patterns from routine activities. *Journal of Biomechanics*, 34(7), 859–871.
6. Heller, M., Bergmann, G., Deuretzbacher, G., Dürselen, L., Pohl, M., Claes, L., Haas, N., and Duda, G. (2001). Musculo-skeletal loading conditions at the hip during walking and stair climbing. *Journal of Biomechanics*, 34(7), 883–893.
7. Cardona, M., and Garcia Cena, C. (2019). Biomechanical analysis of the lower limb: a full-body musculoskeletal model for muscle-driven simulation. *IEEE Access*, 7, 92709–92723.
8. Delp, S., Anderson, F., Arnold, A., Loan, P., Habib, A., John, C., Guendelman, E., and Thelen, D. (2007). OpenSim: open-source software to create and analyze dynamic simulations of movement. *IEEE Transactions on Biomedical Engineering*, 54(11), 1940–1950.
9. Seth, A., Sherman, M., Reinbolt, J., and Delp, S. (2011). OpenSim: a musculoskeletal modeling and simulation framework for in silico investigations and exchange. *Procedia IUTAM*, 2, 212–232.

10. John, C., Anderson, F., Higginson, J., and Delp, S. (2013). Stabilisation of walking by intrinsic muscle properties revealed in a three-dimensional muscle-driven simulation. *Computer Methods in Biomechanics and Biomedical Engineering*, 16(4), 451–462.

11. Delp, S., Loan, J., Hoy, M., Zajac, F., Topp, E., and Rosen, J. (1990). An interactive graphics-based model of the lower extremity to study orthopaedic surgical procedures. *IEEE Transactions on Biomedical Engineering*, 37(8), 757–767.

12. Weinhandl, J., and Bennett, H. (2019). Musculoskeletal model choice influences hip joint load estimations during gait. *Journal of Biomechanics*, 91, 124–132.

13. Mathai, B., and Gupta, S. (2019). Numerical predictions of hip joint and muscle forces during daily activities: a comparison of musculoskeletal models. *Proceedings of the Institution of Mechanical Engineers, Part H: Journal of Engineering in Medicine*, 233(6), 636–647.

14. Arnold, E., Ward, S., Lieber, R., and Delp, S. (2009). A model of the lower limb for analysis of human movement. *Annals of Biomedical Engineering*, 38(2), 269–279.

15. Shelburne, K. B., and Pandy, M. G. (1997). A musculoskeletal model of the knee for evaluating ligament forces during isometric contractions. *Journal of Biomechanics*, 30(2), 163–176.

16. Modenese, L., Phillips, A., and Bull, A. (2011). An open source lower limb model: Hip joint validation. *Journal of Biomechanics*, 44(12), 2185–2193.

17. Roelker, S. A., Caruthers, E. J., Baker, R. K., Pelz, N. C., Chaudhari, A. M. W., and Siston, R. A. (2017). Interpreting musculoskeletal models and dynamic simulations: causes and effects of differences between models. *Annals of Biomedical Engineering*, 45(11), 2635–2647.

18. Hamner, S. R., Seth, A., and Delp, S. L. (2010). Muscle contributions to propulsion and support during running. *Journal of Biomechanics*, 43(14), 2709–2716.

19. Lin, Y. C., Dorn, T. W., Schache, A. G., and Pandy, M. G. (2011). Comparison of different methods for estimating muscle forces in human movement. *Proceedings of the Institution of Mechanical Engineers, Part H: Journal of Engineering in Medicine*, 226(2), 103–112.

20. Modenese, L., and Phillips, A. T. M. (2012). Prediction of hip contact forces and muscle activations during walking at different speeds. *Multibody System Dynamics*, 28, 157–168.

21. Modenese, L., Gopalakrishnan, A., and Phillips, A. (2013). Application of a falsification strategy to a musculoskeletal model of the lower limb and accuracy of the predicted hip contact force vector. *Journal of Biomechanics*, 46(6), 1193–1200.

13 Design optimization of 3D printed shoe sole using lattice structures

Kshitij Ambadas Jungare, Dhakshain Balaji V, Monitakshaya Kar, and Radha R

School of Mechanical Engineering, Vellore Institute of Technology, Chennai, Tamil Nadu India

Abstract: The 3D printing industry has reinvented product manufacturing. It provides for fast, easy and low-cost fabrication of products. Recent trends in the footwear industry adopt 3D printing technologies for footwear and other related apparel. The additive manufacturing of shoes has revolutionized the shoe manufacturing industry. Major footwear labels have now turned to rapid manufacturing methods such as stereolithography, selective laser sintering, Multi Jet Fusion and fused deposition modelling, etc. In this paper, the shoe sole is designed using CAD software and Fused Deposition Modelling technique is used to produce the shoe sole. The model is sliced using a slicing software and parametric study is conducted. A flexible polymer, thermoplastic polyurethane, is used to fabricate the shoe sole. Support structures made of QSR material aid the structural integrity during part building. Lattice structures are introduced to optimize the entire shoe design.

Keywords: Additive manufacturing, fused deposition modelling, lattice structures, shoe sole design.

Introduction

3D printing or additive manufacturing involves the usage of a CAD model to fabricate a three-dimensional object. Many such technologies were invented in the late 1900s and after. In 1986, Chuck Hull invented the stereolithography process and in 1988, Scott Crump invented the FDM process. These two processes laid the foundation for the now booming additive manufacturing industry.

Additive manufacturing processes have revolutionized the footwear industry. Nowadays, 3D foot model extracted from anthropometric data can be converted to design personalized footwear [1,2]. Topological optimization of sole is done to reduce material wastage. Moreover, feet planar pressure map is taken to account for weight distribution analysis [3].

Various methods are in use for shoe manufacturing using 3D printing techniques. A few include Carbon's Direct Light Synthesis, High Speed Sintering, Selective Laser Sintering, Stereolithography, Fused Deposition Modelling and Multi Jet Fusion. Fused Deposition Modelling technique with ABS or TPU as material is used to produce various fashion product [4]. 3D printing has also impacted the medical footwear industry. Foot orthoses for rheumatoid arthritis can be manufactured using Selective Laser Sintering [5].

Lattice structures are introduced in shoe soles to control shoe stiffness. Various topologies are utilized to design the curved surface of shoe sole. These structures are a measure to reduce the stress on the wearer's foot [6,7]. The comfort is a very important aspect when it comes to footwear, so a lot of tests should be conducted for finding the optimal design for a specific person [8]. When multi lattice structures are incorporated in the sole, the weight of the whole shoe is reduced considerably while keeping the strength [9]. The design of shoe with curved sole and flat sole have different pressure distribution and it significantly affects the comfort level as well as strength [10]. The lattice structures have shown similar compressive properties as conventional foam and are very helpful when it comes to absorb the compressive stresses [11,12].

Materials and Methods

Material and Process Selection

The sole is printed using thermoplastic polyurethane (TPU) which is a flexible and abrasion resistant thermoplastic. It is used in many 3D printing processes such as FDM, SLS and SLA. It can withstand ambient temperatures up to 80 degree Celsius. Few properties of TPU include high durability and strength, very high flexibility, high abrasion resistance, medium high chemical resistance and medium water resistance. The support material is QSR, which is water-soluble. This aids in easy hands-free removal. FDM process is chosen. TPU filament can be utilized.

Component Design and Calculations

The design of a product decides if the product will have proper functionality and if the people will like it or not. In this case where we design a shoe sole, it is very important that we consider all the factors affecting the comfort and functionality of the sole. The main purpose of the sole is to support the weight applied by its user, but we went a few steps ahead of it. By using modern design techniques and additive manufacturing we accomplished making a complex design which not only supports the weight of a user according to their needs but also reduces weight. By using lattice structures, it is possible to accomplish such a goal where less amount of material is used without affecting its functionality. Then by taking pressure distribution data design is made according to it. Also using the data of a person's feet, making of custom sole according to the person's needs is possible. After collection of the design data next step towards the design phase is taken. Now it is needed to consider how weight is distributed around the feet while walking or running, then that data can be used to make a rough design of the sole. That rough design can be modified to get finer details sorted. These fine details will affect the aesthetics of the sole and it will also determine the comfort level of the sole.

Lattice Structures

Lattice structures are bio-inspired configurations, it consists of webs, and trusses with a defined pattern. It is usually used for weight reduction. Conventional manufacturing methods make it very difficult to manufacture a product containing lattice structures. But additive manufacturing has made it possible to make all kinds of complex structures. In this design a software is used for latticing according to few defined parameters. This software converts a solid body into lattice structures resulting in reduction in mass. Latticing also helps in giving a lot of cushioning and distributing the force exerted by foot evenly. After latticing the sole, expected design is made. But it also includes some errors like open surfaces, bad edges or cut pipe structures. It is a must to repair those errors and get a proper design. It might also be needed to edit the lattice features like its thickness or angle and for mesh adjusting the density of triangles to get a finer design is needed.

Slicing Parameters

Slicing is a very important part of any product which is to be manufactured by using additive manufacturing method. After exporting the sole with lattice structure in STL format, file is imported into a slicer. As printing of the shoe sole is done, by using FDM technique few things like support structure should be taken care of, the material that is used is TPU 95A [13] and for support QSR support material is used. This support material is water soluble, so it is easy to use, and the support removal process can be hands-free. Basic print settings are:

Layer thickness	0.15 mm
Extrude Temperature	225–250 Deg. Celsius
Print bed Temperature	50
Fan cooling is recommended	
Build plate adhesion	Kapton tape (PEI), Blue painter's tape.
Print speed	15–20 mm/s
Retraction is required	

After setting these things in the slicer, slicing of the STL file of shoe sole is done. The slicer will slice the model according to the settings and provide a preview of the results. If the results are as expected, then its G-code is made by saving the file in proper format. This G-code then can finally be provided to the 3D printing machine to print the shoe sole.

Results and Discussion

Usually, there are three parts in a sole (1) Outsole, (2) Midsole and (3) Insole. But nowadays they are integrated with each other to obtain maximum comfort and efficiency. To compare the results obtained by different types of sole design, two designs were analysed for the stress distribution and reaction forces. One of them was a solid structure and the other one has lattice structure in place of midsole. The lattice used was Octahedral with FCC.

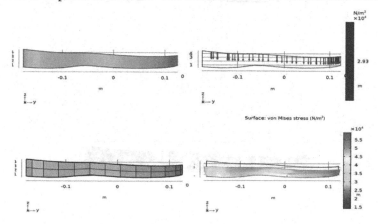

Figure 13.1 (top left) Sole geometry, (top right) boundary loads, (bottom left) mesh, (bottom right) stress distribution for shoe sole before latticing.

The geometry is a solid structure without any gaps in between them to compare it with latticed design. The boundary load of 800N was applied on the top face and the fixed constraint was given on the bottom face.

Table 13.1: Mesh Statistics of the shoe sole before latticing.

Description	Value
Minimum element quality	3.449E-4
Average element quality	0.2413
Prism	20
Hexahedron	104
Triangle	20
Quad	176
Edge element	76
Vertex element	6

Shoe Sole Before Latticing

The sole is printed using thermoplastic polyurethane (TPU) which is a flexible and abrasion resistant thermoplastic. It is used in many 3D printing processes such as FDM, SLS and SLA. It can withstand ambient temperatures up to 80 degree Celsius. Few properties of TPU include high durability and strength, very high flexibility, high abrasion resistance, medium high chemical resistance and medium water resistance. The support material is QSR, which is water-soluble. This aids in easy hands-free removal. FDM process is chosen. TPU filament can be utilized.

Shoe Sole After Latticing

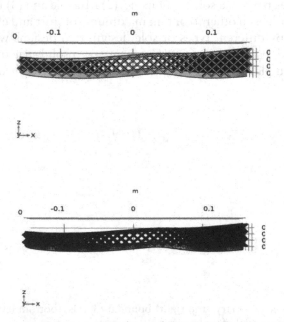

Figure 13.2 (top) Geometry and (bottom) Meshing of shoe sole after latticing.

The lattice structure used was Octahedral with FCC. To keep the results comparable, same material with same properties was used. Here also boundary load of 800N was applied on the top face and the bottom face was given a fixed constraint.

Table 13.2: Mesh statistics of the shoe sole after latticing.

Description	Value
Minimum element quality	5.469E-5
Average element quality	0.6083
Tetrahedron	1095981
Triangle	331706
Edge element	83503
Vertex element	17571

Comparison Between Various Designs

The material used for finite element analysis was Polyurethane with Young's modulus = 15 MPa and Poisson's ratio = 0.25.

The equation used in Solid Mechanics in FEA software is:

$$0 = \nabla \cdot S + F_v$$

The equations governing the elastic material in FEA software are:

$$0 = \nabla \cdot S + F_v$$
$$S = S_{ad} + C : \epsilon_{el}, \; \epsilon_{el} = \epsilon - \epsilon_{inel}$$
$$\epsilon_{inel} = \epsilon_0 + \epsilon_{ext} + \epsilon_{th} + \epsilon_{hs} + \epsilon_{pl} + \epsilon_{cr} + \epsilon_{vp}$$
$$S_{ad} = S_0 + S_{ext} + S_q$$
$$\epsilon = \frac{1}{2}\left[(\nabla u)^T + \nabla u\right]$$
$$C = C(E, v)$$

Note: COMSOL Multiphysics was used for finite element analysis

Table 13.3: Reaction forces after applying the force on a normal Sole.

Total reaction force, x component (N)	Total reaction force, y component (N)	Total reaction force, z component (N)
1.4884E-13	-8.5626E-15	800.00

Table 13.4: Reaction forces after applying the force on latticed sole.

Total reaction force, x component (N)	Total reaction force, y component (N)	Total reaction force, z component (N)
157	157	486

After analysing the results, it was found that after latticing the weight of the sole was reduced by about 40%. It also helped in reducing the reaction forces, which means forces acting on the legs of a person wearing the shoe will also reduce. The octahedral design of the lattice evenly distributes the load across different axes.

Conclusion

The results showed us that latticing a shoe sole can help in reducing the overall weight of the sole as well as making the shoe better for our legs by distributing the reaction forces along X and Y axes. The compressive stress acting on the lattice is bending stress and the elastic nature of material helps the sole to get in its previous position after removing the force.

Additive manufacturing is very essential in these kinds of designs as they cannot be manufactured using traditional manufacturing techniques like injection moulding or casting. As the reaction forces are reduced, the shoe becomes more comfortable and becomes good aesthetically. The paper also showed us that by latticing we use less material, making the shoe more environment friendly by reducing its carbon footprint. This helps in making the design more practical and sustainable.

References

1. Spahiu, T., Politeknik, U., Piperi, E., and Politeknik, U. (2016). 3D printing as a new technology for apparel designing and manufacturing. *International Textile Conference*, 1–5.
2. Spahiu, T., Grimmelsmann, N., Ehrmann, A., Shehi, E., and Piperi, E. (2016). On the possible use of 3D printing for clothing and shoe manufacture. In Proceedings of the 7th International Conference of Textile, (pp. 1–7).
3. Spahiu, T., Almeida, H., Ascenso, R. M. T., Vitorino, L., and Marto, A. (2020). Advanced technologies for shoe sole production. In MATEC Web Conference, (Vol. 318), 01012. https://doi.org/10.1051/matecconf/202031801012.
4. Kim, S., Seong, H., Her, Y., and Chun, J. (2019). A study of the development and improvement of fashion products using a FDM type 3D printer. *Fashion and Textiles*, 6. https://doi.org/10.1186/s40691-018-0162-0.
5. Pallari, J. H. P., Dalgarno, K. W., and Woodburn, J. (2010). Mass customization of foot orthoses for rheumatoid arthritis using selective laser sintering. *Transactions on Biomedical Engineering*, 57, 1750–1756. https://doi.org/10.1109/TBME.2010.2044178.
6. Dong, G., Tessier, D., and Zhao, Y. F. (2019). Design of shoe soles using lattice structures fabricated by additive manufacturing. In Proceedings of International Conference on Engineering Design ICED, 2019 August, (pp. 719–728). https://doi.org/10.1017/dsi.2019.76.
7. Tang, Y., Kurtz, A., and Zhao, Y. F. (2015). Bidirectional evolutionary structural optimization (BESO) based design method for lattice structure to be fabricated by additive manufacturing. Computer-Aided Design (CAD), 69, 91–101. https://doi.org/10.1016/j.cad.2015.06.001.
8. Franciosa, P., Gerbino, S., Lanzotti, A., and Silvestri, L. (2013). Improving comfort of shoe sole through experiments based on CAD-FEM modeling. Medical Engineering and Physics, 35, 36–46. https://doi.org/10.1016/j.medengphy.2012.03.007.
9. Kang, D., Park, S., Son, Y., Yeon, S., Kim, S. H., and Kim, I. (2019). Multi-lattice inner structures for high-strength and light-weight in metal selective laser melting process. *Materials and Design*, 175, 107786. https://doi.org/10.1016/j.matdes.2019.107786.
10. Stewart, L., Gibson, J. N. A., and Thomson, C. E. (2007). In-shoe pressure distribution in "unstable" (MBT) shoes and flat-bottomed training shoes: a comparative study. Gait and Posture, 25, 648–651. https://doi.org/10.1016/j.gaitpost.2006.06.012.

11. Zhang, G., Ma, L., Wang, B., and Wu, L. (2012). Mechanical behaviour of CFRP sandwich structures with tetrahedral lattice truss cores. *Composites Part B: Engineering*, 43, 471–476. https://doi.org/10.1016/j.compositesb.2011.11.017.

12. Brennan-Craddock, J., Brackett, D., Wildman, R., and Hague, R. (2012). The design of impact absorbing structures for additive manufacture. *Journal of Physics: Conference Series*, 382, 1–7. https://doi.org/10.1088/1742-6596/382/1/012042.

13. Ultimaker TPU 95A TDS. (2020). Ultimaker Support. https://support.ulti maker.com/hc/en-us/articles/360012664440-Ultimaker-TPU-95A-TDS.

14 Quantum computing by quantum dot-a review

P. Kour[1,a], Nidhi Priya[2], S. K. Pradhan[3], and Amit Kumar[4]

[1]Department of Physics, BIT, Mesra, Patna Campus, India

[2]Department of Computer Science Engg, BIT, Mesra, Patna Campus, India

[3]Department of Mechanical Engg, BIT, Mesra, Patna Campus, India

[4]Department of Mechanical Engg, NIT, Patna, India

Abstract: With the constant progress in technologies, each day we develop a new approach to overcome the limitations of their classical counterparts. Quantum computing is one such path breaking step, which is expected to overcome all limitations of conventional computers utilizing quantum mechanics, computer science, and mathematical models. Qubit remain an integral part of the quantum computation, and for better performance of quantum computers, the material used in qubit remains a center of consideration. Semiconductors qubits serve most of the purpose, and is marked as a principal material in the information processing. In this study, we have reviewed different types of qubit realizations comparing them on various parameters like their control, readout, materialistic approach, limitations, and recent advancements made, highlighting different opportunities in this field. Quantum scalability proposals have also been studied. We have systematically discussed different properties of semiconductor quantum dots, their applications, and the challenges like decoherence, and the opportunities that underlie, in order to determine a better material, and possibly a satisfactory reason for using semiconductors in quantum computing.

Keywords: Decoherence, quantum dot, quantum scalability, spin qubits.

Introduction

It was in 1982 when researcher [1–3] envisaged the idea to stimulate nature's laws on scalable individual particles, which gave rise to quantum computing concept. Powerful classical computers or even the Supercomputers hitherto, comprising of several processors is still not capable of encoding problems like a myriad combination of real world problems and others, which influences the need for quantum computers. Quantum computers are based on the concept of reversible computing, proposed by Deutsch. Reversible computing can probably provide us chance to go beyond the Von Neumann Landauer limit in a more effective way, compared to the theoretical limit. Present day processors use million times more energy than Von Neumann Landauer limit, whereas applying reversible computing concept to this, would help in reducing the limit. Qubits are the fundamental units of quantum computers capable of storing information in quantum form. Different prognosticate properties of qubits have influenced researchers and scientific groups to put forward ways for its realization to develop realizable quantum computing technologies. Ion trap

[a]paramjit.kour@bitmesra.ac.in

quantum computing [4] (in which ions can be manipulated using visible and infrared light), heteropolymer quantum computers, [5] superconducting circuits (responsible for reducing decoherence), semiconductor quantum computing [6], ionized donors in silicon [7], qubits composed of holes (for faster and reduced decoherence quantum computing), are few realizations techniques. Among these semiconductor quantum dots, shows promising results in designing quantum devices. Revolutionizing technologies in the last decade, has led to many breakthroughs in semiconductor quantum information processing. Semiconductor devices possess a favoring position over others, since the science used for their fabrication makes them eligible for getting interspersed along the real hardware. Pair of quantum dots act as a basic quantum logic unit, which is the qubit [8]. Quantum dots are a nanometer—solid state framework, produced in the laboratories, in which the motion of the charge carriers is constrained along all the three dimensions. Quantum dots are also called artificial atoms due to their singular and discrete nature, similar to atoms or molecules. It was also shown experimentally with ref. to [9], that both the quantum dots and real atom electronic wave functions show the resemblance. Moreover, coupling two or more quantum dots can successfully make artificial molecule exhibit hybridization even at room temperatures [10].

Semiconductor quantum dots comprise of confined electrons which work like electrons in individual atoms. Principally, electrons attain discrete energy level, which forces it to attain ground state of quantum dot, provided the temperature is smaller than the energy level gap [11]. Spin of electron stored in dot couples weakly with surrounding, thereby nominating it as a realizable option for qubit. Generally, material having non zero spin is considered apt for physical realization of qubits. Spin and charge degrees of electron or donor nucleus can potentially act as qubits [12–13]. Spin based qubits like spin 1/2, singlet-triplet (provides a system similar to double QD), and exchange only qubit (solely influenced by exchange interaction, providing universal qubit control), test less charge noise and have a longer decoherence time. However, for scalability purpose, qubits can be realized electrically enabling charge, exchange and hybrid qubits a better option, although they suffer limited fidelity. Another qubit called Hybrid qubit realizes the longer coherence time of spin qubit, and manipulative times of a charge qubit into it, which increases its productivity and scalability.

Quantum Scalability Proposals

DiVincenzo criteria for a Scalable Quantum Computing

DiVincenzo proposed idea of quantum computing to be based on the circuit model [14–15], which employs basic criteria, that ought to be satisfied by different proposals coming up to bring about an efficient computation. This theory introduces the presence of well characterized qubits with scalable physical system, the ability to compute states of the qubit to simple fiducial state, long decoherence time comparable to gate operation time, and to set a universal quantum gates. Well characterized qubit allows qubits interaction, thereby, protecting the system against different forms of decoherence. Having a fleet initialization mechanism is imminent for the quantum error correction. Longer decoherence time can be implied by encoding information into the logical quantum bits involving many single quantum bits. Threshold value generally ranges from 10^{-5} to 10^{-3}, implying decoherence time to range from 10^3 to 10^5 times longer than gate operation duration. Universal set of quantum gates is important to realize generic quantum computing.

Loss-DiVincenzo's Proposal

In Loss-DiVincenzo's proposal [16] qubit was realized as a spin of electron on single electron quantum dot system, focused on localized electrons in electrically gated semiconductor quantum dot. According to the proposal, localized gating implemented by applying the method of exchange coupling, can potentially provide scalability. Initially the realization allowed the electron count of every dot, cut to the single electron system, ensuring isolated spin system. Initializing every qubit to Zeemann ground state, by allowing spin to achieve thermal equilibrium at temperature T in presence of strong magnetic field B (provided $|g\mu_BB| > k_BT$), followed up next. Proposal discerns spin ½ of single electron being intrinsic regime, which protects qubit-state against turbulence.

Semiconductor Quantum Dots

Semiconductor quantum dots (QDs) are nanoscale materials and their size ranges from ~1nm to ~10nm, however, their linear size is bound to 1μm along any dimension [17–19]. This boundness creates a potential barrier of size proportional to the size of quantum dot and depth measure of the confinement potential, called quantum confinement, which highly influences optical and electronic properties of quantum dots. Quantum dots having larger size ~5nm–6nm in diameter, exhibit light of the red end of spectra; while those with smaller diameter ~2nm–3nm exhibit light near the blue end of the spectra. However, the light emitted is dependent on the exact composition of the dots.

Among different quantum dots, electrostatic or gate controlled QDs are pre-eminent because of the possible logic implementations performed using quantum logic gates. They comprise of sequential layer which develops a single or multi-level, potential wells and barriers. Electrostatic dots prove to be good performers in quantum computation, owing to their realizable electronic properties by an appropriate pick of parameters, and adjustability with regard to alternating external voltage give to electrodes. This property helps in achieving the desired design of quantum states and logical control operations performed in these states [20]. Structurally, pillar shaped dot devices are more efficient, with an extra gate electrode present on its cylindrical surface, which adds to its capability of tuning electrostatic field in electrostatic dots.

Quantum dots find various applications in different fields. Vertically gated quantum dot devices act as a prototype of transistor with a single electron, which has great futuristic advantages over others, because of its high performance in electronic devices at reduced power consumption. Quantum dots are also expected to replace phosphor with white LED, because of its narrow and tunable emission bandwidth. In display technologies, QDs attains an active interest owing to the increased energy efficiency and quality control over the color spectrum. The Dots also find application in the medical field; they can replace dyes used in cancer detection owing to their slow migrating tendency.

Provocations and Opportunities

Although advances in quantum computing like its scalability, and high fidelity single and two-qubit have laid possibility of fault tolerant quantum computation, still some challenges like quantum decoherence, qubit scalability and readout, extreme temperature conditions, algorithmic development, weak measurement, need to be addressed.

Semiconductor qubit readout depends on charge sensor; change in charge-state of qubit alters charges sensor's resistance. Different approaches have come forward to improving charge sensor performance, which monitors the readout speed and fidelity of qubits. Developments in the charge sensitivity for particular bandwidths are being brought up by different researchers, with constant improvement in charge sensitivity [21].

Completely isolating the quantum system can provide good quantum coherence, however it lends us measurement difficulties. Thereby at time of measurement, the system loses coherence on interaction with surrounding, which is called decoherence. It is a hedge in path of scalability and realizability of qubits. Low coherence or high fidelity, can reduce overhead error correction, avoiding the need of energy intensive noise reduction methods [22]. Input state and gate type influence decoherence of the system. Lower depth realization of quantum function helps us in attaining a lower decoherence, improving the fidelity by almost 20%. Other than this, Quantum control and Encoding have also been believed to potentially control decoherence in the system [23]. Quantum control theory aims at developing structured methodologies to achieve dynamic manipulation and control over quantum system. Attaining controllable quantum system is a major concern, which is desired to achieve universal quantum computation, and atomic-scale transformation. The finite level quantum system can provide a facile mathematical solution to the issue, in case of isolated systems. However, for larger dimensions, a method based on graph has been observed to promise a controllable quantum system with easily certified controllable criteria [24]. Open loop coherent control is another method which promises to meet the desired control on a quantum system [25]. Lyapunov based control method of open loop coherent control is an excellent tool which makes an artificially-closed loop system and then stimulates open loop methodology (since information cannot be perceived from quantum states without distracting them), debarring classical limitations. Although it is a good approach, still it is limited to individual transformations. Some uncontrollable system cannot be controlled by coherence control method. Incoherent control has been proposed for such systems, which allows quantum measurement and incoherence to employ quantum control task, thereby enhancing the system productivity [15]. Encoding the information lowers the possibility of decoherence in the system. Classical information processing involved messages, transfer in the form of bits through the channels, prone to noise which demolishes message. Encoding the information by using Modular Arithmetic, different Logic Gates, or OSR Algebra will ensure information to be protected against the noise by preserving parity, and error free information can be received on the other end.

Semiconductors have been marked to be highly scalable concerning to their use of computers, mobile phones, electronic devices, lasers, solar cell, and others. However, maximum possible qubit control in device has been still restricted to barely four qubits [26]. Therefore, properties of substrate majorly determine qubit's fabrication capability. GaAs/AlGaAs-heterostructure is a quality substrate, offering high mobility; however, high fidelity is not promised [27]. Silicon on the other hand, offers high fidelity for electron; however, it still suffers certain valley degeneracy complexity. Substantial drop in spin lifetime can be brought by tunable spin-valley mixing in spin ½ qubit [28], while inter valley splitting might limit dephasing time. In case of singlet-triplet-qubits or exchange-only qubits, readout fidelity reduces lifting the spin blockade for low valley [29]. While in Silicon hybrid qubits, controllable valley splitting in range has been ministered, leading to a rise in its demand. Besides Silicon, and Germanium as host material for spin qubit, hole spin is also rising as a potential host which would provide high fidelity qubit-control which is free from valley degeneracy, and can be purified to improve coherence time. After dealing the fidelity and readout issue, the next concern lies in scaling qubits on tens to hundreds scale. Crossbar-network in spin ½ qubits in Si-MOS QD, 2D Si donor qubit lattice, flip flop qubit and donor dot structures, proposals have come up to support scalability issue, however these register an issue of wiring strategy and balance between feasibility and fine performance.

Conclusion

Quantum computers would not only overcome the limitations of the classical computers but are also expected to show better performance. Although qubit is a major parameter for measuring performance of quantum computers having several advantages, however, some of its disadvantages when addressed would improve the performance of the quantum computers. Scalable- and well characterized qubits idea was proposed by DiVincenzo and Loss-DiVincenzo respectively, for achieving a scalable quantum computer. The drawback of spin and charge qubit to be overcome by implementing hybrid quantum circuits which would use advantage of each qubit in its implementation. Evolution in the quantum computer's performance is highly dependent on the quality of material used. Scalability and fabrication of qubits depend on the properties of semiconductors. The progress of the semiconductor qubit system is dynamic, and not static, and the development in this area of science is growing with a fast pace with the introduction to new types of qubits which are under development. Although the study is still underway, the near future, applications of the semiconductor quantum dots can bring a whole lot of changes in the computational and technological sectors.

References

1. Deutsch, D. (1985). Quantum theory, the church-turing principle and the united quantum computer. *Royal Society of London Series A*, 400, 97–212.
2. Feymann, R. P. (1982). Simulating physics with computers. *International Journal of Theoretical Physics*, 21, 467–488.
3. Feymann, R. P. (1986). Quantum mechanical computers. *Foundations of Physics*. 16, 507–531.
4. Brown, K. R., Wilson, A. C., Colombe, Y., Ospelkaus, C., Meier, A. M., Knill, E., Leibfried, D., and Wineland, J. (2011). Single-qubit-gate error below 10-4 in a trapped ion. *Physical Review A*, 84, 1–5.
5. Barends, R., Kelly, J., Megrant, A., Veitia, A., Sank, D., Jeffrey, E., White, T. C., Mutus, J., Fowler, A. G., Campbell, B., Chen, Y., Chen, Z., Chiaro, B., Dunsworth, A., Neill, C., O'Malley, P., Roushan, P., Vainsencher, A., Wenner, J., Korotkov, N., Cleland, A. N., and Martinis, J. M. (2014). Logic gates at the surface code threshold: superconducting qubits poised for fault-tolerant quantum computing. *Nature*, 508, 500–503.
6. Veldhorst, M., Hwang, J. C. C., Yang, C. H., Leenstra, deRonde, B., Dehollain, J. P., Muhonen, J. T., Hudson, F. E., Itoh, K. M., Morello, A., and Dzurak, A. S. (2014). An addressable quantum dot qubit with fault-tolerant control fidelity. *Nature Nanotechnology*, 9, 981.
7. Saeedi, K., Simmons, S., Salvail, J. Z., and Duluhy, P. (2013). Room-temperature quantum bit storage exceeding 39 minutes using ionized donors in silicon-28. *Science*, 342, 830–833.
8. Lucjan, J., Pawel, H., and Arkadiusz, W. (1998). Quantum Dots. Berlin: Spinger.
9. Zak, R. A., Rothlisberger, B., Chesi, S., and Loss, D. (2009). Quantum Computing with electron spins in quantum dots. *La Rivista del Nuovo Cimento*, 33, 1–60.
10. Kastner, M. A. (1993). Aritificial atoms. *Physics Today*, 46, 24–31.

11. Joos, E., and Zeh, H. D. (1985). The emergence of classical properties through the interaction with the environment. *Zeitschrift für Physik B Condensed Matter*, 59, 223–243.

12. Zang, X., Li, H. O., Cao, G., Xiao, M., Guo, G. C., and Guo, G. P. (2019). Semiconductor quantum computation. *National Science Review*, 6(1), 32–54.

13. Zhang, X., Li, H. O., Wang, K., Cao, G., Xiao, M., and Gou, G. P. (2017). Qubits based on semiconductor quantum dots. *SCIENTIA SINICA Informatics*, 47, 1255–1276.

14. DiVencenzo, D. P. (2000). The physical implementation of quantum computation. *Fortschritte der Physik*, 48, 771.

15. Chuang, I. L., and Nielsen, M. A. (2000). Quantum Computation and Quantum Information. New York: Cambridge University Press.

16. Loss, D., and DiVincenzo, D. P. (1998). Quantum computation with quantum dots. *Physical Review*, 57, 120.

17. Bacon, D. (2001). Decoherence, Control, and Symmetry in Quantum Computers. Berkeley: University of California.

18. Banin, U., YunWei, C., David, K., and Millo, O. (1999). Identification of atom—like electronic state in Indium arsenide nanocrystals quantum dots. *Nature*, 400, 542–544.

19. Adamowski, J., Bednarek, S., and Szafran, B. (2005). Quantum computing with quantum dots. *Physical Review A*, 14, 95–110.

20. Yanofsky, N. S. (2007). An Introduction to Quantum Computing. Cambridge University Press, (pp. 1–34).

21. Stehlik, J., Liu, Y. Y., Quintana, M., Eichler, C., Hartke, T. R., and Petta, J. R. (2015). Fast charge sensing of a cavity-coupled double quantum dot using a Josephson parametric amplifier. *Physical Review Applied*, 4, 014018.

22. Saki, A. A., Alam, M., and Ghosh, S. (2019). Study of decoherence in quantum computers: a circuit-design perspective. arXiv preprint arXiv:1904.04323.

23. Dong, D., and Petersen, I. R. (2010). Quantum control theory and applications: a survey. *The Institution of Engineering and Technology*, 4(12), 2651–2671.

24. Turinici, G., and Rabitz, H. (2001). Quantum wavefunction controllability. *Chemical Physics*, 267, 1–9.

25. Kuang, S., and Cong, S. (2008). Lyapunov control methods of control of uncontrollable quantum system. *Automatica*, 41, 98–108.

26. Ito, T., Otsuka, T., Nakajima, T., Delbecq, M. R., Amaha, S., Yoneda, J., Takeda, K., Noiri, A., Allison, G., Ludwig, A., Wieck, A. D., and Tarucha, S. (2018). Four single-spin Rabi oscillations in a quadruple quantum dot. *Applied Physics Letters*, 113(9), 093102.

27. Hanson, R., Kouwenhoven, L. P., Petta, J. R., Tarucha, S., and Vandersypen, L. M. K. (2007). Spins in few-electron quantum dots. *Reviews of Modern Physics*, 79, 1217–1265.

28. Yang, C., Rossi, A., Ruskov, R., Lai, N. S., Mohiyaddin, F. A., Lee, S., Tahan, C., Klimeck, G., Morello, A., and Dzurak, A. S. (2013). Spin-valley lifetimes in a silicon quantum dot with tunable valley splitting. *Nature Communications*, 4, 2069.

29. Maune, B. M., Borselli, M. G., Huang, B., Ladd, T. D., Deelman, P. W., Holabird, K. S., Kiselev, A. A., Alvarado-Rodriguez, I., Ross, R. S., Schmitz, A. E., and Sokolich, M. (2012). Coherent singlet-triplet oscillation in a silicon-based double quantum dot. *Nature*, 481(7381), 344–347.

15 Barriers to implementation of six-sigma in micro, small and medium enterprises by using DEMATEL method

Ritesh Raj[a] and Sonu Rajak[b]

Department of Mechanical Engineering, NIT Patna, Patna, India

Abstract: Over the last two decades, all over the world, six-sigma has been widely adopted by companies to reduce the variation in the processes drastically and to enhance the effectiveness of the business. It can also be seen as a set of management techniques to uplift businesses by maintaining zero defects. So, six-sigma is a business enhancement strategy. From its inception, six-sigma has mainly been adopted and implemented by only large corporate houses possessing vast resources. However, it can also be used by Micro, small and medium enterprises (MSME) for their enhancement in various aspects. In this context, this paper identifies and analyses the barriers for the implementation of six-sigma in MSME. This paper presents various issues and factors that act as barriers in six-sigma adoption by MSME and the ways to overcome them using the DEMATEL method. The results of this research present a roadmap to engineers, managers, researchers, and practitioners to address the barriers to the implementation of six-sigma.

Keywords: Barriers, decision-making, DEMATEL, MSME, six-sigma.

Introduction

In the 1980s, the Six-Sigma program was developed by Motorola engineer Bill Smith in response to the necessity to eliminate defects and increase the quality of their product. Bob Galvin, the CEO, was positively impressed with the successes achieved ahead of time. Under his leadership, Motorola started implementing six-sigma over the entire company, concentrating on manufacturing processes and systems [10]. The capacity of the six-sigma process in manufacturing or service system methods becomes equal to just 3.4 defects per million possible defects. This has been adopted for world-class manufacturing as a practice. The average level is between three and four sigma, around 6200 and 67000 defects per million, in modern industrial applications. Six-Sigma applies facts and data from the manufacturing and service processes to assess how they can be enhanced.

The organization initially used different methods and strategies of improvement to meet their needs. However, organizations have recently improved their efficiency by using the six-sigma initiatives as the new management technique to bring their traditional quality management methods back to life. Six-Sigma focuses on a glance quality control techniques [17]. Companies should make a continuous effort to improve themselves. The key

[a]riteshr.ug16.me@nitp.ac.in, [b]sonu.me@nitp.ac.in

concern for businesses must be to recognize and eradicate significant defects in the functionality of business implementations. Six-Sigma is known to be a business technique that helps businesses to increase their income by optimizing their operations, improving overall productivity, and eliminating defects. Understanding the customers' needs leads us to attain, sustain and optimize commercial performance. The services, production, product, and management processes need to be improved in all the organizations to survive in the long run in the market. With minimum defects, improved methods, reduced process variability, increased consumer satisfaction, reduce cost, increased profits, improved product quality, and enhanced productivity can be achieved by all the organizations [7]. The Six-Sigma approach has been used to tackle more complex problems, where a thorough investigation is required to fix the problem [8].

Established market improvement programs such as six-sigma should be used by MSME manufacturing sectors to meet global competition. In MSME, there are several misconceptions about six-sigma, such as a perception that six-sigma is adopted and applied only for large organizations with skilled and experienced workforce and adequate resources. There was also a rumor among MSME that six-sigma just adds additional expenses to their costs and does not give them a return on investment. But, in fact adoption of six-sigma will reduce the defects and variability of the product. So, Six-Sigma practices should be discussed to increase the awareness and interest of MSME. Enhancing and maintaining service quality is critical to ensuring customer satisfaction. Furthermore, higher service quality will contribute to a more prosperous local economy and provide a market advantage for MSME [9]. MSME should apply strategies that seek to remove the challenges to the Six-Sigma implementation.

Literature Review

To identify the important barriers to the implementation of six-sigma exhaustive literature surveys have been performed. Tyagi et al. [19], examined the setbacks in Six-Sigma execution in small and medium-sized enterprises (SMEs) and claimed that the lack of resources is a critical barrier to Six-Sigma implementation in SMEs, and lack of knowledge is the most crucial barriers to the enactment of the six-sigma program. However, the issue of leadership is also critical to the failure or success of six-sigma implementation. Improving all processes/services by reducing defects can increase the profits of an organization. The prime barrier in the implementation of six-sigma is the lack of training and guidance for successful completion of projects but training workshops are expensive for MSME. Training workshops must be organized by the enterprises for all their employees to overcome this barrier. Snee and Hoerl [18] study shows that nowadays, it is considerably easier for SMEs to obtain better external resources for guidance without paying higher costs. Kumar and Antony [5] showed that in SMEs, the organization had reduced various problems to make the framework more suitable, easily accessible, and applicable, which were incorporated into each phase of the six-sigma methodology. Furthermore, the dominant barrier was a lack of knowledge about implementing the six-sigma program.

Eckes [2] claimed that in SMEs the implementation of Six-Sigma has become poor due to leadership; six-sigma seemed like just a passing management fad. An organization must be created that has top management who not only knows the processes but also has the experience to create an organization that works together. Kumar and Antony [6] studied small and medium-sized enterprises in the UK and concluded that lack of resources was a critical barrier to implement six-sigma in SMEs. This was a result of other factors that play an essential role in Six-Sigma implementation in SMEs such as lack of training, inadequate employee participation, and knowledge, internal resistance, lack of commitment from managers, etc. Rotondaro [16] proposed that organizations require a change in organizational culture to overcome the barriers to the implementation of six-sigma. It is essential to consider six-sigma as a component of overall quality management for SMEs. There should be no cultural conflict arising within the company. A good manager should know how to tackle the issues causing trouble. Henderson and Evans [4] described that organizations should include the details that define and quantify the exact needs and requirements of consumers and the necessary improvement in the value currently offered to their customers. The implementations required cover how customer needs and expectations are interrelated. Only the implementation team will efficiently execute operations in a way that leads to satisfied products and better services. Brue [1] proposed that the type or size of the company in which the six-sigma methodology is applied does not matter. Whether there are 500 employees in a company or it is a family business of 15 employees, six-sigma will work effectively in both as long as the processes are followed. Lande et al. [11] studied that Indian MSME are in a rare instance in the new era of innovative technologies since they seek to increase the quality of their products/service to assure customer happiness while investing as little as possible in customization, automation, competitiveness, and responsiveness. Ghobakhloo [12]; Ghobakhloo and Fathi, [13] stated that MSME must implement higher quality improvements and sustainability strategies in order to take advantage of the new era, which would assure their sustainability in the global market. Further, a list of barriers to the implementation of six-sigma practices in MSME is given in Table 15.1. A literature review clearly illustrates that research has been carried out by researchers in the context of SMEs and very few studies are available for barriers to implementation of six-sigma practices in MSME. And there is no concrete research found by authors for the implementation of six-sigma practices in Indian MSME. In this context, this paper identifies and analyses the barriers to the implementation of six-sigma practices in an Indian MSME manufacturing firm. The following issues are answered in this research study.

• What are the barriers to the implementation of six-sigma in MSME?
• What are the causalities and dependencies between these barriers?
• What is the cause and effect relationship between these barriers?

Table 15.1: List of barriers for implementation of six-sigma practices in MSME.

Barrier	Code	Description	Source
Lack of resources	F1	In any project, there will be a lot of company resources consumed during the implementation of Six-Sigma, such as financial, human resources, time, etc.	[6, 14, 19];
Internal resistance	F2	There will be resistance to reform unless the advantages are visible. Employees get used to the method they have been using for a long time.	[6, 14, 14];
Lack of leadership from top executives	F3	Six-sigma is considered a passing management fad due to poor leadership. An organization should have innovative top management who have an understanding of the process and also know how to build an organization that works together to properly execute Six-Sigma.	[2, 6, 14],

Barrier	Code	Description	Source
Lack of knowledge about six-sigma	F4	Six-sigma needs good experience and proficiency on the part of its experts to successfully manage ventures. For individuals at various levels of an organization, proper knowledge of Six-Sigma strategies is very important.	[5, 14, 19];
Insufficient organizational alignment	F5	A crucial element in its success is the integration of the entire company with the adoption of Six-Sigma.	[14]
Cultural barriers	F6	Cultural tension must not exist inside an organization. A good manager needs to know well how to handle the issues causing trouble tactfully.	[2, 14, 16]
Poor training	F7	If companies recruit professional Six-Sigma consultants, companies should recruit the brightest, not anyone who touts gimmicks. The organizations will have to recognize that their expenses would not be waived by successful consultants and take a proportion of the resulting cost savings. With proven track records and skills, companies can go for the best consultants.	[14, 18],
False notion that Six-sigma is too complex to use	F8	Many companies avoid Six-Sigma, thinking too much work is needed for mathematics and statistics. With modern PC-based software, control charts can be easily created.	[4, 14]
Wrong identification of process parameters	F9	Critical to Process (CTP) is the main input process factors. The Six-Sigma at the highest level of the organization should increase the productivity and effectiveness of the organization. However, companies are involved in improving productivity at the expense of effectiveness. This arises because, after Six-Sigma implementation, the management needs a fast return on investment. Instead of recognizing the costs associated with the current level of ineffectiveness, by concentrating on the current level of inefficiencies, businesses are trying to benefit from short-term costs. As a result, companies consider inaccurate process parameters such as system downtime on critical dimensions instead of incorrect tolerance. Eventually, this leads to higher costs of production.	[14]
Lacunae in data collection	F10	Crucial factors play a key role in determining what information is needed and gathering the full data set in a six-sigma project. The quality and quantity of data collected for analysis depend on the solutions for services and goods that do not meet the needs of the customers. This collection of too much or too little information may be harmful to the process's improvement.	[14]
Poor six-sigma project selection	F11	The selection of proper projects is among the most critical considerations in the efficiency in the implementation of six-sigma. If the projects underway had little to no effect on the result, proper preparation and management of the projects would not matter. It is critically important to pick initiatives that have an impact on the organization's net profit.	[14]

Case Illustration

A case study has been conducted in an automotive component manufacturing MSME firm located in Jharkhand, India. The ISO 9000 quality management system, ISO 14000 environment management system, and total productive maintenance (TPM) are already implemented by the case organization. And, now the management is keen to implement Six-Sigma practices in their organization. In this context, this paper identified the barriers to implementation of six-sigma concept in the MSME. A list of barriers for the implementation of Six-Sigma practices in MSME is shown in Table 15.1. The barriers are analyzed by the DEMATEL (decision-making and evaluation laboratory) method. Also, using the DEMATEL approach, we evaluated the relationships between the barriers and validated the current literature and the findings of the experts.

Data Collection

The study begins with the documentation of barriers that had been identified by the literature survey as well as discussion with experts and decision-makers. To evaluate the relevant barriers for MSME, the relevant literature was checked. Much of the literature reported was about the challenges faced by MSME in India. The list of barriers was sent to experts to elicit their input through a direct relationship matrix that defines inter-relationships among various barriers. As the DEMATEL method can be used with a small sample size, we used three experts [15]. All three experts have rich experience and knowledge in the automotive sector in India. A detailed background of all experts is shown in Table 15.2. These experts had solved many Industrial and manufacturing problems raised in their respective industry over the years. Hence, these experts were considered to provide their opinions for our DEMATEL model. Each expert was provided with a detailed description of the 11 barriers. And, for these barriers, the initial three (11 × 11) direct-relationship matrices were formed.

Table 15.2: Experts' background.

Experts	Position	Area expertise	Experience (years)
E1	General manager	Operations management, six-sigma	20
E2	Quality head	Total quality management, six-sigma	15
E3	Supply chain manager	Supply chain management, logistics management, lean six-sigma	13

DEMATEL Method

The Battelle Memorial Institute Geneva Research Center has proposed a DEMATEL method [3]. DEMATEL is investigated as an efficient tool for defining the components of a complex system's cause-effect chain. The DEMATEL method has been opted for as it provides the overall charge of control of factors [3]. The DEMATEL process procedure is presented in Figure 15.1.

Below are discussed the computations of the DEMATEL method steps to research: Following steps are taken in the DEMATEL method:

Step 1: Find "X" (Direct-relation matrix).

Input data provided by experts upon inter-relationships among factors in the direct-relation matrix "X". This matrix shows how the other factors involved in the analysis are affected by each factor and is shown below:

$$X = \begin{bmatrix} 0 & x_{12} & \ldots & x_{1j} & \ldots & x_{1n} \\ x_{12} & 0 & \ldots & x_{2j} & \ldots & a_{2n} \\ & & \ldots & \ldots & & \\ x_{i1} & x_{i2} & \ldots & x_{ij} & \ldots & x_{in} \\ & & \ldots & \ldots & & \\ x_{n1} & x_{n2} & \ldots & x_{nj} & \ldots & 0 \end{bmatrix} \tag{1}$$

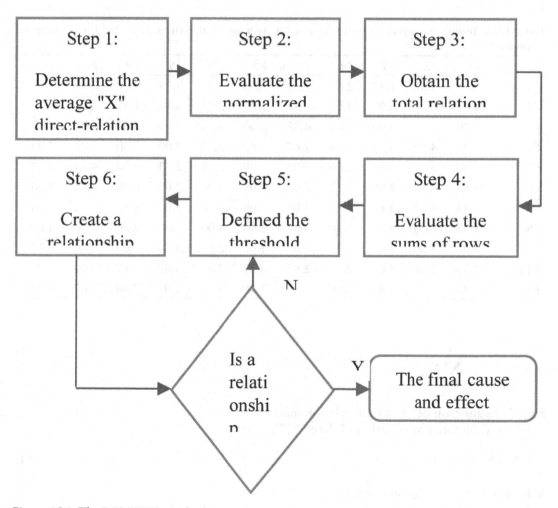

Figure 15.1 The DEMATEL method.

Obtain the direct relation matrix "X". First, the expert used a 6-point linguistic scale to look for the interrelationships between barriers (0: No influence, 1: Very low influence, 2: Low influence, 3: Medium influence, 4: High influence, 5: Very high influence). The ratings show the impact of one barrier on another (e.g., the impact of barrier F1 on barrier F2 on a scale of 0–5). The final direct relation matrix "X" is computed by taking an average of all three matrices provided by three experts and experience-based linguistic value. The "X" matrix entries are the average of the three sets of inputs given by three experts and are shown in Table 15.3. Step 2: Determination of "Y" (Normalized direct-relation matrix).

"Y" is determined using equations (2) and (3). In this matrix, every entry lies between 0 and 1.

$$Y = k \cdot X \tag{2}$$

Where, = Normalization factor and X = Initial relation matrix

Table 15.3: Initial direct-relationship taken as the average of all matrix given by three experts of response "X".

	F1	F2	F3	F4	F5	F6	F7	F8	F9	F10	F11
F1	0.00	3.67	1.33	2.67	3.33	1.33	3.33	4.33	3.00	3.67	4.67
F2	4.00	0.00	3.00	3.00	4.67	1.67	3.33	4.33	1.67	2.00	3.67
F3	1.00	3.00	0.00	4.00	4.33	2.00	4.33	4.00	3.00	3.00	4.33
F4	2.00	4.00	3.00	0.00	2.67	0.67	5.00	5.00	3.00	4.00	4.00
F5	4.00	3.33	2.00	2.00	0.00	0.00	3.00	2.33	4.00	4.33	4.00
F6	0.33	3.67	2.00	2.33	2.00	0.00	1.00	2.00	0.00	1.00	1.00
F7	3.33	4.33	3.67	3.33	3.00	1.00	0.00	3.00	1.67	3.33	4.00
F8	3.33	3.00	4.67	5.00	3.67	2.00	4.00	0.00	2.00	3.00	4.00
F9	4.00	3.00	2.33	4.00	2.67	1.00	2.33	3.00	0.00	3.00	2.67
F10	2.33	2.67	1.67	2.00	2.67	2.67	3.00	3.00	3.67	0.00	3.00
F11	2.33	3.00	2.67	4.00	3.67	2.00	3.00	2.00	2.00	1.67	0.00

$$k = \frac{1}{max\left(\sum_{j=1}^{n} a_{ij}\right)} (i, j = 1, 2, ..., n) \tag{3}$$

Step 3: Estimation of "T" (Total relation matrix)
   "T" is calculated by equation (4) from "Y".

$$T = Y (I - Y)^{-1} \tag{4}$$

Where, I denote the Identity matrix.
   Step 4: Evaluate the total sum of rows and columns in the "T" matrix.

Total sum of rows and columns are determined by equations (5) and (6) respectively.

$$R_i = \left[\sum_{j=1}^{n} t_{ij}\right]_{n\times1} = [t_i]_{n\times1} \ , i = 1, 2, \cdots, n \tag{5}$$

$$C_i = \left[\sum_{i=1}^{n} t_{ij}\right]_{1\times n} = [t_j]_{n\times1} \ , j = 1, 2, \cdots, n \tag{6}$$

Where, R denotes the total sum of rows, and C represents the total sum of columns in the total relation matrix "T".

Step 5: Defining the threshold value (α)

The threshold value is determined by average of total matrix "T" elements using equation (7).

$$\alpha = \frac{\sum_{i=1}^{n}\sum_{j=1}^{n} [t_{ij}]}{N} \tag{7}$$

Step 6: Formation of the casual diagram

At the final step, the values of $(R_i + C_i)$ and $(R_i - C_i)$ are used to construct a causal diagram. For each factor, a graph is plotted on the x-axis and y-axis using the values of $(R_i + C_i)$ and $(R_i - C_i)$.

Total relation matrix and degree of prominence and net cause/effect values for barriers has been shown in Table 15.4 and Table 15.5 respectively. And, DEMATEL prominence-casual diagram is shown in Figure 15.2.

Results and Discussion

The result and findings of this study are as follows. Table 15.5 demonstrates the values for each barrier. If the value for a particular barrier is positive, the barrier is considered to be in the causal group, and if the value for a particular barrier is negative, the barrier is considered to be in the effect group. The calculation of the threshold using is done by equation (7) and found to be 0.535844 that is highlighted in bold in Table 15.4. The threshold value is compared with the entries of the total relation matrix "T". If the values of the total relation matrix "T" are more than the threshold value, α = 0.535844, then there is an interrelation between them; otherwise, there is no interrelation between them. After that, a prominence-causal is drawn. The illustration of DEMATEL prominence-causal can be seen in Figure 15.2.

This research shows that the categories of the effect group include barriers, such as 'Internal resistance' (F2), 'Insufficient organizational alignment' (F5), 'Poor training' (F7), 'Lacunae in data collection' (F10), and 'Poor six-sigma project selection' (F11). Their values are negative and rely on other barriers. And, 'The lack of resource' (F1), 'Lack of leadership from top executives' (F3), 'Lack of knowledge about Six-Sigma' (F4), 'Cultural barriers' (F6), 'The false notation that the six-sigma is too complex to use' (F8) and 'Wrong identification of process parameters' (F9) are listed as causal barriers. These barriers influence other barriers and should be taken into account and eventually eliminated. They have

Table 15.4: Total relation matrix.

	F1	F2	F3	F4	F5	F6	F7	F8	F9	F10	F11
F1	0.474472	0.651951	0.495708	0.61692	0.642389	0.287339	0.643766	0.664143	0.503149	0.593898	0.723942
F2	0.581248	0.562966	0.541244	0.630074	0.681493	0.29608	0.651489	0.671205	0.475648	0.560827	0.708761
F3	0.524997	0.663849	0.481306	0.676364	0.692989	0.31445	0.696602	0.682518	0.523568	0.602849	0.744625
F4	0.565261	0.703193	0.57535	0.589651	0.670239	0.290551	0.730069	0.724676	0.535716	0.641285	0.756013
F5	0.550668	0.608292	0.479675	0.565112	0.520145	0.237898	0.600882	0.58247	0.505361	0.581079	0.671531
F6	0.253365	0.383239	0.291473	0.342631	0.342084	0.12823	0.317369	0.341377	0.219322	0.284333	0.34008
F7	0.555247	0.66419	0.548869	0.628223	0.632464	0.277309	0.554975	0.631098	0.468281	0.582616	0.705679
F8	0.605495	0.696596	0.625716	0.730755	0.709526	0.32933	0.722939	0.6144	0.524546	0.632644	0.773804
F9	0.535958	0.588607	0.479144	0.60159	0.577993	0.256137	0.574229	0.588772	0.389368	0.537574	0.62532
F10	0.460904	0.540891	0.430819	0.514291	0.537277	0.281509	0.545527	0.544259	0.451087	0.418568	0.58571
F11	0.464029	0.555365	0.459996	0.566939	0.568164	0.264212	0.554973	0.528043	0.41581	0.472825	0.515117

Table 15.5: Degree of prominence and net cause/effect values for barriers.

Barrier	R_i	C_i	$(R_i + C_i)$	$(R_i - C_i)$	Identify
F1	6.297677	5.571644	11.86932	0.726033	Cause
F2	6.361036	6.61914	12.98018	–0.2581	Effect
F3	6.604117	5.4093	12.01342	1.194817	Cause
F4	6.782002	6.46529	13.24729	0.316712	Cause
F5	5.903115	6.574763	12.47788	-0.67165	Effect
F6	3.243503	2.963046	6.206549	0.280457	Cause
F7	6.248951	6.59282	12.84177	–0.34387	Effect
F8	6.965752	6.572961	13.53871	0.392791	Cause
F9	5.754692	5.011856	10.76655	0.742836	Cause
F10	5.310843	5.928497	11.21934	–0.59765	Effect
F11	5.365472	7.150582	12.51605	–1.78511	Effect

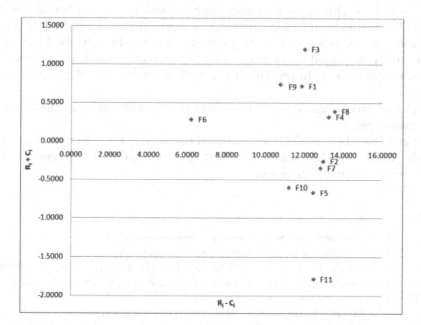

Figure 15.2 DEMATEL prominence-casual diagram.

a more influential effect on than an influential impact on. The larger the value, the higher its effect. High prominence barriers have a beneficial impact on the other barriers and are often influenced by other barriers. These are the barriers that are to be considered as the main barriers in the short term. Barrier F8 'False notion that six-sigma is too complex to use' indicates the highest significance and hence the highest association exists between this barrier and other factors F4 'Lack of knowledge about Six-Sigma' and F2 'Internal resistance'. Barrier F8 'False notation that six-sigma is too complex to use' and F9 'Wrong identification of process parameters' rank after F4 'Lack of knowledge about Six-Sigma'.

The results presented in the study make several contributions and also provides implications for the Six-Sigma practices. The barriers are identified in this research may guide to managers to the implementation of Six-Sigma practices in the industry. Quality managers may first concentrate on these barriers for the fast implementation of the Six-Sigma practices in the industry. This study identified the barriers to the implementation of Six-Sigma in MSME. This study identifies the cause and effect relationship between these barriers. This study identifies the hierarchical level among these barriers.

Conclusions

Nowadays, the market changes influence the definition of quality for consumers. Therefore, there is tough competition among the manufacturers in the market. It thus leads to the rise of many challenges to meet consumers' demands, primarily through the quality of goods having zero defects. This highlights the importance of the six-sigma methodology in organizations, mainly in MSME. This study describes and evaluates the barriers to Six-Sigma's implementation in MSME, using the DEMATEL method. Eleven potential barriers were identified based on an exhaustive literature review as well as a discussion with experts and decision-makers. This study examines the relationship between barriers by adopting the DEMATEL Method. DEMATEL is used to detect the cause and effect relationship among factors taken into account in the analysis and to produce the total impact of every factor. This study may guide managers and practitioners to the fast implementation of six-sigma practices in the industry. This framework is flexible enough to apply any type of organization.

The developed model has eleven barriers and may be extended beyond eleven barriers for better evaluation.

Fuzzy or grey based DEMATEL may be used to improve the effectiveness of the result. Also, a dedicated knowledge-based expert system may be developed to enhance the effectiveness of computation and analysis.

References

1. Brue, G. (2006). Six Sigma for Small Business. Entrepreneur Press.
2. Eckes, G. (2003). Six Sigma for Everyone. John Wiley and Sons.
3. Gabus, A., and Fontela, E. J. B. G. R. C. (1972). World problems, an invitation to further thought within the framework of DEMATEL. *Battelle Geneva Research Center, Geneva, Switzerland,* 1(8), 12–14.
4. Henderson, K. M., and Evans, J. R. (2000). Successful implementation of Six Sigma: benchmarking general electric company. *Benchmarking: an International Journal,* 7(4), 260–282.
5. Kumar, M., and Antony, J. (2008). Comparing the quality management practices in UK SMEs. *Industrial Management and Data Systems,* 108(9), 1153–1166.
6. Kumar, M., and Antony, J. (2009). Multiple case-study analysis of quality management practices within UK Six sigma and non-six sigma manufacturing small-and medium-sized enterprises. *Proceedings of the Institution of Mechanical Engineers, Part B: Journal of Engineering Manufacture,* 223(7), 925–934.
7. Mandahawi, N., Fouad, R. H., and Obeidat, S. (2012). An application of customized lean six sigma to enhance productivity at a paper manufacturing company. *Jordan Journal of Mechanical and Industrial Engineering,* 6(1), 103–109.
8. Hudnurkar, M., Ambekar, S., and Bhattacharya, S. (2019). Empirical analysis of Six Sigma project capability deficiency and its impact on project success. *The TQM Journal,* 31(3), 340–358.

9. Syapsan (2019). The effect of service quality, innovation towards competitive advantages and sustainable economic growth: marketing mix strategy as mediating variable. *Benchmarking*, 26(4), 1336–1356.
10. Montgomery, D. C., and Woodall, W. H. (2008). An overview of six sigma. *International Statistical Review/Revue Internationale de Statistique*, 329–346.
11. Lande, M., Shrivastav, R. L. and Seth, D. (2016). Critical success factors for lean six sigma in SMEs (small and medium enterprises). *The TQM Journal*, 28(4), 613–635.
12. Ghobakhloo, M. (2018). The future of manufacturing industry: a strategic roadmap. *Journal of Manufacturing Technology Management*, 29(6), 910–936.
13. Ghobakhloo, M., and Fathi, M. (2020). Corporate survival in Industry 4.0 era: the enabling role of lean digitized manufacturing. *Journal of Manufacturing Technology Management*, 31(1), 1–30.
14. Raghunath, A., and Jayathirtha, R. V. (2013). Barriers for implementation of six sigma by small and medium enterprises. *International Journal of Advancements in Research and Technology*, 2(2), 1–7.
15. Raj, A., Dwivedi, G., Sharma, A., de Sousa Jabbour, A. B. L., and Rajak, S. (2020). Barriers to the adoption of industry 4.0 technologies in the manufacturing sector: an inter-country comparative perspective. *International Journal of Production Economics*, 224, 107–546.
16. Rotondaro, R. G. (2002). Six Sigma. São Paulo: Atlas, (pp. 375).
17. Schroeder, R. G., Linderman, K., Liedtke, C., and Choo, A. S. (2008). Six sigma: definition and underlying theory. *Journal of operations Management*, 26(4), 536–554.
18. Snee, R. D., and Hoerl, R. W. (2003). Leading Six Sigma: A Step-by-Step Guide Based on Experience with GE and other Six Sigma Companies. Ft Press.
19. Tyagi, D., Soni, V. K., and Khare, V. K. (2014). A review on issues for implementation of six sigma by small and medium enterprises. *International Journal of the Latest Trends in Engineering and Technology*, 3(4), 94–98.

16 Determining safe limit for lifting task parameters in asymmetric posture during manual lifting

Anurag Vijaywargiya[a], Mahesh Bhiwapurkar[b], and A. Thirugnanam[c]

[1]Mechanical Engineering Department, OP Jindal University, Punjipathra, Raigarh, India

[2]Biotechnology & Medical Engineering, National Institute of Technology, Rourkela, India

Abstract: Building construction workers are prone to having musculoskeletal disorder problems while working with a bent and/or twisted trunk, which increases the risk of a back injury. The aim of the paper is to quantify the influence of task variables; lifting weight and destination height on instantaneous loading rate and subjective estimate by the subject during manual lifting task. Twelve male subjects age between 21 to 26 years lifted 5 weights (10–20 kg) to various vertical heights (below knee-ear level) of the subject. The subjects adopted free-style lifting techniques while lifting weights in 135-degrees asymmetric direction to the right. The subjective estimate was assessed by rating the overall workload and also analyses the discomfort perceived of each body part while performing lifting task. The vertical GRF recorded from force plate was used to calculate loading rate. The result shows that lifting weight in elevated destination height increases the both loading rate and overall workloads. When the 15 kg moderate load is raised above the knee level, the task becomes much more stressful. The present study proposes to keep 12.5 kg as the highest lifting weight that is considered suitable from knee to ear level.

Keywords: loading rate, lower back disorders, MMH, subjective workload.

Introduction

In this era of extensive use of automation, manual lifting tasks are still inherent in many different jobs in industry. Manual lifting can require the worker to perform lifting tasks in awkward *postures such as frequent bending,* twisting, *lifting,* pushing, pulling or performing combinations of these activities. The mechanical loading of the low back in manual material handling (MMH) has been identified as a high risk activity for low back pain [1]. The low back, hip, and knee joints are principally responsible for sharing the external force generated by lifting the weight. Mechanical loading of the spine is linked to a higher *risk of developing* low back pain [2, 3], which inflicts a severe global health/economic burden [4]. Therefore, several studies have explored the ergonomic risk factor for mechanical back loading, by experimental investigations using force plates and motion capture systems.

Indian labourers frequently use two-handed lifting techniques while twisting their torsos in the building construction industry, especially when loading and unloading in confined

---

[a]anurag@opju.ac.in, [b]mahesh.bhiwapurkar@opju.ac.in, [c]thirugnanam.a@nitrkl.ac.in

workspaces. Workplace restrictions could result in low back pain. For approximately three decades, researchers have utilized psychophysical methods to develop guidelines (weights, forces and frequencies) for MMH tasks [5, 6]. Many authors [6, 7] *recommended* that the psychophysical approach is a *credible technique* to evaluate the perceived exertion during MMH tasks in low and moderate frequencies (4 lifts/min). Although physiological measures of workload may be more precise, subjective measures are also more practical. Subjective reactions to physical work, such as perceived effort or perceived workload, have often been found to correlate with work intensity and work performance. It was also noted that earlier research on asymmetric lifting was restricted to asymmetry angles of 90 degrees. A void in the body of knowledge exists in that none of these research examined lifting task variable in 135-degree asymmetric angle.

With this goal in mind, the purpose of the current experimental investigation is to use subjective and biomechanical loading evaluation method to assess the risk associated with lifting in asymmetrical posture. The lifting load and destination heights are used as two independent variables for lifting task. The findings may contribute to the advancement of knowledge and the development of new lifting recommendations that relaxes the allowable lifting weight due to higher degree of asymmetry associated with the lift.

Subject and Method

In the current study, twelve young Indian male subjects, aged between 21 and 26 years' institute students have participated. The participants' mean age was 23.5 ± 1.78 years, weight was 70.67 ± 2.57 kg, and height was 1.76 ± 0.027 m. Each subject is required to lift a 'tasla', a container used in field work to lift different construction materials (cement, sand and grit). The lifting cycle started from the below knee height to a given destination level, rest for 3 s. The lifting task required the subjects to lift containers weighing 10, 12.5, 15, 17.5 and 20 kg on five vertical lifting heights: at below knee, at the knee, at the waist, at the shoulder, and at the ear level of the individual. This MMH task is considered suitable for the Indian building construction sector. It was further assumed that subject's foot was fixed on the ground.

Experimental Setup

The experiment was conducted in the NIT Rourkela's Biomechanics Lab [8]. The multi-axial force platform of Kistler made (Model AA9260) was used to record ground reaction forces (GRF). The size of the force plate used is $500 \times 590 \times 50$ mm. The force platform's analogue output is received by the data acquisition system after being amplified internally and sampled at a frequency of 1000 Hz to create a digital signal. Finally, QTM software record the data.

Biomechanical Evaluation

The effect of lifting task variables was analyzing from GRF and subjective workload using a body discomfort chart. A force platform was used to measure GRF *in* vertical direction during the lifting cycle. Figure 16.1 depicts the setup configuration. The rate at which the GRF applied in vertical direction was recorded while being lifted. The peak vertical GRF (loading rate) can lead to stress related injuries, while performing manual lifting task [9].

Loading rate (LR) is calculated from difference in GRF between start and end of lift, and the corresponding time span, as shown in Figure 16.2. Mathematically,

$$LR = \frac{F_{zmax} - F_{zmin}}{t_2 - t_1}$$

Initiation of lift from origin (Fz_{min}) to deposit of container (F_{zmax}) at destination height is represented as the maximum and minimum value of F_z of lift cycle and ($t_2 - t_1$) is the corresponding time span, Figure 16.2.

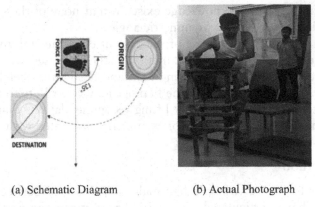

| (a) Schematic Diagram | (b) Actual Photograph |

Figure 16.1 Experimental set-up at biomechanics lab.

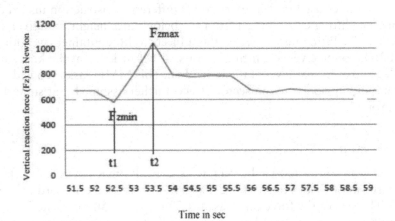

Figure 16.2 Vertical GRF graph for lifting 17.5 kg load from below knee to shoulder height.

The subjective estimates were obtained immediately after the subject performed the lifting task from the ratings of overall workload in body discomfort chart. A questionnaire was given to each respondent as part of the subjective evaluation as shown in Figure 16.3 [10]. The questionnaire of body discomfort chart assessing the subject's degree of discomfort while lifting the weight. This chart looks at the neck, shoulder, upper back, arm, elbow, lower back, forearm, wrists, hands, and fingers as well as the hip, thighs, knees, and ankle. Each subject performed lifting task for a total 25 test conditions (5 weights × 5 destination heights). The subjective responses were recorded for the level of pain in each listed body segments and overall workload. Both the subjective responses have a five-point scale as shown in Figure 16.3.

Test Procedure

Before starting the experiment, a trial was taken from each subject by lifting a container of known weight from below knee level (symmetric sagittal plane) to the various destination heights in one uninterrupted motion. The lifting exercise required each subject to complete 2 to 3 trials, was correctly explained to each participant. Each subject lifted weight in 135-asymmetrical direction, while standing on the force plate with free style lifting as per test conditions. After each lifting, an adequate rest period was provided to relieve from fatigue. No movement of the feet was allowed during the lifting cycle. The Fz was recorded for each test condition for all the twelve subjects.

Response Data Analysis

To assess the subject's response, an analysis of variance (ANOVA) was carried out. The outcomes of the statistical studies are presented in terms p-value and data R^2. All the data were subjected to a non-parametric Wilcoxon matched-pairs signed ranks test to see if the independent variables (weights and heights) significantly affected the dependent variables (LR and overall workload). 5% threshold of statistical significance was accepted. ($p < 0.05$) using two-tailed test and the. For all statistical analyses, SPSS Inc., Chicago, USA, version 16 of its statistical software for social sciences was used. In a within-subjects design of ANOVA, *each* participant is tested under *all* conditions. Given that each subject's judgement was repeatedly recorded for all test situations, the repeated-measures design is ideal.

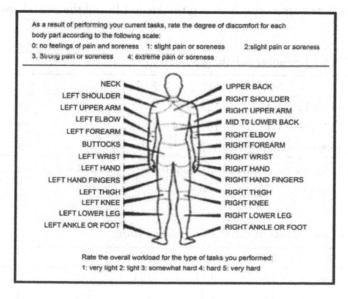

Figure 16.3 The subjective response questionnaire [10].

Results and Discussions

In the within-subject test, general effects *of* two independent variables; weight and destination height on both the dependent variables; LR and overall workload were found (Tables 16.1 and 16.2). All of the major parameters are significantly relevant for the evaluation of responds as shown by the significant interaction effect in this table ($p<0.5$). For all the independent and interaction factors, the observed power often has high values. The results show that all the independent variables and their interaction are equally affecting both the subjective and LR responses during lifting task. The table indicates least contribution from interaction variable for overall workload.

All 12 subjects have had their vertical reaction forces evaluated across all trials. The mean of 12 subjects are plotted for LR and overall workload for lifting five different weights at five destination heights are displayed in Figures 16.4 and 16.5. The graph clearly demonstrates that the LR rises practically linearly as the weight to be lifted increases and also found increases with destination heights. The LR was shown to be identical between waist and knee height irrespective of the load lifted ($p>0.05$). When the lifting heights rise over this level to ear level, there is a noticeable increase in LR ($p<0.05$) that indicates that people performing dynamic lifting tasks are exerting themselves more. Additionally, it can be seen from both plots that the LR for lifting the lightest weight, regardless of destination height, does not significantly differ heights ($p>0.05$). The results are also found to be in line with subjective evaluation.

Lifting weight and lifting heights had a considerable influence on LR, thus different combinations of these variables were employed to determine the smallest and greatest strenuous work conditions. From Figure 16.4, Irrespective of destination height, it was found that when the weight was increased from 15 to 20 kg, the mean LR increased by about 17 to 20%. The LR also increased by about 12 percent when the destination height was raised from waist to shoulder level, irrespective of lifting load. Beyond shoulder level, the LR was found to increase by 20–25 percent.

Figure 16.5 shows the mean overall workload rating for the lifting task. It's interesting to note that there is some correlation between the mean LR and the overall workload rating.

Table 16.1: ANOVA for dependent variable LR.

Source	SS	DF	MS	F	p	η^2	O.P
Weight (W)	1692632	1.3	1238039	5572	.00	.99	1.0
Height (H)	1559302	1.4	1090164	10013	.00	.99	1.0
W * H	460016	2.2	202228	860	.00	.98	1.0

SS: Sum of square; DF: degree of freedom; MS: Mean square; *p*: Significance level; η^2: Partial eta Squared; O.P: Observed power

Table 16.2: ANOVA for dependent variable overall workload.

Source	SS	DF	MS	F	p	η^2	O.P
Weight (W)	256	1.8	139	163	.00	.93	1.0
Height (H)	20	3.0	6.81	47.7	.00	.81	1.0
W * H	20	53	3.94	7.5	.00	.40	0.99

Both responses are based on the notion that subjects in dynamic lifting exercises must exert more effort. According to the subjective ratings of discomfort, it was found that lifting a 15 kg weight up to the shoulders height caused strong pain (rating 3) in both shoulders, both upper arms, both upper arms and mid to lower back and extreme pain in both upper arm (rating 4) at ear level. An increase in weight causes excruciating discomfort in the ear height. Therefore, there is need to established the safe lifting limit in 135-degree asymmetric angle to prevent/reduce injuries to workers involved in construction fields.

Based on LR and subjective rating data, the safe limits have been proposed, assuming alarming levels of both subjective responses; perceived difficulty and workload as rating "2,". For instance, Weights should weigh no more than 12.5 kg when being lifted from knee to ear level. The maximum weight restriction of 15 kg must also be limited to below the knee. The safe upper limit of LR for asymmetric angles from 0 to 60 degrees was proposed by Singh et al. (2014). These finding was based on the regression equation developed from GRF, heart rate and oxygen uptake measurement [11]. From the Figure 16.5, the present study proposed 220 N/s as the safe LR for lifting weight at an asymmetric angle of 135 degrees, assuming that a person engaged in the construction industry has an overall workload rating of 2, which is considered to be less stressful.

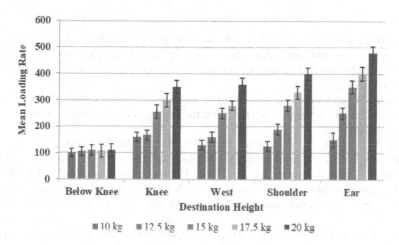

Figure 16.4 Mean LR while lifting weights to destination height.

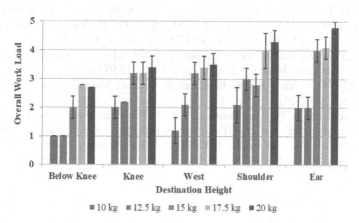

Figure 16.5 Mean overall workload while lifting weights to destination height.

Conclusion

The primary goal of the current study is to decrease worker effort and the risk of subsidence by employing the most suitable ranges of lifting task variables during manual lifting activity. It was seen that increases in lifting loads with elevating destination heights, increases the overall workloads. When the heavy (15 kg) load is raised beyond the knee height, the tasks become significantly more difficult. As a result, it is suggested that 12.5 kg remain the maximum weight that can be lifted safely from knee to ear level.

Based on the *biomechanical* and *subjective* responses in the present study, the safe limit for LR is proposed to be established at 220 N/s for lifting weight in an asymmetric angle of 135-degree. If the LR is increased beyond this safe level, one may feel pain and discomfort, which could cause anxiety about the workload. The findings of this research will be useful for improving lifting guidelines in asymmetric direction during manual material handling. The current study was restricted to the constrained age group of novice male handlers. Adding information about the approach used by expert handlers with a wide age group of both male and female provides a new avenue for reducing the worker's physical effort during manual material handling.

References

1.   Das, B. (2015). An evaluation of low back pain among female brick field workers of West Bengal, India. *Environmental Health and Preventive Medicine,* 20, 360–368.
2.   Maher, C., Underwood, M., and Buchbinder, R. (2017). Non-specific low back pain. *The Lancet,* 389, 736–747.
3.   Da Costa, B. R., and Vieira, E. R. (2010). Risk factors for work-related musculoskeletal disorders: a systematic review of recent longitudinal studies. *American Journal of Industrial Medicine,* 53, 285–323.
4.   Knutsson, B., Sandén, B., Sjödén, G., Järvholm, B., and Michaëlsson, K. (2015). Body mass index and risk for clinical lumbar spinal stenosis, a cohort study. *Spine,* 40(18), 1451–1456.
5.   Alferdaws, F. F., and Ramadan, M. Z. (2020). Effects of lifting method, safety shoe type, and lifting frequency on maximum acceptable weight of lift, physiological responses, and safety shoes discomfort rating. *International Journal of Environmental Research and Public Health,* 17, 3012.
6.   Fernandez, J. E., and Marley, R. J. (2014). The development and application of psychophysical methods in upper-extremity work tasks and task elements. *International Journal of Industrial Ergonomics,* 44(2), 200–206.
7.   Widia, M., Md. Dawal, S. Z., and Yusoff, N. (2019). Maximum acceptable frequency of lift for combined manual material handling task in Malaysia. *PLOS One,* 14(5), e0216918.
8.   Jena, S., Sakhare, G. M., Panda, S. K., and Thirugnanam, A. (2017). Evaluation and prediction of human gait parameters using univariate, multivariate and stepwise statistical methods. *Journal of Mechanics in Medicine and Biology,* 17(5), 1750076.
9.   Nigg, B. M. (1994). Force in Biomechanics of the Human Musculoskeletal System. Chichester: John Wiley, (pp. 200–224).
10.   Sauter, S. L., Schleifer, L. M., and Knutson, S. J. (1991). Work posture, work station design and musculoskeletal discomfort in a VDT data entry task. *Human Factors,* 33, 151–67.
11.   Singh, R. P., Batish, A., and Singh, T. P. (2014). Determining safe limits for significant task parameters during manual lifting. *Workplace Health Safety,* 62(4), 150–160.

17 Performance of 2-lobe non-recessed worn out hybrid bearing operating with non-Newtonian lubricant

Prashant B. Kushare[1,a] and Satish C. Sharma[2,b]

[1]Department of Mechanical Engineering, K.K. Wagh Institute of Engineering Education and Research, Nashik, India

[2]Department of Mechanical and Industrial Engineering, Tribology Laboratory, Indian Institute of Technology Roorkee, India

Abstract: In this study, influence of worn out of bearing surface on a 2-lobe non-recessed non-Newtonian lubricated bearing has been investigated analytically. Worn bearing surface analysis is accounted by Dufrane's abrasive model. The change in lubricant behavior is accounted by cubic shear stress law model. Finite element Galerkins technique is used to get the solution of the Reynold's equation. The results of study show that the value of $\bar{h}_{min}$ decreases as the wear of bearing surface increases. It also has effect of non-linearity factor. The wear defect significantly affects the performance of a 2-lobe non-recessed journal bearing. A maximum percentage rise in the value of direct stiffness coefficient ($\bar{S}_{11}$ & $\bar{S}_{22}$) is found of the order of 37.57% and 19.71 % respectively in unworn condition of 2-lobe journal bearing at offset factor δ = 1.25 as compared to unworn non-recessed circular hybrid bearing. The stability point of view 2-lobe journal bearing perform better than the circular journal bearing.

Keywords: Hybrid bearing, 2-lobe, non-Newtonian, wear.

Introduction

The design of 2-lobe hybrid journal bearing has become quite complex because of technological advancements in high speed precession machines and related engineering applications during the last few decades. Now days, many researchers [1–4, 11, 13–15] have given their attention towards a 2-lobe journal bearing configurations because of excellent dynamic characteristics and shaft stability. Owing to present technological advancement, the machines are to be used for high load and high speed applications, results in to wear. Due to the continuing operation of the machinery, a bearing is subjected to go for a number of cycles during the operational time of a machine and is subjected to quite a lot of start and stop operations. These start and stop operation period of the bearing system initiates the progressive wear of bush due to rubbing. As a result of this, wear occurs, which ultimately change the fluid film distribution with bearing performance. The experimental and analytical studies of influence of worn out bearing surface of journal bearings have been carried by many researchers [5–11]. Dufrane et al. [5] studied and presented the process of occurrence of wear because of start/stop operations in journal bearings and developed

---

[a]pbkushare@gmail.com, [b]sshmefme@iitr.ernet.in

a model of worn segment. Dufrane model has been used by several researchers [6, 7, 11, 14, 15] in their study. As reported in the literature, a very few investigators [8–11, 14] have studied the impact of worn out surface on the functioning of the hybrid journal bearings. Furthermore, performance of bearing lubricant is improved by adding high molecular weight polymers. Because of this, these lubricants conform to a nonlinear behavior like non-Newtonian fluid. Therefore, the analysis of bearing based on Newtonian fluid hypothesis, does not give a realistic bearing performance data. Recent years have witnessed the focus of many researchers in the studies of the non-Newtonian lubricated bearings performance in the literature [12–15]. Many researchers have employed cubic shear stress law to take into account the nonlinear rheology of lubricant for the analysis of circular journal bearing systems [6, 12, 13]. A detailed survey of the available literature reveals that a very few studies, which deals with the influence of wear and nonlinear rheology of lubricant on the operation of the non-circular journal bearing reported in the literature [14, 15]. Thus, theoretical investigation of CFV compensated symmetric 2-lobe non-recessed journal bearing is proposed. The geometry of symmetric 2-lobe (2-Row and 12 Holes per row) journal bearing is shown in Figure 17.1 (a&b).

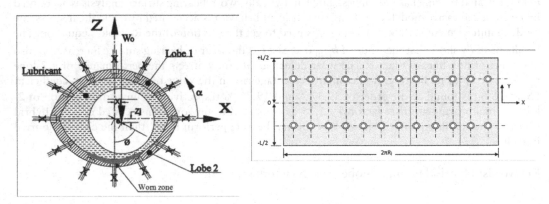

Figure 17.1 (a) Symmetric 2-lobe hybrid (b) Geometry of symmetric journal bearing system 2-lobe bearing

Analysis

Reynolds Equation

The equation of non-Newtonian lubricant flow field in the gap (clearance) of the shaft and bush is expressed as a generalized Reynolds equation [14, 16]:

$$\frac{\partial}{\partial \alpha}\left(\frac{\bar{h}^3}{\bar{\mu}}\frac{\partial \bar{p}}{\partial \alpha}\right) + \frac{\partial}{\partial \beta}\left(\frac{\bar{h}^3}{\bar{\mu}}\frac{\partial \bar{p}}{\partial \beta}\right) = \frac{\Omega}{2}\frac{\partial \bar{h}}{\partial \alpha} + \frac{\partial \bar{h}}{\partial \tau} \tag{1}$$

Minimum Fluid Film Thickness

The worn out bearing surface region geometry and coordinate system is shown in Figure 17.1 (a&b). The expression of the worn zone of a bearing surface is defined by Dufrane et al. [5] the fluid-film thickness expression of a 2-lobe worn bearing is given as [4, 11, 14, 17]:

$$\bar{h} = \frac{1}{\delta} - \left(\bar{X}_j + \bar{x} - \bar{X}_L^i\right)\cos\alpha - \left(\bar{Z}_j + \bar{z} - \bar{Z}_L^i\right)\sin\alpha + \partial\bar{h} \tag{2}$$

Equation of Non-Newtonian Fluid Flow Through a Constant Flow Valve Restrictor

The non-Newtonian lubricant flow equation of a CFV restrictor is written as [15, 16]:

$$\bar{Q}_R = \bar{Q}_C \tag{3}$$

Cubic Shear Stress Law Model

Using cubic fluid law model [12–16], the apparent viscosity $(\bar{\mu}_a)$ of non-linear behavior fluid is defined in terms of shear stress and shear strain arte $\dot{\gamma}$ as

$$\bar{\mu}_a = \bar{\tau}/\dot{\bar{\gamma}} \tag{4}$$

where,

$$\bar{\gamma} = \left[\left(\frac{\bar{h}}{\bar{\mu}}\frac{\partial\bar{p}}{\partial\alpha}\left(\bar{z} - \frac{\bar{F}_1}{\bar{F}_0}\right) + \frac{\Omega}{\bar{\mu}\,\bar{h}\,\bar{F}_0}\right)^2 + \left(\frac{\bar{h}}{\bar{\mu}}\frac{\partial\bar{p}}{\partial\beta}\left(\bar{z} - \frac{\bar{F}_1}{\bar{F}_0}\right)\right)^2\right]^{1/2} = \bar{\tau} + \bar{K}\bar{\tau}^3 \tag{5}$$

Finite Element Formulation

The global system of equations is obtained using Galerkin's orthogonality conditions and can be written as [13]:

$$\left[\bar{F}\right]\ \{\bar{P}\} = \{\bar{Q}\} + \Omega\ \{\bar{R}_H\} + \dot{\bar{X}}_J\ \{\bar{R}_x\} + \dot{\bar{Z}}_J\ \{\bar{R}_z\} \tag{6}$$

Numerical Solution

The steady state pressure has been obtained by solving governing equation of fluid flow field (Eq. 1). Newton–Raphson method is used to obtain the solution of Eq. (3) and (4) for the values of apparent viscosity. The change in the value of viscosity has been utilized further to solve the Eq. (1). The associated boundary conditions reported in reference [14, 17] have been applied to solve the modified form of system equation (6). These boundary conditions are described as [14, 15, 17]:

1. The hole positioned nodes have the same pressure.
2. The compensating element flow is equal to the bearing entry flow at hole position.
3. Zero pressure is assumed on external boundary nodes. $\bar{p}\,|_{\beta\,=\,\mp1.0}\,=\,0.0$.
4. $\bar{p} = \dfrac{\partial\bar{p}}{\partial\alpha} = 0.0$

Results and Discussion

A numeric model has been developed in order to assess the performance of non-Newtonian lubricated symmetric 2-lobe bearing system. To check the authenticity and validity of the developed model and methodology used in the analysis, the results are compared with the available published results [12]. It shows quite a good agreement (Figure 17.2). A 2-lobe bearing performance characteristic obtained from the simulated results are plotted and presented (Figures 17.3–9) for various values of non-circularity, non-linearity factors and wear depth parameter.

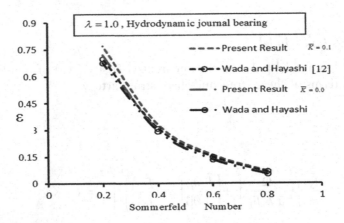

Figure 17.2 Eccentricity ratio (ε) versus Sommerfeld number (S_0).

Maximum Fluid Film Pressure ($\bar{P}_{max}$)

Figure 17.3 depicts the variation $\bar{P}_{max}$ in as a function of external load ($\bar{W}_0$) for 2-lobe journal bearing. The value of $\bar{P}_{max}$ increases with increase in the value of $\bar{\delta}_w = 0.5$.The maximum increase in fluid film pressure is observed in the order of 72 % at $\bar{\delta}_w = 0.5$ for an external load $\bar{W}_0 = 1.5$ at $\delta = 1.25$ with respect to the circular unworn bearing ($\delta = 10$ and $\bar{\delta}_w = 0.0$). The worn out region reduces the bearing clearance and increases the converging area. As a consequence of this, value of $\bar{P}_{max}$ gets increased to bear the applied load. Further, it may be noted that the increase in the value of δ influences the value of $\bar{P}_{max}$. It enhances the value of $\bar{P}_{max}$ for both worn $\bar{\delta}_w = 0.5$ and non-damaged ($\bar{\delta}_w = 0.0$) condition.

Minimum Fluid Film Thickness ($\bar{h}_{min}$):

The variation in $\bar{h}_{min}$ for 2-lobe bearing is shown in Figure 17.4. From Figure 17.4, it is observed that the value of $\bar{h}_{min}$ is reduced for increased value of $\bar{W}_0$. Further, it may be noticed that the bearing with $\delta = 1.25$ and $\bar{\delta}_w = 0.5$ runs at a lesser value of $\bar{h}$. The results also shows the nonlinear behavior of the lubricant increases ($\bar{K} = 1.0$), the value of $\bar{h}_{min}$ decreases. This is happened due to viscosity drop effect of the lubricant. The wear defect ($\bar{\delta}_W$) and viscosity change behavior of the lubricant ($\bar{K}$) significantly affects the value of $\bar{h}_{min}$ and the 2-lobe bearing runs at a lesser value of $\bar{h}_{min}$. Thus, there is a need to take care while choosing the appropriate bearing geometries.

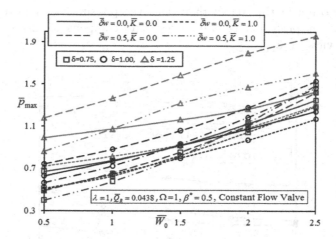

Figure 17.3 Maximum fluid film Pressure ($\bar{P}_{\max}$).

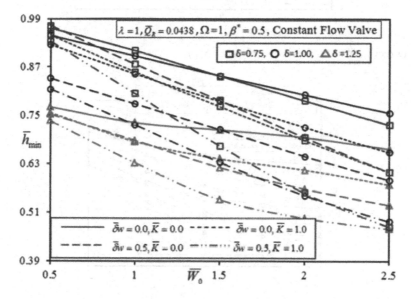

Figure 17.4 Minimum fluid film thickness ($\bar{h}_{\min}$).

Direct Stiffness Coefficient ($\bar{S}_{11}, \bar{S}_{22}$)

The variations of $\bar{S}_{11}$ & $\bar{S}_{22}$ are shown in Figures 17.5 and 17.6. From Figure 17.5, The reduced value of $\bar{S}_{11}$ is observed owing to ruin bearing surface. However; change in non-circularity factor (δ) from 0.75 to 1.25 compensates the loss of $\bar{S}_{11}$ under condition of worn and unworn journal bearing. An increase in the value of $\bar{\delta}_W$ decreases the value of $\bar{S}_{11}$ & $\bar{S}_{22}$ for all the values $\bar{W}_0$ of for 2-lobe non-recessed bearings. The influence of $\bar{\delta}_W$ and $\bar{K}$ decreases the value of $\bar{S}_{11}$ & $\bar{S}_{22}$ in the order of 12.40% and 20.57% at $\delta = 1.25$ when compared with unworn circular non-recessed hybrid bearing operating at $\bar{K} = 0.0$. The combined influence of $\bar{\delta}_W$ and $\bar{K}$ lowers the value of $\bar{S}_{11}$ & $\bar{S}_{22}$ at $\bar{W}_0 = 1.5$. However, a change in bearing profile improves the value of $\bar{S}_{11}, \bar{S}_{22}$ under regular and worn condition of bearing for the offset factor greater than one. An apposite selection of non-circularity factor may enhance the performance of CFV restrictor compensated 2-lobe hybrid bearing from the point of view of $\bar{S}_{11}, \bar{S}_{22}$.

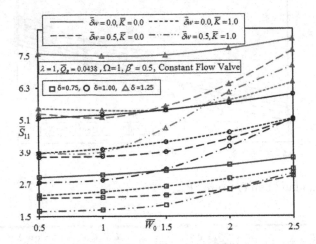

Figure 17.5 Direct fluid film stiffness coefficient ($\bar{S}_{11}$).

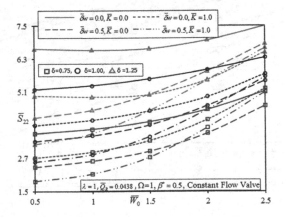

Figure 17.6 Direct fluid film stiffness coefficient ($\bar{S}_{22}$).

Direct Damping Coefficient ($\bar{C}_{11}, \bar{C}_{22}$)

Figures 17.7 and 17.8 clarifies influence on the values of direct fluid film damping coefficient ($\bar{C}_{11}, \bar{C}_{22}$).the values of $\bar{C}_{11}$ & $\bar{C}_{22}$ reduces as value of $\bar{\delta}_w$ increases, whereas; increased in the value of non-circularity factor improves the value of $\bar{C}_{11}$ & $\bar{C}_{22}$. This results in the enhanced value of $\bar{C}_{11}$ & $\bar{C}_{22}$ to damp out the oscillations. The variation in the value of $\bar{C}_{22}$ is being more as compared to $\bar{C}_{11}$ for the chosen values of external load ($\bar{W}_0$). The wear of the bearing surface lies in the perpendicular direction, thus, anticipating great fall in the value $\bar{C}_{22}$ of that $\bar{C}_{11}$.

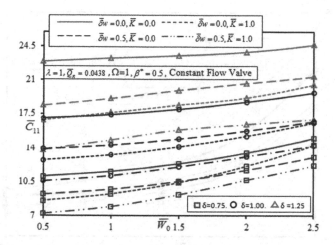

Figure 17.7 Direct fluid film damping coefficient ($\bar{C}_{11}$).

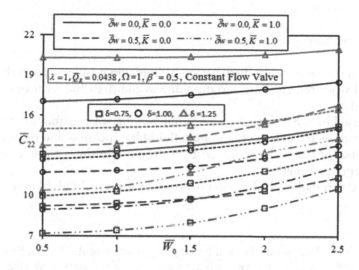

Figure 17.8 Direct fluid film damping coefficient ($\bar{C}_{22}$).

Threshold Speed Margin Stability Parameter ($\bar{\omega}_{th}$)

Figures 17.9 depicts the change in threshold speed margin ($\bar{\omega}_{th}$) w.r.t to load for 2-lobe bearing. The decrease in the value of is observed for an increase in $\bar{\delta}_W$ and non–linearity factor ($\bar{K}$). The maximum reduction in the value of $\bar{\omega}_{th}$ at $\delta = 1.0$ for $\bar{\delta}_W > 0.0$ is 19.13%. Further, shows that the bearing with non-circularity factor ($\delta \geq 1.0$) recompense the diminishing occurred due to wear damage and behavior of the lubricating fluid and provide a larger value of $\bar{\omega}_{th}$. The maximum enhancement in $\bar{\omega}_{th}$ is observed by 16% when 2-lobe bearing operates at $\delta = 1.25$ for the same operating condition.

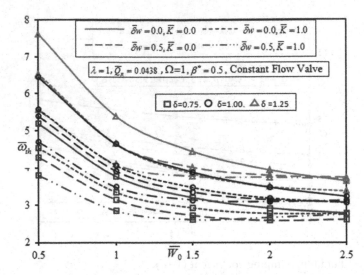

Figure 17.9 Threshold speed margin ($\bar{\omega}_{th}$).

Conclusion

The following conclusion can be drawn:

- The fluid film pressure and its distribution or fluid film profile of 2-lobe hybrid bearing gets affected when the viscosity variation due to additives and wear of bearing surface is considered
- The loss caused by the influence of worn out bearing surface and non-linear behavior of the lubricant on the values of minimum fluid-film thickness and rotor dynamic coefficients can be substantially improved by a proper selection of non-circularity factor and wear depth parameter.
- An improved threshold speed margin is observed for noncircular bearing ($\delta \geq 1.0$)

References

1. Pinkus, O. (1956). Analysis of elliptical bearing. *Transaction of ASME,* 78, 956–973.
2. Malik, M. (1983). A comparative study of some two lobed journal bearing configurations. *ASLE Transaction,* 26(1), 118–124.
3. Schuller, F. T. (1971). Experiments on the stability of water-lubricated three-lobe hydrodynamic journal bearings at zero load. *NASA Technical Note,* 6315, 1–23.
4. Ghosh, M. K., and Satish, M. R. (2003). Rotordynamic characteristics of multilobe hybrid bearings with short sills-part I. *Tribology International,* 36(8), 625–632.
5. Dufrane, K. F., Kannel, J. W., and McCloskey, T. H. (1983). Wear of steam turbine journal bearings at low operating speeds. *ASME Journal of Lubrication Technology,* 105, 313–317.
6. Hashimoto, H., Wada, S., and Nojima, K. (1986). Performance characteristics of worn journal bearings in both laminar and turbulent regimes. Part I: steady-state characteristics. *STLE Tribology Transaction,* 29(4), 565–571.
7. Bouyer, J., and Fillon, M. (2007). Influence of wear on the behavior of a two-lobe hydrodynamic journal bearing subjected to numerous startups and stops. *ASME Journal of Tribology,* 129, 205–208.

8. Laurant, F., and Childs, D. W. (2002). Measurement of rotordynamic coefficients of hybrid bearings surface, with (a) a plugged orifice (b) a worn land. *ASME Journal of Engineering for Gas Turbines and Power*, 124, 363–368.

9. Scharrer, J. K., Hecht, R. F., and Hibbs, Jr. R. I. (1991). The effects of wear on the rotordynamic coefficients of hydrostatic journal bearing. *ASME Journal of Tribology*, 113, 210–213.

10. Soulas, T., and Andres, L. S. (2003). Performance of damaged hydrostatic bearings predictions Vs experiments. *ASME Journal of Tribology*, 125, 451–456.

11. Sharma, S. C., Phalle, V. M., and Jain, S. C. (2012). Performance of a noncircular 2-lobe multirecess hydrostatic journal bearing with wear. *Industrial Lubrication and Tribology*, 64(3), 171–181.

12. Wada, S., and Hayashi, H. (1971). Hydrodynamic journal bearings by pseudo-plastic lubricants, part-I theoretical Studies. *Bulletin of the JSME*, 14(69), 268–278.

13. Tayal, S. P., Sinhasan, R., and Singh, D. V. (1982). Finite element analysis of elliptical bearings lubricated by a non-newtonian fluid. *Wear*, 80, 71–81.

14. Kushare, P. B., and Sharma S. C. (2013). A study of two lobe non recessed worn journal bearing operating with non-newtonian lubricant. *Processor of the Institution of Mechanical Engineers, Part J: Journal of Engineering Tribology*, 227 (12), 1418–1437.

15. Kushare, P. B., and Sharma, S. C. (2014). Nonlinear transient stability study of two lobe symmetric hole entry worn hybrid journal bearing operating with non-newtonian lubricant. *Tribology International*, 69, 84–101.

16. Garg, H. C. (2011). Influence of non-newtonian behavior of lubricant on performance of hole-entry hybrid journal bearings employing constant flow valve restrictors. *Industrial Lubrication and Tribology*, 63(5), 373–386.

17. Kushare P. B., and Sharma S. C. (2015). Influence of wear on the performance of 2-lobe slot entry hybrid journal bearings. *Mechanics and Industry*, 16(5), 14.

18 Film condensation in a vertical tube with pure steam: a numerical investigation

Amit Kumar[a], Dipak Chandra Das[b], and Pritam Das[c]

Department of Mechanical Engineering, National Institute of Technology, Agartala, India

Abstract: This research article presents a numerical analysis focuses on the laminar film condensation occurring within a vertically oriented, isothermal tube of pure saturated vapor, focusing on the analysis of heat transport phenomena within the heat exchanger architecture. The study employs the volume of fluid (VOF) approach to examine the influence of velocity and wall temperature on the thickness of the condensate film heat transfer coefficient and the thickness of the condensate film. The model's accuracy is validated by comparing its predictions with published experimental data, demonstrating satisfactory agreement between the numerical solution and experimental results. This research sheds light on the critical role of film condensation in the heat exchanger's performance and provides valuable insights for the design and optimization of such systems.

Keywords: Film condensation, heat transfer calculation, numerical simulation, pure vapor, vertical tube.

Introduction

Three-dimensional alterations in heat transfer occurring during film-wise condensation processes on external surfaces can yield advantageous outcomes in engineering systems. These benefits encompass energy conversion and thermal management systems, chemical refining, cooling, air-conditioning, and the generation of thermal energy.

Many theoretical, experimental, and computational studies have been performed to enhance the understanding of complex transport and characterize the efficiency of condensation. Nusselt [1] pioneered the condensation research of pure vapor inside a vertical plate and found that heat transport during condensation primarily relies on the local liquid film thickness. Shekriladze et al. [2] shows that the heat transfer rate over a part of surface behind the separation point drops significantly as the boundary layer separates. Minkowycz et al. [3] illustrate decisive effect on the heat-transfer rate due to small concentrations of the non-condensable gas at lower pressure ranges. Panday [4] analyzed numerically the film condensation of vapor within a channel. Groff et al. [5] proposed a computational model based on actual two-phase parabolic boundary conditions for turbulent flow region, in axi-symmetric flows inside a vertical tunnel. Lee et al. [6] suggested

---

[a]amitkumar17788@gmail.com, [b]dchdas12@gmail.com, [c]pritamdas.nit@gmail.com

a reliable, approximate closed-form solution using a novel density function for laminar condensation. A universal method was suggested by Kim and Mudawar [8] to estimate the HTC for mini channel and micro channel flow. This study presents the derivation of two distinct correlations, one for annular flow and another for bubbly flow. Additionally, Li [9] introduces a numerical simulation approach for investigating condensation in the presence of non-condensable gas under turbulent flow conditions within a vertically oriented tube. Ganapathy et al. [10] used the VOF technique to perform numerical simulation of flow condensation in micro channels. The findings in Wu et al. [11] show that air volume has an important impact on heat flux. Gun et al. [12] performed an experimental analysis on air-steam condensation over vertical tube exterior surfaces under normal convection. Le et al. [13] introduced systematic solutions to characterize the condensation process of a laminar film on surface of a vertically oriented cylinder. These solutions were based on the film thickness, providing valuable insights into the condensation behavior on the cylindrical surfaces. Kumar et al. [15–17] has conducted a study on flow boiling phenomena at various saturation temperatures.

In this context, it can clearly be established that as the study on the heat and mass transfer mechanism is largely restricted to the experimental approach, and the results vary greatly under various experimental conditions. Although substantial work on forced convective condensation within the tube has been published, research on forced convective fluid flow condensation inside the vertical tubes are limited. The key objective of the present study is to build a condensation model using Fluent to determine the effect of vapor on the ranges of the characteristics of velocity, interface, and temperature in addition to the HTC.

Numerical Model

The current study focuses on the laminar film condensation occurring within a vertically oriented, isothermal tube. The tube is subjected to a downward-flowing vapor, as depicted in Figure 18.1. Figure 18.2 show mesh diagram for the respective computational domain. Both the vapor and the interface between the vapor and the solvent are assumed to be saturated. The available boundary conditions include the temperatures of the inner surface of the tube wall and the incoming steam mass flow rate. In a vertical tube with a diameter of 20 mm, we consider vapor flowing, retaining a length of 3 m. The saturation vapor density rises on the basis of the thermal behavior of saturation vapor, and with the saturation temperature, the saturation liquid density decreases.

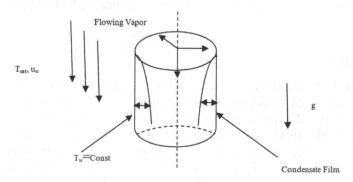

Figure 18.1 Schematic of film wise condensation inside a vertical tube.

In order to predict the single-phase convective HTC, the flow configuration is crucial. Three of the well-established parameters for any downward flow arrangements have been defined to assess the heat transfer. For surface condensation, the basic heat-transfer model, proposed by Nusselt [1], explains how well a pure saturated vapor condenses inside a vertical wall, creating a thin condensate film to allow downward flows caused by gravity. The operating conditions of actual condensers could likely be different from the assumptions. The temperature within the tube is below the saturation temperature of the vapor. The liquid inertia is ignored.

The momentum equation is simplified as

$$Du/Dt = \nabla\sigma/\rho + g \tag{1}$$

By disregarding the surface tension (σ) effect

$$(\mu/r)(\partial/\partial r)(r(\partial u/\partial r)) + (\rho - \rho_v)g = 0 \tag{2}$$

The energy equation represents.

$$(1/r)(\partial/\partial r)(r(\partial T/\partial r)) = 0 \tag{3}$$

Boundary conditions are

$$U = 0 \qquad |(r = ro) \tag{4}$$
$$T = T_w \, |(r = ro) \tag{5}$$

Following the Shekriladze and Gomelauri approach [2], shear stress at the interface ($r = r + \delta$) can be expressed as

$$\mu(\partial u/\partial r) = \int'(u_\infty - u)v \tag{6}$$
$$T = T_{sat} \tag{7}$$

Following the Minkowycz and Sparrow [3], the film temperature is assumed as

$$T_f = T_w + 0.31(T_{sat} - T_w) \tag{8}$$

The thermo physical properties have been obtained using a physical properties database [14].

Numerical Analysis

In this study, the ANSYS 2019 is utilized analyze the laminar filmwise condensation. The solution configuration solver is pressure-based, transient, recognizing gravitational acceleration, with average velocity formulations. The energy and viscous model (standard k-epsilon method) with the standard wall function is considered. The multiphase model, such as water vapor, water liquid and air, has phase content. The pressure-velocity bonding utilizes the PISO algorithm in this study. The pressure discretization is achieved using the PRESTO scheme, while the VOF method is employed for energy calculation, and the second-order upwind method is implemented for momentum calculations. Convergence is attained through careful adaptation of relaxation factors. In addition, until the weighted average of wall surface HTC becomes constant and also the volume proportion of the flowing phase become constant, the simulation is converged.

Problem Description

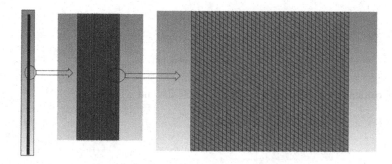

Figure 18.2 Mesh diagram for the respective computational domain.

Figure 18.1 displays the schematic view of the topic being considered. With the balanced flow of mass and stable temperature inside the vertical tunnel, mixture of air-steam comes downwards. The temperature of the wall is held below the boiling point for a specified partial pressure. The steam thus continues to condense on the walls in such a way that in the near-tube wall area, the steam mass fraction declines. A thin film layer along the tube is shaped due to steam condensation. To verify this numerical model condensation flow, the simulation of the Kim et al. [7] was pushed through. The condenser tube possesses an internal diameter of 0.02 m and extends for a length of 3 m., where the steam leads to downward direction. By defining the front wall to be symmetric, the 2D axis symmetric computing domain was implemented.

Results

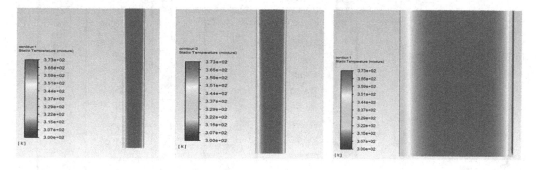

Figure 18.3 Distribution of temperature profile.

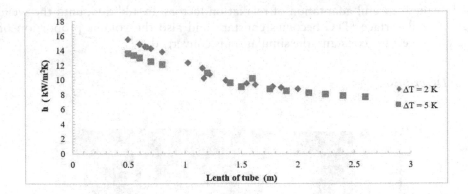

Figure 18.4 Condensation HTC distribution at $\Delta T = 2k$ and $\Delta T = 5k$.

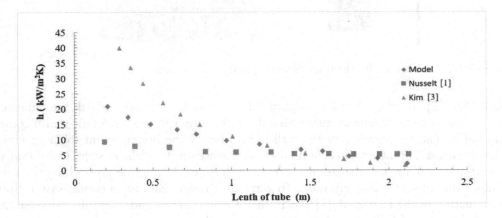

Figure 18.5 Distribution of HTC for the experimental Nusselt model [1], Kim [7] and the proposed model.

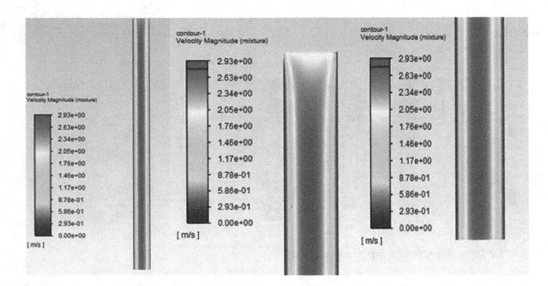

Figure 18.6 Distribution of velocity profile.

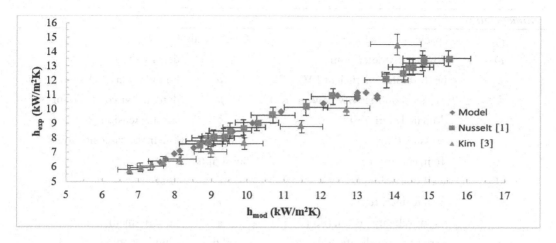

Figure 18.7 Comparison between HTC from the present work (h_{calc}) and experiment (h_{exp}).

Vapor condensation is responsible for the sharp and smooth concentration gradient near the film and the main flow respectively. The temperature distribution within the liquid film follows a linear pattern and aligns closely with the observations outlined by the Nusselt principle [1]. The numerical results are provided through Figures 18.3–18.7. Figure 18.4 shows the variation of HTC along the length of tube with ΔT = 2K and ΔT = 5K. The relation of the current findings with the experimental evidence from Nusselt [1] and Kim et al. [7] is presented in Figure 18.5. The variance of the HTC along the length of the tube is described. The drop in vapor velocity along the wall and near to the wall is shown in Figure 18.6. The condensate layer becomes thinner and the wall heat flux increases as the inlet vapour velocity increases. The analysis of the experimental HTC obtained from previous correlation and the HTC from the current model is shown in Figure 18.7. The examination of heat transfer focuses on analyzing variations in wall temperature. The findings reveal that higher cooling water velocities lead to a decrease in the HTC during the condensation process. Furthermore, as velocity and the difference in saturation wall temperature increase, heat dissipation also experiences an upsurge. For Kim et al. [7] and Nusselt [1], the Model's average absolute error is 25% and 20%, respectively. The variance can be caused by convective heat transfer.

Conclusion

We proposed a numerical model for determining the condensation HTC in an isothermal vertical tube. While reasonably relative large deviation value for the high mass dissipation conditions, the Kim et al. [7] model and experimental results Nusselt [1] agrees well. The variance can be caused by convective heat transfer. The Model's average absolute error is 25% with Kim et al. [7] and 20% with Nusselt [1]. It is possible to use the model outcome as a guide to optimize the heat transfer model condensation process. This model also needs to be weighed by more experimental evidence and the effects of with force flow in dry saturated vapor can also be addressed.

Nomenclature

c_p	specific heat (kJ kg^{-1} K^{-1})	**Greek symbols**		
D	domain diameter (mm)	ρ	density (kg m^{-3})	
h	heat transfer coefficient(kW/m²K)	δ	boundary layer thickness, m	
k	thermal conductivity (kW m^{-1} K^{-1})	μ	dynamic viscosity (kg m^{-1} s^{-1})	
L	domain length (m)	σ	surface tension (Nm^{-1})	
p	pressure (bar)	ϑ	kinematic viscosity, m²/s	
T	Temperature (k)	**Subscripts**		
ΔT_w	wall subcooling (k)	f	fluid	
T_w	wall temperature (k)	i	inlet	
U	vapor velocity vector, (u,v)	g	gravity (m/s²)	
x	cartesian coordinates(m)	l,v	liquid, vapor	
HTC	heat transfer coefficient	s	liquid-vapor interface	
FVM	finite volume method	sat	saturation	
		w	wall	

Acknowledgement

This work is ostensibly supported by the Chemical Engineering Departement, NIT, Agartala for their ANSYS 2019 package.

References

1. Nusselt, W. (1916). Die oberachen kondensation des wasserdampfes, the surface condensation of water. *Zeitschrift Vereines Deutscher Ingenieure*, 27, 541–446.
2. Shekriladze, I. G., and Gomelauri, V. (1966). Theoretical study of laminar film condensation of flowing vapor. *International Journal of Heat and Mass Transfer*, 9(6), 581–591.
3. Minkowycz, W. J., and Sparrow, E. M. (1966). Condensation heat transfer in the presence of non –condensables, interfacial resistance, super heating, variable properties and diffusion. *International Journal of Heat and Mass Transfer*, 9, 1125–1144.
4. Panday, P. K. (2003). Two-dimensional turbulent film condensation of vapours flowing inside a vertical tube and between parallel plates: a numerical approach. *International Journal of Refrigeration*, 26(4), 492–503.
5. Groff, M. K., Ormiston, S. J., and Soliman, H. M. (2007). Numerical solution of film condensation from turbulent flowof vapor–gas mixtures in vertical tubes. *International Journal of Heat and Mass Transfer*, 50, 3899–3912.
6. Lee, C., Park, K., Jung, D., and Kim, J. (2008). Condensation heat transfer coefficients of HFC245fa on a horizontal plain tube. In International Refrigeration and Air Conditioning Conference (IRACC), (p. 905).
7. Kim, D. E., Yang, K. H., Hwang, K. W., Ha, Y. H., and Kim, M. H. (2011). Pure steam condensation model with laminar film in a vertical tube. *International Journal of Multiphase Flow*, 37, 941–946.
8. Kim, S. M., and Mudawar, I. (2013). Universal approach to predicting heat transfer coefficient for condensingmini/micro-channel flow. *International Journal of Heat and Mass Transfer*, 56, 238–250.

9. Li, J. D. (2013). CFD simulation of water vapour condensation in the presenceof non-condensable gas in vertical cylindrical condensers. *International Journal of Heat and Mass Transfer*, 57, 708–721.

10. Ganapathy, H., Shooshtari, A., Choo, K., Dessiatoun, S., Alshehhi, M., and Ohadi, M. (2013). Volume of fluid-based numerical modeling of condensation heat transfer and fluid flow characteristics in microchannels. *International Journal of Heat and Mass Transfer*, 65, 62–72.

11. Wu, X. M., Li, T., and Chu, Q. L. (2017). Approximate equations for film condensation in the presence of no condensable gases. *International Communications in Heat and Mass Transfer*, 85, 124–130.

12. Jang, Y. J., Lee, Y. G., and Choi, D. J. (2017). An experimental study of air–steam condensation on the exterior surface of a vertical tube under natural convection conditions. *International Journal of Heat and Mass Transfer*, 104, 1034–1047.

13. Le, Q. T., Ormiston, S. J., and Soliman, H. M. (2017). A closed-form solution for laminar film condensation from downward pure vapour flow in vertical tubes. *International Communications in Heat and Mass*, 81, 141–148.

14. Lemmon, E. W., Huber, M. L., and McLinden, M. O. (2010). NIST Standard Reference Database 23. Reference fluid thermodynamic and transport properties (REFPROP), 9.

15. Kumar, A., Das, D. C., and Das, P. (2022). Significance of surface morphology of materials on flow boiling heat transfer using R-407c. *Materials Today: Proceedings*, 62(6), 3122–3128.

16. Kumar, A., Das, D. C., and Das, P. (2023). Parametric variation studies of experimental flow boiling heat transfer phenomena using R407c inside an enhanced tube. *Heat and Mass Transfer*, 59, 1353–1363.

17. Kumar, A., Deb, S., Das, D. C., and Das, P. (2023). Comparative experimental studies of flow boiling heat transfer phenomena in smooth and enhanced tubes using R407C. *Heat and Mass Transfer*, 1–17. https://doi.org/10.1007/s00231-023-03379-3.

19 Design of an adaptive neuro-fuzzy expert system for predicting surface roughness in laser direct metal deposition

Ananya Nath[1,a], Shibendu Shekhar Roy[1,b], Anirban Changdar[c], and Aditya Kumar Lohar[2,d]

[1]Department of Mechanical Engineering, NIT Durgapur, West Bengal, India

[2]Surface Engineering & Tribology Laboratory, CSIR-CMERI Durgapur, West Bengal, India

Abstract: Surface roughness is a significant design specification that is known to have considerable influence on properties such as wear resistance and fatigue strength. Laser based Direct Metal Deposition (LDMD) is a metal additive manufacturing process of increasing interest due to its applicability in maintenance and repair of critical high-cost products in the aerospace and automotive industry. Therefore, modeling and prediction of surface roughness of a printed parts play an important role in the modern manufacturing industry. In the present study, an attempt is made to design an optimized fuzzy expert system using adaptive neural networks, so that the surface roughness in LDMD process can be modeled for set of input process parameters, namely laser power, powder flow rate and scan velocity. Three different membership functions, triangular, trapezoidal and gaussian shaped, were adopted during hybrid learning (i.e. combination of the least squares method and back-propagation gradient descent method) of first order Sugeno-type fuzzy system in order to compare the prediction accuracy of surface roughness by three membership functions. The predicted surface roughness values obtained from proposed adaptive neuro-fuzzy expert system were compared with the experimental data. The comparison indicates the gaussian-shaped membership function in proposed system achieves higher prediction accuracy than triangular and trapezoidal membership function.

Keywords: Adaptive neuro-fuzzy, additive Manufacturing, expert system, laser direct metal deposition, surface roughness.

Introduction

Additive manufacturing (AM) has been spread out in all over the globe with its multifunctional techniques. Previously it used to form only protypes, but nowadays this multilayer deposition technique can be utilized in all kind of fields for making any type of complex shaped objects. Among all seven kinds of AM techniques, Laser based direct metal deposition has become a very popular and trustworthy methodology for it's better performance than other AM techniques, where laser beam is used as the power source to melt the powdered material to fuse [1]. This laser-based AM technology is used to make parts which can be easily joined by welding or forming to make the complex shaped object.

[a]an18u10057@btech.nitdgp.ac.in, [b]shibendu.roy@me.nitdgp.ac.in, [c]anirbancmeri04@gmail.com, [d]aklohar@cmeri.res.in

In any kind of manufacturing techniques, surface roughness is a very important primary indicator of good metal deposition. Many process parameters are involved for getting a good surface finish as a response parameter in a laser direct metal deposition process (LDMD). It is a very common problem for any kind of machining or manufacturing process. To get perfect shape and surface finish, a constant layer height is to be maintained [2, 3]. Good surface finish (less surface roughness) results in good fatigue strength, lubricating property, friction, wear and creep life and corrosion resistance. There are a number of parameters are present on which the surface roughness value depends [4]. So, to get the desired surface finish, some input parameters have been chosen in our study, such as laser power (W), scan speed (mm/s) and powder flow rate (g/min) [2, 5, 6]. This surface roughness prediction has been carried out using various modelling previously such as fuzzy logic-based modelling, neural network-based modelling, mathematical modelling (physics based), statistical modelling (by regression equations) etc. In this paper the adaptive neuro fuzzy based modelling has been performed.

Experimentation

The experiment of Laser direct metal deposition is executed in a Laser based direct energy deposition system, developed at CSIR-CMERI Durgapur. Laser has been used of wave length approximately 976 nm±10 nm and the maximum LASER power that can be obtained is 1.2KW. Three input parameters are taken i.e., Laser power (W), scan velocity (mm/s) and powder flow rate (g/min). In Table 19.1 the details of respective input parameter are given. The response parameter is surface roughness, measured in μm. Argon has been used as a carrier gas of flow rate 10 l/min. Helium has been used as shielded or nozzle gas of flow rate is 10 l/mm. Bead overlap ratio has been set in 50%.

In this direct metal deposition technique, the metal powder will be fed on to the bed with e LASER beam. The developed experimental setup, as shown in Figure 19.1, is a 5-axis control system. The bed can move in 3 direction i.e., x, y, and z, but the nozzle can move in only two direction. The chamber is closed at the time of operation. Separate chambers for LASER beam production, power storage, cooling system etc. are present. Two metal powder chambers are present. From these two chambers the metal powder will come and mix in the welding torch. The experimentation is carried out with a focused LASER beam at exact 10mm apart from the print bed.

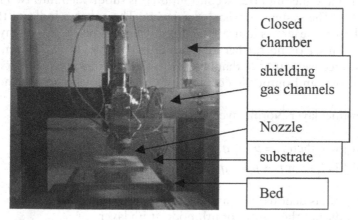

Figure 19.1 Experimental setup of laser direct metal deposition.

The material used here is SS316L in a powder form to deposit [7,8]. Argon has been used as a shielded gas. SS316L is well known for its good corrosion resistance. Stainless steels are very frequently used material for any deposition technics due to their low maintenance. However, SS316L is the mostly used grade of stainless steel. Due to low carbon content, welding of many parts of SS316L can results into good part's strength without any carbon precipitations [7]. Mostly in petrochemical, chemical sectors, wastewater treatment sectors stainless steels are used. Nowadays, in medical world also stainless steel has its many applications. Also the parts made up of stainless steel are very attractive in outer look. That's why SS316L stainless steel in more advantageous to use in deposition processes than other materials. However the powder form of SS316L is used as it has some direct impact on the mechanical properties of manufactured parts. The process is also become more efficient using powdered form of material as the powdered form of metal makes the process easy and more volumetric using less Laser power.. In Laser metal deposition (LMD) and in many more AM technologies, SS316L has performed great. SS316L is made of Cr (18.0), Ni(14.0), Mo(3.0), Mn(2.0), Si(0.75), C(0.03), N(1.0), P(0.045), S(0.03). The substrate material is Mild steel.

Adaptive Neuro Fuzzy Inference System (ANFIS)

The Adaptive Neuro Fuzzy Expert System is also known as Adaptive Neuro Fuzzy Inference System. It is a type of artificial multilayer feed forward neural networking system, based on Takagi-Sugeno's approach, was developed by Jang in 1995. Fuzzy logic operates in "IF-THEN" rule based system. Fuzzification is done to the input parameters before entering into the system. The output is then converted into crisp set by defuzzification. Fuzzy logic based structures are made up of some decision making rule base system. However, to get most accurate and desired output result, selection of membership function, number of neurons per layer, type and number of rules are important.

Adaptive Neuro-Fuzzy Architecture

To understand the detailed illustration of Adaptive Neuro Fuzzy Inference System architecture, let's take two input parameters with each of two levels.

First input is I and the second input is J, the first input (I) is subdivided into two further levels i.e., low (I1) and medium (I2) second input(J) is subdivided into two levels i.e., low (J1) and high (J2). The system consists of 5 layers. The squares indicate the chances of optimization whereas the circles indicate the fixed value i.e., no chance of any optimization or change in parameter or no further adaptation is possible. In each layer, each input is converted into respective output of that layer with the help of some membership functions. Input is denoted by I and output is denoted by O. The whole system is shown in a flow diagram below.

Fuzzy laye Product layer Norm. layer De-fuzzy layer Output

$$\text{If I is } I_1 \text{ and J is } J_1 \text{ then } f_1 = p_1 I + q_1 J + r1 \tag{1}$$
$$\text{If I is } I_2 \text{ and J is } J_2 \text{ then } f_2 = p_2 I + q_2 J + r_2 \tag{2}$$

In each layer input is denoted by I and output is denoted by O.
Now the $O_{i,n}$ indicates the output of nth node at ith layer.

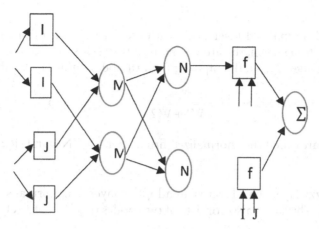

Figure 19.2 Adaptive neuro fuzzy expert system architecture.

Layer 1
This layer is also called fuzzy layer. In this layer adoption or optimization can take place.
   So, in layer1 i = 1.
   And in j = 1, two nodes are present.
   If the membership function (MF) for the node 1 in layer 1 μ is I, n then

$$O_{I1,n} = \mu_{I,n}(I) \qquad n = 1, 2 \tag{3}$$

Nad if the MF for the second node of layer1 is J,n then

$$O_{J1,n} = \mu_{J,n}(J) \qquad n = 1, 2 \tag{4}$$

   Now, the MFs can be of many type, for example trapezoidal, triangular log sigmoid, tan sigmoid etc.
   For trapezoidal, the MF the μ value will be

$$\mu(I) = max\left[min\left\{ \frac{I-a}{b-a}, \frac{c-I}{c-b} \right\} \right]$$

a, b and c is the parameter of the membership function of trapezoidal rule. This parameters will be optimized at the end of the loop.
   The above mentioned Adaptive Neuro Fuzzy Expert System architecture is shown in Figure 19.2.

Layer 2
This layer is also known as product layer. In this layer no modification can take place. Input-output values are fixed. In layer two also there are two nodes (M,M). One output from the first node of first layer and one output from the second node of the first layer will go into the first node of second layer as an input. So,
   If the output of each of the node of second layer is $O2,n = W_n = \mu_{I,n}(I)\,\mu_{J,n}(J)$ n=1,2.
   This layer output is called "firing strength".

Layer 3

This layer is called normalised layer deals with fixed value that means no chance of any adoption. First the firing strengths are normalised between some fixed range. In this layer also there are wo nodes (N,N). the input of the nth node of this layer will be

$$O_{3,n} = \overline{Wn} = \frac{Wn}{W1 + W2} \qquad n = 1, 2$$

This layer outputs are called the "normalized firing strength". N stands for normalised layer.

Layer 4

This layer is de fuzzy layer. This layer is an adoptive layer. The normalised firing strengths are optimised here. This layer also consists of two nodes (f1, f2) where $O_{4,n}$ is the respective layer output.

$$O_{4,n} = \overline{Wn} f_n = \overline{Wn}(p_{nI} + q_{nJ} + r_n) \qquad n = 1, 2$$

p, q, r is the parameters of MF of the neuron of 4th layer. De-fuzzification of the fuzzy set outputs are taken place in this layer.

Layer 5

This is the last single circular nodded layer is called total output layer which deals with the fixed values, means no chance of adoption. Output of this layer is denoted as $O_{5,n}$

$$O_{5,n} = \sum_n \overline{Wn}\, \text{fn} = \frac{\sum_n Wn\text{fn}}{\sum_n Wn}$$

Learning Algorithm

The main aim is to learn algorithm is to make the Adaptive Neuro Fuzzy Logic based Expert System architecture in such a way that the output response after a number of adaptions (to modify the unknown parameters like p1, q1, r etc.) will match with the experimental data. The hybrid learning algorithm of a Fuzzy logic based neural network is shown in Figure 19.3.

In this paper the ANFIS is operated with three input parameters i.e., Laser power, scan speed and flow rate of the powder, and one output i.e., surface roughness. ANFIS model is a very easy interpretable model to the user because of the contribution of fuzzy logic in it, in terms of input variables. In this paper the ANFIS architecture is of five layers. Each layer has multiple nodes, and these nodes have their specific membership functions to govern the input output relationship. In input layer, the each input is also divided in three levels. Here, output is varied in different segment of layers. Different membership functions like triangular, trapezoidal and Gaussian functions are used for testing of three levels of input variable.

Results and Discussion

A number of process parameters are involved to get a desired surface finish. But among them three parameters have been chosen for this study, such as Laser power, scan velocity

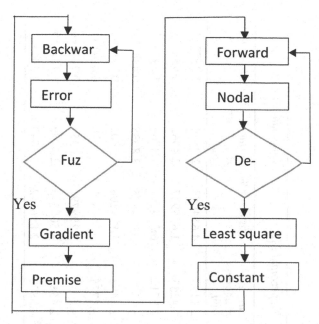

Figure 19.3 Flow diagram of a hybrid learning of adaptive fuzzy logic based expert system.

and flow rate of the powder. The selected input parameters are very much important to measure the surface roughness in which laser power has the most contribution for getting low surface roughness in LDMD process. Selection of the right range of these parameters is also very important. After few literature survey, the range of input parameters is set, given in Table 19.1. Here, the response or output parameter is surface roughness, measured in μm. Using the experimental input-output sets of data [5], the second order regression equation is formed which is as follows,

$$R_{avg} = -22.27 + 0.1612\ P + 2.035\ V - 1.833\ F - 0.000246\ P \times P + 0.02524\ V \times V - 1.433\ F \times F - 0.009729\ P \times V + 0.02202\ P \times F$$

After a careful study, the training data set is generated using these above-mentioned regression equations to model the ANFIS architecture.

First these generated training data sets are fed to the system to model the ANFIS architecture. After that, neuro-fuzzy system generates random parameters (parameters of input membership functions used in Fuzzy logic) and by grid partition, the learning and training

Table 19.1: Input parameters in LDMD process.

Process Parameters and Responses	Symbols	Units	Ranges Considered	
			Max	Min
Laser power	P	Watt (W)	300	200
Scan velocity	V	mm/s	8	4
Flow rate of the powder	F	g/min	0.848	2

Table 19.2: Comparison between the experimental data and ANFIS predicted data.

Laser power (W)	Scan velocity (mm/s)	Flow rate of the powder (g/min)	Surface roughness (μm) Experimental results	Triangular MF		Trapezoidal MF		Gaussian MF	
				Surface roughness (μm)	Abs error%	Surface roughness (μm)	Abs error%	Surface roughness (μm)	Abs error%
200	4	0.848	2.023333	2.1774	7.164515	2.1175	4.654053	2.0504	1.337743
200	6	1.375	2.383333	2.3878	0.187427	2.4233	1.676937	2.4062	0.959455
200	8	2	1.853333	2.1206	14.42088	1.9712	6.359731	1.8764	1.244623
250	4	1.375	3.786667	3.779	0.202474	3.7599	0.706875	3.8035	0.444533
250	6	2	2.77	2.9743	7.375451	2.8492	2.859206	2.7902	0.729242
250	8	0.848	3.153333	3.2307	2.453499	3.1775	0.766395	3.1686	0.484154
300	4	2	4.603333	4.6765	1.589435	4.7092	2.29979	4.6359	0.707466
300	6	0.848	2.543333	2.7062	6.403684	2.658	4.508533	2.5799	1.437759
300	8	1.375	2.32	2.4485	5.538793	2.3872	2.896552	2.3549	1.50431

starts, then by applying the aforesaid hybrid algorithm the optimization begins. In each iteration of input to output, response is validated with the testing data or the real experimental data which are also fed to system thereafter, and the error is evaluated by the system itself to update the architecture according to the need. This process repeats in a number loop. In the adaptive layers of Fuzzy system, the adoption of parameters of MFs occurs and the error between the system output and test output (fed to the system) get minimised with each loop. Here, in the developed ANFIS architecture of this paper, each response is fed and the error get minimized in 1000 iterations.

In Table 19.2 the comparison between the test or experimental output and the predicted response by ANFIS is shown. The absolute percentage error of the experiment value and predicted value is measured. Furthermore, it is calculated that in triangular membership function (MF) the average absolute percentage error is 2.944556 % where in trapezoidal MF the error came out as 2.894778 % and in Gaussian MF it is 2.851778 %. And among these three MF, Gaussian distribution has shown slightly more accurate response than other two. This is because the surface roughness exhibits non-linear relationship with its process parameters.

Conclusion

Surface roughness is a very important feature of any kind of manufacturing process to be measured which indirectly effects the process efficiency. In this study the adaptive fuzzy logic based expert system has been used to predict the surface roughness of a SS316L powdered material, deposited by laser based direct metal deposition onto the mild steel substrate. Three main input parameters are being used to predict the response such as laser power, scan velocity and flow rate of the powder. In order to get the most accurate response, the output predicted by Fuzzy logic baser neural network is compared with the measured surface roughness value of few test data. There a number of membership functions are present in fuzzy logic system among which, triangular, trapezoidal and Gaussian membership functions have been used for this experiment. However, the error percentage of the predicted surface roughness with the testing data came out to be 2.944556 %, 2.894778 % & and 2.851778 % for triangular, trapezoidal and Gaussian membership functions respectively where Gaussian membership function has shown the least error value among the three distribution functions due to its nonlinear behaviour. So, it can be concluded that using Gaussian distribution MF the surface roughness prediction in this Laser based direct deposition technique is done most accurately.

References

1. Shah, K., Pinkerton A. J., Saloman, A., and Li, L. (2010). Efects of melt pool variables and process parameters in laser direct metal deposition of aerospace alloys. *Materials and Manufacturing Processes,* 25(12), 1372–1380.
2. Peng, L., Shengqin, J., Xiaoyan, Z., Qianwu, H., and Weihao, X. (2007). Direct laser fabrication of thin-walled metal parts under openloop control. *International Journal of Machine Tools and Manufacture,* 47(6), 996–1002.
3. Thompson, S. M., Bian, L., Shamsaei, N., and Yadollahi, A. (2015). An overview of direct laser deposition for additive manufacturing; part I: transport phenomena, modeling and diagnostics. *Additive Manufacturing,* 8, 36–62.

4. Hofman, J. T., Pathiraj, B., Van Dijk, J., De Lange, D. F., and Meijer, J. (2012). A camera based feedback control strategy for the laser cladding process. *Journal Materials Processing Technology,* 212(11), 2455–2462.

5. Pant, P., Chatterjee, D., Nandi, T., Samanta, S. K., Lohar, A. K., and Changdar, A. (2019). Statistical modelling and optimization of clad characteristics in laser metal deposition of austenitic stainless steel. *Journal of the Brazilian Society of Mechanical Sciences,* 41(7), 2–83.

6. Fayazfar, H., Salarian, M., Rogalsky, A., Sarker, D., Russo, P., Paserin, V., and Toyserkani, E. (2018). A critical review of powder-based additive manufacturing of ferrous alloys: process parameters, microstructure and mechanical properties. *Mater Design,* 144, 98–128.

7. Boisselier, D., and Sankaré, S. (2012). Infuence of powder characteristics in laser direct metal deposition of SS316L for metallic parts manufacturing. *Physics Procedia,* 39, 455–463.

8. Zhang, K., Wang, S., Liu, W., and Shang, X. (2014). Characterization of stainless steel parts by laser metal deposition shaping. mater desi one of the most accurate AM methodology for getting very near shape of an object in laser metal deposition or Laser direct energy deposition technique. *Glomerulonephritis,* 55, 104–119.

20 Numerical modeling of the thermal phenomenon in wind turbine disc brake

M. Nithin Kumaar, J. Aishwarya, R. Vaira Vignesh,
S. Sai Nischay, B. V. Likhith, and M. Govindaraju

Department of Mechanical Engineering, Amrita School of Engineering, Coimbatore, Amrita Vishwa Vidyapeetham, India

Abstract: This study aims to understand the thermal profile developed in the course of braking operation for a wind turbine using COMSOL multiphysics software. The finite element method is used to obtain the thermal characteristics of the conventional brake pad material. The results show that high-temperature levels at the edges of the brake pads. This increase in temperature eventually favors the fading behavior of the brake pads, which necessitates frequent replacement. It is also noted that the rate of heat dissipation was much lesser than the rate of heat generation.

Keywords: Brake pad, COMSOL multiphysics, Thermal profile.

Introduction

Global energy demand is projected to double by 2050 regard to the increase in the global population and economic status [1–3]. However, more than 80% of the currently consumed energy is obtained from fossil fuels. The consumption of fossil fuels has increased at a rapid pace over the years [4]. With the current trend of consumption of fossil fuels, most of the fossil fuel resources are expected to exhaust by the next century. Besides, fossil fuels are associated with non-eco-friendly characteristics like polluting the environment, creating unsustainable development [5, 6]. Renewable energy sources are now preferred by all countries to battle global climate change.

There are mainly five major renewable sources of energy: wind, water, geothermal, solar, and biomass. Hydroelectricity and wind power accounts for the majority of the total renewable energy generation. Hydropower is more efficient and has an efficiency of 90%. But the world is moving to wind energy as it can conserve water to a greater extent [7]. It is predicted that by 2050 wind energy will satisfy one-third of the world's electricity needs. Harnessing wind energy is becoming more inevitable in recent times in India [8]. India's topography, with its 7600 km long coastline, allows for maximum utilization of wind [9, 10]. Wind turbines are machines that harness the kinetic energy of the wind for the generation of electricity. A typical wind turbine has an electrical system, transmission system, braking system, and supports structures [11].

These systems include the low-speed shaft, high-speed shaft, control unit, sensors, hydraulics, brakes, gearbox, generator, and drivetrain. The braking system consists of aerodynamic (primary) and mechanical (secondary) brakes. Brakes are designed to operate at 10% overspeed of the low-speed shaft and during the maintenance period [12, 13]. For a 750-kW wind turbine, the nominal speed of the low-speed shaft is 18 rpm and the high-speed shaft is 1500 rpm. The wind turbine blades start to rotate at a wind speed greater than 2 m/s. When the wind speed reaches the cut-in speed of 3–4 m/s, wind turbines start to generate power. In the eventuality of winds crossing the cut-out speed of 25 m/s, the brakes are applied.

When the wind turbine blades are parallel to the wind flow, the rotor seizes to rotate, which is aerodynamic braking. After the aerodynamic brakes are applied, the high-speed shaft speed reduces to 350–450 rpm and then the mechanical brakes are applied. Figure 20.1 shows the schematic of a mechanical braking system of a wind turbine [14, 15]. At the instance of mechanical braking, the brake pads are pressed against the brake rotor disc (steel). This creates friction between the brake pad and the brake disc due to their relative motion. The frictional energy thus produced is converted into heat energy. The frictional resistance decelerates the rotor disc and ultimately stops the generator and wind turbine.

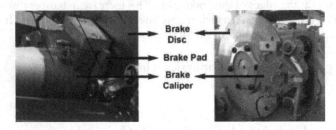

Figure 20.1 Schematic of a mechanical braking system [14,15].

Investigations by Jiguang and Fei [16] showed that the pad volume significantly influences the brake pad temperature profile. As the brake pad volume increases, the temperature is lower as the heat capacity is higher. Belhocine et al. [17]found that the thermal and mechanical stresses developed during the braking operation cause wear of the brake pad and disc through cracks. Borawski's [18] research showed that the friction coefficient depends on the interface temperature as the pad deforms at a high temperature. A review of Thakur and Dhakad [19] on the thermal properties of brakes stated that the brake pad material should have effective thermal conductivity to dissipate the heat and superior damping properties.

In the course of braking, temperature increases at the contact surface [20, 21]. The increase in temperature eventually fades the properties of the brake pad and hence mandates frequent replacements [22]. Hence, it is desirable to understand the thermal profile in the course of the braking phenomenon to design new composite materials. In the current work, brake pads were designed according to the design standards for a 750-kW capacity wind turbine. The thermal phenomenon in the course of the application of mechanical brakes in wind turbines was simulated using COMSOL Multiphysics software. The simulation was performed to study the temperature profile/thermal characteristics developed at the brake pad and brake disc interface on braking [23, 24].

Methodology

All engineering problems can be converted into a mathematical model which is then solved and the solution obtained is studied further to understand its behavior. In real-world applications, these models are usually governed by Partial Differential Equations (PDE), aided with relevant boundary conditions. FEM is one of the numerical methods used to solve PDEs [25]. COMSOL Multiphysics is an interactive user-friendly physics-based solver designed to solve real-life engineering problems [26, 27]. The model designed in the software is a 3D solid model and simulated based on heat transfer equations using a time-dependent solver.

3D Modeling

The brake pad is designed for a wind turbine capacity of 750 kW. The torque of the shaft can be calculated from the equations (1) to (4):

$$P = T \times \omega \, (W) \tag{1}$$

$$T = \frac{P}{\omega} \tag{2}$$

$$T = \frac{P \times 60}{2 \times \pi \times N} \, (Nm) \tag{3}$$

$$\omega = \frac{2 \times \pi \times N}{60} \, (rad/s) \tag{4}$$

The torque developed in the high-speed shaft with a factor of safety 2 is 9554.14 Nm which is calculated from equation (3). As per the design guidelines of PN-M-85000: 1998, the value of shaft diameter to sustain the high-speed shaft torque is 110 mm. A 10% increase in the low-speed shaft with 1.1 factor for possible loss of caliper spring force results in 22 rpm and correspondingly speed of the high-speed shaft increases to 1650 rpm. The angular velocity of the high-speed shaft at 1650 rpm is found to be 172.7 rad/s. Considering the supercritical sliding velocity to be 40 m/s, equation (5) gives the diameter of the brake disc. The brake disc diameter is 0.46 m.

$$V = r \times \omega \, (m/s) \tag{5}$$

$$Q_b = T \times \omega \, (kW) \tag{6}$$

$$A_{bc} = \frac{n \times \pi \times D^2}{4} \, (m^2) \tag{7}$$

The power dissipation (Q_b) at the instance of braking in the high-speed shaft is 1649.28 kW which is calculated using (6). The thermal design constraint of the brake pad is the maximum power dissipation per unit pad area (Q) which should be less than 11.6 MW/m^2. The ratio between Q_b and Q gives the total area of the brake pads covered by the four calipers. The total area of the pads is 0.142 m². Therefore, the area of pads (A_{bc}) in each brake caliper is 0.0355 m². The diameter and number of brake pads are calculated from equation (7). The dimension constraint of the brake pad should be greater than 25 mm. Therefore, each caliper has 18 brake pads of diameter 50 mm as shown in Figure 20.2. The brake pad was modeled in AutoCAD Inventor 2020.

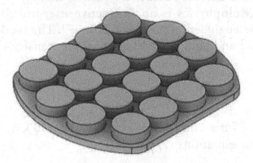

Figure 20.2 Brake pad model.

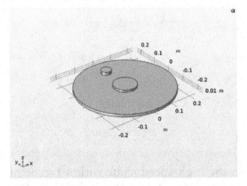

Figure 20.3 Schematic of model geometry.

The simulation was done for a single brake pad pressed against the brake disc. The brake pad and disc were 3D modeled using COMSOL Multiphysics 5.5. A brake disc of diameter is 0.46 m and a shaft of diameter 0.11 m to sustain the high-speed shaft torque is designed. A brake pad of diameter 50 mm is placed at 150 mm from the shaft axis. The thickness of the pad and the disc are 15 mm and 10 mm respectively. Figure 20.3 depicts the model used for the simulation.

Governing Equations

The governing equation is a set of simultaneous algebraic equations that describes the physics of the problem. The problem's physics describes the mass, energy, and momentum of the body. It also describes how one variable influences the other. The heat produced during braking can be studied using the heat transfer equations. This model focuses on the study of the thermal characteristics of the brake pad and hence, the equations governing this model are based on heat transfer in solids. Since the simulation involves all modes of heat transfer, they are specified under different governing equation sections.

Thermal Insulation

Heat flux is defined as the flow of heat from one object to the other. When the heat flux across the boundary is zero, the body is said to be thermally insulated. There is no heat transfer across the boundary of the pad and disc before braking, as the temperature gradient is zero. This is given by equation (8).

$$-n \cdot q = 0 \tag{8}$$

where, n – normal vector on the boundary, and q – heat flux (W/m³)

In the instance of braking, the kinetic energy of the high-speed shaft is converted into heat energy. The converted heat energy would be liberated to surroundings through different modes of heat transfer like conduction, convection, and radiation. Heat transfer from the brake pad to the disc happens through conduction whereas the heat transfer to the external surroundings happens through convection and radiation.

Conductive Heat Transfer

At the interface of the pad and disc, conductive heat transfer takes place. The conductive heat transfer in the course of braking is governed by the equation (9) and (10).

$$\rho C_p \frac{\partial T}{\partial t} + \rho C_p u \cdot \nabla T + \nabla.(q + q_r) = Q + Q_{ted} \tag{9}$$

$$Q_{ted} = -\alpha T \frac{dS}{dt} \tag{10}$$

Where ρ – density (kg/m³), u – velocity vector (m/s), C_p – specific heat capacity at constant pressure (J/kgK), T – Absolute temperature (K), q – heat flux by conduction (W/m²), q_r – heat flux by radiation (W/m²), α – coefficient of thermal expansion (1/K), S – second Piola-Kirchoff stress tensor (Pa), Q – additional heat sources (W/m³), Q_{ted} – thermoelastic damping heat source (W/m³)

The term q_r in the heat, the transfer equation is approximated to zero since there is no heat flux by radiation in the model. The heat flux due to conduction is given by equation

$$q = -k\nabla T \tag{11}$$

Where k – thermal conductivity (W/mK)

Heat Flux

The heat transfer rate can be calculated on the surface from the heat flux equation, a boundary condition. Equation (12) gives the specified heat flux boundary condition across the disc. The term inward heat flux accounts for all modes of heat transfer.

$$-n \cdot q = q_0 \tag{12}$$

Where q_0 – inward heat flux (W/m²)

Convective heat transfer happens on the interface of the disc and the surroundings. External forced convective heat transfer (q_0) over the length of the plate is considered. Considering the temperature difference to be proportional to the convective heat transfer, the equation is given as (13).

$$q_0 = h \cdot (T_s - T_\infty) \tag{13}$$

Where q_0 – convective heat transfer (W/m²), h – convection heat transfer coefficient (W/m² · K), T_s – surface temperature (K), T_∞ – temperature far away from the model (K)

For a flow over a horizontal plate, the convective heat transfer coefficient can be calculated based on the range of Reynold's number. Since the heat interactions take place between a moving fluid and solid, the Prandtl number plays a critical role in calculating the convective heat transfer coefficient. Prandtl number gives the relationship between momentum and thermal diffusivity which can be found from the equation (15). Reynolds number is used to classify the fluid flow pattern and is given as equation (16). The convective heat transfer coefficient can be thus calculated using the equation (14).

$$h = \begin{cases} 2\dfrac{k}{L}\dfrac{0.3387 Pr^{1/3} Re_L^{1/2}}{\left(1+\left(\dfrac{0.0468}{Pr}\right)^{2/3}\right)^{1/4}} & if\ Re_L \leq 5 \cdot 10^5 \\[20pt] 2\dfrac{k}{L} Pr^{1/3}\left(0.037 Re_L^{4/5} - 871\right) & if\ Re_L > 5 \cdot 10^5 \end{cases} \tag{14}$$

$$Pr = \frac{\mu C_p}{k} \tag{15}$$

$$Pr = \frac{\mu C_p}{k} \tag{16}$$

Where L – Length of the plate (m), Pr – Prandtl number, Re_L – Reynolds number, μ – Dynamic viscosity of the fluid (Pa-s), U – stream velocity (m/s).

Convective heat transfer depends on many dimensionless numbers like the Nusselt number, Prandtl number, and Reynolds number. These numbers are used to nondimensionalize the equations governing the system and help in reducing the total number of variables to study. With the help of these dimensionless numbers, flow properties can be obtained which makes the study much easier. Since convection also happens at the instance of braking, it becomes necessary to analyze these parameters.

Thermal Contact

The conduction between the two contact surfaces, the pad, and the disc, can be given using the thermal contact function. The heat generated by thermal friction between the surfaces is distributed according to their material properties. Constriction conduction (h_c) is calculated using the Cooper-Mikic-Yovanovich (CMY) correlation for rough surfaces considering the plastic deformation of the surface abrasives. Gap conductance and radiative conductance are negligible.

$$-n_d \cdot q_d = -h(T_u - T_d) + rQ_b \tag{17}$$

$$-n_u \cdot q_u = -h(T_d - T_u) + (1-r)Q_b \tag{18}$$

$$r = \frac{1}{1+\xi}, \quad \xi = \sqrt{\frac{\rho_u C_{p,u}(k_u n_u) \cdot n_u}{\rho_d C_{p,d}(k_d n_d) \cdot n_d}} \tag{19}$$

$$h_c = 1.25 k_{contact} \frac{m_{asp}}{\sigma_{asp}} \left(\frac{p}{H_c} \right)^{0.95} \tag{20}$$

$$\frac{2}{k_{contact}} = \frac{1}{(k_u n_u) \cdot n_u} + \frac{1}{(k_d n_d) \cdot n_d} \tag{21}$$

$$h = h_c + h_g + h_f \tag{22}$$

Where d – downside of the contact surface, u – upside of the contact surface, r – heat partition coefficient, Q_b – heat produced due to friction (W/m²), h_c – conductance due to construction (W/m²K), $k_{contact}$ – harmonic mean of conductivities of surfaces in contact (W/mK), m_{asp} – average of asperities slopes (m), σ_{asp} – average height of asperities (m), p – pressure at the contact surface (Pa), H_c – microhardness of the softer material (Pa), h – joint conductance (W/m²K), h_g – gap conductance (W/m²K), h_f – radiative conductance (W/m²K). The heat generated by thermal friction is calculated from equation (23) where the heat rate is given by equation (24).

$$Q_b = \frac{P_b}{A} \tag{23}$$

$$P_b = -m_{disc} \cdot v(t) \cdot a(t) \tag{24}$$

Where P_b – heat rate (W), A – boundary area (m²), m_{disc} – mass of the brake disc (kg), $v(t)$ – velocity of the disc (m/s), $a(t)$ – acceleration of disc (m/s²). Though most heat loss is through conductive and convective heat transfer, all surfaces are admitted to radiative heat transfer too.

Surface to Ambient Radiation

Transfer of heat from the surface of the pad and the disc to the ambient surrounding through radiation is defined using surface-to-ambient radiation conditions. The heat flux is given by equation (25).

$$-n \cdot q = \varepsilon \sigma \left(T_{amb}^4 - T^4 \right) \tag{25}$$

Where ε – emissivity of the surface, σ – Stefan Boltzmann constant, T_{amb} – Ambient temperature (K). Room conditions are usually taken as ambient conditions. The emissivity is taken for the brake pad and disc material and radiation is calculated accordingly.

Materials and Parameters

The materials used for the brake pad and the brake disc are Cu and stainless steel respectively. Cu-based brake pads are desirable as they eliminate common spallation and detachment problems associated with brakes [28]. The material properties are shown in Table 20.1 [29–31]. The parameters used for the simulation are given in Table 20.2. The efficiency of the braking depends on the friction coefficient of the brake pad. The friction coefficient must be in an optimum value (in the range 0.3–0.4) for effective braking. If the friction coefficient is higher, the wear of the brake pad will be higher and the effectiveness of the pad is degraded.

If the coefficient of friction is lower, the braking will not be effective. Hence an optimum value of 0.35 is taken. The braking system is programmed to function during maintenance and at Overspeed. At 10% over-speed of the LSS, aerodynamic brakes initially reduce the speed of HSS to 350–450 rpm. Mechanical brakes are then applied to the high-speed shaft which eventually halts the motion of the turbine. From the design considerations, the optimum sliding velocity is chosen as 5.5 m/s. Figure 20.4 shows how the sliding velocity varies with time.

Table 20.1: Brake pad and brake disc material property.

Description	Variable	Disc (Steel)	Pad (Copper)
Emissivity	ε	0.3	0.8
Microhardness (MPa)	H_v	550	800
Density (kg/m³)	ρ	8050	8933
Heat capacity (J/kg °C)	C_p	530	385
Thermal conductivity (W/ mK)	K	17.1	401

Table 20.2: Parameters used in the simulation.

Description	Variable	Value
Friction coefficient	Mu	0.35
Sliding velocity speed	v0	5.5[m/s]
Rotor acceleration	a0	–2.5[m/s²]
Disc radius	r_disc	0.23[m]
Disc mass	m_disc	10000[kg]
Braking time (start)	t_brake_start	2[s]
Braking time (end)	t_brake_end	4[s]
Temperature of air	T_air	300[K]

An increase in sliding speed increases the wear rate and increases the temperature of both materials. A higher braking load aggravates the wear of the brake pad. Considering the clamping force to be 100 kN, mass can be determined as 10,000 kg. The brake pads generate friction when pressed against the rotor disc which is liberated as heat. The deceleration of the wind turbine for the defined sliding velocity is 2.5 m/s². Figure 20.5 represents the deceleration acting to stop the motion of the rotor disc on braking.

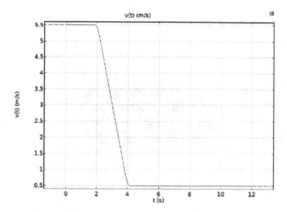

Figure 20.4 Velocity vs Time graph.

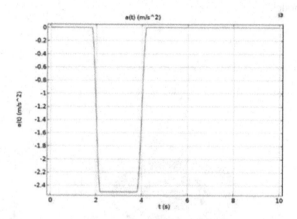

Figure 20.5 Acceleration vs Time graph.

Meshing

After modeling the component, FEM discretizes the body into sub-domains. The geometry is split into smaller and simpler finite units. This is called discretization. In COMSOL Multiphysics, meshing is used to discretize the model. Figure 20.6 depicts the model which is meshed in COMSOL Multiphysics. These smaller units are called the mesh elements. Meshing is critical for smoother visualization of the results. The mesh determines the preciseness, convergence, and speed of the simulation. Free triangular mesh along the boundary which is then swept along the remaining domain is used for meshing in this model. The predefined size of the extra fine is used. The completed mesh had 4860 domain elements and 2194 boundary elements.

The physics which the model uses is heat transfer in solids whose parameters may vary with time. This necessitates the use of a solver to study the change in variables throughout the time domain which is the primary objective of the model. The Time-Dependent solver is the best solver that can be used to study time-dependent problems. With the initiation of the solver, the solver determines the timestep depending upon the total time of the simulation. The model uses the Backward Differential Formula (BDF) method with intermediate steps for solving and determining the solution.

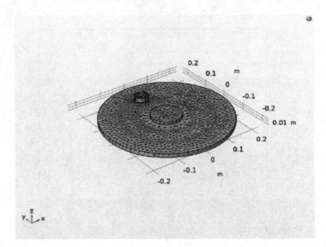

Figure 20.6 Meshed model.

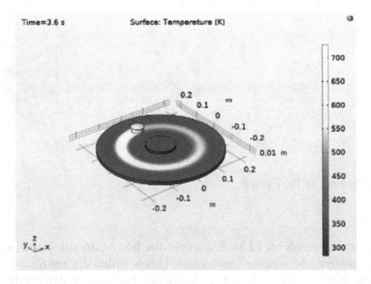

Figure 20.7 Surface temperature variation at the interface.

Results and Discussion

The thermal phenomenon during braking of the brake disc with the brake pad was studied using COMSOL using the finite element method. The pure Cu brake pad on the steel disc is conventionally used for braking purposes as Cu effectively enables heat transfer. On braking, heat is generated at the interface of the brake pad and the brake disc which eventually increases the temperature on the surface of the pad and the disc. Heat is conducted throughout the brake pad and is liberated to the atmosphere by convection and radiation. The temperature profile can be visualized in the form of plots and contours. Figure 20.7 shows the temperature profile developed on braking at 3.6 seconds at the interface of the brake pad and the brake disc.

At the instance of braking, there is a momentary increase in temperature due to the friction between the brake pad and rotor. As the heat is dissipated through convection and radiation, the temperature on the disc and pad decreases and stabilizes. From Figure 20.8, it can be observed that the rate at which heat is dissipated is much lower than the rate of heat produced [16]. The remaining energy is conducted throughout the disc and pads internally during braking and heats the brake components. This inability of the brake pad to dissipate the heat it produces leads to the storage of heat in it. This storage of heat results in deterioration of the properties of the brake pad as well as the brake disc [32]. This demands the need for an additional cooling system. Studies done by Belhocine et al. [33] showed that ventilation plays an important role in the cooling system of the brakes and ventilated brake pads provide better thermal characteristics. It may also lead to the wearing of the brake pad. Wear heavily influences the temperature distribution on the pad and the disc [34].

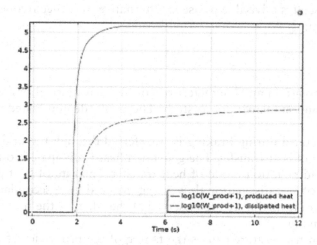

Figure 20.8 Graph comparing the produced and the dissipated heat.

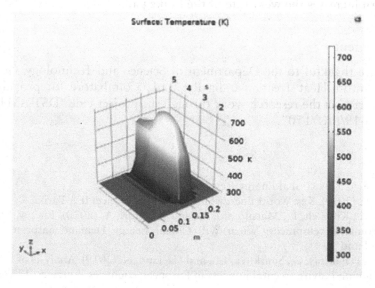

Figure 20.9 Temperature distribution at the interface.

High temperature is observed at the edge of the pad. This creates a possibility for overheating which may lead to plastic deformation of the pad [35]. This results in the fading of the pads and necessitates frequent maintenance and replacement of the pads. Figure 20.9 depicts the temperature distribution at the developed interface during the braking period. It can be seen that the temperature rises to 700 K which lies within the permissible limits. According to Masamichi Mitsumoto's research on Cu-based brake pads, typical Cu-based brake pads can attain a temperature of 500°C on braking [36]. The braking load has a drastic effect on the brake pads during braking. When there is an increase in braking load, the normal force from the brake pad acting against the rotor increases to decelerate the rotor motion. Due to this increase in force, there will be more amount of friction acting on the disc. This friction causes rapid wear of the brake pad surface and generates tremendous heat. Thus, causing a step in the temperature profile plot due to the effect of the braking load. From the thermal phenomenon observed through the simulation, it can be inferred that the poor heat dissipation of the brake pad weakens the surface properties of both the brake pad and disc. It is advisable to use an alternative to reduce frequent maintenance of the braking system.

Conclusion

The thermal phenomenon and characteristics of the conventional brake pad during braking were studied and simulated in this study. The results demonstrated that:

1. The heat generated during braking is transferred through the brake pad by the conduction mode of heat transfer. The generated heat is dissipated to the atmosphere by convection and radiation mode of heat transfer. The rate of heat generation is more than the rate of heat dissipation thus giving the need for a secondary cooling system.
2. Higher levels of temperature are observed at the edges of the brake pad, thus causing plastic deformation.
3. An increase in temperature causes the fading of the properties of the brake pad and mandates frequent replacements. An increase in the braking load generates more heat that in turn increases the wear rate of the brake pad.

Acknowledgement

The authors are thankful to the Department of Science and Technology, Government of India and Quantum Heat Treaters India Pvt. Ltd., Coimbatore for providing financial assistance to carry out the research work through the project vide "DST-AMT-DEVLP, FP, PP & TD/2018-19/R/C /150".

References

1. Agency, I. E. (2020). Global Energy Review 2020. IEA Paris.
2. Agency, I. E. (2019). Key World Energy Statistics 2019. France: IEA Paris.
3. Bogmans, C., Kiyasseh, L., Matsumoto, A., and Pescatori, A. (2020). Energy, Efficiency Gains and Economic Development: When Will Global Energy Demand Saturate?, International Monetary Fund.
4. Martins, F., Felgueiras, C., Smitkova, M., and Caetano, N. (2019). Analysis of fossil fuel energy consumption and environmental impacts in European countries. *Energies*, 12, 964.

5. Wood, N., and Roelich, K. (2019). Tensions, capabilities, and justice in climate change mitigation of fossil fuels. *Energy Research & Social Science*, 52, 114–122.
6. Johnsson, F., Kjärstad, J., and Rootzén, J. (2019). The threat to climate change mitigation posed by the abundance of fossil fuels. *Climate Policy*, 19, 258–274.
7. Shoaib, M., Siddiqui, I., Rehman, S., Khan, S., and Alhems, L. M. (2019). Assessment of wind energy potential using wind energy conversion system. *Journal of Cleaner Production*, 216, 346–360.
8. Sharma, S., and Sinha, S. (2019). Indian wind energy & its development-policies-barriers: an overview. *Environmental and Sustainability Indicators*, 1, 100003.
9. Chaurasiya, P. K., Warudkar, V., and Ahmed, S. (2019). Wind energy development and policy in India: a review. *Energy Strategy Reviews*, 24, 342–357.
10. Rehman, S., Natarajan, N., Vasudevan, M., and Alhems, L. M. (2020). Assessment of wind energy potential across varying topographical features of Tamil Nadu, India. *Energy Exploration & Exploitation*, 38, 175–200.
11. Wang, X., Zeng, X., Li, J., Yang, X., and Wang, H. (2018). A review on recent advancements of substructures for offshore wind turbines. *Energy conversion and Management*, 158, 103–119.
12. Rajesh Kannan, K., Govindaraju, M., and Vaira Vignesh, R. (2021). Development of fly ash based friction material for wind turbines by liquid phase sintering technology. *Proceedings of the Institution of Mechanical Engineers, Part J: Journal of Engineering Tribology*, 235(7), 1463–1469.
13. Kannan, K. R., Vignesh, R. V., Kalyan, K. P., Murugesan, J., Megalingam, A., Padmanaban, R., and Govindaraju, M. (2019). Tribological performance of heavy-duty functionally gradient friction material (Cu-Sn-Fe-Cg-SiC-Al2O3) synthesized by PM route. In: AIP Conference Proceedings, (pp. 020004). AIP Publishing.
14. Sha, Z., Yin, J., Zhang, S., Liu, Y., and Ma, F. (2015). Structure topology optimization design of brake pad in large megawatt wind turbine disc brake. In First International Conference on Information Sciences, Machinery, Materials and Energy. Atlantis Press.
15. Caliper Brake Solutions from Altra Provide Reliable Performance On Wind Turbines Around The World. https://www.twiflex.com/en/newsroom/2016/02/is caliper brake solutions for wind turbines (accessed Nov. 23, 2023).
16. Jiguang, C., and Fei, G. (2015). Temperature field and thermal stress analyses of high-speed train brake disc under pad variations. *Open Mechanical Engineering Journal*, 9(1), 371–378. doi: 10.2174/1874155X01509010371.
17. Belhocine, A., and Abdullah, O. I. (2020). Thermomechanical model for the analysis of disc brake using the finite element method in frictional contact. *Multiscale Science and Engineering*, 2, 27–41.
18. Borawski, A. (2021). Impact of operating time on selected tribological properties of the friction material in the brake pads of passenger cars. *Materials*, 14, 884.
19. Thakur, A. S., and Dhakad, P. (2016). Thermal analysis of disc brake using ANSYS: a review. *Int. J. Adv. Eng. Res. Dev*, 3(03). doi: 10.21090/ijaerd.030359
20. Govindaraju, M., Megalingam, A., Murugasan, J., Vignesh, R. V., Kota, P. K., Ram, A. S., Lakshana, P., and Kumar, V. N. (2020). Investigations on the tribological behavior of functionally gradient iron-based brake pad material. *Proceedings of the Institution of Mechanical Engineers, Part C: Journal of Mechanical Engineering Science*, 234(12), 2474–2486.
21. Keshav, M. G., Hemchandran, C., Dharsan, B., Pradhin, K., Vignesh, R. V., and Govindaraju, M. (2020). Manufacturing of continuous fiber reinforced sintered brake pad and friction material. *Materials Today: Proceedings*, 46, 4493–4496.
22. Petinrin, M. O., and Oji, J. O. (2012). Numerical simulation of thermoelastic contact problem of disc brake with frictional heat generation. *New York Science Journal*, 5, 39–43.
23. Bashir, M., Qayoum, A., and Saleem, S. (2019). Analysis of frictional heating and thermal expansion in a disc brake using COMSOL. In Journal of Physics: Conference Series, (pp. 012094). IOP Publishing.

24. Xiao, Y., Zhang, Z., Yao, P., Fan, K., Zhou, H., Gong, T., Zhao, L., and Deng, M. (2018). Mechanical and tribological behaviors of copper metal matrix composites for brake pads used in high-speed trains. *Tribology International*, 119, 585–592.

25. Pakowski, Z. (2007). Fundamentals of heat and mass transfer. In Incropera, F. P., DeWitt, D. P., Bergman, T.L., and Lavine, A.S. editors. Hoboken NJ: J. Wiley & Sons, (p. 997). Elsevier.

26. Vignesh, R. V., Padmanaban, R., Arivarasu, M., Thirumalini, S., Gokulachandran, J., and Ram, M. S. S. S. (2016). Numerical modelling of thermal phenomenon in friction stir welding of aluminum plates. In IOP Conference Series: Materials Science and Engineering, (pp. 012208). IOP Publishing.

27. Vignesh, R. V., and Padmanaban, R. (2019). Modelling corrosion phenomenon of magnesium alloy AZ91 in simulated body fluids. *Advances in Mathematical Methods and High Performance Computing*, 471–486. Springer

28. Cøengel, Y. (2007). Heat and Mass Transfer: A Practical Approach. New York: McGraw-Hill.

29. Zhang, P., Zhang, L., Wei, D., Wu, P., Cao, J., Shijia, C., and Qu, X. (2020). A high-performance copper-based brake pad for high-speed railway trains and its surface substance evolution and wear mechanism at high temperature. *Wear*, 444, 203182.

30. Sarkar, S., Rathod, P. P., and Modi, A. (2014). Research paper on modeling and simulation of disc brake to analyse temperature distribution using FEA. *International Journal for Scientific Research & Development*, 2, 491–494.

31. Zhang, S., Hao, Q., Liu, Y., Jin, L., Ma, F., Sha, Z., and Yang, D. (2019). Simulation study on friction and wear law of brake pad in high-power disc brake. *Mathematical Problems in Engineering*, 2019, 2019. doi: 10.1155/2019/6250694.

32. Mačužić, S., Saveljić, I., Lukić, J., Glišović, J., and Filipović, N. (2015). Thermal analysis of solid and vented disc brake during the braking process. *Journal of Serbian Society for Computational Mechanics*, 9, 19–26.

33. Belhocine, A., and Bouchetara, M. (2014). Structural and thermal analysis of automotive disc brake rotor. *Archive of Mechanical Engineering*, 61(1), 89–113.

34. Wu, Y., Jin, H., Li, Y., Ji, Z., and Hou, S. (2014). Simulation of temperature distribution in disk brake considering a real brake pad wear. *Tribology Letters*, 56, 205–213.

35. Hatam, A., and Khalkhali, A. (2018). Simulation and sensitivity analysis of wear on the automotive brake pad. *Simulation Modelling Practice and Theory*, 84, 106–123.

36. Mitsumoto, M. (2017). Copper free brake pads with stable friction coefficient. *Tribology Transactions*, 59, 23–24.

21 Numerical modeling of compaction stresses in powder metallurgy processing of fly-ash based ceramic tiles

K. Balasubramanian, R. Vaira Vignesh[a], and R. Padmanaban[b]

Department of Mechanical Engineering, Amrita School of Engineering, Coimbatore, Amrita Vishwa Vidyapeetham, India

Abstract: Flyash, a residual product resulting from coal combustion, is produced extensively within global thermal power plants. Consequently, development of technology for the efficient utilization of flyash is a tenacious requirement. Powder metallurgy (PM) processing technology is one of the operative methodologies to fabricate fly ash based ceramic tiles. However, the properties of the PM components depend on the effective compaction of flyash powders. Hence, the COMSOL Multiphysics model was developed to understand the evolution of pores and stresses in the course of compacting fly ash powders. The results indicate that the fillet in the plunger suggestively influences the void volume fraction, volumetric plastic strain, and stresses that are induced in the course of compacting flyash powders.

Keywords: Ceramic, compaction, flyash, green pellet, stress.

Introduction

Waste can be any kind of substance that may be of any form solid, liquid, or even gas. The municipal wastes are generated from houses, gardens, street sweeping, office buildings, institutions, and other market wastes. These wastes are collected and treated by local municipalities. Plastic wastes often do not degrade easily which takes much longer years and these plastics are very much harmful to the environment [1, 2]. Industrial wastes are very much common all over the world [3]. Industrial wastes are in turn can be categorized as hazardous and non-hazardous waste [4]. The waste generated by industries and manufacturing units includes dirt, gravel, concrete, scrap metals, trash, oil, solvents, chemicals, fly ash, plastics, and wood.

Flyash, a by-product of the combustion of coal, is one of the major industrial wastes that is generated in large quantities across the thermal power plants in the world. Typically, Indian coals are low grade and have around 30–45% ash content [5, 6]. The coal consumed was 406.91 million tons in the first half-year of 2019–2020 in India. And consequently, the fly ash generation was 129.09 million tons in the first half-year of 2019–2020 in India. The percentage of fly ash content was 31%, as shown in Table 21.1 [7]. About 316 individual minerals and 188 mineral groups are present in the fly ash. The elements

[a]r_vairavignesh@cb.amrita.edu, [b]dr_padmanaban@cb.amrita.edu

which are found in the fly ash around the world are silica (SiO_2), alumina (Al_2O_3), ferric oxide (Fe_2O_3), lime (CaO), potassium oxide (K_2O), magnesium oxide (MgO), sulfur trioxide (SO_3), titanium dioxide (TiO_2), sodium oxide (Na_2O), phosphorus oxide (P_2O_5), and manganese oxide (MnO) [8, 9].

The major content of fly ash is silicon dioxide, aluminum dioxide, calcium oxide, and sulfur. Fly ash is classified based on its chemical composition. The fly ash generated from anthracite and bituminous coals are classified as "Class F" [10]. Class F flyash predominantly comprises of alumina and silica. Class F flyash displays a reduced calcium proportion compared to Class C flyash, the latter being produced from sub-bituminous coals and characterized by its composition of calcium alumino-sulfate glass and free lime (CaO). Class C, often referred to as high calcium flyash, contains over 20% CaO. The structure of flyash encompasses spherical fine powder particles, which vary between solid and hollow forms, predominantly glassy in nature. Flyash exhibits distinctive characteristics of high specific surface area and low bulk density [11].

Table 21.1: Fly ash generation in India [7].

Indian fly ash generation	(1st half-year 2018–2019)	(1st half-year 2019–2020)
Number of Thermal Power Stations from which data was received	156	194
Coal consumed (Million tons)	295.42	406.91
Fly Ash Generation (Million tons)	93.26	129.09
Percentage Average Ash Content (%)	31.57	31.73

Around 20% to 40% of the generated fly ash is utilized by the cement factories, construction sites, as a soil amendment [12]. On the word of National Thermal Power Corporation Limited and Central Electricity Authority, fly ash is used as supplementary material for producing Portland cement. Also, it is used as filler materials in cement, concrete, ceramics, construction fills, and road fills [13]. However, the remaining fly ash is usually dumped in the lands near the thermal power plants. The contamination of groundwater resources, water bodies, soil fertility, and air are the direct consequences of dumping fly ash. Fly ash is now used to produce fly ash based bricks which can be a good replacement for the normal chamber bricks. Fly ash is mixed with sand cement to form fly ash based bricks. The fly ash brick has good properties similar to the normal bricks. The breakage is less than 1% compared to clay brick, and the compressive strength for the clay brick is about 50–60 kg/cm^2 and for the fly ash-based brick it about 75–100 kg/cm^2.

The fly ash-based ceramic tiles and bricks are gaining popularity, as it uses a major proportion of fly ash [14]. Conventionally, the fly ash in the ceramic tile varies from 40–60%. These fly ash based tiles will exhibit better properties than granites and other types of wall tile. The utilization of the powder metallurgy processing method is predominantly favored for fabricating ceramic tiles based on fly ash. Powder metallurgy processing technique is a process in which powders (matrix, sintering agents) are used for ease of mixing and effective densification [15]. The reinforcement and powder are mixed, compacted, and sintered. In this process, powders are blended to form a homogenous mixture. Then, the blended mixture is compacted to the required shape (green pellet) [16]. Normally the green pellets are sintered without applying pressure [17].

The sintering process is diffusion-assisted [18]. Blending or mixing of powders is a process of mixing all the ingredients to make a desired homogenous mixture for the production of the green component [19]. Subsequently, the homogeneously blend of fine-sized powders is compacted under high pressure to make it into the desired component (green pellet) [20, 21]. This compaction process helps in the strength of the green pellet. Hao et al., [22] integrated fly ash and glass to form a foam/dense bi-layered insulated ceramic tile. The material was pressed to produce the green specimen which was sintered around 1200°C. The dense bi-layered insulated ceramic tile showed good mechanical bending strength of 31.5 MPa and good compressive strength of 12.9 MPa. Ji et al., [23] introduced an innovative approach for manufacturing ceramic tiles. This method involved utilizing fly ash as the primary raw material, which exhibited a substantial concentration of alumina, quartz, clay, and feldspar. The result indicates the sintering temperature range from 1150 to 1200°C, and 50–70% fly ash was a good mixture for the production of ceramic tile. The rupture modulus was about 51.28 MPa which was 47% greater than the standard requirement. The surface and overall texture were very much acceptable.

Hao et al., [24] synthesized using high alumina fly ash 70%, clay 15%, and quartz 15%. The samples were prepared and sintered at1200°C. It was found that the impurities in the high alumina fly ash lowered the densification transition temperature and reduced the bloating temperature. When the sintering temperature was increased a good densification behavior with high flexure strength of 67 MPa. Zimmer et al. [25] used the mineral coal fly ash to develop a ceramic tile. In this work, they have used different batches of fly ash as a main raw material mixed with other materials like clay, feldspar, and limestone to develop a ceramic tile. These specimens were sintered at 1300°C. The specimen that had limestone exhibited lower flexural strength and higher water absorption. The feldspar 60wt% with fly ash 40wt% specimen showed good flexural strength.

Jian Zhang et al., [26] fabricated a glass-ceramic from the fly ash which was about 90% and the rest 10% was Na_2O which was added for alkali activation. The powdered constituents were blended and subjected to a high temperature of approximately 1350°C, resulting in melting. Subsequently, the sample was subjected to a heat treatment at 770°C for a duration of 2 hours. This process led to the formation of a glass-ceramic specimen exhibiting favorable mechanical characteristics and robust chemical resistance. Kockal [27] evaluated the influence of fly ash on the sintered properties of ceramic tiles. The study indicated that sintering ceramic tiles with a high fly ash content at high temperatures decreased the properties. Bou et al., [28] produced a conventional ceramic tile using pulverized fuel ash with 92% PFA, 5% sodium bentonites, and 3% talc. The powders were mixed, pressed, and sintered when the green specimen was fired. The increased quantity of flocculants led to significant shrinkage during the firing process.

The heightened sensitivity of shrinkage to temperature variations resulted in pronounced distortion of the tiles. Furthermore, the sintered specimens with a water absorption rate of less than 2% exhibited considerable bloating, contributing to an undesirable decline in surface quality. The stresses are developed in the green pellet that affects the effective sintering of specimens. Hence, a study on analyzing compaction stresses in powder metallurgy processing of fly-ash based ceramic tiles becomes mandatory. Chung-man et al., studied different flow stress models that were used to simulate the powder compaction process of a cylindrical object (synchronizer hub).

Diarra et al., [29] simulated powder compaction in the course of hot isostatic pressing. They developed three models known as CAM-Clay, Drucker-Prager, and modified Drucker-Prager with creep. The Drucker-Prager model, combined with a creeping constitutive model, demonstrated remarkable precision in predicting the final geometric outcomes, yielding a relative error ranging from approximately 1.5% to 4.8%. Peng Han et al., simulated the Fe-Al powder compaction model using the multi-particle finite element method. The model was developed with different packing structures using the discrete element method. The compaction process was recreated by the macro and micro properties of the compact. During the compaction process, large stresses were concentrated in the neighboring Fe particles that in turn allowed densification. Abedinzadeh [30] studied the densification behavior of the aluminum powder using microwave hot pressing, both numerically and experimentally. The numerical study was carried by the 3D finite element method. The result showed that the samples right edge corners had a high density compared to bottom corners. The center region was consolidated uniformly.

Abedinzadeh et al. [31] conducted an investigation involving finite element modeling of microwave hot pressing within a multimode furnace. Gurson–Tvergaard–Needleman model was employed to create porous compacts. Calculations were made for the temperature fluctuations of the samples, considering dielectric and thermal properties. The powder within the sample needed a certain amount of time to attain the necessary temperature during the hot pressing procedure. Additionally, stress levels at the corners of the alumina ceramic compact subjected to hot pressing were found to be approximately 45–50% of the applied pressure. Tikare et al., [32] simulated a model on microstructural evolution for the single state solid compaction for complex 3D powers. The model simulated all the stages of sintering from neck growth. The model also demonstrated the microstructural evolution and compared it with the experimental.

The result showed the sliced edges were at the corner of the compacted object and showed porosity. The pores were continuous which percolated through powder.

In this research work, the computational work was carried out using COMSOL Multiphysics software to understand the stresses in the course of compacting fly-ash powders to form a green pellet. Also, volumetric plastic strain, von-mises stress, contact pressure, frictional force (per unit area) in the course of compaction was investigated.

Computational Methodology

The following sections describe the methodology for simulating the compaction process in COMSOL Multiphysics.

3D Modeling

A 2D model was developed using the Geometry section. Figure 21.1 shows 2D Schematic of compaction process. The model consists of a plunger that compacts the fly ash powder. The plunger (rectangle 2) consists of a small fillet radius to reduce the high-stress concentration. Though beneficial for the plunger life, this has a negative influence on the green pellet. The fly ash particles are present in the region of compaction (rectangle 1). The die component (rectangle 3) helps to form the green pellet to the required shape.

Material Properties

In the developed model the plunger (compressing equipment) and the die were assumed to be rigid components. Hence, material properties were neglected for both the die and plunger. The fly ash properties were not available in the open literature. One of the major components of the fly ash includes mullite (ceramic phase of SiO_2 and Al_2O_3). Hence, mullite properties were assigned for the fly ash in the computational work. The properties of the material were assigned as follows: density of 3030 kg/m^3, young's modulus of 224.9 GPa, the initial yield stress of 138 MPa, and Poisson's ratio of 0.24 [33].

Boundary Conditions

In COMSOL Multiphysics, solid mechanics was selected for analyzing the model. Afterward, a linear elastic material was chosen for the simulation, utilizing a porous plasticity model. The fundamental principle underlying porous plasticity models revolves around the progression of relative density. Relative density pertains to the proportion of solid volume within a porous blend and is intricately linked to the porosity (or void volume fraction, φ) by equation (1).

$$\rho_{rel} = 1 - \varphi \tag{1}$$

During the process of compacting a particle mixture, the porosity gradually diminishes towards zero, and concurrently, the relative density progressively approaches unity. Numerous porous plasticity models have been documented in the existing literature to elucidate the intricacies of the compaction mechanism. In this study, the Fleck-Kuhn-McMeeking (FKM) model, often referred to as FKM-GTN, and the Gurson-Tvergaard-Needleman (GTN) model were specifically chosen to facilitate the simulation and analysis of the compaction process. The plastic potential within the GTN criterion is mathematically defined by equation (2).

$$Q_p(\sigma) = \left(\frac{\sigma_e}{\sigma_0}\right)^2 + 2q_1\varphi_e \cosh\left(\frac{3q_2 p_m}{2\sigma_0}\right) - \left(1 + q_3\varphi_e^2\right) \tag{2}$$

Where σ_e is the equivalent stress, σ_0 is the initial yield stress, p_m is the pressure, and φ_e is the effective void volume fraction (effective porosity). The value of effective void fraction is determined by various factors including the present porosity and additional material parameters, and its relationship is expressed through equation (3).

$$\varphi_e = \begin{cases} \varphi, & for\, \varphi < \varphi_c \\ \varphi_c + \dfrac{\varphi_m - \varphi_c}{\varphi_f + \varphi_c}\left(\varphi - \varphi_c\right) & for\, \varphi_c < \varphi < \varphi_f \end{cases} \tag{3}$$

Where φ_c represents the critical void volume fraction, denoting the porosity threshold at which void coalescence initiates, while φ_f signifies the void volume fraction at the point of failure. As porosity progressively escalates towards the failure limit, the effective porosity achieves its peak value denoted as φ_m. At this juncture, the material becomes incapable of sustaining stress. The determination of the maximum porosity φ_m is contingent on various other parameters, as outlined by *equation (4)*.

$$\varphi_m = \frac{q_1 + \sqrt{q_1^2 - q_3}}{q_3} \tag{4}$$

The Fleck-Kuhn-McMeeking criterion, commonly known as the FKM criterion, was formulated to capture the plastic yielding behavior of metal aggregates characterized by substantial porosity. The yield function and the corresponding plastic potential were derived based on formulations applicable to randomly dispersed particles. The criterion is particularly applicable to aggregates featuring porosity levels ranging from 10% to 35%. The expression defining the plastic potential within the FKM criterion is represented by equation (5). Additionally, the material's flow strength under conditions of hydrostatic loading, denoted as , is mathematically defined by equation (6).

$$Q_p = \left(\frac{5\sigma_e}{18p_f} + \frac{2}{3} \right)^2 + \left(\frac{5p_m}{3p_f} \right)^2 - 1 \tag{5}$$

$$p_f = 2.97(1-\varphi)\frac{2\varphi_m - \varphi}{\varphi_m}\sigma_0 \tag{6}$$

The FKM-GTN criterion is an amalgamation of the previously mentioned criteria designed to encompass a broader spectrum of porosity variations. Specifically, the GTN model is employed when dealing with lower void volume fractions (porosity below 10%), whereas the FKM criterion is applied for higher void volume fractions exceeding 25%. Within the transitional zone falling between these ranges, a linear combination of both criteria was adopted to provide a comprehensive characterization.

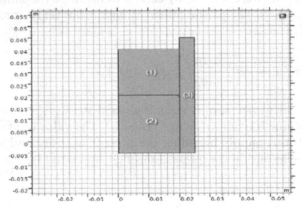

Figure 21.1 2D Schematic of compaction process.

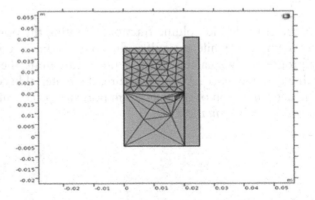

Figure 21.2 Meshing scheme.

Tvergaard correction coefficients were $q_1 = 1.5$, $q_2 = 1$, and $q_3 = 2.25$ [33]. The maximum void volume fraction was assumed to be 0.36, with an initial void volume fraction of 0.28 [14]. The plunger and die are selected as rigid domains. A prescribed displacement boundary condition for the upper and lower face of the powder was given. The displacement of the plunger in the z-direction is controlled by the parameter "para (z)" that varies from 0 (beginning of compaction) to 0.02 (end of compaction).

Meshing Scheme

A customized mesh was applied to improve the convergence and quicken the computations. The straight edges in the rigid domains (die and compaction zone) were set using coarse mesh. The compaction zone near the curved boundary was joined with fine triangular mesh. The plunger with fillet radius was resolved using fine triangular mesh. The meshed model is shown in Figure 21.2.

Solver Configuration

Direct stationary solver (Comsol Multiphysics) was selected for solving the numerical model.

Results and Discussion

This section describes the volumetric plastic strain, void volume fraction, von-mises stress, contact pressure, and frictional force developed during the compaction process are discussed.

Volumetric Plastic Strain

The volumetric plastic strain is the distortion or deformation in the volume in the course of the compaction process. The volumetric plastic strain in the course of compacting the fly-ash powders is shown in Figure 21.3. Figure 21.4 shows void volume fraction (a) Before compaction (z = 0); (b) Middle of compaction (z = 0.001); (c) End of compaction (z – 0.002). In the middle of the compaction, a maximum volumetric plastic strain of 0.035 was observed in the middle zone. At the end of compaction, a maximum volumetric plastic strain of 0.07 was observed in the middle zone. The volumetric plastic strain was high near the ends of the fillet. The volumetric plastic strain of about 0.10 at the corner points is attributed to friction, that was produced during the compaction.

Von Mises Stress

Figure 21.5 represents the stress developed during the compaction process. It was observed that stresses increased with an increase in the displacement of powder particles. At the end of compaction, a small portion of the green pellet was subjected to low stress of magnitude 4×10^7 N/m². The low stress was attributed to the non-contact of the plunger with the powder particles (a direct consequence of fillet). All other regions were subjected to a maximum stress of magnitude $\sim 9 \times 10^7$ N/m²

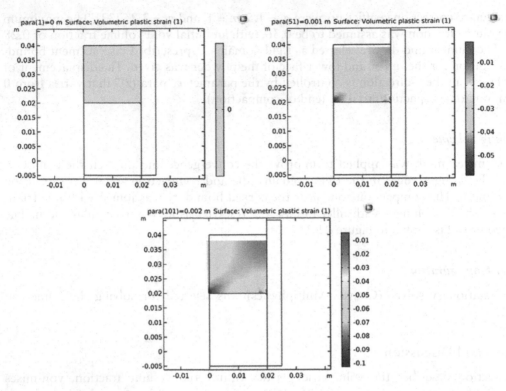

Figure 21.3 Volumetric plastic strain (a) Before compaction (z = 0); (b) Middle of compaction (z = 0.001); (c) End of compaction (z = 0.002).

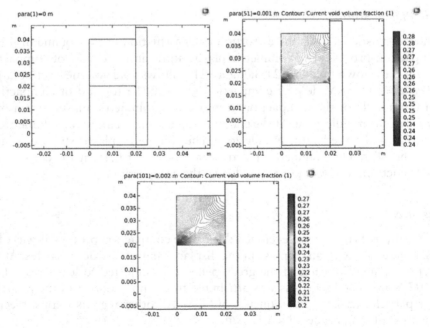

Figure 21.4 Void volume fraction (a) Before compaction (z = 0); (b) Middle of compaction (z = 0.001); (c) End of compaction (z = 0.002).

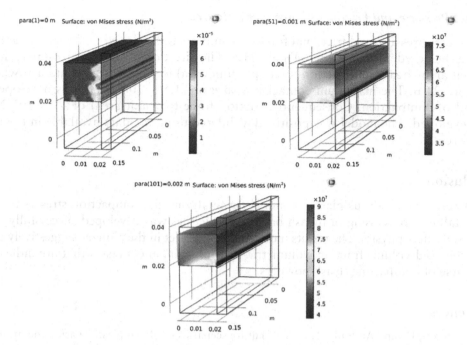

Figure 21.5 Von mises stress (a) Before compaction (z = 0); (b) Middle of compaction (z = 0.001); (c) End of compaction (z = 0.002).

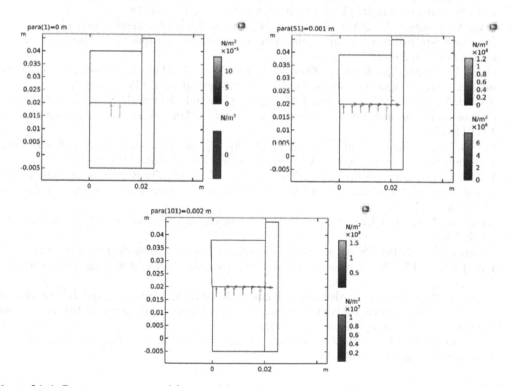

Figure 21.6 Contact pressure and frictional force (per unit area) (a) Before compaction (z = 0); (b) (z = 0.002) Middle of compaction (z = 0.001); (c) End of compaction.

Contact Pressure and Frictional Force (per unit area)

The contact pressure and frictional force (per unit area) developed in the course of compacting the powders are shown in Figure 21.6. The direction of contact pressure is shown as green arrowheads and friction force (per unit area) is shown as magenta arrowheads in Figure 21.6. The maximum contact pressure of ~1.5 × 108 N/m² was developed at the end of compaction. Consequently, friction force (per unit area) of ~1 × 107 N/m² was developed. The model was positively validated using the data available in the open literature.

Conclusion

In this research, a porous plasticity model for analyzing the compaction stresses in powder metallurgy processing of fly-ash based ceramic tiles was developed successfully using COMSOL multiphysics. The results indicate that the fillet in the plunger suggestively influences the void volume fraction, volumetric plastic strain, and stresses that are induced in the course of compacting flyash powders.

References

1. Lebreton, L., and Andrady, A. (2019). Future scenarios of global plastic waste generation and disposal. *Palgrave Communications*, 5, 1–11.
2. Borrelle, S. B., Ringma, J., Law, K. L., Monnahan, C. C., Lebreton, L., McGivern, A., Murphy, E., Jambeck, J., Leonard, G. H., and Hilleary, M. A. (2020). Predicted growth in plastic waste exceeds efforts to mitigate plastic pollution. *Science*, 369, 1515–1518.
3. Liao, M.-I., Chen, P.-C., Ma, H.-W., and Nakamura, S. (2015). Identification of the driving force of waste generation using a high-resolution waste input–output table. *Journal of Cleaner Production*, 94, 294–303.
4. Duan, H., Huang, Q., Wang, Q., Zhou, B., and Li, J. (2008). Hazardous waste generation and management in China: a review. *Journal of Hazardous Materials*, 158, 221–227.
5. Yousuf, A., Manzoor, S. O., Youssouf, M., Malik, Z. A., and Khawaja, K. S. (2020). Fly ash: production and utilization in India-an overview. *Journal of Materials and Environmental Science*, 11, 911–921.
6. Dhadse, S., Kumari, P., and Bhagia, L. (2008). Fly ash characterization, utilization and Government initiatives in India—A review. *J. Sci. Ind. Res. (India)*, 67(1), 11–18.
7. F. L. Y. A. S. H. Generation, I. T. S. Utilization, and I. N. (2018). The Report on At Coal / Lignite Based Thermal Power Stations and Its Utilization in the Country for 1 St Half of the Year 2017–18. 18(June).
8. Kutchko, B. G., and Kim, A. G. (2006). Fly ash characterization by SEM–EDS. *Fuel*, 85, 2537–2544.
9. Giergiczny, Z. (2019). Fly ash and slag. *Cement and Concrete Research*, 124, 105826.
10. Fox, J. M. (2017). Fly ash classification–old and new ideas. In World of Coal Ash Conference, Lexington, KY.
11. Bhatt, A., Priyadarshini, S., Mohanakrishnan, A. A., Abri, A., Sattler, M., and Techapaphawit, S. (2019). Physical, chemical, and geotechnical properties of coal fly ash: a global review. *Case Studies in Construction Materials*, 11, e00263.
12. Ahmaruzzaman, M. (2010). A review on the utilization of fly ash. *Progress in Energy and Combustion Science*, 36, 327–363.

13. Alam, J., and Akhtar, M. (2011). Fly ash utilization in different sectors in Indian scenario. *International Journal of Emerging Trends in Engineering and Development*, 1, 1–14.
14. Luo, Y., Zheng, S., Ma, S., Liu, C., and Wang, X. (2017). Ceramic tiles derived from coal fly ash: preparation and mechanical characterization. *Ceramics International*, 43, 11953–11966.
15. Samal, P. K., and Newkirk, J. W. eds., (2015). Powder metallurgy methods and applications. In ASM Handbook of Powder Metallurgy. ASM international (Vol. 7).
16. Govindaraju, M., Megalingam, A., Murugasan, J., Vignesh, R. V., Kota, P. K., Ram, A. S., Lakshana, P., and Kumar, V. N. (2020). Investigations on the tribological behavior of functionally gradient iron-based brake pad material. *Proceedings of the Institution of Mechanical Engineers, Part C: Journal of Mechanical Engineering Science*, 234, 2474–2486.
17. Jagadeep, R., Vignesh, R. V., Sumanth, P., Sarathi, V., and Govindaraju, M. (2021). Fabrication of fly-ash based tiles using liquid phase sintering technology. *Materials Today: Proceedings*, 46, 7224–7229.
18. Kannan, K. R., Vignesh, R. V., Kalyan, K. P., Murugesan, J., Megalingam, A., Padmanaban, R., and Govindaraju, M. (2019). Tribological performance of heavy-duty functionally gradient friction material (Cu-Sn-Fe-Cg-SiC-Al2O3) synthesized by PM route. In AIP Conference Proceedings, 2128(1), 020004. AIP Publishing.
19. Keshav, M. G., Hemchandran, C., Dharsan, B., Pradhin, K., Vignesh, R. V., and Govindaraju, M. (2020). Manufacturing of continuous fiber reinforced sintered brake pad and friction material. *Materials Today: Proceedings*, 46, 4493–4496.
20. Govindaraju, M., Vignesh, R. V., and Padmanaban, R. (2019). Effect of heat treatment on the microstructure and mechanical properties of the friction stir processed AZ91D magnesium alloy. *Metal Science and Heat Treatment*, 61, 311–317.
21. Kannan, R. K., Govindaraju, M., and Vignesh, R. V. (2021). Development of fly ash based friction material for wind turbines by liquid phase sintering technology. *Proceedings of the Institution of Mechanical Engineers, Part J: Journal of Engineering Tribology*, 235(7), 1463–1469.
22. Wang, H., Sun, Y., Liu, L., Ji, R., and Wang, X. (2018). Integrated utilization of fly ash and waste glass for synthesis of foam/dense bi-layered insulation ceramic tile. *Energy and Buildings*, 168, 67–75.
23. Ji, R., Zhang, Z., Yan, C., Zhu, M., and Li, Z. (2016). Preparation of novel ceramic tiles with high Al2O3 content derived from coal fly ash. *Construction and Building Materials*, 114, 888–895.
24. Wang, H., Zhu, M., Sun, Y., Ji, R., Liu, L., and Wang, X. (2017). Synthesis of a ceramic tile base based on high-alumina fly ash. *Construction and Building Materials*, 155, 930–938.
25. Zimmer, A., and Bergmann, C. P. (2007). Fly ash of mineral coal as ceramic tiles raw material. *Waste Manag*, 27, 59–68.
26. Zhang, J., Dong, W., Li, J., Qiao, L., Zheng, J., and Sheng, J. (2007). Utilization of coal fly ash in the glass-ceramic production. *Journal of Hazardous Materials*, 149, 523–526.
27. Kockal, N. U. (2012). Utilisation of different types of coal fly ash in the production of ceramic tiles. *Boletín de la Sociedad Española de Cerámica y Vidrio*, 51, 297–304.
28. Bou, E., Quereda, M. F., Lever, D., Boccaccini, A. R., and Cheeseman, C. R. (2013). Production of pulverised fuel ash tiles using conventional ceramic production processes. *Advances in Applied Ceramics*, 108, 44–49.
29. Diarra, H., Mazel, V., Busignies, V., and Tchoreloff, P. (2017). Comparative study between drucker-prager/cap and modified cam-clay models for the numerical simulation of die compaction of pharmaceutical powders. *Powder Technology*, 320, 530–539.
30. Abedinzadeh, R. (2018). Study on the densification behavior of aluminum powders using microwave hot pressing process. *The International Journal of Advanced Manufacturing Technology*, 97, 1913–1929.

31. Abdelhafeez, A., and Essa, K. (2016). Influences of powder compaction constitutive models on the finite element simulation of hot isostatic pressing. *Procedia CIRP*, 55, 188–193.
32. Tikare, V., Braginsky, M., Bouvard, D., and Vagnon, A. (2010). Numerical simulation of micro-structural evolution during sintering at the mesoscale in a 3D powder compact. *Computational Materials Science*, 48, 317–325.
33. Pabst, W., Gregorová, E., Uhlířová, T., and Musilová, A. (2013). Elastic properties of mullite and mullite-containing ceramics–Part 1: theoretical aspects and review of monocrystal data. *Journal Ceramics-Silikáty*, 57, 265–274.

22 Design and ergonomic evaluation of a new cashew nut shelling workstation

Krishna Chaitanya Mallampalli[1,a], Debayan Dhar[2,b] and Swati Pal[3]

[1]Vellore Institute of Technology, Vellore, Tamil Nadu, India
[2]Indian Institute of Technology (IIT) Guwahati, Guwahati, Assam, India
[3]Indian Institute of Technology (IIT) Bombay, Powai, Maharashtra, India

Abstract: Over the recent decades, many small-scale industries in India are showing significant interest in ergonomics to improve productivity and workers' health. Re-design of workplace for the worker's safety and comfort is one of the initiatives taken by Indian small-scale industries to prevent various occupational injuries such as musculoskeletal disorders (MSDs). In this context, the focus of this present case study was to design and ergonomic evaluation of a new cashew nut shelling workstation in comparison with the existing cashew nut shelling workstation in a small-scale cashew nut processing industry. A participatory ergonomics approach was adopted for designing the new cashew nut shelling workstation. Based on Indian anthropometric data, a generic 50th percentile Indian female manikin was developed using CATIA V5 software. Assessment of working posture was carried out using the rapid upper limb assessment (RULA) method for analyzing both existing and newly designed workstations. Results showed that the comfort level of trunk, neck, and upper limb postures were remarkably improved with the new cashew nut shelling workstation. The RULA action level was 2 for new design whereas action level was 4 for the existing one. Based on these findings, it is concluded that implementation of newly designed cashew nut shelling workstation in a pragmatic way could minimize MSDs among female cashew nut shelling workers.

Keywords: Cashew nut shelling, ergonomics, posture analysis, shelling workers, shelling workstation.

Introduction

Musculoskeletal disorders (MSDs) are significant health problems in almost all occupations and a major reason for disability in both industrially developed and developing countries [1–3]. These MSDs significantly impact not only the workers' health but also reduce productivity as well as workers' performance [3, 4]. Therefore, identifying MSDs risk-related factors and developing ergonomic measures for preventing MSDs among workers is a challenging area for ergonomists and occupational health researchers.

The Cashew industry in India is one of the oldest and largest industries in the small-scale agro-processing sector. According to the Nuts and dried fruits statistical report [5], India is the second top-ranked country in cashew production in the world. Further, this industry is

[a]mkchaitanya555@gmail.com, [b]debayan@iitg.ac.in

contributing significantly to the economic development of the country in terms of export earnings and employment. According to the Cashew Export Promotion Council of India [6], the export earnings reached USD 911 million during the year 2017–18. Further, half-million workers were employed in this industry [7].

Cashew-nut shelling is the main activity in cashew nut processing industries in India, where the majority of the female workers can be seen in performing shelling operations [7–9]. Cashew nut shelling mainly deals with the shelling and extraction of cashew kernels. For this purpose, a traditional cashew nut shelling workstation is widely used. While using this existing workstation, shelling workers are exposed to ergonomic risk factors. Awkward working postures are one of the important ergonomic risk factors for cashew nut shelling workers [8, 10, 11]. The existing workstation used for cashew nut shelling is operated manually in a standing position throughout the day. This design lacks some ergonomics requirements—for example, provision for sitting and height adjustability of the table. So far, these requirements have not been addressed in India. Some authors described that the shelling workstation design has a significant relationship with MSDs risk [9]. The use of the existing shelling workstation for a prolonged duration is more likely to cause pain in different body regions among shelling workers. Knee (63.7%), lower back (62.7%), and shoulder (58.8%) pains were most prevalent among female cashew nut shelling workers [9].

Previous studies showed that the task performed in awkward standing posture is associated with MSDs in upper body parts in the workplace [2]. At the same time, evidence suggests that inappropriate workplace design is associated with occupational health issues among workers [3,4,12]. Many studies have also shown that poorly designed tools or workstation affect workers' physical health and workplace well-being [12]. Therefore, the physical workload must be considered in occupational ergonomics research to develop ergonomic interventions from a physical ergonomics perspective. So far, several previous studies have reported the positive outcomes of ergonomic design interventions in small-scale industries [4, 3, 12]. Further, some of these studies have also shown the re-design of a workstation and evaluation using digital human modeling (DHM) technology to prevent MSDs in the workplace [4, 12]. DHM was found to be a reliable posture assessment tool for computer-aided ergonomics investigations [13].

However, to date, researchers have not focused on the problems faced by workers in shelling workstation with regard to physical workload. Therefore, the main objective of present study is to identify the postural stress among female cashew nut shelling workers using DHM simulation. The aim was to design and ergonomically evaluate newly developed workstation in comparison with existing shelling workstation design in the small-scale cashew nut processing industry.

Materials and Methods

Assessment of Posture in Existing Shelling Workstation

The assessment of posture in the existing workstation was carried out via the following steps:

Existing cashew nut shelling workstation model. A CAD model of existing shelling workstation was developed using CATIA V5 software. The created CAD model was according to the original dimensions collected from the field. Firstly, the individual elements of the existing workstation were modeled separately and then assembled using the 'Assemble'

Figure 22.1 The existing cashew nut shelling workstation.

feature of CATIA V5 software. Figure 22.1 shows the CAD model of the existing shelling workstation according to actual dimensions.

Digital human model. Due to wide variability in body measurements, a generic 50th percentile Indian female digital human model was created based on an Indian anthropometric database developed by Chakrabarti [14]. For creating this human model, CATIA DELMIA software was used. The human model prepared has the necessary characteristics such as scalability, flexibility, and body parts movement. Using the 'Posture edit' feature, the digital human model was manipulated at free joints and postured as closely as possible to replicate the actual postures of female shelling workers.

RULA

Rapid upper limb assessment (RULA) [15] technique was used for assessing selected activity postures of female shelling workers. The RULA method allows quick evaluation of working postures and suggests distinct action levels for improving working postures. The researchers Kushwaha et al., [4] and Sanjoget al. [12] have also used RULA analysis for optimizing the working postures and ensured better product design and a safer workplace.

In the RULA method, the human body is divided into two groups (A and B). Group A includes shoulders, elbows, and wrists; group B includes neck, trunk, and legs. According to this method, a score (1 for neural posture; 4 for worst posture) is calculated for the working posture of each body segment. The combined individual scores of group A and group B obtain two separate scores (i.e., Score A and Score B). Muscle use and force exertion scores are added to the scores A and B, which obtains two new scores (Score C and Score D). Again the combination of scores C and D gives a final or grand score (ranges from 1 to 7) which indicates the musculoskeletal loading on workers' posture. Finally, four action levels and corresponding actions to be taken are suggested to avoid the risk of musculoskeletal disorders or injuries.

Recording of working posture and interfacing
Video recording and still photography techniques were used to record the shelling activity of female workers. The criteria taken for identifying the most critical posture was based on maximum flexion, deviation, and rotation of the hand and wrist regions. The critical posture was determined using recorded images and direct observations as the shelling operators perform the manual shelling activity.

Interfacing and simulation of shelling activity posture with the interaction of existing shelling workstation were performed using CATIA DELMIA software. This interfacing was performed in full scale (1:1), which is helpful for accurate and precise ergonomic evaluation. Figure 22.2 shows the interfacing of digital human model in existing shelling workstation.

Vision Analysis

(a)

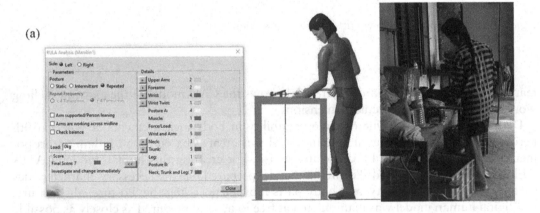

(b)

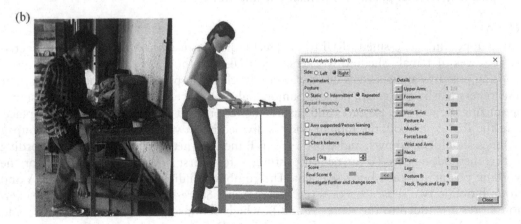

Figure 22.2 Interfacing of digital human model in the existing shelling workstation. (Note: a: top; b: bottom).

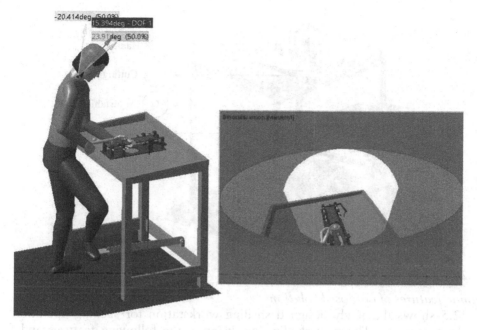

Figure 22.3 Visibility analysis in the existing shelling workstation.

To ensure safety and good control over shelling operation, vision is very important [4]. In order to have continuous observation and good visibility at shelling operation, workers generally bend their head forward direction, which is harmful to the neck region. To assess the risk associated with head flexion, vision analysis was performed using CATIA DELMIA software. Figure 22.3 shows the visibility analysis in existing cashew nut shelling workstation.

Participatory Ergonomics Approach

Contextual knowledge is very important for ergonomic design intervention, especially while designing new workstations for small-scale industries [12]. Understanding the users' expectations thoroughly could help to devise context-specific ergonomic design interventions. For this purpose, in the present study, on-site focus group meeting were conducted. A total of ten female workers in shelling operation were volunteered for free discussions. The average experience of these participants was more than five years in the shelling operation. Initially, the participants were informed about the aim and objective of the present study. During the discussions, they were asked about ergonomic issues faced while operating the existing shelling workstation. Based on their feedback, initially, few preliminary concepts were sketched by the first author in this study. These concepts were shown to the participants on-site. Again, the opinion of participants on preliminary concepts was taken, and refinement of initial sketches was carried out. As a final outcome of this process, the conceptualization of the new cashew nut shelling workstation was done. Figure 22.4 shows the new cashew nut shelling workstation.

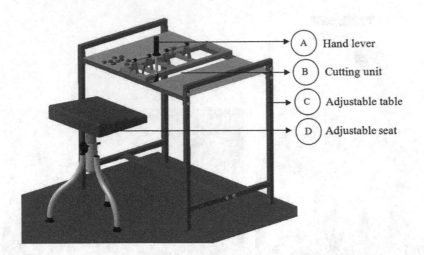

Figure 22.4 The new cashew nut shelling workstation.

Ergonomic features of proposed solution

Figure 22.5 shows the newly designed shelling workstation for shelling operation and adjustable parameters. The new shelling workstation the following features and functionalities: (1) A seat with height adjustability was designed for comfortable sitting. This adjustment of height feature would allow wide range of users to sit. (2) A non-integrated seat was provided with the table. Hence users can adjust their seat position based in their comfort. (3) Handle was designed such a way that users can apply required pressure by pulling it backward direction. (4) Hand lever play (angular adjustment of ±20 deg with vertical line) has been provided, which allows flexible hand movement of users. (5) Height adjustable shelling table. This adjustment feature provides adjustment of table height relative to user sitting position. (6) The shelling table has been designed with space on either side of the shelling area for storing cashew nuts to be shelled.

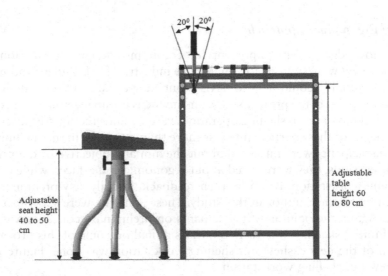

Figure 22.5 Adjustable parameters in new shelling workstation.

Effectiveness of Proposed Solution

The posture analysis in new shelling workstation was performed using CATIA DELMIA software. The female digital model of 50th percentile Indian manikin was used. Figure 22.6 shows the interfacing of digital human model in new shelling workstation.

DELMIA feature 'Biomechanical single angle action analysis' was used to assess compression (N) and moment (N-m) at L4-L5 segment of the lumbar region of digital human models with the interaction of new shelling workstation.

In shelling operation vision plays an important role in accomplishing the shelling operation safely. The vision analysis in new shelling workstation was carried out in CATIA DELMIA software. Figure 22.7 shows the visibility analysis in new cashew nut shelling workstation.

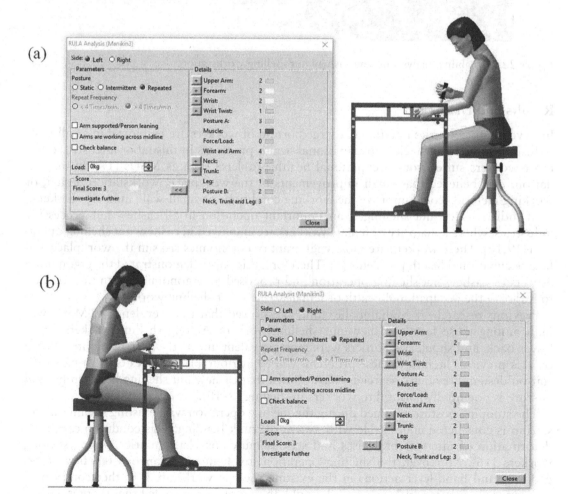

Figure 22.6 Interfacing of digital human model in new shelling workstation. (Note: a: top; b: bottom).

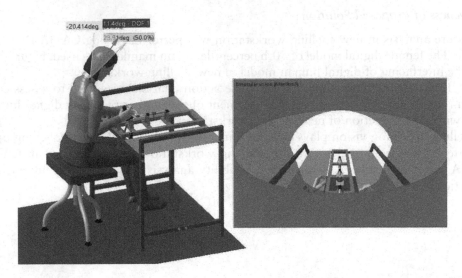

Figure 22.7　Visibility analysis in new cashew nut shelling workstation.

Results and Discussion

In developing countries like India, a large portion of workers are employed in small-scale industries. Due to the lack of occupational health programs in this labor-intensive sector, most workers suffer from occupational health problems such as MSDs [17]. In this situation, it is believed that small improvements in the workplace, workstation design, or working methods could improve the working conditions of small-scale industry workers.

In India, cashew-nut shelling is an important operation in all cashew nut processing industries, where a majority of female workers are employed in cashew nut shelling operations [9,10]. These workers are mostly ignorant of ergonomics risks in the workplace and face occupational health problems [8]. Therefore, this paper demonstrated the ergonomics aspects of cashew nut shelling operation and proposed an ergonomic design intervention to improve the occupational health of female cashew nut shelling workers.

Previous research by Mallampalli et al., [9] showed that the prevalence of MSDs was high among these female cashew nut shelling workers. Among shelling workers, knee, lower back, and shoulder disorders were most prevalent during the past 12 months. Based on this previous literature, these problems can be attributable to issues of existing workstation design. Previous researchers have shown that cashew nut shelling in an un-natural posture is potentially harmful to workers in shelling [8–10].

The stooping posture assumed during the shelling operation with existing shelling workstation is considered undesirable with awkward trunk bending. This condition can cause deformation of intervertebral discs and serious musculoskeletal injuries among shelling workers [16]. Moreover, the shelling operation involving simultaneous repetitive foot-pedaling and hand-lever actions in the existing shelling workstation further contributed to awkward posture and MSDs risk. It could, therefore, be concluded that there is a great need of re-designing existing shelling workstation for improved postures.

Therefore, this study focused on re-designing the existing shelling workstation based on focus group meetings held with end-users. During these discussions with end-users, two major ergonomics risk factors were identified. One is the non-adjustable table height or mismatch between table height and worker's body dimensions. Moreover, users felt that this is the primary reason for their poor trunk posture while shelling. The second reason is that long hours of foot-pedaling in standing position. This feedback collected from the users is very important for designing a new workstation [12]. Some previous studies also demonstrated the use of a participatory approach in designing new workstations [3,12]. In present study, based on the users' feedback and literature study, a new shelling workstation (see Figure 22.4.) was designed to solve the ergonomics issues of female shelling workers.

From the postural analysis, it was observed that the new shelling workstation had improved the working postures noticeably. Results showed that the comfort level of trunk, neck, and upper limb postures were remarkably improved with the new cashew nut shelling workstation. The RULA final score for the new cashew nut shelling workstation was 3, whereas the existing shelling workstation was 7 (Table 22.1). The RULA recommend action level was dropped from 4 in existing shelling workstation to 2 in the new shelling workstation. The reduction in RULA action level for new shelling workstation may be attributed to seated position shelling operation. The new seating arrangement with height adjustability has caused less deviation of upper body parts from neutral position. The biomechanical stress on L4-L5 joint also reduced from 1204 N to 777 N. The results of visibility analysis also revealed that the head posture is satisfactory in the new shelling workstation. The forward head flexion in the new shelling workstations was 11.4 deg, which is an acceptable value according to Karmakaret al. [13].

Table 22.1: RULA, biomechanical stress and visibility analyses in existing and new shelling workstations.

	RULA score	Risk level	Action level	Biomechanical stress at L4-L5 segment		Head flexion (deg)
				Compression (N)	Moment (N-m)	
Existing workstation	7	Very high	4	1204	39	15.3
New workstation	3	low	2	777	31	11.4

Finally, the use of digital human modelling for ergonomics analysis was beneficial in this study. It allowed rapid identification of ergonomics problems and evaluation of developed solutions in simulated environment and also reduced developmental costs. To the best of our knowledge, this is the first study on ergonomic evaluation of cashew nut shelling operation using free modeling of digital human. According to Karmakaret al. [13], proactive ergonomics investigations using DHM at initial design stages were helpful in early estimation of ergonomic risk. Also, there is a high possibility for exercising a broad spectrum of variations at low cost. In present study, therefore, ergonomic evaluation of body postures in both existing and new shelling workstations using simulation-based approach was very much beneficial for promoting occupational design ergonomics research in small-scale industries in developing countries.

Conclusion

In this case study, the working postures while interacting with existing cashew nut shelling workstation in small-scale cashew nut processing industry were found to be poor and subjected to a high risk of MSDs. This finding highlighted the need for improvement in existing shelling workstation. Considering the importance for design of new cashew nut shelling workstation, the present study attempted to design a new cashew nut shelling workstation using participatory ergonomics approach. The ergonomic evaluation of new ergonomic shelling workstation was carried out through a computer-aided ergonomics perspective. Based on the results, it could be concluded that new shelling workstation improved the body posture noticeably. Furthermore, it is suggested that the new shelling workstation can be implemented in a pragmatic way.

References

1. Candan S. A., Sahin, U. K., and Akoğlu, S. (2019). The investigation of work-related musculoskeletal disorders among female workers in a hazelnut factory: prevalence, working posture, work-related and psychosocial factors. *International Journal of Industrial Ergonomics*, 74, 102838.
2. Sakthi Nagaraj, T., Jeyapaul, R., and Mathiyazhagan, K. (2019). Evaluation of ergonomic working conditions among standing sewing machine operators in Sri Lanka. *International Journal of Industrial Ergonomics*, 70, 70–83.
3. Sanjog, J., Patel, T., and Karmakar, S. (2019). Occupational ergonomics research and applied contextual design implementation for an industrial shop-floor workstation. *International Journal of Industrial Ergonomics*, 1(72), 188–198.
4. Kushwaha, D. K., and Kane, P. V. (2016). Ergonomic assessment and workstation design of shipping crane cabin in steel industry. *International Journal of Industrial Ergonomics*, 52, 29–39.
5. Home - International Nut & Dried Fruit Council. http://www.nutfruit.org/wp-continguts/uploads/2019/05/Global-Statistical-Review-2018-2019.pdf (accessed Nov. 15, 2023).
6. The Cashew Export Promotion Council of Indialcashew supplier in Indialcashew producer. http://cashewindia.org/uploads/userfiles/Annual Report.pdf (accessed Nov. 15, 2023).
7. Mohod, A., Jain, S., Powar, A., Rathore, N., and Kurchania, A. (2010). Elucidation of unit operations and energy consumption pattern in small scale cashew nut processing mills. *Journal of Food Engineering*, 99(2), 184–189.
8. Borah S., Chetia, D., Marak, T. R., Chauhan, N. S., and Kumar, A. (2019). Musculoskeletal disorder faced by women workers in cashew nut processing industries of north-east India. *Anthropology*, 38(1–3), 1–8.
9. Mallampalli, K. C., Dhar, D., and Pal, S. (2021). Prevalence of musculoskeletal disorders among female cashew nut shelling workers. In Joint Conference of the Asian Council on Ergonomics and Design and the Southeast Asian Network of Ergonomics Societies. (pp. 281–91). Springer, Cham.
10. Girish, N., Ramachandra, K., Maiya, A. G., and Asha, K. (2012). Prevalence of musculoskeletal disorders among cashew factory workers. *Archives of Environmental and Occupational Health*, 67(1), 37–42.
11. Mallampalli, K. C., Dhar, D., and Pal, S. (2020). Prevalence and risk factors associated with musculoskeletal disorders among cashew-nut shelling workers in India. In International Conference on Applied Human Factors and Ergonomics. (pp. 79–86). Springer, Cham.
12. Sanjog, J., Patnaik, B., Patel, T., and Karmakar, S. (2016). Context-specific design interventions in blending workstation: an ergonomics perspective. *Journal of Industrial and Production Engineering*, 33(1), 32–50.
13. Karmakar, S., Pal, M. S., Majumdar, D., and Majumdar, D. (2012). Application of digital human modeling and simulation for vision analysis of pilots in a jet aircraft: a case study. *Work*, 41(Supplement 1), 3412–3418.

14. Chakrabarti D. (1997). Indian Anthropometric Dimensions for Ergonomic Design Practice. National institute of Design.
15. McAtamney, L., and Corlett E. N. (1993). RULA: a survey method for the investigation of work-related upper limb disorders. *Applied Ergonomics*, 24(2), 91–99.
16. Chaffin, D. B., Andersson, G. B., and Martin B. J. (2006). Occupational Ergonomics. New York: Wiley & Sons.
17. Kawakami, T., Batino, J. M., and Khai, T. T. (1999). Ergonomic strategies for improving working conditions in some developing countries in Asia. *Industrial Health*, 37(2), 187–198.

23 Flexural strength analysis of hybrid composites using different stacking sequence of carbon/glass fiber layers in polymer composites

Ankit Dhar Dubey[1], Kishore Debnath[2,a], and Rajesh Kumar Verma[1,b]

[1]Materials and Morphology Laboratory, Department of Mechanical Engineering, Madan Mohan Malaviya University of Technology, Gorakhpur, India

[2]Department of Mechanical Engineering, National Institute of Technology Meghalaya Shillong, India

Abstract: This article highlights the stacking effect and examines the polymer composite's flexural properties reinforced by carbon fiber and glass fiber. The specimen was developed using the hand layup method with different sequential staking of carbon and glass fiber. Different fiber layers were laid up to observe new changes in the property of the proposed hybrid composites. A positive hybridization effect was observed for different combinations of composite materials and stacking sequences to be an important factor in the flexural properties. The proposed investigation checked the mechanical properties of polymer composites influenced by fiber sequence and stacking configuration. Comparative outcomes recommend other sequences of stacking composite study, and flexural strength gets enhanced with a configuration of C6G2.

Keywords: Carbon, fiber, flexural strength, glass, polymer.

Introduction

In the current scenario, polymer matrix composites demand materials for various manufacturing sectors due to their better engineering properties. It has ascertained high specific stiffness, specific strength, improved dimensional stability, energy absorption, resistance to corrosion and lower processing costs. Composite materials use the qualities of the different parts to make the parent material better. With cost-effective preparation, high tensile, young module, ultimate strain, and impact properties, hybrid composites have great attainment for study. Different reinforcing materials, such as carbon, glass, and aramid fibre, would give the end composite different properties. In the case of carbon fibre, it has a high modulus strength, so we need another fibre with a low modulus, like glass fibre, to balance it out. This introduces properties that may be good for the final composite structure, which is provided by the combination of reinforced fibre (CF/GF). Owing to the alteration of the fabric sequencing of different layers of fiber overlays and the change in the volumetric ranges of different applications of engineering properties of a hybrid composite change a great deal. Very few research works have been done to understand the mechanical properties of the hybrid composite material and its effects on the stacking sequence. This can be used as a test study to find the most efficient proper stacking sequence. Previously published article has been demonstrated on different reinforcement

[a]debnathiitr@gmail.com, [b]rkvme@mmmut.ac.in

materials and their potential advantages. Among the various reinforcements, the use of carbon fiber is widely used due to better mechanical properties [1]. In the automotive industry, carbon fiber is considered a significant weight reduction for vehicles to achieve higher speed [2]. The properties of composite laminates with various stacking sequences and the effect of the composite structure were studied by Zhang et al. [3]. A combination of various reinforcements such as carbon, basalt, kenaf, flax fibers, etc., in laminate form has been developed for structural applications [4–6]. Hybridization effects due to different reinforcement applied to the matrix materials will enhance the composite laminate properties and stacking sequences will provide the structures with the durable application [7, 8]. Kannan et al. [9] influenced the stacking sequence on the composite material properties made up of glass fiber, basalt, jute, etc. and its effect on the tensile and flexural strength was studied. Several other reinforcements were used for research purposes, such as jute fibre, in addition to glass fiber [10]. The hybrid composite's mechanical properties depend on the proportion of the composition and arrangement of different laminates' reinforcements [3, 11, 12]. High-module fibres such as carbon have excellent rigidity and load carrying capacity, while low-module fiber such as glass has better tolerance for damage and lower stiffness with higher elongation [13,14]. Lopresto et al. [15] conducted research on the impact and flexural behavior of carbon/glass fiber using a variety of stacking sequences. The effect of configuration on modification was revealed by Kalantari et al [16] on the flexural properties of C/G fiber-reinforced composites and demonstrated that a positive mixture effect was produced for overlay composites. As described in the literature review, a hybrid sequence on the mechanical properties of the developed composite was enhanced the composite structure. Another technique that has been further established is the enhancement of flexural characteristics in composite materials through the incorporation of fibre reinforcement. A proposal has been taken to combine the effect of two reinforcements to form a hybrid carbon/glass fiber epoxy composite with different laminate sequences and volumetric proportions and to research improvements in its mechanical structure.

Experimental

Materials and Sample Preparation

In this study, the fabrication technique involved the utilization of carbon fiber (400 GSM, plain weave, Bi-directional sheet) and Glass fibre (400 GSM, plain weave) as reinforcing materials, together with epoxy resin as the matrix material. To create the hybrid composite, an epoxy (resin) and hardener are used to prepare the matrix combination. The components are blended at a weighted ratio of 100:10 (Epoxy: Hardener). The composites were made using a hand layup technique. Each fibre was individually set up in the various configurations after sheets of fibre were cut to the requisite diameters (26 cm × 17 cm). Various stacking sequences were generated in the various specimens in an effort to determine the optimal sequence necessary for enhanced mechanical qualities. Specimen detail (in coded form) are displayed in Figure 23.1 as the layup sequences followed by the layers of carbon fiber and glass fiber in the matrix.

Flexural Testing

Samples were prepared with ASTM D7264 standard (dimensions 70 mm × 13 mm × 3mm) for analysis of flexural strength (three-point bending test). A minimum of two sets of specimens was performed of each stacking sequence to avoid error, and their various

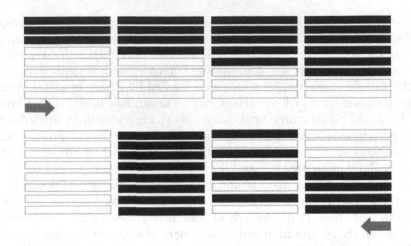

Figure 23.1 Stacking sequence of composite materials blue-Carbon, white-glass as (C3G5), (C4G4), (C5G3), (C6G2), (G4C4), (CG), (C8), (G8) respectively.

parameters were recorded for further analysis. For calculating the flexural modulus and flexural strength of the composites, the following mathematical expression is used [17].

$$\sigma_f = \frac{(3PL)}{(2bd)^2} \tag{1}$$

$$e_f = \frac{(L^3 m)}{(4bd^3)} \tag{2}$$

Here, L is the length of specimen span, b is the width, d is the thickness of the sample, m is the initial slope of L-D curve and P is the applied load. Figure 23.2 shows the flexural testing apparatus and being bending being performed on the workpiece.

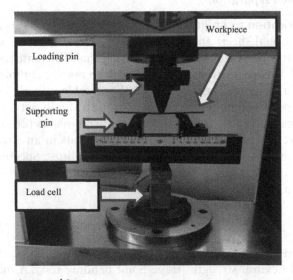

Figure 23.2 Flexural testing machine.

Results and Discussion

Flexural strength was studied by subjecting all composite samples with varying stacking sequences to a Three-point bending test. From Figure 23.3, L-D curves of the hybrid configuration C6G2, G4C4, CG depict a higher load bear among them C6G2 having the highest load caring i.e., highest flexural strength. On the other hand, low displacement value shows the presence of brittle nature. However, G4C4 offers very good displacement value, i.e., good ductility, perhaps due to the presence of elongation properties of glass fiber and position on the compressive side.

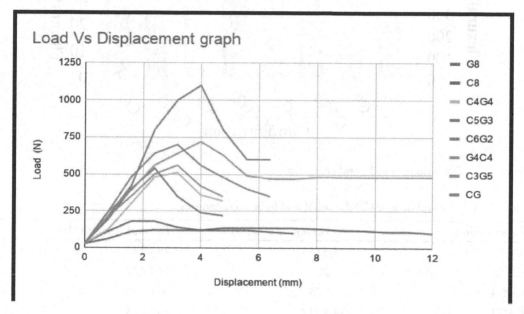

Figure 23.3 Load displacement curve.

From Equations 1–2, flexural strength, and flexural modulus of each composite with different stacking sequence was calculated. Table 1 shows the value of flexural properties of each composite structure. Figure 23.4 shows both flexural strength and flexural modulus with respect to different stacking sequences. In addition, a comparison is made between C4G4 and G4C4, both of which are made up of the same number of layers and proportions of carbon and glass fibre and are designed to withstand a high load with minimal deformation. C4G4, which has four carbon layers at the compressive side, has a lower load value than G4C4, which has 4 glass layers. It has been proven that the characteristics of a sample can vary depending on the reinforcement ratio and the number of fibre layers employed in the stacking sequences. This finding is consistent with the findings of prior researchers like Subagia et al. [17]. In a similar configuration with C3G5 over C5G3, higher values in the L-D curves revealed the increase in the composites' maximum load capacity due to the addition of the higher glass fiber layer with the Carbon fiber layer. Expansion of glass filaments with Carbon strands expands the composites' flexural strength but diminishes the flexibility esteem as C8 composite. It has demonstrated the most elevated dislodging indicating the great ductility nature of carbon fiber. According to Table 24.1, the composite with C6G2 sequencing has the highest flexural strength.

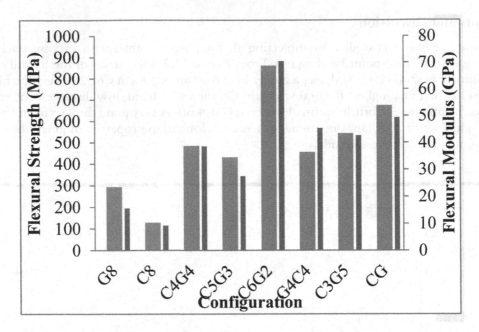

Figure 23.4 Flexural data of hybrid composites.

Table 23.1: Flexural data of CFRP, GFRP, and hybrid composites.

Configuration	Flexural modulus (GPa)	Flexural strength (MPa)
G8	23.5253	194.444
C8	10.289	114.722
C4G4	38.775	481.444
C5G3	34.4937	343.388
C6G2	68.8952	882
G4C4	36.4033	567
C3G5	43.361	531.22
CG	53.814	616.7778

C6G2, G4C4 and CG hybrid structures display a rapid and steep rise in load value, while C4G4, C3G5 shows slight deviation from linear characteristics. C6G2 has six stacked layers of carbon sheet at the compressive face and has the largest flexural strength value. Similar flexural strengths are shown by C4G4 and C3G5. It has four and three carbon sheet stacking on the compressive face, giving it better flexural property than that of C5G3. Glass fibre on the compressive side has been shown to boost flexural strength. In C6G2, maximum carbon layers have increased load value. Chen et al. [18] found that glass fibre layers increase flexural strength over Basalt fibre layers. Only the C6G2 structure has better value than the configuration CG sequenced composite with four layers of each in alternative sequence. The prepared sample's strength order: C6G2, CG, G4C4, C3G5, C4G4, C5G3. This shows how sequencing affects composite structure flexuralness. Bending tests have the greatest effect on the composite laminates' compressive region. Its great bending strength is improved by adding a carbon layer on top. It resists bending loads well. Carbon fibre in the middle layer increases flexural modulus, as shown by Jesthi et al. [19]. Unfortunately, it reduces composites' flexural extension. The carbon fiber-topped composite may shatter due to load resistance. The carbon layer on top minimizes cracking and supports the glass layer below in CG hybrid composites with carbon and glass layers alternately organized. It showed that CG composites had good flexural characteristics. For a stacking sequence with a 50% ratio, namely C4G4, G4C4 and CG. It has differed in strength values due to unlike stacking sequences. On the contrary, in C6G2 hybrid structures, the hybrid ratio is only 25% structure shows the highest value of flexural strength and modulus. Due to the presence of many carbon layers on the compressive side with a glass layer beneath it to further resist the crack initiation during the bending test. It has been concluded that the addition of the glass layer has certainly enhanced the flexural performance. From Figure 23.5, observed that there has a very slight variation of peak strain with respect to different stacking sequences. This indicates that the presence of carbon fiber has been the dominating factor in the failure mode of the composite structure [18].

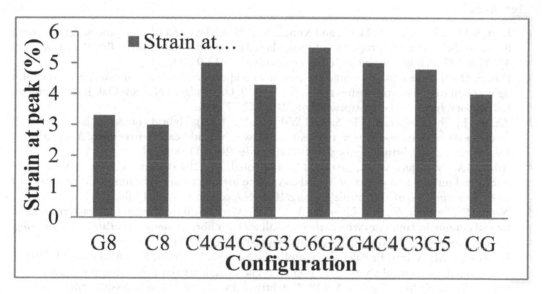

Figure 23.5 Strain at peak value plot.

Conclusion

Various configuration hybrid composite constructions with different stacking sequences were tested for flexural capabilities to determine how fibre affected them. The investigation study shows the following.

- The C6G2 hybrid composite has the maximum flexural strength at 882 MPa. CG, G4C4, and C4G4 configurations have next-best flexural strength values.
- Composite structures with the same percentage of C4G4, G4C4, and CG reinforcements differ in flexural properties. It shows the stacking sequence is crucial to composite structure performance.

- The composite structure's flexural strength and modulus increase with the hybrid ratio. Strain at peak ratio values vary little, suggesting carbon fibre causes failure.
- The proposed novel material with desired (tensile and flexural) properties can be recommended for end product like water pipe, water tank and energy storage tank etc.

Acknowledgments

The authors would like to acknowledge the Uttar Pradesh Council of Science and Technology, Lucknow, India.

Funding

This presented work is financially supported by the Uttar Pradesh Council of Science and Technology, Lucknow, India, under the research project scheme [R&D project ID-UPCST/D-2491].

References

1. Han, S. H., Oh, H. J., Lee, H. C., and Kim, S. S. (2013). The effect of post-processing of carbon fibers on the mechanical properties of epoxy-based composites. *Composites Part B: Engineering*, 45, 172–177. https://doi.org/10.1016/j.compositesb.2012.05.022.
2. Das, S. (2001). The cost of automotive polymer composites: a review and assessment of DOE's lightweight materials composites research (Vol. 47). Oak Ridge, TN, USA: Oak Ridge National Laboratory. https://doi.org/https://doi.org/10.2172/777656.
3. Zhang, J., Chaisombat, K., He, S., and Wang, C. H. (2012). Hybrid composite laminates reinforced with glass/carbon woven fabrics for lightweight load bearing structures. *Materials and Design*, 36, 75–80. https://doi.org/10.1016/j.matdes.2011.11.006.
4. Atiqah, A., Maleque, M. A., Jawaid, M., and Iqbal, M. (2014). Development of kenaf-glass reinforced unsaturated polyester hybrid composite for structural applications. *Composites Part B: Engineering*, 56, 68–73. https://doi.org/10.1016/j.compositesb.2013.08.019.
5. Nisini, E., Santulli, C., and Liverani, A. (2017). Mechanical and impact characterization of hybrid composite laminates with carbon, basalt and flax fibres. *Composites Part B: Engineering*, 127, 92–99. https://doi.org/10.1016/j.compositesb.2016.06.071.
6. De Rosa, I. M., Marra, F., Pulci, G., Santulli, C., Sarasini, F., Tirillò, J., and Valente, M. (2011). Post-impact mechanical characterisation of E-glass/basalt woven fabric interply hybrid laminates. *Express Polymer Letters*, 5, 449–459. https://doi.org/10.3144/expresspolymlett.2011.43.
7. Sevkat, E., Liaw, B., Delale, F., and Raju, B. B. (2009). Drop-weight impact of plain-woven hybrid glass-graphite/toughened epoxy composites. *Composites Part A: Applied Science and Manufacturing*, 40, 1090–1110. https://doi.org/10.1016/j.compositesa.2009.04.028.
8. Sayer, M., Bektaş, N. B., Demir, E., and Çallioğlu, H. (2012). The effect of temperatures on hybrid composite laminates under impact loading. *Composites Part B: Engineering*, 43, 2152–2160. https://doi.org/10.1016/j.compositesb.2012.02.037.
9. Amuthakkannan, P., Manikandan, V., Jappes, J. T. W., and Uthayakumar, M. (2012). Influence of stacking sequence on mechanical properties of basalt-jute fiber-reinforced polymer hybrid composites. *Journal of Polymer Engineering*, 32, 547–554. https://doi.org/10.1515/polyeng-2012-0063.
10. Bunsell, A. R., and Harris, B. (1974). Hybrid carbon and glass fibre composites. *Composites*, 5, 157–164. https://doi.org/10.1016/0010-4361(74)90107-4.

11. Song, J. H. (2015). Pairing effect and tensile properties of laminated high-performance hybrid composites prepared using carbon/glass and carbon/aramid fibers. *Composites Part B: Engineering*, 79, 61–66. https://doi.org/10.1016/j.compositesb.2015.04.015.

12. Abd El-baky, M. A. (2017). Evaluation of mechanical properties of jute/glass/carbon fibers reinforced hybrid composites. *Fibers and Polymers*, 18, 2417–2432. https://doi.org/10.1007/s12221-017-7682-x.

13. Czél, G., and Wisnom, M. R. (2013). Demonstration of pseudo-ductility in high performance glass/epoxy composites by hybridisation with thin-ply carbon prepreg. *Composites Part A: Applied Science and Manufacturing*, 52, 23–30. https://doi.org/10.1016/j.compositesa.2013.04.006.

14. Rahman, M. M., Zainuddin, S., Hosur, M. V., Malone, J. E., Salam, M. B. A., Kumar, A., and Jeelani, S. (2012). Improvements in mechanical and thermo-mechanical properties of e-glass/epoxy composites using amino functionalized MWCNTs. *Composite Structures*, 94, 2397–2406. https://doi.org/10.1016/j.compstruct.2012.03.014.

15. Lopresto, V., Langella, A., and Papa, I. (2016). Residual strength evaluation after impact tests in extreme conditions on CFRP laminates. *Procedia Engineering*, 167, 138–142. https://doi.org/10.1016/j.proeng.2016.11.680.

16. Kalantari, M., Dong, C., and Davies, I. J. (2016). Numerical investigation of the hybridisation mechanism in fibre reinforced hybrid composites subjected to flexural load. *Composites Part B: Engineering*, 102, 100–111. https://doi.org/10.1016/j.compositesb.2016.07.012.

17. Ary Subagia, I. D. G., Kim, Y., Tijing, L. D., Kim, C. S., and Shon, H. K. (2014). Effect of stacking sequence on the flexural properties of hybrid composites reinforced with carbon and basalt fibers. *Composites Part B: Engineering*, 58, 251–258. https://doi.org/10.1016/j.compositesb.2013.10.027.

18. Chen, D., Sun, G., Meng, M., Jin, X., and Li, Q. (2019). Flexural performance and cost efficiency of carbon/basalt/glass hybrid FRP composite laminates. *Thin-Walled Structures*, 142, 516–531. https://doi.org/10.1016/j.tws.2019.03.056.

19. Kumar Jesthi, D., Mandal, P., Rout, A. K., and Nayak, R. K. (2018). Effect of carbon/glass fiber symmetric inter-ply sequence on mechanical properties of polymer matrix composites. *Procedia Manufacturing*, 20, 530–535. https://doi.org/https://doi.org/10.1016/j.promfg.2018.02.079.

24 Machining performance evaluation and parametric optimization during drilling of bio-nano composite

Umang Dubey[a] and Rajesh Kumar Verma[b]

Materials & Morphology Laboratory, Department of Mechanical Engineering, Madan Mohan Malaviya University of Technology, Gorakhpur, India

Abstract: PMMA bone cement assumes the leading part in orthopedic application areas because of its magnificent biocompatibility and mechanical properties. The paper examines the effect of process parameters on drilling performances of Polymethyl methacrylate/hydroxyapatite nanocomposites using Grey theory. The varying parameters are considered, tool material (TOOL) such as TiAlN, Carbide and HSS, speed of spindle (SPEED), and hydroxyapatite weight percentage (Wt.%). The desired value of drilling performances viz. Surface roughness (Ra) and Material removal rate (MRR). Utilising GRA as W-10%, N-1428 rpm, and an HSS drilling tool, the optimal setup was achieved. A confirmatory test that reveals an improvement in Grey performance value has confirmed the predicted outcomes. It was also discovered via interaction plots that both MRR and Ra exhibit their best values at a value of 10 Wt.%.

Keywords: Drilling, machining, Nanocomposite, optimization, PMMA.

Introduction

Since 1930, bone surgery applications have used polymethyl methacrylate (PMMA) as a bone concrete material [1]. Hydroxyapatite (HA) is an osteoconductive and biocompatible nanomaterial, showing an extracellular response to bone-filler minerals [2]. The investigation of Kwon and collaborators demonstrated that consolidation of 30 wt. % HA in bone cement and hydroxyapatite composite surge the interfacial shear strength at the prosthetic embed interface following a month and a half of implantation [3]. A few researchers have examined and contemplated HA strengthened PMMA as a potential bone cement [5, 6]. When the weight percent of HA expanded over 20, high mechanical properties were not achieved, and the cement's mechanical properties decreased [4]. With the addition of HA up to 15 wt percent in the powder component, approximately 2 to 2.5 GPa expansion of flexural modulus was seen. In addition to the unreinforced PMMA cement with 25 percent HA additive, augmentation in the flexural modulus was additionally seen [7]. While making prosthesis utilizing PMMA bone cement, drilling on the prosthesis is likewise needed to put bolts for fiber wires [8].

[a]aaron9396ud@gmail.com, [b]rkvme@mmmut.ac.in

Hydroxyapatite (HA) is found to be the most preferred additive in Polymethyl methacrylate bone cement [6] but the impact of HA on the drilling of PMMA-HA bone cement is less studied. As per the application capability of PMMA bone cement, drilling is a very important phenomenon to be studied. To obtain an ideal parametric setting for machining constraints with desired responses in the current work, the Grey relational analysis (GRA) optimisation method is used. To provide a perfect parametric setting for drilling restrictions with desired response values, PMMA enhanced with HA is employed. The three experiment parameters that were picked in this examination are HA wt.%, speed of the spindle, and drilling tool material. The experimental parameters' impact on the response attribute is MRR and Ra was analysed. ANOVA was implemented to decide the effect of experiment factors on drilling performance.

Materials and Methods

PMMA bone cement from Emsurg health care Kolkata, India, and hydrox-yapatite from Nano Exploration Lab in Jharkhand, India, were both used in the production of nano-composites. Composites containing different amounts of HA were added. After adding HA powder to a concentrated NH_4OH solution, powdered PMMA was substituted for an amount of HA that was equal in weight. 4, 10, or 16% of the PMMA powder was replaced by hydroxyapatite nanoparticle fillers. Although the amount of powder contained both PMMA and hydroxyapatite particles, the ratio of the PMMA powder to the fluid monomer was P/M = 2. Sample preparation was done by adding the methyl methacrylate monomer liquid with the HA-PMMA powder. Three sorts of samples were created (PH1, PH2, PH3) with different HA nanoparticles' proportions in the PMMA lattice (Table 24.1). The prepared sample is shown in Figure 24.1.

Figure 24.1 Prepared sample.

Table 24.1: Classification of specimens according to weight percentage.

Sample	PMMA	HA
PH1	38.4 gm	1.6 gm
PH2	36 gm	4 gm
PH3	33.6 gm	gm

Experimental

The drilling PMMA-HA nanocomposite experiment was executed on a PUSCO Vertical drilling machine (Figure 24.2) at three different levels of drilling parameters appeared in Table 24.2 like HA strengthened rate, Shaft Speed, and tool material (TiAlN, carbide covered, and HSS).

Figure 24.2 PUSCO vertical drilling machine.

Design of experiment using Taguchi method

In this investigation, the settings of drilling boundaries were controlled by utilizing the Taguchi trial plan strategy. Symmetrical varieties of Taguchi, the signal to noise (S/N) proportion, the examination of fluctuation (ANOVA) is utilized to analyze the impact of the drilling boundaries on Ra and MRR. To diminish time and cost, tests are done using the L9 orthogonal array. To notice the level of impact of the cutting conditions in drilling cycle three components (spindle speed, Wt. % and drill bit material), the obtain array of the test has been mentioned in Table 24.3.

Surface Roughness Calculation

The mean surface roughness (Ra) is estimated utilizing a Mitutoyo Surftest SJ-201 roughness test instrument. The cut-off and testing lengths for every estimation are taken as 0.8 and 5mm, separately.

Material Removal Rate (MRR) Calculation

The MRR has been calculated by using the standard relation:

$$MRR = \frac{\left[(W_i) - (W_f) \right]}{\left[(\rho) * (t) \right]} \, mm^3 / sec \tag{1}$$

W_i = Initial weight (gram), W_f = Final weight (gram), t = time in seconds and ρ = Density (gram/mm^3) and.

Table 24.2: Machining parameters at different levels.

Machining Parameters	L 1	L 2	L 3
Speed (S)	318(rpm)	865(rpm)	1428(rpm)
Tool Material (T)	HSS (1)	Carbide (2)	TiALN (3)
WT.%	4	10	16

Table 24.3: Outline of experiment and obtained data.

Ex. No.	Speed (RPM)	Weight %	Tool material	Initial weight	Final weight	Density	Time	Roughness (Ra)
1.	318	4	HSS	41.07	41.00	0.00126	15.19	2.336
2.	318	10	Carbide	34.93	34.86	0.00127	15.07	2.923
3.	318	16	TiAlN	39.82	39.68	0.00133	28.61	2.536
4.	865	4	Carbide	41.00	40.84	0.00126	35.72	3.361
5.	865	10	TiAlN	34.86	34.74	0.00127	19.33	3.245
6.	865	16	HSS	40.00	39.91	0.00133	14.50	3.145
7.	1428	4	TiAlN	40.84	40.72	0.00126	19.52	4.198
8.	1428	10	HSS	35.00	34.93	0.00127	8.64	4.335
9.	1428	16	Carbide	39.91	39.82	0.00133	18.45	4.667

ANOVA Study

To identify the important process boundaries and examine the effects of process factors on reactions by developing a mathematical model, analysis of variance was expressly required. Every investigation was carried out under the assumption that the optimal yield would be obtained with a 94% confidence level. ANOVA results summary of the Ra and MRR is shown below.

The Material Removal Rate Regression Equation is listed below.

$$4.4977 - 0.8340\ S\_318 - 0.1290\ S\_865 + 0.9630\ S\_1428$$
$$- 0.4683\ WT.\ \%\_4 + 0.4760\ WT.\ \%\_10 - 0.0077\ WT.\ \%\_16$$
$$+ 0.4023\ T\_1 - 0.3853\ T\_2 - 0.0170\ T\_3 \tag{2}$$

R-sq(adj) = 97.56%, R-sq = 99.39%, R-sq(pred) = 87.64%
The Surface roughness regression Equation is listed below.

$$3.42067 - 0.8090\ S\_318 + 0.8297\ S\_865 - 0.0207\ S\_1428$$
$$- 0.1223\ WT.\ \%\_4 + 0.0937\ WT.\ \%\_10 + 0.0287\ WT.\ \%\_16$$
$$- 0.1487\ T\_1 + 0.2430\ T\_2 - 0.0943\ T\_3 \tag{3}$$

R-sq(adj) = 99.91%, R-sq = 99.98%, R-sq (pred) = 99.53%

Analysis of S/N ratio

Taguchi utilizes the S/N ratio to gauge the quality characteristic straying from the ideal worth. The S/N ratio is estimated based on the desired characteristics of the responses such as lowest the best, nominal the best and higher the better [9, 10]. The lowest best characteristic (Eq. 4) is utilized for surface roughness, and the highest the best (Eq. 5) is utilized for material removal rate.

Smaller the Best:

$$S/N = -10 * \log\left(\Sigma\left(Y^2\right)/n\right) \qquad (4)$$

Higher the Best:

$$S/N = -10 * \log\left(\Sigma\left(1/Y^2\right)/n\right) \qquad (5)$$

Observations are denoted by n, while the data collected is denoted by Y.

Optimization Method

The choice of ideal parametric combination is a demanding assignment in drilling as it bargains an enormous number of information measure parameters and responses gathered through different experiments. To facilitate the above approach, Grey relational analysis (GRA) is classified to execute.

Grey relational Analysis (GRA)

The GRA is a multiple response optimization procedures. In GRA, black corresponds to having no data and white represents having all data. Between black and white, a grey system has a degree of detail. In other words, some information is identified in a grey structure and some information is unknown [11]. In the subsequent pages, the stepwise process for GRA is explained.

Steps in the GRA Method

Step-1: To prevent variability, the experimental data must first be normalised. A value between 0 and 1 generated from the initial data is to be assigned to each value. Since the reaction to surface roughness is reduced, the following formula uses smaller the better features for normalisation.

$$x_i^* = \frac{x_i(k) - \min x_i(k)}{\max x_i(k) - \min x_i(k)} \qquad (6)$$

If the goal is to maximise the response of the material removal rate, smaller the better features are chosen for normalisation using the formula below.

$$x_i^* = \frac{\max x_i(k) - x_i(k)}{\max x_i(k) - \min x_i(k)} \qquad (7)$$

Where, I = 1… ., m; k = 1… ., n, m is the quantity of trial information and n is the number of reactions. Xi(k) indicates the first succession, de-noticed the grouping after the information pre-handling, max Xi(k) means the biggest estimation of Xi(k), min Xi(k) signifies the smallest estimation of Xi(k), and x is the ideal worth.

Step-2: The Grey Relational Coefficient Determined,

$$\gamma_i(k) = \frac{\Delta_{min} + \gamma\Delta_{max}}{\Delta_{oi}(k) + \gamma\Delta_{max}} \tag{8}$$

Δoi, is the deviation sequence and

$$\Delta_{oi} = |x_o(k) - x_i(k)| \tag{9}$$

Where (k) indicates the reference arrangement and (k) named as likeness succession. also, are the base and maximum estimations of the total contrasts of all looking at arrangements.

Step-3: To calculate the grey relational grade (GRG), one must do the following

$$G_i = \frac{1}{n}\sum_{k=1}^{n}\gamma_i(k) \tag{10}$$

where G_i are the required grey relational grade and n = number of response characteristics.

Results and Discussion

S/N Ratio Plot

Using the above-presented data with the selected formula stated in Eq. 5 for obtaining S/N ratio for material removal rate and eq. 4 for calculating S/N ratio for Ra, the Taguchi experiment results As shown in Figure 24.3 for MRR and Figure 24.4 for surface roughness, which was obtained employing MINITAB 13 statistical software.

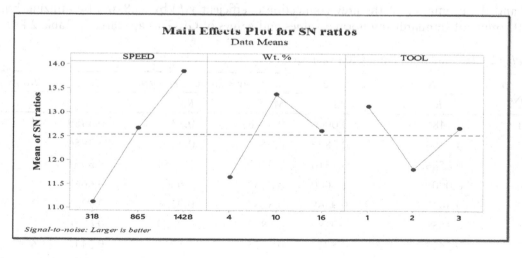

Figure 24.3 S/N curve for the MRR.

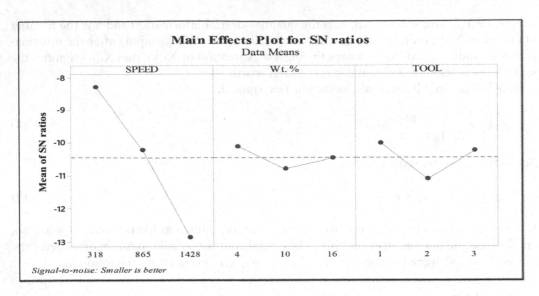

Figure 24.4 S/N curve for the Ra.

Both Figures 24.3 and 24.4, show the impact of drilling parameters on the MRR and Ra, respectively.

The spindle speed appears to be the most important element and Wt.% is least in determining MRR in the S/N responses.

The spindle speed appears to be the most important element and tool type is least in determining Ra in the S/N responses.

GRA Method

At first, standardization of noticed information according to desired quality attributes, for example, the highest, the better or lowest, the better by utilizing equation 6 and equation 7 and the estimation of the grey connection coefficient $\gamma_i(k)$ by eq.8 and by utilizing Eq. 10 counts of standardizing normal values and values of GRG, as appeared in Table 24.4.

Table 24.4: Standardized and GRG values.

Exp. No.	Normalized sequence		Grey relation coefficient		GRG	Rank
	MRR	Ra	MRR	Ra		
1.	0.0489	0.0000	0.3446	0.3333	0.3389	9
2.	0.0484	0.3809	0.3444	0.4468	0.3956	7
3.	0.0586	0.1316	0.3469	0.3654	0.3561	8
4.	0.0000	1.0000	0.3333	1.0000	0.6667	4
5.	0.5447	0.9568	0.5234	0.9205	0.7219	2
6.	0.4652	0.9186	0.4832	0.8600	0.6716	3
7.	0.5412	0.5031	0.5215	0.5016	0.5115	6
8.	1.0000	0.5703	1.0000	0.5378	0.7689	1
9.	0.6266	0.7223	0.5725	0.6430	0.6077	5

The results of the GRA module under ideal conditions are noted as 1428 rpm speed, 10 wt. % and the tool material as HSS with a maximum value of grey relation grade (0.7689).

Optimal Drilling Parameters and Combination of Levels

Ideal parameters setting acquired by the GRA method obtained with the help of Table 24.4 which reveals that HA (10 wt%), speed (1428 rpm), and tool material (HSS) gives optimum machining performance. Higher spindle speed produces optimum results while drilling Polymethyl methacrylate (PMMA) based composite [12]. Hydroxyapatite (HA) content of 10% in polymer composite gives better drilling experimentation results [13]. Surface roughness (Ra) improves at increasing spindle speed drilling [14].

Table 24.5: EA versus GRA optimization result.

Optimal setting	Taguchi orthogonal array design SPEED-865, Wt. %-10, TOOL-Carbide	GRA design SPEED-1428, Wt. %-10, TOOL-HSS	%improvement
Surface roughness	4.5870	3.3350	27.29
Material removal rate	6.3390	6.3780	0.61
Overall assessment	0.7219	0.7689	6.11

Confirmatory Test

The affirmation examination was performed by the best parameter setting to sort out the activity response in the drilling of PMMA-HA nanocomposites. Table 24.4 shows that experiment no. 8 made sure about the most elevated value of the grey-relation grade (0.7689), demonstrating the ideal parametric setting as Wt.%-10/TOOL-HSS/SPEED-1428, which provides the best machining qualities among the executed experiments. Using Taguchi L9 orthogonal array, the optimal solution is/Wt.%-10/ TOOL-2/SPEED-865 obtained. The comparative investigation outcome of both design at ideal settings are portrayed in Table 24.5. Operation response and common performance index value have essentially enhanced in Ra and MRR using the GRA method. The average surface roughness diminishes from 4.5870 µm to 3.3350 µm, and the average MRR increases from 6.3390 mm³/sec to 6.3780 mm³/sec. The relating % improvement in surface roughness is 27.29% and the material removal rate is 0.61%.

Conclusion

This examination utilized the Grey Relational Analysis (GRA) method for drilling PMMA-HA Nanocomposites. The utilization capability of GRA has been analysed and contrasted. The GRA method has been productively used to choose control factors, level, and optimum machining environment, which provides diminished Ra and MRR. The conclusions drawn from the examination are:

- The collection of various responses like Ra and MRR into a solitary objective function, common performance index has been productively obtained with the use of GRA strategy. The ideal union of machining parameters assessed using GRA is established as 10 wt.%, 1428 rpm, and HSS Tool material.

- Analysis of variance results distinguishes the critical machining parameters and their impacts on response parameters, i.e., MRR and Ra.
- GRA fundamentally improves the estimations of Ra up to 27.29% and 0.61% of MRR than expected average results. A near report shows the higher application capability of the GRA technique.
- After drilling on a conventional vertical drilling machine in a workshop, the obtained Ra and MRR value shows that the PMMA-HA nanocomposite can be machined in a commercial workshop and does not require any special arrangements.

Acknowledgment

The authors would like to acknowledge the kind support of o/O DC (Handicrafts), Ministry of Textiles, Govt. of India, New Delhi, INDIA for extending all possible help in carrying out this research work directly or indirectly.

Funding

This research work is financially supported by o/O DC (Handicrafts), Ministry of Textiles, Govt. of INDIA, Under Project ID: K-12012/4/19/2020-21/R&D/ST.

References

1. Anusavice, K. J. (2003). Phillips' science of dental materials. 11th Edition, WB Saunders, Philadelphia.
2. Marino, X. (2005). Functional Fillers for Plastics. Weinheim: WILEY-VCH Verlag GmbH & Co. KGaA.
3. Kwon, S. Y., Kim, Y. S., Woo, Y. K., Kim, S. S., and Park, J. B. (1997). Hydroxyapatite impregnated bone cement: in vitro and in vivo studies. *Bio-Medical Materials and Engineering*, 7, 129–140.
4. Sogal, A., and Hulbert, S. (1992). Mechanical properties of a composite bone cement: polymethylmethacrylate and hydroxyapatite. In Yamamuro, T., Kokubo, T., Nakamura, T., eds. Bioceramics. Japan: Kobunshi Kankokai,. (Vol. 5, pp. 225–232).
5. Walsh, W. R., Svehla, M. J., Russell, J., Saito, M., Nakashimac, T., Gilliesa, R. M., Brucea, W., and Hori, R. (2004). Cemented fixation with PMMA or bis-GMA resin hydroxyapatite cement: effect of implant surface roughness. *Biomaterials*, 25(20), 4929–4934.
6. Itokawaa, H., Hiraideb, T., Moriyaa, M., Fujimotoa, M., Nagashimaa, G., Suzukia, R., and Fujimoto, T. (2007). A 12 month in vivo study on the response of bone to a hydroxyapatite polymethylmethacrylate cranioplasty composite. *Biomaterials*, 28(3), 4922–4927.
7. Vallo, C. I., Montemartini, P. E., Fanovich, M. A., Lopez, J. M. P., and Cuadrado, T. R. (1999). Polymethylmethacrylate-based bone cement modified with hydroxyapatite. *Journal of Biomedical Materials Research: An Official Journal of The Society for Biomaterials, The Japanese Society for Biomaterials, and The Australian Society for Biomaterials*, 48(2), 150–158.
8. Varma, A. K., Varma, V., Mangalandan, T. S., Bal, A., and Kumar, H. (2014). Use of polymethyl methacrylate as prosthetic replacement of destroyed foot bones—clinical audit. *The Journal of Diabetic Foot Complications*, 6(1), 1–12.
9. Tong, L., Wang, C., and Chen, H. (2005). Optimization of multiple responses using principal component analysis and technique for order preference by similarity to ideal solution. *The International Journal of Advanced Manufacturing Technology*, 27, 407–414.

10. Saha, A., and Mondal, S. C. (2017). Multi-objective optimization of manual metal arc welding process parameters for nano-structured hard facing material using hybrid approach. *Measurement*, 102, 80–89.
11. Singh, S., Singh, I., and Dvivedi, A. (2012). Multi objective optimization in drilling of Al6063/10% SiC metal matrix composite based on grey relational analysis. *Journal of Engineering Manufacture*, 227, 1767–1776.
12. Pant, P., Mishra, S., and Mishra, P. (2021). *Advances in Mechanical Processing and Design.* Springer.
13. Mehar, A. K., Kotni, S., Mahapatra, S. S., and Patel, S. K. (2020). A comparative study on drilling performance of hydroxyapatite-polycarbonate and hydroxyapatite-poly-sulfone composites using principal component analysis methodology for orthopaedic applications. *Materials Today: Proceedings*, 33, 5174–5178.
14. Kamaraj, M., Santhanakrishnan, R., and Muthu, E. (2018). Investigation of surface roughness and MRR in drilling of Al2O3 particle and sisal fibre reinforced epoxy composites using TOPSIS based Taguchi method. *In IOP Conference Series: Materials Science and Engineering*, 402(1), 012095. IOP Publishing.

25 Electric discharge machining of inconel X-750 and parametric appraisal using grey theory

Puranjay Pratap[1,a], Radhey Lal[2], and Rajesh Kumar Verma[1,b]

[1]Materials and Morphology Laboratory, Department of Mechanical Engineering, Madan Mohan Malviya University of Technology, Gorakhpur, India

[2]Council of Science and Technology, Lucknow, India

Abstract: This paper emphasizes the Grey theory for parametric appraisal and multiple criteria optimizations during the Electric Discharge Machining (EDM) of Inconel X-750 using the copper tool. Taguchi concept-based L_{16} orthogonal array was selected as an experimental design to execute the EDM tests. EDM is widely employed for cutting harder materials with higher accuracy, tolerance, improved surface finish. Effects of process parameter viz. Input Current (I_p), Pulse on time (T_{on}), Pulse off time (DC) and voltage (V) on material removal rate (MRR), tool wear rate (TWR), and surface roughness (Ra) have been studied. The contribution of process parameters was determined by performing the analysis of variance (ANOVA) test. The optimized condition was determined by the hybrid module of Taguchi based GRA approach. The study revealed that the preferred solution value for Taguchi-based GRA obtains improvement % over the experimental orthogonal array (OA condition) value of 5.54%. Also, ANOVA demonstrates that the input current is the most influencing factor to affect the machining efficiency over the process parameters.

Keywords: EDM, inconel X-750, surface roughness, Taguchi.

Introduction

In manufacturing industries, component quality and productivity play a very prominent role in fulfilling customers' varying needs. The varying strength materials consist of exceptional properties. The carbides, ceramics, Inconel, diamonds, and others are becoming more common in aerospace, nuclear engineering, structural functions, etc. Despite modern engineering innovations from the economic perspective of productivity, traditional machining like turning, grinding, drilling, broaching, milling, etc. are seems ineffective to machine these materials. In addition, machining these materials in a complicated way is challenging and these issues have been overcome. Nowadays, several applications are subject to non-traditional machining processes with widely varying capabilities and requirements. Electrical discharge machining (EDM) played a significant role in spark machining, eroding sparks, sinking die, corrosion of metal burning wire. The considerable improvements in terms of the production process are achieved by electrical discharge. The material is separated from the workpiece by quickly occurring current discharge between two electrodes,

[a]puranjay116@gmail.com, [b]rkvme@mmmut.ac.in

separated by a dielectric liquid and electrical voltage. In this process, tool-electrode generate a spark at 8000 to 12000°C temperature. It is a non-traditional technique of removing the material from the workpiece via thermal energy. Gowthaman et al. [1] performed an experimental investigation using EDM of Monel-super alloy by selecting input parameters such as, gap voltage, Pulse off time, Pulse on time, and discharge current to examines the rate of material removal (MRR) and surface roughness (Ra). Experiments were designed by using L_{27} Taguchi's orthogonal array and machined with die sink EDM. The findings demonstrated that the GRA identified current influence nearly 82% then T_{off} about 12%, T_{on} 4%, and gap voltage. Lin et al. [2] carried out the study on EDM in gas using Taguchi's L_{18} array experimental design. The process parameters were selected as gas pressure, peak current, grain size, machining polarity, and servo reference voltage to determine optimal MRR, EWR and Kumar et al. [3] researched the EDM process to analyze MRR and Ra of titanium alloy (Ti-6Al-4V). In the proposed work of experiment investigation, the process parameters, namely, current, voltage & Pulse, are simultaneously optimized using GRA. They were concluded that the lower current intensity produced a good quality of surface finish and vice versa, at a higher current value possible to maximize the MRR. The optimized conditions were found to be 18A, 100 µs and 40 Mlynarczgkcy ct al. [4] performed micro-electric discharge alloying (EDA) using tungsten on aluminum. EDA analysis was used to find alloying elements up to which diffusion takes place. It has found best voltage condition results in discharge over a large area. However, an increase in voltage did not increase the transfer of the electrode. Kumar et al. [5] carried out a machining process on the SS304 workpiece sample using an electro-discharge machine. They used five levels to design the experiment, and ANOVA was performed to check to influence factors affecting the MRR and TWR characteristics. Finally, it has been concluded that T_{off} has a significant factor affecting the circulation and minimum tool wear rate. Bobbili et al. [6] approached the GRA with Taguchi on wire EDM of ballistic grade Al-Mg-Si alloy 6063 material. They had selected input parameters as T_{on}, peak current, T_{off} and voltage. The machining characteristics of MRR and Ra have been chosen as the performance study. They were confirmed from analysis of variance test T_{on}, I_p & SV had the most significant factor. Uhlmann et al. [7] studied the applications of tungsten carbide electrodes in EDM. They were claimed the SLM (selective laser machining) was used for complex electrode geometry alternative for conventional tool electrode manufacturing.

Based on the literature survey, a lot of work performed on electric discharge machining (EDM) using different type of tool for machining operation of the various type of materials and consider output response parameter like material removal rate (MRR), tool wear rate (TWR), thrust force, torque force, and surface roughness, etc. But EDM of Inconel is passing through initial phases and it needs more attention from eminent scholars and industries for its proper utilization. For optimizing conflicting responses, different types of optimization approaches such as MOORA, Taguchi, GRA, ANN, PCA, VIKOR, TOPSIS, and Utility, etc. This article works on Inconel X-750 composite material by electric discharge machining (EDM) using a copper tool based on Taguchi L_{16} orthogonal array design. An attempt has been proposed to overwhelm the limitations of the traditional Taguchi approach.

Experimentation

In this research, a hybrid module uses as a mathematical technique for evaluating machining variables. In this experimentation, a Taguchi-based GRA method is used. Inconel X-750 has been chosen as work material and copper works as tool material due to high conductivity, better machining capability, and low Tool wear rate as displayed Figure 25.1. The surface roughness was measured by using a roughness tester, Handy Surf Tokyo Seimitsu Model No E-MC-S24B as displayed in Figure 25.2. The MRR and tool wear rate (TWR) is measured from Eq. (1) and the density of the material 8.276 g/cm³.

$$MRR/TWR = \frac{Initial\ weight\ of\ the\ workpiece - final\ weight\ of\ the\ workiece}{Density\ of\ the\ workpiece \times machaning\ time}\ mm^3/min\ (1)$$

The domain of the experiment has four factors and four distinct levels to optimize the response and machining operation during performed operation on Inconel X-750 workpiece using the copper tool, as shown in Table 25.1.

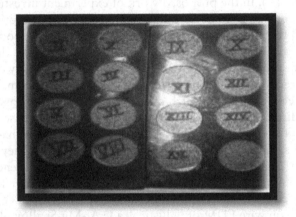

Figure 25.1 Inconel X-750 workpiece.

Figure 25.2 Inconel X-750 workpiece and computation of surface roughness.

Results and Discussion

The experimental performance prediction was evaluated based on Taguchi's integrated Grey concept. In multivariate machining problems, Grey relational analysis (GRA) methods are used efficiently to convert multiple responses into a single aggregated function [8].

Table 25.1: L$_{16}$ Orthogonal array and Response parameters.

Exp. No	I$_p$	T$_{on}$	DC	V	MRR	TWR	R$_a$
1	8	200	8	50	0.1380	0.0019	3.4
2	8	300	9	60	0.1508	0.0010	3.6
3	8	400	10	70	0.2112	0.0048	3.5
4	8	500	11	80	0.1843	0.0033	3.7
5	12	200	9	70	0.2585	0.0041	5.5
6	12	300	8	80	0.1996	0.0027	5.8
7	12	400	11	50	0.3017	0.0041	6.8
8	12	500	10	60	0.3205	0.0088	6.7
9	16	200	10	80	0.3911	0.0132	7
10	16	300	11	70	0.3727	0.0050	6.9
11	16	400	8	60	0.2868	0.0091	7.3
12	16	500	9	50	0.2402	0.0027	7.1
13	20	200	11	60	0.1830	0.0033	7.7
14	20	300	10	50	0.2225	0.0009	9.2
15	20	400	9	80	0.4109	0.0037	9.4
16	20	500	8	70	0.4174	0.0028	9.6

To avoid inconsistencies, this method first normalizes the quality properties of calculated data from 0 to 1. This approach is effectively applied to aggregate multiple conflicting responses into a single objective function known as GRG. GRA method considers the following step:

Normalized the function value

Higher-the-better

$$Y_i^*(j) = \frac{Y_i(j) - \min Y_i(j)}{\max Y_i(j) - \min Y_i(j)} \qquad (2)$$

Lower-the-better

$$Y_i^*(j) = \frac{\max Y_i(j) - Y_i(j)}{\max Y_i(j) - \min Y_i(j)} \qquad (3)$$

The deviation sequence are determined

$$Di(k) = 1 - \frac{dj(k) - d}{\max Di(k) - Di} \qquad (4)$$

Where Di (k) represents deviation value.

Calculate Gray relation coefficient

$$\xi_i(k) = \frac{\Delta_{min} + \xi\Delta_{max}}{\Delta oi + \xi\Delta_{max}} \tag{5}$$

Here, ξ = identical coefficient and considering $\xi = 0.5$ [9], while Δ_{max} and Δ_{min} are the maximum and minimum absolute differences.

Calculate Grey relation grade

$$GRG = \frac{1}{t}\Sigma \in (k) \tag{6}$$

Where t = number of process performance. The Grey relation grade (GRG) computed from Eq. (6) is based on response parameters such MRR, Ra, and TWR, as depicted in Table 25.2.

Table 25.2: Normalized data, GRC and GRG corresponding experiment.

Normalized value			GRC			GRG
MRR	TWR	R_a	MRR	TWR	R_a	
0	0.9259	1	0.3333	0.8710	1	0.7347
0.0457	0.9963	0.9677	0.3438	0.9927	0.9393	0.7586
0.2618	0.6890	0.9838	0.4038	0.6165	0.9687	0.6630
0.1657	0.8059	0.9516	0.3747	0.7204	0.9117	0.6689
0.4313	0.7435	0.6612	0.4678	0.6610	0.5961	0.5750
0.2203	0.8581	0.6129	0.3907	0.7790	0.5636	0.5777
0.5857	0.7458	0.4516	0.5468	0.6630	0.4769	0.5622
0.6533	0.3623	0.4677	0.5905	0.4395	0.4843	0.5048
0.9059	0	0.4193	0.8416	0.3333	0.4626	0.5459
0.8401	0.6678	0.4354	0.7578	0.6008	0.4696	0.6094
0.5327	0.3341	0.3709	0.5169	0.4288	0.4428	0.4628
0.3655	0.8575	0.4032	0.4407	0.7783	0.4558	0.5583
0.1607	0.8086	0.3064	0.3733	0.7232	0.4189	0.5051
0.3022	1	0.0645	0.4174	1	0.3483	0.5885
0.9770	0.7731	0.0322	0.9560	0.6878	0.3406	0.6615
1	0.8485	0	1	0.7675	0.3333	0.7002

Analysis of Variance (ANOVA)

ANOVA is a set of logical modeling techniques to evaluate the variations between means in a data set and their related predicting techniques. ANOVA has been shown to control each input parameter and evaluate the most dominant constraint contribution factor. In this study, Pulse on time (I_p) achieved a higher contribution (86.06%), as shown in Table 25.3. The same finding was concluded by Raja et al. and Kumar et al. [10, 11]automobile and electronics industries. The profile manufactured by EDM process should be dimensionally and geometrically accurate apart from good finish. This expectation is very much important as the die manufactured from EDM process is subjected to subsequent mass production. The material normally selected for die making will be superior in quality and hence time and cost of production will also be high. Selection of optimum EDM parameters may reduce the machining time along with maintaining required surface finish and dimensional accuracy. So there is a need to develop a technique for selecting the optimal EDM parameters to achieve the desired performance measures. In the present work, a recently developed Firefly Algorithm (FA. The preferred solution value on $I_p 8$, $T_{on} 300$, DC9, V70 orthogonal setting obtained predicted value has 0.8007 for Grey relational analysis (GRA) shown in Figure 25.3 and achieves optimal machining condition.

Table 25.3: Analysis of variance of GRG.

Source	DF	Seq SS	Contribution	Adj SS	F-Value	P-Value
Regression	4	5.79592	97.20%	5.79592	103.98	0.000
I_p	1	5.13208	86.06%	0.00270	0.19	0.668
T_{on}	1	0.35388	5.93%	0.00735	0.53	0.482
DC	1	0.26299	4.41%	0.04748	3.41	0.090
V	1	0.04697	0.79%	0.04697	3.37	0.091
Error	12	0.16722	2.80%	0.16722		
Total	16	5.96314	100.00%			

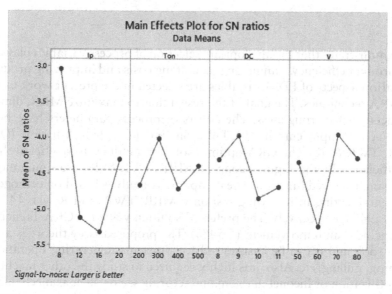

Figure 25.3 S/N ratio of GRG.

Confirmatory Test

The comparative study of Inconel X-750 machined workpiece based on the orthogonal and experimental setting concerning the different considered responses, MRR, TWR, and Ra. In terms of performance and optimal conditions, the GRG values are presented.

Figure 25.3 represents the S/N ratio obtained from Grey theory. From Table 25.2, the IP8, Ton300, DC9 and V60 (experiment number 2) setting at a higher GRG value is achieved. The optimal condition of the process parameters is displayed in Figure 25.3. Table 25.4 presents the corresponding OA value of the MRR, TWR and Ra values of 0.1508 mm³/min, 0.0010 mm³/min and 3.6 µm, respectively. It has found the percentage enhancement value of response such as material removal rate (MRR), tool wears rate (TWR), and surface roughness (Ra) as 14.85%, 10%, and 6.94%, respectively, when contrasting this value from optimal setting. Also, the preferred solution value for Taguchi embedded GRA is 0.8007, resulting in a 5.54% improvement over the OA condition (Table 25.4). Finally, the EDM performances' desired value is acquired through the hybrid Grey module for a product manufacturing environment.

Table 25.4: Confirmatory test for grey relational analysis (GRA).

Response	GRA Taguchi		Experimental	Imp. %
	OA-setting $I_p8T_{on}300DC9V60$	Predicted setting $I_p8T_{on}300DC9V70$		
MRR	0.1508		0.1732	
TWR	0.0010		0.0009	
R_a	3.6		3.35	
Preferred solution value	0.7586	0.8007	0.8007	5.54%

Conclusion

This study demonstrates that the appropriate selection of process variables plays a vital role in ensuring product efficiency, minimizing machining costs and improving production rate. The optimization aspects of EDM variables are selected in the present work using Taguchi embedded GRA techniques. The goal of the research is to maximize MRR, diminish TWR and the reduced surface roughness. The effects of process parameters have been defined by ANOVA such as input current (I_p), Pulse on time (T_{on}), pulse off time (DC) Voltage (V) on MRR, TWR & R_a. The GRA optimal solution modules have effectively evaluated machining efficiency. By the Taguchi concept, MRR, TWR & Ra are simultaneously optimized into a single desired function. The comparative study is based on orthogonal setting and experimental setting in following response MRR, TWR, and R_a via. 14.85%, 10%, and reduced 6.94%, respectively. The preferred solution value for GRA Taguchi is 0.7586 and 0.8007, achieve an improvement (5.54%). The proposed Grey theory is a generalized optimization tool, and it can be customized for other manufacturing operations such as drilling, turning, milling, etc. Also, it is highly required to study the other machining aspect for efficient utilization of Inconel in the manufacturing sector, such as mechanical material removal, interface temperature, tool materials, safety, and health during machining tests.

Acknowledgments

The authors would like to acknowledge the kind support and sincere gratitude to the Madan Mohan Malaviya University of Technology, Gorakhpur India.

References

1. Gowthaman, S., Balamurugan, K., Kumar, P. M., Ali, S. K. A., Kumar, K. L. M., and Ram Gopal, N. V. (2018). Electrical discharge machining studies on monel–super alloy. *Procedia Manufacturing*, 20, 386–391. doi: 10.1016/j.promfg.2018.02.056.
2. Lin, Y. C., Hung, J. C., Lee, H. M., Wang, A. C., and Fan, S. F. (2018). Machining performances of electrical discharge machining combined with abrasive jet machining. *Procedia CIRP*, 68(April), 162–167. doi: 10.1016/j.procir.2017.12.040.
3. Kumar, R., Roy, S., Gunjan, P., Sahoo, A., Sarkar, D. D., and Das, R. K. (2018). Analysis of MRR and surface roughness in machining Ti-6Al-4V eli titanium alloy using EDM process. *Procedia Manufacturing*, 20(2017), 358–364. doi: 10.1016/j.promfg.2018.02.052.
4. Młynarczyk, P., Krajcarz, D., and Bańkowski, D. (2017). The selected properties of the micro electrical discharge alloying process using tungsten electrode on aluminum. *Procedia Engineering*, 192, 603–608. doi: 10.1016/j.proeng.2017.06.104.
5. Ravi Kumar, M., Krishnaiah, A., and Shankar Kalva, R. (2018). Experimental study on micro machining of SS304 by using electric discharge machining. *Materials Today: Proceedings*, 5(13), 27269–27276. doi: 10.1016/j.matpr.2018.09.043.
6. Bobbili, R., Madhu, V., and Gogia, A. K. (2015). Multi response optimization of wire-EDM process parameters of ballistic grade aluminium alloy. *Engineering Science and Technology, an International Journal*, 18(4), 720–726. doi: 10.1016/j.jestch.2015.05.004.
7. Uhlmann, E., Bergmann, A., Bolz, R., and Gridin, W. (2018). Application of additive manufactured tungsten carbide tool electrodes in EDM. *Procedia CIRP*, 68(April), 86–90. doi: 10.1016/j.procir.2017.12.027.
8. Talla, G., Sahoo, D. K., Gangopadhyay, S., and Biswas, C. K. (2015). Modeling and multi-objective optimization of powder mixed electric discharge machining process of aluminum/alumina metal matrix composite. *Engineering Science and Technology, an International Journal*, 18(3), 369–373. doi: 10.1016/j.jestch.2015.01.007.
9. Kumar, J., and Verma, R. K. (2021). Experimental investigation for machinability aspects of graphene oxide/carbon fiber reinforced polymer nanocomposites and predictive modeling using hybrid approach. *Defence Technology*, 17(5), 1671–1686. doi: 10.1016/j.dt.2020.09.009.
10. Bharathi Raja, S., Srinivas Pramod, C. V., Vamshee Krishna, K., Ragunathan, A., and Vinesh, S. (2013). Optimization of electrical discharge machining parameters on hardened die steel using firefly algorithm. *Engineering with Computers*, 31(1), 1–9. doi: 10.1007/s00366-013-0320-3.
11. Kumar, R., Pandey, A., and Sharma, P. (2019). Investigation of surface roughness for inconel 718 in blind hole drilling with rotary tool electrode. *Journal of Advanced Manufacturing Systems*, 18(3), 379–394. doi: 10.1142/S0219686719500203.

26 Evaluation of fly ash bricks for water absorption and compressive strength

Vinay Singh[1,a], Vishal Behl[2], and Vijay Dahiya[2]

[1]Department of Mechanical Engineering, Jagan Nath University, Bahadurgarh, Jhajjar, India

[2]Department of Civil Engineering, Jagan Nath University, Bahadurgarh, Jhajjar, Haryana, India

Abstract: In general, fly ash bricks possess similar texture as that of clay bricks, but contain some additive with spherical fly ash particles. This research paper deals with the properties, process material, methodology, inspection, and quality control as per the provisions of Indian Standard Code. The bricks' density was reduced by the addition of fly ash particles, and their durability was much increased. The use of mixture having (50% ash + 20% lime + 20 % sand+10 % gypsum) also resulted in cost cutting. The bricks so formed were found to have appropriate compressive strength and water absorption capacity. The physical composition of the fly ash bricks provided good results in terms of compressive strength, density, water absorption, color and as well as dimensional stability in comparison with the traditional clay bricks. Present study concludes that in future, fly ash bricks will become an alternative to conventional clay bricks with considerable economy and ease in construction.

Keywords: Clay bricks, compressive strength, fly ash bricks, physical properties, water absorption.

Introduction

In India, major source of energy needed for sustainable development of the country are mainly coal based thermal power plants [1]. The coal quality obtained from different sources is variable with their different burning properties and resultant fly ash properties. Fly ash, also called flue ash, is fine powder of silica powder and metallic oxides along with the traces of heavy metals [2]. The silica particles are very small (5.8 μm) spherical and their color may vary as per coal quality from greyish white to bit darker color and is classified into different classes in view of its physic chemical properties. In India, annual production of fly ash is 320 MT which is likely to increase to 1000 MT in future depending upon expansion in thermal power plants in different parts of the country.

Huang et al. [3] investigated the manner in which nanosilica (NS) affected high-volume fly ash cement HVFAC pastes' compressive strength and water absorption. Additionally, mercury intrusion porosimetry (MIP), scanning electron microscopy (SEM) and X-ray diffraction (XRD) were used to examine the pore structure, microstructure, and phase

---

[a]arijit.majumder28@gmail.com

composition of the pastes. Results revealed that adding NS to HVFAC pastes increased their compressive strength and decreased their water absorption.

Fantu et al. [4] investigated the effect of fly ash on high-strength concrete properties by varying the cement replacement percentage starting from 0% to a maximum of 30% with 5% increment. The study's findings reveal that replacing cement with fly ash increases the slump value up to 10%. However, replacing more than 10% of the cement with fly ash reduces the fresh concrete workability.

Hosan et al. [5] investigated the influence of nano-$CaCO_3$ (NC) on the durability properties along with compressive strengths of high-volume slag fly-ash (HVS-FA) blended and high volume slag (HVS) concretes. The results demonstrate that adding 1% NC increased the compressive strengths of HVS-FA and HVS concretes by an amount of 28% and 43%, respectively.

Fly ash possesses several distinct advantages as compared to clay bricks such as durability, compressive strength, uniformity, hydration and thermal properties be-side lower cost of production [6–8]. However, the quality of fly ash bricks varies with fly ash produced from different qualities of coal. The present study deals with the evaluation of fly ash bricks for various parameters variable to civil engineers and construction industry such as compressive strength, water absorption, dimensional stability etc.

Materials and Methods

Fly ash bricks were prepared from fly ash obtained from Kasnia Associates located near Suratgarh thermal power plant at Suratgarh, District Sri Ganganagar, and Rajasthan, India. The commonly used raw material for production of Fly ash bricks are: (a) Fly ash, (b) Sand Stone, (c) Gypsum, (d) Lime and (e) Water

Methodology

The various methods employed for producing mass fly ash bricks involves mixing of the raw materials and burning them to solidify or mixing soil, fly ash, POP and water, and then allowing the mixture to sun dried to avoid emissions. In present study, we produced fly ash bricks by mixing 50% fly ash, 20% lime, 10% gypsum, 20% sand and the mixer was put into the machine and water was added in sufficient quantities to make the mixture slurry. The slurry was passed through the molds to obtain bricks 230 x 100 x 75 mm and allowed to sun dry on soil surface. The properly dried bricks were evaluated for different physical parameters.

Procedure

Water Absorption Test

Devices: 1) Ventilated Oven and 2) Sensitive Mass Balance

Pre Conditioning:
It includes drying of fly ash bricks in ventilated oven in temperature ranging from 100–115°C until it acquires constant mass. Afterwards, specimens were allowed to cool at room temperature to measure the weight (M1).

Follow Up:
After the specimens were completely dried, they were dipped in to water at a temperature of $25 \pm 4°C$ for about a day. Then the specimens were wiped out with cloth to avoid any water traces followed by weighing of the specimen. Again specimens were weighed after removing from the water. Water absorption, by mass percentage can be derived by using the given formula:

$$\frac{M2 - M1}{M1} \times 100$$

Compressive Strength Test
Devices: Surface Grinder, Measuring tape, Scale, Compression testing machine

Calculations:
Following formula was used to determine the strength:

$$Compressive\ Strength\ \left(N / mm^2\right) = \frac{Maximum\,load\,at\,failure\,in\,N}{Average\,area\,of\,bed\,face\,in\,mm2}$$

Results

The fly ash bricks produced in these sizes (4 Inches: $-230 \times 100 \times 75$ mm) were tested for compressive strength and water absorption. The salient results are pre-scented in Tables 26.1 and 26.2. Average 6.4% water absorption was observed in the specimen.

Table 26.1: Result for water absorption.

SPECIMEN	Dry Weight (in gms)	Wet Weight (in gms)	Absorption (percentage)
A	3550	3825	7.746
B	3722	3942	5.910
C	3740	3953	5.695
	AVERAGE		6.450

Table 26.2: Result for compressive strength.

Specimen	Fly ash Brick (Dimensions) (in mm)			Average area (Surface) (mm²)	Failure Load (in kN)	C. Strength (in N/ mm²)
	L	B	H			
A	230	100	118	23000	118	5.152
B	230	100	104	22800	104	4.580
C	230	100	132	22800	132	5.813
				AVERAGE		5.181 51.81(kgf/cm²)

Table 26.3: Comparison between clay brick and fly ash brick.

Physical Properties	Fly Ash Brick	Clay brick	Remarks
Compressive strength	55–100kg/cm²	25–40 kg/cm²	Good load bearing capacity
Color	Uniformity in color	Variations in color	Good in Appearance
Density	1710–1860kg/cm³	1610–1760 kg/cm³	Good load bearing capacity
Plastering	Provides even surface	Thickness varies	Plaster savings
Transit wastage	< 2%	≤ 12%	Cost savings upto 10%
Water absorption	10–12%	16–25%	Low dampness
Dimensional Stability	High Tolerance	Low Tolerance	Mortar Savings

Table 26.2 shows the compressive strength observed for different brick dimensions with different impact loadings. The comparison between the physical properties of fly ash bricks and clay bricks has been depicted in tabular form in Table 26.3. Fly ash bricks showed superiority over clay bricks in terms of compressive strength, color uniformity, density and water absorption characteristics. Other features like dimensional stability and transit wastage were also find quite appropriate in comparison with clay bricks.

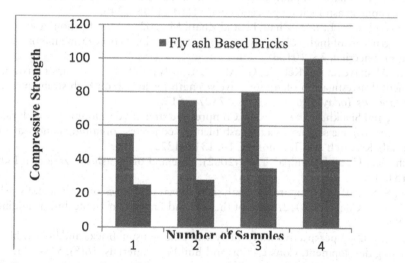

Figure 26.1 Comparative analysis of compressive strength of fly ash based bricks and clay based bricks.

The prepared fly ash bricks were made using the given composition (50% fly ash, 20% lime, 10% gypsum, 20% sand). The comparative analysis of the fly ash brick and clay brick has been shown in Figure 26.1. In all the samples, the fly ash bricks showed better compressive strength in all samples. The study further revealed that fly ash bricks were uniform having high density in comparison with clay bricks. Water absorption capacity in fly ash brick was found to be very low when compared with clay bricks. Also, fly ash brick were of high tolerance nature and with good dimensional stability.

Conclusion and Future Scope

Fly ash can be profitably used to develop bricks. The average compressive strength obtained was $5.725N/mm^2$. Clay (60%) and fly ash (40%) can become a better option. Fly ash bricks are a useful building material made from previously discarded trash from coal-fired thermal power plants. Since the production of fly ash bricks is increasing, its use in brick production can provide effective solutions to manage this potential resource for sustainable development while avoiding its negative environmental impacts. More developments are required to produce fly ash bricks with more favorable physicochemical qualities that are beneficial in the construction business.

While making fly ash bricks utmost care should be given to the aspects of workshop testing in terms of flexural test, compressive strength test, and water absorption properties, smoothness of the fly ash brick surfaces for better settlement in buildings or civil structures. This will enable the strategy to utilize fly ash, lime, gypsum and sand in making fly ash bricks with greater strength, longer durability, less environmental pollution and economize the fly ash bricks production.

References

1. Gupta, P., and Gupta, A. (2014). Biogas production from coal via anaerobic fermentation. Fuel, 118, 238–242.
2. Yoshiie, R., Nishimura, M., and Moritomi, H. (2002). Influence of ash composition on heavy metal emissions in ash melting process. Fuel, 81(10), 1335–1340.
3. Huang, Q., Zhu, X., Liu, D., Zhao, L., and Zhao, M. (2021). Modification of water absorption and pore structure of high-volume fly ash cement pastes by incorporating nanosilica. Journal of Building Engineering, 33, 101638.
4. Fantu, T., Alemayehu, G., Kebede, G., Abebe, Y., Selvaraj, S. K., and Paramasivam, V. (2021). Experimental investigation of compressive strength for fly ash on high strength concrete C-55 grade. *Materials Today: Proceedings*, 46, 7507–7517.
5. Hosan, A., and Shaikh, F. U. A. (2021). Compressive strength development and durability properties of high volume slag and slag-fly ash blended concretes containing nano-CaCO3. Journal of Materials Research and Technology, 10, 1310–1322.
6. Bhattacharjee, U., and Kandpal, T. C. (2002). Potential of fly ash utilization in India. Energy, 27(2), 151–166.
7. Bansode, S. S. (2012). Comparative analysis between properties of steel slag, fly ash, and clay bricks. In Geo Congress 2012: State of the Art and Practice in Geotechnical Engineering (pp. 3816–3825).
8. Kumar, S. (2002). A perspective study on fly ash–lime–gypsum bricks and hollow blocks for low cost housing development. Construction and Building Materials, 16(8), 519–525.

27 Modelling and simulation of squeeze casting process

Rupanshu Singh[a] and Ravi Butola

Department of Mechanical Engineering, Delhi Technological University, New Delhi, India

Abstract: The aluminium composite matrices are used extensively due to their preferable properties. These composite matrices are formed and processed by using different manufacturing processes like squeeze casting, stir casting etc. Squeeze casting is used to produce material with enhanced mechanical properties and finer grain size. The application of constant pressure on the composite, solidifies and forms the desired component without any gas and shrinkage porosities. This research attempts to simulate the stress-strain generation and deformation occurrence in different composite materials (Al, Al 2024, Al 6061, Al 7075) during the squeeze casting process, by conducting analysis on ANSYS software. The results are compared, and it was observed that maximum deformation of 7.81 mm occurred in the Al 6061 and highest stress of 74.9 MPa was generated in Al 7075 material. The results also revealed that explicit dynamics simulation is a reliable method to predict the events undergoing nonlinear, transient forces.

Keywords: Aluminium metal matrix composite, mechanical properties, simulation, squeeze casting.

Introduction

Composites are composed of two or more materials, exhibiting different forms and properties. The usage of composites in various manufacturing industries is extending quite rapidly and is being utilized in various industries like aviation and automobile industry. The tribological and wear properties of a base material can be enhanced by adding a suitable reinforcement material and developing it into a fabricated composite [1]. MMCs manufacturing can be classified into three types-solid, liquid, and vapor. Powder blending and powder metallurgy comes under solid-state methods squeeze casting, stir casting, etc. comes under liquid state methods and the vapor state method includes physical vapor deposition [2]. Out of these squeeze casting is a unique kind of casting process in which the forging and casting processes are combinedly used. Mechanical properties that are obtained with this process are of the highest kind. Squeeze casting is termed as liquid metal forging process. It is a mixture of closed die forging and gravity die casting [3]. Pressure is applied to solidify the metal, for obtaining the minimal gas porosity the heat transfer occurs when pressure interacts through the die surface. Continuous pressure application

[a]singhrupanshu@gmail.com

results in the solidification of the metal, producing the desired output. Squeeze casting is used to produce finely detailed parts, at higher casting yields as no riser and gating is required. The molten metal is solidified by applying continuous pressure, this results in fine grain size with enhanced mechanical properties [4]. The shrinkage defects are not observed and high-quality surfaces can be generated. The applied pressure is less than that in forging. While squeeze casting produces the desired output but the complex tooling results in high cost, thus, to justify the equipment investment, high volumes of output should be produced. The process works on the punching mechanism, the tool is allocated to a specific component, no flexibility is observed. The operational time and cost increases as the process requires to be accurately controlled [5]. Squeeze casting can be classified into subcategories, (Figure 27.1) depicts the various categorization of squeeze casting. There are several experimental research conducted on the squeeze casting process. Sun el al. [6] conducted squeeze casting process of A443 in five steps and determined the heat transfer coefficient by applying the inverse algorithm, the peak value obtained was 9419 W/m²K and the graph results were plotted. Li et al. [7] analyzed the shrinkage defects in the automobile control arm by developing a model and simulating the squeeze casting process on MAGMASOFT software. You et al. [8] presented a frictional and lubrication model of the injection process of squeeze casting technique, the friction's influence was observed and the simulation results were compared by the experimental ones. Shi et al. [9] determined the flow stresses and microstructure of the Al alloy and developed different plots on stress-strain variation, flow stresses behavior and microstructures. This current research work demonstrates the explicit dynamics simulation of the squeeze casting process using ANSYS software. The deformation occurrence and stress-strain generation in different aluminium composites were obtained and compared. A brief overview of squeeze casting and the methodology involved in simulation process is discussed in this paper. This would help the researchers to visualize and understand the stress and strain generation in different materials, when the squeeze casing process is conducted.

Squeeze Casting Setup

The experimental setup comprises of the pathway preheater with maximum capacity of 400°C, stirrer unit, a degasser is used to remove impurities from the material, composite

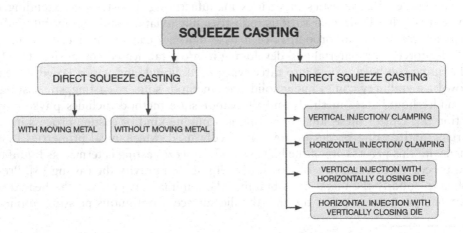

Figure 27.1 Classification of squeeze casting.

preheater which generates temperature up to 500°C, pouring furnace used for melting the metal and a die with punch attached to the hydraulic unit. The liquid metal is placed in the mould passing through the preheated pathway to obtain the desired initial temperature. The die punch surface exerts the squeeze pressure or injects the metal using a piston with the punch mechanism, for obtaining the required properties and avoiding losses. The solidification time is in the range of 15–45 seconds [10].

The parameters governing the process of squeeze casting are Casting temperature depends on the alloys and the geometry, Tooling temperature ranges between 190–315°C, the duration between the instant the punch contacts the pool of molten metal and starts pressurizing it and actual pouring, is known as time delay [11], pressure levels ranging between 50Mpa to 140Mpa, pressure duration, lubrication, a colloidal graphite spray lubricant is used on warm dies. Figure 27.2 shows the sequential squeeze casting process to produce the desired output.

In squeeze casting, pressure application enhances the bonding force and the wettability between reinforcing agent and the matrix [12]. The pressure applied results in reduction of gas porosity and 10% increase in tensile strength. After After observing the of ZnAl alloy it was concluded that their mechanical properties like ductility and high tensile strength are enhanced. Also because micropores in the alloy are eliminated because of high pressure which then showed enhanced improvement in mechanical properties [13,14].

The mechanical properties of aluminium composite varies with the casting temperatures. For obtaining desired results, metal is preheated before getting transferred to the die. Temperatures of 660 or 690°C are considered as the optimal temperatures for the aluminium alloys undergoing squeeze casting. The temperature of die ranges around 200°C–250°C, a rapid increase in elongation and strength is observed, while only a slight increment occurs when the increase in temperature is in range of 250–300°C [15]. While high temperature might affect the die's life, cause shrinkage pores, delays in solidification, resulting in lower strength [4].

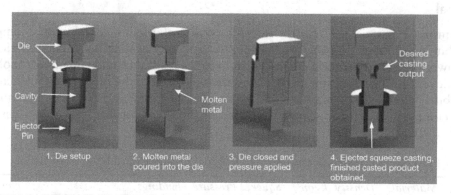

Figure 27.2 Squeeze casting process.

Simulation Methodology

ANSYS operates on a mathematical model based on the governing equations & boundary conditions and generates the numerical solution at different nodes and elements. The structural analysis of deformation and stress strain variation over the whole setup and dynamics can be predicted by using the explicit dynamics analysis [5]. Analysis software develops the result contours and graphs by applying the boundary conditions on a virtual model. This can help in optimizing the design and help researchers to analyze different characteristics of simulation.

Design and Modelling of Setup

A model of squeeze casting experiment was generated to analyze the stress development and other properties of the composite material. The molten material was filled in the cavity. The die was punched over the setup, applying a constant pressure resulting in the solidification of the material. For better analysis and understanding of squeeze casting, the experimental process was simulated using solidworks 2020 and ANSYS. The model of a die had two parts which were assembled together in which a closed die of radius 100mm with a cavity of radius 65mm and height 70mm was generated and filled with the liquid aluminium composite, considering the allowances, shown in Figure 27.3. The die was developed by revolved feature and the composite was filled by using the fill feature in design modeler.

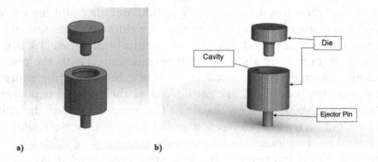

Figure 27.3 a) Model setup in Solidworks'20 b) rendered image of the setup.

Material

The reinforced material used for the research work is aluminium metal matrix composite. It exhibits good thermal properties, higher strength, easy availability, higher elastic modulus, lower weight ratio, good corrosion and fatigue resistance and enhanced tribological properties [16]. When multiple particulates are reinforced in the metal, it is called aluminium hybrid metal matrix composite, while a single reinforcement is called aluminium metal matrix composite. For this research, the material considered for the simulation is described in Table 27.1.

Table 27.1: Aluminium matrix composites and its properties developed by squeeze casting.

S. No.	Material	Density (kg/m³)	Squeeze Pressure	Findings	Reference
1	Al	2770	30, 50 & 70 MPa	Composite showed increase in tensile strength	[17]
2	Al 2024	2785	75 MPa	High mechanical strength	[18]
3	Al 6061	2703	40MPa	Excellent joining properties, good workability	[19]
4	Al 7075	2804	100 MPa	Exceptional fatigue strength, machinability, toughness, and weldability	[20]

Meshing

Meshing is done for calculating the governing equations at the required points and defining the domain [21]. Mechanical properties were analyzed by calculating the mathematical model at different nodes, at each node the governing equations were implemented. For this analysis, a fine mesh of tetrahedron shape was generated with 14168 nodes and 74495 elements, shown in Figure 27.4.a).

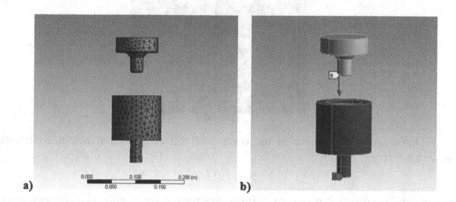

Figure 27.4 **a)** Refined mesh generated on the model **b)** velocity applied to the punch in negative y-direction.

Boundary Conditions

For obtaining optimal results, specific boundary conditions were applied on model of the setup. The Finite element analysis and explicit dynamics provide a method to simulate the structural mechanics branch containing physics interfaces for analyzing deformations, stresses, and strains of solid structures. The analysis types include stationary, eigenfrequency, transient, frequency response, and linear buckling for simulation [22]. Explicit dynamics analysis was conducted for simulating the squeeze casting process. The time step was 10^{-3} sec, the bottom surface was fixed, max no. of cycles were 10^7 beam time safety factor was 0.5, the punch was allowed to fall over the fixed die, as shown in Figure 27.4.b).

Result

The normal stress and elastic strain generated in molten aluminium composite, filled inside the die cavity were analyzed using the ANSYS software. The regions of high stress variation were obtained. Deformations occurring in the composite when the punch closed the die, were observed and critical points were obtained. The explicit dynamics shows that compression in the composite occurs at the interaction of punch and composite, when the die is closed. The simulation analysis was repeated for different material inputs and comparative results were obtained.

Deformation

When the punch interacts with the composite material, the mesh compression occurs as shown in Figure 27.5. The grain size is observed to be reduced, thus resulting in faster

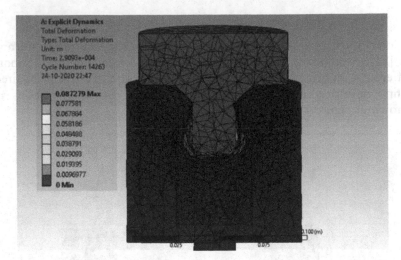

Figure 27.5 Deformation occurrence in the composite material, when punch is completely submerged in composite.

solidification under constant pressure, it increases tensile strength and hardness, maximum deformation of 7.81 mm was developed in Al-6061 material. The sudden impact on the molten composite developed a significant deformation.

Stress and Strain

The stress contours show that high stress is generated near the surface in contact with the die. The stress strain variation in the composite is depicted in Figure 27.6. Maximum strain is developed near the die punch-composite interaction. As the strain increases, stress increases linearly till the yield point. Further, only a slight increase in stress is observed. Table 27.2. Shows the comparison in deformation, stress and strain generation in different composites. The results obtained from the simulation are identical to the result graphs obtained by Shi et al. [9].

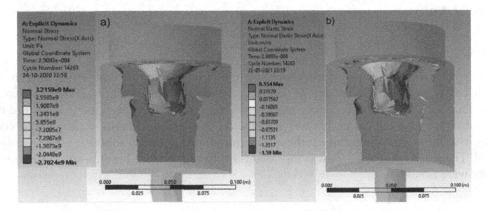

Figure 27.6 a) Stress b) Strain contours, when punch is completely submerged in composite.

Table 27.2: Result deformation and stress strain obtained from simulation.

S. No.	Material	Deformation (mm)	Total Stress (MPa)	Maximum strain
1	Al	7.65	73.9	0.561
2	Al 2024	7.62	74.3	0.554
3	Al 6061	7.81	73.5	0.567
4	Al 7075	7.54	74.9	0.549

Concluding

This article summarizes the effect on the aluminium composite using squeeze casting technique, accomplished by various authors. A model of squeeze casting was virtually setup and simulated on ANSYS software. The total deformation and normal stress, strain variation occurred in different aluminium composites, while undergoing squeeze casting process were analyzed and compared. The maximum deformation of 7.81mm occurred in the Al 6061 and highest stress of 74.9 MPa was generated in Al 7075 material. Stress and strain contours were developed and results of various aluminium composites were compared.

Acknowledgments

The authors would like to thank Department of Mechanical Engineering, Delhi Technological University for providing the workspace and necessary equipment to carry out the required research.

References

1. Coyal, A., Yuvaraj, N., Butola, R., and Tyagi, L. (2020). An experimental analysis of tensile, hardness and wear properties of aluminium metal matrix composite through stir casting process. *SN Applied Sciences*, 2(5), 1–10.
2. Mishra, A., Kumar, V., and Srivastava, R. (2014). Optimization of tribological performance of Al-6061T6-hybrid metal matrix composites using taguchi method & grey relational analysis. *Journal of Minerals and Materials Characterization And Engineering*, 02(04), 351–361.
3. Natrayan, L., and Senthil Kumar, M. (2018). Study on squeeze casting of aluminum matrix composites—a review. *Advanced Manufacturing and Materials Science*, Selected Extended Papers of ICAMMS 2018, 75–83.
4. Srivastava, N., and Anas, M. (2020). An investigative review of squeeze casting: processing effects & impact on properties. *Materials Today: Proceedings*, 26, 1914–1920.
5. Soundararajan, R., Ramesh, A., Sivasankaran, S., and Sathishkumar, A. (2015). Modeling and analysis of mechanical properties of aluminium alloy (A413) processed through squeeze casting route using artificial neural network model and statistical technique. *Advances in Materials Science and Engineering*, 2015, 1–16.
6. Sun, Z., Hu, H., and Niu, X. (2012). Experimental study and numerical verification of heat transfer in squeeze casting of aluminum alloy A443. *Metallurgical and Materials Transactions B*, 43(6), 1676–1683.
7. Li, Y., Yang, H., and Xing, Z. (2016). Numerical simulation and process optimization of squeeze casting process of an automobile control arm. *The International Journal of Advanced Manufacturing Technology*, 88(1–4), 941–947.

8. You, D., Wang, X., Cheng, X., and Jiang, X. (2017). Friction modeling and analysis of injection process in squeeze casting. *Journal of Materials Processing Technology*, 239, 42–51.
9. Shi, Y., Liu, L., Zhang, L., Zhang, L., Zheng, L., Li, R., and Yu, B. (2017). Effect of squeeze casting process on microstructures and flow stress behavior of Al-17. 5Si-4Cu-0.5Mg alloy. *Journal of Iron and Steel Research International*, 24(9), 957–965.
10. Ghomashchi, M., and Vikhrov, A. (2000). Squeeze casting: an overview. *Journal of Materials Processing Technology*, 101(1–3), 1–9.
11. Arunkumar, S., Chandrasekaran, M., Vinod Kumar, T., and Muthuraman, V. (2018). Properties and behavior of squeeze pressure on aluminum molybdenum DI-supplied composite. *International Journal of Engineering and Technology*, 7(3.3), 496.
12. Seyed Reihani, S. (2006). Processing of squeeze cast Al6061–30vol% SiC composites and their characterization. *Materials and Design*, 27(3), 216–222.
13. Li, R., Li, R., and Bai, Y. (2010). Effect of specific pressure on microstructure and mechanical properties of squeeze casting ZA27 alloy. *Transactions of Nonferrous Metals Society of China*, 20(1), 59–63.
14. Yang, L., Mori, K., and Tsuji, H. (2008). Deformation behaviors of magnesium alloy AZ31 sheet in cold deep drawing. *Transactions of Nonferrous Metals Society of China*, 18(1), 86–91.
15. Dhanashekar, M., and Kumar, V. (2014). Squeeze casting of aluminium metal matrix composites-an overview. *Procedia Engineering*, 97, 412–420.
16. Samal, P., Vundavilli, P., Meher, A., and Mahapatra, M. (2020). Recent progress in aluminum metal matrix composites: a review on processing, mechanical and wear properties. *Journal of Manufacturing Processes*, 59, 131–152.
17. Hajjari, E., Divandari, M., and Mirhabibi, A. R. (2010). The effect of applied pressure on fracture surface and tensile properties of nickel coated continuous carbon fiber reinforced aluminum composites fabricated by squeeze casting. *Materials & Design*, 31(5), 2381–2386.
18. Tian, K., Zhao, Y., Jiao, L., Zhang, S., Zhang, Z., and Wu, X. (2014). Effects of in situ generated ZrB2 nano-particles on microstructure and tensile properties of 2024Al matrix composites. *Journal of Alloys and Compounds*, 594, 1–6.
19. Butola, R., Pratap, C., Shukla, A., and Walia, R. (2019). Effect on the mechanical properties of aluminum-based hybrid metal matrix composite using stir casting method. *Materials Science Forum*, 969, 253–259.
20. Butola, R., Ranganath, M. S., and Murtaza, Q. (2019). Fabrication and optimization of AA7075 matrix surface composites using Taguchi technique via friction stir processing (FSP). *Engineering Research Express*, 1(2), 025015.
21. Zhenglong, L., and Qi, Z. (2018). Simulation and Experiment Research on Squeeze Casting Combined With Forging of Automobile Control Arm. Volume 2: Advanced Manufacturing.
22. Butola, R., Murtaza, Q., and Singari, R. (2021). An experimental and simulation validation of residual stress measurement for manufacturing of friction stir processing tool.

28 Thermal analysis of a steam turbine unit of namrup thermal power plant, Assam, India

Kankan Kishore Pathak[1,a], Asis Giri[2,b], Rwittik Barkataki[1,c], Mujakkir A. Saikia[1,d], and Surajit Mahanta[1,e]

[1]Department of Mechanical Engineering, Girijananda Chowdhury Institute of Management and Technology, Guwahati, India

[2]Department of Mechanical Engineering, North Eastern Regional Institute of Science and Technology, Itanagar, India

Abstract: A thermal analysis has been done to rate the efficacy of a steam turbine unit of Namrup thermal power plant situated in Assam, India. Modeling of each component and a detailed worksheet of energetic and exergetic losses are presented. The largest energetic losses are associated with the condenser which alone covers 56.31% of entire cycle loss. Again, the greatest losses of exergy take place in the boiler system covering almost 66.49% of the entire exergy loss in the plant. The energetic and exergetic efficiencies are found to be 20.65% and 23.87% respectively for the chosen plant which are relatively lower for a standard steam powered cycle. Irrespective of the dead states temperature, a large share of the exergy is destructed in the boiler of the plant. A slight increment of exergy losses in the boiler with the rise in the dead state temperature is also observed.

Keywords: Energetic efficiency, energetic loss, exergetic efficiency, exergetic loss, steam turbine.

Introduction

Since the dawn of industrialization, steam has been a widely used method of transferring energy for power generation as well as in process industries like refineries, petrochemicals, etc. Majority of the steam produced originates from fossil fuels such as natural/ coal gas. By assessing performances of thermal power plant, it becomes easy to pinpoint components where energy conversion is poor and where improvement is required, thereby increasing the efficiency; minimizing operating expenses and increasing profitability of industry. Electricity generation in the most parts of the world deeply depend on fossil fuels, especially natural gas and coal. Though the clean energy technologies and resources are rapidly growing; the status of these technologies has not advanced to a phase where they can minimize the fossil fuel dependency. Thus, efficient utilization of the fossil fuels is important and desirable. For resolving the aforesaid complexities, there exist two indispensable techniques accessible: first/second laws analysis (energy/ exergy analysis). Second law based exergy examination has recently been a salient point in furnishing a better knowledge of the procedure, to count the inefficiency sources and to extricate the standard of energy consumed. Some of the published literature related to

[a]kankankishore@gmail.com, kankan_me@gimt-guwahati.ac.in, [b]Asisgiri1467@gmail.com,
[c]Barkataki.rwittik@gmail.com, [d]Saikiamujakkir@gmail.com, [e]Surjitm@gmail.com

the thermal performance analysis of power plants has been briefly discussed as follows: Habib and Zubair [1] have carried out a thermal analysis of reheat regenerative Rankine power plant cycle based on the examination of second law. This report describes the irreversible losses in a reheat-regenerative Rankine power plant cycle and to compare the cycles with closed and open feed water heaters. Dincer and Al-Muslim [2] have examined the consequence of different operational conditions on the energy/exergy efficiencies evaluated through changing the boiler pressure, boiler temperature, mass fraction of the steam and work output. A study based on exergy analysis of a coal based 210 MW thermal power plant under varying working states has been performed by Sengupta et al. [3]. At all loads, the results of the investigation elucidate that almost 60% of exergy demolition occurs in the boiler. The report of Ameri et al. [4] has found that major share of loss of energy takes place in the condenser (70.5%) whereas the second major energy loss occurs in the boiler (15.5%). However, the maximum destruction of exergy takes place in the boiler (81%). Aljundi [5] demonstrated an energy and exergy based thermal analysis of a steam power plant in Jordan. The primary exergy demolition occurs in the boiler system is reported. Mitrovic et al. [6], after performing a first law (energy)/second law (exergy) analysis of a Serbian steam power plant, reports that major portion of the energy is consumed in the condenser. Energetic and exergetic efficiencies were found out to be 37% and 36% respectively. A comprehensive survey on energy-exergy analyses of thermal power producing plants is carried out by Kaushik et al. [7]. The report of Ahmadi and Toghraie [8] examines the energy-exergy performance of Iran's Montazeri steam plant facility. In this article, it is informed that the major heat loss (69.8%) in the chosen thermodynamic cycle occurs in the condenser while analysis of exergy reported that the highest exergy loss occurs in the boiler (85.66%).

It may be possible that in the calculation of the thermal efficiency there exist indispensable dissimilarities between the results available in the literature and the corporations' data as corporations have developed power plants using gross power of electricity instead of net power. In fact, this subject needs to be verified very minutely so that in the near future the thermal power plants may obtain maximum benefits. In this report an attempt has been made to examine a steam turbine unit of Namrup thermal power plant situated in Assam, North-Eastern region of India through energy and exergy approach and recommendations are provided based on the results obtained. Description of the plant is described next.

Description of the Plant

The schematic diagram of 30 MW steam turbine units of Namrup thermal power station (NTPS) is shown in the Figure 28.1. Steam from the gas fired boiler at high pressure (58.62 bar) and temperature (482°C) enters the turbine. The properties of the natural gas used in the NTPS's steam turbine unit and the detailed working conditions are shown in Tables 28.1 and 28.2 respectively.

Analysis

Energy Analysis

For each standard volume, conservation of mass and energy equations are applied, and the status of each component in terms of energy efficiency is calculated, after that the values of uncertain parameters are evaluated through EES (Engineering Equation Solver) software.

Table 28.1: Properties of natural gas used in the plant.

Various Properties	Value	Units
Density at 15°C	0.712	kg/m³
Total Sulfur	4.28	Wt%
FlashPoint	221	°C
Pour Point	−57 to 32	°C
Ignition Point	537	°C
Boiling Point	−164	°C
Melting Point	−182	°C
Kinematic viscositv at 100°C	25.89	cSt
Gross Calorific Value	52.3	MJ/kg
Net Calorific Value	47.3	MJ/kg

Table 28.2: Operating conditions of the plan.

Operating Conditions	Value	Units
Power Produced	30	MW
Mass flow rate of fuel	9	tons/hr
Steam flow rate, main line	150	tons/hr
Steam temperature, main line	482	°C
Steam pressure, main line	58.62	bar
Feed water temperature to boiler	182	°C
Furnace temperature	1840	°C
Mass flow rate of cooling water to condenser	6895	tons/hr
Number of Induced Draft and Forced Draft fans	4	-
Combined pump/motor efficiency		80

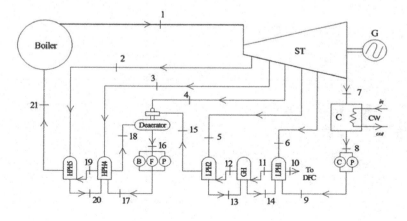

Figure 28.1 Schematic diagram of the steam turbine unit of NTPS.

The generalized equation for energy balance applied to any control volume may be given as follows (Kaushik et al. [7])

$$\sum \dot{Q}_k + \sum \dot{m}\left(h_i + \frac{c_i^2}{2} + gZ_i\right) = \sum \dot{m}\left(h_e + \frac{c_e^2}{2} + gZ_e\right) + \sum \dot{W} \tag{1}$$

For any control volume in steady state with minimal kinetic/potential energy fluctuations, the energy as well as mass balances may be represented as follows (Aljundi [5])

$$\sum \dot{m}_i = \sum \dot{m}_e \tag{2}$$

$$\dot{Q} - \dot{W} = \sum \dot{m}_e h_e - \sum \dot{m}_i h_i \tag{3}$$

It should be mentioned that the energy produced by fuel is computed (Aljundi [5]) as

$$\dot{Q} = \dot{m}_f \times LHV_f \tag{4}$$

Again, the pump power input is estimated following [5] as $\dot{W}_{pump} = \dfrac{\dot{m}(h_{e,s} - h_i)}{\eta_{combined}}$, where, $\eta_{combined} = 0.80$, is the motor/pump combined efficiency. To obtain an operation of steady state and selecting each unit of Figure 28.1 as a control volume, the energy balance/energy efficiency equations are shown in the Table 28.3.

Table 28.3: Energy balance and efficiency for each component of the plant.

Component	Energy Balance	Energy Efficiency
Boiler	$\dot{m}_1 h_1 + \dot{m}_{25} h_{25} = \dot{m}_2 h_2 + \dot{Q}l_{boiler}$	$\eta_{boiler} = \dfrac{\dot{m}_2 h_2 - \dot{m}_1 h_1}{\dot{m}_f h_f}$
Steam Turbine	$\dot{W}_{st} = \dot{m}_1(h_1 - h_2) + (\dot{m}_1 - \dot{m}_2)(h_2 - h_3) + (\dot{m}_1 - \dot{m}_2 - \dot{m}_3)(h_3 - h_4) +$ $(\dot{m}_1 - \dot{m}_2 - \dot{m}_3 - \dot{m}_4)(h_4 - h_5) + (\dot{m}_1 - \dot{m}_2 - \dot{m}_3 - \dot{m}_4 - \dot{m}_5)(h_5 - h_6) +$ $(\dot{m}_1 - \dot{m}_2 - \dot{m}_3 - \dot{m}_4 - \dot{m}_5 - \dot{m}_6)(h_6 - h_7) - \dot{Q}l_{st}$	$\eta_{st} = \dfrac{\dot{W}_{st}}{\dot{m}_1 h_1 - (\dot{m}_2 h_2 + \dot{m}_3 h_3 + \dot{m}_4 h_4 + \dot{m}_5 h_5 + \dot{m}_6 h_6 + \dot{m}_7 h_7 + \dot{m}_8 h_8)}$
Condenser	$\dot{Q}l_c = \dot{m}_7 h_7 - \dot{m}_8 h_8$	$\eta_{cd} = \dfrac{\dot{m}_8 h_8}{\dot{m}_7 h_7}$
Deaerator	$\dot{Q}l_{Da} = \dot{m}_{11} h_{11} + \dot{m}_{12} h_{12} + \dot{m}_{13} h_{13} - \dot{m}_{14} h_{14}$	$\eta_{Dea} = \dfrac{\dot{m}_{14} h_{14}}{\dot{m}_{11} h_{11} + \dot{m}_{12} h_{12} + \dot{m}_{13} h_{13}}$
Condensate Pump	$\dot{Q}l_{cp} = \dot{m}_8 h_8 - \dot{m}_9 h_9 + \dot{W}_{cp}$	-----
Boiler Feed Pump	$\dot{Q}l_{BFP} = \dot{m}_{14} h_{14} - \dot{m}_{15} h_{15} + \dot{W}_{BFP}$	-----
Low Pressure Heater (LPH1)	$\dot{Q}l_{LPH1} = \dot{m}_9 h_9 + \dot{m}_6 h_6 + \dot{m}_{21} h_{21} - \dot{m}_{10} h_{10} - \dot{m}_{22} h_{22}$	-----
Low Pressure Heater (LPH2)	$\dot{Q}l_{LPH2} = \dot{m}_{10} h_{10} + \dot{m}_5 h_5 - \dot{m}_{11} h_{11} - \dot{m}_{21} h_{21}$	-----
Gland Heater (GH)	$\dot{Q}l_{GH} = \dot{m}_{13} h_{13} + \dot{m}_{23} h_{23} - \dot{m}_{12} h_{12} - \dot{m}_{24} h_{24}$	-----
High Pressure Heater (HPH4)	$\dot{Q}l_{HPH4} = \dot{m}_{16} h_{16} + \dot{m}_3 h_3 + \dot{m}_{19} h_{19} - \dot{m}_{17} h_{17} - \dot{m}_{18} h_{18}$	-----
High Pressure Heater (HPH5)	$\dot{Q}l_{HPH5} = \dot{m}_{15} h_{15} + \dot{m}_2 h_2 - \dot{m}_{16} h_{16} - \dot{m}_{19} h_{19}$	-----

Exergy Analysis

Exergy analysis, which determines the level of the destruction of exergy, identifies the location, extent and causes of thermal inefficiencies within a thermal system [2]. Exergy analysis comprises of thermal exergy, chemical exergy and exergy associated with kinetic and potential energy of a system. In this study, only thermal exergy associated with the components of the power plant have been taken into account. The status of the all the components is evaluated in respect of exergy efficiency by applying exergy equations for each standard control volume obtained from the diagram shown in the Figure 28.1. Following is the general equation for exergy balance for any control volume [7].

$$\sum \left(1 - \frac{T_0}{T}\right) \dot{Q}_k + \sum (\dot{m}_i \psi_i) = \sum \psi_w + \sum (\dot{m}_e \psi_e) + \dot{I}_{destroyed} \qquad (5)$$

Exergy balance for any control volume in steady state with minor potential/kinetic energy fluctuations, may be given (Aljundi [5]) as

$$\dot{X}_{heat} - \dot{W} = \sum (\dot{m}_e \psi_e) - \sum (\dot{m}_i \psi_i) + \dot{I}_{destroyed} \qquad (6)$$

Here, the net transfer of exergy by heat ($\dot{X}_{heat}$) at temperature T is given by

$$\dot{X}_{heat} = \sum \left(1 - \frac{T_0}{T}\right) \dot{Q} \qquad (7)$$

where subscript '0' denotes the dead state condition. The specific exergy is given by

$$\psi = h - h_0 - T_0(s - s_0) \qquad (8)$$

The overall rate of exergy related to a fluid flow thus becomes

$$\dot{X} = \dot{m}\psi = \dot{m}[h - h_0 - T_0(s - s_0)] \qquad (9)$$

For a steady-state process, and picking each component on the schematic diagram of steam turbine unit of NTPS as a control volume, as shown in Figure 28.1, their rate of exergy destruction and exergy efficiency equations are written below.

To compute fuel specific exergy, the following equation is used (Kotas [9]):
$\psi_f = \xi_f \times LHV_f$, where $\xi_f = 1.03$, exergy factor ξ_f is based on the fuel's lower heating value.

Results and Discussion

Thermal study of a 30 MW steam turbine unit of Namrup Thermal Power Station (Assam, India) has been performed through energy-exergy approach. 1.01325 bar and 298.15 K are chosen as the reference environment condition of pressure and temperature for the analysis. With this initial assumption, specific exergy and exergy associated with the flowing fluid stream for all the indicated nodes as shown in the Figure 28.1 are evaluated and listed in Table 28.5. The extent of heat loss, heat drop in percentage and first law efficiency for all other important equipment are reported in Table 28.6. From this table, it is observed that the thermal efficiency for the thermodynamic cycle is 20.65 %, which is relatively lower than the efficiency of the modern power producing thermodynamic cycles. In the

Table 28.4: Exergy balance and efficiency for each component of the plant.

Component	Exergy Balance	Exergy Efficiency
Boiler	$I_{Boiler} = X_f + X_{21} - X$	$\eta_{x,Boiler} = \dfrac{X - X_{21}}{X_f}$
Steam Turbine	$I_{ST} = X_1 - (X_2 + X_3 + X_4 + X_5 + X_6 + X_7) - W_{ST}$	$\eta_{x,ST} = \dfrac{W_{ST}}{X_1 - (X_2 + X_3 + X_4 + X_5 + X_6 + X_7)}$
Condenser	$I_c = X_9 - X_8$	$\eta_{x,c} = \dfrac{X_8}{X_9}$
Deaerator	$I_{DA} = X_4 + X_{15} + X_{18} - X_{16}$	$\eta_{x,DA} = \dfrac{X_{16}}{X_4 + X_{15} + X_{18}}$
Condensate Pump	$I_{CP} = X_8 - X_9 + W_{CP}$	$\eta_{x,CP} = \dfrac{X_9 - X_8}{W_{CP}}$
Boiler Feed Pump	$I_{BFP} = X_{16} - X_{17} + W_{BFP}$	$\eta_{x,BFP} = \dfrac{X_{17} - X_{16}}{W_{BFP}}$
Low Pressure Heater (LPH1)	$I_{LPH1} = X_6 + X_9 + X_{14} - X_{15} - X_{11}$	$\eta_{x,LPH1} = \dfrac{X_{15} + X_{11}}{X_6 + X_9 + X_{14}}$
Low Pressure Heater (LPH2)	$I_{LPH2} = X_5 + X_{12} - X_{13} - X_{14}$	$\eta_{x,LPH2} = \dfrac{X_{13} + X_{14}}{X_5 + X_{12}}$
Gland Heater (GH)	$I_{GH} = X_{10} + X_{13} - X_{12} - X_{14}$	$\eta_{x,GH} = \dfrac{X_{12} + X_{14}}{X_{10} + X_{13}}$
High Pressure Heater (HPH4)	$I_{HPH4} = X_3 + X_{17} + X_{19} - X_9 - X_{19}$	$\eta_{x,HPH4} = \dfrac{X_{18} + X_{19}}{X_3 + X_{17} + X_{19}}$
High Pressure Heater (HPH5)	$I_{HPH5} = X_2 + X_{19} - X_{20} - X_{21}$	$\eta_{x,HPH5} = \dfrac{X_{20} + X_{21}}{X_2 + X_{19}}$

Table 28.7, analysis of exergy for the overall thermodynamic cycle and the constituent components are reported. One of the major causes of high loss of exergy in the boiler is the process of combustion. Entropy generation is what causes actual loss in the boiler. Unlike the first law analysis, it is observed that substantial improvements may be possible in the system of boiler than that in condenser. The cycle's exergy efficiency is calculated to be 23 % which is lower than that of the appropriate efficiency for a steam cycle which exemplifies that lots of opportunities are available for improvement.

In Figure 28.2a, the variation of exergy losses or exergy destruction with the changes in the reference environment temperature is shown for the major components. From Figure 28.2a, it is clear that irrespective of the dead state's temperature, the chief part of the exergy is destructed/lost in the boiler of the plant. With increasing dead state temperature, there is a minor rise in boiler exergy losses. The pattern of the fluctuation of exergy losses with respect to reference dead state temperature associated with the steam turbine is comparable to that of the boiler, although the magnitude is significantly smaller. The exergy destruction/loss associated with the condenser shows there is a slight decrease of the losses with the rise in dead state temperature. Usually, the higher is the temperature difference with the reference environment; the lower is the effect of the environment temperature on the exergetic efficiency. The influence of reference environment temperature on the exergetic efficiency is shown in the Figure 28.2b. It is evident from Figure 28.2b that the exergetic efficiency of the boiler system and steam turbine do not alter notably with the change in the reference temperature of the environment but its effect is evidently observable in the case of the condenser.

Again, if concentration is now given to the Figure 28.3a–b, it can be seen that increasing the steam temperature boiler exergy destruction decreases (while exergy efficiency

Table 28.5: Thectnodynamic properties at each state point as shown in the Figure 28.1.

Node	T(°C)	P (bar)	ṁ(kg/s)	h (kJ/kg)	s (kJ/kgK)	ψ(kJ/kg)	X(MW)
1	482	58.615	41.5	3379.72	5.9010	1625.727	67.467
2	206	17.559	4.5	2795.60	6.2865	926.728	4.17
3	181	10.398	3.8	2778.59	6.5714	824.828	3.134
4	154	5.297	4.2	2750.05	6.8011	727.827	4.366
5	135	3.041	3.2	2726.99	6.9870	649.369	2.077
6	86	0.588	2.8	2652.99	7.5379	411.201	1.151
7	50	0.114	23	2593.35	7.9012	243.298	5.595
8	47	0.114	23	210.90	0.6591	18.990	0.436
9	53	8.295	23	226.51	0.6904	25.276	0.581
10	45	0.515	6	205.81	0.6386	13.705	0.082
11	70	0.32	23	293.66	0.9549	20.005	0.46
12	82	0.781	23	343.30	1.0990	20.305	0.467
13	90	0.824	3.2	377.48	1.1926	26.590	0.085
14	62	0.413	3.2	259.90	0.8562	21.258	0.068
15	115	1.569		472.82	1.4735		0.879
16	133	4.821	41.5	634.27	1.6654	142.486	5.913
17	135.8	92.426	41.5	652.32	1.6867	154.186	6.398
18	138	6.311	8.3	671.20	1.7182	163.628	1.358
19	154	83.21	41.5	734.50	1.8736	180.672	7.497
20	158	6.341	41.5	758.34	1.9227	182.881	0.809
21	182	80.442	41.5	874.18	2.1493	238.194	9.885
Dead State	25	1.013	-	104.92	0.3672	-	-

Table 28.6: Heat loss and energy efficiency of the power plant components.

Component	Heat Loss, Ql(MW)	Percentage Ratio	Energy Efficiency, η_z (%)
Boiler	14.64	14.78	87.65
Steam Turbine	4.677	4.72	84.18
Condenser	55.796	56.31	8.1
Condensate Pump	0.128	0.13	-
Boiler Feed Pump	0.186	0.19	-
Deaerator	6.041	6.10	81.33
Heater-	17.62	17.77	-
Cycle	99.082	100	20.65

Table 28.7: Exergy destruction rate and exergy efficiency of the power plant components.

Component	Energy Destruction Rate, I(MW)	Percentage Ratio	Energy Efficiency, η_{zz}(%)
Boiler	64.598	66.49	47.12
Steam Turbine	17.490	18	62.75
Condenser	5.159	5.31	7.79
Condensate Pump	0.343	0.35	29.71
Boiler Feed Pump	0.454	0.47	51.81
Deaerator	0.790	0.82	88.21
LPH1	1.258	1.29	30.11
LPH2	1.580	1.63	37.89
Gland Heater	0.010	0.01	98.16
HPH4	4.486	4.62	83.63
HPH5	0.973	1.01	91.66
Cycle	97.141	100	23.87

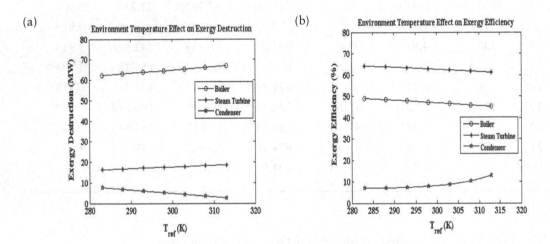

Figure 28.2 Dependence of reference environment temperature of the main plant components on (a) rate of total exergy destruction; (b) exergy efficiency.

increases) but turbine exergy destruction increases (while exergy efficiency decreases) equally. Therefore, not appreciable gain is reported increasing the temperature of the steam from 470°C to 520°C. It may also be understood that more the exergy efficiency more is the system sustainability due to the superior energy quality. On the contrary, lower the exergy efficiencies lower is the sustain ability score due to the energy losses and internal irreversible reactions. It is also noted that the energy-exergy losses and their efficiencies for all the power devices varies as their definitions and designs are different. To get a comparative understanding of the energy-exergy losses and energy-exergy efficiencies of different power equipment available in the chosen plant, Figure 28.4a–b are drawn. One may get a lucid picture of this analysis from these figures.

(a)

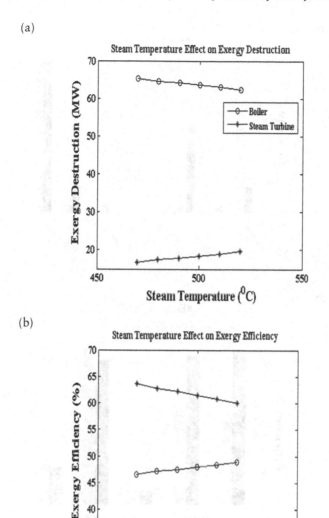

(b)

Figure 28.3 Dependence of steam temperature of the main plant components on (a) rate of total exergy destruction; (b) exergy efficiency.

Conclusions and Recommendations

In this study, an attempt has been made to investigate the thermal performance of a steam turbine unit of Namrup Thermal Power Station (NTPS). Heat loss, exergy destruction, energy-exergy efficiencies are evaluated for the chief equipment of the plant. Moreover, the effect of change in environmental reference temperature and steam temperature on exergy destruction and efficiency is reported. Based on the investigations, the condenser loses the most energy to the environment, losing 56.31% of the energy that is fed into it. The calculated thermal efficiency for the plant is about 20.65%, which is relatively lower than the

(a)

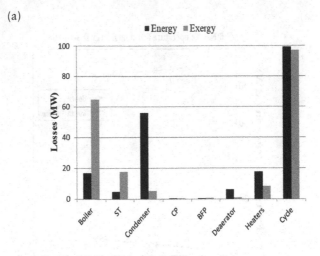

(b)

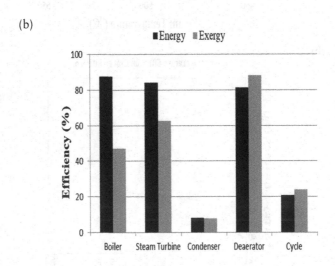

Figure 28.4 (a) The energy and exergy losses of power plant cycle and main components; (b) energy and exergy efficiencies of power plant cycle and main components.

efficiency of the modern power plant cycles. The boiler system is reported to be the source of major exergy destruction, where 66.49% of the input fuel exergy is destroyed. The efficiency of exergy for the present thermodynamic cycle is calculated to be 23 % which is also lower than that of the appropriate efficiency for a steam cycle which exemplifies the scope for improvement. The cycle performance may be enriched through proper preheating the air used for combustion, producing appropriate fuel-air mixture, minimizing the variation in temperature between the steam at each sage and combustion products involved in heat transfer process, properly isolating the boiler parts, etc., which may eventually be effective to control exergy losses.

Acknowledgement

The authors are very much thankful to the management and staff of Assam Power Generation Corporation Limited (APGCL) working in Namrup Thermal Power Station (NTPS), Assam (India) for their assistance and helpful suggestions.

References

1. Habib, M. A., and Zubair, S. M. (1992). Second-law based thermodynamic analysis of regenerative-reheat-rankine cycle power plants. *Energy*, 17(3), 295–301. https://doi.org/10.1016/0360-5442(92)90057-7.
2. Dincer, I., and Al-Muslim, H. (2001). Thermodynamic analysis of reheat steam cycle power plant. *International Journal of Energy Research*, 25, 727–739. https://doi.org/10.1002/er.717.
3. Sengupta, S., Datta, A., and Duttagupta, S. (2006). Exergy analysis of a coal based 210 MW thermal power plant. *International Journal of Energy Research*, 31(1), 14–28. https://doi.org/10.1002/er.1224.
4. Ameri, M., Ahmadi, P., and Hamidi, A. (2008). Energy, exergy and exergoeconomic analysis of a steam power plant: a case study. *International Journal of Energy Research*, 33, 499–512. https://doi.org/10.1002/er.1495.
5. Aljundi, I. (2009). Energy and exergy analysis of steam power plant in Jordan. *Applied Thermal Engineering*, 29, 324–328. https://doi.org/10.1016/j.applthermaleng.2008.02.
6. Mitrovic, D., Zivkovic, D., and Lakovic, M. S. (2010). Energy and exergy analysis of 348.5 MW steam power plant. *Energy Sources, Part A*, 32, 1016–1027. https://doi.org/10.1080/15567030903097012.
7. Kaushik, S. C., Reddy, V. S., and Tyagi, S. K. (2011). Energy and exergy analyses of thermal power plants: a review. *Renewable and Sustainable Energy Reviews*, 15, 1857–1872. https://doi.org/10.1016/j.rser.2010.12.007.
8. Ahmadi, G., and Toghraic, D. (2015). Energy and exergy analysis of montazeri steam power plant in Iran. *Renewable and Sustainable Energy Reviews*, 56, 454–463. https://doi.org/10.1016/j.rser.2015.11.074.
9. Kotas, T. J. (1985). The Exergy Method of Thermal Plant Analysis. London: Butterworth. https://doi.org/10.1016/C2013-0-00894-8.

29 Continuous helical baffle heat exchanger: an experimental investigation

Vivek S.[1,a], Vishnu V.[2], Dipin Kumar R.[3], and Jithu J.[3]

[1]Department of Mechanical Engineering, Indian Institute of Technology, Guwahati, India

[2]Mechanical Engineer Project, Rubfila International Ltd., Dindigul, India

[3]Department of Mechanical Engineering, NSS College of Engineering, Palakkad, India

Abstract: Nowadays, shell and tube heat exchangers are used in automobiles and aerospace applications as it is best suited for applications with high pressure. The use of baffles in heat exchangers helps in supporting the tubes and helps to prevent vibrations from flow induced eddies. Heat transfer is also improved through the proper arrangement of baffles. In this work, a working model of a continuous helical baffle heat exchanger (helixchanger) was made and the effect of mass flow rates (MFR) of each fluid (water and calcium chloride brine solution) on the overall heat transfer coefficient (OHTC) was studied. In both cases, the overall coefficient of heat transfer steadily increased with the flow rate of each of the fluids. After a certain value of MFR, the OHTC tends towards constancy in the case of calcium chloride brine solution as the hot fluid. The pressure drop in the shell increases continuously with increment in the MFR of the cold fluid.

Keywords: Calcium chloride brine solution, helical baffles, helix changer, OHTC.

Introduction

Heat exchangers are a vital part of the operations of many systems. They find a variety of applications in petroleum refineries, chemical plants, natural gas processing, refrigeration, air conditioning, and automotive applications. Heat is efficiently transferred over a solid surface from one fluid to another. As a result, the temperatures of both the fluids as well as the solid surface change over the length of the heat exchanger.

The most commonly used type of heat exchanger is the shell and tube heat exchanger consisting of a cylindrical shell that encloses tubes with one fluid flowing through the tubes and the other fluid outside the tubes. Shell and tube heat exchangers provide high heat transfer efficiency due to their high surface area-to-volume ratios. Heat exchangers with segmental baffles have low coefficient of heat transfer. They pass through the heat transfer surface causing high leakage flow. This along with a high pressure drop increases the pumping costs in industries. The use of helical baffles is a preferred alternative to overcome these limitations.

The helical baffle heat exchanger offers numerous advantages over other types of heat exchangers. It is very effective when the heat exchanger is subjected to vibrations.

[a]vivekskutty10@gmail.com

The flow of the shell side fluid is guided through a helical path by the baffles over the tube bundle. Near plug flow conditions are produced due to a reduction in flow dispersion. Other advantages of helixchangers are enhancement in the rate of heat transfer, decreased bypass effects, reduction in fouling factor at the shell, less vibrations due to flow, low pumping cost, etc.

According to the experimental study by Wen et al. (2015) [1], the usage of ladder type fold baffles in the conventional Helical baffle heat exchanger blocks the triangular leakage zone. Their experimental result shows that the temperature and shell side velocity distribution is more uniform and the circular leakage is eliminated in the ladder type fold baffle heat exchanger. Zhang et al. (2009) [2] observed that for the same volume flow rate, the shell side heat transfer coefficient for the STHXsHB is lower than that of STHXsSB. According to their experiment, for a unit pressure drop, the helixchanger has a higher heat transfer coefficient for the same volume flow rate. And among the four STHXsHB, the one with a 40° helix angle shows the best performance. The experimental analysis by Gao et al. (2015) [3] showed that the pressure drop rises with the increase in volume flow rate at the shell side for all the STHXsHB and the increase is quite evident in the one with a small helix angle. Also with increasing volume flow rate, the shell side coefficient of heat transfer increases and is inversely proportional to the baffle helix angle. However, according to Jayachandriah and Vinay Kumar [4], the effect of pressure drop is minimal when the helix angle is greater than 18°.

Jian et al. [5] conducted experimental analysis with different spiral wound heat exchangers and concluded that the increase in heat transfer coefficient with the decrease in winding angle is due to the larger curvature of the tube which enhances the secondary flow. According to Rasheed et al. [6], helical microtube curvature swirls play a vital part in improving heat transfer. Maghrabie et al. [7] conducted an experimental study on a shell and helically coiled tube heat exchanger (SHCTHE) and found that the OHTC is higher for SHCTHE in the vertical direction by 10% than that of horizontal direction with coil Dean no 1540 and 6.2% more with coil Dean no 3860.

Wang et al. [8] observed that the shell side coefficient of heat transfer increased by 18.2–25.5% and the OHTC went up by 15.6–19.1% when sealers are installed on the shell side. This was due to the blockage in the gap between the baffle and the shell which resulted in the decrease of circular leakage. Jamshidi et al. [9] conducted an experimental analysis to improve the heat transfer in shell and helical tube heat exchanger. They concluded that the maximum heat transfer coefficient occurs at higher coil diameter, coil pitch, and MFR. Bartuli et al. [10] did a trial with shell and tube type polymeric hollow fibre heat exchanger (PHFHE). They found that the overall coefficient of heat transfer with the same shell water flow velocity is 6 times more in the one with cross wound PHFHE than the one with parallel fibres.

Jayachandriah and Vinay Kumar (2015) [4] designed Shell and tube two-pass heat exchanger with a helical baffle and compared with a segmental baffle heat exchanger using the Kern method. They found that the heat transfer coefficient is much better in Heat exchanger with helical baffles than that of segmented baffles. Also, the increment in tube inlet velocity resulted in the enhancement in the coefficient of heat transfer. An experimental study by Peng et al (2007) [11] shows that at a higher shell side flow rate, the heat exchanger with a side-in-side-out shell design has a 2% higher heat transfer coefficient than that with a middle-in-middle-out shell design. The Helical baffle heat exchanger with a side-in-side-out shell design has a 10% more heat transfer coefficient than the segmental baffle heat exchanger. The design and analysis of HE with different baffles by Gireesh and

Rao (2017) [12] showed that among the normal segmental baffle, inclined baffle, and helical baffle, the one with helical baffle gave effective pressure drop.

Kumar and Arasu [13] conducted an experimental investigation using nanocomposite fluid (Ti O2-Ag/distilled water) and found that the rate of heat transfer for the nanocomposite fluid is more than that of water by 18.4%. The study done by Vishwakarma and Jain (2013) [14] showed the advantages of helixchanger over the segmented baffle heat exchanger. The advantages are low bypass effect, lesser fouling in the shell side, prevention of vibration due to flow, and lower maintenance. The experimental analysis by Gowthaman and Sathish (2014) [15] indicated that the helixchanger reduces pumping cost and weight fouling. The lesser mass flux throughout the shell due to the higher cross flow area resulted in a lower pressure drop in the Helical baffle than that of the segmental baffle. Zhang et al. [16] also validated that the pressure drop is minimal in helixchanger than in the segmental baffle heat exchanger. The experimental investigation of shell and tube heat exchanger by Wang et al. (2011) [17] showed that the Nusselt number is 50% more for Shell and tube heat exchanger with flower baffles than that with segmental baffles for the same Reynolds number. And the pressure drop of the one with the flower baffle is just 30% of the segmental baffle.

In the present work, a helical baffle heat exchanger was fabricated. Experiments were performed on it to study the effect of the flow rate of each of the fluids on the OHTC. Experiments were performed for two hot fluids namely water and calcium chloride brine solution. The paper is divided into the following sections namely helixchanger design, experimental setup, results and discussions, conclusions, and future scope.

Design of Helixchanger

Table 29.1 gives the fluid properties and the tube material properties.

Table 29.1: Properties of the working fluid and the tube material.

Properties	Water	Calcium chloride brine	Copper
Density, kg/m^3	1000	1129.959	8978
Dynamic viscosity (hot), kg/ms	0.00051	0.0014	
Dynamic viscosity (cold), kg/ms	0.00083		
Thermal conductivity, W/mK	0.6	0.60031	387.6
Specific heat, J/kgK	4182	3378.102	381

Table 29.2 shows the measured and calculated parameters. The geometrical parameters of the heat exchanger were calculated using the equations given below [4]. The values are given in Table 29.3.

$$\text{Tube clearance, } C = P_t - d_o \tag{1}$$

$$\text{Baffle spacing, } L_b = \pi * D_i * \tan \Phi \tag{2}$$

$$\text{Cross flow area, } A_s = \frac{D_i C L_b}{P_t} \tag{3}$$

Equivalent diameter, $D_E = 4\left[\dfrac{P_t^2\sqrt[3]{3}}{4} - \dfrac{\pi d_o^2}{8}\right] / \left[\dfrac{\pi d_o}{2}\right]$ \hfill (4)

Maximum velocity, $v_{max} = \dfrac{m_c}{\rho_s A_s}$ \hfill (5)

Reynold's number, $R_{es} = \dfrac{v_{max} D_E}{\vartheta_S}$ \hfill (6)

Heat transfer coefficient, $h_o = \dfrac{0.36 * k_s * R_e^{0.55} * P_r^{1/3}}{D_E}$ \hfill (7)

Number of baffles, $N_b = \dfrac{L}{L_b + \Delta_{BT}}$ \hfill (8)

Velocity of tube fluid, $v_t = \dfrac{m_h}{\rho_{t} A_t}$ \hfill (9)

Reynolds's number, $R_{et} = \dfrac{\rho_t v_t d_i}{\mu_t}$ \hfill (10)

Prandtl number, $P_{rt} = \dfrac{\mu_t C_t}{k_t}$ \hfill (11)

Nusselt number, $N_u = 0.023 * R_{et}^{0.8} * P_{rt}^{0.3}$ \hfill (12)

Heat transfer coefficient, $h_i = \dfrac{N_u k_t}{d_i}$ \hfill (13)

OHTC, $U_o = \left[\dfrac{1}{h_o} + \dfrac{d_o}{d_i h_i} + \dfrac{d_o}{2k}\ln\left(\dfrac{d_o}{d_i}\right)\right]^{-1}$ \hfill (14)

Shell side pressure drop, $\Delta p = \dfrac{f_s G_s^2 (N_b + 1) D_i}{(2\rho_s D_E \phi_s)}$ \hfill (15)

where, $f_s = \exp(0.576 - 0.19 ln(Re_s))$ \hfill (16)

$G_s = \dfrac{m_s}{A_s}$ \hfill (17)

Table 29.2: Measured and calculated parameters.

Abbreviation	Name
m_c	MFR of cold fluid
m_h	MFR of hot fluid
v_{max}	Velocity of shell fluid
Re_s	Shell fluid Reynolds number
Pr_s	Shell fluid Prandtl number
h_o	Shell side convective coefficient of heat transfer
Re_t	Reynolds number of tube fluid
Pr_t	Prandtl number of tube fluid
Nu	Nusselt number
U_o	OHTC

Table 29.3: Dimensions and values of the parameters.

S. No.	Description	Notation	Value	Unit
1	Length of heat exchanger	L	600	mm
2	Inner diameter of shell	D_i	135	mm
3	Outer diameter of shell	D_o	140	mm
4	Tube inner diameter	d_i	25.4	mm
5	Tube outer diameter	d_o	31	mm
6	Triangular tube pitch	P_t	40	mm
7	Number of tubes	N_t	5	-
8	Baffle thickness	Δ_{BT}	1	mm
9	Helix angle	Φ	8	deg

Experimental setup

The experiment consists of a heater used to heat the hot fluid as shown in Figure 29.1. The fluids were forced to flow through the shell/tube using a pump from the tank. Copper tubes were used for hot fluid flow. Baffles were made of copper and the shell was made of steel. The MFR of one of the fluids was measured using a rotameter and that of the other fluid by noting the time taken to fill the tank.

Experiments were performed with water as the cold fluid flowing through the shell side of the helixchanger. Two different fluids namely water and calcium chloride brine solution were used as the hot fluid flowing through the tubes. Calcium chloride brine solution was chosen as it has a large heat capacity and thermal conductivity. The MFR of one of the fluids was kept constant (m_h=0.056 kg/s, m_c=0.023 kg/s for water and m_h=0.0157 kg/s, m_c=0.0173 kg/s for calcium chloride brine solution), and values of varied MFR of the other fluid. The properties of calcium chloride brine solution at 15% weight concentration, 900F are chosen for the analysis as it suits the ambient temperature conditions.

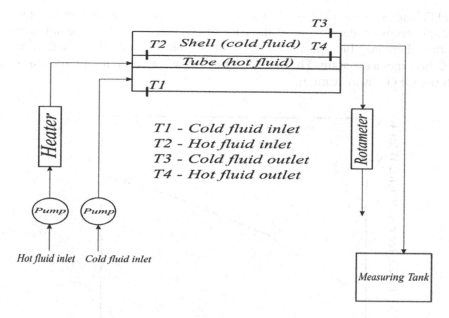

Figure 29.1 Schematic diagram of the experimental setup.

Results and Discussions

The OHTC rises with increment in MFR of each fluid for both water and calcium chloride brine solution. When the flow rate of the hot fluid is kept constant, the plot between OHTC and the MFR of the cold fluid has a small slope as shown in Figures 29.2(a) and 29.3(a). The slope of the plot between OHTC and the MFR of the hot fluid for a constant MFR of the cold fluid is comparatively higher as shown in Figures 29.2(b) and 29.3(b).

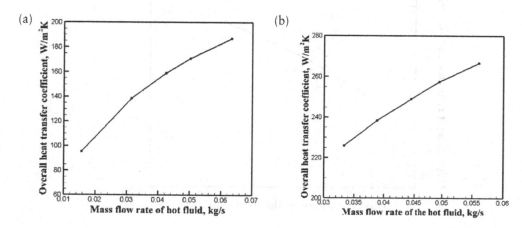

Figure 29.2 OHTC vs MFR for water as the hot fluid (a) Constant MFR of hot fluid (b) Constant MFR of cold fluid.

The OHTC increases almost linearly for water as the hot fluid as shown in Figure 29.2. But the slope keeps on decreasing in the case of calcium chloride brine solution which indicates that there may be a particular MFR of the hot fluid and the cold fluid at which the OHTC becomes a constant. This indicates that there may be a minimum value of MFR for which the OHTC is maximum.

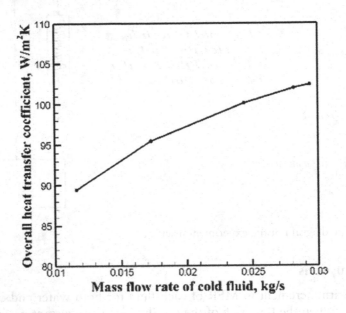

(a)

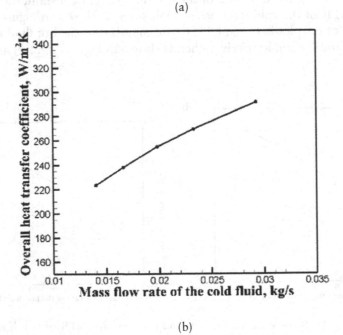

(b)

Figure 29.3 OHTC vs MFR for calcium chloride brine solution as the hot fluid (a) Constant MFR of hot fluid (b) Constant MFR of cold fluid.

Shell side pressure drop computed using the Kern method is plotted against the MFR of cold fluid as shown in Figure 29.4. It can be seen that the pressure drop increases with the MFR for both water and calcium chloride brine solution.

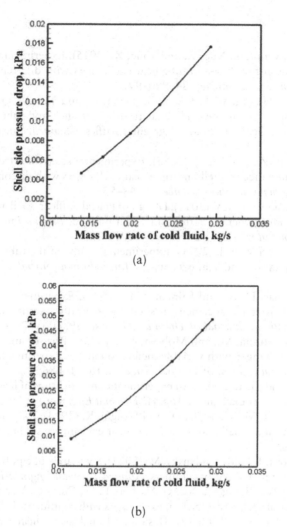

(a)

(b)

Figure 29.4 Shell side pressure drop vs MFR of the cold fluid (a) Water as the hot fluid (b) Calcium chloride brine as the hot fluid.

Conclusions and Scope for Future Work

The OHTC increases with an increase in the MFR of both fluids (one constant and the other varying). Another interesting result obtained from the experiment is that after a particular stage, the OHTC becomes a constant and doesn't increase further with the increase in MFR for calcium chloride brine solution as the hot fluid. Thus, it can be concluded that there is an optimum value for the MFR of both hot and cold fluids for which the OHTC is maximum. Pressure drop in the shell side of the heat exchanger also shows an increasing trend with an increase in the MFR of the cold fluid.

As we have observed that the OHTC tends towards constancy with the increase in the MFR of each of the fluids in the case of calcium chloride brine solution, the optimum MFR at various concentrations can be studied. Computational Fluid Dynamics tools can be used to analyze the problem at hand and then validated using experiments.

References

1. Wen, J., Yang, H., Wang, S., Xue, Y., and Tong, X. (2015). Experimental investigation on performance comparison for shell-and-tube heat exchangers with different baffles. *International Journal of Heat and Mass Transfer*, 84, 990–997.
2. Zhang, J. F., Li, B., Huang, W. J., Lei, Y. G., He, Y. L., and Tao, W. Q. (2009). Experimental performance comparison of shell-side heat transfer for shell-and-tube heat exchangers with middle-overlapped helical baffles and segmental baffles. *Chemical Engineering Science*, 64(8), 1643–1653.
3. Gao, B., Bi, Q., Nie, Z., and Wu, J. (2015). Experimental study of effects of baffle helix angle on shell-side performance of shell-and-tube heat exchangers with discontinuous helical baffles. *Experimental Thermal and Fluid Science*, 68, 48–57.
4. Jayachandriah, B., and Vinay, V. (2015). Design of helical baffle in shell and tube heat exchanger and comparing with segmental baffle using kern method. *International Journal of Emerging Technology in Computer Science & Electronics*, 13(2).
5. Jian, G., Wang, S., and Wen, J. (2021). Experimental study on the tube-side thermal-hydraulic performance of spiral wound heat exchangers. *International Journal of Thermal Sciences*, 159, 106618.
6. Rasheed, A. H., Alias, H. B., and Salman, S. D. (2021). Experimental and numerical investigations of heat transfer enhancement in shell and helically microtube heat exchanger using nanofluids. *International Journal of Thermal Sciences*, 159, 106547.
7. Maghrabie, H. M., Attalla, M., and Mohsen, A. A. (2021). Performance of a shell and helically coiled tube heat exchanger with variable inclination angle: Experimental study and sensitivity analysis. *International Journal of Thermal Sciences*, 164, 106869.
8. Wang, S., Wen, J., and Li, Y. (2009). An experimental investigation of heat transfer enhancement for a shell-and-tube heat exchanger. *Applied Thermal Engineering*, 29(11–12), 2433–2438.
9. Jamshidi, N., Farhadi, M., Ganji, D. D., and Sedighi, K. (2013). Experimental analysis of heat transfer enhancement in shell and helical tube heat exchangers. *Applied Thermal Engineering*, 51(1–2), 644–652.
10. Bartuli, E., Kůdelová, T., and Raudenský, M. (2021). Shell-and-tube polymeric hollow fiber heat exchangers with parallel and crossed fibers. *Applied Thermal Engineering*, 182, 116001.
11. Peng, B., Wang, Q. W., Zhang, C., Xie, G. N., Luo, L. Q., Chen, Q. Y., and Zeng, M. (2007). An experimental study of shell-and-tube heat exchangers with continuous helical baffles. 1425–1431.
12. Gireesh, D., and Rao B. B. J. (2017). Design and analysis of heat exchanger with different baffles. *International Journal and Magazine of Engineering, Technology, Management and Research*, 4(5), 625–636.
13. Kumar, D. D., and Arasu, A. V. (2021). Experimental investigation on dimensionless numbers and heat transfer in nanocomposite fluid shell and tube heat exchanger. *Journal of Thermal Analysis and Calorimetry*, 143, 1537–1553.
14. Vishwakarma, M., and Jain, K. K. (2013). Thermal analysis of helical baffle in heat exchanger. *International Journal of Science and Research (IJSR)*, 2(7).
15. Gowthaman, P. S., and Sathish, S. (2014). Analysis of segmental and helical baffle in shell and tube heat exchanger. *International Journal of Current Engineering and Technology*, 2, 625–628.
16. Zhang, J. F., He, Y. L., and Tao, W. Q. (2010). A design and rating method for shell-and-tube heat exchangers with helical baffles. *Journal of Heat Transfer*, 132(5), 051802.
17. Wang, Y., Liu, Z., Huang, S., Liu, W., and Li, W. (2011). Experimental investigation of shell-and-tube heat exchanger with a new type of baffles. *Heat and Mass Transfer*, 47(7), 833–839.

30 Decision making in a supply chain for optimization of transportation cost

Vamsikrishna A.[a], Manoj Adiga M., Anurag V Rao, and Divya Sharma S. G.[b]

Department of Mechanical Engineering, Amrita School of Engineering, Amrita Vishwa Vidyapeetham, India

Abstract: Decision making plays a critical role in every supply chain. A small mistake in decision making can lead to a drastic change in the functioning of the organization. In this paper, a supply chain of a chemical industry was considered. For reduction of the transportation cost, two systems with four and five warehouses were considered and cost evaluation was done in both the cases. Evaluation was done using Excel Solver and LPP algorithm. The new network proved to reduce the cost of transportation by Rs 8,36,780/month. A payback period of 1 year and a comparative payback period of 2 years 5 months were observed.

Keywords: Decision making, payback period, transportation cost.

Introduction

Supply chain engineering is a very complex discipline. A small change at one end can create a big issue on the other end. Hence correct decision-making plays an important role in any supply chain. Factors influencing the decisions may include cost, time, movement, efforts etc and may vary from decision to decision. Scott J. Mason et al. examined the total cost that can be gained, while performed by an integrated system [1]. Runhaar et al. explores how transport costs regulate freight transport demand in supply chains [2]. Mohamed M. Naim et al. justified transport flexibility into numerous types and then developed a framework for it [3]. Wilson et al. used system dynamics simulation to find the effect of transportation disruption [4]. Potter et al. used spreadsheet simulation to measure the impact on transport [5]. Rodrigues et al. established a supply chain model that transport operations [6]. Özcan et al. has found the solution to choose the best warehouse location among other alternatives by making use of the TOPSIS and ELECTRE methods [7]. Rao et al. found that when the number of cities are increased the solution obtained by GA method outperformed the NNA method [8]. Garvey, et al. have made use of Bayesian Network (BN) approach to evaluate the risk propagation in a supply chain network [9]. Rao et al. designed the shortest route to transport the material in using annealing technique [10]. Emeç et al. have generated a solution to the warehouse location problem in a stochastic

[a]vamsikrishna.ananth@gmail.com, [b]sg_divya@blr.amrita.edu

environment by making use of the SAHP method and fuzzy VIKOR approach [11]. Lee, et al. have generated multi-objective evolutionary algorithms (MOEAs) to find the solution to minimize the travel expenses for two storage assignments. [12]. Fartaj et al. have made use of the BMW-RSR framework to know the factors for the disruption of the transportation network [13]. Veeresh Kumar et al. used the vacuum assisted stir casting technique to perform the study on Al6061-Si3N4 composite. Wear property and loss in ductility was observed [14].

In this paper, a supply chain of a particular product of a chemical industry was considered. It was a 2-stage supply chain system where the products were transported from manufacturing units to warehouses and from the warehouses to the markets. Two networks, with four warehouses and with five warehouses, were considered for the evaluation.

Problem Understanding

This network consisted of two plants, four warehouses and thirty customers. The products were delivered to the customers from the warehouses within the stipulated time. The company had an idea of investing in a new warehouse in a new location in order to reduce the cost of transportation. Hence an evaluation was necessary to get clarity to figure out if there was a reduction in the transportation cost in the network consisting of five warehouses and what kind of investment was necessary on the 5th warehouse.

Methodology

The transportation cost for all required cases were collected. The functions and constraints were formulated based on the collected data. After the analysis with the mathematical model, we determined the decision that the company should take in order to reduce the transportation cost. A methodology flowchart is depicted in Figure 30.1.

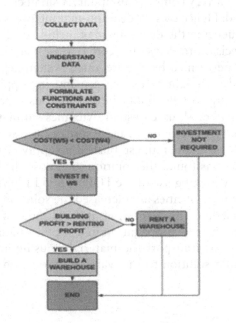

Figure 30.1 Methodology flowchart.

It is very well known that LPP technique is widely used to model and solve real time industrial problems. Hence, the modelling and solving technique adopted in order to analyze and take decision was again Linear Programming. As we are aware that there are many algorithms for solving an LPP, simplex algorithm was chosen as it is known to be the best. All the variables were given non-negative constraint along with other constraints discussed in the results and discussions section of this paper.

Results and Discussions

In the formulation we found that there were three types of constraints: plant capacity, market constraints and warehouse constraints Plant capacity constraint ensures that each plant manufactures a number of units of the products within its manufacturing capacity. Market constraints ensures that the number of units delivered to the customers be equal to forecasted demand. Warehouse constraints confirms that the number of units entering the warehouse be equal to that of leaving the warehouse. Figure 30.2 shows five warehouses Vs four warehouses.

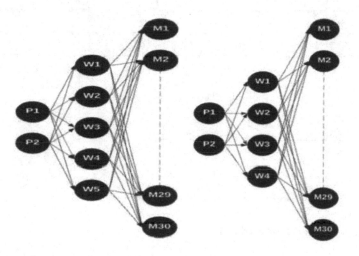

Figure 30.2 Five warehouses Vs four warehouses.

In the tables given below (Table 30.1 and 30.2), the cost of distribution for various cases has been shown. P, W and M stands for plants, warehouses and markets respectively. In this table cells corresponding to a plant and a warehouse represents the transportation cost per unit from that particular plant to particular warehouse similarly the same is applicable for transportation cost per unit from any warehouse to any customer. Table 30.1 corresponds to the existing distribution system i.e., the system with four warehouses, while Table 30.2 corresponds to the new proposed system i.e., with five warehouses.

Initially, to figure out the expected demand of every market/customer, monthly demand of 15 previous months of every market was collected. A demand forecasting was done using an in-built Excel function in order to find out the demand. The forecast graph can be seen in the figure and the forecasted values can be seen in Table 30.3.

Further LPP formulation was done for both the cases. For the 1st case with four warehouses, the formulation consisted of 128 variables, 36 constraints and one objective function. While in the case with the formulation for five warehouses, the formulation consisted

Table 30.1: Distribution cost of the product (per unit) in Rs for four warehouses.

	P1	P2	M1	M2	M3	M4	M5	M6	M7	M8	M9	M10	M11	M12	M13	M14	M15	M16	M17	M18	M19	M20	M21	M22	M23	M24	M25	M26	M27	M28	M29	M30
W1																																
W2																																
W3																																
W4																																

Table 30.2: Distribution cost of the product (per unit) in Rs for five warehouses.

	P1	P2	M1	M2	M3	M4	M5	M6	M7	M8	M9	M10	M11	M12	M13	M14	M15	M16	M17	M18	M19	M20	M21	M22	M23	M24	M25	M26	M27	M28	M29	M30
W1																																
W2																																
W3																																
W4																																
W5																																

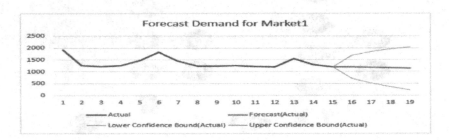

Figure 30.3 Forecast demand for market 1.

Table 30.3: Demand forecast values for the markets per annum.

Markets	Demand	Markets	Demand
1	1194	21	6431
2	3238	22	5115
3	1547	23	7094
4	9393	24	9296
5	7300	25	10401
6	12327	26	9834
7	13144	27	11014
8	7690	28	9116
9	3855	29	6099
10	10190	30	3330
11	13633		
12	5255		
13	4035		
14	7057		
15	2404		
16	9479		
17	8676		
18	9995		
19	6373		
20	6843		

of 160 variables, 37 constraints and one objective function. The inputs were given in Excel solver and the optimal distribution strategy was obtained for both the cases as shown in Tables 30.4 and 30.5. Along with the optimal distribution strategy, the total cost of transportation was also obtained. An estimated savings of about Rs 8,36,780 per month could be seen if there existed a fifth warehouse in the specified location. Figure 30.3 shows forecast demand for market 1.

Table 30.4: Optimal distribution strategy for four warehouses.

Total cost: 3857670

Table 30.5: Optimal distribution strategy for five warehouses.

Total cost: 3020890

Decision on whether a new warehouse should be rented or built is still unknown and has to be evaluated. Cost of renting a warehouse is Rs 350000/ month while building a new warehouse is Rs 9300000 as quoted by a vendor. If a warehouse was rented, we can observe only Rs 486000/month as profit. In case, we build a new warehouse we can't observe any profit for some duration until we get back the money invested on building and later, we can observe Rs 836000/month as profit. Figure 30.4 shows a cumulative profit comparison for the company between building and renting the warehouse.

If we build a warehouse, for the first 12 months we won't observe any profit. Hence, the payback period is one year. If it was rented, we can observe a profit of Rs 486000/ month. Correctly after 29 months (2 year 5 months), profit observed in both cases will be equal and further on, the profit observed will be more in case if the new warehouse is built instead of renting. Hence, the comparative payback period is 2 years 5 months. Since the time period is not too long, inventing a new warehouse will prove to be a wise decision

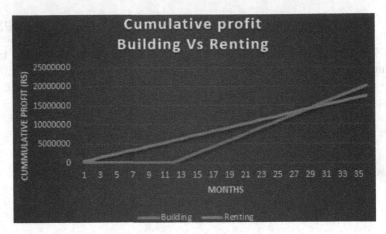

Figure 30.4 Cumulative profit comparison plot.

Conclusions

A comparative study and evaluation for two network designs were developed. Initially, from the past 15 months data of the customer demand, the expected demand was forecasted using excel solver built-in function. Further, for the existing network (two plants, four warehouses and 30 customers), a LPP formulation was developed, optimal distribution strategy was evaluated and optimal distribution cost was observed to be Rs 3857670/month and the similar thing was carried out for the proposed case (two plants, five warehouses and 30 customers) which resulted in optimal distribution cost of Rs 3020890/month. Hence a savings of Rs 8,36,780/month could be observed from the transportation cost stand-point. If invested in building a new warehouse, a payback period of 1 year and a comparative payback period of 2 years 5 months could be observed.

References

1. Mason, S. J., Ribera, P. M., Farris, J. A., and Kirk, R. G. (2003). Integrating the warehousing and transportation functions of the supply chain. *Transportation Research Part E: Logistics and Transportation Review*, 39(2), 141–159.
2. Runhaar, H., and van der Heijden, R. (2005). Public policy intervention in freight transport costs: effects on printed media logistics in the Netherlands. *Transport Policy*, 12(1), 35–46.
3. Naim, M. M., Potter, A. T., Mason, R. J., and Bateman, N. (2006). The role of transport flexibility in logistics provision. *The International Journal of Logistics Management*, 17(3), 297–311.
4. Wilson, M. C. (2007). The impact of transportation disruptions on supply chain performance. *Transportation Research Part E: Logistics and Transportation Review*, 43(4), 295–320.
5. Potter, A., and Lalwani, C. (2008). Investigating the impact of demand amplification on freight transport. *Transportation Research Part E: Logistics and Transportation Review*, 44(5), 835–846.
6. Rodrigues, V. S., Stantchev, D., Potter, A., Naim, M., and Whiteing, A. (2008). Establishing a transport operation focused uncertainty model for the supply chain. *International Journal of Physical Distribution & Logistics Management*, 38(5), 388–411.
7. Özcan, T., Çelebi, N., and Esnaf, Ş. (2011). Comparative analysis of multi-criteria decision making methodologies and implementation of a warehouse location selection problem. *Expert Systems with Applications*, 38(8), 9773–9779.

8. Tatavarthy, S. R., and Sampangi, G. (2015). Solving a reverse supply chain TSP by genetic algorithm. *Applied Mechanics & Materials*, 813, 1203–1207.

9. Garvey, M. D., Carnovale, S., and Yeniyurt, S. (2015). An analytical framework for supply network risk propagation: a bayesian network approach. *European Journal of Operational Research*, 243(2), 618–627.

10. Rao, T. S. (2017). A comparative evaluation of GA and SA TSP in a supply chain network. *Materials Today: Proceedings*, 4(2), 2263–2268.

11. Emeç, Ş., and Akkaya, G. (2018). Stochastic AHP and fuzzy VIKOR approach for warehouse location selection problem. *Journal of Enterprise Information Management*, 31(6), 950–962.

12. Lee, I. G., Chung, S. H., and Yoon, S. W. (2020). Two-stage storage assignment to minimize travel time and congestion for warehouse order picking operations. *Computers and Industrial Engineering*, 139, 106129.

13. Fartaj, S. R., Kabir, G., Eghujovbo, V., Ali, S. M., and Paul, S. K. (2020). Modeling transportation disruptions in the supply chain of automotive parts manufacturing company. *International Journal of Production Economics*, 222, 107511.

14. Kumar, G. B. V., Ulhas Krishna Rao, A., and Srinivas Rao, T. (2020). Fabrication and investigation of mechanical and dry sliding wear characteristics of Al6061-Si3N4 Composites Published by ASTM. 9(1), 20200037. doi:10.1520/MPC20200037.

31 Evaluation and optimization of processing time for construction permits

Vivek Sawant[1,a] and Shilpa Kewate[2,b]

[1]PG Student, Department of Civil Engineering, Pillai HOC College of Engineering and Technology, Rasayani, India

[2]Associate Professor, Department of Civil Engineering, Pillai HOC College of Engineering and Technology, Rasayani, India

Abstract: This research paper includes a detailed database on the approval process for various categories of buildings by evaluating and optimizing the process time for construction permits of buildings in the Mumbai metropolitan region. The investigation has identified technical advisors as useful for monitoring the online system, and automated compliance as well as technical consultant report as per ease of doing business of permitting process in Mumbai as a viable solution to many of the existing challenges. These recommendations are studied keeping in view the global perspective so that infrastructure can be developed in our India through smart urban management application. Three-tier construction permitting processes are presented as a projection of a fully integrated planning system to shape the future of our housing industry by replacing traditional manual documentation permitting methods. In this research, detailed information about the construction activity of tower-like structures in the city of Mumbai are obtained and their permitting process evaluated. A number of challenges with the current scenario are identified and recommendations for improvement are provided. The time optimization evaluated by using the primavera software Network to streamline the construction permits considering the probability of new building proposal approvals are being delayed at the Mumbai municipal corporation jurisdiction based on the number of compliance during various construction stages of the project.

Keywords: Authority, evaluation, permit processing time, technical consultant.

Introduction

Building permits are legal approvals issued by Municipal Corporations, and the other building proposal planning bodies in accordance with the Town Planning Act & Development Planning Regulations to carry out proper construction activities in consonance as per the suitability of the environment and awareness. Mumbai is one of the fastest-growing a city in the metropolitan jurisdiction of India and the area around Mumbai and its suburbs attracts many builders to invest in various construction projects for their livelihood development. There is an option to develop vertically in the form of high rises buildings for houses and as per the demand of the residential cum commercial users. Due to sudden changes in the government planning rules, circulars, laws, and policies, the building permit

[a]sawant_vd@rediffmail.com, [b]shilpakewate@mes.ac.in

process is becoming more complicated. Therefore, frequent changes have to be made in various structures as well as a lack of an understanding of legal rules, and sometimes awkward policy taken by the planning authorities. As a result, the cost of new construction units puts a huge strain on the developer and affects the public. With this in a mind, the purpose of this framework is to modify the conventional permitting process by an evaluation and optimization of processing time for construction permits through digital submissions for online permits and to complete automated IS code checks for building permits as well as for permits issued by private consultant reports. In this research papers studied processing time for optimization for obtaining remark from hydraulic engineers department and remarks for parking layout has been discussed.

Problem Statement

There are major obstacles to the growth of the construction sector in India. One reason for the development of construction projects is the complex permitting process. Due to sudden changes in government planning laws and policies, the permitting process is becoming more and more complicated. Due to the rapidly growing population, the permits required by the relevant regulatory departments to complete the building are becoming more valuable for the development of the construction industry, and it will be more difficult to proceed without planning and management to speed up the construction activities. Permit planning can start smoothly during the entire construction period. Most of the studies follow the prevailing methods of obtaining permission from the planning authority in the jurisdiction of the corporation and the government. However, this research work has developed the direct authority to submit the NOC Completion Report of the permit through the relevant technical consultant and their important role as per the existing government regulations and policy.

Accordingly, the following studies for construction permit procedure are evaluated herewith in this research paper.

Study I: Processing time optimization for obtaining remark from Hydraulic Engineer Department.

Study II: Processing time optimization for obtaining report of parking layout

Assessment Study

This research paper study aims to evaluate the permitting the process in various intervals, and investigates various factors that have an influence thereon. Sara (2012) [1] has concluded that a well-defined service guarantee has been contributed the improving quality of the services. It helps the planning and building department of municipality to increase the customer satisfaction and attract more building developer to invest in the municipality region. Baria et al. (2018) [2] have studied that sound construction regulation will save human lives improve health and safety and support a prosperous and property building sector and economy, it's progressing to facilitate doing business by safeguarding remunerative investments, strengthening property rights, and protecting the overall public from faulty building practices. Personal sector involvement at intervals the human activity of building laws has shown positive finally lands up in achieving restrictive goals. Davit (2017) [3] has studied that the adorned review of the problems the amount of necessary legislative or sensible short-comings unit noted that need improvement and regulation. Implementation of the higher than is unbelievably necessary thus on avoid chaotic architectural-building development once the urban designing and construction change issue on the one hand and forestall any violation of universally recognized human rights secured by the constitution of the country on the choice hand. It is a necessity to eliminate the foremost deficiencies incontestable supported the problems reviewed on the far side for the effective and comprehensive fulfillment of the functions allotted to them by the native autonomy bodies. Hammad and Radhlinah (2015) [4] recommendation are made for improvement that include the need for practicing managers to professionalize project planning and scheduling based on a more proactive and knowledge-based planning approach, which is supported by management. Igor (2010) [5] has concluded to introduce additional anticorruption measures at intervals the scrutiny processes rationalize and consolidate post-completion inspections reduced the time and worth burden on entrepreneurs, introduce multifunctional inspections. Nawari and Adel (2017) [6] concluded that this analysis work seeks to seem at the role of BIM in up the allowing procedure and tackle the defragmentation of the event trade. The investigation additionally provides a prime level read of existing frameworks for automatic code verification in building vogue and notably branch of knowledge and structural vogue. Peter and Darco (2016) [7] has analyzed that the design change mechanism has been adopted to balance the advantages between landowners then the selection stakeholders concerning swish & quick construction development.

Objectives

This research is initiated by the study of existing academic literature. The following objectives are derived based on the current performance of the construction permitting the process in the city of Mumbai where need to be aligned existing the practice systematically with an advance technology which are being implemented through qualified technical experts in the architecture an engineering field as well as reputed builders and developers.

1. To evaluate the construction permitting procedures considering the challenges with the present permit practice.
2. To investigate the permit process by the time optimization for obtaining requisite compliance of building projects.
3. To assess a technical documentation to simplifies and expedite the construction activities.

Statutory Planning Authorities

This planning a regulation embodies those obligatory by municipal and provincial governments as control bodies like MCGM, MHADA, MMRDA, MIDC, MBPT and special planning cell for government buildings. Specific laws for the downtown coring of the Mumbai jurisdiction were additionally enclosed. The Table 31.1 shows that various planning authorities of building permits from the date they came into an existence.

Construction Permit Stages

The building permits are divided into Mumbai-based the metros. Building permits for the residential, commercial, and industrial use is granted to the developer or owner by the concerned planning authority according to their respective area of the responsibility. The building permits are issued in three stages namely IOD/IOA (intimation of disapproval/

Table 31.1: Summary of statutory planning authorities.

Jurisdiction Planning Authorities	Year of Establishment	Act & Regulations				
Municipal Corporation of Greater Mumbai (MCGM)	1967	The Maharashtra Regional and town planning Act	The Mumbai Municipal Corporation Act,	The Mumbai municipal corporation development control & promotion regulation 2034	CRZ Regulation 1991 CRZ Notification 2011	Maharashtra Fire Act
Slum Rehabilitation Authority (SRA)	16 December 1995	The Maharashtra Co-Operative Societies Act. Maharashtra Act	Maharashtra Cooperative Societies Rules.	The Multi-State Cooperative Societies Rules,	-	-
Maharashtra Housing & Area Development Authority (MHADA)	5 December 1977	Maharashtra Housing And Area Development Authority (MHADA) Act 1976	-	The Development Control & Promotion Regulation 2034	-	-
Mumbai Metropolitan Region Development Authority (MMRDA)	1 December, 1999	The Mumbai Metropolitan Region Development Authority Act	-	The Development Control & Promotion Regulation 2034	-	-
Maharashtra Industrial Development Corporation (MIDC)	1 August 1962	Maharashtra Industrial Development Act	-	The Development Control & Promotion Regulation 2034	-	-

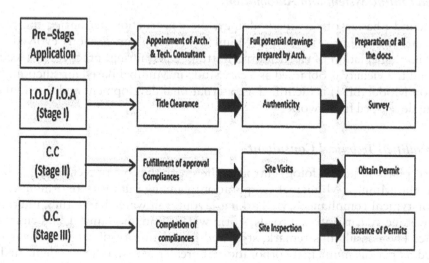

Figure 31.1 Schematic of permitting stages.

approval), commencement certificate and occupation certificate while pre-construction, during construction and after completion of construction project.

Accordingly, it is mandatory to fulfill the requirement before initiate the permitting stages at the pre-application level as follows as per shown above in Figure 31.1. In the pre-application stage, the owner/developer shall confirm the title clearance and authenticity of the plot, and then apply to assistant engineer (survey) for getting elaborate remarks for the property. The architect/license surveyor shall prepare the Building Proposal plans for full potential projected to be utilized by the owner/developer on the property/ land underdevelopment. The owner/developer shall appoint qualified consultants for issuance the remark and completion of various compliance of construction project as a permit. Accordingly, technical consultant should have submitted acceptance letter as such, his appointment is accepted by the planning authority or the concerned department of the Mumbai Municipal Corporation through a letter of acceptance. Architects will be appointed at the three stages of IOD CC and OC respectively. The building permits to be obtained for fulfillment of the above stages from the concerned department through a qualified technical consultant appointed by the owner or developer.

Three Tier Construction Permitting Process

The three-tier permitting process are presented as under for obtaining various construction permits while phases of preconstruction, during construction and after completion of construction activity.

Conventional Method

In this process the manually permit to be obtained after going through various steps based are evaluated as under. As per the studied several permit processes, the optimization and evaluation of the permitting process of some buildings under the case study as compliance where the permits required for commencement and getting occupation certificate of construction activity.

Web-Based Online System and Automation

For the second phasing of this methodology, the construction permitting the process of MCGM planning authority in MMR Mumbai Metropolitan Regions evolved, and the successful implementation of electronic permitting (e permitting) practices in specific locations around the vicinity is obtained as a case study in Mumbai Based jurisdiction. Building Information Model (BIM) is identified as international development and essential component of the developed framework.

Role of Qualified Technical Consultant

In the third phase of methodology introduce herewith the role of technical qualified consultant. In the Mumbai Municipal Corporation planning authority reforming the permit process for typical compliance's viz Parking layout, rainwater harvesting, internal water works, sewerage system, internal roads, storm water drain, mechanical, and electrical ventilation etc. The consultant's remark, style and completion certificate vide their role recommended as per the municipal corporation of greater Mumbai has implemented ease of doing business method.

Methodology

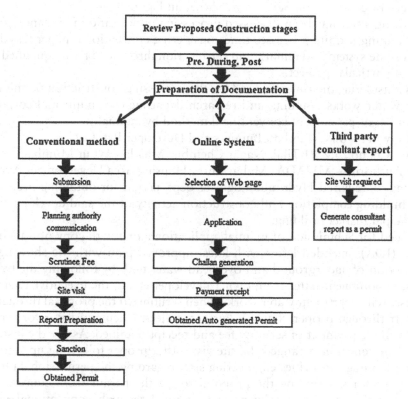

Figure 31.2 Permitting process flow chart.

In 2014, the Government of India launched an ambitious program of regulatory reforms aimed at making it easier to do business in India. Program represents a great deal of construction permits to create a more construction business in peaceful atmosphere. Municipal Corporation of greater Mumbai has initiated various reforms to facilitate the work of construction permits.

The particular reforms are undertaken as reengineered the process along with the approval process is defined with predictability. The official fee design an acceptance is to be made at one window system. The third party verification and compliance does accept for particular no objection certificate (N.O.C.) and approval. The all proposal can be accepted through online. The role of concern planning authority are clearly mentioned as per their standard practice according to manual is available. The parameters wise reforms are introduced by numbers for construction permit reduced from 42 to 8 numbers are proposed. The processing times for getting construction permit are reduced up to 60 days from 164 days earlier. The regulation quality has enhanced by developing auto scrutinize plan with digitally sign approval and completion certificates are issued. The entire construction permit process now is in public domain also instant online approval for miscellaneous as well as addition alteration work are initiated. There are two-way communication facilities by email and SMS are available throughout the construction activity of the particular building. Videos and photos can be uploaded to observe the status of present construction works. Reform impact can be improved the efficiency to simplify standardize uniform and time bound procedure transparently by reducing the time at all levels i.e. Intimation of Disapproval (IOD), Commencement Certificate (CC) and Occupation Certificate (OC). Various stages of construction permits are shown in Figure 31.2.

The following study has been discussed processing optimization for planning and evaluation of planning permits processes time considering conventional paper-based practice, web-based online system and remark and certification through a private qualified technical consultant of particular project.

Study I: Considering the time required for the processing construction permit obtaining NOC from water works planning and research department of municipal corporation of greater Mumbai as per current conventional method by manually.

Name of Owner – M/s Manthan Builders and Developer Pvt. Ltd.

Site Area – Cts no.25,25/1 Tilak Nagar Chembur Mhada Layout Mumbai

Planning Authority – MHADA (Maharashtra Housing Area Development Authority)

Permit Stage – required Hydraulic Engineer dept. permit for obtaining occupation certificate for building completion and construction activity of the said development of proposed Municipal market building.

First proceed for Submission of manual application along with copy of IOD/intimation of approval (IOA)/Amended letter with latest approved plan with the showing location and cross section of underground and overhead water tank by following the bylaws, and regulations, appointment letter from owner developer etc. to the executive engineer planning and research department water works. Then scrutinized the proposal through concern engineer staff through proper channel. Further, collected demand letter for scrutiny fee. Accordingly, made payment of scrutiny fee and receipt obtained. As per the instruction of a junior and sub engineer arranged the site visit. After proper following up, and convince the concern planning authorities engineering staff regarding the permission. Subsequently, prepared report for sanctioning the proposal to get the permit from concern hydraulic engineering department, and finally permit obtained through conventional method by manually

Study II: There is introduced a parking layout remark, as a permitting process required for building construction in Mumbai metropolitan area is now facilitated through trained technical consultants.

Name of owner – M/s Pioneer India Development Pvt. Ltd

Name of architect – M/s Bidco Engineering division

Site Area – Sale bldg. No. D on sub-plot D, FP 16B, TPS VI, Santacruz (W), Mumbai

Planning authority – SRA (Slum Rehabilitation Authority)

Permit stage – Required parking layout permit for obtaining commencement certificate for further construction activity of the said development of proposed residential building.

Initially collected relevant data and documentation required to get the report for obtaining concerned parking layout remark accordingly prepared report as per the plan issued by client subject to following EODB circular of Municipal Corporation of greater Mumbai. Then uploaded the report copy of the particular parking layout remark along with marking of manuring system of vehicles as per development control regulations 2034 on our letter headed as a technical consultant work as known as permit.

The Limitation of this Study

- By an evaluation and an optimization of the processing time for construction permits, it is observed that roles of the concern planning authorities are very vital to provide the construction permit within a time frame.

- The mechanism and an approach should be prompt of a concern architect and a technical consultant whoever is communicating and following up for the construction permit throughout the stages of the construction activity to expedite the permit process.

Results and Discussions

In this context, we hereby presented the comparatively statement of three tier permitting process as adopting conventional paper based a method, the web based online system with an automation and a qualified technical consultant report for obtaining the construction permit by an evaluation and an optimization of the processing time as per Table 31.2 shown. Following observation can be drawn from Table 31.2.

1. **Conventional method** – In this process, it is observed that, planned days are 9 and actual days 18 i.e. delayed by 9 days, due to absence of concerned person, due to

Table 31.2: Study-I, details of permit obtaining for hydraulic engineering dept.

Sr. no	Construction Permit processing steps to be followed	Conventional Method		Web based online system with automation		Third party Qualified technical consultant report	
		Processing time		Processing time		Processing time	
		Planned Days	Actual Days	Planned Days	Actual Days	Planned Days	Actual Days
1	Submission	1	2	NA	NA	NA	NA
2	Scrutinee	1	3	NA	NA	NA	NA
3	Demand Letter	1	2	NA	NA	NA	NA
4	Scrutinee fee payment	1	2	NA	NA	NA	NA
5	Site Visit	1	2	NA	NA	NA	NA
6	Follow up for report	2	4	NA	NA	NA	NA
7	Sanctioned Report	1	2	NA	NA	NA	NA
8	Permit Obtain	1	1	NA	NA	NA	NA
	Subtotal	9	18				
9	Select Official web page	NA	NA			NA	NA
10	Filled Application	NA	NA			NA	NA
11	Generate Challan	NA	NA	2	3	NA	NA
12	Online Payment	NA	NA			NA	NA
13	Receipt Generation	NA	NA			NA	NA
14	Displayed Permit	NA	NA			NA	NA
	Subtotal			2	3		
15	(I) Collection of relevant data, (II) Prepared report according to plan and drawings, (III) Upload the Technical Consultant Report as a permit	NA	NA	NA	NA	2	Half (1/2)
	Subtotal					2	Half (1/2)

computer system slow, due to delayed done in DD making from client etc, delayed process as shown in Table 31.2.

2. **Web based online system with automation** – In this process, it is observed that, By optimizing the Permit took 2 days, it was issued in 3 day which was expected to be completed within 2 day as per planned, it means that the permit received after schedule time by losing 1 day.

3. **Third party Qualified technical consultant report** – In this process, it is observed that, By optimizing the Permit took half day, it was issued in 2 days which was expected to be completed within 2 day as per planned, it means that the permit received before schedule time by saving valuable 1 and 1/2 day.

Table 31.3: Study-II, details of permit obtaining-parking layout remark.

Sr. no	Construction Permit processing steps to be followed	Conventional Method		Web based online system with automation		Third party Qualified technical consultant report	
		Processing time		Processing time		Processing time	
		Planned Days	Actual Days	Planned Days	Actual Days	Planned Days	Actual Days
1	Collection of relevant data	NA	NA	NA	NxA	NxA	NxA
2	Prepared report according to plan and drawings	NA	NA	NA	NxA	2	1
3	Upload the Technical Consultant Report as a permit	NA	NA	NxA	NxA	NxA	NxA
	Subtotal					2	1
4	Submission	1	1	NA	NxA	NxA	NxA
5	Scrutinee	1	3	NxA	NxA	NxA	NxA
6	Demand Letter	1	1	NxA	NxA	NxA	NxA
7	Scrutinee fee payment	1	1	NxA	NxA	NxA	NxA
3	Site Visit	NA	NA	NxA	NxA	NxA	NxA
9	Follow up for report	3	5	NxA	NxA	NxA	NxA
10	Sanctioned Report	2	3	NxA	NxA	NxA	NxA
11	Permit Obtain	1	1	NxA	NA	NA	NA
	Subtotal	10	15				
12	Select Official web page	NA	NA	2	1	NxA	NxA
13	Filled Application	NxA	NxA			NxA	NxA
14	Generate Challan	NA	NA			NxA	NxA
15	Online Payment	NA	NA			NxA	NxA
16	Receipt Generation	NA	NA			NxA	NxA
17	Displayed Permit	NA	NA			NxA	NxA
	Subtotal			2	1		

Following observation can be drawn from Table 31.3.

1. **Conventional method** – In this process, it is observed that, planned days are 10 and actual days 15 i.e. delayed by 5 days, due to absence of concerned engineer busy in other departmental work, Delayed due to verification and scrutinize by executive engineer.
2. **Web based online system with automation** – In this process, it is observed that, planned days 2 and actual day 1 delayed by 1 days, due to client issue for delaying in verification and scrutinize by executive engineer as shown in Table 31.3.
3. **Third party Qualified technical consultant report** – In this process, it is observed that, actually the permit took 1 day, it was issued in 1 day which was expected to be completed within 2 day as per planned, it means that the permit received before schedule time by saving valuable 1 day.

Conclusions

*   The building permit process in the Mumbai metropolitan area is available online and as an automated computer system only for the proposals of the Mumbai Municipal Corporation in a modern way. But, rest of the planning authority should be applied online and auto generated based system, for issuance of digital permit instead of present conventional huge paperwork practice to avoid unnecessary storage of file paper by saving valuable time and cost of construction projects.
*   The permitting process can be expedite by obtaining permit of various departments through private technical consultant in Mumbai Municipal Corporation Jurisdiction planning authority.
*   Assessment of the knowledge of planning regulations by the laws act and planning circulars and notification is very essential to frame the proper permitting process to expedite the construction activities.

References

1. Sara, A. (2012). A Study of Optimizing the Processing Time for Building Permits Study Case: Tyresö Municipality. Sweden: Royal Institute of Technology.
2. Baria, D., Marie, D., Imane, M., Keiko, S., Jayashree, S., and Yelizaveta, Y. (2017). Dealing with Construction Permits Private Sector Participation in Construction Regulation. The World Bank.
3. Churghulia, D. (2017). Problematic issues of protection of the rights of owners of neighboring property in issuing construction permits and construction planning. *TSU Journal of Law*, 235.
4. Hammad, N., and Radhlinah, A. (2015). Assessing understanding of planning and scheduling theory & practice on construction projects. Engineering Management Journal, 27(2), 58–72.
5. Igor, A. (2010). Construction Permits Process in Poland. Investment Climate Regulatory Simplification—CICRS. World Bank Group.
6. Nawari, N. O., and Adel, A. (2017). The role of BIM in simplifying construction permits in Kuwait. *In Architectural Engineering Conference*, 855–866.
7. Wong, P. S., and Maric, D. (2016). Causes of disputes in construction planning permit applications. *Journal of Legal Affairs and Dispute Resolution in Engineering and Construction*, 8(4), 04516006.

32 Case study on implementation of 5S in an industry

Vamsikrishna A.[a], Y Harsha Chaitanya, and Divya Sharma S. G.[b]

Department of Mechanical Engineering, Amrita School of Engineering, Amrita Vishwa Vidyapeetham, Bengaluru, India

Abstract: In this modern busy world, many of us forget the fundamentals of arranging things in the right place and in the right manner. In industries, messy surroundings can bring down the productivity of the workers largely. Hence, implementation of lean manufacturing practices is essential to have a productive and efficient environment. In this paper, a case study about implementation of 5S in a company is explained. The nomenclature and strategies adopted in arranging things in a comfortable manner is briefed. A comparison on the basis of space occupied, time taken to pick the objects, money spent and distance moved etc. before implementation of 5S and after implementation of 5S is shown in the paper. The techniques used to sustain such an arrangement are explained.

Keywords: cost reduction, floor space reduction, 5S methodology, 30 sec checks.

Introduction

Modern management in the present world needs to enhance their financial strength and show continuous improvement at the same time. This aspect attracts the management to incorporate quality improvement techniques like 5S, six sigma, sustainable manufacturing, kaizen method, lean manufacturing, and many other concepts, technologies and approaches. The 5S technique is a good way to obtain continuous improvement. The 5S philosophy has its roots in Japan. The above list describes the way to organize our workplace in an effective and an efficient way. R. Shah et al. examined the effects of the three factors, plant size, plant age and unionization status on implementing several manufacturing practices [1]. A. P. Brunet et al. have studied kaizen as practiced in some Japanese companies, taking Nippon Steel Corporation (NSC) as a base model [2]. Michalska et al. have introduced the way of implementing 5S methodology in the company [3]. Gapp et al. have worked on developing 5S techniques into an integrated management system [4]. Saleeshya et al. have made a case study on the application of lean manufacturing methodology in textile industries. [5]. Wahab et al. have developed a conceptual model for the measurement of leanness in the manufacturing industry [6]. Jaca et al. have developed a multi-case study analysis and identified the important aspects for the successful implementation of 5S methodology [7]. Khandelwal et al. have analyzed the time and energy expenditure by

[a]vamsikrishna.ananth@gmail.com, [b]sg_divya@blr.amrita.edu

a case study and analysed the effectiveness of the methodology with a IDEEA device [8]. Randhawa et al. have worked on 5S methods and suggested the ways to implement sustainable 5S techniques [9]. Helleno et al. have proposed a conceptual model to integrate the VSM tool with new sustainability indicators [10]. Durakovic et al. have discussed the pros and cons of the implementation of lean and quoted that, 'The biggest threat in implementing lean is lack of understanding the concept but those who engage consultants were more successful' [11]. Sreedharan et al. reviewed on the Critical Success Factor (CSF) of various continuous Improvement techniques and performed a content analysis [12]. Yadav et al. have developed a framework for the effective adoption of lean manufacturing in companies using FAHP-DEMATEL hybrid [13]. Palange et al. have reviewed the effects of implementation of lean manufacturing techniques in different manufacturing sectors [14]. In this paper, implementation of 5S in an engineering store is considered. The importance of 5S was quantified based on space, time, distance moved, cost etc. after some simple experimentation of a 30 seconds checks.

Methodology

The case study began with a visit to the engineering store for understanding their practice and their working. Then sorting of items by discarding unwanted ones was done. A list of useful items in the store was made and classified them based on weight and frequency of usage of objects. Heavy objects were to be placed in the bottom and frequently used ones were kept in the front. All the objects were arranged following a particular nomenclature. Further, shine and standardize steps were carried out by training the technicians and by cleaning the stores. Figure 32.1 shows methodology flowchart.

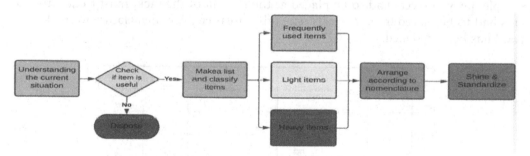

Figure 32.1 Methodology flowchart.

Results and Discussions

In the company to carry on with the day-to-day activity's employees need some equipment and materials for maintenance, service, etc. The company management organizes all these materials and equipment in three locations namely, Engineering store, Workshop and Tool kit Room. Engineering department will take care of these three locations. The main job of this department is maintaining the equipment and materials, to avoid breakdowns and to reduce downtime during the manufacturing. Whenever there is a maintenance issue with a machine, it is immediately informed to the department manager or any responsible person in the department and a work order is released. Then the maintenance department gets the information and takes action to solve the problem. For these activities, lots of tools & materials are necessary to resolve the problem in least time. Hence, it becomes essential that the engineering materials and tools reach the right place at the right time. Amongst the three locations i.e., engineering store, workshop and tool kit room, most of the materials and tools are placed in the engineering store. However, the Engineering store is not organized in a proper way as seen in Figure 32.2. Hence, it was decided to implement 5S methodology in this location for efficient and effective working.

Figure 32.2 Store before implementing 5S.

If 5S could be successfully implemented in this location, the time taken to fetch an item would be reduced, more floor space would be available, movement of the service men could be reduced, downtime of the equipment to be serviced could be reduced which directly increases the efficiency of the all maintenance works and increasing the efficiency of the total working process of the company. To start with the implementation of the 5S in the Engineering store, there should be a good plan first. Initially, the list of items kept in the Engineering store was taken. Then it was decided to classify the setup as rack 1 and rack 2 where rack 1 is smaller than rack 2. Then, items are classified in many aspects. For example, heavy objects had to be placed at the bottom of the rack, most frequently used items had to be placed in more easily accessible areas etc. A nomenclature for rack 1 and rack 2 has been designed.

R1R	R1F
R2R	R2F
R3R	R3F
R4R	R4F
R5R	R5F

R1C1	R1C2	R1C3	R1C4	R1C5	R1C6
R2C1	R2C2	R2C3	R2C4	R2C5	R2C6
R2C3	R3C2	R3C3	R3C4	R3C5	R3C6
R4C4	R4C2	R4C3	R4C4	R4C5	R4C6
R5C1	R5C2	R5C3	R5C4	R5C5	R5C6

C1R	C2R	C3R	C4R	C5R	C6R
C1F	C2F	C3F	C4F	C5F	C6F

Figure 32.3 Rack nomenclature.

The above (Figure 32.3) shows the designed nomenclature. There were 3 columns and 5 rows in rack 1. Each column is divided into two parts. Therefore, the number of columns became 6. Now, the depth is divided into two parts i.e., front and rear part. The above (Figure 32.3), shows the front, top and right-side view of the rack after assigning the desired nomenclatures. Categorization of the items based on many categories such as heavy items, light items, frequently used items etc. was done. Then, design of their locations accordingly was made. For example, all the heavy oil cans and lubricants to be placed in

Figure 32.4 Store after implementing 5S.

the bottom. More frequently used parts like filters, service equipment etc. to be placed in rack 1 which is in the front so that an employee could access it easily.

The above picture (Figure 32.4), shows the result of implementing 5S methodology. The improvement in the look is clearly seen. Similarly for rack 2, classification of the rows and columns with new nomenclature was done. For analysis of the effect of 5S, the data of time taken to fetch a particular item from the store before and after implementing 5S for 8 items, number of technicians, salary pay etc. was collected as shown in Tables 32.1 and 32.2.

Table 32.1: Calculations table.

Items	Time taken(in seconds)	
	Before 5S	Alter 5S
Diffusers-1	70	20
Diffusers-2	81	28
FH hose	79	26
SS ball valves (Vaas)-1	55	26
SS ball valves (Vaas)-2	59	29
Inverter 15 Amp 800 VA portable	38	26
Stirrer hydraulic oil	22	16
ETP blower oil	29	17
Average Time	54.125	23.5
Factor	*Before 5S*	*After 5S*
Effective time to pick the objects (in sec)/day	52393	22748
Effective time to pick the objects (in sec)/month	1571790	682440
Cost calculation per month (in rupees)	18192.01389	7898.611111
Cost calculation per annum (in rupees)	218304.1667	94783.33333
Effective time to pick the objects per technician (in sec)/day	1746.433333	758.2666667
% time wasted in picking the items from store	9.702407407	4.212592593
Floor space occupied (in sq ft)	1362	884
% floor space occupied	82.64563107	53.6407767
Distance moved(in m)/day	73350.2	31847.2
30 seconds check	Failed	Passed

Table 32.2: Data table.

Avg no of times stores visited as per register/day	968
Avg salary of a technician(in rupees) month	30000
Avg salary of a technician(in rupees)/day	1000
Avg salary of a technician(in rupees)/second	0.01157407407
Effective working hours in a day/technician	5
Effective working seconds in a day/technician	18000 sec
number of technicians	30

The 30-seconds check (Figure 32.5) has been conducted and the time taken to fetch each item is noted. The data on the number of visits to the store in a day was collected. With the data, the floor space, distance moved per day, Cost per month etc. has been calculated as shown in the Table 32.3.

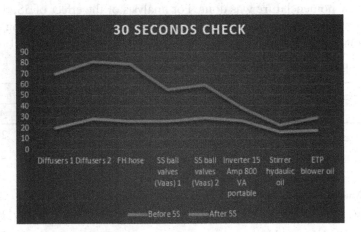

Figure 32.5 30-seconds check plot.

Table 32.3: Comparison table.

Factor	*Before 5S*	*After 5S*
30 seconds check	Failed	Passed
Effective time to pick the objects (in sec)/day	52393	22748
Effective time to pick the objects (in sec)/month	1571790	682440
Cost calculation per month (in rupees)	18192.01389	7898.611111
Cost calculation per annum (in rupees)	218304.1667	94783.33333
Effective time to pick the objects per technician (in sec)/day	1746.433333	758.2666667
% time wasted in picking the items from store	9.702407407	4.212592593
Floor space occupied (in sq ft)	1362	884
% floor space occupied	82.64563107	53.6407767
Distance moved(in m) day	73350.2	31847.2

From the results in Table 32.3, it can be observed that the effective time to pick the object in store per day had reduced from 14.55 hrs to 6.32hrs. The cost spent on fetching objects per month had reduced from Rs 18192 to Rs 7898.6. Occupied floor space had reduced from 1362 sq. ft to 884sg.ft. As there are appreciable changes, the impact of implementing 5S methodology was understood. The last and most important step for the further smooth run of the store will be sustainability. The arrangement sequence should not be changed and should be maintained properly by the engineering department. They can incorporate new technologies like assigning barcodes for each item and its location, connecting to ERP systems such as SAP, introduction of advanced CCTV systems etc.

Conclusion

Implementation of 5S has given positive results. Thirty seconds check was successful. Effective time spent to pick the items per day has decreased by nearly 8 hours. Money invested on picking objects from the store has reduced by almost Rs.1,21,000/annum. Percentage time wasted in picking the items from the store has decreased by nearly 5%. Percentage of floor space occupied has decreased by 29%. Effective distance moved/day has decreased by 41 Km. To sustain these improvements, it is necessary to follow the nomenclature and abide by the 5S and lean manufacturing practices. These improvements were observed merely by arranging things in a systematic manner and by implementation of 5S. Therefore, 5S is a very powerful lean manufacturing practice and further scope of research in this field will prove to be useful as implementation of 5S will give fruitful results.

References

1. Shah, R., and Ward, P. T. (2003). Lean manufacturing: context, practice bundles, and performance. *Journal of Operations Management*, 21(2), 129–149.
2. Brunet, A. P., and New, S. (2003). Kaizen in Japan: an empirical study. *International Journal of Operations and Production Management*, 23(12), 1426–1446.
3. Michalska, J., and Szewieczek, D. (2007). The 5S methodology as a tool for improving the organization. *Journal of Achievements in Materials and Manufacturing Engineering*, 24(2), 211–214.
4. Gapp, R., Fisher, R., and Kobayashi, K. (2008). Implementing 5S within a Japanese context: an integrated management system. *Management Decision*, 46(4), 565–579.
5. Saleeshya, P. G., Raghuram, P., and Vamsi, N. (2012). Lean manufacturing practices in textile industries–a case study. *International Journal of Collaborative Enterprise*, 3(1), 18–37.
6. Wahab, A. N. A., Mukhtar, M., and Sulaiman, R. (2013). A conceptual model of lean manufacturing dimensions. *Procedia Technology*, 11, 1292–1298.
7. Jaca, C., Viles, E., Paipa-Galeano, L., Santos, J., and Mateo, R. (2014). Learning 5S principles from Japanese best practitioners: case studies of five manufacturing companies. *International Journal of Production Research*, 52(15), 4574–4586.
8. Khandelwal, A., Prathik, R., Kikani, R. P., and Ramesh, V. (2014). 5S Implementation and its effect on physical workload. *International Journal of Research in Engineering and Technology*, 3(9), 437–440.
9. Randhawa, J. S., and Ahuja, I. S. (2017). 5S–a quality improvement tool for sustainable performance: literature review and directions. *International Journal of Quality and Reliability Management*, 34(3), 334–361.

10. Helleno, A. L., de Moraes, A. J. I., and Simon, A. T. (2017). Integrating sustainability indicators and Lean Manufacturing to assess manufacturing processes: Application case studies in Brazilian industry. *Journal of Cleaner Production*, 153, 405–416.
11. Durakovic, B., Demir, R., Abat, K., and Emek, C. (2018). Lean manufacturing: trends and implementation issues. *Periodicals of Engineering and Natural Sciences (PEN)*, 6(1), 130–143.
12. Sreedharan, V. R., and Sunder, M. V. (2018). Critical success factors of TQM, Six sigma, lean and lean six sigma: a literature review and key findings. *Benchmarking: An International Journal*, 25(9), 3479–3504.
13. Yadav, G., Luthra, S., Huisingh, D., Mangla, S. K., Narkhede, B. E., and Liu, Y. (2020). Development of a lean manufacturing framework to enhance its adoption within manufacturing companies in developing economies. *Journal of Cleaner Production*, 245, 118726.
14. Palange, A., and Dhatrak, P. (2021). Lean manufacturing a vital tool to enhance productivity in manufacturing. *Materials Today: Proceedings*, 46, 729–736.

33 Comprehensive review on minimum quantity lubrication (MQL) using nano particles in machining operation

Suhas Rewatkar[1], Bhushan Mahajan[1], Prashant Maheshwary[1], Kailash Nemade[2], and Sanjay Dhoble[3]

[1]Department of Mechanical Engineering, JDCOEM, Nagpur, India

[2]Department of Physics, Indira College, Kalamb, Yavatmal, India

[3]Deapartment of Physics, R.T.M.N.U, Nagpur, India

Abstract: During the high-speed machining operation there is a heat generation between tool and work piece. Heat generated deteriorates the quality of finish product and also effect the tool life. Conventional flood cooling technique is not eco-friendly and economical. Hence MQL (Minimum Quantity Lubrication) emerge as an alternative to flood cooling. In the current study, basic machining processes such as drilling, turning, milling, grinding has been discussed where MQL system implemented. Furthermore, effect of different Nano Fluid (NF) along with MQL system has been studied and mentioned in the paper. It can be concluded from the comprehensive study that MQL along with NF's shows much better results on response parameters such as surface roughness, tool wear. However, sedimentation of Nano particles is more challenging. Use of surfactant along with base fluid is very much essential to make fluid stable. Interestingly, it has been found that along with percentage of Nano particles, amount of surfactant in base fluid also has effects on response parameters. Study Reveal that nMQL (Nano Minimum Quantity Lubrication) could be much better option compare to pure MQL, flood cooling and dry machining to improve surface quality, reduce tool wear and cutting forces.

Keywords: Machining, minimum quantity lubrication, nano fluid.

Introduction

Turning operation is most widely used in the field of machining. Heat generation may lead to deterioration of finished product and tool wear [1]. Flood cooling is one of the conventional options to reduce the heat but it leads to wastage of coolant consumption is around 1200 L/hr. Hence MQL is one of the best replacements to flood cooling where consumption reduced drastically to around 60 ml/hr [47–49]. Pervaiz et al. [34] mentioned in their review that physical scarcity of water on earth is more concern as approximately 500million people will suffer from water scarcity in future. Furthermore, total cost contribution of Flood cooling is around 16–20% of the total cost of manufacturing [7]. In flood cooling, the cutting fluid applied fails to enter at the chip-tool-interface [25]. However, dispersing the heat is difficult [2]. National Institute for Occupational Safety and Health (NIOSH) estimated that one million workers had been exposed to metal working fluid (MFW) which may leads to cancer [13] hence minimum quantity lubrication (MQL) has evolved as an option for conventional flood cooling system [28]. MQL set up consist of different components such as pneumatic compressor, pressure regulator, oil reservoir, mixing chamber for oil. Compressed air mix with oil used in mixing chamber and comes out of spray nozzle attached to MQL box. Gaurav G et al. [53] perform his experiment on high-speed precision lathe (NH22, HMT, INDIA). He used Dropsa–UK base MQL model. Flow rate of cutting fluid was 60 ml/h and regulate the compressed air at 6 bar. B. Boswell et al. [46] observed that study of MQL is evenly distributed among the all four basic machining operations i.e, grinding (29%), milling (26%), turning (24%), drilling(21%). Hence effect of nMQL in all four basic machining operations explained in this paper. Many researchers have made a comparison between MQL and conventional way of cooling in different machining operation. Gajrani et al. [3] found that, green cutting fluids gives much better results in MQL system compared to Flood cooling and dry machining [9]. Diego Carou et al. [21] focused his study on effect of MQL and concluded that MQL system improves the turning operation in all type of materials in terms of tool life or surface finish compared to dry machining and wet machining. Zeilmann et al. [4] found that internal drilling with MQL is effective compared to external drilling. Tasdele et al. [5] concluded that MQL stand better in all respect while drilling compared to dry compressed air. E.A. Rahim et al. [6] proposes two lubricants along with MQL i.e palm oil and synthetic ester. Author concluded that palm oil can be used as an alternative lubricant in place of ester oil as it gives much better result in MQL system. Paturi et al. [26], while turning operation of Inconel 718 along with solid lubricant concluded that MQL system improves surface finish by 35%.

MQL with Nano Coolant

Nano particle size are in the range of 1 to 100 nano meter [31, 44–45]. On the basis of research work nanoparticles are broadly classified metallic, non-metallic, metal-oxides, carbide, and ceramics, carbonic and hybrid nano particles [33]. Vasu et al. [42] study the effect of aluminum oxide on turning Inconel 600 alloy. Cetina et al. [37], done milling process using Nano silver particles with different basic fluids such as Boron oil, ethylene glycol, borax. Both authors found Nano MQL (nMQL) much effective compare to pure MQL. Asthana [50] done the comparative study in between copper oxide and aluminum oxide. His study reveals that CuO nano fluids are better than Al_2O_3 in every aspect including Heat transfer coefficient Author's analysis further proves that percentage of surfactant plays crucial role in stability of nano particles. Among all the Nano fluids, the fluid with the base composition of (60:40) EG:water with 5% surfactant shows best results. Sahakian et al. [43] proved in his experimentation that composition of base fluid too have major role in performance as it provide stability to fluid. Fluid with composition of (60:40) EG:water showed best results in that respect along with CuO nano particles [34]. further he mentions that zinc oxide particles are found to be toxic which may be harmful to human being and the environment even though it gives much better results in order to reduce cutting force and temperature. Khedkar et al. studies the thermal conductivity of CuO nanoparticles in the base fluids like mono ethylene glycol and water. Thermal conductivity was checked by varying types of base fluid, sonication and settlement time. It was found that thermal conductivity increased with increase in sonication time. CuO with water as a base fluid found to give much better thermal conductivity. Thermal conductivity enhances with increase in percentage of concentration. Khatai [48] presented comparison of all metal oxide nanoparticles and mention that ZnO has highest thermal conductivity among all with 29.8 W/m.K while aluminum oxide next to it with value of 40 W/m.K. Karlsson et al. [46] revealed in his comparative study of toxicity that carbon nano tubes (CNT) are less toxic in nature compare to all metal oxides While copper oxide nano particles are having most toxicity [39]. Li et al. [22] did experimentation on grinding operation by using six different nano particles. It includes MoS2, ZrO_2, CNT, PCD, Al_2O_3, SiO_2 along with palm oil as a base fluid and found that CNT among all gives highest heat transfer coefficient. MQL with CNT Nano fluid also found to have large contact angle and low surface tension. So it becomes important to study the MQL system with CNT. Das et al. [51] processed hardened steel by using ZnO, CuO, Fe_2O_3, and Al_2O_3 in deionized water and got best result from Cuo followed by ZnO.

MQL with Carbon Nanto Tube

Carbon nano tubes have maximum thermal conductivity (3000 W/m-K) than any other nano particles used so far [29, 35, 41]. Sharmin et al. [8] evaluates carbon nano tube (CNT)-by mixing it with water. He synthesizes four different samples of Carbon nano-tubes. CNT with volume percentage 0.2%, 0.3%, 0.4% and 0.5% were synthesize and concluded that 0.3% volume CNT gives best results. Author mentioned that increase in percentage, increases thermal conductivity too. Raju et al. [11] uses sodium dodecyl sul-phate as surfactant (0.5%) in MWCNT (0.2%) for turning operation on EN 31 material and found to be best effective even compared to Pure mineral oil MQL operation. Hegab et al. [34] done comparison of MWCNT and aluminum oxide (Al_2O_3) while turning Inconel 718. Two different weight percentages of both nano particles have been used i.e., 2% and 4%. He found that MWCNT has better effectiveness compared to Al_2O_3.

MQL with Graphene Nano Particles

Graphene is two dimensional thin layered structures hence more convenient to use in machining process as it enters more easily into the gap of work piece-tool-chip interface. Graphene oxide, graphene nano platelets are other forms of graphene [14, 15]. Graphene oxide has more thermal conductivity of 5800 W-m-1 K-1 [18]. Graphite as a nanoparticle is much better option in machining operation as it provides lubricating as well cooling effect [23] also graphene has Young's modulus of about 1.0 Tera pascals [30], which is greater than diamond. Amrit Pal et al. [16] aims his work on drilling operation AISI 321 stainless steels using vegetable-oil-based cutting fluid. He concluded that drilling with 1.5 % of graphene nano particles gives best machining results compare to 0.5 & 1%. Shuang Yi et al. [17] done his investigation on milling operation using graphene oxide as a Nano particle with different weight percentage and got best result with 0.3 wt. %. Some high-lighted research work has mentioned in Table 33.1 along with its references.

MQL with Hybrid Nano-particles

Hybrid nano coolant found to be potent [12]. Yuanzhou Zheng [37] has done investiga-tion on thermal conductivity (TC) of hybrid nano-additives consist of zinc oxide and silica nanoparticles in base fluid of water/EG (70:30) with different samples of volume fraction. Rabesh Kumar Singh et al. [20] mixed alumina-based nanofluid with graphene nanoplate-lets (GnP) in the volumetric concentrations of 0.25, 0.75 and 1.25 vol %. Gugulothu and V. Pasam [27] done synthesis of two hybrid nano coolant one—carbon nanotubes mixed

Table 33.1: Summary of comparison of review of different published papers on graphene.

Ref.	Machining Operation	Nano Particles	Base fluid	% of Nano Particles in base Fluid	Outcomes
[17]	Milling	Graphene Oxide	mixed mineral oil and saponified nature oil	0.1 wt.% GO, 0.3 wt.% GO and 0.5 wt.% GO	a) Cutting force decreased with the increase in GO concentration. b) 0.3 wt. % showed the best performance on tool wear.
[19]	Turning	Graphene	canola oil	0.2, 0.5, 1.0, and 2.0%) wt %	1% wt of graphene gives best results among all in turning operation
[24]	Turning	Nano Graphite	Soluble oil	0.1, 0.3 and 0.5wt%.	Minimum tool wear & Cuttting force obtained at 0.5wt% nanographite-soluble oil graphite-LB2000
[26]	Turning	Graphene Oxide Nano sheets	LB2000 vegetable-based oil	0.1 and 0.5 wt% of Nano Graphite	Nano fluid reduces cutting force and zone temperature in best way.

with boric acid and other CNT mixed with molybdenum disulfide (CNT/MoS2). sesame, neem, and mahua oils used as base fluid for both hybrid fluids. Zhang, Y et al. [32] choose different concentration of nano fluids, include molybdenum disulfide (MoS2), carbon nanotubes (CNTs), and their mixtures (MoS2-CNTs) while grinding on Ni-based alloy. Results indicate that hybrid nano fluid proves to be potential compare to others. T.M. Duc [10] has done turning operation on 90CrSi steel (62HRC) by using hybrid nanofluid of Al2O3 and MoS2 nanoparticles in the base fluid of soybean oil and water-based emulsion and found that cutting force and surface roughness improved by using hybrid nano cutting fluid. Muhammad Jamil et al. [24] has done comparison between cryogenic CO2 cooling and hybrid nano MQL technique for turning operation of Ti–6Al–4V. Lv, T et al. [26] developed water-based GO/SiO2 hybrid nano cutting fluid and compare the result with pure GO and SiO2 water based MQL system. Kishor Kumar Gajrani et al. [36] developed three nano cutting fluid i.e MoS2, CaF2 and hybrid of both by using 0.3% concentration of CaF2 and 0.3% concentration MoS2. Results shows that hybrid anno coolant perform better compare to pure MoS2, CaF2. Jahar Sarkar et al. [38] mentioned study of different hybrid nano fluids in his review and concluded that due to synergistic effect, hybrid nano fluid provides all properties. Different authors have done comparative studies of nano particles with different base fluid. Anthony Chukwu jekwu Okafor et al. [40] investigated— high oleic soybean oil (HOSO), refined low oleic soybean oil (LOSO), acculube LB2000 (LB2000) oil, and mineral oil–based emulsion cutting fluid (EC). Author concluded that HOSO shows enhanced viscosity compare to rest base fluids at 300c whereas soyabene oil

& LB2000 has similar thermal conductivity. Surjeet Singh Bedi et al. [52] found rice Brine oil is much better option as a base fluid compare to coconut oil.

Conclusion

MQL found to be effective compare to flood cooling as it reduces consumption of cutting fluid and gives better machining performances' system with addition of nano particles (n-MQL) is proposed and mentioned in this paper. n-MQL results shown to be more effective compare to conventional cooling and pure MQL system. However, it is found by many researchers that synthesis of nano particle effects the quality nano particles. Sonication and settlement time for nano particles do have effect on response parameters. Size and concentration of nano particles in base fluid effect the quality of machining process. More work needs to be done on effect of base fluid. Blended oil mixed with nano practical is also the field to be investigated more.

References

1. Rahman, M. H., and Das, D. (2017). A review on application of nanofluid MQL in machining mustafa rifat. In AIP Conference Proceedings, 1919, 020015; https://doi.org/10.1063/1.5018533.
2. Joshi, K. K., and Kumar, R., and Anurag, (2018). An experimental investigations in turning of incoloy 800 in Dry, MQL and flood cooling conditions. In 2nd International Conference on Materials Manufacturing and Design Engineering. *Procedia Manufacturing*, 20, 350–357.
3. Gajrani, K. K., Suvin, P. S., Kailas, S. V., and Sankar, M. R. (2018). Hard machining performance of indigenously developed green cutting fluid using flood cooling and minimum quantity cutting fluid. *Journal of Cleaner Production*, 206, 108–123. doi: 10.1016/jjclepro.2018.09.178.
4. Zeilmann, R. P., and Weingaertner, W. L. (2006). Analysis of temperature during drilling of Ti6Al4V with minimal quantity of lubricant. *Journal of Materials Processing Technology*, 179(1–3), 124–127. https://doi.org/10.1016/j.jmatprotec.2006.03.077.
5. Tasdelen, B.., Wikblom, T., and Ekered, S. (2008). Studies on minimum quantity lubrication (MQL)and air cooling at drilling. *Journal of Materials Processing Technology*, 200, 339–346.
6. Rahim, E. A., and Sasahara, H. (2011). A study of the effect of palm oil as MQL lubricant on high speed drilling of titanium alloys. *Tribology International*, 44, 309–317.
7. Sharma, A. K., Tiwari, A. K., and Dixit, A. R. (2016). Effects of minimum quantity lubrication (MQL) in machining processes using conventional and nanofluid based cutting fluids: a comprehensive review. *Journal of Cleaner Production*, 127, 1–18.
8. Sharmin, I., Gafur, M. A., and Dhar, N. R. (2020). Preparation and evaluation of a stable CNT-water based nano cutting fluid for machining hard-to-cut material. *SN Applied Sciences*, 2, 1–18.
9. Shokoohi, Y. (2018). Application of superabsorbent coolant as a novel approach to semi-dry machining. Doctoral dissertation.
10. Duc, T. M., Long, T. T., and Chien, T. Q. (2019). Performance evaluation of MQL parameters using Al2O3 and MoS2 nanofluids in hard turning 90CrSi steel. *Lubricants*, 7(5), 40. https://doi.org/10.3390/lubricants7050040.
11. Raju, R. A., Andhare, A., and Sahu, N. K. (2017). Performance of multi walled carbon nano tube based nanofluid in turning operation. *Materials and Manufacturing Processes*, 32(13), 1490–1496.
12. Eltaggaz, A., Hegab, H., Deiab, I., and Kishawy, H. A. (2018). Hybrid nano-fluid-minimum quantity lubrication strategy for machining austempered ductile iron (ADI). *International Journal on Interactive Design and Manufacturing (IJIDeM)*, 12, 1273–1281.

13. Pal, A., Chatha, S. S., and Sidhu, H. S. (2020). Experimental investigation on the performance of MQL drilling of AISI 321 stainless steel using nano-graphene enhanced vegetable-oil-based cutting fluid. *Tribology International*, 151, 106508.

14. Yi, S., Li, J., Zhu, J., Wang, X., Mo, J., and Ding, S. (2020). Investigation of machining Ti-6Al-4V with graphene oxide nanofluids: tool wear, cutting forces and cutting vibration. *Journal of Manufacturing Processes*, 49, 35–49.

15. Hand book of Graphene Volume 5: Graphene in Energy, Health Care, and Environmental Applications. Scrivener Publishing Wiley (P.N 27 to 46.)

16. Singh, R., Dureja, J. S., Dogra, M., Gupta, M. K., Mia, M., and Song, Q. (2020). Wear behavior of textured tools under graphene-assisted minimum quantity lubrication system in machining Ti-6Al-4V alloy. *Tribology International*, 145, 106–183.

17. Sharma, A. K., Dixit, A. R., Tiwari, A. K., Pramanik, A., and Mandal, A. (2017). Performance evaluation of alumina-graphene hybrid nano-cutting fluid in hard turning. *Journal of Cleaner Production*, 162, 830–845.

18. Li, G., Yi, S., Li, N., Pan, W., Wen, C., and Ding, S. (2019). Quantitative analysis of cooling and lubricating effects of graphene oxidenanofluids in machining titanium alloy Ti6Al4V. *Journal of Materials Processing Technology*, 271, 584–598.

19. Shen, B., Shih, A. J., and Tung, S. C. (2008). Application of nanofluids in minimum quantity lubrication grinding. *Tribology Transactions*, 51, 730–737.

20. Amrita, M., Srikant, R., Sitaramaraju, A., Prasad, M., and Krishna, P. V. (2013). Experimental investigations on influence of mist cooling using nanofluids on machining parameters in turning AISI 1040 steel. *Proceedings of the Institution of Mechanical Engineers, Part J: Journal of Engineering Tribology*, 227, 1334–1346.

21. Carou, D., Rubio, E. M., and Davim, J. P. (2015). A note on the use of the minimum quantity lubrication (MQL) system in turning. *Industrial Lubrication and Tribology*, 67(3), 256–261. ISSN: 0036-8792. Publication date: 13 April. https://doi.org/10.1108/ILT-07-2014-0070.

22. Li, B., Li, C., Zhang, Y., Wang, Y., Jia, D., Yang, M., and Sun, K. (2017). Heat transfer performance of MQL grinding with different nanofluids for Ni-based alloys using vegetable oil. *Journal of Cleaner Production*, 154, 1–11. https://doi.org/10.1016/j.jclepro.2017.03.213.

23. Su, Y., Gong, L., Li, B., Liu, Z., and Chen, D. (2016). Performance evaluation of nanofluid MQL with vegetable-based oil and ester oil as base fluids in turning. *International Journal of Advanced Manufacturing Technology*, 83, 2083–2089. DOI 10.1007/s00170-015-7730-x.

24. Jamil, M., Khan, A. M., Hegab, H., Gong, L., Mia, M., Gupta, M. K., and He, N. (2019). Effects of hybrid Al2O3-CNT nanofluids and cryogenic cooling on machining of Ti–6Al–4V. *International Journal of Advanced Manufacturing Technology*, 102, 3895–3909. doi:10.1007/s00170-019-03485-9.

25. Gugulothu, S., and Pasam, V. (2020). Testing and performance evaluation of vegetable-oil–based hybrid nano cutting fluids. *Journal of Testing and Evaluation*, 48(5), 3839–3854. https://doi.org/10.1520/JTE20180106.

26. Paturi, U. M. R., Maddu, Y. R., Maruri, R. R., and Narala, S. K. R. (2016). Measurement and analysis of surface roughness in WS2 solid lubricant assisted minimum quantity lubrication (MQL) turning of inconel 718. *Procedia CIRP*, 40, 138–143. doi: 10.1016/j.procir.2016.01.082.

27. Liu, Y., Wang, X., Pan, G., and Luo, J. (2013). A comparative study between graphene oxide and diamond nanoparticles as water-based lubricating additives. *Science China Technological Sciences*, 56, 152–157.

28. Lv, T., Huang, S., Hu, X., Ma, Y., and Xu, X. (2018). Tribological and machining characteristics of a minimum quantity lubrication (MQL) technology using GO/SiO2 hybrid nanoparticle water-based lubricants as cutting fluids. *The International Journal of Advanced Manufacturing Technology*, 96(5–8), 2931–2942. doi:10.1007/s00170-018-1725-3.

29. Lee, C., Wei, X. D., Kysar, J. W., and Hone, J. (2008). Measurement of the elastic properties and intrinsic strength of monolayer graphene. *Science*, 321(5887), 385–388.

30. Pervaiz, S., Kannan, S., and Kishawy, H. A. (2018). An extensive review of the water consumption and cutting fluid-based sustainability concerns in the metal cutting sector. *Journal of Cleaner Production*, 197, 134–153. doi: 10.1016/j.jclepro.2018.06.190.

31. Sharma, A. K., Tiwari, A. K., and Dixit, A. R. (2015). Improved machining performance with nanoparticle enriched cutting fluids under minimum quantity lubrication. *Materials Today: Proceedings*. 2, 3545–3551.

32. Gajrani, K. K., Suvin, P. S., Kailas, S. V., and Mamilla, R. S. (2019). Thermal, rheological, wettability and hard machining performance of MoS2 and CaF2 based minimum quantity hybrid nano-green cutting fluids. *Journal of Materials Processing Technology*, 266, 125–139.

33. Zhang, Y., Li, C., Jia, D., Li, B., Wang, Y., Yang, M., Hou, Y., and Zhang, X. (2016). Experimental study on the effect of nanoparticle concentration on the lubricating property of nanofluids for MQL. *Journal of Materials Processing Technology*, 232, 100–115.

34. Hegab, H., Umer, U., Soliman, M., and Kishawy, H. A. (2018). Effects of nano-cutting fluids on tool performance and chip morphology during machining Inconel 718. *The International Journal of Advanced Manufacturing Technology*, 96(9–12), 3449–3458. doi:10.1007/s00170-018-1825-0.

35. Rad, S., Goodarz, A., Hussein, T., Mahidzal, D., Salim, N. K., Emad, S., and Nashrul, Z. (2014). An experimental study on thermal conductivity and viscosity of nano fluids containing carbon nanotubes. *Nanoscale Research Letters*, 9(151), 1–16. doi: 10.1186/1556-276X-9-151.

36. Khedkar, R. S., Sonawane, S. S., and Wasewar, K. L. (2012). Influence of CuO nanoparticles in enhancing the thermal conductivity of water and monoethylene glycol based nanofluids. *International Communications in Heat and Mass Transfer*, 39(5), 665–669. doi: 10.1016/j.icheatmasstransfer.2012.03.012.

37. Cetin, M. H., and Kilincarslan, S. K. (2020). Effects of cuttingfluids with nano-silver and borax additives on milling performance of aluminium alloys. *Journal of Manufacturing Processes*, 50, 170–182. Accepted 20 December 2019.

38. Sarkar, J., Ghosh, P., and Adil, A. (2015). A review on hybrid nanofluids: recent research, development and applications. *Renewable and Sustainable Energy Reviews*, 43, 164–177. doi: 10.1016/j.rser.2014.11.023.

39. Prabu, L., and Saravanakumar, N. (2019). Experimental study on the anti-wear and anti-corrosive properties of the water-soluble metalworking fluid dispersed with copper and aluminium oxide nanoparticles. *Materials Research Express*, 6(12), 125022.

40. Okafor, A. C., and Nwoguh, T. O. (2020). A study of viscosity and thermal conductivity of vegetable oils as base cutting fluids for minimum quantity lubrication machining of difficult-to-cut metals. *The International Journal of Advanced Manufacturing Technology*, 106, 1121–1131. https://doi.org/10.1007/s00170-019-04611-3.

41. Asthana, S., Rattan, S., and Das, M. (2013). Comparative studies of copper oxide with aluminium oxide nanoparticles in conventional thermal fluids for its enhanced efficiency as coolant. *Proceedings of the National Academy of Sciences, India Section A: Physical Sciences*, 83, 73–77.

42. Vasu, V., and Pradeep Kumar Reddy, G. (2011). Effect of minimum quantity lubrication with Al2O3 nanoparticles on surface roughness, tool wear and temperature dissipation in machining Inconel 600 alloy. *Proceedings of the Institution of Mechanical Engineers, Part N: Journal of Nanoengineering and Nanosystems*, 225(1), 3–16. DOI: 10.1177/1740349911427520.

43. Sahakian, M. (2011). Machining and toxicological performance of a zinc oxide metalworking nanofluid. https://ir.library.oregonstate.edu/concern/graduate_thesis_or_dissertations/bc386m742.

44. Tschätsch, H., and Reichelt, A. (2009). Cutting fluids (coolants and lubricants). In Applied Machining Technology. ed: Berlinm Heidelberg: Springer, (pp. 349–352).

45. Boswell, B., Islam, M. N., Davies, I. J., Ginting, Y. R., and Ong, A. K. (2017). A review identifying the effectiveness of minimum quantity lubrication (MQL) during conventional machining.

International Journal of Advanced Manufacturing Technology*, 92, 321–340. DOI 10.1007/ s00170-017-0142-3.

46. Karlsson, H. L., Cronholm, P., Gustafsson, J., and Möller, L. (2008). Copper oxide nanoparticles are highly toxic: a comparison between metal oxide nanoparticles and carbon nanotubes. *Chemical Research in Toxicology*, 21, 1726–1732.

47. Zheng, Y., Firouzi, M., Manafi, S., and Rostami, S. (2020). Improving the thermal conductivity of an antifreeze by suspending the hybrid nano-additives consist of zinc oxide and silica nanoparticles. *International Communications in Heat and Mass Transfer*, 116, 104649. https:// doi.org/10.1016/j.icheatmasstransfer.2020.104649.

48. Khatai, S., Kumar, R., Sahoo, A. K., Panda, A., and Das, D. (2020). Metal-oxide based nano fluid application in turning and grinding processes: A comprehensive review. *Materials Today: Proceedings*, 26(Part 2), 1707–1713. https://doi.org/10.1016/j.matpr.2020.02.360.

49. Okafor, A. C., and Nwoguh, T. O. (2020). A study of viscosity and thermal conductivity of vegetable oils as base cutting fluids for minimum quantity lubrication machining of difficult-to-cut metals. *The International Journal of Advanced Manufacturing Technology*, 106, 1121–1131. https://doi.org/10.1007/s00170-019-04611-3.

50. Asthana, S., Rattan, S., and Das, M. (2013). Comparative studies of copper oxide with aluminium oxide nanoparticles in conventional thermal fluids for its enhanced efficiency as coolant. *Proceedings of the National Academy of Sciences, India Section A: Physical Sciences*, 83, 73–77.

51. Das, A., Patel, S. K., Arakha, M., Dey, A., and Biswal, B. B. (2020). Processing of hardened steel by MQL technique using nano cutting fluids. *Materials and Manufacturing Processes*, 36(3), 316–328. doi:10.1080/10426914.2020.1832688.

52. Bedi, S. S., Behera, G. C., and Datta, S. (2020). Effects of cutting speed on MQL machining performance of AISI 304 stainless steel using uncoated carbide insert: application potential of coconut oil and rice bran oil as cutting fluids. *Arabian Journal for Science and Engineering*, 45, 8877–8893. doi:10.1007/s13369-020-04554-y.

53. Gaurav, G., Sharma, A., Dangayach, G. S., and Meena, M. L. (2020). Assessment of jojoba as a pure and nano-fluid base oil in minimum quantity lubrication (MQL) hard-turning of Ti–6Al–4V: A step towards sustainable machining. *Journal of Cleaner Production*, 272, 122553.

34 Experimental evaluation of mechanical properties of fused deposition printed thermoplastics materials

Debashis Mishra[a] and Anil Kumar Das[b]

Department of Mechanical Engineering, National Institute of Technology, Patna, India

Abstract: The FDM process named as fused deposition modeling technique is widely used and well-known for the fabrication of multi-dimensional solid objects or parts and has become popular than conventional techniques in terms of quick and ease in inventions. The layer-by-layer deposition is the idea of printing solid parts by using various thermoplastic materials. This experimental investigation was an effort to print the specimens of ABS, Carbon fiber PLA and PETG thermoplastics. The specimens of hardness, tensile and flexural were prepared as per the ASTM standards. The process parameters were like infill density, specimen printing speed and the layer thickness. From the testing, the highest strength of printed ABS was identified as 22.62MPa, flexural strength 76MPa, and hardness 71BHN. In the case of the printed carbon fiber PLA specimen it was reported as 26.31MPa, flexural strength 153MPa, and hardness 76BHN. The maximum strength of printed PETG specimens was found to be 34MPa, flexural strength 37MPa, and hardness 69BHN. This relative investigation and analysis will be helpful to know the detailed perceptive of the technique and the various properties of chosen printing materials and the process is useful for different applications.

Keywords: Acrylonitrile butadiene styrene (ABS), carbon fiber polylactic acid (CF-PLA), flexural strength, hardness value, fused deposition modeling (FDM/3D printing), polyethylene terephthalate glycol (PETG), tensile strength.

Brief About the Fused Deposition Modeling Technique and Related Literature Studies

The concept of manufacturing the complex and composite products or parts by the fused deposition modeling technique lies with an addition of the thermoplastic materials by multiple layers over a substrate. Many researchers used the fused deposition technique and presented their research work in many different ways.

In one of the review articles, it was noted that the process was a useful technique in comparison with the traditional manufacturing techniques in terms of numerous possibilities. The possibilities were like high quality and difficult to produce products that can be manufactured in less time and cost. The design of experimental techniques was largely used in preparing and formulating the experiments to be executed and to identify the influential

[a]debashism.ph21.me@nitp.ac.in, [b]akdas@nitp.ac.in

factors [1]. The additive manufacturing process was pointed as a joining process of materials by layer upon layer.

Carbon fiber reinforced composites were produced by the addition of carbon fiber in With the different percentages such as 0, 3, 5, 7.5, 10, 15, and ABS thermoplastics. The tested results were shown as 5% addition of carbon fiber produced the desired properties [2]. The mechanical properties of PLA were improved with the addition of carbon fiber. The tensile performance was reported to be improved by 47.1%. The flexural strength and stiffness were increased by 89.75% and 230.95% respectively. Dimensional accuracy was not disturbed and up-gradation in surface roughness was reported [3]. Structural strength was reported to be enriched with the addition of carbon fiber and other materials into the polymer matrix [4]. Various case studies and reviews were discussed and the results were reported as the production of composites is better supported by the additive manufacturing processes by making ease to the fixation, positioning, and processing. The additive manufacturing process has a lot of potentials to make it a suitable one than the other various conventional methods for the biomedical, aerospace, mechanical, and electronics applications [5–8]. The impact of different factors such as layer height, raster angle, and width on the flexural strength of fused filament printed PLA was investigated. The printing factors, layer height was observed to be most influential, and then the raster angle. The enhanced mechanical properties were reported by the fused deposition modeling technique than the conventional injection molding technique [9]. The development of smart materials and the industrial revolution brought up futuristic innovation in design and manufacturing were reported by introducing the internet of things, sensors, virtual realities, and robotics [10]. The hardness and tensile strength of printed thermoplastics were impacted by building direction [11]. The 3D printing process was optimized and the conclusion was made suitable melting of filament and controlled solidification depends upon appropriate printing speed and temperature of the nozzle [12]. The strength and weaknesses of PETG with and without the reinforcement of glass fiber was reported [13, 14]. The FDM process was optimized and the effective parameters were reported as fill density, feed rate and layer thickness. The optimization was helpful for various commercial applications in research and industrial areas [15–17]. It was reported that the raster angle and build orientations had insignificant impacts on the Youngs modulus of tensile specimens in ABS material [18]. The factors like deposition positioning, layer thickness, raster width and gap have major influence on strength of FDM fabricated parts [19]. The tensile strength of FDM printed specimen were increased with the reinforcement of continuous way of the carbon fiber from 5 to 40% respectively [20, 21]. In general, the interest of researchers is to present the theories and advancements of fused deposition modeling techniques and experimental investigations on different mechanical and micro-structural properties of printed thermoplastic specimens. This experimental examination is meant to provide a relative and proportional study of the hardness, tensile strength and flexural strength of FDM printed ABS, carbon fiber PLA and PETG thermoplastic materials. The investigation will be beneficial to get a methodical insight into the productivity and performance of chosen thermoplastics for fruitful applications in various industries.

Printing of Various Thermoplastics Test Specimens

The chosen technique named as the fused deposition modeling (FDM/3D printing) was used to fabricate the required numbers of hardness, tensile and flexural strength testing specimens of various thermoplastic materials such as ABS, Carbon fiber PLA and PETG.

The different printing process parameters were used such as printing speed, in-fill density, and layer thickness. The Figure 34.1 is the representation of the printing setup for the experiments. The Guider II Flash-forge 3D printer and the brass nozzle were employed for the fabrication of the required number of specimens. The preferred printing factors and their working settings are given in Table 34.1. The CAD drawings and printed specimen images are shown in Figures 34.2, 34.3, and 34.4 respectively. The internal pattern of infill density pattern of the printed thermoplastics is the honeycomb structure.

Figure 34.1 FDM set-up which is used for printing of the specimens.

Table 34.1: Chosen process parameters and their operational limits for the making of required specimens.

FDM printing process parameters	FDM Printing limits of chosen thermoplastics		
Infill density (%)	40	60	80
Printing speed (m/sec)	0.06	0.08	0.1
Layer thickness (microns)	100	200	300

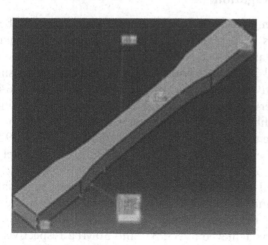

Figure 34.2 The CAD drawing and prepared tensile specimens respectively (ASTM D638).

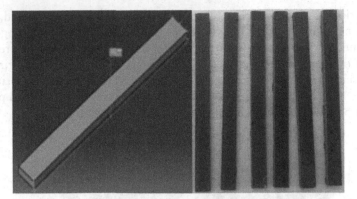

Figure 34.3 The CAD drawing and prepared flexural specimens respectively (ASTMD 790).

Figure 34.4 The CAD drawing and prepared hardness specimen respectively.

Outcomes From the Experimental Investigation

The hardness value, tensile and flexural strength of FDM or 3D printed specimens of ABS, carbon fiber PLA and PETG thermoplastics were reported in this experimental investigation. The obtained hardness value of ABS was followed as 68, 71, and 70BHN. The tensile strength was obtained as 12.37, 14.16, and 22.62MPa with the observed ultimate tensile load, 1.160, 1.560, and 1.320 in kN respectively. The flexural strength was achieved as 64, 76, and 73MPa. The force applied was observed as 80, 86.5, and 85.5N respectively. Table 34.3 shows the tested results of the carbon fiber PLA thermoplastic material. The obtained hardness values are 64, 76 and 74BHN. The tensile strength are measured as 17.12, 16.92, and 26.31MPa concerning the applied tensile load, 1.56, 1.36, and 1.84 in kN respectively. The flexural strength was obtained as 142, 153, and 135MPa for the force applied, 0.24kN respectively. Table 34.4 expresses the tested results of the PETG thermoplastic material. The obtained hardness value was 62, 69 and 63 in BHN. The tensile strength was measured as 25, 20, and 34 in MPa concerning the applied tensile load, 2.28, 1.68, and 2.88 in kN respectively. The flexural strength was achieved as 35, 37, and 30MPa respectively. Table 34.2 shows different tested results of ABS thermoplastic material.

Table 34.2: Different tested results of ABS thermoplastic material.

Test piece number	Hardness value (BHN)	Ultimate tensile load (kN)	Ultimate Tensile strength (MPa)	Force (N)	Flexural strength (MPa)
1	68	1.160	12.37	80.0	64
2	71	1.560	14.16	86.5	76
3	70	1.320	22.62	85.5	73

Table 32.3: Tested results of carbon fiber PLA thermoplastic.

Test piece number	Hardness value (BHN)	Ultimate tensile load (kN)	Ultimate Tensile strength (MPa)	Force (N)	Flexural strength (MPa)
1	64	1.56	17.12	240	142
2	76	1.36	16.92	240	153
3	74	1.84	26.31	240	135

Table 32.4: Tested results of PETG thermoplastic.

Test piece number	Hardness value (BHN)	Ultimate tensile load (kN)	Ultimate Tensile strength (MPa)	Force (N)	Flexural strength (MPa)
1	62	2.28	25	60	35
2	69	1.68	20	57.5	37
3	63	2.88	34	57	30

Conclusion

The results were reported as follows; the maximum hardness in the case of ABS was obtained as 71 BHN, 76BHN in carbon fiber PLA, and 69BHN in PETG thermoplastics. The ultimate tensile strength was attained as 22, 26, and 34MPa in ABS, carbon fiber PLA and PETG thermoplastics respectively. The flexural strength was found as 76, 153, and 37 in MPa in ABS, carbon fiber PLA and PETG thermoplastics respectively. Thus, it was well observed from this experimental exploration that the hardness, tensile and flexural properties of FDM printed materials were self-reliant. These properties are likely to vary with the variation in the input limits of chosen factors or processed parameters. The maximum fill density and low printing speed produce good tensile strength but the combination of both the factors is required to be identified by performing rigorous experimentations. The comparative examination will be supportive to acquire a careful and methodical perceptive about the process and also about the mechanical properties of FDM printed ABS, carbon fiber PLA and PETG thermoplastic material. The workshop is meant for the sharing of the recent progress in various streams of mechanical engineering. The workshop is giving an insight into research activities and fundamentals of various theoretical concepts and their applications to a wider perspective.

Future Scope of Work

The performed experimental exploration can be promoted in examining of the microstructures of the FDM/3D printed ABS, carbon fiber PLA and PETG thermoplastic materials.

The micro-structural examination can be carried and analyzed with the before and after the testing to observe the development of stress and strain under the applications of load. The other composite materials can be prepared and chosen for printing and their properties can be examined. The different filaments can be used in the printing of the specimens and can be compared to identify the most suitable one. The simulation and optimization of the FDM process can be performed to obtain an optimum and preferred printing conditions. The finite element analysis can be performed to predict the printing behavior of thermoplastics.

References

1. Sheoran, A. J., and Kumar, H. (2020). Fused deposition modeling process parameters optimization and effect on mechanical properties and part quality: Review and reflection on present research. *Elsevier Materials Today: Proceedings*, 21(Part 3), 1659–1672. https://doi.org/10.1016/j.matpr.2019.11.296.
2. Ning, F., Cong, W., Qiu, J., Wei, J., and Wang, S. (2015). Additive manufacturing of carbon fiber reinforced thermoplastic composites using fused deposition modeling. *Elsevier, Composites: Part B*, 80, 369–378. http://dx.doi.org/10.1016/j.compositesb.2015.06.013.
3. Reverte, J. M., Caminero, M. Á., Chacón, J. M., García-Plaza, E., Núñez, P. J., and Becar, J. P. (2020). Mechanical and geometric performance of PLA-based polymer composites processed by the fused filament fabrication additive manufacturing technique. *Materials (Basel)*, 13(8), 1924. DOI: 10.3390/ma13081924. PMID: 32325825; PMCID: PMC7215744.
4. Dickson, A. N., Abourayana, H. M., and Dowling, D. P. (2020). 3D Printing of fibre-reinforced thermoplastic composites using fused filament fabrication-a review. *Polymers (Basel)*, 12(10), 2188. DOI: 10.3390/polym12102188. PMID: 32987905; PMCID: PMC7601740.
5. Turk, D. A., Kussmaul, R., Zogg, M., Klahn, C., Leutenecker-Twelsiek, B., and Meboldt, M. (2017). Composites part production with additive manufacturing technologies. *Elsevier, Procedia CIRP*, 66, 306–311.
6. Black, S. (2015). A growing trend: 3D printing of aerospace tooling. *Composites World*, 1(7), 22–31.
7. Singamneni, S., Yifan, L. V., Hewitt, A., Chalk, R., Thomas, W., and Jordison, D. (2019). Additive manufacturing for the aircraft industry: a review. *Journal of Aeronautics and Aerospace Engineering*, 8(1), 351–371. doi:10.4172/2329-6542.1000214.
8. Wang, X., Jiang, M., Zhou, Z., Gou, J., and Hui, D. (2017). 3D printing of polymer matrix composites: a review and perspective. *Composites Part B: Engineering*, 110, 442–458. ISSN 1359–8368. https://doi.org/10.1016/j.compositesb.2016.11.034.
9. Rajpurohit, S. R., and Dave, H. K. (2018). Flexural strength of fused filament fabricated (FFF) PLA parts on an open-source 3D printer. *Advanced Manufacturing*, 6, 430–441. https://doi.org/10.1007/s40436-018-0237-6.
10. Kumar, A. (2018). Methods and materials for smart manufacturing: additive manufacturing, internet of things, flexible sensors and soft robotics. *Manufacturing Letters*, 15(Part B), 122–125. ISSN 2213–8463. https://doi.org/10.1016/j.mfglet.2017.12.014.
11. Calignano, F., Lorusso, M., Roppolo, I., and Minetola, P. (2020). Investigation of the mechanical properties of a carbon fibre-reinforced nylon filament for 3D printing. *Machines*, 8(3), 52. https://doi.org/10.3390/machines8030052.
12. Abeykoon, C., Sri-Amphorn, P., and Fernando, A. (2020). Optimization of fused deposition modeling parameters for improved PLA and ABS 3D printed structures. *International Journal of Lightweight Materials and Manufacture*, 3(3), 284–297. https://doi.org/10.1016/j.ijlmm.2020.03.003.

13. Szykiedans, K., Credo, W., and Osiński, D. (2017). Selected mechanical properties of PETG 3-D prints. *Procedia Engineering*, 177, 455–461. doi: 10.1016/j.proeng.2017.02.245.
14. Durgashyam, K., Reddy, M. I., Balakrishna, A., and Satyanarayana, K. (2019). Experimental investigation on mechanical properties of PETG material processed by fused deposition modeling method. *Materials Today: Proceedings,* 18, 2052–2059. https://doi.org/10.1016/j.matpr.2019.06.082.
15. Padhi, S. K., Sahu, R. K., Mahapatra, S. S., Das, H. C., Sood, A. K., Patro, B., and Mondal, A. K. (2017). Optimization of fused deposition modeling process parameters using a fuzzy inference system coupled with Taguchi philosophy. *Advanced Manufacturing*, 5(3), 231–242.
16. Gordeev, E. G., Galushko, A. S., and Ananikov, V. P. (2018). Improvement of quality of 3D printed objects by elimination of microscopic structural defects in fused deposition modeling. *PLoS One*, 13(6), e0198370. https://doi.org/10.1371/journal.pone.0198370.
17. Dey, A., and Yodo, N. (2019). A systematic survey of FDM process parameter optimization and their influence on Part characteristics. *Journal of Manufacturing and Materials Processing,* 3, 64. doi:10.3390/jmmp3030064.
18. Cantrell, J. T., Rohde, S., Damiani, D., Gurnani, R., DiSandro, L., Anton, J., Young, A., Jerez, A., Steinbach, D., and Kroese, C. (2017). Experimental characterization of the mechanical properties of 3D-printed ABS and polycarbonate parts. *Rapid Prototyping Journal,* 23, 811–824. https://doi.org/10.1108/RPJ-03-2016-0042.
19. Liu, X., Zhang, M., Li, S., Si, L., Peng, J., and Hu, Y. (2017). Mechanical property parametric appraisal of fused deposition modeling parts based on the gray Taguchi method. *International Journal of Advanced Manufacturing Technology,* 89, 2387–2397. https://doi.org/10.1007/s00170-016-9263-3.
20. Luo, H., Tan, Y., Zhang, F., Zhang, J., Tu, Y., and Cui, K. (2019). Selectively enhanced 3D printing process and performance analysis of continuous carbon fiber composite material. *Materials,* 12, 3529. doi:10.3390/ma12213529.
21. Bade, L., Hackney, P. M., Shyha, I., and Birkett, M. (2015). Investigation into the development of an additive manufacturing technique for the production of fiber composite products. *Procedia Engineering,* 132, 86–93.

35 An optimal inventory policy for deteriorating products considering carbon emissions and carbon tax

Rajesh Kumar Mishra[a] and Vinod Kumar Mishra

Department of Mathematics and Scientific Computing, Madan Mohan Malaviya University of Technology, Gorakhpur, UP, India

Abstract: In this research paper, we examine an optimal inventory model for deteriorating products considering carbon emissions. In current scenario, consideration of carbon emissions in an inventory model has become essential. So, this proposed model incorporates the carbon emissions emitted from activities: storage and deterioration. Due to consideration of carbon emissions, this model involves carbon emission tax. In this model, total cost function and total carbon emission function are derived. Further, order quantity and cycle time are determined by minimizing total cost and total carbon emission. A solution procedure is provided to obtain the optimal solution. The convex nature of total cost function is shown by analytical approach as well as graphical approach. Finally, a numerical analysis is set to validate the model and moreover an analysis of sensitivity is carried out to investigate the effect of key model parameters on optimal policy.

Keywords: Carbon emissions, carbon tax, deterioration, optimal inventory model.

Introduction

In an inventory problem, mitigating carbon emissions has become an important issue for business organizations. Nowadays, business organizations are actively pursuing ways to reduce carbon emissions because of the carbon policies implemented by different regulatory bodies. ISO certification, market competitiveness, and consumer loyalty not only encourage policies but also encourage companies to implement the green practices in their supply chain. Regulatory authorities are taking steps to minimize carbon emission in several countries. So, in many countries, carbon taxation and cap-and-trade schemes are two implemented mechanisms [17]. Moreover, as the world economy relies on fossil fuels for industrial activities, emissions of GHG (greenhouse gas) cannot be halted. The primary task of mitigating global warming is to substantially lower the rate of emissions. In fact, the primary cause of GHG emissions, which are the main explanation for environmental pollution, is carbon emissions. Currently, organizations are forced to follow the green practices in production, supply chain, and storage process to mitigate carbon emissions. In this proposed article, we study an inventory model for deteriorating products considering carbon emissions, which includes inventory cost components: ordering cost,

[a]rkmishra1019@gmail.com

holding cost, purchasing cost, deterioration cost, and carbon emission tax and objective of this model is to minimize the total inventory cost and also to determine optimal order quantity and cycle time. Economic Order Quantity (EOQ) developed by Harris [1] is the first inventory model that has been thoroughly studied to date. Most items deteriorate during storage process [2]. In literature, there are available many review papers on inventory model with deterioration [35]. Further, many researchers studied different types of deteriorating inventory models considering realistic phenomena such as nonlinear holding cost and stock-dependent demand [6], price-dependent linear demand [7], varying deterioration rate, price-dependent demand, and shortage [8], rework and stochastic preventive maintenance time [9], multiple prepayments [10], expiration date and advance payments [11], and substitutable products with cost of substitution [12]. With increasing awareness of the environmental issue, many researchers incorporate the carbon emissions into their inventory model [13–15]. Recently, by considering carbon emission, Taleizadeh et al. [16] considered pricing and inventory decision jointly, partial backordering and planned discounts, and Mishra et al. [17] studied a sustainable inventory management considering deterioration and backorder. The structure of this article is as follows: section 2 describes notations, section 3 describes model development and solution approach, section 4 describes numerical example with sensitivity analysis, and lastly section 5 ends with conclusions.

Notations

To build the proposed inventory model, the following notations are used:

Notations	Notation description	Unit
$I(t)$	Level of inventory at time t in $[0, T]$	Units
D	Demand Rate	units/year
A	Ordering cost	$/order
q	Ordering quantity (decision variable)	unit
T	Cycle time (decision variable)	year
C	Purchasing cost per unit	$/unit
h	Holding cost per unit per unit time	$/unit/year
θ	Deterioration rate ($0 \leq \theta \leq 1$	-
d	Deterioration cost	$/unit
E_s	Average storage energy consumption per unit item	kwh/unit/year
C_{ES}	Carbon emission from electricity generation	ton-CO_2/kwh
C_{ED}	Carbon emission from deterioration	ton-CO_2/unit
T_x	Carbon emission tax	$/ton-$CO_2$
q^*	Optimal ordering quantity	units
T^*	Optimal cycle time	years
TC	Total cost per cycle	$/cycle
TCU	Total cost per unit time	$/year
TCE	Total carbon emission per unit time	ton-CO_2/year

Model Development and Solution Approach

Model Development

This model has been developed under the following assumptions:

1. Single product is assumed, replenishment is instantaneous, and lead time is taken to be zero.
2. Items are deteriorating.
3. Demand rate and deterioration rate are deterministic and constant.
4. Carbon emissions and carbon tax are considered.
5. Carbon emissions are due to storage and deteriorating items.
6. Shortage is not allowed.

At the initial time, the retailer received q items from a supplier, after that inventory level depletes due to the combined impact of deterioration and demand and reaches to zero at cycle time T. Inventory diagram of deteriorating items is depicted in Figure 35.1.

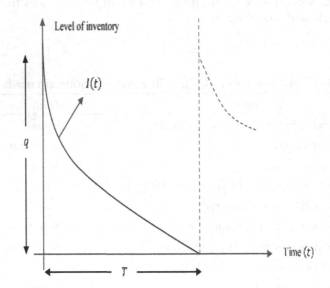

Figure 35.1 Inventory diagram of deteriorating items.

Thus, inventory level over the period [0,T] is described by the following linear differential equation:

$$\frac{dI(t)}{dt} + \theta I(t) + D = 0 \tag{1}$$

with conditions $I(0) = q$, $I(T) = 0$
After solving above boundary value problem, we get

$$I(t) = \left(q + \frac{D}{\theta}\right)e^{-\theta t} - \frac{D}{\theta}, \quad \text{in } [0,T] \tag{2}$$

since $I(T) = 0$ which gives cycle time in terms of order quantity as

$$e^{\theta T} = \frac{\theta q + D}{D}$$

$$T = \frac{1}{\theta} \ln\left(\frac{\theta q + D}{D}\right) \tag{3}$$

To find holding cost and carbon emissions emitted from storage, firstly average inventory will be calculated.

Average inventory per cycle is given by

$$AI = \int_0^T I(t)\, dt$$

$$AI = \int_0^T \left(\left(q + \frac{D}{\theta}\right) e^{-\theta t} - \frac{D}{\theta}\right) dt$$

$$AI = \frac{1}{\theta^2}\left\{(\theta q + D) - D\theta T - (\theta q + D)\, e^{-\theta T}\right\} \tag{4}$$

By equation (3), average inventory (AI) can be expressed as:

$$AI = \frac{1}{\theta^2}\left\{\theta q - D\ln\left(\frac{\theta q + D}{D}\right)\right\} \tag{5}$$

The total cost per cycle is calculated as:

Total cost = Ordering cost + Holding cost + Purchasing cost + Cost of deterioration + Carbon emission tax.

Carbon emission tax is calculated as:

$$\text{Carbon emission tax} = \frac{T_X}{\theta^2}\left\{\begin{array}{l} E_S C_{ES}\left(\theta q - D\ln\left(\frac{\theta q + D}{D}\right)\right) \\ + \theta C_{ED}\left(\theta q - D\ln\left(\frac{\theta q + D}{D}\right)\right) \end{array}\right\} \tag{6}$$

Using above cost components of this model, total cost per cycle (TC) is given as:

$$TC = \frac{1}{\theta^2}\left[\begin{array}{l} A\theta^2 + h\left(\theta q - D\ln\left(\frac{\theta q + D}{D}\right)\right) + Cq\theta^2 + d\theta\left(\theta q - D\ln\left(\frac{\theta q + D}{D}\right)\right) \\ + T_X\left(E_S C_{ES} + \theta C_{ED}\right)\left(\theta q - D\ln\left(\frac{\theta q + D}{D}\right)\right) \end{array}\right] \tag{7}$$

Finally, total cost per unit time is determined by:

$$TCU = \frac{TC}{T}, \text{ which gives as:}$$

$$TCU = \cfrac{1}{\theta\ln\left(\dfrac{\theta q + D}{D}\right)} \left[\begin{array}{l} A\theta^2 + h\left(\theta q - D\ln\left(\dfrac{\theta q + D}{D}\right)\right) + Cq\theta^2 + d\theta\left(\theta q - D\ln\left(\dfrac{\theta q + D}{D}\right)\right) \\ + T_X\left(E_S C_{ES} + \theta C_{ED}\right)\left(\theta q - D\ln\left(\dfrac{\theta q + D}{D}\right)\right) \end{array} \right] \quad (8)$$

Moreover, total carbon emission per unit time (*TCE*) is

$$TCE = \cfrac{1}{\theta\ln\left(\dfrac{\theta q + D}{D}\right)} \left[\left(E_S C_{ES} + \theta C_{ED}\right)\left(\theta q - D\ln\left(\dfrac{\theta q + D}{D}\right)\right) \right] \quad (9)$$

Solution Approach

To check unique optimality, we have shown the convexity of the total cost function per unit time analytically.

Taking first and second order derivative of equation (7), we get

$$\frac{d^2 TC}{dq^2} = \frac{D\left(\theta(d + T_X C_{ED}) + h + T_X E_S C_{ES}\right)}{(\theta q + D)^2} > 0 \, \forall$$

parameters because concerning all parameters are positive. Hence, *TC* is convex for all parameters. Further, by the result: a positive convex function/a positive concave function [18], total cost function *TCU* is a pseudo-convex function.

Algorithm to determine optimal policy.

Using the following algorithm, we have to calculate the optimal ordering quantity, optimal cycle time, minimal total cost, and minimal total carbon emission.

Step 1: Find q such that min *TCU* such that $q \geq 0$

Step 2: Optimal value of T i.e., T^* is calculated from equation (3).

Step 3: Minimal total carbon emission is denoted by TCE^* and is obtained from equation (9) by using optimal ordering quantity q^*. Table 35.1 shows input parameters.

Numerical Example and Sensitivity Analysis

In this section, in order to validate this model, a numerical example is provided.

Table 35.1: Input parameters.

Parameters of the model	Values (with proper units)
Ordering cost, A	200
Demand rate, D	100
Purchasing cost, C	50
Holding cost, h	0.05
Deterioration rate, θ	0.03
Cost of deterioration, d	0.01
Average storage energy consumption, E_s	10.50
Carbon emission from electricity generation, C_{ES}	0.0005
Carbon emission from deterioration, C_{ED}	0.0006
Carbon emission tax, T_X	80

As can be seen from above, pseudo-convexity of total cost, which ensures unique optimality (minimality) has been shown for all parameters. Using above algorithm, minimal total cost is $TCU^* = 5282.83$, optimal ordering quantity is $q^* = 143.44$, and optimal cycle time is $T^* = 1.40$. Thus, minimal solution is $q^* = 143.44$, $T^* = 1.40$, $TCU^* = 5282.83$. Moreover, minimal total carbon emission is $TCE^* = 0.38$. The graphical verification of convexity is given by Figure 35.2.

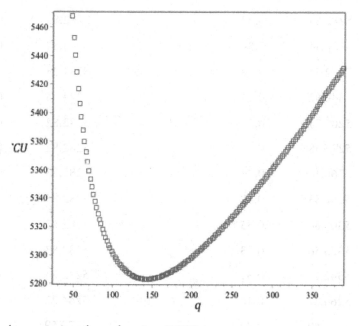

Figure 35.2 Pseudo-convexity of cost function (*TCU*).

Now, we conduct a sensitivity analysis of minimal total cost as well as total carbon emission with variations in all parameters. An analysis of sensitivity with increment and decrement by 20% in parameters is given by Table 35.2 and is graphically given by Figure 35.3.

Table 35.2: Sensitivity analysis in respect of parameters.

Parameter	TCU*	TCE*	Parameter	TCU*	TCE*
A	5178.41	0.23	D	2179.60	0.24
	5218.73	0.29		3219.53	0.29
	5252.78	0.33		4253.18	0.34
	5282.83	0.38		5282.83	0.38
	5310.03	0.41		6309.63	0.41
	5335.08	0.44		7334.29	0.44
	5358.42	0.47		8357.23	0.47

(continued)

Table 35.1: Continued

Parameter	TCU*	TCE*	Parameter	TCU*	TCE*
C	2209.04	0.50	h	5280.69	0.38
	3236.23	0.45		5281.40	0.38
	4260.58	0.41		5282.12	0.38
	5282.83	0.38		5282.83	0.38
	6303.44	0.34		5283.54	0.37
	7322.73	0.33		5284.25	0.37
	8340.92	0.31		5284.96	0.37
θ	5207.75	0.51	d	5282.82	0.38
	5235.38	0.45		5282.82	0.38
	5260.16	0.41		5282.83	0.38
	5282.83	0.38		5282.83	0.38
	5303.86	0.35		5282.83	0.38
	5323.56	0.33		5282.83	0.38
	5342.17	0.31		5282.84	0.38
E_S	5264.27	0.16	C_{ES}	5264.27	0.16
	5270.60	0.24		5270.59	0.24
	5276.78	0.31		5276.78	0.31
	5282.83	0.38		5282.83	0.38
	5288.75	0.44		5288.75	0.44
	5294.55	0.50		5294.55	0.50
	5300.24	0.56		5300.24	0.56
C_{ED}	5282.76	0.37	T_X	5264.20	0.40
	5282.79	0.37		5270.55	0.39
	5282.81	0.37		5276.76	0.38
	5282.83	0.38		5282.83	0.38
	5282.85	0.38		5288.77	0.37
	5282.87	0.38		5294.58	0.36
	5282.89	0.38		5300.30	0.35

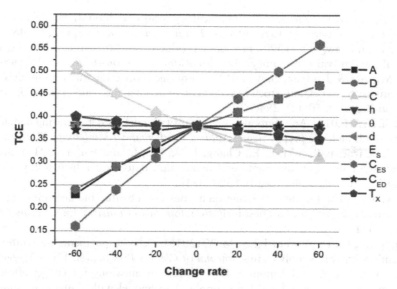

Figure 35.3 Sensitivity of carbon emissions concerning all parameters.

Conclusions

This model studies the optimal inventory and cycle time decisions for an inventory model of deteriorating products with carbon emissions and carbon tax. The aim of this study is to minimize the total inventory cost including ordering cost, holding cost, purchasing cost, deterioration cost, and carbon emission tax and total carbon emissions including carbon emission from storage and deterioration. Here, optimal value of ordering quantity and cycle time is determined. Convexity of objective function have been shown both analytically and graphically. Further, a numerical example is presented and using maple software, an analysis of sensitivity is carried out. The purpose of sensitivity analysis is to investigate the sensitivity of optimal policy with respect to input parameters. Future inventory research is needed to generalize for multi products. Also, it may be extended as probabilistic model by considering some stochastic parameters.

References

1. Harris, F. W. (1915). Operations and Costs. Chicago: A. W. Shaw Company, (pp. 48–54).
2. Ouyang, L. Y., Wu, K. S., and Cheng, M. C. (2005). An inventory model for deteriorating items with exponential declining demand and partial backlogging. *Yugoslav Journal of Operations Research*, 15(2), 277–288.
3. Goyal, S. K., and Giri, B. C. (2003). Recent trends in modelling of deteriorating inventory. *European Journal of Operational Research*, 134(1), 1–16.
4. Bakker, M., Riezebos, J., and Teunter, R. H. (2012). Review of inventory systems with deterioration since 2001. *European Journal of Operational Research*, 221(2), 275–284.
5. Janssen, L., Claus, T., and Sauer, J. (2016). Literature review of deteriorating inventory models by key topics from 2012 to 2015. *International Journal of Production Economics*, 182, 86–112.
6. Pando, V., José, L. A. S., Garc ía-Laguna, J., and Sicilia, J., (2018). Optimal lot-size policy for deteriorating items with stock-dependent demand considering profit maximization. *Computers and Industrial Engineering*, 117, 81–93.

7. Mahmoodi, A. (2019). Joint pricing and inventory control of duopoly retailers with deteriorating items and linear demand. *Computers and Industrial Engineering*, 132, 36–46.
8. Wee, H. M., and Law, S. T. (1999). Economic production lot size for deteriorating items taking account of the time-value of money. *Computers and Operations Research*, 26, 545–558.
9. Wee, H. M., and Widyadana, G. A. (2012). Economic production quantity models for deteriorating items with rework and stochastic preventive maintenance time. *International Journal of Production Research*, 50(11), 2940–2952.
10. Taleizadeh, A. A. (2014). An economic order quantity model for deteriorating item in a purchasing system with multiple prepayments. *Applied Mathematical Modelling*, 38(23), 5357–5366.
11. Teng, J. T., Cárdenas-Barrón, L. E., Chang, H. J., Wu, J., and Hu, Y. (2016). Inventory lot-size policies for deteriorating items with expiration dates and advance payments. *Applied Mathematical Modelling*, 40(19–20), 8605–8616.
12. Mishra, V. K. (2017). Optimal ordering quantities for substitutable deteriorating items under joint replenishment with the cost of substitution. *Journal of Industrial Engineering International*, 13(3), 381–391.
13. Aljazzar, S. M., Gurtu, A., and Jaber, M. Y. (2018). Delay-in-payments—a strategy to reduce carbon emissions from supply chains. *Journal of Cleaner Production*, 170, 636–644.
14. Bazan, E., Jaber, M. Y., and Zanoni, S. (2017). Carbon emissions and energy effects on a two-level manufacturer-retailer closed-loop supply chain model with remanufacturing subject to different coordination mechanisms. *International Journal of Production Economics*, 183, 394–408.
15. Hammami, R., Nouira, I., and Frein, Y. (2015). Carbon emissions in a multi-echelon production-inventory model with lead time constraints. *International Journal of Production Economics*, 164, 292–307.
16. Taleizadeh, A. A., Hazarkhani, B., and Moon, I. (2020). Joint pricing and inventory decisions with carbon emission considerations, partial backordering and planned discounts. *Annals of Operations Research*, 290, 95–113.
17. Mishra, U., Wu, J. Z., and Sarkar, B. (2021). Optimum sustainable inventory management with backorder and deterioration under controllable carbon emissions. *Journal of Cleaner Production*, 279, 123699.
18. Cambini, A., and Martein, L. (2009). Generalized Convexity and Optimization: Theory and Application. Berlin Heidelberg, USA: Springer-Verlag.

36 CFD-based numerical investigation of heat transfer augmentation and fluid flow characteristics in an artificially roughened SAH

Mukesh Kumar[a] and Arun Kumar[b]

Department of Mechanical Engineering, National Institute of technology Patna, India

Abstract: In this study, fluid flow parameters and heat transmission within a surface absorber and heater (SAH) with intentionally roughened surfaces are numerically analyzed. The study makes use of 2-Dimensional Computational Fluid Dynamics (CFD) with a variety of rib shapes, including rectangular, square, rectangle with a triangular end, and rectangle with a semicircular end, placed in different combinations. The RNG k-turbulence model and ANSYS—15 software were also used in the analysis. The 'second order upwind' method was used to solve the momentum and energy equations, and the SIMPLE algorithm was used to solve the coupling of velocity and pressure equations. A modest convergence threshold was applied to every residual in order to guarantee accurate parameter predictions. The study concentrated on turbulent flow conditions between 4000 and 16000 Reynolds numbers (Re). In particular, the results showed that the rectangular ribs with a triangular end had the greatest Nu, reaching 58.35 at a Reynolds number (Re) of 16000.

Keywords: Friction factor (f), Nusselt number (Nu), Reynolds number (Re), ribs, Solar air heater (SAH).

Introduction

One of the major parts of global energy consumption is the production of heat. Typically, heat is generated through the combustion of fossil fuels, leading to environmental issues such as global warming, acid rain, and ozone layer depletion. To address these concerns, the development of alternative energy sources becomes imperative. Among the environmentally friendly and carbon-neutral options, solar energy stands out as a readily accessible and abundant resource [1]. Solar energy serves as a means of generating heat in various ways. During winter, for instance, solar collectors can harness sunlight to produce hot air, which effectively warms buildings and indoor spaces [2]. In India, numerous small-scale food processing industries dealing with fruits and spices have begun to shift away from traditional electric air heaters for crop drying. Instead, they are increasingly adopting SAH as an alternative method [3]. In recent years, numerous researchers have focused on improving the thermo-hydraulic performance of flat plate SAHs due to their low thermal efficiency [4]. Komolafe et al. [5] work on an experimental investigation on SAH thermal evaluation with rectangular roughness on the absorber plates. They used locally fabricated

[a]mukesh.nitpatna@gmail.com, [b]arun@nitp.ac.in

SAH during the day time span of 09:00–18:00 h activity, the maximum strength of solar radiation 827.87 W/m², solar air heater and ambient temperature was, 112.0 and 33.77°C respectively. The thermal efficiency values were measured from 14.0 to 56.5 percent. The simulated maximum and minimum SAH temperature were 21 and 127°C, respectively, which were in fair accordance with the 20 and 112°C results of the experiment. Patel et al. [6] experimentally performed the effects on the efficiency of the SAH by the novel roughness part type in the form of NACA 0040 profile ribs in reverse position for the Re ranging from 6000 to 18000. At Reynolds number of 6000, the thermo hydraulic efficiency consideration for this experiments were get to be 2.53. The numerical finding was confirmed by the experimental result, and the optimum deviations were observed to be 4.84 %. Sarvankumar et al. [7] performed thermal and thermo-hydraulic examine of arc created rib on the SAHs absorber plate attached with baffles and fins. They presented variation of flow factors are provided with regard to baffle design parameters i.e Re and temperature raises parameters. Compared to the arc shapes rib roughened SAHs, the proposed SAH increases energy and performance by 28.3 percent and 27.1 percent. Wang et al. [8] carries out experimentally work on the study of the collector efficiency and pressure drop characteristics of solar air heater's absorber plate with multiple s-shaped ribs with gaps. They found thermal efficiencies of the SAH with artificials roughness were increased as a result of 13 percent to 48 percent under dissimilar conditions match with the smooth plate. Manjunath et al. [3] work on numerical analysis of heat transfer enhancement and consequent thermo hydraulics performances of sinusoidal profiled SAHs absorber plates for flow Re ranging from 4000 to 24000 using 3-Dimentional CFD simulation. Compared to the smooth channels for the design with a non-dimensional wavelength of 10.0 and an spect ratio 1.5, the average improvement in thermal efficiency is around 12.5 percent. Kumar et al. [9] numerically computed the heat transfer improvement and frictional flow character of triangular duct solar air heater having square parts ribs on absorber plates using CFD for Re varies between 3900 and 17900. The result showed that the highest heat transfers are seen to be in the order of 97 percent with a P/e of 10 and an e/D of 0.05 with a Re of 179000. Srivastava et al. [4] performed numerical study of different forms of repeated ribs roughness V shape with gaps and arc shape with gaps with comparable shape and size on the absorber plates of the SAHs for Re varies between 3000 and 15000 using CFD. The result showed that the maximum enhancement in Nusselt number (Nu) is 3.4 for V shape ribs with frictional penalty of 4.05 at Re 3000. Patel and Lanjewar [10] conducted an experimental study aimed at enhancing the thermo-hydraulic performance of solar air heater by employing a V-geometry with a similar gap roughened absorber plate. Their investigation involved varying roughness parameters, including a P/e value of 10, an angle of attack 60°C, and an e/D ratio of 0.043. The Reynolds number (Re) range investigated ranged from 4000 to 15000. The experiments aimed to assess the impact of these parameters on frictional flow and the rate of heat transfer. Remarkably, the maximum enhancements observed were noteworthy, with the Nusselt number (Nu) and friction factor achieving values 2.51 and 2.7 times higher, respectively, compared to those of a plain surface. Azad et al. [11] conducted experimental research to investigate the features of pressure drops and collector efficiency of SAHs with a discrete symmetrical arc geometry plate. The Reynolds number (Re) considered in their study ranged from 3000 to 14000. Various parameters, such as g/e, P/e, and e/Dh were examined during the experiments. The findings revealed that the Nusselt number (Nu) experienced a significant increase, reaching 3.88 times that of a plain surface, particularly at a g/e value of 4. This demonstrates the potential of the discrete symmetrical arc geometry plate in enhancing the performance of

SAHs. Dong et al. [12] used computational fluid dynamics (CFD) to conduct a numerical inquiry to examine the flow characteristics and improvement of heat transmission of solar air heater with a unique inclined groove ripple surface. The experiments involved varying roughness parameters, such as array number, attack angle, and groove amplitude, while considering Reynolds numbers (Re) within the range of 12000 to 24000. The aim was to assess the impact of these parameters on flow-friction and the heat transfer rate. The results demonstrated that SAH equipped with the ripple surface exhibited excellent thermo-hydraulic performance. The Nusselt number (Nu) experienced a notable increase, reaching 1.04 to 1.94 times that of a plain surface. These findings highlight the potential of the inclined groove ripple surface in enhancing the heat transfer efficiency of SAH.

2-D CFD Simulation

Problem Definition

This study's goal is to perform 2-D simulations in a synthetically roughened solar air heater that circulates air. Thin ribs were added to the inner surface of the absorber plate on solar heaters. For several types of ribs, a numerical analysis of heat transfer augmentation and frictional flow characteristics was conducted. The analysis was carried out using the ANSYS-15 program and the RNG k-turbulence model. It was decided to use the "second order upwind" technique for both momentum and energy equations. Velocity and Pressure were coupled using the SIMPLE method.

Computational Domain

This analysis takes into account a rectangular segment with (L) 640 mm dimensions, as shown in Figure 36.1. The segment has a length of 145 mm for the entrance, 245 mm for the test section, and 115 mm for the exit. The width (w) is 100 mm wide, while the height (H) is 20 mm. This geometry was adopted from Tanda's experimental work [13]. To ensure that the flow in the test section reaches a fully-developed state, an entrance length is provided. The simulation was conducted using ANSYS 15 software. In a study by Chaube et al. [14], they performed numerical simulations on their SAH and compared 3D results with 2D results, finding no significant differences. Consequently, in the present analysis, 2D simulations were carried out for the same reason.

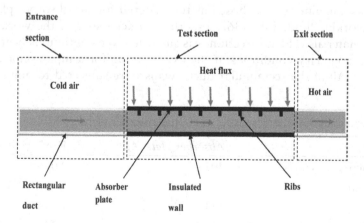

Figure 36.1 Illustration of the computational domain featuring a transverse square rib geometry.

Mess Generation

ANSYS software was used to mesh the geometry once it had been created. The "bias" form was used to meticulously divide each edge into the necessary number of parts. In order to guarantee a perfect rectangular mesh with excellent perpendicularity, the mapped face option was turned on. Finally, a desirable and well-structured mesh was generated by using the "Generate Mesh" option in the ANSYS-15 software.

Grid Independence Test

To attain accurate results, selecting the optimal grid dimension is of utmost importance. Hence, in this study, Grid independence tests were conducted at a Re of 4000 to determine the appropriate grid size for conducting the numerical simulations. Figure 36.2 illustrates the variation in average Nu (Nusselt number) across different grid elements.

Governing Equations

Different conservations equations for turbulent fluid flow with heat transfer are as follows:
Continuity equation

$$\frac{\partial}{\partial x_i}(\rho u_i) = 0 \tag{1}$$

Momentum equation

$$\frac{\partial}{\partial x_i}(\rho u_i u_j) = -\frac{\partial P}{\partial x_i} + \frac{\partial}{\partial x_j}\left[\mu\left(\frac{\partial u_i}{\partial x_j} + \frac{\partial u_j}{\partial x_i}\right)\right] + \frac{\partial}{\partial x_j}\left(-\overline{\rho u_i' u_j'}\right) \tag{2}$$

Energy equation

$$\frac{\partial}{\partial x_i}(\rho u_i T) = \frac{\partial}{\partial x_j}\left[(\Gamma + \Gamma_t)\frac{\partial T}{\partial x_j}\right] \text{ Where } \Gamma = \frac{\mu}{Pr} \text{ and } \Gamma_t = \frac{\mu_t}{Pr_t} \tag{3}$$

Boundary Conditions

For this analysis, aluminum was chosen as the material for the absorber plate, while air served as the working fluid. Table 36.1 lists the characteristics of the working fluid and absorber plate materials. The test section was subjected to a continuous heat flux of 1100 W/m2, and the boundary conditions at the exit were uniform input velocity and atmospheric pressure. All of the rectangular duct's walls were subjected to non-slip boundary criteria.

Table 36.1: Thermo-physical characteristics of air and absorber plates.

Properties	Absorber plate (Al.)	Working fluid (air)
Thermal conductivity, (W/m-K)	202.4	0.0242
Viscosity (kg/m-s)	-	1.7894e-05
Specific heat (J/kg-K)	871	1006.43
Density (kg/m³)	2719	1.225

Equations for Data Deduction

The Nusselt number (Nu) and friction factor (f) can be determined using the following equations.

$$\text{Reynolds number,} \quad Re = \frac{\rho v D_h}{\mu} \tag{4}$$

Following methods are used to obtained average heat transfer coefficient

$$h_{avg} = \frac{\dot{m} C_P (T_o - T_i)}{A_s \left(T_{avg,sur} - T_{avg,air}\right)} \quad \text{Where } T_{avg,air} = (T_i + T_o)/2 \tag{5}$$

$$\text{Friction factor } f = \frac{2(\Delta P) D_h}{4\rho L v^2} \tag{6}$$

$$\text{Average Nusselt number } Nu_{avg} = \frac{h_{avg} D_h}{k} \tag{7}$$

Result and Discussions

Validation of the Smooth Channel's Results

The Nusselt number (Nu) was used to validate the numerical results concerning heat transmission and frictional pressure decrease in a plain rectangular channel under uniform heat flux conditions. Figure 36.3 shows the comparison between the turbulent flow values from Tanda's experimental work [13] and the Nu obtained from the numerical simulations for the plane channel. It was observed that the Nusselt values obtained from the experimental results were considerably higher than those obtained from the smooth channel in the numerical simulations.

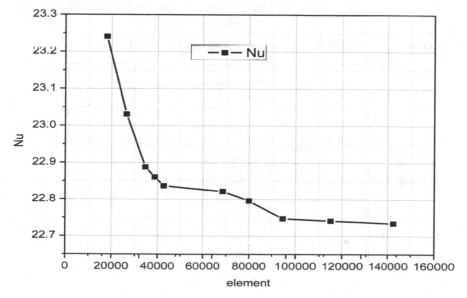

Figure 36.2 Nu changes with grid element at Re = 4000.

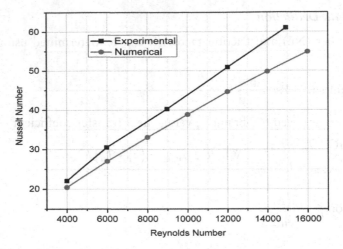

Figure 36.3 Plain channel validation.

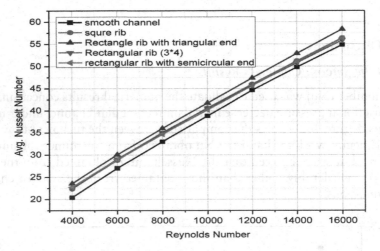

Figure 36.4 Mean Nusselt number changes with Re for different rib shape.

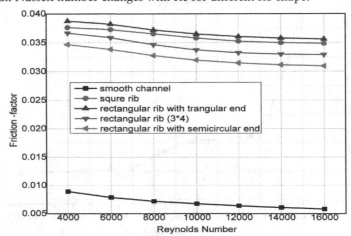

Figure 36.5 Friction factor changes with Re for different rib shape.

Variation in the Nusselt Number with Reynolds Number

In the same range of Reynolds numbers (Re), Figure 36.4 compares the mean Nusselt numbers (Nu) for various rib designs. The heat transmission, represented by Nu, demonstrates a significant increase as the Reynolds number rises. A thorough examination demonstrates that the rectangular rib with a triangular end has a higher Nusselt number than the plain channel, with a maximum Nu value of 58.35 at Re = 16000. These findings are consistent with those of earlier studies by Tanda [15].

Variation in the Friction Factor with Reynolds Number

The fluctuation of the friction factor with Reynolds number (Re) for various rib designs is shown in Figure 36.5. It is clear that using different rib geometry results in a markedly higher friction factor when compared to the plain channel. Due to the greater flow path created by the radial swirl flow and the longer contact time between the working fluid and the channel surface, the rectangular rib with a triangular end exhibits the highest friction factor. As a result, for the rectangular rib with a triangular end, the friction factor achieves its maximum value of 0.03575 at Re = 4000. These findings are in line with Tanda's research.

Conclusions

In order to explore the properties of air flows through roughened rectangular ducts in turbulent flow regions, a 2-Dimentional CFD analysis was conducted. According to CFD research, the observations show that the triangular-ended rectangular rib performs better than other rib shapes in terms of pressure drop and heat transfer rate. Notably, the rectangular rib with a triangular end's maximum Nu is 58.35 at Re = 16000, demonstrating its superior ability to transport heat compared to other configurations.

Nomenclature			
f	Friction factor	P	Rib pitch
Nu	Nusselt number	D_h	Hydraulic diameter
Q	Heat flux in W/m²	e/D	Relative roughness heights
μ	Viscosity	e	Rib element height
SAH	Solar air heater	Re	Reynolds number
A	Angle of attack	h	Heat transfer coefficient in W/m²-K

References

1. Yadav, S., and Saini, R. P. (2020). Numerical investigation on the performance of a solar air heater using jet impingement with absorber plate. *Solar Energy*, 208, 236–248.
2. Wang, D., Liu, J., Liu, Y., Wang, Y., Li, B., and Liu, J. (2020). Evaluation of the performance of an improved solar air heater with "S" shaped ribs with gap. *Solar Energy*, 195, 89–101.
3. Manjunath, M. S., Karanth, K. V., and Sharma, N. Y. (2018). Numerical investigation on heat transfer enhancement of solar air heater using sinusoidal corrugations on absorber plate. *International Journal of Mechanical Sciences*, 138, 219–228.
4. Srivastava, A., Chhaparwal, G. K., and Sharma, R. K. (2020). Numerical and experimental investigation of different rib roughness in a solar air heater. *Thermal Science and Engineering Progress*, 19, 100576.

5. Komolafe, C. A., Oluwaleye, I. O., Awogbemi, O., and Osueke, C. O. (2019). Experimental investigation and thermal analysis of solar air heater having rectangular rib roughness on the absorber plate. *Case Studies in Thermal Engineering*, 14, 100442.
6. Patel, Y. M., Jain, S. V., and Lakhera, V. J. (2020). Thermo-hydraulic performance analysis of a solar air heater roughened with reverse NACA profile ribs. *Applied Thermal Engineering*, 170, 114940.
7. Saravanakumar, P. T., Somasundaram, D., and Matheswaran, M. M. (2019). Thermal and thermo-hydraulic analysis of arc shaped rib roughened solar air heater integrated with fins and baffles. *Solar Energy*, 180, 360–371.
8. Wang, D., Liu, J., Liu, Y., Wang, Y., Li, B., and Liu, J. (2020). Evaluation of the performance of an improved solar air heater with "S" shaped ribs with gap. *Solar Energy*, 195, 89–101.
9. Kumar, R., Kumar, A., and Goel, V. (2019). Performance improvement and development of correlation for friction factor and heat transfer using computational fluid dynamics for ribbed triangular duct solar air heater. *Renewable Energy*, 131, 788–799.
10. Patel, S. S., and Lanjewar, A. (2021). Heat transfer enhancement using additional gap in symmetrical element of V-geometry roughened solar air heater. *Journal of Energy Storage*, 38, 102545.
11. Azad, R., Bhuvad, S., and Lanjewar, A. (2021). Study of solar air heater with discrete arc ribs geometry: experimental and numerical approach. *International Journal of Thermal Sciences*, 167, 107013.
12. Dong, Z., Liu, P., Xiao, H., Liu, Z., and Liu, W. (2021). A study on heat transfer enhancement for solar air heaters with ripple surface. *Renewable Energy*, 172, 477–487.
13. Tanda, G. (2004). Heat transfer in rectangular channels with transverse and V-shaped broken ribs. *International Journal of Heat and Mass Transfer*, 47(2), 229–243.
14. Chaube, A., Sahoo, P. K., and Solanki, S. C. (2006). Analysis of heat transfer augmentation and flow characteristics due to rib roughness over absorber plate of a solar air heater. *Renewable Energy*, 31(3), 317–331.
15. Tanda, G. (2011). Performance of solar air heater ducts with different types of ribs on the absorber plate. *Energy*, 36(11), 6651–6660.

37 Impact of diamond shaped cut-out on buckling nature of composite carbon/epoxy plate

Vineet Sinha[a], Rutul Patel[b], Keval Ghetiya[c], Mikhil Nair[d], Tarun Trivedi[e], and Lokavarapu Bhaskara Rao[f]

School of Mechanical Engineering, Vellore Institute of Technology, Chennai Campus, Vandalur Kelambakkam Road, Chennai, Tamil Nadu, India

Abstract: The present study deals with buckling analysis of composite plate with and without cut out. Composite plates containing layers of carbon and epoxy material were analysed using finite element method. The analysis was done for four different stacking sequences. To study the impact of length on buckling load, composite plates having constant width and thickness, with varying length were analysed. Composite plates of constant length and width, with varying thickness were analysed to study the impact of thickness on buckling load. Plates of constant length and width, having diamond shaped cut out of different sizes were analysed to examine the impact of cut out size on the buckling load values. Also, composite plates having a diamond shaped cut out with seven different orientations with the horizontal were analysed to examine the impact of cut out orientation on buckling load. All the finite element analysis were performed in ANSYS 19.2 software.

Keywords: Buckling, carbon, composite, epoxy, FEA, perforation.

Introduction

Composite plates are vastly used in spacecraft and aircraft industries as they provide very good mechanical properties and light weight. However, proper arrangement of fibers within the materials is required for better result. For many purposes like weight reduction, aesthetics, cooling, etc cut outs of different shapes are introduced to the plates. Thereby making the study of the plates with and without cut out for their buckling strength necessary. In the field of structural engineering many studies have been done on the buckling of plates. Hamani et al. [1] analyzed the buckling of antisymmetric composite laminate plate and found increment in critical buckling load with increase in angle, maximum values of buckling load are obtained when fibers are parallel to the applied stress and minimum values when fibers are perpendicular. Sa'el Sahel and Albazzaz [2] examined buckling of composite plates with central elliptic cut out at different orientation angle and found that the buckling load increased when the elliptical hole orientation angle increases. Mallikarjuna et al. [3] analyzed the composite plates with constant thickness,

[a]vineetsinha99@gmail.com, [b]rrutulpatel@gmail.com, [c]ghetiya.keval.j@gmail.com, [d]nairmikhil@gmail.com, [e]tarunktrivedi98@gmail.com, [f]bhaskarbabu_20@yahoo.com.

varying the number of layers up to 16 and found that maximum critical buckling load obtained at 7 layers for uniaxial compressive load. They examined the effect of aspect ratio, side to thickness ratio, modulus ratio, fiber orientation on the maximum critical buckling load for different materials. Geng et al. [4] examined shear deformable functionally graded GPL-RC plates with a circular hole under buckling load. They employed three different boundary conditions including free edges, simply-supported and clamped. They observed that increment of hole radius leads to increment in the buckling load parameters and increasing GPLs volume fraction made the plate stronger. Husam et al. [5] evaluated the influence of cut out size on load needed to buckle and found reduction in buckling load with the increment in cut out size when uniaxial or biaxial load were applied on simply supported panels. Whereas the buckling load decreases with the increment of cut out size when under uniaxial load and increased for large cut out sizes in the case of biaxial load for clamped supported panels. Gaira et al. [6] evaluated the influence of aspect ratio on the buckling load factor finding that the buckling loads are lower if cut outs are present. Balcıoğlu and Aktaş [7] examined the impact of thickness, cut out diameter, distance between cut out and orientation angle on the buckling load and found that the thicker plates are stronger under buckling load. Falkovicz [8] studied that buckling behavior of composite plate for various ply orientation and calculated the change in critical load by changing the layout of laminate layers. Sonawane [9] examined how varying cut out size affects the plates in terms of overall deformation for different materials. Wysmulski and Debski [10] discussed the effect of eccentric load on a composite column for buckling from different direction. Erklig and Yeter [11] examined the impact of perforation on buckling characteristics for different stacking sequences with three different cut out shapes and found that for all cut out types, critical buckling load decreased when cut out area was increased. Achour et al. [12] performed analysis on plates made of hybrid composite material with and without cut out, and found that the buckling loads increments exponentially with the increment in orientation angle. Thaier et al. [13] investigated the effect of nano carbon weight fraction on buckling of composite plates and found that the critical load increased when the weight ratio of nano carbon tubes was increased. Fernandes and Mirje [14] observed that increasing the number of layers increases the critical buckling load. It has been observed from the mentioned literature that many studies have been performed on composite plates with varying thickness, cut out size, number of layers, orientation, etc. to determine their buckling characteristics However, studies with different combinations can be performed for further understanding. In this study composite plates having different stacking sequences are examined. The effect of their length and thickness on buckling strength is examined. The influence of a diamond shape cut out with varying area and orientation is also performed. Finally, the comparison of buckling nature of plates for different stacking sequences is done.

Finite Element Analysis

Eigen value buckling analyses were executed using ANSYS 19.2 [15]. Models were made in Design Modeler of ANSYS, boundary conditions were defined in ANSYS Mechanical and meshing was also done with help of ANSYS Mechanical. The model was assigned the material properties (of carbon and epoxy) as shown in Table 37.1. The boundary condition of the geometry is shown in Figure 37.1, the left and right edges are clamped, and the top and bottom edges are free.

Table 37.1: Properties of materials.

Material	Young's Modulus (GPa)			Shear Modulus (GPa)			Poisson's Ratio		
Epoxy	3.78			1.4			.35		
Carbon	E_X	E_Y	E_Z	G_{XY}	G_{YZ}	G_{XZ}	v_{XY}	v_{YZ}	v_{XZ}
	230	23	23	9	8.2143	9	0.2	0.4	0.2

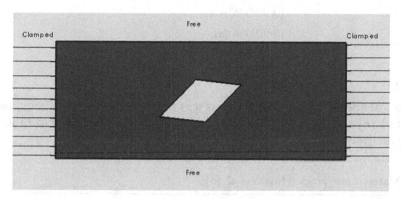

Figure 37.1 Boundary conditions.

Fixed support was assigned on left side of the plate and a compressive load of 1 N/mm was applied to the right side. The top and bottom sides were free. The first five modes were evaluated using eigen value buckling analysis. The following equation was used to evaluate critical buckling load.

$$Critical\ Buckling\ load = Load\ multiplier \times Applied\ load \qquad (1)$$

In Figure 37.2, 'L', 'W' and 't' denote the length, width and thickness of plate respectively, 'OA', 'a' and 'b' denote the orientation, larger diagonal and the smaller diagonal of the cut-out respectively.

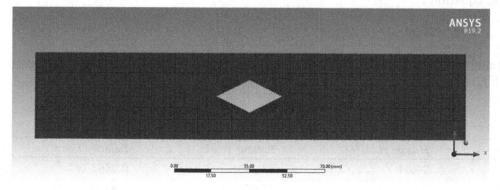

Figure 37.2 Model of plate.

Four different stacking sequences as shown in Table 37.2 were used.

Table 37.2: Stacking sequence for different samples.

Sample	Stacking Sequence
C1	[0]
C2	[90]
C3	[0/90/0/90]s
C4	[0/90/0/90]as

Deliberations

In order to authenticate the method, analysis was performed on models based on the work of Husam et al. [6] and the mechanical properties used by them as given in Table 37.3 were used for the analysis.

Table 37.3: Material properties (Husam et al. [6]).

Properties	Material 1	Material 2	Material 3	Material 4
E_1(GPa)	130	174.6	181	127.8
$E_2 = E_3$(GPa)	10	7	4.5	11
$G_{12} = G_{13}$(GPa)	5	3.5	2.7	5.7
G_{23} (GPa)	3	1.4	2.26	3
$v_{12} = v_{13}$	0.35	0.25	0.25	0.35
v_{23}	0.49	0.49	0.49	0.49

The critical buckling load (N_{cr}) was obtained for the boundary condition CCFF (clamped-clamped-free-free) of laminated composite panels with circular perforations subjected to load along an axis. And the non-dimensional buckling load (λ_u) was calculated using equation 2.

$$\lambda_u = N_{cr}b^2/E_2h^3 \qquad (2)$$

Table 37.4 depicts the comparison between the non-dimensional buckling loads obtained with Husam et al. [6].

Table 37.4: Comparison with Husam et al. [6]

d/b	Husam et al. [6]	Present Work	Relative Error
0	36.64	35.302	0.0365
0.1	35.32	34.986	0.0094
0.2	32.40	33.854	0.0448

Influence of Length

The variations in buckling load with increment in length are displayed in the Table 37.5 and Figure 37.3. It is recorded that buckling load of the plate for all the samples reduces with the increment in the length. Hence, shorter plates are stronger in buckling. Sample C1 and C4 are strongest and sample C2 is weakest in buckling.

Table 37.5: Effect of length.

Boundary condition	Length (mm)	Critical buckling load (N/mm) for Sample			
		C1	C2	C3	C4
C-C	50	4280.7	1558.2	4112.6	4280.7
C-C	100	2628.6	821.34	2481.7	2628.6
C-C	200	1056.8	353.64	988.09	1056.8
C-C	400	704.32	90.07	655.3	704.32

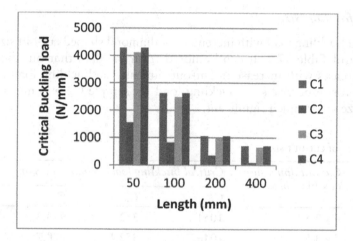

Figure 37.3 Effect of length.

Influence of Thickness

The variations in buckling load with increment in thickness are given in the Table 37.6 and Figure 37.4. With increment in thickness an increment in buckling load is noted. Hence, thicker plates are stronger in buckling. Sample C1 and C4 are strongest, and sample C2 is weakest in buckling.

Table 37.6: Effect of thickness.

Boundary condition	Thickness (mm)	Critical buckling load (N/mm) for Sample			
		C1	C2	C3	C4
C-C	2	139	44.849	139.30	139.05
C-C	4	1056.	353.64	988.09	1056.8
C-C	6	3291	687.02	3165.5	3291.1

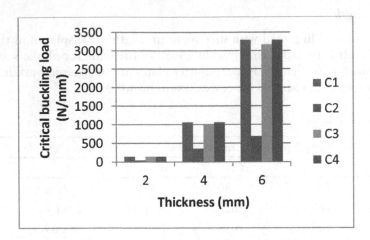

Figure 37.4 Effect of thickness.

Influence of Perforation Size

The variations in buckling load with increment in diamond shaped cut-out size are depicted in Figure 37.5 and Table 37.7. It can be noted from the data that buckling load for all the samples decreases with increase in cut-out dimensions. Buckling load for sample C1 and C4 is same for respective size. Buckling load for sample C2 was much lower and its decrease with size is not much significant.

Table 37.7: Impact of cut-out size.

Boundary condition	Cut-out dimension (a × b) (mm × mm)	Critical buckling load (N/mm) for Sample			
		C1	C2	C3	C4
C-C	5 × 2.5	1051.7	352.8	983.3	1051.8
C-C	7 × 3.5	1046.7	352.1	978.7	1046.7
C-C	10 × 5	1036.4	351.2	968.9	1036.6
C-C	15 × 7.5	1012.9	348.9	947.1	1013
C-C	20 × 10	983.43	345.9	919.8	983.29
C-C	25 × 12.5	949.84	342.0	888.8	949.45
C-C	30 × 15	912.03	337.2	853.3	912.04

Influence of Cut-Out Orientation

Figure 37.6 and Table 37.8 show the effect of cut-out orientation for different samples with cut-out size 30 x 15. It can be noted that the critical buckling load for all the samples is minimum at 90° and maximum at 15° except for C2. However, after 15° there is reduction in buckling load with increment in orientation. Buckling load for sample C2 is found to be much lower and its decrease with orientation is not significant. Also, there no significant difference between the buckling loads of sample C1 and C4 for respective angles.

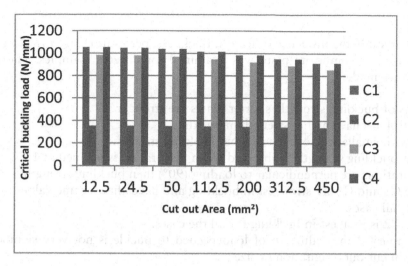

Figure 37.5 Impact of cut-out size.

Table 37.8: Effect of orientation.

Boundary condition	Cut-out orientation (deg)	Critical Buckling Load (N/mm) for sample			
		C1	C2	C3	C4
C-C	0	912	337.95	853.31	912.04
C-C	15	914.83	336.74	856.36	914.71
C-C	30	901.58	333.03	843.27	901.45
C-C	45	850.26	327.15	796.18	850.2
C-C	60	795.1	319.74	745.19	795.16
C-C	75	751.96	318.01	705.49	752.2
C-C	90	735.7	314.24	690.93	735.6

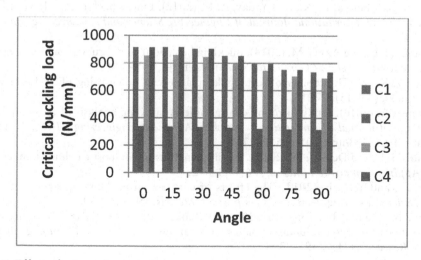

Figure 37.6 Effect of orientation.

Inferences

In the present study the influence of length, thickness, perforation size orientation on the buckling nature of composite plate was examined for different samples. The following conclusions are made:

- In terms of buckling strength, shorter plates are stronger
- Thicker plates have higher strength in buckling.
- Plates with smaller cut-out have higher buckling strength.
- Highest buckling strength is observed when perforation is inclined at 15°
- If perforations are perpendicular to loading (90°) then buckling strength is reduced
- Sample C1 and C4 are strongest in buckling and have almost same values of buckling load for all cases.
- Sample C2 is weakest in buckling for all the cases.
- For sample C2 the reduction of load needed to buckle is not very significant with change in cut-out orientation or size.

References

1. Hamani, N., Ouinas, D., Benderdouche, N., and Sahnoun, M. (2012). Buckling analyses of the antisymetrical composite laminate plate with a crack from circular notch. In: *Advanced Materials Research*. (vol. 365, pp. 56–61), Switzerland: Trans Tech Publications.
2. Al-Jameel, S. S., and Albazzaz, R. K. (2014). Buckling analysis of composite plate with central elliptical cut out. *Al-Rafidain Engineering*, 22(1).
3. Reddy, K. M., Reddy, B. S., and Kumar, R. M. (2013). Buckling anlysis of laminated composite plates using finite element method. *International Journal of Engineering Sciences and Research Technology*, 2(11), 3281–3286.
4. Geng, X., Zhao, L., and Zhou, W. (2021). Finite-element buckling analysis of functionally graded GPL-reinforced composite plates with a circular hole. *Mechanics Based Design of Structures and Machines*, 49(7), 1028–1044. DOI: 10.1080/15397734.2019.1707688.
5. Qablan, H. A., Rabab'ah, S., Alfoul, B. A., and Hattamleh, O. A. (2020). Semi-empirical buckling analysis of perforated composite panel. *Mechanics Based Design of Structures and Machines*, 50(8), 2635–2652. DOI: 10.1080/15397734.2020.1784198.
6. Gaira, N. S., Maurya, N. K., and Yadav, R. K. (2012). Linear buckling analysis of laminated composite plate. *International Journal of Engineering Science and Advanced Technology*, 2(4), 886–891.
7. Balcıoğlu, H. E., and Aktaş, M. (2014). Buckling behaviour of pultruded composite beams with circular cutouts. *Steel and Composite Structures*, 17(4), 359–370.
8. Falkowicz, K. (2017). Stability of rectangular plates with notch using FEM. In: *ITM Web of Conferences*, (vol. 15), 07013.
9. Sonawane, A. (2016). Study of round central hole in buckling analysis of cross ply laminates. *International Journal of Innovative Research in Advanced Engineering (IJIRAE)*, 5(10), 169–175. doi: 10.15623/ijret.2016.0510027..
10. Wysmulski, P., and Debski, H. (2019). Stability analysis of composite columns under eccentric load. *Applied Composite Materials*, 26, 683–692.
11. Erklig, A., and Yeter, E. (2012). The effects of cut-outs on buckling behaviour of composite plates. *Science and Engineering of Composite Materials*, 19, 323–330.
12. Achour, B., Ouinas, D., Touahmia, M., and Boukendakdji, M. (2018). Buckling of hybrid composite carbon/epoxy/aluminum plates with cutouts. *Engineering, Technology and Applied Science Research*, 8(1), 2393–2398.

13. Ntayeesh, T. J., Ismail, M. R., and Saihood, R. G. (2019). Buckling analysis of reinforced composite plates with a multiwall carbon nanotube (MWCNT). *Periodicals of Engineering and Natural Sciences*, 7(3), 1275–1285.

14. Fernandes, R. J., and Mirje, K. S. (2018). Buckling analysis of laminated composite plate using finite element software. *International Journal of Civil Engineering Research*, 9(1), 11–19.

15. ANSYS, Inc (2013). ANSYS Fluent User's Guide Engineering, (Vol. 1), ANSYS, Inc. Southpointe Technology Drive Canonsburg, USA, Release 15.0, November 2013.

38 Experimental study of domestic solar water heating process by using parabolic solar accumulator through propylene fluid charged wickless heat pipes

Harendra Kumar Jha[a] and Arun Kumar[b]

Department of Mechanical Engineering, NIT, Patna, India

Abstract: In this new effort, parabolic sun rays accumulator was used for solar water heating process. Accumulator is going to move with movement according to sun by using sprocket—chain mechanism. After obtaining sunlight at focus of accumulator temperature difference in create with respect to atmosphere, which is transfer to cold water by service of wickless heat pipes. The temperature difference found at focus of accumulator in the range of 25 to 44°C For this purpose wickless heat pipes are designed, constructed and tested at temperatures of 25°C, 35°C, and 45°C solar heat source. The working fluids employed was propylene with filling ratio (FR) 45%. Maximum heat transfer capacity of each pipe was 250W. The outcomes show that proficiency of wickless heat pipes analyzed in range of 35% to 45%. Experimentally found that the perpendicular position is the best performance of wickless heat pipes for extreme mass flow rate of fluid. Four propylene charged wickless heat pipes are hold together at the focus area of parabolic accumulator and transfer heat energy to cold water. Maximum temperature of warm water found in the stretch of 40°C to 45°C during winter season at Patna (November to March). Finally, this experimental effort proves that propylene wickless heat pipes with parabolic accumulator can be used in solar water heating process, because it will give improved performance in low temperature ranges.

Keywords: Parabolic accumulator, propylene, solar water heater, wickless heat pipes.

Introduction

Curiosity in solar energy as a most probable contributor to the world energy economy has risen again mainly on the basis of awaiting reductions in the supplies of economically recoverable conventional fuels and on the awareness that the employment of solar energy leads to a minimum of environmental effect [1–4]. Slim driven two-stage frameworks give huge points of interest over single-stage warm exchange frameworks. With the normally expanded warm limit related with stage change of a employed liquid, impressively littler mass flow rates are essential to transport proportionate sums than in two-stage fluid or gas frameworks for a specified temperature to go besides, warm exchange factors of two-stage frameworks are significantly more prominent than in two-stage streams and outcome in upgraded warm exchange Lesser mass stream rates and improved warm qualities give the repayment of littler framework size and weight while giving expanded execution. The

[a]harendraj.phd19.me@nitp.ac.in, [b]arun@nitp.ac.in

warm limit of a solitary stage framework relies upon the temperature change of the working liquid; consequently, a vast temperature slope or a high mass stream rate is required to exchange a lot of warmth. Then again, a two-stage framework can give basically thermal process paying little respect to varieties in the warmth stack. Moreover, single stage frameworks require the utilization of power-driven pumps and fans to course the working liquid, while narrow driven two-stage frameworks have no outside power prerequisites, which make such frameworks more solid and vibration free. The real parts of a warmth pipe are a fixed holder, a wick mess structure, and a working liquid [4]. The wick mess structure is put on the internal face of the warmth pipe divider and is immersed with the fluid working liquid and gives the structure to build up the fine activity for fluid coming back from the condenser heat releaser to the evaporator heat absorber area. With evaporator warm expansion, the working liquid is dissipated as it assimilates a measure of warmth comparable to the inactive warmth of vaporization, although in the heat releaser condenser area, the working liquid vapor is dense [6–13]. The mass expansion in the vapor centre of the heat absorber evaporator segment and mass dismissal in the heat rejector condenser final products in a weight slope along the vapor network which drives the comparing vapor stream. Arrival of the fluid to the heat absorber evaporator from the heat condenser is given by the wick and mess structure. As soon as vaporization happens in the heat evaporator, the fluid going to start above correspondingly into the wick structure [5].

Experimental Setup

The method of experimental facilities outlined and developed at laboratory. The lengthy evaporative segment wickless warmth tube is built of copper container of 500 mm distance of inner distance across 12.2 mm and 1 mm thick. The employed fluid utilized is propylene with satisfying proportion 45%. The lengths of heat absorber, fluid travel length and heat rejector segments are 200, 100 and 200 WS individually. A most extreme warmth exchange limit of wickless warmth channels is 250 W. The pipe was at first emptied utilizing emptiness pumps (rotational) after a progression of washing procedures to expel conceivable impurities, which can influence the execution and lifecycle of wickless warmth channels. First of all, pipe is pumped downgrade at the encompassing temperature after that, the pushing is proceeded while the tube is warmed [3, 4]. Since high emptiness were compulsory, it was a tedious procedure. Resulting clearing, the working liquid was sip into the tube through an extraordinary valving course of action and the satisfying pipe connected at the higher end was levelled by creasing to a core of 0.1 to 0.2 mm. This procedure mandatory up and about to an hour for tube. The parabolic accumulator was used for solar energy collection. All the incident rays from sun reflected by the fragments of mirror which are attached with the surface of parabolic accumulator after that all the reflected rays accumulate on the focus of parabolic dish. The temperature at the focus was measured by digital thermometer. The total heat available at the focus were transferred by evaporative section of four propylene wickless heat pipes. The adiabatic section of wickless heat pipes was completely covered by cotton ropes to prevent heat leakages. The condensing section of pipes were inserted in the cold water of the tank after releasing heat from condensing section the cold water becomes warm in the range of 35 to 45°C. The divider temperature dispersion lengthways the pipe was calculated utilizing 10 associated thermocouples (K-type). The temperature measuring tool are embedded in 1 mm trenches, drilled in the external face of the water dissemination coat tube divider. A precise watt energy meter is associated with water warmer to calculated correct the power provided. Temperature was perused straightforwardly beginning at advanced show. The water Flow frequency of the hot and chilly liquid was dictated by calculating the measure of the liquid over an interim of the time. Water inlet and outlet temperatures were calculated by digital thermometers. The genuine exploratory set up picture and schematic diagram are appeared in Figures 38.1 and 38.2.

Figure 38.1 Actual test setup of solar water heater.

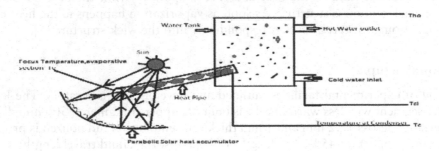

Figure 38.2 Schematic diagram of test setup.

Thermal Analysis and Calculations

The following calorimetric equations were used for calculating the performance of solar water process after obtaining experimental values [5].

$$Q = m.c.\Delta t \tag{1}$$

$$\Delta t = t_0 - t_i \tag{2}$$

$$R = T_{hot} - T_{cold} \big/ Q_{in} \tag{3}$$

$$\mathrm{Re} = \frac{\rho V D_h}{\mu} \tag{4}$$

$$\mathrm{Pr} = \frac{\mu c_P}{k} \tag{5}$$

$$Nu = \frac{h_i d_i}{k} \tag{6}$$

The temperature difference, overall thermal resistance, mass flow rate of water and specific heat capacity of water are denoted by Δt, R, m & C. Re, Pr, Nu are the Reynolds number, Prandtl number and Nusselt number. Reynolds number is determined to predict type of fluid flow. Total heat available at evaporative section of wickless heat pipes is given by

$$Q_h = m_h Cp_h (Thi - Tho)$$ (7)

Total heat available at Condenser section of wickless heat pipes is given by

$$Q_c = m_c Cp_c (Tci - Tco)$$ (8)

Result and Discussion

The maximum temperature variation found at focus from 25 to 44°C. Maximum average temperature at focus found 44°C in the month of March 2021 at Patna region. Superior value of temperature at focus will gives higher efficient heat source for wickless heat pipes. Figures 38.3 and 38.4 indicates the average temperature variation with time morning 08:00 am to evening 04:00 pm for the month of November 2020 to January 2021 at Patna region. Minimum average temperature at the focus found 36°C in the month of January 2021 at Patna region. Figure 38.5. Shows the average temperature variation at the focus with Time for the month of November 2020 to March 2021. The maximum propylene fluid filled wickless warmth pipes efficiency initiate 44.91% at 45°C solar warmth source temperature at 90° position. Faultless employed arrangement of wickless heat pipe is vertical. Wickless heat pipe operates on maximum performance and highest mass flow rate in this situation. For 90° position of wickless heat pipe efficiency is higher than the other inclination the lower inclination reduces the efficiency due to the obstruction of vapour with condensate return from the condenser. From experimental measuring performance of wickless heat pipe is creating graphic dependences usual values of thermal performance from working position of thermosyphon. Figures 38.6 to 38.8 shows average efficiency with different positioning (0°,15°,30°,45°,60°,75°,90°) angle of wickless heat pipes at 25°C, 35°C, and 45°C solar heat source. During experimentation 37.65% proficiency at 45°C also create in inclined 45°angle of location for propylene fluid wickless heat pipes. Propylene fluid filled wickless heat pipes at average focus temperature ranges from 25 to 45°C and utilised in domestic solar water heating method. After testing of wickless heat pipes, design, construct and trial an effectual, less cost flat parabolic solar rays accumulator through solar heating of water for domestic purpose. The major importance was to develop a 24 × 20 cm parabolic collector for the solar heating application. The design of accumulator was directed with the help of mathematical models of the heat transfer method in the collector. The parabolic accumulator design has an aluminium plate and fragments of mirror stick on the parabolic plate were gives reflection of solar rays on the focus.

Conclusions

The present new study involves investigation procedure about the water heating process through wickless heat pipes with solar rays through parabolic accumulator. The average temperature variation range found at focus from 25°C to 45°C in Patna region. Maximum average temperature at focus found 44°C in the month of March 2021 at Patna region.

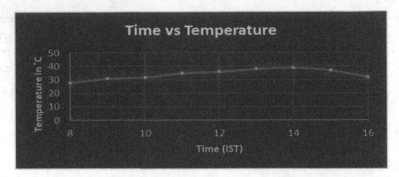

Figure 38.3 Time Vs temperature graph for the month of November 2020 at Patna.

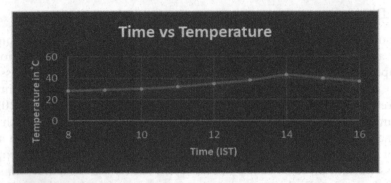

Figure 38.4 Time Vs temperature graph for the month of January 2021 at Patna.

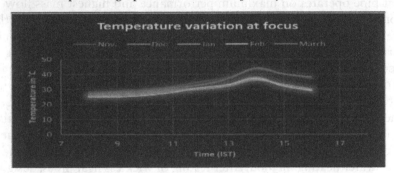

Figure 38.5 Average temperature variation at focus with time for the month of November 2020 to March 2021 at Patna.

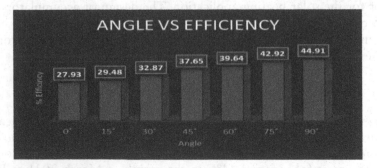

Figure 38.6 Wickless heat pipes Average efficiency with different inclination angle at 25°C solar heat source.

Figure 38.7 Wickless heat pipes Average efficiency with different inclination angle at 35°C solar heat source.

Figure 38.8 Wickless heat pipes Average efficiency with different inclination angle at 45°C solar heat source.

Superior value of temperature at focus will gives higher efficient solar heat source for wickless heat pipes. The maximum propylene fluid filled with no wick structure heat pipes efficiency observed 44.91% at 45°C focus temperature of solar energy at an angle of 90°. Throughout experimentation 37.65% efficiency at 45°C also found at an angle of 45° orientation for propylene fluid wickless heat pipes. Finally, this experimental search gives Propylene fluid filled wickless heat pipes with parabolic solar rays accumulator can be used in domestic solar water heating process.

Nomenclature			
C	Specific heat	Q_h	Heat available at evaporative section
Δt	Temperature difference	Q_C	Heat available at condenser section
R	Over all thermal Resistance	Th_i	Inlet temperature at evaporative section
m	Mass flow rate	Th_o	Outlet temperature at evaporative section
Re	Reynold's number	Tc_i	Inlet temperature at condenser section
Pr	Prandtl number	Tc_o	Outlet temperature at condenser section
Nu	Nusselt number		

References

1. Bhat, A. (1982). Performance investigation of a long, slender heat pipe for thermal energy storage applications. *Journal of Energy,* 6(6), 361–367.
2. Kang, H. M., Kim, H., Lee, C. H., Lee, C. K., and Choi, S. I. (2017). Changes and development plans in the mountain villages of South Korea: Comparison of the first and second national surveys. *Journal of Mountain Science,* 14(8), 1473–1489.
3. Chi, S. W. (1976). Heat Pipe Theory and Practice. New York: McGraw-Hill.
4. Dunn, P., and Reay, D. A. (1978). Heat Pipes. New York: Pergamon Press,.
5. Nemec, P., Caja, A., and Lenhard, R. (2010). Influence Working Position of Heat Pipe on their Thermal Performance, Experimental Fluid Mechanics. TU Liberec.
6. Sukchana, T., and Jaiboonma, C. (2013). Effect of filling ratios and adiabatic length on thermal efficiency of long heat pipe filled with R-134a. *Energy Procedia,* 34, 298–306.
7. Manimaran, R., Palaniradja, K., Alagumurthi, N., and Hussain, J. (2012). Factors affecting the thermal performance of heat pipe—a review. *Journal of the Entomological Research Society (JERS),* III(II/April-June), 20–24.
8. Ahmad, H. H., and Yousif, A. A. (2013). Comparison between a heat pipe and a thermosyphon performance with variable evaporator length. *Al-Rafidain Engineering,* 21(2).
9. Kate, A. M., and Kulkarni, R. R. (2010). Effect of pipe cross section geometries and inclination angle on heat transfer characteristics of wickless heat pipe. *International Journal of Engineering Research and Technology,* 3(3), 699–710. ISSN 0974-3154.
10. Azad, E. (2008). Theoretical and experimental investigation of heat pipe solar collector. *Experimental Thermal and Fluid Science,* 32, 1666–1672.
11. Bong, T. Y., Ng, K. C., and Bao, H. (1993). Thermal performance of a flat-plate heat-pipe collector array. *Solar Energy,* 50, 491–498.
12. Beckman, W. A., Klein, S. A., and Duffie, J. A. (1977). Solar Heating Design by the F-Chart Method. USA: John Wiley & Sons Inc..
13. Budihardjo, I., Morrison, G. L., and Behnia, M. (2007). Natural circulation flow through water-in-glass evacuated tube solar collectors. *Solar Energy,* 81, 1460–1472.

39 Drilling performance investigation of biopolymer nanocomposite modified by graphene nanoplatelet

Virat Mani Vidyasagar[a] and Rajesh Kumar Verma[b]

Materials & Morphology Laboratory, Department of Mechanical Engineering, Madan Mohan Malaviya University of Technology, Gorakhpur, India

Abstract: PMMA-based materials have been used in biomedical applications since the 1930s. PMMA was first used in dentistry before being used in orthopaedics. Various studies had been done on the mechanical properties of PMMA bone cement. This paper explores the developed PMMA/GNP nanocomposite materials and machining of nanocomposite materials to achieve the machining parameter's optimal condition using the TOPSIS method. The machining characteristics considered are Material removal rate (MRR) and Surface roughness (Ra). The most prominent factors are affecting objective function spindle speed (s), Wt % of GNP. The parameter's optimal setting is found as 865 rpm (Level-2), 4 Wt. % 0f GNP (Level-3), and with material from HSS Tool (Level-3), which has been validated by a confirmatory test. The outcomes reveal satisfactory assessments with actual ones. The approach can be recommended for quality and productivity control during the manufacturing of polymer components.

Keywords: drilling, GNP, PMMA bone cement, TOPSIS.

Introduction

Polymethylmethacrylate (PMMA) is a widely used biomaterial to produce orthodontic retainers, dentures, and plastic teeth. Before clinical use, in appliances are cleaned in the dental laboratory to ensure clean, polished, and comfortable surfaces. PMMA is also used as in situ cured bone cement to anchor orthopaedic implants to bone, cranial implants, and acquired and congenital bone defects. For the treatment of broken bones, a wide range of synthetic materials are available, including metals, ceramics, polymers, and cement. Among these items, bone cement plays a unique role in this regard. Calcium phosphate cement and polymethylmethacrylate (PMMA) cement are the two primary types of bone cement. Zafar et al. [1] stated that PMMA is one of the most widely used bone cement in the medical field. This form of cement is a biocompatible polymer that provides excellent initial bone-to-implant fixation. PMMA bone cement has a high polymerization temperature, producing heat between 80° and 100° during the polymerization process. Ballo et al. [2] studied and stated that clinical use of PMMA cement is difficult due to its poor mechanical properties and bone compatibility. Zebarjad et al. [3] studied the hydroxyapatite-doped

[a]vmsagar98@gmail.com, [b]rajeshverma.nit@gmail.com

PMMA nanocomposite material's mechanical properties, and they tried to overcome the limitation of the PMMA bone cement. They concluded that the nanofiller redesign matrix improved thermal, mechanical, and other properties. In this series, GNP frequently has been used in recent years due to its outstanding thermal and mechanical, and biocompatibility properties. Pahlevanzadeh et al. [4] discussed G-PMMA and GO-PMMA bone cement' and stated that mechanical performance improved at low loadings (≤0.25 wt.%), especially the fracture toughness and fatigue performance.

As per the product's manufacturing, the composite material undergoes various machining processes such as milling, drilling, turning, etc., in their specific orthodontic retainers and dentures. According to the most current biomedical field, drilling is the most common machining process for the prosthesis is likewise needed to put bolts for fiber wires. Kharwar and verma [5] studied drilling parameters and stated that unexpected major problems were observed during a drilling operation, such as producing burr, increasing surface roughness, and low material removal rate. The non-homogeneous and multi-phase structure and anisotropic nature of nanocomposite lead to a different failure. There are various multi optimization techniques available such as GRA, MOORA, TOPSIS, etc. The previous research has been used for the optimization of machining parameters. The TOPSIS optimization hypothesis is used in the current job to achieve an optimal parametric arrangement for ideal machining constraints with desired output [6]like force cuts on the drilling of these composites. The focus is to drill on hybrid laminates by changing the cutting conditions like the speed and feed and the selection of optimum machining process parameters. Nine experimental cycles have been performed, and five various attributes were analyzed, for example, torque, tangential force, thrust force, and also the factor of delamination (Enter & Exit.

According to the literature survey, very few studies have been performed on drilling the GNP modified PMMA nanocomposite. It is vital to analyze the MRR and surface roughness during composite drilling in order to accomplish exemplary structural part assembly. This study aims to fill a gap in the literature by exploring nanofiller materials' addition to PMMA. PMMA was doped with GNP at three different weight percentages (1 %, 2.5 %, and 4 %) in the current sample. A different drill tool has been used to machine these GNP-doped PMMA samples. This experiment aims to determine how various machining parameters, such as spindle speed (S), weight wt. % of GNP, and three different tools, affect MRR and surface roughness (R). Taguchi L_9 and ANOVA were used to determine and optimize drilling parameters.

Material and Equipment

Fabrication of Composite Material

The solution blending and casting process was used for the fabrication of composite materials. A mixture of PMMA and monomer is used as a matrix, and GNP has been used as a nanofiller material. PMMA powder of the different weight % of the GNP was substituted. The weight parts of PMMA subbed by the GNP filler were 1, 2.5, 4 Wt. % within the wake of mixing the PMMA powder with the GNP, the squares of concrete were set up by the standard method proposed by the producer. The extent of liquid monomer (M) to the PMMA (P) was M/P= 0.5, the mass of the powder joined the PMMA and GNP particles. The blending of PMMA and GNP powder was exhausted during a ball process for one hour. The test was set up by mixing the MMA monomer with the powder. The mix was allowed to settle at room temperature for one hour, and re-established materials with a noticeable homogeneous scattering of powder were obtained by mixing the powder in an antacid course of action. The fabricated sample is shown in Figure 39.1.

Figure 39.1 Prepared sample.

Experimental Works

Description of Experimental Setup and Measurements

The drilling tests are carried out on a PUSCO vertical drilling machine under dry conditions (Figure 39.2). For the drilling operation of composite material, the three different drill tools (HSS, Carbide, and TiAlN) with the 5 mm drill bit diameter have been used throughout the experimental work. The MRR and average surface roughness (Ra) are considered as responses for this study. The MRR has been calculated using Eq.1. The Ra is measured at three positions spaced around the whole circumference. The Ra of each hole is taken as the mean of three circumferential readings.

Plan of Investigation

Taguchi-based L_9 orthogonal array (O.A.) has been used to conducted experimentation. Table 39.1 shows the variables and their levels that were considered in this study. In general, spindle speed, Wt. % and Tool of the machining parameters effect on drilling quality. However, the present work aims to increase the MRR and reduce surface roughness, therefore increase work quality. The design of the experiment (DOE) and their machining response are listed in Table 39.2.

Table 39.1: Drilling parameters and level.

Parameters	Unit	Level 1	Level 2	Level 3
Speed	RPM	318	865	1428
Wt. (%)	-	1	2.5	4
Tool	-	1 (HSS)	2 (Carbide)	3 (TiAlN)

Figure 39.2 Drilling operation.

Table 39.2: Input parameters and response.

S. No	Input Parameters			Response	
	Speed (rpm)	Wt. %	Tool	MRR (mm³/sec)	Ra (µm)
1	318.00	1.00	1	5.42	1.69
2	318.00	2.50	2	6.9	0.81
3	318.00	4.00	3	7.4	0.71
4	865.00	1.00	2	5.55	1.57
5	865.00	2.50	3	6.22	1.02
6	865.00	4.00	1	7.43	0.65
7	1428.00	1.00	3	5.88	1.86
8	1428.00	2.50	1	6.34	1.02
9	1428.00	4.00	2	7.87	0.85

$$MRR = \frac{Initial\ weight - Final\ weight}{\rho \times Time\ taken}\ mm^3 / sec \tag{1}$$

The mean of the S/N ratio plot has been obtained employing the Minitab analytical tool. Figures 39.3, and 39.4 shows the mean plot of the variation of mean values for MRR and Ra with respect to variables, i.e., Spindle speed, weight % of GNP, and tool selection. From Figure 39.3 can be revealed that MRR obtained increases trend with respect to change in speed, wt. % and Tool changes. Figure 39.3 shows the values of MRR increases as the tool changes from HSS to carbide (2). MRR shows an increasing trend with respect to the change in weight of reinforcement materials. It clearly shows that higher MRR is obtained at higher speed (1428), 4 wt.%, and carbide tool. Figure 39.4 demonstrated that maximum Ra is observed at the lower spindle speed (318 rpm), 4 wt. % of GNP and at HSS tool. And the decreasing trend and smooth surface is observed with respect to the change Spindle speed at (1428 rpm), 1 Wt. % of GNP and TiAlN Tool.

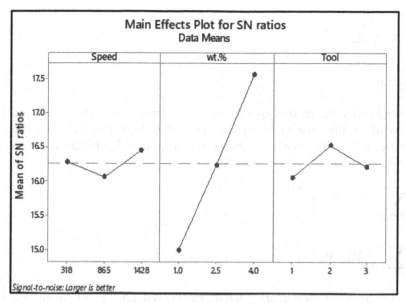

Figure 39.3 Mean effect plot for S/N ratio of MRR.

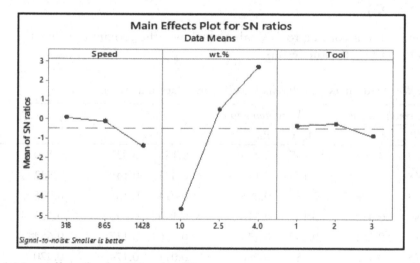

Figure 39.4 Mean effect plot for S/N ratio of Ra.

Result and Discussion

TOPSIS

In this article, a technique for order preference by similarity to the ideal solution (TOPSIS) method was developed in 1995 by Hwang and Yoon. Implementing TOPSIS, the initial step is to normalize the obtained values using Eq. 2 and afterward compute the weighted normalized value using Eq. 3. In this work, an equivalent weightage of 50% is expected for all the two reaction factors: material removal rate and surface roughness.

$$A_{ij} = X_{ij} / \sqrt{\sum\nolimits_{i=1}^{n} x_{ij}^2} \tag{2}$$

$$V_{ii} = A_{jj} \times W_i \tag{3}$$

When the weighted standardized qualities are determined, the ideal best and ideal most noticeably awful qualities for each response are determined. From the obtained ideal best and most noticeably awful qualities, the closest and after Euclidean distances are determined utilizing Eq. 4 and Eq. 5 individually.

$$S_i^+ = \left[\sum\nolimits_{j=1}^{m} \left(V_{ij} - V_j^+ \right)^2 \right]^{0.5} \tag{4}$$

$$S_i^- = \left[\sum\nolimits_{j=1}^{m} \left(V_{ij} - V_j^- \right)^2 \right]^{0.5} \tag{5}$$

The value of the common performance index (Pi) is determined by using Eq. 6.

$$P_i = \frac{S_i^-}{\left(S_i^- + S_i^+ \right)} \tag{6}$$

Score the option according to C_i's value, the lower the proximity value, the higher the score, and thus the better the results.

Table 39.3: Weighted values and closeness coefficient values and ranking.

Exp No.	Normalised value		Weighted values		S_i^+	S_i^-	Pi	Rank
	MRR	Ra	MRR	Ra				
1	0.273	0.466	0.137	0.233	0.156	0.023	0.130	8
2	0.348	0.223	0.174	0.112	0.033	0.149	0.832	4
3	0.373	0.196	0.187	0.098	0.014	0.166	0.925	2
4	0.280	0.433	0.140	0.216	0.140	0.040	0.223	7
5	0.314	0.281	0.157	0.141	0.066	0.117	0.654	6
6	0.375	0.179	0.187	0.090	0.011	0.174	0.970	1
7	0.297	0.513	0.148	0.256	0.174	0.012	0.064	9
8	0.320	0.281	0.160	0.141	0.064	0.118	0.657	5
9	0.397	0.234	0.198	0.117	0.028	0.152	0.848	3

The enhanced results obtained by TOPSIS (Table 39.3) demonstrate that for the directed examination, the ideal mix of the information boundaries Inside the chose test range is 318 spindle speed, 4 wt. % and TiAlN Tool with the most significant estimation of Pi (0.970). Ideal parameters are setting a quired by the TOPSIS method obtained with the help of Table 39.3, which reveals that Spindle Speed (865-RPM, i.e., Level-2), GNP (4 wt%, i.e., Level-3), and Tool (HSS, i.e., Level-1) gives optimum machining performance. The MRR analysis has been performed to use ANOVA and check the significance of the obtained result. It shows the results of the analysis of variance with the machining response for both MRR and Ra. If P values ≤ 0.05 as significant, otherwise, it is non-significant. It is observed that Wt.%, have statistical and physical significance, and speed and Tool are insignificant parameters in the drilling of PMMA/GNP nanocomposite. The ANOVA model also demonstrates an MRR, R^2 value of 97.53 % confidence level, which shows a very high correlation between the proposed method and the results.

The surface roughness analysis has been performed using ANOVA and checks the significance of the obtained result, which is revealed that Wt.% of GNP have significant parameters and the rest are insignificant. It is also observed that R^2 values are 99.16 %, which indicates that there is a strong correlation between the proposed model and the experimental results.

Confirmatory Test

The confirmation test has been performed to confirm the above enhancement technique applied to the machining boundaries. From Table 39.4, it can be observed that experiment no. 6 has the highest Pi value (0.970), indicating that the optimum setting for drilling operation to be performed on the composite is N2-W4-T1. As the drill diameter and spindle speed increased, the precision of the drilled holes decreased but improved as the feed rates increased. Taking the initial experimental one value, the optimal condition for drilling came out to be N1-W1-T1. With reference to the TOPSIS experiments, it can observe that the value of MRR increases up to a significant amount, and the smoothness of the drilled surface improved. Hence it can drown that TOPSIS optimization has significantly improved the working condition. Compared to the optimal and confirmatory test values, the MRR significantly increased and improved surface roughness (Ra), showing the effectiveness of the proposed TOPSIS approach in the machining environment.

Table 39.4: Experimental array versus TOPSIS optimization result.

Optimal setting	Taguchi orthogonal array design N1-W1-T1	TOPSIS design N2-W4-T1	% improvement
MRR	5.42	7.43	37..08
Ra	1.69	0.65	61.53

Conclusion

The optimization of the drilling performance behavior of PMMA/GNP nanocomposites using TOPSIS techniques is investigated effectively. The experimental results are concluded as follows:

- The optimal combination of TOPSIS-assessed process parameters is found as 865 rpm (Level-2), 4 Wt. % 0f GNP (Level-3), and with material from HSS Tool (Level-3). Results indicate that an increase in spindle speed (S), increases the MRR using the carbide tool.
- It is also observed that filler material (GNP) improved the surface roughness (Ra). From the investigation, proper process parameters are required to improve the drilling hole quality on developed composites.

- Results indicate that, with an increase in Tool materials, the surface roughness decreased. Analysis of variance results distinguishes the critical machining parameters and their impacts on response parameters, i.e., material removal rate and surface roughness.
- The confirmatory results revealed that the TOPSIS approach improves the estimations of MRR up to 37.08 % and 61.53 % of surface roughness (Ra) than expected average results. A near report shows the higher application capability of the TOPSIS technique.
- ANOVA results distinguish the critical machining parameters and their impacts on response parameters, i.e., MRR and Ra.
- This methodology is more suitable for optimizing drilling parameters in drilling of GNP doped PMMA nanocomposite at the levels investigated. A higher MRR and minimal Ra can be obtained at high speed in drilling GNP-modified PMMA nanocomposite.

Acknowledgment

The authors would like to acknowledge the kind support of o/O D.C. (Handicrafts), Ministry of Textiles, Govt. of India, New Delhi, INDIA for extending all possible help in carrying out this research work directly or indirectly.

Funding

This research work is financially supported by o/O D.C. (Handicrafts), Ministry of Textiles, Govt. of INDIA, Under Project ID: K-12012/4/19/2020-21/R&D/ST.

References

1. Zafar, M. S., and Ahmed, N. (2014). Nanoindentation and surface roughness profilometry of poly methyl methacrylate denture base materials. *Technology and Health Care,* 22, 573–581. https://doi.org/10.3233/THC-140832.
2. Ballo, A. M., Akca, E. A., Ozen, T., Lassila, L., Vallittu, P. K., and Närhi, T. O. (2009). Bone tissue responses to glass fiber-reinforced composite implants—a histomorphometric study. *Clinical Oral Implants Research,* 20, 608–615. https://doi.org/10.1111/j.1600-0501.2008.01700.x.
3. Zebarjad, S. M., Sajjadi, S. A., Sdrabadi, T. E., Sajjadi, S. A., Yaghmaei, A., and Naderi, B. (2011). A study on mechanical properties of PMMA/Hydroxyapatite nanocomposite. *Engineering,* 03, 795–801. https://doi.org/10.4236/eng.2011.38096.
4. Pahlevanzadeh, F., Bakhsheshi-Rad, H. R., and Hamzah, E. (2018). In-vitro biocompatibility, bioactivity, and mechanical strength of PMMA-PCL polymer containing fluorapatite and graphene oxide bone cements. *Journal of the Mechanical Behavior of Biomedical Materials,* 82, 257–267. https://doi.org/10.1016/j.jmbbm.2018.03.016.
5. Kharwar, P. K., and Verma, R. K. (2020). Machining performance optimization in drilling of multiwall carbon nano tube/epoxy nanocomposites using GRA-PCA hybrid approach. *Measurement,* 158, 107701. https://doi.org/10.1016/j.measurement.2020.107701.
6. Gokulkumar, S., Thyla, P. R., ArunRamnath, R., and Karthi, N. (2020). Acoustical analysis and drilling process optimization of camellia sinensis/ananas comosus/GFRP/epoxy composites by TOPSIS for indoor applications. *Journal of Natural Fibers,* 00, 1–18. https://doi.org/10.1080/15440478.2020.1726240.

40 Performance of heat pipe with nanorefrigerant at different inclination angles

Aruna Veerasamy[a] and Kanimozhi Balakrishnan[b]

Department of Mechanical Engineering, Sathyabama Institute of Science and Technology, Chennai, Tamilnadu, India

Abstract: In this study is a goal aimed at improving the efficiency of the heat pipe with nano refrigerant is presented. The circular heat pipe is conceived and manufactured with a length and diameter of 200 mm and 8mm respectively. Refrigerants such as R-410a and graphene nanoparticles are used in this study. Graphene nanoparticles are mixed with the base fluid by the two-step method, then the refrigerant and nanofluid are directly injected into the heat pipe. The effects of heat pipe, input power and angle of inclination were taken into consideration. The thermal resistance, thermal efficiency and temperature distribution in the evaporator area were evaluated. The experimental results clearly show that heat pipe with Nano refrigerant as a working fluid, charge amount of 50%, tilt angle 900, input power 85W gives the highest efficiency and low thermal resistance. Increasing input power increases HP performance by approximately 47.4%.

Keywords: Heat pipe, nanorefrigerant, thermal efficiency, thermal resistance.

Introduction

Eliminating heat from devices and chilling components is a big challenge in many applications. To recuperate this problem the heat pipes (HP) are invented. Heat pipes absorb surface heat and cool the equipment by releasing latent heat through the condenser. In order to improve the thermal conductivity of the heat pipes, the working fluid in an evaporator section is essential. Because HP performance depends primarily on the working fluid, material and their input parameters. HP's and its applications have been the subject of a great deal of research for several decades.

Amin et al. [1] analyze the thermal performance of gravity assisted heat pipe with zirconia—acetone nanofluids at various input power, tilt angle, filling ratios and concentrations. They concluded that an increased concentration of nanoparticles in the base fluid can reduce the temperature profile but increase the heat transfer coefficient. Filling ratio and heat transfer affect the heat transfer rapidly. Aboutalebi et al. [2] experimentally investigate the performance of novel rotating closed loop pulsating heat pipe at various input parameters. According to their research, the increase in inlet power and rotation speed reduces thermal resistance. Even a low filling ratio increases thermal performance in RCLPHP. Ebrahim et al. [3] also modified an L-shaped heat pipe structure and investigated

[a]arunaveer@gmail.com, [b]kanihwre@gmail.com

performance at different input parameters. The change in the size of the nanoparticles, the concentration and the filling ratio affects the conductivity of HP as well as the inclination angle of HP and the input power also have an influence. Their results showed that increasing the angle and concentration of nanoparticles in the working fluid (SiO_2) decreases the thermal resistance considerably.

Emad et al. [4] analyzed the thermal performance of thermosyphon type HP by implementation of covalent functionalized graphene nanofluids. They altered the concentration, fill ratio, input power and tilt angle. From their experimental result at 5% of concentration improve thermal conductivity and reduce thermal resistance. They analyze graphene structure after a covalent functionalization using diazonium (DS) with XPS. Kerim et al. [5] apply two step method to prepare aqueous Fe-CuO hybrid nanofluid in order to increase thermal performance of plain HP. Thermal resistance reduced and thermal efficiency improved by aqueous hybrid nanofluids than distilled water. Torri [6] experimentally studied a performance and thermal characterization of HP with aqueous graphene oxide. He concluded that viscosity, pressure loss, heat transfer coefficient were increased with increasing volumetric concentration than pure water. Increasing temperature also improved heat transfer performance, heat flux and heat transfer coefficient except viscosity for all filling ratio of working fluid.

Kamlesh Mehta et al. [7] also concluded operational parameters and geometrical parameter affects closed loop flat plate oscillating heat pipe(CLFP-OHP). When the heat pipe charged with acetone, it provides lowest thermal resistance and highest thermal conductivity. In the present study, performance of HP is experimentally evaluated. The main novelty of the study is working fluid. Nano refrigerant (R-410a/Graphene) is used as a working fluid. Thermal resistance, thermal efficiency, overall heat transfer coefficient and temperature difference between evaporator and condenser have been validated.

Experimental Setup

Figure 40.1 illustrates the schematic diagram of experimental setup used in this study. HP has fabricated at copper material with a length of 200 mm and inner diameter of 8mm. HP has made up of three parts named an evaporator, adiabatic and condenser section. Evaporator and adiabatic section has a length of 60 mm each and condenser has 80 mm length. An electric heater was attached at an evaporator to provide a heat supply. Condenser section is connected to water bath, which is used to cool the condenser section. K-type thermocouple were attached to HP at various parts to predict a temperature distribution throughout the HP. Inside an evaporator section nano refrigerant was injected after setting the HP. Heat input has fixed at 75 and 85W. Flow rate of water has fixed as 1 L/hr. Tilt angle of HP and heat power are changed and experimental date were observed.

Initially, working fluid is injected to an evaporator section. Working fluid was prepared by two step method and their properties are already discussed by the same author and given elsewhere [8]. The process which is involved in preparation of nano refrigerant has shown in Figure 40.2. First, DI water is directly mixed with graphene and its stirred for 30 mins. Then stirred solution is subjected to ultrasonication for 2 hrs (50 Hz, 10 mins time interval). Finally, the nano solution and refrigerant is directly injected to heat pipe evaporator section. Once the heat pipe is ready the heat power is applied to evaporator section. When the evaporator wall has heated, inside the working fluid absorb heat power from evaporator wall and change its phase from liquid to vapor. Phase change vapor travels to

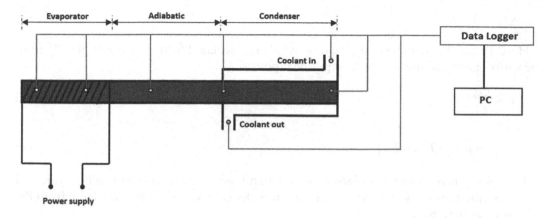

Figure 40.1 Schematic diagram of heat pipe.

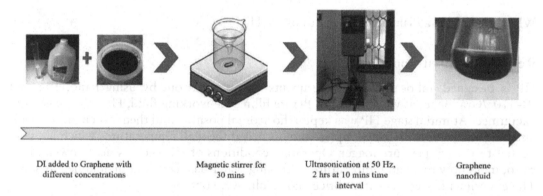

DI added to Graphene with different concentrations   Magnetic stirrer for 30 mins   Ultrasonication at 50 Hz, 2 hrs at 10 mins time interval   Graphene nanofluid

Figure 40.2 Preparation of working fluid by two step method.

condenser section in order to release the latent heat. Water is used as a cooling medium in condenser section. By releasing the latent heat, working fluid back to the liquid state and moves to an evaporator section due to gravity or capillary action. The experiment is done for different parametric inputs, such as power input and inclination angle.

Data Processing

Thermal properties such as thermal resistance, thermal efficiency and overall heat transfer coefficient were used to investigate the thermal performance of HP. Thermal resistance can be calculated from the temperature difference between evaporator and condenser to input heat power.

$$R_{th} = \frac{T_e - T_c}{Q_{in}} \tag{1}$$

$$T_e = \frac{T_1 + T_2}{2} \tag{2}$$

$$Q_{in} = V \times I \tag{3}$$

Here T_1 and T_2 are evaporator reading which is calculated from thermocouples. T_c from the ratio of output heat power to input heat power

$$\eta = \frac{Q_{out}}{Q_{in}} \tag{4}$$

$$Q_{out} = mC_p(T_{out} - T_{in}) \tag{5}$$

T_{in} and T_{out} is represent a coolant water inlet and outlet water. m is a mass flow rate and Cp is specific heat capacity. The overall heat transfer coefficient can be calculated from the following equation.

$$U = \frac{Q_{in}}{\pi Dl(T_e - T_c)} \tag{6}$$

Where D and l is a diameter and length of the HP.

Result and Discussion

HP is designed, fabricated and experiments were carried out by using nanorefrigerant (R-410a/Graphene) as working fluid. Before filling the working fluid, HP has degassed or vacuumed. At initial stage HP was kept in horizontal position and then it is changed to 45° and 90° orientation. Thermocouples were recorded the wall temperatures which are connected to the temperature logger. Operating conditions of HP, such as heat power, filling ratio, mass flow rate and temperature of cooling water has fixed and maintained as same. The experiments were repeated thrice and results were obtained.

Figure 40.3 represents the thermal resistance of HP with different tilt angle at two input powers (75 W and 85W). Initially 75W input power was fixed and HP positions were changed from 0° to 45° and 90°. 0.65°C/W thermal resistance was observed at 0° tilt angle. Increasing inclination angle decreases thermal resistance, at the same time improve thermal performance. 90° tilt angle, thermal resistance was 0.61°C/W. For the same tilt angle but 85W input power, thermal resistance was observed as 0.54°C/W. Thermal resistance readings clearly indicate that increasing input power and tilt angle improve thermal performance of HP. 85W input power decreases thermal resistance up to 10.8% than 75W input power. Dispersed graphene nanoparticles in DI water could notably reduced thermal resistance. In general, nanoparticles has an ability to increase the thermal conductivity in heat pipe [9–11]. At high heat input power, nanoparticles absorb the heat in the particle surface and moves along the heat pipe while carrying the heat. At higher heat power, nanoparticles moves rapidly inside the pipe and makes liquid slug and vapor plug. Due to working fluid movement evaporation and condensation process happened rapidly at the time of increasing input power from low to high. Further increasing heat load would dry out the HP leads to improper thermal conductivity and thermal resistance Thus thermal resistance decreased while increasing heat power and inclination angle [12heat pipe inclination angles, and input heating power. Distilled water (DW–14]. Figure 40.3 clearly displays increasing inclination angle decreasing thermal resistance. Here gravity affects the fluid flow and performance the pipe.

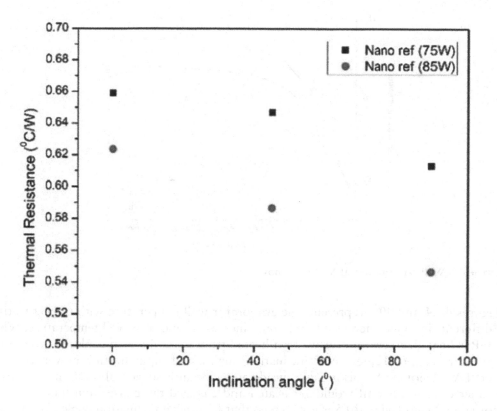

Figure 40.3 Thermal resistance of HP with respect to inclination angles.

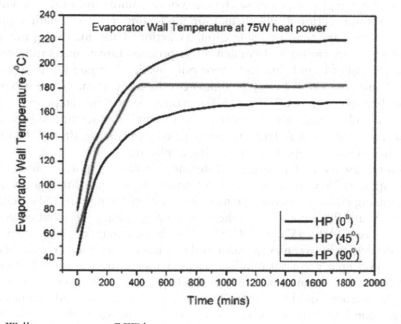

Figure 40.4 Wall temperature at 75W heat power.

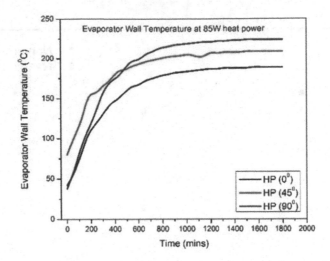

Figure 40.5 Wall temperature at 85W heat power.

Figures 40.4 and 40.5 represents the evaporator wall temperature with respect to time for different tilt angle. Increasing time could increase evaporator wall temperature. When the HP is heated, an evaporator wall get heated from the coil which is placed around the evaporator section [12]heat pipe inclination angles, and input heating power. Distilled water (DW. Amount of working fluid inside an evaporator section absorbs the heat from the inner evaporator wall would get heated and changed the phase from liquid state to vapor state. From Figure 40.5 it is observed that increasing inclination angle can increase evaporator wall temperature significantly. This happened because of gravitational force and capillary action. Gravity is one of the reason to improve evaporation and condensation process. Due to gravity, a phase change working fluids from the condenser moves rapidly to evaporator section at 90° inclination angle. But at 45° inclination angle, liquid droplets reaches to evaporator by using capillary action. Thus, increasing the inclination angle accelerated evaporation and condensation process. Lower inclination angle slows down the movement of liquid slug and vapor plug inside the pipe following performance of heat pipe. Here position of heat pipe influences the performance of heat pipe. Further increasing the heat power from 75W to 85W increases evaporator temperature. Therefore, higher heat load and tilt angle effectively improve thermal conductivity and decreases thermal resistance. Evaporator wall temperature reached steady state after 300 mins, and the steady state time is key to implement in cooling applications.

Figure 40.6 represents the temperature difference between evaporator and condenser at various tilt angles. At 75W input power, 0° tilt angle temperature difference was observed as 41°C. Increasing tilt angle increases temperature difference rapidly and reaches at 46°C and 53°C for 45° and 90° tilt angle. For the same tilt angles but 85W input power temperature differences were 44°C, 55°C and 81.5°C. The above statement clearly mentioned that temperature difference between evaporator and condenser keep on increasing with respect to tilt angle of HP. Input power also influences the temperature variation. Increasing input power at evaporator section could increase boiling point of working fluid and help to change the phase from liquid to vapor. Phase changed working fluid reaches condenser section and get condensed by releasing latent heat. Coolant water absorbs the heat power and help to change the phase of working fluid from hot vapor to cool liquid. Thus higher

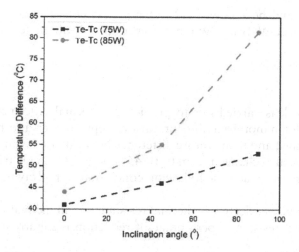

Figure 40.6 Temperature difference between evaporator and condenser sections with inclination angles.

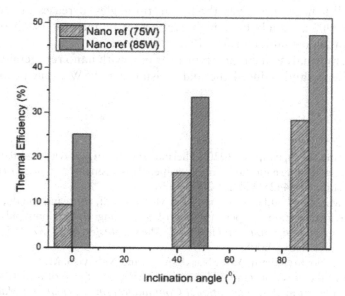

Figure 40.7 Thermal efficiency of HP with respect to respect to different inclination angles.

input power increased temperature at evaporator section within a short period of time as well as improve the performance of HP [15]. An inclination angle also played a significant role to improve the temperature difference in HP's.

Figure 40.7 represents the variation of HP efficiency with heat pipe tilt angle at 75 W and 85 W input powers. Figure 40.7 clearly shows increasing tilt angle significantly increase the thermal efficiency for all input power. Flowing of working fluid between evaporator and condenser become high due to gravitational force. Faster flowing of working fluid can hike heat transfer rate of HP. However, at 75 W input power, thermal efficiency were 9.4%, 16.7% and 28.4%. Further increasing input power from 75 W to 85 W, thermal efficiency were 25.1%, 33.4% and 47.4%. Therefore heat pipe efficiency seems to increase with increasing tilt angle and heat power [12heat pipe inclination angles, and input heating power. Distilled water (DW, 16]. 85W input power and 90° tilt angle shows higher thermal efficiency than 75W.

Conclusion

An experimental work is carried out to predict the thermal efficiency and temperature distribution of HP for nanorefrigerant for various input power and tilt angle. Different input power has applied and temperature difference has evaluated from the thermocouples readings. HP position also changed from 0° to 45° and 90° tilt angle thermal performance were evaluated. From this study the following conclusion were drawn.

- Increasing input power decreases thermal resistance at all inclination angles. Although it reduces thermal resistance, percentage of reduction is slightly more in 90° than 0° tilt angle.
- Decreasing thermal resistance is a key for higher efficiency of HP. Thermal efficiency become higher with high input power and tilt angle.
- Evaporator wall temperature got higher at high heat load. Inclination angle also influences the wall temperature. Even though all tilt angles increases the evaporator temperature, but 90° tilt angle have high evaporator temperature with respect to time. Here time is a important factor for HP applications.
- From the overall analysis a circular type heat pipe with nano refrigerant showed better thermal efficiency and reduced thermal resistance at 85W input power and 90° tilt angle.

References

1. Zadeh, A. A., and Nakhjavani, S. (2020). Thermal analysis of a gravity-assisted heat pipe working with zirconia-acetone nanofluids: an experimental assessment. *Archives of Thermodynamics*, 41(2), 65–83. doi: 10.24425/ATHER.2020.133622.
2. Aboutalebi, M., Moghaddam, A. M. N., Mohammadi, N., and Shafii, M. B. (2013). Experimental investigation on performance of a rotating closed loop pulsating heat pipe. *International Communications in Heat and Mass Transfer*, 45, 137–145. doi: 10.1016/j. icheatmasstransfer.2013.04.008.
3. Khajehpour, E., Noghrehabadi, A. R., Nasab, A. E., and Nabavi, S. M. H. (2020). Experimental investigation of the effect of nanofluids on the thermal resistance of a thermosiphon L-shape heat pipe at different angles. *International Communications in Heat and Mass Transfer*, 113, 104549. doi: 10.1016/j.icheatmasstransfer.2020.104549.
4. Sadeghinezhad, E., Akhiani, A. R., Metselaar, H. S. C., Latibari, S. T., Mehrali, M., and Mehrali, M. (2020). Parametric study on the thermal performance enhancement of a thermosyphon heat pipe using covalent functionalized graphene nanofluids. *Applied Thermal Engineering*, 175, 115385. doi: 10.1016/j.applthermaleng.2020.115385.
5. Martin, K., Sözen, A., Çiftçi, E., and Ali, H. M. (2020). An experimental investigation on aqueous Fe–CuO hybrid nanofluid usage in a plain heat pipe. *International Journal of Thermophysics*, 41(9), 135. doi: 10.1007/s10765-020-02716-6.
6. Torii, S. (2020). Enhancement of heat transfer performance in pipe flow using graphene-oxide-nanofluid and its application. *Materials Today: Proceedings*, 35, 506–511. S2214785320326936. doi: 10.1016/j.matpr.2020.04.078.

7. Mehta, K., Mehta, N., and Patel, V. (2021). Experimental investigation of the thermal performance of closed loop flat plate oscillating heat pipe. *Experimental Heat Transfer*, 34(1), 85–103. doi: 10.1080/08916152.2020.1718802.

8. Veerasamy, A., Balakrishnan, K., and Razack, S. A. (2021). Statistical optimization of closed loop pulsating heat pipe parameters with R-410a and nanorefrigerant in air conditioning applications. *Energy Sources, Part A: Recovery, Utilization, and Environmental Effects*, 1–18. doi: 10.1080/15567036.2021.1916130.

9. Veerasamy, A., Balakrishnan, K., Surya, T., and Abbas, Z. (2020). Efficiency improvement of heat pipe by using graphene nanofluids with different concentrations. *Thermal Science*, 24(1 Part B), 447–452. doi: 10.2298/TSCI190415358V.

10. Arya, A., Sarafraz, M. M., Shahmiri, S., Madani, S. A. H., Nikkhah, V., and Nakhjavani, S. M. (2018). Thermal performance analysis of a flat heat pipe working with carbon nanotube-water nanofluid for cooling of a high heat flux heater. *Heat Mass Transfer*, 54(4), 985–997. doi: 10.1007/s00231-017-2201-6.

11. Tharayil, T., Asirvatham, L. G., Ravindran, V., and Wongwises, S. (2016). Thermal performance of miniature loop heat pipe with graphene–water nanofluid. *International Journal of Heat and Mass Transfer*, 12.

12. Esmaeilzadeh, A., et al. (2020). Thermal performance and numerical simulation of the 1-Pyrene carboxylic-acid functionalized graphene nanofluids in a sintered wick heat pipe. *Energies*, 13(24), 6542. doi: 10.3390/en13246542.

13. Chidambaranathan, S., and Rangaswamy, S. (2020). Experimental investigation of higher carbon alcohols with low latent heat of vaporization as self rewetting fluids in closed loop pulsating heat pipes. *Thermal Science*, (00), 347–347. doi: 10.2298/TSCI200509347C.

14. Mehta, K., Mehta, N., and Patel, V. (2021). Experimental investigation of the thermal performance of closed loop flat plate oscillating heat pipe. *Experimental Heat Transfer*, 34(1), 85–103. doi: 10.1080/08916152.2020.1718802.

15. Huang, Q., Li, X., Zhang, G., Zhang, J., He, F., and Li, Y. (2018). Experimental investigation of the thermal performance of heat pipe assisted phase change material for battery thermal management system. *Applied Thermal Engineering*, 141, 1092–1100. doi: 10.1016/j.applthermaleng.2018.06.048.

16. Sözen, A., Gürü, M., Khanlari, A., and Çiftçi, E. (2019). Experimental and numerical study on enhancement of heat transfer characteristics of a heat pipe utilizing aqueous clinoptilolite nanofluid. *Applied Thermal Engineering*, 160, 114001. doi: 10.1016/j.applthermaleng.2019.114001.

41 An empirical study of sorting algorithms under Skew't' input

Priyadarshini

Assistant Professor, Department of IT/IS, International School of Management, Patna, India

Abstract: This paper presents the statistical analysis of the performance of four sorting algorithms namely quicksort, heap sort, merge sort, and k sort, each exhibiting the same average-case complexity of O(NlogN)). The array for sorting has been randomly generated from Skew't' distribution. The investigation is done for studying the effect of the factor, responsible for introducing 'skewness' in the symmetrical Student's' distribution, on the performance of sorting algorithms. The sorting algorithms have been compared based on the statistical bounds and parametric complexity. In statistics and probability theory, skew t distribution is used to model many real data sets arising in many fields containing moderate to strong asymmetry. The purpose of this paper is to check which sorting algorithm is best suited for data that follows skew t distribution.

Keywords: Parametric complexity, skew t distribution, sorting algorithms, statistical bound.

Introduction

A lot of work has been done related to the parametric complexity analysis of sorting algorithms. Arrays are randomly generated from different probability distributions such as uniform, binomial, Poisson, exponential, normal, truncated, etc [1–4]. These arrays are sorted using sorting algorithms under study and their execution times are recorded. The computer experiments [5] are repeated by changing the distribution parameters and the effect of the parameters of the input distribution on the sorting efficiency of the algorithms are analyzed by using Factorial experiments [7]. This paper is aimed at investigating the effect of the parameters of the skew t distribution on the sorting efficiency of the algorithms under study. The optimal values of the parameters of the distribution are also obtained for various sorting algorithms. A brief introduction of the properties of Skew t distribution has been given below.

Skew-t Distribution

The Student t distribution is more efficient in producing values that fall far from its mean and capturing heavy tail data sets. However, it is a symmetric distribution that cannot capture asymmetry. To accommodate asymmetry and long tailed data, different scientists have proposed different versions of skew t distribution [10,12]. In this paper skew t distribution proposed by Azzalini and Capitanio [13] has been used. The authors defined a skew t variate as a scale mixture of skew-normal and chi-squared variables [11]. C program has been used for generating arrays from skew t distribution using the appropriate logic given for random number generation which can't be shown here due to limitation of the paper. A laptop computer (Intel (R) Core ™ i3-4005U Processor @ 1.70GHz, 8GB RAM, Windows 7 operating system) is used to do serial computer experiments. Execution times of the sorting algorithms are computed by executing codes written in Dev C++ using C language and different comparative graphs are prepared using Minitab 17 Statistical Package.

Relative performance analysis of different algorithms for skew–t input [6].

The relative performance of sorting algorithms was carried out by varying the array size from 1 lakh to 10 lakhs for different pairs of values of the input parameters (ν, α). The results are shown in the tables to follow.

From Table 41.1 and Figure 41.1, it is noticeable that the performance of merge sort and heap sort is more or less the same up to N = 400000, but for N>400000, they have changed their paths. After N = 700000, there is a drastic decrease in the heap sort execution time. K-sort and quick sort approximately showed the same pattern of execution time with the change in the array size.

Table 41.1: Shows the comparison of various sorting algorithms keeping ν = 10 and α = 5.

	$\nu = 10, \alpha = 5$			
N	*Heapsort*	*K-sort*	*Quicksort*	*Mergesort*
100000	0.0798	0.0343	0.0359	0.0736
200000	0.1563	0.0625	0.083	0.1705
300000	0.2609	0.1062	0.1139	0.2394
400000	0.3578	0.1499	0.1628	0.3389
500000	0.4266	0.1796	0.2125	0.339
600000	0.5452	0.2391	0.2468	0.3564
700000	0.6546	0.2626	0.3266	0.3922
800000	0.336	0.3251	0.3454	0.4716
900000	0.0405	0.3734	0.4719	0.5282
1000000	0.0437	0.3876	0.4437	0.6171

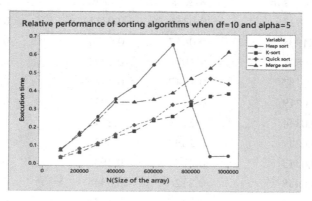

Figure 41.1 Relative performance of sorting algorithms when ν = 10 *and* α = 5.

Table 41.2: Shows the comparison of various sorting algorithms keeping $v = 10$ and $\alpha = 50$.

	$v = 10, \alpha = 50$			
N	Heapsort	K-sort	Quicksort	Mergesort
100000	0.0815	0.0295	0.0467	0.0703
200000	0.1685	0.0685	0.0795	0.1548
300000	0.2578	0.1028	0.1218	0.2233
400000	0.3532	0.1436	0.1689	0.3015
500000	0.4498	0.1923	0.211	0.4093
600000	0.5608	0.2204	0.292	0.4966
700000	0.6656	0.2641	0.3141	0.5811
800000	0.7703	0.3189	0.3298	0.6359
900000	0.9069	0.3656	0.4001	0.7453
1000000	0.819	0.3812	0.4653	0.5736

It is observable from the above Table 41.2 and Figure 41.2 that by increasing the value of the shape parameter from 5 to 50, there is little change in the performance of k-sort and quicksort. Whereas heap sort and merge sort have shown a decrement in the execution time for N> = 900000. It is visible that heaps sort has shown maximum complexity.

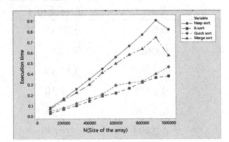

Figure 41.2 Relative performance of sorting algorithms when $v = 10$ and $\alpha = 50$.

Table 41.3: Shows the comparison of various sorting algorithms keeping $v = 40$ and $\alpha = 5$.

	$v = 40, \alpha = 5$			
N	Heap sort	K-sort	Quicksort	Mergesort
100000	0.0923	0.0358	0.0311	0.067
200000	0.1905	0.0657	0.0734	0.1329
300000	0.0531	0.1079	0.1094	0.1843
400000	0.0187	0.1547	0.1483	0.1999
500000	0.0219	0.1902	0.2031	0.2548
600000	0.0344	0.2208	0.2345	0.3093
700000	0.028	0.2392	0.2671	0.3654
800000	0.0343	0.2781	0.2907	0.4077
900000	0.0438	0.3126	0.3346	0.5275
1000000	0.0548	0.3477	0.4481	0.5275

From Table 41.3 and Figure 41.3, the center of attraction is again the heap sort, whose execution time has decreased a lot from N> = 200000, as the degree of freedom is increased from 5 to 40.

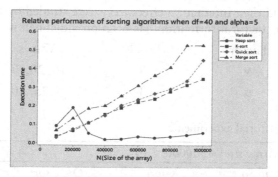

Figure 41.3 Relative performance of sorting algorithms when ν = 40 and α = 5.

Table 41.4: Shows the comparison of various sorting algorithms keeping ν = 40 and α = 50.

	ν = 40, α = 50			
N	Heapsort	K-sort	Quicksort	Mergesort
100000	0.025	0.0281	0.0438	0.0655
200000	0.0078	0.0624	0.0887	0.1562
300000	0.0155	0.1001	0.1344	0.2171
400000	0.0219	0.1359	0.1845	0.2985
500000	0.0248	0.1673	0.2265	0.2749
600000	0.033	0.2015	0.2732	0.3554
700000	0.0421	0.2373	0.5482	0.4062
800000	0.04	0.2735	0.2937	0.4468
900000	0.0341	0.275	0.336	0.5123
1000000	0.0484	0.3081	0.3687	0.5892

Heapsort is greatly affected by increasing the degree of freedom and shape parameter alpha, which is evident from Table 41.4 and Figure 41.4. The rest of the sorting algorithms have performed similarly as in the previous cases. The only difference is shown in the case of quicksort for N = 700000, which may occur due to the random data. The above discussion can be further supported by the parametric complexity in the next section.

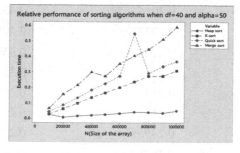

Figure 41.4 Relative performance of sorting algorithms when ν = 40 and α = 50.

Parametric Complexity

Table 41.5 presents the summary of the results obtained in a 3^2-factorial experiment for various sorting algorithms. In the experiment, the effects of two factors namely degree of freedom (v) and (α) and their interaction effect (αv) were studied on the response variable i.e. execution time. The factor information has been provided below.

Factor	Levels	Values
v	3	10, 20, 30
α	3	5, 25, 50

Table 41.5: Results of 3^2-ftorial experiment.

Skew-t distribution (N(Array size) = 500000, v, α)									
		Heap Sort		K-sort		Quicksort		Merge sort	
Sources	D.f.	F	P	F	P	F	P	F	P
α	2	6.32	0.003	1.61	0.207	1.13	0.330	5.71	0.000
v	2	88.00	0.000	2.73	0.072	16.66	0.000	127.58	0.005
αv	4	104.85	0.000	2.91	0.027	3.23	0.017	7.14	0.000

As far as the parameter alpha(α) is concerned, it shows a significant effect on the execution times of heap sort and mergesort, but it is found to have an insignificant effect on the running times of k-sort and quick sort algorithms. In skew-t distribution, alpha (α) models the skewness in the distribution. The results given in Tables 41.6 and 41.7 below could be a support to the results obtained in the factorial experiment by providing an idea about the effect of skewness (α) in the data on the running times of sorting algorithms. The second parameter, degree of freedom (v), is found to be significant in the case of heapsort, quick sort, and merge sort, but insignificant for k-sort which can be further explained by the results obtained in Tables 41.8 and 41.9 given below. Though for some algorithms the skewness parameter α and degree of freedom (v) are singularly non-significant but their joint effect αv is found to be significant for each of the sorting algorithms [8].

Table 41.6: Optimal value of the degree of freedom for $\alpha = 5$.

	$\alpha = 5, N = 500000$		
v	Heap sort	Quicksort	Mergesort
10	0.4172	0.1999	0.4673
20	0.425	0.2016	0.4085
30	0.3686	0.2016	0.4352
40	0.0191	0.2045	0.4054
50	0.0235	0.2203	0.3306
60	0.022	0.2198	0.3119
70	0.0202	0.2125	0.2771
80	0.0215	0.2155	0.281
90	0.0204	0.2499	0.2792
100	0.0203	0.1989	0.2708

From Table 41.6 and 41.7 it is evident that irrespective of the value of α heap sort gives the best performance for the value of dfv > = 40. Thus, it can be concluded that any value of the degree of freedom(df) after 40 i.e. ν > = 40 can be preferred in the case of heapsort. In the case of quicksort, for a smaller value of the shape parameter α, a high value for d.f.(ν) is preferable (when α = 5, ν = 100). For high values of the skew parameter (α = 50), the stability in execution time is observed for any value for the d, f.(ν).

As such, for the value of α as high as 50, any value for the degree of freedom (ν) can be taken as the optimal value. As far as the merge sort is concerned, the optimal value of the degree of freedom (ν) for merge sort in both cases is found to be between 60 and 100.

Table 41.7: Optimal value of the degree of freedom for α = 50.

ν_{df}	α = 50, N = 500000		
	Heap sort	*Quicksort*	*Mergesort*
10	0.4171	0.1783	0.4188
20	0.4844	0.177	0.3986
30	0.2688	0.1835	0.4147
40	0.0236	0.1817	0.3456
50	0.0267	0.1781	0.2845
60	0.0282	0.181	0.2814
70	0.0233	0.1814	0.2755
80	0.028	0.1813	0.2771
90	0.0237	0.1783	0.2733
100	0.025	0.1862	0.2751

Another parameter of skew-t distribution i.e., α is found to be significant for heap and merge sort. Thus, the optimal values of α for the two sorting algorithms are tried to obtain as follows.

Table 41.8: Optimal value α of when ν = 10.

α	ν = 10, N = 500000	
	Heap sort	*Merge sort*
5	0.4218	0.3797
25	0.4470	0.3609
45	0.4466	0.3752
65	0.4220	0.3639
85	0.4626	0.3720
105	0.4310	0.3687
125	0.4248	0.4076
145	0.4186	0.4140
165	0.4446	0.3765
185	0.4188	0.3673

To obtain the optimal value of α for heap sort and merge sort, for which α is found to be significant, the data in Tables 41.8 and 41.9 can be considered for fixed values of degree of freedom $v = 10$ and $v = 40$ respectively. The array size is fixed at $N = 500000$. If we consider the case of $v = 10$, heap sort has not shown systematic performance. But optimal value of α, i.e. minimum execution time of heap sort is found at $\alpha = 145$.

Table 41.9: Optimal value of α when $v = 40$.

	$v = 40, N = 500000$	
α	*Heapsort*	*Merge sort*
5	0.4094	0.3812
25	0.2094	0.3452
45	0.0192	0.2688
65	0.022	0.2578
85	0.022	0.2626
105	0.0188	0.2576
125	0.0184	0.2563
145	0.0222	0.2656
165	0.0250	0.2547
185	0.0186	0.2624

On the other hand, how heap sort has performed is quite interesting in case of $v = 40$. There is a drastic decrement in its execution time after $\alpha = 45$ and further, it remains stable for higher values of alpha [9]. Thus, any value of α that is greater than or equal to 45 can be seen as preferable. If we consider merge sort, then for $v = 10$, it has also performed similarly as the heap sort with a change of alpha. Though the execution time of merge sort is minimum at $\alpha = 165$, but after $\alpha = 65$, it is more or less the same. Thus any value that is greater than or equal to 65 can be considered here.

Conclusion

As far as the relative performance of sorting algorithms is concerned, heap sort can be seen as the best choice for sorting skewed data. The parametric complexity analysis confirms that the shape parameter or the factor responsible for skewness in the data (α) has a significant effect on the execution times of heap sort and merge sort, but it is found to have an insignificant effect on the running times of k-sort and quick sort algorithms. The second parameter degree of freedom (v) is found to be significant in the case of heapsort, quick sort, and merge sort, but insignificant for the k-sort algorithm. It was also observed from the analysis that any value of the degree of freedom after 40 i.e., $v > = 40$ can be preferred in the case of heapsort. Any value of α that is greater than or equal to 45 is preferable for heap sort and for merge sort any value that is greater than or equal to 65 can be considered as preferable. The above discussion will help select an appropriate sorting algorithm in case of skewed data when the values of the parameters of the input distribution are known.

The results obtained in this analysis are quite useful for further research work in this area. This result can be compared with the results of 't' distribution. This is an attempt to add up something to the research work done in this area.

References

1. Priyadrashini, and Kumari, A. (2020). A peek into sorting complexity analysis for skewed normal parameters. *Journal of Information and Computational Science,* 10(8), 113–125. DOI: 10.12733.JICS.2020.V10I8.535569.12764.

2. Priyadrashini, and Kumari, A. (2020). A comparative performance of sorting algorithms: statistical investigation In: Soft Computing: Theories and Applications, Advances in Intelligent Systems and Computing, (vol. 1154, pp. 367–379), Springer Nature Singapore Pte Ltd. https://doi.org/10.1007/978-981-15-4032-5_8.

3. Priyadarshini, and Kumari, A. (2018). Parameterized complexity analysis of heap sort, K-sort and quick sort for binomial input with varying probability of success. *International Journal of Emerging Technologies and Innovative Research,* 5(7), 491–495. (www.jetir.org I UGC and ISSN Approved), ISSN:2349-5162, Available from http://www.jetir.org/papers/JETIR1807424.pdf

4. Priyadarshini, Chakarborty, S., and Kumari, A. (2018). Parameterised complexity of quick sort, heap sort and k-sort algorithms with quasi binomial input. *International Journal on Future Revolution in Computer Science and Communication Engineering (IJFRSCE),* 4(1), 117–123.

5. Kumari, A., Singh, N. K., and Chakraborty, S. (2015). A statistical comparative study of some sorting algorithms. *International Journal in Foundations of Computer Science and Technology,* 5, 21–29. 10.5121/ijfcst.2015.5403.

6. Singh, N. K., Chakarborty, S., and Malick, D. K. (2014). A statistical peek into average case complexity. *International Journal on Recent Trends in Engineering and Technology,* 8(1), 64–67.

7. Kumari, A., and Chakraborty, S. (2013). Parameterized complexity, a statistical approach combining factorial experiments with principal component analysis. *International Journal of Computer Science Engineering,* 5, 166–176.

8. Kumar, P., Kumari, A., and Chakraborty, S. (2012). Parameterised complexity on a new sorting algorithm, a study in simulation. *Annals Computer Science Series,* VII. arXiv preprint arXiv:1202.5957.

9. Kumari, A., and Chakraborty, S. (2007). Software complexity: A statistical case study through insertion sort. *Applied Mathematics and Computation,* 190(1), 40–50.

10. Basalamah, D. (2017). Statistical inference for a new class of skew t distribution and its related properties. Phd Thesis, College of Bowling Green State University.

11. Azzalini, A., and Capitanio, A. (2014). *The Skew-Normal and Related Families.* Cambridge University Press.

12. Kim, H. J. (2001). On a Skew-t distribution. *The Korean Communications in Statistics,* 8(3), 867–873.

13. Azzalini, A., and Capitanio, A. (2003). Distributions generated by perturbation of symmetry with emphasis on a multivariate skew t-distribution. *Journal of the Royal Statistical Society Series B: Statistical Methodology,* 65(2), 367–389.

42 Investigative studies on damage mechanics of base iron metals

Chethan S[1], Ravikumar S[1], Hemaraju[2,a], Jayashree M[3], and Santhosh Kumar T C[4]

[1]Department of Mechanical Engineering, ATME College of Engineering, Mysuru, India

[2]Department of Mechanical Engineering, BGSIT, Adhichunchanagiri University, BG Nagara, India

[3]Department of Electrical & Electronics Engineering, The National Institute of Engineering, Mysore, India

[4]Department of Mechanical Engineering, BGSIT, Adhichunchanagiri University, BG Nagara, India

Abstract: Damage of the base iron metals were happened due to the elastic and plastic deformation due to unidentified nature of stresses and forces. In many applications gradual failure of the materials will takes place due to rubbing nature of two or more elements. These rubbing nature will produce incremental co-efficient of friction cause progressive loss of material results in permanent damage of the element. As many elements in many applications were made from base iron materials present investigation was made on the same. In this investigation three base metals such as high carbon high chromium steel, heat resistant steel, mild steel, and cast iron was considered for finding the damage behaviour when they are subjected to surface abrasion. Surface abrasion test was conducted using dry sand abrader as per G-65 standards. Abrasive of different sizes were used for conduction of study. Varying normal load and time was used conducted for the study. Surface mechanics and sub surface mechanics was studied to identify the inelastic deformation and damage of the materials.

Introduction

Tribology is the investigation of two rubbing bodies. Elements, in any machinery, are used to transfer either force or motion from one location to other locations. Materials are also transferred from one place to other locations using machineries. These elements may be structures like belt conveyors or pipes through which material is delivered using fluids. The pulverized coal in thermal plants is conveyed from one location to other locations using compressed air as conveyor. In all such machineries relative motions are involved between components. Velocity and displacement discontinuities result from the relative motions. The velocity and displacement discontinuities manifest themselves into friction and wear. These friction and wear will be reliant on mating pairs and their surface morphology. Friction and wear were found to be system dependent parameter which makes difficulty in designing of machine elements.

It was found from different scientists all over the world who conducted experiments simulating field conditions that the wear rate was drastically changed in the range of 10^{-15} to 10^{-1} mm^3/ N-m. These wear rates were discovered to be affected by the material pairs and operational circumstances [17]. The operating conditions and materials for a given

[a]hemarajucrp@gmail.com

machine element pair should be known to the design engineer in advance. The complexity of wear phenomenon was quantified by defining parameters like wear rates and modes of wear [8, 9]. Abrasive wear was caused by many mechanisms either by plastic deformation or by brittle fracture. The mechanisms found are cutting mode wear, ploughing mode wear and wedge mode wear in base metals such as ductile cast iron, grey cast-iron, heat-treated steels and cold working metals [10, 11].

Stanisla Verichev et al. [12] summarized the impact of hydrostatic pressure on wear behaviour of structural materials using with very high level of hydrostatic pressure of 250 atm. The target materials used were steel and tungsten carbide. The SEM study revealed the peeling of layers and chipping of hot particles. A reasonable co-relation was obtained for wear data with a deformation factor that incorporated the friction of the abrasive on plastic deformation. Less resistance was observed for materials which were easily deformed plastically. Xu et al. [13] studied the effect of subsurface deformation and wear behaviour of construction steels with the help of multiple dual indenters. Scratch tests were performed on wide range of steels with different microstructures. Based on the experimentation it is identified that the nature of work hardened layer and its thickness formed beneath the abrader surface has more influence on abrasion resistance of a material. Bakshi et al. [14] conducted experimentation using abrasion test rig with standard rubber wheel on different samples of same steel differentiated based on metallurgical structures and discrete hardness. He noticed the impact of micro-structure on wear behaviour by the experimentation and summarized that test results did not exhibit any significance in wear co-efficient but three different micro structure of steels yielded wear loss in different wear mechanisms. It was believed that wear map is most convenient and suitable for choice of material under an extensive range of operating conditions. Designing a tribo-system and selecting material pairs based on wear maps further requires the understanding of wear rates, different wear forms and mechanisms involved.

In the present investigation aims at understanding the role of hardness, micro-structural features, abrader attribute and load levels on inelastic deformations and damage to the base metals selected such as heat resistant steel, mild steel cast iron and high carbon chromium steel.

Materials Selection

There are four base iron metals that were selected for the purpose of study followed with as High carbon chromium steel, heat resistant steel, mild steel, and cast iron.

- **Mild steel:** A base mild steel having 0.25% C, 0.2% Cu, 98.0% Fe, 1.03% Mn, 0.040% P, 0.280% Si, and 0.050% S was used for the test.
- **Heat Resistant Steel:** En 1.47 Category heat resistant steel having 0.12%, 6–8% Cr 0.5–1.0 % Al and 0.015% S, and 0.04% S was used for the test.
- **High carbon High Chromium steel:** .1.40–1.60% C, 11.0–13.00 % Cr, 0.70–1.20% Mo, 1.10% V, 0.030% P, 0.280% Si, and 0.03% S was used for the test.
- **Cast iron:** A base CI having a chemical composition of 2.5–4.0% carbon, 1–3% silicon, and the remainder is iron grey cast iron was used for the test.
- **Abrasive materials:** Natural sand of known size distribution is used for conducting the test.

Experimentation

The experimentation was carried out according ASTM G65 using dry sand abrasion test rig. The test rig is represented in Figure 42.1 which is installed at Department of Industrial Production Engineering, The NIE Mysore.

The specimens were cut according ASTM G65 with specimen size of 75 mm × 25 mm × 12 mm. The specimens were machined to required size by conventional milling process. Natural silica is used as abrasive agent for conducting the wear test and it acts as tertiary body, while the wheel act as secondary body in the three-body abrasive test. The wheel is rotated at 200 rpm against the specimen with abrasive agent in between specimen and wheel.

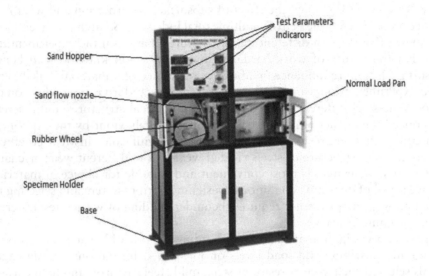

Figure 42.1 Dry sand abrasion test rig.

The load applied during the wear test comprises to variations namely 53.2N and 102.4N. During the wear test, Volume loss occurs in the test specimen and thus the Volume loss of the specimen is computed before and after the test.

Results and Discussions

Tibological Wear Analysis

Experimentation is carried out to determine the wear behaviour of different metals with variation in the hardness and microstructure. The different metals with hardness used for wear test are heat resistant steel of 155.6 BHN, Mild steel of 130.9 BHN, cast iron of 159.3 BHN and High carbon high chromium steel of 158.2 BHN. The different metals with respective hardness are subjected to three body dry sand abrasion test at two different loads of 52.3N and 102.4N respectively. The volume loss occurred due to wear phenomena is represented in Table 42.1.

Figure 42.2 depicts the comparison of wear loss at two different loads of 52.3N and 102.4N for different materials. It is observed that the % volume loss is more at 102.4N

Table 42.1: Volume loss of base iron materials subjected to wear load.

Sl No.	Material	volume loss for 52.3 N load in %	volume loss for 102.4 N load in %
01	Heat resistant steel	0.29	0.57
02	Mild steel	0.41	0.82
03	Cast iron	0.04	0.06
04	High carbon chromium steel	0.07	0.15

load compared to 52.3N for all the materials subjected to wear test. However, cast iron exhibit maximum wear resistance compared to other materials while mild steel experienced maximum wear loss compared to other materials. It is evident from the experimentation that % volume loss of 90% is observed by subjecting mild steel for dry sand abrasion test when compared to cast iron at 52.3N load.

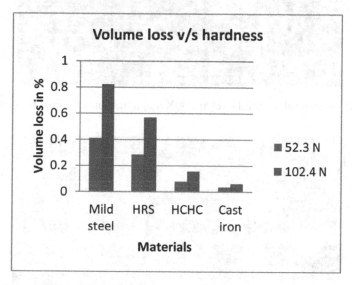

Figure 42.2 %volume loss for different base iron metals at two variant loads.

It is also observed that the increase in percentage of volume loss is found to be approximately double for 102.4N compared to that of percentage of volume loss for 52.3N for Heat resistant steel, Mild steel and High carbon high chromium steel, while cast iron show less wear rate for 102.4N compared to load of 52.3N.

Scanning Electron Microscope

Morphology studies was carried out utilizing SEM, Model: JEOL JSM-6400lv situated at UVCE, Bangalore for characterizing the reliance of volume loss under different loading conditions.

Figure 42.3a represents the SEM micrograph of worn surface of Mild steel recorded at 500X magnification when subjected to wear load of 52.3N. It is observed from the SEM micrograph that profound groove is visible at centre on the wear track and the profound depression raised and clear cut on the ridges. The edges exhibit a measure of plastic distortion where the ridges appear to be detached and also it can be observed that groove appears only till the midway and vanishes towards the end representing plastic deformation.

Figure 42.3b depicts the worn surface micrograph of Mild steel visualized at magnification of 1500X subjected to wear load of 102.4N. It is evident from the SEM micrograph that there is profound notch at the centre of the wear track. It can also be noticed from the micrograph that the wear track appears to be flat without raise in the ridges when compared to the wear track recorded at 52.3N. Hence it can be summarized from Figure 42.3a and 42.3b that the wear track of is flat for higher load in contrast to lower load conditions. However, the mode of abrasive wear indicates severe cutting on the wear surfaces.

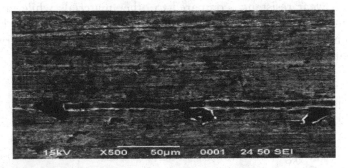

Figure 42.3a SEM micrograph of Mild steel at 500X magnification subjected to wear load of 52.3N.

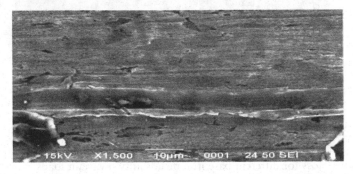

Figure 42.3b SEM micrograph of Mild steel at 1500X magnification subjected to wear load of 102.4N.

The SEM visuals of wear track of Heat resistant steel recorded at magnification of 1000X for wear load of 52.3N and 2000X magnification for wear load of 102.4N is illustrated in Figure 42.4a and 42.4b respectively. It can be noticed from the SEM images that numerous poorly characterized grooves are observed representing a smooth worn surface and it is indicative from the micrograph that mild ripping abrasive wear phenomenon can be observed while experimenting three-body abrasive wear on Heat resistant steel. The presence of poorly defined grooves at 102.4N represent more wear loss in contradiction to wear at 52.3N.

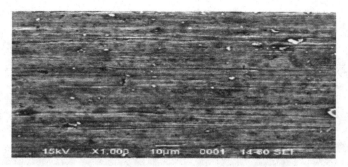

Figure 42.4a Heat resistant steel SEM micrograph at 1000X magnification subjected to wear load of 52.3N.

Figure 42.4b Heat resistant steel SEM micrograph at 2000X magnification subjected to wear load of 102.4N.

Figure 42.5a represents the SEM micrograph of worn area of High Carbon Steel recorded at 2000X magnification when abraded at wear load of 52.3N. It is evident from the SEM images that high carbo steel with better BHN value exhibit smooth surface of wear track with clearly defined grooves. This represents less percentage of wear loss compared to wear behaviour at higher loads.

Figure 42.5b represents the SEM images of High Carbon Steel recorded at 4000X magnification when abraded at wear load of 102.4N. It is clear form SEM images that groove is found on the wear track with no ridges. The presence of groove indicate damage on the worn surface with crack present on it. The feature of the micrographs denotes abrasive wear with wedge mechanism.

The SEM images of heat resistant steel wear track recorded at magnification of 1500X for wear load of 52.3N and 2000X magnification for wear load of 102.4N is illustrated in Figure 42.6a and 6b respectively. It can be noticed from the wear track of Cast iron material, when subjected to 52.3N, that surface is clear with no grooves or ridges resulting in more wear resistance for the applied load in the abrasive environment. It is evident from the SEM images represented in Figure 42.6b that well defined groove is clearly visible at the central portion and no ridges are found towards the end due to greater wear resistance offered by Cast iron material. The higher BHN of Cast iron material, in contrast to other materials offer significant wear resistance towards abrasion.

Figure 42.5a High carbon chromium steel SEM micrograph at 2000X magnification subjected to wear load of 52.3N.

Figure 42.5b High carbon chromium steel SEM micrograph at 4000X magnification subjected to wear load of 102.4N.

It can be summarized from the wear analysis and SEM morphological analysis that increase in hardness yields to more wear resistance thus Cast iron with 159.3 BHN offers maximum wear resistance leading low percentage volume loss in contrary to high carbon chromium steel, heat resistant steel and mild steel, mentioned in the order of decrease in wear resistance.

It can also be noticed that increase in wear load leads to increase in percentage volume loss, however the contribution against wear loss is high for mild steel followed by high carbon chromium steel, heat resistant steel and cast iron.

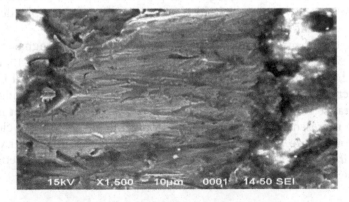

Figure 42.6a Cast iron SEM micrograph at 1500X magnification subjected to wear load of 52.3N.

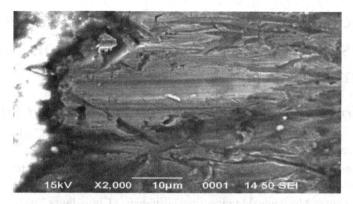

Figure 42.6b Cast iron SEM micrograph at 2000X magnification subjected to wear load of 102.4N.

Conclusions

1. The deformation was happened sequentially in the nature of abrasive mode.
2. Ploughing mode and cutting mode of abrasive wear was found both in my steel and HR steel.
3. Wedge mode and groove mode was found in high carbon steel and cast iron
4. Weight loss of mild steel is slightly more compared to heat resistant steel and cast iron.
5. Weight loss of high carbon steel is slightly more compared to cast iron and comparable with hardness.
6. Weight loss of HCHC steel and cast iron is less compared to mild steel and HR Steel and found to be comparable with hardness.

References

1. Archard, J. J. (1953). Contact and rubbing of flat surfaces. *J. Appl. Phys.*, 24(8), 981–988.
2. Bhansali, K. J, Wear control hand book, Peterson M.B and Winer, W.O, (Eds), 1980, ASME, 373–383
3. Johnson, K. L. (1995). Contact mechanics and the wear of metals, *Wear*, 190, 162–170.
4. Hokkarigawa,K., Wear maps of ceramics, *Ceramics Japan* 32, 19–25.
5. Holm, R., Electric contact, Almquist and Wiksells, Stockhelm, 1946, section 40.
6. Lancaster, J. K. (1978). Wear mechanisms of metals and polymers, *Trans. Inst. Metal Finish*, 56, 4, 145.
7. Rabincowicz, E. (1980). Wear control hand book, Peterson M.B and Winer, W.O, (Eds), ASME, 475.
8. Hokkirigawa, K. (1988). Theoretical Estimation of Abrasive Wear Resistance Based on Microscopic Wear Mechanism of Steel, *Wear*, 123, 241–251
9. Lim, S. C. and Ashby, M.F. (1987). Overview no. 55 Wear-Mechanism maps, *Acta Metallurgica*, 35, 1–24.
10. Zum Gahr, K. H. (1987). Microstructure and wear of materials, *Tribology series*, 10, 132–148.
11. K Srinivasa, S Chethan, N Yathisha, MS Arjun. (2020). A study of physical, mechanical and tribological properties of a biomaterial using magnesium alloy, *AIP Conference Proceedings* 2274.
12. Verichev, S. N., Mishakin, V. V., Nuzhdin, D. A., and Razov, E. N. (2015). Experimental study of abrasive wear of structural materials under the high hydrostatic pressure, *Ocean Engineering*, 99, 9–13.
13. Xu, X., Zwaag, S., and Xu, W. (2015). Prediction of the abrasion resistance of construction steels on the basis of the subsurface deformation layer in a multi-pass dual-indenter scratch test, *Wear*, 338–339, 47–53.
14. Bakshi, S. D., Shipway, P. H., and Bhadeshia, H. K. D. H. (2013). Three-body abrasive wear of fine pearlite, nanostructured bainite and martensite, *Wear*, 308, 46–53.

43 Experimental study of free convective heat transfer from shrouded finned horizontal channel

Juri Sonowal[1,2,a], Kankan Kishore Pathak[3,b], and Biplab Das[4,c]

[1]School of Energy Science and Engineering, Indian Institute of Technology, Guwahati, Guwahati, Assam, India

[2]Department of Mechanical Engineering, North Eastern Regional Institute of Science and Technology, Itanagar, India

[3]Department of Mechanical Engineering, Girijananda Chowdhury Institute of Management and Technology, Assam, India

[4]Department of Mechanical Engineering, National Institute of Technology Silchar, Assam, India

Abstract: An experimental investigation of free convective heat transfer from a shrouded horizontal fin array has been performed. For this purpose, two heat sinks of different types are used with varying fin spacing. A clearance gap of 10 mm to 30 mm is maintained between the fin tip and an adiabatic shroud. A customized test bed is designed and built accordingly. A series of tests are conducted using these heat sinks. The effect of spacing between fins and clearance between the shroud and the fin tip on the total heat dissipation is measured. Results indicate that the widely spaced heat sink shows higher surface temperature than that of the narrowly spaced heat sink. For the narrowly spaced heat sink with clearance 30 mm shows nearly 10.54% higher heat transfer than that of the narrowly spaced heat sink with 20 mm clearance. The same is 13.93% when compared with 10 mm clearance. For the widely spaced heat sink also, similar results are reported but 30 mm clearance shows nearly 11.27% and 6.12% higher heat transfer than that of 10 mm and 20 mm clearance respectively.

Keywords: free convection, shrouded finned horizontal channel.

Introduction

Undesirable heat generation prompting failure of system and operation of many new compact electronic devices is a matter of major concern. From the inception of the extended surfaces/fins, the speedy removal of heat from the various heat generating machineries become important in many engineering applications. In other words, fin heat transfer becomes one of the classical methods of rapid heat removal from the heated source. Adequate cooling of electronic devices, solar PV panels, car radiators, and transformers has recently invigorated experiments on the use of fins to improve natural convective heat transfer. In this essence, it may be conferred that free convection heat transfer is inherently present in all transport phenomenon and accurate analysis of this phenomenon is always desirable. Natural convective transfer of heat from the fin arrays attached to a base oriented vertically is listed as follows: The report of [1] examines the performance of four

[a]jurysonowal@gmail.com, [b]kankankishore@gmail.com, [c]biplab.2kmech@gmail.com

geometrical sets of arrayed rectangular fin are tested in different orientations, namely, vertical, 45° with the horizontal, and horizontal to catch the influences of spacing and height of fins on heat transfer. The experimental results of the investigation [2] show that heat transfer from fin lies in the range between the results of the parallel plate channel and vertical plate. Interference of boundary layer causes lower heat transfer for narrowly spaced fins, which involves reduction of air motion over the surface of the fins. The experimental study of Fisher and Torrance [3] has provided the first experimental evidence of chimney effect which is due to the presence of a vertical shroud before the tip of the fin array on a vertical base undergoing free convection.

Prior investigations on pure natural convective heat transport from the arrays of fins glued to a horizontal base examined are as follows: The results of experimental study of Harahap and McManus [4] reveals that shorter length fins causes single chimney flow persuading air from the ends which enhanced heat transfer. Experimental investigation for optimum fin spacing was noted by Jones and Smith [5]. Interferometer is used to measure the gradient of temperature of the fluid and then the local heat transfer coefficient is determined. A correlation based on the value of properties estimated at the average temperature was provided. Experimental study by Leung et al. [6] varied fin height in a range from 32 mm to 90 mm and observed the elevated base temperature variations in a range from 40°C to 80°C. The influence of fin heights on heat transfer is very small as compared to the fin spacing for the vertical rectangular fins attached to the vertical and horizontal base. Yuncu and Anbar [7] experimentally highlights the fact that coefficient of convective heat transfer of "fin base" system is influenced by fin-height, fin-spacing and excess temperature between ambient and base.

From the existing literature, it is found that numerous studies on fin heat transfer from a fin or fin array mounted on a base plate oriented vertically, horizontally and inclined to the horizontal are available. It is also reported in literature that in natural convection, placing of a shroud in front of the fin tip of a fin array may enhance the effect of buoyancy which in turn augments the heat transfer process. This shrouding effect in free convection remains to be seen for heat sinks (fins) mounted on horizontal base. Since there are extensive applications of this type of fin configuration in engineering and industrial applications, it cannot be overlooked. In this essence, an experimental investigation is performed to rate the shrouded horizontal fin array under free convection

Experimental Setup

A customized test bed is designed and built in the departmental heat and mass transfer laboratory of North Eastern Regional Institute of Science and Technology (NERIST), Arunachal Pradesh, India as in Figure 43.1. Two sets of heat sinks are prepared from highly polished plate of Aluminium alloy with a thermal conductivity of $k = 202.4$ W/m-K at and an emissivity (ε) of 0.04, featuring different fin spacing are tested to calibrate the test-bed against the existing theoretical model. The base plate dimensions and heights of both the fins are fixed at 40 mm each while the spacing of the fins is varied in a range from 6 mm to 8 mm. The base of the fin is 10 mm each. The fin channels are continuous and rectangular. The design dimensions of the heat-sinks are given in Table 43.1. An insulated glass case of thickness 10 mm is used to reduce the thermal losses from the fin. Insulation paper lined with thermocol of 15 mm were used for insulation. A heater made up of nichrome wire, having capacity of 450 W and length 650 mm, laid uniformly on an asbestos plate was used to heat the fin array. This is then covered by mica sheets to provide electrical insulation and asbestos sheet to provide both thermal and electrical insulation. The entire setup is then placed on a glass case to prevent base plate heat losses. T type thermocouples are used to record the temperature variations at various fin positions and on the base-plate.

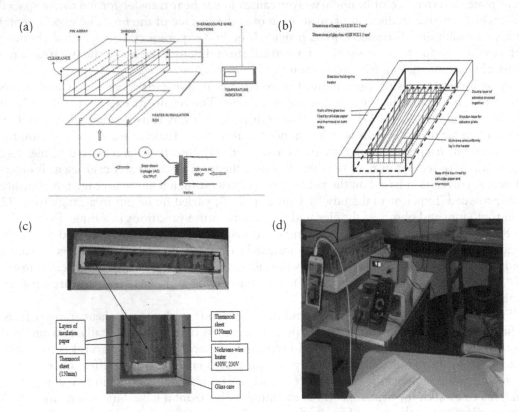

Figure 43.1 (a) Schematic diagram of the test bed layout for the experimental procedure; (b) The assembled test bed; (c) Schematic diagram of the insulation box with heater; (d) The customized heater and the glass box for insulation.

Experimental Procedure and Data Collection

The setup is placed in a windowless room and constant A.C. power is supplied at 120 volts to the heater using a variac throughout the experiments. The flow inside all the channels is assumed to be identical in nature and temperature measurements are taken for only one (middle) channel. At steady state, the base-plate temperature T_b, fin temperature T_f, ambient temperature T_a, power input to heater in terms of voltage is recorded at time intervals of 5 minutes. Table 43.2 shows the readings of this observation. Steady state is considered to be achieved after 45 minutes has elapsed from the start of the test. Heat losses during the experiment are mainly through radiation and convection. Electrical losses are very small and hence neglected. Radiation losses are computed analytically. To measure the convection losses the fins maybe inverted and used to heat water. This loss is calculated as follows:

Power supplied – Power consumed by the fins = Convection losses   (1)

Table 43.1: Design parameters.

Parameter	Height of fin (mm)	Length of fin (mm)	spacing (mm)	clearance (mm)	No. of Fins
	Heat sink dimensions for *test series 1*				
Heat sink 1	40	600	8	30	8
Heat sink 2	40	600	6	30	9
	Heat sink dimensions for *test series 2*				
Heat sink 1	40	600	8	20	8
Heat sink 2	40	600	6	20	9
	Heat sink dimensions for *test series 3*				
Heat sink 1	40	600	8	10	8
Heat sink 2	40	600	6	10	9

Table 43.2: Observation of temperature.

Test series no.	Time (min)	Average fin temperature (T_f)	Average base temperature (T_b)	Ambient temperature (T_a)
Test series 1				
Heat Sink 1	45	62	62	35
Heat Sink 2	45	62	62	36
Test series 2				
Heat Sink 1	45	77	77	45
Heat Sink 2	45	78	78	41
Test series 3				
Heat Sink 1	45	80	80	42
Heat Sink 2	45	78	78	41

Data Reduction

Empirical co-relation used for calculation of heat transfer from base plate Chen et al. [8].

$$Nu = \left\{ \left(\frac{Ra_b}{1500} \right)^{-2} + \left(0.081 Ra_b^{0.39} \right)^{-2} \right\}^{0.5} \tag{2}$$

Here Ra_b is the Rayleigh number of base-plate
Natural convective heat transfer rate Q_{NC}

$$Q_{NC} = Q_{total} - Q_{rad} \tag{3}$$

Rate of heat transfer from fins (Q_{fins})

$$Q_{fins} = Q_{NC} - Q_{base} \tag{4}$$

Where (Q_{base}) is rate of heat transfer from base plate.

Rate of radiation heat transfer (Q_{rad})

$$Q_{rad} = \sigma\varepsilon(T_W^4 - T_a^4) \tag{5}$$

Total area of heat transfer

$$A_{Total} = A_{fin} + A_{base} \tag{6}$$

Total heat input

$$Q_{total} = Power = \frac{V^2}{R} \times power\ factor \tag{7}$$

Here, V = voltage (volts), R = resistance of the heater (Ω)
Heat transfer coefficient

$$h = (Q_{total} - Q_{rad}) / A\Delta T \tag{8}$$

The total heat input to the heater is calculated as 55.13 W from A.C. supply. Temperature variations on the various fin positions due to free convection heat transfer are observed for all the cases until required temperature of fins are obtained and recorded. It is observed that for the cases with higher spacing and clearance the required temperature was achieved at a lesser time than those with less fin spacings and clearance values.

Uncertainty Analysis

Following the Ref. [9], the associated uncertainty with the measurement of voltage and current may be considered as the least count of the respective calculating devices as ±0.1 V and ±0.01A. The uncertainty estimation in evaluating the derived power (product of voltage and current) is therefore calculated as:

$$\sigma_{p1} = \pm\sqrt{\left(\left(\frac{\partial P}{\partial V}\sigma_V\right)^2 + \left(\frac{\partial P}{\partial I}\sigma_I\right)^2\right)} \tag{9}$$

Considering the power level of 64.26 W, the uncertainty in power is obtained as,

$$\sigma_{p1} = \pm\sqrt{\left((I\sigma_V)^2 + (V\sigma_I)^2\right)} \tag{10}$$

$$\sigma_{p1} = \pm\sqrt{\left((0.54\times0.1)^2 + (119\times0.01)^2\right)} = 1.19W \tag{11}$$

$$\frac{\sigma_p}{P} = \pm\frac{1.19}{64.26}\times100 = \pm1.85\%$$

Thermocouple Positions and Calibration

In order to capture the variation of temperature, T-type copper-constantan thermocouples are attached at various positions on the base-plate and the fin surface. For all the experiments, the middle channel is selected and three thermocouples are attached to the base plate. Six thermocouples are attached to the surface and tip of the fins as shown in the Figure 43.2a–b. Figure 43.2 (a) shows positions of thermocouple for the sample 1 and Figure 43.2 (b) shows positions of thermocouple for the sample 2. Based on calibration, temperature measurements with T-type thermocouples used for the experiment are accurate up to ±2.2°C.

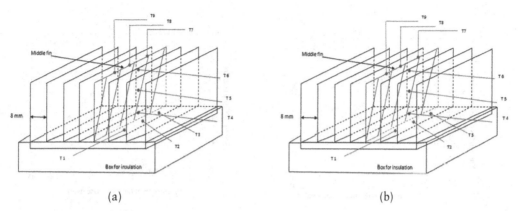

(a)   (b)

Figure 43.2 (a) Positions of thermocouple for the sample 1; (b) Positions of thermocouple for the sample 2.

Results and Discussion

Temperature Variation Over the Fin Surface

The temperature variation over the non-isothermal surface of the fin is shown in the Figure 43.3a–c. These variations of temperature across the base-plate and fin surface are plotted for a particular interval of time. Figure 43.2a–b shows the positions of the thermocouples over the fin surface. From the Figure 43.3a–c, it is observed that the temperature at a particular instant of time after a fixed interval shows variation across the fin. Highest and lowest temperatures are recorded at the base-plate and at the tip of the fins respectively. The temperature variation is of similar nature for both the fin array of different spacing (heat sink 1 and 2). The fin with higher spacing (heat sink 1) reveals higher surface temperature than that of the fin of smaller spacing (heat sink 2) as there is faster heat absorption by the fins due to larger area available for convection. Similarly, effect of the clearance (C) between the fin tip and the shroud on the fin surface temperature distribution may be obtained comparing the Figure 43.3a–c. Similar variation is obtained for all cases across the fin surface from base plate to the fin tip. It is noted that the temperature gradient in case of higher clearance (C = 30 mm) is lesser than that of the chosen lower clearance (C = 10 and 20 mm). This is attributed to the larger volume of circulating air inside the heat sink with increase in fin tip to shroud clearance. This may cause a rise in total heat transfer for the sink resulting in a lesser gradient of temperature. Due to the aforementioned reason temperature of the heat sink at higher clearance is low.

Heat Transfer

Results of the heat transfer for the present experimental investigation are shown in Table 43.3. From the tabulated data, it has been observed that heat sink 2 with clearance spacing

30 mm shows higher heat transport. However, this degree of rate of heat transfer reduces with the decrease in clearance spacing. For heat sink 2, clearance 30 mm shows nearly 10.54% higher heat transfer than that of the clearance 20 mm. The same is 13.93% while compared this heat sink with 10 mm clearance. For heat sink 1 also, similar results are persisted but 30 mm clearance shows nearly 11.27% and 6.12% higher heat transfer than that of 10 mm and 20 mm clearance respectively.

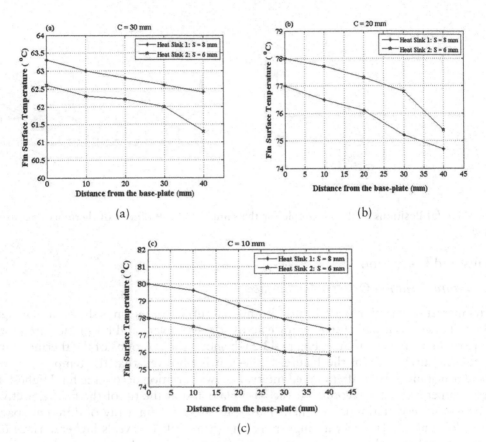

Figure 43.3 Variation of fin surface temperature (a) clerance (C) = 30 mm; (b) Clerance (C) =20 mm; (c) clerance (C) =10 mm.

Table 43.3: Observation of heat transfer.

Test series	Time (min)	Spacing (mm)	Clearance (mm)	Q_{fins} (W)	Q_{base} (W)	Q_{total} (W)
heat sink 1	45	6	30	24.34	22.62	46.97
heat sink 2	45	8	30	31.97	15.25	47.22
heat sink 1	45	6	20	17.44	26.82	44.26
heat sink 2	45	8	20	21.01	21.70	42.72
heat sink 1	45	6	10	19.92	22.29	42.21
heat sink 2	45	8	10	20.42	21.03	41.45

Conclusions

The conclusions of the present study may be drawn as follows: Highest temperature is recorded at the base-plate and lowest temperature at the fin tip. Almost similar nature of temperature distribution is observed for both the heat sinks. The fins with larger spacing (Heat Sink 1: S = 8 mm) shows higher surface temperature than that of the fins with lower spacing (Heat Sink 2: S = 6 mm). This is attributed to the faster heat absorption by the fins due to larger area available for convection. The temperature gradient in case of higher clearance (C = 30 mm) is lesser than that of the chosen lower clearances because of the larger volume of circulating air inside the heat sink with increase in tip of the fin to shroud clearance.

Heat transfer coefficient for heat sink 2 is obtained to be greater as compared to the heat sink 1 for all the chosen clearances. As future scope for this study, effect of variation of the fin heights on heat transfer may also be examined.

References

1. Starner, K. E., and McManus, H. N. (1963). An experimental investigation of free-convection heat transfer from rectangular-fin arrays. *Journal of Heat Transfer, 85,* 273–278.
2. Welling, J. R., and Wooldridge, C. V. (1965). Free convection heat transfer coefficients from rectangular vertical fins. *ASME Journal of Heat Transfer*, 87, 439–444.
3. Fisher, T. S., and Torrance, K. E. (1999). Experiments on chimney-enhanced free convection. *Transactions of the ASME, Journal of Heat Transfer*, 121, 603–609.
4. Harahap, F., and McManus, H. N. (1967). Natural convection heat transfer from horizontal rectangular fin arrays. *Journal of Heat Transfer*, 89, 32–38.
5. Jones, C. D., and Smith, L. F. (1970). Optimum arrangement of rectangular fins on horizontal surfaces for free convection heat transfer. *Journal of Heat Transfer-Transactions ASME*, 92, 6–10.
6. Leung, C. D., Probert, S. D., and Shilston, M. J. (1985). Heat exchangers: optimal separation for vertical rectangular fins protruding from a vertical rectangular base. *Applied Energy*, 19, 77–85.
7. Yuncu, H., and Anbar, G. (1998). An experimental investigation on performance of rectangular fins on horizontal base in free convection heat transfer. *Heat and Mass Transfer*, 33, 507–514.
8. Chen, H., Lai, C., Lin, T., and He, G. (2014). Estimation of natural convection heat transfer from plate-fin heat sinks in a closed enclosure. *International Journal of Mechanical, Aerospace, Industrial and Mechatronics Engineering*, 8(8), 1398–1403.
9. Srikanth, R., and Balaji, C. (2017). Experimental investigation on the heat transfer performance of a PCM based pin fin heat sink with discrete heating. *International Journal of Thermal Sciences*, 111, 188–203.

44 Effect on micro hardness and micro-structure of thin film coated titanium alloy substrate

K. Chandrappa[1,a] *and Sanjay T. Setty*[2,b]

[1]Associate Professor, Department of Mechanical Engineering, Siddaganga Institute of Technology, Tumkur, Karnataka, India

[2]Undergraduate, Department of Mechanical Engineering, Siddaganga Institute of Technology, Tumkur, Karnataka, India

Abstract: Thin film barrier coating on titanium alloy substrate has a vast range of application among which aeronautic and marine application are significant. Talking about 2 very different aspects, whose requirements and applications are way apart. Titanium has always been beneficial in both these product's Application. Introduction of titanium in aerospace and marine locomotive parts will have a significant importance on parameters like pressure withstanding capacity and corrosion resistance. Our immense perseverance is to fulfill the above intent by giving a thin coating on titanium alloy substrate. These coated titanium alloy will be tested for micro-structure and micro-hardness and it is hoped will be in line with the requirement Four different specimen of same material that is OT 4-1 titanium alloy will be coated with TiN, AlCroNa pro, DLC (diamond like carbon) and WCC (Tungsten carbide carbon coating) all the mentioned are coated for about 2micron to 4 microns over the substrate titanium alloy. TiN and AlCroNa pro are carried out in arc spray method and DLC, WLC are carried out in sputtering process.

Through micro structure analysis we will be able to find the surface contour of both uncoated and coated samples. Surface contour plays a major role in determining the corrosive property of material. Notches, cracks, dendrites are few properties which help in initiating corrosion faster, with the samples being coated with a superficial layer will cover all the above mentioned defects hence enhancing the service time of the material. Micro hardness refers to the surface hardness of the material, harder the surface lower the wear rate of the material.

Keywords: Arc spray, microhardness, micro-structure, sputtering, titanium alloy.

Introduction

It has been always aluminum the first preferences for any aeronautic or submarine application. The inherent property of Al like pressure withstanding capacity, corrosion resistance and high strength makes the material most ideal for its usage. The direct use of uncoated Al may not be suitable for use in aeronautic, marine or automotive industry as it fails to fulfill certain atmospheric threats, hence the coating is preferential. In an effort to improve the end requirements fulfilled by coated aluminum, coated titanium [10] alloy seems to be most ideal as it also has the property of durability [3].

[a]kchandrappa@sit.ac.in, [b]sanjay.1si17me104@gmail.com

Further in this article we will be discussing about PVD (physical vapor deposition) and sputtering process, the coated samples will be subjected to microstructure and micro hardness test and the recorded results will be used for further discussion regarding the material property enhancement observed and effect of thin film coating over the base metal.

Having said above information our immense effort is to prove that Ti alloy provides better quality, efficiency and durability when it is coated for its applicability in aerospace and submarine industry or any automotive sector. In connection with this Ti alloy has been subjected to TiN, AlCroNa Pro, DLC (diamond like carbon), WCC (tungsten carbide carbon) coating, and the end requirements will be recorded [11].

Methodology

Physical Vapor Deposition Coating

Any material can be passed into vapor phase if sufficient energy is passed to the surface. However, this process is easiest to apply for metals with low melting point and high vapor pressure. Hence aluminum is the most widely used material for industrial applications. Deposition rate attained can be as high as 1000 nm/sec which is 2 mils/min can be achieved. Refractory materials such as tungsten metal can be effectively evaporated but deposition rate is found to be comparatively low i.e. 10–15 nm/sec [1].

Alloy constituent's similar vapor pressure can be passed into the vapor with little change in composition, particularly when the rod feed electron beam is used. Thus alloys like Ni-Cr, Ag-Cu, Ti-Al-V and Ti-Cr-Al-V can be easily deposited using evaporation method. Problems are encountered when the alloys which contains constituents of widely different vapor pressure.

Sputtering

There are three distinct ways where the sputtered coating structure differ from evaporation process. First the sputtered atoms may be scattered by working gas so that they approach substrate at very low oblique angles thus producing zone 1 type structure. Second, significant number of ions can be neutralized and reflected back to the target surface and then again back to the substrate surface with an energy as high as several thousand Ev (electron volt). At very low working pressure the atoms will reach substrate and subject them to bombardment which promotes dense fine grained, dense zone T structure. Third the substrate may be in contact with plasma. Plasma ion bombardment tends to suppress zone 1 structure, even at elevated Ar pressures and to produce dense fine grained structure of zone T type. Ion bombardment on uncooled substrate also promotes high temperature structures. Thus bombardment has effect on producing dense, high quality deposits.

Experimental Procedure

For Physical Vapor Deposition Coating

For PVD coating or evaporation coating the apparatus typically consists of vacuum chamber, pumping system, pressure measuring instrument evaporation source and power supplies, deposition rate detectors. Pumping system combines high vacuum pump with mechanical backing and pumping systems. High vacuum pumps are generally of oil diffusion type.

Coating flux from an evaporation source is in general non uniform over most substrate surface shapes. The problem is eased for many applications by arranging a array of low cost resistant heated wire sources surrounding the object to be coated. The rate of evaporation is exponential function of the source temperature. Coating thickness deposited from the simple resistive heated sources are generally established by placing a fixed amount of coating material and evaporating all of it. However, the deposition rate monitors must be generally used to control the rate of evaporation and to determine the accumulated coating thickness.

Thanks to Oerlikon Balzers Coating India Pvt. Ltd. Bommasandra, IV Phase Industrial Area,560 099 Bangalore. Who helped me in proceeding with PVD coating as well as sputtering and for the support I received from the company for my research work has been remarkable.

For Sputtering Process

The vacuum chambers and pressure pumps used in the sputtering process is almost similar to that used in evaporation process. In addition, this has sputtering source and feed and control system for the working gas. Chambers with vacuum interlocks are used which permits substrate metal to be introduced without exposing the chamber walls and sputtering source to the atmosphere. The rate of sputtering from a given target is proportional to the power given to the material which can be controlled accurately. Thus deposition thickness is often deposited by time rather than using rate monitor.

One of the electrode monitor is planar diode. Although limited to moderate deposition rates they are used widely in electronic industry to deposit a range of material onto flat wafer substrates using dc or RF power. RF sputtering is generally done in the industrially allowed frequency of 13.56 Mhz. Power levels are 500 W to 5Kw range [2]. Voltages are 2–4 KV, in case of magnetron source magnetic field are used to confine intense plasma ring or sheets containing circulating currents, over planar or cylindrical target.

Result and Discussion

Microstructure

Figure 44.1 OT 4-1 substrate coated with AlCrONa.

Figure 44.2 OT 4-1 substrate coated with TiN.

Figure 44.3 OT 4-1 substrate coated with WCC material.

Figure 44.4 Base metal OT 4-1.

Figure 44.1 is substrate material coated with AlCrONa material to a thickness of 4 micron magnified to 1000x. The coating material is an alloy of aluminum, chromium and nitride. Unevenness of substrate material, scratches and coating conditions count on the surface quality after coating. This material has coefficient of friction of 0.55 [4]. wear resistance of this coating is directly proportional to nitrogen pressure [5].

Figure 44.2 shows substrate material coated with TiN material magnified to 1000x under scanning electron microscope. TiN coating has adhesion of 3.0 [6] Kgf on hard steel substrate, poisons ratio of 0.25, melting temperature is as high as 2950°C, thermal expansion coefficient 9.4×10^{-6}/°C.

WCC coating is well known for its temperature property. WCC abbreviates to Tungsten carbide carbon, this coating has a temperature withstanding capacity of up to 950°C [8], hardness number of 1250 to 1350 Hv [7] which server the requirement for aeronautic application [9]. Figure 44.3 is OT 4-1 substrate coated with WCC for 4 microns magnified to 1000x. Figure 44.4 shows base metal OT 4-1, Figure 44.5 shows micro hardness v/s time graph of base metal, Figure 44.6 shows hardness v/s displacement into surface graph of AlCrONa coated specimen, Figure 44.7 shows hardness v/s displacement into surface graph of TiN plated specimen, and Figure 44.8 shows hardness v/s displacement into surface graph of WCC coated specimen.

Micro-Hardness

Base metal titanium alloy (OT 4-1)

Table 44.1: Micro hardness of base metal in HV.

Temperature in C(across), Time in min (Down)	1000	850	600
90	386.32	380.97	357.96
60	367.14	331.74	329.08
30	320.44	316.34	309.49

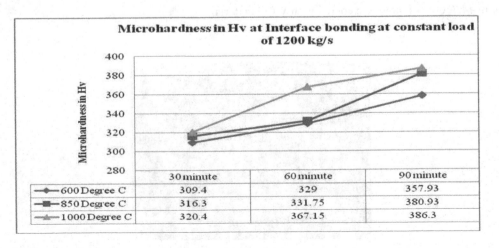

Figure 44.5 Micro hardness v/s time graph of base metal.

Table 44.2: Nano indenter test observations of AlCrONa coated specimen.

Test	Average modulus [200–400 nm] GPa	Avg hardness [200–400 nm] GPa	Temperature C
1	****	****	25.6
2	118.3	8.70	25.7
3	171.4	16.05	25.8
4	180.0	17.37	25.8
5	169.7	15.63	25.8
6	164.4	15.15	25.9
Mean	160.7	14.58	25.8

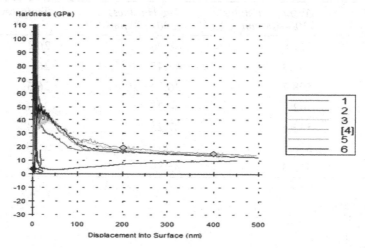

Figure 44.6 Hardness v/s displacement into surface graph of AlCrONa coated specimen.

The micro hardness values of the base metal are shown in the above Figure 44.4 and Table 44.1, shoes that the strength remains the same through the material since this is an alpha alloy. This value can be improved further for many corrosion and erosion application by coating different materials on the surface varying from 2–4 microns. The improvement of this property and microstructure is shown in the following graphs. Table 44.2 shows nano indenter test observations of AlCrONa coated specimen, Table 44.3 shows nano indenter test observations of TiN coated specimen, and Table 44.4 shows nano indenter test observations of WCC coated specimen.

AlCrONa Pro
The AlCrONa coated sample when subjected to Nano indenter testing above results were observed. The specimen is tested for surface hardness in 6 regions as mentioned in the graph and observed for hardness value in each region. Specimen will be laced in a specified region and will be subjected to loading maintaining same atmospheric conditions. Possible reason for the variation in hardness value is dependent on many factors like loading conditions, substrate evenness, coating quality, variation in thickness of coated material. Titanium material is found to be difficult material to be coated, which requires special consideration during pre-coating surface preparation, hence this counts on the quality of coating which

in turn reflects in non-consistent hardness value over same specimen area. Hence the test will be conducted for different instances and average value is considered which is found to be 14.58 Gpa for AlCrONa coating.

TiN

The specimen is tested for micro hardness in 6 different regions and is found that the average hardness of the sample is 35.86 Gpa. As we see dendritic surface in the magnified view of coating in Figure 44.2, plays a major role in variation of micro hardness of the surface in point 6 as shown in the graph. With working temperature up to 600°C and hardness value of 35.86 Gpa which in terms of hardness number is approximately 2500 Hv can be used for aeronautic application as it surpasses aluminum material surface properties.

Table 44.3: Nano indenter test observations of TiN coated specimen.

Test	Average modulus [200–400 nm] GPa	Avg Hardness [200–400 nm] GPa	Temperature C
1	611.8	38.96	26.2
2	702.8	46.45	26.3
3	508.7	26.72	26.3
4	562.8	34.25	26.4
5	556.4	37.25	26.4
6	256.9	7.11	26.5
Mean	577.3	35.86	26.4

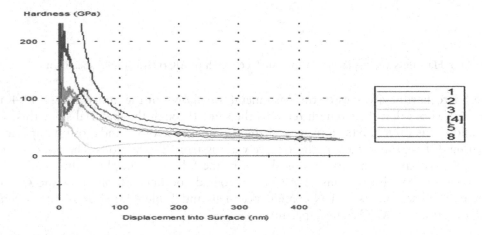

Figure 44.7 Hardness v/s displacement into surface graph of TiN plated specimen.

WCC

WCC which abbreviates to Tungsten carbon carbide has the most satisfactory hardness value. Since the coating is even, the material doesn't show much variation in hardness value irrespective of instances over the surface. We find that average hardness of the specimen to be 51.79 Gpa. Knowing that WCC coated titanium alloy has working temperature of up to 950°C and hardness value 51.79 Gpa which in terms of hardness number is found to be 5285 Hv is a most ideal material for aeronautic application.

Table 44.4: Nano indenter test observations of WCC coated specimen.

Test	Average modulus [200–400 nm] GPa	Avg Hardness [200–400 nm] GPa	Temperature C
1	854.9	56.88	25.9
2	661.5	49.01	25.9
3	836.5	50.61	25.9
4	802.3	58.30	25.9
5	649.5	48.93	26.0
6	1153.9	64.42	26.0
Mean	819.0	51.79	26.0

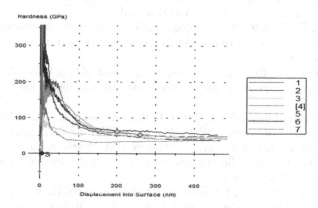

Figure 44.8 Hardness v/s displacement into surface graph of WCC coated specimen.

Conclusion

The above tests clearly indicate "coated surface has a phenomenal enhancement in mechanical property ". The comparative analysis of surface contour via SEM (Scanning electron microscopy) test indicates a drastic reduction in dendritic pattern between uncoated to coated surface. notches, cracks, dendrite, porosity are the initiating point of corrosion. Such substrate when covered with suitable coating, will enhance corrosion resistance.

Comparing the micro hardness results of base metal and coated samples we find significant improvement in the hardness value. Micro hardness of AlCrONa coating is found to be 1487Hv, TiN coating with 2500Hv and WCC having 5285Hv whereas the hardness value of base metal is 400Hv. Among all the above results, it is clearly evident WCC finds to be highly preferential. With the above observed test results we can conclude that OT-41 base with thin film coating can be beneficial and more efficient than conventionally used aluminum alloy which in turn would reflect in better airplanes and submarine performance.

I pay my sincere gratitude to Oerlikon Balzers Coating India Pvt. Ltd. Bommasandra, IV Phase Industrial Area,560 099 Bangalore, for assisting in the selection of coating material as well as selection of coating procedure i.e. Sputtering and PVD methods which reflected in achieving great experimental results. CMTI (Central Manufacturing Technology Institute) who also supported me in testing the coated parts. I wish to convey my sincere thanks to CMTI.

References

1. Thornton, J. A. (1983). Plasma-assisted deposition processes: theory, mechanisms and applications. *Thin Solid Films*, 107(1), 3–19.
2. Sheppard, K. (1984). 3. Metal Surface Preparation. *Electroplating Engineering Handbook*, 58.
3. Gomez-Gallegos, A., Mandal, P., Gonzalez, D., Zuelli, N., & Blackwell, P. (2018, August). Studies on titanium alloys for aerospace application. *In Defect and Diffusion Forum* (Vol. 385, pp. 419–423). Trans Tech Publications Ltd.
4. Cadena, N. L., Cue-Sampedro, R., Siller, H. R., Arizmendi-Morquecho, A. M., Rivera-Solorio, C. I., & Di-Nardo, S. (2013). Study of PVD AlCrN coating for reducing carbide cutting tool deterioration in the machining of titanium alloys. *Materials*, 6(6), 2143–2154.
5. Warcholinski, B., Gilewicz, A., Myslinski, P., Dobruchowska, E., & Murzynski, D. (2020). Structure and properties of AlCrN coatings deposited using cathodic arc evaporation. *Coatings*, 10(8), 793.
6. Ming'e, W., Guojia, M., Xing, L., and Chuang, D. (??). Morphology and mechanical properties of TiN coatings prepared with different PVD methods.
7. Jun, Z., Hui, Z., Zhi-hua, W., & Rui-peng, S. (2011). Structure and mechanical properties of tungsten-containing hydrogenated diamond like carbon coatings for space applications. *Physics Procedia*, 18, 245–250.
8. Esteve, J., Zambrano, G., Rincon, C., Martinez, E., Galindo, H., & Prieto, P. (2000). Mechanical and tribological properties of tungsten carbide sputtered coatings. *Thin solid films*, 373(1–2), 282–286.
9. Chandrappa, K., Sumukha, C. S., Sankarsh, B. B., & Gowda, R. (2020). Superplastic forming with diffusion bonding of titanium alloys. *Materials Today: Proceedings*, 27, 2909–2913.
10. Poondla, N., Srivatsan, T. S., Patnaik, A., & Petraroli, M. (2009). A study of the microstructure and hardness of two titanium alloys: Commercially pure and Ti–6Al–4V. *Journal of Alloys and Compounds*, 486(1–2), 162–167.
11. Kuruvilla, M., Srivatsan, T. S., Petraroli, M., & Park, L. (2008). An investigation of microstructure, hardness, tensile behaviour of a titanium alloy: Role of orientation. *Sadhana*, 33, 235–250.

45 Identifying transport problems in mauritius and proposing some solutions to IT

Smita Barla[1,a], Santaram Venkannah[1,a], Mahendra Goroochurn[1,a], and Avi Agarwal[2,b]

[1]University of Mauritius, Mauritius
[2]Amity School of Engineering and Technology, India

Abstract: Mauritius is a small island, which has recently joined the high-income group as per World Bank classification. The road network in Mauritius is highly developed, with public buses serving almost all regions of the island. But still, the country is facing major traffic jams due to an ever-increasing number of people investing in personal means of transport. Mauritius is facing a lot of transportation related problem such as road congestion, air pollution and road accidents which are considered too high for a small island developing state with less than 1.3 million inhabitants. A lot of research is going on around the world in order to improve public transport, from improving indoor bus conditions to changes in designs. So Mauritius must adopt a holistic approach to solve the transport problems. This paper outlines the current transportation issues in Mauritius, and tries to propose some solutions to alleviate the problems.

Keywords: Comfort, mauritius, public buses, transportation.

Introduction

Bus travel is widely acknowledged as a highly economical, sustainable, and secure means of transportation especially in Mauritius. Due to these advantages, authorities often support it by providing full or partial subsidies, with the aim of encouraging the general public to utilize it frequently. An esteemed and commendable public bus service assumes a pivotal role in fostering remarkable economic growth, accommodating the burgeoning population, and propelling the comprehensive development of both urban and rural areas [1]. A successful public bus service should prioritize accessibility to ensure a reliable, safe, and intelligent transportation system that offers convenience and effectiveness to all its users [2]. In the realm of public transport, the essence of passenger comfort stands as a vital beacon, illuminating the path towards superior service quality. This cherished facet exerts a profound influence on residents' transportation choices, wielding its captivating allure to guide them towards the embrace of a truly delightful travel experience [3]. Various factors, including technical, physiological, and psychological elements, play a role in influencing passenger comfort [4]. Comfort, being a subjective experience, can be influenced by the vehicle's design and ambiance, such as the placement of handrails, levels of noise and

[a]barlasmita@gmail.com, [b]sv@uom.ac.mu, [c]m.goroochurn@uom.ac.mu, [d]99aviagrawal@gmail.com

vibration, heating and ventilation systems, as well as the degree of crowding inside the vehicle [5].

Transport trends in Mauritius

In Mauritius the public transport is very structured, but people still prefer to use their own vehicle to travel [6]. There were 77 cars per thousand people car ownership in Mauritius in the year 1995 [7], and it increased to 175 cars per thousand people in the year 2010. Amidst the vibrant pulse of progress, the annual growth rate of registered vehicles throbs relentlessly at an awe-inspiring 4.8%. The World Bank collection of development indicators has borne witness to the automotive symphony, revealing the mesmerizing statistic of 175 motor vehicles per 1000 people in Mauritius during the illustrious year of 2010 [8]. Such data echoes the harmonious fusion of innovation and mobility, composing a grand opus of modernization on the island nation's dynamic stage. Figure 45.1 shows the ownership per thousand populations in Mauritius from year 2002 to 2010.

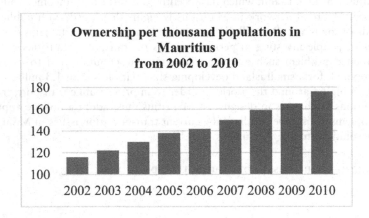

Figure 45.1 Ownership per thousand populations in Mauritius [8].

Transportation Related Problem in Mauritius

In the heart of Mauritius, amidst the splendor of its scenic landscapes, a vibrant and bustling metropolis faces a captivating challenge: significant road congestion. Like an enchanting dance of traffic under the golden sun, the Curepipe to Port Louis corridor experiences a captivating ebb and flow, especially during the enchanting crescendo of peak hours. Yet, amidst the excitement and allure, lies the quest to unveil solutions that shall harmonize the rhythm of movement and transform this captivating challenge into a seamless symphony of mobility. Over the past decade, journey times on this route have increased by 40%, making it a major concern for commuters [6]. Also the traffic jam extends for almost 30km from Terre Rouge on M2 motorway and up to Camp Fouquereaux roundabout on M1 motorway on some days (Figure 45.2). The government has constructed some new infrastructures which have eased the traffic in some regions but transferring the problems to other regions. The government must aim at some sustainable solution and leave the scarce land resources for other more important purposes. The other measure introduced by the Government in 2019 is the Metro Express as an alternative to public buses which has been welcomed by the population. Whereas the bus network provides better connectivity.

Metro Express will cover only a small region of Mauritius and the majority of people will still be using their personal cars or rely on buses. Furthermore, Metro Express itself rely on feeder buses. Also there are lesser number of metro stands, this means that to take the Metro Express, there will be more passengers walking a longer way from home [9]. So the perception of passenger on public transport is critical even for the success of the Metro Express project, as the Metro Express project will still take time to cover a significant portion of the island.

M.P. Enoch [6], found out that the main reason for congestion in Mauritius, is the surge in private vehicle ownership resulting from the country's expanding economy, the rise in personal disposable incomes, and the heightened expectations of its residents. These factors collectively contribute to the increasing number of people opting to own private vehicles to meet their transportation needs and aspirations. The rise in the number of vehicles on the road has resulted in significant road congestion [6]. In Mauritius, the worsening congestion had been calculated to have a yearly economic cost of Rs 200 million (£8 million) in 1991, amounting to 0.5% of the country's GDP [10]. However, by the year 2002, the estimated cost of congestion had soared to Rs 1.2 billion (£30 million), representing nearly 1.1% of the GDP [11]. The increasing number of personal vehicles leads to traffic jams, road rage, pollution and an increase rate of road accidents. Table 45.1 shows the data of road traffic accident causalities.

Within the troves of the World Bank collection of development indicators, a compelling revelation emerges about Mauritius—a significant portion, approximately 25.51% of the total, resonates with carbon dioxide emissions from the domain of transport, standing tall as a figure that beckons for thoughtful consideration, for it stands higher than most. The Figure 45.2 shows the carbon dioxide emission from different sources from the year 2004 to 2014 in Mauritius.

In the ever-evolving realm of road infrastructure, commendable strides have been made to enhance the road network. However, amidst these endeavors, a concerning trend has surfaced—a conspicuous rise in the number of road accidents. A glimpse at Table 45.1,

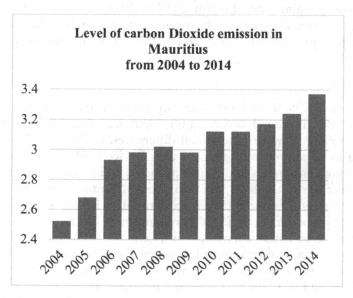

Figure 45.2 Level of carbon dioxide emission in Mauritius [12].

Table 45.1: Road traffic accident and causality [13].

Year	2013	2014	2015	2016	2017	2018
Road Traffic Accident						
NumberRate per 100,000 population	23,563	26,400	28,476	29,272	29,627	29,075
Rate per 1,000 registered Motor Vehicle	1,936	2,165	2,333	2,397	2,425	2,397
Motor Vehicle involved						
Number Rate per 1,000 registered	41,888	51,264	55,617	51,335	58,178	58,962
Motor Vehicle	97	113	117	116	112	105
Casualties						
Total no of casualties	3,610	3,592	3,722	3,862	4,209	3,718
Of which						
Fatal	136	137	139	144	157	143
Seriously Injured	465	505	530	512	560	597
Slightly Injured	3,009	2,950	3,053	3,206	3,492	2,978
Fatality						
Rate per 100,000 population	11.2	11.2	11.4	11.8	12.8	11.7
Rate per 1,000 registered Motor Vehicle	0.3	0.3	0.3	0.3	0.3	0.3
Fatality Index^2	3.8	3.8	3.7	3.7	3.7	3.8

adorned with data spanning from the year 2013 to 2018, unveils the stark reality of this unwelcome development. The numbers bear witness to a narrative that calls for vigilance and concerted efforts to ensure the safety and well-being of all road users.

Literature Review

Existing studies on bus comfort have primarily concentrated on aspects such as noise, thermal comfort, and vibration. However, it is evident that perceived bus comfort is, above all, a subjective notion, subject to individual differences. In 2011, L. Eboli and Mazzulla [14] proposed a novel approach to gauge transit service quality. Their method incorporates input from both passengers and transit agencies, taking into account the key factors that define a transit service. In 2004, A.E. Wahlberg [15] utilized a questionnaire method to explore the correlation between driver acceleration behavior and passenger comfort. According to a study conducted by J.D. Oborne and J.M. Clarke in 2003 [16], when creating a questionnaire to assess comfort, it is essential to prioritize the comfort-related questions and also gather demographic information about passengers, such as age, gender, and physical condition. Nevertheless, relying solely on passengers' feelings was deemed insufficient. K. Shek and W. Chan [17] devised a bus comfort model that considers thermal comfort and air quality as decisive factors in determining the level of comfort. Interestingly, they omitted questions about clothing insulation from their questionnaire, alternatively, in their analysis, they opted to measure the temperature difference inside and outside the bus, assuming that passengers would dress appropriately for the ambient or seasonal conditions to achieve ideal thermal comfort.

In recent times, there has been a growing focus on enhancing thermal comfort for vehicle occupants, making it a key competitive technology and a prominent aspect of automobile design. According to the American Society of Heating Refrigeration and Air Conditioning Engineers (ASHRAE) [18], human thermal comfort is defined as the psychological state that indicates contentment with the surrounding environment. Defining thermal comfort presents challenges due to the need to consider various environmental and personal factors that influence individuals' comfort levels. Determining the optimal temperatures and ventilation that would make everyone feel comfortable is a complex task. Realistically, the aim is to create a thermal environment in workplaces that satisfies the majority of people, often referred to as 'reasonable comfort' [19].

K. Zhang et al. carried out a study in 2014 [20], focusing on public transport buses in Nanjing, China. They assessed the overall comfort of the buses, taking into account various factors such as noise, thermal comfort, vibration, and acceleration. To gauge thermal comfort, they utilized Fanger's predicted mean vote (PMV) and calculated the predicted percentage of dissatisfied (PPD). Their results revealed that passengers considered the bus to be comfortable when the PPD value was equal to or less than 23%.

In 2010, a study conducted by T.P. Lin et al. [21] delved into the assessment of thermal comfort in both short and long-haul vehicles. Through interviews with passengers, the researchers gathered valuable insights into thermal sensation, preference, and acceptance regarding key factors such as temperature, humidity, air movement, and solar radiation. The findings unveiled that the ideal temperature range for short-haul vehicles was between 22.4 and 28.9°C, while for long-haul vehicles, slightly higher temperatures ranging from 22.4 to 30.1°C were preferred.

In 2008, an investigation into bus comfort in Hong Kong was carried out by K. Shek and W. ChanShek. Utilizing interviews with bus passengers and conducting correlation analyses, they constructed a comprehensive comfort model that could forecast the predicted percentage of dissatisfied (PPD) value by considering both thermal comfort and air quality. Their results pointed out that thermal sensation had a more substantial impact than air quality when predicting the PPD [22]. In 2015, N.E.A. Shafie et al. conducted a meticulous study involving Computational Fluid Dynamics (CFD) simulation to examine the airflow dynamics within a passenger bus compartment. Their primary objective was to assess temperature variations at various locations, such as the diffuser, seat, and floor. Through their investigation, they discovered that achieving a diffuser air velocity of 3.1 m/s resulted in a uniform velocity distribution. Fascinatingly, this optimal velocity also led to a noticeable temperature decrease of 0.3 °C in the seat area [23]. This study provides valuable insights into enhancing passenger comfort and thermal regulation inside the bus.

In 2017, S. Khatoon and H. Kim [24] conducted a study utilizing CFD simulation in the ANSYS Fluent module to enhance thermal comfort in a passenger car. Their numerical simulations revealed that the choice of ventilation variant significantly impacts human thermal comfort within the car cabin, as well as the energy efficiency of the vehicle. In 2012, D. Morcotet [25] conducted a study on thermal comfort in public transport, focusing on the summer period when thermal comfort tends to decrease. His research findings indicate that the design and constructional features of the analyzed vehicle play a crucial role in determining the level of thermal discomfort experienced by passengers.

In the year 2008 K.Shek and W. Chan [26] in their research found out that comfort inside the buses can be one of the reasons why people avoid public bus transport. In their research they showed that by improving the thermal comfort and air quality inside the buses there was an increase in the number of passenger travelling by public buses [28–30].

To attract more passengers and alleviate traffic congestion, bus operators and authorities have been actively prioritizing the enhancement of the passenger experience on buses [30]. Recognizing the factors that influence bus comfort levels can prove invaluable to policymakers, as it enables them to implement targeted improvement strategies effectively. Lai and Chen's [31] research in 2011 highlighted a positive correlation between perceived value based on service quality and overall satisfaction, involvement, and behavioral intentions. Comfort, being a vital aspect contributing to service quality, plays a significant role in influencing passenger satisfaction with bus transits. In a separate study conducted by Paul Chang et al., the implementation of expanded high occupancy vehicle (HOV) lanes and the introduction of new coaches resulted in a significant increase in ridership, approximately 5.5%, when commute times were reduced by approximately 2.5 minutes, as revealed by the findings.

The research by W.P. Bradshaw and M.V. McGreevy [32] revealed that customers are willing to pay a premium for improved public services. This finding suggests that providing better-quality public services can lead to increased customer satisfaction and a willingness to invest more in utilizing those services.

Recommended Solutions

Currently, Mauritius is grappling with severe road congestion and a concerning increase in carbon dioxide levels. In this context, public transportation can play a significant role in alleviating congestion and its associated problems, while also combating the rising trend of road accidents and injuries. Public buses in Mauritius have demonstrated a commendable safety record with very few fatalities, making them a relatively safe mode of transport.

Moreover, an efficient public bus service brings various benefits, including improved personal economic opportunities, fuel and cost savings, and reduced environmental impacts. However, it is crucial to ensure the quality of service remains high and reliable, as a poorly managed public bus service can exacerbate transportation system challenges. Therefore, promoting and enhancing public transportation in Mauritius can be a promising solution to address congestion and environmental concerns while providing a safer and more efficient means of travel for the public.

A lot of research and experiments are going on around the world for improving the comfort level of passenger inside the bus, but it's not necessary that each and every place will need the same condition in order to maintain comfort inside the bus. Mauritius being a small island, the ambient conditions varies from the central region to the coastal region. The temperature difference from Curepipe to Port Louis is in the range of 4–50C at all times and the microclimate affect the comfort of passengers in a trip. Moreover, the heat load on the vehicle along the route will vary and the same system will no longer be comfortable at the end of the trip.

Transport systems play a substantial role in environmental impact, contributing to approximately 20% to 25% of global energy consumption and carbon dioxide emissions

(Rodrigue 2020). Given this, even small measures towards sustainable development can have significant benefits, including reducing greenhouse gas emissions, enhancing road safety, and improving overall transport system efficiency.

By embracing sustainable practices, we can promote the adoption of greener transportation options, encouraging people to opt for alternatives like walking, cycling, or using public transport. These steps collectively contribute to creating a more sustainable and eco-friendly transportation landscape, benefiting both the environment and the well-being of individuals.

Therefore a lot of factors and research methods are to be considered and tests are to be conducted in order to find the best comfort level for passenger inside public buses in Mauritius. Also the bus construction in Mauritius is regulated by the regulations and only the dimensions and sizes are taken into account, comfort in terms of air quality is not considered [27], so a lot of research is also to be done in the field of design as well.

Conclusion

The transportation-related problems Mauritius is currently dealing with have been discussed. Its main causes have been highlighted. Some solutions have been proposed, based on current researches that are going on around the world. Emphasis is given more on using public transport and improving the comfort level of public transport, this may increase the passenger's number using public transport.

Students and elders (above 60 years) are already benefitting from free transport and government is providing subsidies to bus operators to invest in semi low floor buses. The country is on the right track but the authorities must take that little step to attract the working class to public transport by providing a comfortable, stress free and hassle free public transportation system.

References

1. Bachok, S., Osman, M. M., and Ponrahono, Z. (2014). Passenger's aspiration towards sustainable public transportation system: Kerian District, Perak, Malaysia. *Procedia—Social and Behavioral Sciences,* 153, 553–565.
2. Amiril, A., Nawawi, A. H., Takim, R., and Latif, S. (2014). Transportation infrastructure project sustainability factors and performance. *Procedia—Social and Behavioral Sciences,* 153, 90–98.
3. Dell'Olio, L., Angel, I., and Patricia, C. (2011). The quality of service desired by public transport users. *Transport Policy,* 18, 217–227.
4. Oborne, J. D., and Clarke, J. M. (2003). Questionnaire survey of passenger comfort. http://dx.doi.org/10.1016/0003-6870(75)90302-6.
5. Bird, R., and Quigley, C. (1999). Assessment of Passenger Safety in Local Service psv's. DfT Report; PPAD 9/33/24. Technical Report.
6. Enoch, M. P. (2003). Transport practice and policy in Mauritius. *Journal of Transport Geography,* 11, 297–306.
7. International Road Federation (1995). World Road Federation 1990–1994. Geneva, Switzerland: International Road Federation.
8. Trading Economics (2019). Mauritius motor vehicle per 1000 people. https://tradingeconomics.com/mauritius/motor-vehicles-per-1-000-people-wb-data.html. (accessed Nov. 23, 2023).
9. Dhookit, L. J. (2017). Our traffic problem: why we should not discard the second harbor option. https://www.lexpress.mu/idee/303753/our-traffic-problem-why-we-should-not-discard-second-harbour-option. (accessed Nov. 23, 2023).

10. Ministry of Land Transport, Shipping and Public Safety (1997). National Transport: Tomorrow's Transport, Port Louis, Mauritius. http://nao.govmu.org/English/Pages/%E280%A2Mimistry-of-Land-Transport,-Shipping-and-Public-Safety.aspx. (accessed Nov. 23, 2023)

11. Rahman, H., Chin, H. C., and Seebaluck, N. (2012). Urban transport sustainability in Mauritius: A balanced scorecard. *OIDA Int. J. Sustain. Dev.*, 5(11), 83–104.

12. Trading Economics (2019). World Bank Indicators. https://tradingeconomics.com/mauritius/co2-emissions-metric-tons-per-capita-wb-data.html. (accessed Nov. 23, 2023).

13. Statistics Mauritius (2018). Road transport and road traffic accident statistics (Island of Mauritius). March, p. 14, 2019.

14. Eboli, L., and Mazzulla, G. (2011). A methodology for evaluating transit service quality based on subjective measures from the passenger point of view. *Transport Policy*, 18(1), 172–181.

15. Wahlberg, A. E. (2004). The stability of driver acceleration behavior, and a replication of its relation to bus accident. *Accident Analysis and Prevention*, 36, 83–92.

16. Oborne, D. J., and Clarke, M. J. (1975). Questionnaire survey of passenger comfort. *Applied Ergonomics*, 6(2), 97–103.

17. Shek, K. W., and Chan, W. T. (2008). Combined comfort model of thermal comfort and air quality on buses. *Science of the Total Environment*, 389(2–3), 277–282.

18. Standard, A. S. H. R. A. E. (1992). Thermal environmental conditions for human occupancy. *ANSI/ASHRAE*, 55, 5.

19. Parsons, K. C. (2000). Environmental ergonomics: a review of principles, methods and models. *Applied Ergonomic*, 31(6), 581–594.

20. Zhang, K., Zhou, K., and Zhang, F. (2014). Evaluating bus transit performance of Chinese cities: developing an overall bus comfort model. *Transportation Research Part A*, 69, 105–112.

21. Lin, T. P., Hwang, R. L., Huang, K. T., Sun, C. Y., and Huang, Y. C. (2010). Passenger thermal perceptions, thermal comfort requirements, and adaptations in short- and long-haul vehicles. *International Journal of Biometeorol*, 54(3), 221–230.

22. Shek, K., and Chan, W. (2008). Combined comfort model of thermal comfort and air quality on buses. *Science of the Total Environment*, 389(2–3), 277–282.

23. Shafie, N. E. A., Kamar, H. M., and Kamsah, N. (2015). CFD Simulation of Air Temperature inside a Bus passenger Compartment. Switzerland: Transformative Technologies, (pp. 85–90).

24. Khatoon, S., and Kim, H. (2017). Human thermal comfort and heat removal efficiency for ventilation variants in passenger cars. *Energies*, 10. 11.

25. Morcotet, D. (2012). The thermal comfort in the surface public transport from Bucharest during the summer period. *Aerul si Apa. Componente ale Mediului*, 367–374.

26. Shek, K., and Chan, W. (2008). Combined comfort model of thermal comfort and air quality on buses. *Science of The Total Environment*, 389(2–3), 277–282.

27. Mauritius Meteorological Services. http://metservice.intnet.mu/ (accessed Nov. 23, 2023).

28. Department of Transport, Road traffic: construction and use of vehicles (Amendment). Regulation 2013, no. 53, 2013.

29. Mauritius (Republic) (2001). Integrated National Transport Strategy Study–Integrated Strategy Study Report. Halcrow Fox in Association with MDS Transmodal for Government of Mauritius.

30. Eboli, L., and Mazzulla, G. (2007). Service quality attributes affecting customer satisfaction for bus transit. *Journal of Public Transportation*, 10(3), 21–136.

31. Chang, P., LaRiviere, J., and Thorpe, J. (2021). Time, Comfort and Preferences for Shared Transit. University of Chicago Press Journal, (pp. 345–348).

32. Bradshaw, W. P., and McGreevy, M. V. (2021). Buses, Passenger Transport after 2000 A.D. (p. 33–50). Routledge.

33. das Neves Almeida, M., de Paula Xavier, A. A., and Michaloski, A. O. (2020). A review of thermal comfort applied in bus cabin environments. *Applied Sciences*, 10(23), 8648.

46 Metal-free terahertz absorber/sensor using patterned graphite

Anil Kumar Soni[1,a], Shrawan Kumar Patel[1,b], and Gaurav Varshney[2,c]

[1]ECE Department, School of Studies in Engineering & Technology, Guru Ghasidas Vishwavidyalaya Bilaspur, India

[2]ECE Department, National Institute of Technology Patna, India

Abstract: In this paper, metal-free absorber is implemented at terahertz (THz) frequency for generation of multiband resonance peaks. Absorber is designed by introducing a semicircle notch and rectangle slot in graphite sheet. Effect of the dimension of notches and slots are analysed in details for selecting the appropriate dimension of the absorber. It is found that a proposed absorber can generate multi band resonance with higher absorption efficiency. By varying the dimension and thickness absorber can also provide merging and separation of resonance peaks. Further, Polarization insensitivity of the proposed absorber is also studied for the wide range of incidence angle (θ) for both TE and TM mode. Performance sensitivity of the proposed absorber is analyzed to shows the use in sensing devices.

Keywords: Absorber, absorption enhancement, graphite, metal-free, multiband, terahertz.

Introduction

In recent year, miniaturization has pushed technology to a new era of optical devices such as biosensor [1, 2], active plasmonic devices [3, 4], modulators [5, 6], filters [7, 8], couplers [9], delay line [10, 11], oscillators[12], antenna [13–16], reflectors [17, 18], phase shifters [11] and absorber [19, 20] at THz frequency band. Electro-magnetic spectrum lies between the frequency ranges 0.1 THz to 10 THz is called the THz frequency band. Due to the sandwiched in between the microwave and the far infrared frequencies, THz band shares the mid-characteristics of the two bands. These mid-characteristic properties of the THz band attracted in-creasing attention due to the growing number of practi-cal applications in astronomy, communication, biomedical imaging, spectroscopy, etc [21–24].

But at the same time controlling and guiding the light at nanometre scale has challenges for the optical devices. In view of these, the sp^2 family of carbon material such as graphite and graphene is one of the best solutions for the implementation of different devices at optical frequency. This is possible because of the non-degradable electrical and optical properties of carbon materials at high frequency ranges which has a capability to guide light in form of plasmons [25–28]. Graphite and graphene also has tuning property which

[a]anilsoni@ggu.ac.in, [b]shrawan.patel@ggu.ac.in, [c]gaurav.ec@nitp.ac.in

is beneficial for the nanoscale devices where manufacturing modifications in the structures are difficult [27–33].

For diagnoses of the human body and material [23, 24], optical sensor is the essential device which uses the absorption of light. Hence, the study to design of optical device such as absorber at THz frequency band is essential for absorption of light at low absorption loss. Researcher utilized various shapes for designing different types of THz absorber with the use of metals/dielectrics and metamaterial structures for reducing the absorption losses [19, 34–39] Shen *et al.* proposed a THz broadband (frequency range of 0.6-1.254 THz with the absorption above 80%) metamaterial absorber, which consists of four sub-cells with multiple metal rings and a metal ground plane separated by a dielectric layer [36]. Cheng *et al.* proposed the concept of sectional asymmetric structures (metal resonator/dielectric/metal ground) which shows the THz broadband absorption spectrum [19]. Zohreh designed a THz absorber by use of metal mirror thin film and a single lossy dielectric buffer which are polarization-independent absorber as a glucose sensor in Food Industry applications. Author also demonstrate the ability of multiband absorption with tuning two bands by changing the reflective index [21]. Based on the application growth towards the implementation of multi-band absorbers a number of dual and multiband absorbers have been implemented [20, 40–44]. Sun et.al. designed a graphene-based dual-band independently tunable infrared absorber by varying the Fermi energy level of the graphene nano-disks and of the graphene layer with nano-holes [22]. Islam et.al. proposed a Plasmon induced tunable meta-surface (gold-silicon-graphene sheet-Zeonex-graphene) based multiband absorber for the variable refractive index based surrounding environment application [30].

Use of metal restricts the absorber at lower frequency band, because of the conductivity of the metals reduces at higher frequency [45]. Natural material such as graphene provides low absorption efficiency when it is used alone for designing absorber. Whereas, the high absorption efficiency obtained by the combination of graphene with metal and dielectric [26, 46, 47]. This required to designing of an absorber which reduces the absorption losses in material. Based on the input gained from the literature, the requirements for optical devices are: (1) design an absorber for THz frequency band (2) design a multiband absorber with high absorption efficiency and (3) design a metal-free absorber with low absorption loss.

In view of the requirements our research work is carried out. In this paper, metal-free graphite-based absorber is designed and analysed. Whereas, few researcher designed a microwave metamaterial absorber with the use of graphite sheet [28, 39, 48, 49]. In this paper, metal-free absorber is proposed by the combination of graphite and dielectric (Silicon dioxide), which has concentric square loop shape with four semicircle notches for generation of multiband resonance with enhanced absorption efficiency. Also the effect of the notches in metal free structure is analyzed and based on the response multi band absorber is finalized.

Principle and Design

Figure 46.1 shows the unit cell geometry of the proposed THz metal-free absorber, which consists of two-layer graphite sheet separated by a dielectric cavity. Top layer of the absorber are designed by a concentric square loop shape with four semicircle notches. For the implementation of the top and bottom layer in proposed absorber, the stack of sp^2 carbon nano-material also known as graphite is selected with the thickness t_g and relaxation time $\tau = 1 \times 10^{-12}$ at room temperature $T = 300K$ [50]. Graphite thickness (t_g) is selected after the parametric study and the final reported value of $t_g = 0.3$ μm, which is equivalent to the stack of around 900 two-dimensional nano-sheets of planar carbon material. The electrical and physical properties of graphite have been selected as reported in [16, 51]. Silicon dioxide ($\varepsilon_s \approx 3.8$) has been used as the dielectric having the thickness $(t_s) = 15.6$ μm and the final optimized dimensions of the proposed absorber are given in Table 46.1.

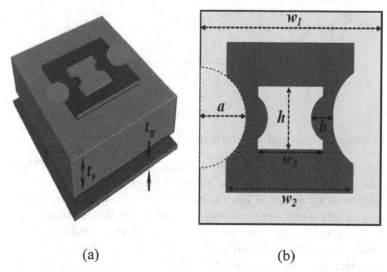

(a)   (b)

Figure 46.1 Proposed graphite-based THz absorber (a) 3D view and (b) top view.

Table 46.1: Optimized dimensions of the proposed absorber.

Parameters	w_1	w_2	a	w_3	h	b	t_s	t_g
Dimensions (µm)	85	60	22	30	25	8	15.6	0.3

After simulation of the designed structure, refection, transmission and absorption are calculated. The absorption coefficient (A) is obtained by $A = 1 - |S_{11}|^2 - |S_{21}|^2$, where $R = |S_{11}|^2$ is reflection coefficient and $T = |S_{21}|^2$ is transmission coefficient. In absorber, the transmission coefficient is approximately zero ($T \cong 0$) because of the large thickness of the ground plane than its skin depth. Figure 46.2 shows the response (A, R and T coefficients) of the proposed absorber. Figure 46.2 depict the three absorption peaks at resonance frequencies 2.88, 3.18 and 3.67 THz with absorption efficiency of 99.9, 99.8 and 99.9% respectively obtained by the proposed absorber for TE mode. Relative frequency interval between two adjacent resonant frequencies $(f_{i+1} - f_i/f_{i+1})$ is calculated for the obtained resonance frequency at absorption peak and 13.35% largest relative frequency interval is attained between two adjacent resonant frequencies.

For better understanding the physical mechanism and operational principle of the proposed multiband metal-free absorber the electric (E) and magnetic (H) field distribution are observed. Figure 46.3 shows the electric (E) and magnetic (H) field distribution on the absorber in $z = t_s$ plane at resonance frequencies 2.88, 3.18 and 3.67 THz. The field distribution for TE mode is illustrated on surface of the designed structure to get insight

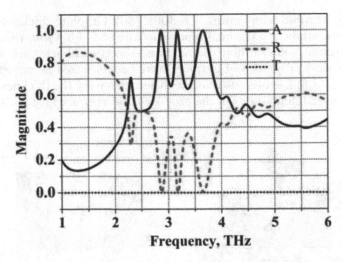

Figure 46.2 The response (A, R and T coefficients) of the proposed absorber.

view of the absorption behaviour. In Figure 46.3a, at frequency 2.88 THz, it can be seen that E-Fields are concentrated to the fine edges of inner and outer graphite structure and in the middle region of structure. While the magnetic field (H-field) are centred to the inner notches of the graphite sheet orthogonally. In Figure 46.3b, at frequency 3.15 THz the E-field of the outer graphite structure is rotated in quadrature and inner structure remains unchanged. Fields are distributed uniformly in the upper section of the structure. At the same time magnetic fields are concentrated 90 degree out of phase in graphite structure and distributed in the lower section of the structure. In Figure 46.3c at, frequency 3.67 THz the E-fields of the outer graphite structure is shifted 90 degree and more fields are distributed uniformly in the outer section of the semicircle notches and higher numbers of TE mode in H-field are also excited. In all cases the fields are concentrated in the structure which is the desirable property on any absorber.

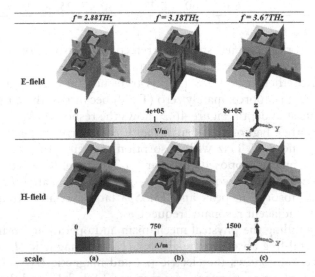

Figure 46.3 E-field and H-field distribution in absorber for TE at resonance frequencies 2.88, 3.18 and 3.67 THz.

Absorber Evaluation

For designing of the proposed structure comprises of five steps of evolution as shown in Figure 46.4 and the response of the absorber (Abs) structures for TE and TM modes are shown in Figure 46.5. In the first step(Abs-1), simple sandwich structure of two graphite sheets and a dielectric layer of silicon dioxide is used as shown in Figure 46.4a. Dimension of the top layer is consider as w_2, while the bottom layer graphite and dielectric are same in size (w_1). As reported in the literature, multiband resonance can obtains by use of an appropriate size of graphite at lower THz band [51]. Therefore, Abs-1 structure is used for generation of multiband resonance and it is clearly seen in Figure 46.5 that three resonances are obtained by Abs-1 at frequency 2.60, 3.62, and 4.49 THz with A $\approx$ 50%. Because of the symmetrical structure of Abs-1 it also shows the same response for TE and TM mode.

If a semicircle notch of radius, a is cut in one side of the top graphite sheet (Abs-2) in E-field direction as shown in Figure 46.4b, the absorption characteristic in TE mode gives higher number of harmonics due to the slight discontinuities in E direction. From the Figure 46.5a, it is found that the lower order resonance is obtained with absorption efficiency around 90% at frequency range 3.1-3.7 THz for TE mode due to the presence of semicircle notch in top graphite sheet. But TM mode results are almost similar as Abs-1. If one more symmetrical notch about y-axis is cut to other side of the top graphite sheet (Abs-3) as shown in Figure 46.4c, then the higher harmonics due to the second notch neutralize the effect higher harmonic in Abs-3 for TE mode as seen in Figure 46.5a. Abs-3 is Asymmetrical in structure, hence it provide the different response for TE and TM mode. From the Figure 46.5a, it is visible that the TE response for Abs-3 neutralized the higher harmonics response but also provide the resonance with higher absorption efficiency (84%) at 3.70 THz. In case of TM mode, absorption efficiency for first peak enhanced to 84% and other peaks are similar as Abs-1.

In the next step of evaluation, a rectangle slot of size ($h \times w_3$) is cut over the centre of the top graphite sheet (i.e. Abs-4), as shown in Figure 46.4d. In Abs-4, the effective E-field and H-field are changing due to number of discontinuities in structure, which provides the change of response in TE and TM mode. The change in TE and TM mode sift the resonance peaks of the Abs-4 more closer and achieve higher frequency band (2.80-3.88 THz) as well as higher absorption efficiency ($\approx$ 100%) as shown in Figure 46.5. It can be seen from the result, Abs-4 provides the three peaks with the absorption efficiency 98.8, 99.4, and 99.1% at frequency 2.89, 3.22, and 3.67 THz respectively in TE mode. Abs-4 also provides the three resonance peaks in TM mode with enhanced absorption efficiency 99.3, 80.5, and 98.2% at frequency 2.36, 2.68, and 3.34 respectively.

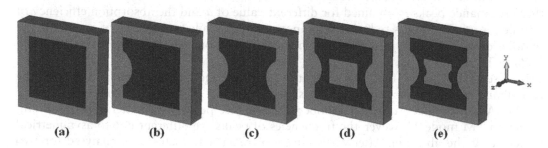

Figure 46.4 Evolution of absorber (a) Abs-1, (b) Abs-2, (c) Abs-3, (d) Abs-4, and (e) Abs-5.

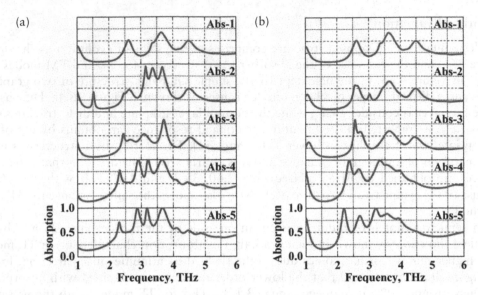

Figure 46.5 The response of absorbers with (a) TE and (b) TM modes.

Further, in Abs-4 two symmetrical semicircle notches of radius b are added in E-field direction in both sides of the rectangle slot (Abs-5) as shown in Figure 46.4e. Addition of semicircle notches in Abs-5 provides another important modification in the response. It is seen in Figure 46.5b, first and third resonance peaks are sifted towards the origin and higher absorption efficiency (i.e. 99.3, 83.9, and 99.6%) is obtained for three resonance peaks in TM mode. From the Figure 46.5 it is also notice that by changing structure from Abs-1 to Abs-5 range of frequency for resonance peaks are reduces.

Absorber Performance Analysis

Selection of the appropriate dimension of the graphite sheet used at the top of the absorber is required to obtain the multiple resonance peaks with higher absorption efficiency. However, in Abs-1 obtained three resonance peaks with absorption efficiency around 50%. For improvement of the absorption efficiency of Abs-1 various modifications in structure is performed and discussed in section 3. First, a semicircle notch with radius a, is cut from Abs-1 and the influence of radial parameter 'a' on the absorption efficiency is analysed in Abs-2. For this study, value of a varying from 18 to 30 μm in the interval of 4 μm is considered and results are shown in Figure 46.6. From the Figure 46.6a, it is found that the three resonance peaks is obtained for different value of a and the absorption efficiency of middle resonance peak is reducing with increase in a. While for the case of $a = 22\mu m$, the middle resonance are divided into three higher order resonance with absorption efficiency 90% in TE mode.

In case of TM mode, first two resonance peaks are merging together and proving the broadband resonance response as increase the value of a as shown in Figure 46.6b. For $a = 22\mu m$Abs-2 provides a greater number of resonance peaks with better absorption efficiency in TM mode. However, the frequencies of peaks are different due to asymmetrical structure of the absorber. Afterwards effects of redial parameter a, is analysed for two

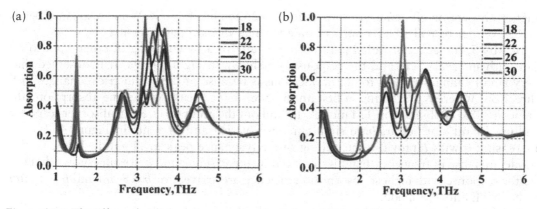

Figure 46.6 The effect of radial parameter *a* in Abs-2 for (a) TE and (b) TM mode.

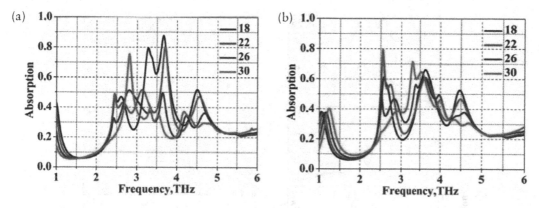

Figure 46.7 The effect of radial parameter *a* in Abs-3 for (a) TE and (b) TM mode.

semicircle notches in top of the graphite sheet (i.e. Abs-3) and results are shown in Figure 46.7. The better response is obtained for the lower value of *a* (i.e. 18 and 22 μm). As increase the value of *a*, the performance is degrading for both TE and TM mode, because the closer the radial notches, reduce the total absorption efficiency.

A rectangular slot is cut in the top graphite sheet for Abs-4 and the influence of slot dimension on response of Abs-4 is analyzed. For this analysis, variation in edges *h* and w_3 are carried out simultaneously ($h = w_3$) from the range 15 to 35 μm and the response is shown in Figure 46.8a. It is seen from the figure, as increases the edges of slot from 15 to 25 μm, the absorption efficiency as well as resonance peaks are increases. Hence, the study also provides a technique for separation and merging of adjacent resonance peaks. Further increase of edges dimensions degrades the performance of Abs-4. In case of TM mode, it is also seen that the first resonance peaks are shifting towards the lower frequency by increasing the value of edge dimension. Afterward the analysis are carried out by changing the height, *h* (20 μm < *h* < 30 μm) with w_3 = 30 μm of the rectangle slot and the response are given in Figure 46.8b. Result shows that the more number of absorption peaks with higher absorption efficiency are arrived for $h = 25 \ \mu m$ in TE mode. While resonance peaks in TM modes provides better absorption efficiency with proper diversity.

In Abs-5, the effect of radial parameter b is analysed by changing in range 6μm to 10 μm with interval 2 μm and the response are shown in Figure 46.9a. It is observed that very small changes arrived for TE mode. This is because of the symmetrical changes in E direction. Whereas, the sifting in first and last resonance peaks with same absorption efficiency (98%) is observed in TM modes. To find out the accurate effect in absorber response radius, b and slot height, h are simultaneously varied and presents in Figure 46.9b. By increasing h with b provides the nulling of the resonance peak in TE mode. While in TM mode, increase in frequency band is observed by increasing h and b. Results shows that the better response with highest absorption efficiency are arrived for $b = 8 \ \mu m$ and $h = 25 \ \mu m$ in both TE and TM mode.

Thickness of substrate and graphite sheet is other important physical parameters which influence the response of the absorber. By change in substrate thickness tuning in resonance bandwidth can achieve, because of the merging and separation of the resonance peaks. For

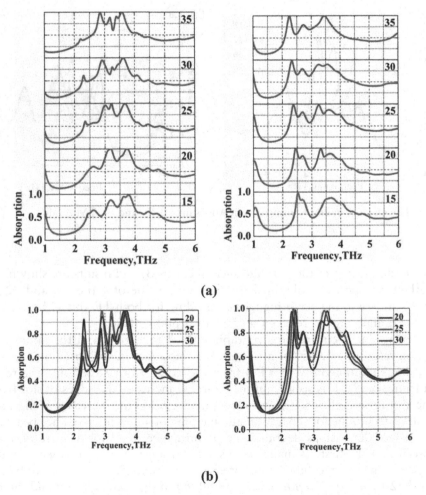

(a)

(b)

Figure 46.8 The effect of rectangle slot in Abs-4 for TE and TM mode by variation in (a) $h = w_3$ and (b) h with $w_3 = 30 \ \mu m$.

this study, substrate thickness (t_s) varied from 9 μm to 21 μm with the interval of 3 μm and results are shown in Figure 46.10. By increasing t_s, separation in resonance peaks for both TE and TM mode is observed in Figure 46.10. From the results it is observed that as increase the t_s (9 μm to 15 μm) number of resonance peaks are increasing with sufficient separation up to the thickness $t_s = 15$ μm, then further increase of degrades the response of the absorber in terms of resonance peaks as well as absorption efficiency. After the precise analysis of thickness $t_s = 15.6$ μm, is found best and considered for final absorber structure (Abs-5).

However, the thickness of the graphite sheet is also an important parameter because graphite is responsible for multiband resonance. To determine the effect of the graphite thickness (t_g), different value of t_g ranging from 0.1 μm to 1.2 μm with variation of 0.3 μm are analysed and response are shown in Figure 46.11. The increasing the value of t_g provides the shifting of resonance peak towards the higher frequency. However, the sifting in reverse direction (itowards the lower frequency) is observed with change in other parameters. By changing the graphite thickness, improvement of absorption efficiency is illustrated. From the results it is also found that the relative frequency interval is reducing due to the reduction of bandwidth of resonance peaks which shows the frequency selectivity quality of the proposed absorber. In Figure 46.11, $t_g = 0.3$ μm shows the enhanced response for TE and

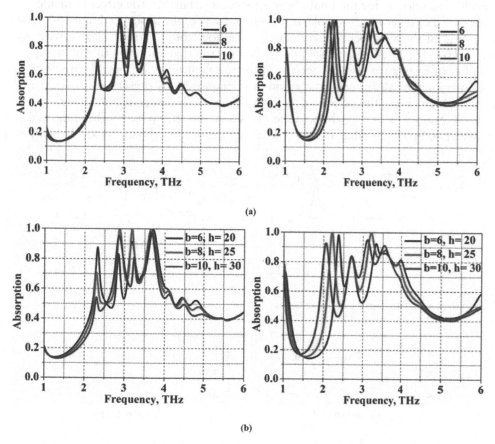

Figure 46.9 The effect of slot dimension in Abs-5 for TE and TM mode by variation in (a) *b*, and (b) *b, h*.

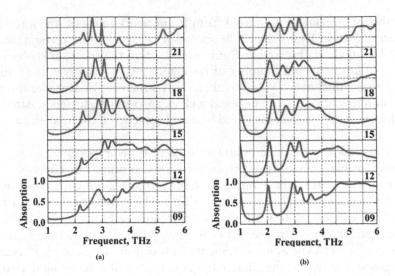

Figure 46.10 The effect of the change in t_s for (a) TE mode, and (b) TM Mode.

TM mode and selected for the final absorber structure.Further, the effect in incident angle (θ) is analyzed to determine the polarization sensitivity of the absorber. Figure 46.12 shows the response of the absorber for wide angle variation from 0° to 90° degree with various stages of θ (i.e. 0°, 25°, 55°, and 85°) and it is observed that variation of θ although effects the number of resonance peaks as well as absorption efficiency for both TE and TM mode. But performance of absorber at the resonance frequencies is preserved for wide angle of incidence which shows the polarization insensitivity of the proposed absorber.

Table 46.2 show the comparison of the proposed absorber with another reported multi-band absorber. From the table, it can be seen that the most of the multi-band absorbers are

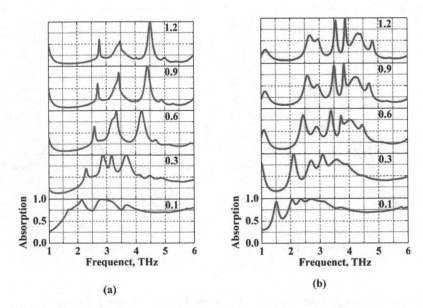

Figure 46.11 The effect of the change in t_g for (a) TE mode, and (b) TM Mode.

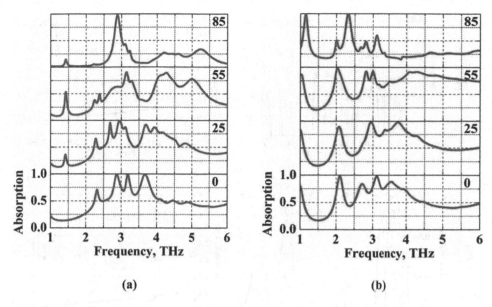

Figure 46.12 The effect in incident angle (θ) for (a) TE mode, and (b) TM Mode.

Table 46.2: Comparison with the other multi-band absorbers.

Ref.	Techniques	Peaks	Frequencies (THz)	Absorption efficiency (%)	Polarization insensitive Allowed θ
[20]	Metal/dielectric/ELC resonator	2	1.41–3.02	≈ 99	Not reported
[22]	gold/insulator/ graphene/ insulator/ graphene	2	20.6–25.2	≈ 100	0–65°
[41]	gold/dielectric/ SGP	2	0.9–3.12	91.5–98.2	0–50°
[40]	Au/Sio$_2$/graphene	3	10.15–14.86	≥ 98	0–80°
[43]	Au/dielectric/Au / dielectric/Au/ dielectric/Au	3	2.06–2.51	≈ 100	Not reported
[44]	Cu/Sio$_2$/graphene	3	4.98–12.97	99.5–99.9	Not reported
[52]	Al/silicon/Al	4	148.9–269.8	79–99	±15°
[53]	Au/Polyimide/Au	4	0.81–3.5	86.4–99.3	0–45°
[54]	Au/quartz dielectric/Au	5	0.37–1.77	94.1–99.6	Not reported
[55]	Copper/Polyimide/Copper	5	0.22–0.96	90.4–99.8	0–60°
[56]	Silicon/copper/polyimide/ copper	5	0–4.5	99–99.8	0–90°
Proposed absorber	Graphite/Sio$_2$/Graphite	3	2.88–3.67	99.8–99.9	0–85°

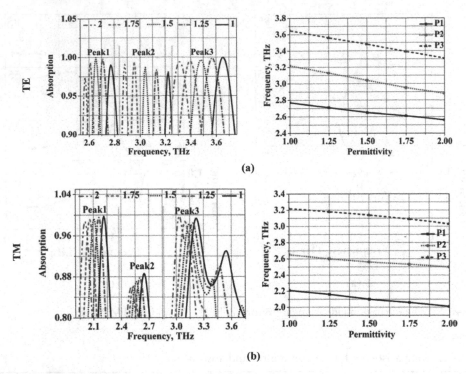

Figure 46.13 Sensing performance of the proposed absorber (a) response and (b) resonance frequency *vs* permittivity of analyte.

designed by the use of metamaterial structure. Also depict that multiband absorbers are developed at the cost of absorption efficiency. Furthermore, it is also observed that most of the absorbers are polarisation sensitive or insensitive for the lower range of allowed incidence angle.

Sensing Performance

To investigate the sensing performance of the proposed absorber in environment an analyte of thickness 10 µm is placed over the top graphite sheet. The relative permittivity (ε) of the analyte is varying from the range of 1 to 2 with the variation of 0.25. Figure 46.13a shows the response of the proposed absorber with varying permittivity of the analyte and depict that as increases ε, resonance frequency is shifting toward the lower frequency with slightly change in absorption efficiency. However, the variation in absorption efficiency is different for the resonance peaks in TE and TM mode. Absorption efficiency of Peak1 in TE mode and peak 1, 2 in TM mode are slightly reducing with increasing ε, whereas Peak 3 in TE mode are in increasing in nature. Further, the relation between the ε of analyte and resonance frequency are studied and plotted in Figure 46.13b. From the result it is observed that the resonance frequency generated due to the graphite sheet resonator is decreases linearly with increase in ε of analyte. Hence, the results it confirms the high sensitivity of the absorber which change the response by small change in relative permittivity (i.e. 0.25).

Conclusion

In conclusion, graphite-based metal-free THz absorber is designed and its response are numerically investigated. Proposed absorber is designed by three layers graphite sheet / Sio2/ graphite sheet structure with the insertion of notches and slot in top layer graphite sheet. The proposed absorber provides multiple resonance peaks for TE and TM mode having ≈100% absorption efficiency with proper diversity. The electric and magnetic field is analyzed for understanding the operation. Effect of the dimension changes of the absorber is also studied and found that the resonance peaks generated by the absorber can control by varying the dimensions of the absorber. Moreover, absorber also preserved better absorption efficiency for wide angle of incidence and serve polarization insensitivity. Highly sensitive nature of the proposed absorber shows the application use in sensing devices for sensing electrical and physical variations in the surroundings. Hence, the proposed metal-free THz has a wide application in biochemical sensing and optical sensing devices.

The structure of the proposed absorber is simple so easily can be fabricated. The large thickness of the carbon material layer will not restrict the fabrication. Furthermore, the parametric analysis is reported which can provide the view from the experimental aspects. The parametric study shows that the proposed absorber will be less sensitive to the variations in the physical dimension which may happen during the fabrication.

References

1. Weisenstein, C., Schaar, D., Wigger, A. K., Schäfer-Eberwein, H., Bosserhoff, A. K., and Bolívar, P. H. (2020). Ultrasensitive THz biosensor for PCR-free cDNA detection based on frequency selective surfaces. *Biomedical Optics Express*, 11, 448. https://doi.org/10.1364/boe.380818.
2. Vafapour, Z., and Ghahraloud, H. (2018). Semiconductor-based far-infrared biosensor by optical control of light propagation using THz metamaterial. *Journal of the Optical Society of America B*, 35, 1192. https://doi.org/10.1364/josab.35.001192.
3. Yi, Z., Chen, J., Cen, C., Chen, X., Zhou, Z., Tang, Y., Ye, X., Xiao, S., Luo, W., and Wu, P. (2019). Tunable graphene-based plasmonic perfect metamaterial absorber in the THz region. *Micromachines*, 10, 194. https://doi.org/10.3390/mi10030194.
4. Gupta, B. D., Pathak, A., and Semwal, V. (2019). Carbon-based nanomaterials for plasmonic sensors: a review. *Sensors (Switzerland)*, 19(16), 3536. https://doi.org/10.3390/s19163536.
5. Liu, M., Yin, X., Ulin-Avila, E., Geng, B., Zentgraf, T., Ju, L., Wang, F., and Zhang, X. (2011). A graphene-based broadband optical modulator. *Nature*, 474, 64–67. https://doi.org/10.1038/nature10067.
6. Correas-Serrano, D., Gomez-Diaz, J. S., Sounas, D. L., Hadad, Y., Alvarez-Melcon, A., and Alu, A. (2016). Nonreciprocal graphene devices and antennas based on spatiotemporal modulation. *IEEE Antennas and Wireless Propagation Letters*, 15, 1529–1533. https://doi.org/10.1109/LAWP.2015.2510818.
7. Yao, Y., Cheng, X., Qu, S. W., Yu, J., and Chen, X. (2016). Graphene-metal based tunable bandpass filters in the terahertz band. *IET Microwaves, Antennas & Propagation*, 10, 1570–1575. https://doi.org/10.1049/iet-map.2016.0335.
8. Varshney, G., Gotra, S., Pandey, V. S., and Yaduvanshi, R.S. (2019). Proximity-coupled graphene-patch-based tunable single-/dual-band notch filter for THz applications. *Journal of Electronic Materials is Springer*, 48, 4818–4829. https://doi.org/10.1007/s11664-019-07274-8.
9. Meng, Y., Hu, F., Shen, Y., Yang, Y., Xiao, Q., Fu, X., and Gong, M. (2018). Ultracompact graphene-assisted tunable waveguide couplers with high directivity and mode selectivity. *Scientific Reports*, 8, 13362. https://doi.org/10.1038/s41598-018-31555-7.

10. Conteduca, D., Dell'Olio, F., Ciminelli, C., and Armenise, M. N. (2015). Resonant graphene-based tunable optical delay line. *IEEE Photonics Journal*, 7(6), 1–9. https://doi.org/10.1109/JPHOT.2015.2496245.

11. Chen, P. Y., Argyropoulos, C., and Alu, A. (2013). Terahertz antenna phase shifters using integrally-gated graphene transmission-lines. *IEEE Transactions on Antennas and Propagation*, 61, 1528–1537. https://doi.org/10.1109/TAP.2012.2220327.

12. Rana, F. (2008). Graphene terahertz plasmon oscillators. *IEEE Transactions on Nanotechnology*, 7, 91–99.

13. Varshney, G., Gotra, S., Pandey, V. S., and Yaduvanshi, R. S. (2018). Inverted-sigmoid shaped multiband dielectric resonator antenna with dual-band circular polarization. *IEEE Transactions on Antennas and Propagation*, 66, 2067–2072. https://doi.org/10.1109/TAP.2018.2800799.

14. Varshney, G., Gotra, S., Chaturvedi, S., Pandey, V. S., and Yaduvanshi, R. S. (2019). Compact four-port MIMO dielectric resonator antenna with pattern diversity. *IET Microwaves, Antennas Propagation*, 13, 2193–2198. https://doi.org/10.1049/iet-map.2018.5799.

15. Varshney, G. (2020). Reconfigurable graphene antenna for THz applications: a mode conversion approach. *Nanotechnology*, 31, 135208. https://doi.org/10.1088/1361-6528/ab60cc.

16. Varshney, G. (2020). Ultra-wideband antenna using graphite disk resonator for THz applications. *Superlattices and Microstructures*, 141, 106480. https://doi.org/10.1016/j.spmi.2020.106480.

17. Deng, G., Xia, T., Yang, J., Qiu, L., and Yin, Z. (2016). Tunable terahertz metamaterial with agraphene reflector. *Materials Research Express*, 3, 1–9. https://doi.org/10.1088/2053-1591/3/11/115801.

18. Keshavarz, A., and Vafapour, Z. (2019). Sensing avian influenza viruses using terahertz metamaterial reflector. *IEEE Sensors Journal*, 19, 5161–5166. https://doi.org/10.1109/JSEN.2019.2903731.

19. Gong, C., Zhan, M., Yang, J., Wang, Z., Liu, H., Zhao, Y., and Liu, W. (2016). Broadband terahertz metamaterial absorber based on sectional asymmetric structures. *Scientific Reports*, 6, 1–8. https://doi.org/10.1038/srep32466.

20. Tao, H., Bingham, C. M., Pilon, D., Fan, K., Strikwerda, A. C., Shrekenhamer, D., Padilla, W. J., Zhang, X., and Averitt, R. D. (2010). A dual band terahertz metamaterial absorber. *Journal of Physics D: Applied Physics*, 43, 0–5. https://doi.org/10.1088/0022-3727/43/22/225102.

21. Vafapour, Z. (2019). Polarization-independent perfect optical metamaterial absorber as a glucose sensor in food industry applications. *IEEE Transactions on Nanobioscience*, 18, 622–627. https://doi.org/10.1109/TNB.2019.2929802.

22. Sun, P., You, C., Mahigir, A., Liu, T., Xia, F., Kong, W., Veronis, G., Dowling, J. P., Dong, L., and Yun, M. (2018). Graphene-based dual-band independently tunable infrared absorber. *Nanoscale*, 10, 15564–15570. https://doi.org/10.1039/c8nr02525h.

23. Yang, X., Zhao, X., Yang, K., Liu, Y., Liu, Y., Fu, W., and Luo, Y. (2016). Biomedical applications of terahertz spectroscopy and imaging. *Trends in Biotechnology*, 34, 810–824. https://doi.org/10.1016/j.tibtech.2016.04.008.

24. Sim, Y. C., Maeng, I., and Son, J. H. (2009). Frequency-dependent characteristics of terahertz radiation on the enamel and dentin of human tooth. *Current Applied Physics*, 9, 946–949. https://doi.org/10.1016/j.cap.2008.09.008.

25. Ye, L., Sui, K., Liu, Y., Zhang, M., and Liu, Q. H. (2018). Graphene-based hybrid plasmonic waveguide for highly efficient broadband mid-infrared propagation and modulation. *Optics Express*, 26, 15935. https://doi.org/10.1364/OE.26.015935.

26. Xiong, F., Zhang, J., Zhu, Z., Yuan, X., and Qin, S. (2015). Ultrabroadband, more than one order absorption enhancement in graphene with plasmonic light trapping. *Scientific Reports*, 5, 1–8. https://doi.org/10.1038/srep16998.

27. Park, M. S., Lee, J., Lee, J. W., Kim, K. J., Jo, Y. N., Woo, S. G., and Kim, Y. J. (2013). Tuning the surface chemistry of natural graphite anode by H 3PO4 and H3BO3 treatments for improving electrochemical and thermal properties. *Carbon*, 62, 278–287. https://doi.org/10.1016/j.carbon.2013.05.065.

28. Ansari, A., and Akhta, M. J. (2017). Co/graphite based light weight microwave absorber for electromagnetic shielding and stealth applications. *Materials Research Express*, 4(1), 016304. https://doi.org/10.1088/2053-1591/aa570c.

29. Varshney, G., Rani, N., Pandey, V. S., Yaduvanshi, R. S., and Singh, D. (2021). Graphite/ graphene disk stack-based metal-free wideband terahertz absorber. *Journal of the Optical Society of America B*, 38, 1–9.

30. Islam, M. S., Sultana, J., Biabanifard, M., Vafapour, Z., Nine, M. J., Dinovitser, A., Cordeiro, C. M. B., Ng, B. W. H., and Abbott, D. (2020). Tunable localized surface plasmon graphene meta-surface for multiband superabsorption and terahertz sensing. *Carbon*, 158, 559–567. https://doi.org/10.1016/j.carbon.2019.11.026.

31. Hanson, G. W. (2008). Dyadic green's functions for an anisotropic non-local model of biased graphene. *IEEE Transactions on Antennas and Propagation*, 56, 747–757.

32. Najafi, A., Soltani, M., Chaharmahali, I., and Biabanifard, S. (2020). Reliable design of THz absorbers based on graphene patterns: Exploiting genetic algorithm. *Optik (Stuttgart)*, 203, 163924. https://doi.org/10.1016/j.ijleo.2019.163924.

33. Jafari Jozani, K., Abbasi, M., Asiyabi, T., Biabanifard, M., and Biabanifard, S. (2019). Multi-bias, graphene-based reconfigurable THz absorber/reflector. *Optik (Stuttgart)*, 198, 163248. https://doi.org/10.1016/j.ijleo.2019.163248.

34. Soheilifar, M. R. (2019). The wideband optical absorber based on plasmonic metamaterials for optical sensing. *Optik (Stuttgart)*, 182, 702–711. https://doi.org/10.1016/j.ijleo.2019.01.090.

35. Mahmud, S., Islam, S. S., Mat, K., Chowdhury, M. E. H., Rmili, H., and Islam, M. T. (2020). Design and parametric analysis of a wide-angle polarization-insensitive metamaterial absorber with a star shape resonator for optical wavelength applications. *Results in Physics*, 18, 103259. https://doi.org/10.1016/j.rinp.2020.103259.

36. Shen, X., Yang, Y., Zang, Y., Gu, J., Han, J., Zhang, W., and Jun Cui, T. (2012). Triple-band ter-ahertz metamaterial absorber: Design, experiment, and physical interpretation. *Applied Physics Letters*, 101(15). https://doi.org/10.1063/1.4757879.

37. Fann, C.-H., Zhang, J., ElKabbash, M., Donaldson, W. R., Campbell, E. M., and Guo, C. (2019). Broadband infrared plasmonic metamaterial absorber with multipronged absorption mechanisms. *Optics Express*, 27, 27917. https://doi.org/10.1364/oe.27.027917.

38. Fang, J., Huang, J., Gou, Y., and Shang, Y. (2020). Research on broadband tunable metama-terial absorber based on PIN diode. *Optik (Stuttgart)*, 200, 163171. https://doi.org/10.1016/j. ijleo.2019.163171.

39. Chen, X., Jia, X., Wu, Z., Tang, Z., Zeng, Y., Wang, X., Fu, X., and Zou, Y. (2019). A graph-ite based metamaterial microwave absorber. *IEEE Antennas and Wireless Propagation Letters*, 18, 1016–1020. https://doi.org/10.1109/LAWP.2019.2907780.

40. Wu, J. (2019). Tunable multi-band terahertz absorber based on graphene nano-ribbon meta-material. *Physics Letters, Section A: General, Atomic and Solid State Physics*, 383, 2589–2593. https://doi.org/10.1016/j.physleta.2019.05.020.

41. Huang, M., Cheng, Y., Cheng, Z., Chen, H., Mao, X., and Gong, R. (2018). Based on graphene tunable dual-band terahertz metamaterial absorber with wide-angle. *Optical Communications*, 415, 194–201. https://doi.org/10.1016/j.optcom.2018.01.051.

42. Zhou, Q., Zha, S., Liu, P., Liu, C., an Bian, L., Zhang, J., Liu, H., and Ding, L. (2018). Graphene based controllable broadband terahertz metamaterial absorber with transmission band. *Materials (Basel)*, 11(12), 2409. https://doi.org/10.3390/ma11122409.

43. Wang, B. X., Tang, C., Niu, Q., He, Y., and Chen, T. (2019). Design of narrow discrete distances of dual-/triple-band terahertz metamaterial absorbers. *Nanoscale Research Letters*, 14, 1–7. https://doi.org/10.1186/s11671-019-2876-3.

44. Cen, C., Yi, Z., Zhang, G., Zhang, Y., Liang, C., Chen, X., Tang, Y., Ye, X., Yi, Y., Wang, J., and Hua, J. (2019). Theoretical design of a triple-band perfect metamaterial absorber in the THz frequency range. *Results in Physics*, 14, 102463. https://doi.org/10.1016/j.rinp.2019.102463.

45. Walther, M., Cooke, D. G., Sherstan, C., Hajar, M., Freeman, M. R., and Hegmann, F. A. (2007). Terahertz conductivity of thin gold films at the metal-insulator percolation transition. *Physical Review B: Condensed Matter and Materials Physics*, 76, 1–9. https://doi.org/10.1103/PhysRevB.76.125408.

46. Wang, K. X., Yu, Z., Liu, V., Cui, Y., and Fan, S. (2012). Absorption enhancement in ultrathin crystalline silicon solar cells with antireflection and light-trapping nanocone gratings. *Nano Letters*, 12, 1616–1619. https://doi.org/10.1021/nl204550q.

47. Wan, M., Li, Y., Chen, J., Wu, W., Chen, Z., Wang, Z., and Wang, H. (2017). Strong tunable absorption enhancement in graphene using dielectric-metal core-shell resonators. *Scientific Reports*, 7, 1–10. https://doi.org/10.1038/s41598-017-00056-4.

48. Gogoi, J. P., Bhattacharyya, N. S., and Bhattacharyya, S. (2014). Single layer micro-wave absorber based on expanded graphite-novolac phenolic resin composite for X-band applications. *Composites Part B: Engineering*, 58, 518–523. https://doi.org/10.1016/j.compositesb.2013.10.078.

49. Borah, D., and Bhattacharyya, N. S. (2017). Design and development of expanded graph-ite-based non-metallic and flexible metamaterial absorber for x-band applications. *Journal of Electronic Materials*, 46, 226–232. https://doi.org/10.1007/s11664-016-4918-2.

50. Chamorro-Posada, P., Vázquez-Cabo, J., Rubiños-López, Ó., Martín-Gil, J., Hernández-Navarro, S., Martín-Ramos, P., Sánchez-Arévalo, F. M., Tamashausky, A. V., Merino-Sánchez, C., and Dante, R. C. (2016). THz TDS study of several sp2 carbon materials: graphite, needle coke and graphene oxides, *Carbon*, 98, 484–490. https://doi.org/10.1016/j.carbon.2015.11.020.

51. Varshney, G. (2020). Wideband THz absorber: by merging the resonance of dielectric cav-ity and graphite disk resonator. *IEEE Sensors Journal*, XX, 1–1. https://doi.org/10.1109/jsen.2020.3017454.

52. Mulla, B., and Sabah, C. (2017). Multi-band metamaterial absorber topology for infrared fre-quency regime. *Physica E: Low-Dimensional Systems and Nanostructures*, 86, 44–51. https://doi.org/10.1016/j.physe.2016.10.003.

53. Bakshi, S. C., Mitra, D., and Minz, L. (2018). A compact design of multiband terahertz metama-terial absorber with frequency and polarization tunability. *Plasmonics*, 13, 1843–1852. https://doi.org/10.1007/s11468-018-0698-2.

54. Sabah, C., Mulla, B., Altan, H., and Ozyuzer, L. (2018). Cross-like terahertz metamate-rial absorber for sensing applications. *Pramana - Journal of Physics*, 91, 1–7. https://doi.org/10.1007/s12043-018-1591-4.

55. He, Y., Wu, Q., and Yan, S. (2019). Multi-band terahertz absorber at 0.1–1 THz frequency based on ultra-thin metamaterial. *Plasmonics*, 14, 1303–1310. https://doi.org/10.1007/s11468-019-00936-7.

56. Meng, T., Hu, D., and Zhu, Q. (2018). Design of a five-band terahertz perfect metamate-rial absorber using two resonators. *Optics Communications*, 415, 151–155. https://doi.org/10.1016/j.optcom.2018.01.048.

47 Electrodeposition studies of zinc-nickel-alumina nanocoating on mild steel—AISI 1144 using hull cell

M. Kannan[1,a], Naveen Rana[2,b], Mukesh Kumar[1,c], and Satyendra Singh[3,d]

[1]Department of Mechanical Engineering, Quantum University, Roorkee, India

[2]Mechanical Engineering Department, MMEC, Maharishi Markandeshwar (Deemed to be University), Mullana-Ambala, Haryana, India

[3]Department of Mechanical Engineering, B.T. Kumaon Institute of Technology, Dwarahat, Uttarakhand, India

Abstract: In this paper, Mild Steel—AISI 1144 specimens have been fabricated by electrodepositing zinc-nickel-alumina nanocoating from Watt's bath. A Comparative study has been conducted by using two different baths. Two baths can be differentiated on the bases of base salt used i.e., zinc sulphate hexahydrate ($ZnSO_4.6H_2O$) is used in bath1 and zinc bromide ($ZnBr_2$) is used in bath 2 as base salt. The impact of different salt concentrations on the characteristics of the obtained deposit is investigated through the utilization of the Hull cell, ultimately determining the ideal electrolyte bath concentration.

Keywords: Electrodeposits, hull cell, mild steel—AISI 1144, nanocoating, nanocomposites, zinc-nickel-alumina.

Introduction

Electrodeposition stands as one of the most extensively employed techniques for applying coatings onto substrates to prevent corrosion. Process parameters in electrodepositing have been widely researched with different substrates and salt solutions. Optimized combinations of parameters are essential for better inhibition of corrosion. O.S.I. Fayomi et al. carried out electrodeposition of zinc-alumina (Al_2O_3)—silicon carbide (SiC) coating on a mild steel substrate. Different bath compositions were used by varying the quantity of alumina and silicon carbide with variable currents ranging from0.5-1.0 A/cm². It was observed that the hardness of the substrate increased with the addition of ceramic composite particles like alumina (Al_2O_3)—silicon carbide (SiC) and with the increasing concentration of the additives [1].

Sekhar et al. have discussed various coating techniques and have found electrodeposition is the best method to obtain a uniform surface structure. A sulphate bath containing zinc sulphate ($ZnSO_4.7H_2O$) and magnesium sulphate ($MgSO_4.7H_2O$) with varying composition was used as the electrolyte. Zinc oxide particles were dispersed in the electrolyte and stirred for a period of 30 minutes. It was concluded that the surface properties and morphology of the substrate surface change with the addition and changing the concentration of additives [2]. D. Mukherjee et al. deposited zinc and alumina on copper and steel substrates. Chloride bath containing zinc chloride and sodium chloride with varying concentrations was

[a]mkananm@gmail.com, [b]er.naveenrana@live.com, [c]mkamboj67@gmail.com, [d]ssinghiitd@rediffmail.com

used as an electrolyte. The coated samples have comparable properties with those obtained by the hot-dip galvanizing process [3]. In the study conducted by Paulo Sergio da Silva [4], zinc particles obtained from a depleted battery were applied to coat a mild steel (AISI-1018) substrate. Both galvanostatic (–10 mA/cm2) and potentiostatic (–1.2V and –1.6V) approaches of electrodeposition were used in the analysis. The results have shown that the deposit becomes fine with the addition of PEG and a smooth coating obtained at –10 mA/cm2 current density [4]. Chandran et al used a Hull Cell for optimizing the bath parameters. For this objective, a conventional 267 ml hull cell was employed, and the optimization trials were conducted over a 10-minute period at a temperature of 30°C, utilizing a direct current of 1 ampere. Hull cell studies provided the effects of varying concentrations of different components in the electrolyte. Throwing power, cathode and anode current efficiency were measured for current density ranging 0.5 to 2.0 A/dm^2 [5]. Shivakumara et al carried-out hull cell (267 ml) experiments with a sulphate bath containing heptahydrate zinc sulphate (ZnSO$_4$.7H$_2$O), hexahydrate nickel sulphate (NiSO$_4$.6H$_2$O) and boric acid (H$_3$BO$_3$) as the basic electrolyte. The effect of varying concentration of condensation product was studied using a hull cell. As a conclusion concentrations of various constituents and current density were optimized. The nature of the deposit was reported to be good and the throwing power of the electrolyte under varying current densities was also found to be fairly good [6]. Maniam [8] researched on the improvements in deposition of zinc and zinc-nickel alloys in ionic liquids [7]. Kumar et al compiled the work on the influence of bath concentration and current density to the process current efficiency using watts bath and proposed optimum parameters for electro-deposition process [8]. Asmae et al, investigated a zinc-coated electro-deposition on a mild steel substrate within an acidic bath, employing cyclic voltammetry and chronopotentiometry methodologies. The impact of the current density on the rate of deposition was examined [9]. Shams Anwar et al investigated the electrochemical characteristics of both pure zinc and zinc-nickel alloy coatings, derived from both citrate and non-citrate baths. and the stability of the electrolyte baths was investigated [10]. Tolumoye et al found that changes in bath temperature had a good influence on the Ni content and form of the deposits [11]. Thangaraj et al. carried out electrodeposition using additives glycineand gelatinin a chlorine bath. The effect of bath concentration and operation parameters on the chemical composition and the nature of deposit was investigated and revealed that a Zinc-Nickel deposit having 13.6% Nickel showed good resistance against corrosion [12].

Experimental Details

Electrolytic Bath

In the galvanization process types of electrolytic baths used are Watts' bath, sulphamate bath, alkaline bath and cyanide bath. Each type of bath has its own advantages and disadvantages. Watts' bath and sulphamate bath are used for experimental purposes whereas zinc cyanide bath is used in commercial applications.

Watt's bath has been included. All the chemicals utilized are specifically intended for laboratory and research purposes. For the purpose of comparative analysis of the samples prepared, two distinct chemical baths have been employed. Hereafter, these two baths will be referred to as bath1 and bath2.

Table 47.1 provides the chemical composition for preparing 1 liter of electrolyte in both bath1 and bath2. In bath1, the foundational salt used is zinc sulphate Hexahydrate (ZnSO4.6H2O), while Nickel chloride hexahydrate (NiCl2.6H2O) is introduced to enhance the deposit's mechanical characteristics within the electrolyte.

Table 47.1: Chemical composition of bath1 for 1 litre of electrolyte.

Chemical Composition in BATH 1	Chemical Composition in BATH 2
Zinc sulphate hexahydrate ($ZnSO_4.6H_2O$)	Zinc bromide ($ZnBr_2$)
Nickel chloride hexahydrate ($NiCl_2. 6H_2O$)	Nickel chloride hexahydrate ($NiCl_2. 6H_2O$)
Boric acid (H_3BO_3)	Boric acid (H_3BO_3)
Glycerine	Glycerine
Sodium lauryl sulphate	Sodium lauryl sulphate
Alumina nanoparticles (α-Al_2O_3)	Alumina nanoparticles (α-Al_2O_3)

The difference between the composition of bath1 and bath2 is that of the base salt. Bath2 employs zinc bromide as the fundamental salt. Boric acid (H_3BO_3) is used as a buffer in electrolyte. It maintains the pH of the electrolyte by controlling the process of hydrogen evolution at the cathode surface. It does not let the hydrogen produced due to electrolysis of water deposit at the cathode surface. The deposition of hydrogen leads to the phenomenon called hydrogen embrittlement later on resulting in pitting and cracking on the deposited surface. Sodium lauryl sulphate is used in electrolyte as a surfactant or a dispersant. It is basically a foaming agent that disperses the salt in the solution. It does not let the salt deposit in the suspension and provides a conductive medium for the ions. Glycerine in the electrolyte acts as a brightener for the deposited coating. It tends to improve the quality of the deposit. Other brighteners, however, have been identified as glycine-glycol, polyethylene glycol.

Experimental Setup and Apparatus

Figure 47.1 illustrates the actual experimental arrangement for the procedure. Anodes consist of pure zinc rods with a purity of 99%. The cathode specimens are fashioned from mild steel, taking the form of both rods and plates. To facilitate the process, a mechanical stirrer operated by a 3-volt DC motor has been employed. This stirrer ensures continuous agitation of the solution during both electrolyte preparation and the procedure itself. By doing so, the mechanical stirrer prevents the settling of alumina nanoparticles, thus ensuring uniform electrolyte density throughout the electrolytic tank. The entire arrangement is linked to a DC power supply, as depicted in Figure 47.2. A schematic representation of this setup is detailed in Figure 47.3.

Figure 47.1   Actual experimental setup.

Figure 47.2 DC power supply source.

   To facilitate the electrolytic procedure, a direct current supply unit (ranging from 0 to 10 Amps at 0 to 10 Volts) has been employed. This power supply device is equipped with constant current (CC) regulation, enabling operation within differing voltage levels. The instrument is configured to sustain a steady current anywhere between 0 and 10 Amperes, while operating within a voltage span of 0 to 10 volts. A picture of the same is shown in Figure 47.2.

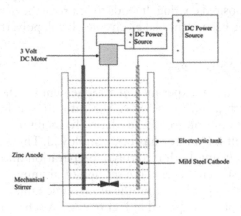

Figure 47.3 Schematic of the electroplating setup view.

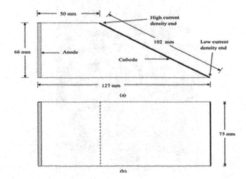

Figure 47.4 Schematic of trapezoidal hull cell (a) Top view (b) Front view.

Process Parameters Optimization-Hull Cell

The electrodeposition process encompasses several factors, including deposition time, current density, and chemical concentration within the electrolyte. To investigate these parameters, a conventional Trapezoidal Hull Cell has been integrated into the setup. Figure 47.4 presents both the top and front views of a representative 267 ml trapezoidal Hull Cell. The Hull cell, a uniquely engineered trapezoidal testing chamber, serves the purpose of conducting practical and qualitative plating evaluations on electroplating solutions. Its design is such that the cathode maintains a predetermined angle to the anode, generating a spectrum of diverse current densities. These current densities, both higher and lower than those typically employed in regular production settings, enable the identification of potential issues within the plating solution prior to the commencement of the actual coating procedure.

Calculation

The current distribution is based upon the formula:

$i = I (5.1 - 5.24 \log10 X)$, where i = current density (A/cm^2),
I = total current (ampere) and X = distance from high current density end of panel.

This configuration permits the test panel to be oriented at an angle relative to the anode, resulting in the deposition occurring at various current densities, which can be quantified using the previously mentioned formula.

Results and Discussion

Hull Cell Patterns for bath1

In Hull cell analysis mild steel plates used as cathode panels were plated at different values of current and varying concentrations. The effect of each component of the electrolyte on the quality of deposit has been studied with the help of legends. Legends for Hull cell is not a standard rule and may vary according to simplicity. These have been depicted in Figure 47.5 and the results would be based on them.

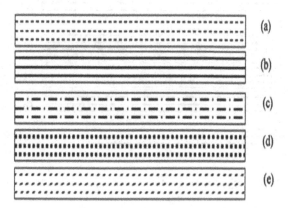

Figure 47.5 Hull Cell legends (a) bright deposit (b) milky white deposit (c) gray (Matty) deposit (d) uncoated area (e) burnt area.

Effect of Variation in Concentration of Zinc Sulphate on the Nature of the Deposit

The effect of change in the concentration of zinc sulphate has been presented in Figure 47.6. The solution contains 60 g/l of boric acid. It can be seen that at a concentration of 170 g/l some burnt area is visible at higher current densities (near about 95 mA/cm^2). There is a visible uncoated area at lower current densities and low concentrations (170 and 185 g/l). A grey and milky white deposit is obtained in the middle range of current densities above 50 mA/cm^2. However, a bright deposit is obtained at a concentration of 200 g/l and higher current densities. The area of bright deposit increases as the concentration increases to 220 g/l and there is also a considerable increase in the milky white deposit portion. It can be concluded that 200 g/l is the optimum concentration because a bright deposit begins to appear.

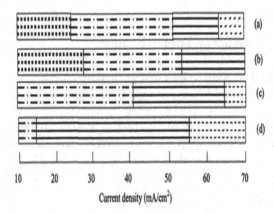

Figure 47.6 Hull cell results for effect of variation in concentration of Zinc sulphate in g/l (a) 170 (b) 185 (c) 200 (d) 220.

Effect of Variation in Concentration of Nickel Chloride on the Nature of the Deposit

The effect of variation in concentration of nickel chloride with 200 g/l of Zinc sulphate and 60 g/l of boric acid at room temperature has been presented in Figure 47.7.

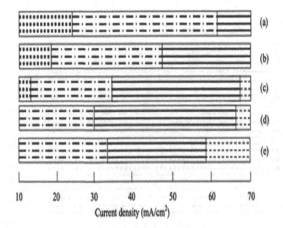

Figure 47.7 Hull cell results for effect of variation in concentration of nickel chloride in g/l (a) 140 (b) 150 (c) 160 (d) 170 (e) 180.

It can be seen that a milky white deposit is visible even at low concentrations of nickel chloride but at higher current density. Bright deposits begin to appear at a concentration of 160 g/l and current density between 90 and 100 mA/cm². A slight uncoated region can also be seen at lower current densities. With further increase in concentration to170 g/l no considerable change is observed in the bright deposit but the milky white deposit region increases with a reduction in the grey deposit area. At a concentration of 180 g/l, a good bright deposit is obtained above a current density of 90 mA/cm². The optimum concentration of nickel chloride can be taken to be 160 g/l at which a bright deposit begins to appear. The other conditions remain unchanged i.e. the concentration of zinc sulphate is 200 g/l, concentration of boric acid is 60 g/l and the experiments are done at room temperature.

Effect of Variation in Concentration of Boric Acid on the Nature of the Deposit

The Hull cell results for change in the concentration of boric acid are shown in Figure 47.8. The concentration of Zinc sulphate and Nickel chloride is kept constant at 200 g/l and 160 g/l respectively at room temperature.

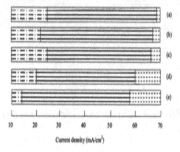

Figure 47.8 Hull cell results for effect of change in concentration of boric acid in g/l (a) 40 (b) 50 (c) 60 (d) 70 (e) 80.

The results have shown that a bright finish can be obtained even at low concentrations of boric acid. But to get a uniform bright finish the optimum concentration of boric acid has been chosen to be 60 g/l. The area of the bright deposit increases with an increase in the concentration of boric acid. Boric acid also acts as a buffer in the electrolyte and regulates the pH by maintaining the concentration of H+ ions in the solution.

Optimized Watts bath Concentration for bath1

The Table 47.2 displays the optimal concentration of the bath for bath1, as determined through Hull cell investigations.

Table 47.2: Optimized watts bath concentration for bath 1

Name of Chemical	Optimum quantity (g/l of water)
Zinc sulphate hexahydrate ($ZnSO_4.6H_2O$)	200
Nickel chloride hexahydrate ($NiCl_2. 6H_2O$)	160
Boric acid (H_3BO_3)	60
Glycerine	5
Sodium lauryl sulphate	0.5
Current density	40–60 mA/cm²
pH	4.5–5.0

Hull Cell Patterns for bath 2

Effect of Variation in Concentration of Zinc Bromide on the Nature of the Deposit

The effect of change in concentration of Zinc bromide from Hull cell studies has been shown in Figure 47.9. The concentration of boric acid is kept constant at 60 g/l and experiments carried out at room temperature. It can be seen from the results that at low concentrations (100 to 120 g/l) and high current density burnt area are predominant due to excessive flow of current. The uncoated area is visible in the region of low current densities. A bright deposit appears at a concentration of 140 g/l with a reduction in uncoated and burnt regions. At higher concentrations of zinc bromide, bright deposits can be seen with some part of the milky white deposit and a grey deposit. 160 g/l can be taken as an optimum concentration for further experiments.

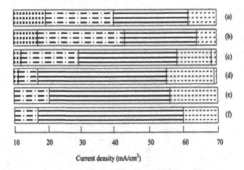

Figure 47.9 Hull cell results for effect of change in concentration of *Zinc bromide in g/l (a) 100 (b) 120 (c) 140 (d) 160 (e) 180 (f) 200.*

Effect of Variation in Concentration of Nickel Chloride on the Nature of the Deposit

The results for the effect of variation in concentration of Nickel chloride have been presented in Figure 47.10.

The concentration of Zinc bromide and boric acid are kept constant at 160 g/l and 60 g/l respectively. Some uncoated region is visible at low concentration of nickel chloride. A good bright deposit is visible at a concentration of 120 g/l and above. At a concentration of 140 g/l and above most of the area is covered by a milky white deposit and some part by grey deposit and a bright deposit. At current densities ranging from 60 to 90 mA/cm^2, a good deposit is obtained. 140 g/l can be taken as the optimum concentration for nickel chloride. The pH of the solution is maintained acidic between 4 and 5.

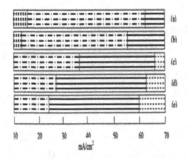

Figure 47.10 Hull cell results for effect of change in concentration of Nickel chloride in g/l (a) 100 (b) 110 (c) 120 (d) 140 (e) 160.

Effect of Variation in Concentration of Boric Acid on the Nature of the Deposit
The results for the change in the concentration of boric acid have been presented in Figure 47.11.

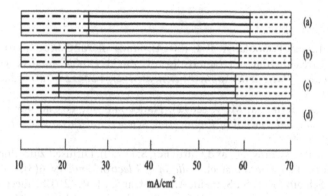

Figure 47.11 Hull cell results for effect of change in concentration of Boric acid in g/l (a) 100 (b) 110 (c) 120 (d) 140 (e) 160.

It can be seen that even at low concentrations of boric acid a bright deposit is obtained at a high current density end but accompanied by a milky white deposit and some grey deposit. On increasing the concentration There is no considerable change in the bright deposit area but most of the region is covered by the milky white deposit (50 to 90 mA/cm²). 60 g/l can be taken as the optimum concentration for boric acid when the concentrations of zinc bromide and nickel chlorides are 160 g/l and 140 g/l respectively. Bright deposits are visible at low concentrations but it is suitable to use high concentrations of boric acid because it acts as a buffer and tends to regulate the pH of the solution.

Optimized Watts bath Concentration for bath2
The optimized watts bath concentration for bath2 have been presented in Table 47.3.

Table 47.3: Optimized watts bath concentration for bath 2.

Name of Chemical	Optimum quantity (g/l of water)
Zinc bromide ($ZnBr_2$)	160
Nickel chloride hexahydrate ($NiCl_2.\ 6H_2O$)	140
Boric acid (H_3BO_3)	60
Glycerine	5
Sodium lauryl sulphate	0.5
Current density	40–60 mA/cm²
pH	4.5–5.0

Conclusion

The Hull cell studies performed using a conventional 267 ml Hull cell have yielded satisfactory results. Effects of variation in concentration of zinc bromide, boric acid and Nickel chloride on the nature of deposit has been studied and reported. The obtained results yield the optimal concentration for the electrolyte bath, revealing that the deposition can be conducted across a range of varying current densities. The hull cell patterns have shown how the appearance of the deposited coating changes from grey to bright with a change in concentration of the electrolyte, current density and time of deposition.

References

1. Fayomi, O. S. I., and Popoola, A. P. I. (2014). Study of Al2O3/SiC particle loading on the microstructuralstrengthening characteristics of Zn-Al2O3-SiC matrix composite coating. *Egyptian Journal of Basic and Applied Sciences*, 1(2), 120–125. http://dx.doi.org/10.1016/j. ejbas.2014.05.003.

2. Sekhar, M. C., and Ramana, M. V. (2012). Nanocomposite coating for improved corrosion resistance. *International Journal of Engineering Science and Technology (IJEST)*, 4(07), 3118–3123. ISSN: 0975-5462.

3. Mukherjee, D., Palaniswamy, N., and Guruviah, S. (1990). Diffused zinc-alumina electrodeposited on cold worked copper and steel. *Bulletin of Electrochemistry*, 6(3), 380–381.

4. da Silva, P. S., Schmitz, E. P. S., Spinelli, A., and Garcia, J. R. (2012). Electrodeposition of Zn and Zn–Mn alloy coatings from an electrolytic bath prepared by recovery of exhausted zinc-carbon batteries. *Journal of Power Sources*, 210, 116–121.

5. Chandran, M., and Ramesh Bapu, G. N. K. (2013). Electrodeposition of nano Zinc—nickel alloy from bromide basedElectrolyte. *Research Journal of Chemical Sciences*, 3(8), 57–62.

6. Shivakumara, S., Manohar, U., Naik, Y. A., and Venkatesha, T. V. (2007). Influence of additives on electrodeposition of bright Zn–Ni alloy on mild steel from acid sulphate bath. *Bulletin of Material Science*, 30(5), 455–462.

7. Maniam, K. K., and Paul, S. (2020). Progress in electrodeposition of zinc and zinc nickel alloys using ionic liquids. *Applied Sciences*, 10, 5321. doi:10.3390/app10155321.

8. Kumar, S., Pande, S., and Verma, P. (2015). Factor effecting electro-deposition process. *International Journal of Current Engineering and Technology*, 5(2), 700–703.

9. El Fazazi, A., Ouakki, M., and Cherkaoui, M. (2019). Electrochemical deposition of Zinc on mild steel. *Mediterranean Journal of Chemistry*, 8(1), 30–41.

10. Anwar, S., Zhang, Y., and Khan, F. (2018). Electrochemical behaviour and analysis of Zn and Zn–Ni alloy anti-corrosive coatings deposited from citrate baths. *RSC Advances*, 8, 28861.

11. Tuaweri, T. J., and Gumus, R. (2013). Zn-Ni electrodeposition for enhanced corrosion performance. *International Journal of Materials Science and Applications*, 2(6), 221–227. doi: 10.11648/j.ijmsa.20130206.18.

12. Thangaraj, V., Udayashankar, N. K., and Hegde, A. C. (2008). Development of Zn-Co alloy coatings by pulsed current from chloride bath. *Indian Journal of Chemical Technology*, 15, 581–587.

48 Single stage and multi stage hole flanging using single point incremental forming

Rudreshkumar Makwana[a], Bharat Modi[b], and Kaushik Patel[c]

Mechanical Engineering Department, Institute of Technology, Nirma University, Ahmedabad, Gujarat, India

Abstract: Single Point Incremental Forming is technique of incremental forming without use of die or support tool. Hole flanging is a forming process to form flange on a sheet metal with precut hole on it. In this work, the principle of single point Incremental Forming with single stage and multistage strategies has been used to form hole flange. Experiments are performed on sheet metals of Aluminium 1050 to form flange with straight wall successfully without failure. Variation in the limit forming ratio, thickness distribution and geometric accuracy on the formed flange obtained by single stage and multistage strategies are compared and analyzed. A new methodology to check geometric accuracy considering local deformation of hole flange is proposed.

Keywords: Geometric accuracy, hole flanging, limit forming ratio, multistage strategy, single point incremental forming, single stage strategy.

Introduction

Incremental forming (IF) is a die less forming technique. The early research work on IF shows that it offers high formability as compared to conventional forming which remains the key aspect of IF. So many researchers are attracted to explore the process further. The different methods of IF includes a method known as single point incremental forming (SPIF), the method is exhaustively studied and analyzed by Jesweit et al. [1]. The study shows the capability of SPIF technique to form asymmetric parts with improved formability. To further explore the fundamentals of SPIF, Silva et al. [2] developed an analytical model of the process by membrane analysis. They claimed that cracks in SPIF occur because of meridional tensile stresses. To explore use of multi stage strategy in SPIF, Skjoedt et al. [3] obtained a circular cylindrical cup of equal height and radius by five stage forming strategy of SPIF. They demonstrated that it is possible to obtain a 90° wall angle cup with multi stage strategy. Apart from, formability of SPIF; many researchers have worked on geometric accuracy of formed part using IF [4–6]. It is noted from these studies that geometric accuracy is considered as deviation between intended and obtained cross sectional profile. In much research works, methods for improving the geometric accuracy are suggested.

---

[a]rudresh.makwana, [b]bharat.modi, [c]kaushik.patel@nirmauni.ac.in

Hole flanging operation is used to form flange on a sheet metal with precut hole on it. In Automobile and Aerospace applications, flanging is required to increase strength of edge and to provide support to join other part or for aesthetics. Cui et al. [7]which use dies and punches, are not cost competitive to make prototype parts or batches of small quantities. Since incremental forming does not require dedicated dies, it has shown promise to reduce cost and cycle time to manufacture such parts. In the present study, incremental forming with three forming strategies was investigated to produce prototype parts with hole-flanges. Results indicate that the forming strategy by increasing the part diameter in small steps during the forming process reaches the final optimum part geometry to improve the formability while it can produce a relatively higher neck height, maximum forming limit ratio (LFRmax performed experimental investigation on effect of different multistage strategies on formability. Here the measure of formability is taken as limit forming ratio (LFR), which is ratio of final flange diameter (d_f) to the minimum pre-cut hole diameter (d_i) The work of Centeno, G. et al. [8] is focused on deformation mechanics and failure in hole flanging using SPIF. Their work involving use of multistage strategy to form conical flange and cylindrical flange concludes that the high formability in hole flanging using SPIF (compared to conventional hole flanging) is because of necking suppression before fracture. However, Silva, M.B. et al. [9]and the strain loading paths resulting from each forming process are determined by circle grid analysis. Results in Aluminium AA1050-H111 and Titanium (grade 2 showed that the process operating window in hole flanging depends on material also. In their work, they demonstrated that the LFR obtained by conventional hole flanging is higher than the LFR obtained by multistage SPIF for Titanium material.

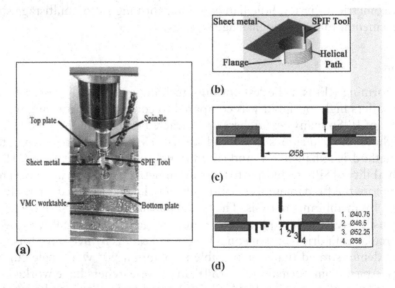

Figure 48.1 (a) Experimental set up (b) Tool path (c) single stage strategy (d) multi stage strategy.

Borrego et al. [10] explored the single stage SPIF technique for hole flanging as multistage SPIF technique increases the time of operation. In their detailed study using Aluminium 7075-O sheet metal, they introduced some useful parameters. The study also analyzed effect of spindle speed and tool diameter and concluded that larger tool gives high LFR whereas spindle speed does not affect formability but results into better surface quality. Morales-Palma et al. [11] proposed a strategy of two stage incremental forming for hole flanging based on the results of thickness distribution obtained in [10], the study included optimization of thickness distribution using FEA simulations. Praveen et al. [12] obtained hole flange with straight 90° wall with multi stage strategy and presented study of flange height, thickness, vertical force and surface roughness. The effect of feed and step depth in single stage hole flanging by SPIF is studied by Makwana et al. [13]. Han et al. [14] investigated the deformation mechanism in straight flanging using incremental forming, it was noted that as the depth of flange increases the geometric accuracy decreases.

Many individual studies are done on hole flanging using single stage SPIF or multi stage SPIF, which used different materials. However, it is required to compare thickness distribution, geometric accuracy and height of straight flange obtained by both the strategies (single stage and multistage) in a common material. The aim of present work is to achieve the comparative results by experimental work. Moreover, the geometric accuracy in most of the studies has been considered as deviation between intended profile and obtained profile. However, it is to be noted that the outer surface and inner surface of the formed part may have local deviations; in such case even if obtained profile matches with intended one, the overall geometry can be inaccurate. So, in the work presented here, a new methodology to check geometric accuracy considering local deviations on hole flange surfaces (both outer and inner) is proposed.

Methodology

Aluminium 1050 sheets with 1.5 mm thickness are used for the experimental study. For performing experiments, sheets of 100 × 100 mm were cut from the as received rolled sheet. The first step is to cut a hole on the sheet followed by forming of flange using the principle of single point incremental forming. The process is performed on a Vertical Machining Center with experimental set up shown in Figure 48.1(a).

To form the sheets, a hemispherical headed tool with 10 mm diameter is used. The tool follows helical path while forming the flange as shown in Figure 48.1(b). The hole flanging operations are carried out using single stage strategy and multistage strategy as shown schematically in Figure 48.1(c) and 48.1(d) respectively. In single stage strategy, the tool is located above the sheet metal at the inner diameter of the flange to be formed and then it is traversed in helical path up to the full height of the flange. In multistage, the final flange diameter is achieved through forming intermediate flanges as shown in Figure 48.1(d). The multi stage strategy used in this research work is the best strategy among the three as per [7]. To find Limit Forming Ratio of the material in the hole flanging process, experiments were performed by varying the diameter of precut hole, the minimum diameter which gives a straight flange without crack is considered for the calculation of LFR. To understand failure of the flange, it is required to measure strain along the flange height at various places. For measurement of strain values, circular grid with 1 mm diameter circles is engraved (using laser engraving) on the sheets. The deformed circles after flanging operation are visible in Figure 48.2. The percentage change in diameter along major axis and minor axis of ellipses represents major strain and minor strain respectively. Portable microscope with image analyzer software is used to measure major and minor axis length of deformed circles.

The hole flanging operations were intended to form straight flanges. So, to check how accurate the obtained shape is, it was required to check the geometric accuracy of flange.

For the same purpose, the outer and inner diameter of the flanges were measured at marked point along the flange height, this gives the deviations on the flange profile. The methods discussed above are applied for both single stage and multistage strategies.

Figure 48.2 Deformed circles on the flanges formed by single stage and multi stage strategies.

Results and Discussion

Limit Forming Ratio, Thickness of Flange and Height of Flange

For single stage strategy it was observed that straight flanges are successfully formed for all the precut hole diameters more than 33 mm. For the diameter 33 mm, the straight flange could not be formed. For multi stage strategy, the minimum precut hole diameter which gave straight flange is 28 mm. For the diameter 27 mm, the straight flange could not be formed. The obtained LFR is 1.70 and 2.07 for single stage and multi stage strategy respectively. The formed flanges are shown in Figure 48.3.

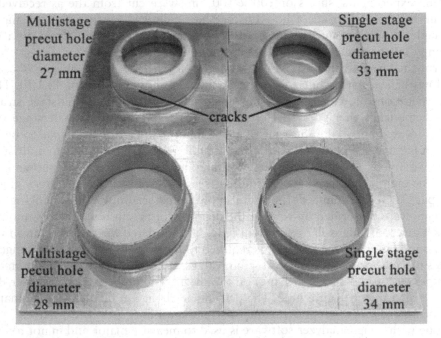

Figure 48.3 Flanges formed by single stage approach and multistage approach.

The strain values observed from the circle grid analysis are shown in Table 48.1. It is observed that in multistage strategy during the intermediate stages, the material is stretched from the intermediate roots towards the edge which reduces the large amount of material being stretched in the final stage.

Table 48.1: Strain measured on formed flanges.

Single stage			Multistage	
Sr. No.	Major strain	Minor strain	Major strain	Minor strain
1	0.66	0.09	0.70	0.1
2	0.53	0.38	0.49	0.28
3	0.55	0.24	0.30	0.08

It is observed that biaxial stretching takes place near the root of the flange in single stage strategy and it remains in biaxial mode until reaching upper zone. In the upper half of the flange, it starts converting into uniaxial stretching and can be clearly observed that near the edge the deformation of circles shows uniaxial strain. In multistage strategy, if every intermediate stage is thought of as a single stage, then it can be understood that all the intermediate edges are stretched under uniaxial stress mode. The mid zone of the flange in multi stage shows bi directional stretching but less than that of mid zone of flange in single stage. This shows that material at the mid zone is not stretched beyond a limit. This helps the tool in forming flange from sheet metal having smaller precut hole diameter as compared to single stage approach. Hence, LFR could be improved by approximately 18 % with multistage approach as compared to single stage approach.

For comparison of thickness distribution and height of the flanges obtained by both strategies, flanges were formed with precut holes of same size (diameter = 35 mm). Thickness on the obtained flanges is measured by pointed anvil digital micrometer with least count of 0.001 mm. The measurements are made at different points at the interval of 1 mm from root towards the flange edge. The maximum and minimum thicknesses on flanges obtained by single stage are 1.243 mm and 0.791 mm respectively. In case of multi stage they are 1.487 mm and 0.789 mm respectively. The average flange thickness is measured to be 0.982 mm and 1.203 mm respectively and the flange height is measured to be 16.20 mm and 12.46 mm respectively.

The distribution of thickness is shown on the plot in Figure 48.4(a). It is observed from the plot that the thickness along the flange formed by single stage strategy decreases from root up to mid distance and then increases towards the edge of the flange, the distribution is similar to what is described by [10]. As more material is stretched from root, thickness in the lower zone of flange keeps on decreasing and as the material is stretched towards the edge, it remains thicker. In case of multistage strategy, the thickness decreases from root towards the edge of flange and thickness distribution remains more uniform. It is also observed that due to multiple stages, thickness bands are generated on the flange. These bands can be observed on the plot shown in Figure 48.4(a) and the image shown in Figure 48.4(b). The reason behind formation of these bands is bending during the intermediate stages. As here in multistage strategy four stages are used, bending of sheet takes place four times at four different places. The material is stretched from all these four roots towards

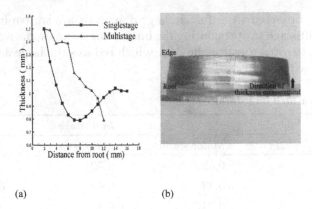

(a)   (b)

Figure 48.4 (a) Thickness distribution along the flange height (b) flange geometry.

the edge. Ultimately, the four roots create steps on the final straight flange which are visible in Figure 48.4(b). These steps also result into thickness bands.

Figure 48.5 shows flange (single stage) cross-section profile obtained by 3D scanning. The obtained results show that reduction in thickness in single stage SPIF and multistage SPIF is 34.54% and 19.8% respectively. It is clear from this result that more stretching takes place in single stage strategy which can also be related to the height of the flange. The measured height of the flange is as shown in Table 48.2. It is observed that although the diameter of the flange is fixed, different height of flanges is achieved. It is observed that the multistage strategy gives shorter flange. By correlating the height, the thickness and the strain values; it is clear that the multistage flange is short because of less stretching and same is the reason behind its higher thickness. The height of single stage flange can be correlated with its lower thickness; and the reason behind it is the high amount of stretching. These results can be correlated to the major strain values, and it is visible that strain values in multistage strategy are smaller than what is obtained in the single stage strategy.

Table 48.2: Height of formed flanges.

Pre-cut hole diameter (mm)	Height of flange (mm)	
	Single stage approach	*Multistage approach*
35	16.12	12.46
28	Failure	20.18

To find the geometrical deviations along the flange height, straight profile is imposed on actual flange profile as shown in Figure 48.6 and it is measured at outer and inner side of the profile. On both sides (inner and outer), deviation from the straight profile towards outer side is considered to be positive and towards inner side is considered to be negative. Measurements are taken on the profile of both the flanges obtained by single stage strategy and multistage strategy using CAD software.

Table 48.3: Maximum and minimum deviations on flange obtained by single stage approach.

Δ_{osmax} *(mm)*	Δ_{osmin} *(mm)*	Δ_{ismax} *(mm)*	Δ_{ismin} *(mm)*
0.24	−0.45	0	−0.37

Table 48.4: Maximum and minimum deviations on flange obtained by multistage approach.

Δ_{ommax} (mm)	Δ_{ommin} (mm)	Δ_{immax} (mm)	Δ_{immin} (mm)
0.55	−0.37	0.21	−0.20

Figure 48.5 Flange (single stage) cross-section profile obtained by 3D scanning.

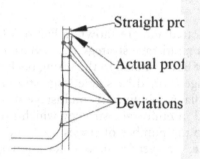

Figure 48.6 Schematic of deviation measurement.

Tables 48.3 and 48.4 shows deviations on outer and inner surfaces of flanges obtained by single stage strategy and multistage strategy respectively.

The total deviation on the outer and inner surface is calculated by equations (1) to (4). The obtained values are shown in Table 48.5.

Total deviation on outer surface – single stage $\delta_{os} = \Delta_{osmax} - \Delta_{osmin}$ (1)

Total deviation on inner surface – single stage $\delta_{is} = \Delta_{ismax} - \Delta_{ismin}$ (2)

Total deviation on inner surface – multi stage $\delta_{om} = \Delta_{ommax} - \Delta_{ommin}$ (3)

Total deviation on inner surface – multi stage $\delta_{im} = \Delta_{immax} - \Delta_{immin}$ (4)

Table 48.5: Total Deviations on outer and inner surface of the flanges.

δ_{os} (mm)	δ_{is} (mm)	δ_{om} (mm)	δ_{im} (mm)
0.69	0.37	0.92	0.41

The total error is calculated by equation (5) and (6),

$$\varepsilon_s = \delta_{os} - \delta_{is}\, \delta_{is} \qquad (5)$$
$$\varepsilon_m = \delta_{om} - \delta_{im}\, \delta_{is} \qquad (6)$$

The calculated errors are $\varepsilon_s = 0.32$ mm and $\varepsilon_m = 0.51$ mm.

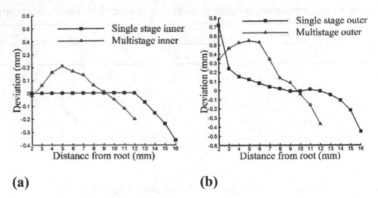

(a)   **(b)**

Figure 48.7 Deviations on flange formed by single stage approach on (a) outer surface and (b) inner surface.

The deviation values are plotted and are shown in Figure 48.7(a) and (b), it is observed that in single stage as well as multistage strategy the deviations in the top portion of the flange are negative which may be attributed to the spring back effect. It is observed further that inner surface of the flange formed by single stage strategy is closer to the required geometry as compared to the flange obtained by multistage strategy. This can be attributed to the multiple steps involved in multistage strategy which involves bending of the sheet at multiple locations equal to the number of stages used and visualized as bands on the flange. In case of outer surface, the overall range of deviation appears more in single stage but that is because of the two extreme values occurring on root of the flange and edge of the flange (can be observed in Figure 48.7 (b), this is due to the high bending stresses at the root and high spring back at the edge. In multistage the bending takes place at different places and the spring back is less because of multiple tool passes at the edge. Overall, the local deviations are less in flange formed by single stage strategy.

Conclusions

The experimental study is carried out on hole flanging using single stage and multistage strategies of single point incremental forming (SPIF) using a common material (Aluminium AA1050 sheet metal). Following points are concluded from the study.

Higher limit forming ratio (LFR) is obtained by multistage SPIF as compared to single stage SPIF. The higher formability is attributed to less stretching of material in the process which is due to forming of the sheet by bending at more than one place. So, multi stage strategy should be used to obtain a thick flange. In single stage strategy, the sheet is stretched from a single root towards the edge which results in high strain values and high amount of stretching; the strategy should be used when thin and long flange without bands on the flange wall is required. It is also useful to decrease the forming time.

The shape of the flange is not enough to define geometric accuracy as there can be small local deviations on the surface of the flange. So, a new method of determining geometric accuracy of flange is proposed. Using the proposed method, measurements are taken on the flanges obtained by single stage and multi stage strategies. The results show that single stage strategy gives high geometric accuracy in terms of local deviations where as there are more deviations on the flange obtained by multistage because of bands created by more than one number of stages.

Funding

This work is supported by Nirma University, Ahmedabad, Gujarat, India. (vide office order no: NU/Ph.D/MRP/IT/2018-19/6439)

References

1. Jeswiet, J., Micari, F., Hirt, G., Bramley, A., Duflou, J., and Allwood, J. (2005). Asymmetric single point incremental forming of sheet metal. *CIRP Annals—Manufacturing Technology*, 54(2), 88–114. doi: 10.1016/S0007-8506(07)60021-3.
2. Martins, P. A. F., Bay, N., Skjoedt, M., and Silva, M. B. (2008). Theory of single point incremental forming. *CIRP Annals—Manufacturing Technology*, 57(1), 247–252. doi: 10.1016/j.cirp.2008.03.047.
3. Skjoedt, M., Bay, N., Endelt, B., and Ingarao, G. (2008). Multi stage strategies for single point incremental forming of a cup. *International Journal of Material Forming*, 1(Suppl. 1), 1199–1202. doi: 10.1007/s12289-008-0156-3.
4. Lu, H., Kearney, M., Wang, C., Liu, S., and Meehan, P. A. (2017). Part accuracy improvement in two point incremental forming with a partial die using a model predictive control algorithm. *Precision Engineering*, 49, 179–188. doi: 10.1016/j.precisioneng.2017.02.006.
5. Besong, L. I., Buhl, J., Ünsal, I., Bambach, M., Polte, M., Blumberg, J., and Uhlmann, E. (2020). Development of tool paths for multi-axis single stage incremental hole-flanging. *Procedia Manufacturing*, 47, 1392–1398. doi: 10.1016/j.promfg.2020.04.290.
6. Vahdati, M., Sedighi, M., and Khoshkish, H. (2010). An analytical model to reduce springback in incremental sheet metal forming (ISMF) process. *Advanced Materials Research*, 83–86, 1113–1120. doi: 10.4028/www.scientific.net/AMR.83-86.1113.
7. Cui, Z., and Gao, L. (2010). Studies on hole-flanging process using multistage incremental forming. *CIRP Journal of Manufacturing Science and Technology*, 2(2), 124–128. doi: 10.1016/j.cirpj.2010.02.001.
8. Centeno, G., Silva, M. B., Cristino, V. A. M., Vallellano, C., and Martins, P. A. F. (2012). Hole-flanging by incremental sheet forming. *International Journal of Machine Tools and Manufacture*, 59, 46–54. doi: 10.1016/j.ijmachtools.2012.03.007.
9. Silva, M. B., Teixeira, P., Reis, A., and Martins, P. A. F. (2013). On the formability of hole-flanging by incremental sheet forming. *Proceedings of the Institution of Mechanical Engineers, Part L: Journal of Materials: Design and Applications*, 227(2), 91–99. doi: 10.1177/1464420712474210.
10. Borrego, M., Morales-Palma, D., Martínez-Donaire, A. J., Centeno, G., and Vallellano, C. (2016). Experimental study of hole-flanging by single-stage incremental sheet forming. *Journal of Materials Processing Technology*, 237, 320–330. doi: 10.1016/j.jmatprotec.2016.06.026.
11. Morales-Palma, D., Borrego, M., Martínez-Donaire, A. J., Centeno, G., and Vallellano, C. (2018). Optimization of hole-flanging by single point incremental forming in two stages. *Materials (Basel)*, 11(10), 2029. doi: 10.3390/ma11102029.
12. Kumar, A., Gulati, V., Kumar, P., Singh, V., Kumar, B., and Singh, H. (2019). Parametric effects on formability of AA2024-O aluminum alloy sheets in single point incremental forming. *Journal of Materials Research and Technology*, 8(1), 1461–1469. doi: 10.1016/j.jmrt.2018.11.001.
13. Makwana, R., Modi, B., and Patel, K. (2020). Effect of feed and step depth in hole flanging using single point incremental forming. *Journal of Physics: Conference Series*, 1706(1). doi: 10.1088/1742-6596/1706/1/012177.
14. Han, K., et al. (2019). Experimental and numerical study on the deformation mechanism of straight flanging by incremental sheet forming. *International Journal of Mechanical Sciences*, 160(June), 75–89. doi: 10.1016/j.ijmecsci.2019.06.024.

49 Characterizing single point tool for incremental die less forming of thin titanium sheet

S. S. Adewar[1,a] and A. K. Bewoor[2,b]

[1]Mechanical Engineering Department, Zeal COER, Pune, India

[2]Mechanical Engineering Department., Cummins C.O.E. for Women's, Pune, India

Abstract: Incremental forming procedure has a capacity to deliver complex shapes and can be utilized for small batch production. The product which has better surface finish, great quality and formability has more importance in industrialized and car application on pure titanium sheet, which has a high potential and multifaceted nature within the ordinary framing process, the single point incremental forming procedure is performed. Through the test plan, parameters such as tool diameter, feed rate, spindle velocity, and step depth were used to characterize performance parameters such as surface roughness. The findings could help to confirm the result characteristics viz, tool size and vertical step size, have the biggest impact on Single to noise ratio fluctuation. The confirmation test also revealed that Taguchi analysis and ANOVA response tables and graphs are effective and efficient approaches for establishing each design parameter's ideal level in order to obtain the Single to noise ratio minimal value.

Keywords: Die less process, single point incremental forming, tool selection.

Introduction

The experimental findings have introduced a novel concept regarding the significance of tool size in single point incremental forming (SPIF). This concept also paves the way for establishing a generalized criterion for selecting the appropriate tool size. The innovation presented in this study is referred to as incremental sheet metal forming (ISMF) [18]. This new approach offers valuable insights into optimizing the tool size selection process, which can lead to improved outcomes and efficiency in the SPIF technique [4]. This procedure has ability to deliver complex parts, no need of pass on and restrictive device set up [10]. Also, ISMF can be performed on a material, for example, steels, polymers, titanium, thermoplastics and so forth. Despite the fact that the procedure requires higher assembling time when contrasted with ordinary framing process, it is reasonable for delivering complex parts in little clusters, for example, model parts for aeronautical, car and modern applications [2]. In this procedure a little tool moves in a progression around external fringe of the part, venturing down between each pass [1]. This procedure requires tool and fixture. This process is perfect for prototyping and low batch creation. The procedure is appeared in Figure 49.1.

---

[a]shraddha.adewar@zealeducation.com, [b]dranandbewoor@gmail.com

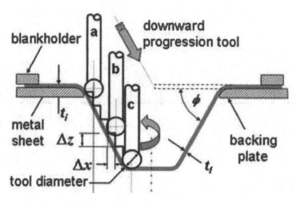

Figure 49.1 SPIF process [3].

In this work, process parameters are investigated to experimentally evaluate their effect on SR variation [7]. A predictive model including the optimum parameter levels that decrease developed parts' SR has been developed and validated. Each process parameter was investigated at four different levels [8]. Surface roughness measurement techniques. The measurement of the forming angle in this context is based on a tangent line drawn from the unformed blank to the deformed part of the surface. This measurement is determined using the sine law. The forming angle provides valuable information about the deformation of the material during the incremental sheet forming process and helps in understanding the extent of shape change achieved in the final formed part.

$$tf = ti \sin (\pi/2-) ti \sin \alpha, \qquad (1)$$

where ti and tf are the initial and final thicknesses of the part, respectively, and α is the semi cone angle [17]. Several industries utilize simple parameters such as SR (surface roughness) to assess surface quality. In Figure 49.2, the average roughness is graphically represented as the area between the irregular roughness profile and its mean value [11]. Another way to describe this is by integrating the absolute value of the roughness height over the sampling length. Mathematically, Ra, which stands for arithmetic average roughness, is defined as:

$$Ra = \frac{1}{L}\int_0^L Y(x)dx \quad [18] \qquad (2)$$

where
Ra is the arithmetic average roughness
Y is co-ordinate of the profile curve
L is the sampling length
Surface roughness (SR) can be estimated or approximated on specimens through various means, including visual inspection and tactile examination using fingertips. Additionally, specialized instruments such as microscopes, profile tracing instruments, or stylus instruments can be employed to obtain more precise measurements of surface roughness. These instruments provide quantitative data and detailed analysis of surface texture, enabling a more accurate characterization of the surface roughness of the material or component being evaluated.

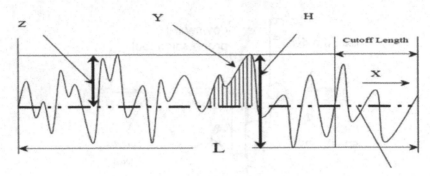

Figure 49.2 SR profile [16].

SPIF Control Parameter

These SPIF parameters and their qualities are appeared Table 49.4. Improvement of parameters is finished with sign to-noise (S/N) proportion according to Taguchi strategy [5]. The bigger the rate association of the parameter to the absolute contrast, bigger is the capacity of that parameter to impact S/N proportion [12]. The ideal blend of factors is acquired from greatest estimation of mean S/N proportion for separate parameters. More the S/N proportion better is the presentation [6].

Table 49.1: SPIF parameter.

Sr no	Factors	Tool Diameter	Spindle Speed	Table Feed	Step down depth
1	Unit	mm	rpm	mm/min	mm
2	Symbol	d	N	f	z
3	Low	1	1000	1000	0.5
4	Medium	2	2000	1500	0.75
5	High	3	3000	2000	1

Design of Experiment

The dimensions of all proposed info factors have been chosen to make process progressively sensible, possible and significant. Considering 4 variables and 3 levels, consequently L9 orthogonal array is chosen. The test is completed to check the impact of spindle speed (N), table feed (f), step down depth (z), tool diameter (d) on surface roughness [13, 14].

Table 49.2: Level of control parameter.

Parameters	Notation	Unit	Original L1	L2	L3	Code L1	L2	L3
Tool Diameter	A	mm	1	2	3	1	2	3
Spindle Speed	B	rpm	1000	2000	3000	1	2	3
Table Feed	C	mm/rev	1000	1500	2000	1	2	3
Step down depth	D	mm/min	0.5	0.75	1	1	2	3

Figure 49.3 shows the experimental setup used in the study. The sheet metal used was Ti-0.05 mm. In these experiments, a forming tool made of cylindrical high-speed steel with a hemispherical head was used. The surface roughness of the formed parts was measured using a "Mar Surf M 400" machine. The Mar Surf M 400 is a specialized instrument designed for accurately measuring surface roughness, providing precise and reliable data for evaluating the quality and characteristics of the formed surfaces in the study.

Figure 49.3 Experimental set up.

Data Analysis

Surface roughness variation in incremental sheet forming (ISF) is analyzed using three distinct statistical criteria: (i) ANOVA (analysis of variance), (ii) general response table with an estimated effect graph, and (iii) main effect plot. These statistical methods aid in understanding and quantifying the influence of various factors on surface roughness during the ISF process. By employing these criteria, researchers can gain valuable insights into the key factors affecting surface roughness and make informed decisions to optimize the ISF process for improved surface quality. Table 49.5 showcases the correlation between surface roughness, maximum forming angle, and their respective signal-to-noise ratios concerning the manipulation of various parameters, including tool diameters, table feeds, step depths of incrementally formed shaped sheets, and spindle speeds [9]. The table provides essential data for evaluating how these parameters influence surface roughness and forming angles during the incremental forming process. By analyzing the signal-to-noise ratios, researchers can gain insights into the experimental results' variations and trends, enabling the optimization and enhancement of the forming process to suit specific material and geometric requirements. It is observed that as the step depths and table feed rates increase, there is a noticeable decrease in surface roughness [5]. The feed rate is that per unit time (mm/min) the distance travelled by shaping the instrument over the sheet surface. Additional Taguchi analysis for surface roughness has been done. Table 49.3 depicts the variation of surface roughness [15]. Excellent signal to noise ratios for various table feeds, step depths, progressively created shaped sheets, spindle speeds, and tool diameters. It is noted that with increasing depth measures and increasing table feed rates, there is a decrease in roughness. The feed rate is that per unit time (mm/min) the distance travelled by shaping the instrument over the sheet surface. Additional Taguchi analysis for surface roughness has been done [19].

Table 49.3: Experimental data.

Expt No.	Tool diameter	Spindle Speed	table feed	Step down depth	Average surface roughness	S/N ratio (Surface roughness)
1	1	1000	1000	0.5	3.1	−9.82723
2	1	2000	1500	0.75	1.172	−1.37855
3	1	3000	2000	1	1.286	−2.18482
4	2	1000	1500	1	2.364	−7.47295
5	2	2000	2000	0.5	1.44	−3.16725
6	2	3000	1000	0.75	2.352	−7.42875
7	3	1000	2000	0.75	1.268	−2.06239
8	3	2000	1000	1	1.12	−0.98436
9	3	3000	1500	0.5	1.633	−4.25972

Response Table and Response Graph of Surface Roughness

Tables are used to make the interpretation of the experimental findings simpler. The table displays a range of values for each factor, from lowest to highest, and then reveals how that factor influenced the final result. As a measure, SR. The response graph, as seen in Figure 49.4a, is used in combination with the response table study to help set the appropriate levels for process parameters.

Effect of Process Variables on Surface roughness

Table 49.4: Signal to noise ratios for surface roughness.

Level	Tool diameter	Speed	Feed	Step Depth
1	−4.464	−6.454	−6.080	−5.751
2	−6.023	−1.843	−4.370	−3.623
3	−2.435	−4.624	−2.471	−3.547
Delta	3.587	4.611	3.609	2.204
Rank	3	1	2	4

Table 49.5: Response parameters for surface roughness.

Level	Tool diameter	Speed	Feed	Step Depth
1	1.853	2.244	2.191	2.058
2	2.052	1.244	1.723	1.597
3	1.340	1.757	1.331	1.590
Delta	0.712	1.000	0.859	0.468
Rank	3	1	2	4

Surface roughness exhibits a decreasing trend whenever the spindle speed is increased from 1000 to 2000 rpm, but it then increases again beyond 2000 rpm. Additionally, an increase in tool diameter from 1 mm to 2 mm leads to an increase in surface roughness, while further increasing the tool diameter from 2 mm to 3 mm results in a reduction of surface roughness.

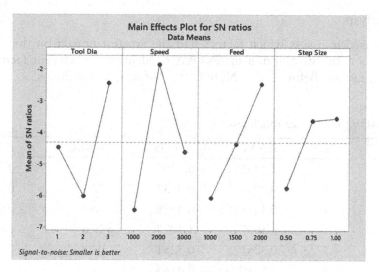

Figure 49.4a S/N& Means graph for average surface roughness.

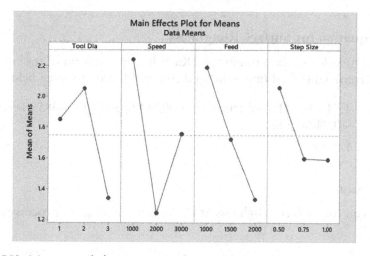

Figure 49.4b S/N& Means graph for average surface roughness.

In Figure 49.4b, the relationship between surface roughness and spindle speed is depicted. The graph shows that as the spindle speed is increased from 1000 to 2000 rpm, there is a notable decrease in surface roughness. However, beyond 2000 rpm, the surface roughness starts to increase. This trend suggests that there exists an optimal spindle speed range where surface roughness is minimized, and exceeding this range leads to a rise in surface roughness. By identifying and understanding this relationship, manufacturers can optimize the spindle speed during the machining process to achieve the desired surface quality for the material being worked on. Furthermore, an increase in tool diameter from 1 mm to 2 mm results in higher surface roughness, while a further increase in tool diameter from 2 mm to 3 mm leads to a reduction in surface roughness. The graph clearly depicts these trends in surface roughness with varying spindle speed and tool diameter.

ANOVA Analysis

ANOVA was used to assess which factors had the greatest impact on the final SR. The F-test is used as a cut-off criterion in ANOVA to identify the relevant factors from the non-significant factors. Below is an ANOVA study of surface roughness.

Table 49.9: ANOVA for surface roughness.

Source	DF	Adj SS	Adj MS	F-Value	P-Value
Regression	4	2.1852	0.5463	1.31	0.399
Tool Dia	1	0.3937	0.3937	0.95	0.0386
Speed	1	0.3558	0.3558	0.85	0.008
Feed	1	1.1077	1.1077	2.66	0.0178
Step Size	1	0.3281	0.3281	0.79	0. 425
Error	4	1.6650	0.4163		
Total	8	3.8503			

Regression Equation for Surface Roughness

Regression equation for surface roughness (Ra value) in the form of tool diameter (mm), spindle speed (rpm), table feed (mm/min) and step size (mm) are given below.

$$\text{Ra Values} = 4.74 - 0.331 \text{ Tool Dia.} - 0.0000244 \text{ Speed} - 0.000859 \text{ Feed} - 0.94 \text{ Step Size (3)}$$

$$= 1.0402$$

Confirmation Run

Confirmation run for surface roughness at the optimal settings of process parameters are A3B2C3D3.

Table 49.10: Confirmation run.

Sr No	Response Parameters	Optimum Conditions	Optimal set of uncoded parameter	Predicted optimal value	Exp. Nr	Actual value	Avg. Result	Error %
1	Surface Roughness	A3B2C3D3	3, 2000, 2000, 1	0.826	1	1.14	1.15	10.3
					2	1.16		
					3	1.15		

Confirmation Test

The last phase of the experimentation process involves confirmation testing, where the predicted results and the average of the experimental findings at specific parameter levels are validated and verified. This step ensures the reliability and accuracy of the obtained data, thus establishing a solid basis for drawing conclusions and making informed decisions based on the experimental outcomes. During the confirmation run, the actual experiments are carried out, and the results are compared to the predicted values. The outcome of the confirmation run indicated a remarkable accuracy of 90% in matching the predicted results, signifying the reliability and effectiveness of the experimental findings.

Conclusion

The main aim of this paper is to achieve optimization in both surface roughness and forming angle of pure Titanium sheets, which have a thickness of 0.05 mm, through the utilization of the Single Point Incremental Forming (SPIF) process. The investigation focuses on analyzing the impact of several parameters, such as spindle speed, tool diameter, stepdown depth, and table feed, in order to identify the optimal combination that produces the desired results. By studying these parameters, the paper seeks to enhance the quality and efficiency of the SPIF process for forming Titanium sheets with improved surface characteristics and forming angles. The experimental design follows an orthogonal array (L9) to conduct the process. The findings indicate that an increase in tool diameter enhances formability, but it negatively impacts surface quality when incremented from 1 to 2 mm. However, when the diameter of tool is further increased to 2 mm, surface quality improves. Regarding speed of spindle, an increase leads to reduced formability. When the spindle speed is increased from 1000 to 2000 rpm, there is an improvement in the quality of the surface. This means that increasing the spindle speed within this range leads to a smoother and better-finished surface for the material being processed. The higher speed likely results in reduced imperfections and improved surface characteristics, which can be beneficial in various manufacturing and machining applications. However, when the spindle speed is further increased beyond 2000 rpm, surface quality starts to deteriorate. The observation suggests that an increase in feed rate results in decreased formability. Conversely, a further increase in feed rate improves the surface quality of the sheet. Lastly, it is concluded that increasing the step size enhances formability in the SPIF process. Overall, the study aims to identify the optimal parameter levels to achieve the desired surface roughness and forming angle for pure Titanium sheets during the SPIF process.

References

1. Verbert, J., Behera, A. K., Lauwers, B., and Duflou, J. R. (2011). Multivariate adaptive regression splines as a tool to improve the accuracy of parts produced by FSPIF. *Key Engineering Materials,* 473, 841–846.
2. Leszak, E. (1967). Apparatus and process for incremental die less forming. Patent US3342051A.
3. Jeswiet, J., Micari, F., Hirt, G., Bramley, A., Duflou, J., and Allwood, J. (2005). Asymmetric single point incremental forming of sheet metal. *CIRP Annals—Manufacturing Technology,* 54, 623–649.
4. Marabuto, S., Afonso, D., Ferreira, J., Melo, F., Martins, M. A., and de Sousa, R. (2011). Finding the best machine for SPIF operations—a brief discussion. In Key Engineering Materials. Trans Tech Publication, pp. 861–868.

5. Duflou, J. R., Verbert, J., Belkassem, B., Gu, J., Sol, H., Henrard, C., Adewar, S., and Bewoor, A. (2008). Process window enhancement for single point incremental forming through multi-step toolpaths. *CIRP Annals—Manufacturing Technology, 57*, 253–256.
6. Malhotra, R., Xue, L., Belytschko, T., and Cao, J. (2012). Mechanics of fracture in single point incremental forming. *Journal of Materials Processing Technology, 212*, 1573–1590.
7. Fang, Y., Lu, B., Chen, J., Xu, D., and Ou, H. (2014). Analytical and experimental investigations on deformation mechanism and fracture behavior in single point incremental forming. *Journal of Materials Processing Technology, 214*, 1503–1515.
8. Duchêne, L., Guzmán, C. F., Behera, A. K., Duflou, J. R., and Habraken, A. M. (2013). Numerical simulation of a pyramid steel sheet formed by single point incremental forming using solid-shell finite elements. In Key Engineering Materials. Trans Tech Publication, (vol. 549, pp. 180–188).
9. Behera, A. K., Lauwers, B., and Duflou, J. R. (2014). Tool path generation framework for accurate manufacture of complex 3D sheet metal parts using single point incremental forming. *Computers in Industry, 65*, 563–584.
10. Behera, A. K., Verbert, J., Lauwers, B., and Duflou, J. R. (2013). Tool path compensation strategies for single point incremental sheet forming using multivariate adaptive regression splines. *Computer-Aided Design, 45*, 575–590.
11. Araujo, R., Teixeira, P., Montanari, L., Reis, A., Silva, M. B., and Martins, P. A. (2014). Single point incremental forming of a facial implant. *Prosthetics and Orthotics International, 38*, 369–378.
12. Essa, K., and Hartley, P. (2011). An assessment of various process strategies for improving precision in single point incremental forming. *International Journal of Material Forming, 4*, 401–412.
13. Martins, P., Kwiatkowski, L., Franzen, V., Tekkaya, A., and Kleiner, M. (2009). Single point incremental forming of polymers. *CIRP Annals—Manufacturing Technology, 58*, 229–32.
14. Darpanah Martins, P., Bay, N., Skjødt, M., and Silva, M. (2008). Theory of single point incremental forming. *CIRP Annals—Manufacturing Technology, 57*, 247–252.
15. Silva, M., Alves, L., and Martins, P. (2010). Single point incremental forming of PVC: experimental findings and theoretical interpretation. *European Journal of Mechanics A/Solids, 29*, 557–566.
16. Jackson, K., Allwood, J., and Landert, M. (2008). Incremental forming of sandwich panels. *Journal of Materials Processing Technology, 204*, 290–303.
17. Mohammadi, A., Vanhove, H., Attisano, M., Ambrogio, G., and Duflou, J. R. (2015). Single point incremental forming of shape memory polymer foam. In MATEC Web of Conferences, EDP Sciences, 04007.
18. Echrif, S. B. M., and Hrairi, M. (2014). Significant parameters for the surface roughness in incremental forming process. *Materials and Manufacturing Processes, 29*, 697–703.
19. Hera, A. K., Lauwers, B., and Duflou, J. R. (2013). Tool path generation for single point incremental forming using intelligent sequencing and multi-step mesh morphing techniques. *Key Engineering Materials, 554–557*, 1408–1418. http://dx.doi.org/10.4028/www.scientific.net/KEM.554-557.1408.

50 Structural design optimization using grey wolf optimizer

Bhavik D. Upadhyay[1,a], Sunil S. Sonigra[2,b], and Sachin D. Daxini[3,c]

[1]Department of Mechanical Engineering, Shantilal Shah Engineering College, Bhavnagar, Gujarat, India

[2]Department of Mechanical Engineering, Government Polytechnic, Rajkot, Gujarat, India

[3]Department of Mechanical Engineering, Babaria Institute of Technology, Vadodara, Gujarat, India

Abstract: Selection of appropriate optimization technique for structural optimization problems plays vital role as these problems involve nonlinear objective function, complex behavior constraints and many design variables. Metaheuristics have become popular in solving such complex and real world optimization problems in last two decades. In the present study, the promising features of the gray wolf optimizer (GWO), a metaheuristic approach, inspired by the social hierarchy and hunting behavior of the grey wolves, are discussed by solving several benchmarked constrained structural optimization problems. To handle behavior and geometric constraints, static penalty method has been used. The performance of the GWO algorithm is compared with the results obtained by the various other meta-heuristic optimization algorithms. In most of the cases, the solutions obtained by GWO are better or equally competent in reference to other methods.

Keywords: GWO, meta-heuristics, structural optimization.

Introduction

Structural optimization has remained an active research field since last five decades as it contributes in developing improved product, reduced cost of the material and saving product development time. Different structural optimization problems deal with different aspects of product like size, shape and topology. Typical structural optimization problem involves nonlinear objective function, nonlinear constraints and many design variables. The constraints are problem specific and defined in terms of allowable stress, displacement, loading and geometric configurations etc. Such nonlinear problems are difficult to deal with traditional gradient-based local search algorithms [1]. Modern metaheuristic algorithms have proved quite useful in solving such complex optimization problems. Metaheuristics are developed based on the successful biological systems and the features of nature. The objective of these algorithms is to find near global optimal solution within reasonable time. Since last two decades, these algorithms are becoming popular in solving nonlinear structural optimization problems [1–3]. Metaheuristics can be categorized in many ways. For instance, based on working they are categorized into single solution based approach and population based approach while on the basis of concept they can be classified into evolutionary algorithms, swarm intelligence based and process based algorithms.

[a]bhavikadit@yahoo.co.in, [b]sssonigra@gmail.com, [c]sachin_daxini@rediffmail.com

GWO is a relatively new swarm intelligence based algorithm proposed by Mirjalili in 2014 [4]. The algorithm mimics intelligent behavior of grey wolves in searching, encircling and hunting the prey. GWO algorithm has several interesting features like ease of implementation, simple concept and less number of algorithm specific parameters to tune [14]. The improved versions of GWO has been presented in some recent articles which improves balancing between local and global search through dimension learning based hunting (DLH) and memory based techniques [5, 6].

The present work employs standard GWO algorithm to solve some benchmark constrained structural optimization problems and results are compared with other stochastic optimization techniques of past literature. For handling behavior constraints e.g. stress and deflections, static penalty approach has been adopted for simplicity and ease of use. The rest of the presentation of an article is organized as follows: Section 2 presents gray wolf optimizer. Section 3 presents numerical examples followed by concluding remarks in Section 4.

GWO

Grey wolves are the apex predators and considered at the upper level of food cycle. They live in a pack and the group size varies from 5 to 15. In a pack, there exist a hierarchy which include α, β, δ and ω wolves. The hierarchy helps in maintaining overall discipline and other activities like hunting. α wolves are the leaders and decision makers in the pack. The second category of wolves in sequence is β wolves that help α wolves in their task of decision making and other activities. The ω wolves are ranked lowest in the pack and accordingly get minimum importance. The wolves which are not α, β or ω, are categorized as δ wolves. They are superior to ω but ranked after α and β. The optimal solution to the underlying problem is determined by GWO through the hunting mechanism of grey wolves. The leading aspects of their hunting behavior include:

- Tracking, chasing and approaching the prey
- Encircling and harassing the prey until it stops its movement
- Attacking the prey

Working of GWO

Like other population based stochastic algorithms, the search process in GWO starts with initial random population of grey wolves with their random positions. Each wolf represents the target solution and their positions are updated by measuring the distances from the target prey position. In their mathematical model, Mirjalili et al. [4] considered the fittest solution as α, while second and third best solutions are referred as β and δ respectively. The mathematical model for working of GWO is briefly explained below:

Encircling the Prey

As mentioned in the hunting behavior above, grey wolves encircle the prey. Mathematically, this behavior is modeled as below:

$$\vec{D} = \left| \vec{C}\,\vec{X}_p(t) - \vec{X}(t) \right| \tag{2.1}$$

$$\vec{X}(t+1) = \vec{X}_p(t) - \vec{A}.\vec{D} \tag{2.2}$$

where, $\vec{X}_p$ and $\vec{X}$ are the position vectors for the prey and grey wolf respectively, $\vec{A}$ and $\vec{C}$ are the coefficient vectors and t represents the current iteration. The values of vectors $\vec{A}$ and C are computed as below:

$$\vec{A} = 2\vec{a}\vec{r}_1 - \vec{a}$$
$$\vec{C} = 2\vec{r}_2 \tag{2.3}$$

where, r_1 and r_2 are the two random vectors in [0, 1] while vector a linearly decrease from 2 to 0 over the iterations.

Hunting

α wolves guide the pack for hunting while β and δ help in decision making. Mathematically, the hunting behavior is simulated by considering the α as best solution, and β and δ as the other two best candidate solutions respectively. Accordingly, the first three best solutions obtained are memorized and other wolf positions are renovated as per the position of best search agents. Following expressions are used to mathematically model the same:

$$\vec{D}_\alpha = \left| \vec{C}_1 \vec{X}_\alpha - \vec{X} \right|$$
$$\vec{D}_\beta = \left| \vec{C}_2 \vec{X}_\beta - \vec{X} \right| \tag{2.4}$$
$$\vec{D}_\delta = \left| \vec{C}_3 \vec{X}_\delta - \vec{X} \right|$$

$$\vec{X}_1 = \vec{X}_\alpha - \vec{A}_1 \vec{D}_\alpha$$
$$\vec{X}_2 = \vec{X}_\beta - \vec{A}_2 \vec{D}_\beta \tag{2.5}$$
$$\vec{X}_3 = \vec{X}_\delta - \vec{A}_3 \vec{D}_\delta$$

The other wolves' positions are updated based on the position of best search agents as follows:

$$\vec{X}(t+1) = \frac{\vec{X}_1 + \vec{X}_2 + \vec{X}_3}{3} \tag{2.6}$$

Attacking the Prey

The hunting operation ends with attacking the prey when it stops its movement. It is approached mathematically by decreasing the value of $\vec{a}$ and accordingly $\vec{A}$ will also decrease whose value varies [$-a$, a]. For exploration, the grey wolves must diverge from each other. Mathematically it is performed with $\vec{A}$ value greater than 1 or less than -1. Another parameter which assists exploration in GWO is $\vec{C}$ whose value varies [0, 2].

General structure of GWO algorithm is as follows:

> Create the pack of grey wolf
> Initialization of the wolves position
> Calculate the fitness of the each wolf
> Record the position of α-wolves, β-wolves and δ-wolves

Repeat
   Update the random parameters
   Calculate the fitness of all the wolves
   Update the position of wolves:
      α-wolves (the best solution),
      β-wolves, (the second best solution)
      δ-wolves (the third best solution)
Until
   Convergence criterion reached

Numerical Examples

Case: I Tubular Column Design

Problem statement: Figure 50.1 shows tubular column section which carries a load of 2500 kgf compressive in nature. The column is joined by means of hinged joints at both the ends. The column length is 250 cm. The mean diameter of the column is kept between 2 cm to 14 cm and the thickness ranged between 0.2 cm to 0.8 cm. The induced stress in the tubular column should be less than the buckling stress and the yield stress. The total cost of the column considering material cost and construction cost can be taken as 5W + 2d, where W is weight in kgf and d is mean diameter of the column in cm. To minimize the cost of the component is taken as objective of the problem.

The material property of the column is given below:

Yield stress = 500 kgf/cm^2;
Elastic Modulus = 0.85×10^6 kgf/cm^2;
Weight density = 0.0025 kgf/cm^3;

Objective Function: Minimization

$$f(x) = f(x_1, x_2) = 9.82\, x_1 x_2 + 2x_1 \tag{3.1}$$

Where,

$$X = \begin{Bmatrix} x_1 \\ x_2 \end{Bmatrix} = \begin{Bmatrix} d \\ t \end{Bmatrix} \tag{3.2}$$

Constraints:

$$c_1 = \frac{2500}{\pi x_1 x_2} - 500 \leq 0 \tag{3.3}$$

$$c_1 = \frac{2500}{\pi x_1 x_2} - \frac{\pi^2 \left(0.85 \times 10^6\right)\left(x_1^2 + x_2^2\right)}{8(250)^2} \tag{3.4}$$

$$c_3 = \frac{2}{x_1} - 1 \leq 0 \tag{3.5}$$

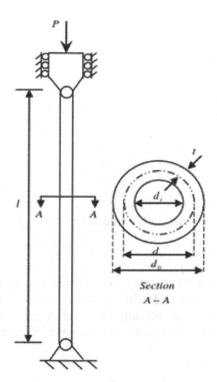

Figure 50.1 Tubular column design.

$$c_4 = \frac{x_1}{14} - 1 \le 0 \tag{3.6}$$

$$c_5 = \frac{0.2}{x_2} - 1 \le 0 \tag{3.7}$$

$$c_6 = \frac{x_2}{0.8} - 1 \le 0 \tag{3.8}$$

Range:

$$2 \le x_1 \le 14$$
$$0.2 \le x_2 \le 0.8$$

Table 50.1: Comparison of the results for tubular column problem.

	Fuzzy PD [7]	*Rao [8]*	*CS [1]*	*Present Study*
d (mean diameter)	5.4507	5.44	5.45139	5.4511
t (thickness)	0.292	0.293	0.29196	0.29204
$F_{min.}$	25.5316	26.5323	26.53217	26.5351

The convergence study for the optimum solution for the tubular column structure is shown in Figure 50.2 and Table 50.1.

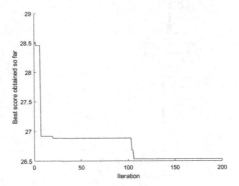

Figure 50.2 Convergence diagram for tubular column structure.

Case: II Cantilever Beam

Problem statement: Figure 50.3 presents stepped cantilever beam (hollow with constant thickness) of square cross section and subjected to a vertical load at free end of the beam structure. The problem is of weight minimization and design variables are height of different beam sections. The thickness of the beam elements is fixed as t = 2/3.

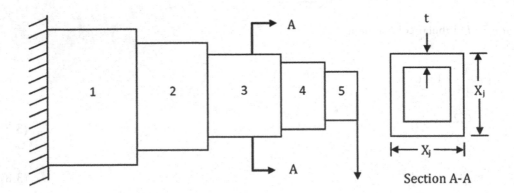

Figure 50.3 Cantilever beam.

Objective Function: Minimization

$$f(x) = f(x_1, x_2, x_3, x_4, x_5) = 0.0624(x_1 + x_2 + x_3 + x_4 + x_5) \tag{3.9}$$

Constraints:

$$C = \frac{61}{x_1^3} + \frac{37}{x_2^3} + \frac{19}{x_3^3} + \frac{7}{x_4^3} + \frac{1}{x_5^3} - 1 \leq 0 \tag{3.10}$$

Range:

$$0.01 \leq x_j \leq 100$$

Table 50.2: Comparison of the results for cantilever beam problem.

	GCA [9]	MMA [10]	CS [1]	Present Study
X_1 (Element – 1)	6.0100	6.0100	6.0089	6.0465
X_2 (Element – 2)	5.3040	5.3000	5.3049	5.2891
X_3 (Element – 3)	4.4900	4.4900	4.5023	4.4449
X_4 (Element – 4)	3.4980	3.4900	3.5077	3.5429
X_5 (Element – 5)	2.1500	2.1500	2.1504	2.1544
$F_{min.}$	1.3400	1.3400	1.33999	1.3402

The convergence study for the optimum solution for the cantilever beam is shown in Figure 50.4 and Table 50.2.

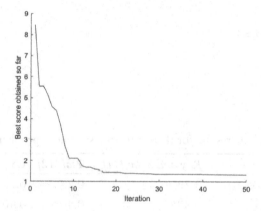

Figure 50.4 Convergence diagram for cantilever beam.

Case: III Three Bar Truss

Problem statement: Figure 50.5 presents three bar truss. The objective function is to find the optimum cross sectional area to minimize the volume of the three bar truss structure. Each member length of the structure is 100 cm. The structure is subjected to load (F) of 2 kN/cm² and the permissible stress (σ) is not exceed to 2 kN/cm².

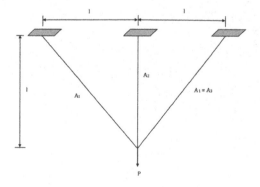

Figure 50.5 Three bar truss.

Objective Function: Minimization

$$f(x) = f(x_1, x_2) = \left(2\sqrt{2}x_1 + x_2\right) \times l \tag{3.11}$$

Where,

$$X = \begin{Bmatrix} x_1 \\ x_2 \end{Bmatrix} = \begin{Bmatrix} A_1 \\ A_2 \end{Bmatrix} \tag{3.12}$$

Constraints:

$$c_1 = \frac{\sqrt{2}x_1 + x_2}{\sqrt{2}x_1^2 + 2x_1x_2} F - \sigma \leq 0 \tag{3.13}$$

$$c_2 = \frac{x_2}{\sqrt{2}x_1^2 + 2x_1x_2} F - \sigma \leq 0 \tag{3.14}$$

$$c_3 = \frac{1}{x_1 + \sqrt{2}x_2} F - \sigma \leq 0 \tag{3.15}$$

Range:

$$0 \leq x_1, x_2 \leq 1$$

Table 50.3: Comparison of the results for three bar truss structural problem.

	Ray and Saini [11]	*Tsai [12]*	*CS [1]*	*Present Study*
A$_1$(C/S area of member 1 & 3)	0.795	0.788	0.78867	0.78776
A$_2$(C/S area of member 2)	0.395	0.408	0.40902	0.41085
F$_{min.}$	264.3	263.68	263.9715	263.897

The convergence study for the optimum solution for the three bar truss is shown in Figure 50.6 and Table 50.3.

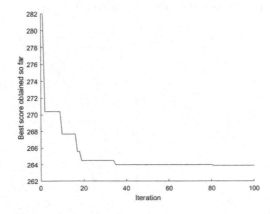

Figure 50.6 Convergence diagram for three bar truss.

Case: IV Corrugated Bulkhead Design

The problem of corrugated bulkhead [13] for a tanker is a minimum weight design problem in which the variables are width, height, length and thickness.

Objective Function: Minimization

$$f(x) = f(x_1, x_2, x_3, x_4) = \frac{5.885x_4(x_1 + x_3)}{x_1 + \sqrt{x_3^2 - x_2^2}} \tag{3.16}$$

Where,

$$X = \begin{Bmatrix} x_1 \\ x_2 \\ x_3 \\ x_4 \end{Bmatrix} = \begin{Bmatrix} b \\ h \\ l \\ t \end{Bmatrix} \tag{3.17}$$

Constraints:

$$c_1 = 8.94\left(x_1 + \sqrt{(x_3^2 - x_2^2)}\right) - x_4 x_2\left(0.4x_1 + \frac{x_3}{6}\right) \le 0 \tag{3.18}$$

$$c_2 = 2.2\left(8.94\left(x_1 + \sqrt{(x_3^2 - x_2^2)}\right)\right)^{\frac{4}{3}} - x_4 x_2\left(0.2x_1 + \frac{x_3}{12}\right) \le 0 \tag{3.19}$$

$$c_3 = 0.0156x_1 - x_4 + 0.15 \le 0 \tag{3.20}$$

$$c_4 = 0.0156x_3 - x_4 + 0.15 \le 0 \tag{3.21}$$

$$c_5 = 1.05 - x_4 \le 0 \tag{3.22}$$

$$c_6 = x_2 - x_3 \le 0 \tag{3.23}$$

Range:

$$0 \le x_1, x_2, x_3 \le 100$$
$$0 \le x_4 \le 5$$

Table 50.4: Comparison of the results for corrugated bulkhead problem.

	CS [1]	Present Study
b (width)	37.1179498	37.7407
h (depth)	33.0350210	33.1913
l (length)	37.1939476	37.7456
t (plate thickness)	0.7306255	0.739053
$F_{min.}$	5.894331	5.8928

The convergence study for the optimum solution for the corrugated bulkhead design is shown in Figure 50.7 and Table 50.4.

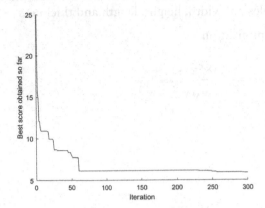

Figure 50.7 Convergence diagram for corrugated bulkhead problem.

Case: V Spring Design

Problem statement: Figure 50.8 presents axially loaded tension/compression helical spring. The spring is subjected to varied constraints like surge frequency, deflection and shear stress. The design variables are wire diameter d, mean coil diameter D and number of active turns N. The objective of the problem is to minimize the weight of the spring.

Objective Function: Minimization

$$f(x) = f(x_1, x_2, x_3) = (x_3 + 2) x_1 x_2^2 \qquad (3.24)$$

Where,

$$X = \begin{Bmatrix} x_1 \\ x_2 \\ x_3 \end{Bmatrix} \begin{Bmatrix} D \\ d \\ N \end{Bmatrix} \qquad (3.25)$$

Constraints:
For deflection-

$$c_1 = 1 - \frac{x_1^3 x_3}{71785 x_2^4} \le 0 \qquad (3.26)$$

For shear stress-

$$c_2 = \frac{4 x_1^2 - x_1 x_2}{12566 \left(x_1 x_2^3 - x_2^4 \right)} + \frac{1}{5108 x_2^2} - 1 \le 0 \qquad (3.27)$$

For frequency-

$$c_3 = 1 - \frac{140.45 x_3}{x_2 x_3} \le 0 \qquad (3.28)$$

For outer diameter-

$$c_4 = \frac{x_1 + x_2}{1.5} - 1 \leq 0 \qquad\qquad (3.29)$$

Range:

$0.25 \leq x_1 \leq 1.3$
$0.05 \leq x_2 \leq 2$
$2 \leq x_3 \leq 15$

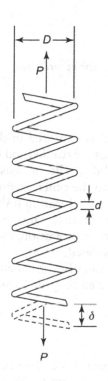

Figure 50.8 Coil Spring.

Table 50.5: Comparison of the results for coil spring problem.

	SA [11]	*Belegundu [15]*	*Coello [16]*	*Arora [17]*	*Present Study*
D (Coil dia.)	0.3215	0.315900	0.351661	0.399180	0.39261
d (wire dia.)	0.0504	0.05	0.05148	0.053396	0.05314
N (No. of turns)	13.979	14.25	11.63220	9.185400	9.4767
$F_{min.}$	0.01306	0.0128334	0.012704	0.127302	0.01272

The convergence study for the optimum solution for the coil spring is shown in Figure 50.9 and Table 50.5.

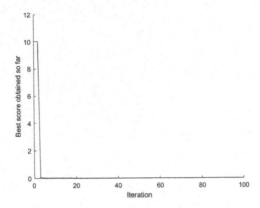

Figure 50.9 Convergence diagram for coil spring problem.

Conclusion

In the present study, GWO algorithm is explored in the field of structural optimization. Various structural benchmark problems are studied and GWO algorithm has been employed to validate its capability and compatibility. The analysis reveals that the GWO algorithm is found competent when compared with the other existing optimization algorithms. This is mostly due to the factors considered as exploration ($|A|>1$) to find the global optimum and exploitation ($|A|<1$) towards the best solution. During exploration, the wolves are forced to diverge from the prey and during the exploitation, the wolves attack the prey. The search mechanism of the prey is also well modeled.

The capability of GWO algorithm can also be enhanced in several ways to improve the results in global means and also to be employed in case of unconstrained structural optimization problems.

References

1. Gandomi, A. H., Yang, X. S., and Alavi, A. H. (2013). Cuckoo search algorithm: a metaheuristic approach to solve structural optimization problems. *Engineering with Computers*, 29(1), 17–35.
2. Durgun, İ., and Yildiz, A. R. (2012). Structural design optimization of vehicle components using cuckoo search algorithm. *Materials Testing*, 54(3), 185–188.
3. Perez, R. L., and Behdinan, K. (2007). Particle swarm approach for structural design optimization. *Computers and Structures*, 85(19–20), 1579–1588.
4. Mirjalili, S., Mirjalili, S. M., and Lewis, A. (2014). Grey wolf optimizer. *Advances in Engineering Software*, 69, 46–61.
5. Nadimi-Shahraki, M. H., Taghian, S., and Mirjalili, S. (2021). An improved grey wolf optimizer for solving engineering problems. *Expert Systems with Applications*, 166, 113917.
6. Gupta, S., and Deep, K. (2020). A memory-based grey wolf optimizer for global optimization tasks. *Applied Soft Computing*, 93, 106367.
7. Hsu, Y. L., and Liu, T. C. (2007). Developing a fuzzy proportional–derivative controller optimization engine for engineering design optimization problems. *Engineering Optimization*, 39(6), 679–700.
8. Rao, S. S. (2019). Engineering Optimization: Theory and Practice. 14th ed. John Wiley & Sons.

9. Chickermane, H., and Gea, H. C. (1996). Structural optimization using a new local approximation method. *International Journal for Numerical Methods in Engineering*, 39(5), 829–846.

10. Fleury, C. (1993). Sequential convex programming for structural optimization problems. In Optimization of Large Structural Systems (pp. 531–553). Dordrecht: Springer.

11. Ray, T., and Saini, P. (2001). Engineering design optimization using a swarm with an intelligent information sharing among individuals. *Engineering Optimization*, 33(6), 735–748.

12. Tsai, J. F. (2005). Global optimization of nonlinear fractional programming problems in engineering design. *Engineering Optimization*, 37(4), 399–409.

13. Kvalie, D. (1967). Optimization of plane elastic grillages. PhD Thesis, NorgesTekniskNaturvitenskapeligeUniversitet, Norway.

14. Ravindran, A., Ragsdell, K. M., and Reklaitis, G. V. (2006). Engineering Optimization: Methods and Applications. 2nd ed. NJ: John Wiley & Sons.

15. Belegundu, A. D. (1983). A Study of Mathematical Programming Methods for Structural Optimization. Internal report, Dept. of Civil and Environmental Engg., University of Iowa.

16. Coello, C. A. C. (1999). Self-adaptive penalties for GA based optimization. In Proceedings of the Congress on Evolutionary Computation. IEEE Service Center: Washington D.C. (Vol. 1, pp. 573–580).

17. Arora, J. S. (2004). Introduction to Optimum Design. 3rd edn. Academic Press Publications, Elsevier.

51 Aerodynamic investigation over class 8 heavy vehicle using CFD approach

Satyavan P. Digole[a], Virendra Talele[b], Gaurav Bhale[c], Amit Bhirud[d], Suyash Hulji[e], and Vikrankt Patil[f]

Department of Mechanical Engineering, MIT School of Engineering, MIT ADT University, Loni Kalbhour, Maharashtra India

Abstract: Across the years, the automotive industry showed the range of harmful emission which directly causes a harmful effect to the environment and human health factor. Over the range, heavy utility vehicles contribute around 25–28 % of harmful gas emissions due to the bad practice of vehicle modification which causes more amount of drag generation. Improving vehicle consumption is directly depends on the practice to reduce the amount of drag in the vehicle. In the present investigation watertight class 8 model benchmarked as a TATA LPT 2518, which widely driven for luggage transport in Asia pacific region has been investigated using the CFD approach to predict drag characteristics over the vehicle for four different practical cases such as 1) Vehicle with equal cab and trolley length, 2) Vehicle having larger trolley, 3) Vehicle with Larger Cab and deflector, 4) Vehicle with equal length of Cab and trolley with Deflector. A mesh independence numerical study is also been proposed to suggest standard guidelines in selecting optimum skewness factors for the validation of numerical study.

Keywords: Aerodynamic, CFD, class 8, deflector, mesh study.

Introduction

Heavy commercial vehicles are considered aerodynamically inefficient compared to other ground vehicles due to their un-streamlined body shapes. Unlike many developed countries, in some south Asian countries, India is still using many discontinued model trucks. The automobile industry is currently seeing an increase in demand for fuel-efficient automobiles. The advent of increasingly fuel-efficient cars has exacerbated increased gasoline prices and the growth of This problem. In contrast to other land vehicles, heavy commercial vehicles have inadequate fuel consumption owing to a solid aerodynamic drag. In general, numerous variables influence a vehicle's fuel usage. Aerodynamic resistance is a significant one. Indeed, above 80 kilometers per hour, aerodynamic drag is the primary source of resistance for a truck traveling at a steady altitude [1]. Aerodynamic improvement is a critical technology for fuel economy. A substantial commercial class 8 vehicle running at 100 km/h approximately consumes 50% of the overall fuel efficiency to generate the required strength to counteract aerodynamic drag [2–4]. According to research, the total annual

[a]satyavan.digole@mituniversity.edu.in, [b]virendratalele1@gmail.com, [c]gaurav.g.bhale@gmail.com, [d]bhirud.amit07@gmail.com, [e]virendratalele48@gmail.com, [f]virendratalele49@gmail.com

mileage of a large commercial vehicle is between 130000 and 160000 kilometers [5,6]. Since heavy trailers are transportation medium through thee link of intercity and state, it significant cause increasing of pollution by hampering human health condition. The use of fuel and diesel by road vehicles is estimated to be over 1.3 trillion liters. It often applies to areas with elevated emission levels of (CO_2), a byproduct of the combustion of fossil fuels [7]. Nowadays, a standard benchmark truck comes with the pre-implemented drag reduction devices, as significantly less research is found on an existing scale of models. In the Asian continent, a truck user has found out that a wide variety of customization takes place without knowing its importance on how it causes the fuel economy and mileage of trucks [8–11].

The aerodynamic Investigation to estimate drag coefficient is a helpful tool to predict truck performance under different airspeed scenarios. The modern tool engrossed, such as external flow investigation, using commercial coded CFD software is a primarily great example to understand aerodynamic resistance. Some recent CFD investigations are done to maximize the fuel performance of trucks in proportion to the reduction of drag [12]. They created advanced computational models that incorporate both LES and RANS modeling techniques and examined the tractor-trailer difference's impact [13]. Theoretical fluid dynamics to evaluate various aerodynamic trailer devices and aerodynamically shaped trailers (CFD), to determine their effect on the flow across the truck. Additionally, the drag contribution from multiple regions was measured to determine where the most drag can be gained and devices deployed in the undercarriage and base area produced the most robust performance, indicating that these regions are the most amenable to drag reduction. Further, they demonstrated the critical nature of tractor-trailer integration. The Asian content countries like India where transportation hub through road depends on heavy utility vehicles the demand for increasing aerodynamics is getting more critical. TATA LPT 2518 Truck is widely used in transportation facilities. Still, the fact of its lower benchmarking on aerodynamics efficiency it emits more pollution, In present numerical study class 8 truck model is investigated for four types of cases such as 1) Vehicle with equal cab and trolley height, 2) Vehicle having larger trolley, 3) Vehicle with Larger Cab and deflector, 4) Vehicle with equal length of cab and trolley with deflector to predict the creation of drag in every particular possible modification.

Modeling and Simulation

Pre-Processing

The initial stage of CFD investigation states the creation of a watertight working fluid domain where the enclosure as wind tunnel is created against the solid vehicle with allocated domain condition as shown in Figure 51.1, in which inlet type is kept as pressure-based velocity inlet, the outlet is kept as an ambient outlet with the pressure of 10^5 Pa at 25°C. It is assumed that the vehicle is running at speed of 25 m/s. The Symmetric and lateral boundary of the wind tunnel enclosure is kept as a Solid wall with the no-slip condition [14].

Discretization of Domain

Meshing is one of the important steps of CFD investigation since it proposed the accuracy of the solution. In the present investigation mesh parametric study has been presented with the selection of optimum mesh based on criteria of skewness factor to verify the

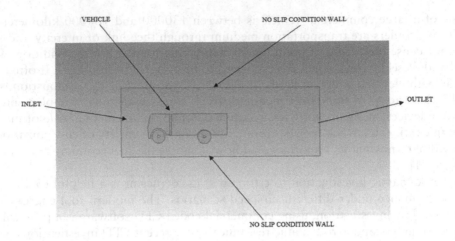

Figure 51.1 Creation of water-tight geometry.

results. A correlative study between program-controlled tetrahedral and polyhedral mesh is proposed to suggest the best guidelines on simulating external flows with less computational power and maintained accuracy. Table 51.1 presents a parametric mesh study on selecting the optimum element based on skewness factor where results are tabulated for case 1) Vehicle with equal cab and trolley height. Table 51.2 shows the comparative results between computational time for Tetrahyderal and Polyhedral mesh, where polyhedral mesh consumes less amount of time to converge solution for same time scale.

Table 51.1 represents the parametric mesh study in a correlation of skewness factor, which verifies the output quality of the solution. The targeted skewness factor was kept as 0.98 for tetrahedral mesh, whereas archived skewness is 0.9538 for element size as 10 mm. Element size 10 mm was selected because it satisfies the condition of maintaining a skewness factor below 0.98. In comparative results with mesh size 5 mm, 10 mm selected

Table 51.1: Mesh parametric study on the selection of optimum mesh size with a relation of skewness factor.

Sr no	Element size (mm)	Total Nodes	Total elements	Skewness Factor	CD
1	40	7093	34187	1	196.32342
2	30	14334	71753	0.99995	87.618695
3	20	35683	181433	0.99703	45.549613
4 (Selected)	10	164238	857969	0.9538	37.656167
5	5	769659	4115680	0.9454	37.952328

Table 51.2: Comparative mesh study between Tetrahedral and polyhedral mesh.

Sr no	Case under Investigation	Type of mesh	Element count	Converged Timescale	Time for convergence
1	Case 1	Tetrahedral	164238	600 Iteration	326 S
2	Case 1	Polyhedral	42368	96 Iteration	96 S

although it has low skewness factor as compare to 5 mm because 5 mm element size creates more amount of denser mesh in the enclosure region due to which solution takes more time to converge as part of output monitored drag characteristics it doesn't affect the solution, based on parametric correlation 10 mm element size is proposed.

The above-presented Table 51.2 determines a detailed comparative study between computational timescale analysis for tetrahedral vs polyhedral mesh criteria in which polyhedral mesh having polygonal faces seems more robust and cost-effective, bringing convergence to the residuals of the solution in less time. The polyhedral mesh covers a 3-Dimensional polygonal shape that occupied the maximum adaptive ratio in a working domain compared to the tetrahedral mesh, due to which polyhedral mesh creates less amount of element count in solver by maintaining the same quality of the solution. Figure 51.2 represents adaptive refine polyhedral mesh around the vehicle.

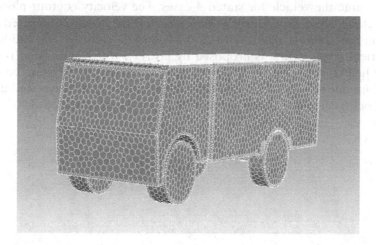

Figure 51.2 Polyhedral mesh.

Numerical Scheme

The presented CFD investigation is performed by using commercial coded Ansys software having Ansys Fluent, where the convergence scheme for implicit solution works on the gauss serial method. The present numerical scheme is based on the external flow around a vehicle in which a pressure-based solver scheme is used along with the Momentum for a type of inlet flow kept is as a velocity inlet. The energy equation is kept off for the solution. The solution is conserved for Mass and Momentum as the Mass of inlet air is assigned as pressure-based, and Momentum is conserved as Velocity-based 25 m/s Inlet condition.

Turbulence Modeling

The standard K-Epsilon turbulence model has been selected for the presented solution because the type of inlet velocity is tabulated as for high-velocity flow of inlet at a maximum achieved the speed of the vehicle in turbulent condition. The standard guideline suggests kinetic dissipation rate of the K-Eps model works better for high-velocity inlet flow [15]. Equation 1 represents mathematical modeling for the K-Epsilon model.

$$\frac{\partial(\rho k)}{\partial t} + \frac{\partial(\rho k u_i)}{\partial x_i} = \frac{\partial}{\partial x_j}\left[\frac{\mu_t}{\sigma k}\frac{\partial k}{\partial x_j}\right] + 2\mu_t E_{ij}E_{ij} - \rho\varepsilon$$

$$\frac{\partial(\rho\varepsilon)}{\partial t} + \frac{\partial(\rho\varepsilon u_i)}{\partial x_i} = \frac{\partial}{\partial x_j}\left[\frac{\mu_t}{\sigma\varepsilon}\frac{\partial\varepsilon}{\partial x_j}\right] + C_{1\varepsilon}\frac{\varepsilon}{k}2\mu_t E_{ij}E_{ij} - C_{2\varepsilon}\rho\frac{\varepsilon^2}{k} \qquad (1)$$

Where,
u_i = Velocity component in the desired direction
E_{ij} = Confined rate of deformation in Cartesian coordinate
μ_t = Eddy viscosity for vortices confinement

Post-Process Validation

The post-processing criteria are set in the CFX post-process capability of Ansys. The post-process scheme targeted to visualize certain creation of overpressure and under pressure regions around the vehicle for stated 4 cases. The velocity contour plots present the region on attachment and detachment of flow which is the need to optimized for reducing pressure region and indirect quantity of drag around the vehicle. The validation for the present numerical Investigation is proposed by [16,17], which states that the practice of extending the height of trolley is directly created on increasing pressure contours in the working medium due to which the quantity of drag suddenly increases and more number of empty air pockets generate at trailing section as shown in Figure 51.3.

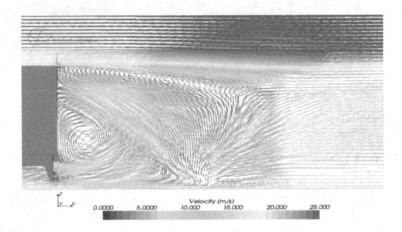

Figure 51.3 Trailing edge air pockets.

Results and Discussion

The CFD investigation has been proposed for the following four types of configurations, the base models for 4 configurations are prepared using inbuilt Ansys design modeler software as shown in Figure 51.4.

Case 1) Vehicle with equal cab and trolley length
In the configuration of case 1, the slant deviation at the front area of the vehicle denoted by subscript "A" is the cause of fluid flow operation and the creation of the overpressure region. Annotation "B" indicates the velocity surface contour at the trailing section for empty air pockets. The contours obtained for the above case are shown in Figure 51.5.

1) Vehicle with equal cab and trolley length

2) Vehicle having the larger trolley

3) Vehicle with cab deflector.

4) Extended trolley with cab deflector

Figure 51.4 Four configurations under CFD investigation.

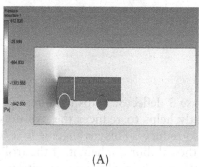

(A)

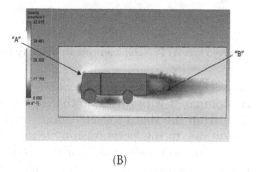

(B)

Figure 51.5 A) Pressure contour B) Velocity contour.

Case 2) Vehicle having the larger trolley

In a typical class 8 LPT 2518 truck, the load-carrying capacity is limited to up 25,000 Kg, but in considering the scenario of the extended load-carrying capacity height of the trolley is mean to be increased. In Figure 51.6, the present pressure and velocity contour show how the extended trolley affects the aerodynamic characteristics of the truck. The increasing drug resistance is the quantity of increasing pressure coefficient around the extended surface of a trolley. As shown in Annotation "A" and "B" shows the early flow separation takes place around the corner of the bluff and subscript "C" shows the increased under pressure Vacuum zone behind the truck, which pulls it backward causing empty air pockets creating uneven downforce. As compared to case 1 case 2 causes more amount of empty air pockets which pull the truck backward.

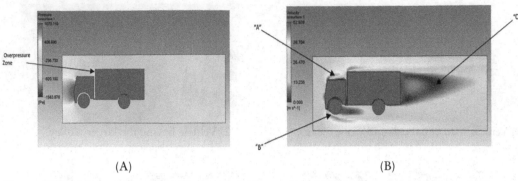

(A)

Figure 51.6 A) Pressure contour.   *Figure 51.6* B) Velocity contour.

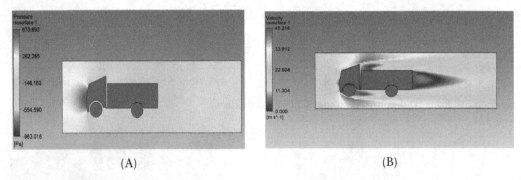

(A)   (B)

Figure 51.7 A) Pressure contour B) Velocity contour.

Case 3) Vehicle with Cab deflector.

As compared with case 2, in the configuration of case 3 deflector helps to re-match inlet flow from the sharp corner region which significantly helps to reduces the overpressure region due to this phenomenon creation of drag also reduces. This arrangement is based on modification where the deflector is deployed on the cab but the height of the trolley is not increased. This type of arrangement decreases the net of local velocity contours and overpressure regions as compared to case 2 which can be seen from Figures 51.6 and 51.7.

Case 4) Extended trolley with CAB deflector.

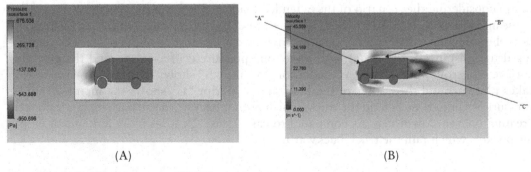

(A)   (B)

Figure 51.8 A) Pressure contour B) Velocity contour.

In the configuration of case 4 (Figure 51.8) CAB has been associated with deflector and the trolley is having a standard height equal to the cab with deflector arrangement. In this the creation of under pressure with void zone package region is significantly reduce as compare to case 3). The Annotation "A" suggests detachment of the flow and Annotation "B" suggests reattachment of the flow over a top surface area of the vehicle. Annotation "B" shows the void space which significantly reduces as compared to the case 2 and case 3 arrangement. Below Figure 51.9. Shows a comparative result of drag coefficient with different cases, the x-axis shows the cases under Investigation and the y-axis shows the value of drag coefficient.

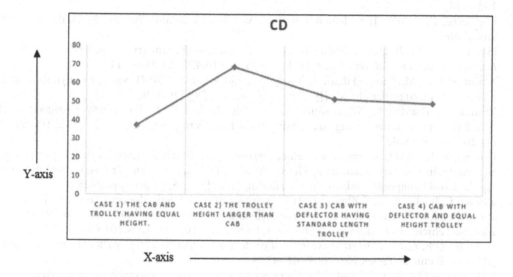

Figure 51.9 CD vs Cases.

Conclusion

In the region of Asia and subcontinents Class 8, LPT 2518 Trucks are widely used to transport the goods by road, but due to over language above the capacity of a truck it causes the uneven increasing height of trolley. The present Investigation is performed by using CFD software to suggest standard guidelines on how the creation of drag resistance changes by changing the arrangements. The following points can be drawn from the present Investigation:

The creation of Coefficient of Drag is correlated with the quantity of increasing pressure in the grid as showing in comparative case 1), case 2), case 3) and case 4)

The creation of drag forces tends to increase more in case 2) where the height of the trolley was extended in considering extra load capacity. This extended shape of the trolley causes early detachment of air molecules by causing a more number overpressure region due to which case 2 contributes the highest number of a drag coefficient.

The Drag coefficient tends to decrease in case 3 and case 4 due to the attachment of the deflector which lowers the contour value for the overpressure region near the oblique surface.

1. Investigation suggests standard guideline that when a trolley is operating on extra payloads over the limit, incorporation of the deflector is necessary to lower the drag

coefficient and increase the fuel economy of a vehicle because the creation of drag is directly proportionate to the cube of locally generated velocities which means that around 10% reduction in drag can contribute to the increasing fuel economy by 5 % in cursing mode.

References

1. Howell, J. P. (2012). Aerodynamic Drag Reduction for Low Carbon Vehicles [M]// Sustainable Vehicle Technologies, Driving the Green Agenda. Woodhead, Elsevier Publication, (pp. 145–154).
2. Lögdberg, O. (2008). Turbulent Boundary Layer Separation and Control. Ph.D. thesis, KTH, Stockholm.
3. Hallqvist, T. (2009). The cooling airflow of heavy trucks—a parametric study. *SAE International Journal of Commercial Vehicles,* 1(1), 119–133. doi:10.4271/2008-01-1171.
4. Mahmoodi-K, M., Davoodabadi, I., Višnjić, V., and Afkar, A. (2014). Stress and dynamic analysis of optimized trailer chassis [J]. *Technical Gazette,* 21, 599–608.
5. Yuniar, D., Djakfar, L., Wicaksono, A., and Efendi, A. (2020). Truck driver behavior and travel time effectiveness using smart GPS. *Civil Engineering Journal,* 6, 724–732. 10.28991/cej-2020-03091504.
6. Cooper, K. R. (2004). Commercial vehicle aerodynamic drag reduction: Historical perspective as a guide. In The Aerodynamics of Heavy Vehicles: Trucks, Buses, and Trains. Lecture Notes in Applied and Computational Mechanics. Berlin, Heidelberg: Springer, (pp. 9–28).
7. Zhao, H., Burke, A., and Zhu, L. (2013). Analysis of class 8 hybrid-electric truck technologies using diesel, LNG, electricity, and hydrogen, as the fuel for various applications. In 2013 World Electric Vehicle Symposium and Exhibition, EVS, 2014. 10.1109/EVS.2013.6914957.
8. Schoon, R. E. (2007). On-road evaluation of devices to reduce heavy truck aerodynamic drag [R]. SAE Technical Paper, No. 2007–01–4294.
9. CD-adapco (2014). CFD and CAE Engineering Simulation Software. http://www.cd-adapco.com.
10. Gullberg, P., Löfdahl, L., Adelman, S., and Nilsson, P. (2009). An investigation and correction method of stationary fan CFD MRF simulations. SAE Technical Paper, 2009-01-3067. doi:10.4271/2009-01-3067.
11. Dominy, R. G., Corin, R. J., and He, L. (2008). A CFD investigation into the transient aerodynamic forces on overtaking road vehicle models [J]. *Journal of Wind Engineering and Industrial Aerodynamics,* 96, 1390–1411.
12. Englar, R. J. (2001). Advanced aerodynamic devices to improve the performance, economics, handling and safety of heavy vehicles [R]. SAE Technical Paper, No. 2001–01–2072.
13. Håkansson, C., and Lenngren, M. J. (2010). CFD Analysis of Aerodynamic Trailer Devices for Drag Reduction of Heavy Duty Trucks [D]. Göteborg Sweden: Chalmers University of Technology.
14. Talele, V., Mathew, V. K., Sonawane, N., Sanap, S., Chandak, A., and Nema, A. (2021). CFD and ANN approach to predict the flow pattern around the square and rectangular bluff body for high Reynolds number. *Materials Today: Proceedings,* 47, 3177–3185. ISSN 2214-7853. https://doi.org/10.1016/j.matpr.2021.06.285.(https://www.sciencedirect.com/science/article/pii/S2214785321046757)
15. ANSYS, INC (2011). ANSYS 14.0 FLUENT User's Guide [CP]. Pennsylvania USA. 10, 813–818.
16. Srivastava, S. (2018). Numerical simulation of supersonic over expanded Jet from 2-D convergent-divergent nozzle. *International Journal of Integrated Engineering,* 10(8), 195–201.
17. Hariram, A., Koch, T., Mårdberg, B., and Kyncl, J. (2019). A study in options to improve aerodynamic profile of heavy-duty vehicles in europe. *Sustainability,* 11(19), 5519. doi:10.3390/su11195519.

52 Conventional material challenges in dental implants

Sambhrant Srivastava[1] and Saroj Kumar Sarangi[2]

[1]Research Scholar, Department of Mechanical Engineering, NIT Patna, India

[2]Associate Professor, Department of Mechanical Engineering, NIT Patna, India

Abstract: Dental implant materials presently available need to be studied, their compatibility and recent developments are required to be examined. This review is aimed to examine the applications of modern dental implant materials such as titanium alloys, zirconia and polyether ether ketone (PEEK) on the mechanical characteristics and biocompatibility issues. Finally, a conclusive remark is suggested for future research direction in the field of dental implants.

Keywords: Dental implant, titanium alloy, zirconia, PEEK.

Introduction

Nowadays living quality of human beings is increasing which demands progress in technologies. Due to the development in social economy, people have started spending more on their looks and aesthetics. Loosing of teeth due to old age, accidents and any chronic disease is one of the major outlooks. Tooth Loss may affect aesthetic look in case of front teeth, chewing problems in case of molar teeth and dysfunction of temporomandibular joint (connecting jawbone to the skull) in case of adjacent teeth. Various solution is there to above problems but to restore the masticatory functions the best way in opinion of stomatology community is dental implant. Due to excellent fit of aesthetics, and zero handling because of permanent fixation dental implant procedure is becoming popular day by day and as per new research its steadily increasing and reached one million implants in a year [1]. The very foundation of dental implantology was initiated by Prof P-I Brånemark over 70 years ago but dental implants are complex to function in some environments, so to overcome this subgingival part (beneath the gums) is implanted into alveolar bone which would help to form a strong biological connection. However, subgingival part bears the oral environments moisture, acidity, bacterial circumstance and resist the stress loading. Earlier clinical trials in dental implants early failure were investigated to be 4% [2] but in the long span of 12, 36 or 60 months 2.6%, 3.6%, and 4.2% failure is noticed [3]. However, even though specific situation of patients and treatment methods failure due to implant material can't be ignored. This emerges a need to select a suitable material to improve the success rate of long-term implantation. To review the major advancements in dental implant material is the main objective of this review. Beginning with titanium alloys as they are the main focus of research in these many years. Then moving to other promising interests in dental implants are like ceramics, polymers. At the end, some focus on zirconia and polyether ether ketone (PEEK) are summarized. Finally, prospects of future dental implants materials conclusive remarks are outlined.

Dental Implantation Components and Methods

In dental implant to support dental prosthesis a surgical component implanted to the alveolar bone divided in three parts namely implant fixture, abutment and Implant crown [4] as described in Figure 52.1.

According to the mode of connectivity between implant body and abutment three types of structures are classified i.e., one-, two-, and three-piece (with cross-core screw) implants. One-piece implant is considered to be to be simple and quicker in clinical procedure and as it uses no thread to connect the abutment so there are be no gaps between the implant and abutment. Due to this the bacterial infection is least. Moreover, after implantation, artificial crown can be installed to abutment very soon. In small diameter implant it is observed high stress in two-piece implant material compared to one-piece implants [5, 6].

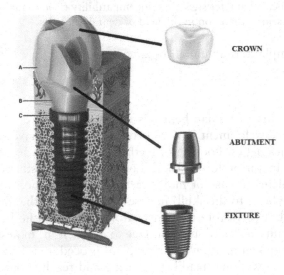

Figure 52.1 Components of restored implant. (A-implant crown, B-abutment, C-implant fixture).

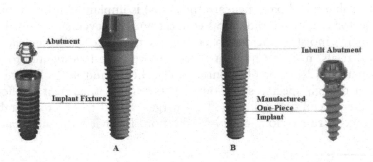

Figure 52.2 Computer aided design model with difference: (A) Two-piece implant and (B) One-piece implant.

Challenges in Materials of Implant Body

Osseointegration is the very first and important step of dental implantation which means structural and functional connection between the living bone and surface of dental implant to withstand occlusal forces. Discovery of osseointegration occurred accidently when Swedish researchers placed chambers into the bone of animals to examine the vasculature of the bone [4]. This process needs fixing and healing time of utmost 3 to 4 months to implants without stress loading before installing abutment [7]. To select a proper material for implants it is likely to have optimized structural design and biocompatibility for long term stability of dental implants. Although dental implants are very common nowadays there are still some problems related to materials of implants that are needed to be resolved. Considering the challenges as

- Difference in elastic modulus of implants and bones is general problem. According to Hooke's law equal elastic modulus will result equal amount of deformation. Elastic Modulus of Implants material like titanium alloy [8] is almost 8 to 10 times then elastic modulus of trabecular bone (10–20 GPa) and cortical bone (1.5–2.5 GPa) [9]. High elastic modulus material will bear more stress thus results loosening of implant after a while.
- Oral environment exposure causes bacterial infection around dental implants and bacterial plaques will destroys the tissue around the implants thus this loosens the Osseointegration and result to implant failure [10].

Suitable Materials Properties for Dental Implants

Metals

Due to excellent mechanical properties like good processability, easy to manufacture, resistant to stress and high melting point, metals are popular in clinical treatment. However, metals are biological inert [11]. In metals, stainless steel (316L) used in implant material for bone plates and screws as they are cheaper as well as machinable but its corrosion properties are much inferior to titanium, this makes stainless steel unsuitable for dental implants. Despite good occlusal forces resistant and better biodegradable properties then titanium, cobalt-based alloy is not as resistant in corrosion as titanium [12, 13]. Both titanium and surgical steel are almost inert to degradation. However, titanium is capable to bear more load than surgical steel but both found to be biologically inert [14]. Titanium alloy (Ti-6Al-4V) and Cobalt-Chromium alloy are Osseointegration in rabbit model after investigation titanium alloy is proved to be more corrosion resistant, better mechanical properties and biocompatible then Cobalt-Chromium alloy [15]. Despite of above properties the problem of toxicity because of Vanadium in Ti-6Al-4V emerged need of research for low modulus and least toxicity in dental implant.

Then β type Titanium alloys are investigated i.e., much better in biocompatibility, and low elastic modulus of about 55 to 85 GPa which is apparently much smaller than α type and α + β type (Like Ti-6Al-4V alloys) (Figure 52.3) but reasonably higher than that of bone [8].Target to attain least toxicity and reasonably low moduli, many new low moduli and non-toxic elements like Nb, Ta, Zr etc. are composed which leads to the development of β type titanium alloys like T-6Al-7Nb and Ti-5Al-2.5Fe which are better in biocompatibility, toughness and strength and low elastic modulus than α + β type titanium alloys (like Ti-6Al-4V) [8]. Yet the moduli are more than that of bone so further investigation by an idea of controlling the young modulus of β type titanium alloys done by aging treatment

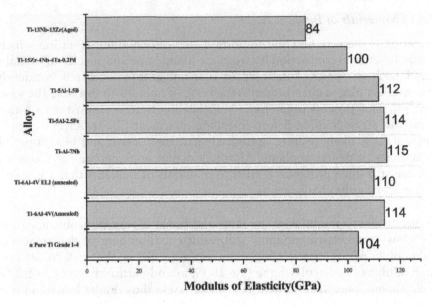

Figure 52.3 Comparison of elastic modulus of biomedical titanium alloy [8].

[16]. To continue with idea, Niinomi in 2003, decreased the Young modulus of Ti-13Nb-13Zr to 44 GPa by water quenching and cold working which is lowest moduli reported [16]. The main drawback of the metal dental implant is releasing of metal ions in the body and can harm various organs. During an experiment some ions are investigated in rabbit tissue by Lugowski et al [17].

Ceramics and Polymers

Discussion with titanium alloy left with corrosion and release of metal ions in the tissue due to wear. Ceramic implants introduced to overcome the grey appearance of Titanium and biological inert issues. The very first ceramic dental implant was made from alumina in 1960s.

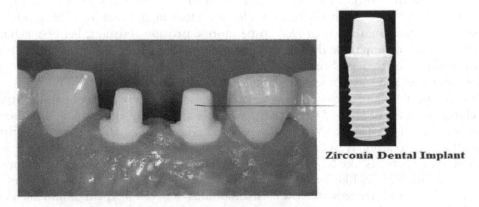

Figure 52.4 Zirconia implant before cementation of crown.

However, due to low fracture toughness it is replaced with zirconia (Zro2), which is the only alternative material to titanium for dental implants in ceramics [18]. In zirconia, osse-ointegration is not a problem and stress simulation into bone is similar to titanium alloy [19]. However, zirconia (Figure 52.4) is considered metal free, so release of metal ions in tissues are not a problem but ceramic materials are brittle, hard, strong in compression and weak in shear and tension which lacks strength compared to titanium alloy [11, 20]. Thus, zirconia is still not the first choice in dental implant over titanium. Ceramics are generally bio-inert type material but do not release metals ions. So, after implant a fibrous capsule surround the implant that do not bind with tissue thus fibrous capsule continues to thickens and block blood supply. Thus, dental implant after accumulation of waste around the tissue loosen due to unbalanced stress [11]. Another idea of using polymer due to limitation in ceramics of low ductility and high brittleness. Polymers elastic modulus much lower than metals and they can be attached to tissues [12]. An early investigation of polymer-coated implants are reported in 1970. After so many researches a high-performance engineering plastic, PEEK (polyether ether ketone) has been proposed as a substitute for metals in biomaterial [20]. The young modulus of PEEK ranges from 3 GPa to 4GPa i.e., much closer to the moduli of human bone [12]. PEEK exhibit excellent injection molding and 3D printing which makes it excellent processable material [21]. PEEK reinforcement phases were investigated but a balance point of modulus and brittleness need to be found [12].

Conclusions

Concluding on restriction of clinical trial results, titanium implants are still the mainstream and gold standard of dental implants even the modulus mismatching is still unaddressed. Whereas, ceramic and polymer-based implants like zirconia and PEEK demonstrated better aesthetic, biocompatibility and corrosion resistant than titanium, which can be another descent choice of dental implant but need more clinical trials. The future challenge is a need of composite material with lowest elastic modulus like PEEK and high strength like Titanium along with excellent biocompatibility of ceramic.

References

1. Le Guehennec, L., Soueidan, A., Layrolle, P., and Amouriq, Y. (2007). Surfacetreatments of titanium dental implants for rapid osseointegration. *Dental Materials*, 23, 844–854.
2. Baqain, Z. H., Moqbel, W. Y., and Sawair, F. A. (2012). Early dental implant failure: Risk factors. *The British Journal of Oral & Maxillofacial Surgery*, 50, 239–243.
3. Lee, C. T., Tran, D., Jeng, M. D., and Shen, Y. T. (2018). Survival rates of hybrid rough surface implants and their alveolar bone level alterations. *Journal of Periodontology*, 89, 1390–1399..
4. Hupp, J. R., DMD, MD, JD. (2017). Introduction to implant dentistry: a student guide. *Journal of oral and Maxillofacial Surgery*, 75(2).
5. Wu, A. Y., Hsu, J. T., Chee, W., Lin, Y. T., Fuh, L. J., and Huang, H. L. (2016). Biomechanical evaluation of one-piece and two-piece small-diameterdental implants: In-vitro experimental and three-dimensional finite element analyses. *Journal of the Formosan Medical Association*, 115,794–800.
6. Cehreli, M. C., Akca, K., Iplikcioglu, H. (2004). Force transmission of one- and two-piece morse-taper oral implants: a nonlinear finite element analysis. *Clin Oral Implants Research*,15, 481e9
7. Albrektsson, T., Brånemark, P. I., Hansson, H. A., and Lindström, J. (1981). Osseointegrated titanium implants: Requirements for ensuring a longlasting, direct bone-to-implant anchorage in man. *Acta Orthopaedica Scandinavica*, 52, 155–170.

8. Niinomi, M. (1998). Mechanical properties of biomedical titanium alloys. *Materials Science and Engineering A*, 243, 231–236.
9. Bayraktar, H. H., Morgan, E. F., Niebur, G. L., Morris, G. E., Wong, E. K., and Keaveny, T. M. (2004). Comparison of the elastic and yield properties of human femoral trabecular and cortical bone tissue. *Journal of Biomechanics*, 37, 27–35.
10. Zhao, L., Wang, H., Huo, K., Cui, L., Zhang, W., Ni, H., … and Chu, P. K. (2011). Antibacterial nano-structured titania coating incorporated with silver nanoparticles. *Biomaterials*, 32, 5706–5716.
11. Wang, S.-F., Yang, C.-K., Lee, S.-Y., Yang, J.-C., and Ho, I.-L. (2013). *Composite Bio-Ceramic Denatal Implant and Fabrication Method Thereof* (U.S. Patent No. US 2013/0150227 A1).
12. Xunyan, J., Yitong, Y., Weiming, T., Dongmei, H., Li, Z., Ke, Z., Shuanjin, W., and Yuezhong, M. (2019). Design of dental implants at material level: an overview. *Journal for Biomedical Material Research*, 2020, 108, 1634–1661.
13. McCracken, M. (1999). Dental implant materials: Commercially pure titanium and titanium alloys. *Journal of Prosthodontics*, 8, 40–43.
14. Rahim, M. I., Ullah, S., and Mueller, P. P. (2018). Advances and challenges of biodegradable implant materials with a focus on magnesium-alloys and bacterial infections. *Metals*, 8, 532.
15. Jinno, T., Goldberg, V. M., Davy, D., and Stevenson, S. (1998). Osseointegration of surface-blasted implants made of titanium alloy and cobalt-chromium alloy in a rabbit intramedullary model. *Journal of Biomedical Materials Research*, 42, 20–29.
16. Hao, Y., Yang, R., Niinomi, M., Kuroda, D., Zhou, Y., Fukunaga, K., and Suzuki, A. (2003). Aging response of the Young's modulus and mechanical properties of Ti-29Nb-13Ta-4.6Zr for biomedical applications. *Metallurgical and Materials Transactions A: Physical Metallurgy and Materials Science*, 34, 1007–1012.
17. Lugowski, S. J., Smith, D. C., McHugh, A. D., and van Loon, J. C. (1991). Release of metal ions from dental implant materials in vivo: Determinationof Al, Co, Cr, Mo, Ni, V, and Ti in organ tissue. *Journal of Biomedical Materials Research Part A*, 25, 1443–1458.
18. Wenz, H. J., Bartsch, J., Wolfart, S., and Kern, M. (2008). Osseointegration and clinical success of zirconia dental implants: a systematic review. *International Journal of Prosthdont*, 21, 27–36.
19. Nicola, M., Francesco, M., and Santo, C. (2016). Ceramic Material as an alternative to titanium for dental implant fabrication, INTECH Open Science.
20. Rahmitasari, F., Ishida, Y., Kurahashi, K., Matsuda, T., Watanabe, M., and Ichikawa, T. (2017). PEEK with reinforced materials and modifications for dental implant applications. *Dental Journal*, 5, 35–42.
21. Panayotov, I. V., Orti, V., Cuisinier, F., & Yachouh, J. (2016).Polyetheretherketone (PEEK) for medical applications. *Journal of Materials Science*, 27, 118–128.

53 Low temperature combustion engines for future transport—a review

Satendra Singh[a], D. Ganeshwar Rao[b], and Manoj Dixit[c]

Department of Mechanical engineering, Faculty of Engineering, DEI, Dayalbagh, Agra, India

Abstract: The internal combustion engines are still a primary source for transportation, goods carriers, public transport and power generation units. The curiosity in compression ignition engines increases constantly due to their many advantages like high efficiency, reliability, durability, and low-operating cost. But these engines produce lot of harmful engine emissions creating pollution problems globally and causing several health problems in humans. Over the period many rules and regulations have been imposed to control engine emissions. Many aftertreatment emission control technologies have been used but none of them have effectively controlled engine emissions without effecting the performance of the engine. One effective solution to this problem may be to use low temperature combustion strategies which not only would control the engine emissions but simultaneously increase the efficiency of the engine. Low temperature combustion engine as compared to conventional diesel engines reduces particulate matter and nitrogen oxides to nearly zero-level, reduces heat losses and carbon dioxides by 2% and 20% respectively. Increase in carbon monoxide and hydrocarbons emission was observed by 7.8 times while increase in thermal efficiency observed is around 18%.

Keywords: Diesel engine, Emission, HCCI, LTCE, PCCI, RCCI.

Introduction

The world has more than 4.8 billion vehicles including passenger cars, commercial vehicles, transport vehicles, railways and marine ships in the world which are uninterruptedly increasing. More than 99.9% are powered by internal combustion engines which use liquid hydrocarbon (HC) fuel for power production. As per new findings, the retrieval rates for global crude oil supply size have been growing rapidly and it is projected that the demand for oil will remain increasing and will be at peak by 2040 [1, 2]. The total number of vehicles in the world are expected to reach 2440 billion in 2050 [3] and only one-third of light duty vehicles would run on electricity and natural gas. The majority of vehicles will continue to run on gasoline and diesel fuel due to many reasons like infrastructure, cost, climate, and geography. Fossil fuel-based energy accounts for 82.67% of total energy consumption globally. Also implies that internal combustion engines are the primary energy conversion devices as on date [4].

[a]satendrasingh.metech@gmail.com, [b]dgra°.dei@gmail.com, [c]manojdixit@dei.ac.in

Diesel engines are widely used compared to gasoline engines due to their high durability, energy efficiency, low-operating costs and reliability [5]. Nevertheless, due to various factors like fuel and air ratio, air–fuel concentration, combustion temperature, combustion form, turbulence, ignition timing etc. affects the ideal combustion process thus, harmful exhaust emission is produced during combustion [6]. In India approximately 72% of transportation vehicles are used diesel as fuel [7]. Various types of renewable sources of energy were used to meet energy requirement but did not get much success due to various problems. Biodiesel are emerged as the supplementary of diesel fuel among all other alternates [8]. Various types of biodiesel fuels like jatropha, neem, mahua etc. have been also used and they are giving good results but till now did not commercialized because of different problems [9].

Although diesel engine has many advantages, but they have also produced significant amount of engine emission which contains mainly four major types of pollutant especially CO, HC, NOx and PM. Particularly, contains higher amount of NOx emissions (more than 50 %) and particulate matter (second highest proportion) and low amount of CO and HC emissions. Depending upon the specifications and quality of fuel little amount of SO2 is produced due to the presence of sulphates in diesel fuel [10]. The engine emissions emitted by diesel engines have an adverse effect on both environment and human health. The problems created to environment are acid rains, air pollution, ground-reduce, Climate change, acidifying of ocean, extreme heat waves, drought affecting food crop, greenhouse effect, global warming and visibility level ozone. For human life it causes respiratory problems, lung damage and even cancer [11, 12]. So, due to these various rules and regulation are imposed on internal combustion engines time to time to reduce the engine emission. On such regulation i.e., Euro VI engine emission norm for heavy-duty vehicle are 130, 1500, 10 and 400 mg/kWh for HC, CO, PM and NOx emissions [13].

Emission Control Systems used in Diesel Engines

To reduce the engine emissions various methods are being used like improved fuel properties, engine modifications and electronic controlled fuel injections systems but all these remedial steps have been unsuccessful to attain desired emission decrement. After that various after-treatment emission control systems used in which the emission is treated just before it leaves the tailpipe of engine [10, 11]. To eliminate the NOx emissions exhaust gas recirculation (EGR), lean NOx trap (LNT), and SCR are among the majorly focused technologies used so far by the researches. In EGR, exhaust gas from the tailpipe of engine is recirculated back and again combined with fresh air in different proportions into the combustion chamber at intake stroke. This reduced the temperature inside the combustion chamber hence reduces the NOx emission but it lead to degraded the combustion efficiency and results in to the increased HC and CO emissions [11]. NOx-storage-reduction (NSR) or LNT or NOx adsorber catalyst (NAC), operates during lean engine conditions to reduce NOx emissions. LNT is working in a way that it stores NOx on the catalyst wash coat and after that, under the rich fuel burning environments it releases the NOx and reacts by the traditional three-way type reactions but are insufficient to reduce NOx emissions. SCR technology attracts many researchers it includes DOC, DPF and SCR exhaust treatment devices are used in all new vehicles for wayfaring cars, light and heavy-duty working diesel engines [14].

But the detrimental effects of these engine exhaust emissions on both the environmental condition and human health the currently used emission control technologies and methods are proven insufficient. Every method used to control emission have some disadvantages and effect the engine performance, engine efficiency and also not able to reduce engine emission completely. In fact, the emission reduction efficiencies of these technologies are between 40–90% but are vary with the temperature variation greatly because each of these devices provides maximum efficiency at a certain specific temperature thus not only affects the performance of after treatment device efficiency but also reduce the engine efficiency, increase fuel consumption [4, 12, 14].

So, more emphasize is required to the technological improvements, advancement and developments, making numerous legal arrangements, such as development of control systems, using alternate fuels, creating several types of model structures, and shaping the basic structure of traffic. One such solution of these problem discussed in above sections is to adapt low temperature combustion engine (LTCE) which not only reduce the emission level to EURO VI norms but at the same time increase the thermal efficiency, can use any type of fuel for their power generation etc.

Low Temperature Combustion Engines

In a LTCE's the combustion temperatures inside the cylinder and fuel-air equivalence ratio are well below 2200 K and 2 respectively due to this nearly zero level NOx and soot engine exhaust emissions were formed [15]. LTC engines results in, lower in-cylinder peak temperature, uniform temperature distribution, higher thermal efficiency and lower heat losses due to lean combustion and homogeneous fuel air mixture [16]. Even though LTC engines have various advantages but commercial implementation in automotive industries need to resolve various problems associated with LTC like to controlling the timing of ignition in engine, lower operating load range extension to wider range and to minimizing CO and HC engine exhaust emissions [17].

Homogeneous Charge Compression Ignition Engine

HCCI is one of the first type of LTC engine instigated by Shigeru Onishi et al. 1979 with gasoline injected through port of the engine due to the gasoline fuel characteristics such as higher volatility in comparison to diesel. After that it is also very effectively employed to the diesel engines by various investigators [18, 19]. But as the properties of diesel are different form gasoline some problems are created and to address these problems various methods like early direct injection (DI) into the combustion chamber with single spray that is targeting the piston crown, direct injection with low pressure to combustion chamber with a narrow spray cone angle and external mixture preparation with the help of a fuel vaporizer are endeavoured [16]. HCCI engine has an advantage like high fuel flexibility but the considerable challenges associated with the HCCI strategy are include higher pressure rise rate, higher unburned emissions, misfire problem at low loads, narrow load range and improper combustion phasing [20]. To overcome the difficulties related to load range with HCCI engine various approaches like mixed mode combustion, variable compression ratio utilization during engine operation, exhaust gas recirculation (EGR), utilizing supercharger and water injection technique is used. But precise ignition timing cannot be achieve hence cycle-to-cycle variation occurred [19].

Premixed Charge Compression Ignition Engine

PCCI engine was introduced in which instead of external mixture preparation, the mixture of fuel and air is prepared inside the cylinder hence mixture stratification concept is used [16]. In PCCI, shorter ignition dwell is used compared to HCCI strategy so, that charge stratification can be obtained because of partially premixed fuel-air mixture. Nevertheless, due to early fuel injection timing compared to HCCI some control over combustion phasing is achieved hence results in maximum thermal efficiency with lower combustion noise [21]. To overcome the increased emissions in PCCI two strategies are used by which NOx and PM emissions are reduced. First method is to use EGR increases ignition dwell period which provide more time to vaporize fuel in combustion chamber and results in cooling effect. Second method is very late fuel injection obtained with the help of a narrow cone angle injector near the Top Dead Centre (TDC). But many studies revealed that PCCI engine also have numerous limitations like higher rate of unburned emissions, higher rate of fuel consumption, misfire at low loads and knocking at higher loads [22].

Reactivity Charge Compression Ignition Engine

In RCCI combustion a fuel with low reactivity like gasoline is injecting and mixed with air in the intake manifold and a fuel with high reactivity like diesel is directly injected into the combustion chamber of engine during the compression stroke [23]. By varying the quantity of low and high reactive fuels, at varying loads optimal working conditions can be achieved in RCCI which in comparison to HCCI and PCCI leads to better combustion phasing, better ignition timing control and having wide operating load range [24].

Kokjohn SL and Reitz R., (2013) [25] determined that RCCI engine could able to achieve 4% higher fuel economy and total fluid consumption improved by 7.3% along with very low NOx and smoke engine exhaust emissions with comparison to CDC. Jesus Benajes et al. (2016) [26] at reduced compression ratio experimentally found that RCCI engine produce lower pressure rise rate along with nearly-zero level NOx and smoke engine exhaust emissions.

A comparative assessment between HCCI, PCCI and RCCI strategies using different fuels has been attempted. The results of these investigations proposed that HCCI and PCCI engine suffer form few limitations including that the operating load range is narrow, higher cycle by cycle combustion variation and higher combustion noise whereas RCCI engine is suffer only from higher CO and HC emissions which can be treated effectively with the help of catalytic converter with minimum cost and higher life. So, this comparison suggests that RCCI engine can be one of the engines that can be used in future to power IC engines, especially for farming purposes and small-scale electricity production with nearly zero-level emission and high thermal efficiency of more than 50%.

Comparison between Conventional CI Engine and low Temperature Combustion Engine

From Figure 53.1, on comparing CDC and LTCE it can be clearly seen that on CDC engines are producing more emission because of the high operating temperature i.e., 2500 K and fuel-air equivalence ratio while LTCE works under low temperature range i.e., below 2000 K and an equivalence ratio below 2. In CDC engine NOx versus PM trade-off relationship has been seen while in LTCE NOx versus (CO + UHC) trade-off has been seen [27]. It has been observed that over 11 g/kWh NOx emission and very low PM emissions are produced under CDC engine operation that are more than 28 times when compared with EU-VI emission norms while in case of LTCE ultra low NOx and PM exhaust emissions are produced which are much below than EU-VI emission norms. LTCE produced approximately 20% less CO_2 emissions compared to CDC engine without any after treatment

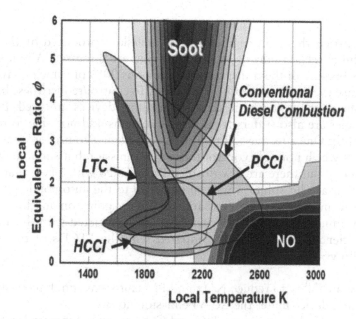

Figure 53.1 HCCI, PCCI, LTC and CDC operating T-Φ range [29].

devices. One of the biggest disadvantages of LTCE is that they produce around 7.8 times more THC and CO engine exhaust emissions as compared to the CDC engine operation which can be treated by some catalytic converter effectively to a well below level of EU-VI emission norms [28].

Since invention, LTCE shows great ability and potential not only it produces high thermal efficiency but at the same time also reduce engine emissions specially NOx, PM and CO2 [28]. However, LTC engines have some disadvantages also like low load capacity, cycle-to-cycle variation in combustion, low control over combustion etc. which affects the production of LTC engines commercially. But due to its enormous advantages and huge emphasis on research and development by researchers and organisations very soon these LTC engines will replace conventional IC engines in near future. A comparison between CI engine and LTC engine on various factors are shown in Table 53.1.

Table 53.1: Comparison between CDC and LTC engine on major factors [30].

Factors	CI Engine	LTC Engine
Major Emissions	Low CO & HC, High PM & NOx	CO, HC
Exhaust gas treatment	Required, thus increase cost.	Do not require, so economical.
Net work output	44%	45–53%
Heat transfer losses	16%	14%
Temperature	High (2700 K)	Low (below 2100 K)
Fuel air ratio (Φ)	Rich (fuel rich regions present)	Lean (fuel rich regions are absent)
Thermal efficiency	30–35%	50%

Conclusion

In this research article the main engine exhaust emissions produced by the diesel engine and their harmful effects on human and environment are discussed. About 60–70% local pollution occurs because of the transportation sector as 90% of vehicles have ICE as their power unit and are expected to remain so till 2040. To control/reduce these harmful engine emissions various methods techniques and controlling devices are used. But these after treatment strategies are also suffers from some problems and not able to reduces engine emissions efficiently. One effective solution of this problem is to use low temperature combustion strategies which not only control the engine emission but simultaneously increase the engine efficiency. But there are few problems associated with LTCE that are needed to be resolved so that a solid solution will be provided to the automobile sector. For those further continuous intensified researches and studies are going on and hopefully they will provide solution in near future. The decisive findings associated with the aspects of CDC combustion, efficiency, and emissions compared with the LTCE's of this study may be expressed as follows:

- LTCE's have an ability to reduce NOx and PM emissions simultaneously without any after treatment device under the EU-VI emission norms.
- LTCE's produces 7.8 times more THC and CO emissions compared to CDC engine.
- LTCE's reduces CO2 emissions by 20% as compared to CDC engine.
- Thermal efficiency of LTCE's is 53% which is higher as compared to CDC i.e., 35%.
- Heat transfer losses are around 2% less in LTCE compared to CDC.
- LTCE's has combustion stability and load carrying capacity problems which affects its commercialisation.
- A dual fuel based RCCI engine is the most promising among all the LTCE's due to its advantages like combustion stability, varying load carrying capacity, any types of fuel can be used etc.
- More deep and intense research are required for LTCE's in order to replace current IC engine technology.

References

1. Kalghatgi, G. (2018). Is it really the end of internal combustion engines and petroleum in transport? *Applied Energy*, 225, 965–974. https://doi.org/10.1016/j.apenergy.2018.05.076
2. Sioshansi, F., and Webb, J. (2019). Transitioning from conventional to electric vehicles: the effect of cost and environmental drivers on peak oil demand. *Economic Analysis and Policy*, 61, 7–15. https://doi.org/10.1016/j.eap.2018.12.005.
3. EIA, U. (2019). International energy outlook 2019 (IEO2019) reference case. US Energy Information Administration.
4. Wu, Y., Wang, P., Farhan, S. M., Yi, J., and Lei, L. (2019). Effect of post-injection on combustion and exhaust emissions in DI diesel engine. *Fuel*, 258, 116131. https://doi.org/10.1016/j.fuel.2019.116131.
5. Lao, C. T., Akroyd, J., Eaves, N., Smith, A., Morgan, N., Nurkowski, D., Bhave, A., and Kraft, M. (2020). Investigation of the impact of the configuration of exhaust after-treatment system for diesel engines. *Applied Energy*, 267, 114844. https://doi.org/10.1016/j.apenergy.2020.114844.
6. Lheywood, J. B. (2010). Internal combustion engine gasket. John Wiley & Sons.
7. Dash, S. K., Lingfa, P., and Chavan, S. B. (2020). Combustion analysis of a single cylinder variable compression ratio small size agricultural DI diesel engine run by Nahar biodiesel and

its diesel blends. *Energy Sources, Part A: Recovery, Utilization and Environmental Effects*, 42, 1681–1690. https://doi.org/10.1080/15567036.2019.1604878.

8. Hina, P. N., Pradip, L., and Dash, S. K. (2021). An experimental investigation on the combustion characteristics of a direct injection diesel engine fuelled with an algal biodiesel and its diesel blends. *Clean Technologies and Environmental Policy*, 23(6), 1769–1783. https://doi.org/10.1007/s10098-021-02058-3.

9. Papu, N. H., Lingfa, P., and Dash, S. K. (2020). Euglena Sanguinea algal biodiesel and its various diesel blends as diesel engine fuels: a study on the performance and emission characteristics. *Energy Sources, Part A: Recovery, Utilization and Environmental Effects*, 00, 1–13. https://doi.org/10.1080/15567036.2020.1798566.

10. Prasad, R., and Bella, V. R. (2010). A review on diesel soot emission, its effect and control. *Bulletin of Chemical Reaction Engineering and Catalysis*, 5, 69–86. https://doi.org/10.9767/bcrec.5.2.794.69-86.

11. Olabi, A. G., Maizak, D., and Wilberforce, T. (2020). Review of the regulations and techniques to eliminate toxic emissions from diesel engine cars. *Science of the Total Environment*, 748, 141249. https://doi.org/10.1016/j.scitotenv.2020.141249.

12. Apicella, B., Mancaruso, E., Russo, C., Tregrossi, A., Oliano, M. M., Ciajolo, A., and Vaglieco, B. M. (2020). Effect of after-treatment systems on particulate matter emissions in diesel engine exhaust. *Experimental Thermal and Fluid Science*, 116, 110107. https://doi.org/10.1016/j.expthermflusci.2020.110107.

13. Hoekman, S. K., and Robbins, C. (2012) Review of the effects of biodiesel on NOx emissions. *Fuel Processing Technology*, 96, 237–249. https://doi.org/10.1016/j.fuproc.2011.12.036.

14. Bauner, D., Laestadius, S., and Iida, N. (2009). Evolving technological systems for diesel engine emission control: Balancing GHG and local emissions. *Clean Technologies and Environmental Policy*, 11, 339–365. https://doi.org/10.1007/s10098-008-0151-x.

15. Dec, J. E. (2009). Advanced compression-ignition engines—understanding the in-cylinder processes. *Proceedings of the Combustion Institute*, 32(II), 2727–2742. https://doi.org/10.1016/j.proci.2008.08.008.

16. Reitz, R. D., and Duraisamy, G. (2015). Review of high efficiency and clean reactivity controlled compression ignition (RCCI) combustion in internal combustion engines. *Progress in Energy and Combustion Science*, 46, 12–71. https://doi.org/10.1016/j.pecs.2014.05.003.

17. Agarwal, A. K., Singh, A. P., and Maurya, R. K. (2017). Evolution, challenges and path forward for low temperature combustion engines. *Progress in Energy and Combustion Science*, 61, 1–56. https://doi.org/10.1016/j.pecs.2017.02.001.

18. Gan, S., Ng, H. K., and Pang, K. M. (2011). Homogeneous charge compression ignition (HCCI) combustion: implementation and effects on pollutants in direct injection diesel engines. *Applied Energy*, 88, 559–567. https://doi.org/10.1016/j.apenergy.2010.09.005.

19. Fathi, M., Saray, R. K., and Checkel, M. D. (2011). The influence of exhaust gas recirculation (EGR) on combustion and emissions of n-heptane/natural gas fueled homogeneous charge compression ignition (HCCI) engines. *Applied Energy*, 88, 4719–4724. https://doi.org/10.1016/j.apenergy.2011.06.017.

20. Bendu, H., and Murugan, S. (2014). Homogeneous charge compression ignition (HCCI) combustion: mixture preparation and control strategies in diesel engines. *Renewable and Sustainable Energy Reviews*, 38, 732–746. https://doi.org/10.1016/j.rser.2014.07.019.

21. Lu, Y., Yu, W., and Su, W. (2011). Using multiple injection strategies in diesel PCCI combustion: Potential to extend engine load, improve trade-off of emissions and efficiency. In SAE 2011 World Congress & Exhibition. https://doi.org/10.4271/2011-01-1396.

22. Kulkarni, A. M., Stricker, K. C., Blum, A., Shaver, G. M. (2010). PCCI control authority of a modern diesel engine outfitted with flexible intake valve actuation. *Journal of Dynamic Systems, Measurement and Control, Transactions of the ASME*, 132, 1–15. https://doi.org/10.1115/1.4002106.

23. Kokjohn, S., Hanson, R., Splitter, D., Kaddatz, J., and Reitz, R. (2011). Fuel reactivity controlled compression ignition (RCCI) combustion in light- and heavy-duty engines. *SAE International Journal of Engines*, 4(1), 360–374. https://doi.org/10.4271/2011-01-0357.

24. Splitter, D., Wissink, M., Kokjohn, S., and Reitz, R. (2012). Effect of compression ratio and piston geometry on RCCI load limits and efficiency. SAE Technical Papers. https://doi.org/10.4271/2012-01-0383.

25. Kokjohn, S. L., and Reitz, R. D. (2013). Reactivity controlled compression ignition and conventional diesel combustion: a comparison of methods to meet light-duty NOx and fuel economy targets. *International Journal of Engine Research*, 14, 452–468. https://doi.org/10.1177/1468087413476032.

26. Benajes, J., García, A., Monsalve-Serrano, J., and Martínez-Boggio, S. (2019). Optimization of the parallel and mild hybrid vehicle platforms operating under conventional and advanced combustion modes. *Energy Conversion and Management*, 190, 73–90. https://doi.org/10.1016/j.enconman.2019.04.010.

27. Simescu, S., Fiveland, S. B., and Dodge, L. G. (2003). An experimental investigation of PCCI-DI combustion and emissions in a heavy-duty diesel engine. SAE Technical Papers, 2003. https://doi.org/10.4271/2003-01-0345

28. Shim, E., Park, H., and Bae, C. (2020). Comparisons of advanced combustion technologies (HCCI, PCCI, and dual-fuel PCCI) on engine performance and emission characteristics in a heavy-duty diesel engine. *Fuel*, 262, 116436. https://doi.org/10.1016/j.fuel.2019.116436.

29. Neely, G. D., Sasaki, S., Huang, Y., Leet, J. A., and Stewart, D. W. (2005). New diesel emission control strategy to meet US Tier 2 emissions regulations. *SAE Transactions*, 2005, 512–524. https://doi.org/10.4271/2005-01-1091.

30. Seong, H., Lee, K. O., Choi, S., Adams, C., and Foster, D. E. (2012). Characterization of particulate morphology, nanostructures, and sizes in low-temperature combustion with biofuels (No. 2012-01-0441). SAE Technical Paper.

54 Inventory cost optimization in small and medium enterprises: a case study

P. V. Joshi[a], G. B. Narkhede[b], and N. R. Rajhans[c]

Department of manufacturing Engineering and Industrial Management, College of Engineering Pune, Savitribai Phule Pune University, India

Abstract: In SMEs, lack of purchase strategies leads to either higher inventories or stockout situations. A case study is conducted on a small-scale manufacturing firm that is engaged in manufacturing transmission parts for different customers to develop an integrated purchase policy. In the existing policy, inventory managers place the orders as per their requirements separately from different sources which resulted in high-cost fluctuations. The current research is aimed to use the concept of rank order clustering in grouping identical components followed by lot-sizing techniques to manage inventories effectively and to reduce the total cost of the product. A systematic replenishment strategy is developed to minimize the inventory cost. Insights derived from the case study showed a substantial cost saving of up to 25% (2142672 Rs. per annum).

Keywords: Cost optimization, Inventory management (IM), Lot sizing techniques, rank order clustering (ROC), small and medium-sized Enterprises (SMEs).

Introduction

This paper presents work on the identification and explanation of cost optimization approaches used by small and medium-sized Enterprises (SMEs). This paper aims to open up a new window in the industrial sector for the development of a cost management framework for SMEs.

Literature

There is a need for cost savings for SMEs when and whenever possible [1]. Some of the factors behind the slower development of small and medium enterprises are i) inadequate access to finance, ii) shortage of a database, iii) weak R&D expenses, iv) poor supplier selection strategy, v) weak inventory management, and vi) a low level of finance [2]. Bad inventory management leads to a lot of cash losses which can lead to higher inventory costs for manufacturers [3]. Our objective is to minimize the overall cost of inventory. Harris introduced the first inventory model [4]. The well-known economic order quantity (EOQ) inventory model has been developed which considers demand to be constant and known.

[a]joshipv19.prod@coep.ac.in, [b]ganesh.narkhede@viit.ac.in, [c]nrr.mfg@coep.ac.in

Harris is considered to be the founding father of inventory theory [4]. In this direction, Gupta developed an inventory model for multi-items, taking stock-dependent consumption rates into account [5]. In our study we are also focusing on the inventory by considering quantity and cost of the component.

Methodology

Data Collection

The questionnaire was distributed via self-directed surveys mostly through e-mail and through face-to-face surveys.

Irregular Ordering

It is observed that there is irregular ordering of components. Which disturbs the SMEs economically and inventory management becomes complicated.

Rank Order Clustering

Rank Order Clustering is a technique used to group components and machines. The same technique is used in this case to group components and assemblies. A matrix is generated for components and products. Two clusters of component and assembly highlighted in Table 54.1.

Table 54.1: Rank order clustering.

	Spring Washer M10	Drain Plug 1/2" BSP	Breather Plug 1/2" BSP	Level Plug 1/2" BSP	Spring Do well Pin 4x30	Internal Circlip B 62	Allen Bolt 6x70	Internal Circlip B 47	External Circlip A 35	Ball Brg. 6005
2380	1	1	1	1	1					
1280	1	1	1	1						
3380	1	1	1	1	1					
3340	1	1	1	1	1					
4300	1	1	1	1						
2340	1	1	1	1	1					
3300	1	1	1		1					
3260	1	1		1	1					
2260	1		1		1					
2240	1				1					
4240	1				1		1			
1340		1	1	1						
3130			1			1		1	1	1
4190							1			

(continued)

Table 54.1: Continued

	Spring Washer M10	Drain Plug 1/2" BSP	Breather Plug 1/2" BSP	Level Plug 1/2" BSP	Spring Do well Pin 4x30	Internal Circlip B 62	Allen Bolt 6x70	Internal Circlip B 47	External Circlip A 35	Ball Brg. 6005
3160						1	1	1	1	
2160						1				
1130						1	1			
2130						1		1	1	1
4130						1	1	1	1	1
2095						1		1	1	1
3095						1		1	1	1
1095						1		1	1	1

Existing Cost (Considering Varying Cost)

The components are ordered as and when required. This leads to variation in cost. The components were ordered as and when required. For example, as in collected data, Ball Bearing 6013 has large variation in the cost in the month say January and March. In both the months quantity ordered is same but the cost is Rs. 445 In January while in Rs. 1204 in March. This results in high inventory costs.

Calculations

Total inventory cost was calculated using the following formula. The costs were calculated for each component and total cost was calculated by summing up individual costs.

$$\text{Existing cost of each component} = \text{Quantity ordered in that month} \times \text{Cost of component in that month} \quad (1)$$

$$\text{Total existing cost} = \sum_{1}^{48} \text{Existing cost of each component} \quad (2)$$

The total cost of purchased items was calculated as Rs. 85,31,871.

New Cost (Considering Minimum Cost)

As in the existing cost calculations, varying cost is considered, as per that for new cost consider only the minimum cost of component in that year.

Using equation (1) and (2), new cost was calculated. The cost per item here is considered as the minimum cost incurred during the year. No quantity discounts are considered.

$$\text{Cost Saving} = \text{Total existing cost} - \text{Total new cost}$$
$$\%\,\text{Saving} = (\text{Cost saving} \div \text{Total existing cost}) \times 100$$

Where, minimum cost value gives Rs. 65,01,978. Which gives almost Rs. 20,30,893 savings over existing expenditure.

Lot Sizing Techniques

Current study aims at seven lot sizing techniques for inventory management, which are Lot for lot, Economic Order Quantity (EOQ), Periodic Order Quantity (POQ), Least Total Cost, Least Unit Cost, Least Period Cost, Wagner-Whitin Model (WW).

All these lot sizing methods are applied on the data, an example of application of one of them is shown in Table 54.2.

After implementing all the seven algorithms to all components, calculated new cost using equation (2)

$$\text{Cost Saving} = \text{Total existing cost} - \text{Total new cost}$$
$$\% \text{ Saving} = (\text{Cost saving} \div \text{Total existing cost}) \times 100$$

After studying all these techniques, algorithm is prepared in excel for implementation of methods on data, as shown in Table 54.2. The table shows the small part of the calculation, similar to that algorithm were applied to all components. Comparison of results of algorithm is shown in Table 54.3.

Validation of Work

An algorithm was developed for selected techniques to optimize the inventory cost.

New data of another SME was collected to validate the current method. The current cost of inventory in this case is Rs. 96,32,510. Using the derived algorithm sheet of lot sizing technique, which is shown in Table 54.4, costs were recalculated. From the results of algorithm, it is observed that it gives substantial saving on inventory. And also gives idea, which technique is best suited for lot sizing. It was observed that lot sizing and inventory cost optimization were best suited ubiquitously. Results of algorithm and percentage savings are shown in Table 54.5.

Result and Discussion

Selection of inventory management policy influences productivity and profitably of any organization. In this paper, an integrated multi-period inventory management technique with stochastic demand is developed to reduce the inventory cost in SMEs. ROC and lot sizing techniques are coupled together to take maximum advantage of quantity discounts. The practicality of developed approach is tested with a case study conducted on a SME which shows substantial amount of cost saving. The computational evidence indicates that this developed approach is a viable alternative to other approaches found in the literature to tackle inventory management issues of SMEs. Also, this approach can become a practical tool for inventory managers working in SMEs who do not possess the resources necessary to use specialized algorithms.

Conclusion

The results of numerical example illustrate that proposed method has satisfactory performance in managing inventories in SMEs. Application of ROC resulted in identifying the

Table 54.2: Application of lot sizing techniques on data (Wagner-Whitin).

Inventory cost/Period/ Component	21.55	21.55	21.55	21.55	21.55	21.55	21.55	21.55	21.55	21.55	21.55	21.55
Ordering cost/Period (At)	150	150	150	150	150	150	150	150	150	150	150	150
Demand/Period (Dt)	0	0	0	0	5	5	10	10	10	0	0	25
Period (t)	1	2	3	4	5	6	7	8	9	10	11	12
1	150	150	150	150	581	1119.75	2412.75	3921.25	5645.25	5645.25	5645.25	11571.5
2		300	300	300	623.25	1054.25	2131.75	3424.75	4933.25	4933.25	4933.25	10320.75
3			300	300	515.5	838.75	1700.75	2778.25	4071.25	4071.25	4071.25	8920
4				300	407.75	623.25	1269.75	2131.75	3209.25	3209.25	3209.25	7519.25
5					300	407.75	838.75	1485.25	2347.25	2347.25	2347.25	6118.5
6						450	665.5	1096.5	1743	1743	1743	4975.5
7							557.75	773.25	1204.25	1204.25	1204.25	3898
8								707.75	923.25	923.25	923.25	3078.25
9									857.75	857.75	857.75	2474
10										1007.75	1007.75	2085.25
11											1007.75	1546.5
12												1007.75
Min costs =	150	150	150	150	300	407.75	557.75	707.75	857.75	857.75	857.75	1007.75

Table 54.3: Comparison sheet of algorithms and percentage savings.

Part No.	Part Name	Annual Requirement	Price	Qty x Price	LFL	EOQ	POQ	LTC	LUC	LPC	W-W	Min cost	Best
40	Internal Circlip B 140	250	97.1	24275	24575	31788.49	25603.25	24655.92	24655.92	24655.92	24655.92	24575	LFL
41	Internal Circlip B 110	900	17.6	15840	16440	20806.53	16524.67	16363.33	16448	16363.33	16363.33	16363.33	LTC/LPC/W-W
42	Internal Circlip B 90	1100	38	41800	42850	49287.22	42921.67	42835	43025	42858.33	42835	42835	LTC/W-W
43	Oil seal 55x72x10	1340	24.9	33366	34566	36991.47	34696.4	34277.25	34355.5	34323.5	34277.25	34277.25	LTC/W-W
44	Oil seal 85x120x12	630	50	31500	32850	35739.43	32691.67	32566.67	32566.67	32566.67	32416.67	32416.67	W-W
45	Level Plug 1/4"BSP	2550	12	30600	31350	35216.83	32040	31400	31510	31400	31400	31350	LFL
46	Drain Plug 1/4"BSP	5503	9	49527	50277	53780.06	50652	50352	50772.9	50352	50352	50277	LFL

Total cost 63,90,199

TOTAL SAVING 21,42,672

% SAVING 25.1108

Table 54.4: Validation of algorithms (Wagner-Whitin).

Period (t)	1	2	3	4	5	6	7	8	9	10	11	12
Inventory cost/Period/Component (ht)	11.45	11.45	11.45	11.45	11.45	11.45	11.45	11.45	11.45	11.45	11.45	11.45
Ordering cost/Period (At)	150	150	150	150	150	150	150	150	150	150	150	150
Demand/Period (Dt)	20	22	5	0	10	20	65	240	40	25	35	10
1	150	401.9	516.4	516.4	974.4	2119.4	6584.9	25820.9	29484.9	32061.15	36068.65	37328.15
2		300	357.25	357.25	700.75	1616.75	5338	21826	25032	27322	30928.75	32073.75
3			450	450	679	1366	4343	18083	20831	22834.75	26040.75	27071.25
4				507.25	621.75	1079.75	3312.5	14304.5	16594.5	18312	21117.25	22033.25
5					507.25	736.25	2224.75	10468.75	12300.75	13732	16136.5	16938
6						657.25	1401.5	6897.5	8271.5	9416.5	11420.25	12107.25
7							807.25	3555.25	4471.25	5330	6933	7505.5
8								957.25	1415.25	1987.75	3190	3648
9									1107.25	1393.5	2195	2538.5
10										1257.25	1658	1887
11											1407.25	1521.75
12												1557.25
Min costs =	150	300	357.25	357.25	507.25	657.25	807.25	957.25	1107.25	1257.25	1407.25	1521.75

Table 54.5: Comparison sheet.

Part No.	Part Name	Annual Requirement	Price	Qty x Price	LFL	EOQ	POQ	LTC	LUC	LPC	W-W	Min cost	Best
95	Oil Seal 20X42X7 Neoprine	170	12.85	2184.5	2784.5	4164.976	2743.642	2572.308	2593.642	2518.767	2518.767	2518.767	LPC/WW
96	Ball Bearing 6205 FAG	14	29	406	556	955.3435	562.7667	610.1333	610.1333	706	610.1333	556	LFL
97	Copper Washer Ø8	1000	3	3000	3150	4791.169	3300	3150	3150	3150	3150	3150	LFL/LTC/LUC/LPC/WW
98	Ball Bearing 16007 FAG	270	161	43470	43920	50339.9	44740.83	19620	43920	43920	43920	19620	LTC
99	Grub Screw M6X10	400	2.23	892	1492	1460.013	1053.15	1142.35	1142.35	1218.017	1142.35	1053.15	POQ
100	Key 6x6x30 (OER)	100	8.5	850	1150	1319.02	1014.167	1134.583	1134.583	1171.25	1134.583	1014.167	POQ
												cost after algorithm	4875581
												Existing cost	9632510
												savings	4756929
												% savings	0.493841

common components from different assemblies, which resulted into aggregation of quantities and application of lot sizing techniques enabled ordering in large quantities. This integrated approach saved 25 percent of total cost. The above method suggests that using an integrated approach, ordering policies can be derived to generate substantial savings. SME's can work in a systematic way to save inventory costs.

Grouping of items and identification of correct technique leads to substantial savings. The method is applied to a small data set. If applied to the complete organization, savings will be much more than what is being shown in the current study. This approach can be easily adopted by purchasing managers working in SMEs. There can be an extension of this work which might be interesting research problems. An incremental discount scheme can be applied to this problem to evaluate the cost savings and can be studied further for the best combination with ROC.

Acknowledgments

We would like to thank all the members who directly or indirectly helped us while carrying out this work.

Declaration

The authors report to conflicts of interest. The authors are alone responsible for the content and writing of this article.

References

1. Mani, V., Jabbour, C. J. C., and Mani, K. T. (2020). Supply chain social sustainability in small and medium manufacturing enterprises and firms' performance: empirical evidence from an emerging asian economy. *International Journal of Production Economics,* 227, 1–13.
2. Crawford, R. L., and Ibrahim, A. B. (1985). A strategic planning model for small business. *Journal of Small Business & Entrepreneurship,* 3(1), 45–53.
3. Mayer, K. B., and Goldstein, S. (1961). The first two years: problems of small firm growth and survival. *Small Business Administration,* 67(5).
4. Gupta, R. V. P. (1986). Inventory model with multi-items under constraint systems for stock dependent consumption rate. *Operations Research,* 24(1), 41–42.
5. A. O. D. A. H. M. Elaswad (2019). Cost optimization approach in input-output of manufacturing smes growth towards sustainability: a review. *International Journal of Engineering Research and Management (IJERM),* 6(5), 2058–2359.
6. T. C. and Bressan, A. (2018). A systematic review of future research challenges and prospects of organizational learning research in small medium size enterprises. *Journal of Small Business and Enterpreneurship,* 30(2), 175–191.
7. N. Y. a. F. Taghizadeh-Hesary (2016). Major challenges facing small and medium-sized enterprises in asia and solutions for mitigating them. *ADBI Working Paper,* no. 564.
8. F. J. and M. H. B. M. Sopha (2020). Analysis of the uncertainty sources and SMEs' performance. *Journal of Small Business and Entrepreneurship,* 37(1), 187–212.

55 Investigation on the effect of diethyl ether (DEE) with waste fish oil biodiesel in direct injection (DI) diesel engine

Ambeprasad Kushwaha[a], Gaurav Bhawde[b], Mohammad Arif I. Upletawala[c], and Ajoy kumar[d]

St. John College of Engineering and Management, Palghar

Abstract: The study deals with the test conducted on direct injection CI engine to obtain characteristic different proportions of diethyl ether in waste Fish Oil biodiesel liquid blends. The addition of DEE (Diethyl Ether) additive in biodiesel is limited to 5% to enhance the properties of biodiesel and to control the emissions. The DEE 5% is blended with 5%, 10%, 15%, 20%, 25%, 30% Fish oil to obtain different liquid blends (B00, B05, B10, B15, B20, B25, B30). The trials were conducted in a Four-Stroke single-cylinder direct injection CI Engine at a constant speed of 1500 rpm. The results show reduction in NOx due to the addition of DEE and CO emission for B10, B15 Biodiesel blend. Drop in specific fuel consumption is reflected and 8% increase in brake thermal efficiency when engine is fuelled with Biodiesel.

Keywords: Biodiesel, brake specific fuel consumption, diethyl ether, emission, fish oil.

Introduction

The largest source of non-renewable energy is going to be reduce in the near future, due to excessive consumption in different sector. The consumption has tremendously increased in the 19th and 20th century after the start of Industrial revolution due to its reliability, better combustion efficiency and better output of power, diesel engines are used in agriculture, industry and transportation sectors. Biodiesel is a sustainable, biodegradable, and non-toxic fuel with properties similar to diesel. As a result, it can be used in diesel engines when combined with diesel [1]. Diesel engines with diesel-biodiesel blends produce lower soot, HC and CO emissions [2]. Biodiesel has gained wide acceptance as another energy supply despite its high value of production [3]. In the study conducted by Farid et al. [4], Glycerol etherification is a good way to transform glycerol into fuel additives, and the TGME additive is a crucial one. Engine performance parameters, brake thermal efficiency, and induced power cost can all be improved by adding oxy-genated additives to diesel fuel. For biodiesel transesterification, a base catalyst method was used, and up to 50% transesterification was achieved. ASTM criteria were used to classify the biodiesel (ASTM D 6751-09) [5]. Dennis Y. C Leung et al. [6], investigated that fish oil biodiesel has the property to oxidize during storage, 440ppm of Ethoxyquin, an antioxidant reduced oxidation.

[a]ambeprasad88@gmail.com, [b]gauravb@sjcem.edu.in, [c]arifupletawala@gmail.com, [d]drajoy@gmail.com

In the research conducted by Limin Geng et al. [7], The spray characteristics were analysed using a high-speed camera and Mal-vern optical system analysis, while the combustion and emission efficiency of a turbocharged common rail direct injection (CRDI) ICE was tested. The findings reveal that adding ethanol to biodiesel increases the spray cone angle (SCA) and decreases spray tip penetration (STP). Addition of ester to ethanol–diesel blends cause an improvement in the stability of fuel blend with desirable effect of reduced Nox, low UBHC [8]. The Experimentation conducted by Oyetola Ogunkunle et al. [9], Biodiesel made from Parinari polyandra oil extracted via alkali catalysed methanolysis The results of varying Parinari polyandra biodiesel blends revealed that B10 was the best blend for improving the engine's overall performance in terms of speed, energy, and thermal efficiency, while B30 showed stable performance characteristics. There is recorded decreased in emission compared to diesel fuels. Mohankumar Subramaniam et al. [10], This research attempted to acquire algae fuel and blend it with diesel on a volume basis, namely A10, A20, A30, A40, and A100, with testing carried out in single cylinder direct injection diesel engines. Closer calculation with diesel results in higher thermal efficiency, lower HC, CO, Smoke, and particulate matter emissions among the blends evaluated. Canola oil up to 5% in the diesel fuel blend shows less CO emissions [11]. Hazelnut oil diesel blend can be used as an alternative fuel in prechambered diesel engine without any modification to give better performance and emission parameter [12]. A.J Torregrosa et al. [13], analyse the performance, emission and noise quality using Fischer-Tropsch and biodiesel fuels, results show less variation in the noise quality, decreasing soot level and increase in NOx emissions. H.K Rashedul et al. [14], reviewed the effect of various additives on the performance of biodiesel fuelled C.I Engine which shows additives to be useful for improving both the properties, Engine performance and emission control. The engine output and exhaust emissions of an HINO H07C DDF engine running on diesel, biodiesel, diesel–CNG, and biodiesel–CNG were studied in the lab, and it was found that biodiesel could be used as an alternative and environmentally friendly fuel without any engine modifications, particularly in heavy transportation fleets. [16]. Fish oil extracted shows the presence of palmitic acid, oleic acid and linolenic acid as a desirable biodiesel component with zero traces of glycerol [17].

Biodiesel Preparation

Fish oil is extracted from the waste fish slurry obtained from the local fish market in Palghar and then filtration process is done so that some solid particles are separated. This fish oil is poured in the beaker and it is heated with the help of mechanical heater and the stirrer is rotating at 600 rpm. Fish oil is heated at 110°C for 20 minutes and this process is called as DE moisturization. This process helps in taking out the moisture content in the oil. Fish oil is allowed to cool for some time and Acid esterification is done. As the Free Fatty Acid (FFA) content is very high in fish oil, this process is done 2 times. Acid esterification is done by adding 0.5% of sulphuric acid which works as a catalyst with 0.7% of methanol as a solvent. This process is done for 45 mins and temperature is maintained 55–60°C. 0.4% of Potassium Hydroxide crystals are dissolved in 7% of methanol and this mixture is added in the oil and heated for 20 mins by maintaining the temperature 55–60°C. This process is called as Transesterification. This mixture is allowed to cool for about 8–10 hours and settling process takes place. Solid glycerine (20%) and biodiesel (80%) is obtained and separated. For purification, hot distilled water is added and this mixture is heated for 20 minutes at about 110°C.

One litre of each blend was used in the experiment and the analysis was done. The produced Biodiesel was compared to the American Society for testing and Materials standards before its testing and analysis on the Diesel Engine.

B00–100% Pure Diesel
B05–5% Fish oil Biodiesel + 5% additives + 90% Pure Diesel
B10–10% Fish oil Biodiesel + 5% additives + 85% Pure Diesel
B15–15% Fish oil Biodiesel + 5% additives + 80% Pure Diesel
B20–20% Fish oil Biodiesel + 5% additives + 75% Pure Diesel
B25–25% Fish oil Biodiesel + 5% additives + 70% Pure Diesel
B30–30% Fish oil Biodiesel + 5% additives + 65% Pure Diesel

Properties of Waste fish oil Biodiesel were obtained as shown in Table 55.1.

Table 55.1: Properties of the fuel and DEE.

Property	Diesel	DEE	Fish oil B20
Viscosity(cSt)	2.75	0.23	2.96
Density (Kg/m³)	0.83	0.713	0.837
Cetane number	45	>125	50.13
Calorific Value (kj/Kg)	43000	33,900	41900
Auto Ignition temperature	250°C	160°C	102°C

Experimental Setup

The test rig used for the experimental evaluation is shown in Figure 55.1. It is a single cylinder, 4-strokes, water cooled CI engine with cylinder bore of 87.50 (mm), stroke length of 110 (mm), connecting rod length of 234 (mm) with variable compression ratio and its Swept volume is 661.3 (cc). The load on the engine is varied by using an eddy current dynamometer.

Figure 55.1 Experimental setup.

Results and Discussion

The aim of this study was to look into the performance parameters and exhaust emissions of a single cylinder diesel engine that ran on both standard diesel (SD) and biodiesel fuel. This research was carried out to see whether the exhaust emission and engine efficiency could be improved. Pollutant gases such as nitrogen oxides (NOx), carbon monoxide (CO) are examined. Mechanical Efficiency, Brake Thermal Efficiency, and Specific Fuel Consumption were the calculated parameters to reflect engine output. The results were obtained by using AVI smoke meter and Exhaust gas Analyzer.

Engine Performance and Emissions

Figure 55.2 shows how the presence of biodiesel decreases mechanical efficiency at full load. However, at 50% of applied load, the Mechanical Performance of Different Blend Biodiesel shows little difference. Close observation is obtained at SD and B30 Blend at full operating load. Brake thermal efficiency increases for all biodiesel blends from 50% to 100% of operating load as shown in Figure 55.3. Brake thermal efficiency is maximum for B25 blend at full load compared to SD blend showing an increase of 8%. Figure 55.4 shows the results of the BSFC graph, that as the percent of waste fish oil fuel increases, so does specific fuel consumption. Such a finding may be explained by the influence of Biodiesel's lower heating value [15]. At 75% of operating load, all blends have virtually the same SFC. As the biodiesel content increases, CO concentration decreases, as seen in Figure 55.5. There is a significant difference between the SD and B20 blends. At 50% of load B05, B10, B15, B20, B25 shows reduced CO emission. At full load B05 and B10 shows lowest CO emissions. Further increase in biodiesel content from B15 to B30, increase in CO Emission is observed.

In Figure 55.6, B30 show lower NOx emission compared to SD B00. NOx is formed when the combustion temperature in the engine reaches 1100°C. Addition of biodiesel lowers the Calorific value of fuel which results in longer ignition delay, restricting the combustion chamber temperature to reach above 1100°C and thus reduces the NOx.

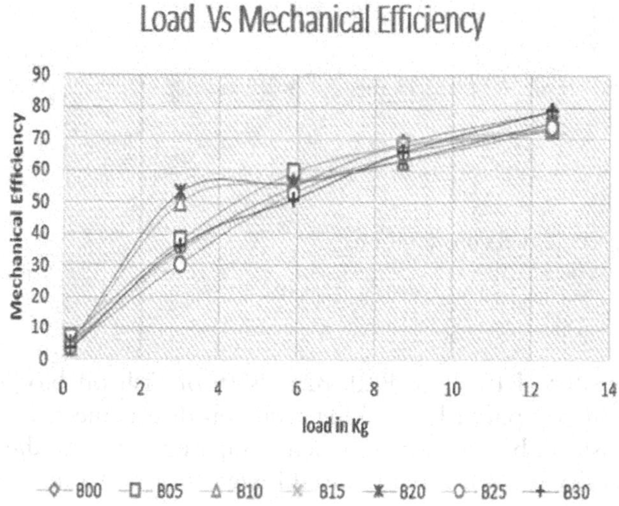

Figure 55.2 Load vs mechanical efficiency.

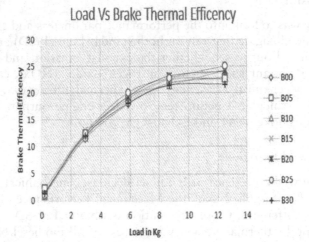

Figure 55.3 Load vs brake thermal efficiency.

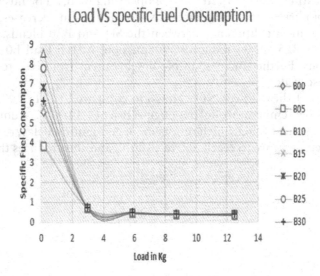

Figure 55.4 Load vs specific fuel consumption.

Conclusion

Six different blends (B05, B10, B15, B20, B25, B30) of Fish oil biodiesel and Standard Diesel B00 have been compared by running trials on direct injection diesel engine. The property of Waste Fish oil biodiesel used, when compared with Standard fuel, has a comparatively higher density (+7%), and a reduced lower calorific Value (–2.5 %).

1. Drop in Specific fuel consumption is obtained at 25% of load for all biodiesel Fuel Blend.

Load vs CO Emission

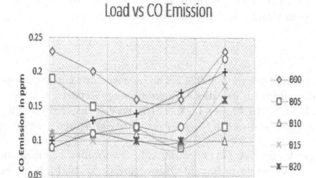

Figure 55.5 Load vs CO emission.

Load vs NOx Emission

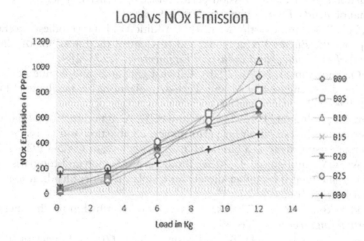

Figure 55.6 Load vs NOx emission.

2. At 25% of operating load blend B10 shows the highest rise in Mechanical efficiency compared to standard diesel (SD).
3. Brake thermal Efficiency increases for B25 blend at full load by 8%, and at 75% of operating load Brake thermal efficiency remains same and shows similar pattern.
4. The blends of B10 and B15 has considerably high CO reduction as compared to the standard diesel due to the complete combustion Process [2].
5. In contrast to standard Diesel (SD) at 75 percent of load, a pattern of NOx emission reduction was observed for all forms of Fish oil biodiesel blends. The maximum reduction of NOx was obtained at B30 blend. Addition of 5% of DEE helps in NOx reduction as it has high cetane number and low boiling point.

6. The current work can be used to compare the emission and performance parameter by varying the injection timing and the fuel injection pressure.

References

1. Ma, Q., Zhang, Q., and Liang, J. (2021). The performance and emissions characteristics of diesel/biodiesel/alcohol blends in a diesel. *Engine Energy Reports*, 7, 1016–1024.
2. Chen, H., Ding, M., Li, Y., Xu, H., and Li, Y. (2014). Feedstocks, environmental effects and development suggestions for biodiesel in China. *Journal of Traffic and Transportation Engineering*, 7(6), 791–807.
3. Adekunle, A. S., Oyekunle, J. A. O., Oduwale, A. I., Owootomo, Y., Obisesan, O. R., Elugoke, S. E., Durodola, S. S., Akintunde, S. B., and Oluwafemi, O. S. (2020). Biodiesel potential of used vegetable oils transesterified with biological catalysts. *Engine Energy Reports*, 6, 2861–2871.
4. Shoar, F. H., Najafi, B., and Mosavi, A. (2021). Effects of triethylene glycol mono methyl ether (TGME) as a novel oxygenated additive on emission and performance of a dual-fuel diesel engine fueled with natural gas-diesel. *Engine Energy Reports*, 7, 1172–1189.
5. Lhawanga, T., Sirigeri, S., Menona, S., and Vadiraj, K. T. (2021). Azadirchta excelsa seed oil, a potential non-conventional biodiesel feedstock. *Environmental Challenges*, 3, 100057
6. Leiung, D. Y. C., Wu, X., and Leung, K. H. (2010). A review based on biodiesel production using catalyzed transesterification. *Applied Energy*, 87(4), 1083–1095.
7. Geng, L., Bi, L., Li, Q., Chen, H., and Xie, Y. (2021). Experimental study on spray characteristics, combustion stability, and emission performance of a CRDI diesel engine operated with biodiesel–ethanol blends. *Energy Reports*, 7, 904–915.
8. Jagadish, D., Puli, R. K., and Murthy, M. (2011). Addition of Ester (Biodiesel) to Ethanol-Diesel Blend to Improve the Engine Performance and to Control the Emissions of Nitrous Oxides. *Jordan Journal of Mechanical and Industrial Engineering*, 3, 275–284.
9. Ogunkunle, O., and Ahmed, N. A. (2020). Exhaust emissions and engine performance analysis of a marine diesel engine fuelled with Parinari polyandra biodiesel–diesel blends. *Energy Reports*, 6, 2999–3007.
10. Subramaniam, M., Solomon, J. N., Nadanakumar, V., Anaimuthu, S., and Sathyamurthy, R. (2020). Experimental investigation on performance, combustion and emission characteristics of DI diesel engine using algae as a biodiesel. *Energy Reports*, 6, 1382–1392.
11. Roy, M. M., Wang, W., and Bujold, J. (2013). Biodiesel production and comparison of emissions of a DI diesel engine fuelled biodiesel–diesel and canola oil–diesel blends at high idling operations. *Applied Energy*, 106, 198–208.
12. Çetin, M., and Yüksel, F. (2007). The use of hazelnut oil as a fuel in prechamber diesel engine. *Applied Thermal Engineering*, 27, 63–7.
13. Torregrosa, A. J., Broatch, A., Plá, B., and Mónico, L. F. (2013). Impact of fischere–tropsch and biodiesel fuels on trade-offs between pollutant emissions and combustion noise in diesel engines. *Biomass and Bio Energy*, 52, 22–33.
14. Rashedul, H. K., Masjuki, H. H., Kalam, M. A., Ashraful, A. M., and Rahman, S. M. A. (2014). The effect of additives on properties, performance and emission of biodiesel fuelled compression ignition engine. *Energy Conservation and Management*, 88, 348–364.
15. Mattarelli, E., Rinaldini, C. A., and Savioli, T. (2015). Combustion analysis of a diesel engine running on different biodiesel blends. *Energies*, 8(4), 3047–3057.
16. Mohsin, R., Majid, Z. A., Shihnan, A. H., Nasri, N. S., and Sharer, Z. (2014). Effect of biodiesel blends on engine performance and exhaust emission for diesel dual fuel Engine. *Energy Conversion and Management*, 88, 821–828.
17. Kara, K., Ouanji, F., Lotfi, E. M., El Mahi, M., Kacimi, M., and Ziyad, M. (2018). Biodiesel production from waste fish oil with high free fatty acid content from moroccan fish-processing industries. *Egyptian Journal of Petroleum*, 27, 249–255.

56 Experimental investigation on feasibility of using stokes settling law for batch settling at different temperatures and performance analysis of magnetorheological fluids using magnetorheological damper

Ashok Kumar Kariganaur[a], Hemantha Kumar[b], and Arun Mahalingam[c]

Department of Mechanical Engineering, NITK Surathkal, Mangalore, India

Abstract: Settling is a major phenomenon in systems where particles consist of a certain mass is dispersed in the dielectric base fluid. To increase the settling time of the particles some amount of additive is added. In this study, the characterization of silicone oil and fork oil for its density and viscosity on basic Redwood viscometer at different temperatures is carried out along with the characterization of pre-pared magnetorheological (MR) fluids using silicone oil as the carrier fluid. The properties of both the carrier fluids were used to calculate the stokes settling velocity and it is compared with the sedimentation rate of the magnetorheological fluids (batch settling). The results show that stokes law cannot be used for MR fluid batch settling evaluation. The characterization of silicone-oil based MR flu-id sample at 0A,0.25A,0.5A,0.75A,1A and 2A currents is also carried out for their flow characteristics and observed that MR fluid produce high yield behavior at higher currents. And the performance analysis of the prepared MR fluid is characterized using MR damper has one of the applications at different ampli-tudes, frequencies and currents.

Keywords: HB model, MR damper, MR fluids, redwood viscometer, rheometer, stokes law.

Introduction

The subclass of smart fluids is magnetorheological fluids where magnetic particles are dispersed in the dielectric base fluid. The peculiar ability of these fluids is they can change their yield point on the application of an external magnetic field [1]. Some of the applications where magnetic fluids were used are magnetorheological damper, magnetorheological brake, magnetorheological clutch, prosthetics etc. Using hydrophilic treated magnetic particles in the preparation of MR fluids, sedimentation can be reduced drastically with similar flow behaviour as that of the commercially available MR fluids [2]. Adding polymethyl methacrylate (PMMA) particles to MR fluids can increase the stability and also loss modulus and shear modulus increases with an increase in volume fraction of PMMA par-ticles compared to other surfactant addition [3]. Higher particle size increases the sedimentation constant and particle increase increases the yield stress of the prepared sample [4]. Particle is with lower diameter having a higher magnetic saturation compared high particle size and also increase in volume fraction in-creases yield stress of the MR fluid [5, 6]. Oleic acid and aluminum stearate give better dispersibility stability compared

[a]ashokk.nitk@gmail.com, [b]hemanta76@gmail.com, [c]m.arun1978@gmail.com

to silica gel but the latter give better sta-bility at static position losses its stability when some amount of shear is applied [7]. Coating of particles and plasma-treated particles will increase the sedimentation behaviour of MR fluids and yield point can be increased [8, 9]. Nanostructures with wire-like shape are mixed into conventional magnetic particles de-creases the sedimentation rate and increases the yield stress of MR fluid [10]. Bidisperse magnetic particles will give enhanced sedimentation stability and yield stress of MR fluids can be increased by increasing the applied magnetic field [11, 12]. Magnetic saturation will decrease with an increase in temperature of the particles and yield stress of the samples will also decrease at higher temperatures [13]. Particle size and volume fraction play a signifi-cant role in enhanced flow behaviour of magnetorheological fluids and also non-spherical particle will also improve the sedimentation and flow behaviour of the MR fluids [14]. Graphene oxide additive improves sedimentation stability and also gives better flow prop-erties [15]. Optimal design of the MR damper is considered through finite element (FE) analysis using the golden section algorithm and parametric language [16]. Using different piston profiles optimisation of double coil MR damper dimensions is obtained for better performance [17]. MR fluid performance analysis is carried out in damper working with different currents and shear rates using Bingham fluid model. Pressure drop is reduced due to an increase in the fluid flow gap [18]. Experimental and numerical study has been car-ried out using the FE 2D model and the results show that a close relationship between the numerical study can be reliable [19].

The study consists of the preparation of two MR fluid samples of silicone oil and hydrau-lic oil as the carrier fluid. After preparation, the samples were employed for sedimentation study for some amount of time then the sedimentation rate is compared with that of the stoke's settling velocity for application of this law to batch settling of particles and finally characterization of the is carried out for their flow properties.

Methodology

Particle Characterization, Preparation, Stability Analysis and Characterization of MR Fluid

Before going to find the velocity using stokes law, the properties of carrier fluid from which MR fluids are prepared are obtained at various temperatures using a redwood vis-cometer, by which kinematic viscosity and density variation are obtained as a function of temperature. Further by using the properties of carrier fluids, stokes velocity is calculated. Preparation of MR fluid is carried out by dis-persing a 30% volume fraction of magnetic particles in the carrier fluid with 1.5% additive is added and made to stir for some time. Figure 56.1a shows the red-wood viscometer to measure the density and viscosity of sil-icone oil and fork oil. Figure 56.1b shows the complete methodology flowchart of the present study. After preparation, the samples were tested on the rheometer for their flow behaviour by maintaining a constant magnetic field and giving a variable shear rate. The characterized MR fluid is fitted with the Herschel-Bulkley model [19] to see the yield stress of both samples

$$\tau = \tau_o + K\dot{\gamma}^n \qquad\qquad 1)$$

τ = shear stress [Pa] $_o$ = yield stress [Pa], K = constitutive index, $\dot{\gamma}$ = shear rate [s^{-1}], n = flow index. For n<1 the fluid is shear-thinning, whereas for n>1 the fluid is shear-thicken-ing. If τ_o = 0 and n = 1, this model reduces to the Newtonian fluid.

Stoke's velocity can be calculated by obtaining the properties of the base fluid and carbonyl iron powder. Redwood viscometer is used to obtain the properties such as kinematic viscosity, density and dynamic viscosity of the carrier fluids. The equation used to measure the stokes velocity is given below.

$$V = (d^2 * (\rho_p - \rho) * g)/18\mu \qquad 2)$$

where d = diameter of the particle(m) ρ_p = density of the particle (kg/m³), ρ = density of carrier fluid (kg/m³), g = acceleration due to gravity(m/s²), μ = dynamic viscosity (kg/m-s).

Kinematic Viscosity is measured by filling the sample cup with a carrier fluid and heating the cup until the required temperature is reached, immediately opening the orifice hole

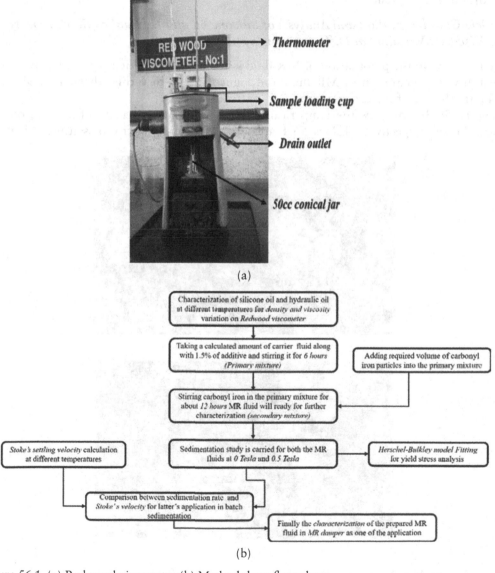

(a)

(b)

Figure 56.1 (a) Redwood viscometer (b) Methodology flow chart.

under the cup and sample fluid is collected in a *50cc* conical flask and time taken to fill the conical flask is noted. Finally, the collected fluid is weighed to measure the mass of fluid by subtracting the mass of the empty conical flask, the equation below gives the kinematic viscosity of the carrier fluid.

$$\text{Kinematic Viscosity (K.V.)} = [0.26^*t - 179/t]; \text{ when } t < 100 \text{ seconds (cSt)} \tag{3}$$

$$= [0.247^*t - 50/t]; \text{ when } t > 100 \text{ seconds (cSt)} \tag{4}$$

$$\text{Dynamic Viscosity} = K.V.^* \rho_{oil} \tag{5}$$

where t is time taken to fill the 50CC conical jar.

Results and Discussion

Particle Characterization and Analysis of Silicone oil and Hydraulic Oil for Density and Viscosity Variation at Different Temperatures

The first step in the preparation of MR fluid is to verify the particle size and shape for its feasibility in preparation of MR fluid and Figure 56.2(a–b) verifies through SEM image and particle size of the carbonyl iron (CI) particles.

Figure 56.3(a–b) gives the temperature versus kinematic viscosity of silicone oil and hydraulic oil respectively. Tables 56.1 and 56.2 gives the velocity of settling calculated

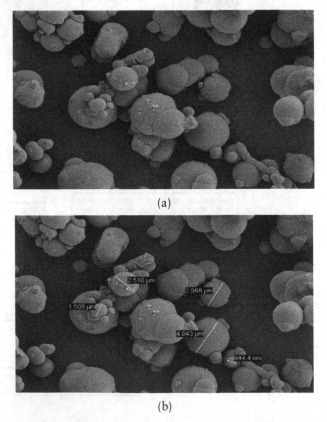

(a)

(b)

Figure 56.2 (a–b) SEM image and particle size of the particles used for MR fluid preparation.

from stoke's law. From the results it is evident that hydraulic oil-based MR fluid will settle at a faster rate when compared to silicone oil-based MR fluid. The viscosity and density of silicone oil are more than the hydraulic oil which will give more advantage in reducing settling and yield behaviour of the silicone oil-based MR fluid. Generally, oil viscosity decreases by the exponential pattern with an increase in temperature [20]. The MR fluid analysis plays an important role when it comes to the application part for their stability, dispersibility and flow characteristics which decide the performance of the system for long term operations and also a life of the MR fluid along with the system.

Sedimentation Rate of MR Fluids

By the definition, sedimentation ratio is the particles settle forming a layer leaving behind clear fluid [21]. The sedimentation rate is calculated by noting down, the time taken by the

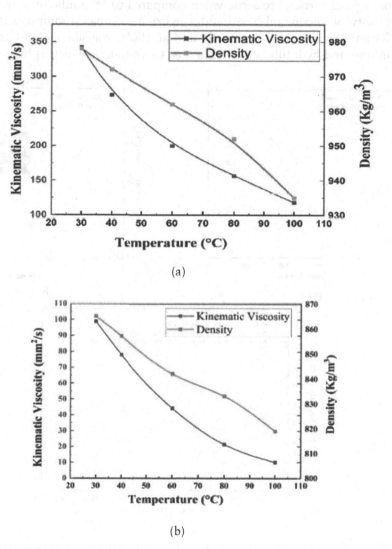

(a)

(b)

Figure 56.3 (a–b) Kinematic viscosity, density vs temperature variation of silicone oil and hydraulic oil.

fluid layer to settle at individual temperatures maintained in the incubator after the incubator reaches a steady state. The carrier fluid has a major effect on the sedimentation rate of the MR fluid than the particles because the effect of a decrease in viscosity and density with an increase in temperature is seen in base fluid and as the temperature increases the particles settle at a faster rate. At 30°C and 100°C, the velocity of hydraulic oil-based MR fluid is two times faster than silicone oil-based MR fluid. From the results, it is also evident that the time taken by the fluid samples at 30°C is to attain a steady state for the silicone oil-based MR fluid sample is 100 hours and that of the hydraulic oil-based sample is 400 hours. And at 100°C the time for the silicone oil-based fluid to attain a steady state was 2 hours and for the hydraulic oil-based sample is 12 hours. From the analysis, it is observed that silicone oil-based sample is settling at a slower rate and takes fewer hours to attain steady-state compared to hydraulic oil-based sample which faster than the latter and takes 400hours to attain the steady-state and it can be observed that silicone oil-based sample is not allowing the particles to settle when compared to hydraulic oil sample. The stokes settling velocity of silicone oil-based and a hydraulic oil-based sample is 0.0395mm/min and 0.1574mm/min at 30°C respectively and at 100°C the silicone oil-based sample has 0.1204mm/min and hydraulic oil sample has 1.62mm/min which is higher than that of

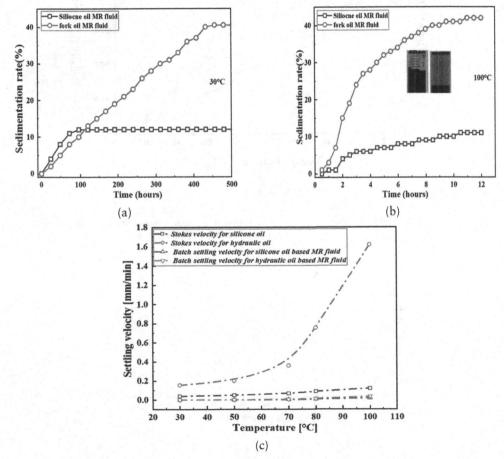

Figure 56.4 (a–b) Sedimentation rate of MR fluid at 30°C and 100°C, (c) Stoke's velocity comparison with batch sedimentation.

the batch sedimentation velocities. Figure 56.4(a–b) shows the settling velocity of both the samples at 30°C and 100°C. The detailed behaviour of MR fluids sedimentation rate and stokes velocity comparison for both the samples is depicted in Figure 56.4(c) and decay is fitted with the exponential fit which is having an $R^2 > 0.95$ which shows that the results are feasible.

Table 56.1: Velocity measurement for hydraulic oil.

Temperature (°C)	Kinematic Viscosity($m^2/s * 10^{-3}$)	Density (Kg/m^3)	Dynamic viscosity (Pa-s)	Velocity (mm/min)
30	0.09885	865	0.085505	0.1574
40	0.0779	857	0.0667603	0.2018
60	0.04431	842	0.03730902	0.36185
80	0.02146	833	0.01787618	0.7562
100	0.0101	819	0.0082719	1.62

Table 56.2: Velocity measurement for silicone oil.

Temperature(°C)	Kinematic Viscosity($m^2/s * 10^{-3}$)	Density (Kg/m^3)	Dynamic viscosity (Pa-s)	Velocity(mm/min)
30	0.34206	978	0.3345347	0.03957
40	0.27356	972	0.2659003	0.04982
60	0.2	962	0.1924	0.06897
80	0.1567	952	0.1491784	0.0891
100	0.11832	935	0.1106292	0.1204

Characterization of Silicone Oil-Based MR Fluid

The important parameter in the analysis of the MR fluids is their flow behaviour. Shear rate versus viscosity and shear rate versus shear stress is shown in Figure 56.5(a–b).

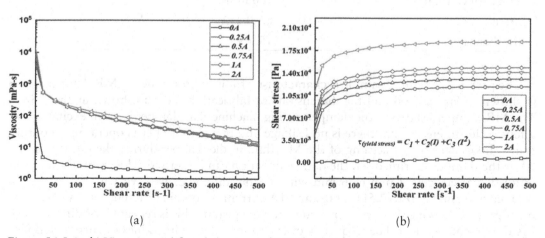

(a)   (b)

Figure 56.5 (a–b) Viscosity and flow behaviour of MR fluid at various currents.

The shear rate is kept at 500s⁻¹ throughout the tests and measured at 0A,0.25A,0.5A,0.75A,1A,2A currents. The viscosity of MR fluid samples increases drastically for higher currents making the yield point of the MR sample increase. By fitting the HB model to the flow characteristics, the yield stress is obtained. The yield stress obtained at 0A for the sample was 40Pa. And at 2A, the yield stress obtained was 8000Pa. The results indicate that MR fluid has better yield characteristics at higher currents.

Characterisation of MR Damper

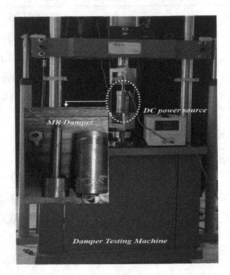

Figure 56.6 MR damper testing facility and its geometric dimensions.

Table 56.3: Dimensions of MR.

Flange length	12mm
Fluid gap	2mm
Number of coils turns	100 turns
Currents	0A,0.25A,0.5A,0.75A,1A
Core diameter	15mm

The prepared MR fluid sample is characterised for its performance in MR damper which is shown in Figure 56.6 and the dimensions of fabrication are also shown in Table 56.3. The MR damper is fitted to the damper testing machine. The damper is given excitation for a few cycles to ensure that there is no leakage and that the damper is operating smoothly. It also takes care of any settling of the MR fluid in the damper during the idle condition. After the damper is primed set, the data acquisition can be started. The amplitude of excitation is varied from 2 mm to 6 mm with a 2mm difference with varying the excitation frequency from 1.5Hz to 2.5Hz at 0A and 1A currents to obtain different force vs displacement graphs showing the behavior fluid inside the piston. This is repeated for different values of current supplied. The frequency of excitation along with values of current supplied

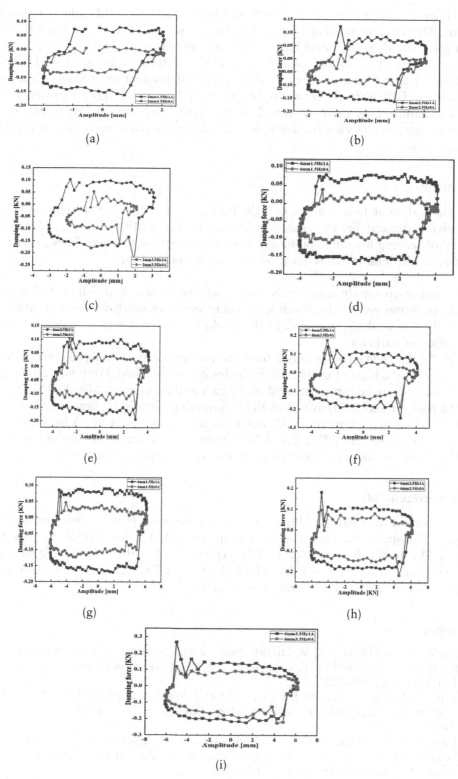

Figure 56.7 (a–i) Force vs Displacement curves at different currents, amplitudes and frequencies.

is provided. Figure 56.7(a–i) shows the force obtained at lower and higher amplitudes and currents. The force obtained at 2mm and 1.5Hz frequency at 0A and 1A is 8.94N and 72N which gives the dynamic range 8.056. The force obtained at 2mm and 2.5Hz frequency at 0A and 1A is 18.77N and 84.17N which gives a dynamic range of 4.325. From the above analysis, it is evident that dynamic range is dependent on frequencies at constant amplitude and currents. At 6mm amplitude and 1.5Hz and 2.5Hz frequency at 0A and 1A the dynamic range obtained was 2.96 and 1.631 respectively. From both amplitude and frequency point of view the dynamic range decreases and at constant current dynamic range increases.

Conclusions

From the sedimentation analysis of MR fluids, it is evident that silicone oil-based MR fluids give better stability than that of hydraulic oil-based MR fluid. When it comes to the velocity of settling hydraulic oil-based MR fluid sample settles two times faster than the silicone oil MR fluid which is not feasible for MR fluid stability.

- A comparative study between the Stokes settling velocity and sedimentation rate evaluation shows velocity for batch sedimentation is very much slower than stoke's velocity. By this analysis, it is evident that stoke's velocity law cannot be used for batch settling of particles.
- Finally, the yield stress obtained from the rheometric study was, silicone oil MR fluid gives 15.92% higher than that of the hydraulic oil MR fluid. From flow characteristics, silicone oil can be further studied as the carrier fluid in many MR fluid applications.
- The performance analysis of MR fluid shows that there is 3.2 times increase in damping force at no current and a 20% increase in damping force at 1A between 2mm and 6mm amplitude. At 1.5Hz and 2.5Hz frequency between 2mm and 6mm, there is a 75% increase and 53% increase at 1A and zero currents respectively.

Acknowledgements

The authors acknowledge the Ministry of Human Resource Development and the Ministry of Road Transport and Highways, Government of India, for supporting this research through IMPRINT Project No. IMPRINT/2016/7330 titled "Development of Cost-Effective Magnetorheological (MR) Fluid Damper in Two wheelers and Four Wheelers Automobile to Improve Ride Comfort and Stability".

References

1. Phulé, P. P., and Ginder, J. M. (2019). Synthesis and properties of novel magnetorheological fluids having improved stability and redispersibility. *International Journal of Modern Physics B*, 13(14n16), 2019–2027.
2. Park, J. H., Chin, B. D., and Park, O. O. (2001). Rheological properties and stabilization of magnetorheological fluids in a water-in-oil emulsion. *Journal of Colloid and Interface Science*, 240(1), 349–354.
3. Iglesias, G. R., Roldán, A., Reyes, L., Rodríguez-Arco, L., and Durán, J. D. G. (2015). Stability behavior of composite magnetorheological fluids by an induction method. *Journal of Intelligent Material Systems and Structures*, 26(14), 1836–1843.

4. Gorodkin, S. R., Kordonski, W. I., Medvedeva, E. V., Novikova, Z. A., Shorey, A. B., and Jacobs, S. D. (2000). A method and device for measurement of a sedimentation constant of magnetorheological fluids. *Review of Scientific Instruments*, 71(6), 2476–2480.

5. Acharya, S., Saini, T. R. S., and Kumar, H. (2019). Determination of optimal magnetorheological fluid particle loading and size for shear mode monotube damper. *Journal of the Brazilian Society of Mechanical Sciences and Engineering*, 41(10), 1–15.

6. Kittipoomwong, D., Klingenberg, D. J., and Ulicny, J. C. (2005). Dynamic yield stress enhancement in bidisperse magnetorheological fluids. *Journal of Rheology*, 49(6), 1521–1538.

7. López-López, M. T., Zugaldía, A., González-Caballero, F., and Durán, J. D. G. (2006). Sedimentation and redispersion phenomena in iron-based magnetorheological fluids. *Journal of Rheology*, 50(4), 543–560.

8. Cheng, H., Wang, M., Liu, C., and Wereley, N. M. (2018). Improving sedimentation stability of magnetorheological fluids using an organic molecular particle coating. *Smart Materials and Structures*, 27(7), 075030.

9. Jiang, W., Zhang, Y., Xuan, S., Guo, C., and Gong, X. (2011). Dimorphic magnetorheological fluid with improved rheological properties. *Journal of Magnetism and Magnetic Materials*, 323(24), 3246–3250.

10. Wereley, N. M., Chaudhuri, A., Yoo, J. H., John, S., Kotha, S., Suggs, A., Radhakrishnan, R., Love, B. J., and Sudarshan, T. S. (2006). Bidisperse magnetorheological fluids using Fe particles at nanometer and micron scale. *Journal of Intelligent Material Systems and Structures*, 17(5), 393–401.

11. Wang, X., and Gordaninejad, F. (2006). Study of magnetorheological fluids at high shear rates. *Rheologica Acta*, 45(6), 899–908.

12. Arief, I., Sahoo, R., and Mukhopadhyay, P. K. (2016). Effect of temperature on steady shear magnetorheology of CoNi microcluster-based MR fluids. *Journal of Magnetism and Magnetic Materials*, 412, 194–200.

13. De Vicente, J., Klingenberg, D. J., and Hidalgo-Alvarez, R. (2011). Magnetorheological fluids: a review. *Soft Matter*, 7(8), 3701–3710.

14. Zhang, W. L., and Choi, H. J. (2012). Graphene oxide added carbonyl iron microsphere system and its magnetorheology under applied magnetic fields. *Journal of Applied Physics*, 111(7), 2012–2015.

15. Nguyen, Q. H., and Choi, S. B. (2008). Optimal design of a vehicle magnetorheological damper considering the damping force and dynamic range. *Smart Materials and Structures*, 18(1), 015013.

16. Hu, G., Xie, Z., and Li, W. (2015). Optimal design of a double coil magnetorheological fluid damper with various piston profiles. *In World Congress on Structural and Multidisciplinary Optimisation*, 2–7.

17. Bajkowski, J., and Skalski, P. (2012). Analysis of viscoplastic properties of a magnetorheological fluid in a damper. *Acta Mechanica et Automatica*, 6(3), 5–10.

18. Mangal, S. K., and Kumar, A. (2015). Geometric parameter optimization of magneto-rheological damper using design of experiment technique. *International Journal of Mechanical and Materials Engineering*, 10(1), 1–9.

19. Acharya, S., Saini, T. R. S., and Kumar, H. (2019). Determination of optimal magnetorheological fluid particle loading and size for shear mode monotube damper. *Journal of the Brazilian Society of Mechanical Sciences and Engineering*, 41(10), 1–15.

20. Qin, Y., Wu, Y., Liu, P., Zhao, F., and Yuan, Z. (2018). Experimental studies on effects of temperature on oil and water relative permeability in heavy-oil reservoirs. *Scientific Reports*, 8(1), 12530.

21. Zhang, Y., Li, D., and Zhang, Z. (2020). The study of magnetorheological fluids sedimentation behaviors based on volume fraction of magnetic particles and the mass fraction of surfactants. *Materials Research Express*, 6(12), 126127.

57 Thermal performance of artificially packed bed solar air heater

Ajoy Kumar[1,a], Mohammad Arif Iqbal Upletawala[1], Ambeprasad Kushwaha[1], Kumar Abhinandan[2], and Gaurav Bhawde[1]

[1]Department of Mechanical Engineering, St. John College of Engineering and Management, Palghar, India

[2]Department of Mechanical Engineering, National Institute of Technology, Patna, India

Abstract: For use at low temperatures, the standard flat-type solar air heaters are utilized extensively. Due to its low convective heat transfer coefficient between the absorber plate and air, its efficiency is found to be low. Several methods have been proposed to enhance its efficiency. Many designers are drawn to packed bed solar air heaters because they have some unique advantages over the traditional approach. In this paper, an experimental investigation has been done on the performance of a solar air heater packed with scrap iron turnings carried out at varying mass flow rates under actual outdoor conditions. During the investigation, it has been found that the thermal efficiency of such solar air heaters varied between 40–80% compared with plane one having an efficiency value ranging between 20–30 % under the same mass flow rate and intensity of solar radiation available. Rating criteria including the overall heat loss coefficient, effective absorption coefficient, heat removal efficiency factor, and plate efficiency factor have all been recorded.

Keywords: Efficiency factor for heat removal, plate efficiency factor, total heat loss coefficient.

Introduction

Typically, solar air heaters have a rectangular duct with a high aspect ratio that is made up of a top absorber plate, a bottom plate, two side walls, and a glass cover or covers. Air conventionally flows below the absorber plate.

Many applications that call for low and moderate temperatures can benefit from using solar air heaters. Some of these include air conditioning and heating in buildings, drying in agriculture and industry, seasoning wood, and producing electricity. The efficiency of flat plate collectors is often low, which results in larger heat losses to the surrounding environment due to the low value of convective heat transfer coefficient between the absorber plate and the moving air. In this regard, various solar air heater designs have been created in the past to enhance their functionality. The flow duct of the solar air heater has been designed to be the hot spot of interest in order to increase the rate of heat collection. In addition to the methods used for increasing the intensity of radiation and reducing thermal losses from the absorber plate to ambient air, different packing media (wire mess, expanded metals, stone pebbles, etc.), extended surfaces, and artificial roughness have been

[a]drajoyk@gmail.com

incorporated with the flow duct to enhance the heat collection rate, resulting in a relatively low absorber plate temperature. Hence, it leads to an increase in the thermal efficiency of solar air heaters.

By using irregular shapes like ribs, grooves, baffles, winglets, and twisted tapes in an absorber plate, passive techniques are used to improve heat transfer in the solar duct [1–3]. Many authors [4–7] have carried out experimental and analytical studies using passive methods in absorber plates that resulted in improved heat transfer.

The exergy analysis of various solar collector types with corrugated and reverse trapeze, corrugated base flat plate collector was suggested by Huseyin Benli et al. [8].

The thermo hydraulic performance evaluation of rib roughness under experimentation, V-down ribs with gap, and similar reported roughness geometries used in solar air heater duct were presented by Sukhmeet Singh et al. [9].

Karwa et al. [10], experimented by using different rib arrangements, such as transverse, angled, continuous, and discrete, in v-pattern for ribs of different shapes, and expanded metal wire mesh. Due to the creation of local wall turbulence between artificial roughness of ribs, friction factor and heat transfer coefficient vary with the roughness type.

The present study examines an experimental investigation on a packed-bed absorber plate with black-painted scrap iron turnings. In order to compare the performance of the two collectors, a packed bed collector and an aircraft collector, tests have been carried out under identical real-world outdoor settings. These tests produce valuable information that can be used to rate these sort of solar air heaters according to their thermal performance.

Due to several inherent benefits over flat plate heaters, packed-bed air heaters have garnered a lot of interest. According to the packing density of iron turnings, which has a higher ratio of heat transfer area and results in a relatively low absorber temperature, the insolation penetrates to a greater depth and is gradually absorbed.

As a result, the absorber plate will lose less heat to the surrounding air, increasing the heater's thermal efficiency.

Experimental Set-up

The used experimental set-up's schematic diagram is shown in Figure 57.1. A blower powered by a single-phase D.C. motor suctions air into the duct in the directions indicated by arrows. The temperature of the air in the flow duct was measured using thermometers that were built into the duct's sidewalls.

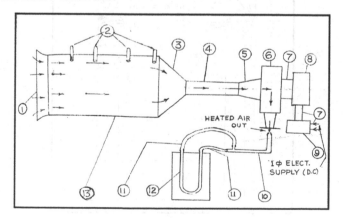

Figure 57.1 Schematic diagram of experimental set up.

Figure 57.2 Experimental set-up.

1. Bell mounted entry section 2. Thermometer 3. Diverging section 4. Flow pipe 5. Blower Hopper 6. Blower 7. Transmission wire 8. Electric motor 9. Auto variac 10. Pitot tube 11. Connecting tubes 12. U-Tube manometer 13. Solar air duct

A variac has been used to control the power input to the blower and hence to vary the mass flow rate of air. The pitot provided with an inclined U-tube manometer measured the flow rate through the duct. The photo of the experimental setup is shown in Figure 57.2.

Solar Air Heater Duct

Figure 57.3 shows the cross section of the experimental solar air heater duct. A wooden piece of 1.51 M × 0.36 M and 25 mm thickness formed the bottom plate of the duct. Air gap of 50 mm has been provided. Saw dust of 10 mm thickness has been filled in the bottom of the absorber plate to check bottom heat losses. Wooden side-walls suffice to check the side heat losses. A bell-mounted entry section of dimensions 380 mm × 155 mm as shown in Figure 57.1 allowed smooth entrance of air in the duct.

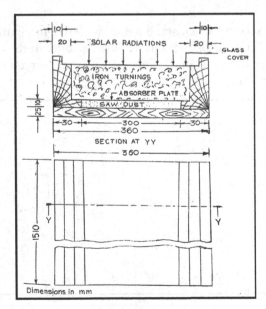

Figure 57.3 Cross section of the experimental solar air heater duct.

Black painted 20 SWG G.I. sheet measuring 1.51 M in length and 30 cm in breadth was utilized as the absorber plate. Scrap and black painted iron turnings were filled in the air gap of the flow duct up to 40 mm depth only 4 mm thick window glass plate formed the cover of the duct. Special care was taken to prevent any leakage in the duct.

An electric motor with a blower–A blower powered by a 1-0, 0.6 A, 220 volt D.C motor pulled ambient air through the air heater system.

Flow Control Setup–A 1-Φ variac linked in line with the blower motor was used to control the air flow rate through the duct. The variac's voltage regulation enabled a controlled power supply to the motor, resulting in the required blower speeds.

Instrumentation

Air flow Measurement–A pitot tube inserted at mid depth of the exit section of flow pipe and which was fed to an inclined U-tube water manometer of least count 0.627 mm measured the air flow rate through the duct. Inclination of one vertical to 7.67 horizontal was kept in the U-tube manometer. Pressure difference in the manometer was noted at the corresponding power delivered to the blower-motor from the variac. To determine the discharge, first, velocity of hot air inside the duct was calculated by the help of pressure difference using the relationship:

$$V = C_v \sqrt{2g\Delta P} \tag{1}$$

Where, ΔP = pressure head by Pitot tube
C_v = Co-efficient of Pitot tube
The mass flow rate could thus be determined as

$$\dot{m} = \rho A_s V \tag{2}$$

Where ρ = specific density of air at mean temperature.
A_s = Cross sectional area of the duct.

Temperature Measurement–To gauge the temperature of the air inside the duct, four mercury thermometers were placed into it at various points. At every 15 minutes interval, temperatures were noted from 11 A.M.to 1.30 P.M. temperatures were taken for 5 mass flow rates each for one day. Ambient temperature was also measured by mercury thermometer. Every thermometer was having the same least count of 0.1°C.

Experimental Procedure

Figure 57.1 shows the schematic of the experimental set-up. All the components of the set-up were checked for their proper positioning and functioning. The joints between the entry and exit sections and other points of possible leakage were properly sealed. All measuring instruments i.e.–tube manometer, Volt meter and mercury thermometer were properly checked. Test runs were done in a controlled mass flow rate environment. The test data were gathered daily between the hours of 11:00 and 1:00 at quarter-hour intervals.

The following parameters were measured:

1. Temperature of air in the duct
2. Pressure drop across the duct
3. Ambient temperature

Insolation was measured using a pyranometer Model No.0062, having calibration constant of 129.3 wm^{-1} (mV^{-1}).

Thermal Performance Equations

Solar air heaters are operated in the two ways:

1. With air cycling
2. Without air recycling

In the first type of solar air hater, heated air is recycled resulting in variation of air inlet temperature T_1. Therefore, the conventional thermal performance equation of Hottel-Whillier-Bliss written as:

$$\eta = F_R(\tau\alpha) - U_L\left(\frac{T_i - T_a}{I}\right) \tag{3}$$

Above equation represents the thermal performance of such solar air heaters.

But in the second type of solar air heater without air recycling. T_i is always equal to T_a resulting in $(T_i - T_a) = 0$ i.e. the conventional Eq.3 reduces to

$$\eta = F_R(\tau\alpha) \tag{4}$$

Equation 4 is meaningless. It is therefore, suggested that adopted in practice also by so many authors [11–19] represent the thermal performance of such solar air heaters by the modified equation in terms of a collector performance factor F_O, based on outlet air temperature instead of conventionally collector heat removal factor F_R based on air inlet temperature T_i, which is written here as

$$\eta = F_o(\tau\alpha) - F_o U_L\left(\frac{T_0 - T_a}{I}\right) \tag{5}$$

As $T_i = T_a$ in such solar air heaters Eq.5 can be written as:

$$\eta = F_o(\tau\alpha) - F_o U_L\left(\frac{T_0 - T_i}{I}\right) \tag{6}$$

Equation 6 has been utilized to represent the thermal performance of solar air heaters which will give the two parameters when Plotted ($\frac{T_0 \quad T_i}{I}$) on abscissa and on ordinate which are $F_o(\tau\alpha)$ and $F_o U_L$. These two values of intercept and slope respectively are not to be referred as performance $F_R U_L$ and $F_R(\tau\alpha)$ respectively, the slope and intercept values can be used to determine the values of $F_R U_L$ and $F_R(\tau\alpha)$ of the conventional thermal performance curve by the following equation [11]:

$$F_R(\tau\alpha) = F_o(\tau\alpha)\left[\frac{\dfrac{\dot{m}Cp}{A_c}}{\dfrac{\dot{m}Cp}{A_c} + F_o U_L}\right] \tag{7}$$

$$F_R U_L = F_o U_L \frac{\dfrac{\dot{m} Cp}{A_c}}{\dfrac{\dot{m}p}{A_c} + Fo U_L} \tag{8}$$

Results

To determine the values of necessary parameters, experimental data have been minimized, which have been further utilized to represent the experimental results in the form of thermal performance curve. Additionally, plots have been created to show how different operating settings affect both the plate temperature and thermal performance. Discussion on the results of famous parameters have also been taken up. Table 57.1 represent the values of different important parameters like $\dot{m}$, F_R, U_L and F′ .

A sample calculation for mass flowrate, $\dot{m} = 8.83 \times 10$ kgs^{-1} corresponding to a set of readings recorded at 12.00 P.M.is as follows:

Table 57.1: Values of Heat removal factor (F_R), overall loss coefficient (U_L) and collector efficiency factor (F′).

Particulars	$\dot{m}_1 = 7.25 \times 10^{-3}$ Kg/s	$\dot{m}_2 = 8.05 \times 10^{-3}$ kg/s	$\dot{m}_3 = 8.83 \times 10^{-3}$ kg/s	$\dot{m}_4 = 9.96 \times 10^{-3}$ kg/s	$\dot{m}_5 = 11.29 \times 10^{-3}$ kg/s
F_R	0.451	0.600	0.746	0.780	0.810
U_L	8.093	8.116	8.150	8.200	8.660
F′	0.513	0.704	0.830	0.842	0.853

Mass Flow Rate-Through Solar Air Duct (m)-

Pressure drop (Δ P) across the air duct was noted as 1.905 cm (vertical).
Velocity of air inside the duct was calculated as follows:

$$V = C_v \sqrt{2g\,\Delta P}$$
$$= 0.952\sqrt{2 \times 9.81 \times 1.905 \times 10^{-2}}$$
$$= 0.580 \, m/s$$

Where C_v = Co-efficient of pitot tube
Specific density at mean temperature of air in the duct was taken as 1.016 kgm^3.
Cross sectional area of duct As = (30 × 5) = 150 cm^2
Mass flow rate of air $\dot{m} = \rho A_s V = 1.016 \times 150 \times 10^{-4} \times 0.58 = 8.83 \times 10^{-3}$ kg s^{-1}

Collector Efficiency (η)

Useful heat gain (q_u) was calculated using relationship

$$q_u = \dot{m} C_p (T_{f4} - T_{f1}) \tag{9}$$

Where C_p = Specific heat = 1.009 kJ kg^{-1} k^{-1}

T_{f4} = Outlet temperature of air in the duct
T_{f1} = Inlet temperature of air in the duct
$q_u = 8.83 \times 10\ 0 \times 1.009(367 - 330.5)$
Therefore, collector efficiency

$$\eta = \frac{q_U}{I\,A_C} \tag{10}$$

Where I = 1004.17 Wm^{-2}
A_c = Collector area of the duct = (151 × 30) cm = 453cm² = 0.453 m²

$$\eta = \frac{0.3251 \times 103}{1004.17 \times 0.453} \times 100 = 71.46\%$$

Thermal Performance Curve

Taking the values of $\left(\dfrac{T_f - T_a}{I}\right)$ on abscissa and thermal efficiency η on the ordinate, Figure 57.4 displays the thermal performance curves that were created for a range of mass flowrates.

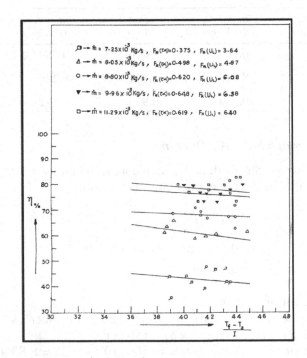

Figure 57.4 Thermal performance of solar air heater.

Air Temperature Curve

Figure 57.5 has been drawn to represent the air temperature at various sections along the particular mass flow rate of 7.25×10^{-3} kgs^{-1} through duct length rates at varying intensity of solar radiation.

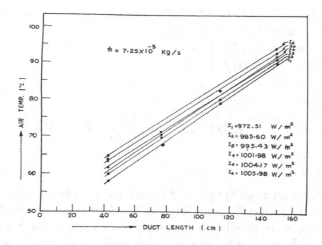

Figure 57.5 Temperature distribution through duct.

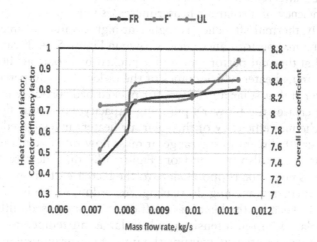

Figure 57.6 Mass flow rate effects on total loss coefficient, heat removal factor, and collector efficiency factor.

Results From Thermal Performance Curve

The experimental slope and intercept values $F_o U_L$ and $F_o (\tau\alpha)$ obtained from the thermal performance curves shown in Figure 57.4; have been utilized to calculate for the conventional performance parameter $F_R U_L$ and $F_o (\tau\alpha)$ by the Eqns.7 and 8. In order to calculate the values of the collector heat removal factor FR, an appropriate value of the absorptivity-transmissivity product $(\tau\alpha)$ equal to 0.83 was used. By replacing the experimental data of the performance parameter F_R with the calculated value of F_R, the value of U_L has been derived. The values of U_L obtained are taken to calculate for the values of collector efficiency factor F′ by the following equation [18]:

$$F' = \frac{\dot{m}Cp}{A_c} \ln \frac{F_o U_L}{F_R U_L} \Big/ U_L \tag{11}$$

To demonstrate how to calculate the collector heat removal factor and collector efficiency factor, a sample calculation is shown.

From Figure 57.4, for $\dot{m} = 7.25 \times 10$ kgs^{-1}

Experimental values of $F_R (\tau \alpha) = 0.375$

$F_R U_L = 3.64$

Taking $(\tau\alpha) = 0.83$

$$F_R = \frac{0.375}{0.83} 0.451 \ and \ U_L = \frac{3.64}{0.451} = 8.093 \ Wm^{-2}$$

Discussion on Results

Artificial roughness put on absorber plate of the collector leads to break the boundary sub-layer thereby increasing heat collection. Singh & Singh [20] conducted CFD based analysis of nonuniform square wave profile rib as an artificial roughness on the absorber plate of such type of solar air heaters and compared with similar uniform transverse rib arrangements. Through the use of flow visualization, they attempted to show how the roughness and flow parameters affected the thermal and hydraulic performance of the solar air heater.

The thermal efficiency of a solar air heater increases with an increase in mass flow rate, or more specifically, thermal efficiency is higher at higher values of mass flow rate as can be seen from the thermal performance curve shown in Figure 57.4. It can also be attributed from this figure that thermal performance of a packed bed solar air heater is superior to that of a plane solar air heater which comes of the order of 20–30 percent only. However, it has been discovered that solar air heaters for packed beds produce extremely high fluid pressure, which increases the blower's pumping capacity. From this figure it can be seen that the value of thermal efficiency of the solar air heaters investigated varies from about 40 percent to about 80 percent in the range of mass flow rate taken and intensity of solar radiation available. It can also be seen from Figure 57.4 that the rate of increase of thermal efficiency, seems not to be proportionate to the increase of mass flow rate, the rate of increase of thermal efficiency being decreasing. According to Table 57.2's tabular values for U_L, this might be because the heat loss coefficient has increased (although marginally). larger values of solar radiation intensity result in larger air temperature rise, as shown in Figure 57.5, which was created to examine the impact of varying solar radiation intensity throughout the flow duct length. Additionally, it demonstrates that the air temperature curve's slope gradually declines as solar radiation intensity rises, resulting in smaller net air temperature rises (To-Ti) at higher solar radiation levels, which could reduce heat absorption for a given mass flow rate. This might be due to increase in temperature of the packed gap of the duct leading to higher top-loss co-efficient. Figure 57.6 illustrates how mass flow rate affects the heat removal factor, collector efficiency factor, and overall loss coefficient. It clearly shows that heat removal factor and collector efficiency factor increase up to mass flow rate of 8.05×10^{-3} kgs1 and almost remain constant where as overall loss coefficient slightly increases up to mass flow rate of 10×10^{-3} kgs1 and then after increases with fast rate. It is good to see that the value of overall loss coefficient comes in the range of 2–10 Wm^{-2} which is safe as per design point of view of flat plate solar collector.

Conclusions

1. The following results have been drawn from the current experimental examination of packed bed solar air heaters:

2. When it comes to heat collection alone, solar air heaters for packed beds are more effective than those for planes.
3. Higher top losses as the flow are above the absorber plate.
4. Higher rate of solar energy collection at higher mass flow rates and intensity of solar radiation.
5. High values of air outlet temperature which may be very much suitable for solar heat storage.
6. At very high mass flow rate such solar air heaters can perform very fast for artificial drying of agricultural farm products.

References

1. Luo, L., Wen, F., Wang, L., Sundén, B., and Wang, S. (2016). Thermal enhancement by using grooves and ribs combined with delta-winglet vortex generator in a solar receiver heat exchanger. *Applied Energy*, 183, 1317–1332.
2. Gupta, A. D., and Varshney, L. (2017). Performance prediction for solar air heater having rectangular sectioned tapered rib roughness using CFD. *Thermal Science and Engineering Progress*, 4, 122–132.
3. Al-Shamani, A. N., Sopian, K., Mohammed, H. A., Mat, S., Ruslan, M. H., and Abed, A. M. (2015). Enhancement heat transfer characteristics in the channel with trapezoidal rib–groove using Nano fluids. *Case Studies in Thermal Engineering*, 5, 48–58.
4. Moon, M. A., Park, M. J., and Kim, K. Y. (2014). Evaluation of heat transfer performances of various rib shapes. *International Journal of Heat and Mass Transfer*, 71, 275–284.
5. Gill, R. S., Hans, V. S., Saini, J. S., and Sukhmeet, S. (2016). Investigation on performance enhancement due to staggered piece in a broken arc rib roughened solar air heater duct. *Renewable Energy*, 104, 148–162.
6. Vipin, B., Gawande, A. S., Dhoble, D. B., and Zodpe, S. C. (2016). Experimental and CFD investigation of convection heat transfer in solar air heater with reverse L-shaped ribs. *Solar Energy*, 131, 275–295.
7. Kumar, K., Prajapati, D. R., and Samir, S. (2017). Heat transfer and friction factor correlations development for solar air heater duct artificially roughened with 'S' shape ribs. *Experimental Thermal and Fluid Science*, 82, 249–261.
8. Huseyin, B. (2012). Experimentally derived efficiency and exergy analysis of a new solar air heater having different surface shapes. *Renewable Energy*, 50, 58–67.
9. Sukhmeet, S., Subhash, C., and Saini, J. S. (2015). Thermo-hydraulic performance due to relative roughness pitch in V-down rib with gap in solar air heater duct—comparison with similar rib roughness geometries. *Renewable and Sustainable Energy Reviews*, 43, 1159–1166.
10. Karwa, R., Sharma, A., and Karwa, N. (2010). A comparative study of different roughness geometries proposed for solar air heater ducts. *International Review of Mechanical Engineering*, 4(2), 159–166.
11. Duffie, J. A., and Beckman, W. A. (2013). Solar Energy Thermal Process. 4th ed. New York: Wiley Interscience.
12. Gupta, D., Solanki, S. C., and Saini, J. S. (1997). Overall performance of solar air heaters with roughened absorber plates. *Solar Energy*, 61, 33–42.
13. Hans, V. S., Saini, R. P., and Saini, J. S. (2010). Heat transfer and friction factor correlations for a solar air heater duct roughened artificially with multiple v-ribs. *Solar Energy*, 84, 898–911.
14. Jaurker, A. R., Saini, J. S., and Gandhi, B. K. (2006). Heat transfer and friction characteristics of rectangular solar air heater duct using rib-grooved artificial roughness. *Solar Energy*, 80, 895–907.
15. Karmare, S. V., and Tikekar, A. N. (2007). Heat transfer and friction factor correlation for artificially roughened duct with metal grit ribs. *International Journal of Heat and Mass Transfer*, 50, 4342–4351.

16. Kumar, S., and Saini, R. P. (2009). CFD based performance analysis of a solar air heater duct provided with artificial roughness. *Renewable Energy*, 34, 1285–1291.
17. Kumar, A., Saini, R. P., and Saini, J. S. (2012). Experimental investigation on heat transfer and fluid flow characteristics of air flow in rectangular duct with multi-V-shapes rib with gap roughness on the heated plate. *Solar Energy*, 86, 1733–1749.
18. Prasad, B. N., and Saini, J. S. (1988). Effect of artificial roughness on heat transfer and friction factor in a solar air heater. *Solar Energy*, 41, 555–560.
19. Verma, S. K., and Prasad, B. N. (2000). Investigation for the optimal thermo hydraulic performance of artificially roughened solar air heaters. *Renewable Energy*, 20, 19–36.
20. Singh, I., and Singh, S. (2018). A review of artificial roughness geometries employed in solar air heaters. *Renewable and Sustainable Energy Reviews*, 92, 405–425. doi: 10.1016/j.rser.2018.04.108.

58 Flexible automation in public distribution system

Vicky V. Choudhary[a] and Rajeev Srivastava[b]

Motilal Nehru National Institute of Technology Allahabad, Prayagraj, India

Abstract: Automation in the huge distribution channel, such as the Public distribution sys-tem (PDS) can ease the repetitive task of food commodities distribution by the government. However, past experiences have shown that a fully automated sys-tem has not eliminated the issues in a complete way, especially in the rural areas. The fact that the majority of families which are part of this PDS belong to rural areas drives the motivation for a Flexible system requirement. The study of Indian Public distribution system reveals that it is a complex system where different technologies and reforms are incorporated in different states according to the actions taken by the respective governments. However, there lie some serious gaps in the system which question the efficiency of the whole system. The following work addressed mainly the issues of online transaction failure, manual queuing, and online authentication failure and thus provides a solution with the help of modules/solutions created.

Keywords: Automation, fair price shop, information and communication technology, public distribution system, virtual queuing system.

Introduction

The Public Distribution System (PDS) is a food protection system run by the Ministry of Consumer Affairs, Food, and Public Distribution in India. PDS start-ed as a mechanism for dealing with shortages by supplying food crops at fair rates. The PDS is managed jointly by the federal and state governments. The Food Corporation of India (FCI) has taken over the procurement, packaging, transportation, and bulk allocation of food grains to state governments on behalf of the central government. State governments are in charge of duties such as distribution within the state, registration of deserving families, issuance of Ration Cards, and oversight of the service of Fair Price Shops (FPSs), among others. Wheat, rice, sugar, and kerosene are among the resources currently allocated to States/UTs for distribution under the PDS. Pulses, edible oils, iodized salt, spices, and other products of mass consumption are also sold through PDS outlets in some states/UTs.

The government's planning and policy have focused on providing food security. Food safety means access to adequate food grain to satisfy domestic demand and access to adequate food quantities at an affordable price on an individual level. One of the most significant achievements of the country was achieving self-sufficiency in the production of food

---

[a]vickyvc196@gmail.com, [b]rajmnnit@mnnit.ac.in

grain at the national level. The Government has implemented the 2013 National Food Security Act, which provides subsidized food grains to eligible homes, to address the issue of household food safety [5]. While identifying the problems at the local level there is a prevalent issue which is quite common especially in the rural areas is, overcrowding at Fair Price Shops (FPS) when the food grains are distributed to Beneficiaries under the provision of the National Food Security Act (NFSA). Figure 58.1 shows overcrowding at the local Fair Price Shop due to manual queuing.

The issues that the public (Beneficiaries) face is manual queuing issue, no assurance of getting food grains on the same day, loss of work and time, the recurring procedure of queuing for every month, no notification is provided to the people, and quarreling among people to avail the food grains due to queuing. However, these issues are listed as per observation in Fair Price Shops of Thane, Maharashtra. Further study has revealed the overall FPS conditions in the different states.

Figure 58.1 Overcrowding at local fair price shop due to manual queuing.

Literature Review

Lucas S. Dalenogare et al. (2018): discussed the Industry 4.0 technologies based on a study done in Brazil which is a developing country and concluded that the success of an advanced technology depends upon several factors such as area of adoption, adaptability of technologies in that area/sector and feasibility in adoption [1]. Aspire' (2017–18) and Akshaya Patra's Care (2018–19): described the importance of the advanced ICT technologies at every stage of its Mid-day meal schemes which works on a similar aim as for the PDS [2,3]. Prof. Rahul J. Jadhav et al. (2015): The paper gives an understanding of the obstructions in ICT adoptions and the advantages of ICT [4]. Raghav Puri (2017) has discussed the features of NFSA and listed the drawbacks in improper implementation of the act in different states of the country [5]. Varun Chhabra (2017) studied the ICT reforms in the public distribution system of Chhattisgarh, which is one of the best performing states as far as PDS food grain distribution is concerned [6]. Gowd Kiran Kumar (2017) has mainly compared the public distribution systems in Chhattisgarh and Telangana since both these are among the highest-performing states. The advantages, as well as disadvantages of both systems, have been described [7]. Aaditya Verma et al. (2018): have studied the various proposed schemes in PDS by the Government body and commissions in India. The paper lists the whole timeline of all the schemes [8].

The literature study revealed the fact that certain technologies have already been incorporated into the Public distribution system of different states. Some of the technological reforms are the Digitization of ration cards, Computerized allocation to FPS, Smart cards in place of ration cards, GPS technology, SMS-based notification to beneficiaries, and web-based citizens' portal. Unfortunately, every state has adopted different reforms in its way which has resulted in a non-uniform PDS system across the whole country which seems to be the major drawback of the Indian PDS system. However, not all the States and Union Territories have implemented every listed reform. Some states like Tamil Nadu and Chhattisgarh have excelled while some like Maharashtra and Jharkhand have struggled in even adopting one of these reforms [5]. Although the most successful PDS in terms of automation is the Chhattisgarh surveys revealed that the Chhattisgarh model was successful only in urban areas which question the deprived situation of the poor belonging to the rural areas. On a ground reality, in many areas of the Smart card system, FPS owners ask the beneficiaries to leave the cards at the shop giving technical fault/error in the system as an excuse. They distribute the food grains but fill in the details of beneficiaries after they leave the shop. Now this question the sincerity of the FPS owners and the actuality of the details input by the FPS owners on the system [7].

In the areas of Aadhaar authentication if there is a network issue, the beneficiaries have to wait for a long time till the network issue is resolved or revisit the shop another time. FPS owners in the states which have not yet fully adopted the technology are not ready to adopt the system. In some states like Rajasthan and Jharkhand beneficiaries faced problems due to faulty Aadhaar seeding and biometrics data issues.

It is clear that advanced technology already exists in the country. However, there lies some gap in the integration of various tools that can solve many problems as discussed previously. Automation has decreased the load on lots of network channels but not necessarily all the problems in rural areas. Hence the problems stated previously provide an opportunity to study the complexity of the Indian Public Distribution System and thus allows to brainstorm for providing solutions for issues of online transaction failure, manual queuing, and online authentication failure.

Objectives

The objectives of this work are listed as follows.

- To study the fully automated public distribution system
- To design a new fair price shop model based on reforms that have already been incorporated into the existing system
- To design modules for offline transactions
- To design a Virtual queuing system (offline and online adaptable) for the beneficiaries
- To solve online authentication failure issues of beneficiaries.

Methodology Workplan

The Public Distribution System has the following major issues to be solved: online transaction failure, manual queuing, and online authentication failure. These laid the requirement of multiple modules as stated in the objectives. Figure 58.2 shows the workplan flowchart.

This was facilitated firstly through the data collected from the official food distribution web portal of the state (Maharashtra under consideration as an example: www. mahaepos. gov.in)) and people working in local fair price shops and the beneficiaries in the targeted region (Thane FPS). Appropriate tools for solving each issue were to be identified by the

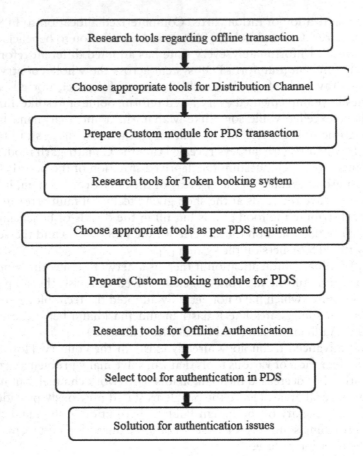

Figure 58.2 Work plan flowchart.

research study. This was followed by a custom modules-making process. These modules are the Offline Point of Sales module, Online and offline adaptable Virtual queuing system, and module/solution for authentication.

Modules Details

All the modules or solutions to the aforesaid issues are discussed further.

Offline Point of Sales Tool (Flexible FPS)

A tool 'Point of Sales (POS) System Offline' enables offline transaction. It is basically a billing system. Thus, this tool was used to add an offline POS module. Figure 58.3 shows the Offline Point of Sales Tool. A custom module 'Flexible Fair Price Shop (Flexible FPS)' was created. For the development of this module, different commodities were added to the module. Further, all the basic details such as the quantity and price per unit was added for a particular commodity. Following work includes rice and wheat as the added examples of PDS commodities. Figure 58.4 shows the addition of rice as one of the commodities. Now, for proper accountability, inventory details were added. Thus, initial inventory for both, rice and wheat were entered as 1500 kg as an example. The quantity sold is reflected in the bill report and the same is reflected in the final left inventory whenever any transaction takes place. As a result, it becomes easy to maintain accountability. This module facilitates the following features:

- Category Item Management: This feature allows the management of different categories of business/ services on a single platform. Each category has an option to add different commodities/ products. Details such as product's name, price, image, quantities number along with discount and tax settings can be added in the respective category. Figure 58.4 shows the category management for rice.

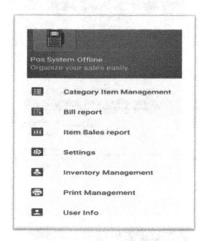

Figure 58.3 Offline point of sales tool.

Figure 58.4 'Flexible FPS' module addition.

- Bill report: This feature allows the creation of individual bill receipts as well as the whole bill report. This eases the task of individual bill calculation. Figure 58.5 shows the bill report option.

- Inventory Management: This feature helps to manage all the inventories. The inventory details can be added initially so that it is possible to manage the inventories according to the sales. After each sale, the quantity data is fetched automatically and thus inventory number reduces in the report which helps to order inventories if required. Figure 58.6 shows the inventory management option.
- Item Sales report: This feature allows to track sales of each product on a particular date or within a particular time range as per the requirements.
- Print Management: This option enables you to take the print of the bill receipt through the option of Bluetooth scan or direct print.

Figure 58.5 Bill report.

Figure 58.6 Inventory details after the transaction.

Virtual Queuing System

Token/Slot booking system can eliminate the long queue formation by the public during monthly food grain distribution. A Virtual Queuing System, 'PDS booking' was created for slot booking of the beneficiaries for the food grain distribution using the software called Calendly which is automated scheduling software. Following are the features of the PDS booking system as discussed.

- Slot booking option: Appointments can be booked according to the slots and working days set by the service provider. If slots are allotted to the beneficiaries the unwanted manual queuing can be avoided which ensures no unfair means are adopted by people to get the food grains quickly and thus there will be no quarreling. It will be a first come-first serve the type of service. This can reduce the load on the public and save their time too. Figure 58.7 shows the PDS slot booking option.

- No Cost: The beneficiary does not require to pay any money for booking a slot.
- Details of Beneficiary: For the slot to be booked, it is required to get the details of the beneficiary so that the authorities identify the real service beneficiary for a particular booked slot. Although these details can be simply the name, phone number, and email address just for slot booking, the detailed questionnaire can be added and actual authentication is done again during food grain collection. Figure 58.8 shows the beneficiary details.
- Confirmation of slot booked to the beneficiary: After the slot is booked, the beneficiary receives the confirmation for the booked slot. It shows details of date, location and time. Figure 58.9 shows the slot booking confirmation message.
- Proper details of slot booked available to the authority: After the slots booked by different beneficiaries the authorities are able to access the details of slots booked along

Figure 58.7 Slot booking for beneficiaries.

Figure 58.8 Beneficiary details.

Figure 58.9 Slot booking confirmation.

Figure 58.10 Slot booking details available to authorities.

with beneficiary details so that real benefiter can be identified. Figure 58.10 shows the slot booking details available to the authorities.

• Rescheduling and Cancellation: In case, a beneficiary is absent even after booking the slot, there are two options available. The slot can either be cancelled or rescheduled for a later time slot, as per the availability. Figure 58.8 shows the 'Reschedule' and 'Cancel' option for the adjustment of a booked slot.

• Slot limit: In order to avoid overbooking of the slots, a limit can be set for the total number of slots that can be booked in a day.

• No multiple bookings for the same slots: The system takes care that the same slots are not booked multiple times by different users.

• Offline and online adaptability: One can integrate offline booking with the online system using the integration option. Thus, it provides offline and online modes of slot booking.

The Solution to Online Authentication Failure

It is clear from the literature study that smart cards can be a solution to Aadhaar authentication failure. However smart cards had the disadvantage of being used unfairly by the authorities or any other person. But smart card solves the problem of Aadhaar authentication failure (due to poor biometrics data). Since the PDS governing body generates the smart card there can be a (One Time Password) OTP-based authentication using the registered mobile number of the beneficiary. Smart cards are allotted to particular beneficiary only after validating the Aadhaar details. This validation is required only once by the state government with the help of Unique Identification Authority of India (UIDAI). Once the validation is done, the Aadhaar details remains safe with the UIDAI. The smart card carries only the details such as ration card number, the quantity of food grains allotted. Aadhaar details are not reflected in the smart card. Thus, it secures the Aadhaar details and provides a reliable mode of authentication.

Results

The following work focused on the gaps in the present PDS. It provides solution to the issues of online transaction failure, manual queuing, and online authentication failure. Thus, the work tries to provide solutions to these issues with the help of several modules that were developed. It proposes a model for the local fair price shops. In this model, the beneficiary first requires to book the slot through the Virtual Queuing System which replaces the manual queuing in current system. After the slot is booked, the beneficiary arrives at the fair price shop at allotted time. The offline transaction takes place through the Offline POS tool. For authentication, the smart cards can be used to validate the beneficiary's identity. In this way, the proposed model can reduce the difficulties of both, the beneficiaries as well as the authorities.

Conclusion

The results obtained in this work suggest the following conclusions.

* The fully automated system has not eliminated the issues of PDS beneficiaries in a complete way, especially in rural areas. Thus, a flexible Public Distribution System, adaptable to all the regions can be more efficient.
* Technological reforms have already been incorporated in Indian PDS but are non-uniform throughout the different states of the nation.
* The main issues in the current automated system are online transaction failure, manual queuing, and online authentication failure.
* The proposed model tries to provide solutions to these issues. For this, several modules were created. This includes the Flexible FPS module, the Virtual Queuing System and solution to offline Aadhaar authentication.
* The Flexible FPS module facilitates transactions in case the online transaction fails. It also helps the authority to maintain the data/details of all the offline transactions with features of Category Item Management, Bill report, Item sales report, Inventory management, Print management.

- PDS booking–a virtual queuing system (online and offline adaptable) eliminates the long manual queue formation by the beneficiaries during monthly food commodities distribution.
- Usage of Smart cards with OTP authentication instead of Aadhaar authentication can eliminate online authentication failure issues.
- The modules can be used in many distributions and network channels other than Public Distribution Systems such as Civil Hospitals and Municipal Clinics, Uniform Clothes Selling Vendors (During Re-opening of Schools), Registration Office (Stamp Duty, Marriage Registration, Land, etc)

References

1. Dalenogare, L., Benitez, G., and Ayala, N. (2018). The expected contribution of Industry 4.0 technologies for industrial performance. *International Journal of Production Economics*, 204, 383–394.
2. Akshay Patra Foundation (2017–18). Aspire–an Annual Report. Akshay Patra Foundation. India (2017–18).
3. Akshay Patra's Care (2018–19). An Abridged Annual Report. Akshay Patra Foundation. Vol. 19. India, (2018–19).
4. Jadhav, R. L., and Nalawade, K. (2015). Benefits and challenges of ICT adoption in Indian PDS. *International Advanced Research Journal in Science, Engineering and Technology*, 2(12).
5. Puri, R. (2017). India's National Food Security Act (NFSA) Early Experiences—A Report. Levering Agriculture for Nutrition (LANSA) in South Asia.
6. Chhabra, V. (2017). Technology Adoption in the Public Distribution System of Chhattisgarh—A Report. Iowa State University.
7. Kumar, G. (2017). ICT and zero hunger. In 3rd International Conference on Public Policy (ICPP3), June 2017, Singapore.
8. Verma, A., Rathore, A., and Kumari, A. (2018). An Automated approach to public distribution system using internet of things. In International Conference on Computational Intelligence and Data Science.

59 Experimental study of static and dynamic load test of railway wagon bogie

Apurba Das[1,a] and Gopal Agarwal[2,b]

[1]Department of Aerospace Engineering & Applied Mechanics, IIEST, Shibpur, Howrah, India

[2]Department of Mechanical Engineering, Jadavpur University, Kolkata, India

Abstract: This paper is a study representing an experimental approach on static and dynamic load testing on railway wagon bogie. The dynamic load condition are simulated by applying static load in different directions simultaneously. In this experimental work, three load cells (2000 KN) of HBM Quantum X-MX840 are used to apply the static load in three different orientations. The applied loads in different directions during testing are calculated based on EN 13749 standards. 'CATMAN easy' software is used to measure the generated strain in the predefined locations of the bogie. Total 10 numbers of experiments are conducted considering different loading conditions. The steady-state strain at predefined critical locations of the bogie is recorded using a data acquisition system. The bogie flexibility in terms of deflection produced under different loading conditions is measured. The performance of the bogie springs in terms of stiffness is also checked. In this work a test methodology is established properly to conduct the static load test of wagon bogie. The strain values and the deflection of the bogie measured bolster under different loading conditions are inspected for the bogie specific requirements. It is noted that the measured strain and the bogie deflection during different loading conditions are within the permissible design condition of this bogie.

Keywords: Bogie load test, dynamic test, static test, strain gauge.

Introduction

Transportation serves as a necessity in everyone's day to day life. Railways are one of the elements that help us fulfill the need for transportation. Railways are majorly used for long-distance transportation of goods and passengers [1]. This technology apart from being cost-effective is comparatively safe and environ-mentally friendly. The design and the parameters of testing the design play a crucial role in its safe operation. So, in order to validate a wagon design, one has to comply with the design standards [2]. Every newly designed railway wagons and its components must be within the standard codes before it is commissioned [3]. Modal analysis of bogie with its study of static behavior has been done by Claus and Schielen [4]. Hwa et al. [5] used finite element methods to study the bogie frame in tilted condition for fatigue strength and weight. Lu et al. [6] through modal analysis have analyzed his study by replacing a rigid bogie frame from a flexible one.

[a]apurba.besu@gmail.com, [b]gopalagarwal1259@gmail.com

Wei et al. [7] have worked on a model of a linear metro system and analyzed it through dynamic simulations. Vatulia et al. [8] with the help of software packages presented work on tank wagon structural improvements. Virtual prototyping computer tools were used to design open type freight wagon and its structural analysis was done using FEM by Harak [9]. Experimental data are basically the key for every new design of a wagon or its components. Some sensational works were carried out by Wickens [10] and Andersson et al. [11] with respect to the dynamics of rail vehicles. Both static and dynamic tests on flat wagons were carried out by Sandu and Zaharia [12]. The test was carried out on chassis of different sizes (20 feet, 30 feet, 40 feet). Rezvani and Mazraeh [13] has worked on analysis of stability and dynamics of a freight wagon with different railway track and wheelset operational conditions. Das & Agarwal [14, 15] has established experimental work and explanation on compression test, tension test and lifting test, which highlights the various critical spots of a wagon. In conclusion from the mentioned works, it is evident that a good knowledge of wagon's behavior is to firstly required before one approaches to solve the increasing problem on safety and security of railway vehicles. The only available work to determine structural integrity of a modified freight wagon bogie frame was done by Dižo et al. [16] and was limited to the numerical determination of the developed stress in the bogie frame, where no experimental data and experimental techniques were established. An important work on experimental test of rail vehicle's suspension monitoring system of a prototype wagon was reported by Mełnik and Kostrzewski [17]. As wagons are mainly used as the means for the transportation of the goods, the experimental dynamic, and static load testing is a very important criterion to comply with as per the standard codes. The experimental work carried out in the present research is based on standard EN 12663-2:2010 [18]. The aim of this paper is to study the dynamic and static loading of a wagon bogie. Several tests were carried out on the wagon bogie with the load ap-plied on the x, y and z directions. The Stresses were recorded with the help of strain gauges located at the several critical locations of the bogie. With the help of the stress-strain methodology, the critical locations of the wagon bogie are identified as per the loads applied. This could be guideline for the prototype wagon before mass production and this test results could serve as ready reference for future need.

Bogie Specification and Testing Load Calculation

The loading capacity of the bogie is generally specified by axle load in Ton. A bogie consists of two axles. Hence the total loading capacity of a bogie is twice the axle load. The present bogie is having 17T/axle load and capable of carrying 34T load in normal operating conditions. A wagon generally consists of two bogies unless it is an articulated wagon. A typical nomenclature of different key parameters of a bogie and loading pattern for present flat wagon are shown in Fig-ure 1(a) and 1(b), respectively. The parametric data used for testing the present bogie is tabulated in Table 59.1. The total weight of the present bogie(Wm) is 3.5T with wheelbase(b) = 1.720m. The bogie is determined for a freight wagon with a total weight MW of 68T. The flexibility of the bogie is designed as 1.3 mm/Ton. Figure 59.1(b) represents the loading pattern of the flat wagon containing the ISO container. The different variations and combinations of the ISO 20 feet or 40 feet container can be used for transportation, while the case 3 loading pattern will experience the loading on one side of the bogie. For the worst case one bogie will bear the total load and the other side of the bogie will act as support through a pivot pin. Load point application in three-axis

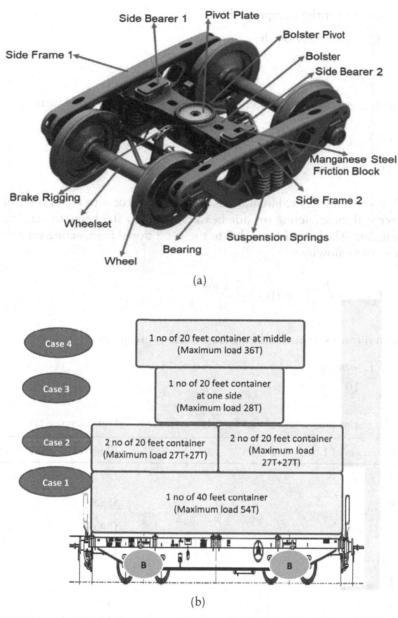

Figure 59.1 (a) Typical configuration of bogie, (b) Loading pattern for flat wagon.

directions for bogie experiencing static and dynamic conditions is shown in Figure 59.2. The total load of the wagon is transmit-ted through the bogie and each bogie is experienced the vertical force [3].

$$F_z = \left(0.5W_m - W_b\right)g \qquad (1)$$

The vertical direction of the exceptional load is represented as,

- if vertical forces act only on the center pivot:

$$F_{zpmax} = 2F_z \qquad (2)$$

- if vertical forces act on the center pivot and also on the one side bearer:

$$F_{z1max} = F_{z2max} = 1.5F_z\beta \qquad (3)$$

$$F_{zp} = 1.5F_z(1-\beta) \qquad (4)$$

where F_z is the total vertical load, F_{zp} is the vertical force acting on center pivot, F_{z1} and F_{z2} are vertical forces acting on side bearers, with $\alpha = 0.3$ being the coefficient for the body swinging. The lateral force due to the exceptional load, acting on every wheelset, is given as the following:

$$F_{y1max} = F_{y2max} = \frac{F_{ymax}}{2} = 10 + \left(\frac{F_y + mg}{6}\right) \qquad (5)$$

The longitudinal force strains the bogie frame and is given as:

$$F_{x1max} = \left(\frac{F_y + mg}{10}\right) \qquad (6)$$

Table 59.1: The testing load is calculated for a bogie frame based on reference [16].

Sl. No	Direction of Load	Theoretical Value (kN)
1	F_z	299.2
2	F_{zpmax}	598.4
3	$F_{z1max} = F_{z2max}$	134.64
4	F_{zp}	314.17
5	$F_{y1max} = F_{y2max}$	65.59
6	F_{x1max}	33.35

As previously stated, this work focuses on the unfavorable load. Cases that combine two or more elements the specified standard include these load combinations. They essentially reflect various combinations of estimated loads from Table 59.1 and Figure 59.2 depicts individual load applied on the sites and its directions.

Experimental Setup and Strain Gauge Position

The geometrical dimension of the bogie is determined as per its purpose and the railway of track gauge. These class of bogies are general purpose bogies, basically used for transporting solid, liquid or bulk products in loose or within a container. The designed bogie consists of 17Ton/axle load and is a T17APB Bogie. The major components of the bogie are made of 280-480-M2 cast steel grades for general purposes as per NFA32051-81 standard. The ultimate strength of the 280-480-M2 cast steel grades is 480 MPa. The structural requirements of the flat wagon (for which the bogie is supposed to design and tested) are considered based on EN 12663-2:2010 standard [18].

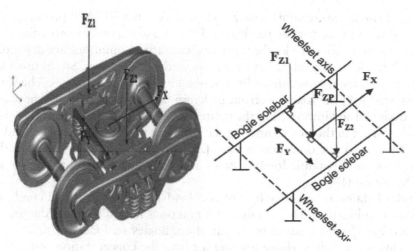

Figure 59.2 Load point application for bogie test [3].

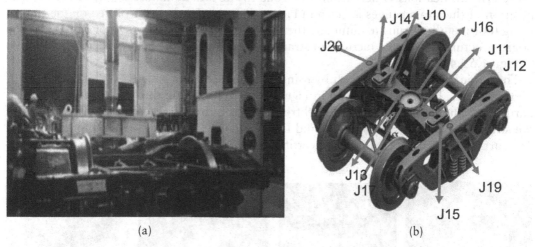

(a)	(b)

Figure 59.3 (a) Bogie test rig with load sensor and instrumentation (b) strain gauge location.

For the experiment, three load cells (2000 kN) were used of HBM Quantum X-MX840. The system was connected with CATMAN easy software. The software and load cells are used to apply the static load in different orientations and to measure the generated strain in the predefined locations of the bogie. The bogie at various critical locations is equipped with strain gauges to determine the stress developed, as derived from the reference [16]. The critical locations, in accordance with the developed stress, serves as the basis to select strain gauges location to capture the measured strain from the experiments process. Figure 59.3(a) shows the bogie placed on the test rig for carrying out the static and simulated dynamic testing by applying a combination of loads in different orientations as described in Table 59.2. A total of 10 numbers of experiments are conducted considering different combinations of loads in three different axes.

Analysis of Static and Dynamic Load Test

The static and simulated dynamic tests are carried out on a T17APB prototype bogie. The bogie is lifted with the strings as per Figure 59.3(a) and carried towards the test rig. The frame is loaded by a variety of loads; the actual operating conditions are dependent on the loading level, track quality, ideal load position, and other factors. So, in this inconsistent loading operation, the results cannot be described by load dependencies. The way the new bogies are tested [19] the load spectrum is determined according to it. The wagon bogie frame was designed and improved in the matter of strength for the load testing. This testing is done with respect to the norms and regulations as stated by the UIC [20] and AAR [21] which can be carried out on a new wagon [22]. The bogie frame is tested with the help of derived static and dynamic loading cases and the calculations done are based on the European Standard [18].

The standard states the combination of two loads that are static load combination and dynamic load combination. The specified combinations not only specify designs and frame optimizations but also the basis of test verification. Bogies are loaded with a large dynamic load of varying sizes. This variable loading leads to the concentration of loads in critical locations. In the present work, the static load combinations are considered to determine those critical locations which are more prone to the fatigue failure of the structure. The position of the strain gauges is given in Figure 59.3 (b). The bogie is placed on the test rig using four metallic chains to eliminate the pre-deformation and any type of damage on the bogie structure. The actual pictures of strain gauge pasted on the bogie structure are shown in Figure 59.4.

The bogie is constructed for test by using 10 numbers strain gauges (120 ohms, length of 10 mm: TML, Japan) as displayed in Figure 59.3(b) following the ASTM E251-20a standard. A 60 channel DAQ–MGC PLUS from HBM, Germany records the deformation on the strain gauges. The strain gauges and DAQ system used for testing is calibrated with a 3rd party agency and the calibration certificates are also provided ensuring the correctness

Figure 59.4 Pictures of the strain gauges J10, J11, J12, J13.

of the measured value. The tests are carried out at room temperature (270C) and humidity was recorded 60%. For different testing conditions the force is applied through the load cells in the defined direction as per the test number mentioned in Table 59.3.

Once the applied force is reached the defined value and reached in steady state condition the strain values at different locations are measured and recorded in the CATMAN easy software. After every test, the bogie was inspected to check whether any cracks or deformation formed or not. After confirming there is no crack or deformation is formed the next set of tests is conducted to ensure the accuracy of the test results. The performance of the bogie springs in terms of deflection under each loading condition is also checked to measure the overall spring flexibility of the bogie.

Table 59.2: The applied load is given for the different testing condition.

Test No	Applied Load in kN						
	F_{zpmax}	F_{zp}	F_{z1max}	F_{z2max}	F_{y1max}	F_{y2max}	F_{x1max}
1	598.4	-	-	-	-	-	-
2	-	314.2	134.64	-	-	-	-
3	-	314.2	-	134.64	-	-	-
4	-	314.2	134.64	-	65.59	-	-
5	-	314.2	134.64	-	-	65.59	-
6	-	314.2	-	134.64	-	65.59	-
7	-	314.2	134.64		65.59	-	33.35
8	-	314.2	-	134.64	65.59	-	33.35
9	-	314.2	134.64	-	-	65.59	33.35
10	-	314.2	-	134.64	-	65.59	33.35

Table 59.3: Strain gauge values captured during different loading condition test.

Test No	Measured Strain (μm/m) at different gauge location on the bogie									
	J10	J11	J12	J13	J14	J15	J16	J17	J19	J20
1	855	880	869	883	840	843	875	891	509	561
2	516	782	423	749	494	466	777	759	495	652
3	437	782	865	749	457	854	772	756	495	663
4	720	798	536	761	745	575	788	787	497	627
5	731	801	592	756	757	556	775	753	505	642
6	486	794	856	745	485	864	788	788	507	623
7	741	810	532	757	765	546	799	792	494	663
8	472	815	845	753	498	865	806	787	502	629
9	724	829	559	756	758	35	787	781	505	640
10	715	827	817	758	749	855	808	804	499	623

Results and Discussions

The strain values are recorded in the data collecting system in the various selected sites for each experimental test case. During the static and simulated dynamic bogie tests, when all of the values are stable then strain values are measured and recorded each time. The strains measured at various places on the wagon bogie are reported in tabular form with regard to the several test situations. Table 59.3 represents the strain valies of the prototype bogie for different test numbers considering static and simulated dynamic testing. One can derive the corresponding stress values from the Hooke's law using the measured strain value (σ = ε *E), as the deformations are small in which E is Young's modulus in GPa, ε is a measured strain in μm/m and σ is stress in MPa. The measured strain and the derived stress results at different critical locations are analyzed and discussed from the bogie design and safety point of view. The bogie flexibility in terms of deflection produced under different loading conditions is also measured. The performance of the bogie springs in terms of stiffness is also checked for every case and it is found that the average flexibility of bogie is 1.28mm/T, which is as per the design parameter of the bogie. The first load scenario, as shown in Figure 59.2, occurs when the maximum exceptional load acts on the frame (F_{zpmax} = 598.4 kN) on the center pivot. The maximum stress values are observed on the center pivot region under the effect of this load. The maximum stress is observed at location J17 (316.3 MPa) for the load test condition 1.

On the other hand, the minimum stress is observed at location J12 (150.2 MPa) for the load test condition 2. Intermediate stresses are observed for other load test conditions at different locations as described in Table 59.3. This in general indicates the limit of acceptability of the frame structure in this region. The maximum stress developed for the worst-case loading condition is 316.3 MPa which is well below the stress limit (480 Mpa) of the bogie cast grade steel material. It is to be noted that the location at the side frame of the bogie (J19, J20) is experiencing the lowest strain rate while the critical locations at the bolster are experiencing the higher strain. Hence, from a design and manufacturing point of view special care has to be taken for the bolster during series production. The prototype bogie testing given in this work stands as a basic guideline for the series production of bogies. It can be clearly noted that the measured strain is max for test number 1, where the maximum static loading considering the worst-case loading is given on the bogie. The maximum strain of 891.0μm/m is measured at the J17 location for test number 1. The bogie is made of 280-480-M2 cast grade steel material with a maximum allowable stress value of 480 MPa as per NFA32051-81 standard. These strain gauges show which sections are more load sensitive and hence more likely to fail due to fatigue. However, in real case scenario, the wagons are not operated considering the worst-case load where the max payload (54T) of the wagon will be experienced by one side of the bogie. As per Figure 59.2, the maximum load on one side of the bogie will experience 28T, which is much below the worst case testing condition. As a result, even after considering numerous unknown characteristics during manufacture, the bogie component will withstand the worst-case load scenario.

Conclusion

A conventional methodology is established to conduct the static load test of a wagon bogie and this methodology can be extended for various other type of wagon. The following are concluded based on the present work.

1. The test condition 1 predicts the maximum strain values of 891.0μm/m at location J17. The derived maximum stress at that is noted as 316.3 MPa.
2. The test condition 2 shows the minimum strain of 423.1μm/m at location J12. The developed least stress is observed at location J12 as 150.2 MPa.
3. The worst-case static condition load case is vulnerable compared to the combined static and dynamic loading conditions. Hence, it is not recommended to load the wagon where the full load of the container will be experienced by one side of bogie.

Uniformly distributed loading pattern is recommended, even if it is required to carry an uneven load on one side of the wagon load should be 50% of the total load. The maximum developed stress in the worst-case loading condition is 34% lower than the maximum permissible stress.

References

1. Kendra, M., Babin, M., and Barta, D. (2012). Changes of the infrastructure and operation parameters of a railway line and their impact to the track capacity and the volume of transported goods. *Procedia-Social and Behavioral Sciences*, 48, 743–752.
2. Funfschilling, C., Perrin, G., Sebes, M., Bezin, Y., Mazzola, L., and Nguyen-Tajan, M. L. (2015). Probabilistic simulation for the certification of railway vehicles. *Proceedings of the Institution of Mechanical Engineers, Part F: Journal of Rail and Rapid Transit*, 229(6), 770–781.
3. EN 13749 (2011). Railway Applications—Wheelsets and Bogies—Method of Specifying the Structural Requirements of Bogie Frames. Brussels, Belgium: European Com. for Stand.
4. Claus, H., and Schiehlen, W. (1998). Modeling and simulation of railway bogie structural vibrations. *Vehicle System Dynamics*, 29(S1), 538–552.
5. Hwa Park, B., Po Kim, N., Seok Kim, J., and Yong Lee, K. (2006). Optimum design of tilting bogie frame in consideration of fatigue strength and weight. *Vehicle System Dynamics*, 44(12), 887–901.
6. Lu, Y. H., Zeng, J., Wu, P. B., and Guan, Q. H. (2009). Modeling of rigid-flexible coupling system dynamics for railway vehicles with flexible bogie frame. In Fourth International Conference on Innovative Computing, Information and Control (ICICIC), Kaohsiung, (pp. 1355–1360).
7. Wei, Q., Wang, Y., Zhang, Y., and Deng, Y. (2007). A dynamic simulation model of linear metro system with ADMAS/rail. In International Conference on Mechatronics and Auto, Harbin, (pp. 2037–2042).
8. Vatulia, G., Falendysh, A., Orel, Y., and Pavliuchenkov, M. (2017). Structural improvements in a tank wagon with modern software packages. *Procedia Engineering*, 187, 301–307.
9. Harak, S. S., Sharma, S. C., and Harsha, S. P. (2014). Structural dynamic analysis of freight railway wagon using finite element method. *Procedia Materials Science*, 6, 1891–1898.
10. Wickens, A. H. (2003). Fundamentals of Rail Vehicle Dynamics. Lisse, Netherlands: Swets & Zeitlinger.
11. Andersson, E., Berg, M., and Stichel, S. (2007). Rail Vehicle Dynamics. Centre for Research and Education in Railway Engineering. Railway Group KTH.
12. Sandu, N., and Zaharia, N. L. (2014). Static and dynamic tests performed on a flat wagon. *Problemy Kolejnictwa*, 163, 67–77.
13. Rezvani, M. A., and Mazraeh, A. (2017). Dynamics and stability analysis of a freight wagon subjective to the railway track and wheelset operational conditions. *European Journal of Mechanics-A/Solids*, 61, 22–34.
14. Das, A., and Agarwal, G. (2020). Investigation of torsional stability and camber test on a meter gauge flat wagon. In Maity, D., Siddheshwar, P., and Saha, S. eds. Advances in Fluid Mechanics and Solid Mechanics. Lecture Notes in Mechanical Engg. Singapore: Springer.

15. Das, A., and Agarwal, G. (2020). Compression, tension & lifting stability on a meter gauge flat Wagon: an experimental approach. *Australian Journal of Mechanical Engineering*, 20(4), 1113–1125.
16. Dižo, J., Harušinec, J., and Blatnický, M. (2017). Structural analysis of a modified freight wagon bogiframe. In MATEC Web of Conferences, (Vol. 134, p. 00010). EDP Sciences.
17. Mełnik, R., and Kostrzewski, M. (2012). Rail vehicle's suspension monitoring system-analysis of results obtained in tests of the prototype. *Key Engineering Materials*, 518, 281–288.
18. EN 12663-2 (2010). Railway applications—structural requirements of railway vehicle bodies–Part 2: freight wagons.
19. American Railway Encyclopedia (1961). Wagons and their Maintenance. All-Union Publishing and Polygraph Union of Railway Ministry.
20. International Standard UIC Code 517 (2007). Wagons—Suspension Gear—Standardization. 7th ed.
21. Association of American Railroads (1994). Manual of Standards and Recommended Practices, Section C-Part II, Volume 1, Specification M-1001.
22. UIC Leaflet 577 (2005). Wagon Stresses, 4th ed. Paris, France: (UIC) International Union of Railways.

60 Modified ackerman for all terrain vehicles (ATVs)

Siddharth Tripathi, Suhrid Purohit, and Hiren Prajapati[a]

Department of Mechanical Engineering, Institute of Technology, Nirma University Ahmedabad, Gujarat, India

Abstract: In this article modified version of Ackerman steering system is designed for All Terrain Vehicles (ATVs), which adopts rack and pinion mechanism above front axle of the vehicle. The mathematical modeling of the system as well as interdependence between various geometric parameters is presented. Geometric relations are derived for three different conditions of steering Zero-steer, Inner wheel and Outer wheel respectively. These relations are computed using MATLAB code and optimum value of various steering parameters are obtained for Acker-man percentage. At the end, effect of geometric parameters on Ackerman percentage is discussed.

Keywords: Ackerman steering mechanism, all terrain vehicles (ATVs), bump steer.

Introduction

Steering system in a vehicle allows the driver to maneuver vehicle in certain direction. Over the various steering mechanisms, rack and pinion type is considered to be first choice for modern cars manufacturers as well as race car engi-neers because of its simple construction, easy operation, agile response even to small inputs, compact size and less backlash error. There is only one downside with this mechanism in view of race cars. Spatial arrangement of linkages (that are deployed in rack-pinion system) and gear ratio affects steering ratio achingly. Small deviation in linkages leads to steering error that causes difficulties in handling of car during sharp turns.

Since long time, four-bar mechanism has been widely used for steering geometry with rack and pinion. Divergent end behavior has always been an issue for four-bar linkage. However, researchers have developed six-bar and eight-bar linkage concept in order to reduce divergent end behavior. As a steering system eight-member mechanism was first used by Fayeh and Huston [1], which had seven precision points and minute error in an elongated motion. Due to very small pin joint connection between links and frame even a little wear in joint has adverse effect on system. To eliminate this flaw in eight-bar, S. Pramanik [2] suggested six-bar mechanism which had five-point precision and elongated mo-tion of links up to 61°. Later, Gautam et. al. [3] have modified six-bar linkage by adopting seven-point precision which has lesser error compared to previous one. Nalla

---

[a]hiren.prajapati@nirmauni.ac.in

et al. [4] have optimized the geometry of six-bar steering mechanism using pattern search technique and enhanced the cornering performance of race car. Dewei et. al. [5] and Dong et al. [6] have studied numerical analysis to synthesize the six-bar mechanism by reducing the specific input and output dependency. These researchers have provided base to couple six-bar mechanism with Ackerman principle, which is further used by other researchers like Hanzaki et al. [7], Pradhan et al. [8], etc. for further improvement. Hanzaki et al. [7] analyzed planar six-bar linkage for Ackerman principle and developed simulating models. Pradhan et al. [8] developed true Ackerman geometry using synthesis of planar six-bar mechanism considering it as two separate slider crank arrangement.

Various algorithms have been developed to optimism the path generation and make four-bar more efficient. Recursive algorithm is adopted by Chaudhary et. al. [9] for dynamic analyses of four-bar and make kinematic more efficient. Bulatovi´c et. al. [10] as well as Mateka et. al. [11] have developed path synthesis using differential evolution algorithms. Sleesongsom et. al. [12] have used meta-heuristic algorithms for optimum path generation of four-bar which comes out to be more accurate than any. Steering geometry discussed in this article has also deployed four-bar trapezoidal mechanism with rack-and-pinion.

Many parameters affecting the dynamics of the vehicle are neglected (like slip, traction, etc.); while performing the comparison between different steering geometry. Steering geometries are basically classified on the relative turning of inner wheel with respect to outer wheel for the same steering input. There are three types of steering geometries parallel, Anti-Ackerman and Ackerman.

In parallel steer, steering arms are parallel as a result outer and inner wheel turn same angle for a given steering input. Parallel steer induces under steer during turning, which is not desirable for tight turns. This steering mechanism is completely replaced by Anti-Ackerman and Ackerman. [13] With Anti-Ackerman geometry inner wheel steer less then outer wheel and has more grip, due to sudden upheavals and more grip for inner tire. It is more preferable for high-speed racing. Anti-Ackerman should not be used for low speed for that Ackerman is more preferable. The most commonly used steering mechanism in modern vehicle is Ackerman steering where four-bar linkages are used to control the steering action. The kinematic geometry of linkages is usually trapezoidal which pivots all the four wheels about an instantaneous center point and assures zero skidding.

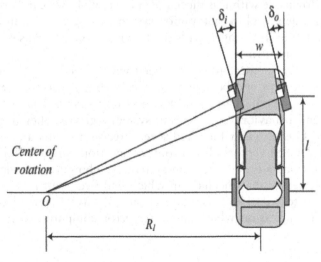

Figure 60.1 Ackerman geometry [14].

In Ackerman steering, inner wheel steers at larger angle than the outer wheel in order to rotate the vehicle about IC (Instantaneous center) and ensure zero skidding of wheels. Accordingly, outer wheel travels faster than inner wheel.

Following condition must be satisfied for Ackerman steering:

$$\cot\delta_o - \cot\delta_i = \frac{w}{l}$$

Here, δo is outer wheel angle, δi is inner wheel angle, l is wheel base of the vehicle and w is track width. Among various steering principle, Ackerman shows ideal steering geometry therefore sufficient work is done on analysis of Ackerman geometry. Zhao et al. [15] proposed steering mechanism which satisfies Ackerman geometry using synthesis of four-bar linkage. Qui et al. [16] extended two-wheeler Ackerman principles for coordinated movement of a four-wheel drive. Yao et al. [17] derived co-relations for trapezium mechanism for the front wheel steering, four-wheel steering and crab shape steering. Deepal Kumar [18] has derived geometric relation among various parameters involved in Ackermann geometry.

In all these articles, Ackerman geometry is described by trapezoidal four-bar linkage behind the front axle. The aim of present work is to develop a criterion for the steering geometry (with rack in front of the axle) based on purely geometric relation. Initially, geometric equations are derived based on Ackerman principle and trapezoidal mechanism. Subsequently, effects of different parameters are studied to develop a criterion.

Rack in Front of Front Axle

Ackerman geometry with rack behind the front axle is better for slow speed turning. Also, length of tie-rod is less as compared to anti-Ackerman geometry and locking of geometry occurs at 45°–50°. Locking of geometry means after a particular angle of turning wheels get locked and not able to steer further. This kind of steering geometry is very common with commercial vehicle. But large locking angle is required in ATV with little bit of speed during turning. Limita-tion of rack behind the axle can be overcome by introducing geometry with rack in front of axle. This geometry is very similar to rack placed behind the front ax le. Trapezoidal arrangement of linkages is shown in Figure 60.2 with straight steering. Where, x is steering arm length, y is tie rod length (in top view), p is rack casing length, p + 2r is rack ball joint center to center length, q is travel of rack, d is distance between front axis and rack center axis, β (beta) is Ackerman angle and a is the distance between spindle center and Tire center.

Geometric relation among various parameters can be deduced for three conditions, 1) Zero-steering 2) inner wheel and 3) outer wheel.

- Zero-steer: $y^2 = \left(\dfrac{B-(p+2r)}{2} + x\sin\beta\right)^2 + (d - x\cos\beta)^2$

- Inner wheel: $y^2 = \left(\dfrac{B-(p+2r)}{2} + x\sin(\delta_i+\beta)\right)^2 + (d - x\cos(\delta_i+\beta))^2$

- Outer wheel: $y^2 = \left(\dfrac{B-(p+2r)}{2} - x\sin(\beta-\delta_o)\right)^2 + (d - x\cos(\beta-\delta_o))^2$

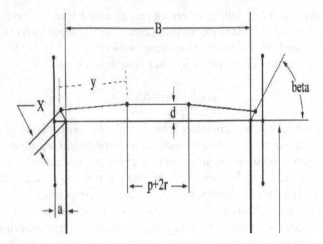

Figure 60.2 Trapezoidal arrangement of linkages in front of the front axle (Zero-steer).

β is defined as variable in iteration for particular track width and wheel base while solving equations for zero steer, inner wheel, outer wheel for particular steering angle you want to keep 100% Ackerman. It gives us more freedom to choose steering geometry. It is impossible to synthesize a mechanism which satisfies exact Ackerman condition for all steering angles. So, we have designed steering geometry which satisfies Ackerman condition, not exactly but close to it, for all steering input angles.

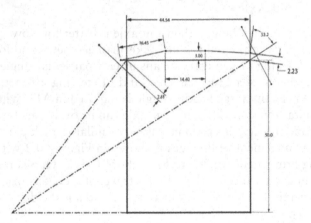

Figure 60.3 Turning left.

Conclusion

We vary values of 100% Ackerman angle (inner wheel angle), distance be-tween rack and front axis, rack length, rack travel for 100% Ackerman angle to get values of steering arm length, b and tie rod length while monitoring values of mean deviation from perfect Ackerman condition.

Track width and wheelbase are decided considering dynamics of the vehicle. B (track width-distance between wheel center and pivot point) is decided from wheel assembly design. So, these two parameters are essentially fixed while per-forming the iterations. We managed to perform iterations using MATLAB. Code performs multiple iterations for given set of inputs and provides us with values of tie rod length, b and steering arm length. b is kept less than 90 degrees to maintain compactness and strength of knuckle. Values obtained from the iterations are also compared with respect to the mean deviation from 100% Acker-man condition. However, 100% Ackerman condition is difficult to satisfy for entire range of wheel turning angles for given set of other parameters.

Mean deviation is calculated using following equation for steering input zero to max steering angle.

$$\Sigma \, (\delta ia - \delta oa)/(\delta i - \delta o)$$

δia is inner wheel angle
δoa is outer wheel angle (outer ackerman angle for different inner angle)
δi is inner wheel angle (same as δia)
δo is outer wheel angle (calculated outer angle for different inner angle)

According to Ackerman condition:

$$y^2 = \left[\frac{B}{2} - \left(\frac{p}{2} + r + q\right) - x\sin(\delta_o - \beta)\right]^2 + \left[d - x\cos(\delta_o - \beta)\right]^2$$
$$c = (A - q)\sin\gamma - d\sin\gamma$$
$$\gamma = \delta_o - \beta$$

Where,

$$c = \frac{\left[y^2 - d^2 - x^2 - (A - q)^2\right]}{2x}$$
$$c = (A - q)\sin\gamma - d\left(1 - (\sin\gamma)^{0.5}\right)^2$$
$$\left[(A - q)^2 + d^2\right] * (\sin\gamma)^2 - 2c(A - q)\sin\gamma + c^2 - d^2 = 0$$
$$\sin\gamma = c * (A - q) + \left[c^2 + (A - q)^2 - \left\{(A - q)^2 + d^2\right\}(c^2 - d^2)\right]^{0.5} = K$$
$$\gamma = \sin^{-1}K; \quad \delta_o = \cot^{-1}\left(\cot\delta_i + \frac{B}{L}\right)$$

Here, dynamic parameters are not considered. However, this method is useful to estimates the size and arrangement of different steering linkages. Bump steer phenomenon is frequently observed in ATV. Therefore, again iteration are performed using MATLAB for minimization of bump steer.

Bump Steer

When vehicle's wheel steer themselves without the input from steering, it is an undesirable condition called bump steer i.e., when tires hit a bump or droop. It also during cornering of vehicle. For Zero bump steer tie-rod must follow the same arc as suspension. As it is impossible to eliminate Bump steer completely in rack and pinion steering. So, to minimize it, the line drawn from upper arm, tie-rod and lower arm must intersect at the same point (IC) as shown in Figure 60.4.

Moreover, steering geometry is two dimensional so we get only x and y coordinates of the steering arm. The z-coordinate can be varied to minimize the bump steer further. This is the most suitable method as it also provides flexibility of setting roll center and other suspension parameters as per our need. Because in the former method RC (Roll center) of vehicle becomes definite due to the inter-section of arms lines and tie-rod line. The result of varying z coordinate can be seen by placing geometry in software like Lotus, ADAMS, etc. Figure 60.5 shows one of the results obtained from Lotus after creating the geometry.

Some of the race car uses bump-steer as dynamic toe in-out. During, turning lateral load transfer takes place due to which toe in-out occurs and helps in handling of vehicle. But for bumpy track bump-steer should be as minimum as possible [19]. For ATV, it must be minimum or zero.

Figure 60.4 Intersection of tie-rod and arm line at IC.

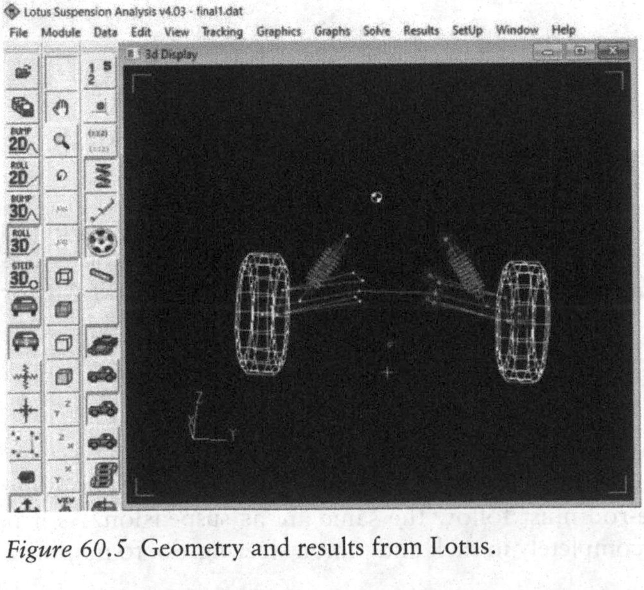

Bump Travel (mm)	Camber Angle (deg)	Toe Angle (deg)
−50.00	0.0169	−0.8264
−40.00	0.0243	−0.5749
0.000	0.0000	0.0000
30.00	−0.0704	0.0753
50.00	−0.1409	−0.0105
80.00	−0.2817	−0.3173
100	−0.3994	−0.6312
130	−0.6138	−1.2580
150	−0.7841	−1.7784

Figure 60.5 Geometry and results from Lotus.

Result and Discussion

Varying parameters like Ackerman angle, length of rack, etc. from low to high to observe the change in X, Y, b and % Ackerman. Summary of this comparison is shown Table 60.1.

Table 60.1: Relation between ackerman percent and various geometric parameters.

Varying Parameter	Steering arm: x	Tie rod: Y	β	%Ackerman
Ackerman angle for 100% Ackerman	Decrease	Decrease	Increase	Decrease
Length of steering rack	Increase	Decrease	Decrease	Increase
Maximum displacement of rack	Increase	Increase	Increase	Increase
Distance between wheel axis and rack	Decrease	Increase	Increase	Decrease

Table 60.2: Observations to ensure minimum bump steer.

Indicator	Correction
Toe in during rebound and out in compression all in one direction	Decrease the wedge on outer end or lower the inner end of tie-rod
Toe in during compression and out in rebound all in one direction	Increase the wedge on outer end or raise the inner end of tie-rod
Toe in during both rebound as well as compression	Increase the length of tie-rod
Toe out during both rebound as well as compression	Decrease the length of tie-rod
Toe out during compression and high rebound but in at time of low rebound	Decrease the wedge on outer end of tie rod and shorten the tie-rod
Toe in during compression and high rebound but out at time of low rebound	Increase the wedge on outer end of tie rod and lengthen the tie-rod

Trend observed from Table 60.1 for different parameters can be used to optimize the design of the steering system. This data will make the process of iteration faster for designers and help them to make an efficient steering system. However, certain other factors must be kept in mind while optimizing the setup like interference, wheel assembly, weight, strength etc.Another trend is observed during iteration (as shown in Table 60.2) which matches general trend of minimizing bump steer [20].

Advantages of Front Rack Geometry

* In front rack geometry locking of geometry occurs at higher angle; locking of geometry means after steering the wheels to particular angle they get locked. At this condition, wheels cannot be retrieving back to their original position.
* Front rack geometry has better performance in high-speed turning.

Limitations of Front Rack Geometry

- For same range of angle, overall Ackerman percentage is less compared to geometry with rack behind the axle.
- Weight of steering column increases as its length is extended, because position of rack is in front of wheel center-line and it also increases weight of tie-rods.
- In this geometry, pick up points for steering arm is on the outboard side of kingpin axis, so at higher angles there may be chance of collision of tie-rod, upright and wheel.

Conclusion

In this article, kinematic equations were derived for four-bar trapezoidal Ackerman geometry (in front of front-axle), and were constrained under geometric parameters. Three equations were derived one based on zero-steer condition, and other two were based on inner and outer wheel during turning (at true Ackerman). These equations were solved using Matlab, parameters like rack length, tie-rod length, steering angle, etc. were included. Moreover, as the model was two dimensional and lacks to incorporate bump steer effect another method was suggested to minimize it.

At the end effect of various geometric parameters on the Ackerman percentage was shown, along with-it certain indicators and corrections were mentioned in order to find out bump steer and correct it. The presented study involves only geometric constraint for Ackerman mechanism. This work can be further extended by including dynamic factors like tire-slip, scrub, caster, etc. and by analyzing components like hub-knuckle assembly.

References

1. Fahey, S. O. F., and Huston, D. R. (1997). A novel automotive steering linkage. *ASME Journal of Mechanical Design*, 119, 481–484.
2. Pramanik, S. (2002). Kinematic synthesis of a six-member mechanism for automotive steering. *Journal of Mechanical Design*, 124(4), 642–645.
3. Gautam, E. N. S., and Awadhiya, P. (2007). Kinematic synthesis of a modified ackermann steering mechanism for automobiles. In 14th Asia Pacific Automotive Engineering Conference Hollywood, California, USA 5–8 August, 2007.
4. Nalla, P., Sarkar, A., and Namaduri, M. S. (2013). Maximization of race car cornering performance through steering geometry optimization. In Proceedings of International Conference on Advances in Mechanical Engineering, AETAME, New Delhi, India, Elsevier.
5. Dewei, Y., and Jiansheng, L. (2003). Study on numerical comparison method for planar six-bar dwell mechanism synthesis. In Proceedings of the 11th World Congress in Mechanism and Machine Science, IFToMM, Tianjin, China, 18–21 August 2003.
6. Dong, H., Wang, D., and China, D. (2007). New approach for optimum synthesis of six-bar dwell mechanisms by adaptive curve fitting. In Proceedings of the 12th World Congress in Mechanism and Machine Science, IFToMM, Besancon, France, (pp. 17–21), 16–21 June 2007.
7. Hanzaki, A. R., Shaha, S. K., and Rao, P. V. M. (2005). Analysis of a Six-Bar Rack and Pinion Steering Linkage. Society of Automotive Engineers Inc.
8. Pradhan, D., Ganguly, K., Swain, B., and Roy, H. (2020). Optimal kinematic synthesis of 6 bar rack and pinion Ackerman steering linkage. *Proceedings of the Institution of Mechanical Engineers, Part D: Journal of Automobile Engineering*, 1–10.
9. Chaudharym H., and Saha, S. K. (2005). Analyses of four-bar linkages through multi-body dynamics approach. In 12th National Conference on Machine and Mechanism, NaCoMM, Guwahati, India, (pp. 45–52), 16–17 December 2005.

10. Bulatovíc, R. R., and Dordevíc, S. R. (2009). On the optimum synthesis of a four-bar linkage using differential evolution and method of variable controlled deviations. *Mechanism and Machine Theory*, 44(1), 235–246.

11. Mateka, S. B., and Gogate, G. R. (2012). Optimum synthesis of path generating four-bar mechanisms using differential evolution and a modified error function. *Mechanism and Machine Theory*, 52, 158–179.

12. Sleesongsom, S., and Bureerat, S. (2018). Optimal synthesis of four-bar linkage path generation through evolutionary computation with a novel constraint handling technique. *Hindawi Computational Intelligence and Neuroscience*, 2018, 16 (Article ID 5462563).

13. Gillespie, T. D. (1992). Fundamentals of Vehicle Dynamics. Pennsylvania: Society of Automotive Engineers, Inc., pp. 27–30.

14. Jazar, R. N. (2008). Vehicle Dynamics: Theory and Application. Springer.

15. Zhao, J. S., Liu, X., Feng, Z. J., and Dai, J. S. (??). Design of an Ackermann-type steering mechanism. *Proceedings of the Institution of Mechanical Engineers, Part C: Journal of Mechanical Engineering Science*, 227(11).

16. Qiu, Q., Fan, Z., Meng, Z., Zhang, Q., Cong, Y., Li, B., Wang, N., and Zhao, C. (2018). Extended ackerman steering principle for the coordinated movement control of a four-wheel drive agricultural mobile robot. *Computers and Electronics in Agriculture*, 152, 40–50.

17. Yao, K., Wang, Y., Hou, Z., and Zhao, X. (2008). Optimum design and calculation of ackerman steering trapezium. In International Conference on Intelligent Computation Technology and Automation.

18. Koladia, D. (2014). Mathematical model to design rack and pinion ackerman steering geometry. *International Journal of Scientific and Engineering Research*, 5(9). ISSN 2229-5518.

19. Milliken W. F., and Milliken, D. L. (1995). Race Car Vehicle Dynamics. SAE International.

20. Longacre Racing (2021). Bump Steer, <https://www.longacreracing.com/technical-articles.aspx?item = 8162>.

61 An investigation of fatigue crack closure on 304LSS & 7020-T7 aluminium alloy

Chandra Kant[a] and G.A. Harmain[b]

NIT Srinagar, Srinagar, Jammu and Kashmir, India

Abstract: Cyclically loaded structures are the most prone to failure hence efficient life prediction of such structures is crucial to ensure the reliable design. In this article, constant amplitude loaded structures at room temperature are studied and fatigue life prediction methods have been revisited. For constant amplitude loading, an investigation of crack closure (CC) effect is carried focusing on the effective resistance of crack propagation and effect of CC with stress ratio (R) is also analyzed. Various models of CC for plasticity-induced CC consideration is taken. Effects of CC are broadly seen in two categories (a) threshold crack propagation region, and (b) Steady crack propagation region. In this article, 304LSS and 7020 T7 Al-alloy material are used for comparison of various CC models and their performance and limitations are categorized based on linear regression parameters.

Keywords: Constant amplitude load, Fatigue crack closure, plasticity-induced crack closure, stress ratio effect.

Introduction

Reliability in life cycle design is dependent on the intrinsic and extrinsic behavior of material and fracture mechanics is one of the way to assess the life cycle of structure. Reliable model is needed for accurate and early fatigue life assessment. The fatigue crack propagation rate (FCPR) (da/dn) can be calculated using Paris model [1] and other models reported in [2]. By integrating the Paris model, remaining life of structure can be determined. Model in [1] is a function of stress intensity factor (ΔK), crack driving force and da/dn (crack growth rate in a cycle). Relation between log(da/dn) and $log(\Delta K)$ is sigmoidal, Stress intensity factor is determined via linear elastic fracture mechanics, which shows stress states at *sharp* crack tip in case of linear elastic condition or negligible plasticity. When material is ductile enough to make significant plastic zone at the crack tip, parameter (ΔK) is not able to capture crack propagation. Due to significant plasticity and plastic flow near crack tip, the process of crack opening and closing is affected which was reported in [3] for austenitic stainless steel 316L(N) Thus, with the inclusion of impact of effective stress intensity in fatigue crack propagation, it becomes crack closure (CC) model. The phenomena of CC is due to (a) plasticity (b) roughness (c) oxide induced (d) phase transformation. Fatigue

[a]chandra07phd17@nitsri.net, [b]gharmain@nitsri.net

crack propagation . modelling is categorized as (i) crack driving force models (ii) crack closure model. After Paris model, one aim of the fatigue crack propagation (FCP) modelling was to condense the effect of load ratio in a single FCPR curve. In this sequence, [4–10] reported various models, with different crack driving parameter and combination of driving parameters ΔK, K_{max}, R, These models have less physical significance than CC models. Thus CC phenomena is cardinal to explicate the impact of fickle loading giving CC model study a paramount role in fatigue crack propagation modelling.

In this article, performance of CC models is critically analyzed for various R and effect of CC is verified with experimental data of the Paris model and merit is quantified by effective stress intensity factor plot with FCPR. Effective stress intensity (K_{eff}) factor calculation is the need of hour after Elber [11] and [12] reported the crack closure phenomena and its influence in negative stress ratio respectively. CC analysis and impact is studied in creep fatigue interaction in [13]. Analysis of CC can be done analytically and numerically [14, 15]. Analytical models are the most widely used for their simplicity and cost effectiveness, but accuracy is compromised due to uniqueness of parameter of experimental data on which the model is based. Numerical analysis is more accurate as their parameter and model selection is based on material behavior replication via FEM type, mesh selection, material deformation model [16]. Using various method and models for effective stress intensity factor, efficient crack growth equation can be established.

This paper pertains a comparative study of various analytical crack closure models in 2 regions of fatigue crack propagation (a) threshold regime (b) steady state regime, analyzing the effect of CC models on varying R for FCP data. This study is based on constant amplitude loaded FCP behavior with or without CC. This article contributes novel insights of life cycle prediction models by incorporating and analyzing the ΔK_{eff} equations thus enabling better understanding of material deformation behavior. This article facilitates in predicting more accurate life cycle design of structure and in choosing an appropriate ΔK_{eff} equation.

FCP in Constant Amplitude Load

Fatigue life prediction models are reliant on loading hence broadly categorized in (a) constant amplitude loading (b) variable amplitude loading. Damage tolerant FCPR prediction models are generally based on stress intensity factor (ΔK). Crack propagation models are of 2 categories: crack driving force and effective crack driving force models (CC models). Crack driving force model [1] does not consider the effect of R and its consequences. However, effect of R on crack propagation is a well-established certitude and hence accounting effect of R is entail [4] has proposed a model with the effect of R and [17] reported a strong correlation of R with FCP.

Few authors found K_{max} as crack driving parameter instead of (ΔK) [5, 6] which shows strong correlation with FCP. These model are free from ascendancy of R. Since K_{max} and ΔK were not able to model FCPR, two parameter based model was proposed by [7, 8] and individual effect on FCP was quantified by [9] and reported that K_{max} has more influence on crack propagation than ΔK. Author [8] proposed a model based on K_{max} and ΔK^+ and observed that compressive load was not contributing for crack propagation concluding the fatigue damage can happen in tension-tension mode.

Effective crack driving force model is based on CC hence called crack closure models. In such models, effective stress intensity is determined for different R and loading conditions based on the models for CC determination. CC can be measured via experimental setup, using cut compliance technique, digital image correlation, in-situ fractography. Details of

CC models are discussed in next section. Experimental CC analysis is costly thus is not feasible always, so categorized and tested models for specific material are important.

Crack Closure Phenomena and Models

Elber in [18] proposed a hypothesis, fatigue crack will propagate under fully open crack only. The stress for crack opening was greater than σ_{op}. Up to a certain value of load, crack flank ΔK is closed and corresponding ΔK is closure stress intensity factor. After Elber model, another correlation is found between R and effective crack driving force and effect of varying R is normalized considering CC, reported in [19]. The consequences of R on CC in [20], reported at higher R (>= 0.7) CC has less influence, while under lower R it influences significantly. CC consideration makes FCP relation close to actual experimental data and independent of R which makes process easy for designer by reducing the involved parameter, R.CC models can simulate the overload induce FCPR retardation, was reported in [21]. Unified 2 parameter based crack propagation equation shows independence of R [8].

$$U = \frac{\Delta K_{eff}}{\Delta K} \qquad [1]$$

[18], proposed a mathematical relation in opening/closure stress with R. The Eq. 2 is established based on Al-alloy2024-T7. CC level is not constant throughout the crack propagation hence its effect is varying in FCPR. CC level is significant in crack initiation zone. In steady state zone, CC depends on R. However, in case of large scale yielding CC load decreases till crack fully opens. CC models [18] are valid under certain range of ΔK, R and material.

$$U = 0.5 + 0.4R \quad 0.7 > R > -0.1 \qquad [2]$$

Another similar relation was proposed by [11] relation is empirically established, based on experimental observation, for $-1 \le R \le 1$.

Another model for CC determination was given by ASTM [22, 23]. Newman Jr. [24] proposed a crack opening load determination model considering stress state involving a stress state variable α, in the model. Relation between α, Poisson ratio ν, was reported by [25] and reported a strong co-relation between them. State of stress affects the plastic zone size, hence in case of plane strain plasticity induced crack closure effect is limited. According to [26, 27] in threshold regime plane strain condition prevails hence negligible effect of plasticity CC. In this article, CC model based on crack tip stress field is also considered. [28, 29] proposed effective stress intensity was given in terms of crack tip tensile stress intensity factor, and defined crack opening when crack plane is under tensile stress field. B&H [30] proposed CC and opening load determination model in both plane stress and plane strain condition, most of the model gives only one parameter either crack closure load or crack opening load, but [30] proposed separate equations. Analogous equation for CC parameter determination is given by [31]. Another CC model proposed by B&C [32, 33] in terms of R, K_{max}.

Material and Methodology

Numerical study on experimental FCP data of [34, 35] is used for this numerical comparative study of CC models for stress ratio, R, 0.1, 0.2, 0.5, 0.6, 0.8 for Al-alloy 7020 T7 and 0.1, 0.2, 0.5, 0.75, 0.85 for 304L stainless steel. Qualitative analysis of models is done via linear regression (LR). Stress ratio limitation of models are categorized based on LR result: coefficient of determination (R^2) and standard error, which together represents dispersion of data, hence quantifies the model output. Effect of R on LR parameters is used to find the limiting R for models.

Results and Discussion

Methodical analysis of crack propagation data for varying R is done for 304Lss and 7020-T7 Al. Elber and Schjive model are found to be in good agreement which can unify the crack propagation data of varying R for 304LSS and 7020T7 Al-alloy shown in Figures 61.1, 61.2, 61.5 and 61.6.

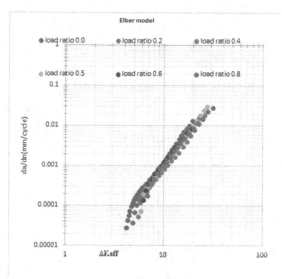

Figure 61.1 Crack closure correction via elber model for Al 7020T7.

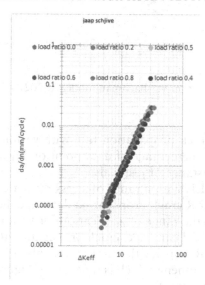

Figure 61.2 Crack closure correction via schijve model Al 7020T7.

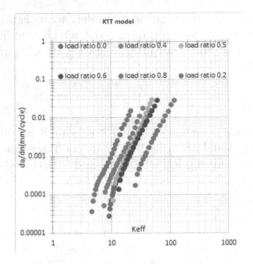

Figure 61.3 Crack closure correction via KTT model Al 7020T7.

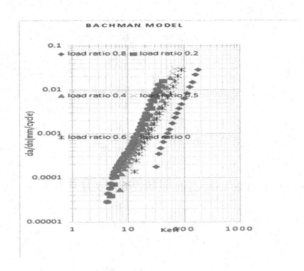

Figure 61.4 Crack closure correction via Bachman model Al 7020T7.

Degree of dispersion of crack propagation rate with various R is determined based on LR result. R^2 Represents the percentage of data point on linearly fitted line and standard error quantified the mean distance of residuals.

A comparison for all models with a range of stress ratio are summarized in Tables 61.1 and 61.2. Effect of particular R crack propagation data on coherence of crack propagation rate is studied which results the range of application for various CC models.

In this study, Bachman model can make crack propagation R independent on range of R 0.0–0.5, further increase of R disperse the data more. Subcritical fatigue crack growth has shown less effect of CC phenomena which is reflected in Figures 61.3 and 61.4 for Al and in Figures 61.7 and 61.8 for 304L SS.

Table 61.1: Comparison of crack closure models for 7020 T7 aluminium alloy.

Linear regression parameter	B &H model (R=0, 0.2,0.4, 0.5,0.6, 0.8)	ASTM model (R=0, 0.2,0.4, 0.5,0.6, 0.8)	KTT (R=0, 0.2,0.4,0.5, 0.6, 0.8)	Bachman (R=0, 0.2,0.4, 0.5,0.6, 0.8)	Schjive (R=0, 0.2,0.4, 0.5, 0.6, 0.8)	Elber (R=0, 0.2,0.4, 0.5, 0.6, 0.8)	KTT (R=0, 0.2, 0.4, 0.5, 0.6)	Bachman (R=0, 0.2, 0.4, 0.5,0.6)
Multiple R	0.91169	0.82595	0.73272	0.74606	0.91356	0.91336	0.83769	0.89197
R Square	0.83119	0.68219	0.53688	0.55661	0.83459	0.83422	0.70173	0.79544
Adjusted R Square	0.82996	0.67926	0.53352	0.55339	0.83339	0.83302	0.69925	0.79373
Standard error	0.00257	0.00349	0.00425	0.00416	0.00254	0.00254	0.00320	0.00265

Table 61.2: Comparison of crack closure models for 304LSS.

Linear regression parameter	B &H model (R=0, 0.1, 0.2, 0.5, 0.75, 0.85)	ASTM model (R=0, 0.1, 0.2, 0.5, 0.75, 0.85)	KTT (R=0, 0.1, 0.2, 0.5, 0.75, 0.85)	Bachman (R=0, 0.1, 0.2, 0.5, 0.75, 0.85)	Schjive (R=0, 0.1, 0.2, 0.5, 0.75, 0.85)	Elber (R=0, 0.1, 0.2, 0.5, 0.75, 0.85)	KTT (R=0, 0.2, 0.4, 0.5)	Bachman (R=0, 0.2, 0.4, 0.5)
Multiple R	0.86948	0.85420	0.58191	0.47429	0.88030	0.87751	0.83387	0.89296
R Square	0.75601	0.72966	0.33862	0.22495	0.77494	0.77004	0.6953	0.79738
Adjusted R Square	0.75513	0.72869	0.33625	0.22218	0.77413	0.76921	0.69361	0.79623
Standard error	0.00014	0.00015	0.00024	0.00026	0.00013	0.00014	0.00019	0.00015

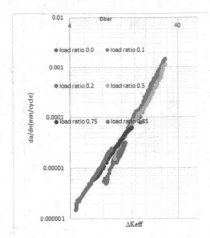

Figure 61.5 Crack closure correction via elber model for 304LSS.

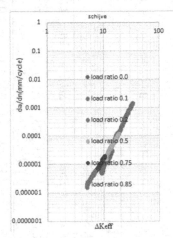

Figure 61.6 Crack closure correction via schjive model for 304LS.

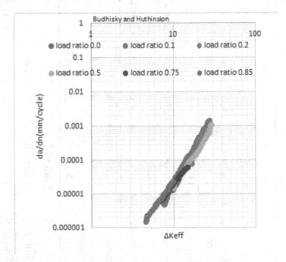

Figure 61.7 Crack closure correction via KTT model for 304LSS.

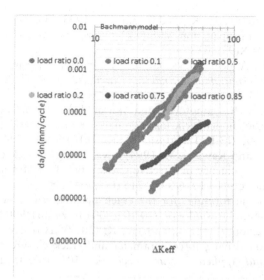

Figure 61.8 Crack closure correction via bachman model for 304LSS.

Conclusion

Fatigue crack closure correction is inevitable in FCP and residual life prediction for Al-alloy 7020 T7 and 304L SS. Elber, Schjive and B&H model performed well for stress range 0.0 to 0.8 and Bachman gave reasonable values up to 0.5 for Al-alloy 7020 T7 and 304L SS. Below Paris zone, the studied CC models are not observed to be much relevant and for full range, KTT model prediction is not able to condense the data for various R but shows relatively better prediction at R<0.5. ASTM model prediction is found to be in good agreement for 304L SS than Al-alloy 7020-T7.

Crack closure correction model needs further attention for better understanding and modelling of crack closure phenomena at various environmental conditions, loads and crack locations.

References

1. Paris, P., and Erdogan, F. (1963). A critical analysis of crack propagation laws. *Journal of Fluids Engineering, Transactions of the ASME*, 85, 528–533. https://doi.org/10.1115/1.3656900.
2. Kant, C., and Harmain, G. A. (2021). A model based study of fatigue life prediction for multifarious loadings. *Key Engineering Materials*, 882, 296–327. https://doi.org/10.4028/www.scientific.net/KEM.882.296.
3. Babu, M. N., and Sasikala, G. (2020). Effect of temperature on the fatigue crack growth behaviour of SS316L(N). *International Journal of Fatigue*, 140, 105815. https://doi.org/10.1016/j.ijfatigue.2020.105815.
4. Walker, K. (1970). The effect of stress ratio during crack propagation and fatigue for 2024-T3 and 7075-T6 aluminum.
5. Beevers, C. J. (1977). Fatigue crack growth characteristics at low stress intensities of metals and alloys. *Materials Science*, 11, 362–367. https://doi.org/10.1179/msc.1977.11.8-9.362.
6. Shahinian, P., and Sadananda, K. (1979). Effects of stress ratio and hold-time on fatigue crack growth in alloy 718. *Journal of Engineering Materials and Technology, Transactions of the ASME*, 101, 224–230. https://doi.org/10.1115/1.3443681.

7. Vasudévan, A. K., Sadananda, K., and Louat, N. (1993). Two critical stress intensities for threshold fatigue crack propagation. *Scripta Metallurgica et Materialia*, 28, 65–70. https://doi.org/10.1016/0956-716X(93)90538-4.

8. Kujawski, D. (2001). A new $(\Delta K + K max)0.5$ driving force parameter for crack growth in aluminum alloys. *International Journal of Fatigue*, 23, 733–740. https://doi.org/10.1016/S0142-1123(01)00023-8.

9. Ueno, A., Kishimoto, H., Kawamoto, H., and Akura, M. A. (1991). Crack propagation behavior of sintered silicon nitride under cyclic loads of high stress ratio and high frequency. *Engineering Fracture Mechanics*, 40, 913-IN12. https://doi.org/10.1016/0013-7944(91)90251-U.

10. Choi, G., Horibe, S., and Kawabe, Y. (1994). Cyclic fatigue crack growth from indentation flaw in silicon nitride: Influence of effective stress ratio. *Acta Metallurgica et Materialia*, 42, 3837–3842. https://doi.org/10.1016/0956-7151(94)90449-9.

11. Schijve, J. (1981). Some formulas for the crack opening stress level. *Engineering Fracture Mechanics*, 14, 461–465. https://doi.org/10.1016/0013-7944(81)90034-5.

12. Zhang, P., Xie, L., Zhou, C., Li, J., and He, X. (2019). Two new models of fatigue crack growth rate based on driving force parameter and crack closure method at negative load ratios. *Theoretical and Applied Fracture Mechanics*, 103, 102315. https://doi.org/10.1016/j.tafmec.2019.102315.

13. Potirniche, G. P. (2019). A closure model for predicting crack growth under creep-fatigue loading. *International Journal of Fatigue*, 125, 58–71. https://doi.org/10.1016/j.ijfatigue.2019.03.029.

14. Huang, X., Cui, W., and Leng, J. (2005). A model of fatigue crack growth under various load spectra. In Proceedings of Sih GC, 7th International Conference of MESO, August. (pp. 1–4).

15. Boljanović, S., and Maksimović, S. (2009). Fatigue life analysis of cracked structural components using crack closure effects. In Proceedings of the 2nd South-East European Conference on Computational Mechanics, SEECCM. (pp. 1–9), Rhodes, Greece.

16. Ohji, K., Keijiogura, and Ohkubo, Y. (1975). Cyclic analysis of a propagating crack and its correlation with fatigue crack growth. *Engineering Fracture Mechanics*, 7, 457–464. https://doi.org/10.1016/0013-7944(75)90046-6.

17. Wu, S. C., Li, C. H., Luo, Y., Zhang, H. O., and Kang, G. Z. (2020). A uniaxial tensile behavior based fatigue crack growth model. *International Journal of Fatigue*, 131, 105324. https://doi.org/10.1016/j.ijfatigue.2019.105324.

18. Elber W. (1971). The Significance of Fatigue Crack Closure. ASTM Special Technical Publication, pp. 230–242. https://doi.org/10.1520/stp26680s.

19. McEvily, A. J., and Ritchie, R. O. (1998). Crack closure and the fatigue-crack propagation threshold as a function of load ratio. *Fatigue & Fracture of Engineering Materials & Structures*, 21, 847–855. https://doi.org/10.1046/j.1460-2695.1998.00069.x.

20. Krenn, C. R., and Morris, J. W. (1999). The compatibility of crack closure and Kmax dependent models of fatigue crack growth. *International Journal of Fatigue*, 21, S147–S155. https://doi.org/10.1016/s0142-1123(99)00066-3.

21. Kant, C., and Harmain, G. A. (2021). Fatigue life prediction under interspersed overload in constant amplitude loading spectrum via crack closure and plastic zone interaction models—a comparative study. In 9th International Conference on Fracture Fatigue and Wear (FFW 2021). Springer (Lecture Notes in Mechanical Engineering), Ghent, Belgium.

22. Amercan Society for Testing and Materials (ASTM) (1999). ASTM E647: standard test method for measurement of fatigue crack growth rates. Annual Book of ASTM Standards, West Conshohocken, Pa, USA.

23. Geerlofs, N., Zuidema, J., and Sietsma, J. (2004). On the relationship between microstructure and fracture toughness in trip-assisted multiphase steels. In Proceedings of the 15th European Conference of Fracture (ECF15'04).

24. Newman, J. C. (1984). A crack opening stress equation for fatigue crack growth. *International Journal of Fracture*, 24. https://doi.org/10.1007/BF00020751.

25. Kirmani, G. A.-H. (1997). Single overload fatigue crack growth retardation: an implementation of plasticity induced closure. http://dspace.library.uvic.ca/handle/1828/8100.
26. Louat, N., Sadananda, K., Duesbery, M., and Vasudevan, A. K. (1993). A theoretical evaluation of crack closure. *Metallurgical and Materials Transactions A*, 24, 2225–2232. https://doi.org/10.1007/BF02648597.
27. Sadananda, K., and Ramaswamy, D. N. V. (2001). Role of crack tip plasticity in fatigue crack growth. *Philosophical Magazine A: Physics of Condensed Matter, Structure, Defects and Mechanical Properties*, 81, 1283–1303. https://doi.org/10.1080/01418610108214441.
28. Sehitoglu, H., and Sun, W. (1991). Modeling of plane strain fatigue crack closure. *Journal of Engineering Materials and Technology, Transactions of the ASME*, 113, 31–40. https://doi.org/10.1115/1.2903380.
29. Sun, W., and Sehitoglu, H. (1992). Residual stress fields during fatigue crack growth. *Fatigue & Fracture of Engineering Materials & Structures*, 15, 115–128. https://doi.org/10.1111/j.1460-2695.1992.tb00042.x.
30. Budiansky, B., and Hutchinson, J. W. (1978). Analysis of closure in fatigue crack growth. *Journal of Applied Mechanics, Transactions ASME*, 45, 267–276. https://doi.org/10.1115/1.3424286.
31. Llorca, J., and Sánchez Gálvez, V. (1990). Modelling plasticity-induced fatigue crack closure. *Engineering Fracture Mechanics*, 37, 185–196. https://doi.org/10.1016/0013-7944(90)90342-E.
32. Bachmann, V., and Munz, D. (1975). Crack closure in fatigue of a titanium alloy. *International Journal of Fracture*, 11, 713–716. https://doi.org/10.1007/BF00116381.
33. Chand, S., and Garg, S. B. L. (1983). Crack closure studies under constant amplitude loading. *Engineering Fracture Mechanics*, 18, 333–347. https://doi.org/10.1016/0013-7944(83)90144-3.
34. Mohanty, J. (2009). Fatigue life prediction under constant amplitude and interspersed mode-i and mixed-mode (I and II) overload using an exponential model. http://ethesis.nitrkl.ac.in/2802/.
35. Kalnaus, S., Fan, F., Jiang, Y., and Vasudevan, A. K. (2009). An experimental investigation of fatigue crack growth of stainless steel 304L. *International Journal of Fatigue*, 31, 840–849. https://doi.org/10.1016/j.ijfatigue.2008.11.004.

62 A CFD investigation to predict delay effect and thermal performance of high heat source electronic IC chip under submerged PCM cooling

Virendra Talele[1,a], Akshay Deshmukh[2,b], Rushikesh Kore[3,c], Saurabh Patrikar[1,d], and Niraj Badhe[1,e]

[1]Department of Mechanical Engineering, MIT School of Engineering, MIT ADT University Pune, India

[2]School of Physics, astronomy, engineering and computer science, University of Hertfordshire, UK

[3]Department of Aerospace Engineering, MIT School of Engineering, MIT ADT University Pune, India

Abstract: This paper deals with the CFD investigation over five unidentical integrated circuit chips (IC), for high modular heat sources to predict their thermal behavior under the passive thermal cooling arrangement. The investigation is proposed based on time-dependent transient scaled analysis with providing standard guidelines on how to predict passive thermal performance by fixing time depend on variables. The comparative results show the case of arrangement with PCM and Without PCM arrangements. The selected PCM for the present case is Paraffin wax which provides a better delay effect causing cooling of IC chips in passive mode. A standard verification guideline is also proposed to select a proper grid while dealing with phase change time-dependent simulations.

Keywords: CFD, IC, mesh, paraffin wax, PCM, simulation.

Introduction

Electronic devices are developing at extremely rapid rate. Modern electronic device is getting more compact and miniaturized in size. Making the ICs smaller gives an advantage to the product for developing portable and compact electronic devices. As the electronic devices are getting compact, the space between the adjacent ICs is getting reduced, giving rise to heating of the ICs embedded in it. Various researchers propose many techniques to cool the ICs. Most of the cooling techniques make use of an active cooling system. Active cooling systems are good and efficient, but they increase the weight of the system. They also require an additional power supply to drive that, which causes decreases in overall performance of the electronic devices and decreases the life of ICs as the appropriate amount of heat is not getting dissipated. So, to increase the IC's life, we have to remove the appropriate amount of heat from IC chips using a passive cooling system. This system does not make use of any additional power supply and the total weight is also maintained. This paper focus on the study of passive cooling arrangement developed for an unidentical

[a]virendratalele1@gmail.com, [b]ajdeshmukh18@gmail.com, [c]rushikeshkore77@gmail.com, [d]saurabhpathrikar6@gmail.com, [e]virendratalele46@gmail.com

IC chip on substrate board. Passive system experimental investigation on aluminum fins was conducted by H.M Ali et al [1], They used square and circular fins in their experiment. They study phase transition materials to cool fins. They used six different types of pcm materials and tested them with and without pcm at 4 to 8 watts. A comparison of the results of various pcm and fin designs was also done. Raising the Rayleigh number sped up cooling. Dehghan and Behnia [2] studied this experiment with many heat sources on different boards. They discovered that as the distance between the heat sources increases, so does the velocity of the cooling fluid, increasing Rayleigh's number. Unfinned triangular cross-section PCM heat sinks were examined by Usman et al [3]. RT-35HC, paraffin wax, and RT-44 were used in the experiment. The heat sink base got 5W-8W of heat. They found that the RT-44 PCM in combination with the inline triangular layout provided the best thermal performance. Cooling high-modularity power electronics was proposed by [4,5]. Several researches have indicated that applying heat management system causes a significant delay effect [6–12]. The main purpose of this experiment was to see how the concentration of nanoparticles affected the results of testing on various PCM-made heat sink designs. Using commercial coded Ansys fluent to the mixed convective heat transfer character is-tics from seven non-similar IC chips (Aluminum) put on a Switch Mode Power Supply (SMPS) board (substrate) have been implemented in a very creative manner, use (glass-reinforced epoxy laminate material). The goal is to find the ideal placement of these ICs on the substrate board. A thermal management setup employing tiny channels embedded with n-eicosane pcm material was investigated [13–15]. [16] On a substrate board, seven non-identical heat sources (Aluminum) were employed to describe three types of conjugate convective heat transfer (Bakelite). Further, temperature is a non-dimensional zed arrangement and prediction using fuzzy logic control. Calculating the temperature distribution of the integrated circuit chips is done using ANSYS Icepak. Mixed convection is preferred since it lowers the temperature of heat sources. Dimension and positioning of heat sources affect heat source temperature. Numerical numbers from ANSYS Icepack correlate well with predicted values from fuzzy logic. [17–21] This article compares the use of mini-channel embedded parafinwax pcm to absorb heat from high heat generating IC chips. The numerical analysis shows that both sets of data match well within the IC chip's 4–5°C temperature range. [22] found that when force convection speed is 5 m/s, the temperature of the board drops suddenly when utilizing copper clad-ding board. The majority of the research in the literature has been done on PCM-based heat sinks using experimental and computational methods. The study focused on heat transfer between the PCM and heat-generating surfaces (IC chips). Accordingly, the present inquiry deals with numerical modelling of high heat source IC chip potentially buried in PCM, which creates delay effect and large quantity of cooling to the heat source. Materials and Methods. The present study is carried out using commercial codded Ansys fluent CFD software v 18.1. The convergence technique of Ansys is based on the gauss serial method. Figure 62.1. Shows methods carried out for the present investigation.

Pre-Processing

The present investigation aims to predict the delay effect of PCM for unidentical power electronic IC chip which are mimic with aluminum material. The selected PCM material is paraffin wax p116 which is submerged filled around 70% to the whole electronic arrangement, property of PCM material is shown in Table 62.1. The geometry creation consists of one extended board on which mounting of a substrate board is presented having a total height of 5 mm with FR-4 material whose property is shown in Table 62.2. The material

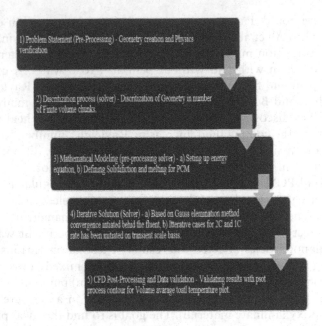

Figure 62.1 Process flowchart.

used for the extended board is Bakelite due to its high thermal insulating property. The electronic element IC chips are placed in a random localized arrangement. Figure 62.2 shows the detail arrangement of the system. The present investigation is carried out for 2 configurations as shown in Table 62.3 where the threshold limit is denoted by 2 C rate.

Table 62.1: PCM Material property.

SI No.	Property	Paraffin wax (P116)
1	Density (kg/m³)	760
2	Thermal Conductivity (W/mK)	0.24
3	Viscosity kg/ms	1.7894e-05
4	Specific Heat (J/Kg K)	2950
5	Latent heat of fusion (KJ/Kg k)	266000
6	Liquids temperature K	43–45 (melting range)
7	Solidus temperature	316

Mesh Convergence Study

Mesh convergence study has been proposed in the present investigation to verify the selection of optimum mesh size concerning convergence time. PCM is the type of material that changes its face from solid to a liquid by absorbing latent heat emitted by an electronic heat source, in between the solid-fluid interaction there might be the creation of floating-point divergence due to the bad selection of skewness factor. Table 62.4 shows the mesh selection of optimum mesh in correlation with the convergence's skewness factor and time-dependent quantity. Figure 62.3 shows the selected mesh.

Table 62.2: Isotropic material property.

SR NO.	Property	Value
Material–FR 4 (Solid for Substrate board)		
1	Density (kg/m3)	1850
2	CP (Specific heat) j/kg-k	1150
3	Thermal Conductivity w/m-k	0.29
Material–Bakelite (Extended Board)		
1	Density (kg/m3)	1250
2	CP (Specific heat) j/kg-k	1380
3	Thermal Conductivity w/m-k	0.2
Material–Aluminium (mimic IC)		
1	Density (kg/m3)	2719
2	CP (Specific heat) j/kg-k	871
3	Thermal Conductivity w/m-k	202.4

Table 62.3: Threshold configuration.

SR no.	Component's	Dimensions (l × w × t)	Heat flux W/cm^2
2 C–Rate (Maximum Threshold limit)			
1	U1	10.6 × 7 × 2.8	3.5623
2	U2	5 × 3 × 2	6.2682
3	U3	10.2 × 7.3 × 3.4	1.9635
4	U4	10.8 × 7.9 × 2.7	2.0354
5	U5	19.95 × 6.4 × 3.2	1.9863
1 C–Rate (low Threshold)			
1	U1	10.6 × 7 × 2.8	1.78115
2	U2	5 × 3 × 2	3.1341
3	U3	10.2 × 7.3 × 3.4	0.98175
4	U4	10.8 × 7.9 × 2.7	1.0177
5	U5	19.95 × 6.4 × 3.2	0.99315

Numerical Scheme

The present investigation runs on enthalpy porosity equation with the kept mushy zone constant as C = 105 which is running in solidification and melting model of commercial coded Ansys software. The phase change transformation in PCM is denoted by liquid fraction rate in between 0 to 1 range. In this range, Table 62.5 suggested for 0 PCM in solid form and 1 suggest PCM is completely liquid formation. This is because PCM absorbs and store the latent heat emitted by the electronic element to transfer its phase from solid to liquid in this case energy equation mode is kept active as shown in equation 1.

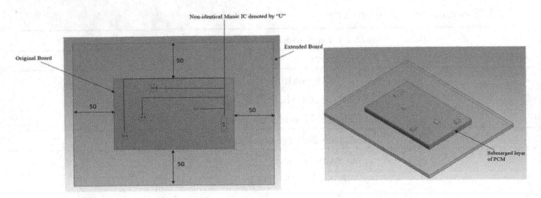

Figure 62.2 Domain creation–a) Geometry creation b) Submerged layer of PCM.

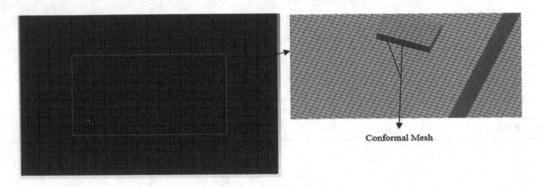

Figure 62.3 Selected mesh size of 1 mm with conformal mesh creation.

Table 62.4: Mesh independence study.

Sr no.	Element size	Element	Nodes	Skewness	Time for Convergence	Total Temp at 2 C (PCM)
1	50 mm	37	148	0.9963	Divergence	–
2	20 mm	215	564	0.9912	Divergence	–
3	10 mm	1840	806	0.9863	Divergence	–
4	5 mm	3108	6644	0.9521	300 S	124.178 °C
5	3 mm	16255	26182	0.9364	395 S	65.194 °C
6 (Selected)	1 mm	427932	515788	0.6589	653S	50.365 °C
7	0.5 mm	1053055	1188272	0.7624	1500 S	50.263

$$H = Sh + Lh \tag{1}$$

The liquid fraction (α) is defined as $\alpha = 0$, if T < T solidus; $\alpha = 1$, if T > T liquidus and $\alpha = $ *T–T solidious T liquidious–T Solidious* if T solidus < T Liquid. Now the latent heat of the material will be written as H = αL. Now for the formulation of solidification and melting numerical scheme energy equation will be rewritten as –

$$\partial(\rho H)t + \nabla\cdot(\rho v H) = \nabla\cdot(k\nabla T) + Sh \tag{2}$$

Validation

The validation for the present investigation is proposed with Mathew V K [21], which suggests a correlation that PCM cooling works on delay effect in time scale analysis. As PCM absorbs the electronic element's latent heat, it delays the threshold limit point of the temperature.

Results

The present investigation is carried out by using time-dependent analysis in which time is taken as a fixed variable in the transient model for comparative configuration between Elements submerged in PCM and Without PCM. The dependence of time is fixed based on temperature achieved at stated 55–56 C then fixed time taken without PCM model to achieve that certain temperature then the model is run with PCM arrangement for the same time which is fixed for the case of without PCM arrangement. Table 62.5 shows the temperature without PCM at different locations after time 200 S and 1600 S.

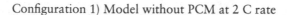

Configuration 1) Model without PCM at 2 C rate

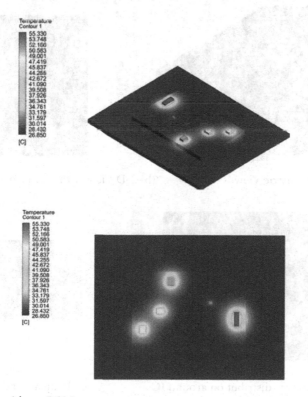

Figure 62.4 **2** C rate without PCM temp contour.

Configuration 2) Model without PCM at 1 C rate

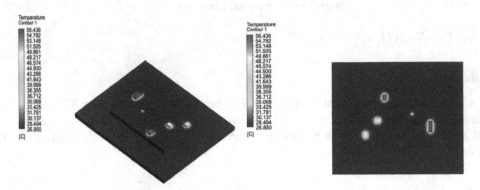

Figure 62.5 1 C rate without PCM temp contour.

Table 62.5: Temperature without PCM.

C rate	u1	u2	u4	u5	u7	board	extended surface	Time
2	322.43781	307.5801	328.0918	322.4263	329.7198	300.6941	300.05239	200 S
1	318.8405	304.8305	324.4897	318.5441	328.6133	302.5013	300.73844	1600 s

Configuration 3) With PCM arrangement at 2–C rate for 200 S

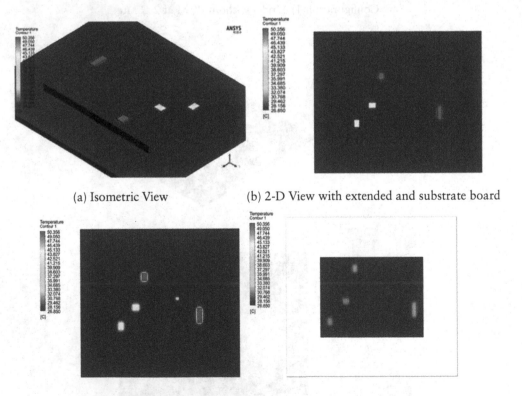

(a) Isometric View   (b) 2-D View with extended and substrate board

(c) Temperature distribution around IC   d) Temp around PCM

Figure 62.6 Temperature contour for 2–C rate with PCM arrangement.

Configuration 4) With PCM arrangement at 1–C rate for 1600 S

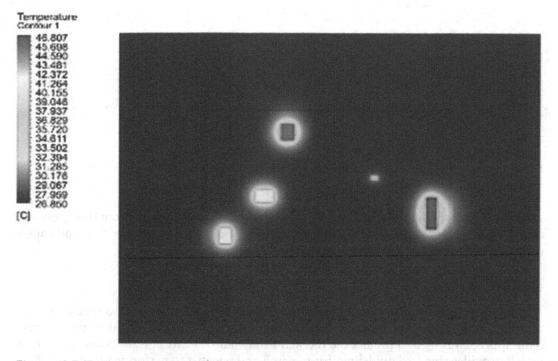

Figure 62.7 Temperature contour for with PCM at 1–C rate arrangement a) Isometric view b) Temperature contour c) Temperature contour at IC.

The time taken for 2 C rate threshold limit for without PCM arrangement to reach 55 °C is 200 S, so 200 S is taken as fixed time quantity then the same model for 2 C rate is run for 200 S in case of arrangement with submerged PCM model which suggests a comparative time delay thermal cooling characteristics. The time taken for the configuration arrangement without PCM 1 crate limit is 1600 S, so 1600 S will be fixed quantity against which below results are plotted to predict thermal cooling concerning delay effect.

The contour plots in Figures 62.4–62.7 suggests there is a drop of total temperature around medium due to the embedded PCM since PCM transfer its phase from solid to a liquid by absorbing stored latent heat. The comparative PCM cooling effectiveness can be predicted between the arrangement of configuration 1 in case of without PCM and configuration 2 in case of with PCM as by below equation 3.

(2 C rate) Total Temperature Drop $= T_{total(without\ pcm)} - T_{total\ (with\ pcm)} = 55.330 - 50.356 = 4.59\ °C$

Same technique can be applied for 1 C rate

(1 C rate) Total Temperature Drop $= T_{total(without\ pcm)} - T_{total\ (with\ pcm)} = 56.436 - 46.807 = 9.269\ °C$

The arrangement of the above plots suggests a strong correlation between working C-rates and impedance of time delay factor which increase passive cooling performance of PCM. The liquid fraction shown in Figure 62.8 are the plot in scale from 0 to 1 in which

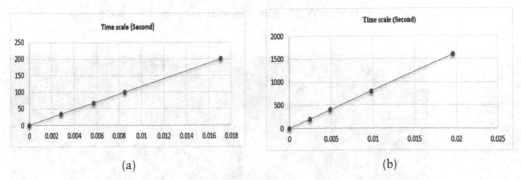

(a)   (b)

Figure 62.8 Time (S) Vs liquid fraction rate of PCM at the 2-C rate at fixed 200 S (a) and Time (S) Vs liquid fraction rate of PCM at the 1-C rate at fixed 1600S.

value near to 0 states the PCM is in the state of solid, increasing value from 0 suggest face transformation which causes a delay in overall threshold limit of temperature and causes thermal cooling around Integrated electronic circuits.

Conclusion

A Numerical investigation had been presented using the CFD approach to predict passive thermal management for high modularity rate Integrated circuits. A novel time depen-dant approach is used to predict the thermal behaviour of PCM in transient scale analysis which provides standard guidelines for research and scientist to performed transient phase change simulation. The following points can be drawn from the present study-

1. To avoid the divergence in the solution an optimum mesh parametric study should be prepared for the selection of mesh because the skewness factor shows a strong amount of correlation with the results.
2. There is a significant drop of temperature had been observed in the comparative case with and without PCM where a conclusion can be drawn as selected Paraffin wax change its states from solid to liquid in minute proportion by absorbing latent heat produce from the electronic element. This phase change characteristic of PCM causes delay effect for the time-scaled factor due to which sudden amount of temperature difference can be observed.

References

1. Ashraf, M. J., Ali, H. M., Usman, H., and Arshad, A. (2017). Experimental passive electronics cooling: parametric investigation of pin-fin geometries and efficient phase change materials. *International Journal of Heat and Mass Transfer*, 115, 251–263.
2. Dehghan, A., and Behnia, M. (1996). Numerical investigation of natural convection in a ver-tical slot with two heat source elements. *International Journal of Heat and Fluid Flow*, 17(5), 474–482.
3. Usman, H., Ali, H. M., Arshad, A., Ashraf, M. J., Khushnood, S., Janjua, M. M., and Kazi, S. N. (2018). An experimental study of PCM based finned and un-finned heat sinks for pas-sive cooling of electronics. *Heat and Mass Transfer*, 54, 3587–3598. https://doi.org/10.1007/s00231-018-2389-0.
4. Srikanth, R., and Balaji, C. (2017). Experimental investigation on the heat transfer perfor-mance of a PCM based pin fin heat sink with discrete heating. *International Journal of Thermal Sciences*, 111, 188–203.

5. El Qarnia, H., Draoui, A., and Lakhal, E. K. (2013). Computation of melting with natural convection inside a rectangular enclosure heated by discrete protruding heat sources. *Applied Mathematical Modelling*, 37(6), 3968–3981.

6. Phan-Thien, Y. L. N. (2000). An optimum spacing problem for three chips mounted on a vertical substrate in an enclosure. *Numerical Heat Transfer, Part A: Applications*, 37(6), 613–630.

7. Zeng, S., Kanargi, B., and Lee, P. S. (2018). Experimental and numerical investigation of a minichannel forced air heat sink designed by topology optimization. *International Journal of Heat and Mass Transfer*, 121, 663–679.

8. Arshad, A., Muhammad, H., Yan, W., and Kadhim, A. (2018). An experimental study of enhanced heat sinks for thermal management using n-eicosane as phase change material. *Applied Thermal Engineering*, 132, 52–66.

9. Farah, S., Liu, M., and Saman, W. (2019). Numerical investigation of phase change material thermal storage for space cooling. *Applied Energy, Elsevier*, 239(C), 526–535.

10. Jouhara, H., Żabnieńska-Góra, A., Khordehgah, N., Ahmad, D., and Lipinski, T. (2020). Latent thermal energy storage technologies and applications: A review. *International Journal of Thermofluids*, 5–6, 100039. https://doi.org/10.1016/j.ijft.2020.100039.

11. Asgharian, H., and Baniasadi, E. (2019). A review on modeling and simulation of solar energy storage systems based on phase change materials. *Journal of Energy Storage*, 21, 186–201.

12. Kothari, R., Sahu, S. K., and Kundalwal, S. I. (2021). Investigation on thermal characteristics of nano enhanced phase change material based finned and unfinned heat sinks for thermal management system. *Chemical Engineering and Processing—Process Intensification*, 162, 108328.

13. Mathew, V. K., and Hotta, T. K. (2018). Numerical investigation on optimal arrangement of IC chips mounted on a SMPS board cooled under mixed convection. *Thermal Science and Engineering Progress*, 7, 221–229.

14. Mathew, V. K., and Hotta, T. K. (2022). Experiment and numerical investigation on optimal distribution of discrete ICs for different orientation of substrate board. *International Journal of Ambient Energy*, 43(1), 1607–1614.

15. Mathew, V. K., and Hotta, T. K. (2019). Role of PCM based mini-channels for the cooling of multiple protruding IC chips on the SMPS board-a numerical study. *Journal of Energy Storage*, 26, 100917.

16. Karvinkoppa, M. V., and Hotta, T. K. (2017). Numerical investigation of natural and mixed convection heat transfer on optimal distribution of discrete heat sources mounted on a substrate. *IOP Conference Series: Materials Science and Engineering*, 263(6), 062066. IOP Publishing.

17. Mathew, V. K., and Hotta, T. K. (2021). Experimental investigation of substrate board orientation effect on the optimal distribution of IC chips under forced convection. *Experimental Heat Transfer*, 34(6), 564–585.

18. Mathew, V. K., Patil, N. G., and Hotta, T. K. (2021). Role of constrained optimization technique in the hybrid cooling of high heat generating IC chips using PCM-based mini-channels. *Constraint Handling in Metaheuristics and Applications*, 231–249.

19. Karvinkoppa, M. V., and Hotta, T. K. (2019). Transient analysis of phase change material for the cooling of discrete heat sources under mixed convection. *International Society for Energy, Environment and Sustainability*, 8, 1–6.

20. Mathew, K., and Patil, N. (2020). Convective heat transfer on the optimum spacing of high heat dissipating heat sources—a numerical approach. In *Emerging Trends in Mechanical Engineering*. Springer, Singapore, 8, 73–84.

21. Karvinkoppa, M., and Hotta, T. (2021). Performance enhancement of high heat generating IC chips using paraffin wax based mini-channels—a combined experimental and numerical approach. *International Journal of Thermal Sciences*, 164, 106865. 10.1016/j.ijthermalsci.2021.106865.

22. Kurhade, A. S., Rao, T. V., Mathew, V. K., and Patil, G. N. (2021). Effect of thermal conductivity of substrate board for temperature control of electronic components: a numerical study. *International Journal of Modern Physics C*, 32(10), 2150132.

63 A review on the effect of different performance parameters of H-darrieus turbines

Amit Kumar Gupta, Abhik Mukhopadhyay[a], Arnab Sarkar, Abhishek Shaw, Nabarun Paul, and Anal Ranjan Sengupta

Department of Mechanical Engineering, JIS College of Engineering, Kalyani, Nadia, West Bengal, India

Abstract: Fuel depletion rate is increasing day by day as population is increasing and the gap of demand of the energy is generating. To sustain in such situation, research-ers are very devoted for the replacement of conventional energy sources with renewable sources of energy. In the recent times, wind has become very famous renewable resource and researchers are focusing particularly on VAWT because of their unique performance characteristics. There are still lots of scopes of re-search in VAWTs. Till date lots of research papers are published on VAWTs. To increase the efficiency of VAWT's, investigations are being performed on various blade shapes and effect of different perfor-mance parameters using experimental and numerical approaches. In this present work, effect of several performance parameters like Blade Shape, Solidity, Tip Speed Ratio, Reynolds Number etc. are studied for H-Darrieus turbines for performance improvement. Moreover, the review paper gives a vast picture of the current improvements of efficiency of wind turbine.

Keywords: blade shape, H-darrieus turbine, solidity, tip speed ratio, wind energy, wind speed.

Introduction

In our day-to-day life, energy plays an important role for survival where there is heavy dependence of modern industries on conventional energy such as coal, natural gas, oil, nuclear power etc. The heavy usage of this conventional energy is the main cause of green-house effect, ozone layer depletion, global warming. Therefore, we have to utilize the energy that we can generate from the renewable energy sources such as solar energy, wind energy, hydro energy etc. The most popular renewable energy is Wind energy. To extract the energy from wind, wind tur-bines were developed. In some countries, there are not enough infrastructures to generate electricity by using conventional fuels. The population in Nigeria is 170 million but only 6 GW electricity is produced. In Brazil, the population is 190 million but 200 GW electricity is generated [1]. In this Scenario, the easiest and comparatively cheapest way to generate electricity is by using Wind Turbines. For this, many researchers around the world are researching for increasing the efficiency of Wind Turbine. In many countries like Taiwan, US, Japan, South Korea, Vietnam, Brazil, Europe,

[a]abhik.muk9827@gmail.com

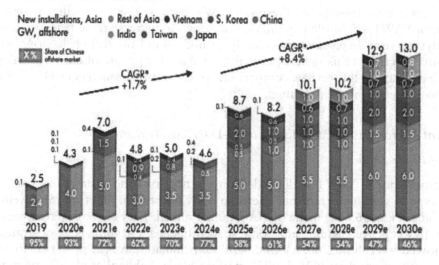

Figure 63.1 Installation growth to 2030 in Asia (a)[2].

Wind Turbine industries are rapidly growing [2]. In 2019, China secured 1st position for establishing 2.4 GW electricity from wind energy. United Kingdom and Germany secured 2nd and 3rd position for installing 1.8 GW and 1.1 GW electricity respectively [2]. It was showed that the installations are increased from 9.14 GW in 2018 to 25.44 in 2018. By seeing the data and graphs in Figure 63.1, it can be said that the usage of wind turbine is increased drastically.

Wind Turbine is a device by which the kinetic energy of wind can be transformed into electrical energy. There are basically two types of Wind Turbines: Vertical Axis Wind Turbine (VAWT), Horizontal Axis Wind Turbine (HAWT), as shown in Figure 63.2(a). VAWT has several advantages like it is omni-directional, can be situated in turbulent air, creates less noise, good self-starting ability, no limitation in size. The disadvantage of VAWT is torque produced varies with time, guy wire is required to stable the whole structure.

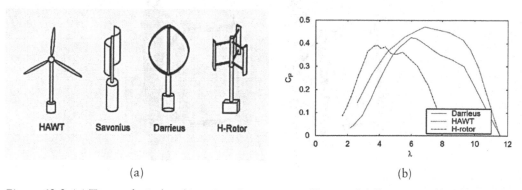

(a)   (b)

Figure 63.2 (a) Types of wind turbines [3], (b) Power coefficient of different Wind Turbine [4].

Objective of the Current Paper

Researchers are focusing particularly on VAWT because there are still lots of scopes of research in VAWTs. Versatile experiments are done to increase the efficiency. Researchers are analysing very deeply and thoroughly to understand the flow behaviour on blades. The main objective of this review paper is to give a wide picture of existing researches on H-Darrieus rotor. In this review, comparisons of different parameters of H-Darrieus rotor with power coefficient are presented.

Different Performance Parameters of H-Darrieus Turbine

Effect of Blade Shape

The performance of VAWT is depended on the flow of wind over airfoil blade. So, the blade shape is very important parameter in VAWT research. Su et al. [5] developed a V shape blade (Figure 63.3a) made of NACA 0021 airfoil. They chose different values of the distance (ΔV) between leading edge of V shaped blade and leading edge of straight blade. They showed for TSR of 2.60, ΔV of 0.6c, V shaped blade gave the power coefficient 0.38, which is 20% more than NACA 0021 straight blade. Sobhani et al. [6] investigated the performance of Darrieus VAWT made of NACA 0021 with circular cavity at different diameter and position in airfoil. They found that the power coefficient of turbine using

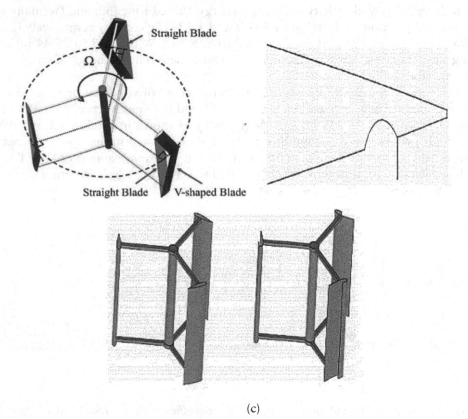

(c)

Figure 63.3 (a) V shape blade [5], (b) Dimple with gurney flap [7], (c) J shape blade [8].

airfoil blade with a cavity is 0.43 for TSR of 2.6. Ismail et al. [7] performed an experiment on a blade NACA 0015 which is having a circular dimple and a Gurney flap (Figure 63.3b) located at the end of trailing edge. They showed the tangential force is increased by 35% than standard NACA 0015, for TSR of 3.50. Zamani et al. [8] performed a 3D simulations of J shaped blade (Figure 63.3c) made of NACA 0015 profile taking different TSR. They observed that at TSR of 1.6, it gives best power coefficient value of 0.3358. Mohamed [9] further studied about the J shaped blade taking three airfoil profiles bladed NACA 0015, NACA 0021, and S1046 at TSR of 2 to 6. It was found at TSR of 5, the power coefficient of NACA 0015 J shaped blade is reduced to 0.25. Kumar et al. [10] used CFD approach to identify the performance of wind turbine made of NACA 0021 airfoil blade with a cavity. It was found that at TSR of 1.3, best power coefficient is 0.16.

Effect of Solidity

Researchers are more focusing on rotor solidity, as better performances are shown for a large range of solidity. Qamar et al. [11] investigated cambered blades airfoil and solidity having variations of NACA 4312 airfoil by CFD analysis to improve the performance of turbines. It was seen that at solidity of 0.9, maximum Cp was 0.38. Singh et al. [12] investigated a rotor equipped with S1210 blade at solidities of 0.8 to 1.2. It was found that maximum Cp of 0.32 was observed for optimum solidity of 1.0. Peng et al. [13] investigated the performance of NACA 0018 blade with solidity of 0.45. It was found that the tangential force is increased and it also has a good self-starting capability. Rezaeiha et al. [14] investigated the impact of solidity within a wide range from 0.09 to 0.36 for 2 bladed, 3 bladed and 4 bladed turbines with NACA 0018 airfoil with varying TSR from 1.5 to 5.5. It was observed that for solidity of 0.30, the maximum Cp obtained for 2 bladed, 3 bladed, 4 bladed turbines are 0.454, 0.450, 0.445 respectively. Sengupta et al. [15, 16] conducted an experiment with three types of blade profiles like S815, EN0005, NACA 0018. They found that at solidity of 0.51, S815 blade showed best power coefficient 0.19 compared to the others, as shown in Figure 63.4. A comparative experiment was done by B. Belabes et al. [17] between small and large H-Darrius VAWTs to investigate the turbulence intensity effect on the power coefficient. It was found that at solidity of 0.10, small H-Darrius VAWT having NACA 0018 blade profile gave power coefficient of 0.34.

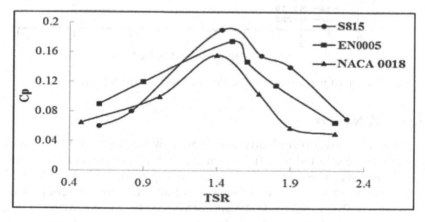

Figure 63.4 Diagram of power coefficient vs TSR of different airfoil blades [15].

Effect of Tip Speed Ratio

The value of TSR predicts the air flow through the wind turbine which decides the performance of wind turbine. So, TSR is the fundamental design parameter of wind turbine. TSRA 2D CFD simulation was done by Mohamed [18] with 5 different series of 20 airfoils by taking TSR range 2–20 and solidity of 0.1. He found that for TSR of 4, S1046 airfoil blade gave maximum Cp of 0.4051. Li et al. [19] experimented the NACA 0021 airfoil profile blade by Field and Wind tunnel test and concluded that for TSR of 2.19, maximum Cp was 0.20. Hashem et al. [20] experimented 2D simulation with 24 different airfoils to find the best efficient blade and also investigated the most suitable wind lanes. They found that for TSR of 4, S1046 showed best Cp of 0.3463 as shown in Figure 63.5.

Qamar et al. [21] have taken 2D CFD approach to simulate airfoil profile, NACA 0012, NACA 4512, and NACA 7512. It was shown that for TSR of 4, NACA 7512, NACA 4512 gave maximum Cp of 0.15. Deshpande et al. [22] used a nu-merical approach to study Giromill-type wind turbine with NACA 0018, NACA 0015, S1210 airfoils by taking two solidities 0.2 and 0.4. They found that at TSR of 3, NACA 0018 gave best Cp of 0.39. Sengupta et al. [23] performed an exper-iment between three airfoil profile blade S815, EN0005, NACA 0018. They observed at optimum rotor height to diameter ratio equals to 1, NACA 0018 gave the power coefficient 0.196 at TSR of 1.43. Q. Li [24] performed an CFD experiment to analyse the performance & amp; vortex characteristic of straight blade VAWT. They found that at TSR of 2.37 and pitch angle of 5 degrees, NACA 0021 blade profile showed Cp of 0.18.

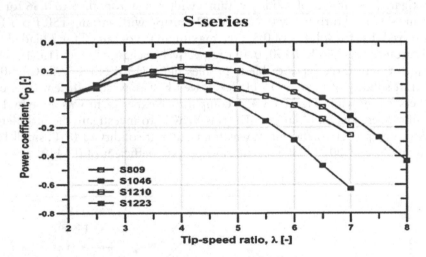

Figure 63.5 Comparison of performances of different S series airfoils [20].

Effect of Reynolds Number

The types of flow of air over rotor blades are decided by Reynolds number. A wide range of Reynolds numbers are selected for experiments to find the performances of wind turbines. Roh et al. [25] showed at Re of 7 × 105 and solidity of 0.0833, NACA 0015 revealed Cp of 0.4. Bogateanu et al. [26] experimented to find the performances of 4 airfoils like NACA 0012, NACA 0015, NACA 0018 and NACA 0021. They concluded that at Re of 106 and TSR of 3.2, NACA 0015 exhibited the highest Cp of 0.5 as shown in Figure 63.6a. Bachant et al. [27] showed at Re of 1.1×106 and TSR of 1.4, NACA0020 showed Cp value of 0.28. Rezaeiha et al. [28] used CFD approach and found that the Cp of NACA 0018, at Re of 105 is 0.47. Figure 63.6b is the contour plot of this CFD investigation. Armstrong et al. [29] experimented to find the performances of straight and canted blade rotor with airfoil NACA 0015 and NACA 0013 respectively. They showed canted blade ro-tor gave best Cp of 0.30 at Re of 5.5×105. Sengupta et al. [30] performed a CFD analysis with two unsymmetrical blade profiles like S815 and EN0005. It was found that S815 blade profile exhibited higher Cp of 0.2.

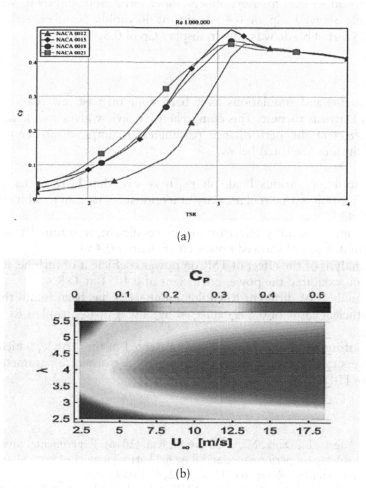

(a)

(b)

Figure 63.6 (a) Cp vs TSR at Reynolds Number 10^6 for different Airfoil [26], (b) Contour plot at Reynolds number 10^5 [28].

Discussions

From the above literature study on different parameters of H-Darrieus turbine, it is seen that:

- From the blade shape analysis is seen that H-Darrieus turbine having NACA0021 V-shaped blade exhibited Cp of 0.38 while NACA0021 blade profile having circular cavity on its pressure side showed Cp of 0.43.

- The investigation of blade solidity for several H-Darrieus turbines showed that NACA4312 blade profile displayed maximum Cp of 0.38 at solidity 0.9. Again, two-bladed H-Darrieus turbine having NACA0018 profile showed Cp of 0.454 at solidity of 0.3 in the TSR range of 1.5 to 5.5.
- From a comparative study of various TSR ranges for different H-Darrieus turbines, it can be observed that S1046 blade showed maximum Cp of 0.4051 at TSR of 4 and solidity 0.1. Again, NACA0018 blade displayed Cp of 0.39 at TSR of 3 and solidity 0.2.
- It has been noticed that for Reynolds Number 7×10^5 and solidity 0.083, NACA0015 blade profile showed Cp of 0.4. Again, at Reynolds Number 10^6 and TSR 3.2, NACA0015 airfoil blade was able to display Cp of 0.5.

Discussions

Lots of experiments and simulations have been done on past few years to improve the efficiency of H-Darrieus turbine. This comprehensive review gives a wide range for understanding the effect of the performance parameters to improve power coefficient. The important conclusions are listed below.

- From the study of various blade shape, it is seen that H-Darrieus turbine having NACA0021 airfoil with circular cavity at its pressure side can exhibit the power coefficient of 0.43.
- After studying the solidity effect on power coefficient, it is found that NACA 0018 airfoil at solidity of 0.3 showed power coefficient of 0.454.
- After the analysis of the effect of TSR on power coefficient of turbine, it is found that, S1046 airfoil exhibited the power coefficient of 0.4051 at TSR 4.
- From the analysis of different Reynolds numbers, it has been found that the highest power coefficient value of 0.5 is exhibited by NACA 0015 airfoil at Re of 10^6.

A plenty of information are collected and presented in this article, which can help the future researchers to understand the effect of several performance parameters to improve the efficiency of H-Darrieus turbines.

References

1. Eboibi, O., Angelo, L., Dana, M., and Howell, R. I. (2016). Experimental investigation of the influence of solidity on performance and flow field aerodynamics of vertical axis wind turbine at low reynolds number. *Renewable Energy*, 92, 474–483.
2. Global Wind Energy Council (GWEC) (2020). Global Wind Report Annual Market Update. Technical Report. https://www.renepoly.com/EN/products-wind (accessed on 19th March, 2021).
3. Eriksson, S., Bernhoff, H., and Leijon, M. (2008). Evaluation of different turbine concepts for wind power. *Renewable and Sustainable Energy Reviews*, 12, 1419–1434.
4. Su, J., Chen, Y., Han, Z., Zhou, D., Bao, Y., and Zhao, Y. (2020). Investigation of V-shaped blade for the performance improvement of vertical axis wind turbines. *Applied Energy*, 260, 114326.
5. Sobhani, E., Ghaffari, M., and Maghrebi, M. J. (2017). Numerical investigation of dimple effects on darrieus vertical axis wind turbine. *Energy*, 133, 231–241.

6. Ismail, M. F., and Vijayaraghavan, K. (2015). The effects of airfoil profile modification on a vertical axis wind turbine performance. *Energy*, 80, 20–31.

7. Zamani, M., Nazari, S., Moshizi, S. A., and Maghrebi, M. J. (2016). Three-dimensional simulation of J-shaped darrieus vertical axis wind turbine. *Energy*, 116, 1243–1255.

8. Mohamed, M. H. (2019). Criticism study of J-shaped darrieus wind turbine: performance evaluation and noise generation assessment. *Energy*, 177, 367–385.

9. Kumar, Y., Sengupta, A. R., Biswas, A., Mazarbhuiya, H. M. S. M., and Gupta, R. (2021). CFD analysis of the performance of an H-darrieus wind turbine having cavity blades. In Recent Advances in Mechanical Engineering, Lecture Notes in Mechanical Engineering (pp. 711–719).

10. Qamar, S. B., and Janajreh, I. (2017). A comprehensive analysis of solidity for cambered darrieus VAWTs. *International Journal of Hydrogen*, 42, 19420–19431.

11. Singh, M. A., Biswas, A., and Misra, R. D. (2015). Investigation of self-starting and high rotor solidity on the performance of a three S1210 blade H-type darrieus rotor. *Renewable Energy*, 76, 381–387.

12. Peng, Y. X., Xu, Y. L., Zhu, S., and Li, C. (2019). High-solidity straight-bladed vertical axis wind turbine: numerical simulation and validation. *Journal of Wind Engineering & Industrial Aerodynamics*, 193, 103960.

13. Rezaeiha, A., Montazeri, H., and Blocken, B. (2018). Towards optimal aerodynamic design of vertical axis wind turbines: impact of solidity and number of blades. *Energy*, 165, 1129–1148.

14. Sengupta, A. R., Biswas, A., and Gupta, R. (2017). Investigations of H-darrieus rotors for different blade parameters at low wind speeds. *Wind and Structures*, 25, 551–567.

15. Sengupta, A. R., Biswas, A., and Gupta, R. (2017). The aerodynamics of high solidity unsymmetrical and symmetrical blade H-darrieus rotors in low wind speed conditions. *Journal of Renewable and Sustainable Energy*, 9, 043307.

16. Belabes, B., and Paraschivoiu, M. (2021). Numerical study of the effect of turbulence intensity on VAWT performance. *Energy*, 233, 121139.

17. Mohamed, M. H. (2012). Performance investigation of H-rotor darrieus turbine with new airfoil shapes. *Energy*, 47, 522–530.

18. Li, Q., Maeda, T., Kamada, Y., Murata, J., Yamamoto, M., Ogasawara, T., Shimizu, K., and Kogaki, T. (2016). Study on power performance for straight-bladed vertical axis wind turbine by field and wind tunnel test. *Renewable Energy*, 90, 291–300.

19. Hashem, I., and Mohamed, M. H. (2018). Aerodynamic performance enhancements of H-rotor darrieus wind turbine. *Energy*, 142, 531–545.

20. Qamar, S. B., and Janajreh, I. (2017). Investigation of effect of cambered blades on darrieus VAWTs. *Energy Procedia*, 105, 537–543.

21. Deshpande, P., and Li, X. (2013). Numerical study of giromill-type wind turbines with symmetrical and non-symmetrical airfoils. *European International Journal of Science and Technology*, 2(8), 195–208.

22. Sengupta, A. R., Biswas, A., and Gupta, R. (2016). Studies of some high solidity symmetrical and unsymmetrical blade H-darrieus rotors with respect to starting characteristics, dynamic performances and flow physics in low wind streams. *Renewable Energy*, 93, 536–547.

23. Li, Q., Cai, C., Maed, T., Kamada, Y., Shimizu, K., Dong, Y., Zhang, F., and Xu, J. (2021). Visualization ofaerodynamic forces and flow field on a straight bladed vertical axis wind turbine by wind tunnel experiments and panel method. *Energy*, 225, 120274.

24. Roh, S. C., and Kang, S. H. (2013). Effects of a blade profile, the Reynolds number, and the solidity on the performance of a straight bladed vertical axis wind turbine. *Journal of Mechanical Science and Technology*, 27, 3299–3307.

25. Bogateanu, R., Dumitrache, A., Dumitrescu, H., and Stoica, C. I. (2014). Reynolds number effects on the aerodynamic performance of small VAWTs. *UPB Scientific Bulletin, Series D: Mechanical Engineering*, 76(1), 25–36.

26. Bachant, P., and Wosnik, M. (2016). Effects of reynolds number on the energy conversion and near-wake dynamics of a high solidity vertical-axis cross-flow turbine. *Energies*, 9, 73.

27. Rezaeiha, A., Montazeria, H., and Blocken, B. (2018). Characterization of aerodynamic performance of vertical axis wind turbines: impact of operational parameters. *Energy Conversion and Management*, 169, 45–77.

28. Armstrong, S., Fiedler, A., and Tullis, S. (2012). Flow separation on a high reynolds number, high solidity vertical axis wind turbine with straight and canted blades and canted blades with fences. *Renewable Energy*, 41, 13–22.

29. Sengupta, A. R., Biswas, A., and Gupta, R. (2019). Comparison of low wind speed aerodynamics of un-symmetrical blade H-darrieus rotors-blade camber and curvature signatures for performance improvement. *Renewable Energy*, 139, 1412–1427.

30. Sengupta, A. R., Biswas, A., and Gupta, R. (2017). Investigations of H-darrieus rotors for different blade parameters at low wind speeds. *Wind and Structures*, 25, 551–567.

64 Development of prototype for offshore integrated wave and wind energy platform for power generation

Rushikesh Chinnari[a], A Ramesh Kumar[b], and Y P Deepthi[c]

Department of Mechanical Engineering, Amrita School of Engineering, Amrita Vishwa Vidyapeetham, Bengaluru, Karnataka, India

Abstract: The offshore wave and wind energies have the potential to address the needs of the people and the industries in the coastal areas which can meet the increased dependence on the renewable energy. In this paper hybrid wind and wave energy system at different offshores are explained and highlights the various types of structures, configurations. A prototype of hybrid wind and wave energy system was developed. Power developed across wind turbine, wave turbine and hybrid system are calculated and found out that hybrid system showed a higher value. The hybrid project can also improve the economic viability relative to the individual projects of wave energy and also wind energy and enhancing the offshore renewable energy.

Introduction

Renewable energy development attracts more people to paying attention to the success of the revolution in energy. They may be given the opportunity to assist in the resolution of challenges such as fossil fuel exhaustion, greenhouse effect and contamination of the environment. Wind energy is the most commonly used renewable source of energy. In the recent years, the development over the off-shore energy has been increased a lot since off-shore has the enormous potential to supply the sustainable energy for the long time. There is major construction, installation, grid connectivity and service and repair difficulties in accessing the shallow waters, which greatly raise costs. However, if prices can be reduced to a sustainable standard, there is a significant opportunity for deep water wind farms. One potential way to reduce offshore wind costs is to harness synergies with other green maritime technology [1, 2]. In the off-shore energy industry, these two types of energy, wave and wind, already play a significant role.

Background and Resource

Wind energy is one of the world's most advanced green energies, since the real concept of wind energy has been revealed. The two well-developed forms of wind concepts are based on the axis direction. HAWT (Horizontal axis wind turbine) are preferred for higher power coefficient and axis of rotation is parallel to air stream whereas VAWT

[a]deepthi.prajna@gmail.com, [b]BLENU4MEE17012@blr.amrita.edu, [c]blenu4mee17006@blr.amrita.edu

(Vertical axis wind turbine) has the lower power coefficient but receives wind from any direction. Among the marine energy, wave energy converters are the well-developed form of green energy. The two main types under which most of the available wave energy converters can be classified are tidal kinetic energy and tidal potential energy [3]. And relative density is the one of the important physical factor that should be considered in the wave energy converter (WEC).

Two aspects will be taken into account in the union of wave and wind energy. One of them is the generation of electricity in a farm that includes wave energy equipment and wind turbines such as autonomous WEC's [4]. Whereas the second part, such as floating wind wave energy converters or bottom fixed hybrid wind wave energy converters, is the so-called hybrid system where the wave absorbers and wind power equipment are combined onto a single base.

The one objective of adding the wave energy converters to the off shore wind turbines is to extract the wave energy as WEC that act as the dampers to the structure thus increases the stability. The WEC absorbs the incoming wave energy, reducing the buoyant wind turbine's movement and increasing overall output [5]. With the wave energy absorption, the waves at the rear end are calmer and hence the boat landing and maintenance work can be monitored easily as in Figure 64.1. On the other side, addition of the wind turbines to the floating wave energy device makes it more stable and also enhances power production.

Figure 64.1 Rear side calmer water compared to front side [5].

Wind Turbines have to be positioned with a certain distance between each other to enhance effects of the air turbulence effect in the wind farm, in large scale wind turbines the distance can be in the range of half a kilometer depends on the rotor diameter [6]. Wave plants, on the other hand, are densely spaced between wind turbines, resulting in more electricity in the ocean in less space.

Synergies in Hybrid System

The mixture of wave and wind energy converters or systems has increased the renewable energy production per area, Utilized the both wave and wind energy simultaneously, smooth output power and results in overall improvement. It also minimizes the uncertainty of the electricity produced.

These hybrid systems focusing on marine resources have generated four synergies, according to the researchers, which could increase growth [7, 8].

- For the creation of such hybrid systems, a standard regulation is required. This includes a legal administrative system, marine spatial planning, a streamlined licensing process and the planning of grid and auxiliary infrastructures.
- Synergies in process engineering: marine fuel-oriented hybrid systems may be paired with other marine projects, such as oil and gas production, hydrogen processing, breakwater, aquaculture, desalination etc.
- Cooperation-sharing facilities, construction and equipment: this form of synergy applies to machinery, logistics, grid connectivity, storage and service and maintenance. Costs may be minimized for any of these products by integrating multiple forms of suppliers.
- Synergies-sharing areas: between solar energy projects and additional installations (oil and gas production, desalination, hydrogen processing etc.). Sharing areas enables the densification of ocean usage to be enhanced, the output provided by the nearby operations to be exchanged and studies to be limited to a single venue.

Stability of Platform

Stability of base is one of the aspect where the base will be in the equilibrium, which can affect the efficiency of the hybrid system and also the reliability of the structure. On the availability of the different concepts that decides the stability in the floating condition, the elemental three concepts are shown in the Figure 64.2. The Ballast stabilized is the first one which has low center of gravity. Spar is the best suited element in this category. Mooring stabilized is the second type that has mooring line tension for example tension leg platform. Buoyancy or water-plane stabilized (distributed buoyancy) e.g., semisubmersible [9].

Arrangement of Systems

There are two possible types of systems depending on their arrangement of wind and wave energy systems:

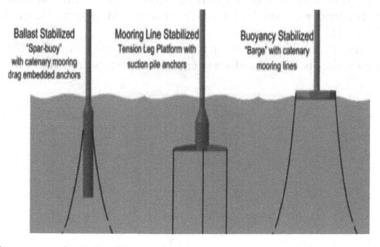

Figure 64.2 Three types of stability [9].

Co-Located Systems

At the current point, this may be the easiest and most natural alternative. Basically, these systems compound an offshore wind farm with a line-up of converters for wave energy. In other words, these processes are composed of disconnected systems that are brought together. These networks share a range of assets such as maritime zones, grid links, facilities for operation and repair etc. These systems either bottom-fixed or floating are typically relying on an offshore wind farm. It is possible to further classify these systems as follows.

Independent Array

These systems have different wave and wind farms inland and cover distinct marine regions. However along with other facilities, they will share the same power grid link. The schematic representation of an isolated device is shown in Figure 64.3 [10, 11].

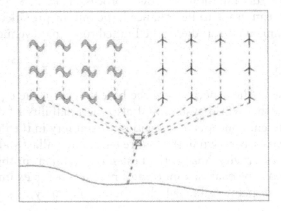

Figure 64.3 Independent array [11].

Combined Array

In this situation, the offshore wave and wind systems are built in the same marine environment, unlike separate arrays, and they share different kinds of infrastructures, in essence, making them function as a single, interconnected hybrid system. Hence it is named as single array [10, 11].

Combined array is classified into three types as uniformly, non-uniformly and peripherally depending on the distribution of wave energy converters and wind turbines as shown in Figure 64.4 [7].

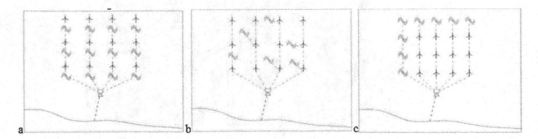

Figure 64.4 (a) Uniformly distributed array; [11] (b) Non-uniformly distributed array; [11] (c) Peripherally distributed array [7].

Hybrid Systems

This incorporates a wind turbine and a wave energy converter on the same platform. Hybrid systems possibly be further divided as bottom-fixed systems suitable for deep waters, useful for deeper or transitional waters and floating form.

Floating Hybrids

With developers developing floating platforms that compoud wave energy converters with offshore wind turbines, the floating hybrid is a relatively recent idea that has come into perspective due to the development of floating offshore wind prototypes [12]. Two EU sponsored initiatives, MARINA and TROPOS, have been applied in the real world, dealing specifically with these hybrid networks [13]. Hybrid solutions in general are based on the idea of multipurpose offshore platforms [14].

Bottom-Fixed Hybrids

Bottom-fixed hybrids, in effect, naturally follow from the latest structures used by the offshore wind plants. In order to put up a wave energy converter, these devices may be changed. Essentially, by installing a converter of wave energy onto an existing offshore wind plant, a bottom-fixed hybrid can be created.

Case Study

W2Power, floating hybrid wave and wind concept that combines wind turbines assisted with WECs (point absorber type) on a semi-submersible structure, as seen in Figure 64.5. A semi-submersible triangle support system is part of the W2Power, floating wind, wave power plant. The Pelton turbine is accompanied by two wind turbines mounted in both corners and the third corner, powered by three channels of wave-actuated pumps assembled on the sides of the structure. Two 3.6 MW conventional wind turbines, e.g. The 3.6-107 Siemens (80–85 m hub height and 107 m rotor diameter) can be used. In offshore region with a heavy wave environment, the marine platform can be valued at greater than 10 MWferrography ... [1]. The side-way powers can be efficiently minimized with the use of counter-rotating wind turbines. WT's thrust and gyro-force can help stabilize the platform.

Figure 64.5 (a) W2Power schematic configuration, a floating hybrid wave-wind concept based on an integrated semi-submersible wind turbine with WECs point absorber [15]. (b) Combined Wind Wave energy by W2Power [15].

Experimentation

An attempt was done to understand the output of the hybrid system with comparison to the individual operation of the wind and wave energy systems. A small scale prototype was made using Poly Vinyl Chloride pipes and 4V DC Generators and deployed in a water pool as shown in Figure 64.7. The prototype has two wind turbines with four floater type wave energy converters which works on the principle of electromagnetic induction as shown in Figure 64.6. The experimentation is done in an opened place where the waves are generated manually and the wind turbines are operated naturally. The generated voltage and resistance was measured through multimeter.

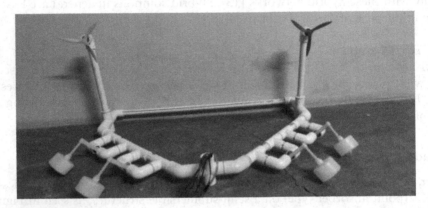

Figure 64.6 Prototype of hybrid wind and wave energy system.

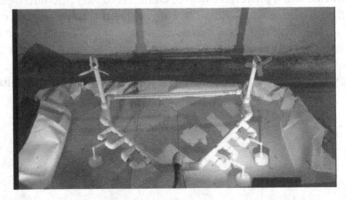

Figure 64.7 Hybrid wind and wave energy system deployed in water pool.

Results and Discussion

The wave energy converters were operated and a varying output of 2mV–3mV was observed at each converter whereas the wind turbines were operated and a varying output of 60mV–110mV was noticed. The hybrid system consists of wind turbines and wave energy converters that were connected in series. When operated simultaneously an output of 135mV–141mV is observed. The resistance offered by the hybrid system found to be

$6.5\,\Omega$ as measured using the multimeter. The trials were repeated numerous times to ensure that the systems were repeatable, and average voltage was used to calculate power(P) as shown in the equation 1

$$P = \frac{V^2}{R} \tag{1}$$

The power produced by wind system, wave energy system and hybrid system are 0.0009W, 0.0002 W and 0.00132W respectively. Results shows that the output of the hybrid system is higher than the individual wind and wave energy systems.

Conclusion

Ocean has the immense energy that can be used sustainably for the long time. The importance of this paper is the compounding of the wave energy and the wind energy to meet the growing needs. The hybrid system has different advantages and synergies. The different arrays of co-located systems and the hybrid systems explains about the space utilization. Overall power output will be increased in the lesser area compared to the plants deployed individually. The experimentation shows that hybrid system has the greater output than the individual systems which shows the feasibility of the better output when applied in real time. This ocean energy can replace the conventional energy market like natural gas, coal and other fossil fuels.

References

1. Li, L., Gao, Z., and Moan, T. (2015). Joint distribution of environmental condition at five european offshore sites for design of combined wind and wave energy devices. *Journal of Offshore Mechanics and Arctic Engineering*, 137(3), 031901.
2. Butterfield, S., Musial, W., and Jonkman, J. (2007). Engineering challenges for floating offshore wind turbines. In 2005 Copenhagen Offshore Wind Conference, Copenhagen, Denmark.
3. Roberts, A., Thomas, B., Sewell, P., Khan, Z., Balmain, S., and Gillman, J. (2016). Current tidal power technologies and their suitability for applications in coastal and marine areas. *Journal of Ocean Engineering and Marine Energy*, 2, 227–245.
4. Subekti, M., Parjiman and Hanifah, N. (2020). Design of sea wave power hybrid power generation through utilization of wave and wind energy as renewable electric energy sources for leading, outermost and disadvantaged areas. In Proceedings of the International Joint Conference on Science and Engineering (IJCSE 2020).
5. Li, H. (2020). Wave Energy Potential, Behavior and Extraction. MDPI.
6. Choi, N. J., Nam, S. H., Jeong, J. H., and Kim, K. C. (2013). Numerical study on the horizontal axis turbines arrangement in a wind farm: Effect of separation distance on the turbine aerodynamic power output. *Journal of Wind Engineering and Industrial Aerodynamics*, 117, 11–17.
7. Roy, A., Auger, F., Dupriez-Robin, F., Bourguet, S., and Tran, Q. T. (2018). Electrical power supply of remote maritime areas: a review of hybrid systems based on marine renewable energies. *Energies*, 11(7), 1904.
8. Clemente, D., Rosa-Santos, P., and Taveira-Pinto, F. (2021). On the potential synergies and applications of wave energy converters: a review. *Renewable and Sustainable Energy Reviews*, 135, 110162.
9. Tillenburg, D. (2021). Technical challenges of floating offshore wind turbines. *EGU Journal of Renewable Energy Short Reviews*, 3, 13–18.
10. Iglesias, G., and Carballo, R. (2009). Wave energy potential along the death coast (Spain). *Energy*, 34(11), 1963–1975.

11. Kesari, D. J., Gupta, A., Shukla, K., and Garg, P. (2019). A review of the combined wind and wave energy technologies. *International Journal of Engineering Trends and Technology (IJETT),* 131–136.

12. Le, C., Ren, J., Wang, K., Zhang, P., and Ding, H. (2021). Towing performance of the submerged floating offshore wind turbine under different wave conditions. *Journal of Marine Science and Engineering*, 9(6), 633.

13. Pérez-Collazo, C., Greaves, D., and Iglesias, G. (2015). A review of combined wave and offshore wind energy. *Renewable and Sustainable Energy Reviews*, 42, 141–153.

14. Crowle, A., and Thies, P. (2021). Installation innovation for floating offshore wind. In Maritime Innovation and Emerging Technologies Conference.

65 Stiffness improvement using leaf vein structure in additively manufactured components

Jyothirmay Bhattacharjya[1,a], Meenakshi Devi Parre[2,b], and Vamsi Krishna Pasam[3,c]

[1]Department of Mechanical Engineering, RGUKT, RK Valley, Idupulapaya, YSR Kadapa (Dt), Andhra Pradesh, India

[2]Department of Mechanical Engineering, National Institute of Technology, Warangal, Telangana, India

Abstract: In this present work stiffness of the additive manufactured components is improved by using bionic leaf vein structures. CREO parametric is used to model various vein structures over rectangular bars and analyzed using ANSYS Workbench to check the ability of the vein structures to improve stiffness. Finite element analysis of the parts is done for poly lactic acid. These are fabricated using fused deposition modeling technology. Three-point bending test was conducted on the fabricated parts in a universal testing machine. Simulation results are compared with experimental results to check the validity of the FE model developed. It is concluded that adoption of leaf vein structures improved the stiffness and flexural strength. Parallel leaf vein structures are having more stiffness and flexural strength compared to structure with no vein structure.

Keywords: Additive manufacturing, leaf vein structure, stiffness, three-point bending test.

Introduction

Additive manufacturing (AM) is a technology in which products are made in a layer upon layer process from the 3D CAD data. It is more popularly known as 3D printing or Rapid Prototyping (RP) in the commercial world. According to ASTM standard this technology is classified as Vat photopolymerization, powder bed fusion, material extrusion, material jetting, binder jetting, sheet lamination and direct energy deposition. In this present work, a sub classification of material extrusion process called as fused deposition modelling (FDM) is used. The main advantage of AM technology over other manufacturing technologies is, it eliminates all the constraints regarding the complexity in design of a part and reduces the wastage of material to a large extent. But like all other manufacturing technologies it also has both pros and con. The con part includes the limitation in materials that can be used by AM machines and the limited or inferior mechanical properties of the components fabricated by using AM technologies. Though an optimum value of these properties can be obtained by changing the machine parameters like layer height/thickness, scanning speed, scanning pattern, part building direction etc., still it is lower than that obtained from conventional manufacturing technologies. Antonio et al. [1] evaluated the role of different variables on properties of additive manufactured PLA components. It was observed that

[a]jyotirmoy3411@gmail.com, [b]pmeenakshidevi@gmail.com, [c]vamsikrishna@nitw.ac.in

as the infill orientation approaches the 90° the strength of the parts decreases and similarly strength decrease when layer thickness increases the value 0.18 mm. Swayam Bikash Mishra et al. [2] studied the external perimeter and its effect on the flexural strength of the components fabricated by FDM. It was identified that adding external perimeter increase the strength of the build parts as the stress concentration is shifted to the center due to the addition of contours. The raster angle of 30° reduces the anisotropy of the build parts. Ozgur Keles et al. [3] studied the impact of build orientation on the mechanical properties of 3D printed ABS parts. Tensile test was performed on ABS parts with hole and without hole at the center. The parts were built in XY, XZ and C+45 orientation. It was observed that the parts produced in XZ orientation possessed highest strength in terms of fracture and the parts with C+45 has the lowest value. Wear behaviour of powder bed fusion made Titanium alloy was studied by Hua Li et al. [4]. The results indicated that the manufacturing techniques had no influence on the wear rate. However, the deformation and fracture behaviours differed significantly.

Nature always inspires scientist and engineers to solve the real-life engineering problems. Distribution of leaf veins and distribution of tree roots under the soil are some of the examples. Leaf veins are self-growth structures in the leaf to provide the required stiffness so that they can steadily expose to the sun rays for photosynthesis. Similarly roots spread under the soil in search of water and minerals and at the same time it provides stability to the tree to withstand the self-weight and other external factors. The engineering problems can be solved using these bionic structures. Baotong Li et al. [5] carried out topological optimization using bionically inspired structures for improving the stiffness of the column. Three levels of analogical analysis should be satisfied for a bionic structure to be ulilized to solve the engineering problems namely topology or shape, loading and functional similarity. Adam Runions et al. [6] mentioned about different leave vein structures and also their distribution. The veins present on a leaf are divided into three type namely primary veins, secondary veins and tertiary veins. The distribution and differentiation of the leaf veins on the leaf blade is controlled by the Canalization hypothesis. Mahbub Ahmed et al. [7] mentioned that the relatively low stiffness of different thermoplastics can be increased by introducing carbon fibers into it. PLA and ABS were studied in this work. In the experimental works by Zhi Sun et al. [8] a sandwich structure odelledg two thin, stiff face sheet of carbon fiber reinforced polymer and a thick low density porous structure was used to improve the stiffness. A noticeable improvement was observed in stiffness with this biologically inspired model. The chances of local failure of the sandwich structure were avoided in the model by incorporating structures like leaf veins. An approach for improving the mechanical properties of FDM components was given by Jianlei Wang et al. [9]. In this work, microspheres that are thermally expandable are used into the matrix during manufacturing and exposed to temperature raise later. The microsphere expanded due to raise in temperature and occupy the voids in the parts. Positive results in terms of improved tensile strength were reported. The wear resistance of different bionic structures was investigated by Mengyao Wu et al. [10]. The study involved coating of Titanium alloy with bionic structures with powder reinforced composite. The surface damage behaviour and microstructure were studied. The bionic structure proved to have better wear resistance.

In this work, various vein structures are odelled using CREO parametric on a rectangular bar of dimensions according to ASTM D790 standard [11] and analyzed using ANSYS in order to identify the ability of the vein structures to increase stiffness. Parts are fabricated with PLA material using a Flash Forge machine which works on Fused Deposition

Modelling (FDM) technology. Three-point bending test is carried out in a UTM to study the load vs. deformation behaviour. The results from simulation are compared with experimental results in order to validate the FE model.

Design and Simulation for 3 Point Bending Test

Leaf vein structures (LVS) studied in this work were Cross-venulate, Parallel, and Rotate [12] and are shown in the Table 65.1. These leaf vein structures were analysed to improve the stiffness of parts fabricated by AM. These are modeled in CREO Parametric. The veins were taken as circular cross-section with a diameter of 1.5 mm, which converges to 1.2 mm along the length for primary veins and with a diameter of 0.8 mm, which converges to 0.6 mm along the length for secondary veins. Similarly, in case of tertiary veins, the cross section has a diameter of 0.6 mm which converges to 0.4 mm along the length of the veins. These vein designs were made on a rectangular bar of 127 × 12.7 × 3.2 mm according to ASTM D790 standard [11]. The diameter of the veins considered are directly proportional to the thickness of the rectangular bar.

The governing equation for deflection of curve is [13],

$$EI\frac{d^2y}{dx^2} - M(x) = 0 \tag{1}$$

Force equilibrium equation used is [13],

$$[F] = [k] \times [X] \tag{2}$$

Where k is the stiffness matrix for the beam element [13].

To get an overhanging length according to ASTM D790 standard, deflection of the beam in perpendicular direction is considered as zero between support point of 12.5 mm and 114.5 mm from one side. Load was applied at the middle of the beam and deflection was allowed only in y direction at the centre of the beam. The material chosen for simulation was PLA. The simulation was done for a series of load conditions starting from 0 to 30 N for all the 3 leaf vein structures. The span length used in the Three-point bending simulation according to ASTM D790 standard was 102 mm. These are compared with the standard specimen i.e. the rectangular bar without any leaf vein structure and the most suitable vein structures to improve the stiffness were chosen. The properties of PLA used in the study was tabulated in Table 65.2.

Table 65.1: Three Leaf Vain Structures.

S. No	Name of the structure	Structure
1.	Cross-Venulate	
2.	Parallel	
3.	Rotate	

Table 65.2: Properties of PLA.

S.No	Properties	PLA
1	Youngs modulus	3500 MPa
2	Ultimate tensile strength	50 MPa
3	Yield strength	30 MPa
4	Density	1.3 g/cm^3
5	Poissons ratio	0.36

Mesh Independency Checking

The meshed model is shown in Figure 65.1. Mesh independency checking was done to ensure that the results or output obtained is independent of the element size of the mesh. Mesh independency checking was done and the result obtained is tabulated in Table 65.3.

Figure 65.1 Meshed model of the sample.

Table 65.3: Mesh independency checking for 3-point blending simulation.

S. No	Element size	Nodes	Number of elements (Tetragonal)	Deformations in 'mm'
1	Default	5926	966	8.9605
2	1 mm	33599	6656	8.9746
3	0.5 mm	212008	46410	8.9829
4	0.25 mm	1450732	337467	8.9902
5	0.22 mm	2141935	502860	8.9915

Fabrication and Testing

After simulation, parts were fabricated in a Flash Forge printer. Parts with leaf vein structure having more stiffness based on simulation results were fabricated along with standard specimen without any vein structure as shown in Figure 65.2 and tested in 3 point bending test. Three samples of each specimen were fabricated according to ASTM D790 standard i.e. 127 × 12.7× 3.2 mm. The material used during fabrication was PLA, which is used in modelling and simulation of the components. The specifications of the 3D Printer were given in Table 65.4.

Table 65.4: 3D Printer and its specifications.

S.No	Properties	Magnitude
1	Printing speed	40–60 mm/s
2	Layer resolution	100–300 µm
3	Temperature of nozzle	220 °C
4	Build volume	140 × 140 × 140 mm³
5	Diameter of nozzle	0.4 mm
6	Extruder	Single
7	Diameter od filament	1.75 mm
8	Supported filament	PLA

The flexural strength of the components was calculated from the data obtained from the experiment. The sample was placed horizontally on an adjustable support and loaded with the help of a cross head. The cross head speed was 5.4 mm/min, which was obtained using the equation (1) and standard support span length used was 102 mm. The deflection of the sample was allowed till the rapture takes place. The average depth and width of the samples were 3.2 mm and 12.7 mm respectively. For all the three types of samples, experimentation was carried out in triplicate. The average value of deformation from the three values for each sample was taken and load vs. deformation graph was drawn.

The relations used to evaluate the flexural strength of the components from the experimental data of 3-point bending test are mentioned in equation (4) to (7).

Loading rate of the machine or the cross-head speed is calculated using the equation,

$$R = zL^2/6d \tag{4}$$

In this experiment the rate of straining of outer fiber is taken as 0.01 mm/mm/min and the samples are deflected till the rapture.

Flexural stress is calculated using the equation,

$$\sigma_f = 3PL/2bd^2 \tag{5}$$

Figure 65.2 Fabricated samples with and without leaf vein structures.

Flexural strain is calculated from the equation,

$$\varepsilon_f = 6Dd/L^2 \tag{6}$$

Flexural modulus is obtained using the equation,

$$E_f = L^3m/4bd^3 \tag{7}$$

Figure 65.3 Three-point bending test in UTM.

Results and Discussion

Simulation Results for 3 Point Bending Test

All are according to Figure 65.3. Simulation results for total deformation due to the applied external load on PLA specimens with and without leaf vein structures are shown in Figures 65.4–65.7.

Figure 65.4 Cross venulate vein structure.

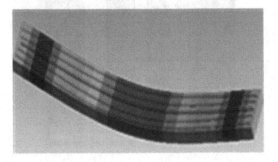

Figure 65.5 Parallel vein structure.

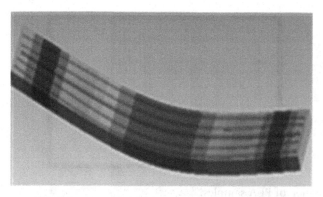

Figure 65.6 Rotate vein structure.

Figure 65.7 Sample without vein structure.

Table 65.5: Summarized Deformation results for eight vein structures.

S.No	Samples with leaf vein structure	Deformation of 'mm'
1	Cross–Venulate	7.0294
2	Parallel	5.5382
3	Rotate	6.5732
4	Samples without vein structure	8.9915

The standard specimen has shown a deflection of 8.9915 mm whereas with parallel leaf vein structures the deflection was 5.5382 mm This indicates stiffness improvement in design with LVS. The LVS act as an aid to endure the externally applied load and hence the resistance to bending increases for the component. The deformations of the samples were plotted in Figure 65.8 and Load vs. deflection curve is shown in the Figure 65.9. Values are varying according to Table 65.5.

It can be seen that slope of load-deflection curve for parallel leaf vein structures surpasses the standard specimen. Higher the slope, more is the stiffness. Hence, it is observed that use of LVS on the rectangular components improved the stiffness. parallel structures can improve the stiffness more effectively among all the LVS. The applied LVS acted as the support for the samples when external load is applied and hence the improvement in stiffness is achieved. Moreover, the difference in the deformation of the samples with

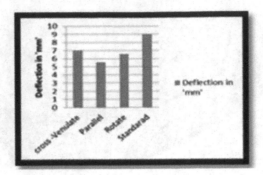

Figure 65.8 Deformation of PLA samples.

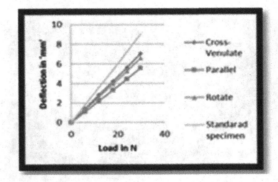

Figure 65.9 Load vs deflection responses for different leaf vein structures.

different LVS under the application of similar amount of load is due to the distribution pattern of the leaf veins over the surface of the sample. In parallel vein structure only one primary vein is present. The secondary veins are in parallel and symmetrically distributed with the primary veins. This may be attributed to provide more support to the applied load and stiffness improvement is more. In rotate vein structure, more than one primary vein is present but they are not distributed uniformly over sample's surface. Hence, parallel vein structure also helped to improve stiffness.

Results of 3-Point Bending Test

Load vs. displacement graph for sample with parallel leaf vein structures and samples without any vein structure is plotted in Figure 65.10. It can be observed that the average value of deflection for the sample having leaf vein structure is less than the one which is having no leaf vein structure. Moreover, the maximum load that can be withstood by the samples is also more in case of the samples having leaf vein structure on it. The average value of deflection for a particular value of load is less in case of sample having parallel vein structure compared to the standard sample i.e., the one having no vein structure. similarly, the maximum value of load is more in case of sample with parallel vein structure than sample without any vein structure. The average values for mechanical properties of the samples were given in Table 65.6. Value of flexural stress, flexural strain and flexural modulus are evaluated using the equations 4 to 7. It is observed that flexural modulus of the samples is improved from 0.764 to 1.38 because of the implementation of leaf vein structures.

Average flexural modulus value for standard sample is 0.764 and it is 1.38 for sample with parallel leaf vein structures. The average flexural strength is about 23.57 MPa for standard specimen. An increase of about 22.61 MPa or 57.52% is observed with parallel vein structure. The simulation and experimental results were compared at a load of 30 N and error is calculated as given in Table 65.7. The error obtained in case of parts with parallel leaf vein structure is 3.23% and in the case of parts without any vein structures with a value of 2.17%. The difference is attributed to the value of Young's modulus which is taken as 2.4 GPa, whereas in actual conditions, it ranges between 2 to 3 GPa. Furthermore, the element size for the fine node was taken as 0.22 mm during the analysis. This is obtained from the mesh independency of the components without any LVS. This element size was adopted for the other two samples also. Comparison between experimental and simulation results of deformation bending test is shown in Figure 65.11.

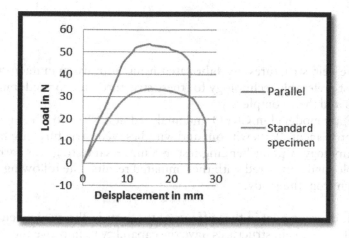

Figure 65.10 Load–Displacement responses of samples.

Table 65.6: Average mechanical properties of all samples.

S.No	Specimen type	Avg. max. load (N)	Avg. Deflection at the center of the sample (mm)	Avg. flexural stress	Avg. flexural strain (mm/mm)	Avg. flexural modulus (GPa)
1	Sample without vein structure	33.415	27.85	39.31	0.0514	0.764
2	Sample with parallel vein structure	52.630	24.36	61.92	0.0449	1.38

Table 65.7: Comparison of simulation and experimental values.

S.No	Sample	Deformation in structure with parallel veins	Deformation in structure without veins
1	Simulation	5.5382	8.9915
2	Experimental	5.7233	9.191
3	% Error	3.23	2.17

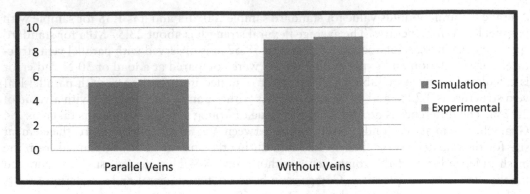

Figure 65.11 Comparison of deformation between experimental and simulation results for Three-point bending test.

Conclusions

In this work, the vein structures are fabricated using additive manufacturing technology which is the most preferable technology for these structures due to randomness in distribution of the veins and their complexity.

Different LVS are modeled in CREO and analyzed in ANSYS for PLA material to study the stiffness improvement. Parts without and with best vein structures are fabricated using 3D printing technology. 3 point bending test in a universal testing was carried out for the fabricated samples and compared with the simulated results. The following are the conclusions were drawn from the study:

1. A significant improvement in the stiffness of the parts is observed when vein structures are used. All type of vein structures have the capability to improve the stiffness. Among all the vein structures, most preferable vein structures are parallel veins.
2. The experimental and simulation outcomes are in good agreement with each other proving that the developed model is reliable.
3. The Flexural modulus value of sample with parallel leaf vein structure is higher than the sample without any vein structure. The increase of about 22.61 MPa or 57.52% is observed with parallel vein structure.

References

1. Lanzotti, A., Grasso, M., Staiano, G., and Martorelli, M. (2015). The impact of process parameters on mechanical properties of parts fabricated in PLA with an open-source 3-D printer. *Rapid Prototyping Journal*, 21(5), 604–617.
2. Mishra, S. B., Malik, R., and Mahapatra, S. S. (2017). Effect of external perimeter on flexural strength of FDM build parts. *Arabian Journal for Science and Engineering* 42(11), 4587–4595.
3. Keleş, Ö., Blevins, C. W., and Bowman, K. J. (2017). Effect of build orientation on the mechanical reliability of 3D printed ABS. *Rapid Prototyping Journal*, 23(2), 320–328.
4. Li, H., Ramezani, M., and Chen, Z. W. (2019). Dry sliding wear performance and behaviour of powder bed fusion processed Ti–6Al–4V alloy. *Wear*, 440, 203103.
5. Li, B., Hong, J., and Liu, Z. (2014). Stiffness design of machine tool structures by a biologically inspired topology optimization method. *International Journal of Machine Tools and Manufacture*, 84, 33–44. Retrieved from http://dx.doi.org/10.1016/j.ijmachtools.2014.03.005.

6. Runions, A., Fuhrer, M., Lane, B., Federl, P., Rolland-Lagan, A.G., and Prusinkiewicz, P. (2005). Modeling and visualization of leaf venation patterns. ACM SIGGRAPH 2005 Papers, pp. 702–711.

7. Ahmed, M., Islam, M. R., Vanhoose, J., Hewavitharana, L., Stanich, A., and Hossain, M. (2016). Comparisons of bending stiffness of 3D printed samples of different materials. In ASME 2016 International Mechanical Engineering Congress and Exposition (Vol. 50633, p. V009T12A023). American Society of Mechanical Engineers Digital Collection.

8. Sun, Z., Li, D., Zhang, W., Shi, S. and Guo, X. (2017). Topological optimization of biomimetic sandwich structures with hybrid core and CFRP face sheets. *Composites Science and Technology*, 142, 79–90.

9. Wang, J., Xie, H., Weng, Z., Senthil, T., and Wu, L. (2016). A novel approach to improve mechanical properties of parts fabricated by fused deposition modeling. *Materials and Design*, 105, 152–159.

10. Wu, M., Zhan, X., Bu, H., Liu, L., Song, Y., and Li, Y. (2020). Wear resistance of different bionic structure manufactured by laser cladding on Ti6Al4V. *Metals and Materials International*, 27, 2319–2327.

11. ASTM Subcommittee D20. 10 on Mechanical Properties, 1997. Standard test methods for flexural properties of unreinforced and reinforced plastics and electrical insulating materials. American Society for Testing Materials. Retrieved from ASTM International. https://www. astm.org/Standards/D790.htm.

12. Rebello, A. S., Moore, B. W., Sochacki, D., and Alkhaledi, F. (2019). Leef Jerky™: A Sustainable Meat Product for a Better Future. Retrieved from https://www.thoughtco.com/ id-trees-using-leaf-shape-venation-1343511.

13. Perry Jr, W. H. (1985). Finite element analysis of polymer flows (Doctoral dissertation, Ohio University).

66 Sustainable design innovation & analysis of customized solar frame bicycle

Suraj[1,a], Arun Kumar[1,b], Tameshwer Nath[2,c], and Bikas Prasad[3,d]

[1]National Institute of Technology, Patna, Bihar, India
[2]Department of Mechatronics, IIIT Bhagalpur Bihar, India
[3]Govt. Polytechnic, Dheri-On-Sone, Rohtas Bihar, India

Abstract: Though we are improving our living standards with technology, we are neither bothered about our fitness nor our environment. The article focuses on both, this research will not only help to keep an individual fit but also help in improving air pollution. In this article using the concept of seven concern of innovation process, a customized Bicycle frame (Photo Voltaic (PV) frame) is selected after overcoming the flaws in many designs and final conception is analyzed for maximum deformation at different loads. In this article five concerns out of seven is considered to achieve final Design. Flexible Solar Cell (mentioned in a magazine created by Massachusetts Institute of Technology, Energy Initiative) on frame will help to charge the battery and that battery will operate other devices like Air Purifier, Mobile Charger etc. This article only focuses on innovative design analysis of customized solar frame but not the working of solar frame or air purifier. This innovative design is going to change the concept of cycling in future.

Keywords: 7C's of innovation, customized bicycle frame, FEM, deformation, sustainable design, sustainable transportation.

Introduction

Innovation should not compromise future generation and should include development of environment indeed. In this article five concerns out of seven concerns of innovation is used keeping sustainable design in mind. Seven concerns of in-novation are Cause, Context, Comprehension, Check, Conception, Crafting, and Connection. Five concerns of innovation used for this design are Cause considered is that humans nowadays are being affected with fatal disorders like heart syndromes and cancer. The main cause of this is just breathing polluted air of ur-ban cities and not keeping them fit. Context considered is that cycling is one of the most effective ways by which the problem of increasing pollution as well as obesity of an individual can be controlled to some extent. Comprehension & Check concluded that in today's market there is no such product targeted for sports enthusiast who live in polluted cities and have potential for a very bright future but are unable to achieve. Therefore, not doing the sports activity like cy-cling due to the concern of air pollution that may cause irreversible damage to their body, this design will solve the problem for those. Conception gave multi-ple ideas to address the problems listed in check. As clear

[a]suraj.ph21.me@nitp.ac.in, [b]arun@nitp.ac.in, [c]tnath.mea@iiitbh.ac.in, [d]bikasprasad@yahoo.com

from the problem as-sociated with air quality has increased with the increase in vehicles running on fossil fuels. As an alternative way of transport, Ministry of Housing and Urban Affairs, Government of India promotes bicycle. It will help in reduction in air pollution, energy use, high traffic, noise pollution and allowing at same time a healthier lifestyle for riders. In addition to this, use of bicycle is one of the inex-pensive and adequate modes of transportation. Using the design process shown in Figure 66.1 final design was customized.

Figure 66.1 Design process.

In designing a bicycle, a designer has to work on structure of frame and as from previous work on frame design and analysis, diamond type frame has been found highest rigidity [1]. To manufacture the frame, materials is selected in such a way that it should be light weight, good strength and mostly safe [2]. Aesthetics is also considered to improve the quality of riding and increase the acceptance by the individuals as customer into design process [3, 4]. Just the Design, material and strength of the frame is not enough and so design of fitting system carried by some laboratory test and experiments to investigate the aspects of cycling comforts for pilot were also considered [5]. Innovative and integrated procedure for optimizing the road bicycle frame, we also have to give effort to reduce the deformation of bicycle frame permanently and lessening the mass from the bicycle frame [6]. With the growth of renewable energy and sustainability research on power as-sisted solar bicycle (battery get charged by using of solar energy were solar panel is placed on carriage, solar plate externally placed) came into action [7–9]. In this article, a frame is designed in such a way that the Flexible Solar Cell is on the frame of bicycle. Customized Frame design is finalized after overcoming the flaws in many designs using the concept of 7 concerns of innovation. In final customized frame design, ANSYS software is used for the structural analysis of the bi-cycle frame and ANSYS software analyze the problems based on the principle of "Finite Element Method" (FEM).

Design Concepts

The main aim of proposed design is to integrate solar panel, battery, and air purifier on to the bicycle frame rather than adding them separately. Adding a solar plate separately increases weight due to addition of standoffs and mounting brackets that are needed to hold it in place. This reduces the ergonomics of cycle due to all the wire management issues, reduces ease of maintainability due to all the wires tangling on the frame. It also reduces effectiveness of air purifier if it is placed at the back of the bi-cycle frame. Based on their modification, 4 designs are given:

Construction and Drawbacks of Design Concept 1

Main construction of the Bicycle frame, as shown in Figure 66.2a, incorporates housing for battery and an inverter to charge electronics, give power supply to the air purifier and get charged from a solar panel. The solar panel acts as an en-closure to the housing on the bicycle. This enclosure does not need any wires to be separately connected since there are contact pads below the panels that directly connect to the inverter. The solar panel can be easily removed to service or replace the batteries and inverter in case of any problem.

Drawbacks in design 1 was found as it was not perfect and had some flaws like complicated mechanism to clamp a carrier base that would be able to mount the air purifier and the analysis of the cycle showed a heavy deformation of bicycle on the front end of the frame. To overcome flaws in design-1, Design-2 was conceived.

Construction and Drawbacks of Design Concept 2

It overcame some problems faced in design-1 like the front-end frame deformation was reduced and focus was on ergonomics and ease of accessibility and maintainability as shown in Figure 66.2b. Same solar panel from design-1 was used to enclose the battery and inverter. Other replacement in this design was the clamp mechanism for the carrier was removed and the frame was extended backwards to act as a carrier. This extended carrier would be used to attach all the standoffs and mounting brackets to mount the Air purifier of 220×600 mm^2. The extended backside frame has an opening that can be used as a storage and to upgrade battery storage if required.

The biggest drawback of this design was that it increased the weight of the bicycle and thus making it hard for the cyclist. The other one was the space that could be utilized for storage was extremely limited. To overcome flaws in design-2, Design-3 was conceived.

Construction and Drawbacks of Design Concept 3

Construction of design concept 3 was meant to be a 2-sided exoskeleton de-sign that would require forging of a thick sheet for both the sides and then sandwiched with cycle parts like seat, steering handle etc. This design as shown in Figure 66.2c solved many previous problems and included some extra features like a USB outlet to charge your device.

However, the main drawback of this design was that it was too bulky and boxy. In addition, in deformation analysis there was a sulk deformation observed where the cyclist would sit. The design was rejected due to high deformation from the middle of the frame. To overcome flaws in design-3, Design-4 was conceived.

Construction of Final Design, Design Concept 4

It is a complete redesign of the other bicycle design variants which are dis-cussed above. This design as shown in Figure 66.2d focuses on sleekness, lightness, easy to manufacture and handle. The frame is a bare bone double triangle frame with a sheet metal frame enclosure that encloses an air purifier, battery system and an inverter to charge your electronics for some time. The frame enclosure and front mud guard has integrated flexible solar cells with solar frame exposer area of 240499.52 millimeter2 that can charge the batteries and power the air purifier. The air purifier has 3 stages of air filters. The filters can be easily changed by unscrewing the front frame enclosure and the air purifier can be easily accessed. The lower frame enclosure houses battery and inverter. The air purifier is mounted on the front head of the cycle frame.

Figure 66.2 Design concept phases a) Design concept 1, b) Design concept 2, c) Design concept 3, d) Final design.

One end of hose pipe is connected to purifier and another end is connected to a custom-ized mask, self-designed as shown in Figure 66.3, which will help cyclist to breathe fresh air while cycling. It is designed to be effective in improving health of the cyclist keeping their lungs healthy in polluted cities, efficient in purifying air and elegant to catch the eyes of the buyers.

Analysis of Final Designed Frame

Design made on CAD software as shown in Figure 66.4a is transferred to ANSYS, material (Stainless Steel grade 304(SS 304)) is selected and meshing in static structural system is done as shown in Figure 66.4 b.

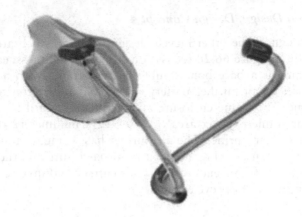

Figure 66.3 Customized mask.

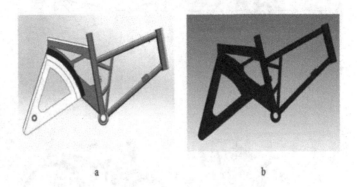

a   b

Figure 66.4 Frame design & meshing a) Design on CAD software b) Design meshing in static structure.

Stainless Steel grade 304(SS 304) was the material selected in-stead of other raw materials namely titanium (Ti-64), Aluminum (Al6061-T6) for the analysis of the bi-cycle frame since:

- It is readily available from the market unlike Aluminum and titanium.
- SS 304 has ultimate tensile stress of 505MPa.
- Yield Tensile Strength is around 215Mpa.
- Its elongation at break is 70 %
- Modulus of elasticity is 193Gpa to 200Gpa
- Has good weldability and machinability characteristics
- SS 304 costs around Rs.170/kg, whereas Aluminum and titanium costs Rs. 450/kg and Rs. 4000/kg respectively.

In view of factors above, SS 304 was selected for being the material of choice for the frame despite its disadvantage of being heavy. Analysis of deformation is done by giving the boundary conditions and applying different load where the cyclist will be seated. The weights applied on the frame are 800N, 1000N, 1500N and 2000N.

Results and Discussion

The results obtained for the bi-cycle frame for the given boundary conditions are as follows, when force of 800N or 80kg was applied on the frame, the total deformation (max) is 0.024 as can be observed in the Figure 66.5a, Similarly appli-cation of force 1250 N or 125kg was applied on the frame, which gave deformation (Max) on the frame was around 0.0386mm, as can be observed from the Figure 66.5b, When the force of 1500 N or 150 kg was applied on the frame, the total deformation (Max) on the frame was around 0.0463mm, as can be observed from the Figure 66.5c, When the force of 2000 N or 200 kg was applied on the frame, the total deformation (Max) on the frame was around 0.0617mm, as can be observed from the Figure 66.5d. It is analyzed that maximum Deformation when maximum load of 2000N is applied is 0.0617mm.

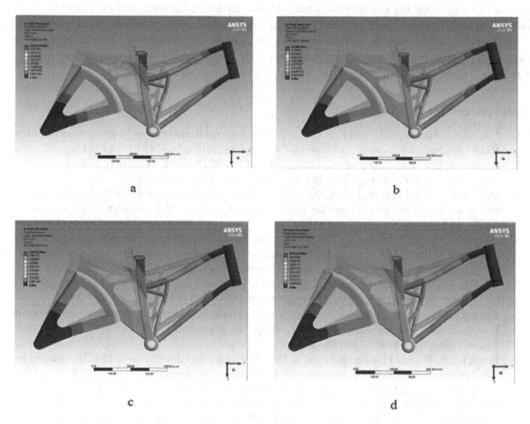

Figure 66.5 Deformation analysis at different loads.

Table 66.1: Deformation in mm at different weight in kg and deformation difference.

Body Weight	Deformation of frame	Deformation Difference
80kg	0.024	-
125kg	0.0386	0.014
150kg	0.0463	0.0077
200kg	0.0617	0.0154

By the report of National Institute of Nutrition in year 2020 the average weight of man is 65kg and average weight of women is 55kg in India. In this article weight of the cyclist is considered much more than the average weight of the man and women across the world. As it can be observed from Table 66.1 that when weight of the cyclist is increased from 80kg to 125kg, 125kg to 150kg and 150kg to 200kg deformation changes by 0.014mm, 0.0077 and 0.0154kg respectively which can be acceptable.

Conclusion

Material used for the frame of solar bicycle is Stainless Steel grade 304(SS 304), the frame of the bicycle is rugged. Analysis of the customized frame concludes that an economical weight of the cyclist is 125kg. This design is made only for a single person cycling. The final design (design-4) of the solar framed air purifying cycle is finalized because in this design we have integrated the components namely solar panels, battery, and air purifier on to a bicycle frame, it is less bulky than the initial designs. This design will be less cumbersome. The de-sign will be easy to handle by the end user. It will be easy to manufacture, maintain and replace faulty parts. It will also improve the health of the end user by letting him breathe. This Design will change the concept of cycling in future.

References

1. Lin, C. C., Huang, S. J., and Liu, C. C. (2017). Structural analysis and optimization of bicycle frame designs. *Advances in Mechanical Engineering,* 9(12), 1687814017739513.
2. Rontescu, C., Cicic, T. D., Amza, C. G., Chivu, O., and Dobrota, D. (2015). Choosing the optimum material for making a bicycle frame. *Metalurgija,* 54(4), 679–682.
3. Siddique, Z., and Ninan, J. A. (2006). Modeling of modularity and scaling for integration of customer in design of engineer-to-order products. *Integrated Computer-Aided Engineering,* 13(2), 133–148.
4. Ninan, J. A., and Siddique, Z. (2007). Internet-based framework to support integration of the customer in the design of customizable products. *Proceedings of the Institution of Mechanical Engineers, Part B: Journal of Engineering Manufacture,* 221(3), 529–538.
5. Christiaans, H. H., and Bremner, A. (1998). Comfort on bicycles and the validity of a commercial bicycle fitting system. *Applied Ergonomics,* 29(3), 201–211.
6. Cheng, Y. C., Lee, C. K., and Tsai, M. T. (2016). Multi-objective optimization of an on-road bicycle frame by uniform design and compromise programming. *Advances in Mechanical Engineering,* 8(2), 1687814016632985.
7. Bachche, A. B., and Hanamapure, N. S. (2012). Design and development of solar assisted bicycle. *International Journal of Engineering and Innovative Technology (IJEIT),* 2(6).
8. Abagnale, C., Cardone, M., Iodice, P., Marialto, R., Strano, S., Ter-Zo, M., and Vorraro, G. (2016). Design and development of an innovative e-bike. *Energy Procedia,* 101, 774–781.
9. Apostolou, G., Reinders, A., and Geurs, K. (2018). An overview of existing experiences with solar-powered E-Bikes. *Energies,* 11(8), 2129.

67 Sliding mode control design for MR damper in vehicular applications

Pinjala Devikiran[a], Puneet NP[b], and Hemantha Kumar[c]

Department of Mechanical Engineering, National Institute of Technology Karnataka Karnataka, India

Abstract: In present days MR dampers are widely used for vibration isolation applications. These semi active systems are advantageous when compared to passive dampers in controllability and adaption to vibrations at variable frequencies. One major issue with MR dampers which limits its performance is the need of proper control logic with instantaneous control signal from the controller to enhance the ride comfort of vehicle/system. In this paper, simulation of sliding mode control logic has been done to improve the ride comfort of the vehicle using MR damper(s). The simulation results obtained show considerable reduction in sprung mass acceleration. The simulation is performed in MATLAB Simulink.

Keywords: Magneto-rheological damper, modified bouc-wen model, sliding model control.

Introduction

As technology advanced, suspension with the use of smart fluids providing variable viscosity to effectively control the damping force, have taken the spotlight replacing the traditional hydraulic valves. Smart fluid dampers which are feasible to design practically are Electro-rheological fluid (ER) and also the Magneto-rheological fluid (MR) dampers. These kinds of fluids are a compound mixture of elements like silicon oil, iron particles, grease etc., and so can change its state from a semi-solid state to a free-flowing state and vice-versa. Current being varied in MR Damper using a current controller passing through the coil wound to the piston of the damper and the maximum current handled by the coil depend on the number of turns wound to it.

Magnetorheological dampers (MRD) are quite effective in controlling the vehicular vibration although it could not completely isolate it. The use of MR fluids produces controllable damping force providing both stability and comfort thus having both the facility of active control systems and reliability of passive systems with few watts of electrical power supply. Due to their reduced power, simplicity, higher dynamic range robustness and larger force capacity, offer an attractive means of vibration protection.

[a]devikiranpro@gmail.com, [b]puneetnp12@gmail.com, [c]hemantha@nitk.edu.in

Literature Review

Many researchers have contributed their time an effort in the improvement of MR technology. Some of them were referred in bringing this study so far. Choi et al. have provided a brief explanation on different control strategies which could be applied on MRD in different applications [1]. Chen proposed and implemented a skyhook surface sliding mode control for a quarter car model to improve ride comfort [2]. Goldasz et al. described about the modelling of different types in modelling of MRD [3]. Yao et al. presented a sliding mode control considering an appropriate ideal skyhook control as a reference model [4]. Choi et al. developed a hysteresis model for a magneto rheological damper [5]. Spencer et al. provided with a review of several nonlinear mechanical models for MR dampers and implemented them for the prototype damper for better accuracy [6]. Kafafy et al. implemented a basic sliding mode control for a particular modified Bouc-Wen model for MRD in a two degree of freedom quarter car model [8]. Rajendran et al. implemented a fractional order fuzzy sliding mode control for considering a quarter car model with dual actuators. The model also considers the sprung mass variation [9]. Jhin et al. provided a comparative study of two different MRD for a better ride comfort. The study included a skyhook control strategy under bump and random road profiles [10].

MR damper Suspension-Based Quarter Car Model (QCM)

The Figure 67.1 describes a two degree of freedom mechanical suspension system which represents the quarter car model with both passive and MR damper. The vehicular body mass is represented by M_s, typically the sprung mass and the wheel mass including all the fixtures and the other components associated with it is considered as the unsprung mass and is represented by M_{us}. The vertical displacements are represented with Z_s and Z_{us} respectively for both the sprung and unsprung masses. The road undulation or profile or simply the vertical disturbance is indicated with Z_r, the spring stiffness and tire stiffness are labelled as K_s and K_{us} Equations of motion for sprung and unsprung masses are as follows:

$$M_s \ddot{Z}_s + k_s(Z_s - Z_{us}) + F_d = 0 \tag{1}$$

$$M_{us} \ddot{Z}_{us} + k_s(Z_s - Z_{us}) - k_{us}(Z_r - Z_{us}) - F_d = 0 \tag{2}$$

The force produced by the MR Damper is indicated as F_d. In the passive suspension system quarter car model F_d is replaced by $C_s(\dot{z}_s - \dot{z}_{us})$.

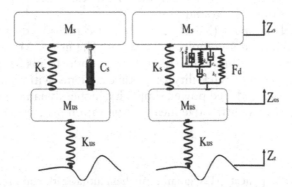

Figure 67.1 Quarter car model representing both passive and semi active suspensions.

The mathematical model for MR damper using the modified Bouc-Wen model in presence of the hysteresis behaviour is considered from Spencer et al [6] for the simulation. The modified Bouc-Wen parameters as specified in the Table 67.1.

Table 67.1: The modified Bouc-Wen model parameters of considered MRD.

Parameter	Value of the parameter	Parameter	Value of the parameter
C_{ia}	14649 Ns/m	K_o	3610 N/m
C_{ib}	34622 Ns/Vm	α_a	12441 N/m
C_{oa}	784 Ns/m	α_b	38430 N/Vm
C_{ob}	1803 Ns/Vm	n	2
β	2059020 1/m²	σ	190 1/s
δ	58	γ	136320 1/m²
K_i	840 N/m		

$$F_{mr} = C_i(\dot{y} - \dot{Z}_{us}) + k_i\{(Z_s - Z_{us}) - Z_0\} \tag{3}$$

$$\dot{y} = \frac{1}{C_o + C_i}\{\alpha Z + C_o \dot{Z}_s + C_i \dot{Z}_{us} + K_o(Z_s - y)\} \tag{4}$$

$$\dot{Z} = -\gamma|\dot{Z}_s - \dot{y}|Z|Z|^{n-1} - \beta(\dot{Z}_s - \dot{y})|Z|^n + \delta(\dot{Z}_s - \dot{y}) \tag{5}$$

$$\alpha = \alpha_a + \alpha_b V \tag{6}$$

$$C_i = C_{ia} + C_{ib}V \tag{7}$$

$$C_o = C_{oa} + C_{ob}V \tag{8}$$

$$\dot{V} = -\sigma(V - v) \tag{9}$$

Here the parameters, α and δ are the functions related to the height width and slope of the applied magnetic field hysteresis loop, the viscous damping coefficient is represented by C_i. This component at lower velocities produces rollouts. x_o is considered to be the

Table 67.2: Parameters of the quarter car model.

Quarter car Parameters	Value of the parameter
M_s	240 Kg
M_{us}	36 Kg
C_s	1500 Ns/m
K_s	16000 N/m
K_{us}	160000 N/m

deflection of the damper in prior (initial deflection) for the damper accumulator which is represented by the stiffness k_i. Z is the Bouc-Wen variable governed by equation (5). As the damper force F_{mr} is dependent on the input current which is internally dependent on the voltage signal input from the control logic, linear relations are used as specified in equations (6) to (9). All value of the parameters are taken from Table 67.2.

Sliding Mode Control Strategy

Sliding mode control (SMC) is a nonlinear control methodology which alters the system dynamics of the specified system by applying a discontinuous system that makes the system to slide and to be retained on a specified surface called the sliding surface within the state space. Advantage with the sliding mode control is its immunity to the parametric variations and disturbances once the system is in the sliding mode, thus clearly eliminates the requirement of exact modelling [9]. The error equation for the sliding surface (S) can be in different ways provided it follows the condition:

$$S\dot{S} < -\eta|S| \tag{10}$$

Where η is a positive real number. As the quarter car suspension is of second order, the sliding surface can be designed by obtaining either suspension deflection or sprung mass velocity or both as error.

$$S = \dot{x}_2 + \lambda x_1 \tag{11}$$

$$\dot{S} = \dot{x}_2 + \lambda \dot{x}_1 \tag{12}$$

Here in equations (11) and (12), x_1 is the relative suspension deflection $Z_s - Z_{us}$ and x_2 is the vertical velocity of the vehicular body (sprung mass velocity) and λ is the sliding surface gain

The condition to drive trajectory towards sliding surface is

$$\dot{S} = 0 \tag{13}$$

Which on substitution from equation 12

$$0 = \ddot{x}_2 + \lambda \dot{x}_1 \tag{14}$$

$$0 = \ddot{Z}_s + \lambda(\dot{Z}_s - \dot{Z}_{us}) \tag{15}$$

Substituting the sprung mass acceleration from the equation (1) then the equation turns out to be

$$0 = \frac{1}{M_s}\left\{-K_s(Z_s - Z_{us}) - C_s(\dot{Z}_s - \dot{Z}_{us}) - F_d'\right\} + \lambda(\dot{Z}_s - \dot{Z}_{us}) \tag{16}$$

Solving the above equation, we have the equivalent control force F_d' as

$$F_d' = (\dot{Z}_s - \dot{Z}_{us})(\lambda M_s' - C_s) - K_s(Z_s - Z_{us}) \tag{17}$$

The control force is thus formulated as

$$F_{cs} = F_d' - K_o M_s' sign(S) \tag{18}$$

The damping force from the MR damper is taken as a feedback and the error signal is produced by comparing with the desired force generated by the SMC control [7]. As we know that the MR dampers do not generate energy the control signal is to be applied only when the desired damping force produced by SMC and the direction of the error are in the same direction.

From Figure 67.3 the control loop can thus be simulated as an if else control as shown below

$$\begin{aligned}
&if : G\left(F_{cs} - F_{mr}\right)sign\left(F_{mr}\right) > V_{max} \Rightarrow v = V_{max}\\
&elseif : G\left(F_{cs} - F_{mr}\right)sign\left(F_{mr}\right) < V_{min} \Rightarrow v = V_{min}\\
&else : v = G\left(F_{cs} - F_{mr}\right)sign\left(F_{mr}\right)
\end{aligned} \tag{19}$$

Where v is the input signal to the MR Damper with V_{max} and V_{min} are the maximum and minimum values of the signal (here considered as 2 and 0 respectively) and G is the gain of the drive circuit.

Simulation Results

The Figure 67.2 is the Simulink block diagram of sliding mode control strategy applied on quarter car model. The quarter car model with passive and semi active suspension system is subjected to sine input with 0.012m amplitude and frequency 5Hz and continuous pulsed input with 0.02 m amplitude and frequency 5Hz.

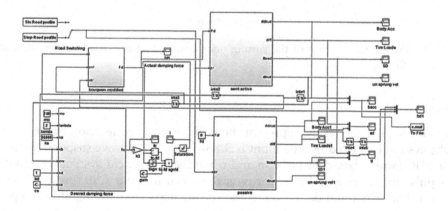

Figure 67.2 Simulink block diagram of sliding mode control.

The Figure 67.4 shows semi active suspension has significant reduction in sprung mass acceleration providing effective ride comfort and tire loading compared to passive suspension.

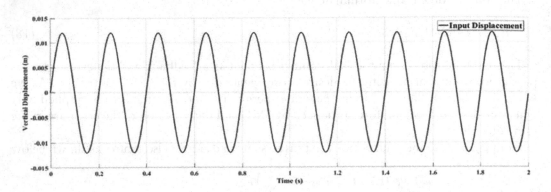

Figure 67.3 Sine Road profile and Continuous pulse road profile input.

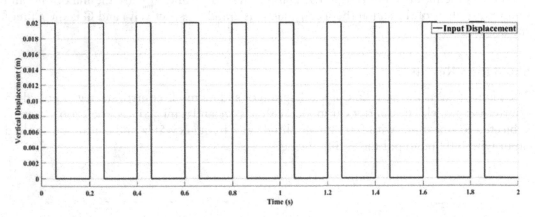

Figure 67.4 Vertical acceleration of the sprung mass considering sine road profile and continuous pulse road profile.

Discussions

From Figure 67.5 as the results represent the peak vertical acceleration of the sprung mass when the system is excited by sine input is 3.8387 m/s^2 for passive suspension and 2.9192 m/s^2 for MR damper-based suspension and when the quarter car system is excited by continuous pulse input are 6.344 m/s^2 for passive suspension and 5.805 m/s^2 for semi active suspension. The maximum deflection of the sprung mass from the equilibrium position is 0.0068m for the passive suspension and 0.0067m for the semi active suspension under sine input excitation and sprung mass from the equilibrium position is 0.01255m for the passive suspension and 0.01242m for the semi active suspension under continuous pulsed

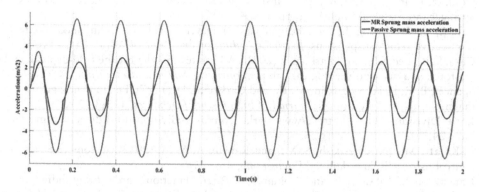

Figure 67.5 Suspension deflection considering sine road profile and continuous pulse road profile.

input excitation. The RMS value of sprung mass acceleration for sine input is 2.7215 m/s2 for passive suspension and 1.8861 m/s2 for semi active suspension. When the system is excited by continuous pulse input, sprung mass accelerations are 3.8387 m/s2 for passive suspension and 3.4294 m/s2 for semi active suspension.

Conclusion

This study simulates a quarter car suspension model with both passive suspension and MR damper based semi active suspension with sliding mode control. The RMS value of sprung mass acceleration for sine input is 2.7215 m/s^2 for passive suspension and 1.8861 m/s^2 for semi active suspension. When the system is excited by continuous pulse input, sprung mass accelerations are 3.8387 m/s^2 for passive suspension and 3.4294 m/s^2 for semi active suspension. The results specify a 30.69% reduction of sprung mass RMS acceleration for sine road profile and 10.66% reduction for continuous pulsed road profile. This specifies the effectiveness of semi active suspension over passive suspension system for the QCM. Thus, the sliding mode control strategy applied to MR damper provides better ride comfort and stability in comparison with the traditional passive suspension.

Acknowledgement

This work was supported by the Ministry of Human Resource Development and Ministry of Road Transport and Highways, Govt. of India under IMPRINT project titled with "Development of Cost Effective Magneto-Rheological (MR) Fluid Damper in Two wheelers and Four Wheelers Automobile to Improve Ride Comfort and Stability" [No. IMPRINT/2016/7330].

References

1. Choi, S. B., and Han, Y. M. (2012). Magnetorheological Fluid Technology: Applications in Vehicle Systems. CRC Press. pp. 17–30.
2. Chen, Y. (2009). Skyhook surface sliding mode control on semi-active vehicle suspension systems for ride comfort enhancement. *Engineering*, 1(1), 23–32.

3. Gołdasz, J., and Sapiński, B. (2015). Insight Into Magnetorheological Shock Absorbers. Cham: Springer International Publishing, pp. 1–224.

4. Yao, J., Shi, W., Zheng, J., and Zhou, H. (2013). Development of a sliding mode controller for semi-active vehicle suspensions. *Journal of Vibration and Control,* 19(8), 1152–1160.

5. Choi, S. B., Lee, S. K., and Park, Y. P. (2001). A hysteresis model for the field-dependent damping force of a magneto rheological damper. *Journal of Sound and Vibration,* 245(2), 375–383.

6. Spencer, B. F., Dyke, S. J., Sain, M. K., and Carlson, J. D. (1997), Phenomenological model for magneto rheological dampers. *Journal of Engineering Mechanics,* 123(3), 230–238.

7. Lord Division. (2004). https://www.parker.com/us/en/divisions/lord-division.html (accessed Nov. 24, 2023).

8. El-Kafafy, M., El-Demerdash, S. M., and Rabeih, A. A. M. (2012). Automotive ride comfort control using MR fluid damper. *Engineering,* 4(4). 179–187.

9. Rajendiran, S., Lakshmi, P., and Rajkumar, B. (2016). Fractional order fuzzy sliding mode controller for the quarter car with driver model and dual actuators. *IET Electrical Systems in Transportation,* 7(2), 145–153.

10. Park, J. H., Kim, W. H., Shin, C. S., and Choi, S. B. (2016). A comparative work on vibration control of a quarter car suspension system with two different magneto-rheological dampers. *Smart Materials and Structures,* 26(1), 015009.

68 Ergonomic approach to whole body vibration-a review

Mohammad Raza[a] and Rajesh Kumar Bhushan[b]

Mechanical Engineering, N.I.T Manipur, Imphal, Manipur, India

Abstract: The drivers of light and heavy vehicles with longer shifts usually of 8 hours are subjected to different degrees of Whole Body Vibration (WBV) which depends upon the nature of the vehicle, it's weight, velocity and type of road. In most of the vehicles due care is not taken for absorption of vibration especially to the driver through it's seat. Vibration in forward, backward and sideway is less harmful than the vertical direction which are denoted in the text as x, y and z axis. The standard procedure has been employed by different authors in their respective studies, although the low back pain was the most common side effect but the beneficial effect were no less especially when WBV was applied to patent's suffering from neuromuscular disorders. Nowadays topic of research is mainly focused on the modeling and measurement of biodynamic response of seated human body which employs an extensive ergonomics.

Keywords: benefits, diseases, light vehicle, heavy vehicle, whole body vibration.

Introduction

Human body is subjected to various types of vibrations under various situations. The vibration spectrum is very broad varying from unperceivable subatomic particles to perceivable situation like whole body vibration. Magnetic resonance imaging (M.R.I) is based upon the exploitation of magnetic property of hydrogen ion (proton) which is a classical example of subatomic vibration. Hydrogen ions are present in the aqueous medium in the human body. Under strong magnetic field the nucleus of hydrogen atom is excited from ground state and thereby releasing energy in the form of electromagnetic radiation i.e. processed into an image, it is due to the intense vibration of hydrogen ion [1]. On the other hand the extreme cases of whole body vibration a person is jolted strongly which may invariably lead to different types of aches and pains and neuromuscular disorders. However in certain situations the WBV is also of therapeutic importance like in cases of complete or partial weakness of neuromuscular system [2–6]. Unlike pain or temperature sensation the perception of vibration is not carried by a special set of nerve fibers but is the perception of touch, is readily applied. Broadly speaking the epicritic sensory modalities enters the nervous system via largemyelinated peripheral nerve fibers for clinical purpose vibration and position sense are conveyed to central nervous system and ultimately the

[a]mohammadrazaamu01@gmail.com, [b]rkbnitm@gmail.com

sensory cortex through this route thereby an individual senses the vibration [7]. It is also worth mentioning that vibration sense is impaired in old age above 70 years especially at ankle and lower limb [8]. In between these situations of vibration lies the sonographic induced vibrations in the form of ultrasound which may be of diagnostic and therapeutic value. In cases of arthritis and musculoskeletal disorders the sound wave (ultrasound) in the range of 1Megahertz to 5Megahertz are used. These waves when enter the body apart from producing imaging effect also have heating affect by virtue of increase circulation of blood which helps in tissue repair. Different studies with different settings of experiments have suggested the positive and negative effect of vibrations on human body. Some common symptoms of vibrations with relation to the frequency are given in the following Table 68.1 [9].

Table 68.1: Symptoms and related frequency.

Symptoms	Frequencies(Hz)
A general feeling of discomfort	4–9
Head symptoms	13–20
Low jaw symptoms	6–8
Influence on speech	13–20
Lump in throat	12–16
Chest pain	5–7
Abdominal Pain	4–10
Urge to urinate	10–18
Increased muscle tone	13–20
Influence on breathing movement	4–8
Muscle contractions	4–9

M.Cardinale and M.H. Pope [10] have reported in their review paper that short term exposure to whole body vibration has been shown to increase the serum level of testosterone and growth hormone and the combine effects on the neuromuscular system and endocrine system seem to suggest its effectiveness as a therapeutic approach for sarcopenina and possibly osteoporosis.

As regards the WBV is concerned it has both advantageous and disadvantageous outcome. The common sources of WBV are driving automobiles, piloting aircrafts, heavy vehicles, earthmovers etc. Earlier the scientists and researchers focused only on the adverse or disadvantageous side effects [11] like Raynaud's phenomenon in the persons whose vocation required the use of vibrating hand tools such as chain saws or jackhammers and keyboard operators [12]. Other common harmful effects were low back ache, slip disk etc. However in recent years the WBV is also applied for the management and treatment of neuromuscular diseases in addition to conventional training and therapy [13,14].

Literature Review

For review the studies were searched using the keywords Whole Body Vibration, light vehicle, heavy vehicle, diseases, and benefits on websites like PubMed, Google scholar and

Science direct. In all 80 papers were selected from which 51 research papers and standards were finally reviewed. Three books of medical stream were also taken into account for the respective study. Based on the review, researchers aimed to specify the WBV due to light and heavy vehicles. Human body receives vibration which may be locally or to the whole body which is called Whole Body Vibrations (WBV). The prototype of local vibrations called hand arm vibrations is due to the road driller power tools etc. however in general day to day life WBV is mainly due to light motor vehicles. On the other hand vibrations due to heavy vehicles are due to cargo vehicles, busses, trucks etc.

WBV *Induced by Light Vehicles*

Degan et al. [15] studied on two vehicles, one a light vehicle designated as model vehicle while the other one was reinforced with heavy armored shield. Both the vehicles had the similar driver seat but different total weight and the work was as per the recommendation by ISO 2631-1(1997) for the computation of WBV. It was observed that the vibration level at driver seat was less than the armored vehicle when compared to model vehicle. It is because of different interaction between seat and vehicle chassis. The special material for ballistic protection for armored vehicle had absorption effect due to its increased absorption effect. So the modification of the model car reduced the WBV exposure and the reduction was from 5% to 11%. He suggests that the study sample was small further research is needed on large no of vehicles. In another study carried out by Derakhshanjazari et al. [16] measured the WBV in the taxi drivers operating Peugeot 405. They observed that vibration transmitted to body was related to speed of the vehicle i.e. lower the speed lower was the amount of WBV. It was their experimental study in which apart from speed other factors were constant like tire pressure, engine speed, road gradient etc. They employed ISO 2631-1 measurement as standard. As the speed of vehicle increase from 20 to 80 Km/hr. increased vibration was recorded at z axis (Vertical axis). At 80 Km/hr. and beyond the average vibration limit exceeded health limit by ISO 2631-1(1997). They concluded that z axis vibration was most injurious to health as compared to x and y axis. Kumaresh et al. [17] studied the WBV on three different models of cars. The vibration level in x, y and z axis at the seat and the back rest were measured. All the three models were run at the speed of 40, 60, 90 and 110Km/hr. and the vibration dose was evaluated as per ISO 2631-1 and BS 6841. Their objective was discomfort and health hazard of the passengers. Ultimately they concluded that vibration decreases from low to medium speed whereas it increases from medium to high speed. The vibration at the seat was not attenuated much. At normal operation there was insignificant health hazard to the passengers in all the three cars.

WBV *Induced by the Heavy Vehicles*

Sharma et al. [18] in their review paper have highlighted the effect of heavy vehicles which usually ply in unpaved and undulated surfaces induced WBV which depends upon the sitting posture, duration of operation and transmission of vibration to the operator.

Bortolini et al. [19] studied the vibration effect of heavy trucks used for all tests within their manufacture not lasting more than 12 months. Measurements of vibration were recorded on a standard seat pad to which data acquisition system was attached. The trucks were operated on three types of tracks i.e. dirt, parallelepiped and asphalt. In their experiment drivers were different and one truck was loaded with 30 tons and another was empty. Their vibration measurement standard was as per ISO 2631-1(1997). Daily exposure to

the driver was 8 hours. Ultimately it was concluded that parallelepiped track is most hazardous for both loaded and unloaded cargo trucks and a large difference was observed in
perceiving the WBV by the drivers of unloaded and loaded trucks.

Kumar [20] carried study on empty truck meant for transportation of overburden material with the aim to work out the vibration magnitude in orthogonal axis of seat pan and
vibration exposure to the driver. He observed that magnitude of vibration in vertical direction was always harmful be the tuck empty or overloaded. He suggested multiple cushioning of the seat and control of the movement.

Wolfgang and Limerical's [21] study was based on haul truck operation and WBV in
different road conditions which were mainly rough in moist weather, graded surface and
also rough and smooth road of coal mine in New South Wales and concluded that vibration magnitude increase with load carrying capacity with higher vibrations in small size
dumpers.

Aye et al. [22] measured the whole body vibrations mainly daily exposure and concluded
that exposure action value exceeded for 50% of the vehicle, and the exposure limit was
less than the 90% of the vehicle. They concluded that vibration magnitude is directly proportional to the condition of the road, age and quality of machine and experience and age
of the operators.

Diseases Caused by WBV due to Light and Heavy Vehicles

The effect of WBV induced by moving vehicle has been studied by many researchers and
the injury to low back was found to be most significant by the operators of all type of
vehicles [23] which is due to continuous exposure to WBV [24]. One study in Great Britain
showed that over 9 million people every week are exposed to occupational WBV. Out of
these 374000 exceeded the recommended vibration value and they were mostly truck drivers [25]. The bus drivers were also the sufferers of low back pain due to WBV, mechanical
shocks and awkward postures on the seat [26, 27]. In their studies the third group suffered
from low back pain (LBP) were policemen who were exposed to WBV and at times they
had permanent injuries due to combat and arrest of the culprit [28]. The other risk of
occupational driving was pain in the shoulder, hand and wrist both with standard petrol
cars and armored cars [29]. Apart from WBV low frequency vibrations have adverse physiological conditions like degenerative diseases of the vertebrae, motion sickness, tiredness
and fatigue [30], low back pain [31], digestive problems, impairment of vision and an
increased risk of certain cancers [32]. Reduction in human reaction time and impairment
of balance are some other short term effects of WBV [33, 34]. According to Virtanen et al
most common illness linked to WBV is low back pain caused by intervertebral disc disease
[35]. Global Burden of Disease Study 2010 low back pain was also ranked first in 291
subject studies [36]. The other conditions causing low back pain due to WBV include trunk
posture [37–40], body mass distribution [41], muscular response to the machinery suspension [42] and occupants age [43]. Even with the assist vibration dosage below the standard
threshold level drivers working in transport, construction and agricultural industries also
have higher rate of lower back disorders [44–46]. However, In the review article published
by Gallais and Griffin [47] published a conflicting report that there was no significant
correlation of low back pain in car drivers subjected to whole body vibrations and they
emphasized further study in this field. Their study was carried out between 1975 to 2005.
Bovenzi et al [48] studied the effect of whole body vibrations on professional drivers and

concluded that assessment of internal spinal dose is better than the external dose for the prediction of low back pain.

WBV Benefits

As mentioned earlier in the text WBV has therapeutic implications also. Therefore it has gained popularity in physical exercise and rehabilitation centers. It is because it has improve muscle strength and powers [13,14], adds flexibility [49,50]. These improvements are due to the exposure of the individual to low frequency-low amplitude mechanical stimuli through a vibrating platform. The muscle spindles are stimulated and send nerve impulse leading to muscle contraction according to tonic vibration reflex [51]. WBV is also effective in balanced training, enhancing the strength and preventing falls [52,53]. The evaluation improvement was assessed on the basis of surface electromyography (sEMG). Mukhtar et al [54] in their review paper have concluded that there is significant improvement in muscle power after WBV. This improvement depended upon the range of amplitude and frequency, type of vibration and exercise protocol. According to them WBV also improves the vertical jumping height. At the cellular level they also concluded that WBV with frequency between 20 to 50Hz improves EMG activity, thereby increasing muscle strength.

Conclusion

From the above review it is concluded that the vibration exposure to human being can be of diagnostic and therapeutic value. When the whole body is subjected to vibration it has harmful as well as useful effect. The harmful effect is mainly concerned with the drivers operating light and heavy vehicle. The degree of vibration depends mainly upon the weight of the vehicle, type of the road, condition of the tire and transmissibility of vibration from chassis to drivers and passenger's seat. It also related to speed of the vehicle and the vibrations are maximum experienced during acceleration or retardation. Obviously the WBV was less on the smooth road. The most common health hazard is low back pain experienced by the professional driver's. Other health hazards have been shown in the table with relation to different frequencies. The measurement of vibration was carried out according to ISO 2631-1(1997) and in some studies BS 6841. The detailed explanation and methodology to measure the vibration level is beyond the scope of this paper for which it can be referred separately. In our opinion the beneficial effect of WBV outweighs its hazards. It is so because improvement in neuromuscular functions is a very significant achievement in a paralyzed or person with muscle disease. The vibration enhances the muscle strength, power, flexibility and degree of joint movement. The data of regarding the beneficial effect is far less than the data of its hazardous effects. Therefore it is suggested that application of WBV should be more extensively studied for the amelioration of neuromuscular disorders.

References

1. Goldberger, A. L. (2008). Harrison's Principles of Internal Medicine. 17th ed. (Vol. II, p. 1401).
2. Lundeberg, T., Nordemar, R., and Ottoson, D. (1984). Pain alleviation by vibratory stimulation. *Pain, Elsevier*, 20, 25–44.
3. Stillman, B. C. (1970). Vibratory motor stimulation. *The Australian Journal of Physiotherapy*, 16(3), 118–23.

4. Hagbrath, K. E., and Eklund, G. (1968). The effects of muscle vibration in spasticity, rigidity, and cerebellar disorders. *Journal of Neurology, Neurosurgery, and Psychiatry*, 31(3), 207–213.
5. Johnston, R. M., Bishop, B., and Coffey, G. H. (1970). Mechanical vibration of skeletal muscles. *Physical Therapy*, 50(4), 499–505.
6. Green, J. G., and Stannard, S. R. (2010) Active recovery strategies and handgrip performance in trained vs. untrained climbers. *Journal of Strength and Conditioning Research*, 24(2), 494–501.
7. Pradat, P. F., and Bede, P. (2020). *Biomarkers and Clinical Indicators in Motor Neuron Disease*. Lausanne: Frontiers Media SA. doi: 10.3389.
8. Alagappan, R. (2018). Manual of Practical Medicine. JP Medical Ltd.
9. Rasmussen, G. (1983). Human body vibration exposure and its measurement. *The Journal of the Acoustical Society of America*, 73(1), 1–27.
10. Cardinale, M., and Pope, M. H. (2003). The effect of whole body vibration on humans: dangerous or advantageous. *Acta Physiologica Hungarica*, 90(3), 195–206. DOI:10.1556/aphysiol.90.2003.3.2.
11. European Committee for Standardization (1996). Mechanical vibration-guide to the health effects of vibration on the human body. CR Report 12349. Brussels: CEN.
12. Woehrle, E. (2019). Is Professional Breath-Hold Diving Associated with Endothelial Dysfunction? (Doctoral dissertation, The University of Western Ontario (Canada)).
13. Roelants, M., Delecluse, C., and Verschueren, S. (2004). Whole-body vibration training increases knee-extension strength and speed of movement in older women. *Journal of the American Geriatrics Society*, 52(6), 901–908.
14. Rees, S., Murphy, A., and Watsford, M. (2008). Effects of whole-body vibration exercise on lower-extremity muscle strength and power in an older population: a randomized clinical trial. *Physical Therapy*, 88(4), 462–470.
15. Degan, G. A., Coltrinari, G., Lippiello, D., and Pinzari, M. (2017). Effects of Ground Conditions on Whole-Body Vibration Exposure on Cars: A Case Study of Drivers of Armored Vehicles. Urban Transport XXIII, (Vol. 176, pp. 431–437). doi:10.2495/UT170371.
16. Derakhshanjazari, M., Monazzam, M. R., Hosseini, S. M., and Poursoleiman, M. S. (2018). One-way ANOVA model to study whole-body vibration attributes among taxi drivers—an emphasis on the role of speed. *Journal of Low Frequency Noise, Vibration and Active Control*, 37(1), 156–164. DOI:10.1177/1461348418757907.
17. Kumaresh, S. A., and Aladdin, M. F. (2019). A study of vibration transmission on seated person in passenger vehicle. *AIP Conference Proceedings*, 2137, 040001. https://doi.org/10.1063/1.5120999. Published Online: 07 August (2019).
18. Sharma, A. S., Mandal, S. K., Suresh, G., Oraon, S., and Kumbhakar, D. (2020). Whole body vibration exposure and its effects on heavy earthmoving machinery (HEMM) operators of opencast mines—a review. *Journal of Mines, Metals and Fuels*, 68(7), 221–228.
19. Bortolini, A., Miguel, L. F. F., and Becker, T. (2018). Measurement and evaluation of whole-body vibration exposure in drivers of cargo vehicle compositions. *Human Factors and Ergonomics in Manufacturing & Service Industries*, 29(3), 253–264. DOI: 10.1002/hfm.20780.
20. Kumar, S. (2004). Vibration in operating heavy haul trucks in overburden mining. *Applied Ergonomics*, 5(6), 509–520.
21. Wolfgang, R., and Burgess-Limerick, R. (2014). Whole-body vibration exposure of haul truck drivers at a surface coal mine. *Applied Ergonomics*, 45(6), 1700–1704.
22. Aye, S. A., and Heyns, P. S. (2011). The evaluation of whole-body vibration in a South African opencast mine. *Journal of the Southern African Institute of Mining and Metallurgy*, 111(11), 751–757.
23. Porter, J. M., and Gyi, D. E. (2002). The prevalence of musculoskeletal troubles among car drivers. *Occupational Medicine*, 52(1), 4.
24. Zhao, X., and Schindler, C. (2014). Evaluation of whole-body vibration exposure experienced by operators of a compact wheel loader according to ISO 2631-1: 1997 and ISO 2631-5: 2004. *International Journal of Industrial Ergonomics*, 44(6), 840–850.

25. Palmer, K. T., Griffin, M. J., Bendall, H., Pannett, B., and Coggon, D. (2000). Prevalence and pattern of occupational exposure to whole body vibration in Great Britain: findings from a national survey. *Occupational and Environmental Medicine*, 57(4), 229–236.

26. Okunribido, O. O., Shimbles, S. J., Magnusson, M., and Pope, M. (2007). City bus driving and low back pain: a study of the exposures to posture demands, manual materials handling and whole-body vibration. *Applied Ergonomics*, 38(1), 29–38.

27. Velmurugan, P., Kumaraswamidhas L. A., and Sankaranarayanasamy, K. (2014). Whole body vibration analysis for drivers of suspended cabin tractor semitrailer. *Experimental Techniques*, 38(2), 47–53.

28. McKinnon, C. D., Callaghan, J. P., and Dickerson, C. R. (2011). Field quantification of physical exposures of police officers in vehicle operation. *International Journal of Occupational Safety and Ergonomics*, 17(1), 61–68.

29. Donnelly, C. J., Callaghan, J. P., and Durkin, J. L. (2009). The effect of an active lumbar system on the seating comfort of officers in police fleet vehicles. *International Journal of Occupational Safety and Ergonomics*, 15(3), 295–307.

30. Mabbott, N., Foster, G., and McPhee, B. (2002). Heavy vehicle seat vibration and driver fatigue. Department of Transport and Regional Services, Australian Transport Safety Bureau, Technical Report. CR 203.

31. Bovenzi, M., and Hulshof, C. T. J. (1999). An updated review of epidemiologic studies on the relationship between exposure to whole-body vibration and low back pain (1986–1997). *International Archives of Occupational and Environmental Health*, 72(6), 351–365.

32. Nadalin, V. (2012). Prostate cancer and occupational whole-body vibration exposure. *Annals of Occupational Hygiene*, 56(8), 968–974.

33. Newell, G. S., and Mansfield, N. J. (2008). Evaluation of reaction time performance and subjective workload during whole-body vibration exposure while seated in upright and twisted postures with and without armrests. *International Journal of Industrial Ergonomics*, 38(5–6), 499–508.

34. McPhee, B., Foster, G., and Long, L. (2009). Bad Vibrations: A Handbook on Whole-Body Vibration Exposure in Mining. Coal Serv. Heal. Safety Trust, Queensland Min. Safety, Sydney, NSW, Australia.

35. Virtanen, I. M. (2007). Occupational and genetic risk factors associated with intervertebral disc disease. *Spine*, 32(10), 1129–1134.

36. Hoy, D., March, L., Brooks, P., Blyth, F., Woolf, A., Bain, C., Williams, G., Smith, E., Vos, T., Barendregt, J., and Murray, C. (2014). The global burden of low back pain: Estimates from the global burden of disease 2010 study. *Annals of the Rheumatic Diseases*, 73(6), 968–974.

37. Milosavljevic, S., Bagheri, N., Vasiljev, R. M., McBride, D. I., and Rehn, B. (2012). Does daily exposure to whole-body vibration and mechanical shock relate to the prevalence of low back and neck pain in a rural workforce. *Annals of Occupational Hygiene*, 56, 10–17.

38. Gooyers, C. E., McMillan, R. D., Howarth, S. J., and Callaghan, J. P. (2012). The impact of posture and prolonged cyclic compressive loading on vertebral joint mechanics. *Spine*, 37, E1023–E1029.

39. Bovenzi, M., Pinto, I., and Stacchini, N. (2002). Low back pain in port machinery operators. *Journal of Sound and Vibration*, 253, 3–20.

40. Morgan, L. J., and Mansfield, N. J. (2014). A survey of expert opinion on the effects of occupational exposures to trunk rotation and whole-body vibration. *Ergonomics*, 57(4), 563–574.

41. Bovenzi, M. (2006). An epidemiological study of low back pain in professional drivers. *Journal of Sound and Vibration*, 298, 514–539.

42. Seidel, H. (2005). On the relationship between whole-body vibration exposure and spinal health risk. *Industrial Health*, 43, 361–377.

43. Ayari, H., Thomas, M., Doré, S., and Serrus, O. (2009). Evaluation of lumbar vertebra injury risk to the seated human body when exposed to vertical vibration. *Journal of Sound and Vibration*, 321, 454–470.

44. Kowalska-Koczwara, A., and Stypula, K. (2016). Assessment of the vibration influence on humans in buildings in the standards of different countries. *Procedia Engineering*, 161, 970–974.
45. Rantaharju, T., Mansfield, N. J., Ala-Hiiro, J. M., and Gunston, T. P. (2015). Predicting the health risks related to whole-body vibration and shock: a comparison of alternative assessment methods for high-acceleration events in vehicles. *Ergonomics*, 58(7), 1071–1087.
46. Zhao, X., and Schindler, C. (2014). Evaluation of whole-body vibration exposure experienced by operators of a compact wheel loader according to ISO 2631-1: 1997 and ISO 2631-5: 2004. *International Journal of Industrial Ergonomics*, 44(6), 840–850.
47. Gallais, L., and Griffin, M. J. (2006). Low back pain in car drivers: a review of studies published 1975 to 2005. *Journal of Sound and Vibration*, 298, 499–513.
48. Bovenzi, M., Schust, M., Menzel, G., Prodi, A., and Mauro, M. (2015). Relationships of low back outcomes to internal spinal load: a prospective cohort study of professional drivers. *International Archives of Occupational and Environmental Health*, 88(4), 487–499 https://doi.org/10.1007/s00420-014-0976-z.
49. Kinser, A., Ramsey, M., O'Bryant, H., Ayres, C., and Sands, W. (2008). Vibration and stretching effects on flexibility and explosive strength in young gymnasts. *Medicine and Science in Sports & Exercise*, 40(1), 133–140.
50. Cochrane, D., and Stannard, S. (2005). Acute whole body vibration training increases vertical jump and flexibility performance in elite female field hockey players. *British Journal of Sports Medicine*, 39(11), 860–865.
51. Cardinale, M., and Bosco, C. (2003). The use of vibration as an exercise intervention. *Exercise and Sports Sciences Reviews*, 31, 3–7.
52. Rees, S., Murphy, A., and Watsford, M. (2008). Effects of whole-body vibration exercise on lower-extremity muscle strength and power in an older population: a randomized clinical trial. *Physical Therapy*, 88(4), 462–470.
53. Bautmans, I., Van, H. E., Lemper, J. C., and Mets, T. (2005). The feasibility of whole body vibration in institutionalized elderly persons and its influence on muscle performance, balance and mobility: a randomized controlled trial [ISRCTN62535013]. *BMC Geriatr*, 5, 17–24.
54. Alam, M. M., Khan, A. A., and Farooq, M. (2018). Effect of whole-body vibration on neuromuscular performance: a literature review, *Work*, 59, 571–583. DOI:10.3233/WOR-182699 IOS Press.

69 Experimental and numerical analysis of vehicle grille nozzle air flow for engine bay cooling

Vinay D. Patel[a], Gaurav Gawad[b], Mitesh Shinde, Aashish Gawande, Vinayak Amle, and Ashish Chaudhari[c]

Department of Mechanical Engineering, Vidyavardhini's College of Engineering and Technology, Vasai (W), Maharashtra, India

Abstract: New technologies are emerging in order to improve engine power with efficient methods of heat rejection through radiator. The vehicle under-hood temperature is a major concerned of engine performance. The vehicle grills commercially available are aesthetically designed. However, grille opening shape and air mass flow rate are important in engine cooling of the vehicle. In this paper, an attempt has been made to reduce the temperature of engine bay using newly designed vehicle front grille. The vehicle grill is designed and developed with a convergent nozzle shape such that larger cross section acts as inlet of air and smaller one would be exit towards the engine bay. This causes the pressure difference with rise in velocity, causing the efficient circulation of air in the engine bay with substantial decrease in temperature. Further, the air striking the radiator and entering the engine bay leads to a considerable increase in performance and efficiency of the engine. The numerical and experimental performance of various cross section nozzles both individual and as entire grill were performed and analyzed to predict front end airflow pattern. The nozzle with rectangular inlet and elliptical outlet reported to be the most optimum design attaining the maximum temperature drop.

Keywords: convergent nozzle, engine bay cooling, vehicle grill.

Introduction

In automotive vehicle, the internal combustion engine cooling is important to extract the waste heat from the engine bay. Cooling is needed because hood under high temperatures tend to damage engine components. Cooling becomes most important when the climate becomes very hot especially in the Indian subcontinent region. Therefore, engine cooling system is employed to remove heat quickly and to keep temperatures low so the engine can survive at elevated temperature. The conventional grille available in today's vehicles is aesthetically designed without much engineering constraint. As a large portion of air coming in contact with vehicle enters the engine bay through grille. This air entered inside engine bay improve the heat carrying capacity of radiator as well as keeps engine cool with natural draught. Many researchers were worked in this area of engine bay cooling in which Kim et al. stress upon key design geometric parameters like vertical height, horizontal width, size, linear deformation, position, blockage, aerodynamic drag and grille inlet flow rate for cooling performance of the car [1]. Kim et al expressed about front-end airflow

[a]vinay.patel@vcet.edu.in, [b]gawadgaurav18@gmail.com, [c]ashish.chaudhari@vcet.edu.in

pattern for engine cooling [2]. D,Hondt performed study on flow around engine compartment and commented about lower aerodynamic drag to decrease CO_2 and greenhouse gas emissions. Author corelates the aerodynamic drag with the cooling air flow and suggested method of reducing cooling drag and increasing air flow rate [3]. Importance of internal airflow to improve engine cooling performance by using scale model was investigated by the author [4]. The grill and engine bay are modelled and analyzed to determine the suitable grille shape. The author reported that the grille blade setting at 45 degree with highest flow velocity into engine bay gives insignificant effect on vehicle speed [5]. Engine cooling system maintaining the operating temperature of engine in control plays a vital role in improving the vehicle fuel economy and meeting the stringent emission norms as well as in determining the aerodynamic drag of vehicle [6]. The authors Naraki et al, Kumar et al., Ali et al,Chavan et al., worked on application of nanofluids in automobile radiator in order to improve the heat transfer coefficient and thereby a have good engine performance. Some compositions are CuO/water nanofluid, Al_2O_3-water nanofluid wherein nanoparticle's dispersion into base fluids improve the thermal properties of the fluid and hence, thermal performance of the thermal system or device improve [7–13]. Kartik et al performed experimental and parametric study of louvered fin and reported that the increase in air velocity has significant influence for heat transfer coefficient [14]. Pang et al. elaborate about the numerical modelling of engine cooling system, heat transfer at radiator [15]. Lin et al. presented different cooling methods, flow field and thermal management in designing engine cooling systems and discuss about the flow field and reduction in thermal resistance and its effect on cooling performance [16]. Mason et al. discusses 2-D, C-D nozzles with suitable geometries and tested in static-test facility to determine the effects on internal performance like discharge coefficients, internal thrust ratios and contouring the nozzle throat by varying throat radius [17]. The exhaustive literature suggest the importance of engine bay cooling for improvement of overall thermal gradient and engine performance. There are few researchers who commented about the role of vehicle grill for air circulation inside engine bay cooling. To focus in depth of the engine bay cooling, the study has performed with various geometries of grill experimentally and numerically to optimize the shape of grill for efficient engine performance.

Modelling and CFD Analysis on Individual Nozzle

The 3-Dimensional modeling of various nozzles and complete grill are done using Solidworks16. All nozzles were modeled with same below mentioned dimensions. Area at inlet 3692 mm², area at outlet 509 mm² and length of nozzle 100 mm² with parabolic

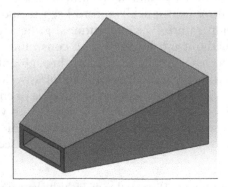

Figure 69.1 Simple conical nozzle.

convergence from inlet to outlet is maintained for all nozzles. Modelling of entire grille for optimum nozzle cross section. Figures 69.1–69.4 below shows the different shapes of nozzle with inlet and outlet.

The analysis of convergent nozzle consist of meshing done using ANSYS Fluent software. The mesh was created of trigonal elements with element size 6.4844e-002mm. near the wall of the nozzle. The inlet and outlet named selections are defined and the meshing is updated. The Pre-processing of the nozzle was done in ANSYS Fluent. Double precision

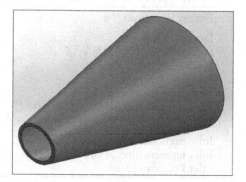

Figure 69.2 Rectangular nozzle.

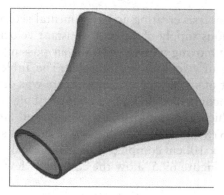

Figure 69.3 Nozzle with elliptical inlet and outlet.

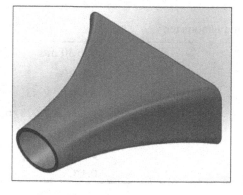

Figure 69.4 Nozzle with rectangular inlet and elliptical outlet.

settings were used while reading the mesh. The mesh was scaled since all dimensions were initially specified in mm. The mesh was checked in fluent and no critical errors were reported. The absolute criterion for convergence of the scaled residuals for x-velocity, y-velocity and z-velocity are 0.001. The following parameters are selected for simulation as shown in Table 69.1.

Table 69.1: Parameters for simulations.

General	Solver type: Pressure based
Models	Energy equation: On
Materials	Density: Ideal gas Specific heat: 1006.43J/kg.K Viscous model: Laminar Gamma: 1.4 Viscosity: 1.789×10^{-5} kg/m.s Thermal conductivity: 0.0242 W/m.K Mean molecular mass: 28.966 kg/kmol
Boundary Conditions	Inlet pressure: 1.01325bar Inlet temperature: 300K Inlet velocity: 15m/s

Experimental Analysis

To validate results, CFD analysis creating an experimental setup close to CFD condition is must necessary. A continuous supply of air with constant velocity blower is require simulating condition of vehicle moving on road and the air passing through grille and nozzle. A centrifugal blower with air velocity 15m/s selected. The Table 69.2 shows specifications of anemometer. The outlet of air blower is open so there was drastic velocity drop and the flow of the air was also uncontrolled. In CFD a counter of air passage was used, and grille was placed in contour which gave uniform air velocity and laminar air flow. To achieve same condition as of ANSYS a conduit was built with help of sheet metal. Dimensions of conduit are 35cm × 27cm × 100cm complete closure and leak proof was main aim while constructing conduit. The Figure 69.5 show the complete view of experimental setup test rig of grille testing.

Provision was provided to measure velocity and temperature of air at inlet and outlet by providing holes at distance of 5cm from outlet of grille. During design phase a complete

Table 69.2: Specifications of anemometer.

Air Velocity Range	0 to 30 m/s
Air Velocity Resolution	0.1 m/s
Power Supply	AAA 1.5V x3
Dimensions	(160 x 50 x 28) mm
Temperature Range	-10 to 50°C
Temperature Accuracy	±2°C
Temperature Resolution	0.1°C

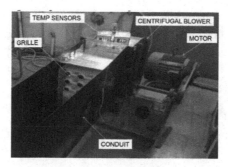

Figure 69.5 Experimental setup test rig of grille testing.

Figure 69.6 3D Printed grille section (rear section).

grille was designed with standard dimensions of swift grille. Size of the blower was major constrain. 1/3rd grille was printed which represent a prototype of complete model.

Grille was printed using 3D printing as shown in Figure 69.6 as it is most economical of single object. Wax coat applied on inner surface of nozzle, and it is covered with aluminum tape to make surface smoother. Anemometer monitoring is used to measure air velocity. The specification of Anemometer is as follows: RTD sensor is used for monitoring temperature. RTD with resolution of 0.1°C and its range is 0 to 200°C. To calculate the outlet temperature of nozzle the following assumptions are made: i) No heat transfer within the nozzle. ii) The flow of air is assumed to be streamlined. By using continuity equation for compressible flow.

Results and Discussion

CFD results of individual nozzles are depicted below from which the best nozzle which suits for purpose of maximum temperature drop will be selected. Figure 69.7 the temperature distribution drop over the conical nozzle was 2.3 degrees. In While in Figures 69.8 and 69.9 for the rectangular nozzle and elliptical nozzle it was 0.9 degrees and 1.8 degrees respectively for the same conditions. From above results, the nozzle with rectangular inlet and elliptical outlet was selected as it had the highest velocity and maximum temperature drop at the outlet.

By observing the velocity for rectangular inlet and elliptical outlet shape of the nozzle. Exit velocity is less than theoretical calculations, and maximum incurred velocity in this nozzle is 73 m/s. Figure 69.10 shows the temperature contours obtained from the

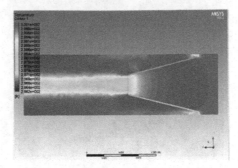

Figure 69.7 Temperature distribution over conical nozzle.

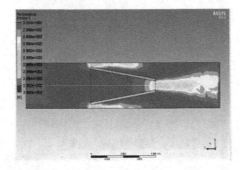

Figure 69.8 Temperature distribution over rectangular nozzle.

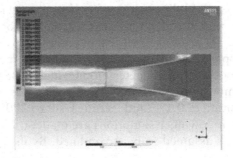

Figure 69.9 Temperature distribution over elliptical inlet outlet nozzle.

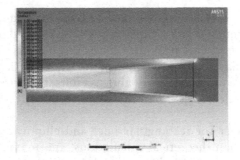

Figure 69.10 Temperature distribution over rectangular inlet and elliptical outlet nozzle.

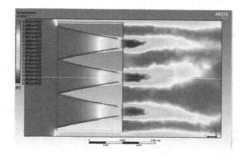

Figure 69.11 CFD analysis on the final grille.

rectangular inlet and elliptical outlet shape of nozzle show the highest temperature drop of 3.8 degrees. For the same inlet and outlet condition and keeping the inlet, exit areas same outlet velocity for the simple circular nozzle is 70 m/s. While for the elliptical nozzle and rectangular nozzle the outlet velocities are 62 m/s and 47 m/s respectively. The Figure 69.11 was selected for CFD analysis on the final grille. The rectangular inlet and elliptical outlet nozzle performed efficiently in terms of temperature drop when the maximum velocity reached. The variation of temperature with distance for various nozzles at velocity of 15 m/s are shown in Figure 69.12 comprises of a superimposed graph for various shapes of nozzles.

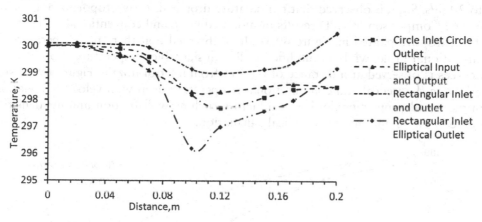

Figure 69.12 Temperature vs distance for various nozzles.

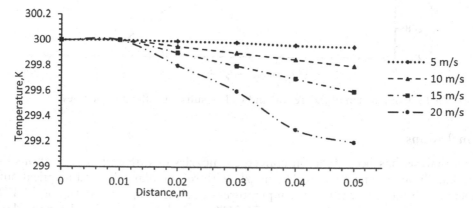

Figure 69.13 Temperature Vs distance graph of modified grille for different velocities.

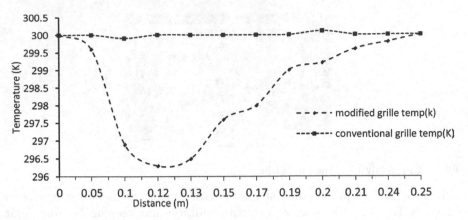

Figure 69.14 Temperature vs distance for conventional and modified grille.

X axis indicates distance from outlet while Y axis shows respective temperature. In conclusion nozzle with rectangular inlet and elliptical outlet gives maximum temperature drop of 3.7 degrees at 0.11m. The temperature variation with distance for different velocities are shown in Figure 69.13 As nozzle with rectangular inlet and elliptical outlet is selected for efficient performance, temperature drop is observed for various velocities ranging from 5 m/s to 20 m/s. So, it is observed that temperature drop is directly proportional to velocity of air. The Comparison of CFD results of modified grille and conventional grille at velocity of 15 m/s) are plotted in Figure 69.14, It is observed that there no temperature drop in conventional grille, while in modified grille considerable temperature drop of around 3.7 degrees is observed at a distance of 0.12 m from inlet of nozzle. Figure 69.15 shows deviations in theoretical and computational temperature drop with velocity. Deviation in graph increased proportionally because factors such as wall friction and air density are considered as assumptions in theoretical calculation.

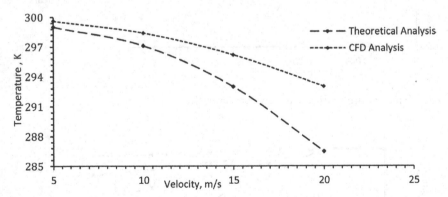

Figure 69.15 Comparison of theoretical and CFD results for different velocities.

Conclusions

CFD analysis has been done on convergent nozzles of different cross sections like rectangular, elliptical, and circular. The nozzle with rectangular inlet and elliptical outlet was selected as it had the maximum temperature drop. The temperature drop is 39.47% more than circular nozzle. It is 52.63% and 76.31% more than elliptical and rectangular nozzle, respectively. The CFD and theoretical results for temperature drop at different velocities show a great resemblance up to 10m/s but as the velocity increases further, after there is deviation from theoretical results due to increased skin friction in inner walls of nozzle which is neglected in theoretical calculations. Comparison of CFD results of conventional and modified grille show a considerable temperature drop of 3.7 degrees. Such engine cooling performance improvement done in various smoke tunnel like light pickup truck, passenger car, and sport car, with decreases by 2°C, 4°C, and 6°C, respectively.

Acknowledgment

Authors are thankful to Vidyavardhini's Trust, Vasai for their support in Experimental and Numerical Analysis of Vehicle Grille Nozzle Air Flow for Engine Bay Cooling experimental setup.

References

1. Kim, J. M., Kim, K. M., Ha, S. J., and Kim, M. S. (2006). Grille design for a passenger car to improve aerodynamics and cooling performance using CFD technique. *International Journal of Automotive Technology*, 17(6), 967–976.
2. Kim, H. J., and Kim, C. J. (2008). A numerical analysis for the cooling module related to automobile air-conditioning system. *Applied Thermal Engineering*, 28, 1896–1905.
3. D'Hondt, M., Gilliéron, P., and Devinant, P. (2011). Experimental investigation on the flow around a simplified geometry of automotive engine compartment. *Experiments in Fluids*, 50, 1317–1334.
4. Hoshino, T., Yoshino, R., and Takada, H. (1981). Improvement of engine cooling performance by cooling airflow visualization. SAE Technical Paper, 811392.
5. Hutacharern, N., and Ridluan, A. (2017). Airflow analysis of engine cooling grille using computational fluid dynamics. In Proceedings of 76th IASTEM International Conference, Seoul, South Korea, 7–9, 2017.
6. Baskar, S., and Rajaraman, R. (2015). Airflow management in automotive engine cooling system-overview. *International Journal of Thermal Technologies*, 5(1), 1–8.
7. Naraki, M., Peyghambarzadeh, S. M., Hashemabadi, S. H., and Vermahmoudi, Y. (2013). Parametric study of overall heat transfer coefficient of CuO/water nanofluids in a car radiator. *International Journal of Thermal Sciences*, 66, 82–90.
8. Kumar, A., and Subudhi, S. (2018). Preparation, characteristics, convection and applications. *Heat and Mass Transfer*, 54, 241–265.
9. Kumar, A., and Subudhi, S. (2019). Preparation, characterization and heat transfer analysis of nanofluids used for engine cooling. *Applied Thermal Engineering*, 160, 114092.
10. Ali, M., El-Leathy, A. M., and Al-Sofyany, Z. (2014). The effect of nanofluid concentration on the cooling system of vehicles radiator. *Advanced Mechanical Engineering*, 6, 962510.
11. Chavan, D., and Pise, A. T. (2013). Performance investigation of an automotive car radiator operated with nanofluid as a coolant. *Journal of Thermal Science and Engineering Applications*, 6(2), 021010.
12. Kılınç, F., Buyruk, E., and Karabulut, K. (2020). Experimental investigation of cooling performance with graphene based nano-fluids in a vehicle radiator. *Heat Mass Transfer*, 56, 521–530.
13. Muruganandam, M., and Kumar, P. C. M. (2020). Experimental analysis on internal combustion engine using MWCNT/water nanofluid as a coolant. *Materials Today: Proceedings*, 21, 248–252.
14. Karthik, P., Kumaresan, V., and Velraj, R. (2015). Experimental and parametric studies of a louvered fin and flat tube compact heat exchanger using computational fluid dynamics. *Alexandria Engineering Journal*, 54(4), 905–915. http://dx.doi.org/10.1016/j.aej.2015.08.003.

15. Pang, S. C., Kalam, M. A., Masjuki, H. H., and Hazat, M. A. (2012). A review on air flow and coolant flow circuit in vehicles cooling system. *International Journal of Heat and Mass Transfer,* *55,* 6295–6306.
16. Lin, W., and Sundén, B. (2010). Vehicle cooling systems for reducing fuel consumption and carbon dioxide: literature survey. In SAE Technical Paper, htpps://doi.org/10.4271/2010-01-1509.
17. Mason, M. L., Putnam, L. E., and Re, R. J. (1980). The effect of throat contouring on two-dimensional converging-diverging nozzles at static conditions, NASA Technical Paper 1704. Langley Research Center Hampton, Virginia, August 1980.

70 Dimension measurement using machine vision and its application in industry 4.0

Labhesh Sanjay Yawalkar[a], Onkar Babaso Chougule[b], Kishore Jawale[c], and Abhishek D. Patange[d]

College of Engineering Pune, Shivajinagar, Pune, India

Abstract: This paper presents a simple yet powerful approach for automated visual inspection. The proposed setup can be recreated at the final step of the manufacturing process that is to perform quality inspection of the samples from the selected lot. Machine vision or computer vision can be used for object detection, segmentation, counting, measurement etc. Here an attempt is made to combine machine vision with sensors for metrology inspection. This application is a potential solution that can take us further ahead in Metrology 4.0. The latest research in dimension measurement using machine vision is focused on using depth vision camera which uses infra-red light to get the depth information. However, in the present study a reference object such as a square of standard size is used, which will always be made available in the frame and the system will be calibrating itself in real-time to find out the real dimensions of the desired object. This results in reducing the cost of the setup and also the computation power requirement. In comparison to conventional optical methods which are used for dimension measurement, machine vision-based systems are cheaper, computationally effective and user friendly. The developed setup presented in this article has shown good potential with error less than 1.7% which can further be reduced. We can use the acquired dimensions data in order to perform different kinds of automated tasks such as inspection, classification, packing, etc.

Keywords: automated dimension measurement, machine vision, metrology 4.0.

Introduction

Integration of the latest technologies, such as machine vision with the manufacturing industry, has helped to automate the process and the technology can be put to use in hazardous manufacturing environment thus, eliminating the need of human intervention. The additional output provided by such system can be then used for various other purposes.

Initially, the concept of machine vision was used predominantly in robotics. Machine vision was used for calibration of robotic mechanism [1], and it was observed to be a very efficient way of calibration for robot with increased position accuracy. Machine vision is slowly becoming popular even for quality inspection as it offers multiple advantages. One of the earlier research has focused on optimizing the resolution, speed, and illumination for their setup which was used for continuous 2D measurement [2].

[a]yawalkarls19.prod@coep.ac.in, [b]chouguleob17.mech@coep.ac.in, [c]kishore.jawale@gmail.com, [d]adp.mech@coep.ac.in

Optical systems have also been employed to aid inspection and quality control. Further the use of Industry 4.0 techniques such as sensors, communication, machine vision, machine learning, etc. have led to a new concept of Metrology 4.0 [3]. A setup developed by [4], on-machine in-process surface metrology which eliminated the need for additional stage of quality inspection post manufacturing. An overview of past utilization of machine vision in the field of metrology can be referred from [5]. Here the authors have focused on assimilating literatures which had used machine vision in inspection and measurement tasks. They had also enlisted the major challenges that industry faces in making machine vision reliable so as to be used for quality control. Another article [6], which is a book chapter, also presents a guideline of how the machine vision system can be used for measurement. In the mentioned chapter the authors have provided flowcharts which could be useful to design a quality inspection setup using the machine vision concept. Machine vision systems have shown to be versatile. They have also been used for quality control inspections of gears [7]. Another example of use of machine vision is to perform in-line quality check for assembly processes [8]. In addition to all the measurement examples mentioned so far, machine vision systems have proven to be useful in identification and classification of materials [9]. The advances are leading to an environment where the quality inspection would be completely automated. A recent article [10] has conceptualized the idea of using machine vision to find the defective product using machine learning. They had focused on comparing various machine learning algorithms and making use of machine learning instead of actual dimension measurement to carry out quality inspection. In the literature it is found the machine vision system is often required to be calibrated if the height/ focus changes. This drawback needs to be scrutinized and the present study tries to address this problem.

The article's primary goal is to measure the dimensions of the manufactured mechanical parts using machine vision, which are usually carried at the final stage of a manufacturing line as a visual inspection task. The algorithm first finds for the reference object whose dimensions are already known to it and then calibrates itself and then looks for other objects. The setup includes a conveyor belt on which the component to be measured is placed. The component is then made to pass under a camera which is a part of machine vision system, which is expected to carry out measurement. In the captured frame only one component and reference object are captured.

To use an optical method for dimension measurement such as CMM trained professional is required while the machine vision system is user-friendly and completely depends on the software rather than skilled operator. The floor space required for optical method setup is relatively more also these instruments are expensive when compared to the machine vision systems. An extra feature that comes with the machine vision systems is that they can be embedded with the artificial intelligence. The current setup automatically detects the objects in the frame without any special arrangement to aid the object location which is not possible in the optical methods-based systems. However, the machine vision systems are prone to manually induced errors in the measurement.

The article structured in following order: setup, working, results and discussion and conclusion.

Setup

The setup (presented in Figures 70.1 and 70.2) is created in such a way that it can be replicated as an add-on to the existing assembly line. If the assembly line already consists of the product on a conveyer belt as most of the production lines do, the setup can be added to the same.

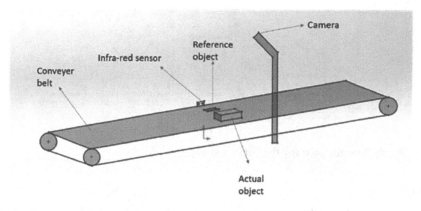

Figure 70.1 Schematic of the proposed setup.

Figure 70.2 Current setup of the method proposed.

A camera is arranged directly above the conveyer belt so that it captures the reference object as well as the main product whose dimension needs to be measured in single frame. The infra-red sensor is used to detect if the actual object is below the camera or not. The light source might be used to increase the contrast of the image which was used while taking the actual readings. The background of the object whose dimensions are to be measured are taken as a white paper to reduce the noise in the image and for better edge detection. The reference object is drawn on the white paper of its bottom left corner whose dimensions are 2cm * 2cm.

Working

The IR sensor is in such a way that it detects if the object whose dimension needs to measured is under the camera so that it is captured in the frame. After detection the IR sensor sends the signal to the Arduino circuit which then turn on the camera and takes the picture and sends it to the raspberry pi computer which has the image processing code loaded in it. The raspberry pi chip processes the image and find out the dimensions of the object which can then further used to perform automated tasks such as stacking, classification etc.

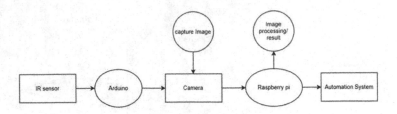

Figure 70.3 Work-flow adopted for the current study.

Image Processing

The setup time is reduced because the user is not required to calibrate or care about the distance aspect as the system auto calibrate itself in real-time. Initially we have tried to carry out the experiments with white background of the image and further make it robust for wild conditions.

For image processing the OpenCV library in python which is an open-source library for image/video processing in python is used.

The image is made up of 3 channels red, green, blue with each of the 3 values for each pixel ranging from 0 to 255. We are capturing such kind of RGB image and using it for further processing.

Preprocessing

Applying gaussian blur convolution to the image in order to reduce false contour detection. Gaussian blur reduces the speckle noise that can be captured in an image and smoothens the image.

Thresholding

First the RGB image is converted into greyscale image so that the grey image now has only one channel with values ranging from 0 to 255, 0 presenting black while 255 representing white. Thresholding is the process of converting a grey image into a binary image in which each pixel can either be white or black that is 0 or 255, which is done using a threshold, the pixels with value less than the threshold in the grey image are converted to 0 while those with more than threshold are converted to 255. Now the image is ready for finding the contours on the basis of Figures 70.3 and 70.4.

Finding Contours

The contour finding algorithm as implemented in OpenCV finds for sudden changes in intensity that is white objects in black background.

Algorithm to Find Dimensions After Finding Contours

Algorithm to find the desired object contour and its dimensions is mentioned below:

Input	:	All the contours found by OpenCV contour finding algorithm
Output	:	Contour of desired object with its dimensions

Calibration Dimension measurement	:	1.	For contour in found contours If contour Area < threshold: Delete contour *//The threshold is found using experimenting which depends on the frame size for frame size of 620*480 it is 1000.This is to ensure that we don't delete the important contours but also screen those which are unnecessary*
		2.	Contours=sorted contours based on the location *// the reference object is located in the frame near the 0-x co-ordinate so we will get the reference object as first element of the sorted contours;*
		3.	Draw a minimum area rectangle around the reference object
		4.	Find the vertices of the minimum area rectangle *//Now we have tl,tr,bl,br points of the reference object:* *//tl: top left point* *//tr: top right point* *//bl: bottom left point* *//br: bottom right point*
		5.	W1=Euclidean distance(tl,tr) *//width of upper edge*
		6.	W2=Euclidean distance(bl,br) *//width of lower edge*
		7.	H1=Euclidean distance(tl,bl) *//height of left edge*
		8.	H2= Euclidean distance(tr,br) *//height of right edge* *//Euclidean distance between two points is the distance between two points in multi-dimensional space so here* *//Euclidean distance=square root((x2-x1)**2+(y2-y1)**2)*
		9.	Pixel_per _mm=(w1+w2+w3+w4)/known dimension of the reference object *//for our purpose we drew a square of 20 mm side so the formula becomes* *//pixel_per_mm=(w1+w2+h1+h2)/80* *//we are taking the average to reduce the calculation error*
		10.	For contour in sorted contours: Find w1,w2,h1,h2 as explained above Measured width=(w1+w2)/(pixel_per_mm*2) Measured height=(h1+h2)/(pixel_per_mm*2)

While calibrating average of the four sides are taken thus, trying to reduce the error while measuring dimensions. Also, for object dimensions average of heights and widths are taken, reducing the average error produced while measurement.As long as the reference object is in the frame, there is no need to calibrate the system as it will auto-calibrate in real-time. Values are taken from Table 70.1.

Results and Discussion

4 slip gauges of standard size namely, 23.5 mm, 24 mm, 25 mm, 50 mm are used in order to check the accuracy and calculate the error produced in dimension measurements. 10 readings of each slip gauge are taken and the absolute error produced by the machine vision system is calculated.

Also, the deviations in the measured dimensions produced by the system during live image capturing using Logitech c270 webcam and producing the dimensions are shown in the graphs below.

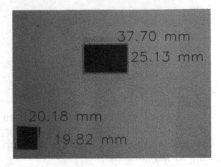

Figure 70.4 Image generated by the proposed setup.

Table 70.1: The actual dimensions measured by the setup for 10 iterations on 4 different sizes of slip gauges.

Actual dimension (mm) / iteration	1	2	3	4	5	6	7	8	9	10
23.5	23.86	23.91	23.71	23.63	23.57	23.72	23.4	24.06	23.68	23.7
24	24.48	24.89	24.76	24.31	24.13	24.31	24.25	24.44	24.18	24.23
25	25.74	25.23	25.13	24.91	25.29	25.63	25.1	25.64	25.5	25.49
50	50.3	50.94	50.74	50.61	50.72	50.2	49.99	49.98	50.64	50.57

The graphs showing the variations in the measurements by the setup are shown in the form of a graph below for a better visualization.

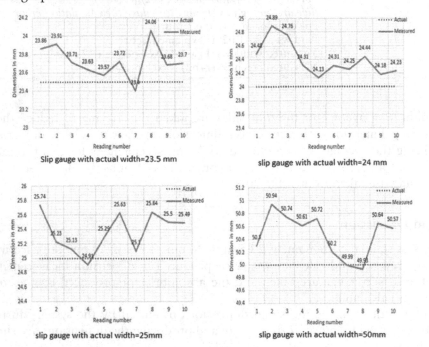

Figure 70.5 Measurements obtained from the setup.

The mean error, variance and the standard deviation for the 10 readings for each slip gauge are calculated and are shown in the Table 70.2.

Table 70.2: Mean dimension measured, mean error, Standard Deviation, Percentage error for the dimensions (in mm) measured by the setup for each slip gauge.

Actual	Mean measured dimension	Mean Error	Variance	Standard Deviation	Percentage error
23.5	23.724	0.224	0.034027	0.184463	0.95 %
24	24.398	0.398	0.062951	0.250901	1.65 %
25	25.366	0.366	0.075182	0.274194	1.46 %
50	50.464	0.464	0.11536	0.339647	0.92 %

Unlike conventional dimension measurement systems like labview or NI vision where the calibration needs to be done manually also the contour whose dimension is to be measured needs to be specified the setup does both the things automatically based on the single image provided to it via camera. As seen in the research for tool measurement using machine vision where the researchers have used pixel values for edge detection and the system needs to be calibrated before taking measurements, they have achieved 1.136% mean error as compared to sophistication Canny edge detection algorithm used in this study and the mean is between 0.95% to 1.65% which is comparable [11].

Further this setup can be connected via cloud to enable remote control of the process. The future of this project extends to using IoT enabled sensors to transmit data to cloud, where the data will be analyzed and presented in a form which will be convenient for the responsible personnel to take required action.

Conclusion

The developed setup focuses on the machine vision-based system rather than the conventional optical methods for dimension metrology in production engineering. The average percentage error is at maximum 1.7%, which can be considered in some applications. The current setup can be further improved by using a stable light source to reduce the shadow effect in the image. Advanced edge/ object detection algorithms can be applied to improve the reliability of the system. The novelty of this setup is its real-time calibration without affecting the process time. The conventional manual methods like micrometer, vernier caliper for dimension measurement are highly dependent on the operator hence they are labor intensive and are subjected to human errors which are eliminated by an automated machine vision-based system.

References

1. Renaud, P., Andreff, N., Lavest, J. M., and Dhome, M. (2006). Simplifying the kinematic calibration of parallel mechanisms using vision-based metrology. *IEEE Transactions on Robotics,* 22(1), 12–22.
2. Maresca, P., Duarte, A., Wang, C., Caja, J., and Gómez, E. (2019). Evaluation of traceability in continuous 2D measurements employing machine vision systems. *Procedia Manufacturing,* 41(1), 922–929.

3. Benitez, R., Ramirez, C., and Vazquez, J. A. (2019). Sensors calibration for Metrology 4.0. In Workshop on Metrology for Industry 4.0 and IoT (MetroInd4.0&IoT) 1(1), pp. 296–299. IEEE.

4. Gao, W., Haitjema, H., Fang, F. Z., Leach, R. K., Cheung, C. F., Savio, E., and Linares, J. M. (2019). On-machine and in-process surface metrology for precision manufacturing. *CIRP Annals,* 68(2), 843–866.

5. Alonso, V., Dacal-Nieto, A., Barreto, L., Amaral, A., and Rivero, E. (2019). Industry 4.0 implications in machine vision metrology: an overview. *Procedia Manufacturing,* 41, 359–366.

6. Kumar, B. S., Vijayan, V., and Davim, J. P. (2019). Machine vision in measurement. *Measurement in Machining and Tribology, Materials Forming, Machining and Tribology,* 113–123.

7. Moru, D. K., and Borro, D. (2020). A machine vision algorithm for quality control inspection of gears. *International Journal of Advanced Manufacturing Technology,* 106(1), 105–123.

8. Frustaci, F., Perri, S., Cocorullo, G., and Corsonello, P. (2019). An embedded machine vision system for an in-line quality check of assembly processes. *Procedia Manufacturing,* 42, 211–218.

9. Penumuru, D. P., Muthuswamy, S., and Karumbu, P. (2020). Identification and classification of materials using machine vision and machine learning in the context of industry 4.0. *Journal of Intelligent Manufacturing,* 31(5), 1229–1241.

10. Benbarrad, T., Salhaoui, M., Kenitar, S. B., and Arioua, M. (2021). Intelligent machine vision model for defective product inspection based on machine learning. *Journal of Sensor and Actuator Networks,* 10(1), 7–25.

11. Mikołajczyk, T., Kłodowski, A., and Mrozinski, A. (2016). Camera-based automatic system for tool measurements and recognition. *Procedia Technology,* 22(1), 1035–1042.

71 Drivers and barriers for implementation of green supply chain management: a review

Gagandeep[1,a], Shilpi Ahluwalia[2,b], Rahul O. Vaishya[1,c], Mohit Tyagi[3,d], and R. S. Walia[1,e]

[1]Department of production and Industrial Engineering, Punjab Engineering College, Chandigarh

[2]Dr. S. S. Bhatnagar University Institute of Chemical Engineering & Technology, Punjab University, Chandigarh

[3]Department of Industrial and Production Engineering, Dr. B R Ambedkar National Institute of Technology, Jalandhar

Abstract: Nowadays, Green Supply Chain Management (GSCM) is an emerging topic due to an upsurge of awareness in customers and government guidelines about the environmental issues. The implementation of GSCM significantly impacts the environmental, social, and economic performance of an organization. Thus, several manufacturing firms have started to implement the same for sustainable future. Several drivers play an important role to motivate the organizations for the implementation of GSCM but at the same time numerous barriers of GSCM are also there that are required to be compensated. The present study aims to review the available literature related to the drivers and barriers of GSCM and identify the various techniques used to prioritize the barriers and drivers of GSCM. Further, it is concluded that 'cost implication' and 'the lack of support from government' are identified as the most critical barriers, whereas 'the support from top management' comes out as the most important driver. The present study may be beneficial for academicians, researchers, and practitioners for the better understanding about the drivers and the barriers faced during the implementation of GSCM in future research.

Keywords: Barriers, drivers, environmental issues, green supply chain management (GSCM).

Introduction

In the era of automation, Supply chain management (SCM) acts as an important part of the strategic planning of any organization for collaborating directly or indirectly between related parties to satisfy the customers' requirements. Supply chain is network of firms wherein the material, information and money flow from supplier to the customer's end [28] and SCM assures delivery of right product to right location at right time. SCM starts from the purchasing of raw materials, followed by manufacturing processes after which final product is delivered to the customer. Generally, conventional SCM approach deals with the economic and business aspects only but does not address the environmental aspects. However, most of the countries have been concerning about environmental issues for last few decades. The environment related issues are gradually increasing due to pollution created by various industries. Thus, the uneven climate changes in environment and depletion of natural resources attract the industrial as well as academic organizations

[a]gagansekhu@gmail.com, [b]shilpi.ahluwalia@gmail.com, [c]rahul_mv@yahoo.com, [d]tyagim@nitj.ac.in, [e]waliaravinder@yahoo.com

to evaluate, understand and manage the environmental issues [47]. To mitigate the environmental issues, the industrial organizations have started adopting Green Supply Chain Management (GSCM). GSCM is well-defined as an integration of environmental concerns with conventional SCM that comprises multiple practices relating to the environmental management [7]. These practices incorporate the environmental concerns into forward as well as reverse logistics in the several phases of supply chain [6]. The consideration of environmental issues distinguishes the GSCM and SCM. GSCM begins from product design to product life ending to recycling or dispose by considering the environmental issues at each stage of SCM. The concept of reverse logistics is effectively growing in past decades due to numerous factors [41]. By adopting GSCM, the environmental performance can be enhanced which helps to improve efficiency and profit of organization, and reduce the cost for leading the competitive market [1, 9, 56]. Therefore, the importance of the GSCM is gradually increasing among the supply chain management experts and researchers [50]. GSCM consist of green procurement, green manufacturing, green delivery and reverse logistics [14]. The model shown in Figure 71.1, as reproduced below, gives information about the stages of GSCM.

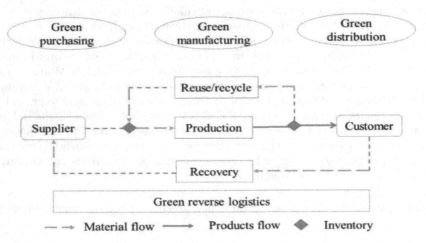

Figure 71.1 GSCM Model [1].

The manufacturing organizations reduce environmental hazards by adopting green practices such green sourcing, green management, green purchasing, green manufacturing, green delivery, and ecological design. To effectively implement GSCM, the organizations must understand about barriers and drivers of GSCM. The present study is aimed to evaluate the various drivers and barriers of GSCM which play an important role to implement the GSCM practices in the manufacturing organizations. The identification of major drivers and barriers of GSCM helps the managerial staff to resolve the issues related to environment. Thus, the purpose of this study is to update the knowledge related to drivers and barriers of GSCM by analyzing of the existing literature. The drivers and barriers as identified in the present study, provides a route map for the prioritization of drivers and barriers, which can be further analyze in future, by using decision making techniques for prioritization to evaluate most critical drivers and barriers. The following section describes the methodology followed for selecting the relevant articles and next section describes the literature analysis of GSCM and critical review of the selected papers. The conclusion points and future directions are summarized in the final section.

Method for Selecting Relevant Papers

The papers related to the barriers and drivers of GSCM were obtained using Google Scholar. Total 61 papers were found by limiting the search results from 2011 to 2020, which are seemed to be relevant to green supply chain. Further, the papers were scrutinized with the purpose exclude the papers related to other green supply techniques rather than evaluation of barriers and drivers of GSCM. Thus, 26 papers were selected as most relevant to the identification of barriers and drivers and, therefore most relevant 26 papers are critical reviewed. The literature summary of the selected papers, including implemented methodology, country, year, and industries of research, is described in Table 71.1. The year wise analysis of the relevant papers, reviewed in the present study, as shown in the Figure 71.2. Most of the study related to GSCM are conducted in the latest years.

Table 71.1: Selected papers related to the drivers and barriers of GSCM.

S. no.	Authors/ year	Methodology	Country	Industry	Drivers	Barriers	Divers & Barriers
1	[16]	ISM	India	Aluminum product manufacturing	✓		
2	[27]	ISM	India	Automobile industry		✓	
3	[8]	ISM	UAE	Construction sector		✓	
4	[10]	Survey	Denmark	Product manufacturing			✓
5	[12]	ISM	India	-		✓	
6	[49]	ISM	India	Manufacturing		✓	
7	[54]	ISM	India	Automobile	✓		
8	[30]	ANOVA and T test	India	Plastic		✓	
9	[15]	Literature Review	India	-			✓
10	[5]	ISM	Gulf countries	Plastic, Petrochemical and Textiles	✓		
11	[46]	ISM	India	Fertilizer	✓		
12	[45]	ISM	India	Fertilizer	✓		
13	[26]	ISM-fuzzy	India	Automobile		✓	
14	[23]	Literature Review	Canada	-		✓	
15	[24]	Pareto analysis	Canada	-		✓	
16	[39]	AHP	India	SMEs		✓	

(continued)

Table 71.1: Continued

S. no.	Authors/ year	Methodology	Country	Industry	Drivers	Barriers	Divers & Barriers
17	[3]	Descriptive statistics	Ethiopia	Leather		✓	
18	[4]	Problem conflict index	Malaysian	Automobile		✓	
19	[53]	Hierarchical cluster analysis	Bangladesh	Textile		✓	
20	[28]	ISM	India	Textile and apparel		✓	
21	[18]	ISM	India	Argo-industry	✓		
22	[44]	ISM	Bangladesh	Chemical	✓		
23	[6]	ISM	Jordan	Manufacturing		✓	
24	[52]	Confirmatory factor analysis	Indonesia	Automobile	✓		
25	[31]	Qualitative study	UK	Restaurants			✓
26	[33]	Delphi	Iran	Oil industry			✓

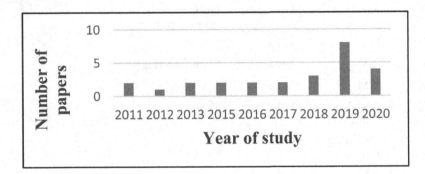

Figure 71.2 Year wise analysis of selected papers.

Literature Analysis

In recent years, several researchers are working on the GSCM as a potential concept due to environmental concern in the manufacturing practices. The consideration of the environmental issues and efficient utilization of the resources distinguishes GSCM from the conventional supply chain management. The manufacturing organizations are trying to take the competitive advantage by putting green labeling on products with adaptation of GSCM. As per the government norms, the manufacturing companies are facing more strict environmental rules for controlling the pollution in the environment. Therefore, many developing nations have just started to implement the GSCM for mitigating the environmental issues. Among the European manufacturing industries, Volkswagen was the first manufacturing firm to adopt environment friendly practices in 1995 [15]. Sundarakani et al. [51] conducted a study using lagrangian transport model to measure carbon footprints, to investigate that the manufacturing and transportation processes consume more energy, generate significant carbon emission, and they suggested some strategies to reduce the carbon emissions. Khan [1] has discussed the green practices such as green material procurement, green market strategies, green management, green delivery and warehousing, green production, eco-friendly design, green transport and reverse logistics, renewable sources of energy, and bio products etc., in supply chain management (SCM) which help in gaining popularity of GSCM for improving the environmental performance. Moreover, Zhu et al. [57] conducted a comparative analysis in manufacturing firms which shows that internal environment management practice is one key practice of GSCM. It was also indicated that big firms may implement the practices of green supply chain by introducing win-win relationships with the small-scale suppliers and customers of the firms. Rao [38] conducted research in SMEs in India that has explored the extent of implementation of GSCM wherein the survey was conducted from environment management representatives (EMR) using a 4-point Likert scale questionnaire.

Literature Review for Barriers of GSCM

Several organizations are widely adopting the GSCM but sometimes it is not easy to take GSCM in practice because of plethora of factors such as government guidelines, high expenses, deficiency of customer awareness, and lack of support from top level superiors etc. These factors which restrict the effective implementation of GSCM, are termed as "barriers". Therefore, there is an immense need to get proper knowledge about these barriers for successful implementation of GSCM. Luthra et al. [27] performed a study in the automotive organization in India to identify the barriers of GSCM and further prioritize these barriers by using Interpretive Structural Modelling (ISM) approach. This research found that the lack of government support system on the lower stage of ISM approach is an important barrier of GSCM, which may further introduce other barriers of GSCM. Ratna and Kumar [39] conducted their research in SMEs and identify ten barriers of GSCM. Further, the Analytical Hierarchy Process (AHP) is implemented to rank the barriers for finding the most critical barriers. Herein, the cost implication was found as major barrier which may restrict the implementation of GSCM. Al-Refaie et al. [6] identified 10 barriers in their research in the manufacturing firms situated in Jordan. A contextual relationship was found between the barriers and ISM tool was used to find most critical barriers. Further, MICMAC analysis was implemented for categorizing the barriers of GSCM into four sections, i.e., dependent, independent, linkage and autonomous variables, in accordance with the driving and dependence power. This study investigated that the lack of government support may be a critical barrier which obstruct the implementation of GSCM. Mathiyazhagan et al. [30] conducted research in the Indian plastic industries and identify 38 barriers of GSCM from available literature and with the help expert opinions. This study prioritizes the top ten important barriers out of 38 barriers based on expectation value. Herein, the Lack of adopting new technology for implementation GSCM was found as the most critical barrier. Moreover, the class having high influence was evaluated through the use of ANOVA and paired T-test. Srivastav and Gaur [49] performed a study in the small-scale manufacturing industries located in north India to identify sixteen barriers of GSCM by analyzing the available literature and consulting to the industrial professionals. Further, ISM and MICMAC analysis was utilized to position the barriers in model and to find the relationship between them. Herein, the Lack of support from government system was found as an important obstacle to implementation of GSCM. Kumar et al. [26] investigated various issues which hinder the implementation of green practices in SCM. This research was performed in automotive manufacturing firms located in the North Capital Region (NCR) to identify eleven issues of GSCM. Further, ISM-fuzzy MICMAC analysis was used for the assessment of issues to sort out most critical barriers. In this study the lack of customer cooperation was identified as the leading barrier that further results in the occurrence of other barriers in the implementation of green practices in SCM. Abebaw and Virdi [3] carried out study in leather product industry in Ethiopia. They identified cost implementation, lack of customer awareness and lack of qualified human resource experts in sustainability as the most critical barriers by conducting descriptive statistics in SPSS 23 software. Afroz et al. [4] examined top barriers of GSCM in automotive industry in Malaysia. In this study, the market competition and uncertainty came out as the main barrier followed by the lack of implementation of GSCM practices. Majumdar and Sinha [28] conducted research to study various barriers in textile and apparel industries located in south-east Asian countries. By concerning with the industrial experts and reviewing the available literature, 36 barriers were identified. Further, these barriers are classified into seven categories such as supplier, strategies of management, technical aspect, information, human resource, external stakeholder and economic considerations. Further, ISM-MICMAC technique was used to develop the hierarchy model for finding inter-relationship among the barriers of GSCM. Thus, this research found the most critical four barriers such as complexity of green process and system design, lack of consumer support, lack of support from regulatory authorities, and high implementation and maintenance cost. Similar study was performed by Tumpa et al. [53] in the textile industries of Bangladesh but by using a different technique i.e. hierarchical cluster analysis although, they use different technique but the most critical barrier identified by them were almost similar with that in the study of Majumdar and Sinha [28]. Balasubramanian [8] conducted a study in construction sector of United Arab Emirates (UAE) in which 32 barriers were identified which were further grouped into 12 criteria. The most critical barriers, obtained using ISM-MICMAC technique, were identified as the lack of skilled sustainability experts, lack of green suppliers and developers, lack of government support, lack of public awareness, and demand and market uncertainty. In 2018, Kaur and Awasthi [23] reviewed 17 papers and identified 54 barriers which were further categorized into six categories such as multiple M's, (man, machine, market, machine, methods, money) supply chain processes, organizational hierarchy, stakeholders, sustainability areas and implementation. According to this review, in 2019, Kaur et al. [24] performed research using Pareto analysis to prioritize the barriers of GSCM. This study found that the lack of awareness regarding to reverse logistics adoption and difficulty in transforming positive environmental attitudes into action are most critical barriers. Dashore and Sohani [12] developed a hierarchical framework to examine the important barriers of GSCM. This research identified 14 barriers of which the dependence and driving power were analyzed by using MICMAC technique, and most influencing barriers were found as uncertainty and competition in market, and poor organizational culture. Bey et al. [10] performed a survey-based study in the product developing and manufacturing industries to identify major current drivers and barriers of GSCM. They identified that 'the lack of information on environmental impacts', 'lack of expert's knowledge' and 'lack of allocated resources such as manpower and time' played role as the main barriers of GSCM. Meager et al. [31] mentioned that a few numbers of studies have been conducted on implementation of GSCM in the restaurants of UK. This study is aimed in same area to find important drivers and barriers of GSCM by in interacting with managerial staff of nine hotels. This was found the 'logistics', 'corporate lying' and 'financial responsibility' as key barriers. Narimissa et al. [33] thought to conduct same research in the oil and gas industry of Iran in which they covered 112 drivers (about which discussion is done in next section) and 41 barriers. The Delphi three-round method was used to identify the essential drivers and barriers of GSCM by taking the opinion of experts working in the oil and gas company. Finally, the high priority barrier was identified as 'the cost implication'.

By reviewing the previous research, various barriers influencing the implementation of GSCM are indentified. Several barriers of GSCM as identified by diffirent authors, are decribed in the Table 71.2.

Table 71.2: List of selected barriers of GSCM.

S.No.	Barriers of GSCM	References
1	Cost implication	[27, 39, 6, 49, 3, 4]
2	Lack of training	[15, 56, 49, 24]
3	Unawareness of customers	[49, 6, 23]
4	Lack of government support	[15, 3, 49, 4, 39, 6, 30]
5	Lack of acceptance of advancement in technology	[56, 25, 6, 27, 12]
6	Lack of support from top management	[41, 28, 12, 32]
7	Unwillingness of supplier to change towards GSCM	[56, 3, 27]
8	Too complex to implement GSCM	[28, 8, 23]
9	Lack of social responsibility	[39, 4, 30]
10	Uncertainty and competition in market	[27, 32, 4, 39, 23]
11	Lack of understanding among stakeholder of Supply chain	[28, 8]
12	Lack of technology infrastructure	[8, 23]
13	Lack of Knowledge and experience among suppliers	[8, 14, 12]
14	Poor organizational culture	[12, 49]
15	Lack of initiatives for Transport and Logistics	[39, 19, 28, 49]
16	Lack of sustainability of certification (ISO 14001)	[26, 4, 8, 6]
17	Resistance from employees	[42]
18	Resource management	[8, 37, 17]
19	Reduce involvement in environmentally related conferences	[4, 23, 24]
20	No specific environmental goals	[28, 19, 23]
21	Fear of failure	[19, 30, 24]
22	High investment and low return on investments	[19, 28, 24]
23	Customer desire for low price	[10, 34]
24	Lack of communication and information sharing among supply chain stakeholders	[8, [23]
25	Slow return on Investments (ROI) after implementing GSCM	[8, 23]
26	Lack of promotion of sustainable products	[53]
27	Lack of skilled human resource professionals in GSCM	[12, 39]
28	Non supporting nature of commercial bank	[48, 23]
29	Lack of strategic planning	[28, 19]

Literature Review for Drivers of GSCM

The factors which motivate to effective implementation of green supply practices for handling the environmental issues in various industries, are identified as drivers of GSCM. Therefore, there is an immense need to get the proper knowledge about these drivers for effective implementation of GSCM. Dhull and Narwal [15] reviewed number of papers related to GSCM to identify the most influencing drivers and barriers which affects the implementation of GSCM. They classified drivers in six categories such as internal, external, competition, market, customers, and suppliers. Tyagi et al. [54] implemented the ISM-MICMAC approach to find the interactions between the drivers of GSCM in automotive industries which are situated in National Capital Region in India. Eleven drivers were identified from the available literature and were leveled using ISM approach. This study found that legislative and regulatory compliance was at base position of model having high driving power, so the proper understanding and strict compliance came out as most important driver. In 2017, Singh et al. [46] also used ISM technique and prioritized eight drivers in fertilizer industry. The top management commitment was found the most important driver for implementation of GSCM in fertilizer manufacturing industry. Then, in 2018, again Singh et al. [45] performed research in fertilizer manufacturing industry in Punjab region of India wherein twelve indicators were found which motivate the industrial staff to adopt GSCM. Herein, ISM and MICMAC analysis was used to develop a model of drivers and to find contextual relationship among the drivers. Finally, 'government regulatory systems' and 'Top Management Support' were identified as the most important drivers of GSCM. Gardas et al. [18] explored the scope of GSCM for agro-industry and developed a model using ISM and MICMAC. Finally, as a result, three drivers, i.e., 'the environmental management', 'competitive pressure', and 'regulatory pressure', were identified as the crucial indicators out of total fourteen performance indicators which were identified through existing literature. Shohan et al. [44] chose chemical industry in Bangladesh to construct a hierarchy model of drivers of GSCM using total interpretive structural modeling, and MICMAC analysis was performed to evaluate the direct and indirect relationship of drivers. Initially, a total of eight drivers were considered out of which two most influencing factors were identified as the supplier pressure and willingness towards GSCM. Thaib [52] performed research in an automotive industry located in Indonesia, and concluded that four drivers such as 'competitor pressure', 'customer pressure', 'regulatory measures (ISO 140001 etc.)' and 'socio-cultural responsibility' were identified as the most important drivres. Furthermore, the confirmatory factor analysis was used to study the hypothesis. Agi & Nishant [5] made their area of research vast, and investigated various influential factors of GSCM in plastic, petrochemical mechanical and textiles industries loacted in the gulf countries. This study found nineteen factors that were further assessed using ISM-MICMAC analysis, and as a result dependence, and 'trust and durability of relationship with supply chain' factors had the highest influence on the implementation of GSCM. Diabat & Govindan [16] also used ISM-MICMAC tool to find key drivers in their research in an aluminium manufacturing industry located in south india. They identified 11 drivers and found that out of these factors two, i.e. 'government regulation & legislation' and 'reverse logistics' had the highest influence in adopting the green practices in supply chain management. Bey et al. [10] identified some most influencing drivers such as 'the legislative demands', 'customer demands' and 'competitive advantages' in manufacturing industries. As per the reported study of Narimissa et al. [33] in the oil and gas industry of Iran in 2020, 'the customer satisfaction' came out as the most affective driver. Meager et al. [31]

was found that the 'company culture', and 'media focus 'were the new drivers of GSCM which are not covered in detailed manner in available literature. By reviewing the selected papers, multifarious drivers motivating the implementation of GSCM are found. Several drivers of GSCM as identified by diffirent authors, are decribed in the Table 71.3.

Table 71.3: List of selected drivers of green supply chain management.

S. N.	Drivers of GSCM	Source
1	Customer awareness	[15, 44]
2	Government rules and legislation	[44, 16, 56]
3	Economic benefits	[44, 46, 45, 57]
4	Stakeholder and investor pressure	[15, 29]
5	Competitors	[44, 55, 52]
6	Top management support	[45, 46, 5]
7	Supplier collaboration and willingness	[18, 54]
8	Eco-friendly design	[16, 40]
9	Green options	[46, 43]
10	Demand from NGOs	[15]
11	Improve quality	[54]
12	Employee involvement	[10, 20]
13	Advance/IT technology implementation	[25, 45]
14	Brand image	[45, 13, 22, 10, 40, 11]
15	Social responsibility	[15, 52]
16	ISO 14001 certification	[54, 52]
17	Desire to reduce cost	[15]
18	Environmental subsidies	[15]
19	Reverse logistics	[16, 40, 43]
20	Knowledge and training	[18, 5]
21	Reducing energy consumption	[16, 35, 51]
22	Global marketing	[40, 11, 43]

Conclusion

This study finds various barriers and drivers faced during the implementation of GSCM in several industries. The present paper briefly reviews few of the studies related to GSCM which have been conducted in various industries of different countries using different techniques like ISM, AHP, and ISM fuzzy MICMAC to prioritize the factors that play important role in GSCM. Based on available literature, 'lack of government support' and 'cost implication' have been found as the most critical barriers in this paper which hinder the implementation of GSCM in industries, and 'support from the top management' has come out as the most influencing driver to motivate the implementation of GSCM. The findings of the present study will effectively contribute to understand the factors affecting the implementation of GSCM. Further, the researchers and practitioners can study these factors and implement GSCM in different industries. The research on GSCM has been getting more attention in different fields for last few years due to an impressive increase in the environmental considerations, but a limited research work has been reported to identify the GSCM practices in the medical equipment manufacturing industries. Therefore, the future research can be focused on medical equipment manufacturing industries to explore the scope of GSCM in this field also. Moreover, the techniques such as, Interpretive structural modelling (ISM) Fuzzy-MICMAC, Analytic Hierarchy Process (AHP), Decision making trial and evaluation laboratory (DEMATEL), fuzzy analytical hierarchy process (FAHP), structural equation modelling (SEM), fuzzy TOPSIS etc., can be used to prioritize the above-mentioned drivers and barriers for finding one or more most critical drivers and barriers.

References

1. Khan, S. A. R. (2019). Introductory Chapter: Introduction of green supply chain management. In Green Practices and Strategies in Supply Chain Management.
2. Khan, S. A. R., and Qianli, D. (2017). Impact of green supply chain management practices on firms' performance: an empirical study from the perspective of Pakistan. *Environmental Science and Pollution Research*, 24, 16829–16844.
3. Abebaw, H., and Virdi, S. S. (2019). Barriers for green supply chain management implementation: in ethiopia leather and leather product industry. *International Journal of Research and Analytical Reviews (IJRAR)*.
4. Afroz, R., Rahman, A., Muhibbullah, M., and Morshed, N. (2019). Malaysian automobile industry and green supply chain management. *International Journal of Recent Technology and Engineering*, 7(6), 158–162.
5. Agi, M. A. N., and Nishant, R. (2017). Understanding influential factors on implementing green supply chain management practices: an interpretive structural modelling analysis. *Journal of Environmental Management*, 188, 351–363.
6. Al-Refaie, A., Al-Momani, D., and Al-Tarawneh, R. (2020). Modelling the barriers of green supply chain practices in jordanian firms. *International Journal of Productivity and Quality Management*, 29(3), 397–417.
7. Badhotiya, G. K., Soni, G., Chauhan, A. S., and Prakash, S. (2016). Green supply chain management: Review and framework development. *International Journal of Advanced Operations Management*, 8(3), 200–224.
8. Balasubramanian, S. (2012). A hierarchical framework of barriers to green supply chain management in the construction sector. *Journal of Sustainable Development*, 5(10), 15–27.
9. Balon, V. (2020). Green supply chain management: pressures, practices, and performance—an integrative literature review. *Business Strategy & Development*, 3(2), 226–244.
10. Bey, N., Hauschild, M. Z., and McAloone, T. C. (2013). Drivers and barriers for implementation of environmental strategies in manufacturing companies. *CIRP Annals—Manufacturing Technology*, 62(1), 43–46.
11. Bhool, R., and Narwal, M. S. (2013). An analysis of drivers affecting the implementation of green supply chain management for the indian manufacturing industries. *International Journal of Research in Engineering and Technology*, 2(11), 2319–1163.
12. Dashore, K., and Sohani, N. (2013). Green supply chain management : a hierarchical framework for barriers. *International Journal of Engineering Trends and Technology*, 4, 2172–2182.
13. Deng, H., Yang, O., and Wang, Z. (2016). Considerations of applicable emission standards for managing atmospheric pollutants from new coal chemical industry in China. *International Journal of Sustainable Development and World Ecology*, 00(00), 1–6.

14. Dhull, S., and Narwal, M. S. (2016a). A state-of-art review on green supply chain management practices. *Accounting*, 2, 129–136.

15. Dhull, S., and Narwal, M. S. (2016b). Drivers and barriers in green supply chain management adaptation: a state-of-art review. *Uncertain Supply Chain Management*, 4(1), 61–76.

16. Diabat, A., and Govindan, K. (2011). An analysis of the drivers affecting the implementation of green supply chain management. *Resources, Conservation and Recycling*, 55(6), 659–667.

17. Dubey, R., Gunasekaran, A., Papadopoulos, T., and Childe, S. (2015). Green supply chain management enablers: mixed methods research. *Sustainable Production and Consumption*, 4, 72–88.

18. Gardas, B., Raut, R., Jagtap, A. H., and Narkhede, B. (2019). Exploring the key performance indicators of green supply chain management in agro-industry. *Journal of Modelling in Management*, 14(1), 260–283.

19. Govindan, K., Kaliyan, M., Kannan, D., and Haq, A. N. (2014). Barriers analysis for green supply chain management implementation in Indian industries using analytic hierarchy process. *International Journal of Production Economics*, 147, 1–14.

20. Hanna, M. D., Newman, W. R., and Johnson, P. (2000). Linking operational and environmental improvement through employee involvement. *International Journal of Operations & Production Management*, 20(2), 148–165.

21. Hsu, C. W., and Hu, A. H. (2008). Green supply chain management in the electronic industry. *International Journal of Environmental Science and Technology*, 5(2), 205–216.

22. Kannan, G., Haq, A. N., Sasikumar, P., and Arunachalam, S. (2008). Analysis and selection of green suppliers using interpretative structural modelling and analytic hierarchy process. *International Journal of Management and Decision Making*, 9(2), 163–182.

23. Kaur, J., and Awasthi, A. (2018). A systematic literature review on barriers in green supply chain management. *International Journal Logistics Systems and Management*, 30(3), 330–348.

24. Kaur, J., Sidhu, R., Awasthi, A., and Srivastava, S. K. (2019). A Pareto investigation on critical barriers in green supply chain management. *International Journal of Management Science and Engineering Management*, 14(2), 113–123.

25. AlKhidir, T., and Zailani, S. (2009). Going green in supply chain towards environmental sustainability. *Global Journal of Environmental Research*, 3(3), 246–251.

26. Kumar, D., Jain, S., Tyagi, M., and Kumar, P. (2018). Quantitative assessment of mutual relationship of issues experienced in greening supply chain using ISM-fuzzy MICMAC approach. *International Journal of Logistics Systems and Management*, 30(2), 162–178.

27. Luthra, S., Kumar, V., Kumar, S., and Haleem, A. (2011). Barriers to implement green supply chain management in automobile industry using interpretive structural modeling technique-an Indian perspective. *Journal of Industrial Engineering and Management*, 4(2), 231–257.

28. Majumdar, A., and Sinha, S. K. (2019). Analyzing the barriers of green textile supply chain management in South-east Asia using interpretive structural modeling. *Sustainable Production and Consumption*, 17, 176–187.

29. Mathiyazhagan, K., Datta, U., Bhadauria, R., Singla, A., and Krishnamoorthi, S. (2017). Identification and prioritization of motivational factors for the green supply chain management adoption: case from Indian construction industries. *Opsearch*, 55(1), 202–219.

30. Mathiyazhagan, K., Haq, A. N., and Baxi, V. (2016). Analysing the barriers for the adoption of green supply chain management—the Indian plastic industry perspective. *International Journal of Business Performance and Supply Chain Modelling*, 8(1), 46–65.

31. Meager, S., Kumar, V., Ekren, B., and Paddeu, D. (2020). Exploring the drivers and barriers to green supply chain management implementation: a study of independent UK restaurants. *Procedia Manufacturing*, 51(2020), 1642–1649.

32. Mudgal, R. K., Shankar, R., Parvaiz, T., and Tilak, R. (2010). Modelling the barriers of green supply chain practices. *International Journal Logistics Systems and Management*, 7(1), 81–107.

33. Narimissa, O., Kangarani-Farahani, A., and Molla-Alizadeh-Zavardehi, S. (2020). Drivers and barriers for implementation and improvement of sustainable supply chain management. *Sustainable Development*, 28(1), 247–258.

34. Orsato, R. J. (2016). Competitive environmental strategies: when does it pay to be green?. *California Management Review*, 48(2), 127–143.

35. Paulraj, A. (2009). Environmental motivations: a classification scheme and its impact on environmental strategies and practices. *Business Strategy and the Environment*, 18(7), 453–468.

36. Pil, F. K., and Rothenberg, S. (2003). Environmental performance as a driver of superior quality. *Production and Operations Management*, 12(3), 404–415.

37. Porter, M., and Van der Linde, C. (1995). Green and competitive: ending the stalemate. *The Dynamics of the Eco-efficient Economy: Environmental Regulation and Competitive Advantage*, 33, 120–134.

38. Rao, P. H. (2019). Green supply chain management : a study based on SMEs in India. *Journal of Supply Chain Management*, 8(1), 15–24.

39. Ratna, S., and Kumar, B. (2019). Evaluating the green supply chain management barriers in a SME: an AHP approach. *International Journal of Mechanical and Production Engineering Research and Development*, 9(2), 271–276.

40. Raut, R. D., Narkhede, B., and Gardas, B. B. (2017). To identify the critical success factors of sustainable supply chain management practices in the context of oil and gas industries: ISM approach. *Renewable and Sustainable Energy Reviews*, 68(June 2016), 33–47.

41. Ravi, V., and Shankar, R. (2005). Analysis of interactions among the barriers of reverse logistics. *Technological Forecasting and Social Change*, 72(8), 1011–1029.

42. Sarkis, J., Zhu, Q., and Lai, K. H. (2011). An organizational theoretic review of green supply chain management literature. *International Journal of Production Economics*, 130(1), 1–15.

43. Sellitto, M. A., Bittencourt, S. A., and Reckziegel, B. I. (2015). Evaluating the implementation of GSCM in industrial supply chains: two cases in the automotive industry. *Chemical Engineering Transactions*, 43, 1315–1320.

44. Shohan, S., Ali, S. M., Kabir, G., Ahmed, S. K. K., Suhi, S. A., and Haque, T. (2019). Green supply chain management in the chemical industry: structural framework of drivers. *International Journal of Sustainable Development and World Ecology*, 26(8), 752–768.

45. Singh, M., Jawalkar, C. S., and Kant, S. (2018). Analysis of drivers for green supply chain management adaptation in a fertilizer industry of Punjab (India). *International Journal of Environmental Science and Technology*, 16(7), 2915–2926.

46. Singh, M., Kant, S., and Jawalkar, C. S. (2017). Development of interpretive structural modeling: an approach for fertilizer industries. *I-Manager's Journal on Mechanical Engineering*, 7(2), 35.

47. Singh, P., and Kumar, V. (2017). Quantitative analysis of drivers affecting green supply chain management in Rajasthan SMEs. *International Journal of Process Management and Benchmarking*, 7(3), 332–353.

48. Soni, G., Prakash, S., Kumar, H., Singh, S. P., Jain, V., and Dhami, S. S. (2020). An interpretive structural modeling of drivers and barriers of sustainable supply chain management: a case of stone industry. *Management of Environmental Quality: An International Journal*, 31(5), 1071–1090.

49. Srivastav, P., and Gaur, M. (2015). Barriers to implement green supply chain management in small scale industry using interpretive structural modelling technique—a north Indian perspective. *Journal of Advances in Engineering and Technology*, 2(2), 6–13.

50. Srivastava, S. K. (2007). Green supply-chain management: a state-of-the-art literature review. *International Journal of Management Reviews*, 9(1), 53–80.

51. Sundarakani, B., De Souza, R., Goh, M., and Shun, C. (2008). Measuring carbon footprints across the supply chain. In 13th International Symposium on Logistics (ISL2008): Integrating the Global Supply Chain, Nottingham University Business School, UK, (pp. 555–562).

52. Thaib, D. (2020). Drivers of the green supply chain initiatives: evidence from indonesian automotive industry. *Uncertain Supply Chain Management,* 8(1), 105–116.

53. Tumpa, T. J., Ali, S. M., Rahman, M. H., Paul, S. K., Chowdhury, P., and Khan, S. A. R. (2019). Barriers to green supply chain management: an emerging economy context. *Journal of Cleaner Production,* 236, 117617.

54. Tyagi, M., Kumar, P., and Kumar, D. (2015). Analysis of interactions among the drivers of green supply chain management. *International Journal of Business Performance and Supply Chain Modelling,* 7(1), 92–108.

55. Uddin, S., Ali, S. M., Kabir, G., Suhi, S. A., Enayet, R., and Haque, T. (2019). An AHP-ELECTRE framework to evaluate barriers to green supply chain management in the leather industry. *International Journal of Sustainable Development and World Ecology,* 00(00), 1–20.

56. Walker, H., Di Sisto, L., and McBain, D. (2008). Drivers and barriers to environmental supply chain management practices: lessons from the public and private sectors. *Journal of Purchasing and Supply Management,* 14(1), 69–85.

57. Zhu, Q., Geng, Y., Fujita, T., and Hashimoto, S. (2010). Green supply chain management in leading manufacturers: Case studies in Japanese large companies. *Management Research Review,* 33(4), 380–392.

72 Utilization of waste HCl for the manufacturing of aluminum chloride from bauxite

Vishal Kumar U. Shah[1,a] and Pratima Wadhwani[2,b]

[1]Chemical Engineering Department, SNPIT & RC, Umrakh, Bardoli, Surat

[2]Chemical Engineering Department, School of Chemical Engg and Physical Sciences, Lovely Professional University, Phagwara, Punjab, India

Abstract: In today's era purification and recovery of waste water is biggest issue. For those various types of chemicals and recovery techniques are used and many more are developing. In the present paper, a new technique for the manufacturing of Aluminum chloride using spent HCl and bauxite has been developed. This process allows using of spent Hydrochloric acid (HCl) and thus large amount of effluent can be retreated. This would be much beneficial for the conservation of environment. The spent HCl contains 35% concentration by volume which is reacting with pretreated Bauxite. The process is carried out at atmospheric pressure and at elevated temperature. One can obtain 30% purity of Aluminum chloride using this process and raw material. The process has been modeled by parameters like time and temperature using conversion as the response using Box-Behnken Design (BBD) to get R-sq. as 99.35% which says that the model is in good agreement with the experimental data.

Keywords: Aluminum chloride, bauxite, polyaluminium hydrochloride, waste hydrochloric acid.

Introduction

In the present work, the process for the preparation of Polyaluminium hydrochloride complexes having the general formula $Al_n(OH)mCl_3nm$ has been discussed in a novel way. The use of aluminum hydroxide with hydrochloric acid at a temperature of 130°C–180°C for 1.5–2 hours has been done under atmospheric pressure [1]. The Polyaluminium hydrochloride product can be used in paper production, water treatment as it provides high coagulation efficiency. The study reports the production of aluminium chloride using HCl. The present methodology is more focused on the methods that reduce the cost of production for aluminium chloride compared to the widely used methods. Treating industrial waste water and removing color from it is a major problem, which can be easily sorted out using Aluminum Chloride as coagulating agent. In some industrial waste water treatment, Polyaluminium Chloride is preferred over Aluminium Chloride, which requires inorganic polymerization using bauxite and spent HCl. For producing aluminium chloride using bauxite and HCl, in autoclave which itself releases heat, care should be taken that the

[a]vishalshah16286@gmail.com, [b]drpratiitriitk@gmail.com

temperature should not cross beyond 1500°C that takes 7–9hrs. Also, during this process, explosive hydrogen gas is released in a continuous manner.

It is well known that a large amount of spent HCl is removed which cannot be utilized. In this regard, the use of waste HCl to produce Aluminium chloride could be a better option. It has been found that highly reactive amorphous powders are obtained by treatment with concentrated hydrochloric acid of metakaolin or kaolin, followed by removal of the liquid by distillation and calcination of the residue. The porous silica was homogeneously mixed with chloride-containing alumina begins to react with gaseous carbon tetrachloride at 450 K. At temperatures above 700K, aluminium chloride formed has high yield and selectivity [2] as Kaolinite clays are potentially a vast domestic resource for aluminum. Utilization of these resources could decrease or eliminate the nearly complete dependence of the United States on foreign raw materials for this important metal. Furthermore, processing of clay to aluminum through anhydrous chloride metallurgy could reduce the high electrical energy requirements of the conventional Hall-Heroult aluminum reduction process. Several anhydrous chloride processes have been proposed; however, unresolved technical problems have prevented their commercialization [3]. An acceptable chemical means has not been found to extract aluminum from clay as a highly pure anhydrous aluminum chloride. The Bureau of Mines report identifies and discusses the important chemical problems involved in achieving an acceptably rapid, self-heating, selective chlorination reaction and the subsequent separation of iron chloride byproduct from the anhydrous aluminum chloride [4–6].

Materials and Method

Bauxite was purchased from the S.D fine chemicals and waste HCl was brought from the industries. Typical yields give 52.60% of conversion when 23 g of bauxite and 60 ml of waste Hydrochloric acid was used. The raw materials used in the present study are shown in Figure 72.2.

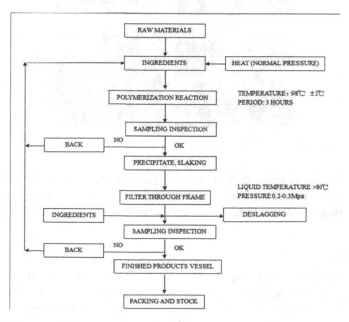

Figure 72.1 Block diagram of the experimental process.

The process takes place at 110°C temperature and 1.3 atm pressure with constant stirring for 3.5 hour. If one go, if 12.78 gm of obtain product the pure product was about 10 gm and other 2.78 gm was product with impurity of HCl with complex structures. One modification in the process can be made with regards to temperature that if the process takes place at 1300°C and 2 atm pressures; we get very high yield of Aluminum Chloride and conversion of raw materials. But at 1500°C temperature and 4.7 atm pressure; we get high purity and basicity of product. It simply shows that elevation in temperature gives higher purity. But the utility required for higher purity is comparatively very high than lower purity product. Rather than manufacturing highly pure product, production and purification of lower quality product is more economical. At lab scale, three different runs were taken as shown in Table 72.1 from Figure 72.1. The process takes place at 1300°C and 2 atm pressure; we get very high yield of Aluminum Chloride and conversion of raw materials. But at 1500°C temperature and 4.7 atm pressure; we get high purity and basicity of product [7–8]. In first run, 75 g/mole of bauxite and 184 ml (excess) of HCl were reacted at atmospheric pressure and at 70°C for a period of 1.5hrs on magnetic stirrer. The whole mass was then filtered to adjust the pH from 3–4 using aluminum hydroxide. It was cooled in ice to obtain 30% conversion.

$$\text{Bauxite} + \text{HCl} \longrightarrow \text{AlCl}_3 + x\text{H}_2\text{O}$$

The above steps were repeated for second run where, 59.987 g/mole of bauxite and 92 ml (excess) of HCl were reacted for 2hrs, keeping temperature of 80°C. The pH was set to 3–4 using aluminum hydroxide to get conversion of 41.675%.

Figure 72.2 Raw material.

Similarly the third run was taken by reacting 23 g/mole of bauxite and 60 ml (excess) of HCl keeping temperature of 110 °C for a period of 3.5hr and keeping pH of 3–4. The product was cooled in ice and here we get conversion of 52.60% in reaction.

Model Summary

The modeling was done using the Box-Behnken Design (BBD) method where the temperature and time were used as the variable from Table 72.1. The obtained R-sq value is 99.35% which says the model is in good agreement with the experimental data. Also Figure 72.3 shows the contour plot of yield with respect to temperature and time and Response plot also is also depicting the same.

Table 72.1: Different practical run table.

Sr. No	Test	Temperature (°C)	Pressure (atm)	Time (hour)	Result (gm)	Conversion (%)
1	Bauxite=1mol=119.975 g/mol HCl = 6 mol = 184 ml(excess)	70	1	1.5	36	30.00
2	Bauxite = 0.5 mol = 59.987 g/mol HCl = 3 mol = 92 ml(excess)	80	1	2	25	41.675
3	Bauxite = 0.191 mol = 23g/mol HCl = 1.959 mol = 60 ml(excess)	110	1	3.5	12.78	52.60

Table 72.2: Model summary.

S	R-sq	R-sq(adj)	R-sq(pred)
0.801858	99.35%	98.88%	95.36%

Table 72.3: Coded coefficients.

Term	Coef	SE Coef	T-Value	P-Value	VIF
Constant	45.200	0.359	26.05	0.000	
Temp	8.866	0.283	31.27	0.000	1.00
time	1.734	0.283	6.12	0.000	1.00
Temp*Temp	−1.925	0.304	−6.33	0.000	1.02
time*time	−0.925	0.304	−3.04	0.019	1.02
Temp*time	−0.900	0.401	−2.24	0.060	1.00

Table 72.4: Analysis of variance.

Source	DF	Adj SS	Adj MS	F-Value	P-Value
Model	5	685.111	137.022	213.11	0.000
Linear	2	652.879	326.440	507.70	0.000
Temp	1	628.828	628.828	978.00	0.000
Time	1	24.051	24.051	37.41	0.000
Square	2	28.992	14.496	22.55	0.001
Temp*Temp	1	25.778	25.778	40.09	0.000
time*time	1	5.952	5.952	9.26	0.019
2-Way Interaction	1	3.240	3.240	5.04	0.060
Temp*time	1	3.240	3.240	5.04	0.060
rror	7	4.501	0.643		
Lack-of-Fit	3	4.501	1.500	*	*
Pure Error	4	0.000	0.000		
Total	12	689.612			

Table 72.5: Regression equation in uncoded units.

C7	=	−53.92 + 1.422 Temp + 10.41 time − 0.004812 Temp*Temp − 0.925 time*time − 0.0450 Temp*time

Result

The synthesis of the aluminium chloride was successfully made which has the capability to remove the color and produce sludge. The yield obtained in this case is 52.60 %. The experimental data was then used for modeling to use BBD response surface method to study the effect of time and temperature on the yield. Table 72.2 shows the coded coefficient where the model variables are in good agreement with the experimental data as the value of P is all the relations is less than 0.05 whereas for Temp.time value is 0.06. Table 72.3 shows the Model summary where the R-sq value is 99.35%, Table 72.4 shows the Analysis of variance data where the two way interaction data for time temp is 0.06 and rest values of P is less than 0.05 which shows that the experimental and model data are in good fit.

The Regression Equation in Uncoded Units is shown in Table 72.5 where the response C7 is the conversion and is related to the variables time and temp by the stated equation. Contour plot of response with respect to time and temperature is shown in Figures 72.3 and 72.4 shows the surface plot of response with respect to time and temperature where it is seen clearly that as the temperature and time, both are increased the response C7 (conversion) is increasing.

Conclusion

The proposed method utilizes the waste HCl and bauxite to convert into aluminum chloride. The Aluminum chloride produced from this method can be used for the treatment of effluent in ETPs. This method would be beneficial if used at large scale. After utilizing this

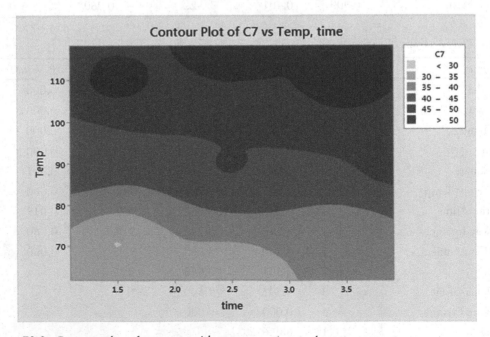

Figure 72.3. Contour plot of response with respect to time and temperature.

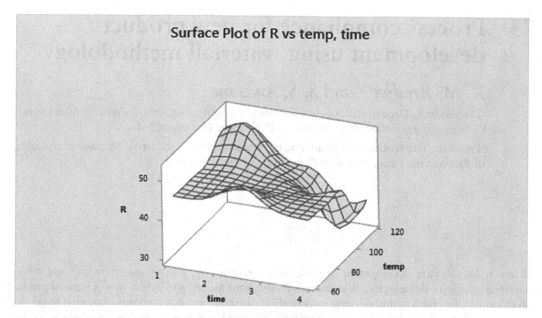

Figure 72.4. Surface plot of response with respect to time and temperature.

methodology, one concludes that using this method one can reduce waste of HCl in industries and can reuse it to produce mentioned product (Aluminum chloride). HCl has detrimental effects on the water bodies. Hence proposed method can be used in those industries where, waste HCl being produced as side product of low concentration.

References

1. Blumenthal, G., and Wegner, G. (1989). Preferential formation of aluminum chloride by chlorination of a highly reactive powder obtained from metakaolin or kaolin by hydrochloric acid treatment. *Reactivity of Solids*, 7(2), 105–113.
2. Landsberg, A. (1983). Aluminum from Domestic Clay Via a Chloride Process: The State of the Art.
3. Luck, R., and Wegner, G. (1980). Untersuchungen zur Chlorierung von Metakaolin im heterogenen System Gas/Feststoff, Dissertation A, Central Institute of Inorganic Chemistry of the Academy of Sciences G. D. R., Berlin.
4. Blumenthal, G. (1987). Zur Bedeutung der anorganischen Chlorierung: Chlorierung von Aluminium-Rohstoffen und-Verbindungen, Dissertation B, Central Institute of Inorganic Chemistry of the Academy of Sciences G.D.R., Berlin.
5. Chao, T., and Sun, S. C. (1967). Trans. Sot. Min. Engin. Dec., 420.
6. Grob, B. (1983). Selektive Chlorierung von Aluminiumoxid in Kaolinit, Dissertation, Eidgenoss. Techn. Hochsch. Zurich.
7. Grob, G., and Richarz, W. (1984). Chlorination of alumina in kaolinitic clay. *Metallurgical and Materials Transactions B*, 15, 529–533.
8. Eisele, J. A., Bauer, D. J., and Shanks, D. E. (1983). Bench-scale studies to recover alumina from clay by a hydrochloric acid process. *Industrial & Engineering Chemistry Product Research and Development*, 22(1), 105–110.

73 Process compliance for new product development using waterfall methodology

K. M. Pandya[1,a] and S. S. Anasane[2]

[1]PG student, Department of Manufacturing Engineering and Industrial Management, College of Engineering Pune, Savitribai Phule Pune University, India

[2]Professor, Department of Manufacturing Engineering and Industrial Management, College of Engineering Pune, Savitribai Phule Pune University, India

Abstract: In software development or information technology, project managers can use project management methodologies like the waterfall model, agile model, and hybrid model for new product development. But for a manufacturing organization, there are no such methods or models suggested. This research has been done for the manufacturing organization for a hardware product. At first, the methodology for new product development using waterfall methodology has identified. After identifying methodology documentation, part of each phase is specified, depending on product and market research. For each stage, separate documents are selected, leading every team member to connect with the new product and fulfill each requirement. The main challenge was the uncertainty of conditions during new product development. This challenge is solved in the case study given in this research paper. The case study used was on the product named cycle controller of electronics company. This research paper focused on the methodology used in new product development from idea generation to commercialization. This study provides theoretical support to new product development using the waterfall methodology. Finally, discussions and future scope of opportunity were discussed.

Keywords: New product development, waterfall methodology.

Introduction

The waterfall model is the plan-driven approach that needs human cooperation, communication, and collaboration for its implementation. A limited basic design was sufficient to bring to the end-user [1]. The scope of this study is to use waterfall methodology for new product development for an organization in the electronics sector. The primary purpose of this project is to generalize the documentation required for new product development. By doing this product development process gets organize. The nature of this Waterfall model is to complete each phase before starting the next step ideally.

Consumers' habits are continually changing due to volatile economies, external uncertainties, and many other reasons. Therefore, NPD firms have two ways to build their core NPD capability by either enhancing their technology or maintaining coordination capability. The central part of new product development to maintain its quality by doing effective resource allocation and cross-functional integration in the process [4].

[a]pandyakm19.prod@coep.ac.in

Understanding the market structure and knowledge of business activities is essential for more efficient manufacturing and distribution of products and services [2]. The mistake involved in any phase tends to delay the project schedule, and so that it is a costly model. Proper documentation is needed to prevent this from developing a new product, which helps maintain partnerships and collaborations with suppliers and helps the team stay on track [5]. Different types of documents are necessary for the smooth process flow of new product development, consisting of pricing and marketing strategy for commercializing products in the market [9]. New product development plays an essential role in building relationship interaction and collaboration with a supplier. Developing new products comes with different kinds of opportunities for suppliers as well. The organization must strengthen and reinforce supplier interaction which implies innovativeness [3]. The authors propose a new model, which allows evaluating every stage of the innovation process, identifying the main problems that improve innovation [6]. Some semi-structured interviews with integration and testing engineers and project team members helped in new product development for root cause analysis. Project managers need to do different feasibility studies for new product development explained in this paper [8]. How this feasibility can be helpful in the process of making a roadmap for product development considering outside and inside changes in the product, adaptability is crucial for better execution [11].

Learning surroundings has a remarkable and positive impact on the knowledge areas and new product development processes. Likewise, knowledge processes have a unique and positive impact on new product development processes [10].

Empirical research comparing agile and waterfall models and their impact on the successful completion of new product development processes must be done [7].

Adopted Methodology

There is always an important discussion topic in meetings for a manufacturing organization about whether to use waterfall or agile for new product development. For a manufacturing organization, it is crucial to choose which method for new product development is more profitable and how that method will impact the organization's financial sheets.

New Product Development using Waterfall Model

Each company's perspective to manufacture products is significantly different from the others from the same domain. In addition, their processes are substantially different from others depending upon the product and many other factors.

There are eight steps for new product development using waterfall methodology identified for the product where scope and requirements are defined. The same is given below:

Idea generation for new products: There are some methods for generating the new idea that includes internal sources, market research, Competitors, SWOT analysis, etc. The main objective of this step is to come up with the idea that separates business from the competition through some competitive advantage. At this stage, there should not be any limit on ideas.

Idea evaluation and feasibility study: The purpose of idea screening is to reduce the number of ideas generated in the previous phase and test them with different organizational capabilities. A project manager should finalize the concept by doing a technical, financial, and operating feasibility study.

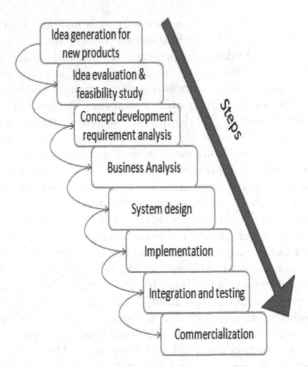

Figure 73.1 Steps for NPD using waterfall methodology.

Business Analysis: This analysis requires constructing a business requirement document to provide information about 'what needs to be done'. Business analysis can guide the organization throughout the roadmap of development and also helps not to distract from the scope.

Concept development and requirement analysis: Concept analysis should define why this product needs to be built and how it can impact internally and externally. Functional requirement documents with many other forms are used for how it needs to be done. Both of these analyses will provide a roadmap for the development of the product.

System Design: This phase consists of designing the architecture, hardware, software, and other environments with a compatibility test of the product to be made. A system design specification (SDS) document has to be made in which there are all possible components and their alternatives if possible. SDS document also provides a better understanding of hardware, software, layout, etc., required for the product to be made.

Implementation: This phase divides the product into small tasks to be performed and small units that can be integrated later. These units need to be tested positive for further process.

Integration and Testing: At this stage, all the units should be tested ok. For integrating these units, firms need to follow specific processes that provide a positive response. Integrating these units requires better cross-functional team involvement that provides a better perspective from all domains. Testing is done on the basis of Figures 73.1–73.3.

Commercialization: This phase consists of bringing the newly developed product into the market and providing all data to the sales and marketing team to be there in the customer's hands with the best possible convenience.

Case Study

Electronics company (EC) provides innovative development and manufacturing solutions for electric vehicle electric control modules and systems. EC has more than 25 clients for BLDC motors and controllers. In 2020, senior authorities had approved the business case for the cycle controller and planned manufacturing as new product development using waterfall methodology. Therefore, our research focused on the cycle controller. The data gathered from structured, semi-structured, and unstructured interviews, observations, market research, and documentation analysis has been written down in transcripts and systematic summaries to be done. Therefore, it was understandable that this project will need a waterfall approach to achieve its goals and objectives. However, some pitfalls in the waterfall approach may be defining practical necessities within the project's scope.

The idea of producing a cycle controller came from a customer requirements and market research. For idea evaluation, EC has used many techniques such as SWOT analysis, five forces, cost-benefit analysis, etc.

Eventually, project team member involvement is crucial to achieving end-user acceptance at the end. The business analysis phase helped the maker to have a separate business model for the cycle controller. A system design specification (SDS) document has been made for the cycle controller, giving an overview of what needs to be built and how it needs to work. In the implementation part, the cycle controller is divided into unit parts which can be developed and tested. Some features required more calibration at the unit stage. By integrating all these units, a cycle controller has been designed. The testing team has tested the controller and found it Ok.

The commercialization phase manufacturer has made more than 20 cycle controllers and dispatched them to the customers for feedback before putting them into production.

Analysis

Here are given project time analysis, project cost analysis, and project quality analysis which gives a better idea about how time, cost, and quality has improved by applying this framework.

Project Time Analysis

By having data of earlier products that have been developed at the company, one of the products was BLDC controller (rated 1kW). The manufacturer took nine months to build a BLDC controller, and the primary reason was they had not followed any specific method. After developing the cycle controller, we can see that the actual time taken by this project is nearly seven months.

Project Cost Analysis

By comparing the cycle controller with the BLDC controller made by the EC, the BLDC controller has developed at the cost of more than 15 lack. In contrast, a cycle controller has developed at the expense of 10 lack. Thus, the cycle controller has grown at a lower price comparing other EC products, which was the main objective of implementing this framework.

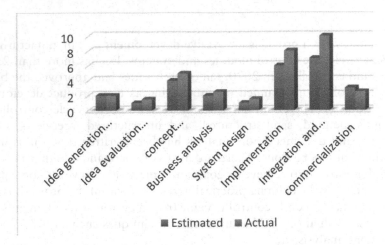

Figure 73.2 Project phase time analysis (estimated and actual).

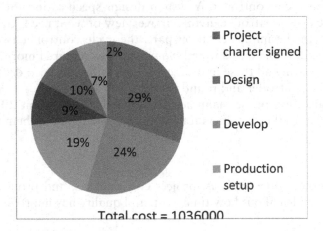

Figure 73.3 Project baseline cost distribution.

Conclusion

The main research questions in this study were: How does a new product development using waterfall methodology work? The data gathered was unstructured but still good enough to gain insights into process flow and product development strategy.

This methodology works better when the matured product needs to be improved. This improved product with the added feature is called a new product. In the EC case, the grown product was a BLDC controller, and by adding some features and improving from the software part, the new controller can be named a cycle controller. The critical factor was getting team members connected with the project as this framework had more documentation. Some documents need approval from all the departments to go into the other process. These phase gates are nothing but meetings with the customer at the end of each phase completed. The phase gates put the customer in the loop, which improves product development. EC used this methodology to develop a cycle controller from the same BLDC controller segment. By applying this methodology, EC has reduced development time by 38 days and cost reduced by 18%. In EC, this framework has turned out to be sufficient and resulted in the successful end-user acceptance of the cycle controller. Interestingly, the impact of requirement uncertainty and modifications in practical situations was relatively less than expected at the start of this project.

References

1. Ruël, H. J. M., and Bondarouk, T. (2010). The waterfall approach and requirement uncertainty: a case study. *International Journal of Information Technology Project Management,* 1(2), 43–60.
2. Petersen, K., Wohlin, C., and Baca, D. (2009). The waterfall model in large scale development. In Product-Focused Software Process Improvement: 10th International Conference, PROFES 2009, Oulu, Finland, June 15-17, 2009. *Proceedings 10,* 386–400. Springer Berlin Heidelberg.
3. Bhuiyan, N. (2011). A framework for successful new product development. *Journal of Industrial engineering and management,* 746–770.
4. Jayaram, J., and Narasimhan, R. (2007). The influence of new product development competitive capabilities on project performance. *IEEE Transactions on Engineering Management,* 54(2), 241–256.
5. Wuang, M. S., and Chiang, S. M. (2011). Activities and problems in new product development process in the networking industry—a case of different business models. *In 2011 IEEE International Conference on Industrial Engineering and Engineering Management,* 664–668. IEEE.
6. Igavens, M., Amantava-Salmane, L., and Silineviča, I. (2016). Research of the new product development processes. *Latgale National Economy Research,* 3.
7. Schuh, G., Rebentisch, E., Riesener, M., Diels, F., Dölle, C., and Eich, S. (2017). Agile-waterfall hybrid product development in the manufacturing industry introducing guidelines for implementation of parallel use of the two models. *In 2017 IEEE International Conference on Industrial Engineering and Engineering Management (IEEM),* 725–729. IEEE.
8. Bausea, K., Radimerskya, A., Iwanickia, M., and Albersa, A. (2014). Feasibility studies in the product development process. In 24th CIRP Design Conference, (pp. 473–478).
9. Huang, S. Y. (2016). Building up new product development strategy by product pricing and marketing analysis chart. In IEEE International Conference IEEE-ICAMSE.
10. Jing, N., and Yang, C. (2010). The Interrelationship among learning environment, knowledge process, and new product development performance. In Proceedings of the 2010 IEEE IEEM.
11. Martinich, L. P. (2016). Excellent execution in new product development: adaptability and resilience. *IEEE Engineering Management Review,* 44(3), 20–22.

74 Applications of composite materials in the field of medical implants: a review

Jitendra Rajput[1,a] and Rajesh Kumar Bhushan[2,b]

[1]Research Scholar, Mechanical Engineering Department, N.I.T Manipur, Imphal, Manipur, India

[2]Professor, Mechanical Engineering Department, N.I.T Manipur, Imphal, Manipur, India

Abstract: Composite material is the combination of two or in excess of two materials having unlike composition which combine at macroscopic scale and are not soluble to each other. This combination of materials develops new properties or improved the mechanical, physical and other properties. These composite materials are combined to achieve distinct applications. Due to singular nature many of composite materials are commercialized some of these materials are under study for the use in medical implant applications like knee implant, heart valve, dental implant, tendon repair, hip implant, bone and artificial skin. So this investigation is the review of composite material application in medical implants. In the recent of advance medical research, the polymer composite and bio composite materials can be used for making artificial human body organs. Composite material coatings are used in medical implants and scaffolds with different aims. These aims are dependent on the purpose and the location of implants. Once implanted in the body, the composite materials are expected that they have properties according to function required from the implant.

Keywords: composite, macroscopic, medical implants.

Introduction

The immediate increase in the population of world over the eras has elevated the requirement of medical transplants for damaged human body parts by definite diseases. In the existing research, initially a bio-composite material premeditated and developed according to the necessity of medical implants [24]. Composite materials are referring to the combination two or more material with different composition at macroscopic level. The composite material has two phases first is continuous phase i.e. matrix phase and second is discontinuous phase i.e. filler material. The two phase have special properties when works together macroscopically. It is the heterogeneous combination of different material having different properties and morphology at macroscopic level [1, 2]. The properties of composite materials are not same as their separate constituents from which they are made. They have excellent corrosion resistance, light weight, higher fatigue strength, extraordinary tensile strength, and great load bearing ability which are the main advantages in different applications [3]. Although, a number of surveys have been done in the area of composite materials metal alloys aimed at biomedical uses such as knee, hip and

[a]jitendrarajput16@gmail.com, [b]rkbnitm@gmail.com

teeth transplants [27]. The purpose of this study is to pull the attention about the use and benefits of polymer composite materials which relate to manufacturing technologies. The polymer composite materials are excellent with biocompatibility so these can be used in various medical applications [4]. The polymer matrix provides the protection to the filaments from mechanical damages and degradation caused by environmental effects. The bio fibers or filaments are the filaments which mainly extracted from biological origin like crops, wood and regenerated cellulose. The bio-fibers are biocompatible and recyclable to the environment, cheaper, recyclable, light weight, and decomposable in nature. There are two types of woods one is hardwood and second one is softwood [5]. The fibers containing cellulose and lignin have better mechanical and physical properties, these fibers are non-wood fibers. Non-wood fibers are used in industrial application purpose but due to −OH group these fibers absorb water and causes swelling of product while contacting with water [6]. Decomposable materials are broadly studied for stents and bone transplants. Biodegradable materials can be totally despoiled in the human body in comparison to permanent implant. Number of biodegradable materials calcium alloys, magnesium alloys, zink alloys and ceramics have been industrialized for biomedical uses. Poly L-lactic acid (PLLA)-magnesium composites have been broadly studied as probable ecological materials for bone implantations [23]. A thermoplastic organic polymer polyether ether ketone (PEEK) is becoming popular because of its applications for manufacturing of human body implants medical tools. These PEEK material implants have wide range of important properties and unique qualities [25]. There is the problem of infection associated with biomaterials. So more thoughts and research have been focused on developing super hydrophobic surfaces of implants over biomaterials to reduce bacterial bio film formation to eliminate the biomaterial associated infection [28]. In recent years the natural polymer widely discovered for their biomedical applications, including cellulose and collagen. The structure features of Bacterial cellulose (BC) offer an ideal environment for developing composites which used for cancer treatment [29]. Ferromagnetic composites using magnetic particles fixed in the polymer matrix are paying attention from scientific viewpoint in the present time. Integration of particles, reinforcements of nanomaterial into polymers permits the creation of polymer matrix composites and these composite shows excellent functionality and mechanical performance [30].

Applications in Dentistry

The rigid part of teeth tissues, mainly enamel and dentine are made of collagen and hydroxyapatite microcrystals which are natural composites. Polymeric matrix composites materials are mainly used in dentistry. Polymer matrix composites are used to replace damaged the tooth, fill cavities and also to restore fractured tooth [1].

Dental Restorative Materials

Composite materials for restoring posterior and anterior of teeth in the clinical practices. Dental composites consist of a methacrylic matrix reinforced with particles of ceramics are used in most applications. Amalgams have excellent compressive strength, used for filling of posterior teeth [1].

The applications of fibrous composite materials are increased widely and has been increasing in volume also. The composite materials needs to be aesthetic in dental application, in addition to biocompatible. There are some major factors which influenced the usage of fibrous composite in dentistry;

- Non-corrosive: In comparison to metal alloy the polymer matrix composites are less corrosion susceptible.
- *Aesthetics:* every type of dental shades and structure can be achieved using the transparent matrix system and variety of ceramic fiber.
- *Toughness:* the nature of polymer matrix is not brittle. So the problem of breakage is reduced.
- *Metal allergy:* due to the presence of metal in the body some people show an allergic responses. But the dental parts made by polymer composites eliminate the allergic reactions [7–9].

The main duty of the dental post is to be responsible for holding to the core of the tooth as shown in Figure 74.1(a).

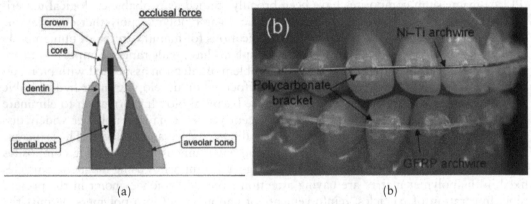

(a)   (b)

Figure 74.1 (a) An endodontically treated teeth using a dental post [7]; (b) Comparison of glass-fiber-reinforced plastic (GFRP) arch wire and nickel-titanium (Ni-Ti) arch wire [11].

The force is conveyed to dentin through post and core while applying the occlusial force to the crown. Tapered and parallel are two main geometries typrs of post. The stiffness decrease, of a post using tapered geometry [10].

Problems of Using Metals as Dental Implants

Metals have excellent mechanical characteristics in comparison to polymeric materials however corrosion resistance, aesthetic and biocompatibility of material are mainly concern. Implants of stainless steel are less corrosion resistive but have better mechanical properties and biocompatibility is also one of the challenges. Titanium and zirconium have better corrosion resistance so these materials are used widely in dental implants but the allergic reaction and biocompatibility is the main concern of titanium. The aesthetic appearance is also one of the concerned parameters in dental implant applications where polymer composites are better. Ni-Ti arch wire and Glass fiber reinforced polycarbonate composites arch wire are compared in Figure 74.1(b) [11].

Applications in Oral and Maxillofacial Surgery

The main causes for maxillofacial surgery are implant of lost teeth, replacement of facial bone, restoration of temporomandibular joint and fixation of broken facial bones.

Dental Implants

The applications of dental implantations are to eliminate the complications related to the bridges and changeable dental prosthesis. Research is going on to develop composite materials for the application as endosseus element. The composite materials which made of silicon carbide (SiC) and carbon fiber reinforced carbon have some benefits over metal alloys and ceramics. These composites combine sufficient strength having low modulus of elasticity, have similarity to natural teeth, effect of stress field minimizes and host response. They have excellent fatigue properties [1].

Cranial Bones Repair

Cranial bone defects are frequently faced in clinical practice and there are various factors that causes for defects, infections, and bone implantations harvesting techniques. Removal of these imperfections necessary for brain safety. Mostly cranial defect restored using original bone flaps, relocated bone flaps cause for rise of infection. So this one of the challenge for scientists that the supreme material for cranioplasty is not existing [12,13]. In present, the standard material for cranioplasty are autologous bone grafts but availability is limited [14].

Applications in Tissue Engineering

Traditionally the damaged tissue were repaired by bone grafting. Plastic polymers are being considered as alternatives to these treatments in tissue engineering. Biodegradable material can be used as temporary substitute material for damaged tissues. The material which shows the ability for bone repair are biodegradable polymers. Polycaprolactone (PCL), poly(glycolic acid) (PGA) and poly(lactic acid) (PLA) are the synthetic biodegradable polymers [15].

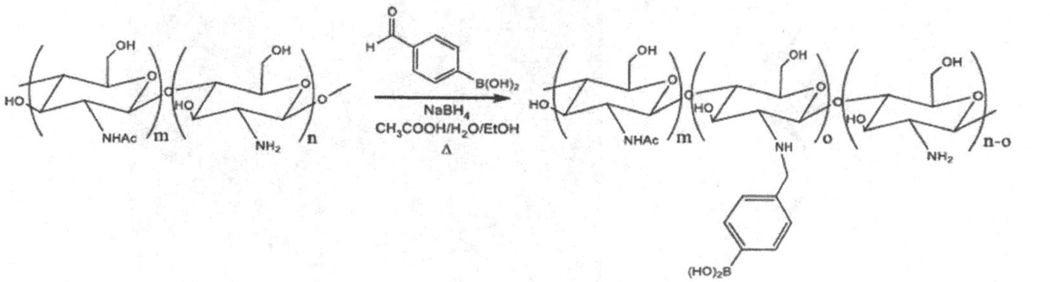

Figure 74.2 Synthesis of chitosan-phenylboronic acid (Chi-PBA) [15].

Chitosan, hyaluronic acid and cellulose are natural polymers. Biological glass or ceramic like calcium phosphate and Bioglass 45S5 used to help the development of biological active layer of hydroxycarbonate apatite. Bioactive glasses have low toughness hence, not easy to use in load bearing situations. There has been a good progress in the use of polymer ceramic composites as frameworks for bone repairmen. But further improvement in mechanical properties and biocompatibility is required [16]. The first polymer/Bioglass 45S5 composites is comprising PBA functionalized chitosan (Chi-PBA) [15].

Orthopedic Applications

According to Figure 74.2 the human bone which provides the support to the body is the composite which contains Hydroxyapatite crystal precipitated in the collagens. The bone have lower shear and tensile strength which caused for fracture when subjected to impact load. In the external fixation method, braces, casts and other fixation systems are used for alignment of the bone. Plaster materials were used for manufacturing of external fixation system which has low wear resistance and strength. So the polymer composite based fixators replace the plaster material fixators. But in early days metals were used in the manufacturing of orthopedic implants. Stainless steel and cobalt-chrome-based alloys have high strength, good corrosion resistance and low cost which were used firstly. The epoxy matrix carbon reinforcement were used in different bone treatments spinal cord plates and rods, total hip replacement and bone plates because these materials provide sufficient strength and hardness [11]. The orthopaedic medical method trusts significantly on the development of biomaterials which are used for the spare of injured bogy parts. The implanted materials in the body are called as biomaterials and should show admirable mechanical properties such as yield strength, ultimate tensile strength and elastic modulus to withstand different impact loads. These implanted materials also have higher corrosion resistance, low density, wear resistance and biocompatibility.

Bioceramics are the materials which are used for making of the orthopedic application devices and implants for the purpose of replacing the diseased and damaged part of the human body such as bone, joints, teeth, skeleton, etc. as shown in Figure 74.3 [17]. Some important alloying elements are found for Mg-based material which are use full for orthopedic applications these elements are Zinc (Zn), Calcium (Ca), and Strontium (Sr). Mainly Zinc is found in the bones and muscles of human body which helps to improve the yield

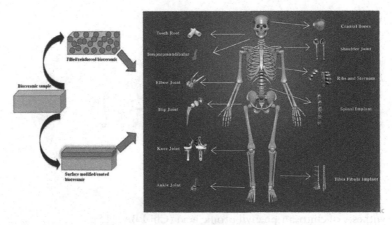

Figure 74.3 Implants and orthopedic application devices for human body [17].

strength of the implant. By refining the grain size, Sr can increase the corrosion resistance and strength [18]. In general, Titanium (Ti) and its alloys, magnesium and its alloys, niobium, tantalum, stainless steel, are used to rebuilding the injured bone by insertion of medical implants [19]. So hydroxyapatite/ poly sorbitol sebacate glutamate (PSSG) composite coating is used on titanium metal transplant orthopedic implantation [26].

In the recent research, in the field of medical, biomedical inserts are developed which are used to substitute a number of parts of human body. In the present, medical implants of Titanium (Ti) composites are getting attention having good corrosion resistance, high specific resistance, and biocompatible properties [20]. Protective bioactive coating of ceramics are used for long term applications of implants. Hydroxyapatite (HAP) has important properties like strong bond forming capability with host bone tissue, osteoinductivity, biocompatibility and non-toxicity [21]. Artificial bone transplant materials should have porosity for nutrient distribution, and bone-like properties. The homogenization of similarities is an important subject between bone and the microstructure of implants. So composites of poly ether-ether-ketone (PEEK) considered in the research to design microstructures for enhancing the compatibility of implant. Porous hip bone implant is made-up by rapid prototyping are recommended to imitate closely natural bone with excellent properties and countless homogenization lattice structures [22].

Future Scope

There is the wide scope of composite materials by enhancing the essential properties intended for biomedical applications like implant interaction with bone, rapid bone healing, etc. by aiming to improve the biocompatibility of medical grafts for long term applications. Antibacterial properties can be improved. The durability and biocompatibility of the present composite material should be improved. So the researcher should pay attention to develop unique composite materials for example zirconia toughened alumina (ZTA), ceria-stabilized zirconia (Ce-TZP), yttria-stabilized zirconia (Y-TZP), etc. with enhanced important properties like toughness, mechanical strength, biocompatibility and biodegradability. Design and development of such material which prevent biomaterial associated infections.

Objective of the Study

The objective of the study is to know the past study of composite materials in the field of medical implants and find out the scope and research gap of composite material in this field.

Conclusion

The use of composite materials offers substitute and different possibilities for the design of implants in biomedical implants. With use of composite materials, the components can be design with the wide range of mechanical properties. In voluminous biomedical applications the study and improvement of composite materials is going on and highly developed. In various biomedical applications metals are replaced by polymer composites because these materials have excellent properties in comparison with metals. This paper reviewed those composite materials utilized in various biomedical applications such as orthopedic, tissue engineering and dentistry. Bio ceramic proved that they are excellent alternative materials, having superior bioactivity biocompatible, for biological load bearing applications. In the medical industry, development of Al and Zirconium (Zr) based ceramic implants is the breakthrough. In the polymer-based biomaterials, the hybrid reinforcement has been increased and the purpose of this reinforcement is to advance the mechanical properties of the polymer matrix.

References

1. Salernitano, E., and Migliaresi, C. (2003). Composite materials for biomedical applications: a review. *Journal of Applied Biomaterials and Biomechanics,* 1, 3–18.
2. Biswal, T., BadJena, S. K., and Pradhan, D. (2020). Synthesis of polymer composite materials and their biomedical applications. *Materials Today: Proceedings,* 30(Part 2), 305–315.
3. Marhoon, I. I. (2018). Mechanical properties of composite materials reinforced with short random glass fibers and ceramics particles. *International Journal of Scientific and Technology Research,* 7, 50–53.
4. Nagavally, R. R. (2017). Composite materials-history, types, fabrication techniques, advantages, and applications. *International Journal of Mechanical and Production Engineering,* 5, 82–87.
5. Thirumalini, S., and Rajesh, M. (2017). Reinforcement effect on mechanical properties of bio-fiber composite. *International Journal of Civil Engineering and Technology,* 8, 160–166.
6. Bavan, S., and Channabasappa, M. K. G. (2010). Potential use of natural fiber composite materials in India. *Journal of Reinforced Plastics and Composites,* 29, 3600–3613.
7. Fujihara, K., Teo, K., Gopal, R., Loh, P. L., Ganesh, V. K., Ramakrishna, S., Foong, K. W. C., and Chew, C. L. (2004). Fibrous composite materials in dentistry and orthopaedics: review and applications. *Composites Science and Technology,* 64, 775–788.
8. Smith, D. C. (1962). Recent developments and prospects in dental polymer. *Journal of Prosthetic Dentistry,* 12, 1066.
9. Goldberg, A. J., and Burstone, C. J. (1992). The use of continuous fiber reinforcement in dentistry. *Dental Materials,* 8, 197–202.
10. Cooney, J. P., Caputo, A. A., and Trabert, K. C. (1986). Retention and stress distribution of tapered-end endodontic posts. *Journal of Prosthetic Dentistry,* 55, 540–546.
11. Krishnakumar, S., and Senthilvelan, T. (2021). Polymer composites in dentistry and orthopedic applications-a review. *Materials Today: Proceedings,* 46, 9707–9713.
12. Zeng, N., van Leeuwen, A. C., Grijpma, D. W., Bos, R. R. M., and Kuijer, R. (2017). Poly(trimethylene carbonate)-based composite materials for reconstruction of critical-sized cranial bone defects in sheep. *Journal of Cranio-Maxillofacial Surgery,* 45(2), 338–346.
13. Neumann, A., and Kevenhoerster, K. (2009). Biomaterials for craniofacial reconstruction. *GMS Current Topics in Otorhinolaryngology, Head and Neck Surgery,* 8.
14. Rogers, G. F., and Greene, A. K. (2012). Autogenous bone graft: basic science and clinical implications. *Journal of Craniofacial Surgery,* 23(1), 323–327.
15. Thibault, M. H., Comeau, C., Vienneau, G., Robichaud, J., Brown, D., Bruening, R., Martin, L. J., and Djaoued, Y. (2020). Assessing the potential of boronic acid/chitosan/bioglass composite materials for tissue engineering applications. *Materials Science and Engineering C,* 110, 110674.
16. Rezwan, K., Chen, Q. Z., Blaker, J. J., and Boccaccini, A. R. (2006). Biodegradable and bioactive porous polymer/inorganic composite scaffolds for bone tissue engineering. *Biomaterials,* 27(18), 3413–3431.
17. Shekhawat, D., Singh, A., Banerjee, M. K., Singh, T., and Patnaik, A. (2021). Bioceramic composites for orthopaedic applications: a comprehensive review of mechanical, biological, and microstructural properties. *Ceramics international,* 47(3), 3013–3030.

18. Chandra, G., and Pandey, A. (2022). Preparation strategies for Mg-alloys for biodegradable orthopaedic implants and other biomedical applications: a review. *IRBM*, 43(3), 229–249.

19. Al Mahmood, A., Gafur, M. A., and Hoque, M. E. (2017). Effect of MgO on the physical, mechanical and microstructural properties of ZTA-TiO2 composites. *Materials Science and Engineering A*, 707, 118–124.

20. Arab, A., Sktani, Z. D. I., Zhou, Q., Ahmad, Z. A., and Chen, P. (2019). Effect of MgO addition on the mechanical and dynamic properties of zirconia toughened alumina (ZTA) ceramics. *Materials* 12(15), 17–19.

21. Dharmaraj, B. M., Subramani, R., Dhanaraj, G., and Louis, K. (2020). Multifunctional halloysite nanotube based composite coatings on titanium as metal implant for orthopedic applications. *Composites Part C: Open Access*, 3, 100077.

22. Oladapo, B. I., Zahedi, S. A., and Ismail, S. O. (2021). Mechanical performances of hip implant design and fabrication with PEEK composite. *Polymer*, 227, 123865.

23. Wang, T., Lin, C., Batalu, D., Zhang, L., Hu, J., and Lu, W. (2021). In vitro study of the PLLA-$Mg_{65}Zn_{30}Ca_5$ composites as potential biodegradable materials for bone implants. *Journal of Magnesium and Alloys*, 9(6), 2009–2018.

24. Singh, G., Sharma, N., Kumar, D., and Hegab, H. (2020). Design, development and tribological characterization of Ti–6Al–4V/hydroxyapatite composite for bio-implant applications. *Materials Chemistry and Physics*, 243, 122662.

25. Haleem, A., and Javaid, M. (2019). Polyether ether ketone (PEEK) and its 3D printed implants applications in medical field: An overview. *Clinical Epidemiology and Global Health*, 7, 571–577.

26. Pan, J., Prabakaran, S., and Rajan, M. (2019). In-vivo assessment of minerals substituted hydroxyapatite/poly sorbitol sebacate glutamate (PSSG) composite coating on titanium metal implant for orthopedic implantation. *Biomedicine and Pharmacotherapy*, 119, 109404.

27. Gautam, C., Gautam, A., Mishra, V. K., Ahmad, N., Trivedi, R., and Biradar, S. K. (2019). 3D interconnected architecture of h-BN reinforced ZrO2 composites: structural evolution and enhanced mechanical properties for bone implant applications. *Ceramic International*, 45(1), 1037–1048.

28. Kavitha Sri, A., Deeksha, P., Deepika, G., Nishanthini, J., Hikku, G. S., Antinate Shilpa, S., Jeyasubramanian, K., and Murugesan, R. (2020). Super-hydrophobicity: Mechanism, fabrication and its application in medical implants to prevent biomaterial associated infections. *Journal of Industrial and Engineering Chemistry*, 92, 1–17.

29. Ul-Islam, S., Ul-Islam, M., Ahsan, H., Ahmed, M. B., Shehzad, A., Fatima, A., Sonn, J. K., and Lee, Y. S. (2021). Potential applications of bacterial cellulose and its composites for cancer treatment. *International Journal of Biological Macromolecules*, 168, 301–309.

30. Xiang, Z., Ducharne, B., Della Schiava, N., Capsal, J. F., Cottinet, P. J., Coativy, G., Lermusiaux, P., and Le M. Q. (2019). Induction heating-based low-frequency alternating magnetic field: high potential of ferromagnetic composites for medical applications. *Materials and Design*, 174, 107804.

75 Automation and digitalization of service sector in India in the wake of COVID-19 crisis

Ramun Prasad[1], Maryam Sabreen[2,a], and Deepak Kumar Behera[1]

[1]Department of Humanities and Social Sciences, NIT, Patna, India

[2]Department of Economics, Oriental College, Patliputra University, India

Abstract: The outbreak of the coronavirus disease (COVID-19) had forced the country's economy to stop its usual functioning and move towards digitalization to work from home through the internet in the tertiary sector and with the help of robots in the secondary sector. As new technologies such as social robots, artificial intelligence, machine learning, cloud computing, and data analytics are adopted and used more often, it is anticipated that this will have a significant impact on employment across a range of economic sectors. In this regard, this paper attempts to explore the impact of mechanization, automation, and digitalization on the employability of the service sector of the Indian Economy. So, the relevant questions are- What is the automation pattern of Indian services market? What role does digitalization play in enhancing the employability for skilled people? How is the adaptation of new technology affecting employability in the services sectors?

Keywords: Automation, digitalization, employment, mechanization, unemployment.

Introduction

The Coronavirus (COVID-19) originated in Wuhan, China, in December 2019. The rapid spread of the disease lead to a country-wide lockdown in India from 25th March 2020. As a result, country's economy stopped usual functioning. Offices were shut and employees were working from home, shops were shut, and online shopping increased, schools were shut, and the classes were running online. Interestingly a major change took place where most of the essential services, even the neighborhood *Kirana* stores started to digitalize. Online orders and digital payments mode were preferred while observing social distancing. This is a positive step coordinating the fourth Industrial Revolution that is digital revolution. In today's times, these technological advancements are impacting more so with emerging artificial intelligence, robotics, the Internet of Things, autonomous vehicles, 3-D printing, nanotechnology, biotechnology, materials science, energy storage, and quantum computing.

The Digital India Campaign had a vision to transform India into a digitally empowered society and knowledge economy. Similar to other countries, India is investing in the adoption of digital technologies like social robots, artificial intelligence, machine learning, driverless automobiles, cyber security to gain from the fourth industrial revolution [4,18].

[a]maryam.sabreen@gmail.com

However, with technological advancements, workers are getting replaced by computers and online shopping and payment are taking place instead of physical shopping and banking activities. Nearly, 69 percent of jobs in India are under threat from automation and this continuing health crisis has speeded the demand for automation in the secondary and the tertiary sector of the Indian economy [4,18]. To sum up, the increased focus on digitalization and the rising demand from users for these services may have an effect on the quantity and quality of jobs available on the labour market, necessitating a response to its economic and social ramifications [8].

Given this background, this study focuses on the impact of digitalization and automation on service sector in India. More specifically, this paper attempts to explore the pattern and resulting impact of mechanization, automation, and digitalization on the employability of the service sector. Finally, to explore the role of digitalization in enhancing the employability for skilled people and the government initiatives to fill the skill gap. This paper has six sections. The first section introduces the study. Section two gives a brief review of literature. Section three and four deals with state of digitalization on employment and post COVID-19 situation. Section five describes government initiatives for skill development and finally, the study concludes with some policy implication in section six.

Review of Literature

Changing technology and advancements in the field of Artificial Intelligence, Blockchain, Big Data Analytics, driverless cars, Cloud Computing, Internet of Things, Cyber Security, Mobile Tech, Robotic Process Automation, Virtual Reality, and 3D Printing are importance in economic development of a country. Innovative technologies like artificial intelligence, sensors, and robotics are gradually being promoted to increase the efficiency of food production with minimum use of resource [7, 1, 20]. Digital transformation using industrial robots are also taking place in every industry [2, 3]. These emerging technologies changes the nature of work, and a massive disturbance is being faced by the information technology (IT) and information technology enabled services (IT-ITES) industry (NASSCOM). On the other hand, automation has the potential to raise overall living standards, while low skill employees would lose, and employers or capital owners would gain [11]. The consequence of automation would manifest, causing in what Keynes [15] termed as "technological unemployment".

Moreover, this automation reduces around 14 percent of the labour share over the last decades in the United States [18]. Some other studies reveal that fourth industrial revolution (4.0) has changed the fundamental structure of work through technological innovations [6,7,10,18] and Global digital companies like Google, Microsoft, APPLE, IBM and Amazons, etc. are forerunner of technological innovations, digitalization and automation [16,17,19]. The need is to understand whether emerging technologies like Artificial Intelligence, Blockchain, Big Data Analytics, driverless cars, Cloud Computing, Internet of Things, Cyber Security, Mobile Tech, Robotic Process Automation, Virtual Reality, and 3D Printing are demanding to re-skilled the labour or retrench them from existing employment [21, 22]. More specifically, this paper is a dedicated enquiry into exploring the pace of digitalization and automation in the tertiary sector of the economy during the COVID period. Further, this study has tried to explore that what kind of reskilling or skill enhancement programmes would be effective to catch up the pace of changes in new-age technologies and make people employable [9, 12, 14].

Tracing the Digitalization and Employment Scenario in Service Sector

The first industrial revolution towards the end of 18th century transformed the agrarian economy into the manufacturing economy through introduction of mechanical production facilities with the help of water propelled steam engines, while the second industrial revolution restructured it into the service-based economy by introducing electrical energy in the beginning of 20th century. The third industrial revolution began with application of electronics and information technology (IT) to automate production and made knowledge-based economy at the beginning of 1970 of 20th century but industry 4.0 is the convergence of IT and operational technology which build up a cyber-physical system. This industry 4.0 provide appearance of smart plants, smart factories, smart cities, smart homes, smart logistics, smart products, smart grids, etc. which demanded employees having new-age technological skills [5].

Currently, the outbreak of COVID-19 has seriously affected humans, economically, mentally, and physically. As it is globally observed that the pandemic has forced government of each country to impose the lock down to keep people indoors make them safe from the pandemic. Due to this lock down most of the individuals in countryside have lost their jobs but circumstances compelled them to search their alternative options for earning their livelihood. Many skilled labours also rendered jobless. Highly skilled persons like teachers, professors, lawyers, and engineers started to work from their home; those who were technically sounds in the field of computers aided work, etc. have got an opportunity to discharge their duty from home. For example, those who were in the field of education started to teach their students through YouTube, Google meets, Zoom, MS-Team, WebEx, and Teachmint to make their students learn and run their livelihood.

Doctors started to treat their patients through video conferencing and those who were in the legal profession started to attend their virtual court. No doubts, a small portion of skilled persons were either unknown about such things, but circumstances compel them to think a new. They also started to learn computer operation, use of internet, etc. with the help of their children and colleagues. Many computer literature persons open their cybercafé in rural areas where villagers were aided with filling a form for government aids, etc. Thus, the outbreak of covid-19 is not only a course for human race, but it is also a boon for the same. Because, it has accelerated the pace of Digital India campaign which had been being launched by the government of India to improved online infrastructure and Internet connectivity to make the country digitally empowered in the field of technology. Moreover, in the present situation women Labour having certain types of skills can utilize it through a digital platform from home. And most of the skilled women have taken an initiative and played a significant role in the constructive and productive work from their homes during pandemic.

Digitalization Across the Country During Covid-19 Pandemic

Covid-19 opened new avenues for those skilled workers who transformed themselves to changing needs. However, on the flip side, it is important to see how the demand side has got affected. Government of India has taken several initiatives to motivate the people for digital transactions across the country. But its pace was not too fat as government expected. While the outbreak of Coronavirus has accelerated it and compel people to the do their transaction in digital mode.

It is important to see the state of digital transactions before pandemic outbreak. Figure 75.1 gives details about the trend of digital payment transactions between 2016 to 2019.

A substantial increase in the digital transaction can be observed that went up from 1013 crore during 2016–17 to 2070.39 during 2017–18 and 3133.58 Crore in 2018–19. Further, the number of digital payment transactions in India was 975.98 crore during 2019–20 (till 11th July 2019). Graph of digital payment transactions revealed that it is increasing with the very rapid rate which probably reduces physical movement of people towards banks for money.

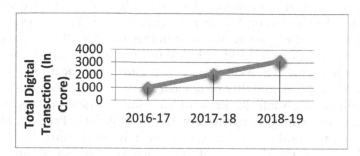

Figure 75.1 Growing digital transactions in India.

Health care services were most affected for patients suffering from other ailments. Nevertheless, Telemedicine emerged as an important tool for patients through which they were able to avail the healthcare facilities though mostly in urban areas. It became a popular fallback option for patients during COVID lockdown. The Health Ministry of India launched the National Telemedicine Taskforce in 2005, and under COVID, its penetration quickly rose. According to Seek Med data, superspecialties like cardiology (150 percent), orthopaedics (167 percent), psychiatry (400 percent), and gastroenterology (344 percent) experienced a triple digit rise in teleconsults during the COVID period compared to the previous quarter of the same year 2020 [5].

Similarly, online shopping in India saw a great surge. In India while the consumers prefer neighborhood brick and mortar shop for daily purchases, the nature of the disease and inherent lockdown has led to a rise in the number of first-time e-commerce-users, who had been so far inhibited to shop online [13]. Not only that, education sector also had transformed, and online mode of education started to increase. As the only option left online classes began for schools, colleges and universities however many online teaching and learning platforms have started that goes beyond normal textbook learning method.

Government Initiatives for Skill Development to Reduce the Skill Mismatch

Ever since development started, labour force needed adjustments and adapting the changing technology. Even during the time when the first Industrial revolution began, the relation of labour to their tasks shifted and craftsmen working with hand started operating machines in factories. Thus, it can be rightly said that with technological advancements, specialized workers are needed to operate the machines and highly skilled workforce would reap the benefits. The existing labour as well as those getting ready to enter the labour market need to acquire new and distinctive skills. In India, the government has undertaken several initiatives to bridge this skill gaps. The Skill development cell has been established to provide youth with skills and enhance their employment opportunities. Several schemes are run under this cell to meet the skill training requirements of Indian Labour force. Pradhan Mantri Kaushal Vikas Yojana for Technical Institutes (PMKVY-TI) is implemented through AICTE approved Colleges to impart Engineering skills to drop-out students and find placement in suitable private sector jobs.

The National Employability Enhancement Mission (NEEM) provides on-the-job training to enhance employability for those who are either pursuing graduation or a diploma in any technical or non-technical stream or have discontinued studies, whereas the Employability Enhancement Training Programme (EETP) focuses on implementation and enhancement of employment opportunities under skill initiatives. The AICTE-Startup policy encourages students of technical institutions to create tech-based start-ups and employment opportunities. Skill Assessment Matrix for Vocational Advancement of Youth (SAMVAY) supports mobility of students between various levels of general education, community colleges and Bachelor of Vocational Courses (B.Voc) courses run by polytechnics and other colleges approved by UGC and AICTE. Pradhan Mantri Kaushal Kendras (PMKK) are model training centers that is an attempt by the government to transform India into the skill capital of the world. Pradhan Mantri YUVA Yojana (PM-YUVA) is the initiative of Ministry of Skill Development and Entrepreneurship (MSDE) on entrepreneurship education and training, advocacy, and easy access to entrepreneurship support network.

Bihar Government has also come up with the ambitious Bihar Skill Development Mission (BSDM) to empower youths by providing them with requisite skills and employment that would ultimately fuel the growth of Bihar. In that, the primary role of BSDM is firstly to establish a wide network of training centers for the youth and further provide employment opportunities to them.

Conclusion and Policy Suggestion

Social robots, artificial intelligence, machine learning, driverless cars, cloud computing, and data analytics are just a few of the developing technologies that are displacing old manual occupations and raising unemployment not just in India but worldwide. Thus, we put forward following policy prescriptions based on our conclusion.

1. To keep people employed and employable, steps should be taken to reskill the professionals, employees, and students according to needs and demand of economy system in this era of 4.0 industrial revolutions and in this technology driven ecosystem to reap the benefit of emerging technology.
2. It is also important catch-up future skill in the emerging technology like cyber security, 3D modeling, block chain, robotic process automation, artificial intelligence, internet of the things, virtual reality, big data analytics, cloud computing, mobile tech, etc. to make himself/herself more employable.
3. An initiative should also be taken by central as well as state governments to establish the Centre of Excellence in every technical as well as non-technical institution of the country which will prepare skilled manpower by providing training for a focus area. It will also support the *Atmanirbhar Bharat* (self-reliant India) initiative of government of India which may be reduced the demand of government job.

References

1. Acemoglu, D., and Pascual, R. (2019). Automation and new tasks: hoTechnology displaces and reinstates labour. *Journal of Economic Perspectives, 33*(2), 3–30.
2. Acemoglu, D., and Pascual, R. (2020). The wrong kind of AI. artificial intelligence and the future of labour demand. *Cambridge Journal of Regions, Economy and Society, Cambridge Political Economy Society, 13*(1), 25–35.
3. Acemoglu, D., and Restrepo, P. (2020). Robots and jobs: evidence from US labor markets. *Journal of Political Economy, 128*(6), 2188–2244.
4. Arntz, M., Gregory, T., and Zierahn, U. (2019). Digitalization and the future of work: macro-economic consequences. In Zimmermann, K. F. (Editor-in-Chief). Handbook of Labor, Human Resources and Population Economics, (pp. 19–24).
5. Awasthi, A. (2020). Digitalization in healthcare, the revolution in India is going to experience post COVID-19. accessed from http://www.businessworld.in/article/Digitization-In-Healthcare-The-Revolution-India-Is-Going-To-Experience-Post-COVID-19/03-09-2020-316145/ on 02/01/2021.
6. Bodrow, W. (2017). Impact of industry 4.0 in service-oriented firm. *Advances in Manufacturing, 5*(4), 394–400.
7. Bayraktar, O., and Ataç, C. (2018). The Effects of Industry 4.0 on Human Resources Management. Globalization, Institutions and Socio-Economic Performance. Peter Lang GmbH, (pp. 337–360).
8. Ernst, E., Merola, R., and Samaan, D. (2018). The economics of artificial intelligence: implications for the future of work. ILO Future of Work Research Paper Series. International Labour Organization, Geneva.
9. Fossen, F. M., and Sorgner, A. (2018). The effects of digitalization on employment and entrepreneurship. In Conference Proceeding Paper, IZA–Institute of Labor Economics.
10. Gera, I., and Singh, S. (2019). A critique of economic literature on technology and fourth industrial revolution: employment and the nature of jobs. *The Indian Journal of Labour Economics, 62*(4), 715–729.
11. Graetz, G., and Michaels, G. (2018). Robots at work. *Review of Economics and Statistics, 100*(5), 753–768.
12. Himanshu, (2011). Employment trends in India: a re-examination. *Economic, and Political Weekly, 46*(37), 43–59.
13. Halan, D. (2020). Lessons for brick-and-mortar Retailers impacted by epidemics and pandemics ET Retail.com. Accessed from https://retail.economictimes.indiatimes.com/rc-tales/lessons-for-brick-and-mortar-retailers-impacted-by-epidemics-and-pandemics/4816 on 06/01/2021.
14. Keune, M., and Dekker, F. (2018). The Netherlands: The Sectoral Impact of Digitalisation on Employment and Job Quality. Work in the Digital Age. London: Policy Network.
15. Keynes, J. M. (1930 [1932]). Economic Possibilities for our grandchildren. In Essays in Persuasion. New York: Harcourt Brace, pp. 358–373.
16. Nathan D., and Ahmed, N. (2018). Technological change and employment: creative destruction. *The Indian Journal of Labour Economics, 61*(2), 281–298.
17. Nagano, A. (2018). Economic growth and automation risks in developing countries due to the transition toward digital modernity. In Proceedings of the 11th International Conference on Theory and Practice of Electronic Governance. pp. 42–50.
18. Nippani, A. (2020). Automation and labour in India: policy implications of job polarisation pre and post COVID-19 crisis. SocArXiv. August, 11.
19. Prettner, K. (2017). A note on the implications of automation for economic growth and the labor share. *Macroeconomic Dynamics, 23*(3), 1294–1301.

20. Rotz, S., Gravely, E., Mosby, I., Duncan, E., Finnis, E., Horgan, M., and Fraser, E. (2019). Automated pastures and the digital divide: how agricultural technologies are shaping labour and rural communities. *Journal of Rural Studies,* 68, 112–122.

21. Sovbetov, Y. (2018). Impact of digital economy on female employment: evidence from Turkey. *International Economic Journal,* 32(2), 256–270.

22. Routray, C., Soma, S., and Danish, M. F. (2021). The Silence behind the Walls-Implication of Maternity Benefit Act 2017. Available at SSRN 3810459.

76 Measuring the violations, errors and lapses using driving behaviour questionnaire (DBQ) on Indian drivers

A. Nageswara Rao[1], R. Jeyapaul[1], B. Chaitanya[2], and G. Manikanta[1]

[1]National Inistitute of Technology Tiruchirapalli, Tamil Nadu, India

[2]Lakireddy Bali Reddy College of Engineering, Mylavram Andhra Pradesh, India

Abstract: India is the leading country in road accident deaths in the world. Human factors play a key role in driving a vehicle. The present study focuses on measuring behaviour of driver with applying the Driver Behaviour Questionnaire (DBQ). The DBQ stands as one among the most extensively utilized tools in order to measure driving behavior, and its factorial validity and reliability have been thoroughly tested among Indian drivers. 506 responses received from participants for 28item DBQ. Subsequently, an Exploratory Factor Analysis (EFA) was employed, utilizing Principal Component Analysis, which disclosed a three-factor structure encompassing errors, lapses and violations. The three-factor formation explained 64.91 percent of the total variance. Factor analysis affirmed the questionnaire's reliability, as evidenced with a high Cronbach's Alpha 0.970. According to the descriptive statistics, the most prevalent driving behaviors were found to be exclusively violations. Violations are potentially dangerous in road accidents so more stringent rules required to make driver behaviour properly on roads.

Keywords: Driver behaviour questionnaire, errors, exploratory factor analysis, lapses, road accident, violations.

Introduction

India accounts for nearly 11%, among all fatalities caused by accidents globally, making it the leading country in terms of road accident fatalities among 199 nations [1,2] ("WHO Global Report on Road Safety 2018, Ministry of Road Transport & Highways, 2018"). Road fatalities may involve three major components: the human driver, the vehicle, and the surrounding environment. However, the human factor was found to a key role in road accidents. Driver error was responsible for 80% of crashes, contributing to 75% of deaths in India [3]. Driving errors can be studied in three ways, firstly Naturalistic studies with the help of instrumented vehicle with all apparatus, secondly simulating environment with a driving simulator, thirdly conducting surveys with driver behavior related questionnaires. Using a driving simulator, researchers in India investigated the impact of mobile phone use on driving [4]. When drivers were engaged in a phone call, the likelihood of an accident increased by three times, and it climbed by four times when drivers were texting. However present study focuses on using the Driver Behaviour Questionnaire(DBQ)

414918051@nitt.edu.

response analyzed. The DBQ holds a prominent position as one of the most widely utilized tools by researchers to assess driving behavior performance. Reason [5] developed questionnaire consist of 50 items most referred DBQ. The number of questions in DBQ varying different authors from different countries. Violations, errors, and lapses are three categories of driving behavior. Violations are described as "purposeful deviations from practices believed necessary to ensure the safe operation of a potentially hazardous system." Errors pertain to "failures in executing planned actions to achieve their intended outcomes," while lapses involve "attention and memory lapses that may cause minor inconveniences but are unlikely to pose a threat to driving safety." Majorly violations and errors cause deviation of driving performance which may lead to an accident.

In the Australia DBQ study, researchers utilized a 28-item four-factor model to examine driving behavior. The results indicated that this model showed gender-invariant properties and displayed robust partial measurement invariance among drivers aged 26 to 64 [6]. The validation process, conducted through confirmatory factor analysis, confirmed the model's gender-invariant nature and provided insights into its varying measurement invariance across different age groups. DBQ tool used [7] for measuring the Violations, Errors and Lapses using Driving Behaviour Questionnaire and with the help of Exploratory Factor Analysis plus Principle Component Analysis were used to extract important factors. In Australia study [8] was conducted a study to examine the driver's relationship between distractions and driving errors. In a French study [9], employing a 41-item DBQ, researchers found that French drivers can enhance their understanding of driving habits and categorizing them into two types of violations and three types of errors. In Finland [10] when using the DBQ to determine the problem areas within a driving population, the study emphasizes the importance of conducting exploratory analysis. The Manchester DBQ applied [11] to assess violent and abnormal driving behavior that result in collisions, statistical analysis of univariate and multivariate logistic regression was performed. The result of EFA was a 65-item scores divided into four factors. Canada study reported that inattention Errors, Age-Related [12] Problems, Distraction and Hurry, and Aggressive Violations were tentatively labelled as the factors that would be predictors of self-report in regression studies. In a UK study [13] Compared to correlated factors and second-order factor systems, a bifactor model with a common factor onto which all items formulated and special factors for ordinary violations, violent violations, slips, and errors fit sample. Researchers in Denmark [14] utilized Confirmatory Factor Analysis (CFA) to examine the fit of the original DBQ as opposed to a shorter version. Moreover, they explored the potential presence of a second-order structure for the DBQ. A study [15] was conducted on elderly drivers using the DBQ to observe changes in driving behavior across three different time-points within a group of elder drivers. In Malaysia, a study [16, 17] investigated the equivalency of DBQ factor structures in elderly vehicle drivers and motorcycle riders. When compared to older automobile drivers, older motorbike riders had a different perception of lapse items. In Kuwait [17], the DBQ applied to examine driver behavior, and the data analysis revealed a three-factor structure.

In Thailand [18], the study combined data on mobile chatting, social media-related driver behavior, and observable actions like alcohol and drug use in driving, applying the DBQ tool for assessment. Research Gap was found after literature review, that in Indian contest no study was reported that with the help of DBQ tool analyze driver behavior. The scope of the present study is to explore different factors that influence to the cause of road accidents using DBQ tool in India. India is the leading country in the world in road accident deaths. So there is much to study in the behavior of driver to reduce road accidents. Research flow followed in present study is shown in Figure 76.1.

The aim of the present study is to examine and critically explore the Driver performance using the Driving Behaviour Questionnaire (DBQ) and measuring the Violations, Errors and Lapses using DBQ. Principal Component Analysis (PCA) was used in Exploratory component Analysis (EFA) to extract component loadings and identify the critical main factors.

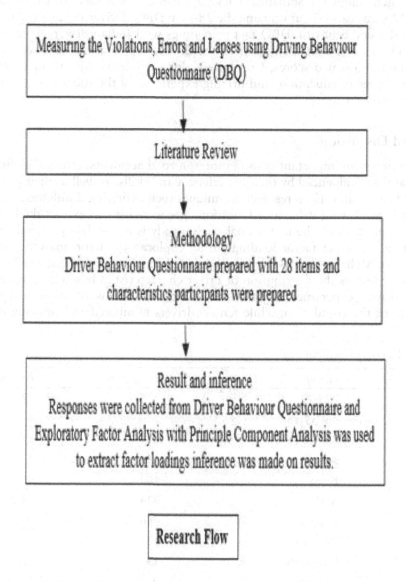

Figure 76.1 Research flow followed in present study.

Methodology

DBQ was employed to measure estimate driver performance of Indian drivers in this study. Response from questionnaire collected through manual and online, in online emails and social media were used. This DBQ [6] have 28 items includes 11 items related to violations,

11 items related to errors, and 6 items related to lapses. The participants were tasked with rating the occurrence of their driving behaviors for every item using a five-point Likert-type scale, varying from "Never" (1) to "Nearly all the time" (5). Initially, 549 questionnaires were collected. However, 43 outlier cases were identified and subsequently removed, leaving 506 usable responses for analysis. The data was examined using SPSS version 26 software, which stands for Statistical Package for Social Sciences. An exploratory factor analysis (EFA) was carried out utilizing the 28-item DBQ (Driver Behavior Questionnaire) to examine drivers' behavior. DBQ factor loadings with Indian drivers were investigated using Principal Component Analysis (PCA) with varimax rotation. To assess the dependability of the DBQ scaled scores, Cronbach's alpha reliability was also determined. The age, gender, degree of education, and driving experience of the subjects were investigated using descriptive statistics.

Results and Discussion

Driver behavior is an important consideration in road accidents. Drivers' abilities exhibit variations and are influenced by their experiences and skills, as well as their physical characteristics. As a result, various research techniques such as driving simulators, instrumented vehicles, and the Driving Behaviour Questionnaire were used to extract the behavior and identify their predictors. The major goal of the study is to use DBQ to evaluate features and determine construct factor loadings using exploratory factor analysis. Additionally, the Cronbach's Alpha value was used to examine the assessment's dependability.

Table 76.1 presents the distribution of driver characteristics based on age, gender, education, and driving experience. Among the respondents, there were 324 male drivers, comprising 64% of the population, while female drivers numbered 182, making up 36% of

Table 76.1: Details of participants.

Variable	Group	Frequency	Percent
Age	<25years	215	42.5
	26–35years	206	40.7
	36–45years	60	11.9
	>46years	25	4.9
Gender	Female	182	36.0
	Male	324	64.0
Education	Degree	466	92.1
	Illiterate	12	2.4
	Intermediate	19	3.8
	Primary	9	1.8
Driving Experience	<2 years	159	31.4
	2–5 years	157	31.0
	5–10 years	101	20.0
	>10 years	89	17.6

the sample. The majority of drivers belonged to the younger age group, with 42.5% of respondents aged below 25 years and 40.7% falling within the 25–35 years age range. The population of response more were education level of graduates with 92%. The driving experience is the key element in driving to avoid accidents, 159 participants with 31.4% were having less than 2 years of experience and 2 to 5 years of driving experience drivers were 157 participants with 31%. 190 drivers with 37% in the study had more than 5 years of experience in driving.

Table 76.2: Driving behavior questionnaire mean scores.

Q No	Statement	Mean	SD
	Lapses		
1	In a parking lot, forget where you parked your car.	1.11	0.482
2	Recognize that you have no idea of the route you've been on.	1.4	0.638
3	While you were intending to utilize another feature, like the wipers, you turned on one, like the headlights.	1.46	0.584
4	While approaching a roundabout or crossroads, get into the wrong lane.	0.99	0.387
5	After setting out to go somewhere, you suddenly notice you're on your way somewhere else.	1.37	0.55
6	Take the incorrect exit from a roundabout after misreading the signals.	0.96	0.446
	Violations		
7	To express your frustration to another road user, use your horn.	1.8	0.835
8	Become frustration with a specific type of driver and express your frustration in whatever manner you can.	1.28	0.609
9	Become frustration at another driver and run after them with the purpose of showing your anger.	1.1	0.481
10	On the left, overtake a slow driver.	1.61	0.689
11	Disregard the speed limit when driving on a motorway or a rural road.	1.2	0.52
12	Breaking the speed limit on a road where homes are located.	1.1	0.521
13	With the purpose of overtaking the driver in front of you, ran away from traffic lights.	0.84	1.408
14	Delay moving to the open lane until the last possible moment, even though you are aware that the current lane will close ahead.	1.39	0.569
15	You should be aware that the traffic signals have already Changed against you as you approach an intersection	1	0.449
16	Drive so close to the vehicle in front of you that braking in an emergency would be tough.	1.13	0.506
17	Even if you believe your blood alcohol level is higher than the legal limit, you should drive.	0.53	1.285

(continued)

Table 76.2: Continued

Q No	Statement	Mean	SD
	Errors		
18	On a slippery road, brake too quickly.	1.23	0.513
19	When passing a moving vehicle, overestimate its speed.	1.28	0.514
20	Failing to look in your rearview mirror before turning, lane-changing, etc.	1.08	0.574
21	Failing to keep an eye out for pedestrians crossing when turning onto a side street from a large thoroughfare.	0.76	0.377
22	You are waiting to turn left into a busy road when you almost strike the car in front of you because you were paying too much attention to the traffic on the main route.	1.01	0.511
23	You hit something you hadn't seen previously while reversing.	0.82	0.346
24	In the wrong gear, try to get away from traffic lights.	0.86	0.377
25	Attempt to approach someone who was signaling a right turn and you hadn't seen.	0.91	0.353
26	Drive far enough into an intersection so the car with the right-of-way must come to end and let you out.	1.47	0.578
27	When turning left, you almost hit with a cyclist who has come up on your left.	1.59	0.484
28	Simply stay out of the path of incoming traffic if you don't see a "Give Way" sign.	1.3	0.652

Table 76.2 displays the descriptive findings that showed three of the most frequent incidents resulted from the driving behaviors that were deemed violations [3]. Among the 'lapses' category, the three most prevalent driving behaviors were "Turn on one feature, such as the headlights, while you meant to use another, such as the wipers." Mean of 1.46 (SD=0.584) followed by "Recognize that you have no idea of the route you've been on" (M=1.4 (SD=0.638)), "After setting out to go somewhere, you suddenly notice you're on your way somewhere else" (M=1.37, SD=0.55).

The error category had only one participant who received a relatively high score for frequently exhibiting driving behavior "When turning left, you almost hit with a cyclist who has come up on your left" came with a mean score of 1.59 and a standard deviation (SD) of 0.484. Followed by "Drive far enough into an intersection so the car with the right-of-way must come to end and let you out." with (M=1.47, SD=0.578), "Simply stay out of the path of incoming traffic if you don't see a "Give Way" sign." (M=1.3, SD=0.652). Because of their potential impact to road accidents, driver errors and violations are of primary interest [2].

Among the 'violations' category, the three most prevalent driving behaviors were "To express your frustration to another road user, use your horn" with a mean score of 1.8 and standard deviation (SD) of 0.835 followed by "On the left, overtake a slow driver" (M=1.61, SD=0.68), "Delay moving to the open lane until the last possible moment, even though you are aware that the current lane will close ahead." (M=1.39, SD=0.569).

Table 76.3: DBQ item factor loading solution.

Item	F1	F2	F3
With the purpose of overtaking the driver in front of you, ran away from traffic lights	0.76		
You should be aware that the traffic signals have already changed against you as you approach an intersection	0.747		
Drive so close to the vehicle in front of you that braking in an emergency would be tough.	0.719		
Attempt to approach someone who was signaling a right turn and you hadn't seen.	0.703		
In the wrong gear, try to get away from traffic lights.	0.702		
Take the incorrect exit from a roundabout after misreading the signals.	0.701		
Become frustration with a specific type of driver and express your frustration in whatever manner you can.	0.684		
Become frustration at another driver and run after them with the purpose of showing your anger.	0.679		
Disregard the speed limit when driving on a motorway or a rural road.	0.679		
When passing a moving vehicle, overestimate its speed.	0.576		
On the left side, overtake a slow driver.	0.572		
Recognize that you have no idea of the route you've been on.	0.547		
Breaking the speed limit on a road where homes are located.	0.519		
When turning left, you almost hit with a cyclist who has come up on your left.	0.433		
Even if you believe your blood alcohol level is higher than the legal limit, you should drive.		0.765	
Failing to keep an eye out for pedestrians crossing when turning onto a side street from a large thoroughfare.		0.74	
You hit something you hadn't seen previously while reversing.		0.708	
While approaching a roundabout or crossroads, get into the wrong lane.		0.706	
You are waiting to turn left into a busy road when you almost strike the car in front of you because you were paying too much attention to the traffic on the main route.		0.699	
After setting out to go somewhere, you suddenly notice you're on your way somewhere else.		0.661	
Failing to look in your rearview mirror before turning, lane-changing, etc.		0.584	
In a parking lot, forget where you parked your car.		0.535	
On a slippery road, brake too quickly.		0.535	
To express your frustration to another road user, use your horn.			0.726
While you were intending to utilize another feature, like the wipers, you turned on one, like the headlights.			0.622

(continued)

Table 76.3: Continued

Item	F1	F2	F3
Drive far enough into an intersection so the car with the right-of-way must stop and let you out.			0.602
Delay moving to the open lane until the last possible moment, even though you are aware that the current lane will close ahead.			0.525
Simply stay out of the path of incoming traffic if you don't see a "Give Way" sign.			0.514

The Driving Behaviour Questionnaire's factorial structure was developed using EFA, PCA, and varimax rotation. Threshold values included factor loadings above 0.4 and Eigenvalues more than 1.0, consider for 64.91% to the total variance. Bartlett's test of sphericity was statistically significant (df = 378, P 0.000), and the Kaiser-Meyer-Olkin (KMO) index demonstrated a high level of adequacy (0.972).

Factor loadings were extracted after the factor analysis, resulting in three groups (as displayed in Table 76.3). The first group exhibited factor loading values ranging from 0.760 to 0.433 "With the purpose of overtaking the driver in front of you, race away from traffic lights" had a high value of 0.76 comes under the violation category. In the second group factor loading values ranging from 0.765 to 0.535, "Even if you believe your blood alcohol level is higher than the legal limit, you should drive" had a high value of 0.765 comes under the violation category. In the third group factor loading values ranging from 0.726 to 0.514, "To express your annoyance to another road user, use your horn" had a high value of 0.726 comes under the violation category. From factor loading analysis all the three maximum values related to violations only means that the driving behaviour of participates has deviated in violations. The proper concern will take place to reduce accidents in terms of violations.

Conclusions

To prevent road accident crashes studying driver behaviour is essential. The study examines Indian drivers' behavior using the Driving Behaviour Questionnaire (DBQ) through factor analysis. The questionnaire comprises of 28 items with a five-point Likert scale, and 506 responses were analyzed. The study identified three important driving behavior factors: violations, errors, and lapses. EFA was applied to confirm the validity and reliability of the DBQ. 'Violations' scored the highest mean, suggesting a need for strict enforcement and public awareness to enhance road safety in India.

The study's limitations include a small number of questionnaire items (28) and only three factors considered. Future research should use a more comprehensive questionnaire to explore driver behavior further. Conducting the survey online might have made it difficult for some drivers to respond, so the next phase will involve preparing a local language questionnaire to address this issue. In the future study, important driving errors will be analyzed with the help of a driving simulator.

References

1. WHO (2017). Road Traffic Injuries. Geneva: World Health Organization.
2. Transport Research Wing (2019). Road Accidents in India. New Delhi: Government of India.
3. Sajan, S., and Ray, G. G. (2012). Human factors in safe driving—a review of literature on systems perspective, distractions and errors. In IEEE Global Humanitarian Technology Conference.
4. Choudhary, P., and Velaga, N. R. (2017). Mobile phone use during driving: effects on speed and effectiveness of driver compensatory behaviour. *Accident Analysis and Prevention*, 106, 370–378.
5. Reason, J., Manstead, A., Stradling, S., Baxter, J., and Campbell, K. (1990). Errors and violations on the roads: a real distinction. *Ergonomics,* 33(10–11), 1315–1332.
6. Stephens, A. N., and Fitzharris, M. (2016). Validation of the driver behaviour questionnaire in a representative sample of drivers in Australia. *Accident Analysis and Prevention*, 86, 186–198.
7. Rosli, N. S., et al. (2017). Testing the driving behavior questionnaire (DBQ) on Malaysian drivers. In The Proceeding of the 12th Malaysian Universities Transport Research Forum Conference (MUTRFC).
8. Young, K. L., and Salmon, P. M. (2012). Examining the relationship between driver distraction and driving errors: a discussion of theory, studies and methods. *Safety Science*, 50(2), 165–174.
9. Guéhoa, L., Graniéa, M. A., and Abric, J. C. (2014). French validation of a new version of the driver behavior questionnaire (DBQ) for drivers of all ages and level of experiences. *Accident Analysis and Prevention,* 63, 41–48.
10. Martinussen, L. M., et al. (2013). Age, gender, mileage and the DBQ: The validity of the driver behavior questionnaire in different driver groups. *Accident Analysis and Prevention,* 52, 228–236.
11. Bener, A., et al. (2019). The effect of aggressive driver behaviour, violation and error on vehicle crashes involvement in Jordan. *International Journal of Crashworthiness*, 25(3), 276–283.
12. Cordazzo, S. T. D., et al. (2016). Modernization of the driver behaviour questionnaire. *Accident Analysis and Prevention,* 87, 83–91.
13. Rowe, R., et al, (2014). Measuring errors and violations on the road: a bifactor modeling approach to the driver behavior questionnaire. *Accident Analysis and Prevention,* 74, 118–125.
14. Martinussen, L. M., et al, (2013). Short and user-friendly: the development and validation of the Mini-DBQ. *Accident Analysis and Prevention,* 50, 1259–1265.
15. Koppel, S., Stephens, A. N., Charlton, J. L., Di Stefano, M., Darzins, P., Odell, M., and Marshall, S. (2018). The driver behaviour questionnaire for older drivers: do errors, violations and lapses change over time. *Accident Analysis and Prevention,* 113, 171–178.
16. Ang, B. H., Chen, W. S., and Lee, S. W. (2019). The malay manchester driver behaviour questionnaire: a cross-sectional study of geriatric population in Malaysia. *Journal of Transport and Health,* 14, 100–573.
17. Matar, H. B., and Al-Mutairi, N. Z. (2020). Examining the factors affecting driver behavior in metropolitan Kuwait. *International Journal of Crashworthiness,* 26(6), 705–710.
18. Jomnonkwao, S., Uttra, S., and Ratanavaraha, V. (2021). Analysis of a driving behavior measurement model using a modified driver behavior questionnaire encompassing texting, social media use, and drug and alcohol consumption. *Transportation Research Interdisciplinary Perspectives,* 9, 100–302.

77 A literature study on the applications of blockchain technology in supply chain management

J Sreejith[1,a] and P. G. Saleeshya[2,b]

[1]Research Scholar, Department of Mechanical Engineering, Amrita School of Engineering, Amrita Vishwa Vidyapeetham, Coimbatore, India

[2]Professor, Department of Mechanical Engineering, Amrita School of Engineering, Amrita Vishwa Vidyapeetham, Coimbatore, India

Abstract: Blockchain technology (BCT) is now gaining more attention beyond the field of cryptocurrency. BCT has been implemented in variety of sectors like finance, information, legal industries etc., Transparency, immutability, and traceability are the major highlights of BCT and still the researches are going on in the field of BCT in Supply chain management. In this research paper, we are trying to explain the advantage of using BCT in the field of Supply chain and Logistics through a systematic literature study. Also, we are trying to identify the current research gap in the mentioned area. More than 30 papers with a systematic study are used in the current research.

Keywords: Agriculture, blockchain technology, supply chain, systematic mapping study.

Introduction

A blockchain is a public distributed ledger and it acts as a decentralized system for recording and documenting transactions [1]. A blockchain is an indigenous way of transferring information and data in a fully automated and safe manner. One party initiates the process by creating a block and this block verified by thousands or millions of another computers distributed around the world. And this verified block also added into the network and hence it creates a chain of blocks.

Ongoing studies are based on the implementation of Artificial intelligence (AI) and Machine Learning (ML) and Information & Communication technology in supply chain management [2–7]. Researches were carried out to strengthen the financial supply chain sector by reducing wastes [8]. BCT can be more effective in the financial sector of the supply chain and it has been proved by studies. Trade has been a vital activity in our day-to-day activities and Barter system was one of the earliest trading methods. But, too much dependency on third parties also not good for a perfect trade and hence the idea of Blockchain is evolved. An efficient consensus protocol can enhance the efficiency of Blockchain technology [9, 10].

The structure of the research paper is as follows: Section 2 consists of the Applied Research Methodology which includes the process of the collection of research papers. The

---

[a]sree8333@gmail.com, [b]pg_saleeshya@cb.amrita.edu

results and discussion is included in Section 3 and Section 4 respectively, from the extracted data. Section 5 discusses the limitations of the study and the conclusion is in Section 6.

Research Methodology

Systematic Mapping Study (SMS) Process

This research study follows the systematic mapping study (SMS) presented by Petersen et al. [11] and to search relevant papers which are covered from all years to 2019, used systematic literature review (SLR) guidelines described by Kitchenham et al [12]. The main objective of using the SMS process is to identify the research areas and research gaps in the field of BCT and supply chain. The SMS process consists of five steps and it is depicted in Figure 77.1.

Formulation of Research Questions

Five research questions were formed for the review based on the SMS process and the research questions are mentioned below:

RQ1: What are the current research areas and topics related to SCM and BCT?
RQ2: What are the pros and cons in the existing researches?
RQ3: Is there any systematic approach or model for the implementation of BCT in SCM?
RQ4: What are the existing research gaps in the field of SCM and BCT?
RQ5: Which sector in the Indian economy is less influenced by the implementation of BCT in SCM?

Detailed Method of Reviewing Process

The review was conducted for the papers published between, January 2017 to December 2019. The primary search for the papers was conducted by using the terms "Supply Chain",

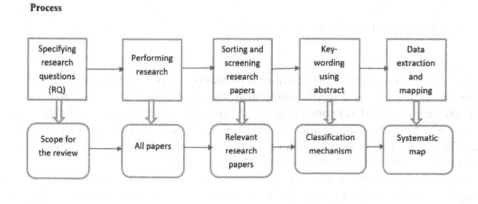

Process

Outcomes

Figure 77.1 Process and outcomes in SMS.

"Logistics" and "Blockchain" in every search. Five filter processes were used for the screening of relevant papers.

1. Duplicates, thesis, and technical reports were removed from the first set of selected research papers.
2. Next filter process was eliminating the non-relevant research papers.
3. Papers were selected which are written and published in English language.
4. The Fourth process was reading the abstract of the second round of filtered papers and remove non-relevant papers if any.
5. Final process is to read full text of all included filtered papers.

After the classification of papers based on key-wording using abstract, the next process is data extraction. The formulated research questions of the current mapping study should be addressed by the data extraction process. The information which has been extracted from each study is as follows:

1. Title of paper
2. Name and country of authors
3. Publication type
4. Abstract and methodology
5. Formulated research questions
6. Limitations and research gap of the existing study

Results

Figure 77.2 represents the results of the search and selection of papers based on the SMS process.

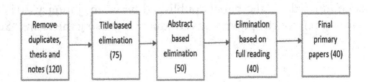

Figure 77.2 Selection process of primary papers.

Figure 77.3 represents the distribution of primary papers based on the publication year. The novelty of the area is directly visible from the chart since the selected papers are published after the year 2015. From Figure 77.3, out of all selected papers, 1 paper (2.50%) and 2 papers (5.00%) were published in 2016 and 2017 respectively. In the years 2018 and 2019, the number of selected papers were increased to 14 (35.00%) and 22 (55.00%) respectively. 1 paper (2.50%) was selected in the year 2020.

Figure 77.4 depicts the percentage distribution of the selected primary papers as per geographical areas. In the current study, the primary papers were selected from the BCT researches conducted in 19 different countries. USA and Germany gained first and second positions respectively with 6 papers (15.00%) from USA-related institutions and with 5 papers (12.50%) for Germany. The third position was occupied by India and Finland with

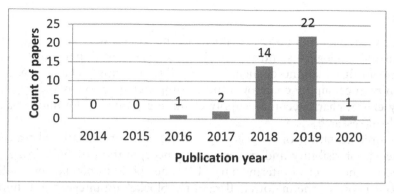

Figure 77.3 Selection of papers based on the publication years.

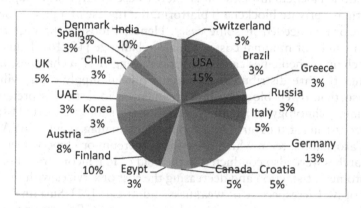

Figure 77.4 Percentage distribution of primary papers based on geographical areas.

4 papers each (10%), followed by Fi Austria with 3 papers (7.50%). The rest of the countries had published one or two papers which have 3% or 5% weightage.

Figure 77.5 represents the publication channel of selected primary papers. Out of 40 papers, 27 papers (67.50%) were published in journals and 10 papers (25.00%) were published through conferences. One white paper (2.50%) were published through World Economic Forum was also considered for the review. Rest of the papers were published in symposium and workshops (1 paper each–2.50%).

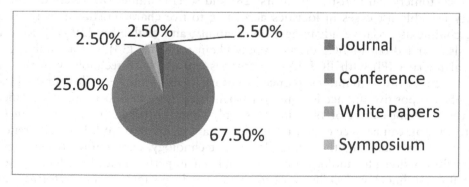

Figure 77.5 Percentage distribution of primary papers based on publication channel.

Discussions

From the segregated journal papers, majority of the studies were carried out in the implementation of BCT in logistics and supply chain management (9 papers). Then, the studies were focused on the risks and challenges of BCT in the supply chain (4 papers). BCT in supply chain finance, supply chain operations, supply chain performance, agriculture, and food supply chain, pharmaceutical supply chain, etc. gained more interest as a result of several kinds of research.

Selective implementation of BCT is required for inter-organizational transactions that need high levels of visibility and transparency. Incorporation of BCT in B2B, B2C, and C2C can reduce the effect of intermediary [13]. The ability of blockchain to displace people from work is one of the major challenges for blockchain integrated technologies [14]. Another challenge regarding BCT in supply chain is the selection of blockchain platform. An open and public blockchain platform is likely to use in large-scale supply chain applications rather than a private blockchain platform. But the use of a private or permissioned blockchain platform is needed in some cases. Hence, choosing the required blockchain platform which can be of minimal cost is a major challenge [15, 16]. Transparency of the official supply chain of medicines can be enhanced using BCT techniques. But tracking of counterfeit drugs that are distributed outside the official supply chain will be a difficult task [17]. Consortium-owed blockchain deployment in industries is preferred instead of public blockchain technology as per the opinion of the subject experts [18]. A decentralized traceability system for the Agri-Food supply chain management using AgriBlockIoT is presented by Caro and the team which use either Ethereum or Hyperledger Sawtooth [19] and other researches were also conducted in the field of IoT incorporated blockchain [20]. Enhancing customer satisfaction and decreasing the cost of services with the help of BCT is the advantage of BCT irrespective of weakness and threats [21]. Sustainability in the supply chain and logistic sector can be greatly enhanced using BCT by tracing purchase orders, order changes, and freight documents [22–24]. The application of BCT in the healthcare and retail sectors are yet to gain proven by researchers and industrial experts [25]. For better implementation of BCT in supply chain, participation of all the different actors is crucial since BCT should be incorporated into the business strategy [23]. Standardized documentation, smoothening logistics, enhanced transparency, gearing up of transactions, etc. are the few benefits of BCT in trade [26]. Integration of smart solutions such as the Internet of Things (IoT), Cloud Computing, Big Data Analytics, and BCT can be an effective method to prevent challenges and threats facing in SCM [27]. Blockchain-based logistics monitoring system (BLMS) using Ethereum enables more transparent operations for the customers and logistic operators [28] and several frameworks were developed to identify possible use cases in logistics according to five characteristics namely compatibility, complexity, relative advantage, observability, and trialability [29–32]. Researches were also carried out in BCT-driven supply chain resilience [33] and a combination of physical internet (PI) with BCT [34]. Industries can use BCT technology in the supply chain to reduce order quantities, reduce inventory levels and lower selling prices [35–37]. A significant positive impact is made on partnership growth and partnership efficiency after the implementation of BCT in the supply chain [38]. Digital supply chain (DSC) transformation can be accelerated using blockchain techniques [39]. Four different identities like information lighthouse, exploration technology, exploitation technology, and relationship-building technology improve supply chain performance but identities such as domination technology, deskiller, and straitjacket reduce supply chain performance [40].

BCT enabled manufacturing industry will be beneficial for different range of users: the consumers can access the data information about the product and on the other hand, the organization can develop their marketing and technological strategies using the information on how the products are used further along the supply chain [41]. Application of BCT in the construction industry logistic sector gained more importance because of the real-time value sharing with the end-users and constructors [42]. BCT is not a silver bullet to solve all the existing problems in SCM. Understanding of the existing issues is required before selecting the tool to eliminate problems; otherwise, it results in a less than optimal task-technology-fit and finally poor results of the project [43] [44]. A circular economy can be enabled with the use of smart contracts which manage the entire supply chain processes by removing intermediaries [45–47]. Securing the origin of multiple identities, improve inventory management, help to detect fraud, fast-tracing of goods and build customer trust, etc. are the value creation activities through BCT in supply chain management [48, 49]. Scalability, implementation of blockchain for precision agriculture and food supply chain network design and interoperability, etc. are the main challenges for the application of BCT in the agriculture supply chain [50]. BCT can be also used as a risk mitigation tool in supply chain management by enhancing the supply chain performance and efficiency [51].

Conclusion and Future Scope of Work

Blockchain technology adoption in supply chain management is one of the promising researches in the current scenario. This research paper focused on the pros and cons of BCT in supply chain management with the aid of a literature mapping study. Blockchain is not a silver bullet technology for all the existing problems in the supply chain even though the technology possesses so many advantages like traceability of different parameters like location, process data, the environment of goods, security, etc.. Researches were carried out in the manufacturing supply chain, construction supply chain, logistics sector, pharmaceutical supply chain, agriculture supply chain, and food supply chain, etc. IoT incorporated blockchain applications in supply chain and logistic sector can be well executed to enhance the supply chain performance and efficiency.

In conclusion, the implementation of BCT in the supply chain is not the final solution for all the existing problems. First, the organization needs to realize the exact issue, carry out root cause analysis and then carefully select the tool to eliminate the problems. Incorporating other smart technologies like IoT, cloud computing, big data analytics with blockchain can be a future technology which acts as an enabler in supply chain management.

The agriculture supply chain is facing many challenges like too many intermediaries and information asymmetry. These mentioned problems have led to the formation of multiple intermediaries, long marketing channels, cost inflation, etc. Hence, implementing BCT in the agriculture supply chain can be a solution to these problems but still, studies should carry out on these issues.

References

1. Fu, D., and Fang, L. (2016). Blockchain-based trusted computing in social network. In 2nd IEEE International Conference on Computer and Communications, (pp. 19–22).
2. Min, H. (2010). Artificial intelligence in supply chain management: theory and applications. *International Journal of Logistics: Research and Applications*, 13(1), 13–39.

3. Bousqaoui, H., Achchab, S., and Tikito, K. (2017). Machine learning applications in supply chains, An emphasis on neural network applications. In 3rd International Conference of Cloud Computing Technologies and Applications (Cloudtech), (pp. 1–7).

4. Priore, P., Ponte, B., Rosillo, R., and de la Fuente, D. (2018). Applying machine learning to the dynamic selection of replenishment policies in fast-changing supply chain environments. *International Journal of Production Research*, 1–15.

5. Carbonneau, R., Laframboise, K., and Vahidov, R. (2008). Application of machine learning techniques for supply chain demand forecasting. *European Journal of Operational Research*, 1140–1154.

6. Soleimani, S. (2018). A perfect triangle with: artificial intelligence, supply chain management and financial technology. *Archives of Business Research*, 6(11), 85–94.

7. Nair, P. R., and Dr. Anbuudayasankar, S. P. (2016). An investigation on the benefits of ICT deployment in supply chain management (SCM). *Indian Journal of Science and Technology*, 9(30), 1–7.

8. Dr. Saleeshya, P. G., Rajan, V., and Premkumar, N. (2019). Optimising the financial performance of supply chain—a case study. *International Journal of Advanced Operations Management*, 11, 232–255.

9. Sankar, L. S., Sindhu, M., and Sethumadhavan, M. (2017). Survey of consensus protocols on blockchain applications. In 4th International Conference on Advanced Computing and Communication Systems, (pp. 1–5).

10. Sajana, P., Sindhu, M., and Sethumadhavan, M. (2018). On blockchain applications: hyperledger fabric and etheruem. *International Journal of Pure and Applied Mathematics*, 118(18), 2965–2970.

11. Petersen, K., Feldt, R., Mujtaba, S., and Mattsson, M. (2008). Systematic mapping studies in software engineering. In Proceedings of the 12th International Conference on Evaluation and Assessment in Software Engineering, (pp. 68–77).

12. Kitchenham, B., Pretorius, R., Budgen, D., Brereton, O. P., Turner, M., Niazi, M., and Linkman, S. (2010). Systematic literature reviews in software engineering—a tertiary study. *Information and Software Technology*, 52, 792–805.

13. Kumar, A., Liu, R., and Shan, Z. (2019). Is blockchain a silver bullet for supply chain management? technical challenges and research opportunities. *Decision Sciences*, 1–30.

14. Boschi, A. A., Borin, R., Raimundo, J. C., and Batocchio, A. (2018). An exploration of blockchain technology in supply chain management. In 22nd Cambridge International Manufacturing Symposium, (pp. 1–12).

15. Litke, A., Anagnostopoulos, D., and Varvarigou, T. (2019). Blockchains for supply chain management: architectural elements and challenges towards a global scale deployment. *Logistics*, 1–17.

16. Treiblmaier, H. (2018). The impact of the blockchain on the supply chain: a theory-based research framework and a call for action. *Supply Chain Management: An International Journal*, 23(6), 545–559.

17. Bryatov, S. R., and Borodinov, A. A. (2019). Blockchain technology in the pharmaceutical supply chain: researching a business model based on hyperledger fabric. In V International Conference on Information Technology and Nanotechnology, (pp. 1–7).

18. Hellwig, D. P. (2019). The feasibility of blockchain for supply chain operations and trade finance: an industry study. *American Council on Germany*, 1–33.

19. Caro, M. P., Ali, M. S., Vecchio, M., and Giaffreda, R. (2018). Blockchain-based traceability in agri-food supply chain management: a practical implementation. IoT Vertical and Topical Summit on Agriculture—Tuscany, (pp. 1–4).

20. Kshetri, N. (2018). Blockchain's roles in meeting key supply chain management objectives. *International Journal of Information Management*, 39, 80–89.

21. Dujak, D., and Sajter, D. (2019). Blockchain Applications in Supply Chain. SMART Supply Network, (pp. 21–46).

22. Tijan, E., Aksentijevic, S., Ivanic, K., and Jardas, M. (2019). Blockchain technology implementation in logistics. *Sustainability*, 11(1185), 1–13.

23. Perboli, G., Musso, S., and Rosano, M. (2018). Blockchain in logistics and supply chain: a lean approach for designing real-world use cases. *IEEE Access*, 6, 62018–62028.

24. Saberi, S., Kouhizadeh, M., Sarkis, J., and Shen, L. (2018). Blockchain technology and its relationships to sustainable supply chain management. *International Journal of Production Research*, 1–20.

25. Blossey, G., Eisenhardt, J., and Hahn, G. J. (2019). Blockchain technology in supply chain management: an application perspective. In Proceedings of the 52nd Hawaii International Conference on System Sciences, (pp. 6885–6893).

26. Norberg, H. C. (2019). Unblocking the bottlenecks and making the global supply chain transparent: how blockchain technology can update global trade. *The School of Public Policy Publications*: SPP Briefing Paper, 12(9), 1–24.

27. El Mesmary, H., and Said, G. A. E. N. A. (2019). Smart solutions for logistics and supply chain management. *International Journal of Recent Technology and Engineering*, 8(4), 2996–3001.

28. Helo, P., and Hao, Y. (2019). Blockchains in operations and supply chains: a model and reference implementation. *Computers and Industrial Engineering*, 136, 242–251.

29. Dobrovnik, M., Herold, D. M., Furst, E., and Kummer, S. (2018). Blockchain for and in logistics: what to adopt and where to start. *Logistics*, 2(18), 1–14.

30. Shamout, M. (2019). Understanding blockchain innovation in supply chain and logistics industry. *International Journal of Innovative Technology and Exploring Engineering*, 8(6C2), 309–315.

31. Hackius, N., and Petersen, M. (2017). Blockchain in logistics and supply chain: trick or treat?. In Proceedings of the Hamburg International Conference of Logistics (HICL)–23, (pp. 1–17).

32. Koh, L., Dolgui, A., and Sarkis, J. (2020). Blockchain in transport and logistics—paradigms and transitions. *International Journal of Production Research*, 58(7), 2054–2062.

33. Min, H. (2018). Blockchain technology for enhancing supply chain resilience. *Business Horizons*, 1–11.

34. Treiblmaier, H. (2019). Combining blockchain technology and the physical internet to achieve triple bottom line sustainability: a comprehensive research agenda for modern logistics and supply chain management. *Logistics*, 3(10), 1–13.

35. Chang, J. A., Katehakis, M. N., Melamed, B., and Shi, J. J. (2018). Blockchain design for supply chain management. *SSRN Electronic Journal*, 1–35.

36. Niu, X., and Li, Z. (2019). Research on supply chain management based on blockchain technology. *Journal of Physics: Conference Series*, 1176, 1–8.

37. Reddy, V. P. V. (2019). Enhancing supply chain management using blockchain technology. *International Journal of Engineering and Advanced Technology*, 8(6), 4657–4661.

38. Kim, J. S., and Shin, N. (2019). The impact of blockchain technology application on supply chain partnership and performance. *Sustainability*, 11(6181), 1–17.

39. Korpela, K., Hallikas, J., and Dahlberg, T. (2017). Digital supply chain transformation toward blockchain integration. In Proceedings of the 50th Hawaii International Conference on System Sciences, (pp. 4182–4191).

40. Hald, K. S., and Kinra, A. (2019). How the blockchain enables and constrains supply chain performance. *International Journal of Physical Distribution & Logistics Management*, 1–25.

41. Abeyratne, S. A., Monfared, R. P. (2016). Blockchain ready manufacturing supply chain using distributed ledger. *International Journal of Research in Engineering and Technology*, 5(9), 1–10.

42. Sivula, A., Shamsuzzoha, A., and Helo, P. (2018). Blockchain in logistics: mapping the opportunities in construction industry. In Proceedings of the International Conference on Industrial Engineering and Operations Management, (pp. 1–7).

43. Verhoeven, P., Sinn, F., and Herden, T. T. (2018). Examples from blockchain implementations in logistics and supply chain management: exploring the mindful use of a new technology. *Logistics*, 2(20), 1–19.

44. Cole, R., Stevenson, M., and Aitken, J. (2019). Blockchain technology: implications for operations and supply chain management. *Supply Chain Management: An International Journal*, 24(4), 469–483.

45. Casado-Vara, R., Prieto, J., De la Prieta, F., and Corchado, J. M. (2018). How blockchain improves the supply chain: case study alimentary supply chain. *Procedia Computer Science*, 134, 393–398.

46. Saini, V. K., and Gupta, S. (2019). Blockchain in supply chain: journey from disruptive to sustainable. *Journal of Mathematical and Computational Science*, 14(2), 498–508.

47. Warren, S., Wolff, C., and Hewett, N. (2019). Inclusive deployment of blockchain for supply chains: part 1-introduction. *World Economic Forum*, 1–26.

48. Koul, S. (2018). Value creation through blockchain technology in supply chain management. *Journal of Information Technology and Software Engineering*, 8(5), 248–249.

49. Mavale, S. J., and Bhosale, J. (2019). A review of 'blockchain for effective supply chain management' current progress and future ahead. *Annual Research Journal of SCMS*, 7, 28–34.

50. Yadav, V. S., and Singh, A. R. (2019). A systematic literature review of blockchain technology in agriculture. In Proceedings of the International Conference on Industrial Engineering and Operations Management, (pp. 973–981).

51. Layaq, M. W., Goudz, A., Noche, B., and Atif, M. (2019). Blockchain technology as a risk mitigation tool in supply chain. *International Journal of Transportation Engineering and Technology*, 5(3), 50–59.

78 A highly efficient tapering bimorph rotational energy harvester subjected to dynamic axial excitation

Rakesh Ranjan Chand[1,a] and Amit Tyagi[2,b]

[1]Department of Mechanical Engineering, C.V. Raman Global University, Bhubaneswar, Odisha, India

[2]Mechanical Engineering Department, IIT (BHU), Varanasi, Uttar Pradesh, India

Abstract: In this research article, the energy generating capability of a piezoelectric tapering bimorph cantilever structure mounted on a rotating hub under an axial dynamic load is studied. The mathematical modelling and the system's motion equations are formulated using Hamilton's equation. Subsequently, the expression for the harvested rotational vibration energy, which is induced in the piezoelectric coupled beam structure because of the rotational motion and the external axial load, is given. Numerical simulations are accomplished using MATLAB software to investigate the system's modal frequencies and the piezoelectric coupled system's power scavenging performance. Effects of tapering, the rotational speed variation, and the piezoelectric patches to the substrate's thickness ratio on the first three modal frequencies and the energy harvesting performance are investigated and presented. The proposed rotational vibration energy scavenger has an improved harvested energy density since the maximum harvested power at both the piezo-patches can be accomplished using the model. Results demonstrate that the suggested rotational vibration energy scavenger can supply sufficient power to low-power wireless transmitters and sensors used for concurrent condition monitoring of smart vehicle wheels, rotating machinery such as propeller blades and tire pressure monitoring system (TPMS), etc., by appropriate dimension and parameter choice.

Keywords: Axial dynamic loading, piezoelectric-composite, rotational system, tapered-bimorph, vibration energy harvesting.

Introduction

Much interest is being shown in the development of a mechanism that can operate remotely and harvest abundant energy and provide power to wireless electronic devices for improved safety, health monitoring, and fault detection mechanism for rotational applications. Sodano et al. [10] and Anton [1] conducted a thorough assessment of cutting-edge methods for power harvesting with piezoelectric materials. Over the years, there has been marginal growth in chemical battery techniques for increased longevity and issues related to maintenance. For power scavenging in remote sites, a variety of power supply technologies have been taken into consideration. To power micro and small electronic devices, however, the piezoelectric harvesting technique has produced high power density utilizing relatively small size systems, as given by Saddon and Sidek [9]. By utilizing

[a]rakeshranjan.chand@cgu-odisha.ac.in, [b]atyagi.mec@iitbhu.ac.in

the gravitational excitation induced on the mass fixed on the tip of a cantilever beam with attached PZT (lead-zirconate-titanate) and (PVDF) polyvinylidene fluoride patches, Khameneifer et al. proposed a theoretical configuration [7] and presented experiments [6] of a PZT based energy harvester for rotary motion excitation. However, following the high rotating frequency of 18 Hz, they could only generate 6.4 mW of power. The total electric charge is directly connected to the amount of mechanical stress applied to the piezoelectric patch. The stress distribution along the substrate geometry must be optimized to enhance the piezoelectric materials' capacity to generate electricity. Due to the underutilization of piezoelectric potential, resulted from the unequal distribution of stress in the substrate beam, i.e., the most significant strain occurred closer to the fixed end, as proposed by Li et al. [8]. Therefore, commonly used piezoelectric energy harvesters perform less efficiently. These factors suggest that a cantilever with a trapezoidal or triangular shape will result in a better distribution of stress and reduce the weight and size of the energy harvester. Xie et al. [11] calculated the RMS power to be 0.1376 W at a rotational frequency of 10 rad/s in their theoretical investigation for a tapering shape cantilever type vibration energy harvester. They also examined the consequence of the tapering parameter, patch widths, etc., on the energy harvesting performance. However, due to the intricate system configuration, experimental analysis of their model is hard to realise.

Most piezoelectric energy harvesters rely on a conventional cantilever structure, and the tapered cantilever harvester proposed is complicated for practical experiments and applications. Therefore, it's the aim to develop a tapered rotating piezoelectric energy scavenger, which will be easy to fabricate and provide a higher power density. With this proposed model, a relatively higher power density is achieved. Various parameters such as the taper parameter, PZT thickness, and rotational frequency on the energy harvesting performance are also investigated.

System Modelling

The main part of the suggested rotational piezoelectric tapered vibration energy harvester (RVEH), shown in Figure 78.1, is a bi-morph piezoelectric coupled tapered beam with a rectangular area of cross-section and linear convergence from fixed to free end. The PZT coupled beam is positioned on a hub with a radius of R_0 and rotates parallel to the surface of the Earth about its Z'-axis. The system is exposed to an axial pulsating forceat $P(t) = P\cos(\omega t)$ the centroid of the cross-section of the free end at $x' = R_0 + l$, where l is the total span of the beam, ω is the excitation frequency, and P is the amplitude of the dynamic load. As shown in the figure, t_p and t_h are the PZT patches and the substrate beam thicknesses, correspondingly, b and $b(x)$ are the widths of the coupled beam at the clamped side and at $x' = R_0 + x$ whereas $A_p(x)$ and $A_h(x)$ are the area of the cross-sections of the PZT patches and the substrate beam respectively. It is assumed that the substrate material is homogeneous and isotropic; extensional deflection is minimal. Euler-Bernoulli's beam theory is applied to the beam. Every point in a cross-section experiences a small and small transverse deflection $w(x,t)$, and the effects of rotational and extensional inertia are minimal. The substrate's width and the patches are considered to be varying according to $b(\phi) = b(1 - \phi \beta)$, where $\phi = x/l$, $\beta = 1 / \left(\dfrac{l}{b} \right)$, is the width taper parameter. Accordingly, $A_p(\phi) = A_p(1 - \phi\beta)$, $A_h(\phi) = A_h(1 - \phi\beta)$, $I_p(\phi) = I_p(1 - \phi\beta)$, and $I_h(\phi) = I_h(1 - \phi\beta)$, are the cross-section and

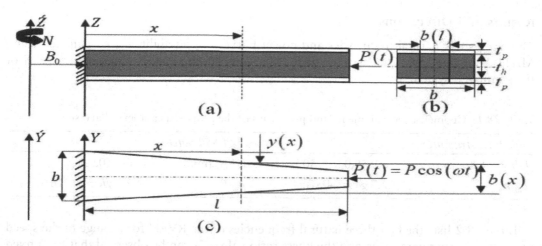

Figure 78.1 Schematic of the proposed RVEH (a) front view(b) side view (c) top view.

the MOI at any standard section. In order to investigate the RVEH's mode shapes and natural frequencies, the numerical formulations are adopted from Chand et al. [2–4]

As proposed by Guan and Liao [5], the approximate formulation to calculate the electrical power generated over an optimal resistive load can be written as

$$P_t = \left(\frac{ma^2}{\omega_n}\right)\left(\frac{1}{(r\Omega + \pi/2)^2}\right)\left(\frac{k_e^2\Omega^2 r}{\left[2\xi + \dfrac{2k_e^2 r}{(r\Omega + \pi/2)^2}\right]^2 \Omega^2 + \left(1 - \Omega^2 + \dfrac{\Omega k_e^2 r}{r\Omega + \pi/2}\right)^2}\right) \quad (1)$$

where a is the base acceleration, in this case, the system operates essentially parallel to the horizontal plane, so its magnitude g always remain constant, m is the modal mass, ω_n is the system's natural frequency, k_e^2 is the electromechanical coupling coefficient, $\Omega(=\omega/\omega_n)$ is the frequency ratio, ξ is the mechanical damping ratio and the normalized resistance is

$$r_n = R_{opt}\omega C_p \quad (2)$$

in which R_{opt} is the optimum resistive load and C_p is the piezoelectric patch's capacitance. The capacitance of the patches can be written as;

$$C_p = \frac{k\varepsilon_0 A}{t_p} \quad (3)$$

where k is the relative dielectric constant of the piezo-material, ε_0 is the permittivity of the space ($\varepsilon_0 = 8.9 \times 10^{-12}$ f/m), A is the surface area of the electrode. The peak output open-circuit voltage response can be related to the peak output power by;

$$V_{oc} = \sqrt{4P_t \Big/ C_p\omega} \quad (4)$$

Results and Discussions

The harvester's natural frequencies and power harvesting capabilities are calculated using MATLAB code. Table 78.1 is a list of the characteristic properties and measurements of the simulated system.

Table 78.1: Geometric measurements and properties of the proposed cantilever harvester.

Substrate structure		PZT-850 patches	
l, b, t_h, R_0 (mm)	60, 50,0.3-1, 50	l, b, t_p (mm)	60, 50,0.2
ρ_h, E_h	2.7 g/cm³, 4.5GPa	ρ_p, E_p	7.6 g/cm³, 63GPa

Table 78.2 lists the first three natural frequencies of the RVEH for a range of the speed of rotation, taper parameter, and thickness ratio values. It can be observed that for a fixed taper parameter $\beta = 0.5$, $t_p = 0.2$ mm, and $t_h = 0.7$ mm, the natural frequencies move toward higher values with an increase in the rotational speed. This is credited to the increase in the centrifugal force, diminishing the consequence of the external dynamic axial load. An increase in the tapering parameter increases the natural frequency values because of the rise in the harvester's flexural stiffness. The PZT-patches to substrate's thickness ratio are varied by changing the substrate structure's thickness and keeping the piezoelectric patch thickness constant. Due to the harvester's decreased stiffness with an increase in the thickness ratio and the system's decreased mass, a fall in the natural frequency values of the system is noticed as the thickness ratio rises.

Table 78.2: Three principal natural frequencies (Hz) of the system for different speeds of rotation, thickness ratio, and tapering parameter.

$N = 2, t_p = 0.2, t_h = 0.7$			$\beta = 0.5, t_p = 0.2, t_h = 0.7$				$\beta = 0.5, N = 2, t_r = (2t_p/t_h)$		
β	ω_1	ω_2	ω_3	N	ω_1	ω_2	ω_3	t_r	ω_2	ω_3
0.3	2.85	16.94	86.48	2	2.99	17.17	85.17	0.4	23.24	115.9
0.5	2.99	17.17	85.17	4	3.18	17.44	85.51	0.58	17.17	85.17
0.7	3.15	17.47	83.62	6	3.47	17.88	86.08	0.8	14.29	70.83
0.9	3.36	17.85	81.76	8	3.81	18.47	86.87	1.3	12.92	64.12

Figure 78.2 represents the effect of the rotational speed and the system's taper parameter on the peak power and the open-circuit voltage generated. The host cantilever and the PZT patches in this simulation have thicknesses; $t_h = 0.7$ mm and $t_p = 0.2$ mm, respectively. Figure 78.2(a) shows that the peak power increase from 0.0020 W to 0.0094 W and decreases from 0.0094 W to 0.0009 W, and again decreases from 0.0020 W to 0.0004 W when ω and α change from (1,0.3) to (4,0.3) and (4,0.3) to (8,0.3) and again (1,0.3) to (1, 1.1), respectively. Figure 78.2(b) shows that the peak open-circuit voltage decreases from 91.6097 V to 43.8622 V and increases from 91.6097 V to 131.1452, and then decreases to 22.6214 V ω and α change from (1,0.3) to (1,1.1) and (1,0.3) to (2,0.3), and then to (8,0.3), respectively. It can also be seen that the peak power at $\omega = 4$Hz and $\beta = 0.3$ is 23.5 times the peak power at $\omega = 1$Hz and $\beta = 1.1$. Similarly, the open-circuit voltage at $\omega = 2$Hz and $\beta = 0.3$ is 7.2 times the open-circuit voltage at $\omega = 8$Hz and $\beta = 1.1$, respectively.

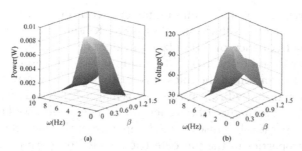

Figure 78.2 (a) Peak electric power and (b) generated voltage versus the rotational frequency and the taper parameter with t_h = 0.7 mm and t_p = 0.2 mm.

The variation of the peak-to-peak open-circuit voltage and the maximum generated power corresponding to the variation of PZT patches' thicknesses is also studied (figure not shown). It is observed that the peak power at t_h = 0.6 mm and t_p = 0.15 mm is almost four times that of the peak power at t_h = 0.9 mm and t_p = 0.1 mm. Similarly, at t_h = 0.6 mm and t_p = 0.15 mm, the open-circuit voltage is almost twice that of t_h = 0.9 mm and t_p = 0.1 mm.

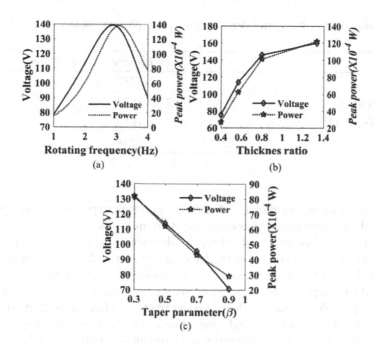

Figure 78.3 Harvested voltage and power for different (a) rotating frequency with β = 0.5 (b) thickness ratio with β = 0.5 (c) taper parameter with t_r = 0.58.

Eqs. (1) and (4) are used to calculate the maximum power and output voltage. It is observed that peak output voltage rises with an increased rotating frequency. Figure 78.3(a) displays the output peak output voltage and maximum harvested power versus the rotating frequency for l = 60 mm, b = 50 mm, t_h = 0.7 mm, and t_p = 0.2 mm. If the proposed rotational energy scavenger is applied to a revolving vehicle wheel, the vehicle's translational speed and tire diameter will be related to the wheel's rotating frequency. For evaluation, a standard motorcycle tire of 431.8 mm diameter is considered, for which 1 Hz rotating frequency corresponds to the vehicle speed of 4.883 Km/hr. The output voltage and peak power for a rotating frequency of 3.05 Hz, equivalent to the vehicle's translational speed of 14.893 Km/hr, are found to be 137.7217 V and 0.0138 W, respectively. Figure 78.3(b) presents the effect of variation in the thickness ratio on the peak power and the peak output voltage of the rotational vibration energy scavenger for N = 2rps, b = 50 mm, l = 60 mm, and β = 0.5. The performance of the energy scavenger is observed to be enhanced with the rise in the thickness ratio because of the dominance of the piezoelectric properties over the host structure properties in the piezoelectric coupled bi-morph cantilever system and also due to the linear decrease of the first modal frequency from 4.06 Hz to 2.24 Hz. The harvester's power-generating performance is diminished with the taper parameter increase for N = 2rps, t_p = 0.2 mm, and t_h = 0.7 mm, as shown in Figure 78.3(c). This is because the total volume of piezoelectric patches decreases with the increase in the taper parameter.

The power density concept, which is equal to the peak power produced by the harvester divided by the system's total volume, is used to compare the present model's performance with the other reported articles. Table 78.3 displays the power density obtained from the current analysis and the results of earlier reported articles. It can be observed that the present model is more efficient in comparison to the previously reported models.

Table 78.3: Comparison of the power density of the bimorph rotating tapered vibration energy harvester.

Articles	Power density ($\mu W/mm^3$)
Present study	5.5757
Khameneifer et al. [6]	3.5833
Glynne-Jones et al. [12]	0.1619
Sodano et al. [10]	4.6236

Conclusions

The power harvesting performances of a novel bimorph tapering width rotating piezo-electric-coupled vibration energy harvesting system under the influence of a dynamic axial load are presented in this article. Analytical solutions and the power harvesting performance of the RVEH have been obtained by developing a MATLAB program. The effects of the rotational speed, tapering parameter, and thickness variation on the system's natural frequencies and the output power responses are studied. The system's stability is improved by increasing the revolution speed, the taper parameter, and decreasing the thickness ratio. Numerical results show an enhanced performance of the harvester with an increase in thickness ratio. The rise in the taper parameter drastically reduces the harvester's performance, but the harvester's power density has not essentially decreased. For a harvester with b = 50 mm, l = 60 mm, t_h = 0.7 mm, t_p = 0.2mm, ω = 3 Hz and β = 0.5, the peak power can reach up to 0.0138 W, which is adequate for charging the batteries or super-capacitors of wireless sensors used for TPMS, smart wheels, robotic manipulators, and real-time status monitoring of turbine and propeller blades. However, for the proper assessment of the performance of the given RVEH model, substantial experimental investigations in realistic operating settings are required.

References

1. Anton, S. R., and Sodano, H. A. (2007). A review of power harvesting using piezoelectric materials (2003-2006). *Smart Materials and Structures*, 16, 09544062231199564. doi: 10.1088/0964-1726/16/3/R01.

2. Chand, R. R., and Tyagi, A. (2022). Investigation of the effects of the piezoelectric patch thickness and tapering on the nonlinearity of a parabolic converging width vibration energy harvester. *Journal of Vibration Engineering and Technologies,* 10, 1–18. doi: 10.1007/s42417-021- 00359-x.

3. Chand, R. R., and Tyagi, A. (2023). Parabolic tapering piezoelectric rotational energy harvester: numerical analysis with experimental validation. *Mechanics of Advanced Materials and Structures,* 30, 3652–3661. doi: 10.1080/ 15376494.2022.2080893.

4. Chand, R. R., and Tyagi, A. (2023). Design and experimental validation of an exponentially tapering width rotational piezoelectric vibration energy harvester. *Journal of Intelligent Materials Systems and Structures,* 34, 15–28. doi: 0.1177/ 1045389X221093315.

5. Guan, M., and Liao, W. H. (2016). Design and analysis of a piezoelectric energy harvester for rotational motion system. *Energy Conversion and Management,* 111, 239–244. doi: 10.1016/j.enconman.2015.12.061

6. Khameneifar, F., Arzanpour, S., and Moallem, M. (2013). A piezoelectric energy harvester for rotary motion applications: design and experiments. *IEEE/ASME Transactions on Mechatronics,* 18, 1527–1534. doi: 10.1109/TMECH.2012.2205266.

7. Khameneifar, F., Moallem, M., and Arzanpour, S. (2011). Modeling and analysis of a piezoelectric energy scavenger for rotary motion applications. *Journal of Vibration and Acoustics*, 133, 011005. doi: 10.1115/1.4002789.

8. Li, H., Tian, C., and Deng, Z. D. (2014). Energy harvesting from low frequency applications using piezoelectric materials. *Applied Physical Review,* 1, 0–20. doi: 10.1063/1.4900845.

9. Saadon, S., and Sidek, O. (2011). A review of vibration-based MEMS piezoelectric energy harvesters. *Energy Conversion and Management,* 52, 500–504. doi: 10.1016/j.enconman.2010.07.024.

10. Sodano, H. A., Inman, D. J., and Park, G. (2004). A review of power harvesting from vibration using piezoelectric materials. *Shock and Vibration Digest,* 36, 197–205. doi: 10.1177/0583102404043275.

11. Xie, X. D., Carpinteri, A., and Wang, Q. (2017). A theoretical model for a piezoelectric energy harvester with a tapered shape. *Engineering Structures,* 144, 19–25. doi: 10.1016/j.engstruct.2017.04.050.

12. Glynne-Jones, P., Beeby, S. P., and White, N. M. (2001). Towards a piezoelectric vibration-powered microgenerator. *IEE Proceedings - Science, Measurement and Technology,* 148, 68–72. doi: 10.1049/ip-smt:20010323.

79 Experimental study and optimization parameters of turning UD-GFRP in cryogenic condition by using PCD tool with Taguchi method

Hazari Naresh[1,a] and Chinmaya Prasad Padhy[2,b]

[1]Research Scholar, Department of Mechanical Engineering, School of Technology, GITAM University, Hyderabad, India

[2]Associate Professor, Department of Mechanical Engineering, School of Technology, GITAM University, Hyderabad, India

Abstract: Cryogenic condition of composites is very difficult for machining due to in homogeneous and anisotropic atmosphere and it require unique cutting tools. At present machining of composite is importance of faster and other light weight application. This work is going to examine in the cryogenic condition machining of tool wear, quality of the surface and forces generated in the various stages of inputs given to the machining of UD-GFRP composite. The optimization technique i.e. Taguchi L9 orthogonal array is used for examinations and find out the targeted. Here considering the input cutting parameters i.e., cutting speed, depth of cut and feed rate, environmental parameters i.e., cryogenic condition, cutting tool parameters i.e., tool geometry, tool material, work piece material parameters i.e., UD-GFRP composite materials. The major focus of the work on the monetary condition for receiving better values based on input parameters given. The investigated values are obtained minimize cutting forces, surface roughness and tool wear. It is believed that the used method provides a robust way of looking at the optimum parameter selection problems.

Keywords: cryogenic condition, cutting parameters, L9 Taguchi optimization and surface roughness, polycrystalline diamond (PCD) tool, UDGFRP composites.

Introduction

Many parts are of made up of carbon fiber reinforced plastics (CFRP) and are regularly delivered close to required shape. Turning, processing and boring, notable from the metal machining, additionally have a place with the much of the time utilized machining forms for CFRP parts. CFRP machining is of absolute significance for beginning sequential creation of high accuracy parts.

Cryogenic machining, as substitute to conventional machining and must enhance machinability of hard-to-cut the materials like assort aluminum, titanium and abrasive composites. Maximum temperatures occurred throughout machining of such materials may manipulate tool wear, surface quality and geometrical accurateness of the machined work piece. One of the aims of the cryogenic machining is to take out heat generated for the period of machining at critical zone and getting well again surface roughness. In cryogenic machining, tremendous cold liquefied gases of oxygen, nitrogen, hydrogen or helium are used as coolants. Among such liquefied gases, liquid nitrogen is the most used one because it is abundant.

---

[a]hazarinaresh@gmail.com, [b]padhy.gitam@gmail.com

Machining of glass fiber composite is a major crisis since high hardness and inert nature. Because of their dissimilar applications, the need for FRP machining has not been fully eliminated. The mechanism of machining GFRP is quite different from metals because of non-homogenous, anisotropic nature.

Literature Survey

Examination and explores the effect of utilizing cryogenic fluid nitrogen and least amount greasing up coolants on the boring quality in GLARE 2B11/10-0.4 fiber metal covers and pneumatic force on the gap quality. The outcomes show that utilizing cryogenic and least amount grease coolants can significantly diminish leave burr arrangement Giasin et al [1]. Furthermore, the pressure of cryogenic coolant on the tensile properties, the microstructure of surface, and machined surface of the CFRP laminates were analyzed with scanning electron microscopy (SEM). Morkavuka et al. [2] The exploratory investigation on front line span of boring tool, external corner wear of boring tool, trust power, force, The discoveries show that cryogenic cooling pro fondly affects diminishing the front lined just in boring tool and external corner wear; it additionally helps upgrading the surface trust worth in qualities of delivered opening Xia1 and Kaynak et al [3]. The cryogenic investigation presents major device math examinations dependent on symmetrical cutting of unidirectional CFRP-material. The examination has demonstrated the influence of the fiber direction and device calculation on an ensuing report breaks down the surface unpleasantness just as the presence of the chip root Subsurface harms are assessed utilizing micrograph segments. Accomplished work piece quality has been broken down in another ensuing investigation Henerichs et al. [4].

Methodology

This paper deals with the output parameter of surface roughness, tool wear and cutting forces, we have considered input parameters as given Table 79.1. The Glass fiber reinforced plastic (GFRP) rods of 20 mm diameter are produced by pultrusion process. The fabricated GFRP composite rods are cut into small segments of length 150 mm and machined in the CNC machine by changing input parameters (speed, feed, and depth of cut) and the cutting forces is estimated by the dynamometer setup. After each trial the wear of the tool is estimated by using the tool maker's microscope. The composite rods are then taken into the surface roughness tester to find out surface roughness values. These resultant parameters are assessed by using the Taguchi and Minitab programming to find economical values with respect to given input values [5–11].

Table 79.1: The fixed Parameters and levels of variable process parameters.

Input parameters	stage 1	Stage 2	Stage 3
Depth of the cut in mm	0.5	1	1.5
Tool feed in mm/rev	0.05	0.1	0.15
Cutting speed in rpm (m/min)	1000 (1047.19)	1200 (1256.63)	1500 (1570.79)
Tool material	PCD	PCD	PCD
Tool rake angle	6	6	6
Cutting Environment	Cryogenic	Cryogenic	Cryogenic

The Parameters are speed of cutting, rate of feed and depth of cut, Environment parameters: Cryogenic condition, Cutting tool parameters: tool geometry, tool material, Work piece material: UD-GFRP. Cutting fluids cryogenic physics is a branch of physics which deals with the production of very low temperatures. In this research to get the behavior of machining of composite under this condition, here liquid nitrogen is used for machining of composite and examine the various parameters. In general liquid nitrogen has –196 °C [–320°F].

Experimentation

Here the glass fiber reinforced composite rod in turning was done with the different machinability trials. Different setup and tools used for the experiments, and machined components are shown in Figurers 1–9. The properties of composite bar is given in Table 79.4. The cutting tool used is PCD inserts shown in Figure 79.5 and its description furnished in Table 79.3. The GFRP bar of 20 mm diameter, 150 mm long jobs prepared and experimented in SUPER JOBBER 500 CNC (SIEMENS 802 D SL) turning center Figure 79.1.

Figure 79.1 CNC super jobber.

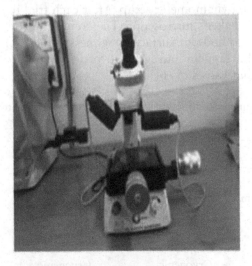

Figure 79.2 Tool maker's microscope.

Figure 79.3 Coolant setup.

Figure 79.4 Surface roughness tester .

Figure 79.5 The PCD insert.

Figure 79.6 CNC lathe dynamometer.

Figure 79.7 CNC machine setup.

The cryogenic machining environment helps in extends tool life and save the UD-GFRP composite from thermal damages. The flow rate maintained at 1 liter per minute. The Mitutoyo make tester employed for measuring surface peaks on machined work pieces. The analysis of variance (ANOVA) Taguchi L9 orthogonal array has been applied to consider the impact of the input parameters on its reaction, surface harshness, Tool wear and cutting forces.

Figure 79.8 Machined Components.

Figure 79.9 Liquid nitrogen.

Table 79.2: Orthogonal array (L9) of Taguchi along with assigned value.

S. No.	Speed (rpm)	Feed (mm/rev)	Depth of cut(mm)	Tool material	Cutting environment	Tool rake angle
1.	1000	0.05	0.5	PCD	Cryogenic	6
2	1000	0.1	1	PCD	Cryogenic	6
3	1000	0.15	1.5	PCD	Cryogenic	6
4	1200	0.05	1	PCD	Cryogenic	6
5	1200	0.1	1.5	PCD	Cryogenic	6
6	1200	0.15	0.5	PCD	Cryogenic	6
7	1500	0.05	1.5	PCD	Cryogenic	6
8	1500	0.1	0.5	PCD	Cryogenic	6
9	1500	0.15	1	PCD	Cryogenic	6

Table 79.3: Specification of PCD tool [1].

Properties of PCD tool	
Clearance angle	6°
Grade	M10
Density	3.80e4.50 g/cm3
Hardness	1600 Vickers kg/mm2
Young's modulus	800e900 GPa
Tool rake angle	6°

Table 79.4: Properties of composite bar [1].

Description of composite	Quantity/specification
Composite's strength on compression	600 N/mm²
Youngs modulus of composite	320 N/mm²
Composite's strength on tensile	650 N/mm²
Agent of reinforcement,	Roving: E- glass

Table 79.5: Experimental observations under cryogenic machining condition.

	Tool wear response				Surface response				Cutting Force(N)			
Exp No.	V_{b1}	V_{b2}	V_{b3}	Average V_b (Avg) mm	R_{a1}	R_{a2}	R_{a3}	Average In μm R_a	F_1	F_2	F_3	Average In F_{Avg}
1	0.115	0.179	0.119	0.138	3.956	5.219	5.698	4.958	15	13	15	14
2	1.022	1.08	0.842	0.981	4.991	5.359	5.458	5.269	13	14	15	14
3	0.161	0.152	0.181	0.165	5.435	5.021	4.922	5.126	15	17	14	15
4	0.065	0.051	0.053	0.056	4.951	5.942	5.644	5.512	16	15	17	16
5	0.189	0.194	0.201	0.195	4.696	4.224	5.024	4.648	18	15	17	17
6	0.688	0.359	0.356	0.468	4.949	6.25	6.651	5.950	17	19	18	18
7	0.059	0.037	0.035	0.044	4.825	5.234	5.245	5.101	19	18	20	19
8	0.277	0.198	0.109	0.195	3.454	4.352	4.988	4.265	19	20	22	20
9	0.895	1.095	0.968	0.986	5.318	5.854	6.532	5.901	18	22	20	20

Results and Discussion

The analysis of variance (ANOVA) Taguchi L9 orthogonal array (shown in Table 79.2) has been applied to consider the impact of the input parameters on its reaction, surface harshness, tool wear and cutting forces (shown in Table 79.5).

Taguchi Analysis: Surface Response (μm) versus Speed, Feed, Depth of cut

The effect of cryogenic condition on surface response (μm) versus speed, feed, depth of cut, The below Figure 79.10 explains the main effecting parameter of surface roughness for different speed, Feed and depth of cut is as Speed for 1200 rpm, feed 0.05 mm/rev and depth of cut 1.0 mm gives economical surface roughness meanwhile it will followed least is better The exact vales are shown in Table 79.6.

Here response Table 79.7 for means of means given according to the Figures 79.11, the three experimental values of surface roughness which are took by only 1000 rpm i.e. surface roughness of all 1000 rpm speed were calculated. The below Figure 79.11 explains the means of mean parameter of surface roughness for different speed, feed and depth of cut is as Speed for 1500 rpm, feed 0.15 mm/rev and depth of cut 0.5 mm gives economical surface roughness meanwhile it will followed by speed 1000 rpm, feed 0.1 mm/rev and depth of cut 1.0 mm gives moderate surface roughness.

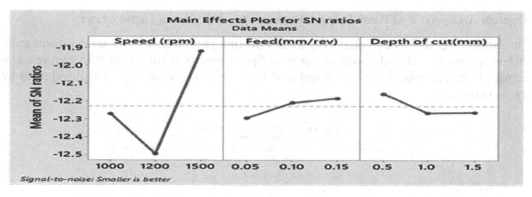

Figure 79.10 Main Effects plot for S/N ratios.

Table 79.6: Response Table for signal to noise ratios.

Level	Speed (rpm)	Feed(mm/rev)	Depth of cut(mm)
1	−12.27	−12.29	−12.16
2	−12.49	−12.21	−12.26
3	−11.92	−12.18	−12.26
Delta	0.58	0.11	0.11
Rank	1	2	3

Table 79.7: Response table for means.

Level	Speed (rpm)	Feed(mm/rev)	Depth of cut(mm)
1	4.107	4.12	4.056
2	4.215	4.079	4.109
3	3.945	4.069	4.102
Delta	0.27	0.05	0.053
Rank	1	3	2

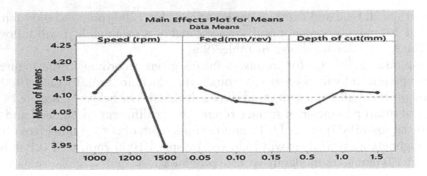

Figure 79.11 Main Effects Plot for Means.

Taguchi Analysis: Tool Wear Response versus Speed, Feed and Depth of cut

The below Figure 79.12 explains the main effecting parameter of tool wear response for different speed, feed and depth of cut is as Speed for 1000 rpm, feed 0.15 mm/rev and depth of cut 1.0 mm gives economical tool wear response meanwhile it will followed by lest is better one.

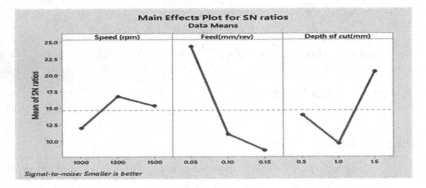

Figure 79.12 Main Effects plot for S/N ratios.

In Figure 79.13. Tool wear response for different speed, Feed and depth of cut is as speed for 1200 rpm, feed 0.05 mm/rev and depth of cut 1.5 mm gives economical tool wear response meanwhile it will followed by speed 1500 rpm, feed 0.1 mm/rev and depth of cut 0.5 mm gives moderate tool wear and speed 1000, feed 0.15 mm/rev and depth of cut 1.0 mm gives very high tool wear, means the above sequence will influences the more.

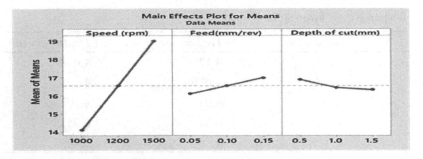

Figure 79.13 Main effects plot for means.

Taguchi Analysis for Cutting Force Taguchi Analysis: Cutting Force (N) versus Speed, Feed, Depth of Cut

The below Figure 79.14 explains the main effecting parameter of cutting force (N) response for different speed, Feed and depth of cut is as Speed for 1500 rpm, feed 0.15 mm/rev and depth of cut 0.5 mm gives economical cutting force response meanwhile it will followed by speed 1200 rpm, feed 0.1 mm/rev and depth of cut 1.0 mm gives moderate cutting force and speed 1500 rpm, Feed 0.05 mm/rev and depth of cut 1.5 mm gives very high cutting force.

The below Figure 79.15 explains the main effecting parameter of cutting force (N) response for different speed, Feed and depth of cut is as Speed for 1000 rpm, feed 0.05 mm/rev and depth of cut 1.0 mm gives economical cutting force response meanwhile it will followed by speed 1200 rpm, feed 1.0 mm/rev and depth of cut 1.0 mm gives moderate cutting force.

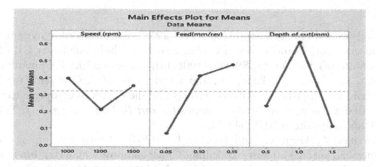

Figure 79.14 Main Effects plot for S/N ratios.

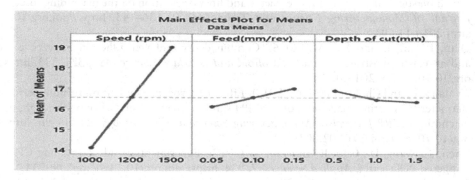

Figure 79.15 Main Effects Plot for Means.

Conclusion

The investigation on some aspects of machinability such as surface roughness and cutting force and tool wear during turning of GFRP composite materials and considering as Polycrystalline diamond tool (PCD) as cutting tool based on the experimental results.

As concern of Surface roughness parameter the better results got when Speed for 1200 rpm, feed 0.05 mm/rev and depth of cut 1.0 mm used. In UG-GFRP composites the surface roughness is highly influenced by Feed rate and it got rank 1 and 2nd depth of cut and 3rd speed parameters. The most helpful parameter for getting less cutting force according to Taguchi L9 orthogonal array is Speed for 1500 rpm, feed 0.15 mm/rev and depth of cut 0.5 mm. The cutting forces are produced high rate when the 1000 rpm feed 0.05 mm/rev and depth of cut 1.5 mm gives very high cutting force.

- Under this cryogenic condition of machining observed that no fibred are pulling out while machining.
- Getting better surface roughness as compare to normal wet machining.
- Tool life increased by using cryogenic machining because of tool maintain less temperature while in operation.

The future scope is to, the number of machining parameters can be extended hence; the data base can be improved by extensive experimentation.

References

1. Giasin, K., Ayvar-Soberanis, S., and Hodzic, A. (2016). The effects of minimum quantity lubrication and cryogenic liquid nitrogen cooling on drilled hole quality in GLARE fiber metal laminates. *Materials and Design,* 89, 996–1006. https://doi.org/10.1016/j.matdes.2015.10.049.
2. Morkavuka, S., Köklüa, U., Bağc, M., and Gemic, L. (2018). Cryogenic machining of carbon fiber reinforced plastic (CFRP) composites and the effects of cryogenic treatment on tensile properties: a comparative study. *Composites Part B: Engineering,* 147, 1–11. https://doi.org/10.1016/j.compositesb.2018.04.024.
3. Xia, T., Kaynak, Y., Arvin, C., and Jawahir, I. S. (2016). Cryogenic cooling-induced process performance and surface integrity in drilling CFRP composite material. *International Journal of Advanced Manufacturing Technology,* 82, 605–616. DOI 10.1007/s00170-015-7284-y.
4. Henerichs, M., Voß, R., Kuster, F., and Wegener, K. (2015). Machining of carbon fiber reinforced plastics: Influence of tool geometry and fiber orientation on the machining forces. *CIRP Journal of Manufacturing Science and Technology,* 9, 136–145. https://doi.org/10.1016/j.cirpj.2014.11.002.
5. Qian, J., Lin, J., and Shi, M. (2016). Combined dry and wet adhesion between a particle and an elastic substrate. *Journal of Colloid and Interface Science,* 483, 321–333. https://doi.org/10.1016/j.jcis.2016.08.049.
6. Sivaiaha, P., and Chakradharb, D. (2018). Effect of cryogenic coolant on turning performance characteristics during machining of 17-4 PH stainless steel: a comparison with MQL, wet, dry machining. *CIRP Journal of Manufacturing Science and Technology,* 21, 86–96. https://doi.org/10.1016/j.cirpj.2018.02.004.
7. Chen, T., Wang, D., Gao, F., and Liu, X. (2017). Experimental study on milling CFRP with staggered PCD cutter. *Applied Sciences,* 7, 934. https://doi.org/10.3390/app7090934.
8. Gupta, M., and Kumar, S. (2015). Investigation of surface roughness and MRR for turning of UD-GFRP using PCA and Taguchi method. *Engineering Science and Technology, an International Journal,* 18, 70–81. http://dx.doi.org/10.1016/j.jestch.2014.09.006.
9. Lee, E. S. (2001). Precision machining of glass fibre reinforced plastics with respect to Tool characteristics. *The International Journal of Advanced Manufacturing Technology,* 17(11), 791–798. DOI: 10.1007/s001700170105.
10. Henerichs, M., Voß, R., Kuster, F., and Wegener, K. (2015). Machining of carbon fiber reinforced plastics: Influence of tool geometry and fiber orientation on the machining forces. *CIRP Journal of Manufacturing Science and Technology,* 9, 136–145. 10.1016/j.cirpj.2014.11. 002; https://dx.doi.org/10.1016/j.cirpj.2014.11.002.
11. Davim, J. P., and Mata, F. (2007). New machinability study of glass fibre rein forced plastics using polycrystalline diamond and cemented carbide (K15) tools. *Master of Design,* 28, 1050–1054.

80 Wear and frictional characteristic of natural fiber reinforced polymer composites: a review

Jiban Jyoti Kalita[1,a] and Kalyan Kumar Singh[2]
[1]Dhemaji Engineering College, Dhemaji, India
[2]Indian Institute of Technology (ISM) Dhanbad, India Dhanbad, India

Abstract: In recent years due to frequent use of fiber reinforced polymer composites in industry (natural as well as synthetic) it is necessary to understand Mechanical as well as tribological (friction and wear) Properties of FRPs. Due to harmful effect of synthetic fiber to the environment, now a day's researchers are looking for some other alternatives that will provide better mmechanical and tribological properties. Light weight, easy availability, cheap is some of the attractive properties of natural fiber reinforced polymer composites. This review paper mainly aims at the study of recent progress in tribological study of different natural fiber reinforced polymer composites in last few decades.

Keywords: Composite material, natural fiber, tribology.

Introduction

Presently friction and wear of machine components are one of the major key factors of production loss. Friction and wear also cause major losses in energy sector, as a major portion of generated energy is utilized to overcome friction [1–4]. Therefore it is of outmost concern to investigate the tribological behavior of materials as well as equipment to improve productivity, maintainability, efficiency and reliability of components. Mostly synthetic fiber composite has high Mechanical strength compared to natural fiber reinforced polymer composite (NFRPCs) but harmful environmental effects of it leads researchers to search for other alternatives such as jute fiber composite, animal fibre Composite, oil palm fibre composites etc. [5–11,41,42]. Most of the automobile manufacturers [59, 60] of developed as well as developing nations are nowadays focusing on the use of natural fiber reinforced polymer composites for vehicle parts. Table 80.1 shows the various advantages and application of natural fiber reinforced polymer composites presently available in the market. It was also observed by researchers that factor like orientation of fiber (parallel to sliding direction, ant parallel to sliding direction, normal to sliding direction), stacking sequence, composition (epoxy etc.), applied load, operating environment, operating temperature, rubbing velocity, types of filler plays an important role in tribological behavior of FRPs. [22–31,39,40,61, 62,64]. In order to identify/evaluate the friction and wear Properties different types of equipment e.g. Block on disc, Block on ring, Pin on disc

[a]kalita.jiban@gmail.com

Table 80.1: Application and advantages of different NFRPCs.

Sl. No.	Fiber	Application	Advantage	Reference
1	Kenf	Automotive industry, construction industry	Low cost, light weight, safety	[12–14]
2	Jute	Household application, Automotive industry	Light weight, ease of availability	[15]
3	Oil-Palm	Packaging Industry, building industry	Higher load sustaining capacity	[16–18]
4	Coir	Rope material, boats	Cheap, waterproof	[19–21]

arrangements are used by various researchers. Losses of mass, specific wear rate, coefficient of friction are the main parameters used to compare tribological characteristic of materials. [32–36,38,40, 59].

Influence of Orientation of Fiber, Applied Load, Sliding Velocity, Chemical Treatment on Friction and Wear behavior of NFRPCs

Depending on direction of sliding effect of variation of three fiber orientation on friction and wear Properties of oil-palm, kenaf fiber, coir fiber etc. is discussed below.

Alshammari et al. [37] in their experimental study investigated the tribological characteristic of Jute fiber under different fiber orientation. Fabrication of the Composite is done by hand layup technique. It was observed that wear rate, weight loss in anti-parallel direction is lowest and in normal orientation it is maximum. Yousif et al. [32] studied the tribological behavior of kenaf fiber with epoxy as matrix and water as surrounding environment. Similar to above mentioned study the wear properties were studied under three conventional orientation, and it was noticed that normal orientation provides better wear resistance characteristic in comparison to other two orientation. It was also observed that for a particular sliding distance, applied load the specific wear rate is more in anti-parallel orientation and least in normal orientation. Friction coefficient in normal orientation is most in normal orientation. It was also concluded in their study that kenaf fiber can be a better replacement of GFRP (glass fiber reinforced polymer) composite under similar operating conditions as shown in Table 80.2.

Table 80.2: Data relevant to reference [37].

Orientation	Friction coefficient
Antiparallel	0.85 to 0.69
Parallel	0.63 to 0.55
Normal	0.51 to 0.29

Narish et al. [24] investment frictional characteristics of kenaf/polyurethane based Composite s. The wear rate in anti-parallel orientation is $2*10^{-8}$(lowest) and in parallel orientation $10*10^{-8}$(Highest). It was also noticed that fiber orientation has less influence in change in friction coefficient in all three orientations. Chin et al. [23] investigated friction and wear characteristic of kenaf/epoxy Composite under three conventional orientation. Three different size (small, medium and large) abrasive sand particles used in an ASTM B 611 machine. It was noticed that wear rate is lowest in normal orientation compared to other two orientations. El. Sayed et al. [35] had taken jute fiber/ unsaturated polyester Composite, and studied friction and wear characteristic under three different orientations. It was observed that with increase in fiber volume fraction wear rate decreases and for a particular volume fraction normal orientation has lowest wear rate. Edeerozey et al. [49] investigated the effect of NaOH treatment (3%,6% and 9%) on kenaf fiber. It was observed that chemically treated kenaf fiber provides better performance than untreated one. It was also observed that at 6% of NaOH/kenaf fibre provides optimum tensile strength. Shuhimi et al. [50] investigated/ compared the tribological characteristic of oil palm and kenaf fiber Composite. Both the composite were chemically treated with NaOH, epoxy was used as matrix. It was noticed that friction coefficient increases as operating temperature increases, but wear rate shows a reverse characteristic. It was also observed that with increase in wt% of kenaf fiber composition tribological property improves. A higher wear rate is observed in kenaf/epoxy Composite compared to oil palm/epoxy Composite. Matějka et al. [51] used Jute as filler and graphite and hazelnut as particulate filler, and studied the wear characteristic of different concentration of Jute fiber. HCl is used for treatment of jute fiber. It was noticed that at 5.6 vol% Jute/ Graphite Composite shows lowest wear rate than Jute/ Hazelnut Composite. Yousif et al. [52] compared the tribological behavior of chemically treated (NaOH) oil palm fiber Composite and untreated Composite. It was noticed that chemically untreated Composite shows poor wear behavior. It was also observed that wear rate of treated composite is 11% less than treated fiber. Table 80.3 shows the Effect of various testing parameters (load, sliding velocity etc.) on different NFRPCs.

Table 80.3: Effect of various testing parameters (load, sliding velocity etc.) on different NFRPCs.

Sl. No	Testing Parameters	Materials	Observations	Reference
1	Machine: ASTM B611 Load: 5 to 20 N	Kenaf fiber/ epoxy and neat epoxy Composite	Low value of friction coefficient in neat epoxy Composite	[23]
2	Machine: TR 600 Load: 5N,20N,30N Sliding velocity:14 m/sec	1. Kenaf/ polyester Composite 2. kenaf/epoxy Composite	1. Specific wear rate decreases until it achieves a steady value 2. At high load loss of mass increase	[34]
3	Two body abrasion test Load:1N, 3N,5N,7N	1. Jute/ polypropylene (untreated, solution treated, melt mixed) 2. Pure polypropylene	1. At 7N wear rate is maximum 2. As applied load increases wear rate increase	[43]
4	Load:10N,20N,30N Sliding distance:0 to 3000m Velocity 1m/sec,2m/sec,3m/ sec	Jute polypropylene Composite	As applied load increases friction coefficient decreases	[31]

(continued)

Table 80.3: Continued

Sl. No	Testing Parameters	Materials	Observations	Reference
5	Load: 20 to 60N	Jute/polyester Composite	Friction coefficient decreases from 1.1 to 0.55 when load increases from 20 N to 60N	[45]
6	Load: 5N to 20N Environment: Dry conditions	1. Banana/ polyester 2. Coconut coir/ polyester 3. Coconut coir/banana/ polyester 4. Coconut/ banana/glass/ polyester	1. As load increases wear rate increase 2. Wear rate is less in material 3 and 4 compared to 1 and 2.	[46]
7	Load: 4 N to 6N	Coir fiber/ polyester	As load increases friction coefficient decreases	[47]
8	Load: 10N-50N Rubbing Speed:1 m/sec to 5m/sec Sliding distance: 1000 to 2000m	1. Jute/Epoxy 2. Flax/Epoxy 3. Hemp/Epoxy 4. Jute/Hemp/ Epoxy 5. Hemp/Flex/ Epoxy 6. Jute/Hemp/ Flax/Epoxy 7. neat epoxy	1. Neat epoxy Composite shows poor tribological characteristic 2. in Jute/Epoxy friction coefficient, wear rate, friction force is most.	[48]
9	1. load (10 N-30N) 2. Rubbing speed(1-3m/s) 3. Rubbing distance (100000cm–300000cm) 4. Dry environment	1. Grewia optiva, nettle, sisal fiber 2. PLA as matrix	1. Significant improvement in wear behavior of PLA matrix composite compared to neat composites. 2. TGA analysis were done on the composites to check thermal stability	[63]
10	1. Variation in fiber length 2. Variation in fiber content (15 wt%-35wt.%)	1. Alkali treated and untreated Lagenaria siceraria fiber. 2. Epoxy as matrix	Alkali treated composites shows better properties compared to untreated composites.	[65]

Prakash et al. [58] have studied the effect of impingement angles and impact velocities on syngonanthus nitens (SN) natural fiber reinforced polymer composites. By varying the weight proportion from 10 wt.%–40 wt.% four different combination of composites were prepared by hand lay-up technique. It can be concluded from their experiment that the at low velocity wear rate maximum at 45° impingement angle.

Ramesh et al. [62] investigated the tribological characteristics of Calotropis gigantea/ Epoxy composites. Chemical treatment of sodium hydroxide and potassium permanganate on the composites were investigated. In order to manufacture the composites compression molding technique was used. It was observed that sodium hydroxide treated composite shows better friction and wear resisting characteristics compared with potassium permanganate treated composites. Further the surfaces obtained from the above experiment were observed under the scanning electron microscope.

Effect of Various Testing Parameters on Tribological Behavior of Animal, Grass, Leaf Fiber etc

Yunhai et al. [53] in their experimental study observed the wear characteristic of wool fiber. Three different types (0wt%, 3wt%, and 4wt%) of composites were prepared. under different temperatures range (100–350°C) among the three, 3wt% wool-fiber shows better wear resisting behavior. During the experiment at high temperature a steady friction coefficient is also observed.

Narendiranath et al. [54] studied the tribological characteristic of angora (animal fiber), kenaf fiber and ramie fiber composite (epoxy is used as matrix). Wear and friction test is conducted on pin on disc apparatus and test pieces are prepared according to ASTM G99 standards. It was observed that kenaf fiber had highest wear and ramie fiber has lowest wear (200 micrometer).

Jyoti et al. [55] done wear test of date palm leaf/ polyvinyl Pyrrolidone Composite on pin on disc apparatus. Six different (0 wt%, 10wt%, 20wt%, 26wt%, 30wt%, and 40wt%) fiber loading is used. It was observed that at 26wt% fiber loading material shows least wear rate as well as weight loss, but 0wt% test specimen shows poorest wear characteristic.

Sneha et al. [56] used bamboo(B) and E-glass(G) as reinforcement and epoxy as matrix. By varying the stacking sequence five different test specimen (BBBB, BGBG, BGGB, GBBG, GGGG) are prepared. Wear rate of different test specimen is tested by various angle of impingement. It was observed that in BBBB Composite erosion wear rate is lowest. By varying wt% of banana fiber [57] seven different composite (0wt%, 5wt%, 10wt%, 15 wt%, 20 wt%, 25 wt%, 30wt%, 35 wt%) were prepared with epoxy as resin. It was observed that erodent impact velocity is most important factors of wear mechanism and angle of impingement has least effect on wear rate.

Conclusions

The following conclusions can be drawn out from the literature review

* It was noticed that chemical treatment on NFRPCs has positive impact on wear characteristic.
* Number of research paper related to effect of surrounding environment (e.g. Inert gas, oil etc) on tribological characteristic on NFRPCs is less.

- Relevant analysis of effect of fiber orientation on oil palm fiber and coir fiber is less.
- Hybridization of NFRPCs improve wear characteristic in most of the cases.

References

1. Holmberg, K., and Erdemir, A. (2017). Influence of tribology on global energy consumption, costs and emissions. *Friction, 5*, 263–284.
2. Bhushan, B. (2013). Introduction to Tribology. 2nd ed. Hoboken, NJ, USA: John Wiley & Sons Ltd., pp. 1–201.
3. Ma, Y., Liu, Y., Gao, Z., Lin, F., Yang, Y., Ye, W., and Tong, J. (2014). Effects of wool fibres on tribological behavior of friction materials. *Journal of Thermoplastic Composite Materials, 27*, 867–880.
4. Alajmi, M., and Shalwan, A. (2015). Correlation between mechanical properties with specific wear rate and the coefficient of friction of graphite/epoxy *Composites Materials, 8*, 4162–4175.
5. Nägeli, C., Camarasa, C., Jakob, M., Catenazzi, G., and Ostermeyer, Y. (2018). Synthetic building stocks as a way to assess the energy demand and greenhouse gas emissions of national building stocks. *Energy Building, 173*, 443–460.
6. Dente, S. M. R., Aoki-Suzuki, C., Tanaka, D., and Hashimoto, S. (2018). Revealing the life cycle greenhouse gas emissions of materials: the Japanese case. *Resources, Conservation and Recycling, 133*, 395–403.
7. Shahinur, S., and Hasan, M. (2019). Natural fibre and synthetic fibre composites: comparison of properties, performance, cost and environmental benefits. *Reference Module in Materials Science and Materials Engineering*, 794–802.
8. Rao, C. H. C., Madhusudan, S., Raghavendra, G., and Rao, E.V. (2012). Investigation in to wear behavior of coir fibre reinforced epoxy composites with the taguchi method. *International Journal of Engineering Research and Applications, 2*, 371–374.
9. Aliyu, I. K., Mohammed, A. S., and Al-Qutub, A. (2018). Tribological performance of ultra-high molecular weight polyethylene nanocomposites reinforced with graphene nanoplatelets. *Polymer Composites, 40*, E1301–E1311.
10. Mohammed, A. S., Ali, A. B., and Nesar, M. (2017). Evaluation of tribological properties of organo-clay reinforced UHMWPE nanocomposites. *Journal of Tribology, 139*, 012001-1–012001-6.
11. Yan, L., Su, S., and Chouw, N. (2015). Microstructure, flexural properties and durability of coir fibre reinforced concrete beams externally strengthened with flax FRP composites. *Composites Part B Engineering, 80*, 343–354.
12. Wang, J., and Ramaswamy, G. N. (2003). One-step processing and bleaching of mechanically separated kenaf fibres: effects on physical and chemical properties. *Textile Research Journal, 73*, 339–344.
13. Akil, H. M., Omar, M. F., Mazuki, A. A. M., Safiee, S., Ishak, Z. A. M., and Bakar, A. A. (2011). Kenaf fibre reinforced composites: a review. *Master of Design, 32*, 4107–4121.
14. Chin, C. W., and Yousif, B. F. (2009). Potential of kenaf fibres as reinforcement for tribological applications. *Wear, 267*, 1550–1557.
15. Acha, B. A., Marcovich, N. E., and Reboredo, M. M. (2005). Physical and mechanical characterization of jute fabric composites. *Journal of Applied Polymer Science, 98*, 639–650.
16. Khalil, H. P. S. A., Alwani, M. S., Ridzuan, R., Kamarudin, H., and Khairul, A. (2008). Chemical composition, morphological characteristics, and cell wall structure of malaysian oil palm fibres. *Polymer—Plastics Technology and Engineering, 47*, 273–280.
17. Valášek, P., Ruggiero, A., and Müller, M. (2017). Experimental description of strength and tribological characteristic of EFB oil palm fibres/epoxy composites with technologically undemanding preparation. *Composites Part B Engineering, 122*, 79–88.

18. Shinoj, S., Visvanathan, R., Panigrahi, S., and Kochubabu, M. (2011). Oil palm fibre (OPF) and its composites: a review. *Industrial Crops and Products*, 33, 7–22.
19. Ayrilmis, N., Jarusombuti, S., Fueangvivat, V., Bauchongkol, P., and White, R. H. (2011). Coir fibre reinforced polypropylene composite panel for automotive interior applications. *Fibres Polymers*, 12, 919–926.
20. Shireesha, Y., and Nandipati, G. (2019). State of art review on natural fibres. *Materials Today: Proceedings*, 18, 15–24.
21. Yousif, B. F. (2009). Frictional and wear performance of polyester composites based on coir fibres. *Proceedings of the Institution of Mechanical Engineers, Part J: Journal of Engineering Tribology*, 223, 51–59.
22. Yallew, T. B., Kumar, P., and Singh, I. (2014). Sliding wear properties of jute fabric reinforced polypropylene composites. *Procedia Engineering*, 97, 402–411.
23. Chin, C. W., and Yousif, F. (2010). Influence of particle size, applied load, and fibre orientation on 3B-A wear and frictional behaviour of epoxy composite based on kenaf fibres. *Proceedings of the Institution of Mechanical Engineers, Part J: Journal of Engineering Tribology*, 224, 481–489.
24. Narish, S., Yousif, B. F., and Rilling, D. (2011). Adhesive wear of thermoplastic composite based on kenaf fibres. *Proceedings of the Institution of Mechanical Engineers, Part J: Journal of Engineering Tribology*, 225, 101–109.
25. Nordin, N. A., Yussof, F. M., Kasolang, S., Salleh, Z., and Ahmad, M. A. (2013). Wear rate of natural fibre: Long kenaf composite. *Procedia Engineerings*, 68, 145–151.
26. Valášek, P., Amato, R. D., Müller, M., and Ruggiero, A. (2018). Mechanical properties and abrasive wear of white/brown coir epoxy composites. *Composites Part B Engineering*, 146, 88–97.
27. Hashmi, S. A. R., Dwivedi, U. K., and Chand, N. (2007). Graphite modified cotton fibre reinforced polyester composites under sliding wear conditions. *Wear*, 262, 1426–1432.
28. Yousif, B. F., and Tayeb, E. N. S. M. (2007). The effect of oil palm fibres as reinforcement on tribological performance of polyester composite. *Surface Review and Letters*, 14, 1095–1102.
29. Rahnejat, H., and Johns-Rahnejat, P. M. (2014). Mechanics of contacting surfaces. In Encyclopedia of Automotive Engineering. Hoboken, NJ, USA: John Wiley & Sons Ltd, pp. 1–9.
30. Sanal, I., and Verma, D. (2019). Construction Materials reinforced with natural products. In Martínez, L. M. T., Kharissova, O. V., and Kharisov, B. I., Eds. Handbook of Ecomaterials. Basel, Switzerland: Springer, pp. 1–24.
31. Dente, S. M. R., Aoki-Suzuki, C., Tanaka, D., and Hashimoto, S. (2018). Revealing the life cycle greenhouse gas emissions of materials: the Japanese case. *Resources, Conservation and Recycling Advances*, 133, 395–403.
32. Yousif, B. F., and Chin, C. W. (2012). Epoxy composite based on kenaf fibres for tribological applications under wet contact conditions. *Surface Review and Letters*, 19, 1250050-1–1250050-6.
33. Ochi, S. (2008). Mechanical properties of kenaf fibres and kenaf/PLA composites. *Mechanics, Materials*, 40, 446–452.
34. Archard, J. F. (1953). Contact and rubbing of flat surfaces. *Journal of Applied Physics*, 24, 981–988.
35. El-Sayed, A. A., El-Sherbiny, M. G., Abo-El-Ezz, A. S., and Aggag, G. A. (1995). Friction and wear properties of polymeric composite materials for bearing applications. *Wear*, 184, 45–53.
36. Yousif, B. F., and El-Tayeb, N. S. M. (2008). Adhesive wear performance of T-OPRP and UT-OPRP composites. *Tribology Letters*, 32, 199–208.
37. Alshammari, F. Z., Saleh, K. H., Yousif, B. F., Alajmi, A., Shalwan, A., and Alotaibi, J. G. (2018). The influence of fibre orientation on tribological performance of jute fibre reinforced epoxy composites considering different mat orientations. *Tribology in Industry*, 40, 335–348.
38. Yallew, T. B., Kumar, P., and Singh, I. (2014). Sliding wear properties of jute fabric reinforced polypropylene composites. *Procedia Engineering*, 97, 402–411.

39. Correa, C. E., Zuluaga, R., Castro, C., Betancourt, S., Vázquez, A., and Gañán, P. (2017). Influence of tribological test on the global conversion of natural composites. *Polimeros*, 27, 339–345.

40. Gupta, M., and Wang, K. K. (1993). Fibre orientation and mechanical properties of short-fibre-reinforced injection-molded composites: simulated and experimental results. *Polymer Composites*, 14, 367–382.

41. Saheb, D. N., and Jog, J. P. (1999). Natural fibre polymer composites. a review. *Advances in Polymer Technology*, 18, 351–363.

42. Agrawal, S., Singh, K. K., and Sarkar, P. K. (2016). A comparative study of wear and friction characteristics of glass fiber reinforced epoxy resin, sliding under dry, oil-lubricated and inert gas environments. *Tribology International*, 96, 217–224.

43. Chand, N., and Dwivedi, U. K. (2006). Effect of coupling agent on abrasive wear behaviour of chopped jute fibre-reinforced polypropylene composites. *Wear*, 261, 1057–1063.

44. Mahapatra, S., and Chaturvedi, V. (2010). Modelling and analysis of abrasive wear performance of composites using Taguchi approach. *International Journal of Engineering, Science and Technology*, 1, 123–135.

45. Dwivedi, U. K., and Chand, N. (2009). Influence of fibre orientation on friction and sliding wear behaviour of jute fibre reinforced polyester composite. *Applied Composite Materials*, 16, 93–100.

46. Kumar, V., Mohan, N., and Bongale, A. K. (2019). Fabrication and tribological investigation of Coconut coir/Banana fiber/Glass fiber reinforced hybrid polymer matrix composites-A Taguchi's approach. *Materials Research Express*, 6(10), 105345.

47. Ibrahem, R. A. (2016). Friction and wear behaviour of fibre/particles reinforced polyester composites. *International Journal of Advanced Materials Research*, 2, 22–26.

48. Chaudhary, V., Bajpai, P. K., and Maheshwari, S. (2018). An investigation on wear and dynamic mechanical behavior of jute/hemp/flax reinforced composites and its hybrids for tribological applications. *Fibre Polymers*, 19, 403–415.

49. Edeerozey, A. M. M., Akil, H. M., Azhar, A. B., and Ariffin, M. I. Z. (2007). Chemical modification of kenaf fibres. *Materials Letters*, 61, 2023–2025.

50. Shuhimi, F. F., Abdollah, M. F. B., Kalam, M. A., Hassan, M., Mustafa, A., and Amiruddin, H. (2016). Tribological characteristics comparison for oil palm fibre/epoxy and kenaf fibre/epoxy composites under dry sliding conditions. *Tribology International*, 10, 247–254.

51. Matějka, V., Fu, Z., Kukutschová, J., Qi, S., Jiang, S., Zhang, X., Yun, R., Vaculík, M., Heliová, M., and Lu, Y. (2013). Jute fibres and powderized hazelnut shells as natural fillers in non-asbestos organic non-metallic friction composites. *Master of Design*, 51, 847–853.

52. Ayrilmis, N. Jarusombuti, S., Fueangvivat, V. Bauchongkol, P., and White, R. H. (2011). Coir fibre reinforced polypropylene composite panel for automotive interior applications. *Fibres Polymer*, 12, 919–926.

53. Ma, Y., Liu, Y., Gao, Z., Lin, F., Yang, Y., Ye, W., and Tong, J. (2014). Effects of wool fibers on tribological behavior of friction materials *Journal of Thermoplastic Composite Materials*, 27, 867–880.

54. Narendiranath Babu, T., Aravind, S. S., Naveen Kumar, K. S., and Sumanth Rao, M. S. (2018). Study on mechanical and tribological behaviour of Angora, kenaf and ramie hybrid reinforced epoxy composites. *International Journal of Mechanical Engineering and Technology*, 9, 11–20.

55. Mohanty, J. R., Das, S. N., and Das, H. C. (2014). Effect of fiber content on abrasive wear behavior of date palm leaf reinforced polyvinyl pyrrolidone composite. *International Scholarly Research Notices*, 2014, 1–10.

56. Latha, P. S., and Rao, M. V. (2017). Erosion wear behaviour of bamboo-glass hybrid. *Fiber Reinforced Polymer*, 10, 2438–2446.

57. Mukti, P. S., and Choubey A. K. (2017). Tribological Characterization of Natural Fiber Composites. *International Journal of Engineering Science and Computing (IJESC)*, 7, 1–12.

58. Prakash, V., Pradhan, S., and Acharya, S. K. (2021). Tribological behaviour of syngonanthus nitens natural fiber reinforced epoxy composites, *Transactions of the Indian Institute of Metals*, 74(7), 1741–1750. 10.1007/s12666-021-02268-9.

59. Prakash, V., Bera, T., and Acharya, S. K. (2019). *Materials Today: Proceedings*, 19, 223.

60. Mohanty, A. K., Misra, M., and Drzal, L. T. (2002). Sustainable Bio-composites from renewable resources: opportunities and challenges in the green materials world. *Journal of Polymers and the Environment*, 10, 19–26.

61. Singh, K. K., Kalita, J. J., and Sharma, N. (2018). Investigation of friction and wear behaviour of MWCNTs filled HFRP composites under different sliding parameters and different environment. *Materials Today: Proceeding 5*, 28347–28353.

62. Ramesh, M., Deepa, C., Rajeshkumar, L., Tamil Selvan, M., and Balaji, D. (2021). Influence of fiber surface treatment on the tribological properties of Calotropis gigantea plant fiber reinforced polymer composites. *Polymer Composites*, 42(9), 4308–4317.

63. Bajpai, P., Singh, I., and Madaan, J. (2013). Tribological behavior of natural fiber reinforced PLA composites. *Wear*, 297, 829–840.

64. Kumar, S., and Saha, A. (2021). Effects of stacking sequence of pineapple leaf-flax reinforced hybrid composite laminates on mechanical characterization and moisture resistant properties. *Proceedings of the Institution of Mechanical Engineers, Part C: Journal of Mechanical Engineering Science*, 236(3), 1733–1750. doi:10.1177/09544062211023105

65. Nagappan, S., Subramani, S. P., Palaniappan, S. K., and Mylsamy, B. (2021). Impact of alkali treatment and fiber length on mechanical properties of new agro waste lagenaria siceraria fiber reinforced epoxy composites. *Journal of Natural Fibers*, 19(13), 6853–6864. DOI: 10.1080/15440478.2021.1932681.

81 Investigation of polycarbonate machining by magnetic field aided electro-chemical discharge machining (ECDM) process

Maneetkumar R. Dhanvijay[1,a] and Vinay A. Kulkarni[2,b]

[1]Department of Manufacturing Engineering and Industrial Management, College of Engineering, Pune

[2]Department of Mechanical Engineering, D.Y. Patil College of Engineering, Akurdi, Pune

Abstract: Electro Chemical Discharge Machining (ECDM) is a hybrid oriented, non-conventional process and vividly used for machining of non-conducting materials like alumina, glass, ceramics, composites etc. This work explores the effect of magnetic field assisted ECDM process on machining performance of polycarbonate material. Magnetic field induces magneto hydrodynamic (MHD) effect in electrolyte to enhance electrolyte circulation in narrow gap of machining zone. Investigation of input parameters like duty factor, electrolyte concentration and voltage were taken into consideration for improved response parameters, mainly Material Removal Rate (MRR), and Diametric Over Cut (DOC) under influence of magnetic field (mf)/nonmagnetic field (nmf). It is observed that MRR is 0.1187 mg/min under no magnetic effect and in contrast 0.22 mg/min of MRR was obtained under the influence of magnetic field. However, an increase in DOC has been observed while machining with magnetic field in the ECDM processes.

Keywords: BBD, ECDM, hybrid process, magnetic field assisted ECDM (MFAECDM), RSM.

Introduction

The electrochemical discharge machining (ECDM) used in machining of hard and non-conductive brittle material such as glass (optical, Plexi, and mainly Pyrex), composites, quartz and ceramics. ECDM process has the ability to perform machining operation on electrically non-conducting materials has attracted many researchers to perform experimental investigation on various non-conducting materials used in MEMS devices, sensors and micro-fabrication.

The ECDM process set up is similar to electrolysis process. The tool electrode is anode and auxiliary electrode having surface area more than fifty times than tool electrode is made anode. Generally these electrodes are arranged in straight polarity. The work-piece and the two electrodes are dipped in an electrolyte solution of NaOH and KOH. The DC supply or pulsed DC supply is provided at the electrodes. When a voltage is applied across the electrodes, electrolysis process commences and thereby evolution of oxygen, hydrogen gas bubbles occurs at tool electrode and auxiliary electrodes. Hydrogen bubbles collapse at critical voltage to form a thin layer round the tool surface. This thin layer act as dielectric

[a]mrd.mfg@coep.ac.in, [b]vakulkarni@dypcoeakurdi.ac.in

medium and thus discharges starts appearing in the layer. The thermal energy provided by such discharges soften and melts the workpiece material and thus material removal by means of discharge takes place.

Discharge performance is influenced by gas film quality, and thus machining quality and geometric accuracy of the process. Kolhekar et al. [1] observed that, low electrolyte concentration and level with higher inner gaps in electrode, contributes to less overcut in optimized conditions, by the study of gas film thickness stability, characterization with mean value of discharge current at critical voltage. Dhanvijay and Ahuja [2] found that continuous electrolyte flow with stainless steel tool yields highest MRR with minimized diametric over cut when stagnant and electrolyte flow is used. Sabahi et al. [3] studied the heat affected zone of workpiece machined by ECDM by using nano indentation test. In presence of magnetic field heat affected zone is reduced. By using KOH electrolyte heat affected zone is increased as compared to NaOH electrolyte. The size of Potassium ion is larger than sodium ion also the electrical conductivity of potassium ion is higher than sodium ions because of more heat power and larger heat affected zone is generated. Higher electrolytic concentration and higher applied voltage also causes increase in heat affected zone. With increase in machining depth of micro hole, there can be difficulty of removing debris from hole which results in less electrolyte circulation and low wettability of the tool. The difficulty of electrolyte circulation in narrow gap between tool and workpiece interface, in micro-fabrication is the concern which requires attention. Many researchers have tried various methods to improve electrolyte circulation by using tool rotation, electrolyte flow through hollow tool, providing ultrasonic vibrations to electrolyte [4, 5], etc. Cheng et al. [6] calculated magnetic field effect on machining performance of ECDM using Pyrex glass as work-piece. They performed drilling operation with and without the use of magnetic field around the tool workpiece interface and found that magneto hydrodynamic convection induced on electrolyte enhances its circulation in narrow gap and thus improves machining efficiency. It also helps in maintaining stability of gas film and thus ensuring stable discharge performance. Rattan et al. [7] performed ECDM by traveling wire (TW-ECDM) in presence of magnetic field. The experimental results showed that there was improvement in MRR if assisted by magnetic field. Circulation of electrolyte roots excess bubble removal and residues from the machining area undergoing machining process. They also found that application of magnetic field reduces discharge current requirement. Hajian et al. [8] deliberated on magnetic field orientation effect on the performance of ECDM. Changing direction of magnetic field causes change in the direction of motion of gas baubles in inter-electrode area. Microchannel depth and surface quality were considered as response parameters. For low concentration of electrolyte, higher voltage value and under the magnetic field assistance, the smoother surface quality of machined surface was found.

With the help of high speed camera images, Xu et al. [9] studied the gas bubble mechanism with formation of gas film, under the influence of magnetic field. They developed the theoretical model of gas film formation under the influence of magneto hydrodynamic effect which was supported by experimental investigation. They also found improvement in machining efficiency due to improved electrolyte circulation. In this study effect of magneto hydrodynamic (MHD) to improve accuracy and efficiency of ECDM processes induced by magnetic field are explored. In 1975 Mukaibo and other researchers originated the study effect of magnetic field on electrolysis process and discovered that Lorentz forces acting on electrolyte ions causes convection in electrolyte under the influence of magnetic field. Antil et al. [10] worked on the precise machining problem of glass fiber reinforced polymer metal matrix composite using ECDM technique. They used the multi objective genetic algorithm (MOGA) to study the effect of the various process parameters on MRR and overcut. MRR decreases at higher electrolyte concentration levels due to the lower degree of dissolution efficiency. Overcut is found to be increasing with the increase in inter electrode gap and voltage. Singh and Dvivedi [11] have adopted a mechanism of titrated flow of electrolyte to machine micro holes in 1350-μm-thick work material. A reduction in machining time by 7 times with an improved hole accuracy (entrance diameter) of 1.25 times was achieved. Saini et. al. [12] classify magnetic field assisted ECDM process as a hybridized ECDM process where the MHD convection effect is utilized for the penetration of electrolyte in narrow machining zones. The present work investigates the machining of polycarbonate material by ECDM process with and without the effect of magnetic field on the electrolyte. An attempt has been made to improve electrolyte circulation in narrow gaps of tool and workpiece using magneto hydrodynamic effect.

Experimental Planning

Insufficient tool wetting occurs with increase in drilling depth. Therefore, stimulation of electrolyte circulation in narrow gap between tool and workpiece, is the deciding factor in precision attainment in ECDM process.

When pulsed current is applied, in the occurrence of magnetic field, electrolyte solution containing positive ions (H^+) and negative ions (OH) acted upon by Lorentz force. The Lorentz force direction is perpendicular to the motion of ions and also perpendicular the magnetic field direction. Thus the Lorentz force influences the direction of motion of charged particles and causes circular convection of electrolyte. The direction of circular motion of electrolyte depends on the direction of applied magnetic field. From Figure 81.1(a) it was observed that when north pole of the magnet is in downward direction, the motion of electrolyte is in counter clockwise direction. Similarly from Figure 81.1(b), when south pole of the magnet is in downward direction electrolyte moves in clockwise direction.

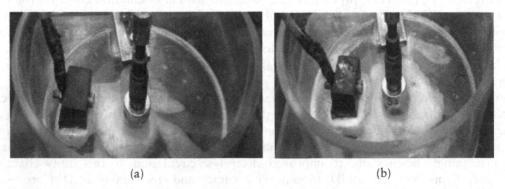

(a)   (b)

Figure 81.1 (a) Electrolyte motion in counter clockwise direction. (b) Electrolyte motion in clockwise direction.

The effect of convective motion of electrolyte due to Lorentz force is called as magneto hydrodynamic (MHD) effect. Such motion of electrolyte causes the flow of fresh electrolyte in the narrow gap of tool workpiece interface and thus improves gas film quality.

Experimental set-up and Material Specification

The representative diagram of ECDM setup is shown in Figure 81.2. It consists of electrolyte chamber, workpiece holder, tool feeding mechanism, power supply unit and two axes work table with manual control.

Material Specifications

For experimental investigation and study of Electrochemical Discharge Machining, polycarbonate sheet of 4 mm thickness is used as a workpiece material. Pure copper tools of diameter 1.2 mm were made as a cathode electrode. The measurements of weights of the workpieces and tools before and after machining were carried out on the electronic balance scale machine with least count of 0.0001 gram. For diametric overcut, diameter of hole machined in workpiece was quantified on Machine vision measuring system unit. The workpiece samples of size (45 × 20 × 4) mm were cut from polycarbonate (Lexan TM 9030) sheet. Polycarbonate ($C_{15}H_{16}O_2$) is naturally transparent amorphous thermoplastic material having very high impact resistance as compared to glass. Many researchers have investigated ECDM process on borosilicate glass or acrylic (Plexi glass). As polycarbonate material has higher impact strength than borosilicate glass and acrylic material, it is widely used where safety is of paramount importance. So it has been decided to study and investigate the machining performance of ECDM process on polycarbonate material. The properties of polycarbonate material are mentioned in Table 81.1.

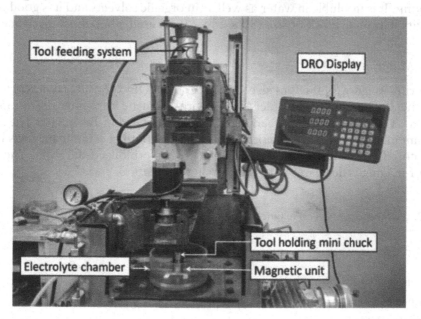

Figure 81.2 Machining set up of ECDM.

Table 81.1: Properties of polycarbonate (Lexan™ 9030).

Property	Test Method	Value	Unit
Density	ISO 1183	1.2	g/cm^3
Poisons ratio	ASTM-D638	0.38	-
Tensile stress at yield 50 mm/min	ISO 527	60	MPa
Hardness H358/30 95	ISO 2039/1	95	MPa
Izod impact, notched 23°C	ISO 180/1A	65	kJ/m^2
Light transmission	ASTM-D1003	89%	-
Thermal conductivity	DIN52612	0.2	W/m.°C
Vicat B/120 softening temperature	ISO306	145	°C

Tool Electrode (Cathode)

A copper tool with 1.2 mm diameter was used and had the same end geometries (cylindrical). Copper as a tool material has been selected because it is a good conductor of electricity, has good chemical inertness, lowest specific heat capacity and higher wear resistance. Higher discharge activity results due to lower specific heat capacity with increase in temperature of tool electrode. The material removal rate in the discharge mechanism is significantly higher for copper tool-electrode as compared to any other materials.

Auxiliary Electrode (Anode)

Graphite was used as auxiliary electrode because it is corrosion resistant and with high melting point. It is insoluble in water as well as in organic solvents and it is good conductor of electricity. Graphite block of size (30 × 25 × 20), is used as auxiliary electrode (anode).

Magnetic Material

Neodium ring magnet (NdFeB) of size (20 × 10 × mm (OD x ID x T) was used to produce magnetic field. The grade of magnet is n35 with magnetic field intensity of 3500 gauss (0.35 Tesla). The properties of NdFeB magnet are mentioned in Table 81.2. The advantage of Neodium magnet is that it has more magnetic strength than ceramic magnets but it also prone to corrosion when exposed to atmosphere for prolonged duration. The magnetic strength of magnet is decreases with increase in temperature.

Table 81.2: Properties of NdFeB magnet.

Property	Value
Grade	35
Magnetic field intensity (B)	3500 Gauss or 0.35 Tesla
Curie temperature	80°C
Relative permeability	1.05

Electrolyte Solution

Sodium hydroxide (NaOH) is most commonly used electrolyte in ECDM process as it has high electrical conductivity. NaOH is a strong electrolyte alkaline base which is available in pellets, flakes. Aqueous solution of NaOH was made with demineralized water of concentration of NaOH by weight percentage. Properties of NaOH are mentioned in Table 81.3.

Table 81.3: Properties of NaOH.

Property	Value
Molar mass	39.99 g/mol
Appearance	white, opaque
Density	2.13 g/cm³
Melting point	318 °C
Boiling Point	1,388° C
Specific heat capacity	59.67 J/mol K

Trial/Screening Experimentations and Experimental Design

Few screening experiments were performed at different levels of voltage and duty cycle to decide the range of levels for voltage and duty factor. Based on literature review, previous researches and screening experimentations levels for voltage, electrolyte concentration and duty cycle were selected for design of experimentation. Statistical design of experiments helps in careful planning of experimental data with valid and fruitful conclusion. Since, Response surface methodology (RSM) uses multivariate statistical technique to optimize the process, where several process parameters significantly influence response and thereby quality characteristics, it is used for further development. RSM is used in many real world applications where more than one response is involved.

a) Response variable parameters
Many studies were carried out to improve material removal rate of ECDM process. As it is also used in the form of micro ECDM to create micro features, diameter overcut (DOC) also needs to be investigated to improve the accuracy of the process. The following response parameters were considered to study their effect on the performance of ECDM process.

b) Material removal rate (MRR)
By measuring the weight of workpiece before and after machining and then dividing by machining time, MRR can be calculated. It can be calculated by taking the ratio of the volume of material removed from the workpiece with respect to the machining time.

$$\left(\begin{array}{c} MRR \\ (mg \: / \: min) \end{array} \right) = \frac{\left(\begin{array}{c} Weight \: of \: workpiece \\ before \: machining \end{array} \right) - \left(\begin{array}{c} Weight \: of \: workpiece \\ after \: machining \end{array} \right)}{Machining \: time}$$

Tool Wear Rate (TWR)

Tool wear rate is computed by measuring the weight of the tool before and after the machining and then dividing it by machining time.

$$TWR(mg\,/\,min) = \frac{(Initial\,weight\,of\,tool) - (Final\,weight\,of\,tool)}{Machining\,time}$$

Diametric Overcut (DOC)

The machined hole diameter may be larger than the tool diameter because of stray discharges in the process. To improve machining accuracy, it is necessary to reduce the diametric overcut. It is calculated using the following equation.

$$DOC = (Diameter\,of\,the\,machined\,hole) - (Diameter\,of\,tool)$$

The machining parameters given in Table 81.4, are set constant for all experimental runs.

Experimental Results and Discussions

After performing experimental runs according to randomized run order as mentioned in design matrix created by 'Design expert 11'software, the values of response parameters were put in the respective columns of design matrix for further analysis. The regression analysis with the help of ANOVA was performed to create mathematical modeling to predict the response value for defined set of input parameters.

Table 81.4: Constant machining parameters for all experimental runs.

Sr. No	Machining Condition	Specification
1	Work-piece material	Polycarbonate (LexanTM 9030)
2	Tool electrode material	Copper tool (1.2 mm Dia)
3	Auxiliary tool material	Graphite
4	Machining time	10 min
5	Level of electrolyte	2 to 5 mm (above w/p)
6	Distance between anode to cathode	40 to 45 mm
7	Pulse on time	1000 µS
8	Tool feed rate	1.67 µm/sec

Experimental Data

The observed data for response parameters MRR in mg/min, TWR in mg/min and DOC in mm, is given in Table 81.5. For convenience, the data is arranged according to standard run order.

Optimization of Process Parameters

Optimization of factors or input process parameters is necessary in order to get desired output or responses. In this section, optimization of input process parameters is conceded by 'Optimization module' available in 'Design expert 11' software.

It searches for a factor level combination, simultaneously satisfying criteria placed by factors and their responses governed by numerical optimization. This will enable computation of best trade-offs to achieve various goals by searching factor space. It has been decided to optimize the parameters for maximum MRR, minimum TWR and minimum DOC.

Setting the Optimization Criterion

Available menu of optimization module are helpful in finalizing desired goal and responses for every factor. Goals may be to minimize, maximize, range, target value settings as desirable. The software restricts factor ranges to factorial levels and response limits default to observed extremes. For any desirable function, weight is allocated to each goal. Importance of each goal can be readjusted with other goals by giving equal importance to all goals with 3 pluses (+++).

For maximization of desirability function, goals are combined from random starting point to steepest slope. Two or more than two maximums, can be found due to curvatures in response surfaces. From several points in the design space, best local maximum can be attained. Initially 50 starting points were chosen.

Desirability

Desirability indicates range from zero to one as objective function. Numerical optimization searchers a point, maximizing popularity function by combining individual and overall greatest desirability function. Derringer and Suich's method, is used by Design-Expert Software.

Desirability uses objective function, D(x) in multiple responses reflecting desirable ranges for every response (di). Geometric mean of all transformed responses defines simultaneous objective function can be given by the formula:

$$D = \left(\prod_{i=1}^{n} d_i \right)^{\frac{1}{n}}$$

Where n is the number of responses in measure.

Table 81.5: Experimental values of MRR, TWR and DOC.

Std	Run	Factor 1 A:vol	Factor 2 B:ele con	Factor 3 C:duty	Factor 4 D:condition	Response 1 MRR	Response 2 TWR	Response 3 DOC
		V	wt%	%	-	mg/min	mg/min	mm
1	12	45	20	65	nmf	0.08	2.51	0.1443
2	14	55	20	65	nmf	0.09	2.46	0.024
3	15	45	40	65	nmf	0.15	2.75	0.2959
4	7	55	40	65	nmf	0.2	2.22	0.3252

(continued)

Table 81.1: Continued

		Factor 1	Factor 2	Factor 3	Factor 4	Response 1	Response 2	Response 3
Std	Run	A:vol	B:ele con	C:duty	D:condition	MRR	TWR	DOC
		V	wt%	%	-	mg/min	mg/min	mm
5	9	45	30	55	nmf	0.08	2.91	0.1981
6	27	55	30	55	nmf	0.07	2.05	0.2238
7	25	45	30	75	nmf	0.09	2.21	0.3724
8	5	55	30	75	nmf	0.1	2.25	0.1597
9	4	50	20	55	nmf	0.04	2.76	0.03
10	24	50	40	55	nmf	0.18	1.69	0.4707
11	1	50	20	75	nmf	0.16	2.2	0.3212
12	3	50	40	75	nmf	0.17	1.9	0.1934
13	11	50	30	65	nmf	0.14	2.16	0.4145
14	8	50	30	65	nmf	0.11	2.51	0.15
15	28	50	30	65	nmf	0.12	3.04	0.0337
16	29	45	20	65	mf	0.12	1.75	0.1672
17	2	55	20	65	Mf	0.13	1.63	0.2507
18	19	45	40	65	Mf	0.19	1.21	0.4956
19	18	55	40	65	Mf	0.22	2.2	0.5941
20	23	45	30	55	Mf	0.09	1.06	0.2834
21	22	55	30	55	Mf	0.08	1.79	0.332
22	10	45	30	75	Mf	0.06	1.15	0.3959
23	16	55	30	75	Mf	0.21	1.09	0.6353
24	26	50	20	55	Mf	0.05	1.99	0.0191
25	20	50	40	55	Mf	0.22	2.2	0.5046
26	6	50	20	75	Mf	0.14	1.82	0.3653
27	21	50	40	75	Mf	0.19	2.44	0.2438
28	13	50	30	65	Mf	0.18	2.3	0.5157
29	30	50	30	65	Mf	0.11	1.46	0.0055
30	17	50	30	65	Mf	0.17	1.85	0.4585

Optimization for no Magnetic Field (nmf)

The goals and ranges set for factors and responses used in analysis are mentioned in Table 81.6. After setting the optimization criterion, the design expert software gives the 52 number of solutions based on starting points. These solutions were arranged according to desirability index and solutions arranged in descending order of desirability index.

Table 81.6: Optimization criterion for factors and responses (nmf).

Name	Goal	Lower limit	Upper limit	Lower weight	Upper weight	Importance
A:vol	is in range	45	55	1	1	3
B:ele con	is in range	20	40	1	1	3
C:duty	is in range	55	75	1	1	3
D:condition	is equal to Nmf	nmf	mf	1	1	3
MRR	maximize	0.04	0.22	1	1	3
TWR	minimize	1.06	3.04	1	1	3
DOC	minimize	0.0055	0.6353	1	1	3

Out of these 52 solutions obtained, solution number 1 and solution number 2 having highest desirability index. Multiple maximum obtained due to curvature in the response surfaces and their combination into the desirability function. Solution number 1 is selected as optimum solution. Thus the optimum input process parameters are voltage at 55 V, electrolyte concentration at 40 wt% and duty cycle at 75%. The response values predicted at these input setting would be MRR of 0.179 mg/min, TWR of 2.265 mg/min and DOC of 0.1 mm. the desirability index is 0.635. Optimal factor settings are shown with red points and optimal response prediction values are displayed in blue. Figure 81.3 shows the optimization plot for all factors and responses.

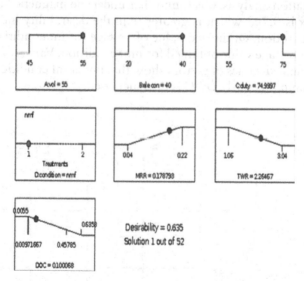

Figure 81.3 Optimal factors and corresponding responses for nmf condition.

Optimization for Magnetic Field (mf)

The goals and ranges set for factors and responses used in analysis are mentioned in Table 81.7. After setting the optimization criterion, the design expert software gives the 42 number of solutions based on starting points. These solutions are arranged according to desirability index in descending order.

Out of these 42 solutions obtained, solution number 1 and solution number 2 having highest desirability index. Multiple maximum obtained due to curvature in the response surfaces and their combination into the desirability function. Solution number 1 is selected as optimum solution. Thus the optimum input process parameters are voltage at 53 V, electrolyte concentration at 40 wt% and duty cycle at 75%. The response values predicted at these input setting would be MRR of 0.220 mg/min, TWR of 1.617 mg/min. DOC of 0.275 mm. the desirability index is 0.743. Figure 81.4 shows the optimization plot for all factors and responses when magnetic field is applied during the ECDM process.

Table 81.7: Optimization criterion for factors and responses (mf).

Name	Goal	Lower limit	Upper limit	Lower weight	Upper weight	Importance
A:vol	is in range	45	55	1	1	3
B:ele con	is in range	20	40	1	1	3
C:duty	is in range	55	75	1	1	3
D:condition	is equal to mf	nmf	mf	1	1	3
MRR	maximize	0.04	0.22	1	1	3
TWR	minimize	1.06	3.04	1	1	3
DOC	minimize	0.0055	0.6353	1	1	3

From the optimization analysis, it is found that under the magnetic field condition value of desirability index is 0.743 which is greater than the desirability index obtained under no magnetic field condition. Optimized value of voltage for mf condition is 53 V which is slightly less than the voltage value obtained for nmf condition. Values of responses parameters obtained at optimized values of factors show that under mf condition there is improvement in MRR but simultaneously DOC is also increased.

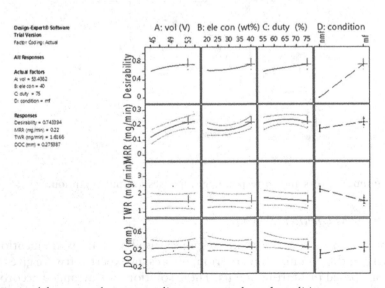

Figure 81.4 Optimal factors and corresponding responses for mf condition.

Multiple response method called as desirability is used to optimize the factors according to desired responses. Optimum solutions for no magnetic field (nmf) condition and magnetic field (mf) conditions were obtained. For nmf condition optimum levels of factors are voltage at 55 V, electrolyte concentration at 40 wt % and duty cycle at 75%. For mf condition optimum values of factors are voltage at 53 V, electrolyte concentration at 40 wt % and duty cycle at 75%.

Conclusions

Application of magnetic field in ECDM induces MHD effect in electrolyte and causes convective motion electrolyte. The effect of magnetic field on electrolyte circulation judged visually. With downward direction of magnetic field, electrolyte circulation takes place in counter clockwise direction and upward magnetic field causes circulation of electrolyte in clockwise direction.

Average MRR for all 15 experiments under no magnetic field condition (nmf) is 0.1187 mg/min and average MRR for all 15 experiments under magnetic field condition is 0.1440 mg/min. Thus there is trend of increase in MRR under magnetic field condition (mf) as compared with no magnetic field (nmf) condition. But there is increase in DOC which is undesirable as it reduces the accuracy.

For no magnetic field condition, optimum solution is obtained for voltage at 55 V, electrolyte concentration at 40 wt% and duty cycle at 75%. The response values predicted at these input setting would be MRR of 0.179 mg/min, DOC of 0.1 mm and the desirability index is 0.635.For magnetic field condition optimum solution is obtained for voltage at 53 V, electrolyte concentration at 40 wt% and duty cycle at 75%. The response values predicted at these input setting would be MRR of 0.22 mg/min, DOC of 0.275 mm. and the desirability index is 0.743.

References

1. Kolhekar, K. R., and Sundaram, M. (2018). Study of gas film characterization and its effect in electrochemical discharge machining. *Precision Engineering*, 53, 203–211. 10.1016/j.precisioneng.2018.04.002; https://dx.doi.org/10.1016/j.precisioneng.2018.04.002.
2. Dhanvijay, M. R., and Ahuja, B. B. (2015). Performance enhancement of ECDM process on Al2O3 ceramics using tubular tool. *International Journal of Scientific and Engineering Research*, 6(7), 369–372.
3. Sabahi, N., and Razfar, M. R. (2018). Investigating the effect of mixed alkaline electrolyte (NaOH+ KOH) on the improvement of machining efficiency in 2D electrochemical discharge machining (ECDM). *The International Journal of Advanced Manufacturing Technology*, 95(1–4), 643–657. 10.1007/s00170-017-1210-4; https://dx.doi.org/10.1007/s00170-017-1210-4.
4. Paul, L., George, B. P., and Varghese, A. (2018). Characterisation of micro channels machined with ECDM for fluidic applications. *Applied Mechanics and Materials*, 877, 82–86. 10.4028/www.scientific.net/amm.877.82; https://dx.doi.org/10.4028/www.scientific. net/amm.877.82.
5. Saranya, S., and Sankar, A. R. (2018). Fabrication of precise micro channels using a side-insulated tool in a spark assisted chemical engraving process, 33(13), 1422–1428. 10.1080/10426914.2017.1401728; https://dx.doi.org/10.1080/10426914.2017.1401728.
6. Cheng, C. P., Wu, K. L., Mai, C. C., Hsu, Y. S., and Yan, B. H. (2010). Magnetic field-assisted electrochemical discharge machining. *Journal of Micromechanics Microengineering*, 7, 20–20.
7. Rattan, N., and Mulik, R. S. (2017). Improvement in material removal rate (MRR) using magnetic field in TW-ECSM process. *Materials and Manufacturing Processes*, 32(1), 101–107. 10.1080/10426914.2016.1176197; https://dx.doi.org/10.1080/ 10426914.2016.1176197.

8. Hajian, M., Razfar, M. R., and Movahed, S. (2016). An experimental study on the effect of magnetic field orientations and electrolyte concentrations on ECDM milling performance of glass. *Precision Engineering*, 45, 322–331. 10.1016/j.precisioneng.2016.03.009; https://dx. doi.org/10.1016/j.precisioneng.2016.03.009.
9. Xu, Y., Chen, J., Jiang, B., Liu, Y., and Ni, J. (2018). Experimental investigation of magneto hydrodynamic effect in electrochemical discharge machining. *The International Journal of Mechanical Sciences*, 1033, 86–96.
10. Antil, P., Singh, S., Kumar, S., and Manna, N. (2019). Taguchi and multi-objective genetic algorithm based optimization during ECDM of SiCp/ glass fiber reinforced PMC's. *Indian Journal of Engineering and Materials Sciences*, 26, 211–219.
11. Singh, T., and Dvivedi, A. (2020). On prolongation of discharge regime during ECDM by titrated flow of electrolyte. *The International Journal of Advanced Manufacturing Technology*, 107, 1819–1834.
12. Saini, G., Kumar, A., Mohal, S., Kumar, R., and Chauhan, A. (2021). Electrochemical discharge machining process, variants and hybridization: A review. *IOP Conference Series: Materials Science and Engineering*, 1033, 012070.

82 Review: solar energy storage materials and technology

Brihaspati Singh, Anmesh Kumar Srivastava, and Om Prakash

Mechanical Engineering Department, NIT Patna Patna, Bihar, India

Abstract: Solar thermal energy storage material can store thermal energy during sunshine and stored energy can be extracted for variety of application during cloudy and off-sunshine time. Solar energy absorption & storage in a medium could be done in different way like sensible heat, latent heat and thermochemical storage technology. Solar energy storage can be done at different temperature and in different quantity that depends on the energy storage technology and type of material used for solar energy storage. In sensible energy storage technology, solar energy can be stored in a storage medium at varying temperature and storage yield depends on specific heat of storage material and volume, while in latent storage technology energy can be stored in a medium at constant temperature. Liquid and solid mediums are being used for sensible heat storage and for latent heat storage technology PCMs are used. The use PCMs for energy storage is limited due to leakage of PCM during melting, high degree of supercooling, lower thermal conductivity, and lower heat storage capacity per unit volume. In this current study a comprehensive review has been done in area of solar energy storage technology. Organic, inorganic, eutectic mixture, nano fluid and carbon-based material is used as solar energy storage options. Encapsulation is also a technique which improves the thermal conductivity, decrease chance phase separation and increases cycle stability of PCMs.

Keywords: Energy storage, latent heat, sensible, solar.

Introduction

In this current time when most of the fossil fuels are about to finish in upcoming years in all around the world, research and innovation must be promoted in solar energy field mainly in that region where there is plethora of solar energy. Solar energy is available in large quantities in India and it is necessary that we must use that free of cost solar energy as an energy source. Efficient and effective ways must be developed for better utilization of solar energy. It is well known fact that the above-mentioned systems functionality is totally based on availability of solar energy radiation but solar energy is available only at the time of sun-light. Therefore, the working devices based on solar energy are intermittent in nature. Solar energy storage technology has been developed to stock the solar radiation energy in form of electricity or heat for application of solar devices in hazy day or night time. In solar thermal systems, solar energy extraction and storage can be done mainly

lncs@springer.com

in three ways, a). Sensible energy storage technology, b). Latent heat storage technology, c). Solar thermal system with thermochemical energy storage unit. This study is mostly focused on two types of solar thermal energy storage system: 1) Sensible solar energy storage technology, 2) Latent solar energy storage technology.

Sensible Solar Thermal Energy Storage System

Sensible heat storage is a technology in which heat energy of can be extracted as well stored in a medium where temperate rise of the medium is directly proportional to the solar energy absorption in the medium. In many of the industrial and domestic applications water is commonly used as a sensible heat storage material. Sensible solar heat energy storage system are mainly two advantages: 1) it is cheaper and it has less danger because of less harmful energy storage materials are used for solar thermal energy storage [1]. The classification of thermal energy storage technology is given below in Figure 82.1. In sensible heat storage system for low and medium temperature (0–100°C) application water is used as an energy storage medium [3]. For high temperature (more than 100°C) energy storage application some other mediums are used like molten salt, oils liquid metals [4]. For high temperature application some solid phase sensible thermal energy storage material can be used the lists of materials are given below in Table 82.1. In a study kanchey marbles has been used as a sensible heat energy storage medium in single basin double slope solar still and with the use of kanchey marble as a sensible heat storage material, the system performance has been enhanced [5]. Packed bed based sensible heat energy storage is too one of the promising sensible heat storage technology for lower temperature applications. Efficient packed bed solar heat energy storage technology should have high energy density, good heat transfer, stability of storage material, complete reversibility of cycle, low thermal loss, economic feasibility and environmentally friendly [6].

Solar energy storage is a technology which store heat energy of sun radiation in a medium and this heat can be utilized in the later time for heating, cooling or power generation in various daily life or industrial application. In another study the conclusion has been drawn that the improved performance of solar air heater was obtained with sensible heat storage medium (high quality synthetic oil) as compare to the conventional solar air heater without energy storage [8]. The main cons of sensible heat energy storage technology are, the lower heat energy storage density per unit volume of heat storage medium and it is shown below in Figure 82.2.

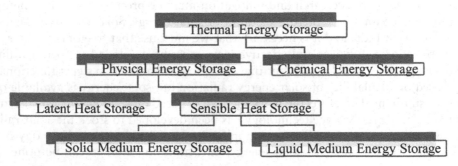

Figure 82.1 Classification of sensible energy storage system [2].

Table 82.1: Solid state sensible thermal energy storage material [7].

Storage materials	Density(kg/m³)	Working temperature (°C)	Specific Heat(kJ/kg-K)	Thermal conductivity(W/m-K)
Sand rock minerals	1700	200–300	1.30	1.0
Reinforced concrete	2200	200–400	0.85	1.5
Cast Iron	7200	200–400	0.56	37.5
NaCl	2160	200–500	0.85	7.0
Cast steel	7800	200–700	0.60	40.0
Silica fire bricks	1820	200–700	1.00	1.5
Magnesia fire bricks	3000	200–1200	1.15	5.0
Waste Glass [9]	2900	-	0.714–1.122	1.16–1.59
Powder Material formed during steel making process in electric arc furnace [10]	3967	-	0.510 @100C	0.7

Apart from this, in a research it is found that thermal energy storage capacity of sensible materials is lesser than the phase change material thermal storage system but for energy storage at more than 150°C in phase change material is costlier [12]. Comparison of storage energy density for sensible storage material with phase change material in heating range from 20 to 26°C is shown below in Figure 82.3.

During review it has been found that the solar energy absorption and storage in sensible heat storage system are limited due to less energy storage density as compare to the other solar energy storage technology like latent heat storage technology and thermochemical energy storage technology and also due to variation in temperature of energy storage material during charging and discharging process. This system is not suitable for constant temperature applications.

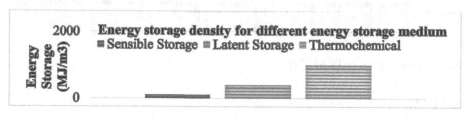

Figure 82.2 Comparative analysis of energy storage density with energy storage technique [11].

Latent Solar Heat Energy Storage System

Phase Change Materials (PCM) are used for the solar energy absorption and storage in latent heat energy storage system. During melting of PCM heat is absorbed from the solar radiation and during solidification heat is released for practical use. Phase change materials have high thermal energy storage capacity. Charging and discharging of heat take place relatively at particular constant temperature.

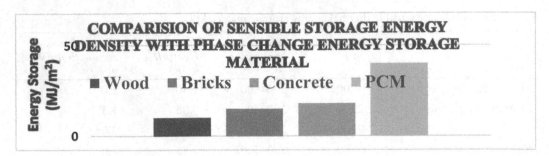

Figure 82.3 Comparative analysis of energy storage density with storage material [13].

High latent heat of fusion of phase change materials provides high thermal energy storage density. A comparative analysis has been carried out to compare the energy storage capacity of paraffin wax with other materials. In the analysis temperature rise has been observed to store 5000 kJ energy. It has been found that least mass of paraffin wax (a phase change materials) is required for same amount of energy storage. It can store 5000 kJ energy at its melting point of 59°C with initial temperature 35°C. it means the size and weight of energy storage will be lesser [14].

Types of Phase Change Material
PCMs are classified in broadly three categories: Organic, Inorganic and Eutectic. The classification of PCMs is shown below in Figure 82.4.

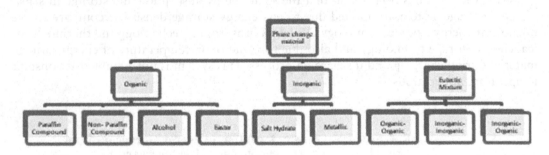

Figure 82.4 Classification of PCM [21].

Organic PCM
The major problems to use organic phase change materials as an energy storage in solar thermal systems are, low thermal conductivity, high-cost fire safety, etc [17, 18]. In past researches organic phase change materials has been successfully used in various applications like in solar space heating, food processing and storage, space industry, electronic devices cooling, refrigeration system, automobiles etc [19]. Examples: Paraffin wax, Al, Na, n-Dodecane (Carbon atoms-12), n-Dodezane (Carbon atoms-22), n-Triacotane(Carbon atoms-30), Caprylic acid (Carbon atoms-08), Lauric acid (Carbon atoms-12), n-Dodezane (Carbon atoms-22), n-Triacotane(Carbon atoms-30), Lauric acid (Carbon atoms-12), Mytristic acid (Carbon atoms-14), Palmitic acid (Carbon atoms-16), Ethylene Glycol distearate, Methyl-12-Hydroxystearate, Poly ethylene glycol etc [14].

Inorganic PCM

Inorganic phase change materials have mainly two parts, salt hydrate PCM and metallic PCM. This phase change materials mainly have high thermal energy storage density, higher thermal conductivity and good cycle stability [21]. The inorganic hydrated salt has revealed some advantages of higher latent heat of fusion, economical, non-inflammable and so on, offering extensive hope for use in solar heat energy storage applications. Nevertheless, these salt hydrate PCMs typically suffer from high degree of supercooling, phase separation and corrosivity problems [20]. In a study it has been found that shape stabilized hydrated salt phase change material will provide higher thermal conductivity, better photothermal efficiency, high latent heat of fusion, less degree of supercooling, less corrosivity and high heat energy storage capacity. Profitably, the shape stabilized SSG (Sodium acetate trihydrate-Sodium dihydrogen phosphate monohydrate-Graphene nano platelets) composites showed shape stability and high thermal reliability together with outstanding photo thermal efficiency of 92.6%. [22]. Examples: $MgCl_2.6H_2O$, Mg $(NO_3)_2$, Ca $(NO_3)_2$, Na_2CO_3, LiOH, NaCl, Na_2SO_4, LiF, NaF, Zn/Mg (53.7/46.3), Zn/Al (96/4), Mg/Ca (84/16), Cu/Zn/Si (74/19/7), Al/Si (87.76/12.24), Si/Mg (56/44), etc [14].

Eutectic PCM

Eutectic PCMs forms with the combination of two or more components and it can be mixture of organic-inorganic, organic-organic or inorganic-inorganic. During freezing and melting these PCMs do not exhibit phase separation [19]. Examples: Lauric acid-capric acid (45/55), Capric acid-Palmitic acid (76.5/23.5), Lauric acid-mytristic acid (66/34), Lauric acid-Palmitic acid (69/31), Lauric acid-stearic acid (75.5/24.5), Mytristic acid-Palmitic Acid (58.0/42.0), Mytristic acid-stearic acid (64.0/36.0), Mytristic acid-stearic acid (65.7/34.3), Palmitic acid-stearic acid (64.2/35.8), $CaCl_2 \cdot 6H_2O/CaBr_2.6H_2O$ (45/55), LiOH/KOH (40/60), KNO_3/KCl (95.5/4.5), $MgCl_2$/KCl (39/61), $BaCl_2$/KCl/NaCl (53/28/19), LiF/MgF_2/KF (64/30/ 6 mol%), LiF/CaF_2(80.5/19.5 mol%), etc [14].

Advantages and Disadvantages of Phase Change Materials
Advantages and disadvantages for different categories of phase change materials are given below in Table 82.2.

Table 82.2: Advantages and disadvantages of organic, inorganic and eutectic PCMs [15].

Phase Change Materials	Advantages	Disadvantages
Organic	Sub cooling degree is lower	In solid state low thermal conductivity.
	Congruent melting	Some are insoluble in water
	Compatible with metals	Flammable
	No phase segregation	Low phase change enthalpy
	Self-nucleating property	Low density
	High heat of fusion	More surface area required
	Available for wide temperature range	Freezing cycle required high heat transfer rates.
	Chemically stable	Latent heat storage capacity is low

(continued)

Table 82.1: Continued

Phase Change Materials	Advantages	Disadvantages
Inorganic	Greater phase change enthalpy	High subcooling
	Large heat storage capacity	Thermally instable
	Low volumetric expansion	Phase segregation
	Higher latent heat of fusion	Incongruent melting
	Less costly	Some have high weight
	Non-flammable	Corrosive to materials
	Sharp phase-change	Prone to degradation
	Thermal conductivity is higher	Chemical instability
Eutectic	Sharp melting	High cost
	High volumetric thermal storage	Latent heat capacity is lower
	No phase segregation	Some PCM have super-cooling
	Congruent phase change	Strong Odor

In many past studies has been found that the lower thermal conductivity, phase separation and cycle stability are the key problems linked with PCMs. Thermal conductivity is directly function of amount of energy stored in the medium in a given time.

Thermal Conductivity Augmentation Technique of Phase Change Materials
Many studies have shown that carbon-based materials have a huge ability to engage with inorganic PCM to enhanced their heat conductivity and also size stability. In addition to that, carbon-based materials have an outstanding capability for solar light absorption which is always a major necessity for solar-thermal energy storage. In other words, the composite PCM materials are not only works as thermal energy storage, but also serve as capture of photon and heater under solar thermal radiation [24]. Between different carbon materials, graphene [27], carbon nanotube [31], expanded graphite [25,26], graphene oxide [28–30], and carbon fiber [32] have been used for composite PCMs in various applications. Encapsulation of PCMs is also a technique to diminish the problem of phase separation and enhance the thermal conductivity of storage medium. Encapsulation is broadly classified into three categories: 1) Micro encapsulation, 2) Macro encapsulation, 3) Nano encapsulation. Encapsulation materials should have following properties: corrosion resistant, strength, thermal stability. Metallic shells are used in most of the microencapsulation while plastic shells are used in the microencapsulation [21]. In nano-encapsulation metallic nano particles have been used as shell material. This technique is successful because it provides increased surface area of heat transfer, reduce wall thickness and increased PCM quantity, due to nano particles [33]. In order to increase the effectiveness of PCMs, it is suggested to encapsulate them prior to application due to liquid-solid transformation of PCM during charging and discharging. Encapsulated phase change materials (EPCMs) consist of inorganic or polymer as shell and phase change materials as core [34]. Addition of nano particles, addition of expended graphite, metallic foam and encapsulation of PCM, are some techniques by which thermal conductivity of PCM can be improved [16].

Solar Energy Storage Technology Applications [23]

Solar heat energy storage technology is being used now a days to enhance the performance of solar thermal systems and provides smooth solar thermal system performance in off sunshine time and hazy days. Solar thermal energy storage technology can be used in various solar systems like solar still, solar water heater, evacuated tube solar water heater, solar air heater, parabolic trough collector, solar cooker, solar PVT systems.

Conclusion

Solar energy is going to be one of the major resources of energy in near future as this energy is free of cost, easily available and pollution free. Major drawbacks of solar energy are, availability of solar energy radiation is intermittent (maximum 8 hour during sunshine out of 24 hours) and performance of solar based devices is lower. In this study sensible and latent solar energy technology have been reviewed and it is found that integration of energy storage material with solar based devices improves the system performance. Thermal properties of storage material have discussed and suitable storage material and technology can be selected as per applications. The criteria to select storage material may be: temperature of application, capacity of energy storage, specific heat for sensible, latent heat of PCMs, conductivity of heat, stability of thermal cycles, phase separation, system size, flammability, economy, availability, etc. Shape stabilization of salt hydrates, micro, macro, nano-encapsulation can also be done to improve the performance of solar energy storage materials.

Future Scope

In the field of solar energy still many more new technologies need to be invented. Being a major source of energy on earth in near future, researcher should think of in two major areas: 1) To improve the performance of any kind of solar based devices, 2) To make balance between supply and demand of energy by inventing new technology.

References

1. Hussain, F., Rahman, M. Z., Sivasengaran, A. N., and Hasanuzzaman, M. (2020). Energy storage technologies, energy for sustainable development. 125–165. Academic Press. https://doi.org/10.1016/B978-0-12-814645-3.00006-7.
2. Khan, M. M. A., Saidur, R., and Al-Sulaiman, F. A. (2017). A review for phase change materials (PCMs) in solar absorption refrigeration systems. *Renewable and Sustainable Energy Reviews*, 76, 105e137. https://doi.org/10.1016/j.rser.2017.03.070.
3. Ayyappan, S., Mayilsamy, K., and Sreenarayanan, V. V. (2016). Performance improvement studies in a solar greenhouse drier using sensible heat storage materials. *Heat and Mass Transfer*, 52(3), 459e467. https://doi.org/10.1007/s00231-015-1568-5.
4. Sarbu, I., and Sebarchievici, C. (2018). A comprehensive review of thermal energy storage. *Sustainability*, 10(1), 1–32. https://doi.org/10.3390/su10010191.
5. Saravanan, N. M., Rajakumar, S., and Moshi, A. A. M. (??). Experimental investigation on the performance enhancement of single basin double slope solar still using kanchey marbles as sensible heat storage materials. *Materials Today: Proceedings*, 1600–1604. https://doi.org/10.1016/j. matpr.2020.05.710.
6. Gautam, A., and Saini, R. P. (2020). A review on sensible heat based packed bed solar thermal energy storage system for low temperature applications. Department of Hydro and Renewable

Energy, Indian Institute of Technology Roorkee, 247667, India. *Solar Energy*, 207, 937–956. https://doi.org/10.1016/j.solener.2020.07.027.

7. Tian, Y., and Zhao, C. Y. (2013). A review of solar collectors and thermal energy storage in solar thermal applications. *Applied Energy*, 104, 538e553. https://doi.org/10.1016/j.apenergy.2012.11.051.

8. Kalaiarasi, G., Velraj, R., Vanjeswaran, M. N., and Pandian, N. G. (2020). Experimental analysis and comparison of flat plate solar air heater with and without integrated sensible heat storage. *Renewable Energy*, 150, 255–265, https://doi.org/10.1016/j.renene.2019.12.116.

9. Gutierrez, A., Gil, A., Aseguinolaza, J., Barreneche, C., Calvet, N., Py, X., Fernandez, ´ A. I., Grageda, M., Ushak, S., and Cabeza, L. F. (2016). Advances in the valorization of waste and by-product materials as thermal energy storage (TES) materials. *Renewable and Sustainable Energy Reviews*, 59, 763–783.

10. Navarro, M. E., Martínez, M., Gil, A., Fernandez, A. I., Cabeza, L. F., Olives, R., and Py, X. (2012). Selection and characterization of recycled materials for sensible thermal energy storage. *Solar Energy Materials & Solar Cells*, 107, 131–135.

11. Tatsidjodoung, P., Pierr`es, N. L., and Luo, L. (2013). A review of potential materials for thermal energy storage in building applications. *Renewable and Sustainable Energy Reviews*, 18, 327–349.

12. Mawire, A., and McPherson, M. (2009). Experimental and simulated temperature distribution of an oil-pebble bed thermal energy storage system with a variable heat source. *Applied Thermal Engineering*, 29(5–6), 1086–1095. https://doi.org/10.1016/j.applthermaleng.2008.05.028.

13. Madad, A., Mouhib, T., and Mouhsen, A. (2018). Phase change materials for building applications: a thorough review and new perspectives. *Buildings*, 8(5), 63. https://doi.org/10.3390/buildings8050063.

14. Regin, A. F., Solanki, S. C., and Saini, J. S. (2008). Heat transfer characteristics of thermal energy storage system using PCM capsules: a review. *Renewable and Sustainable Energy Reviews*, 12(9), 2438–2458.

15. Nazir, H., Batool, M., Osorio, F. J. B., Isaza-Ruiz, M., Xu, X., Vignarooban, K., and Kannan, A. M. (2019). Recent developments in phase change materials for energy storage applications: a review. *International Journal of Heat and Mass Transfer*, 129, 491–523. https://doi.org/10.1016/j.ijheatmasstransfer.2018.09.126

16. Qureshi, Z. A., Ali, H. M., and Khushnood, S. (2018). Recent advances on thermal conductivity enhancement of phase change materials for energy storage system: a review. *International Journal of Heat and Mass Transfer*, 127, 838–856. https://doi.org/10.1016/j.ijheatmasstransfer.2018.08.049.

17. Magendran, S. S., Khan, F. S. A., Mubarak, N., Vaka, M., Walvekar, R., Khalid, M., Abdullah, E., Nizamuddin, S., and Karri, R. R. (2019). Synthesis of organic phase change materials (pcm) for energy storage applications: a review. *Nano-Structures Nano-Objects*, 20, 100399.

18. Zhou, L., Tang, L.-S., Tao, X.-F., Yang, J., Yang, M.-B., and Yang, W. (2020). Facile fabrication of shape-stabilized polyethylene glycol/cellulose nanocrystal phase change materials based on thiol-ene click chemistry and solvent exchange. *Chemical Engineering Journal*, 396, 125206.

19. Sharma, A., Tyagi, V. V., Chen, C. R., and Buddhi, D. (2009). Review on thermal energy storage with phase change materials and applications. *Renewable and Sustainable Energy Reviews*, 13(2), 318–345.

20. Peng, S., Hu, Y., Huang, J., and Song, M. (2020). Surface free energy analysis for stable super-cooling of sodium thiosulfate pentahydrate with microcosmic-visualized methods, *Solar Energy Materials and Solar Cells*, 208, 110390.

21. Javadi, F. S., Metselaar, H. S. C., and Ganesan, P. (2020). Performance improvement of solar thermal systems integrated with phase change materials (PCM), a review. *Solar Energy*, 206, 330–352. https://doi.org/10.1016/j.solener.2020.05.106.

22. Mehrali, M., ten Elshof, J. E., Shahia, M., and Mahmoudi, A. (2021). Simultaneous solar-thermal energy harvesting and storage via shape stabilized salt hydrate phase change material. *Chemical Engineering Journal*, 405, 126624.

23. Kalidasan, B., Pandey, A. K., Shahabuddin, S., Samykano, M., Thirugnanasambandam, M., and Saidur, R. (2020). Phase change materials integrated solar thermal energy systems: global trends and current practices in experimental approaches. *Journal of Energy Storage*, 27, 101118. https://doi.org/10.1016/j.est.2019.101118 Received 28 August 2019.

24. Xiao, Q., Fan, J., Fang, Y., Li, L., Xu, T., and Yuan, W. (2018). The shape-stabilized light-to-thermal conversion phase change material based on ch3coona·3h2o as thermal energy storage media. *Applied Thermal Engineering*, 136, 701–707.

25. Wang, Q., Zhou, D., Chen, Y., Eames, P., and Wu, Z. (2020). Characterization and effects of thermal cycling on the properties of paraffin/expanded graphite composites. *Renewable Energy*, 147, 1131–1138.

26. Zou, T., Fu, W., Liang, X., Wang, S., Gao, X., Zhang, Z., and Fang, Y. (2020). Hydrophilic modification of expanded graphite to develop form-stable composite phase change material based on modified cacl2·6h2o. *Energy*, 190, 116473.

27. Liao, H., Chen, W., Liu, Y., and Wang, Q. (2020). A phase change material encapsulated in a mechanically strong graphene aerogel with high thermal conductivity and excellent shape stability. *Composites Science and Technology*, 189, 108010.

28. Chen, G., Su, Y., Jiang, D., Pan, L., and Li, S. (2020). An experimental and numerical investigation on a paraffin wax/graphene oxide/carbon nanotubes composite material for solar thermal storage applications. *Applied Energy*, 264, 114786.

29. Mehrali, M., Latibari, S. T., Mehrali, M., Metselaar, H. S. C., and Silakhori, M. (2013). Shape-stabilized phase change materials with high thermal conductivity based on paraffin/ graphene oxide composite. *Energy Conversion Manage*, 67, 275–282.

30. Mehrali, M., Latibari, S. T., Mehrali, M., Mahlia, T. M. I., and Metselaar, H. S. C. (2013). Preparation and properties of highly conductive palmitic acid/graphene oxide composites as thermal energy storage materials. *Energy*, 58, 628–634.

31. Zhu, X., Han, L., Lu, Y., Wei, F., and Jia, X. (2019). Geometry-induced thermal storage enhancement of shape-stabilized phase change materials based on oriented carbon nanotubes. *Applied Energy*, 254, 113688.

32. Zhu, X., Han, L., Yang, F., Jiang, J., and Jia, X. (2020). Lightweight mesoporous carbon fibers with interconnected graphitic walls for supports of form-stable phase change materials with enhanced thermal conductivity. *Solar Energy Materials and Solar Cells*, 208, 110361.

33. De Matteis, V., Cannavale, A., Martellotta, F., Rinaldi, R., Calcagnile, P., Ferrari, F., Ayr, U., and Fiorito, F. (2019). Nano-encapsulation of phase change materials: from design to thermal performance, simulations and toxicological assessment. *Energy and Buildings*, 188–189, 1–11.

34. Jamekhorshid, A., Sadrameli, S. M., and Farid, M. (2014). A review of microencapsulation methods of phase change materials (PCMs) as a thermal energy storage (TES) medium. *Renewable and Sustainable Energy Reviews*, 31, 531–42.

83 Analysis of combustion stability and radiation efficiency on porous radiant burners: a mini-review

Kishan Dash[a] and Sikata Samantaray

Department of Mechanical Engineering, ITER, SOA Deemed to Be University, Bhubaneswar, India

Abstract: In the case of a conventional burner the combustion of an air-fuel mixture is less effective because the thickness of the burner is less and the combustion occurs in an open environment. So researchers experimented with porous media combustion technology. Various types of porous radiant burners (PRB) have developed using porous media combustion. This paper analyzes how the combustion stability and radiation efficiency in PRB are affected due to different parameters like porosity, thermal conductivity, geometry, emissivity, and pore sizes of porous media. The paper has discussed how the stable flame is achieved in PRB by varying the optical path length in the preheat zone, introducing the oxygen-added air, supplying more air, preheating the air, etc. This paper also describes the improvement in radiation efficiency of PRB by stabilizing the flame at the center of the burner, keeping downstream section porosities in the range of 0.95–0.97 for any value of upstream section porosities, increasing the equivalence ratio near stoichiometry, decreasing the thermal load, etc. The works that have been published in reputed journals have been examined, summarized, and given appropriate comments and ideas for future research.

Keywords: Combustion stability, porous radiant burner, radiation efficiency.

Introduction

The performance of the conventional burner is ineffective due to free flame combustion and unfinished combustion fuel-air mixture because of the thin combustion zone. Porous media combustion (PMC) has emerged as a novel concept of combusting air-fuel mixture in a porous solid medium that is highly conductive and radiative for improved heat transport. The porous medium has good thermal conductivity and emissivity, so the conduction and radiation mode of heat transport increases. The convection heat transport also increases because of the large inside surface area. Due to better heat transport, the PRB has several advantages like homogeneous temperature distribution, high radiation efficiency, high energy density, high energy modulation, and less emission, etc [1]. PMC is classified into two types: submerged combustion (maximum temperature obtained in the interface of two layers in a porous burner) and surface stabilized combustion (maximum temperature on the surface of the porous burner), both are dependent on working parameters [2, 3, 4].

---

[a]kishandash90@gmail.com

The heat after combustion moves to the upstream region of the PRB and helps in preheating the incoming air-fuel mixture.

PRB is of two types: one-layered and bi-layered PRB. The one-layered PRB is made of a lone layer of large porosity porous media, and the bi-layered PRB is made of two different or the same porous media of dissimilar porosities, as well as dissimilar radiative and thermal properties [5, 6, 7]. The preheat zone (PZ) and combustion zone (CZ) are the main parts of the bi-layered PRB [8, 9]. The bi-layered PRB has more applications than one-layered PRB due to its better flame stability [10]. The position of the flame in PRB depends on material characteristics and porosity [9, 11]. The bi-layered PRB schematic diagram is shown in Figure 83.1.

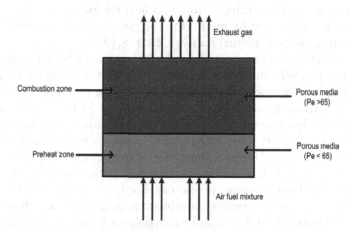

Figure 83.1 A bi-layered PRB schematic diagram.

The stability of flame in the bi-layered PRB is highly dependent on the peclet number (P_e) [12]. In the bi-layered PRB, the peclet number (P_e) must be greater than 65 in the CZ and less than 65 in the PZ for better flame stability [12].

The objective of the paper is to analyze the effect of parameters on combustion stability and radiation efficiency in the PRB. The works that have been published in reputed journals have been examined, summarized, and given appropriate comments and ideas for future research.

Combustion Stability in Porous Radiant Burner

The combustion stability limit (without flashback and blow-off) is the most significant characteristic which describes the performance of the PRB. When the input velocity of the air-fuel mixture is higher compared to the input velocity of the flame, the blow-off occurs. In such a scenario, the flame goes out of the burner surface. In some cases, the input velocity of the air-fuel mixture is less compared to the velocity of the flame, resulting in the movement of the flame towards the upstream region of the PRB, causing a flashback. In the equivalence ratio (φ) range, the burner is stable at a given firing rate, referred to as the range of stable combustion.

A group of researchers observed the stable flame at the interface of PZ and CZ and adjacent to the surface of CZ for methane (CH_4-air) combustion [13]. It has been found that the speed of the flame increased for internal heat circulation properties of porous media. This property helps to achieve stable combustion for various φ and input loads [8, 14, 15]. Some authors found that with the increase in φ the flame speed increased in case of lean combustion in PRB. They also reported with the increase in φ, the lower limit of the flame speed ratio decreases, but the upper limit of the flame speed ratio first decreases and then increases [8]. However, at the same φ, the occurrence of two critical flame speeds has been dismissed by some researchers [16]. They found stable combustion at a specific φ. Another researcher found that a Yttria-stabilized zirconia-aluminum composite (YZA) burner offers better combustion stability than Zirconia-toughened mullite (ZTM) at the φ of 0.65 and a thermal load range of 675–3951 kW/m². So the YZA burner has offered stable combustion for a broad range of thermal loads of 675–3951 kW/m²[17]. It has been found that the heat losses, heat recirculation, and heat release are controlled in the porous burner to get stable combustion by modifying the porous media's porosity, thermal conductivity, emissivity, geometry, air-fuel mixture, etc [18,19].

A new group of authors experimented on the bi-layered PRB by taking three different 10, 30, and 45 ppi Partially stabilized zirconia (PSZ) in CZ, keeping 65 ppi PSZ in PZ [20]. The submerged combustion was achieved by varying the pore size of PSZ. The stable combustion in PRB occurred at the φ of 0.41, 0.44, 0.51 for 10, 30, and 65 ppi PSZ respectively, which are less compared to free flame combustion in the conventional burner at φ of 0.52. Some researchers have developed two porous media burners by keeping 45 psi PSZ in the PZ and taking 10 ppi alumina oxide (Al_2O_3) and 20 ppi Siliconized silicon carbide (SiSiC) in CZ [21]. It has been found that SiSiC offers more stability compared to Al_2O_3 because of better heat transport properties. Kaplan and Hall [22] provided two proposals for the improvement of submerge combustion: a) the path length of PZ may be 0.8 μm whereas for CZ it may be 1.5 μm, and b) turbulence intensities may be minimized for smaller pore material. Some authors [7] found the flashback occurs in the PRB (PZ of 66 ppi cordierite and CZ of 20 ppi cordierite) for higher thermal load (> 300 kW/m²). Some authors found the lower and upper combustion stability of the PRB made of cordierite is at the φ of 0.49 and 0.66 [23]. It has been observed that the lower and upper combustion stability limit increases due to the increase in φ [22, 23]. A new group of researchers developed a new PRB made of Yattria-stabilized zirconia for the burning of propane and methane [24].

In another research, a group of researchers [25] made a bi-layered PRB, consisting of alumina balls of size 3 and 8 mm in PZ and CZ respectively for the combustion of biogas and methane. It has been observed that the PRB offers stable combustion at a higher φ range for biogas than methane, but the flame velocity is less in biogas compared to methane. The combustion stability and flame velocity have reduced in the PRB with the raise of CO_2 percentage in biogas [25]. In another experiment, the same research team tested the methane/(CO_2, Ar, N_2) combustion in the developed PRB and found that the blow-off and flashback limit for methane/CO_2 is lower than for methane/Argon mixture and methane/Nitrogen mixture because of higher heat capacity of CH_4/CO_2 mixture [26]. The same research group also experimented with methane-air combustion in the bi-layered PRB of different materials like Aluminium oxide, Zirconium oxide, Silicon carbide, and Iron-Chromium-Aluminium [27]. It has been found that the stable flame velocity of PRB follows increasing order according to the material used (Aluminium oxide< Zirconium oxide< Silicon carbide < Iron-Chromium-Aluminium) because of the increase in thermal conductivity of the material [27].

Some authors reported that the bilayered-PRB consists of 10 ppi SiSiC in the CZ and Al_2O_3 in the PZ providing large flexibility at lean conditions for excess air (k) of 1.2 [28]. The bi-layered PRB offers more combustion temperature for LPG than methane at k of 1.2–1.8 [28]. The stability of PRB has more due to input load compared to φ [28]. Some authors investigated the combustion of natural gas on the ceramic fiber burner and reticulated ceramic burner with oxygen-added air [29]. It has been found that the blow-off limit and flame velocity improved [29]. Bakry et al. [30] investigated the effect of temperature and pressure of premixed air-methane on the combustion stability of dissimilar porous inert media. It has been observed that the blow-off limit affects due to the temperature, but not the pressure of the premixed mixture.

Radiation Efficiency/ Radiation Output in Porous Radiant Burner

Radiation efficiency is defined as the ratio of the radiative flux leaving the PRBs to the heat produced by the combustion of the fuel-air mixture. The radiation efficiency of different PRBs as a function of thermal load, firing rate, fuel velocity, φ are presented in Table 83.1.

A group of researchers found the maximum radiation efficiency of 40% at φ of 0.5 by stabilizing the flame at the center of the PRB [13]. Kulkarni and Peck [11] studied theoretically on a bi-layered PRB and found that the radiation efficiency, radiation output, and flame speed depend on the upstream (preheat zone) and downstream (combustion zone) porosity. The maximum radiation efficiency, radiation output, and flame speed were found at the downstream section's porosity (ϕ_2) of 0.95–0.97 and the considered upstream section's porosity (ϕ_1) of 0.83, 0.90, 0.95, 0.99. The radiation output was improved to 17% by reducing the ϕ_2 from 0.99-0.95. The increment of ϕ_2 resulted in a decrease in radiation

Table 83.1: Radiation efficiency as a function of several input parameters.

Authors	Porous Media, and Type of fuel	Input Parameters				Outcome
		Firing rate (kW)	Thermal load (kW/m²)	Fuel velocity (m/s)	φ	Radiation Efficiency (%)
Gao et al. [25]	Alumina balls PZ: d= 3mm CZ: d=18mm Fuel: Biogas			0.15–0.6	0.9	~14–28
Bouma and Goey [31]	Ceramic foam poresize:0.4 mm φ: 90% Fuel: CH₄		100–600		0.90	~33-8
Leonardi et al. [32]	Fibrous metal Fuel: CH₄		170–340		0.90–1.1	~18-26
Yoksenakl and Jugjai [33]	Al₂O₃ balls d= 15 mm Fuel: LPG	23–61			2.08–2.5	~23-15
Qiu and Hayden [35]	Ceramic fiber felt, d: 10μm Fuel: Natural gas		200–500		0.60– 0.90	~14-12

(continued)

Table 83.1: Continued

Authors	Porous Media, and Type of fuel	Input Parameters				Outcome
		Firing rate (kW)	Thermal load (kW/m²)	Fuel velocity (m/s)	φ	Radiation Efficiency (%)
Keramiotis et al. [36]	SiSiC Pore density:10 ppi Fuel: Methane (CH_4) Biogas (CO_2:40%, CH_4: 60%) Syngas/Natural gas (CH_4:47.29%, H_2:19.79%, CO: 32.89%) Biogas-Syngas mixture (CH_4:38%, CO: 26%, H_2:16%, , CO_2: 20%) Mixture (CH_4:28%, CO: 20%, H_2:12%, CO_2:40%)		200–800		0.83	~52-25 ~28-25 ~30-65 ~30-55 ~45-25
Abdelaal et al. [37]	Mullite ϕ: 85% Fuel: LPG		94–610		0.7–1.3	~45-13
Devi et al. [38]	PZ: Al_2O_3 Ceramic, t=15mm, ϕ:~7% CZ: Sic Ceramic t=20mm, ϕ:90% Fuel: Biogas (CH_4=43-56%, CO_2=34-38%)	5–10			0.97– 0.75	~33-19

PZ: Preheat zone, CZ: Combustion zone, Al_2O_3: Alumina, SiC: Silicon Carbide, SiSiC: Silicon infiltrated Silicon Carbide, CH_4: methane, CO_2: carbon dioxide, CO: carbon monoxide, H_2: hydrogen, t = thickness, d= diameter, ϕ = porosity, ppi= pores per inch, ppc= pores per centimeter, LPG= Liquefied Petroleum Gas.

efficiency and radiant output due to the less availability of solids to participate in a heat transfer.

Both radiant output and flame speed increased with the reduction of ϕ_1 due to low thermal conductivity does not play a role in the radiative heat transfer in the burner.

Khanna et al. [15] experimented on a double-layered PSZ burner operating on CH_4 at various φ of 0.60, 0.65, 0.70, 0.75, 0.80, and 0.87. It has been found that the radiation efficiency of the PRB decrease due to the increase in flame speeds or firing rates. A similar type of observation, the decrease in radiation efficiency due to the increase in firing rates was observed by various authors [31,33]. Some researchers [34] made a porous media burner, consisting of fibrous ceramics of porosity 90%. The radiation efficiency of the burner increased due to the increase in φ, peaked near stoichiometry, and then decreased. The radiation efficiency was ~14–26% for the thermal loads of 300–450 kW/m² at φ = 0.75–0.95.

Mital et al. [7] built a PRB, made of cordierite. The burner had a fixed thickness of 19 mm and 26 ppc in the PZ. The test was conducted for different combinations of combustion zone thickness: (3.2mm,4ppc), (6.5mm,4ppc), and (3.2mm,8ppc). The radiation efficiency was found to be between ~20–30% depending on the thermal load, equivalence ratio, and thickness of the combustion zone. At a lesser thermal load (<300 kW/m²) and φ of 0.9, the radiation efficiency of the burner was found less for thicker combustion zone (6.5 mm) than thinner (3.2mm), while at a higher thermal load, the trend was reversed.

Qiu and Hayden [35] reported the relationship between radiation efficiency and thermal load for various combustion air temperatures of 25°C, 300°C, and 550°C and found that the radiation efficiency improved with air preheating. In the no-air preheating case, the radiation efficiency decreased from ~14–12% with an increase in the thermal load from 200–500 kW/m².

Gao et al. [25] found radiation efficiency of ~14–28% in the PRB for biogas-air combustion at different CO_2 concentrations of 25%, 30%, 35%,40%, and fuel velocity of 0.15–0.6 m/sec, and φ of 0.9. The CO_2 concentrations were less effective compared to the fuel velocity in the improvement of radiation efficiency. The same research team [26] studied the variation in radiation efficiency due to the influence of three diluents: Ar, N_2, and CO_2 at a fixed φ of 0.9 and a dilution ratio of 30%. The radiation efficiency decreased in the order of CO_2, N_2, and Ar. Abdelaal et al. [37] observed the effect of oxygen concentration on radiation efficiency. The radiation efficiency in the PRB can be improved to ~10% by varying the oxygen concentration from 21–25% at φ of 1.1 and a thermal load of 350 kW/m².

Conclusion

It has been found that the combustion stability and flame velocity improves due to φ in the lean mixture, the internal heat recirculation mechanism, the use of a cooling tube in PRB, and for good hard porous medium, etc. The combustion stability and flame length in PRB highly depend on the porous material's conductivity, emissivity, geometry, porosity, and pore size. The combustion stability in PRB has been enhanced by varying the optical path length in the PZ, providing more oxygen-added air, k > 1.2, heating the supplied air, etc. The radiation efficiency has improved by stabilizing the flame at the center of the burner, keeping ϕ_2 in the range of 0.95–0.97 for any value of ϕ_1, increasing the φ near stoichiometry, decreasing the thermal load, etc. Investigations are needed on surface mode combustion in PRB for domestic purposes. The study is also limited to raw biogas fuel which is a renewable fuel. So the future scope is to analyze the combustion of raw biogas produced by the anaerobic digestion of kitchen waste in the PRB.

References

1. Howell, J. R., Hall, M. J., and Ellzey, J. L. (1996). Combustion of hydrocarbon fuels within porous inert media. *Progress in Energy and Combustion Science*, 22(2), 121–145.
2. Marbach, T. L., and Agrawal, A. K. (2003). Experimental study of surface and interior combustion using composite porous inert media. In Turbo Expo: Power for Land, Sea, and Air (Vol. 36851, pp.549–556).

3. Marbach, T. L., Sadasivuni, V., and Agrawal, A. K. (2007). Investigation of a miniature combustor using porous media surface stabilized flame. *Combustion Science and Technology*, 179(9), 1901–1922.

4. Nakamura, Y., Itaya, Y., Miyoshi, K., and Hasatani, M. (1993). Mechanism of methane-air combustion on the surface of a porous ceramic plate. *Journal of Chemical Engineering of Japan*, 26(2), 205–211.

5. Bubnovich, V. I., Zhdanok, S. A., and Dobrego, K. V. (2006). Analytical study of the combustion waves propagation under filtration of methane–air mixture in a packed bed. *International Journal of Heat and Mass Transfer*, 49(15–16), 2578–2586.

6. Pantangi, V. K., Mishra, S. C., Muthukumar, P., and Reddy, R. (2011). Studies on Porous Radiant Burners for LPG (liquefied petroleum gas) cooking applications. *Energy*, 36(10), 6074–6080.

7. Mital, R., Gore, J. P., and Viskanta, R. (1997). A study of the structure of submerged reaction zone in porous ceramic radiant burners. *Combustion and Flame*, 111(3), 175–184.

8. Barra, A. J., and Ellzey, J. L. (2004). Heat recirculation and heat transfer in porous burners. *Combustion and Flame*, 137(1–2), 230–241.

9. Trimis, D., and Durst, F. (1996). Combustion in a porous medium-advances and applications. *Combustion Science and Technology*, 121(1–6), 153–168.

10. Hsu, P. F., Evans, W. D., and Howell, J. R. (1993). Experimental and numerical study of premixed combustion within nonhomogeneous porous ceramics. *Combustion Science and Technology*, 90(1–4), 149–172.

11. Kulkarni, M. R., and Peck, R. E. (1996). Analysis of a bilayered Porous Radiant Burner. *Numerical Heat Transfer, Part A Applications*, 30(3), 219–232.

12. Babkin, V. S., Korzhavin, A. A., and Bunev, V. A. (1991). Propagation of premixed gaseous explosion flames in porous media. *Combustion and Flame*, 87(2), 182–190.

13. Sathe, S. B., Kulkarni, M. R., Peck, R. E., and Tong, T. W. (1991). An experimental and theoretical study of Porous Radiant Burner performance. *Symposium (International) on Combustion*, 23(1), 1011–1018.

14. Kotani, Y., Behbahani, H. F., and Takeno, T. (1985). An excess enthalpy flame combustor for extended flow ranges. *Symposium (International) on Combustion*, 20(1), 2025–2033.

15. Khanna, V., Goel, R., and Ellzey, J. L. (1994). Measurements of emissions and radiation for methane combustion within a porous medium burner. *Combustion Science and Technology*, 99(1–3), 133–142.

16. Liu, J. F., and Hsieh, W. H. (2004). Experimental investigation of combustion in porous heating burners. *Combustion and Flame*, 138(3), 295–303.

17. Mathis, W. M., and Ellzey, J. L. (2003). Flame stabilization, operating range, and emissions for a methane/air porous burner. *Combustion Science and Technology*, 175(5), 825–839.

18. Du, L., and Xie, M. (2006). The influences of thermophysical properties of porous media on super adiabatic combustion with reciprocating flow. *Heat Transfer—Asian Research: Co-sponsored by the Society of Chemical Engineers of Japan and the Heat Transfer Division of ASME*, 35(5), 336–350.

19. Yoshio, Y., Kiyoshi, S., and Ryozo, E. (1988). Analytical study of the structure of radiation controlled flame. *International Journal of Heat and Mass Transfer*, 31(2), 311–319.

20. Hsu, P. F., Evans, W. D., and Howell, J. R. (1993). Experimental and numerical study of premixed combustion within nonhomogeneous porous ceramics. *Combustion Science and Technology*, 90(1–4), 149–172.

21. Djordjevic, N., Habisreuther, P., and Zarzalis, N. (2012). Experimental study on the basic phenomena of flame stabilization mechanism in a porous burner for premixed combustion application. *Energy and Fuels*, 26(11), 6705–6719.

22. Kaplan, M., and Hall, M. J. (1995). The combustion of liquid fuels within a porous media radiant burner. *Experimental Thermal and Fluid Science*, 11(1), 13–20.

23. Min, D. K., and Shin, H. D. (1991). Laminar premixed flame stabilized inside a honeycomb ceramic. *International Journal of Heat and Mass Transfer*, 34(2), 341–356.

24. Smucker, M. T., and Ellzey, J. L. (2004). Computational and experimental study of a two-section porous burner. *Combustion Science and Technology*, 176(8), 1171–1189.

25. Gao, H., Qu, Z., Tao, W., He, Y., and Zhou, J. (2011). Experimental study of biogas combustion in a two-layer packed bed burner. *Energy and Fuels*, 25(7), 2887–2895.

26. Gao, H. B., Qu, Z. G., Tao, W. Q., and He, Y. L. (2013). Experimental investigation of methane/(Ar, N2, CO2)–air mixture combustion in a two-layer packed bed burner. *Experimental Thermal and Fluid Science*, 44, 599–606.

27. Gao, H. B., Qu, Z. G., Feng, X. B., and Tao, W. Q. (2014). Methane/air premixed combustion in a two-layer porous burner with different foam materials. *Fuel*, 115, 154–161.

28. Keramiotis, C., Stelzner, B., Trimis, D., and Founti, M. (2012). Porous burners for low emission combustion: An experimental investigation. *Energy*, 45(1), 213–219.

29. Qiu, K., and Hayden, A. C. S. (2009). Increasing the efficiency of radiant burners by using polymer membranes. *Applied Energy*, 86(3), 349–354.

30. Bakry, A., Al-Salaymeh, A., Ala'a, H., Abu-Jrai, A., and Trimis, D. (2011). Adiabatic premixed combustion in a gaseous fuel porous inert media under high pressure and temperature: novel flame stabilization technique. *Fuel*, 90(2), 647–658.

31. Bouma, P. H., and De Goey, L. P. H. (1999). Premixed combustion on ceramic foam burners. *Combustion and Flame*, 119(1–2), 133–143.

32. Leonardi, S. A., Viskanta, R., and Gore, J. P. (2003). Analytical and experimental study of combustion and heat transfer in submerged flame metal fiber burners/heaters. *Journal Heat Transfer*, 125(1), 118–125.

33. Yoksenakul, W., and Jugjai, S. (2011). Design and development of a SPMB (self-aspirating, porous medium burner) with a submerged flame. *Energy*, 36(5), 3092–3100.

34. Williams, A., Woolley, R., and Lawes, M. (1992). The formation of NOx in surface burners. *Combustion and Flame*, 89(2), 157–166.

35. Qiu, K., and Hayden, A. C. S. (2006). Premixed gas combustion stabilized in fiber felt and its application to a novel radiant burner. *Fuel*, 85(7–8), 1094–1100.

36. Keramiotis, C., Katoufa, M., Vourliotakis, G., Hatziapostolou, A., and Founti, M. A. (2015). Experimental investigation of a radiant porous burner performance with simulated natural gas, biogas and synthesis gas fuel blends. *Fuel*, 158, 835–842.

37. Abdelaal, M. M., El-Riedy, M. K., and El-Nahas, A. M. (2013). Effect of oxygen enriched air on porous radiant burner performance and NO emissions. *Experimental Thermal and Fluid Science*, 45, 163–168.

38. Devi, S., Sahoo, N., and Muthukumar, P. (2020). Experimental studies on biogas combustion in a novel double layer inert Porous Radiant Burner. *Renewable Energy*, 149, 1040–1052.

84 Students' performance prediction using machine learning algorithms: a comparative study

Eldose K. K.[a], Ravina Agarwal[b], Pranav Dabhade[c], Fathima Raniya A. T.[d], Al Ameen K. P.[e], Sridharan R.[f], Ratna Kumar K.[g], and Gopakumar G.[h]

Abstract: The academic performance of students is of significant importance for every educational institution. Educational data mining has acquired a significant interest in recent years as machine learning applications are growing dramatically in various fields. Students' academic performance has a dependency on various factors, i.e., behavioural, personal, academic and extracurricular. The current research focuses on the academic performance prediction of undergraduate students in an engineering institute in South India. The data were collected using a questionnaire based on Google form and from the institute's academic section. Data cleaning, feature scaling, and data reduction have been conducted on the collected data to eliminate data inconsistencies, decrease data dimensionality, and prepare data for further analysis. The Python programming language libraries are used in the Jupyter notebook for the comparative study of different regression-based algorithms. The analysis of results reveal that linear support vector regression model provides the most reliable prediction with an R-square score of 0.8178.

Keywords: academic performance prediction, data mining, machine learning, regression.

Introduction

Assessment of students' academic performance generates enormous interest and concern of many educational institutions, teachers, parents, and every nation's government. Educational institutions must keep watch on students' academic performance and take necessary steps for its improvement. It is required for educators to monitor students' performance to reach academic goals and create an environment to improve consistently. There are various factors considered for understanding the performance of academic institutions. Depending on the considered factors, institutions need to improve their rankings. A student's academic accomplishment is considered a critical measure in the position of educational institutions [1]. Educational institutions are consistently providing quality education to achieve excellent performance. Educators need to get insights into students' problems in securing better academic performance. College professors can take remedial measures and improve the students' poor performance [2]. Educational data mining is used

[a]eldosekk.me@adishankara.ac.in, [b]ravinaagarwal20@gmail.com, [c]dabhadepranav@gmail.com, [d]alameenkp9068@gmail.com, [e]raniya.fara@gmail.com, [f]sreedhar@nitc.ac.in, [g]ratna_kumar@nitc.ac.in, [h]gopakumarg@nitc.ac.in

to gain important insights and patterns from the information collected from educational settings [3,4]. The present work emphasizes the development of prediction models using data mining techniques.

Motivation for the Research

The primary motivation for the present research is the importance of students' academic performance which can affect the graduation rate of an institute, future of the students and thereby affect the future of the country and academic standards. Academic performance is affected by various factors. Thus, there is a need for higher educational institutions to undertake studies to analyses how the students with particular features are performing academically. Also, it is necessary to predict their future performance and to take proper actions to improve it. The institute considered for the present study is among premier engineering institutions in India and is always progressive in proposing studies to solve real-world problems.

Most of the existing research focuses on past academic performance and some selected demographic features of students. However, personal, behavioural and extracurricular features can also be considered while building academic performance prediction models. Some articles from the literature show the impact of behavioural features, but very few attributes were considered. Thus, there will be limitations to the studies considering only a few attributes. Hence, for the present study, all the possible features of the students are considered to make the study more comprehensive.

Predominantly, classification techniques are used in most of the literature. However, obtaining accurate predictions of semester grade point average (SGPA) of students is more helpful than prediction in terms of performance classification. Hence, the present study uses regression techniques such as support vector regression (SVR), multiple linear regression (MLR), random forest regressor (RFR), stochastic gradient descent regressor (SGDR), gradient boosting regressor (GBR) and light gradient boosting machine regressor (LGBMR). Application of various regression methods allows to identify the best performing algorithm, and thus, comparative analysis can be carried out. Moreover, the correlation among the SGPA and other student features was studied in the present work.

Hence, the main contributions of the paper are as follows:

- The significant factors affecting the student's SGPA were identified and enumerated.
- The relationship between different features and SGPA was determined.
- Prediction models were developed using the selected attributes to evaluate the academic performance of undergraduate students in terms of SGPA.
- Comparative analysis was conducted for the developed performance prediction models.

The rest of the paper is structured as follows. Section 2 provides the salient features of the literature on performance prediction. Section 3 narrates the approach adopted for the development of prediction models. Section 4 highlights the results obtained and their interpretation. Section 5 provides conclusion and scope for future study.

Literature Review

Interesting studies can be found in the literature where researchers have applied machine learning and data mining techniques. Guabassi et al. [5] analysed the students' data and proposed a comparative analysis using a set of supervised machine learning algorithms such as SVR, linear regression (LR), RFR, and decision tree regression. RFR was found the most suitable algorithm for the study. Hussain et al. [6] performed regression analysis using LR and deep learning (DL) models. The DL-based model performed better than the LR model and it was also found that honours/major papers have a substantial impact on students' academic performance. Yang et al. [7] suggested using MLR after the Principal Component Analysis (PCA). A multivariable learning profile collected from the proposed blended calculus course was used while predicting the accuracy of MLR. It was found that measures to improve the accuracy of the MLR using PCA was better than MLR without PCA. Mean square error (MSE) values were 12.33 ad 10.23 before and after PCA, respectively. Ulfa et al. [8] applied MLR to predict students' activities which would improve students' learning output in the online learning management system. The study proved that the use of concept mapping while training was adequate to help the students learn the content. Razak et al. [9] used linear regression and J48 Decision Tree. WEKA and SPSS tools have been employed to determine the relationship between independent variables and dependent variables using regression and decision trees. The analysis revealed a strong relation between cumulative grade point average (CGPA) and SGPA. It also shows that gender does not affect performance.

Kumar et al. [10] determined the most influential features affecting students' academic performance. The regression model predicted the second-semester performance with 15% deviation. Rajalaxmi et al. [11] implemented the multivariate regression technique in the prediction of students' performance based on the independent variables containing the hours spent in internet activities. The researchers predicted the performance based on the CGPA category of the students. The dataset consisted of 150 undergraduate engineering students' data for model study. The error between actual and predicted values was determined for each CGPA category. Adekitan and Salau [12] proposed java-based models consisting of six algorithms. The study predicted the grades of the final year of the course using the performance of the previous three years. Regression models were used for further performance validation. Some researchers have used a nonlinear neural model for forecasting and error measures for evaluation [13]. Chakraborty et al. [14] have used analysis of variance and various measures for model evaluation purpose. Cloud computing environment was suggested by the researchers [15, 16] for securely handling the data.

In the literature, most of the studies have considered only a few attributes for prediction. However, it is required to consider students' background information, extracurricular activities and behavioural features as well. The present study is focused in this direction.

Research Methodology

Students' past academic performance and their personal, behavioural, and extracurricular features have been considered for performance prediction of students of a higher education engineering institution in South India for the current study. The features used in this research are extracted from the relevant past studies. some additional features are added specifically to the students at the chosen institute. For the data collection, a Google form questionnaire-based survey was conducted with 85 questions based on 80 selected features consisting of four major sections, i.e., educational, personal, behavioural and extracurricular features. A brief description of the chosen features is outlined in Table 84.1. A database with 564 students' data has been used based on the responses obtained from the

Table 84.1: Brief description of the categories of features affecting students' studies.

Feature	Description
Personal	Occupation of parents; Demographic details; Interests; Family's annual income; Time spent on entertainment and social media
Educational	Academic details; Entrance exam rank; Reason for choosing engineering; SGPA of previous semesters; career goals; Learning experience; Skills; Details of the internship
Behavioural	Participation in the academic tasks; Effect of group study and friend circle; Type of personality; Interaction with faculty.
Extracurricular	Participation in seminars/talks; Student representatives

Google form circulated among the B. Tech. students of the chosen institute. Other specific academic details like SGPA, 12th class percentage, entrance rank and the admission details have been obtained from the academic division of the institute.

Data Pre-Processing

Data pre-processing involves cleaning of data, feature scaling/normalisation of data, data reduction and factor analysis. Using Microsoft Excel, datasets acquired through the Google form-based survey were merged with that obtained from the academic division of the institute. Python 3 programming language libraries such as pandas, seaborn, and matplotlib were used for data cleaning and data visualisation in jupyter notebook. The data have been modified by removing some unnecessary/irrelevant features and generalising the responses for multiple response type questions, e.g., interests and hobbies. Categorical data comprises nominal data and ordinal data. The data were converted to numerical value using integer for nominal data and one-hot encoding for ordinal data. After encoding, a total of 80 features were converted to 162 input variables and one output variable.

Feature scaling normalises the features present in the data to a specific fixed range. In the current study, robust scalar has been used to scale the data and handle the variability in the values of the variables to obtain a better fit from the algorithms. Data reduction decreases data dimensionality by reducing the number of variables without losing the integrity of the data. Principal Component Analysis (PCA) method is applied for dimensionality reduction and for obtaining a correlation matrix between the variables considered. Faster performance of the machine learning algorithms can be achieved by decreasing the dimensionality of data without losing the data. The dimensionality of the dataset with the 162 variables has been reduced to 15.

Machine Learning Algorithms Adopted

Machine learning is categorized as unsupervised, semi-supervised, supervised and reinforcement learning. Classification and regression algorithms belonging to supervised learning are used for discrete and continuous data respectively. In the present research, SVR, MLR and other regressors are applied to understand the impact of various features on academic performance. The following sub-sections provide a brief description of the algorithms.

Multiple Linear Regression
Linear regression finds a linear relationship between the independent (input) variable and the dependent (output) variable, as follows:

$$y = \alpha + \beta x + e \tag{1}$$

where, y: output value, x: input value, α: intercept, β: slope, and e: error.

Multiple Linear Regression (MLR) defines the linear relationship between two or more independent variables and the dependant variable. The MLR model can be represented as follows:

$$y = \alpha_1 x_1 + \alpha_2 x_2 \dots\dots + \alpha_n x_n + e \tag{2}$$

where, α_1, α_2... α_n are the coefficients of factors x_1, x_2... x_n respectively.

Support Vector Regression
Support Vector Machine (SVM) is an algorithm applied for both classification and regression. SVM applied for machine learning using regression is termed as Support Vector Regression (SVR). For two-dimensional variables, SVR is analogous to linear regression. Here, the equation form of a straight line is represented below:

$$y = \alpha + \beta x \tag{3}$$

In SVR, a straight line partitioning independent variables is known as a hyperplane. A boundary line is drawn with the help of the data points near the hyperplane, and these points are known as support vectors. Regression algorithms try to minimise the error between the predicted and the actual values. In contrast, SVR tries to match the best fit line under a specified limit (distance between the boundary line and the hyperplane). Hence, SVR attempts to meet the boundary constraint $-a < y - \alpha - \beta x < a$. It uses the points along this boundary for prediction.

Random Forest Regression
Random forests (RF) or random decision forests are the ensemble techniques that work by constructing multiple decision trees for training data and output the mean or average of the individual trees used in the training set. The single decision tree is susceptible and may overfit easily. However, the average of such multiple decision trees may reduce the over-fitting and enhance the model's accuracy.

Stochastic Gradient Descent Regression
Stochastic gradient descent (SGD) regression determines the coefficients of functions to minimise the cost/loss function. The SGD regressor fits the regulariser (additional penalty added to the cost function) under the convex cost function like SVM or logistic regression to reduce error and build an estimator. It is mainly used for large datasets, which may be prone to significant errors.

Gradient Boosting Regression
Gradient boosting is the ensemble technique that can boost the performance of weak algorithms, mainly decision trees. It sometimes performs better than random forest. It can be

used for both regression and classification models. The algorithm builds the model step-by-step, the same as other ensemble methods, but it optimises every loss function, resulting in widening its usage scope.

Light Gradient Boosting Machine Regression

Light Gradient Boosting Machine (LGBM) is adopts algorithms based on decision trees. LGBM increases model efficiency along with faster training speed and reduced memory usage. Compared to other tree-based algorithms, it develops decision trees leaf-wise instead of level-wise. Leaf-wise algorithms can minimise more loss than level-wise when growing the same leaf.

Performance Measures for Regression

The performance measures employed to evaluate the regression models in the present study are listed below.

$$\text{Root Mean Square Error (RMSE)}, RMSE = \sqrt{\frac{\sum_{i=1}^{n}(Y_i - \hat{Y}_i)^2}{n}} \tag{4}$$

$$\text{R-Square score } R^2 = 1 - \frac{\sum_i (Y_i - \hat{Y}_i)^2}{\sum_i (Y_i - \bar{Y})^2} \tag{5}$$

where Y_i: actual output, $\hat{Y}_i$: predicted output and, $\bar{Y}$: mean of actual output, n: number of data points.

Results and Analysis

After applying the proposed methodology, 80% of data was used for training purpose, and the rest of the data for testing purpose. The effect of all the considered features and the SGPAs up to the second semester were considered as inputs, and the output was the third-semester SGPA. The combined dataset of students of second, third and fourth-year of the case institution was used for training and testing the model. The training dataset has been modelled using various regressors and predictions were made for the test dataset. The built-in functions available in the scikit-learn library of Python 3 programming language were used to apply various regressors using the Jupyter notebook, and analysis was carried out.

The R-square score and RMSE are used as performance measures. The R-square score shows the proportion for the variance of the dependent variable explained by the independent variables, and it is a relative measure. RMSE shows the closeness between actual and predicted data points, and it provides an absolute measure of fit. RMSE denotes standard deviation of unexplained variance. Hence, a higher R-square score and lower RMSE values are desired for best fit. In the present study, the linear SVR model provides an R-square score of 0.8178 and an RMSE value of 0.629; hence, it performs better than other algorithms and provides the best fit as shown in Figure 84.1. For the random forest algorithm, an R-square score of 0.7238 and RMSE value of 0.7756 shows the least fit for the current model. SGD regressor and MLR also perform well with an R-square score closer to the

SVR model, but with comparatively higher RMSE values. When the test size was increased to 30%, the R-square score reduces to 0.7110, and the RMSE value increases to 0.7120 for the SVR_linear model because of lesser data available for training purpose. Comparison between the fit obtained with different test sizes is shown in Figure 84.2. Table 84.2 shows a comparison of performances metrics for different algorithms.

Table 84.2: Comparison of performance metrics for different algorithms.

Algorithm	R^2 score	RMSE
MLR	0.8028	0.6550
SVR_Linear	0.8178	0.6290
Random Forest Regressor	0.7238	0.7756
SGD Regressor	0.8138	0.6367
Gradient Boosting Regressor	0.7349	0.7598
LGBM Regressor	0.7386	0.7545

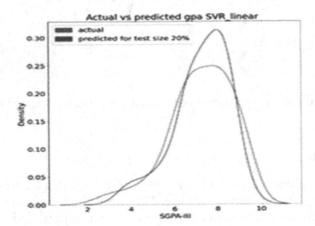

Figure 84.1 Actual vs Predicted performance at 20% test size for SVR_linear.

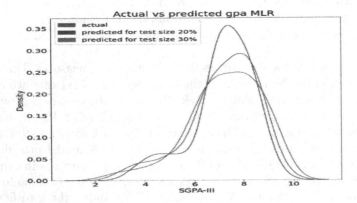

Figure 84.2 Comparison of fit at different test sizes.

Conclusion

In the current work, machine learning algorithms have been applied for predicticting students' academic performance in the undergraduate engineering program of the case institution. The algorithms include support vector regression-linear, multiple linear regression, random forest regression, stochastic gradient descent regression, gradient boosting regression and light gradient boosting machine regression. The required data have been collected from 564 undergraduate engineering students. Data cleaning and visualization were performed. During data pre-processing, the dataset with 162 variables has been reduced to 15 using principal component analysis. The pre-processed data were divided into a testing set and a training set. Machine learning algorithms were applied to the data sets. The linear support vector regression model provides the finest fit by showing an R-square score of 0.8178. A significant positive correlation was observed between past academic performance and the behaviour features such as active participation in academic tasks, regular learning, interest in the engineering degree programme and own branch studies.

Using the results obtained from the analysis, institutions can focus on improving the outcomes. The present study is conducted with the data obtained from a limited number of students. However, there may be variations in the obtained outcomes for a larger dataset since ensemble techniques like random forest work best for a larger dataset. Also, the reason for higher RMSE values can be investigated, and RMSE values can be reduced in order to obtain a better fit. Such enhanced investigation will lead to better prediction. Moreover, artificial neural network and other regression methods can also be used in future work. Based on the results obtained, a recommendation system can be developed so that the administrators and instructors can take corrective action.

Acknowledgements

The authors are most grateful to the reviewers for their constructive comments.

References

1. Yaacob, W. F. W., Nasir, S. A. M., Yaacob, W. F. W., and Sobri, N. M. (2019). Supervised data mining approach for predicting student performance. *Indonesian Journal of Electrical Engineering and Computer Science, 16*(3), 1584–1592.
2. Sultana, J., Rani, M. U., and Farquad, M. A. H. (2019). Student's performance prediction using deep learning and data mining methods. *International Journal of Recent Technology and Engineering, 8*(1 Special Issue 4), 1018–1021.
3. Makombe, F., and Lall, M. (2020). A predictive model for the determination of academic performance in private higher education institutions. *International Journal of Advanced Computer Science and Applications, 11*(9), 415–419.
4. Almasri, A., Celebi, E., and Alkhawaldeh, R. S. (2019). EMT: ensemble meta-based tree model for predicting student performance. *Scientific Programming,* (art. no. 3610248).
5. Guabassi, I. E., Bousalem, Z., Marah, R., and Qazdar, A. (2021). A recommender system for predicting students admission to a graduate program using machine learning algorithms.. *International Journal of Online and Biomedical Engineering, 17*(2), 135–147.
6. Hussain, S., Gaftandzhieva, S., Maniruzzaman, M., Doneva, R., and Muhsin, Z. F. (2021). Regression analysis of student academic performance using deep learning. *Education and Information Technologies, 26*(1), 783–798.

7. Yang, S. J. H., Lu, O. H. T., Huang, A. Y. Q., Huang, J. C. H., Ogata, H., and Lin, A. J. Q. (2018). Predicting students, academic performance using multiple linear regression and principal component analysis. *Journal of Information Processing*, 26, 17–176.

8. Ulfa, S., and Fatawi, I. (2020). Predicting factors that influence students, learning outcomes using learning analytics in online learning environment. *International Journal of Emerging Technologies in Learning*, 16(1), 4–17.

9. Razak, R. A., Omar, M., and Ahmad, M. (2018). A student performance prediction model using data mining technique. *International Journal of Engineering and Technology (UAE)*, 7(2), 61–63.

10. Kumar, A., Eldhose, K. K., Sridharan, R., and Panicker, V. V. (2020). Students academic performance prediction using regression: a case study In The International Conference on Systems, Computation, Automation, and Networking.

11. Rajalaxmi, R. R., Natesan, P., Krishnamoorthy, N., and Ponni, S. (2019). Regression model for predicting engineering students academic performance. *International Journal of Recent Technology and Engineering (IJRTE)*, 7(6S3), 71–75. ISSN: 2277–3878.

12. Adekitan, A. I., and Salau, O. (2019). The impact of engineering students' performance in the first three years on their graduation result using educational data mining. *Heliyon*, 5(2), 1–21. (Article No. e01250).

13. Namasudra, S., Dhamodharavadhani, S., and Rathipriya, R. (2021). Nonlinear neural network based forecasting model for predicting COVID-19 cases. *Neural Processing Letters*, 1–21. https://doi.org/10.1007/s11063-021-10495-w.

14. Chakraborty, R., Verma, G., and Namasudra, S. (2020). IFODPSO-based multi-level image segmentation scheme aided with Masi entropy. *Journal of Ambient Intelligence and Humanized Computing*, 12, 1–19. 10.1007/s12652-020-02506-w.

15. Namasudra, S., Devi, D., Kadry, S., Sundarasekar, R., and Shathini, A. (2020). Towads DNA based data security in the cloud computing environment. *Computer Communications*, 151, 539–547. ISSN 01403664, https://doi.org/10.1016/j.comcom.2019.12.041.

16. Namasudra, S. (2020). Fast and secure data accessing by using DNA computing for the cloud environment. *IEEE Transactions on Services Computing*, 15(4), 2289–2300. doi: 10.1109/TSC.2020.3046471.

85 Comparative numerical investigations on S1223 airfoilfor mixed flow fan applications at high reynold's number

Abhinav Mahajan[1], Anand Mathur, Aakash Goel[1], Akshay Thakur[1], Shivam Wahi[1], Chetan Mishra[2], and Aditya Roy[2,a]

[1]Department of Mechanical and Automation Engineering, Maharaja Agrasen Institute of Technology, GGSIP University, Delhi, India

[2]Department of Mechanical Engineering, Indian Institute of Technology, New Delhi, India

Abstract: Following the individual disadvantages posed by usage of either axial or centrifugal fan in high suction resistance applications such as indoor air purifiers, there has been inclination towards the usage of mixed flow fans. This study has been carried out specifically considering the first staging airfoil profile of a mixed flow fan setup. Detailed study of a high lift low Reynold's number airfoil (S1223) has been conducted using numerical simulations. Three turbulence models namely, Spalart Allmaras, SST k-ω and realizable k-ε were used for determining numerical solutions. These solutions are compared with experimentally obtained results. By producing simulation results at higher Reynold's number, the study provides an additional domain to the viewpoint of S1223 being regarded as a low Reynold's number airfoil. The comparison has been done on the basis of coefficient of lift and drag generated. Among the employed turbulence models, Spalart Allmaras model was found to be the most accurate model for predicting flow phenomena over the S1223 airfoil. Additionally, the S1223 airfoil displayed enhanced flow characteristics at elevated Reynold's number values.

Keywords: Airfoil, angle of attack, S1223, stalling, turbulence modeling.

Introduction

Industrialization is advancing in the present day by leaps and bounds with mankind reaping its benefit to extents unimaginable until a few years back. Our pace to technological advancement has largely been comprehended as a boon by us but little do we realize that a larger picture speaks otherwise. Air pollution is at its peak with innumerable reported cases of respiratory and lung diseases worldwide [1], rendering the health conditions of people living in close proximity to small scale and medium scale industrial areas even more vulnerable. A widely accepted solution has been the usage of indoor air purifiers for gaining respite. Lately, indoor air purifiers have picked up pace by playing a major role in domestic, healthcare and certain industrial applications [2,3]. Till date, air purifiers used for domestic application have been making use of single staging axial or centrifugal fans for clean air delivery. While they both do a decent job in delivering the required clean air, axial and centrifugal fans both have some advantages and disadvantages. As reported by many researchers such as Plint and Martyr [4], axial flow fans can be said to suitable for high-volume, low-pressure system applications whereas vice versa is true for centrifugal fans.

[a]aditya.roy1510@gmail.com

Axial fans perform poorly in cases when there is a large suction resistance such as that posed by filtration media in an air purifier whereas on the other hand, centrifugal fans are unable to impart high flow rate to the clean air at the outlet. A solution is to make use of a mixed flow fan setup where two axial fans with a thick and thin combination of airfoil profiles are used for getting the advantages of both, axial and centrifugal fans. This paper mainly focusses on the first staging fan of a mixed flow fan air purifier for generating high suction pressure. For the first staging, we have chosen a high lift low Reynold's number (Re) airfoil viz., S1223. It was found that S1223 was one among the best airfoils for our application as it acquired highest coefficient of lift when compared with the other airfoils. Given that generation of suction head via first staging of mixed flow fan requires high amount of lift generation in order to overcome pressure drop, choice of S1223 was justified. The chosen airfoil possesses a maximum thickness of 12.1% at 19.8% chord and a maximum camber of 8.1% at 49% chord [5]. Figure 85.1(a) shows a plot of the S1223 airfoil. Another research perspective behind this choice would be highlighted in the following paragraphs.

Selig and Guglielmo [5] were the researchers behind designing of S1223 high lift low Reynold's number airfoil for application in aircrafts carrying increased payload with added objectives of shortening take-off and landing distance. Since then, various researchers have focused on this particular airfoil for usage in varied applications where benefits of high lift at low Reynold's number can be utilized. Quite recently, Oller et al. [6] performed a study involving S1223 airfoil for application in design of hydrokinetic turbine blades. Computational fluid dynamics (CFD) analysis was performed on the same for finding out its suitability in the mentioned application and it was found that the lift and drag characteristics exhibited by S1223 airfoil were well suited for usage. Ma and Liu [7] performed numerical analysis on S1223 at Reynolds number between 10^5 and 10^6 keeping in mind Near Space applications. Another application of S1223 airfoil was seen in stratospheric aircraft propulsion by Liu and He [8] who compared the results produced by Spalart Allmaras turbulence model and XFOIL in designing a high-altitude propeller for stratospheric aircraft propulsion system. Mathur et al. [9] recently studied the application of S1223 airfoil on a Swan-Neck wing for motorsport applications. Speaking of the turbulence models used to study the flow, we found that a set of turbine blades comprising of NACA 0012 (symmetrical) and NACA 4412 (cambered) airfoils was designed and numerically simulated by Costa et al. [20] using the k-ω SST model, for which the least root mean square error was obtained. Oukassou et al. [21] used the NACA 0012 and NACA 2412 airfoils for designing a wind turbine blade and numerically simulated the flow using three models viz., Spalart-Allmaras, k-ε RNG and k-ω SST for evaluating various aerodynamic parameters.

In the present work, numerical investigations using CFD codes have been carried out on the S1223 high lift low Reynold's number airfoil. At first, numerical modeling and meshing techniques have been validated against experimentally available wind tunnel testing data. For this, three different eddy viscosity turbulence models have been used and the model predicting closest results with experimental ones has been pointed out. Subsequently, performance of the airfoil at a higher value of Re has been extensively studied with the help of all the three models, and stalling angle investigations have been performed and compared with the low Re case. As is evident from an aforementioned note, there exists a fair amount of literature on experimental and numerical analyses of flow past an S1223 airfoil. However, in most of the cases, the studies have been concentrated on a low Re regime. The potential of such an airfoil is yet to be seen and realized in applications that might involve a moderately high *Re* flow such as in our envisioned multi-stage mixed flow fan. We aim to fulfil this gap with the help of the present investigation in order to have a better idea on whether the said airfoil could prove to be useful in high *Re* applications. Our work is one among the first ones to explicitly draw comparison between stalling angle and allied resulting flow parameters while operating an S1223 airfoil at low *Re* and high *Re* values with a detailed investigation of the best suited eddy viscosity turbulence model.

Numerical Modeling

The computational domain was created around the airfoil geometry in two-dimensional configuration which also goes on to state the fact that our studies were conducted inside a 2D domain. Innumerable numerical studies have been conducted on various types of airfoils in 2D domains such as the one by Morgado et al. [15] who presented a comparison between prediction capabilities of RANS models and XFOIL for S1223 airfoil. It is a known fact that the requirement of computational resources in numerical simulations scales with the number of cells present in the discretized domain or mesh. Adding on this, a 2D simulation scales with an order of magnitude of L^2 whereas a 3D simulation scales with that of L^3, where L is a characteristic dimension. Given that 3D simulations prove to be highly computationally intensive, a 2D domain was chosen instead of 3D domain due to limitations on the availability of computational resources. There is at least one limitation involved in making such a choice, which is that the induced drag component resulting from circulation differences due to the presence of a finite span dimension is not accounted for in a 2D domain. However, there are theories such as the Prandtl lifting-line theory, a detailed account on which may be found in Anderson [22], which despite being linear and inviscid provide good approximations of 3D effects from 2D data. This adds on to the basis behind why 2D domains are still a well-accepted benchmark in the simulation of external flow past airfoils. The present work strictly reports results and analyses based on 2D numerical calculations. Various types of grids were possible in our case, such as the C-grid, O-grid, etc. Literature was reviewed for making a selection between the two with a critical review of advantages and disadvantages of both. The O-grid was selected based on Lutton's [10] in-depth comparison between usage of O-grid and C-grid for conducting numerical simulations around NACA 0012 airfoil in which, it was communicated that O-grid proved to be better, considering high flow gradients across the trailing edge.

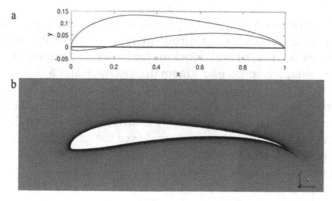

Figure 85.1 (a) Profile and (b) biased-structured mesh of S1223 airfoil with chord length of 1 m.

Figure 85.1(a) and 1(b) show a representation of the airfoil profile and generated mesh respectively wherein, hexahedral structured elements were selected for dividing the computational domain into finite control volumes and solving governing equations in each of them. For similar domain size, tetrahedral mesh is more computationally expensive. Also, storage and running time required for hexahedral mesh is less than that required for tetrahedral mesh as a result of which, hexahedral mesh was preferred over tetrahedral mesh. In turbulence modeling where wall-bounded effects play a major role, the mesh should have a y+ value of nearly equal to or less than 1 that is, within viscous sub-layer. Hence the mesh created in our case had a y+ value of 1. Grid biasing technique was employed for generating very fine elements near the wall for accurately capturing the boundary layer phenomena and coarse ones away from the wall with sizes gradually increasing in order to reduce the number of elements. Skewness between adjacent cells was also reduced to a great extent.

One of the primary objectives of this paper is to compare the numerical results predicted by some of the most widely used eddy viscosity RANS models in the academia and industry. Three such turbulence models used in aerodynamic simulations are Spalart Allmaras (S-A) model, shear stress transport (SST) k-ω model and realizable k-ε model. Governing equations solved in our numerical studies include the continuity equation in two dimensions, x-momentum equation, y-momentum equation and one or two additional equations depending on the turbulence model under consideration for modeling the Reynold's stress terms as per Boussinesq hypothesis and attaining closure of the Navier-Stokes' momentum equations. The S-A model is a one equation model while the other two are two equation models. Governing equations for S-A model, realizable k-ε model and SST k-ω model are shown in equations (1), (2), (3), (4) and (5) respectively.

Additional equations pertaining to the S-A model, realizable k-ε model and SST k-ω model can be referred to from the work of Spalart and Allmaras [14], Launder and Spalding [11] and Shih et al. [12], and Menter [13], respectively. In addition, a detailed treatment of the realizable k-ε model and SST k-ω model turbulence models may also be found in the works of Roy and Dasgupta [16] and Gupta et al. [17, 18, 19], who recently addressed problems in external vehicle aerodynamics and conjugate heat transfer using the same, respectively.

$$\frac{\partial}{\partial t}\left(\rho\tilde{v}\right) + \frac{\partial}{\partial x_i}\left(\rho\tilde{v}u_i\right) = G_v + \frac{1}{\sigma_{\tilde{v}}}\left[\frac{\partial}{\partial x_j}\left\{\left(\mu+\rho\tilde{v}\right)\frac{\partial\tilde{v}}{\partial x_j}\right\} + C_{b2}\rho\left\{\frac{\partial\tilde{v}}{\partial x_j}\right\}^2\right] - Y_v + S_{\tilde{v}} \qquad (1)$$

In the above equation for S-A model, u_i is the velocity component, $\tilde{v}$ is the modified turbulent viscosity, G_v is production of turbulent viscosity, Y_v is destruction of turbulent viscosity which occurs because of wall blocking and viscous damping in near-wall region. $\sigma_{\tilde{v}}$ and C_{b2} are constants, v is molecular kinematic viscosity and S_v is the source term. The transported variable in this model, $\tilde{v}$, is identical to the turbulent kinematic viscosity in all regions other than the near wall region. The S-A model requires the viscosity-affected region within the boundary layer to be properly resolved and is essentially a low Re model. This translates to the requirement that the y+ value of the grid should be close to 1, which is the case in the present work. In addition, there have been previous works which have shown reliable applicability of the model in higher Re regimes such as the one which is being investigated here.

$$\frac{\partial}{\partial t}(\rho k) + \frac{\partial}{\partial x_j}(\rho k u_j) = \frac{\partial}{\partial x_j}\left[\left(\mu + \frac{\mu_t}{\sigma_k}\right)\frac{\partial k}{\partial x_j}\right] + G_k + G_b - \rho\varepsilon - Y_M + S_k \tag{2}$$

$$\frac{\partial}{\partial t}(\rho\varepsilon) + \frac{\partial}{\partial x_j}(\rho\varepsilon u_j) = \frac{\partial}{\partial x_j}\left[\left(\mu + \frac{\mu_t}{\sigma_\varepsilon}\right)\frac{\partial\varepsilon}{\partial x_j}\right] + \rho C_1 S_\varepsilon - \rho C_2\frac{\varepsilon^2}{k + \sqrt{v\varepsilon}}$$
$$+ C_{1\varepsilon}\frac{\varepsilon}{k}C_{3\varepsilon}G_b + S_\varepsilon \tag{3}$$

In the above equations for realizable k-ε model, σ_k and σ_ε represent turbulent Prandtl numbers for k and ε respectively, whereas G_k and G_b denote generation of turbulent kinetic energy due to mean velocity gradient and generation of turbulent kinetic energy due to buoyancy, respectively. μ_t in the effective diffusivity term denotes turbulent viscosity. C_1, C_2 and $C_{1\varepsilon}$ are constants and, S_k and S_ε are user defined source terms. The effect of molecular viscosity is considered negligible in the derivation of the k-ε model. The solution of two separate equations allows the determination of a turbulent length scale as well as a turbulent time scale. The standard k-ε model originally given by Launder and Spalding [11] had some limitations, one of the major ones being that it was only applicable in fully turbulent flows. The realizable k-ε model, given by Shih et al. [12] differs from the standard model in a way such that the turbulent viscosity formulation has been given an alternative treatment, and there is a modified transport equation for dissipation rate, ε. There is, however, one major drawback in this model, which is evident in its application to multiple frames of reference, for example, in computational domains containing both rotating and stationary fluid zones. Since, this is not the case in the present work, the realizable k-ε model has been taken as a candidate turbulence model for our numerical simulations.

$$\frac{\partial}{\partial t}(\rho k) + \frac{\partial}{\partial x_i}(\rho k u_i) = \frac{\partial}{\partial x_j}\left[\left(\mu + \frac{\mu_t}{\sigma_k}\right)\frac{\partial k}{\partial x_j}\right] + G_k + Y_k + S_k \tag{4}$$

$$\frac{\partial}{\partial t}(\rho\omega) + \frac{\partial}{\partial x_i}(\rho\omega u_i) = \frac{\partial}{\partial x_j}\left[\left(\mu + \frac{\mu_t}{\sigma_\omega}\right)\frac{\partial\omega}{\partial x_j}\right] + G_\omega + Y_\omega + S_\omega \tag{5}$$

In the above equations for SST k-ω model, σ_k and σ_ω denote turbulent Prandtl numbers for k and ω respectively. G_ω represents generation term for ω due to mean velocity gradients and dissipation of k and ω is denoted by Y_k and Y_ω, respectively. When compared with the standard k-ω model, the SST modification blends the accurate formulation of the k-ω model in the near wall region with the robust freestream independence outside the shear layer offered by the k-ε model. In short, a product of the transformed k-ε model and standard k-ω model with a blending function is taken separately and added together. The value of this blending function is 1 near the wall, thus activating the standard k-ω model and 0 away from the wall, thus activating the transformed k-ε model. This feature makes the SST k-ω model more reliable for flows exhibiting mild to adverse pressure gradients such as those seen in airfoils.

Standard velocity inlet and pressure outlet condition was specified as boundary condition in each of of the three numerical models. Air flowing at an inlet velocity of 2.922 m/s was considered for validating our model against experimentally available data at $Re = 2 \times 10^5$ while our application specific simulations were run at inlet velocity of 29.22 m/s, corresponding to $Re = 2 \times 10^6$. Density and dynamic viscosity of air were taken as being equal to 1.225 kg/m^3 and 1.79kg/m-s respectively. Coupled solver was employed for achieving pressure-velocity coupling and all the equations were solved using second order discretization. It is to be noted that first 100 iterations in each of the cases were solved using first order discretization for attaining stability in the solution post which, focus was shifted to accuracy. Figure 85.2 shows the results of a grid dependence study that governed the selection of our mesh.

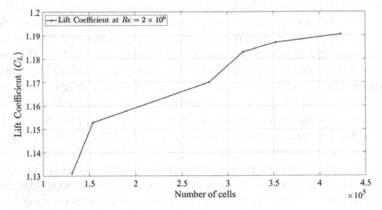

Figure 85.2 Grid dependence test at 0° of angle of attack.

Figure 85.3(a) shows the results of experimental validation carried out against the data of Selig and Guglielmo [5] at $Re = 2 \times 10^5$. It may be observed that the S-A model slightly over predicts lift data whereas the other two models under predict the same. The SST k-ω model predicts closest results up to almost 10° of attack angle after which the deviation starts to creep in. Realizable k-ε model predicts the trend up to low angles of attack satisfactorily but suffers when the flow separation increases. S-A model slightly under predicts the stalling angle but gives the best overall prediction as compared to the other two models. Through experimental validation, it is clear that Spalart Allmaras turbulence model provided the most reliable results as it had a maximum difference in results of 3.15% at stall point as compared to 7.7% and 16.7% at their respective stall points for SST k-ω model and realizable k-ε model, respectively.

Results and Discussion

Simulations were run for attack angles starting from 0° to 18° and results of the same are shown in Figure 85.3(b) which portrays a variation of lift coefficient with angle of attack. In the graphs as shown, the coefficient of lift is plotted against angle of attack (AOA) for the three turbulence models that were used in our study. Lift coefficient with AOA of an airfoil increases up to a certain angle after which it starts decreasing with significant addition in drag. This critical angle is known as stalling angle and the phenomenon is called stalling. It is an important characteristic for determining flow phenomena and allied wake region formations. After stalling, fully separated flow occurs over the geometry of airfoil. This type of flow is highly undesirable as it generates higher drag and low amounts of lift. Therefore, moving beyond stalling angle would not cause any benefit in terms of lift increment. Given that many design decisions such as blade pitch angle, etc. in fan development are based on angle of attack of the airfoil, determination of stalling point is indispensable in our studies. Since coefficient of lift keeps on increasing up to stalling point, higher angle of stall would be desirable.

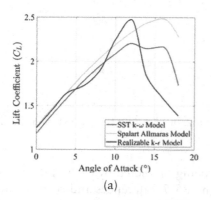
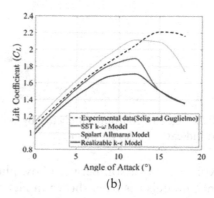

(a) (b)

Figure 85.3 (a) Experimental validation at $Re = 2 \times 10^5$ and (b) Variation of C_L with angle of attack at $Re = 2 \times 10^6$.

It is clearly evident from the results that coefficient of lift increased considerably with transition in flow regimes (Reynolds Number changing from 2×10^5 to 2×10^6). Stalling point also showed an upward trend in high velocity regime as compared to low velocity regime. Results obtained using all the three models are shown here for comparison and reference, but usage of S-A model is recommended because of more accurate predictions as seen in the section of experimental validation. The results of S-A model show that stalling occurred at 12° for $Re = 2 \times 10^5$ which changed to 16° at $Re = 2 \times 10^6$. Furthermore, the maximum Coefficient of Lift also increased by 17.50%. It can therefore be said that in addition to the literature available for S1223 which states the same as being a high lift low Reynolds number airfoil, there is considerable performance improvement of the same at elevated values of Reynolds Number as well. Also, there is significant decrement in drag coefficient by about 50% on transition from low to high Re and this proves the applicability of S1223 airfoil in our specific case. The same factor also happens to be an important finding of the present work. An important observation to be made is that the two equation models provide a more consistent prediction at low Re regime than high Re regime as can be seen from Figure 85.3(b). The variation in trend of C_L values predicted by two equation models at the two considered Re values further highlight the difference in their abilities to capture separated flows at highly turbulent flow regimes. While the new eddy viscosity formulation and modified equation for turbulence dissipation rate enhance the capability of realizable k-ε model for capturing flow separations in comparison to standard and RNG models, maximum amount of variation shown by realizable k-ε model in Figure 85.3(b) points towards the superiority of SST k-ω model over realizable k-ε model, given that similar wall treatment approach has been used while using both the models.

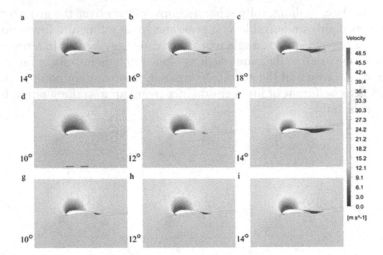

Figure 85.4 Velocity contours predicted by S-A (a,b,c), realizable k-ε (d,e,f) and SST k-ω (g,h,i) models at indicated angles of attack.

An explicit comparison between flow phenomena being predicted by each of the three turbulence models has been shown in Figures 85.4 and 85.5. Inferences and correlations may be drawn with variation of C_L shown in Figure 85.3(b). From Figure 85.4(a), it can be deduced that for Spalart Allmaras turbulence model, the flow separation at 14° AOA is minimum and to its contrary, at 18° AOA, it is maximum. Generation of wake region (blue region) is also biggest at 18° AOA and smallest at 14° AOA. At 14° AOA, the flow separates after the leading edge, generating very small wake region as compared to 16° AOA where the flow separation is almost similar, but the wake region is bigger. Subsequently at 18° AOA, the wake region is biggest as the flow is totally separated. For the same turbulence model from Figure 85.4, it can be determined that at 14° and 18° AOA the negative pressure region is smaller in comparison to the negative pressure region in 16° AOA. Negative pressure at the top surface of airfoil and higher positive pressure at the corresponding bottom surface of the airfoil generate pressure difference and this generates lift. Therefore, maximum lift generated was at 16° AOA following which, it can be said that 16° is the critical angle giving maximum value of coefficient of lift 2.49 in Spalart Allmaras model. In realizable k-ε turbulence model, from Figure 85.4(d,e,f), it can be said that that the flow separation starts on the upper surface much before the trailing edge thus generating a significant wake region. The pattern of flow separation in 12° AOA is similar to that of 14°, but the wake region at 14° AOA is bigger in comparison. This shows that at 14° AOA, the flow is totally separated. From pressure contours (Figure 85.5), it is observed that high pressure region at the bottom of the airfoil is smaller in 12° AOA in comparison to 14° AOA, signifying that the pressure difference at 12° AOA is higher than at 14° AOA. Therefore, the optimum value of coefficient of lift is 2.474 at the stalling angle of 12°.

In k-ω SST turbulence model, from Figure 85.4(g,h,i), it can be inferred that the flow separation starts at 10° AOA and subsequently, the flow is totally separated at 14° AOA. This results in a significant wake region at 14° AOA. In this turbulence model as well, the stalling angle occurs at 12°. From pressure contours it is observed that at the stalling angle, that is, at 12° AOA, the high-pressure region is bigger in comparison to other angles.

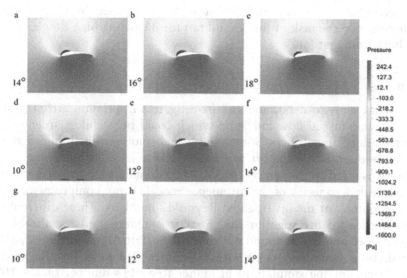

Figure 85.5 Pressure contours predicted by S-A (a,b,c), realizable k-ε (d,e,f) and SST k-ω (g,h,i) models at indicated angles of attack.

Highly negative pressure region (blue zone) is generated at 12° AOA. This signifies that there is larger pressure difference at this angle which generates optimum lift. The optimum coefficient of lift predicted by k-ω SST turbulence model is 2.2033 at 12° AOA. Seeing the pressure contours of Figure 85.5, a note on the negative pressure regions could be made. Beginning at the surface of the wing and travelling up and farther away from the surface, the pressure increases with increasing distance until the pressure reaches the ambient pressure. This creates a pressure gradient where the higher pressures further along from the radius of curvature (surface of airfoil) push inwards towards the center of curvature where the pressure is lower. This provides an accelerating force on the fluid particle. Due to this curved and cambered surface of the wing, there exists a pressure gradient above the wing, where the pressure is lower(negative) right above the surface. This facilitates lift generation.

All the turbulence models were compared to visualize the trend of results prediction by each one of them and it was observed that S-A model predicts the highest value of C_L. This is also along the lines of experimental validation seen above. It can be inferred that in S-A model, the boundary flow separation started significantly at 18° AOA which is higher in comparison to other two models. The size of wake region is also bigger at 18° AOA in comparison to other two models where wake region becomes larger at smaller angles. The difference between the high-pressure region and the high negative pressure region obtained is highest in S-A model in comparison to other two models. This difference of pressure creates suction, which is the primary objective of our application. The highest suction pressure would be obtained at 16° AOA, as predicted by S-A model.

Our application of mixed flow fan for filtration purpose requires a cascaded assembly of two fans. First fan in this cascaded system is used for generating suction and other is used to generate output flow rate. Successful simulation validation at high velocity regimes shows better performance of lift generating airfoil. A well-designed diverging enclosure around the first staging fan, which is a prerequisite in the concept of mixed flow fans, in addition to selection of the investigated airfoil for fan blade profile would enable the system to successfully overcome high pressure drops imposed by filtration layers. This would further ease up the task of the second fan for imparting high flow rate to the clean air approaching from first fan.

Conclusion

The present work aims to determine and investigate an optimum airfoil profile for the first-staging fan in a mixed flow fan setup which could be useful in an envisioned novel application to indoor air purifiers. Some established computational fluid dynamics codes were used for understanding the lift generation and allied flow phenomena. Significant consideration was given to the setup of the meshed domain with focused attention to the grid y+ value, so that accuracy of the turbulence models is not compromised due to insufficient mesh resolution at the investigated Reynold's numbers. Three different turbulence models were chosen for investigation viz., Spalart Allmaras model, realizable k-ε model and SST k-ω model, and it was found out that the Spalart Allmaras model gave the most accurate predictions when compared with experimental data at low Reynold's number of 2×10^5. On conducting simulations at higher Reynold's number of 2×10^6, relevant to our application, it was seen that coefficient of lift at each of the angles of attack had increased with a maximum percentage increase of 17.50% at stalling point with a decrease in maximum drag coefficient by 50%. The Spalart Allmaras, realizable k-ε and SST k-ω models, respectively predicted the stalling angle at this Reynold's number to be 16°, 12° and 12°. This implies that a very high stalling angle of 16° may be utilized for generating a significant lift (corresponding to a lift coefficient of 2.49) using the S1223 airfoil at the said Reynold's number. Furthermore, it was inferred that stalling angle predicted by the most accurate turbulence model (Spalart Allmaras) had shown an increase by almost 4° between low and high Reynold's number regimes. These were the most significant findings of our work. Scope of further research lies in varying the Reynold's number to higher values depending on motor speed of the fan for finding a safe zone of operation in terms of angle of attack within a given range of motor speed. These results can be then used as input conditions for selection of thin airfoil for second-staging fan design. We feel that the academic implication of this work could be in the exploration of the S1223 high lift, low Reynold's number airfoil in applications which require considerably high lift generation at a moderately high Reynold's number regime, such as the case investigated in this work.

References

1. Health Effects Institute (2018). State of Global Air 2018. Special Report. Boston, MA: Health Effects Institute.
2. Oh, H. J., Nam, I. S., Yun, H., Kim, J., Yang, J., and Sohn, J. R. (2014). Characterization of indoor air quality and efficiency of air purifier in childcare centers, Korea. *Building and Environment*, 82, 203–214.
3. Roy, A., Mishra, C., Jain, S., and Solanki, N. (2019). A review of general and modern methods of air purification. *Journal of Thermal Engineering*, 5, 22–28.
4. Martyr, A. J., and Plint, M. A. (2012). Ventilation and air-conditioning in powertrain test facilities. Engine testing: the design, building, modification and use of powertrain test facilities. Elsevier.
5. Selig, M. S., and Guglielmo, J. (2003). Low reynolds number airfoil design. *Journal of Aircraft*, 34, 72–79.

6. Oller, S. A., Nallim, L. G., and Oller, S. (2016). Usability of the selig S1223 profile airfoil as a high lift hydrofoil for hydrokinetic application. *Journal of Applied Fluid Mechanics*, 9, 537–542.
7. Ma, R., and Liu, P. (2009). Numerical simulation of low-reynolds-number and high-lift airfoil S1223. In Proceeding World Congress on Engineering, II, (pp. 1–6).
8. Liu, X., and He, W. (2017). Performance calculation and design of stratospheric propeller. *IEEE Access*, 5, 14358–14368.
9. Mathur, A., Mahajan, A., Aggarwal, A., Mishra, C., and Roy, A. (2021). Numerical study of swan neck rear wing for enhancing stability of ground vehicle bodies. In Das, L. M., Kumar, N., Lather, R. S., and Bhatia, P. (eds.) Emerging Trends in Mechanical Engineering. Lecture Notes in Mechanical Engineering. Singapore: Springer.
10. Lutton, M. J. (1989). Comparison of C- and O-Grid generation methods using a NACA 0012 airfoil. Doctoral dissertation, Air Force Institute of Technology.
11. Launder, B. E., and Spalding, D. B. (1972). Lectures in Mathematical Models of Turbulence. London, England: Academic Press.
12. Shih, T. H., Liou, W. W., Shabbir, A., Yang, Z., and Zhu, J. (1995). A new k-ε eddy viscosity model for high reynolds number turbulent flows. *Computers and Fluids*, 24, 227–238.
13. Menter, F. R. (1994). Two-equation eddy-viscosity turbulence models for engineering applications. *AIAA Journal*, 32(8), 1598–1605.
14. Spalart, P. R., Allmaras, S. R., and Reno, J. (1992). One-Equation Turbulence Model for Aerodynamic Flows Boeing. AIAA, (p. 23).
15. Morgado, J., Vizinho, R., Silvestre, M. A. R., and Páscoa, J. C. (2016). XFOIL vs CFD performance predictions for high lift low Reynolds number airfoils. *Aerospace Science and Technology*, 52, 207–214.
16. Roy, A., and Dasgupta, D. (2020). Towards a novel strategy for safety, stability and driving dynamics enhancement during cornering manoeuvres in motorsports applications. *Scientific Reports*, 10, 1–14.
17. Gupta, A., Roy, A., Gupta, S., and Gupta, M. (2020). Numerical investigation towards implementation of punched winglet as vortex generator for performance improvement of a fin-and-tube heat exchanger. *International Journal of Heat and Mass Transfer*, 149, 1–16.
18. Gupta, S., Roy, A., Gupta, A., and Gupta, M. (2019). numerical simulations of performance of plate fin tube heat exchanger using rectangular winglet type vortex generator with punched holes. *SAE Technical Paper. Series*, 1, 1–14.
19. Gupta, S., Roy, A., and Gupta, A. (2020). Computer-aided engineering analysis for the performance augmentation of a fin-tube heat exchanger using vortex generator. *Concurrent Engineering: Research*, 28(1), 47–57.
20. Rocha, P. A. C., Rocha, H. H. B., Carneiro, F. O. M., da Silva, M. E. V., and de Andrade, C. F. (2016). A case study on the calibration of the k-ω SST (shear stress transport) turbulence model for small scale wind turbines designed with cambered and symmetrical airfoils. *Energy*, 97, 144–150.
21. Oukassou, K., El Mouhsine, S., El Hajjaji, A., and Kharbouch, B. (2019). Comparison of the power, lift and drag coefficients of wind turbine blade from aerodynamics characteristics of Naca0012 and Naca2412. *Procedia Manufacturing*, 32, 983–990.
22. Anderson, J. D. (1995). Computational Fluid Dynamics. McGraw Hill Education. ISBN 978-0070016859.

86 A DFMA approach to reduce the assembly lead time of a vehicle wash machine

Dhananjay Dhandole[1,a] and Inayat Ullah[2,b]

[1]PG Scholar, Department of Mechanical Engineering, G H Raisoni College of Engineering, Nagpur, India

[2]Assistant Professor, Department of Mechanical Engineering, G H Raisoni College of Engineering, Nagpur, India

Abstract: The automatic vehicle wash machine has been developed to increase the rate of cleaning of an automobile, whether it may be a personal car or a commercial mining truck. Although several designs of automatic vehicle wash machines are available commercially. However, when it comes to the manufacturing of this automatic machine, it may take up to months, and most of the customers need their machines to be delivered as soon as possible. Thus, there is a need to minimize the manufacturing lead time for this machine. To this end, the present study attempts to minimize the assembly lead time of an automatic vehicle wash machine by employing the design for manufacturing and assembly (DFMA) framework. The findings of the study report that after incorporating the design modifications, the number of manufacturing processes has been reduced to half. Whereas, the assembly lead time has been reduced from sixteen days to nine days.

Keywords: Assembly lead time, automatic car wash, DFMA, modular structure, sheet metal design.

Introduction

Having a vehicle can either be a necessity or a luxury status for one. Yet, it has its challenges when it comes to vehicle maintenance. Ensuring the cleanliness of the vehicle is one of these challenges. A vehicle can be washed by hand, however, the process may take a significant amount of time and effort, especially in the case of larger vehicles like buses or trucks. The washing efforts and time can be reduced by using an automatic vehicle wash machine. An automatic vehicle wash machine allows cleaning of the vehicle quickly and with little effort. Besides, by using an automatic vehicle wash machine, the undercarriage of a vehicle can be washed with ease that was difficult to perform with hand wash. The advantages of automatic vehicle wash include time savings, reduced physical effort, and fairly thorough cleaning. Owing to the usefulness of automatic vehicle wash machines, several industries across the world produce and deliver these machines to be used commercially. Manufacturing of these machines includes several challenges such as producing high-quality products while keeping the manufacturing costs at a minimum, delivering the machine within the committed delivery time, ease of disassembly for annual maintenance and/or repair work. To meet the above challenges and remain competitive, the organizations need

[a]dhananjay2206@gmail.com, [b]inayatu6@gmail.com

to be adaptive and implement the approaches including modularity in products, design for manufacturing and assembly (DFMA), time-based manufacturing, lean manufacturing, continuous improvement, etc., as the order specifications may vary with customers, or a rapid technology change may result in a complete change in the machine. Also, the mass production of this machine and keeping the finished good inventory in the anticipation of a future order might be disastrous for organizations Senapati et al. [13]. Nowadays, many organizations have adopted a make-to-order strategy and focus on developing the capabilities to produce and deliver the machine in the minimum possible time and costs Ullah and Narain [17]. For a manufacturer, it is becoming extremely difficult to compete in the market since the number of competitors is increasing gradually ever since 1951, when the first fully automatic car wash machine was invented in Seattle, WA. (AUTEC Car Wash n.d.) [3]. Thus, to survive and thrive in today's hypercompetitive environment, a focus on cutting down the assembly lead time, as well as the manufacturing costs is of extreme importance. The present study attempts to optimize the design of automatic vehicle wash machine by applying the principle of DFMA, a well-known approach for design optimization and time compression. By following the guidelines of DFMA, the design has been modified in such a manner that the number of structural components gets reduced considerably, without diminishing the functionality and the strength of the structure.

The entire assembly of this machine consists of several sub-assemblies like left, right, and top structure sub-assembly, transmission system, water, and air inlet system, pneumatic control system, water, and chemical distribution system, chemical containers, electric panel enclosure, pneumatic panel enclosure, horizontal brush lifting arrangements, vertical brush arrangements, wheel wash system, blow-air dryer system, under-chassis wash, and automation interfaces.

By reviewing the existing design, it has been observed that the complete assembly of the machine can be done in a period of sixteen to twenty days, which is considerably larger when there are other sellers available in the market for similar products. Hence, there is a need to introduce significant changes in the design of the structure of the machine so that it could be manufactured and assembled with a comparatively smaller lead time and reduced number of processes. This study attempts to highlight the significance of modular structural design to reduce the assembly lead time by application of DFMA. In this paper, the following tasks have been performed to design a better product:

- Identification and improvement in the current manufacturing methodology.
- Design modifications to reduce the number of manufacturing processes, assembly lead time, and overall cost.

Literature Review

Several researchers have shown that the application DFMA methodology leads to the lead time reduction. In 1788, a Frenchman, LeBlanc, first introduced the idea of interchangeable parts in the manufacturing industry. Eli Whitney, an American inventor of the cotton gin, incorporated and used some techniques of DFM way before the term DFM was derived, for the manufacturing of muskets for the U.S. government in 1801. Sivasankaran et al. [14] redesigned a few of the parts of the system and used DFA software to compute the assembly process time, efficiency, and labor costs. However, it was performed for a system consisting of the mounting base, piston, steel cover, spring, and screws. The authors managed to reduce the lead time almost by half. Alad and Deshpande [2] reviewed that the tools like project planning and control (PPC) and design for production (DFP) can result in the reduction of setup, processing, moving, and waiting time which is beneficial for reduction of cost and lead time and improves product quality. Garza-Reyes et al. [6] investigated the DMAIC process and helped the civil aerospace industry in achieving a thirty percent reduction in assembly lead time of an engine. Makwana and Awasthi, [10] performed a case study on Toyota Production System (TPS) and Value Stream Mapping (VSM) and explained how the in-house production components act as bottlenecks and the parts have to be transferred from machine to machine for completion of the operations which causes a decrease in labor productivity and increases material handling cost. Senapati et al. [13] conducted an exhaustive review and rendered that the reduction of lead time is controllable and is especially beneficial in case of high demand uncertainty. Also, by reducing the assembly lead time, the safety stock can be reduced while minimizing the loss of orders due to stock out, the service level to the customer can be significantly improved, and also, the competitive ability of the organization can be amplified Nahm et al. [11]. In the study carried out by Gokulraju [8], the use of Value Stream Mapping (VSM) resulted in the reduction of lead time from 75 hours to 8 hours, and throughput was gained by 75.23%. Ahmad and Rahman [1] found that the implementation of VSM resulted in the reduction of cycle time by 44.96% and cost-saving up to $1063 per month by labor reduction. Suresha et al. [16] used structural analysis to improve the traditional design of the machine structure by reducing its weight by 41kg. Furthermore, the authors found that due to the weight reduction the impact in terms of safety of the lathe bed was negligible. The research performed by Dejene & Nemomsa [5] showed the three ways to implement DFMA and their results indicated the modified design had a lowered number of parts with enhanced ease of assembly than the earlier design. Harlalka et al. [9] stated that the application of DFMA increased the DFA index from 15.99 to 19.93 for a food processor and which aided in a cost-saving of 1.59% of the total value of the product. The thesis of Gillberg and Sandberg [7] summarized the guidelines for bending sheet metal and showed the advantages of pressing over bending. Besides, the welded components have been found as a crucial place for redesign as they significantly affect the material properties. The concept of the modular car wash machine was introduced in the research conducted by Sabet et al. [12]. Their suggestion of modular rollover design for carwash has helped in reduction of lead time by 30% along with weight reduction up of 35% by reducing its size by 10%. The modified structural design was found to be modular enough to achieve the intended benefits. However, the design contained more structural members that could lead to a longer lead time for manufacturing and assembly.

During the literature review, although very few studies have been found related to the optimization of automatic vehicle wash machines, nevertheless, there is always a window for further development in the design for the reduction in the number of components being used as structural members that can lead to reduced lead time for manufacturing, as well as assembly.

Problem Statement

As per the data collected by interviewing the installation workers of a vehicle wash machine, it has been observed that the current manufacturing process consists of a series of operations that includes cutting the square pipes as per the dimensions and welding them together to manufacture the skeleton of the structure, further, the skeleton is covered with the metallic sheets for protection of internal hardware, as well as for aesthetics. This usually takes more than two weeks, which is considerably long, and also, several workers are needed for the production of a single machine. The structural decomposition of the machine is shown in Figure 86.1. Since the whole work is performed manually, there is a high possibility of human error. In addition, the design needs aesthetical modifications to make it better looking than the rivals. Here, Figure 86.2 represents the CAD model of the old machine structure. (Refer to Table 86.1, for details about parts count and time consumed by different operations before design optimization.)

Structural Decomposition of Automatic Vehicle Wash Machine

Figure 86.1 Members in the old vehicle wash machine.

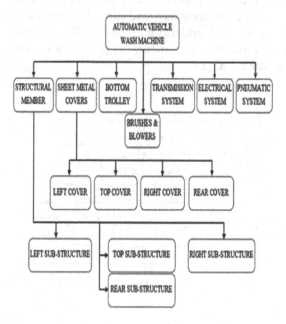

Figure 86.2 CAD model of the old machine structure.

Table 86.1: Scenario before design optimization.

Particulars	Duration
Total No. of components	85 Nos.
Raw material cutting	2
Welding frame	2
Welding of sub-components	2
Laser cutting and bending	3
Grinding and Finishing	2
Painting and Powder coating	2
Assembly of the main structure	2
Total	ays

Research Methodology

The methodology adopted for the present research is shown in Figure 86.3. In the initial phase, technical data related to the features and specifications, dimensions, functionality, bay layout, intakes, and strength were collected from the previous generation machines, and information regarding the number of problems faced during the installation on sites was collected from the installation specialists. Further, this data has been analyzed and using CAD modeling software like SolidWorks and Fusion 360, a new structure has been developed following the DFMA guidelines. To adhere to the guidelines of DFMA, at first, the structural members were identified for elimination by checking the interdependence and their impact on the other components, then by integrating two or more components, it has been investigated whether the condition of minimum part satisfies, as the larger number of parts tends to increase the assembly lead time. The criteria stated that an additional part can exist only if (a) it is a foundational member; (b) different material composition is required; (c) parts have relative movement; (d) part is designed for ease in assembly—Harlalka et al. [9] and Stienstra, [15].

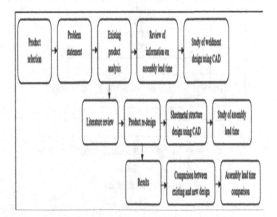

Figure 86.3 The methodology adopted in the research.

Approach Method

No Profile Skeleton

In this approach, the skeleton and the covers have been integrated in such a manner that the new component can be used alone as a structural member for the new machine. Each structural component is made by laser cutting metal sheets and bent into the required shape using a hydraulic bending machine to avoid human errors. Also, the components are designed in such a manner so that they can be assembled with ease. All the components can be assembled by holding them together at their respective location and fastened by using bolts, nuts, or rivets. Doors for the machine are made by following the same practice. Since doors do not have significant load directly associated with them, so material like Carbon Based Material (CBM) can be used to manufacture as they can lead to weight reduction and withstand the higher load, if required by Bera et al.[4]. Hence, manufacturing processes like welding, cutting, and grinding have been eliminated. To carry out DFMA analysis, the machine structure before and after was studied thoroughly and the count of total parts was documented in the form of Bill of Material (BOM).

Findings and Results

The comparison of old and redesigned structures in terms of DFMA can be witnessed by observing the following examples shown in Figures 86.4 and 86.5 are representing the bottom and top of the old machine structure respectively. Further, as the number of parts has been reduced significantly, the new structural design is trouble-free to assemble and also has lesser assembly lead time.

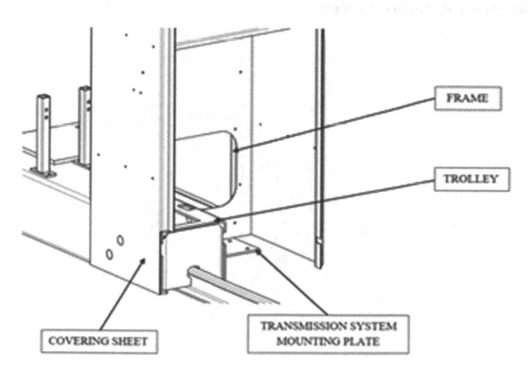

Figure 86.4 Old machine structure (bottom).

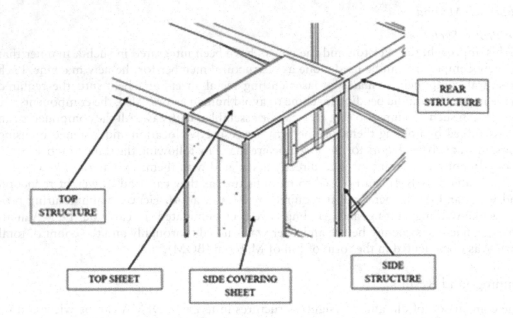

Figure 86.5 Old machine structure (top).

Integration of Structural Members

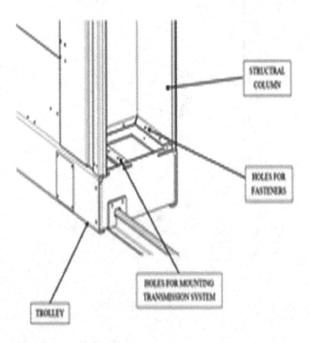

Figure 86.6 New machine structure (bottom).

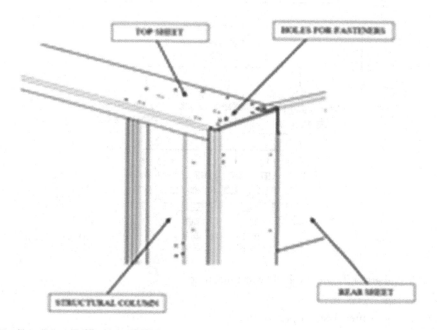

Figure 86.7 New machine structure (top).

The concept of integrating the structural members to reduce the overall part count has become possible due to the use of CAD software. Here, (Figures 86.6 and 86.7 are representing the bottom and top structure of the new machine respectively) the frame has been integrated with the sheet covers, and the new component performs two functions, as a structural member, as well as a cover. It has been performed mainly for the left structure, right structure, top frame, and rear structure. The above exercises resulted in the components such as the left column, right column, top sheet, and rear sheet. Whereas, the trolley which is a base part of the whole assembly, was earlier shorter in terms of width and the rest of the parts were being welded onto it. However, in the new design, the trolley has been widened up and holes are drilled using CNC laser cutting machine for fastening each component directly to their respective position.

In this study, a vertical shell/column has been designed in such a manner that it alone can act as a supporting pillar for the machine structure and no additional member is required. Also, as this pillar is hollow from the inside, so there is sufficient space available and it can accommodate all the necessary hardware by directly fastening over the column. This approach complements the previous approaches in this domain, as previously, the metallic plates were formed to make the structural frame, and additional metal plates are being used as cover [12].

Figure 86.8 shows the structural decomposition of the vehicle wash machine after the design optimization. Whereas Figure 86.9 represents the optimized design for the structural column and Figure 86.10 shows the CAD model of the entire structure. (Refer to Table 86.2, for details about parts count and time consumed by different operations after design optimization.)

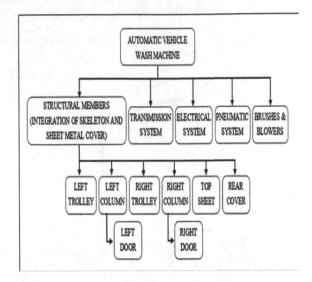

Figure 86.8 Members in the new vehicle wash machine.

Table 86.2: Scenario after design optimization.

Particulars	Duration
Total No. of components	65 Nos.
Laser cutting and Bending	3
Powder coating	2
Assembly of sub-components	2
Assembly of the main structure	2
Total	9 Days

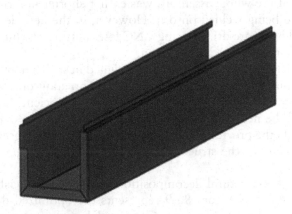

Figure 86.9 New structural column.

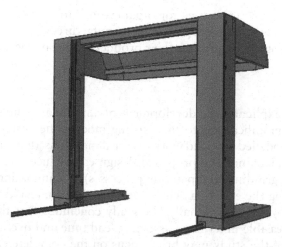

Figure 86.10 CAD model of integrated structure.

Discussion

The main objective of this study was to reduce the assembly lead time for a vehicle wash machine. To achieve the above, a series of changes in the design of the machine structure has been performed. So that, the components can be easily installed with less time, cost, and effort. DFMA principles and guidelines have served the purpose very well in this study. A complete change in the structure has been made by the extensive usage of CAD software from concept generation to final design and also, the domain like Sheet Metal Design was very convenient during the entire work.

The findings of the study show that after changing the structural design, the chances of error while manufacturing has been reduced significantly due to the use of the CNC laser cutting machine in place of manual cutting. The part count has been reduced from 85 to 65 which is around 24%. The significant reduction in the total number of parts provides several benefits such as lesser time for assembly, lesser material requirement, weight reduction, and overall lowered manufacturing costs. Also, the total number of manufacturing processes involved was reduced by 50%, and this succor the assembly lead time to get reduced to 9 days from 16 days, which is nearly 43.75%.

However, the bending process is still performed manually on a hydraulic bending machine due to the extra-large size of some components, which includes the risk of human error. These errors may have an impact on the assembly process, for example, if bending distance on a particular part is having an error of 2 mm and the mating part also have an error of 2 mm then the axes of corresponding holes on both the parts may get deviated up to 4mm, which is not acceptable and it may require additional efforts to rectify the errors.

The study provides several implications. First, the findings of the study can be used as a roadmap to optimize the structural design of different mechanical structures similar to the automatic vehicle wash machine. Second, the managers can get crucial insights by following the steps that have been performed in this study to reduce the number of structural members, the number of manufacturing operations, as well as the overall manufacturing costs. Finally, as it has been well established that to survive and thrive in today's turbulent business environment, it is imperative for organizations to focus on continuous improvements in their products and processes. The manufacturers of automatic vehicle wash machines can use this study to improve their products and stay ahead of their competitors.

Conclusions

This proposed work explicates the development of a modular structure of a vehicle wash machine by making modifications to the existing model. The design of the vehicle wash machine has been modelled in SolidWorks with dimensions that are obtained from the previous generation machine. The proposed design can reduce assembly lead time/labor (for cutting, welding, grinding, and painting processes) by eliminating the number of parts by 24% and increasing the ease of assembly and installation, resulting in the reduction of assembly lead time by 43.75%. Finally, the study concludes that the updated structural design can be more feasible in terms of assembly lead time and overall costing.

The future scope of the study may bring focus on the complete elimination of various human errors performed throughout the current processes by certain design modifications and application of poka-yoke.

References

1. Ahmad, F., and Rahman, M. (2016). Lead time reduction and process cycle improvement of an ice-cream manufacturing factory in bangladesh by using value stream map and kanban board: a case study. *Australian Journal of Basic and Applied Sciences*, 10, 250–260.
2. Alad, A. H., and Deshpande, V. A. (2016). Lead time reduction: a review of various tools and techniques for lead time reduction. *International Journal of Engineering Development and Research*, 2(January), 1159–1164.
3. AUTEC Car Wash. (n.d.). https://autec-carwash.com/blog/news/how-the-first-automatic-car-wash-system-came-to-be/.
4. Bera, M., Gupta, P., and Maji, P. K. (2018). Structural/Load-Bearing Characteristics of Polymer—Carbon Composites Structural/Load-Bearing Characteristics of Polymer—Carbon Composites. Singapore: Springer. https://doi.org/10.1007/978-981-13-2688-2
5. Naol Dessalegn Dejene, Sololo Kebede Nemomsa. (2020). Design for Manufacturing and Assembly. *World Academics Journal of Engineering Sciences*, 7(3), 60–67. https://doi.org/10.1007/978-1-4615-5785-2.
6. Garza-Reyes, J. A., Flint, A., Kumar, V., Antony, J., and Soriano-Meier, H. (2014). A DMAIRC approach to lead time reduction in an aerospace engine assembly process. *Journal of Manufacturing Technology Management*, 25(1), 27–48. https://doi.org/10.1108/JMTM-05-2012-0058.
7. Gillberg, L., and Sandberg, C. (2017). Developing Design Guidelines for Load Carrying Sheet Metal Components with Regards to Manufacturing Method. Master's Thesis, Production Engineering, School of Industrial Engineering and Management (ITM), KTH, Stockholm, Sweden. https://kth.diva-portal.org/smash/get/diva2:1198837/FULLTEXT01.pdf.
8. Gokulraju, R., Vigneshwar, K., and Vignesh, V. (2016). A case study on reducing the lead time and increasing throughput by using value stream mapping. *International Research Journal of Engineering and Technology*, 3(12), 411–424. www.irjet.net.
9. Harlalka, A., Naiju, C. D., and Janardhanan, M. N. (2016). Redesign of an in-market food processor for manufacturing cost reduction using DFMA methodology. *Production and Manufacturing Research*, 3277(November 2017), 0. https://doi.org/10.1080/21693277.2016.1261052.

10. Makwana, K., and Awasthi, S. (2017). A case study on reducing the lead time by using value stream mapping. *International Conference on Ideas, IMpact and Innovaion in Mechanical Engineering*, 5(6), 52–57.

11. Nahm, A. Y., Vonderembse, M. A., Rao, S. S., and Ragu-Nathan, T. S. (2006). Time-based manufacturing improves business performance—Results from a survey. *International Journal of Production Economics*, 101(2), 213–229. https://doi.org/10.1016/j.ijpe.2005.01.004.

12. Sabet, S. M. M., Marques, J., Torres, R., Nova, M., Gomes, L., and Nobrega, J. (2015). Design of a modular rollover carwash machine structure. *Machine Design*, 7(4), 129–136.

13. Senapati, A. K., Mishra, P. C., Routra, B. C., and Biswas, A. (2012). An extensive literature review on lead time reduction in inventory control. *International Journal of Engineering and Advanced Technology*, 1(6), 104–111.

14. Zhai, Y., Sun, Y., Li, Y., and Tang, S. (2023). Design for Assembly (DFA) evaluation method for prefabricated buildings. *Buildings*, 13(11), 2692.

15. Tuvayanond, W., and Prasittisopin, L. (2023). Design for manufacture and assembly of digital fabrication and additive manufacturing in construction: a review. *Buildings*, 13(2), 429.

16. Suresha, D. B., Devendra Reddy, R., and Murali, C. T. (2017). Structural optimization of lathe machine bed. *International Journal of Mechanical and Production Engineering Research and Development*, 7(5), 145–154. https://doi.org/10.24247/ijmperdoct201715.

17. Ullah, I., and Narain, R. (2018). Analysis of interactions among the enablers of mass customization: An interpretive structural modelling approach. *Journal of Modelling in Management*, 13(3), 626–645. https://doi.org/10.1108/JM2-04-2017-0048.

87 Study of surface roughness properties of machined surface on Al based metal matrix composites by near-dry-EDD

Pabitra Kumar Sahu[1,a], Sabindra Kachhap[2,b], Abhishek Singh[3,c], and Pranjali Jain[4,d]

[1]Ph.D. research scholar, NIT Manipur, Imphal, Manipur, India

[2]Assistant Professor, NIT Manipur, Manipur, India

[3]Assistant Professor, NIT Patna, University Campus, Bihar, India

[4]B.Tech. Student, NIT Manipur, Imphal, Manipur, India

Abstract: Electric discharge machining (EDM) is making an impactful appearance in unconventional machining processes in terms of its sustainability and efficiency. This method is commonly being used in defense, dental, medical as well as automotive industries. The Electric discharge drilling (EDD) is a process of ignition erosion for creating holes in conductive materials. To further enhance the process, we use near dry electric discharge drilling (ND-EDD). Near-Dry-EDD is a process which uses a blend of liquid or gas as a dielectric medium which helps in providing a better surface finish to the work piece. In this investigation, the electrode was made of copper and it had external diameter of 9 mm having four openings of inner diameter of 2.5 mm for interior flushing. The current exploration enlightens the examination of ideal degree of input control factors of near-dry electric discharge drilling (ND-EDD) measure in the machining of blind holes of Al based metal composite matrix. The input control factors were considered pulse-on-time, current discharge, duty factor and speed of tool, while surface roughness (SR) was considered as response during the near-dry-EDD experiment. The experimental plan was designed using central composite rotatable design (CCRD) approach of response surface methodology for drilling on work piece. The analysis of variance (ANOVA) was utilized to find the percentage contribution of input parameters. It was found that the lower degree of pulse on time and current value necessary to achieve better surface finish.

Keywords: Dielectric medium, dry EDM, electric discharge drilling, near dry EDM, surface roughness.

Introduction

Electric discharge machining (EDM) is a metal removal process in which utilizes electric spark disintegration to remove metal. This cycle utilizes electric sparks to machine the work piece to produce the ideal shape and measurements. The cycle of EDM does exclude an immediate contact between the copper electrode and the work piece. So mechanical stress and vibration issues during machining measure are eliminated. Even while machining of thin components deformation of work piece does not occur. Therefore, EDM is ideal for machining of materials having high hardness, high strength and high toughness. Since it is versatile and able to cut hardened steels [1]. It has become widely accepted in various manufacturing industries such as die-making, mold manufacturing, aerospace, automotive as

[a]pabitrasahu1994@gmail.com, [b]sabindra.05@gmail.com, [c]abhishek.singh@nitp.ac.in, [d]writetopranjali@gmail.com

well as bio medical [2]. Applications for electric discharge machining (EDM) underway are huge and accordingly become a main-stream process after association, milling and excavation. It is widely used to produce mold and die. It is also used in the manufacture of medical and complementary components in the automotive and aerospace industries [3]. In the EDM cycle, experimental equipment is taken out from the working environment because of erosion. The transformation of electrical into heat energy in the succession of electrical releases via copper electrodes [4]. The dielectric medium is a necessary deciding factor in the EDM process. This method of operation not only acts as a non-conductor, but also contributes to the erosion of the material during release [5]. The recycling gap and waste disposal after disposal have also not been affected [6]. Notwithstanding the high-level material removal framework, EDM has so far not been utilized to certain more noteworthy degree due to the restrictions it showed, especially when surface roughness is considered; impediments, for example, wrong surface quality, surface deformities, layer weakness to thermal heat and residual stress garners poor reviews of the EDM process [7, 8]. Generally, hydrocarbon oils are used as dielectric medium in EDM process [9]. During the machining process, the dielectric medium was used at high temperature (8000–12000°C) leading to its decomposition and hence produces noxious gases [10]. To prevent complications and further enhance the EDM process. Kunieda et al. [11] used various combination of gases in EDM and called it as Dry EDM which ensured severe debris deposits leading to poor surface quality and accuracy. Tanimura et al. [12] discovered an alternative to improve EDM method. This process alternative utilized a liquid-gas mixture as dielectric fluid. This unconventional method is known as near-dry EDM (ND-EDM) process. Near-Dry-EDM comprises of utilizing a combination of air-water or gases (nitrogen, helium, argon etc.) rather than customarily utilized oil containing hydrocarbons by means of a fast pressurized stream. It was found near-dry-EDM effective for finishing process with good machining stability when compared to wet and dry EDM at the same energy level. It is hypothesized that gas medium consisting of dispersed liquid particle has a similar process to promote finishing process as additive powders of PMD EDM. The disposal of dielectric is also environmentally friendly than PM DEDM. Dhakar et al. [13] inferred that glycerin-air blend accompanied a better followed by a combination of oil-air and distilled water-air. It produces surface finish of a level of 0.09 microns with better strength and stability during the experimental process and reduces dielectric consumption and occupies less floor than traditional EDM. Kao and Tao et al. [14] investigated on near-dry- EDM method to improve the surface roughness (SR) and they observed the process of near-dry- EDM had some advantages over dry-EDM on parameters like SR, more sharp cutting edge and decrease in debris deposition. When compared with wet EDM, near-dry-EDM shows a better SR at a lower release energy input and has better outcomes to improve work piece integrity of the surface Srivastava et al. [15] observed that discharge current and pulse-on-time were the most important parameters affecting the surface roughness. Bai et al. [16] found the effect of process parameter such as peak current, pulse on time, pulse off time, flow rate, and concentration of powder particles on material removal rate (MRR), tool wear rate (TWR) and surface roughness (SR) of powder mixed dielectric medium of near dry EDM. It was observed that as the powder concentration and pulse on time rise, the MRR rises at first and then falls. Ravi et al. [17] Studied material removal rate, tool wear rate, and surface roughness by drilling mild steel using various dry EDM process parameters. They found that with increase in current and pulse on time then the MRR increases. Mathias Lorenz et al. [18] Improved results in terms of erosion time, relative electrode wear, as well as bore inlet by switching from standard EDM to near dry EDM. In this investigation nitrogen and oxygen mixed with deionized water and supplied as dielectric medium to near dry EDM.

Sundriyal et al. [19] used Gaussian heat distribution for modelling of material removal rate. The comparison done between modelling result of MRR and experimental result of MRR. It was found that error percentage due to mathematical modelling of MRR and experimental value of MRR was 30%. Yadav et al. [20] investigated to find the capability of near dry EDM. In this investigation, there are five process parameters such as tool rotation speed, current, pulse on time, liquid flow rate and gas pressure were used to perform experiments on AISI M2. The MRR, SR, and hole overcut (HOC) experimental findings were investigated. The response surface technique was used to plan and perform the experiments. Models of regression were also formed. The tool rotation speed has an important impact on MRR, SR, and HOC, according to the findings. FE-SEM micrographs revealed that the RT-ND-EDM machined surfaces have less micro - cracks, debris formation, and craters. Near-dry-EDM has a wide selection palette of gases and liquids along with the greater freedom to adjust the concentration of liquid in gas. The dielectric properties thus can be handpicked in near dry EDM to meet various parameters like a fine surface finish [21]. In study, tubular tools, slotted tools, and helical tools were employed by Yadav et al. [22]. The effect of input parameters such as tool rotation speed, current, pulse on time, gas pressure, and the liquid flow rate on the performance of the RT-ND-EDM process is studied for selected tool electrodes. The results showed that a double-start helical-shaped tool electrode outperformed all other shape electrodes in terms of process performance.

In this investigation, the influence of near dry EDD parameters such as discharge current, pulse on time, duty factor, and tool speed on the surface roughness of an aluminum-based metal matrix composite using a multi-holed copper tool was examined. Central composite rotatable design (CCRD) approach of RSM was taken as main method for experimentation for EDD. There has been less work in near-dry EDM using an aluminum-based metal matrix composite.

Experimental Details

The present examination begins with the study of surface roughness properties of machined surface on Al based metal matrix composites by near-dry-electric discharge drilling (Near-Dry-EDD). The dimension of work piece was 100 mm × 47 mm × 4 mm. To conduct the investigation, copper electrode and air as dielectric medium were preferred. The experiment was performed on the CNC Machine for EDM (Electronica Smart 1 CNC EDM machine). To perform EDD a penetrating contraption was created and was converged with the current EDM arrangement. A CNC Machine for EDM (Electronica Smart 1 CNC EDM machine) was used for the experimental work. The electric discharge drilling setup was connected to component of CNC EDM machine. The copper electrode was pivoted through a connection comprising of an engine, v-belt and a pulley appended in simultaneousness. Control factors were fixed as flushing pressure of 20 psi, gap voltage of 60V and straight electrode polarity. The electrode was made of copper and it had external diameter of 9 mm having four openings of inner diameter of 2.5 mm for interior flushing. The commercially available EDM oil was used as dielectric fluid and tool polarity was positive in nature.

Method Followed

The work consisted of designing of experiment (DOE) and application of response surface method (RSM). Table 87.1 highlights the minute differences among Control Variables of N-D-EDD and EDD. The procedure carried experiments with numerous input control variables with surface roughness as the response variable.

Table 87.1: Input control variables of near dry electric discharge drilling (N-D-EDD) and electric discharge drilling (EDD).

Input control variables	Near Dry EDD	EDD
Dielectric flow rate(ml/min)	60	Submerged in dielectric tank
Air pressure (Bar)	20 psi	
Electrode rotation speed (RPM)	400–1200	400–1200

Input Control Variables

The subsequent variables were used as input control variables on the basis of recommendations of previous literatures. Central composite rotatable design (CCRD) was taken as main method for experimentation for EDD along with Design-Expert statistical software as elucidated in Table 87.2. Discharge current, pulse on time, duty factor and tool speed were taken as input parameters. The experiment consists of four input parameters at five levels. There were 30 experiments conducted.

Table 87.2: Selected input control parameters and their levels.

Symbol	Process parameters	Units	Levels				
			–2	–1	0	+1	+2
A	Discharge current	amp	3	6	9	12	15
B	Pulse on time	μs	50	100	150	200	250
C	Duty factor	-	8	12	16	20	24
D	Tool speed	Rpm	400	600	800	1000	1200

The following Figures 87.1 and 87.2 indicates the experimental set up of near dry EDD and multiple hole cylindrical copper tool respectively.

Figure 87.1 Experimental Setup of Near Dry EDD.

Figure 87.2 Multiple hole cylindrical co.pper tool.

Results

Analysis of Variance

The analysis of variance (ANOVA) technique was used to explore the output variable i.e., SR and in current analysis the SR was collected doing experiments. Table 87.3 shows SR values at different runs and Table 87.4 shows ANOVA for response surface. If the probability value becomes more then 0.1, it is known as insignificant. If probability value becomes less than 0.05 then is known as significant at 95% confidence interval. The investigator wants that lack of fit should be insignificant and model should be significant. Form ANOVA technique, it was found that insignificant lack of fit value is 0.1174 and model is significant. The adjusted R^2 value and predicated R^2 found to be 0.7906 and 0.4426 respectively. The values of adjusted R^2 and predicated R^2 are close for surface roughness. The coefficient of variation and the appropriate precision value are 9.80 and 10.875 respectively, it shows that acceptance of model. Following Table 87.3 presents the various combinations of the levels of the input control variables, obtaining certain response values during N-D-EDD.

Table 87.3: Control log of the experimental trial runs.

Std. order	A: Discharge current (Amp)	B: Pulse-on-time (µs)	C: Duty-factor	D: Tool-speed (Rpm)	SR (µm)
1	6	100	12	600	5.26
2	12	100	12	600	8.51
3	6	200	12	600	4.83
4	12	200	12	600	7.72
5	6	100	20	600	6.41
6	12	100	20	600	7.92
7	6	200	20	600	4.97

(continued)

Table 87.3: Continued

Std. order	A: Discharge current (Amp)	B: Pulse-on-time (μs)	C: Duty-factor	D: Tool-speed (Rpm)	SR (μm)
8	12	200	20	600	7.03
9	6	100	12	1000	4.88
10	12	100	12	1000	7.47
11	6	200	12	1000	7.19
12	12	200	12	1000	7.5
13	6	100	20	1000	4.83
14	12	100	20	1000	7.96
15	6	200	20	1000	4.33
16	12	200	20	1000	6.81
17	3	150	16	800	3.82
18	15	150	16	800	8.84
19	9	50	16	800	5.62
20	9	250	16	800	5.84
21	9	150	8	800	7.95
22	9	150	24	800	5.36
23	9	150	16	400	6.94
24	9	150	16	1200	6.6
25	9	150	16	800	4.96
26	9	150	16	800	5.07
27	9	150	16	800	5.79
28	9	150	16	800	5.63
29	9	150	16	800	5.97
30	9	150	16	800	5.42

Table 87.4: ANOVA for response surface quadratic model.

Source	Sum of Squares	DF	Mean Square	F- Value	Prob > F	
Model	46.33	14	3.31	8.82	< 0.0001	Significant
A-Discharge-current	33.28	1	33.28	88.69	< 0.0001	
B-Pulse-on-time	0.24	1	0.24	0.65	0.4326	
C-Duty-factor	2.86	1	2.86	7.61	0.0146	
D-Tool-speed	0.23	1	0.23	0.62	0.4438	
AB	0.47	1.00E+00	0.47	1.25E+00	2.81E-01	
AC	1.23E-03	1	1.23E-03	3.27E-03	0.9552	

(continued)

Table 87.4: Continued

Source	Sum of Squares	DF	Mean Square	F- Value	Prob > F	
AD	0.09	1	0.09	0.24	0.6314	
BC	1.63	1	1.63	4.33	0.0549	
BD	1.12	1	1.12	2.99	0.104	
CD	0.61	1	0.61	1.62	0.2223	
A^2	1.47	1	1.47	3.92	0.0663	
B^2	0.18	1	0.18	0.49	0.4957	
C^2	2.69	1	2.69	7.16	0.0173	
D^2	3.2	1	3.2	8.53	0.0105	
Residual	5.63	15	0.38			
Lack of Fit	4.83	10	0.48	3.02	0.1174	not significant
Pure Error	0.8	5	0.16			
Cor Total	51.96	29				

Std. Dev.	0.61	R-Squared	0.8917
Mean	6.25	Adj R-Squared	0.7906
C.V. %	9.8	Pred R-Squared	0.4426
PRESS	28.96	Adeq Precision	10.875

Graphical Analysis

Figure 87.3 shows the normal probability plot for surface roughness, reveal that a straight line surrounded with the residuals at a closer distance and errors following the normal distribution trend. Figure 87.4 presents an apt fit between the experimentally obtained values and regression model in the normal probability plot between actual and predicted values.

Three-Dimensional Interaction Plots

Figures 87.5 to 87.8 show the effect of dDielectric medium (mixture of liquid-gas) and input control variables on surface roughness (SR). Increase in discharge current then SR increases. This can be found that more rough surfaces are formed with rise in pulse on time as shown in Figure 87.5. It is further noted that at lower levels of peak current and pulse on time, the better surface finish is produced. Figure 87.6 indicates that, at increase value of the duty factor, the surface finish of the machined surface slightly increases while the pulse on time rises then SR increases steadily. Form Figure 87.7 shows the 3D surface roughness plot with the duty factor and the tool speed. A higher duty factor value and a moderate tool speed value are both indicated in Figure 87.7 for the minimum value of surface roughness. It is explained in Figure 87.8 that the discharge current has a major effect on surface roughness and for good surface roughness, a lower value of peak current

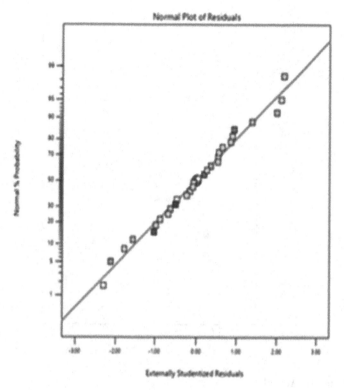

Figure 87.3 Normal plot of residuals.

is desired. It's also found that for better finishing of the work piece, a moderate value of tool speed is necessary.

Scanning Electron Microscopy (SEM) Test

The analysis of SEM was conducted and the images are shown below with distinctive markings. The SEM images were examined at optimal conditions of parameters. The surface characteristics shown below in the SEM images were obtained during EDD with multiple holes tool copper electrode. Figure 87.9 shows SR result, investigated by SEM Morphology; discharge-current is kept at 3 amp, pulse-on-time 150μs, duty factor 16 and tool-speed 800rpm which resulted a lower SR. Figure 87.10 shows SR result investigated by SEM Morphology; discharge-current is kept at 15 amp, pulse-on-time 150μs, duty factor 16 and tool-speed 800 rpm, which resulted a higher SR and the micro crack and long debris particles are found on the surface of work piece. When the supplied current is increased then more amount of energy is generated, which leads to poor surface finish.

Figures 87.11 and 87.12 indicates their dimension profilometer of machined surface at optimum parameter setting. The SR are found more at higher current (I=15 amp) as indicated in Figure 87.12 and surface irregularities are noticed to be less at lower value of current (I=3 amp) as indicated in Figure 87.11.

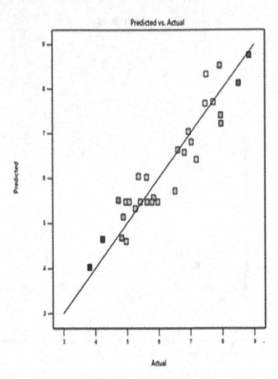

Figure 87.4 Predicted vs. Actual.

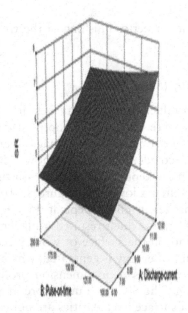

Figure 87.5 Discharge-current vs. Pulse-on-time.

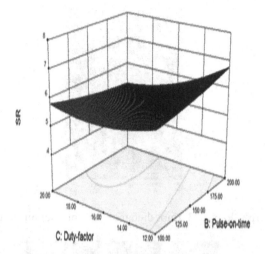

Figure 87.6 Pulse on time vs. Duty-factor.

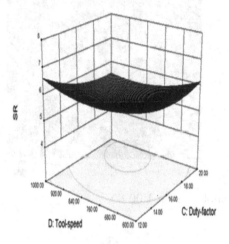

Figure 87.7 Duty factor vs tool speed.

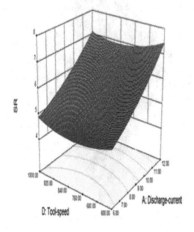

Figure 87.8 Discharge current vs tool speed.

Figure 87.9 Surface Morphology at 3 amp. dicharge-current, magnified at 1.00 KX.

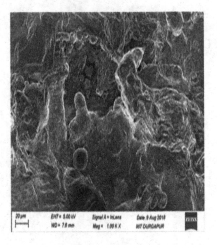

Figure 87.10 Surface Morphology at at 1.00 KX.

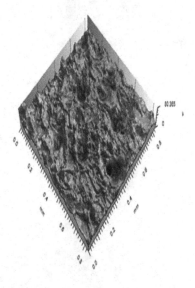

Figure 87.11 Surface profile at 3amp discharge-current.

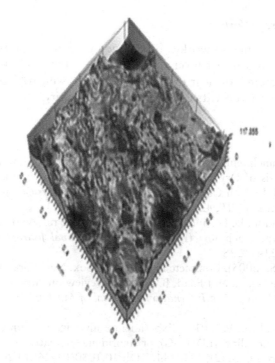

Figure 87.12 Surface profile at 15amp discharge-current.

Conclusions

The effect of each operating parameters on SR is analyzed. ANOVA was also used to investigate the effects of various process parameters on SR. The present investigation for modeling of surface roughness of Al primarily based MMC throughout close to Near Dry EDD has the following conclusions:

1. The PMND-EDM method was found to produce work pieces with higher surface finish, when the input process parameters were optimized.
2. The optimum parameter conditions at discharge-current at 3 amp, pulse-on-time at 150 μs, duty factor at 16 and tool-speed at 800 rpm were found to be the most important in obtaining the minimum surface finish of machined aluminum based composite work piece.
3. Discharge current is the most critical consideration for surface roughness, while other parameters have an almost insignificant effect on surface roughness, i.e., time pulse, instrument speed and pulse duty factor.
4. It is inferred from the experiment that a lower pulse value on time is ideal for lower surface roughness.

   - Furthermore, it is noted that it is necessary to achieve a better surface finish at a lower degree of peak current. In comparison, for minimum surface roughness, a moderate value of instrument speed is required.
   - The P value is less than 0.001 and the F value of model is 8.82, which shows that model is significant.
   - The lack of fit value is 0.1174, which indicates that it is non-significant

Future Scope of Research Work

For future research experimental work can be done by taking response as micro hardness, overcut, residual stress and recast layer thickness of machined workpiece using different parameter. Other parameters of near dry EDD such as power, voltage, servo feed etc. can take as input parameter to know its effect on surface integrity.

References

1. Uhlmann, E., Schimmelpfennig, T. M., Perfilov, I., Streckenbach, J., and Schweitzer, L. (2016). Comparative Analysis of Dry-EDM and conventional EDM for the manufacturing of micro holes in Si3N4-TiN. In 18th CIRP Conferencw on Electro Physical and Chemical Machining (ISEM XVIII). *Procedia CIRP*, 42, 173–178.
2. Abbas, N. M., Solomon, D. G., and Bahari, M. F. (2007). A review on current research trends in electrical discharge machining (EDM). *The International Journal of Machine Tools and Manufacture*, 47, 1214–1228.
3. Carl, S., and Steve, S. (2005). Complete EDM Hand Book. New York: Advance Publishing Inc.
4. Norliana, A., Darius, G. S., and Fuad, B. (2006). A review on current research trends in electrical discharge machining. *The International Journal of Machine Tools and Manufacture*, 47, 1214–1228.
5. Yadav, R. S., and Yadava, V. (2017). Experimental investigations on electrical discharge diamond face surface grinding (EDDFSG) of hybrid metal matrix composite. *Materials and Manufacturing Processes*, 32(2), 135–144. DOI: 10.1080/10426914.2016.1221089.
6. Chengbo, G., Shichun, D., and Dongbo, W. (2017). Study of electrical discharge machining performance in water-based working fluid. *Materials and Manufacturing Processes*, 31(14), 1865–1871.
7. Tao, J., and Shih, A. (2008). Near-dry EDM milling of mirror-like surface finish. *International Journal of Electrical Machining*, 13, 29–33. DOI: 10.2526/ijem.13.29.
8. Yadav, V. K., Kumar, P., and Dvivedi, A. (2017). Investigations on rotary tool near-dry electric discharge machining. In Wang, S., Free, M. L., Alam, S., Zhang, M., and Taylor, P. R. (eds.). Applications of Process Engineering Principles in Materials Processing, Energy and Environmental Technolo-gies. Cham.: Springer (pp. 327–334).
9. Ekmekci, B. (2007). Residual stresses and white layer in electric discharge machining (EDM). *Applied Surface Science*, 253, 9234–9240.
10. Evertz, S., Dott, W., and Eisentraeger, A. (2006). Electrical discharge machining - occupational hygienic characterization using emission based monitoring. *International Journal of Hygiene and Environmental Health*, 209, 423–434. DOI: 10.1016/j.ijheh.2006.04.005.
11. Kunieda, M., and Furuoya, S. (1991). Improvement of EDM efficiency by supplying oxygen gas into gap. *CIRP Annals*, 40, 215–218. DOI: 10.1016/S0007-8506(07)61971-4.
12. Tanimura, T., Isuzugawa, K., and Fujita, I. (1989). Development of EDM in the Mist: Proceedings of Ninth International Symposium of Electro Machining (ISEM IX) Nagoya, Japan, (pp. 313–316).
13. Dhakar, K., Dvivedi, A., and Dhiman, A. (2016). Experimental investigation on effects of dielectric mediums in near-dry electric discharge machining. *Journal of Mechanical Science and Technology*, 30(5), 2179–2185.
14. Kao, C. C., Tao, J., and Shih, A. (2007). Near-dry electrical discharge machining. *The International Journal of Machine Tools and Manufacture*, 47, 2273–2281. DOI: 10.1016/j.ijmachtools.2007.06.001.
15. Srivastava, V., and Pandey, P. M. (2012). Performance evaluation of electrical discharge machining (EDM) process using cryogenically cooled electrode. *Materials and Manufacturing Processes*, 27, 683–688.

16. Bai, X., Yang, X., and Zhang, Q. (2018). Experimental study on the electrical discharge machining with three-phase flow dielectric medium. *The International Journal of Advanced Manufacturing Technology*, 96, 2003–2011.

17. Sundriyal, S., Vipin, V., and Walia, R. S. (2020). Experimental Investigation of the Micro-hardness of EN-31 Die Steel in a powder-mixed near-dry electric discharge machining method, strojniški vestnik. *Journal of Mechanical Engineering*, 66(3), 184–192.

18. Lorenz, M., and Schimmelpfennig, T. M. (2019). Process optimization by transferring conventional electrical discharge machining to near dry proceedings for precise bore hole in CoCrMo. 2019 International Interdisciplinary PhD Workshop.

19. Sundriyal, S., Yadav, J., Walia, R. S., Kumar, V. R. (2020). Thermophysical-based modeling of material removal in powder mixed near-dry electric discharge machining. *Journal of Materials Engineering and Performance (JMEP)*, 29, 6550–6569.

20. Yadav, V. K., Kumar, P., and Dvivedi, A. (2019). Effect of tool rotation in near-dry EDM process on machining characteristics of HSS. *Materials and Manufacturing Processes*, 34(7), 779–790. ISSN: 1042-6914,1532-2475.

21. Singh, N. K., Pandey, P. M., and Singh, K. K. (2017). Experimental investigations into the performance of EDM using argon gas-assisted perforated electrodes. *Materials and Manufacturing Processes*, 32(9), 940–951. ISSN: 1042-6914.

22. Yadav, V. K., Singh, R., Kumar, P., and Dvivedi, A. (2021). Performance enhancement of rotary tool near-dry EDM process through tool modification. *Journal of the Brazilian Society of Mechanical Sciences and Engineering*, 43, 72.

88 Investigation of tensile, hardness and double shear behaviour of basalt aluminium composites

J. Melvin antony[a], D. Kumaran[b], G. Lokeshwaran[c], P. Santhana srinivasan[d], B. Vijaya Ramnath[e], and R. Senthil Kumar[f]

Department of Mechanical Engineering, Sri Sairam Engineering College, West Tambaram, Chennai, Tamil Nadu, India

Abstract: The experimental investigation of a composite laminate comprised of aluminum and basalt fiber is the subject of this paper. The minerals plagioclase and pyroxene make up the majority of the igneous rock known as basalt. Its mechanical qualities are improved when mixed with aluminum, a boron group metal that is soft, non-magnetic, and ductile and can be employed in a variety of technical applications. Using the hand lay-up process, basalt-aluminum laminate is created. The use of silicon carbide improves the material's abrasive and lasting qualities. The ultimate breaking load with ultimate stress is examined using a double shear test, and the ultimate tensile strength of the Basalt-Aluminum laminate is assessed using a tensile test. The conclusions proved that adding SiC increased the ultimate tensile strength and ultimate stress and decrease the hardness of the material. This material can be used in areas where wear is common, such as on conveyor belts, clutch plates, and brake callipers.

Keywords: Aluminium, basalt fiber, double hear test, shore d hardness test, silicon carbide, tensile strength.

Introduction

It is difficult to find materials to replace existing materials like mild steels, cast iron etc., the vast requirement in various fields of engineering as these materials possess great mechanical properties. The usage of natural fibers which can have equivalent properties to that of existing materials, the difficulties and the challenges faced by engineers can be solved. By combining basalt fibre which are extracted from the igneous rocks obtained from the volcanic eruptions, are mainly cost-effective material. Aluminium which is a relatively weak metal, can produce a composite with the addition of varying weight percentage of silicon carbide (SiC) in order to produce mechanical capable laminar composites. The necessity of the replacement and development of existing materials is the way for bringing in a sustainable development and to prevent over usage of well proclaimed materials like iron, steels etc.

In this paper the authors studied the Hand lay-up process which was used to fabricate the fibre metal laminates (FML). They experimentally investigated the blast response of

[a]melvinantony007890@gmail.com, [b]kumaranstudent@gmail.com, [c]lokeshpers10@gmail.com, [d]sansriniana@gmail.com, [e]vijayaramnathb@gmail.com, [f]senthilkumar.mech@sairam.edu.in

the fiber metal laminates and gradient aluminium honeycomb sandwich panels skin FML. Localized failure of sandwich panels can be introduced when the honey comb with large cell side length and thickness is set as the first layer is obtained as result [1].

The methodology of hand-layup process used to combine materials of glass fabric/epoxy laminated composites and matrix reinforcement with pre-dried graphene nanoplatelets (GNPs) up to 30 wt%. E-glass fabrics of two types are being employed in the preparation of the composites. Scanning electron microscopy is used up for microstructural analysis. The result is observed as the number of GNPs increases the fracture varies accordingly [2].

In order to improve the material's tensile, flexural, and impact properties, a study of the fabrication technique employed in hand layup compression moulding has been conducted. The material's hybrid sandwich structure makes it suitable for commercial car applications such as flooring, frames, and bonnets [3].

The Thermal cycling is referred in this paper with increasing temperature, the adhesive becomes more ductile, e Young's modulus and tensile strength decrease while the tensile strain increases. With decreasing temperature, BFRP damage (delamination and fibre breakage) tends to occur and complete delamination failure is more likely to be observed [4, 5].

In this paper tests proves that adding natural basalt fiber improves the tensile strength of the composite. The average tensile strength of the fabricated basalt fibre was 4111 MPa. Formulation optimization is a key component of technological innovation in the field of basalt fiber [6].

Researchers conducted various mechanical experimentations by using abaca, jute, flax and kenaf as both mono fiber and hybrid fiber composite adapting the hand layup method. The results shows that hybrid composites have better performance than mono fiber composite [7].

The paper deals in with the analysis of the alloy of Sn-Pb solder constituting the alloy of Pb in 15, 60 and 80% and are mixed in proper proportions and are subjected to tensile and hardness testing after being heated up to a temperature range of 350°C. The results have shown that the enhancement in the hardness as well as the tensile properties were observed in the composition of Pb-Sn alloy constituting 60% of the alloy in it [8].

Natural fibers namely Jute-Flax, Banana-Flax, abaca–jute, Banana Jute and basalt-banana composites were fabricated and their mechanical behaviours were analysed. It was found that hybrid composite possesses better mechanical behaviour than mono fiber composites [9, 10].

Materials Used

Basalt Fibre

In this work, the basalt fibre has been used as a reinforcement which can be used for numerous applications like corrosion resistant pipelines, non-flammable coverings, brake callipers, clutch plates, etc. The Basalt fibre is an igneous rock which had been formed from the rapid cooling of liquified volcanic rock exposed on the earth's crust. The Basalt fibres are created within the form of chopped fabric, rolls and in powder formats.

Aluminium

The aluminium is a white silvery material that is lite weight, malleable alloy, it is one of the most abundant material on the earth's crust, the composite material is fabricated with 5000 series Aluminium mesh.

Epoxy Resin

In this work epoxy, Araldite LY 556 is used as epoxy resin which is based on Bisphenol-A followed by Aradur HY 951 hardener. It's flexural strength and adhesive characteristics and fatigue resistance is superior.

Silicon Carbide

Crystalline silicon carbide a semiconductor made of silicon and carbon typically has high hardness and wear resistance as well as excellent chemical and thermal shock tolerance. They are used in diodes, MOSFETs, brake pads, grinding wheels, clutch plates, and bullet-proof vests, among other products.

Methodology

The composite material is fabricated with the help of epoxy resin, SiC of 3 and 6wt%, hand layup method and compression moulding respectively.

Hand Layup Method

It is an easy and simple method of fabricating natural fiber composites. Even though the fabrication process is extremely slow, it's appropriate for all kinds of fibres. In this work, the layup method involves stacking up of each basalt fibre ply into shape by hand and then firmly placing aluminium on top of successive basalt layers, where epoxy resin and SiC are mixed together and are used as an adhesive in between the basalt aluminium layers, the compression moulding process is used to finely pack the composite, thus eliminating air pocket between plies.

Testing of Composites

Tensile Test

Tensile test is usually performed in the specimens wherein the central portion is of uniform cross-section, and the length which falls under that uniform cross section is termed out to be as gauge length, this gauge length can be of uniform circular or of rectangular cross-section. The larger ends of the specimen are mounted on the gripper of the universal testing machine and the larger ends are termed out to be as gripping ends.

The specimens are cut into the required number of samples as per the standards ASTM D638as shown in Figure 88.1. The test specimen is mounted over the universal testing machine and the loads are increased cautiously so as to note down the strength of the material accurately.

The specimen after being exposed to such force experiences a change in its initial length to an instantaneous length which denotes that there is an elongation spotted from its initial length to some extent which is the instantaneous length, at a certain point, a sudden non-linearity in the shape of the specimen is observed which leads to the necking at a particular point in the structure of specimen, there is a drastic change in the cross-sectional area of the specimen when it reaches the ultimate tensile stress of the material which leads to the formation of necking over the specimen surface respectively.

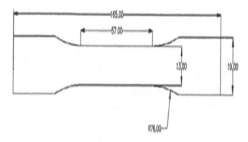

Figure 88.1 Tensile test specimen—ASTM D638.

Double Shear Test

The universal testing machine is used to perform various mechanical testing such as tensile, compression, torsion etc., the 3 and 6wt% SiC specimens are cut into the required dimensions so as to perform the double shear test of the required dimension as per the standard mentioned by the testing laboratory.

The specimen is mounted inside the double shear dies and then the specimen is mounted by the gripper on one end of the attachment from the upper portion followed by the other end which is mounted in a similar manner from the lower end. The dies which carry the specimen to be experimented is inserted over the universal testing machine and the whole apparatus is arranged in an appropriate manner. By applying the loads gradually over the specimen, at a particular point the shear over the specimen can be observed, the load at which the shear is observed over the specimen is noted down respectively.

Shore D Hardness Test

Materials like soft rubbers, polymers, plastics cannot be used for the generally preferred hardness methodologies like Rockwell or Brinell hardness. Instead, these materials are exposed to a unique hardness method known as Shore D Hardness settleability to measure the resistance offered by the material in allowing the penetration of a spring like working needle with a blunted point on it is known as Shore D Hardness test.

The instrument used up in performing the test is known as durometer, this Shore Hardness is of two types which are namely A type hardness and D type hardness. wherein shore A is used up in case of measuring in the softer elastomers by using in a needle with a blunted point whereas, in the case of shore D hardness its being used up in cases wherein harder elastomers are subjected to experimentation and is performed with the help of a needle inclined to 30 degrees at its end.

The material which is to be subjected to hardness testing is prepared and tested as per the ASTM D2240 standard.

Result and Discussion

Result of Tensile Test

The graphs from Figure 88.2 are analyzed, and they are summarized as shown in Table 88.1. The sample with a 6 weight percent SiC content has a tensile strength of 55.94 MPa and a maximum force of 1,800 N, it is discovered. Tensile strength and Fmax values for 3wt% SiC are 64.82 MPa and 1,980 N, respectively.

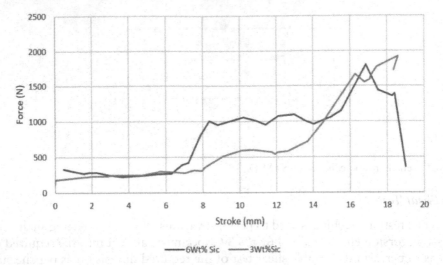

Figure 88.2 Force vs stroke graph for 6 and 3wt% SiC specimen.

Table 88.1: Tensile test results.

Specimen	Fmax (N)	Ultimate Tensile Strength (MPa)
6wt% SiC	1,800	55.94
3wt% SiC	1,980	64.82
Difference	180	8.88
Percentage	110.0%	115.87%

According to the aforementioned result, the sample with 3wt% SiC surpasses the 6wt% SiC composite specimen in terms of Tensile Strength by 8.88 MPa and Fmax of 180 N, respectively. These results indicate that the sample with 3wt% SiC performs 1.1 and 1.158 times better when compared directly with the 6wt% SiC specimen, respectively.

Result of Double Shear Test

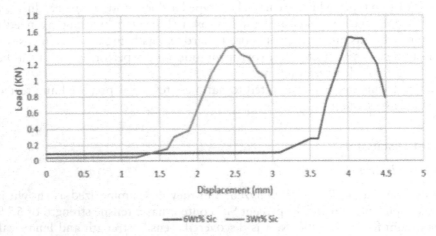

Figure 88.3 Load vs displacement graph of 3 and 6wt% SiC specimen.

Table 88.2: Double shear test results.

Specimen	Ultimate Break Load (N)	Ultimate Stress (N/mm²)
6wt% SiC	1,525	136
3wt% SiC	1,415	126
Difference	110	10
Percentage	107.77%	107.93%

The graphs from Figure 88.3 are analyzed, and they are summarized as shown in Table 88.2. The sample containing 3 weight percent SiC has an ultimate break load of 1,415 N and an ultimate Stress of 126 N/mm2, while the sample containing 6 weight percent SiC has a greater ultimate break load of 1,525 N and an ultimate Stress of 136 N/mm2.

According to the aforementioned result, the sample with 6 weight percent SiC beats the composite specimen with 3 weight percent SiC in terms of ultimate break load and ultimate Stress by 110 N and 10 N/mm2 respectively, which is 1.077 and 1.079 times better when compared directly with the 3 weight percent SiC specimen.

Result of Hardness Test

The values are tabulated as shown in Table 88.3. It is found out that the sample containing 6wt% SiC has hardness averaging 66.233 in the range 64.966 to 68.266, 3wt% SiC has hardness averaging 70.233 in the range of 66.773 to 77.166 respectively.

According to the aforementioned result, the sample with 3 weight percent SiC performs 4 times better than the composite specimen with 6 weight percent SiC in terms of hardness, which is 1.060 times better when compared directly with the 6 weight percent SiC specimen.

Table 88.3: Shore D hardness test results.

Specimen	Trial 1	Trial 2	Trial 3	Average	Average-High	Average-Low
6wt% SiC	67	67.5	64.2	66.233	-1.267	2.0333
3wt% SiC	73.7	73.7	63.3	70.233	-3.467	6.9333

Conclusion

- The tensile strength test values obtained are 55.94 MPa and 64.82Mpa ultimate tensile strength for both 6wt% and 3wt% of SiC respectively. The tensile strength of this aluminium basalt composite is comparable to unreinforced polyester resin and ultrahigh molecular weight polyethylene tensile strength. It is found that adding SiC to the aluminium basalt composite increases its tensile strength.
- The double shear test results are compared and concluded that adding SiC increased the laminate ultimate stress by 7.77%.
- In the case of shore D hardness test, it is found that the material has hardness level compared to that of hard hat and the addition of SiC may decrease the composite materials hardness that making it more flexible.

References

1. Ma, X., Li, X., Li, S., Li, R., Wang, Z., and Wu, G. (2019). Blast response of gradient honeycomb sandwich panels with basalt fiber metal laminates as skins. *International Journal of Impact Engineering*, 123, 126–139. doi: 10.1016/j.ijimpeng.2018.10.003.
2. Seretis, G. V., Kouzilos, G., Manolakos, D. E., and Provatidis, C. G. (2017). On the graphene nanoplatelets reinforcement of hand lay-up glass fabric/epoxy laminated composites. *Composites Part B: Engineering*, 118, 26–32. doi: 10.1016/j.compositesb.2017.03.015.
3. Santhosh, M. S., Sasikumar, R., Thangavel, T., Pradeep, A., Poovarasan, K., Periyasamy, S., and Premkumar, T. (2019). Fabrication and characterization of basalt/kevlar/aluminium fiber metal laminates for automobile applications. [Online]. *International Journal of Materials Science*, 14(1), 1–9. Available: http://www.ripublication.com.
4. Na, J., Mu, W., Qin, G., Tan, W., and Pu, L. (2018). Effect of temperature on the mechanical properties of adhesively bonded basalt FRP-aluminum alloy joints in the automotive industry. *International Journal of Adhesion and Adhesives*, 85, 138–148. doi: 10.1016/j.ijadhadh.2018.05.027.
5. Karvanis, K., Rusnáková, S., Krejčí, O., and Žaludek, M. (2020). Preparation, thermal analysis, and mechanical properties of basalt fiber/epoxy composites. *Polymers*, 12(8), 1–17. doi: 10.3390/polym12081785.
6. Meng, Y., Liu, J., Xia, Y., Liang, W., Ran, Q., and Xie, Z. (2021). Preparation and characterization of continuous basalt fibre with high tensile strength. *Ceramics International*, 47(9), 12410–12415. doi: 10.1016/j.ceramint.2021.01.097.
7. Ramnath, B. V., Kokan, S. J., Raja, R. N., Sathyanarayanan, R., Elanchezhian, C., Prasad, A. R., and Manickavasagam, V. M. (2013). Evaluation of mechanical properties of abaca–jute–glass fibre reinforced epoxy composite. *Materials and Design*, 51, 357–366.
8. Adetunji, O. R., Ashimolowo, R. A., Aiyedun, P. O., Adesusi, O. M., Adeyemi, H. O., and Oloyede, O. R. (2021). Tensile, hardness and microstructural properties of Sn-Pb solder alloys. *Materials Today: Proceedings*, 44, 321–325. doi: 10.1016/j.matpr.2020.09.656.
9. Sakthivel, M., Vijayakumar, S., and Ramnath, B. V. (2018). Investigation on mechanical and thermal properties of stainless-steel wire mesh-glass fibre reinforced polymer composite. *Silicon*, 10(6), 2643–2651.
10. Ramakrishnan, G., Ramnath, B. V., Elanchezhian, C., Arun Kumar, A., and Gowtham, S. (2019). Investigation of mechanical behaviour of basalt-banana hybrid composites. *Silicon*, 11(4), 1939–1948.

89 Design optimization to enhance the structural properties of a vehicle wash machine

Dhananjay Dhandole[1,a] and Inayat Ullah[2,b]

[1]PG Scholar, Department of Mechanical Engineering, G H Raisoni College of Engineering, Nagpur, India

[2]Assistant Professor, Department of Mechanical Engineering, G H Raisoni College of Engineering, Nagpur, India

Abstract: This paper aims to enhance the structural properties of a vehicle wash machine by optimizing the structural design under static loading conditions. The design optimization has been carried out by employing the Design for Manufacturing and Assembly (DFMA) approach. Primarily, the study focuses on the enhancement of the strength and improvement in the factor of safety of the machine structure using Computer-Aided Engineering (CAE). More specifically, the Solidworks and Fusion360 application of Autodesk has been utilized for creating 3D visualization of the structure, and also, for the optimization of design using the simulation module. Besides, an in-depth comparison between the iterations in the designs of structural components has been presented. The outcomes of the study have been shown using graphic images and tables. The efforts employed in the research lead to a 21% reduction in structural weight and 21% changes in factor of safety.

Keywords: Automatic vehicle wash, CAE, finite element analysis, structural optimization.

Introduction

Nowadays, the use of automatic vehicle wash machines is exponentially increasing since they save time, effort, as well as water which is a scarce resource. In addition, the method requires less work than hand-washing because the procedure of cleaning a vehicle by this machine is quite simple and user-friendly. These machines, on one hand, assists the users in maintaining the vehicle in optimum condition, and on the other hand, it allows washing more vehicles with a spotless and glossy appearance in the given time while avoiding the wastage of water. Among others, one of the most important benefits of using these machines is that it helps to maintain the vehicle in proper condition so that it could be sold at a good price if the need arises.

According to the current market research report, despite the Covid-19 pandemic situation around the world, there has been observed a steep growth in the manufacturing segment of these machines. The report indicates that the global market for automatic vehicle wash machines is expected to reach USD 1025.9 million by 2026, increasing from USD 916.3 million in 2020 with a CAGR of 1.9 % between 2020 and 2026. These statistics reflect a huge increase in the market competition in the above segment in near future. The present Indian market is still dominated by large players such as Ezytek Clean, Nissan

[a]dhananjay2206@gmail.com, [b]inayatu6@gmail.com

Clean, Green Motorz, etc. due to their high-quality products and competitive prices. To be able to compete with these giants, it is imperative for small manufacturers to focus on improving the quality of their products in terms of product design, manufacturing lead time, total incurred costs, and the price of the products. To this end, the present study explores the limitations in existing machines available in the market and focuses on the optimization of the structural design to cut down the long manufacturing time lead and costs while maintaining structural integrity. The present study employs finite element analysis (FEA) to improve the design of the automatic vehicle wash machine and also, to investigate whether the design changes have any substantial impact on the machine's strength and performance. The findings show that the strength of the structure of the machine based on the new design is equivalent to that of the old one after design modifications. The design has been updated to suit the Design for Manufacturing and Assembly (DFMA) guidelines, resulting in a significant reduction in the number of structural components while maintaining the structure's strength and functionality.

Literature Review

A large number of researches has been done in the area of product design and optimization which confirms that the application of FEA and DFMA in product design leads to improvement in terms of performance under different conditions, reduction in the number of parts, material requirements, and time-saving due to elimination of redundant manufacturing processes [4–6]. Kaurase & Chopade [7] investigated the desired material properties for bending applications using FEA and the structural steel was found to be the best option amongst the various materials and alloys selected for the study. Mandliya et al. [11] carried out the optimization of the stationary platen of plastic using Computer-Aided Design (CAD) and Computer-Aided Engineering (CAE). In the study performed by Desai et al. [4], the optimal design for flanges has been achieved by using FEA and numerical simulation approaches. Patil and Kulkarni, [12] have conducted experimental validation of the optimization of washing machine crosspieces by utilizing CAE and FEA. Leiva [9] has reviewed various techniques to optimize the design of product structure and stated that dimensional optimization should be preferred to achieve the optimal thickness and to minimize the mass which has been subject to stress and displacement. Jain and Sharma [6] have compared the data from different experimental studies and suggested the use of FEA to optimize the buckling strength adding while bending sheet metal. In their study, Magdum et al. [10] performed the optimization of the weight of a roller bracket up to 27% by adopting the CAE optimization tool from Altair. Antony and Arunkumar [1] performed a case study on the use of DFMA to reduce the number of components in a prosthetic knee. While aiming to simplify the aircraft design by reducing the number of components, Rajamani and Punna [13] has utilized DFMA principles and reported a significant improvement. Sabet et al. [14] focused on the design optimization of automatic vehicle wash machines by introducing the concept of modularity in the structural design. However, the design contained a significantly large number of steel plates that are bolted together to form a column, thus, leading to a very long assembly lead time. In another study, Sabet et al. [15] designed a modular accessory that can be readily put in a vehicle wash machine for drying applications.

During the literature review, it has been observed that while there exist a plethora of studies explaining the use of optimization tools including FEA, CAE, and DFMA, the studies investigating the use of these tools for the optimization of automatic vehicle wash machines are scant. The present study is one among the few that attempts to fill this gap by conducting the structural design optimization of the automatic vehicle wash machine.

Research Methodology

Initially, all the technical data related to features & specifications, dimensions, functionality, bay layout, intakes, and strength of the machine was analyzed, and a new structure was developed using CAD software including SolidWorks and Fusion 360 while following the DFMA guidelines.

To understand the effects of the optimization techniques, a pilot test was first carried out using a specimen sheet metal of equal dimensions. To ensure the structural integrity of the modified design, FEA is carried out and the results were compared with the old structural design. These results are discussed in a further section of this study. The steps involved in optimization by CAE method are as follows:

1. Data collection to create a 3D design.
2. FEA setup for the product.
3. Simulating the part for the applied load.
4. Iteration of simulation by varying the sheet thicknesses.
5. Comparison of results for ensuring maximum safety, minimum cost, and assembly lead time.

The specimen sheet has been modeled using Fusion 360 CAD as shown in Figure 89.1. The CAE module has been used for carrying out the simulation. Static analysis has been performed over the sheet having the dimensions 3620 × 40 mm^2. Further, changes in the specimen sheet have been made and flanges of 50 mm were added on the opposite ends (Figure 89.2) and the simulation of the specimen is again carried out under similar loading and boundary conditions. Figures 89.3 and 89.4 depict the variation of results in both of the specimen sheets.

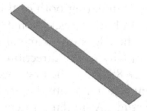

Figure 89.1 Specimen sheet without flanges.

Figure 89.2 Specimen sheet with flanges.

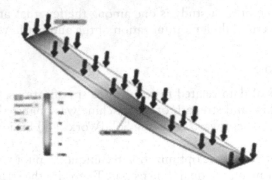

Figure 89.3 Static analysis on the first specimen.

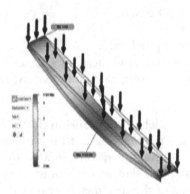

Figure 89.4 Static analysis on the specimen with flanges (50 mm).

After this modification and then simulating both of the specimen sheets, the results are found to have satisfactory values. In both cases, the material used is MS flat (5 mm) with dimensions of 3620 × 440 mm2. The edges of the metal sheet have been fixed and a uniform load of 500N is applied normally in the direction of the ground. The result shows the displacement of 150 mm (Figure 89.3) for the first specimen, whereas, for the specimen with flanges, the total displacement is just 4.5 mm (Figure 89.4). These results were encouraging to proceed further into the study. A similar strategy has been applied for developing the entire structure of the machine, right from the structural column to the top sheet which supports all the major accessories at their respective positions.

Analysis and Results

In Figure 89.5, the old structure of the vehicle wash machine has been presented. This structure is made by the traditional method of manufacturing in which the metal frame is made using square pipes and then metal sheets are welded to cover the frame. Figure 89.6 represents the structure obtained after the design modifications using DFMA guidelines to reduce the number of components. This is a modular structure that is made by using forming processes and metal sheets. To create the structure, the member fastened to each

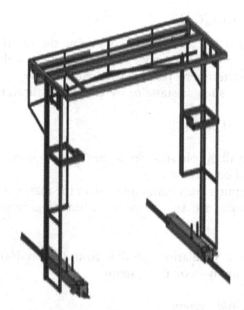

Figure 89.5 CAD model of the old machine structure.

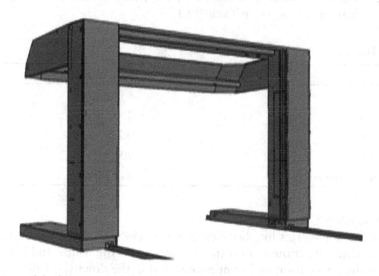

Figure 89.6 CAD model of the new machine structure.

other using fasteners, and no welding operation is performed. An in-depth analysis of the strength and safety of both the structures is carried out further in the study.

Finite Element Analysis

FEM is a mathematical technique to determine the approximate numerical solution to differential equations related to engineering and mathematical modeling [2]. The steps involved in FEM include discretization of the domain; selection of the interpolation function; formulation of the system of equations; solution of the system of equations.

Computer-Aided Engineering (CAE)

Design engineers mostly prefer CAE tools for obtaining the solution of complex engineering problems, as it saves time and money spent on testing actual prototypes. Also, it is a widely used tool in industries (aerospace, marine, automotive, heavy machinery, etc.) for analysis of various systems such as static or dynamic structural, thermal, harmonics, and fluid dynamics.

The steps involved in CAE are: [8]:

1. Preprocessing (where all the physical properties of the system are defined such as material, applied load, and constraints).
2. Solver (here, the computer formulates and solves the mathematical model).
3. Post-processing (Results are shown in the form of graphics and reports) (CAE | Siemens) [3].

In the present study, the "Simulation" module from Fusion 360 software has been used to perform the necessary analysis on the systems.

Case 1: Static Analysis of old System

The system consists of a frame that is made using square pipes joined together using an arc welding process. Visual representation of the frame is shown in Figure 89.7 and the material specifications are given in Table 89.1.

Table 89.1: The material used and its properties.

Material	MS Square 50*50
Thickness	4 mm
Young's Modulus	220 GPa
Poisson's Ratio	0.275
Yield Strength	207 MPa
Ultimate Tensile Strength	345 MPa

Load and Boundary Conditions

Considering the accessories like brushes, blowers, electric cables, air, and water distribution systems which are mounted on the top section of the frame and hence, these accessories causes the application of a continuous load in the downward direction. It has been calculated that a combined load of 5KN is subjected in the direction of the ground. All the square pipes are joined together using fasteners at every adjoining location. Vertical square pipes are fixed at the bottom to restrict the relative motion with the ground. After applying the above loading and boundary conditions, the maximum stress developed in the frame is shown in Figure 89.7 and the safety factor analysis is shown in Figure 89.8.

The results from the study of the old structure show that the total displacement, Von mises stress, and factor of safety are shown in Table 89.2. Here, after the application of 5KN load, the total displacement is observed to be 1.52 mm and the von mises stress as 130 MPa along with the minimum safety factor of 2.2.

Figure 89.7 Von Mises stress analysis on old skeleton/frame.

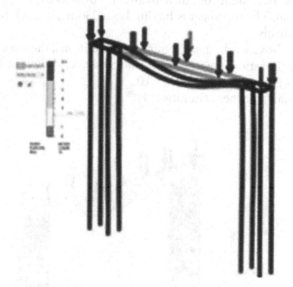

Figure 89.8 FOS analysis on old skeleton/frame.

Table 89.2: Results of case 1.

Variables	Value
Load	5000N
Displacement	1.52 mm
Von mises stress	130 MPa
Factor of safety	2.2 Min

Case 2: Static Analysis of the New System

The new system consists of a frame that is made by forming sheet metal as discussed above. The visual representation of the frame is shown in Figure 89.6 and the material specifications are given in Table 89.3.

Table 89.3: The material used and its properties.

Material	GI Sheet (Side Columns)	MS Flat (Top Sheet)
Thickness	3mm	4mm, 5 mm, 6 mm
Young's Modulus	200 GPa	220 GPa
Poisson's Ratio	0.3	0.275
Yield Strength	207 MPa	207 MPa
Ultimate Tensile Strength	345 MPa	345 MPa

Load and Boundary conditions

In this case, the same accessories which are mounted on the top of the frame and the combined load exerted of 5KN subjected downwards is considered. Both the vertical columns are fixed at the bottom. The top sheet is having bolted joints at each hole with the left and right columns respectively.

After applying the above loading and boundary conditions, the maximum stress is developed in the frame is (Iteration 1 with 4 mm thickness), as shown in Figure 89.9, and the safety factor analysis is shown in Figure 89.10.

For top sheet – 4 mm thickness (Iteration 1)

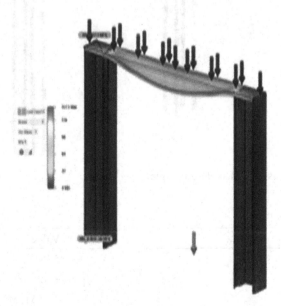

Figure 89.9 Von Mises stress analysis on 4 mm sheet.

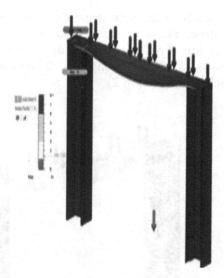

Figure 89.10 FOS analysis on 4 mm sheet.

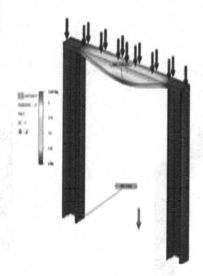

Figure 89.11 Displacement analysis on 4 mm sheet.

Table 89.4: Results of Case 2 (iteration 1).

Variables	Value
Load	5000N
Displacement	5.88 mm
Von mises stress	152.5 MPa
Factor of safety	1.35 Min

The above results for the study of the new structure with 4 mm thickness of top sheet show the total Von mises stress, factor of safety, and displacement (as shown in Table 89.4). Here, after the application of 5KN load, total displacement is observed to be 5.88 mm and the von mises stress as 152.5 MPa along with the minimum safety factor of 1.35.

For top sheet – 5 mm thickness (Iteration 2)

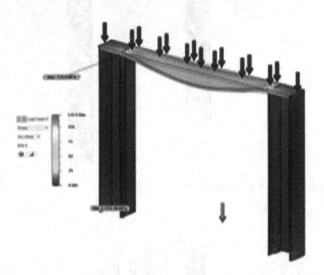

Figure 89.12 Von Mises stress analysis on 5 mm sheet.

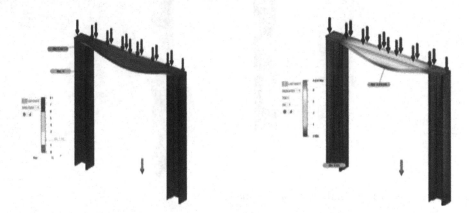

Figure 89.13 FOS analysis on 5 mm sheet.

Table 89.5: Result of case 2 with 5 mm thickness (iteration 2).

Variables	Value
Load	5000N
Displacement	4.82mm
Von mises stress	120.9 MPa
Factor of safety	1.72 Min

The results for the study of the new structure with 5 mm thickness of top sheet show the Von mises stress, factor of safety, and displacement is shown in Table 89.5. Here, after the application of 5KN load, total displacement is observed to be 4.82 mm and the von mises stress as 120.9 MPa along with the minimum safety factor of 1.72. Displacement analysis on 4 mm sheet is shown in Figure 89.11. Von Mises stress analysis and FOS for 5 mm sheet is shown in Figures 89.12–89.14. Von Mises stress analysis and FOS, and Displacement analysis for 6 mm sheet is shown in Figures 89.14–89.16 respectively.

For top sheet – 6 mm thickness (Iteration 3)

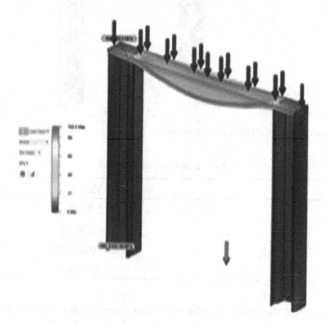

Figure 89.14 Von Mises stress analysis on 6 mm sheet.

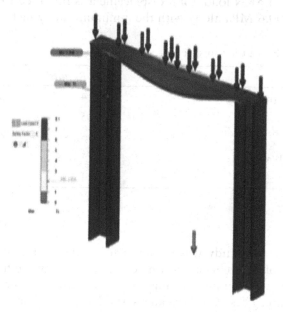

Figure 89.15 FOS analysis on 6 mm sheet.

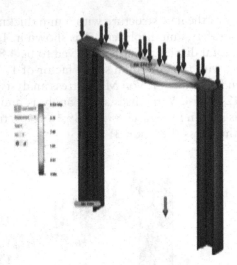

Figure 89.16 Displacement analysis on 6 mm sheet.

Table 89.6: Result of Case 2 with 6 mm thickness (iteration 3).

Variables	Value
Load	5000N
Displacement	4.05mm
Von mises stress	100.6 MPa
Factor of safety	2.05 Min

The above results for the study of the new structure with 6 mm thickness of top sheet show the Von mises stress, factor of safety, and displacement as shown in Table 89.6. Here, after the application of 5KN load, total displacement is observed to be 4.05 mm and the von mises stress as 100.6 MPa along with the minimum safety factor of 2.05.

Table 89.7: Result of FEA of the top sheet for varied thickness.

Variables	Results		
Thickness to top sheet (in mm)	4	5	6
Load (in KN)	5	5	5
Displacement (in mm)	5.88	4.82	4.05
Von mises stress (in MPa)	152.5	120.9	100.6
Factor of safety	1.35	1.72	2.05

Discussion

The main objective of this study was to optimize the design for the enhancement of the structural properties of the vehicle wash machine. CAE software has served the purpose very well in this study. A complete change in the structure has been made by extensive usage of CAD software. These CAD packages also include an analysis module, using their further analysis of the machine structure for safety, and domains like static structure analysis have been performed and the results were found to be astounding during the entire study.

As far as safety is concerned, the old machine structure has been tested several times in the past and met the minimum safety guidelines and it was being produced and sold for the past few decades. During the CAE analysis of the old machine structure, the applied load was kept constant (5 KN) and it was intentionally kept higher than the actual load values, due to the fluctuating changes in the accessories as per the requirements of the customer. For the old structure, the stress value was found to be 130 MPa and for the new structure, the thickness of the top sheet was varied a little, and several tests were performed for each variation so that an optimum value of thickness can be obtained. The optimum thickness value for the top sheet is decided based on the factors of safety. From Table 89.7, it can be observed that for sheet thickness of 4 mm, the safety factor lies below 1.5, this implies that the selected value is less reliable in terms of safety. Whereas, for thickness values 5 mm and 6 mm the safety factor is above 1.5 and these results comply with the safety guidelines. Since the sheet of thickness value of 6 mm is having more safety factors than that of 5 mm thick, it is more likely to be the optimum value. However, since the objective was to keep the machine weight and cost at a minimum, the 5 mm thickness sheet is proved to be more appropriate as it lies in the permissible limit of safety factor as well as the total displacement.

The optimum stress value was found to be 120.9 MPa which lies within the permissible limit. The total displacement was found at 1.52 mm for the old machine structure and total displacement found to be 4.82 mm after design modifications. The reason behind the reduction of stress value and increase in total displacement could be due to the weight reduction, as the number of additional components has been reduced. Concerning the factor of safety, the value for the previous structure was 2.2, whereas, for the new structure, the safety factor is found to be 1.72, which is again within the permissible limits as there is no other dynamic loading condition present in the machine.

The present study renders several implications to both industries and academia. To the industries, the implications include the development of a new optimized structural model of an automatic vehicle wash machine that could save a significant amount of material, costs, as well as time while ensuring optimal performance, therefore, managers could use this study as a roadmap to produce better products at a much faster and cheaper rate that would lead to the enhancement of customer satisfaction, as well as a higher competitive performance in the market.

Owing to the paucity of research for optimizing the structure of automatic vehicle wash machines, the present study fills the research gap and makes significant contributions by using an integrated approach including several techniques for design optimization and lead time reduction. The study also suggests the use of advanced manufacturing processes which could help in eliminating redundant processes and excessive labor. Furthermore, the present study is the first of its kind that effectively utilizes the DFMA approach to reduce the number of components of the structure of an automatic vehicle wash machine without affecting the structural integrity.

Conclusion

This study focuses on the design optimization of automatic vehicle wash machines by investigating the equivalent stress values and performing a safety analysis of the structural components for varied thicknesses. Based on the analysis of the data, the following conclusions can be made. The equivalent stress and total deformation in the new optimized design lie within the acceptable limit as compared to the traditional design. Therefore, the new optimized structure can be accepted as a replacement for the traditional design of the structure. Adding flanges to a metal sheet can boost the overall strength of the component whereas, the total deformation can be reduced by several times that of the original value. The complete construction of a vehicle wash machine can be made of mild steel (MS) and galvanized iron (GI) sheets of the desired thickness depending on the loading conditions, which can serve the purpose as a structural member, instead of a metal frame/skeleton made by welding square pipe.

By introducing the concept of modularity in the structural design, the additional material needs and labor cost can be reduced, since this helps to reduce the design and fabrication requirements. In the future, a detailed analysis of cost and weight optimization is needed to be done to introduce a superior product and to gain higher profit in the business.

References

1. Antony, K. M., and Arunkumar, S. (2020). DFMA and Sustainability analysis in product design. *Journal of Physics: Conference Series*, 1455(1), 12028. https://doi.org/10.1088/1742-6596/1455/1/012028.
2. Bathe, K. J. (1996). Finite Element Procedures. Englewood Cliffs New Jersey. http://www.amazon.com/Finite-Element-Procedures-Part-1-2/dp/0133014584.
3. Charnsamorn, C., and Thanakitirul, P. (2021, November). The Problem-Based Learning by Using Computer-Aided Engineering: The Heatsink Design Competition. *In 2021 6th International STEM Education Conference (iSTEM-Ed)*, 1–4. IEEE.
4. Desai, P. M., Pathak, B. C., Patel, V. A., and Rana, P. B. (2013). Design, analysis and optimisation of body flange and cover flange using finite element analysis. *International Journal of Mechanical Engineering and Technology*, 4(5), 81–87.
5. Elsayed, Y., Vincensi, A., Lekakou, C., Geng, T., Saaj, C. M., Ranzani, T., Cianchetti, M., and Menciassi, A. (2014). Finite element analysis and design optimization of a pneumatically actuating silicone module for robotic surgery applications. *Soft Robotics*, 1(4), 255–262. https://doi.org/10.1089/soro.2014.0016.
6. Jain, V., and Sharma, P. (2016). A study of strength analysis in sheet metal by bending process : a. *Ijerst*, 5(9), 55–60.
7. Kaurase, S., and Chopade, M. P. (2016). Design FE analysis of sheet metal while bending. *International Journal of Engineering Research and Technology*, V5(10), 90–99. https://doi.org/10.17577/ijertv5is100089.
8. Kolbasin, A., and Husu, O. (2018). Computer-aided design and computer-aided engineering. *MATEC Web of Conferences*, 170, 1–6. https://doi.org/10.1051/matecconf/201817001115.
9. Leiva, J. P. (2011). Structural optimization methods and techniques to design efficient car bodies. In International Automotive Body Congress (IABC), 2011. http://bodyandassembly.com/wp-content/uploads/2012/01/G-Edited-Leiva-Paper.pdf.
10. Magdum, P. R., Kamble, V. A., and Balwan, A. R. (2019). Design and development of a bandsaw machine roller bracket for weight optimization. *International Research Journal of Engineering and Technology (IRJET)*, 6(7), 48–51.
11. Mandliya, D., Agrawal, Y., and Seshagiri Rao, G. V. (2014). International journal of engineering sciences & research technology design optimization of stationary platen of plastic injection molding machine using FEA. *International Journal of Engineering Sciences and Research Technology*, 3(5), 633–639.

12. Patil, S., and Kulkarni, S. A. (2013). Optimization of crosspiece of washing machine. *International Journal of Research in Engineering and Technology*, 02(03), 389–392. https://doi.org/10.15623/ijret.2013.0203027.
13. Rajamani, M. R., and Punna, E. (2020). Enhancement of design for manufacturing and assembly guidelines for effective application in aerospace part and process design. SAE Technical Papers, 2020. https://doi.org/10.4271/2020-01-6001.
14. Sabet, S. M. M., Marques, J., Torres, R., Nova, M., Gomes, L., and Nobrega, J. (2015). Design of a modular rollover carwash machine structure. *Machine Design*, 7(4), 129–136.
15. Sabet, S. M. M., Marques, J., Torres, R., Nova, M., and Nóbrega, J. M. (2016). Design of a drying system for a rollover carwash machine using CFD. *Journal of Computational Design and Engineering*, 3(4), 398–413. https://doi.org/10.1016/j.jcde.2016.07.001.

90 Drag and heat reduction on hypersonic re-entry vehicle using combination of spike and counter jet

Kanala Harshavardhan Reddy[a] and Nagaraja S. R.[b]
Department of Mechanical Engineering, Amrita School of Engineering, Bengaluru, Amrita Vishwa Vidyapeetham, India

Abstract: Drag and Heat flux reduction is a major challenge in the design of hypersonic vehicles. Spikes and counter flowing jets have been used to reduce the drag and heat flux. Spikes predominantly reduce the drag, whereas jets reduce heat flux. In this paper a combination of spike and counter flow jet is numerically studied for flow Mach number of 8 using ANSYS Fluent. Spalart-Allmaras model coupled with implicit AUSM solver is used for solving RANS equation. The simulations are done for different jet pressure ratios and jet diameters. The results show the proposed design reduces drag and heat flux over the entire surface. The introduction of jet increases the size of the recirculation zone and moves the separation shock further away from the blunt body. As the jet pressure ratio increases jet boundary expands further and thereby drastically reducing the peak heat flux on the aerodisk and on the blunt body.

Keywords: Blunt body, counter flowing jet, drag, heat flux, Spike.

Introduction

Re-Entry Vehicle (REV) cruising at hypersonic speed experiences high drag and heat flux because of high pressure and aerodynamic heating on the surface of the blunt body. This induces large heating loads and pressure. These heating loads and pressure can damage the vehicle and hence it is required to reduce these loads. Wave drag contributes more to the total drag on the surface of a body and increases as speed of vehicle increases [1]. The heating loads can be reduced using two types of Thermal Protective Systems (TRP)-Passive and Active systems. Passive systems protect surface of REV by way of thermal protective materials, concentrated energy deposition or by ablating on the surface. These passive mechanisms reduce about 20% heat flux, which is comparatively less. These systems cannot be reused and also they are expensive. Some of the Active systems are, spike-a mechanical device attached to the main body and counter flowing jet-a high velocity jet from the stagnation point of main body pushing free stream fluid further away. These methods alter fluid flow and reduces drag forces and heat flux.

[a]kanalahvr@gmail.com, [b]sr_nagaraja@blr.amrita.edu

Spike

Reduction of drag and heat flux on the surface of reentry vehicles is being studied since the 1940s. The heat flux can be reduced by using blunt nose instead of sharp nose, however the drag increases with increased bluntness. Aerospikes have been used to reduce drag on the blunt surface. Crawford [2] in his research studied the shape and nature of flow over spike body with Mach number of 6.8 over blunt surfaces. Variation of flow around a surface is studied by varying spike length to diameter ratio (L/D) up to 3. It is reported that there is no change in flow field for L/D above 2, but aerodynamic heating on the surface reduces. These aerospike designs were further modified to aerodisk where in a disk is attached to the front of aerospike. Rolf Guenther and Peter Reding [3] in their experimental investigation have found that attachment of aerodisk to the front of spike of shorter length than normal aerospike reduces the drag. Kalimuthu et al. [4] have done experimental studies on different aerospikes like conical, hemispherical, flat plate at Mach number 6 for varying angle of attacks and L/D ratios varying from 0.5 to 2. They have found that hemispherical shaped spike with an L/D ration of 1.5 and 2 reduce drag forces on the blunt surfaces. According to them the factors influencing drag forces on the blunt surface are the L/D ratio, shape of spike nose and an angle of attack. G. Gopala Krishnan et al. [5, 6] have analyzed drag and heat flux reduction using single and twin spikes configurations along with counter flowing jets. They have reported maximum reduction of 59% in the drag coefficient and 44% reduction in peak heat flux. Motoyama et al. [7] have found that addition of aerodisk also reduces heat flux on the blunt body. Sebastian et al. [8] in their numerical simulation have found that for increased angle of attack drag coefficient increases. Ahmed and Qin [9] have stated that optimum diameter of aerodisk is inversely proportional to the length of the spike. They found that as the spike length increases, aerodisk size decrease for stable flow on the spike surface. As spike is parallel to the free stream, pressure drag and surface heat flux does not change but the skin friction drag varies. This variation is observed in three phases [10]. Friction drag increases first then decreases and then slowly increases. Skin friction on the spike surface increases as the ratio of diameter of aerodisk to diameter of the main body (d/D) increases, because of the shift in reattachment point. As the d/D increases reattachment point shifts towards the blunt surface. As the d/D increases recirculation zone behind the spike head increases thus causing increase of skin friction drag. Negative shear stress is attained behind the reattachment point thus causing the reduction of drag forces. Third phase is because of the reduction of maximum negative shear stresses. Sai Krishna Mohan et al. [11] have studied flow separation on blunt body with a spike with aerodisk at different Mach numbers. Srikanth et al. [12] have used a secondary spike for drag and heat flux reduction. The effectiveness of secondary spike depends on its location and freestream conditions.

Counter Flowing Jet

Counter flowing jet is a fluid jet injected in the opposite direction to the freestream fluid. Warren et al. [13] have conducted experiments for a flow Mach number of 5.8 with Nitrogen and Helium jets. It was found that swirl injection increases Reynolds number of the flow which is not desirable. The fluid injected acts like an extended spike if the pressure ratio is higher than the critical value. Finley [14] observed flow field around blunt surface with counter flowing jet categorizing the flow field as steady and unsteady flow. P. Visakh et al. [15] have optimized counter flowing jet angle for a blunt body at a particular angle of attack. For a blunt body at 6-degree angle of attack, the maximum heat flux reduction occurs at 9 degrees of counter flowing jet angle. Shen and Liu [16] in their experimental study observed one more regime of steady flow which occurs at low pressure ratio. Three types of flow fields around the blunt body are observed when a counter flow jet is used. Steady SPM at low pressure ratios for hypersonic flows, unsteady LPM and steady SPM at high pressure ratio. The heat flux decreases as the jet injection pressure is above critical value. The maximum pressure and heat flux occurs at the reattachment point. Also as the jet injection pressure increases the shock standoff distance also increases. Robert J Mc Ghee [17] considered nozzle thrust as sum of drag forces and jet forces same is used by Jaecheong Lee et al. [18] to study characteristics of counter flowing jet. In their numerical simulation studied about the effect of nozzle thrust, free stream pressure and Mach number on the drag reduction. For increasing nozzle thrust the drag reduces and then increases. They also studied the variation of total drag with variation of free stream Mach number and the nozzle thrust. Total drag variation follows same trend as of nozzle thrust. M Venkat Prashanth et al. [19] have used counter flowing micro jets to reduce the drag and heat flux using. Z. Eghlima and K. Mansour [20] have used a combination of spike and counter jet to reduce drag of a blunt body. They have studied drag reduction for different L/D ratios of spike and pressure ratios of the jet. Jie Huang and Wei-Xing Yao [21] have used a novel concept of spike and rear opposing jet to reduce drag and heat flux. The rear opposing jet is introduced at the point of attachment of the spike to the spherical body. The rear opposing jet pushes the reattachment shock away from the main body. Recently B. John et al. [22] have studied equivalence between spike and counter flowing jet techniques in reducing drag and heat flux.

From the available literature it can be concluded that spikes and counter flowing jets substantially reduce drag and heat flux respectively. Having a spike on the main body reduces both drag and heat flux on it, however it increases the heat flux on the spike substantially. Introducing a counter flowing jet from the spike head can reduce the heat flux on it. Hence in this study effect of combination of spike and jet are studied for different jet pressure ratios and jet sizes. These numerical studies are done using ANSYS Fluent with flow Mach number of 8.

Computational Methodology

In this work combination of spike and counter-flowing jet is studied at zero angle of attack for drag and heat flux. The base diameter (D) of the blunt body is same as that used by Finley [14] in their experimental study. The simulations are done for a L/D ratio of 1.5, where L is the length of the spike. The spike diameter is 2.6 mm. The diameter (d) of aerodisk is directly proportional to the base diameter (D) and d/D ratio is taken as 0.16. The d/D ratio is chosen such that a low pressure expansion wave is created at the shoulder of the spike and a low reattachment pressure is attained. Reynold's Averaged Navier Stokes equations are solved using ANSYS Fluent. Face meshing is done by maintaining first layer thickness of y+ =1. The density based solver is used for solving RANS turbulent equation. In this study strain/vorticity Spalart–Allmaras one equation turbulence model is used to solve turbulence model as proposed by researchers that for high velocity flows one equation model gives good results and also reduces time for solving. As proved by Ahmed [10] axisymmetric model results are in accordance with full 3D model, which reduces simulation time. In the present study an axisymmetric model is used. The geometry of the complete model is as shown in Figure 90.1. Free stream conditions at an altitude of 20 km are used with a Mach number of 8. The freestream pressure is 5529Pa and temperature 216.7K. The pressure far field conditions are used at the inlet. The opposing jet boundary conditions are Mach number = 0.8, Jet fluid = air, Jet fluid temperature = 300K, and the temperature of the wall and spike is 300K. Counter jet from the aerodisk head is injected against the free stream with different pressure ratios. The jet diameters used are 1 mm, 1.5 mm and 2 mm.

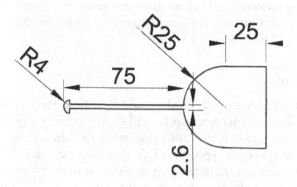

Figure 90.1 Geometry of the computational model.

The simulation model is validated for blunt body with an empirical equation for surface pressure from modified Newtonian (equation 1). Figure 90.2 shows the variation of pressure over blunt body from the present study and from the Newtonian theory. It is observed that the simulation results closely follow the results from the Newtonian theory.

$$\frac{p}{p_{of}} = \cos^2 \theta + \left(\frac{p_\infty}{p_{of}} \right) \sin^2 \theta \tag{1}$$

where p is the surface pressure, p_∞ is the free stream pressure and p_{of} is the freestream pitot pressure.

Proposed model is also validated for peak heat flux on the blunt body using Chapman's [23] empirical equation. Stagnation heat flux found using Chapman equation is 3.2×10^6 W/m². This matches with the value obtained from numerical solution (Figure 90.4), when near first layer thickness is maintained by with y+ = 1.

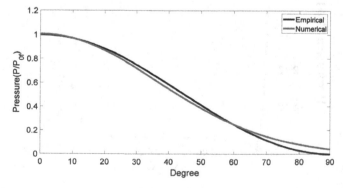

Figure 90.2 Validation of static pressure with modified newtonian empirical equation.

Results and Discussions

This section contains the simulation results are for different jet diameters and jet pressure ratios (PR). The PR is the ratio of stagnation pressure of the jet to the stagnation pressure of freestream fluid i.e. $PR = \dfrac{Stagnation\,Pressure\,of\,Jet\,P_{oj}}{Stagnation\,Pressure\,of\,freestream\,P_{o\infty}}$. The drag coefficient, variation of heat flux and flow field dynamics are studied for two different cases, first case with constant pressure ratio (PR) of 0.1 for different jet diameters of 1 mm, 1.5 mm and 2 mm, and the second case with constant jet diameter of 1 mm for different pressure ratios (PR) of 0.008, 0.01, 0.05, 0.075 and 0.1.

Pressure and Heat Flux

The spike on a blunt body reduces both drag and heat flux. Figure 90.3 shows variation of pressure on the blunt body without spike and on the entire body with spike. It can be observed that there is drastic reduction of pressure on the blunt body and hence decrease in drag. However, pressure on the spike head i.e. aerodisk increases. Since the surface area of aerodisk is not very large compared to that of blunt body, the increase in drag associated with this is negligible. The reduction in drag in this case is 53.22%. Introduction of spike on the blunt body does not decrease the heat flux significantly (Figure 90.4). The reduction of peak heat flux is only 12%. Introduction of spike reduces the drag substantially, but reduction in heat flux is very small. However, heat flux increases drastically on the aerodisk. Hence there is need to reduce the increased heat flux on the aerodisk. Many researchers have used counter flowing jets to reduce the heat flux on the surface of the blunt bodies [15–19]. A counter jet from the aerodisk can reduce the heat flux on the aerodisk and also on the blunt body. So a combination of spike and jet can reduce both drag and heat flux substantially.

Figures 90.5 and 90.6 respectively show pressure and heat flux over the spike and blunt body with counter jet from aerodisk for a PR of 0.075. The size of the jet in this case is 1 mm. The counter flowing jet coming from aerodisk pushes the bow shockwave away from aerodisk and thereby drastically reducing the pressure and heat flux on aerodisk. Due to

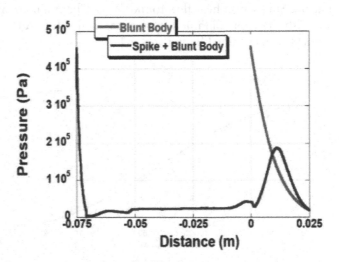

Figure 90.3 Comparison of pressure on spike and Blunt Body without Jet.

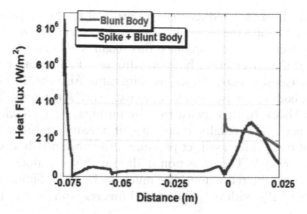

Figure 90.4 Comparison of heat flux on spike and Blunt Body without Jet.

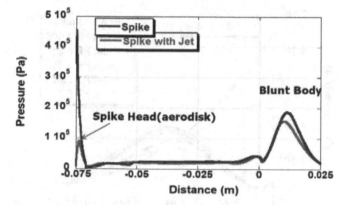

Figure 90.5 Comparison of pressure with and without Jet.

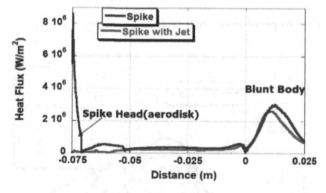

Figure 90.6 Comparison of heat flux with and without Jet.

this re attachment point on the blunt body is also moved further and pressure and peak heat flux on blunt body are also reduced. This results in decrease of drag. Hence using counter flowing jet from the aerodisk not only reduces pressure and heat flux on aerodisk, it also reduces these quantities on the blunt body. The cold air coming from jet completely surrounds the surface of the aerodisk and hence reduces heat flux on it. The heat flux on the aerodisk reduces by 98% for PR of 0.075 compared to the case without jet. The percentage reduction in drag and heat flux are calculated by comparing with corresponding values for blunt body without spike and jet.

The amount of the heat flux reduced is proportional to the jet pressure ratio PR. The simulations are done for different pressure ratios with constant size of the jet. Table 90.1 shows percentage reduction in drag and heat flux. Figures 90.7 and 90.8 show comparison of pressure and heat flux over blunt body for different PRs. It can be observed that the peak heat flux decreases for increased jet pressure ratio. Also the heat flux over the entire surface of the blunt body decreases as the jet pressure ratio increases. As PR increases, the jet pushes the bow shock further away, thereby pushing the reattachment point further away on the blunt body. This results in decrease of pressure and heat flux over the entire surface of the blunt body. Increased jet pressure also substantially decreases the amount of heat flux on the aerodisk. The reduction in drag with spike alone is 53.22%, however with introduction of jet the reduction is as much as 76.85%. Similarly, the reduction in peak heat flux is only 12% with only the spike, however with jet it is as much as 34.5%.

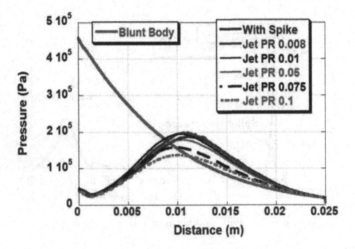

Figure 90.7 Comparison of pressure on blunt body.

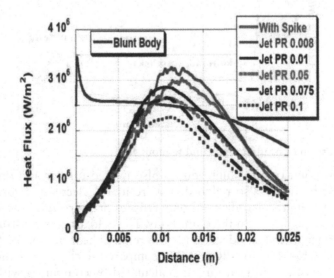

Figure 90.8 Comparison of heat flux on blunt body.

Table 90.1: Percentage reduction of drag and heat flux for different PRs.

Pressure ratio (PR)	only Spike	0.08	0.01	0.05	0.075	0.1
Reduction in Drag (%)	53.22	52.85	52.95	56.61	59.8	76.85
Reduction in Peak Heat Flux (%)	12.00	5.81	17.2	23.66	23.5	34.5

The diameter also affects the drag and heat flux. Simulations were done for varying size of the jet with constant pressure ratio. The results for PR = 0.1 and jet diameters of 1 mm, 1.5 mm and 2 mm are tabulated in Table 90.2. As the size of the jet increase there is substantial decrease in both drag and peak heat flux. As the size of the jet increases the mass flux of coolant gas also increases. The decrease in heat flux is very substantial with increased jet size. For PR = 0.1 and jet of 1 mm diameter the reduction in heat flux is 34.5%, whereas for same PR for jet of 2 mm diameter, it is 82.51%. The reduction is drag for these two cases is 76.85% and 85.69%. Hence the reduction in heat flux is strongly dependent on the mass flux of the counter flowing jet.

Table 90.2: Percentage reduction of drag and heat flux for different jet sizes.

	Only Spike	*1 mm jet*	*1.5 mm jet*	*2 mm jet*
Reduction in Drag (%)	53.22	76.85	80.22	85.69
Reduction in Peak Heat Flux (%)	12.00	34.5	75.40	82.51

Variation of Flow Field

Figure 90.9 represents Mach contours around the blunt body with spike and without the counter flowing jet. It also represents the complete flow domain. Figure 90.10 represents the velocity stream lines for a jet pressure ratio of 0.1. As the PR of the counter jet is above critical value of PR, jet coming from the aerodisk pushes the free stream and changes the shape of the shock wave formed. The size of the recirculation zone increase with introduction of jet. The separation shock on the spike moves further away from the blunt body

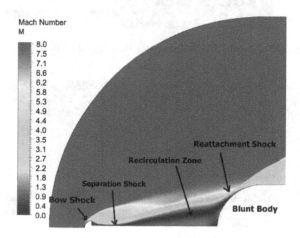

Figure 90.9 Mach contours around spike and blunt body without counter flowing jet.

when counter flowing jet is introduced on the aerodisk. Figures 90.11 and 90.12 represent the aerodisk part of the model considered. The Counter flowing jet injected with different diameters changes the flow structure of free stream. The mass flux of the jet increases for increased diameter of the jet. These figures also show the variation of shape of Mach disk. The Mach disk moves further away from the aerodisk for increased jet pressure ratio and diameter of the jet. The jet boundary expands further and thereby drastically reducing the peak heat flux on the aerodisk and also on the blunt body. From these Mach contours it is observed that as the diameter of the counter jet nozzle increases, the size of the Mach disk increases. Mach disk formed for these conditions is stable Short Penetration Mode (SPM).

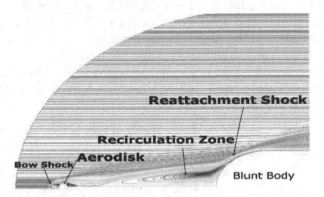

Figure 90.10 Velocity stream lines around spike and blunt body with counter flowing jet.

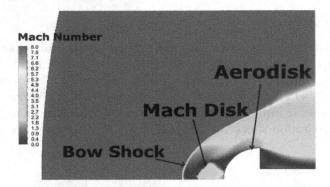

Figure 90.11 Flow field around aerodisk for counter flowing jet of diameter 1 mm.

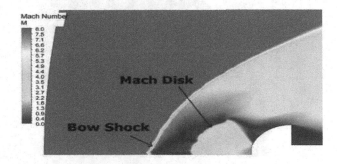

Figure 90.12 Flow field around aerodisk for counter flowing jet of diameter 2 mm.

Conclusions

The effect of combination of spike and counter flowing jets on drag and heat transfer over blunt bodies in hypersonic flow is studied. Simulations were done for varying jet pressure ratios and diameters of the jet. It is found that as the jet pressure ratio increases both drag and heat flux on the blunt body decreases. Having spike alone on the blunt body reduces the drag by 53.22% and peak heat flux by 12%. The counter flowing jet from the aerodisk reduces the drag by 76.85% and peak heat flux on the blunt body by 34.5% for a jet of diameter 1 mm and jet pressure ratio of 0.1. The jet not only reduces the peak heat flux, but it also reduces heat flux over the entire surface of the blunt body. The size of the jet also has an effect on the drag and heat flux. As the size of the jet increases there is substantial decrease in the heat flux. This decrease is up to 82.51% for a jet of 2 mm with jet pressure ratio of 0.1.

References

1. Bushnell, D. M. (2004). Shock wave drag reduction. *Annual Review of Fluid Mechanics,* 36, 81–96.
2. Crawford, D. H. (1959). Investigation of the Flow over a Spiked-Nose Hemisphere Cylinder at a Mach Number of 6.8. National Aeronautics and Space Administration.
3. Guenther, R. A., and Reding, J. P. (1977). Fluctuating pressure environment of a drag reduction spike. *Journal of Spacecraft and Rockets,* 14(12), 705–710.
4. Kalimuthu, R., Mehta, R. C., and Rathakrishnan, E. (2017). Investigation of aerodynamic coefficients at Mach 6 over conical, hemispherical and flat-face spiked body. *The Aeronautical Journal,* 121(1245), 1711–1732. doi: 10.1017/aer.2017.100.
5. Gopala Krishnan, G., Akhil, J., and Nagaraja, S. R. (2017). Drag reduction for hypersonic re-entry vehicles. *International Journal of Mechanical Engineering and Technology,* 8(10), 878–885.
6. Gopala Krishnan, G., Akhil, J., and Nagaraja, S. R. (2015). Heat transfer analysis on hypersonic re-entry vehicles with spikes. In Proceedings of the 23rd National Heat and Mass Transfer Conference and 1st International ISHMT-ASTFE Heat and Mass Transfer Conference IHMTC2015.
7. Motoyama, N., Mihara, K., Miyajima, R., Watanuki, T., and Kubota, H. (2001). Thermal protection and drag reduction with use of spike in hypersonic flow. In 10th AIAA/NAL-NASDA-ISAS International Space Planes and Hypersonic Systems and Technologies Conference, 1828.
8. Sebastian, J. J., Suryan, A., and Kim, H. D. (2016). Numerical analysis of hypersonic flow past blunt bodies with aerospikes. *Journal of Spacecraft and Rockets,* 53(4), 669–677. doi: 10.2514/1.A33414.
9. Ahmed, M. Y. M., and Qin, N. (2010). Drag reduction using aerodisks for hypersonic hemispherical bodies. *Journal of Spacecraft and Rockets,* 47(1), 62–80. doi: 10.2514/1.46655.
10. Ahmed, M. Y. M., and Qin, N. (2011). Recent advances in the aerothermodynamics of spiked hypersonic vehicles. *Progress in Aerospace Sciences,* 47(6). 425–449.
11. Sai Krishna Mohan, K. V. N. B., Bandaru, K., and Prabhudev, B. M. (2018). Numerical studies on effect of changing mach number on aero disk model on flow contours and drag force. In IOP Conference Series: Materials Science and Engineering, (Vol. 377), 012096 doi:10.1088/1757-899X/377/1/012096.
12. Sreekanth, N., Akhil, J., and Nagaraja, S. R. (2016). Design and analysis of secondary spike on blunt head. *Indian Journal of Science and Technology,* 9(45), 1–8. doi: 10.17485/ijst/2016/v9i45/104649.

13. Warren, C. H. E. (1960). An experimental investigation of the effect of ejecting a coolant gas at the nose of a bluff body. *Journal of Fluid Mechanics,* 8(3), 400–417. doi: 10.1017/S0022112060000694.

14. Finley, P. J. (1966). The flow of a jet from a body opposing a supersonic free stream. *Journal of Fluid Mechanics* 26(2), 337–368. doi: 10.1017/S0022112066001277.

15. Visakh, P., Akhil, J., and Nagaraja, S. R. (2018). Effect of counter flowing jet on heat transfer and drag in hypersonic re-entry vehicle. In IOP Conference Series: Materials Science and Engineering, (Vol. 377), 012163. doi:10.1088/1757-899X/377/1/012163.

16. Shen, B. X., and Liu, W. Q. (2020). Thermal protection performance of a low pressure short penetration mode in opposing jet and its application. *International Journal of Heat and Mass Transfer,* 163, 120466.

17. Mc Ghee, R. J. (1971). Effects of a retro nozzle located at the apex of a 140 deg blunt cone at mach Numbers of 3.00, 4.50, and 6.00. Tech. rep., NASA TN-D-6002.

18. Lee, J., Lee, H. J., and Huh, H. (2020). Drag reduction analysis of counter flow jets in a short penetration mode. *Aerospace Science and Technology,* 106, 106065.

19. Prashanth, M. V., and Nagaraja, S. R. (2019). Effect of micro jet on heat and drag in hypersonic flows. In AIP Conference Proceedings. 2200. 020066: https://doi.org/10.1063/1.5141236.

20. Eghlima, Z., and Mansour, K. (2017). Drag reduction for the combination of spike and counter flow jet on blunt body at high Mach number flow. *Acta Astronautica,* 133, 103–110.

21. Huang, J., and Yao, W. X. (2020). Parameter study on drag and heat reduction of a novel combinational spiked blunt body and rear opposing jet concept in hypersonic flows. *International Journal of Heat and Mass Transfer,* 150, 119236.

22. John, B., Bhargava, D., Punia, S., and Rastogi, P. (2021). Drag and heat flux reduction using counterflow jet and spike—analysis of their equivalence for a blunt cone geometry at Mach 8. *Journal of Applied Fluid Mechanics,* 14(2). 375–388.

23. Chapman, G. T. (1964). Theoretical laminar convective heat transfer & boundary layer characteristics on cones at speeds to 24 km/s. NASA TN D-2463.

91 Free vibration analysis of cylindrical shells with reference to the modal orthogonal property using classical shell theories

K. Rajasekhar Reddy[1,a] and N. Srujana[2,b]

[1]Department of Mechanical Engineering, Vardhaman college of Engineering, Hyderabad, India

[2]Department of Civil Engineering, Vardhaman college of Engineering, Hyderabad, India

Abstract: Multiple shell theories were proposed to estimate the natural frequencies of thin circular cylindrical shells based on in-plane displacements, transverse shear components, and moment resultants. Solutions to these domain equations of motion are assumed to satisfy the boundary conditions of the shell members. An attempt has been made in this paper to discover the orthogonality limitations of the modal solutions assumed in the classical shell theories. Fundamental frequencies are estimated using frequency equations of membrane theory, DMV theory, and love theories. Mode shapes are calculated using the eigenvalue problem of the cylindrical shell equations of motion. Orthogonality check has been performed on first two normalized modes of the shell system. A cylindrical shell having simply support on both sides is considered in the study. The fundamental frequencies obtained from the shell theories are validated through the numerical model. The solutions used in classical shell theories confirmed the orthogonal property of modes.

Keywords: Cylindrical shells, Donnell-Mushtari-Vlasov theory, eigenvalue problem, love theory, natural frequency, orthogonal property.

Introduction

The employment of non-linear methods is essential to identify the behavior of cylindrical structures based on various applications. As far as the mechanical applications are concerned, the expected loads are from free vibration and occasionally time-varying random loads. One of the important free vibration characteristics i.e., fundamental frequency, must be estimated for the structure to proceed with further non-linear dynamic analysis to study the behavior and design of the shell structures as well.

Several shell theories are developed over the decades based on Love's classical linear elastic approximations (1888) [8] and (1944) [9] and Kirchhoff's hypothesis. As Love's theory has inconsistencies related to differential operator matrix, the need for a mathematically rigorous two-dimensional set of the shell equations employing the Kirchhoff–Love assumptions led to different versions of the first-order approximation theories. Timoshenko's theory (1921), [13] of thin shells was very close to the Love theory. Higher order approximation has also been developed in which the thinness assumption is delayed in the derivation and established by retained terms of strain constitutive relations related

[a]rajasekhar730@vardhaman.org, [b]n.srujana@vardhaman.org

to thickness coordinate. Flugge (1960) [5], independently developed the second-order approximation theory of circular cylindrical shells because of the application of Kirchhoff hypotheses together with the small-deflection assumption. Donnell et al (1985) [4] independently developed a simplified engineering theory of thin shells of general form. Due to their simplicity, the governing equations of this theory were found to be extremely convenient for solving many engineering shell problems. Apart from the Kirchhoff–Love hypotheses, some additional assumptions that simplify the strain–displacement relations, equilibrium, and compatibility equations were used in deriving these equations.

Based on the classical theories of thin shells, significant research work is available on free vibration analysis of cylindrical shells having various boundary conditions. Sharma et al (1972) [10] and (1974) [11] discussed the rising natural frequency based on shell geometry effects the ring eccentricity of cylindrical shells. The lower natural frequencies are seen for the shells of large length to radius ratios. Xuebin et al (2006) [14] proposed a new method for calculating the natural frequencies based on the Flugge shell theory. Coupled polynomial eigenvalue problem is developed and solved using the classical dynamic approach for the cylindrical shells. Zhang et al (2001) [16] and Xuebin et al (2008) [15] proved to determine natural frequencies of circular cylindrical shells using wave propagation approach based on the Flugge theory. Tang et al (2016) [12] developed an analytical procedure to calculate the natural frequencies using the reverberation-ray matrix based on the Flugge shell. The exact solutions of the waveform function in the axial direction and circumferential directions are estimated. Alujevic et al (2017) [1] studied the natural frequencies and mode shapes of rotating cylindrical shells with both the ends free by considering flexible foundations in the sense of a radial and circumferentially distributed stiffness. Aruna et al (2016) [2] has performed mesh convergence studies on natural frequencies of cylindrical shells for the different length to radius. Cao et al (2018) [3] applied an improved Fourier series method to identify the vibration characteristics of cylindrical shell-circular plate. Hong Liang and Hao-Jie (2013) [7] presented an analytical study on forced vibration of the cylindrical shell to optimize the engineering functionality. Gao et al (2020) [6] presented method to calculate the natural modes and frequencies of the thin orthotropic circular cylindrical shell by considering the phase coincidence condition of the wave modal shapes. Most of the previous research work is limited to the identification of natural frequencies of the cylindrical shell structures using improvised classical cylindrical shell theories. However, the authors emphasize research work related to mode shapes and the modal orthogonality are needed for the effective estimation of the fundamental modes.

The present work has been carried out on the response solutions assumed for various classical shell theories which are assumed to satisfy the boundary conditions of the cylindrical shells. The objective of the study is i) to identify the natural frequency be relevant to the orthogonal modes of the cylindrical shells. ii) to confirm the response solutions assumed in various shell theories must satisfy the orthogonal property of mode shapes in connection with global mass and global stiffness of shell structure. The objectives of the study are accomplished by i) determining the natural frequencies from the frequency equation drawn from domain equations of motion related to various shell theories. ii) identification of mode shapes for the corresponding natural frequencies using eigenvalue solutions. iii) Performing orthogonal property check on two vibrational modes of the cylindrical shell structure in connection with mass and stiffness matrices globally. A sample shell structure of simply supported at both ends is considered in the study to validate the solutions of classical shell theories. The classical theories considered in the present study are membrane theory, Donnell-Mushtari-Vlasov (DMV) theory, and Love theory.

Frequency Solutions of Classical Shell Theories

- Membrane theory: The displacements of the cylindrical shell are defined in u, v and w, in the x, θ and z directions respectively. Governing equation as per the membrane theory is given in Eq.1.

$$\frac{\partial^4 w}{\partial x^4} + 4\beta^4 = \frac{Z}{D} \tag{1}$$

Where $Z = \tilde{m}\dfrac{\partial^2 w}{\partial t^2}$, $D = \dfrac{Eh^3}{12(1-\vartheta^2)}$, $\beta^4 = \dfrac{3(1-\vartheta^2)}{R^2 h^2}$, and w is the transverse displacement

function, $w = \Phi(x)\,f(t)$. $\Phi(x)$ is mode shape in axial x-direction and f(t) is the coordinate in x-direction varying with time. The solution to the domain equation is,

$$\Phi(x) = A \sin (a_n\, x) + B \cos (a_n\, x) + C \sinh (a_n\, x) + D \cosh (a_n\, x) \tag{2}$$

Boundary conditions for simply support at x = 0, deflection in w direction w(0, t) = 0 and the moment $\ddot{w}(0, t) = 0$ and at x = l, deflection in w direction w(l, t) = 0 and the moment $\ddot{w}(l, t) = 0$. Hence the solution to the modal solution is,

$$\Phi(x) = A \sin (a_n\, x) \tag{3}$$

where $a_n = \dfrac{n\pi}{l}$ where n is the nth axial mode where shell tends to vibrate.

$$\text{and } \omega = \sqrt[2]{\frac{\left\{\left(\dfrac{n\pi}{l}\right)^4 + 4\beta^4\right\}D}{\overline{m}}} \tag{4}$$

- Donnell- Mushtari- Vlasov theory (DMV): DMV theory is developed based on the assumptions related to the in-plane displacements of bending strains and shear terms are negligible. Solutions in three directions are assumed to satisfy the boundary conditions of the shell are given in Eq. 5 to 7.

$$u(x,\theta) = \sum_m \sum_n A_{mn} \cos\frac{m\pi x}{l} \cos n\theta \cos \omega t \tag{5}$$

$$v(x,\theta) = \sum_m \sum_n B_{mn} \sin\frac{m\pi x}{l} \sin n\theta \cos \omega t \tag{6}$$

$$w(x,\theta) = \sum_m \sum_n C_{mn} \sin\frac{m\pi x}{l} \cos n\theta \cos \omega t \tag{7}$$

m and n are the number of halfwaves of displacements in axial and circumferential directions of the cylindrical shell. A_{mn}, B_{mn} and C_{mn} are constant. Frequency equation in Eq. 8.

evolved by substituting the assumed solutions in the basic equations of motion proposed by DMV theory. The determinant of the coefficient matrix from the domain equations,

$$\begin{vmatrix} -\lambda^2 - a_1 n^2 + \Omega & a_2 \lambda n & \vartheta \lambda \\ a_2 \lambda n & -a_1 \lambda^2 - n^2 + \Omega & -n \\ \vartheta \lambda & -n & -1 - \lambda^4 \mu - 2\lambda^2 n^2 \mu - n^4 \mu + \Omega \end{vmatrix} = 0 \qquad (8)$$

$$\lambda = \frac{m\pi R}{l}, \ \mu = \frac{h^2}{12R^2}, \ \Omega = \frac{(1-\vartheta^2)R^2}{E}\rho\omega^2, \ a_1 = \frac{1-\vartheta}{2}, \ a_2 = \frac{1+\vartheta}{2}$$

On solving the determinant for ω^2, fundamental frequency is obtained.

- Love theory: Love's assumptions are universally accepted to be valid as a first approximation theory. First approximation defines a thin shell. Second assumption enables us to make all computations in the undeformed configurations of the shell and ensures that the governing differential equation will be linear. Third assumption related the normal stress in z-direction is neglected in the absence of external load. The solutions provided to satisfy the boundary conditions of above are given in Eq. 9 to11.

$$u(x,\theta) = C_1 cos\frac{m\pi x}{l}\cos n(\theta - \Phi_0) \qquad (9)$$

$$v(x,\theta) = C_2 sin\frac{m\pi x}{l}\sin n(\theta - \Phi_0) \qquad (10)$$

$$w(x,\theta) = C_3 sin\frac{m\pi x}{l}\cos n(\theta - \Phi_0) \qquad (11)$$

Where C_1, C_2 and C_3 are the constants and Φ_0 is the phase angle.

The determinant (Eq.12) of the coefficient matrix is obtained on substituting these assumed solutions in domain governing equations of motion.

$$\begin{vmatrix} \rho h\omega^2 - d_{11} & d_{12} & d_{13} \\ d_{12} & \rho h\omega^2 - d_{22} & d_{23} \\ d_{13} & d_{23} & \rho h\omega^2 - d_{33} \end{vmatrix} = 0 \qquad (12)$$

$$d_{11} = C\left(\frac{m\pi}{l}\right)^2 + C\frac{1-\vartheta}{2}\left(\frac{n}{R}\right)^2, \ d_{12} = d_{21} = C\frac{1+\vartheta}{2}\frac{m\pi}{l}\frac{n}{R}, \ d_{13} = d_{31} = C\frac{\vartheta}{R}\frac{m\pi}{l},$$

$$d_{22} = C\frac{1-\vartheta}{2}\left(\frac{m\pi}{l}\right)^2 + C\left(\frac{n}{R}\right)^2 + \frac{D}{R^2}\frac{1-\vartheta}{2}\left(\frac{m\pi}{l}\right)^2 + \frac{D}{R^2}\left(\frac{n}{R}\right)^2$$

$$d_{23} = d_{32} = -\frac{Cn}{R^2} - \frac{Dn}{R^2}\left(\frac{m\pi}{l}\right)^2 - \frac{Dn}{R^2}\left(\frac{n}{R}\right)^2$$

$$d_{33} = \frac{C}{R^2} + D\left(\frac{m\pi}{l}\right)^4 + 2D\left(\frac{m\pi}{l}\right)^2\left(\frac{n}{R}\right)^2 + D\left(\frac{n}{R}\right)^4$$

On solving the determinant for ω^2, fundamental frequency is obtained.

Modal Orthogonality

Since dynamic response is the function of displacement and time, modal analysis can transform the coupled equations to decoupled equations. The shape term is the function of location/ displacement converts the response at any location and time to the function of time alone. In a system of masses, the shape term related to the mode shapes corresponding to natural frequencies should be orthogonal to each other. The characteristic equation of multi degree of freedom system is,

$$|K - \lambda M| \Phi = 0 \tag{13}$$

The orthogonality condition is derived from the Eq. 30 as,

$$\text{For } r \neq n, \ \left(\Phi^{rT} M \Phi^n \right) = 0 \tag{14}$$

$$\left(\Phi^{rT} K \Phi^n \right) = 0 \tag{15}$$

The Eq.s. 14 and 15 shows r^{th} mode shape is perpendicular to the nth mode shape with respect to global mass and global stiffness. In this paper, the rth and n^{th} mode shapes obtained from the solutions of various classical shell theories are determined to check the modal orthogonality.

Estimation of Natural Frequencies and Mode Shapes

Modes shapes and corresponding natural frequencies are estimated for basic classical theories-membrane theory, DMV theory and love theory. A both ends simply supported circular cylinder of length 2000 mm, diameter 250 mm and thickness 2 mm is considered in the study. The physical properties of the material are given in the Table 91.1.

Table 91.1: Physical properties of the material.

Material	Aluminium
Density	652.87 kg/m³
Modulus of elasticity	20E11 Pa
Poisson's ratio	0.3

Membrane Theory of Cylindrical Shells

The profile expression (Eq. 3) of the hollow circular cylinder is derived from the governing equation connected to the bending equilibrium. Mode shapes and natural frequency of the circular cylinder are calculated using the Eq. 3 and Eq. 4. The cylinder is idealized as continuous system having discrete masses at each 100 mm. The mass points are considered as lumped masses of 3.64 kg. Mode shape solution is derived in the axial direction of the cylinder. The mode shapes extracted for the idealized continuous system are shown in Figure 91.1.

Stiffness at each intermediate level is considered as $12EI/L^3$ allowing translational motion to the length axis and by neglecting the rotation. An assembled global stiffness

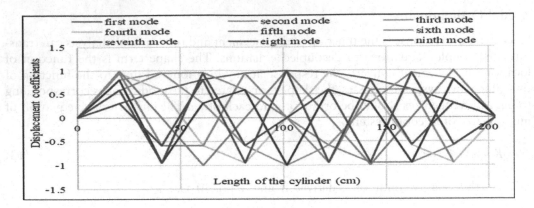

Figure 91.1 Modes shapes of the cylinder using membrane theory.

matrix (Table 91.2) is developed for the transverse vibration of cylinder throughout the length. The normalized mode shapes to global mass matrix given in Table 91.3 is used to perform the orthogonality condition. Effective mass or normalized mass of 18.234 kg is determined using the expression $\Phi^{1T} M \Phi^1$.

Table 91.2: Global stiffness matrix of the cylinder.

2.34E+6	−1.17E+6	0	0	0	0	0	0	0
−1.17E+6	2.34E+6	−1.17E+6	0	0	0	0	0	0
0	−1.17E+6	2.34E+6	−1.17E+6	0	0	0	0	0
0	0	1.17E+6	2.34E+6	−1.17E+6	0	0	0	0
0	0	0	−1.17E+6	2.34E+6	−1.17E+6	0	0	0
0	0	0	0	−1.17E+06	2.34E+6	−1.17E+6	0	0
0	0	0	0	0	−1.11.17E+6	2.34E+6	−1.17E+6	0
0	0	0	0	0	0	−1.17E+6	2.34E+6	−1.17E+6
0	0	0	0	0	0	0	−1.17E+6	1.17E+6

Table 91.3: Normalized modal matrix.

First mode	Second mode	Third mode	Fourth mode	Fifth mode	Sixth mode	Seventh mode	Eight mode	Ninth mode
7.23E−2	1.37E−1	1.89E−1	2.22E−1	2.34E−1	2.23E−1	1.90E−1	1.38E−1	7.27E−2
1.37E−1	2.23E−1	2.23E−1	1.38E−1	3.73E−4	−1.37E−1	−2.23E−1	−2.23E−1	−1.37E−1
1.89E−1	2.23E−1	7.27E−2	−1.37E−1	−2.34E−1	−1.38E−1	7.16E−2	2.22E−1	1.90E−1
2.22E−1	1.38E−1	−1.37E−1	−2.23E−1	−7.46E−4	2.22E−1	1.38E−1	−1.37E−1	−2.23E−1
2.34E−1	3.73E−4	−2.34E−1	−7.46E−4	2.34E−1	1.12E−3	−2.34E−1	−1.49E−3	2.34E−01
2.23E−1	−1.37E−1	−1.38E−1	2.22E−1	1.12E−3	−2.23E−1	1.36E−1	1.39E−1	−2.22E−1
1.90E−1	−2.23E−1	7.16E−2	1.38E−1	−2.34E−1	1.36E−1	7.41E−2	−2.23E−1	1.88E−1
1.38E−1	−2.23E−1	2.22E−1	−1.37E−1	−1.49E−3	1.39E−1	−2.23E−1	2.22E−1	−1.35E−1
7.27E−2	−1.38E−1	1.90E−1	−2.23E−1	2.34E−1	−2.22E−1	1.88E−1	−1.35E−1	6.95E−2

The orthogonal property of mode shapes 1 and 2 connected to the global mass matrix is verified using the Eq.14. $\Phi^{1T} M \Phi^2$ = 3.43E-5 ($\approx$ 0). Hence the condition of perpendicular mode shapes for a dynamic system is verified. In the case of orthogonal property of mode shapes about the global stiffness matrix is partially verified. The result of the expression shown in Eq. 15 is 0.4632. In this case, the condition of perpendicular mode shapes with respect to stiffness matrix for the system is not verified. The natural frequency obtained from the membrane theory is 237217613.61 rad/sec and is observed to be unrealistic.

DMV Theory

The solution based characteristic equation is derived to determine natural frequency from the determinant of Eq. 8. A MATLAB program is used to identify the unknown quantity of the equation ω. The parameters m and n of the Eq. 8 represent the number of half sine waves in the axial and radial/circumferential directions. Fundamental frequency of the cylinder at m = n = 1 is observed as 1062.90 rad/sec. Table 91.4 represents natural frequencies with varying modes in axial and circumferential directions.

Table 91.4: Natural frequencies in axial and radial directions.

Axial mode	Circumferential mode	Natural frequency (rad/sec)
1	1	1062.90
1	2	948.48
1	3	2106.27
2	1	371130
2	2	1613.82
2	3	2220.65

Mode shapes are derived from the eigenvalue expression of classical dynamics using Eq.13. Mode shapes for corresponding natural frequencies (Figure 91.2) are derived from the MATLAB program. Modes at individual frequencies are normalized with global mass and global stiffness is shown in Tables 91.5 and 91.6.

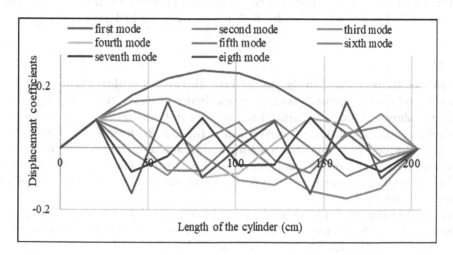

Figure 91.2 Modes shapes corresponding natural frequencies from DMV theory.

Table 91.5: Normalized mode shapes as per DMV theory.

First mode	Second mode	Third mode	Fourth mode	Fifth mode	Sixth mode	Seventh mode	Eight mode	Ninth mode
9.11E–2	9.11E–2	9.11E–2	9.11E–2	9.11E–2	9.11E–2	9.11E–2	9.11E–2	9.11E–2
1.70E–1	1.51E–1	1.23E–1	8.64E–2	4.10E–2	–1.31E–2	–7.60E–2	–1.48E–1	–2.27E–1
2.26E–1	1.59E–1	7.51E–2	–9.01E–3	–7.25E–2	–8.91E–02	–2.74E–2	1.49E–1	4.81E–1
2.53E–1	1.12E–1	–2.14E–2	–9.49E–2	–7.37E–2	2.60E–2	9.90E–2	–9.41E–2	–9.79E–1
2.45E–1	2.73E–2	–1.04E–1	–8.11E–2	3.93E–2	8.53E–2	–5.53E–2	3.89E–3	1.98E+0
2.02E–1	–6.71E–2	–1.19E–1	1.79E–2	9.14E–2	–3.83E–2	–5.27E–2	8.78E–2	–3.98E+0
1.36E–1	–1.38E–1	–5.70E–2	9.81E–2	1.88E–03	–7.98E–2	9.94E–2	–1.46E–1	8.00E+0
4.92E–2	–1.62E–1	4.22E–2	7.52E–2	–9.05E–2	4.98E–2	–3.03E–2	1.50E–1	0.00E+0
–4.37E–2	–1.30E–1	1.14E–1	–2.68E–2	–4.27E–2	7.26E–2	–7.40E–2	–9.72E–2	0.00E+0

Table 91.6: Normalized mode shapes as per DMV theory.

First mode	Second mode	Third mode	Fourth mode	Fifth mode	Sixth mode	Seventh mode	Eight mode	Ninth mode
4.18E–4	4.18E–4	4.18E–4	4.18E–4	4.18E–4	4.18E–4	4.18E–4	4.18E–4	4.18E–4
7.80E–4	6.93E–4	5.65E–4	3.97E–4	1.88E–4	–6.04E–5	–3.49E–4	–6.79E–4	–1.05E–3
1.04E–3	7.30E–4	3.45E–4	–4.14E–5	–3.33E–4	–4.09E–4	–1.26E–4	6.84E–4	2.21E–3
1.16E–3	5.16E–4	–9.84E–5	–4.37E–4	–3.38E–4	1.19E–4	4.55E–4	–4.32E–4	–4.50E–3
1.12E–3	1.26E–4	–4.78E–4	–3.73E–4	1.81E–4	3.92E–4	–2.54E–4	1.79E–5	9.08E–3
9.32E–4	–3.08E–4	–5.48E–4	8.25E–5	4.20E–4	–1.76E–4	–2.42E–4	4.03E–4	–1.83E–2
6.21E–4	–6.36E–4	–2.62E–4	4.51E–4	8.65E–6	–3.67E–4	4.57E–4	–6.73E–4	3.67E–2
2.26E–4	–7.46E–4	1.94E–4	3.45E–4	–4.16E–4	2.29E–4	–1.39E–4	6.89E–4	0.00E+0
–2.00E–4	–5.99E–4	5.24E–4	–1.23E–4	–1.96E–4	3.33E–4	–3.40E–4	–4.48E–4	0.00E+0

The first mode and second mode are perpendicular with reference to global mass matrix $(\Phi^{1T} M \Phi^{2})$ and is verified as 0.035. The orthogonal property of mode shapes with reference to the global stiffness matrix $(\Phi^{1T} K \Phi^{2})$ is partially verified as 0.5104.

Love Theory

Similar procedure is adopted in the case of Love theory. Natural frequencies are identified in the axial and circumferential modes of the cylinder using the frequency equation of the determinant shown in Eq. 12. Natural frequencies in both significant directions of axial and circumferential directions are derived using the MATLAB program. Mode shapes are estimated using the eigenvalue characteristic equation and are shown in Figure 91.3. The fundamental frequencies in the axial and radial direction are identified as 1052.76 rad/sec. The other natural frequencies in various combinations of axial and radial directions are tabulated in Table 91.7.

The normalized modes with respect to global mass and global stiffness (Tables 91.8 and 91.9) have non-zero terms out of diagonal, which has higher displacement coefficients in the 9th mode of the cylinder. The same scenario is observed in the normalized mode shapes

Table 91.7: Natural frequencies in axial and radial directions.

Axial mode	Circumferential mode	Natural frequency (rad/sec)
1	1	1052.76
1	2	751.03
1	3	1874.90
2	1	3711.70
2	2	1506.83
2	3	2000.62

Table 91.8: Normalized mode shapes as per love theory.

First mode	Second mode	Third mode	Fourth mode	Fifth mode	Sixth mode	Seventh mode	Eight mode	Ninth mode
7.43E-2	7.43E-2	7.43E-2	7.43E-2	7.43E-2	7.43E-2	7.43E-2	7.43E-2	7.43E-02
1.42E-1	1.27E-1	1.07E-1	7.76E-2	4.21E-2	−4.77E-4	−5.02E-2	−1.06E-1	−1.71E-1
1.92E-1	1.43E-1	7.67E-2	6.84E-3	−5.03E-2	−7.42E-2	−4.02E-2	8.05E-2	3.11E-1
2.27E-1	1.18E-1	3.57E-3	−7.04E-2	−7.06E-22	9.56E-4	7.75E-2	−9.09E-3	−5.69E-1
2.35E-1	5.82E-2	−7.17E-2	−8.05E-2	1.02E-2	7.42E-2	−1.22E-2	−6.74E-2	9.93E-1
2.22E-1	−1.80E-2	−1.06E-1	−1.36E-2	7.65E-2	−1.43E-3	−6.92E-2	1.06E-1	−1.72E+0
1.86E-1	−8.92E-2	−7.92E-2	6.62E-2	3.32E-2	−7.42E-2	5.91E-2	−8.62E-2	2.98E+0
1.30E-1	−1.33E-1	−7.15E-3	8.28E-2	−5.77E-2	1.91E-3	2.93E-2	1.82E-2	−5.15E+0
6.18E-02	−1.42E-1	6.91E-02	2.03E-2	−6.59E-2	7.43E-02	−7.88E-0	6.01E-2	8.91E+0

Table 91.9: Normalized mode shapes as per love theory.

First mode	Second mode	Third mode	Fourth mode	Fifth mode	Sixth mode	Seventh mode	Eight mode	Ninth mode
5.70E-5	5.70E-5	5.70E-5	5.70E-5	5.70E-5	5.70E 5	5.70E-5	5.70E-5	5.70E-5
1.08E-4	9.76E-5	8.13E-5	5.96E-5	3.24E-5	−3.67E-7	−3.86E-5	−8.24E-5	−1.32E-4
1.49E-4	1.10E-4	5.90E-5	5.26E-6	−3.87E-5	−5.70E-5	−3.09E-5	6.19E-5	2.47E-4
1.74E-4	9.04E-5	2.74E-6	−5.41E-5	−5.43E-5	7.34E-7	5.95E-5	−6.98E-6	−4.37E-4
1.81E-4	4.47E-5	−5.50E-5	−6.18E-5	7.84E-6	5.70E-5	−9.39E-6	−5.18E-5	7.63E-4
1.71E-4	−1.39E-5	−8.12E-5	−1.05E-5	5.88E-5	−1.10E-6	−5.32E-5	8.18E-5	−1.32E-3
1.43E-4	−6.84E-5	−6.08E-5	5.09E-5	2.55E-5	−5.70E-5	4.54E-5	−6.63E-5	2.29E-3
1.00E-4	−1.03E-4	−5.49E-6	6.36E-5	−4.43E-5	1.47E-6	2.24E-5	1.39E-5	−3.96E-3
4.75E-5	−1.08E-4	5.30E-5	1.56E-5	−5.06E-5	5.70E-5	−6.06E-5	4.62E-5	6.84E-3

of DMV theory as well. The first mode and second mode are perpendicular with reference to global mass matrix. Verified using the expression $\Phi^{1T} M \Phi^2 = 0.016$. The orthogonal property of mode shapes with reference to the global stiffness matrix is partially verified as $\Phi^{1T} K \Phi^2 = 0.0103$.

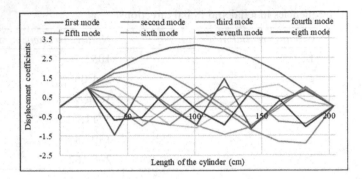

Figure 91.3 Modes shapes corresponding natural frequencies from Love theory.

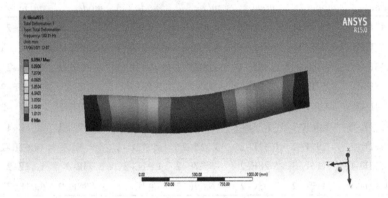

Figure 91.4 First mode of circular cylinder in axial direction.

a. Finite element analysis: Modal analysis is performed on cylindrical shell by considering the support conditions as both ends simply supported. The fundamental frequency of 1110.1 rad/sec in axial direction is observed. The mode shape corresponding fundamental frequency is shown in Figure 91.4. The natural frequencies corresponding axial and radial directions are given in Table 91.10.

Table 91.10: Natural frequencies in axial and radial directions.

Axial mode	Circumferential mode	Natural frequency (rad/sec)
1	1	1132.1
1	2	573..79
1	3	1151.56
2	1	1273.89
2	2	1575.27
2	3	2039.93

Conclusions

In the process of estimating natural frequencies of circular cylindrical shells, the condition of orthogonality is assessed for the solutions assumed in classical shell theories such as membrane theory, DMV theory, and Love theory. The mode shapes corresponding natural frequencies which are obtained from Love theory has satisfied the modal orthogonality with respect to global mass and global stiffness. The membrane theory and DMV theory exhibited the orthogonal property with respect to global mass matrix alone, failed to attain modal orthogonality with the global stiffness matrix. In contrary the fundamental frequencies obtained from the DMV theory and Love theory confirmed the numerical model developed using FEM in Ansys software.

1. The natural frequency obtained using membrane theory is high and not considered. The analytical formula of natural frequency is the function of mass and geometric parameters of the shell structure. Inclusion of moment of inertia in the frequency term can result appropriate natural frequency.
2. Despite the fact that the initial mode shape recovered from the natural frequency acquired from DMV theory is erroneous, the numerical model produced and assessed in Ansys has a convincing fundamental frequency.
3. The shell classical theories attain modal mass orthogonality due to the mass points are assumed at specified intervals. Modal stiffness orthogonality may be achieved by modifying the stiffness matrix.
4. It is observed that the natural frequencies in axial direction derived from the classical theories are accurate and matching with the results of FE model. Hence the condition of perpendicular mode shapes with reference to any one of global mass matrix and global stiffness matrix is validated. Discrepancy is seen in case of other modes of axial and radial directions are not in line with FE results. Since the DMV and Love theories showed accurate results up to first fundamental mode, it can be considered as limitation to these theories.

References

1. Alujevic, N., Campillo-Davo, N., Kindt, P., Desmet, W., Pluymers, B., and Vercammen, S. (2017). Analytical solution for free vibrations of rotating cylindrical shells having free boundary conditions. *Engineering Structures,* 132, 152–171.
2. Aruna, R., Vasanth, M., and Nagpal, A. K. (2016). Finite element analysis of thin circular cylindrical shells. *Proceedings of the Indian National Science Academy,* 82(2), 349–355.
3. Cao, Y., Zhang, R., Zhang, W., and Wang, J. (2018). Vibration characteristics analysis of cylindrical shell-plate coupled structure using an improved fourier series method. *Hindawi Shock and Vibration,* 2018, 1–19.
4. Donnell, L. H., Musthari, K. M., and Vlasov, V. Z. (1985). Chapter 15: donnell-musthari-vlasov theory. *North Holland Series in Applied Mathematics and Mechanics,* 29, 355–362.
5. Flügge, W. (1960). Stresses in Shells. New York (NY) USA: Springer.
6. Gao, R., Sun, X., Liao, H., Li, Y., and Fang, D. (2021). Symplectic wave-based method for free and steady state forced vibration analysis of thin orthotropic circular cylindrical shells with arbitrary boundary conditions *Journal of Sound and Vibration,* 491, 1–15.
7. Hong-Liang, D., and Hao-Jie, J. (2013). Forced vibration analysis for a FGPM cylindrical shell. *Shock and Vibration,* 20(3), 531–550.
8. Love, A. E. H. (1888). The small free vibrations and deformation of thin elastic shell. *Philosophical Transactions of the Royal Society at London,* 179, 491–546.

9.   Love, A. E. H. (1944). A Treatise on the Mathematical Theory of Elasticity. 4th edition. New York (NY) USA: Dover Publications.
10. Sharma, C. B. (1974). Calculation of natural frequencies of fixed free circular cylindrical shells. *Journal of Sound and Vibration*, 35, 55–76.
11. Sharma, C. B., and Johns, D. J. (1972). Free vibration of cantilever circular cylindrical shells—a comparative study. *Hindawi Sound and Vibration*, 25, 433–449.
12. Tang, D., Wu, G., Xiong-liang, Y., and Wang, C. (2016). Free vibration analysis of circular cylindrical shells with arbitrary boundary conditions by the method of reverberation-ray matrix. *Journal of Shock and Vibration*, 2016, 3.
13. Timoshenko, S. P. (1921). On the correction for shear of the differential equation for transverse vibrations of prismatic bars. *Philosophical Magazine LXVI*, 41, 744–746.
14. Xuebin, L. (2006). A new approach for free vibration analysis of thin circular cylindrical shell. *Journal of Sound and Vibration*, 296, 91–98.
15. Xuebin, L. (2008). Study on free vibration analysis of circular cylindrical shells using wave propagation. *Journal of Sound and Vibration*, 311, 667–682.
16. Zhang, X. M., Liu, G. R., and Lam, K. Y. (2001). Vibration analysis of thin cylindrical shells using wave propagation approach. *Journal of Sound and Vibration*, 239(3), 397–403.

Sixth sense for visually impaired: a mobile assistant to aid the blind

Upinder Kaur[a], Himanshu Joshi[b], Mohd. Intesab Khan[c], and Aditya Mishra[d]

Department of Computer Science and Engineering, Lovely Professional University, Phagwara, Punjab, India

Abstract: Vision may be impaired due to multiple reasons. These could be due to eye damage, failure of the brain to receive and read the visual cues sent by the eyes etc. In this research work, we have built a software that uses cell phone camera and sounds to help the blind walking on pedestrians, inside the house or anywhere else. This research paper primarily focuses on three major functions which are providing a description of the object which actually they are seeing, helping them to locate objects around them and give warnings to them about whether an object is will hit or any other kind of collision which can take place. In this mobile application a blind user keeps his/her cell phone in the shirt's pocket or holds it in front of them with the mobile screen facing towards them. The user should be wearing an earphone or a headphone on which he/she will hear the spatial sound produced according to the location of the object. Threat detection will be automatic whereas to access the scene description and locating object feature. User will identify the threats with the help of vibrations or sound alerts. Good quality and up to the mark results are obtained while obtaining the outputs through this application.

Keywords: Blind, GVR, and machine learningzzz, physical disability, vision impaired.

Introduction

"Visual impairment, also known as vision impairment or vision loss, is a decreased ability to see to a degree that causes problems not fixable by usual means, such as glasses. Some also include those who have a decreased ability to see because they do not have access to glasses or contact lenses. Visual impairment is often defined as a best corrected visual acuity of worse than either 20/40 or 20/60." The biggest challenges for visually impaired people, especially the ones with complete loss of the vision (or blind), are related to navigation. We tried to find some existing solutions for this problem of navigation faced by visually impaired users. To the best of our knowledge, most of state-of-the-art works on blind assistance try to produce navigational assistance system or the systems only inform the visually impaired user about the environment or the type of objects. But none of them does both the things at the same time. However, the application designed in this research work has the objective to serve both purposes at the same time. This android application requires camera, to take live imagery of the environment in front of the visually impaired user [1–4].

[a]upinderkaur45@gmail.com, [b]himanshu07joshi@gmail.com, [c]khanintesab07@gmail.com, [d]mishraaaditya29@gmail.com

Literature Survey

Y. Li et al. [5] has mentioned several proposed works that are related to our approach: It is estimated that 1.3 billion people around the world have a kind of vision disability, with the bulk of them being over the age of fifty. For direction, the visually impaired must be linked to others. When walking canes and guide dogs are mixed, they provide little assistance. Canes are being used, but they are unable to assist the individual in assessing what challenge lies ahead with a high degree of certainty. Putting them together isn't especially useful. To make a long story short, it's not a smart idea to deal with those situations merely because of higher body or head level barriers.

Nikhil Thakur et al. [6] proposed new idea contrasting the previous methods that detecting only objects from images, videos or camera feeds do not really count as self-sustaining. Now the whole system is being modulated in three segments; 'Object Detection', 'Depth Estimation or Object Localization' and 'Audio Translation'. Therefore 'You Only Look Once' (YOLO) algorithm 'Version 3' is chosen, implemented, and thus stated in the paper for object detection. In YOLO we reframe object detection as single regression problem straight from image pixel to bounding boxes, coordinates and class probabilities.

Kedar Potdar et al. [7] proposed the traditional 'Sliding Window Method,' which was commonly used before one of the best deep learning algorithms boomed in the atmosphere of image pattern, recognition, and detection techniques, was widely used before one of the best deep learning algorithms boomed in the atmosphere of image pattern, recognition, and detection techniques. The obtained result is reasonably precise, and the detection time on the portable computer is relatively short. It also has been observed that improvements can be made by enhancing the image quality or video picture quality before performing the detection technique even the 'Frames per Second' (FPS) also plays a significant role which mainly takes the consideration of GPU and that is not ubiquitous as all the devices don't have that compatibility and specifications.

M Narendran et al. [8] proposed in his research that this technique will not be that effective unless making the visually impaired ones adaptive to the mapping of the 'Object Detection' with the 'Name of the Object' and its distance from the user himself/herself (object localization) which can be assisted through the output generated viz. the audio engine i.e. SpeechRecogniser (Android API). Although YOLO has a higher rate of localization errors, but is less likely to predict false positives in the past. YOLO is used to process the input image.Since this algorithm is the 'State of the Art,' it is less computationally costly and more efficient than network architectures such as the 'Convolutional Neural Network' (CNN).

Q. Nguyen et al. [9] proposed the most basic means for linguistics information exchange is text. It appears on street signs, shop signs, food labels, restaurant menus, and other places in everyday life. Scene texts are texts that appear in a natural environment. Text identification and text recognition are two steps in an end-to-end text recognition method for associates in nursing. Data regions are detected and labelled with their bounding boxes in text detection. Text information is extracted from detected text regions in text recognition. Text identification is an important step in end-to-end text recognition, which can be done using still images or live video capture. There are many professional applications designed for visually impaired persons in the world. Approaches defined on YOLO algorithms. There is shorter detection time on YOLO with reduced miss rate and error rate compared to discuss traditional methods. Algorithm may also miss small and adjacent objects in an image. Also, Night images require pre-processing C. Frauenberger et al. [10] used YOLO and found that since each grid cell only predicts two bounding boxes and can only have one class, YOLO imposes strict spatial restrictions on bounding box predictions. As a result, the number of neighbouring items that one model can predict is limited.

Chen et al. [11] worked on traditional YOLO Algorithms to detect real time object, which have been used in the past, are used in the approaches described. The detection time of these conventional Algorithms is limited. They're often not very reliable, which can lead to mistakes.

Amir et al. [12] proposed the technique Blind Square. This works in conjunction with the GPS functions on your iPhone to provide you with voiced knowledge about where you are and where you're going. It displays the street you're on, addresses you pass, cross streets, and interesting landmarks. It's great for orienting yourself in a new place.

Google Maps: This will give you voice instructions that will take you step by step. With the three settings: walking, transit, and driving, you can walk, take public transportation, or drive from the back seat of your taxi. In your Safari web browser, type the name of your destination and tap directions.

Jesus Zegerra et al. [13] proposed the technique called Ariadne GPS: Ariadne is used to navigate by tapping on streets. This app then presents voice directions for quick exploration of the area. The app vibrates the user's iPad after a road has been crossed and can announce bus or train stops. It works in several languages for Navigation using Google Maps.

Problem Statement

It has been concerned for quite a long time to improve the way of leading the life by the blind people in terms of eye sight and correspondingly the waking and daily common chores. There have many approaches adopted to implement applications. The earlier approaches defined for blind eye projects have certain limitations in different aspects. Following are few of the limitations, which were faced in the past by the different researchers and developers:

There is no well-known comprehensive software or hardware application for visually impaired persons available having all the features envisioned in our objectives in one package

- Expensive for someone. Also, these approaches rely entirely on ultrasonic sensors and hardware devices.
- Used external embedded sensors which can be expensive for someone.

Therefore, Smartphones came into the picture because smartphones have more usability in regard to these external sensors like watches and bands. Nearly every person have smart-phones as it serves many functions.

Solution lies in perspective of technology. General Technology and Assistive technology. Technology should be more and more used extensively in the area of these visually impaired people by making applications, inbuilt features or websites.

Methodolgy

The android app developed in this research work will aid appropriate solution for the aforementioned problem. This application is mainly used for the blind users. So, the users here are referred to as visually impaired persons. Users will not be required to create new accounts to access the app. They just need to allow the permission to the app to access the device camera and that has been done by user clicking volume button and opens the application hence the camera is turned on and also the users are required to connect either earphones or headphones to the mobile device. The application uses the back camera of the mobile phone. So, the users are required either to keep their mobile phones inside their T-shirt's or shirt's pocket with phone screen facing towards them and back camera popping out of the pocket or they can hold the mobile phone in front of the them facing the phone screen.

The object detection model and the object descriptor model used will detect and describe the object respectively. The models used for these objects which are being detected by the camera lens trained using Convolutional Neural Network and dataset which has been used is 'COCO' dataset which is having 80 distinct classes of real-time entities. Some objects like a speeding car, a person running towards you etc. may have a certain probability of having a collision with user. For this the application has a special feature that is alerting the user of a car or person coming towards them when the distance between them is less than some particular value.

Steps to use application:

- The user clicks volume button and opens the application hence the camera is turned on.
- The camera takes images of the surroundings and sends the images to the model for detecting the presence of objects.
- The model detects objects present in the image and their location. Along with which the audio runs in parallel. This process is continuously repeated until the app is terminated.
- This process can be repeated as long as the user wants
- In the end the application is closed.

The model was built and tested for the testing dataset and a great accuracy was obtained with very less latency of a few milliseconds. 3d sound module was tested to work as per requirements, provided the input object label and location with respect to user. It was found to be working accurately. The process of loading up various media files and 3d rendering and playing up of media sound when the object is detected was found to be taking (approx. 0.2ms) amount of time. It was observed that neither it was too fast to understand, nor too slow taking too much time. It was taking just the right amount of time with sufficient gap between two or more labels.

Figure 92.1 describes the procedural workflow of the process. It gives a detailed explanation of how the application will work starting from opening the application till the end of the process.

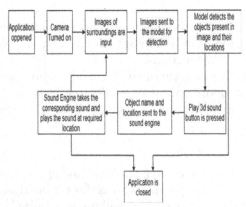

Figure 92.1 Procedural workflow diagram.

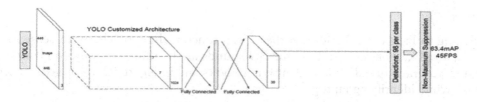

Figure 92.2 Architecture diagram of YOLO.

Figure 92.2 describes the architecture of YOLO. In YOLO we reframe object detection as single regression problem straight from image pixel to bounding boxes, coordinates and class probabilities. YOLO imposes strong spatial constraints on bounding box predictions since each grid cell only predicts two bounding boxes and can have one class, this spatial constraint limits the number of nearby objects than one model can predict.

Tools Used

*   **Android Studio (4.0)**-It is used to speed the development and it also helps to generate best-quality applications for any android device.
*   **Windows 10**-Operating system used to develop the application. All the tools are installed on Windows10.
*   **Python 3.8.2**-Python is a language that is reliable and versatile, and it offers developers with a variety of tools. Python's unique set of characteristics makes it the first option to use it for machine learning.
*   **Android's camera**-It is used to identify and detect the object.
*   **Head-phone**-It is used to signal the user if there is any object. User will hear the sound through the headphones.
*   **Neural Networks:**
*   **TensorFlow**-It is a Python library for rapid numerical computing. It is also used to develop Deep Learning models directly or through wrapper libraries developed on top of TensorFlow.
*   **TFLite**-It enables developers to execute machine-learned models on mobile devices with low latency, allowing them to do classification and regression. It is now available for Android and iOS.

The user interface of this application is the common android app interface, running on smartphone having a camera. 3D Headphone is the major hardware component required, as the application will produce 3D sounds to the user for describing actual scenes around the visually impaired person and alarming him/her for any kind of obstacles around him/her. The UI interface must be easily understandable for the user. The hardware should be lightweight and comfortable to wear. A system should have a minimum of 4GB RAM and 2GB disk space. Also, an Android phone with android OS 4.4 (KitKat) and above with working back camera is required. Other algorithms implemented using programming languages such as Java for android development and Speech Recogniser API is used to produce 3D sounds.

Results

According to Figures 92.3 and 92.4 the crucial functionalities of The Sixth Sense for visually impaired i.e., object detection and 3D sounds are working properly. In Figure 92.3 it is giving accuracy of 0.82 while identifying books and in Figure 92.4 it is giving accuracy of 0.65 while identifying laptop.

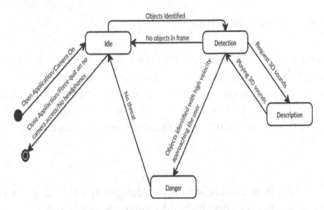

Figure 92.3 Detection of object correctly.

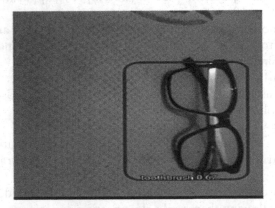

Figure 92.4 Detection of object correctly.

Figure 92.5 Detection of object incorrectly.

The time delay of entire process and speed of speaking was found to be in just right amount in almost all cases but it can give errors also as we can see in Figure 92.5 it is showing spectacle as toothbrush with accuracy of 0.6. (We processed input image by using YOLO. Now for final prediction YOLO combines both confidence score for bounding boxes and class prediction score. If final score is less than 30% then YOLO leaves those bounding boxes). And final testing was done on more than 80 objects and found the similar results as we have mentioned above.

Conclusion

The whole end-to-end implementation of this application helps the visually impaired users to participate in the daily causal activities like walking, roaming around etc. as normal people do. The main objective of this implementation is to provide a proper assistance more technically and in a handy manner rather than any physical assistance. Object detection model embedded with 'Speech Recogniser' android API provides audio assistance as the camera captures the object or person in the frame and respectively providing the audio signal sideways. This model is based on 'YOLO' algorithm and the dataset which is used is 'COCO' dataset consists of 80 distinct classes and then the model is converted into android API using TFLite which is basically a light weight solution for mobile and embedded devices.

Eventually, we are able to succeed in implementation of this project along with proper integration of sound. It also localizes the position of the object detected in the frame like whether it is on the left, on the right or in the middle which helps the blind users to identify the location and perform the next move accordingly as well as collisions can also be avoided. Also, in future we will implement a distance calculation system which could approximately tell how far the object is present and along with that, we tend to implement a scene description feature for describing the whole scenario in front of them could be an extension for the current research work.

References

1. Brunes, A., Hansen, M. B., and Heir, T. (2019). Loneliness among adults with visual impairment: prevalence, associated factors, and relationship to life satisfaction. *Health and Quality of Life Outcomes*, 17(1), 1–7.
2. Loomis, J. M., Golledge, R. G., Klatzky, R. L., Speigle, J. M., and Tietz, J. (1994). Personal guidance system for the visually impaired. In Proceedings of the First Annual ACM Conference on Assistive Technologies, (pp. 85–91).
3. Helal, A., Moore, S. E., and Ramachandran, B. (2001). Drishti: an integrated navigation system for visually impaired and disabled. In Proceedings Fifth International Symposium on Wearable Computers, IEEE, (pp. 149–156).
4. Ćorović, A., Ilić, V., Đurić, S., Marijan, M., and Pavković, B. (2018). The real-time detection of traffic participants using YOLO algorithm. In 2018 26th Telecommunications Forum (TELFOR), IEEE, (pp. 1–4).
5. Dai, J., Li, Y., He, K., and Sun, J. (2016). R-fcn: object detection via region-based fully convolutional networks. *Advances in Neural Information Processing Systems*, 29. arXiv preprint arXiv:1605.06409.
6. Thakurdesai, N., Tripathi, A., Butani, D., and Sankhe, S. (2019). Vision: a deep learning approach to provide walking assistance to the visually impaired. arXiv preprint arXiv:1911.08739.
7. Potdar, K., Pai, C. D., and Akolkar, S. (2018). A convolutional neural network based live object recognition system as blind aid. arXiv preprint arXiv:1811.10399.

8.   Narendran, M., Padhi, S., and Tiwari, A. (2018). Third eye for the blind using arduino and ultrasonic sensors. *National Journal of Multidisciplinary Research and Development*, 3(1), 752–756.
9.   Nguyen, Q. H., and Tran, T. H. (2013). Scene description for visually impaired in outdoor environment. In 2013 International Conference on Advanced Technologies for Communications (ATC 2013), IEEE, (pp. 398–403).
10. Sánchez, J., and Sáenz, M. (2006). Three-dimensional virtual environments for blind children. *Cyber Psychology and Behavior*, 9(2), 200–206.
11. Chen, Z., Khemmar, R., Decoux, B., Atahouet, A., and Ertaud, J. Y. (2019). Real time object detection, tracking, and distance and motion estimation based on deep learning: Application to smart mobility. In 2019 Eighth International Conference on Emerging Security Technologies (EST), IEEE, (pp. 1–6).
12. Zamir, A. R., and Shah, M. (2010). Accurate image localization based on google maps street view. In European Conference on Computer Vision, (pp. 255–268). Berlin, Heidelberg: Springer.
13. Flores, J. Z., and Farcy, R. (2013). GPS and IMU (inertial measurement unit) as a navigation system for the visually impaired in cities. *Journal of Assistive Technologies*, 7(1), 47–56.

93 Simulation of flat plate collector solar water heater in TRNSYS

Subham Show[1,a], Binayak Pattanayak[1,b], and Harish Chandra Das[2,c]

[1]Department of Mechanical Engineering, Siksha 'O' Anusandhan Deemed to be University, Bhubaneswar, Odisha, India

[2]Department of Mechanical Engineering, NIT, Meghalaya, India

Abstract: Increased electricity cost has made solar energy a lucrative alternative source for various household hot water needs. The Earth annually receives about 885 million TWh solar energy. Daily solar energy received by earth is 100 W/m². Hence the use of solar energy to meet household hot water demand is an economic option. The flat plate collectors are the most common types of solar collectors for any solar thermal system. Most water heaters use flat plate collectors being low in cost. The environmental conditions influence the performance of the water heater. The current paper focuses on the simulation of a solar water heating system in TRNSYS. The result for the simulation in various month throughout the year is presented here under Indian climatic conditions. The TRNSYS software simulates transient system with adherence to original results and an error range of 5–10%. Various parameters like the useful solar energy rate, heat transfer rate and hourly plot of system temperature were studied to come to the closure that the south facing solar collectors performs the best.

Keywords: Flat plate collector, simulation, solar water heater, TRNSYS.

Introduction

The requirement of hot water is a common household need for various day-to-day purposes. Mainly this need is fulfilled by electric or fossil fuel-based heaters and boilers. But India, which is already battling increased energy demand, needs to find an alternative to these conventional energy sources wherever feasible. Solar energy is abundant and can be tapped economically. Moreover, it is clean, non-polluting, and is able to provide up to 50–80% of daily domestic hot water needs [10]. Hence it can instrumentally help households to fulfill their desired hot water need without adding to its carbon footprint. Most household solar water heaters use flat plate collector water heaters due to their ability to work with both direct and diffused solar radiation as well as in cloudy conditions [4]. Generally Flat plate collectors work on the principle of thermosiphon in which the working fluid on acquiring heat moves to the top of the storage tank and is replaced by cooler working fluid from the bottom of the tank. The thermosiphon effect is generated by variation in density of water on taking up heat from the solar collector [8] though the flat plate solar collectors can be made to operate with external pump circulating the working fluid to ensure forced convection heat transfer, in turn ensuring better thermal efficiency on a day to day measure [12].

[a]showsubham@gmail.com, [b]pattanayak.binayak@gmail.com, [c]harishdas@nitm.ac.in

The beam radiation or Direct Natural Irradiance (DNI) falling on any solar collector at a right angle to its plane of incidence is the only useful part. DNI is the intensity of solar energy received by the earth at any plane normal to the sun rays [5]. But owing to metrological factors, aerosol content of the air, and air mass, it is transient in nature. Hot, sunny days with clear sky and arid conditions ensure better DNI reception. Hence the regions around the equator and ones between the tropics are best suited for CSP devices operation. So, the study of its transient nature is needed. Most of the time due to the unfeasibility to create the required testing environments, physical experiments cannot always be carried out. Hence simulation is the best choice for performance mapping and design optimization [3]. Ahlgren et al. [1] established the correlation between Direct Natural Irradiance at a place and the annual yield of power. Using this finding and Indian Solar Resource published by National Renewable Energy Laboratory (NREL) we can ascertain the places with the best expected hot water generation. Tiwari et al. [2] provide the inference to use Flat Plate Collectors. Among the available options of solar collectors, the flat plate collectors are the best suited for low temperature and low hot water demand operations. The software TRNSYS is chosen for simulation which according to Messaouda et al. [9] is the most capable and complete software package for solar energy simulation and modeling. This paper is aimed at developing a model for a Flat Plate Collector based solar water heater. The modeling is so done that it can predict the water heater performance throughout the year under changing seasons and climatic conditions. The performance of the Flat Plate Collector-based solar heater is evaluated for the locations in and around Delhi where the supply water temperature is within the range of 21.3–28°C [13].

System Description

The solar water heater operates for eight hours from 8 a.m. to 6 p.m. and is capable of delivering 50 liters of hot water per hour. Water enters the heater at a constant temperature of 20°C and a flow rate of 40 L/h-m². The pump assisted water circulation is chosen over thermosiphon to induce forced convection and ensure better thermal efficiency [12]. The pump used by the heater is 16.7W, 5% of this is gets converted to heat and is absorbed by the water. The model of the collector consists of a 5.6m² collector [2, 13]. An auxiliary heater of 1400W is added to the model so that during times of Direct Natural Irradiance (DNI) unavailability the supply of hot water does not stop.

Methodology

To simulate the water heater it is needed to identify the factors influencing performance and components needed in the simulation. The components used in the simulation are: TYPE 109: Weather data processor which used the weather data in Typical Metrological Year (TMY) format from Meteonorm software to calculate the solar radiation properties on a surface. It takes into account wind speed, outdoor ambient temperature, incident solar radiation as a function of tilt angle and orientation by reading the TMY format weather data file of the simulation site from the weather data bank of TRNSYS. Flat plate collectors yield the best performance at an optimum reflection angle with a similar optimized angle of the top glass cover [11]. The slope is determined using the correlation prescribed by Stanciu and Stanciu [6]. The correlation is $\beta_{optimal} = \phi - \delta \pm$ error where ϕ the latitude and δ is the declination angle of the sun. Using this correlation for our chosen location the slope angle is found to be 30° for the month of December representing the winter months and

0°/horizontal for June representing the summer months. But the flat plate collectors being non-tracking and fixed in nature the slope is set to 30° for its operation annually, rounding of the optimal tilt angle of 27.95° as found by Jamil et al. [7]. Azimuth 0°, 90°, 180°, and 270° meaning the collectors faces towards south, west, north, and east respectively in the northern hemisphere. Ground reflectance is set to 0.2.

TYPE 1b: This TRNSYS component models flat Plate Collector. The solar collector arrays may be connected in parallel or series but the models determine the thermal performance by considering the collectors to be connected in series. The user is needed to input the specific heat capacity of the working fluid. The working fluid temperature can be inlet temperature, outlet temperature, or average temperature. In this component, the second-order quadratic function determines the incidence angle modifier. The rest of the coefficients of the component are of ASHRAE standard. The slope of the flat plate collector is set to 45° and the incidence angle modifier 0.2-0.0. The aperture area of each collector is 5.6 m².

TYPE 6: Auxiliary heater, which is modeled to elevate the temperature of a fluid flow using external or internal control or a combination of both. The heater is so modeled that it can add heat to the fluid flow stream at a user-specified rate to achieve a user-specified maximum temperature. On specifying a constant value of control function of one with an adequate value of heat input rate, the heater will take the control like a routine and function using the internal control to deliver hot water at a set temperature like conventional household water heaters. In the same way, if the heater is a given control function of zero or one from a thermostat or controller, the heater will function like a furnace adding heat at a constant rate so as not to exceed an arbitrary safety temperature limit. For the current application maximum power is set to 5040 KJ/h and the set temperature to which the heater ensures heating is 60°C [13].

TYPE 3b: Single speed pump, modeled to compute a mass flow rate using a variable control function, ranging from 1 to 0. The maximum flow capacity is fixed by the user. The pump power is calculated either as a linear function of mass flow rate or by user-specified relation between them. The amount of heat generated arising from pump power loss is also user-specified. This component sets the flow rate of all other components in the flow loop by multiplying the maximum flow rate and the control signal (parameter 1 and input 3). The maximum flow rate for the current work is set to 50 Kg/h and a constant temperature of 20°C.

TYPE 14h: graphical plug-in for forcing function. This component of the model ensures consistent results for any time step of the simulation. The pattern of the forcing function is user-specified by some discrete data points indicating the value of the function at different times throughout the cycle. The model uses linear interpolation to generate a continuous forcing function from the discrete data. It is also the load profiler describing the water flow rate, duration of working of the solar water heater. The cycle repeats itself every "n" hour where "n" is the last specified time value. The forcing function used in the present simulation runs from 8 in the morning (800 hours) to 6 in the evening (1800 hours) every day for a total of 10 hours.

TYPE 24: It is the integrator; its function is to collect all the results from the working components and assimilating or integrating the results over some time and sending it to the printer. An individual quantity integrator can have up to, but not exceeding 500 inputs.

TYPE 25c: Printer, creates simulation summary and stores all the data and readings after the simulation in text files, and prints the specified system variables at even specified time intervals.

TYPE 65d: Online plotter, which plots the various readings in graphs. It displays selected system variables when the simulation is in progress. It also helps the user to check if the system is functioning in the desired manner by proving valuable variable information.

These components are connected with each other as shown in Figure 93.1 according to related input and output parameters like temperatures, flow rates, Solar radiation, angle of incidence, collector slope. Once desired numbers of variables are set and are named the simulation of the water heater is started and carried out for various values of azimuth angles. The Azimuth angle is the angle between the true north (for southern hemisphere) or south (for northern hemisphere) direction at a location and the direction of sun's projection, taking shift towards west as positive increment. For northern hemisphere the true south is denoted by 0° azimuth angle and moving due west, west, north and east are 90°,180° and 270° respectively.

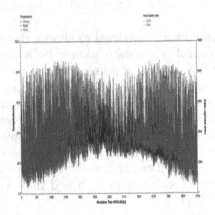

Figure 93.1 TRNSYS schematic of the simulation.

Result and Discussion

In this section the year-round performance of the flat plate collector solar water heater is presented. The aim was to study the performance of flat plate solar water heater in Indian climate. Figure 93.2 shows the outlet temperature of water from the solar collector for azimuth angle 0°. This angle represents that the collector is facing towards south. For the collector facing south the maximum temperature of water delivered is 87°C and 86.8°C for the month of march and October respectively. For this orientation during the peak of

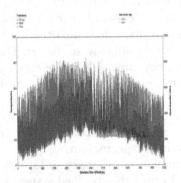

Figure 93.2 The outlet temperature of water from the solar collector for azimuth angle 0°.

summer the highest temperature recorded is 72.65°C. It can be noted that the performance of the collector facing south declines during summer while the winter months performance it surprisingly better. Conversely for the other three orientations the collector performs best only during he months with higher temperature.

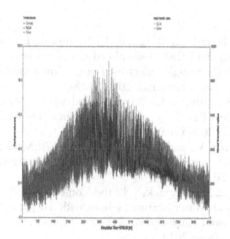

Figure 93.3 The useful solar energy transfer rate to the water by the collector throughout the year.

The Figure 93.3 shows the useful solar energy transfer rate to the water by the collector throughout the year. It is the amount of useful energy provided by the collector or the amount of energy available for thermal operation. It is calculated using the formula $Q_{collector} = \dot{m} * Cp * (Tout - Tin)$ where $\dot{m}$ stands for mass flow rate of water in Kg/s, Cp is the specific heat capacity of water, T_{out} and T_{in} being the outlet and inlet water temperatures respectively. In south wards orientation, where the collector utilizes solar energy consistently throughout the year the with the maximum value of $Q_{collector}$ exceeding 10,000 KJ/hr, and the solar energy transfer rate is consistent. The test location being in the northern hemisphere the results are indicative of the performance of water heaters in northern the hemisphere only.

Conclusion and Future Scope

The present transient analysis of a flat plate solar water heater was conducted for the entire year for the climatic conditions of Delhi. The simulation work was carried out in TRNSYS16. The aim was to predict the best performance time of a domestic water heater. Upon finishing the study, the conclusion drawn are as follows:

The water heater generates more hot water from autumn to spring and during the summer months there is fall in the maximum temperature of delivered hot water only when the collector faces towards south.

During month of lower temperatures, the collector must face towards south.

For the summer month the orientation of the collector should be towards east for maximum hot water delivery.

The effect of transient nature of solar energy on rate of solar energy utilization is seen mostly during rainy season due to occasional cloud cover.

As future scope of work in this concern in the more minute simulation of the water heater taking into account the piping, other type of solar collector and recirculation of water through the collector for better efficiency. Moreover, the polygene ration of multi-disciplinary renewable energy utilization systems is the buzzword for the future.

References

1. Ahlgren, B., Tian, Z., Perers, B., Dragsted, J., Johansson, E., Lundberg, K., Mossegård, J., Byström, J., and Olsson, O. (2018). A simplified model for linear correlation between annual yield and DNI for parabolic trough collectors. *Energy Conversion and Management*, 174, 295–308. https://doi.org/10.1016/j.enconman.2018.08.008.
2. Tiwari, A. K., Gupta, S., Joshi, A. K., Raval, F., and Sojitra, M. (2021). TRNSYS simulation of flat plate solar collector based water heating system in Indian climatic condition. *Materials Today: Proceedings*, 46, 5360–5365.
3. Shrivastava, R. L., Kumar, V., and Untawale, S. P. (2017). Modeling and simulation of solar water heater: A TRNSYS perspective. *Renewable and Sustainable Energy Reviews*, 67, 126–143. doi:10.1016/j.rser.2016.09.005.
4. Ayompe, L. M., Duffy, A., McCormack, S. J., and Conlon, M. (2011). Validated TRNSYS model for forced circulation solar water heating systems with flat plate and heat pipe evacuated tube collectors. *Applied Thermal Engineering*, 31(8–9), 1536–1542. ISSN 1359–4311. https://doi.org/10.1016/j.applthermaleng.2011.01.046.
5. Bijarniya, J. P., Sudhakar, K., and Baredar, P. (2016). Concentrated solar power technology in India. a review. *Renewable and Sustainable Energy Reviews*, 63, 593–603. ISSN 1364–0321. https://doi.org/10.1016/j.rser.2016.05.064.
6. Stanciu, C., and Stanciu, D. (2014). Optimum tilt angle for flat plate collectors all over the World—a declination dependence formula and comparisons of three solar radiation models. *Energy Conversion and Management*, 81, 133–143. ISSN 0196–8904. https://doi.org/10.1016/j.enconman.2014.02.016.
7. Jamil, B., Siddiqui, A. T., and Akhtar, N. (2016). Estimation of solar radiation and optimum tilt angles for south-facing surfaces in humid subtropical climatic region of India. *Engineering Science and Technology, an International Journal*, 19(4), 1826–1835. http://dx.doi.org/10.1016/j.jestch.2016.10.004.
8. Harrabi, I., Hamdi, M., Bessifi, A., and Hazami, M. (2021). Dynamic modeling of solar thermal collectors for domestic hot water production using TRNSYS. *Euro-Mediterranean Journal for Environmental Integration*, 6, 21. https://doi.org/10.1007/s41207-020-00223-6.
9. Messaouda, A., Hamdi, M., Hazami, M., and Guizani, A. A. (2020). Analysis of an integrated collector storage system with vacuum glazing and compound parabolic concentrator. *Applied Thermal Engineering*, 169, 114958. ISSN 1359–4311. https://doi.org/10.1016/j.applthermaleng.2020.114958.
10. Ahmed, S. F., Khalid, M., Vaka, M., Walvekar, R., Numan, A., Rasheed, A. K., and Mubarak, N. M. (2021). Recent progress in solar water heaters and solar collectors: a comprehensive review. *Thermal Science and Engineering Progress*, 25, 100981. ISSN 2451–9049. https://doi.org/10.1016/j.tsep.2021.100981.
11. Thakur, A. (2021). Development and usability of solar thermal collectors in different fields: an overview. *Materials Today: Proceedings*, 46, 6644–6649. ISSN 2214–7853. https://doi.org/10.1016/j.matpr.2021.04.107.
12. Vengadesan, E., and Senthil, R. (2020). A review on recent development of thermal performance enhancement methods of flat plate solar water heater. *Solar Energy*, 206, 935–961. ISSN 0038–092X. https://doi.org/10.1016/j.solener.2020.06.059.
13. Singh, S., Anand, A., Shukla, A., and Sharma, A. (2021). Environmental, technical and financial feasibility study of domestic solar water heating system in India. *Sustainable Energy Technologies and Assessments*, 43, 100965. ISSN 2213–1388. https://doi.org/10.1016/j.seta.2020.100965.

94 Investigation of surface hardness of AISI 316l SS from QPQ complex salt bath treatment process using response surface methodology

Senthil Kumar K. M.[1,a], V. Muthukumaran[1,b], Thirumalai R.[1,c], T. Sathishkumar[2,d], and Santhosh S.[3,e]

[1]Department of Mechanical Engineering, Kumaraguru College of Technology, Coimbatore, Tamil Nadu, India

[2]Department of BioTechnology, Kumaraguru College of Technology, Coimbatore, Tamil Nadu, India

[3]Department of Advanced Manufacturing, SASTRA University, Thanjavur, Tamil Nadu, India

Abstract: The hardness of AISI 316L under Quench Polish Quench (QPQ) surface treated can be examined using an effective mathematical model. This model is formulated to envisage the surface hardness considering the factors like nitrocarburizing bathe chemistry CNO^-%, nitrocarburizing time, nitrocarburizing temperature, oxidizing time and post oxidizing time. An innovative, effective mathematical model has been derived by implementing central composite rotatable second-order response surface. The tests are performed on AISI 316L SS and the hardness of the steel surface is measured. The responses are obtained after varying the factors involved in the process. This in turn facilitated the choice of determinant factors in view of achieving the desired the surface hardness of AISI 316L SS.

Keywords: DOE, hardness, mathematical model, QPQ, response surface.

Introduction

AISI 316L austenitic stainless steel is extensively used for many applications due to its excellent mechanical properties like, sufficient formability, resistance to corrosion, and cost effectiveness. Certain applications of AISI 316L stainless steel in biomedical implants, such as orthopedic, cardiovascular and dental devices are notable [1]. Metin Usta et al. [2] has done considerable research on use of AISI 316L stainless steel for implant fabrication. Although the AISI 316L SS have exceptional bulk properties like ideal strength and flexibility, comparatively lesser surface characteristics are displayed, e.g. low wear resistance. Cui and Luo et al. [3] explored the problems related to wear. Surface modification methods are used to a great extent to increase resistance to corrosion. De Oliveira et al. [4] explored the plasma nitriding on stainless steel. Nickel and aluminum were electroplated in double layers on 316L stainless steel by Kannan et al. [5]. Fossati et al. [6] Studied and characterized the glow discharge nitrided AISI 316L steel in the solution of NaCl. Quench-Polish-Quench complex salt bath nitrocarburizing and post-oxidation process known as QPQ is an advanced technology used to surface strengthening. QPQ process is applied widely to

[a]drkmsenthilkumar2014@gmail.com, [b]vmuthukumaran123@gmail.com, [c]vkrthirumalai@gmail.com, [d]sathishkumar@kct.ac.in, [e]ganesanrtm@gmail.com

raise surface attrition resistance, and to intensify fatigue as well as resistance to corrosion of the treated components [7]. However since additional processes of oxidizing of nitro-carburized components are also applied, it significantly improves the surface properties.

Franjo Cajer et al. [8] have dealt with this in detail. Li et al. [9] applied the QPQ sur-face-treatment process on AISI 1045 steel and studied the nitriding temperature of the nitriding layer. Gui-jiang Li et al. [10] explored the mircrostructure of AISI 316L stainless steel under QPQ process and observed a significant improvement in surface hardness and investigated the effect of QPQ process on surface microstructure of the 17-4PH stainless steel and they found a substantial increase in surface hardness. Gui-jiang Li et al. [10] further made analysis of microstructure 304L austenitic stainless steel by QPQ complex salt bath treatment. A process monitoring system measures the surface hardness in QPQ process. Guemmaz et al. [11] discussed that AISI 316L is largely used in many industrial applications, due to its high resistance to corrosion. As a matter of fact, an increase in hardness is usually correlated with a corresponding improvement in wear resistance. De Oliveira et al. [4] executed plasma nitriding on AISI 316L SS and observed that the treated layers exhibited an accentuated increase in their hardness in relation to the hardness of the substrate. The microhardness distribution from the surface of the plasma nitride 316L SS was measured by Chih-Neng Chang and Fan-shiong Chen et al. [12]. Metin usta et al. [2] observations on hardness measurements indicated that the nitrides have much higher hardness when compared to the base metal in the nitrided samples. The importance of the effect of surface roughness was discussed by Thirumalai et al. [13] in detail. In this work a model is developed for predicting the hardness of AISI 316L SS in terms of QPQ process parameters such as Nitrocarburizing Bath chemistry-CNO⁻%, Nitrocarburizing temp, Nitrocarburizing time, Oxidizing time and post oxidizing time. Maurya et al. [14] developed a system to control flow of a coolant in turning operation.

Experimental Part

Vickers hardness is a measure of the hardness of a material, calculated from the size of an impression produced under load. Micro hardness testing machine is shown in Figure 94.1.

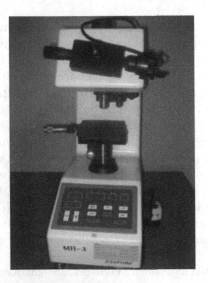

Figure 94.1 Microhardness testing machine.

In the present work, the response variable is surface hardness and the process variables are nitrocarburizing Bath chemistry-CNO⁻%, nitrocarburizing temp, nitrocarburizing time, oxidizing time and post oxidizing time have been considered as the process parameters and surface hardness. Table 94.1. represents the process variables and their factor levels used for this work.

Table 94.1: Process variables.

Process Parameters	Notations	Units	Factor level				
			-2	-1	0	1	2
Nitrocarburizing bath chemistry-CNO⁻%	C	%	25	30	35	40	45
Nitrocarburizing process temp	T	°C	540	565	590	615	640
Nitrocarburizing process time	t1	min	120	150	180	210	240
Oxidizing time	t2	min	10	20	30	40	50
Post oxidizing time	t3	min	10	20	30	40	50

The response surface hardness can be expressed as a function of process parameters. On the precedence of rotatability, central composite rotatable design was considered by Box and Table 94.2 presents the proposed plan of experiment that comprises 32 experiments.

Table 94.2: Experimental design—central composite design matrix.

Expt No	C (%)	-T (°C)	t1 (min)	t2 (min)	t3 (min)	Observed Hardness (HV 0.1)	Predicted Hardness (HV 0.1)	% Error
1	-1	-1	-1	-1	1	762	767	-0.7
2	1	-1	-1	-1	-1	996	976	2.0
3	-1	1	-1	-1	-1	700	676	3.4
4	1	1	-1	-1	1	816	797	2.3
5	-1	-1	1	-1	-1	681	658	3.4
6	1	-1	1	-1	1	849	855	-0.7
7	-1	1	1	-1	1	611	616	-0.9
8	1	1	1	-1	-1	623	627	-0.6
9	-1	-1	-1	1	-1	763	755	1.0
10	1	-1	-1	1	1	800	798	0.3
11	-1	1	-1	1	1	875	868	0.7
12	1	1	-1	1	-1	826	795	3.8
13	-1	-1	1	1	1	485	483	0.5
14	1	-1	1	1	-1	844	840	0.4
15	-1	1	1	1	-1	854	850	0.5
16	1	1	1	1	1	559	584	-4.4

(continued)

Table 94.2: Continued

Expt No	C (%)	–T (°C)	t1 (min)	t2 (min)	t3 (min)	Observed Hardness (HV 0.1)	Predicted Hardness (HV 0.1)	% Error
17	–2	0	0	0	0	329	345	–5.0
18	2	0	0	0	0	487	495	–1.7
19	0	–2	0	0	0	817	817	0.0
20	0	2	0	0	0	735	737	–0.2
21	0	0	–2	0	0	939	979	–4.3
22	0	0	2	0	0	765	749	2.1
23	0	0	0	–2	0	886	925	–4.4
24	0	0	0	2	0	940	925	1.6
25	0	0	0	0	–2	798	828	–3.7
26	0	0	0	0	2	756	726	4.0
27	0	0	0	0	0	809	777	4.0
28	0	0	0	0	0	751	777	–3.4
29	0	0	0	0	0	812	777	4.3
30	0	0	0	0	0	750	777	–3.6
31	0	0	0	0	0	752	777	–3.3
32	0	0	0	0	0	763	777	–1.8

32 sets of coded conditions are involved in this design. The quench polish quench cycle is shown in Figure 94.2. Figure 94.3 shows the nitrocarburizing bath followed next to the Quench process cycle.

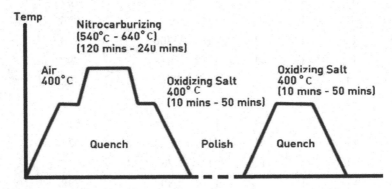

Figure 94.2 QPQ process cycle [15].

Results

The following quadratic polynomial equation brings out the association between response and the process variable.

$$y = b_0 + \sum_{i=1}^{k} b_i x_i + \sum_{i=1}^{k} b_{ii} x_i{}^2 + \sum_{i<j} b_{ij} x_j \tag{1}$$

where b_0 = constant, bi = linear term coefficient, bii = quadratic term coefficient, and bij = interaction term coefficient. Multiple regression method facilitated in calculating the values of the coefficients of the polynomials. A model is developed, and the adequacy of the model is presented in Table 94.3. It is observed that the calculated F ratio of the model lies within acceptable range whereas the calculated R ratio also attains the preferred 95% level of confidence. Table 94.3 illustrates that only less than 5% of error value is found and model is adequate.

Figure 94.3 Nitrocarburizing bath.

Table 94.3: Adequacy of the model.

Parameter	Factors (SS)	DoF	Lack of fit (SS)	DoF	Error Terms (SS)	DoF	F-ratio		R-ratio		Adequacy
							Model	Standard	Model	Standard	
Hardness (HV0.1)	604994.05	13	8085.33	12	5828.83	6	0.69	3.00	47.90	2.92	Adequate

Discussion

The impact of the different process parameters on the surface hardness was envisaged by expounding a mathematical model. The process parameters considered were nitrocarburizing bath chemistry-CNO⁻%, nitrocarburizing temp, nitrocarburizing time, oxidizing time and post oxidizing time. Their direct and indirect influences on the hardness were considered and marked out. The cause and effects were evaluated. The plotted movements of the direct and indirect influences of the parameters on the surface hardness helped in augmenting the surface hardness of AISI 316L SS. Furthermore, many parameters were found to be relatively important. This is due to the fact that every parameter and its indirect factor were approximately equal.

Direct Effect of Variables

In this work, the effects of nitrocarburizing bath chemistry-CNO⁻%, Nitrocarburizing temp, Nitrocarburizing time, oxidizing time and post oxidizing time were experimentally investigated. From Figures 94.4, 94.5, 94.6, 94.7, and 94.8, it is clear that the

846 *Challenges and Opportunities in Industrial and Mechanical Engineering*

nitrocarburizing bath chemistry-CNO%, nitrocarburizing temp, nitrocarburizing time and post oxidizing time have a significant effect on surface hardness of QPQ treated AISI 316L SS. From Figure 94.4, the increase in micro hardness is observed and it attains a maximum value. From Figure 94.5, the increase in nitrocarburizing temperature is noted along with decrease in hardness value. From Figure 94.6, it is observed that the nitrocarburizing time t1 increases with decrease in hardness. From Figure 94.7, it is observed an increase in nitrocarburizing time is noted with decrease in hardness at the low value at oxidizing time t2 and again increasing with nitrocarburizing time t2. From Figure 94.8, it is observed an increase in post oxidizing time t3 as the hardness value decreases.

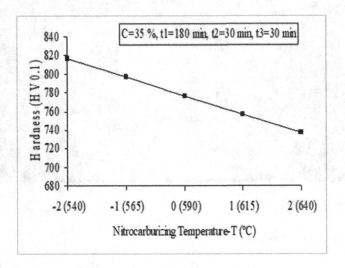

Figure 94.4 Direct effect of nitrocarburizing bath chemistry-C on hardness.

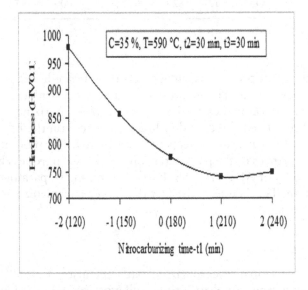

Figure 94.5 Effect of nitrocarburizing temperature-T on hardness.

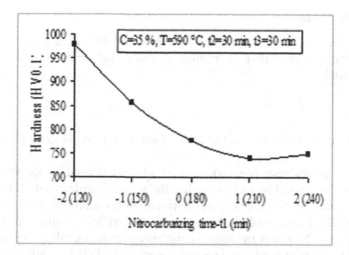

Figure 94.6 Effect of nitrocarburizing time-t1.

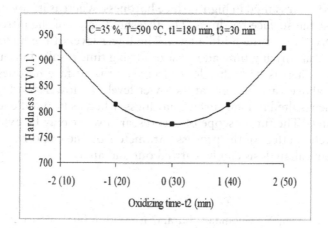

Figure 94.7 Effect of oxidizing time-t2 on hardness. on hardness.

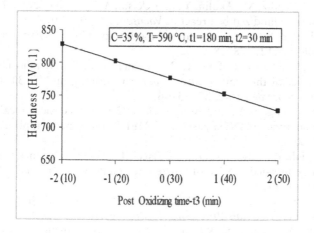

Figure 94.8 Effect of post oxidizing time-t3 on hardness.

Validation of the Model

The validity of the model is checked for the different levels of the parameters. From the validation results, it is seen that experimental value and predicted value are closer and confirms the model.

Conclusions

The findings based on the results of the present investigation are detailed as follows:

1. A central composite response surface methodology is used in the present research to develop a mathematical model to predict the surface hardness of QPQ processed AISI 316L SS in terms of nitrocarburizing bath chemistry-CNO⁻%, nitrocarburizing temp, nitrocarburizing time, Oxidizing time and post oxidizing time. The nitrocarburizing bath chemistry-CNO⁻% is the most significant parameter, which increases the surface hardness. The surface hardness increases from 375 HV0.1 to 800 HV0.1 by increasing the nitrocarburizing bath chemistry. The middle level of the nitrocarburizing bath chemistry CNO⁻% resulted in high surface hardness, whereas its lower level and higher levels decrease the surface hardness. The threshold value of the nitrocarburizing bath chemistry CNO⁻% is at the middle level. The lower level of the nitrocarburizing temperature, nitrocarburizing time and post oxidizing time resulted in increase in the surface hardness, whereas its higher level decreases the surface hardness. The threshold values of the above variables are at its lower level. The lower and higher levels of the oxidizing time resulted in high surface hardness, whereas its middle level decreases the surface hardness. The future scope of this research work may be extended for analyzing the interaction effect of the process parameters on the responses. Also, microstructural and chemical analysis can be carried out and analyzed.

References

1. Chenglong, L., Dazhi, Y., Guoqiang, L., and Min, Q. (2005). Resistance to corrosion and hemocompatibility of multilayered Ti/TiN-coated surgical AISI 316L stainless steel. *Materials Letters*, 59, 3813–3819.
2. Usta, M., Oney, I., Yildiz, M., Akalin, Y., and Ucisik, A. H. (2004). Nitriding of AISI 316L surgical stainless steel in fluidized bed reactor. *Vacuum*, 73, 505–510.
3. Cui, F. Z., and Luo, Z. S. (1999). Biomaterial modification by ion-beam processing. *Surface and Coating Technology*, 112, 278–285.
4. De Oliveira, A. M., Munoz Riofano, R. M., Casteletti, L. C., Tremiliosi, G. F., and Bento, C. A. S. (2003). Effect of the temperature of plasma nitriding in AISI 316L austenitic. *Revista Brasileira de Aplicacoes de vacuo*, 22, 63–66.
5. Kannan, S., Balamurugan, A., and Rajeswari, S. (2005). Electrochemical characterization of hydroxyapatite coating on HNO3 passivated 316L SS for implant applications. *Electrochimica Acta*, 50, 2065–2072.
6. Fossati, A., Borgioli, F., Galvanetto, E., and Bacci, T. (2006). Resistance to corrosion properties of glow-discharge nitrided AISI 316L austenitic stainless steel in NaCl solutions. *Corrosion Science*, 48, 1513–1527.
7. Yeung, C. F., Lau, K. H., Li, H. Y., and Luo, D. F. (1997). Advanced QPQ complex salt bath heat treatment. *Journal of Materials Processing Technology*, 66, 249–252.
8. Cajer, F., Landek, D., and Stupnisek Lisac, E. (2003). Improvement of properties of steels applying salt bath nitrocarburizing with post-oxidation. *Materiali in Technologije*, 37, 333–339.

9. Li, H. Y., Luo, D. F., Yeung, C. F., and Lau, K. H. (1997). Microstructural studies of QPQ complex salt bath heat-treated steels. *Journal of Materials Processing Technology*, 69, 45–49.
10. Li, G. J., Peng, Q., Wang, J., Li, C., Wang, Y., Gao, J., Chen, S. Y., and Shen, B. L. (2008). Surface microstructure of 316L austenitic stainless steel by the salt bath nitrocarburizing and post-oxidation process known as QPQ. *Surface and Coatings Technology*, 202, 2865–2870.
11. Guemmaz, M., Moser, A., Grob, J. J., and Stuck, R. (1998). Sub-surface modification induced by nitrogen ion implantation in stainless steel (SS316L), Correlation microstructure and nanoindentation results. *Surface and Coatings Technology*, 100–101, 353–357.
12. Chang, C. N., and Chen, F. S. (2003). Wear resistance evaluation of plasma nitrocarburized AISI 316L stainless steel. *Materials Chemistry and Physics*, 82, 281–287.
13. Thirumalai, R. Senthilkumaar, J. S., Selvarani, P., Arunachalam, R. M., and Senthilkumaar, K. M. (2012). Investigations of surface roughness and flank wear behavior in machining of Inconel 718. *Australian Journal of Mechanical Engineering*, 10(2), 157–168.
14. Niranjan, M. S., Kumar Maurya, N., and Dwivedi, S. P. (2020). Development of a system to control flow of coolant in turning operation. *Journal of Mechanical Engineering*, 17(1), 17–31.
15. Boopathi, S. (2022). An experimental investigation of Quench Polish Quench (QPQ) coating on AISI 4150 steel. *Engineering Research Express*, 4(4), 045009.

95 Effect of bake hardening on the dent resistance of DP490 steels

Shaik Shamshoddin[1,2,a], Abhishek Raj[1,b], Niranjan Behra[1,c], Rahul Kumar Verma[1,d], and N Venkata Reddy[2,e]

[1]Research and Development, Tata Steel Ltd, India

[2]Department of Mechanical and Aerospace Engineering, IIT Hyderabad, India

Abstract: Dent resistance is the performance requirements of the automotive outer panel. Several steel grades, such as interstitial free steel or Bake hardening steel, are used for manufacturing outer panels of automotive. Further, there are steels that become extra hardened during paint baking, and such behaviour is called the bake hardening effect. The focus of this work is to study the effect of bake hardening on the dent performance of a dual-phase 490 (DP490) steel. A custom die-setup is used for making schedule-A type coupon samples for testing. Sheets are drawn to different depths to study the effect of higher flow strains on the dent performance with and without bake hardening.

Keywords: Bake Hardening, dent resistance, DP490 steel.

Introduction

Usage of high strength steels is increasing in the automotive industry to improve the fuel efficiency and the human safety [11]. The exterior panels often experience impacts from various sources such as hail storm, flying stone chips, impact of luggage and doors of another car in the parking. Such impacts causing permanent deformation of 0.1 mm at the deformation point is defined as a dent [1], and the force to produce such deformation is called the dent resistance. Dents caused by accidents are not covered by the previous definition.

It is reported that sheet metal dent resistance depends upon sheet thickness, yield strength [3, 15], part curvature, supporting conditions and indenter size [6]. Researchers extensively studied the effects of sheet strength and thickness [4, 5, 15]. If other parameters are assumed to be constant, dent resistance is proportional to thickness, yield strength, and panel curvature. However, it decreases with an increase in modulus of elasticity and stiffness [2, 3].

Consideration of dent resistance at the design stage requires good prediction accuracies. Several researchers used FE models with different material models and yield criterion [7, 13] to predict dent resistance. Similarly, there is significant work on analytical models [2] and empirical models [3, 15] to predict the dent resistance of sheet metals.

Since dent resistance is proportional to yield strength, using higher strength steel grades helps in improving the dent resistance [11]. Bake hardening is a phenomenon that increases

[a]shamshoddin@tatasteel.com, [b]abhishek.raj@tatasteel.com, [c]niranjan.behera@tatasteel.com, [d]Rahul.verma@tatasteel.com, [e]nvr@mae.iith.ac.in

the yield strength during paint baking due to accelerating age hardening. There are several steel grades that show bake hardening properties. Typically, the strength increase after paint baking is of the order of 35–40 MPa. [8, 10]. Such an increase in yield strength is termed as the bake hardening effect, and it results in improved dent performance [12]. One of the problems with traditional bake hardening steel is the limited shelf life. Even the nitrogen presence of 5–10 ppm could lead to unwanted strain ageing at room temperature. Hence, researchers are shifting towards other high strength materials with enhanced shelf life. DP490 is one such alternative with similar formability, elongation and higher yield strength. DP steels consist of ferrite-martensitic microstructure, and these steels are specifically developed for higher strength and improved elongation [9].

In this study, we have evaluated the dent resistance of DP490 steel of 0.55 mm thickness before and after the baking process to understand the influence of the bake hardening effect and plastic equivalent strain.

Measurements of Dent Resistance

Dent testing is done as per the standard mentioned in SAE-J2575 using the schedule-A type component. The sample type is chosen based on the product development stage. Assembled die system used to make the schedule-A component is shown in Figure 95.1a, whereas the dimensions of the sample are shown in Figure 95.1. One important feature of the die is the stringer beads to avoid material flow and allow pure stretching without any drawing. DP 490 steel grade sheet is formed to different draw depths of 12 mm, 14.5 mm and, 16.5 mm to achieve different plastic equivalent strains of 2%, 3% and 4%, respectively [14].

The dent test system at Tata Steel is shown in Figure 95.2. The indenter is placed perpendicular to the panel at the sample centre, followed by incremental cyclic loading. Load and displacement are measured. A typical plot of Indenter load versus indenter displacement is shown in Figure 95.2b schematically, which consists of a pre-load, loading and unloading cycles, and a post-load cycle. Values A, B and C are the residual dent depths after each cycle.

The load versus displacement curve in the dent testing process is subdivided into three regions like the initial, secondary and final stiffness regions (Figure 95.3). Initially, the applied load is resisted by bending and compressive membrane stresses. As the applied load increases, local curvature at dent location changes from a convex to a concave shape and the applied load is resisted by a combination of bending and tensile membrane stresses.

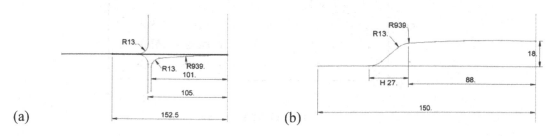

(a) (b)

Figure 95.1 (a) Block diagram of die and punch set used for the forming. (b) Schedule-A sample used in the present study.

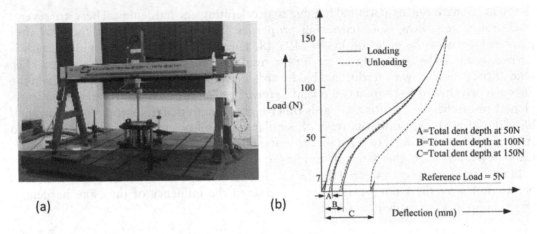

Figure 95.2 (a) Dent measurement system used in the experimentation to quantify the dent resistance (b) Sample dent load-deflection curve (SAE-J2575).

Once, local curvature transforms to a concave shape in the final stage, and panel stiffness increases again as more of the load is carried by membrane tension.

During the experiment, both load and deflection are measured on the indenter. SAE-J2575 standard is followed to measure and characterize the dent.

Dent Behavior of DP490 Before Bake Hardening

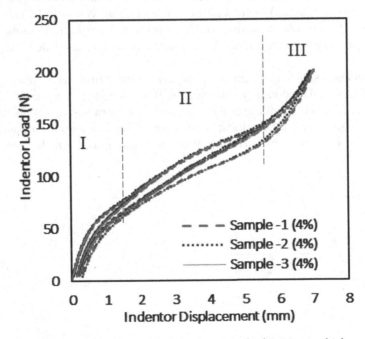

Figure 95.3 Dent load-deflection behavior of the DP490 steel of 0.55 mm thickness. It consists of three zones as indicated by I, II, III.

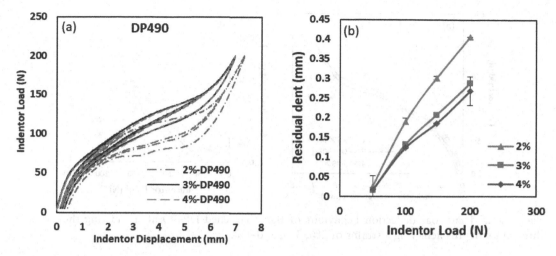

Figure 95.4 Dent load-deflection behavior of the DP490 steel of 0.55 mm thickness at three draw depths 12 mm, 14.5 mm and 16.5 mm corresponding to strains of 2%, 3% and 4%. (b) Residual dent is increasing with the increase in the dent load and draw depth for regular DP490 0.55 mm thickens steel.

Samples are tested for the dent resistance with the 50, 100, 150 and 200N load cycles for draw depths of 12 mm, 14.5 mm and, 16.5 mm corresponding to plastic equivalent strains (vonMises) of 2%, 3% and 4%. Three samples are tested for each strain. DP490mm steel with 0.55 mm thickness samples shows three distinct zones, namely the initial stiffness region, the secondary region and the final region (Figure 95.5) for the 2% strain samples. Stiffness in the initial region is higher than in the other two regions due to the convexity of the surface. The secondary region is a transition from the convex to the concave region. Oil canning is observed in this region. The final stage exhibits a rise in stiffness due to the increase in resistance from membrane stresses and involvement of the larger sample region. Such clear demarcation is not observed for the 3% and 4% strain samples due to work hardening, thus improved flow strength.

Dent resistance is defined as the load corresponding to the residual dent of 0.1 mm (SAE-J2575). Since increments of 50N are taken, 0.1 mm residual dent may not be exactly achieved. Hence, load corresponding to the 0.1 mm could be calculated by linearly interpolating the load corresponding to residual dent immediately after 0.1 mm and immediately before 0.1 mm of depth. Dent resistance values are increasing with the draw depth. It is because of the increase in the flow strength with the work hardening. Higher draw depth causes higher stretch resulting in higher work hardening. The rate of decrease in residual dent is attributed to the rate of increased work hardening.

Dent Behavior of DP490 After Bake Hardening

DP490 sheet of 0.55 mm is formed to draw depths of 12 mm, 14.5 mm and 16.5 mm. Formed samples are air soaked in the closed furnace at 170°C temperature for 20 minutes, after which the samples are allowed to cool to room temperature with natural convection. Such cooled samples are tested for dent resistance. Figure 95.4 shows the load-deflection curve for the DP490 bake hardened steel of 0.55 mm thickness. It could be observed by

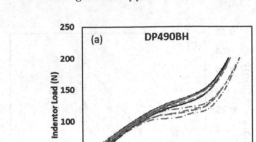

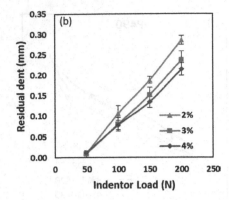

Figure 95.5 Dent load-deflection behaviour of Bake hardened DP490 after drawing the sheet to three draw depths achieving a strains of 2%, 3%, and 4%.

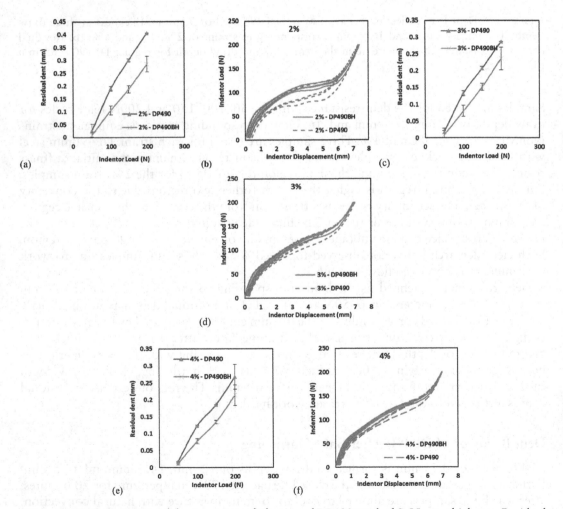

Figure 95.6 Comparison of dent resistance behavior of DP490 steel of 0.55 mm thickness. Residual dents values for Bake hardened and without Bake-hardened DP490 at (a) 2% strain (c) 3% strain (e) 4% strain. Comparison of Load-deflection behavior with and without bake hardening DP490 steel at (b) 2% strain (d) 3% strain (f) 4% strain.

comparing the dent performance of DP490 with and without bake hardening (Figures 95.4 and 95.5) that bake hardened panels are stiffer.

Residual dent values of the bake hardening DP490 steel are compared with the regular DP490 steel at three draw depths, 12 mm, 14.5 mm, and 16.5 mm and is shown in Figure 95.5. Average residual dent values at 100N for three draw depths (12 mm, 14,5 mm, 16.5 mm) are 0.19 mm, 0.13 mm, 0.12 mm for the regular DP490 steel grade and 0.11 mm, 0.08 mm, 0.079 mm for the Bake hardened DP490 grade, which clearly indicates an improvement in dent performance by up to 40% and the improvement is about 20% for the 200N. Similar behavior is observed if draw depth is changed by keeping the load constant (refer Figure 95.6). It indicates that the bake hardening effect will be higher for lower loads and lower strains. As the flow strain increases, the bake hardening effect (percentage) decreases, which in turn reduces the effect on dent performance.

Conclusions

Dent resistance of the DP490 steel grade is evaluated before and after the bake hardening. Samples are drawn to three different draw depths–12 mm, 14.5 mm and 16.5 mm, corresponding to vonMises plastic equivalent strains of 2%, 3% and 4%. Higher draw depth increased the dent resistance, thus lower residual dent values. After bake hardening, there is a further increase in the dent resistance at all the draw depths. It can be concluded that the bake hardening effect of DP490 is significant enough to improve the dent resistance performance. The effect of bake hardening is reducing with the increase in either draw depth or dent load, or both. Thus, at lower strains bake hardening effect is higher and in turn, dent performance improvement is better.

Acknowledgements

We acknowledge the experimental support of the Product Engineering Application Research Center, Tata Steel, India. The authors are also thankful to Tata Steel for permission to publish the results. They also thank Mr Pavithran, Mr Nitish, and Mr Guna for helping in the experimentation and sample collection.

References

1. Surface Vehicle Standard (2004). Standardized Dent Resistance Test Procedure. SAE Standard J2575.
2. Nader, N. (1995). On strength, stiffness and dent resistance of car body panels. *Journal of Materials Processing Technology*, 49, 13–31.
3. DiCello, J. A., and George, R. A. (1974). Design criteria for the dent resistance of auto body panels. *SAE Transactions*, 83, 389–397.
4. Jung, D.-W. (2002). A parametric study of sheet metal denting using a simplified design approach. *KSME International Journal*, 16(12), 1673–1686.
5. Jung, D.-W., and Worswick, M. J. (2004). A parameter study for static and dynamic denting. *KSME International Journal*, 18(11), 2009–2020.
6. Johnson Jr, T., and Schaffnit, W. (1973). Dent resistance of cold-rolled low-carbon steel sheet. *SAE Transactions*, 82, 1719–1730.
7. Lee, J. Y., Lee, M. G., Barlat, F., Chung, K. H., and Kim, D. J. (2016). Effect of nonlinear multi-axial elasticity and anisotropic plasticity on quasi-static dent properties of automotive steel sheets. *International Journal of Solids and Structures*, 87, 254–266.

8. Manikandan, G., Verma, R. K., Lansbergen, M., Raj, A., and Deshpande, D. (2015). Effect of yield strength, pre-strain, and curvature on stiffness and static dent resistance of formed panel. In Advances in Material Forming and Joining: 5th International and 26th All India Manufacturing Technology, Design and Research Conference, AIMTDR 2014 (pp. 47–59). Springer India.

9. Mohrbacher, H. (2013). Advanced metallurgical concepts for DP steels with improved formability and damage resistance. In Proceedings of the International Symposium on New Developments in Advanced High-Strength Sheet Steels. Vail, AIST.

10. Pereloma, E., and Timokhina, I. (2017). Bake hardening of automotive steels. In Automotive Steels. Woodhead Publishing, (pp. 259–288).

11. Sato, K., Tokita, Y., Ono, M., and Yoshitake, A. (2003). Dent simulation of automotive outer panel using high strength steel sheets. SAE Technical Paper, (No. 2003-01-0606).

12. Seel, T. N. (1991). Bake hardening steel application study-key factors of dent resistance improvement. *SAE Transactions*, 100, 283–291.

13. Shen, H., Li, S., and Chen, G. (2010). Numerical analysis of panels' dent resistance considering the bauschinger effect. *Materials and Design*, 31(2), 870–876.

14. Pereloma, E., and Timokhina, I. (2017). Bake hardening of automotive steels. In Automotive Steels. Woodhead Publishing (pp. 259–288).

15. Yutori, Y., Nomura, S., Kokubo, I., and Ishigaki, H. (1980). Studies on the static dent resistance. *Memoires scientifiques de la revue de*, 70, 561–569. IDDRG 1980.

96 A computational analysis and thermal design aspects of twisted tape fitting in circular tube type heat exchanger

Mohan Singh[1], Amit Kumar[1], Arun Kumar[1], Abhijit Rajan[2], and Laljee Prasad[2]

[1]Department of Mechanical Engineering, NIT Patna, India

[2]Department of Mechanical Engineering, NIT Jamshedpur, India

Abstract: This intent presents a comprehensive overview of the numerical analysis and thermal design aspects of twisted tape type heat exchanger fitting in the circular tube. When fluid is flow through the twisted tape then it produces a swirl flow with merge of fluid resulting in advance heat transfer, impulse dispersion and thermal execution as compared to the plain tube. So, now impressive work will be done to raise the heat transfer from the surface work is under consideration. Twisted tape is a great method for heat transfer improvement. Twisted tape ratio (y/w) divergence from $(2 \leq y/w \leq 6)$; in laminar flow regions $(800 \leq Re \leq 2000)$ apply water as working substance. For a laminar flow in the laminar form a fixed volume method was applied. At Reynolds number of 2000, the maximum thermal execution influence of 5.12 has been secure by using the twisted-tape ratio (TR= 2) of the tube in the laminar flow. The study has been proved that the concern of twisted tape is an upgrade approach for laminar convection heat transfer upraise.

Keywords: Computational analysis, heat exchanger tube, laminar flow, swirl flow, twisted tape.

Introduction

A Heat exchanger is expensive equipment built to transfer heat as per our requirement while keeping separation between both the source of heat & the receiver. Heat exchangers are initially used for brewery, pharmaceuticals, dairy and food processing hygienic applications, currently, its applications have been found in modernistic industry also [1]. To raise the area of heat transfer, appliance intimately linked to the primary surface for providing and extending the secondary surface. These extended surface elements are referred to as fins, while, the mechanical design includes the design of tube, baffles, gasket etc. Circular tubes fitting with twisted tape type heat exchangers have rapid response to control action and retain capacity to recover heat from too small temperature differences, thus extend their applications.

Heat transfer amplification fashion could be intimate toward following three categories: active, passive and compound method. Active heat transfer method (HTM) uses exterior energy for the remover of heat for example forced convection heat transfer. Passive HTM is a self-sufficient energy for example fitted tapes. This method is used to distinct heat exchanger systems namely automobiles, thermal power plants, air-conditioning and refrigeration equipment, etc. Whereas compound HTM consists of the compound of both active and passive HTM to increase the thermo-hydraulic execution of heat exchangers. A great variety of interpolate has been used to build up the heat transfer rate through mesh, twisted tape & whatever. Currently, in the heat transfer system twisted tapes are widely accepted due to its manufacturing and installation cost is low. By adding twisted tape in the tube generates swirl flow, which adjusts the velocity profile on the wall side [2–6].

There are several investigations has been done for the up gradation of heat transfer rate along constant and variable twisted tape introduction. Coefficient of heat transfer of flowing water measured by Wang and Sunden [7] in 2000 with the help of twisted tape in a tube. Results show that the gained of heat transfer coefficient 3 and 3.5 times more in the twisted tube than other. Sarmaet et al. [8, 9] correlated transfer of heat and pressure in the fitted tube with twisted tape and conserve the transitional movement from laminar to turbulent regime.

Identical twins swirling flow motion occurs over the entire pipe length fitted with twisted tape for transfer of heat execution since, it possess features like low manufacturing cost, maintenance, easy to installation, feasibility of use and so on. Therefore, twisted tape became an important device to produce uninterrupted swirling flow.

On the minor axis an examination for the transfer of heat execution in the tubular heat exchangers had been done at the combination of the twisted tape and tube as a result laminar regime obtained [10]. Due to the helical swirl flow motion fluid mixing increases in the twisted tapes and their powerful utilization has been generally declared for the heat transfer [11, 12]. Twisted tapes are passive method which applied in the all-water heating systems to raise the transfer of heat rate also. Applying of twisted tape in the solar collector for heat transfer is easier than the plain tube collector for better thermal performance with twining of different twisted tapes reported by Kumar and Prasad [13, 14]. Later, an experimental analysis also given by Jaisankar et al. [15] as for photovoltaic cell through twisted-tape devices for the friction factor, thermal execution and heat transfer.

For a laminar flow, heat transfer, coefficient of friction and thermo-hydraulic execution has been numerically investigated on twisted tapes fitting tube. A computational result show that the heats transfer can be marvelous increases by reducing flow resistance. Conversely, application with a suitable twisted tape ratio (TR) with the fitting twisted tape tube enhances of the heat transfer rate.

Computational Analysis

Physical Pattern

The geometries of a twisted tape fitting in the tube are characterized in Figure 96.1. The entire length of the tube is fitted with the collector area (103×133) in cm^2 and twisted tapes with thickness (t) of 1.5 mm. Length of the tube is 96 cm and diameter is 1.6 cm. Also tubes separation is 10 cm. Twist tape ratio is the ratio of the pitch to its width of a twisted tape. By definition, Twist tape ratio = pitch of the tube (y)/width of the tube (w). As a result, the relative twisted tape ratio (y/w) are 2, 4, 6. An introduction given by Webb [16] for Reynolds number (Re), Nusselt number (Nu), friction factor (f) and thermal execution factor (η) are as follows:

$$Re = \frac{\rho u D}{\mu} \tag{1}$$

$$Nu = \frac{hD}{k} \tag{2}$$

$$f = \frac{\Delta P}{\left(\frac{\rho u^2}{2}\right)\left(L/D\right)} \tag{3}$$

$$\eta = \frac{Nu / Nu_P}{\left(f / f_P\right)^{1/3}} \tag{4}$$

Figure 96.1(a) represents the circular tube along twisted tape and Figure 96.1(b) represents meshing of the tube over the entire length. The working substance is water which entering the tube with initial temperature (T_{in}) and flow through the entire length of the tube owning its diameter. The numerical investigation has been implemented at distinguish values of TR 2, 4 and 6 for laminar flow regions ($800 \leq Re \leq 2000$). The following expectation was done for the calculations:

- At the initial condition the uniform heat flux, velocity and at the outlet condition pressure were relevant to the length of the tube.
- Applying mass flow principle, at constant flow pattern the flow was resolved at each Reynolds number.
- All properties of fluid such as incompressibility, non-viscous, specific gravity and specific weight negligible.
- The wall of twisted tape was considered at constant adiabatic affection.
- As per the analytical value of Reynolds number obtained at constant heat flux was upon the mass flow rate of water.

In this study, we analyzed the laminar flow regime at the mass flow rate at initial and final conditions of the tube with the help of Reynolds number. In the presence of laminar flow in the tube without fitting of twisted tape, the velocity profile is only axial flow but considering twisted tape thickness (t) is 1.5 mm, for the simplification.

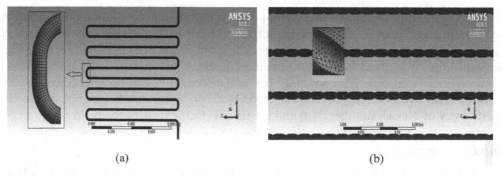

(a)   (b)

Figure 96.1 (a) The arrangement of the circular tube along twisted tape type heat exchanger, and (b) tetrahedral meshing twisted tape.

Methodology and Validation

The three-dimensional model with steady state laminar and non-viscous flow is adopted. For the fluid flow, the equation of continuity, momentum and energy are given below in a tensor form:

Continuity equation,

$$\frac{\partial(\rho u_j)}{\partial x_j} = 0 \tag{5}$$

Momentum equation,

$$\frac{\partial(\rho u_i u_j)}{\partial x_j} = -\frac{\partial p}{\partial x_i} + \frac{\partial}{\partial x_j}\left[\mu\left(\frac{\partial u_i}{\partial x_j} + \frac{\partial u_j}{\partial x_i}\right)\right] \tag{6}$$

Energy equation,

$$\frac{\partial(\rho c_p T u_j)}{\partial x_j} = \frac{\partial}{\partial x_j}\left[k\left(\frac{\partial T}{\partial x_j}\right)\right] \tag{7}$$

On the tube wall a uniform heat flux are utilized. Equations (8) and (9) represent the equation of velocity and temperature respectively at inlet and pressure occurs at outlet of the tube [17].

$$u = u_c\left(1 - \frac{r^2}{R^2}\right) \tag{8}$$

$$T = T_c + \frac{qR}{k}\left[\left(\frac{r}{R}\right)^2 - \frac{1}{4}\left(\frac{r}{R}\right)^4\right] \tag{9}$$

To validate [18] the present model in the plain tube a laminar flow are used as below;

$$Nu = 3.6 \tag{10}$$

$$f = \frac{64}{Re} \tag{11}$$

For this work software, Fluent 18.1 is employed. At the boundary conditions, the above equations (5), (6) and (7) can be solve applying software Fluent 18.1 at a finite volume approach. A single device is used in the numerical form for the momentum, pressure and energy equation. The Semi-Implicit Method for Pressure-Linked Equations (SIMPLE) is used for handling the pressure velocity coupling. For continuity 10^{-5} and velocity 10^{-7} components convergence criteria for energy are used respectively. The theoretical correlation has been employed to check the efficiency of Nusselt number (Nu) and the friction factor (f) with the plain tube.

Boundary Condition

Refer to Figure 96.1 (a), there are two types of boundary surfaces, namely inlet and outlet ends of the tape in the axial direction contacting with the fluid and measured velocity at inlet and pressure at outlet cross-sections of the tape.

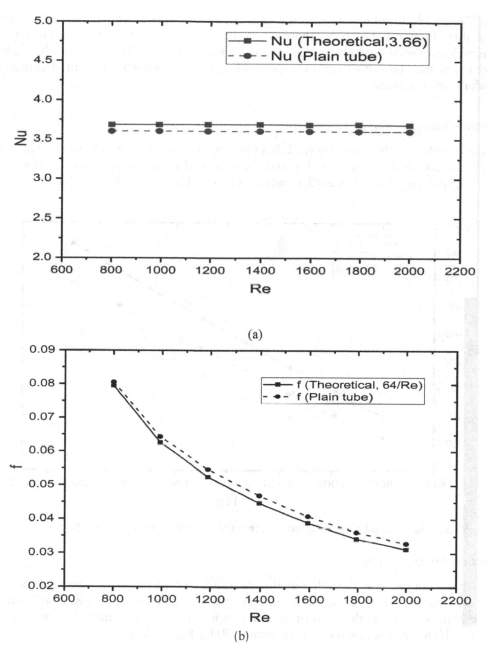

Figure 96.2 (a) & (b) Validation with the plain tube: (a) Nu vs. Re, (b) f vs. Re. [18].

Result and Discussion

In this project, applying the software Fluent 18.1, we have obtained the value of friction factor, transfer of heat rate, flow structure and thermal execution respectively. A proper design of twisted tape, in the circular tube is essential to enhance the transfer of heat rate and the thermal performance. Twisted tape ratio (y/w) divergence from (2 ≤ y/w ≤ 6), in laminar flow regions (800 ≤ Re≤ 2000) had been utilized water as working substance (refer

Figure 96.2). Finite volume approach was applied for a laminar flow in the laminar form. As laminar flow, the maximum thermal execution influence of 5.12 has been obtained by using the twisted-tape ratio (TR=2) at Reynolds number of 2000. The study has been proved that the concern of twisted tape is an upgrade approach for laminar convection transfer of heat upraise.

Effect on Nusselt Number

As shown in Figure 96.3, we observed that for laminar flow Nusselt number (Nu) increases with increment of Reynolds number and decrement of twist ratio (TR). i.e., Nu is maximum 56.50 with TR = 2, Nu = 49.17 with TR = 4 and Nu = 42.93 with TR= 6.

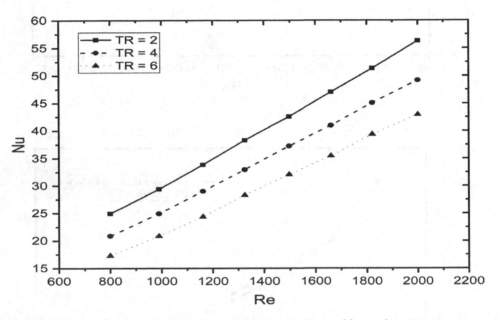

Figure 96.3 Relationship between Nusselt number (Nu) and Reynolds number (Re).

Effect on Friction Factor

It has been noticed that the value of the friction factor (f) increases along decreases the Reynolds number and decrement twisted tape ratio (TR). At the value of Reynolds number, (Re) = 2000 and TR= 6, the value of friction factor is, f = 0.311(minimum). Similarly, at Re = 800 and TR= 2, then, f = 0.612 (maximum). Refer Figure 96.4.

Effect on Thermo-Hydraulic Performance

The thermal performance factor is a vital role of any heat exchanger device. At the same input power, a comparison may be done for coefficients of heat transfer of the tubes with and without expansion devices. Refer to Figure 96.5, we see that the thermal execution factor is above the unity along twisted tape better than other geometry configuration at the same input power. It has been noticed that the thermal performance (η) varies according to the variation of Reynolds number and the twisted tape ratio value (TR). At the minimum,

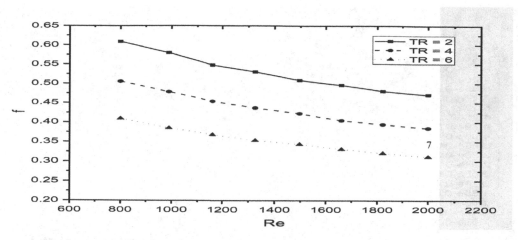

Figure 96.4 Relationship between friction factor (f) and Reynolds number (Re).

(Re) = 800 and TR = 6, the value of thermal performance is, η = 1.9 (minimum). Similarly, at the maximum, (Re) = 2000 and TR = 2, the value of thermal performance is, η = 5.12 (maximum) in the laminar regime. Figure 96.6 represented the stream line of investigated twisted tape geometry fitted in heat exchanger tube with TR = 2 among Re 2000.

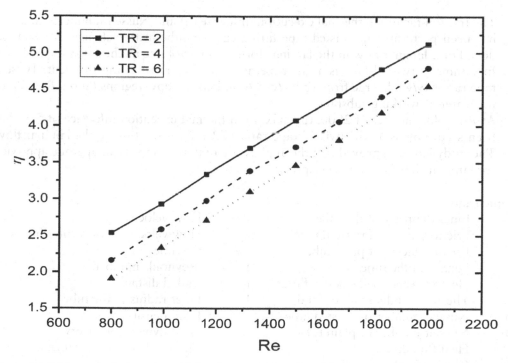

Figure 96.5 Numerical alteration of the thermo-hydraulic performance *(η) vs.* Reynolds number (Re).

Figure 96.6 Streamline for y/w = 2 at Re 2000.

Conclusion

This intent presents a comprehensive overview of the numerical analysis and thermal design aspects of twisted tape type heat exchanger fitting in the circular tube. The overall conclusions are as follows:

*   The friction factor (f), thermal execution influence (η) and Nusselt number evaluation has been performed in twisted tape fitting circular tube heat exchanger for laminar flow. For a laminar flow in the laminar form a finite volume method was applied.
*   In summary, twisted-tape is a great method for heat transfer improvement. Twisted tape ratios (y/w) diverge from (2 ≤ y/w ≤ 6) in laminar flow regions (800 ≤ Re≤ 2000) for water as working substance.
*   At Reynolds number of 2000, the maximum thermal execution influence of 5.12 has been secure by using the twisted-tape ratio (TR = 2) of the tube in the laminar flow. The study has been proved that the application of twisted tape is an upgrade approach for laminar flow heat transfer upraise.

Nomenclature

D	Inner diameter of the tube	u	Flow velocity
f	Friction factor of twisted tape	u_c	Flow velocity at the centerline of
f_p	Friction factor of plain tube		the tube
L	Length of the tube	Re	Reynolds number
h	Heat transfer coefficient of fluid	r	Radial distance
k	Thermal conductivity of fluid	R	Inner radius of the tube
Nu	Nusselt number of twisted tape	ρ	Fluid density
Nu_p	Nusselt number of plain tube	μ	Fluid dynamic viscosity
q	Heat flux density	η	Thermo-hydraulic performance
Δp	Change in pressure difference		factor
T_c	Temperature at the centerline of the tube		

References

1. Singh, K. U. (2014). Video steganography: text hiding in video by LSB substitution. *International Journal of Engineering Research and Applications*, 4(5), 105–108. ISSN:2248-9622 (version 5).
2. Martemianov, S., and Okulov, V. L. (2004), On heat transfer enhancement in swirl pipe flows. *International Journal of Heat and Mass Transfer*, 47(10–11), 2379–2393. https://doi.org/10.1016/j.ijheatmasstransfer.2003.11.005.
3. Smithberg, E., and Landis, F. (1964). Friction and forced convection heat-transfer characteristics in tubes with twisted tape swirl generators. *Journal of Heat Transfer*, 86(1), 39–48.
4. Webb, R., Narayanamurthy, R., and Thors, P. (2000). Heat transfer andfriction characteristics of internal helical-rib roughness. *Journal of Heat Transfer*, 122(1), 134–142.
5. Skullong, S., Thianpong, C., Jayranaiwachira, N., and Promvonge, P. (2016). Experimental and numerical heat transfer investigation inturbulent square-duct flow through oblique horseshoe baffles. *Chemical Engineering and Processing: Process Intensification*, 99, 58–71.
6. Promvonge, P., Khanoknaiyakarn, C., Kwankaomeng, S., and Thianpong, C. (2011). Thermal behavior in solar air heater channel fitted with combined riband delta-winglet. *International Communications in Heat and Mass Transfer*, 38(6), 749–756. https://doi.org/10.1016/j.icheatmasstransfer.2011.03.014.
7. Wang, L., and Sunden, B. (2002). Performance comparison of some tube inserts. *International Communications in Heat and Mass Transfer*, 29(1), 45–56.
8. Sarma, P. K., Kishore, P. S., Rao, V. D., and Subrahmanyam, T. (2005). A combined approach to predict friction coefficients and convective heat transfer characteristics in a tube with twisted tape inserts for a wide range of Re and Pr. *International Journal of Thermal Sciences*, 44(4), 393–398.
9. Agarwal, S. K., and Rao, M. R. (1996). Heat transfer augmentation for the flow of a viscous liquid in circular tubes using twisted tape inserts. *International Journal of Heat and Mass Transfer*, 39(17), 3547–3557. https://doi.org/10.1016/0017-9310(96)00039-7.
10. Khoshvaght-Aliabadi, M., and Feizabadi, A. (2020). Performance intensification of tubular heat exchangers using compound twisted-tape and twisted-tube. *Chemical Engineering and Processing—Process Intensification*, 148, 107799.
11. Bergles, A. E., and Manglik, R. M. (2012). Enhanced heat transfer in single-phase forced convection in tubes due to helical swirl generated by twisted-tape inserts. In ASME 2012 Heat Transfer Summer Conference collocated with the ASME 2012 Fluids Engineering Division Summer Meeting and the ASME 2012 10th International Conference on Nanochannels, Microchannels, and Minichannels, (pp. 77–86). 10.1115/HT2012-58285.
12. Manglik, R. M., Maramraju, S., and Bergles, A. E. (2001). The scaling and correlation of low reynolds number swirl flows and friction factors in circular tubes with twisted-tape inserts. *Journal of Enhanced Heat Transfer*, 8, 383–395.
13. Sadhishkumar, S., and Balusamy, T. (2014). Performance improvement in solar water heating systems—a review. *Renewable and Sustainable Energy Reviews*, 37, 191–198.
14. Kumar, A., and Prasad, B. N. (2000). Investigation of twisted tape inserted solar waterheaters: heat transfer, friction factor and thermal performance results. *Renewable Energy*, 19, 379–398.
15. Jaisankar, S., Radhakrishnan, T. K., and Sheeba, K. N. (2009). Experimental studies on heat transfer and friction factor characteristics of forced circulation solar waterheater system fitted with helical twisted tapes. *Solar Energy*, 83, 1943–1952. https://doi.org/10.1016/j.solener.2009.07.006.
16. Webb, R. L. (1981). Performance evaluation criteria for use of enhanced heat transfer surfaces in heat exchanger design. *International Journal of Heat and Mass Transfer*, 24, 715–726.
17. Holman, J. P. (2002). Heat Transfer. 9th Edn. New York: McGraweHill.
18. Incropera, F. P., DeWitt, D. P., Bergman, T. L., and Lavine, A. S. (2007). Fundamentals of Heat and Mass Transfer. Wiley.

97 3D technology printing: a review

Amit gupta[1,a], P. Sudhakar Rao[1,b], and K. K. Mittal[2,c]

[1]Department of Mechanical Engineering, National Institute of Technical Teachers Training & Research, Chandigarh, India

[2]Department of Mechanical Engineering, Ajay Kumar Garg Engineering College Ghaziabad, U.P., India

Abstract: Additional Production is the contemporary mode in building techniques due to its innumerable convenience. It can be described as the manner of producing parts with the aid of inserting objects in the structure of every layer. It has been the problem of in-depth learn about and overview with the aid of many researchers. In this work, a complete review of the included production has been achieved. The evolution of add-on products such as popular innovation and its different categories is examined. The significance of partial guidance, time measurement, furthermore, cost calculation have likewise been evaluated. A prominent feature of this work is the ID of issues related with various production techniques for supplementation. Due to blemishes in additional production, its integration with different strategies, for example, subtraction, is underscored. This audit will help students comprehend the various viewpoints of additional productivity furthermore, investigate new ways of forthcoming study.

Keywords: Additive manufacturing, cost models, hybrid manufacturing, part orientation, subtractive manufacturing.

Introduction

The emergence of ventures relies upon new jobs what's more, bleeding edge research related to production processes, building materials, and item construction. Notwithstanding low-cost and high-quality customer needs, market rivalry in the current manufacturing industry is connected to the needs of complex products, with short life cycles, short conveyance times, customization, and requiring low-talented specialists. Truth be told, the current type of items is very complex what's more, testing to plan [1–3]. Similarly, there is a solid motivating force for the turn of events, development, what's more, execution of innovative what's more, innovative production measures [4–5]. Production cycles can be divided into five classifications, specifically, subtraction, addition, joining, separation, what's more, conversion, as demonstrated in Figure 97.1.

- Traction technology can be defined as the removal of layers of building material to construct the want geometry. Over the past 20 years, the technology of extraction has undergone major changes.

---

[a]amit.mech19@nitttrchd.ac.in, [b]psrao@nitttrchd.ac.in, [c]mitkamal@gmail.com

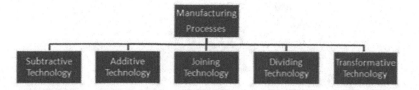

Figure 97.1 Categorization of manufacturing processes.

- The presentation of three-dimensional (3D) faces displaying programming has supplanted conventional code-making, for example, G and M codes. In contrast to the computerized number control system (CNC) of the 1940s, the list of modern CNC mechanism is exceptionally robotized dependent on the coordination of computer-assisted computer assistants (CAD)/production support systems (CAM) systems.
- Additional innovation depends on the expansion of building layers to make a form of your favorite work.
- Joining a innovation such as welding involves physically combining two or more functions to bring out the desired condition.
- Separating technologies such as cutting the fall is against the joining process.
- Flexible technologies, for example, construction, thermal conductivity, and subfreezing chilling use a single material to make something, keeping the size constant.

Additive Manufacturing

Extra production or AM can be characterized as the interaction of bonding, or then again setting substances like fluid gum and powders. Create a component with a layer type using a 3D CAD model. Terms for example, 3D printing (3DP), fast prototyping (FP), digital direct production (DDM), fast production, and strong framing (SFF) can be utilized to portray measures for AM. AM measures create 3D PC information components or Standard Tessellation Language (STL) records, containing details related to the calculation of the article [5,6]. AM is extremely helpful when required for low creation volumes, high plan intricacy, and complexity, and general configuration changes. It provides an opportunity to create complex parts by surviving obstacles to the construction of conventional production strategies. In spite of the fact that AM has numerous advantages, its use is still restricted due to its low exactness along with long construction times contrasted with CNC mechanism. It doesn't have something similar issues with CNC machines since it separates part of the leading areas by a goal equivalent to that of the cycle Figure 97.2.

Various Stages of Development AM

AM, or Additive Manufacturing, had its origins in the 1980s when Kodama introduced the concept through a unit called "Three-Dimensional Data Display by Automatic Preparation of the Three-Dimensional Model." The technology saw significant progress in 1987 with the development of Hurl Structure's STL (courtesy of 3D Frameworks), which played a pivotal role in advancing AM techniques. A cycle that reinforces flimsy layers of bright (UV) light-mirroring light using a laser. It can successfully make complex and modified items with a couple of talented laborers, short conveyance times, and short item alternation-of-generations. The outline in Figure 97.3 portrays the expansion in the number of licenses appended

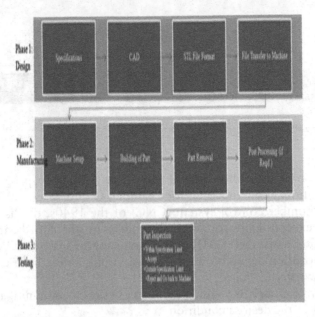

Figure 97.2 Various states in AM operation.

to AM from 1982 to 2012. The innovation dependent on laser sintering (SLS) was earliest protected in 1988 via Carl Deckard at the College of Texas, trailed by the organizing of coordinated documenting (FDM) in 1989 by Scott Crump, originator of Strasays Inc. It very well may be said that the instructive task in 2005 by Dr. Adrian Bowyler at the College of Shower was particular of the incredible ways that arose in the 3DP space. They have made a self-recreating program, known as Rep Rap, which can do whatever it might want to do. From that point forward, numerous organizations, like Bakerbot (2009), Structure Labs (2011), and the HP Combination Stream 3D (2016) printer have emerged [7–9].

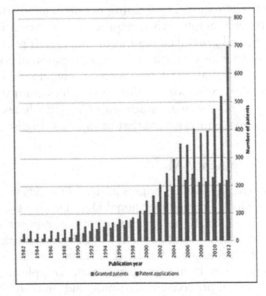

Figure 97.3 Annually examination in the numerous allowed patents.

It tends to be seen that the approach of AM innovation started during the 1980s to deliver 3D items in computer-aided design documents [9]. It tends to be inferred that AM will change and change creation in the coming business.

Demand for AM has grown from automotive and aviation to housing and the clinical, food, and aviation ventures. The Public Air transportation and Space Organization (NASA) has been attempting to track down a more productive and proficient 3DP innovation for Zero-G innovation. Scientists have been attempting to create 3D printed particles at a micron or lower level, to be helpful in delivering little hardware and batteries. Or on the other hand, there is a 3D printer available (graciousness: Photonic Proficient GT) that can make parts that are not more extensive than a human hair. There have been huge upgrades since its commencement, regarding new advances (like hybridization), building materials (e.g., gamma TiAl), items, and AM organizations.

Classification

AM cycles or hardware can be sorted dependent on mechanism size, pipe size, pipe speed, and working region. AM can be arranged from multiple points of view contingent upon the usefulness of the item. Despite the fact that grouping techniques may likewise incorporate download capacities, the way toward making traditional math, the kind of material utilized, and the help interaction. Be that as it may, from a more extensive perspective, AM cycles can be summed up and ordered by the kind of material utilized [10]. Figure 97.4 sums up the current AM strategies dependent on the sort of item.

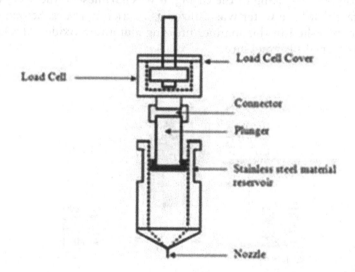

Figure 97.4 Characterization of AM measures relying upon the condition of crude material.

Strong-based AM. AM innovation where the undeniable green info is in a strong state is examined in this sub-area. Among the numerous solid AM-based advancements, FDM, expulsion form (FF), mineral material creation (LOM), and direct ink (DIW) are the most mainstream.

FDM. The FDM execution as demonstrated in Figure 97.5 depends on the rule of concentrated creation innovation. In this innovation, plastic materials (filaments) are removed through a mouthpiece, which warms to liquefy the material. The amplifier head moves toward the instrument, which is delivered by each layer.

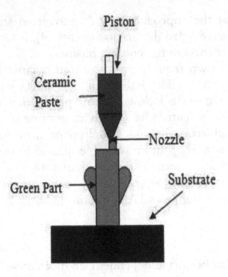

Figure 97.5 Schematic of FDM.

FEF. This cycle was created at the College of Missouri—Rolla, Rolla, Missouri [11], and chipped away at removing fluid earthenware glue. With this innovation, the cement was eliminated from the floor of the structure, which was kept underneath room fundamental quantity to accomplish the icing of the casing. It was harmless to the ecosystem technique in light of the fact that lone water was utilized as restricting issues. Strong heaps of up to 60 volts are accomplished in this manner utilizing aluminum oxide (Al2O3). Figure 97.6 demonstrate the formal FEF measure.

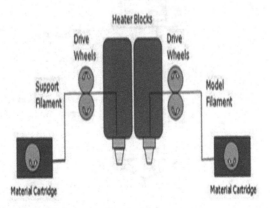

Figure 97.6 Schematic of DIW measure.

LOM. LOM, which stands for Laminated Object Manufacturing, allows for the use of various raw materials such as metal, paper, or different types of polymers in its process. The chosen sheet material is blended using an ultrasonic mixer to achieve the desired shape. In the case of paper, sheets are bonded together and precisely cut to form the required shape. Polymer sheets are joined using heat and pressure [7, 12, 13]. LOM offers

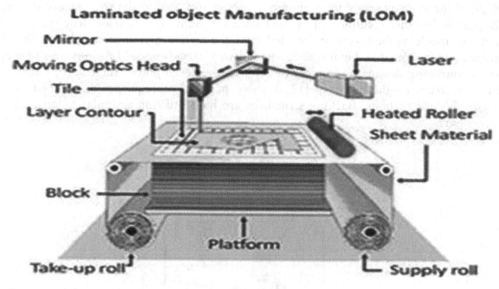

Figure 97.7 Schematic of LOM process [11].

several advantages, including high surface roughness, low material consumption, afford-able machinery and production costs, and excellent durability (as depicted in Figure 97.7).

DIW. DIW (otherwise called "Robocasting", "Robot-Helped Statement", and -Direct Writ Creation") first showed up during the 1990s at Sandia Public Research centers [14, 15]. In this innovation, a bogus impression a plastic-cement blended in with artistic particles was separated to get a segment. The progression of the amplifier was constrained by directions made by the product utilizing information from the computer-aided design model. It is fundamentally the same as FDM, yet its objective depends on mock-versatility, instead of firmness to keep up the state of the parts. 30 mm fine lines are needed for ceramic pastes [16] and 1 mm for polymer glues. Bigger lines (.1 mm) are likewise more agreeable with part fabricating measure quicker than other AM frameworks (Figure 97.8)

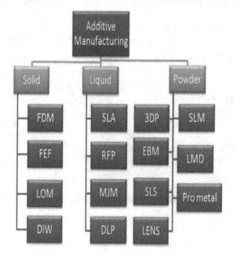

Figure 97.8 FEF process [9].

SLA. In the SLA, a tank of fluid photopolymer gum is used to shape segments as a layer. This part is likewise solidified or treated with UV light. The development stage brings down the article as the layers are framed after the mending of each new layer. The SLA standard can be found in Figure 97.9. The halfway execution of SLA innovation necessitates encouraging designs to connect the unit to the development stage and preserve the item as it coasts on the fluid sap [12, 47]. No material prerequisites are needed for the supporting constructions. Backing structures are built utilizing a similar material as the genuine part.

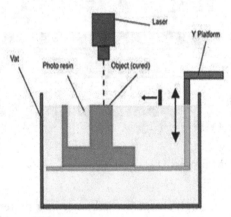

Figure 97.9 Vat Photo polymerization measures conspire [10].

MJM. This cycle utilizes different line planes to convey UV polymer or treatable wax on a case-by-case basis. Exactly when the printed polymer emerges from the head, the UV light enlightens to fix the substance. MJM prints the segments on a versatile stage that goes down after the current layer treatment to continue the process. It is an inexpensive method because it can create short-term components. The MJM cycle is protected and calm and can in this manner be utilized to print polymeric parts in a research facility climate. In any case, in this interaction, the firmness of the units and the naivety of the part is poor and the bottom area often prints the quality parts; the strength of the part is poor and the lower part is frequently relaxed, hardened, or on the other hand lopsided (Figure 97.10) [17].

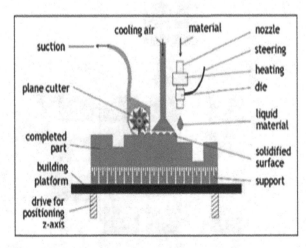

Figure 97.10 MJM interaction [16].

RFP. This interaction shapes part by specifically adding and cooling water in a consecutive way [18]. Two sorts of strategies, for example, persistent establishment and lower interest (Figure 97.11) can be used to introduce water [19]. The structure site is kept at a low fundamental quantity of measure, underneath freezing. In this interaction, unadulterated water or saline solution can be utilized as a help material that is extricated from the pit and set in pre-strong ice. Lines and taking care of lines are kept at a specific temperature so that water does not cool down before the area where the ice is also kept in liquid form to allow the equipment to flow freely and smoothly.

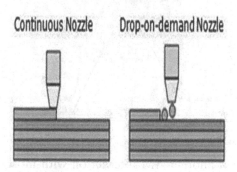

Figure 97.11 Principle of RFP.

DLP. DLP at first appeared in 1987 with "Texas Instruments" This cycle is fundamentally the same as the SLA in that it likewise uses photopolymers [20]. The principle contrast is in the light source. For DLP, a standard light source (circular segment light) what's more a shot (DLP projector) are utilized as demonstrated in Figure 97.12. Resulting in minimal damage and low running costs [21,48,49].

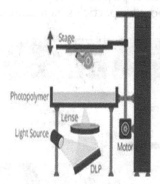

Figure 97.12 Generalisation of DLP20.

3DP. In 3DP, a fluid bond is used to tie fine buildup. The powder layer is as yet appropriated on the development stage what's more the fastener fluid is gone through pipes through a cross-part of the 3D model [22]. This cycle doesn't need extra help, all things being equal, support is given by non-restricting powder particles Figure 97.13 [12,52].

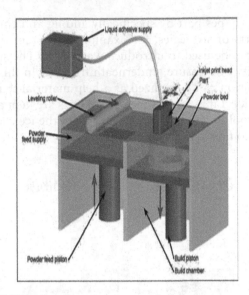

Figure 97.13 Schematic of 3DP10.

EBM. EBM innovation is another AM interaction with most utilization in the clinical what's more aviation industry [23–25] was a licensed innovation by ARPAM for 3DP specifically. the electrons are delivered from the link, which is warmed-over 2500°C. These high-strength electrons are sped up by an anode. The attractive field of the focal point centers around the bar and the gases analyzing the layers. The powdered particles are taken care of by gravity from the containers what's more are partitioned into layers on the design table. The whole interaction is done under a vacuum, which kills contaminations what's more delivers high energy [23, 26, 27].

SLS. In Selective Laser Sintering (SLS), as depicted in Figure 97.14, a laser beam is employed to selectively fuse or melt powder particles in successive layers. The laser beam precisely targets and fuses the powder particles as it scans across the build platform. Once

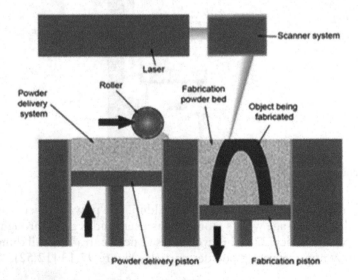

Figure 97.14 Selective laser sintering framework [10].

a layer is scanned, the build platform is lowered, and a new layer of powder is evenly spread over the previous layer. This iterative process continues until the object is partially formed [28,50,51].

SLM. In Selective Laser Melting (SLM), each layer of the component is formed by selectively melting metal powder using a laser. The mechanical properties of components produced through SLM closely resemble those manufactured using conventional manufacturing methods. Previous studies have claimed that SLM can produce parts with a weight of up to 99.9%. Despite its significant potential, the industrial implementation of SLM is hindered by various factors [29–33]. Among the main challenges are the presence of residual stresses and pronounced deformations caused during the process.

LENS or Focal point. The Focal Point process, developed by Sandia National Laboratories and commercialized by Optomec Inc (Albuquerque, NM, USA) since 1997, can be classified as a laser-based additive manufacturing technique [34, 35]. In this method, a laser serves as the heat source to melt powder particles that are deposited onto the substrate. Focal Point is capable of processing various materials, including steel, aluminum, titanium alloys, nickel-based alloys, and metal matrix composites. The rapid cooling during this process results in a thin, fine-grained structure and high ductility (Figure 97.15). The advantage of Focal Point lies in its ability to achieve high tensile strength due to the quick cooling, which produces a refined microstructure [36, 37].

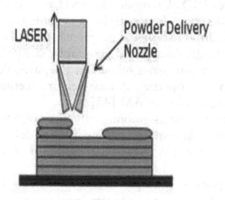

Figure 97.15 Laser-designed net molding chief interaction plot.

Pro metal. Ace metal can be described as a 3D printing (3DP) process used for creating injection molds and dies. It is a powder-based technique predominantly employing stainless steel as the raw material. The printing process involves pumping a water-based binder into a bed of metal powder [38, 39]. The powder bed is controlled using cylinders, and after each layer design is completed, the bed is lowered, and the feed cylinder supplies material for the next layer [40]. In the case of the mold, it doesn't require additional post-processing, but for the active component's construction, processes like compaction, internal penetration, and elimination are needed [41].

LMD. In this technology, the 3D printing (3DP) head is typically integrated with a mechanical arm, equipped with nozzles to deposit metal powder onto the build platform. A laser beam provides the required energy, melting the powder particles to create the desired shape, as illustrated in Figure 97.16 [12].

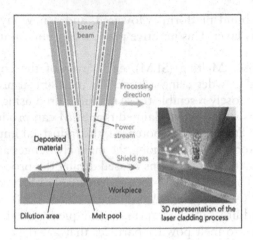

Figure 97.16 Laser metal deposition [10].

The AM design tools can be coordinated into four fundamental classifications. Three classes are strong demonstrating (computer-aided design) frameworks, point cloud preparing, and measure situated plan programming these permit the displaying, representation, transformation, and lattice of 3D plan models into STL records. The STL record change has an issue producing inner designs. The fourth kind of configuration device was acquainted with tackle the issue related to past plan instruments isolated into two sub-classifications Cellular structure design programming and geography streamlining-based programming. There have been a few uncertain issues related to the particular of material distribution and characterization of part properties. A four-venture method that thought about the assembling abilities and imperatives of AM [42].

A four-venture technique for the assembling capability of AM can be recognized as plan procedure, plan investigation, adjusting, and approval of definite math. Some proposed a decision-making framework for a viable determination of creative methodologies in AM strategy.

This framework incorporates multiple decision-making assumptions (MCDAs) and data-opening analysis (DEA) to assist producers in selecting appropriate production strategies among many choices in light of the fact that CNC machines have a few impediments related to shape intricacy, many examination exercises center around the complex plan of AM utilizing non-metallic models, instead of performance. The foundation of uses in metal images was growing at a rapid rate. Great effort is aimed at AM of metal objects based on the operational phase [43]. Figure 97.17 summarizes some of the current AM strategies for performance-based instruments.

There are numerous applications for AM counting lightweight aerospace items, car, clinical, displaying development, and energy ventures. This includes applications where low volume creation is required, high plan intricacy, and the capacity to change plan habitually. Be that as it may, AM actually needs critical improvements in terms of design, equipment, novel strategies, and equipment.

Aviation and Auto Industry. Aviation has shown interest in this innovation in light of its capacity to configuration direct metal parts utilizing materials like titanium (appropriate for planes) and the capacity to make complex, profoundly proficient items with no tool. An

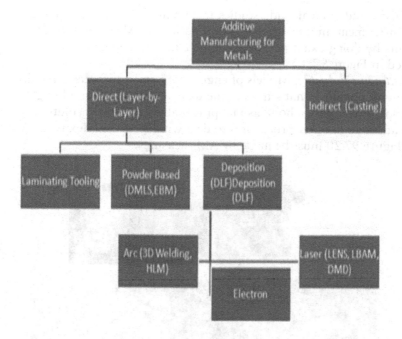

Figure 97.17 Prevailing procedures for added substance assembling of metallic items.

aircraft component with small structures with various interlocking tubes and sells multiple crossings using the AM arc attachment process (Figure 97.18) AM technology has boosted aerospace sector productivity in the accompanying manners:

- Reduce the lead time of the product by 30–70%.
- Reduce the cost of unprocessed products by up to 45%.

A 30–35% reduction in the production costs of low-volume component

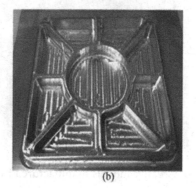

(a)   (b)

Figure 97.18

Clinical Industry. With AM, intricate primary segments can be created straightforwardly from filtered information (registered tomography pictures) that give a better perception of specific life systems. It additionally aids exact pre-medical procedure arranging, asset specialists, and clinical understudies to update different surgeries, and fills in as a specialized instrument among specialists and patients. AM utilized for the detachment of Siamese twins by doing exact pre-medical procedure anticipating AM-molded models, as demonstrated in Figure 97.19.

Design and gems industry. Models of engineering and adornments require a significant level of manual exertion what's more, time as a result of the mind-boggling conditions needed to apply imaginative thoughts in a practical way. The best solution is possible with AM than that found in other procedures to deal with these complex types. Models similar to those in Figure 97.20 must be made by AM measures.

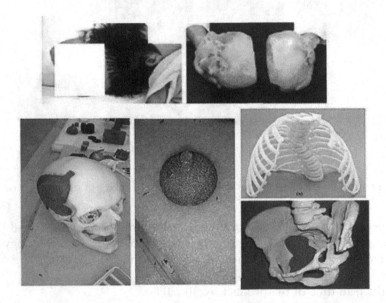

Figure 97.19 AM clinical applications [45].

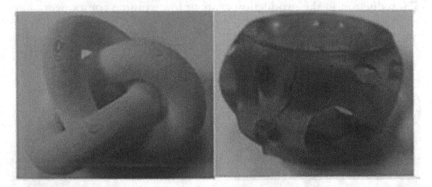

Figure 97.20 Creative models manufactured by AM innovation [46].

Impact on Climate

There are numerous ways AM can influence the climate, however, numerous analysts are zeroing in addition on the utilization of energy and the risks of the climate. *Energy utilization.* Concerning energy utilization, a correlation of different assembling boundaries for the accompanying three frameworks: FDM 3000 (Stratasys), EOSINT M250 Xtended (EOS), and Thermojet (3DS). A bunch of boundaries for lessening electrical energy utilization was recognized. Analyzed energy utilization rates (kW h/kg of part calculation) for LS, FDM, and SLA. The yield of this investigation uncovered that energy utilization was completely reliant on the length of the work. *Ecological perils.* The ecological perils of the AM business are essentially identified with the synthetic compounds (solvents) utilized in help evacuation.

Problems in Added Substance Fabricating

The relationship of various AM advancements regarding their qualities and shortcomings it is seen that none of these innovations are ideal in each measurement. For instance, the energy utilization in FDM is low contrasted with EBM; notwithstanding, EBM outperforms FDM as far as high material strength. Positively, it isn't attainable to look at two totally changed AM cycles like EBM and FDM. The goal here is to underscore the choice of proper AM strategies relying upon the application prerequisite. For example, in few assembling applications or machines, even parts with low strength are adequate to accomplish the ideal component. Around there, it is consistently desirable to utilize an easy FDM part. Depicted as a methodology that coordinates more than one assembling activity from various assembling advancements. Interestingly, sub-half breed producing is the blend of tasks of single assembling innovation. The sub-mixture order is proposed to defeat mathematical issues. The half breed cycle can likewise be characterized relying upon the crude material utilized (like composite items), working standard, (for example, laser-helped processing, and turning), or type of energy. Additionally, contrasted with singular cycles, the mix of various assembling measures has opened new application zones for delivering minimal effort segments [44]. The Global Foundation for Creation Designing (CIRP) proposed the accompanying definitions for characterizing crossover measures:

1. *Open definition.* Hybrid Manufacturing (HM) involves integrating two or more manufacturing methods into a novel and distinct process, allowing the advantages of each individual process to be harnessed synergistically.
2. *Narrow definition.* Hybrid strategies encompass the simultaneous application of various processing principles (chemical, physical, and controlled) within the same processing zone.

A half and half-cycle are characterized as training where at least two assembling strategies from various assembling innovations have blended. The advantages of consolidating numerous procedures like blending of various materials during testimony, liquefying at various warmth conditions, and statement of discrete materials.

AM Challenges

Without a doubt, there are numerous advantages, for example, plan adaptability, the capacity to print complex constructions, convenience, and customization of an item, which can be related to AM. Notwithstanding, AM innovation isn't yet developed enough to be utilized in true applications. There have been difficulties and difficulties that require research and technological advancement. Partial size limit, anisotropic machinery structures, overhead construction, high cost, production efficiency, poor accuracy, bending, spraying, wiring, spaces in upper parts, under extraction, layer defects, overpass, elephant foot, production the quantity and limitations of the materials used are challenges that require further examination and evaluation. A portion of the limits and difficulties associated with AM is portrayed as follows.

Stair-Stepping

Perhaps the greatest test to the AM cycle is the rise of the impact of steps or the situation mistake in the underlying parts. This sort of blunder isn't huge in the inward territories made; be that as it may, it significantly influences the nature of open-air conditions. Albeit, numerous techniques (foundation preparing, for example, sanding can be utilized to lessen or wipe out this issue yet additionally increment the time and cost of the entire cycle.

Anisotropic in mechanical properties and microstructure

Another test that can be seen with AM is the presence of anisotropy in microstructure and mechanical properties. AM advancements produce parts in a layer-by-layer style by restoring the photograph pitch, dissolving the fiber, or liquefying the powder bed, bringing about the age of the warm inclination. The AM parts regularly bring about various microstructure and mechanical properties along fabricate course and different headings for instance; the plates produced by FDM innovation have better strength in x, y-bearing when contrasted with the strength in the z-course (assemble bearing/directions).

Conclusions

After Headways in the assembling business rely upon driving edge research related to assembling cycles, materials, and item plan. As item intricacy builds, there is a requirement for new and inventive assembling measures. AM is a new design in progress cycles on account of the various benefits it gives similarly as the troubles it needs to endure. It has been presented to serious assessment and inside and out audits by the investigation local area. In this work, an exhaustive review related to AM has been completed. The significance of part direction, fabricate time assessment, and cost calculation has been evaluated in detail. The main component of this work is the ID of the issues related to various AM techniques. Among the essential AM difficulties, part size constraint, anisotropic mechanical properties, the structure of shade surfaces, significant expenses, helpless exactness, twisting, layer misalignment, large scale manufacturing, an impediment in the utilization of materials need further examination and examination. In light of this audit, the different parts of AM innovation can be summed up as follows. It tends to be stated that the choice of proper part bearing is pivotal in AM. It helps with improving mathematical also, dimensional blunder, decreases assemble time and limit support volume and part creation costs. The flight of stairs impact has been distinguished as the main factor influencing the part precision. In reality, the flight of stairs impact is straightforwardly corresponding to the layer thickness. It has likewise been seen that from time, which has more noteworthy importance in improving profitability, relies upon machine speed, part size, layer thickness, and fabricate direction It has been seen that there is a necessity for the advancement of a powerful also, inescapable cost model for the AM. Then, the costs of AM can be named far as far as material, machine, assembling, and work costs. The rundown of these expenses addresses the complete expense of the unit.

References

1. Zhu, Z., Dhokia, V., Nassehi, A., et al. (2013). A review of hybrid manufacturing processes—state of the art and future perspectives. *International Journal of Computer Integrated Manufacturing, 26,* 596–615. Gibson, I., Rosen, D. W., and Stucker, B. (2009). Additive Manufacturing Technologies: Rapid Prototyping to Direct Digital Manufacturing. Berlin: Springer. Zeltmann, S. E., Gupta, N., Tsoutsos, N. G., et al. Manufacturing and security challenges in 3D printing. *Journal of Orthomolecular Medicine, 68,* 1872–1881.

2. Wohlers, T., and Gornet, T. (2014). History of additive manufacturing. wohlers report. http://www.wohlersassociates.com/history2014.pdf.

3. Williams, C. B., Mistree, F., and Rosen, D. W. (2011). A functional classification framework for the conceptual design of additive manufacturing technologies. *Journal of Mechanical Design, 133,* 121002.

4. Manogharan, G. (2014). Hybrid Manufacturing: Analysis of Integrating Additive and Subtractive Methods. Raleigh, NC: North Carolina State University.

5. Huang, S. H., Liu, P., Mokasdar, A., et al. (2013). Additive manufacturing and its societal impact: a literature review. *International Journal of Advanced Manufacturing Technology, 67,* 1191–1203.

6. Bikas, H., Stavropoulos, P., and Chryssolouris, G. (2016). Additive manufacturing methods and modelling approaches: a critical review. *International Journal of Advanced Manufacturing Technology, 83,* 389–405.

7. Ahmad, A., Darmoul, S., Ameen, W., et al. (2015). Rapid prototyping for assembly training and validation. *IFAC Papersonline, 48,* 412–417.

8. Yagnik, P. D. (2014). Fused deposition modeling a rapid prototyping technique for product cycle time reduction cost effectively in aerospace applications. *IOSR Journal of Mechanical and Civil Engineering,* 62–68.

9. Huang, V. (2007). Fabrication of ceramic components using freeze form extrusion fabrication. PhD Dissertation, Department of Materials Science and Engineering. University of Missouri-Rolla, Rolla, MO.

10. Mathur, R. (2016). 3D printing in architecture. *International Journal of Innovation Science, 3,* 583–591.

11. Mueller, B., and Kochan, D. (1999). Laminated object manufacturing for rapid tooling and pattern making in foundry industry. *Computers in Industry, 39,* 47–53.

12. Cesarano, J. (1997). Robocasting: direct fabrication of ceramics from colloidal suspensions. *In: Proceedings of Solid Free-Form Fabrication Symposium,* Austin, TX, 11–13, August 1997, (pp. 25–36).

13. Cesarano, J., King, B. H., and Denham, H. B. (1998). Recent developments in robocasting of ceramics and multimaterial deposition. In Proceedings of Solid Freeform Fabrication Symposium, Austin, TX, (pp. 11–13).

14. Li, Q., and Lewis, J. A. (2003). Nanoparticle inks for directed assembly of three-dimensional periodic structures. *Advanced Materials, 15,* 1639–1643.

15. Chua, C. K., Leong, K. F., and Lim, C. S. (2003). Rapid Prototyping: Principles and Applications. 2nd ed. Singapore: World Scientific Publishing Company.

16. Hoffmann, J. (2000). Thermo Jet und medizinische ModelleEin Erfahrungsbericht. RAPROMED. https://tu-dresden. de/ing/maschinenwesen/if/ff/ressourcen/dateien/pazat/for schung/for_ber_ pdf_html/lit00_html/hoff-001.pdf?lang=en.

17. Zhang, W., Leu, M. C., Ji, Z., et al. (1999). Rapid freezing prototyping with water. *Master of Design, 20,* 139–145.

18. Zhang, W., Leu, M. C., Ji, Z., et al. (1997). Method and apparatus for making three-dimensional objects by rapid freezing. Patent US6253116B1, USA.

19. Hong, S. Y., Kim, Y. C., Wang, M., Kim, H. I., Byun, D. Y., Nam, J. D., and Suhr, J. (2018). Experimental investigation of mechanical properties of UV-Curable 3D printing materials. *Polymer*, 145, 88–94.

20. Steenhuis, H. J., and Pretorius, L. (2016). Consumer additive manufacturing or 3D printing adoption: an exploratory study. *Journal of Manufacturing Technology Management*, 27(7), 990–1012.

21. Ameen, W., Al-Ahmari, A., Mohammed, M. K., et al. (2018). Manufacturability of overhanging holes using electron beam melting. *Metals*, 8, 397.

22. Murr, L., Quinones, S. A., Gaytan, S. M., et al. (2009). Microstructure and mechanical behavior of Ti–6Al–4V produced by rapid-layer manufacturing, for biomedical applications. *Journal of the Mechanical Behavior of Biomedical Materials,* 2, 20–32.

23. Murr, L. E., Gaytan, S. M., Ceylan, A., et al. (2010). Characterization of titanium aluminide alloy components fabricated by additive manufacturing using electron beam melting. *Acta Materialia*, 58, 1887–1894.

24. Koike, M., Martinez, K., Guo, L., et al. (2011). Evaluation of titanium alloy fabricated using electron beam melting system for dental applications. *Journal of Materials Processing Technology,* 211, 1400–1408.

25. Karlsson, J. (2015). Optimization of electron beam melting for production of small components in biocompatible titanium grades. Uppsala: Acta Universitatis Upsaliensis.

26. Ameen, W., Ghaleb, A. M., Alatefi, M., et al. (2018). An overview of selective laser sintering and melting research using biblio-metric indicators. *Virtual and Physical Prototyping,* 13, 282–291.

27. Casavola, C., Campanelli, S. L., and Pappalettere, C. (2009). Preliminary investigation on distribution of residual stress generated by the selective laser melting process. *Journal of Strain Analysis for Engineering Design*, 44, 93–104.

28. Vandenbroucke, B., and Kruth, J. P. (2007). Selective laser melting of biocompatible metals for rapid manufacturing of medical parts. *Rapid Prototyping Journal,* 13, 196–203.

29. Yasa, E., and Kruth, J. (2011). Application of laser re-melting on selective laser melting parts. *Advances in Production Engineering and Management*, 6, 259–270.

30. Paul, R. (2013). Modeling and optimization of powder based additive manufacturing (AM) processes. PhD Thesis, University of Cincinnati, Cincinnati, OH.

31. Klingbeil, N. W., Beuth, J. L., Chin, R. K., et al. (1998). Measurement and modeling of residual stress-induced warping in direct metal deposition processes. In: SFF symposium, 1998, (pp. 367–374).. http://sffsymposium.engr.utexas.edu/Manu-scripts/1998/1998-40-Klingbeil.pdf.

32. Zaeh, M. F., and Branner, G. (2010). Investigations on residual stresses and deformations in selective laser melting. *Production Engineering,* 4, 35–45.

33. Clijsters, S., Craeghs, T., and Kruth, J. P. (2011). A priori process parameter adjustment for SLM process optimization. In Proceedings of the 5th International Conference on Advanced Research and Rapid Prototyping: Innovative Developments in Virtual and Physical Prototyping, 28 September–1 October 2011, (pp. 553–560). USA: CRC Press.

34. Abdulhameed, O., Al-Ahmari, A., Ameen, W., and Mian, S. H. (2019). Additive manufacturing: Challenges, trends, and applications. *Advances in Mechanical Engineering*, 11(2), 1687814018822880.

35. Boddu, M. R., Landers, R. G., and Liou, F. W. (2001). Control of laser cladding processes for rapid prototyping: a review. In Proceedings of the 12th Annual Solid Freeform Fabrication Symposium, Austin, TX, 6–8 August 2001, (pp. 460–467).

36. Griffith, M. L., Ensz, M. T., Puska, J. D., et al. (2000) Understanding the microstructure and properties of components fabricated by laser engineered net shaping (LENS). In Proceedings of Materials Research Society symposium. Boston, MA, 27 November–1 December 2000. (Vol. 625). http://mfgshop.sandia.gov/1400 ext/MRS00mg.pdf

37. Hedges, M., and Kiecher, D. (2001). Laser engineered net shaping™—technology and applications. In Proceedings of 3rd National Conference on Rapid Prototyping, Tooling and Manufacturing, High Wycombe, UK, 20–21 June 2001, (pp. 17–23).

38. Hanemann, T., Bauer, W., Knitter, R., et al. (2006). Rapid prototyping and rapid tooling techniques for the manufacturing of silicon, polymer, metal and ceramic microdevices. In Leondes. C. T. (eds.) MEMS/NEMS. Boston, MA: Springer, pp. 801–869.

39. Cooper, K. (2001). Rapid Prototyping Rechnology, New York: Marcel Dekker.

40. Kruth, P. P. (1991). Material incress manufacturing by rapid prototyping techniques. *CIRP Annals: Manufacturing Technology*, 40, 603–614.

41. Xu, F., Loh, H., and Wong, Y. (1999). Considerations and selection of optimal orientation for different rapid prototyping systems. *Rapid Prototyping Journal*, 5, 54–60.

42. Chen, C. C., and Sullivan, P. A. (1996). Predicting total build-time and the resultant cure depth of the 3D stereolithography process. *Rapid Prototyping Journal*, 2, 27–40.

43. Ding, D., Shen, C., Pan, Z., et al. (2016). Towards an automatedrobotic arcwelding-based additive manufacturing system from CAD to finished part. *Computer-Aided Design*, 73, 66–75.

44. Tukuru, N., Gowda, S. K. P., Ahmed, S. M., et al. (2008). Rapid prototype technique in medical field. *Research Journal of Pharmacy and Technology*, 1, 341–344.

45. Petrovic, V., Gonzalez, J. V. H., Fernando, O. J., et al. (2011). Additive layered manufacturing: sectors of industrial application shown through case studies. *International Journal of Production Research*, 49, 1061–1079.

46. Shahrubudin, N., Lee, T. C., and Ramlan, R. J. P. M. (2019). An overview on 3D printing technology: Technological, materials, and applications. *Procedia Manufacturing*, 35, 1286–1296.

47. Martin, J. H., Yahata, B. D., Hundley, J. M., Mayer, J. A., Schaedler, T. A., and Pollock, T. M. (2017). 3D printing of high-strength aluminium alloys. *Nature*, 549(7672), 365–369.

48. Saheb, S. H., and Kumar, J. V. (2020, October). A comprehensive review on additive manufacturing applications. *In AIP Conference Proceedings*, 2281(1). AIP Publishing.

49. Saheb, S. H., and Kumar, J. V. (2020). A Comprehensive Review on Additive Manufacturing Applications. *AIP Conference Proceedings*, 2281, 020024. https://doi.org/10.1063/5.0026202.

50. Singh, T., Kumar, S., and Sehgal, S. (2020). 3D printing of engineering materials: A state of the art review. *Materials Today: Proceedings*, 28, 1927–1931.

51. Rao, P. S., Jain, P. K., and Dwivedi, D. K. (2015). Electro chemical honing (ECH) of external cylindrical surfaces of titanium alloys. *Elsevier Publishers Journal of Procedia Engineering*, 100, 936–945.

52. Rao, P. S., Jain, P. K., and Dwivedi, D. K. (2015). Precision finishing of external cylindrical surfaces of EN8 steel by electro chemical honing (ECH) process using OFAT technique. *Elsevier Publishers, Journal of Materials Today Proceedings*, 2, 3220–3229.

98 Response surface modelling of end milling cooling environment of biocompatible Ti-6Al-4V alloy

Rahul Davis[1,a], Abhishek Singh[1,b], Robson Bruno Dutra Pereira[2,c], and Leonardo Rosa Ribeiro da Silva[3,d]

[1]Department of Mechanical Engineering, National Institute of Technology Patna, Patna, India

[2]Department of Mechanical Engineering, Centre for Innovation in Sustainable Manufacturing—CIMS Federal University of São João del-Rei, São João del-Rei, Minas Gerais, Brazil

[3]Federal University of Uberlandia, School of Mechanical Engineering, Av. João Naves de Ávila, Uberlândia, MG, Brazil

Abstract: Outstanding resistance toward corrosion and strength to weight ratio and outstanding properties of operating effectively even at extremely high temperatures are some of the most remarkable attributes of titanium-based alloys. Moreover, Ti-6Al-4V alloy is the preferable alloy of titanium in the area of biomedical implant applications such as scaffold, bracket, screw, plate, dental implants, etc. Due to its superior biocompatibility to many other biocompatible materials. However, spring back, pressure load, impoverished surface integrity, heat stress, etc. are some of the commonly known issues faced during traditional subtractive manufacturing of Ti-6Al-4V alloy. Therefore, this work demonstrates a comparative study consisting of wet (using conventional cutting oil), cryogenic (using liquid nitrogen), and hybrid (with the aid of conventional cutting oil and liquid nitrogen simultaneously) cooling during end milling of biocompatible Ti-6Al-4V alloy, with an attempt to improve the surface characteristics. Regression models are achieved to predict surface roughness and microhardness. The best performance in terms of surface roughness and microhardness is achieved with the cryogenic milling environment.

Keywords: Coated end mill, End milling, surface microhardness, surface roughness, Ti-6Al-4V alloy, wet, cryogenic, and hybrid cooling.

Introduction

The constant progression of various types of physical sicknesses, viruses including COVID-19, mental traumas, and tissue damage/fracture in the world is causing a continuous deterioration of human health. However, the medical industry and healthcare organizations are also in the process of finding possible solutions to all such issues. In the case of damage of a tissue/bone of the human body, the possible solutions are present in the form of varieties of implantable biomedical devices or implants such as screws, nails, rods, etc. which are made up of a specific class of materials, known as biocompatible materials. The commonly used biocompatible materials are metallic, inorganic, polymeric, and ceramic. The biocompatibility of these materials can be interpreted as their ability through which they can stay with the living tissues inside the human body without causing any toxicity or

[a]rahuldavis2012@gmail.com, [b]abhishek.singh@nitp.ac.in, [c]robsondutra@ufsj.edu.br, [d]leonardo.rrs@gmail.com

immunogenicity [1]. Additionally, with the rise in the cases requiring biomedical implants, the development, remodeling, and application of suitable manufacturing techniques for the same is of great significance. In this connection, the machining (subtractive manufacturing) of biocompatible materials has been spotted as one of the most attractive areas of researchers for the last many decades. A subtractive manufacturing technique can exhibit its efficiency primarily in terms of the tight tolerances of produced geometrical features, improved surface integrity along with tool wear, and considerable material subtraction rate (MSR) [2]. However, while considering the subtractive manufacturing of metallic implant biocompatible materials such as stainless steel 316L, titanium-based alloys, cobalt-based alloys, nickel-titanium shape memory alloys, magnesium alloys, the most commonly implemented traditional subtractive manufacturing techniques are turning, drilling, and milling. Bone screws and plates, prosthetic devices, dental implants, artificial ligaments, and heart valves, etc. are some of the examples of implants manufactured via subtractive manufacturing techniques. Although nowadays the industry 4.0 emphasizes the additive manufacturing (AM) of implants, the role of subtractive manufacturing cannot be ignored due to its proficiency as a post-processing technique [3].

Ti-6Al-4V alloy is one of the most prominent titanium-based alloys which finds its application in the area of pacemaker covering, joints replacement, spine system, etc. However, the traditional subtractive manufacturing of Ti-6Al-4V alloy exhibits various challenges such as serrated chips, burr formation, excessive heat generation and tool wear, residual stress, tool wear, poor surface integrity. Therefore, sustainable manufacturing has provided an effective way to improve the difficult subtractive manufacturing of Ti-6Al-4V alloy by employing a cryogenic coolant such as liquid nitrogen (LN_2). Thus, such a provision terms the process as cryogenic subtractive manufacturing or machining [4, 5]. The extremely low temperature of the cryogenic coolants (around −196°C) not only substantially dampens the adverse effects of the high temperatures during the interaction of cutting tool and workpiece, but also improves all the surface characteristics including the integrity of the surface, tool wear, MSR, etc. [4, 6]. In addition, some recent research also included the application of various types of hybrid cooling such as cryogenic and minimum quantity lubrication (MQL), CO_2 snow-based cryogenic and MQL, LN_2, and cutting oil during (wet machining) the subtractive manufacturing of difficult-to-cut materials and noticed the noteworthy positive impact of the same on the surface characteristics [6–8]. The limited study comprising the subtractive manufacturing of biocompatible Ti-6Al-4V alloy with distinct cooling ambiances and promising findings have motivated the present research to perform the end milling process with the assistance of wet, cryogenic, and hybrid cooling, and thus, to extend the scope of this field.

Experimental Details

The present research was conducted on Ti-6Al-4V alloy of biomedical grade, which had Al (6.21%), V (3.92%), C (<0.082%), Fe (<0.24%), N_2 (<0.052%), and O_2 (<0.21%), as the alloying elements. All the experimental trials were executed on a XL Mill (PC-based 3 axes) milling machine setup using wet, cryogenic, and hybrid cooling environments. An aluminum titanium nitride-coated end mill cutting tool (3700 HV) was chosen for each subtractive operation since such coatings minimize the tool wear considerably [9]. The surface roughness and microhardness were examined using a Precise/TR110 Plus tester and Blue Star Ltd./ BSHT-FHV1-50 Vickers hardness measuring instrument. The surface roughness and microhardness were basically measured at 3 distinct and random locations on each processed surface, followed by averaging the numeric values. The Field emission scanning electron microscope (FESEM)-GeminiSEM 500-Zeiss facilities were availed to examine the produced surfaces of interest via FESEM images. To study the milling process with distinct lubri-colling environments an L27 experimental design was considered with the facing toolpath. The L27 orthogonal array whit 4 factors is a 3^{4-1} fractional factorial design. Considering the 3^{4-1} array coded in 3 levels (−1,0,1) the levels of the 4th factor can be obtained as $x_4 = (x_1 + x_2 + x_3)mod$ (3) − 1. Table 98.1 presents the input variables and their levels where the symbols WME, HME, and CME stand respectively for wet-milling, hybrid-milling, and cryogenic-milling environments. The parameters' range was chosen based on the preliminary test results.

Table 98.1: Input variables and levels.

Input		Measurement unit	Levels		
			−1	0	1
x_1	n	rev/min	1200	2000	2800
x_2	f	mm/rev	0.3	0.6	0.9
x_3	a_p	mm	0.10	0.15	0.20
x_4	env	-	WME	HME	CME

Individual response models, via conventional least-squares, are obtained for each singular response and to the first principal component score. Second-order models were obtained in the light of the significance of categorical input variables of the cooling environment. The significance test is executed with the help of t-test (that pertains to the regression coefficients). Whereas Shapiro-Wilk normality test is employed to examine the normality of the residuals. The significance level adopted for the present analysis was $\alpha = 0.05$. Further, the data was analyzed statistically in R-language [10] through RStudio software. The packages ggplot2 [11] and ggpubr [12] were incorporated to achieve the necessary plots.

Results and Discussion

Table 98.2 displays the L27 experimental-design considering the spindle-speed (n) in rev/min, feed-rate (f) in mm/rev, cutting-depth (a_p) in mm, the cutting environment (env), and the responses R_a and *mhd* results. Average surface roughness results were from 0.34 to 1.43 μm, while microhardness results were from 263.9 to 396.6 HV.

Table 98.3 shows the coded response model coefficients of the responses R_a and *mhd*, while the numerical input variables and the categorical variables are considered. The models are also presented in Equations 1 and 2, respectively. In addition to additive constants (to alter the cooling ambiances), both models also present intercept, linear, quadratic, and interaction coefficients. CME was considered as the reference environment. To change the environment, it is necessary to sum up the related constant presented in Table 98.3.

Both the models demonstrate (in Table 98.3) a significant statistical difference among the environments. For both models, the linear effects of n and f (x1 and x2) are statistically significant. No quadratic or interaction effect of the numerical input variables were observed for both the models. The R_a and *mhd* models present good data variability accounting, with 0.9528 and 0.9967 of explained proportion, respectively. The assumption

Table 98.2: L27 experimental-design and outcomes.

Run	Milling process input variables				Output variables	
	n (rpm)	*f (mm/rev)*	*a$_p$ (mm)*	*env*	*R$_a$ (μm)*	*mhd (HV)*
1	1400	0.2	0.1	WME	0.90	298.71
2	1400	0.2	0.15	CME	0.57	341.32
3	1400	0.2	0.20	HME	0.73	323.32
4	1400	0.4	0.10	CME	0.70	329.42
5	1400	0.4	0.15	HME	0.87	311.33
6	1400	0.4	0.20	WME	1.21	282.51
7	1400	0.6	0.10	HME	0.99	301.46
8	1400	0.6	0.15	WME	1.43	263.95
9	1400	0.6	0.20	CME	0.83	315.82
10	2100	0.2	0.10	WME	0.76	321.43
11	2100	0.2	0.15	CME	0.45	369.31
12	2100	0.2	0.20	HME	0.59	349.43
13	2100	0.4	0.10	CME	0.56	356.22
14	2100	0.4	0.15	HME	0.72	339.32
15	2100	0.4	0.20	WME	0.89	309.21
16	2100	0.6	0.10	HME	0.84	326.42
17	2100	0.6	0.15	WME	1.04	294.34
18	2100	0.6	0.20	CME	0.68	342.92
19	2800	0.2	0.10	WME	0.61	346.81
20	2800	0.2	0.15	CME	0.34	396.63
21	2800	0.2	0.20	HME	0.43	379.74
22	2800	0.4	0.10	CME	0.41	381.92
23	2800	0.4	0.15	HME	0.58	363.52
24	2800	0.4	0.20	WME	0.74	332.41
25	2800	0.6	0.10	HME	0.73	348.33
26	2800	0.6	0.15	WME	0.88	317.44
27	2800	0.6	0.20	CME	0.52	366.64

of normality of the residuals cannot be rejected, as the p-value $\gg 0.05$ of the Shapiro-Wilk normality tests of R_a and *mhd* models.

$$\hat{R}_a = 0.58310 - 1661x_1 + 0.1422x_2 + 0.0067x_3 + 0.0228x_1^2 \qquad (1)$$
$$- 0.0022x_2^2 - 0.0356x_3^2 - 0.025x_1x_2 - 0.020x_1x_3 - 0.0328x_2x_3$$

$$\hat{mhd} = 355.5759 + 25.8667x_1 - 13.8611x_2 - 0.4833x_3 - 0.8889x_1^2 \qquad (2)$$
$$- 0.4389x_2^2 + 1.0278x_3^2 - 0.7167x_1x_2 + 0.8000x_1x_3 + 0.5389x_2x_3$$

Figure 98.1 presents the effects plot of R_a response variable when numerical input variables and environments are taken into consideration. Spindle-speed displays a negative influence for roughness, whereas the feed-rate presents a favorable impact on roughness. Besides, the outcomes of significance test (presented in Table 98.3) nullify the influence of cutting-depth. Further, the cryogenic environment can be identified as the best cooling ambiance.

Table 98.3: Response models of R_a and *mhd*.

Term	R_a			*mhd*		
	Estimate	Pr(>|t|)		Estimate	Pr(>|t|)	
Intercept	0.5831	1.11E-11	***	355.5759	<2.00E-16	***
x_1	-0.1661	1.23E-09	***	25.8667	<2.00E-16	***
x_2	0.1422	1.03E-08	***	-13.8611	2.99E-15	***
x_3	0.0067	0.6055		-0.4833	0.2800	
x_1^2	0.0228	0.3145		-0.8889	0.2520	
x_2^2	-0.0022	0.9205		-0.4389	0.5650	
x_3^2	-0.0356	0.1251		1.0278	0.1890	
env HME	0.1250	0.0020	**	-16.9500	2.22E-10	***
env WME	0.3778	2.04E-10	***	-48.1556	<2.00E-16	***
x_1x_2	-0.0250	0.1271		-0.7167	0.1950	
x_1x_3	-0.0200	0.2159		0.8000	0.1510	
x_2x_3	-0.0328	0.1550		0.5389	0.4820	
R_{adj}^2	0.9528			0.9967		
Shapiro-wilk	0.2204#			0.9142#		

\# p-values
Signif. codes: 0 '***' 0.001 '**' 0.01 '*' 0.05 '.' 0.1 ' ' 1

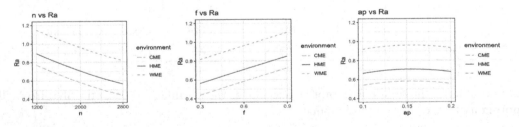

Figure 98.1 Effects plot of R_a.

Figure 98.2 shows how the input variables and the environments cause variation in the obtained microhardness.

The spindle-speed favors microhardness and with the cryogenic milling environment, it achieved the highest microhardness. Whereas the feed-rate deteriorates the microhardness.

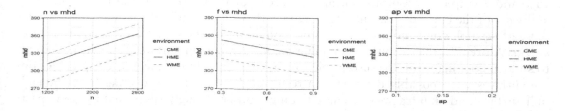

Figure 98.2 Effects plot of mhd with.

These results also assure a high negative correlation between R_a and *mhd*. The models displayed in Table 98.3 can be modified with respect to the cooling ambiance. Therefore, Figure 98.3 shows the contour plot of R_a, whereas Figure 98.4 shows the contour plot of *mhd* when selecting the cryogenic milling environment, which presented the best performance. Thus, the explained effects of the numerical input variables might be substantiated. The lowest value of roughness and the highest value of microhardness were observed when the highest spindle-speed was coupled with the lowest feed-rate (in both combinations of cutting tool-environment).

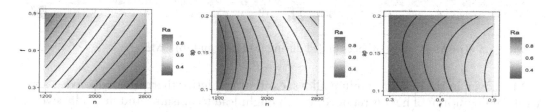

Figure 98.3 Contour plot of R_a with CME.

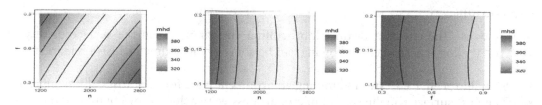

Figure 98.4 Contour plot of *mhd* with CME.

The quality of the surfaces produced using optimal process input variables amid HME and CME can be seen via FESEM images displayed in Figure 98.5 (a) and 98.5 (b), respectively. The surface defects such as abrasion and feed marks, including irregularity and adhered particles have been highlighted. The credible reason for the deeper and more visible marks in Figure 98.5 (a) is the warmer cooling environment of HME compared to CME (Figure 98.5 (b)), which could most likely dissipate lesser amount of generated heat. This would have possibly caused more amount of heat to be retained at end mill cutting edge and workpiece subsurface during HME compared to CME, leading to more melting and wear of the end mill's cutting edge amid HME. Consequently, machining with a worn-out cutting tool would induce feed and abrasion marks on the surface, followed by surface cracks. Moreover, the subsurface retaining a considerable amount of heat, when exposed to sudden cooling, would be susceptible to cracking. In addition, the adhered particles on the surfaces were possibly detected as a result of delamination of coating from the end mill surface and chip fragments. These findings can be strongly supported by some past investigations [13, 14].

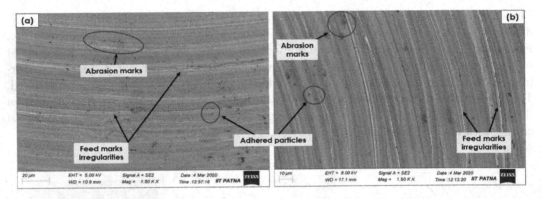

Figure 98.5 Ti-6Al-4V alloy surfaces milled during (a) HME and (b) CME.

Moreover, in the case of titanium alloys, exhibiting inferior thermal conductivity, a relatively high amount of heat is retained, which might lead to mushing and thus adhesiveness. The machining input variables with relatively higher levels and worn-out cutting edge of the cutting tool are mainly liable for the formation and wavy morphology of the surface, as they deform the surface at a high strain rate [14].

Conclusions

In this work, the effect of distinct lubri-colling environments on the surface quality of Ti-6Al-4V alloy was studied. Following can be some of the main conclusive remarks:

- Spindle-speed, feed-rate, and milling environment were found statistically significant.
- The best surface roughness and microhardness were achieved by combining the cryogenic milling environment, 2800 RPM spindle-speed, and 0.3 mm/rev feed-rate.
- FESEM images presented some of the surface defects through visual inspection, which could also demonstrate the favorable influence of CME on the overall surface quality of Ti-6Al-4V alloy.

References

1. Ratner, B. D. (2011). The biocompatibility manifesto: biocompatibility for the twenty-first century. *Journal of Cardiovascular Translational Research*, 4, 523–527. https://doi.org/10.1007/s12265-011-9287-x

2. Axinte, D., Guo, Y., Liao, Z., Shih, A. J., M'Saoubi, R., and Sugita, N. (2019). Machining of biocompatible materials—recent advances. *CIRP Annals*, 68(2), 629–652. doi:10.1016/j.cirp.2019.05.003.

3. Saptaji, K., Gebremariam, M. A., and Azhari, M. A. B. M. (2018). Machining of biocompatible materials: a review. *International Journal of Advanced Manufacturing Technology*, 97, 2255–2292.

4. Sun, S., Brandt, M., Palanisamy, S., and Dargusch, M. S. (2015). Effect of cryogenic compressed air on the evolution of cutting force and tool wear during machining of Ti-6Al-4V alloy. *Journal of Materials Processing Technology*, 221, 243–254.

5. Krishnamurthy, G., Bhowmick, S., Altenhof, W., and Alpas, A. T. (2017). Increasing efficiency of Ti-alloy machining by cryogenic cooling and using ethanol in MRF. *CIRP Journal of Manufacturing Science and Technology*, 18, 159–172.

6. Davis, R., and Singh, A. (2020). Tailoring surface integrity of biomedical Mg Alloy AZ31B using distinct end mill treatment conditions and machining environments. *Journal of Materials Engineering and Performance*, 29, 7617–7635.

7. Scoop, J., Sales, W. F., and Jawahir, I. S. (2017). High speed cryogenic finish machining of Ti-6Al4V with polycrystalline diamond tools. *Journal of Materials Processing Technology*, 250, 1–8. https://doi.org/10.1016/j.jmatprotec.2017.07. 002

8. Al-Ghamdi, K. A., Iqbal, A., and Hussain, G. (2014). Machinability comparison of AISI, 4340 and Ti-6Al-4V under cryogenic and hybrid cooling environments: a knowledge engineering approach. *Proceedings of the Institution of Mechanical Engineers, Part B: Journal of Engineering Manufacture*, 229(12), 2144–2164.

9. Özel, T., Thepsonthi, T., Ulutan, D., and Kaftanoğlu, B. (2011). Experiments and finite element simulations on micro-milling of Ti–6Al–4V alloy with uncoated and cBN coated micro-tools. *CIRP Annals*, 60, 85–88.

10. R Core Team (2020). R: A language and environment for statistical computing. R Foundation for Statistical Computing, Vienna, Austria. URL https://www.R-project.org/.

11. Wickham, H. (2016). ggplot2: Elegant Graphics for Data Analysis. New York: Springer-Verlag.

12. Gray, R. J. (2020). Sorry, we're open: Golden open-access and inequality in non-human biological sciences. *Scientometrics*, 124(2), 1663–1675.

13. McAllister, E. W. (2013). *Pipeline rules of thumb handbook: a manual of quick, accurate solutions to everyday pipeline engineering problems*. Gulf Professional Publishing.

14. Uddin, M. S., Pham, B., Sarhan, A., Basak, A., and Pramanik, A. (2017). Comparative study between wear of uncoated and TiAlN-coated carbide tools in milling of Ti6Al4V. *Advanced Manufacturing*, 5, 83–91. https://doi.org/10.1007/s40436-016-0166-1.

99 Study on drag coefficient for the flow past an aerofoil

Chirag Arya[a], Dinesh Deshwal[b], Monika Deshwal[c], and Ankit Tyagi[d]

Department of Mechanical Engineering, Faculty of Engineering & Technology, Shree Guru Gobind Singh Tricentenary University, Haryana, India

Abstract: Wind energy is one of the practical, sustainable power sources. Here, a wind tunnel is used for the aerodynamic study of an airfoil. Wind speed ought to be in the scope of 5m/s to 15m/s for its viable age. Energy age from wind includes a partial of electrical, mechanical, and wind-related boundaries and Investigation of these boundaries and their impact on research subject for a long time. This exploration paper is centered on the fundamentals of an air stream and its testing. Many driving exploration associations, research labs in colleges and wind energy enterprises throughout the planet are attempting to create small or smaller than usual air streams for taking care of the power needs of metropolitan houses. In this research paper, there will be a comparative study of practical simulation in lab and flow simulation on the Solidworks, and its ease in differentiating the results of physical simulation and virtual simulations

Keywords: Aerofoil, Drag coefficient, dynamic viscosity, operating fuel, Specific heat.

Introduction

Airfoil is similar to the wing of airplanes, and Wind tunnels are installations whose primary purpose is to experiment with aerodynamic characteristics for different models. It's a tool used in aerodynamic research to study the effects of wind moving over the solid object. The wind tunnel was designed and fabricated to reduce the drag and lift forces. Aerodynamic drag offers 50% resistance from net resistance offered on vehicles. A typical open-circuit wind tunnel consists of a motor and fan unit, settling chamber, contraction cone, test section, and diffuser. The main work of the wind tunnel is to improve the design according to the aerodynamic shapes. In many wind tunnel tests, mean velocity in the test section is determined indirectly by measuring stagnation and static pressures. One or several pressure probes can be used in these measurements. Finding the aerodynamic performance of airfoil is a bit typical task since more time is required for it [1].

Many scientists are trying to calculate the value of the Coefficient of drag and minimize it by varying the designing parameters of airfoil on the other hand pressure parameters can be changed, analysis of types of flow is possible using the Reynolds number for the optimization of an airfoil. Here we are trying to do flow simulation on solid works and compare

[a]chiragarya.1000@gmail.com, [b]dinesh_fet@sgtuniversity.org, [c]monika.nain@sgtuniversity.org, [d]ankit.feat@sgtuniversity.org

results with a physical experiment performed in the laboratory. Similarly, as the name recommends, an airstream is a cylinder or passage with man-made windblown through it at a specific speed. Designers and Researchers put a plane, vehicle or wind factory model in the path and afterward study how air moves around the model. Airstreams work on the possibility that a fixed model with air moving around it carries a similar way a real, full-scale object under test traveling through fixed air does [2]. An airfoil or aerofoil is the state of a wing or edge (of a propeller, rotor, or turbine) or sails as found in a cross-area. An airfoil-formed body traveled through a liquid creates a streamlined power. The part of this power opposite to the bearing of movement is called lift. The segment corresponding to the course of movement is called drag [3].

Procedure

Operating Fluid

When the flow simulation is performed, then working fluid creates a huge variation in the reading. Multiple simulations are available on the system at a specific temperature, pressure, or open/close circuit [4].

Type of Flow

The two-dimensional and precarious stream was researched mathematically in the scope of conditions $60 \leq Re \leq 160$ and $0.5 \leq n \leq 2.0$. Over this scope of Reynolds numbers, the stream is occasional on schedule. Laminar and turbulent flow depends on the value Reynolds number (Re) [5].

Need of Wind Tunnel

There are numerous strategies through which test data can be gotten to take care of optimal design issues. Drop tests, water burrows, ballistic reach Flight tests, rocket flights are a portion of the courses through which the crucial conduct could be determined. Wind tunnel is used to study the behavior of solid when it is placed in the stream fluid. Coefficient of drag, lift force, stress generation and research work can be originated with the help of wind tunnel [6].

Angle of Attack-

At the point when you stretch your arm out through window of vehicle which moves at fast, you can feel that your arm pull back while slamming into approaching air and when you hold your hand outside of window in corresponding to the course of street and just slanted it in certain point, you feel like your hand push up it is a result of approaching air strikes at your hand. Approach is the point between the reference line of a body and relative breeze or approaching air is angle of attack [7].

Theory of Aerodynamics

As per this hypothesis, a stream is viewed as compressible if its thickness changes along smooth out. Mach number assumes a vital part in principle of optimal design. If match number is under 1 demonstrate subsonic stream and on the off chance that it is more noteworthy than 1 it is supersonic stream. Mach number to a limited extent or the entirety of the stream surpasses 0.3. The Mach 0.3 worth is somewhat self-assertive, however, it is utilized because gas streams with a Mach number beneath that worth exhibit changes in thickness of under 5%. Moreover, the most extreme 5% thickness change happens at the stagnation point (the point on the item where stream speed is zero).

Geometry

In streamlined features, Airfoil configuration is a significant aspect. Unique flight systems show various outcomes. There are essential dissimilarities among symmetric and lopsided aerofoil like at zero degree approach; Unbalanced airfoils can create lift. At the same time, incessant reversed flights suit symmetric airfoil as on account of an aerobatic plane [8].

Geometry of wind tunnel is optimized by multiple methods to get better results these methods are following [9].

1. Line Search Method
2. Particle Swarm Optimization

Aerodynamic Loads on an Airfoil Section

As per the BEM theory, air loads on an airfoil are relative to the influential pressing factor at just that segment. Lift and drag coefficients are proportionality constants that empower the estimation of streamlined powers, as follows [10]:

$$L = CL \cdot Pd \cdot c \tag{1}$$

$$D = CD \cdot Pd \cdot f \tag{2}$$

$$Pd = 0.5\rho \cdot Vr^2 \tag{3}$$

Where,
 L, lift force per unit span (lb/ft)
 D, drag force per unit span (lb/ft)
 CL, lift coefficient of the section
 CD, drag coefficient of the section
 Pd, dynamic pressure (lb/ft2)
 ρ, air density (slugs/ft2)
 C, chord length (ft)

Applications of Wind Tunnels

- Aerodynamic application –
- Non-aerodynamic application-
  a. Civil engineering
  b. Automobile engineering
  c. Calibration of instruments
- Civil engineering—important class of civil engineering applications in which wind tunnel tests wind effect on structure.

- Automobile engineering—wind tunnel testing was applied to automobiles, to determine aerodynamics forces and reduces the power required move vehicle on road at uniform speed.
- Calibration of instrument—The calibration of a wind tunnel consists in determining the mean values and uniformity of various flow parameters in the region to be used for model testing. The parameters basic to any wind-tunnel calibration are stagnation pressure and temperature, velocity or Mach number, and flow angularity.

Drag Coefficient

Drag coefficient is used to find out resistance which is offered by the fluid on the object. Understanding the drag qualities of articles in fluid stream is fundamental for designing plan angles, for example to lessen the drag on autos and airplanes for speed and mileage. Coefficient of drag is definitely not a steady yet differs as an element of speed, stream bearing, object position, object size, liquid thickness and liquid consistency. The obstruction of a body as it travels through a liquid is of extraordinary specialized significance in hydrodynamics and streamlined features [11].

Airfoil is placed in the wind tunnel's test section and fixed properly with the help of threads in the arrangement so that stream velocity (v) can also be calculated [12].

Pressure Variation

The pressing factor minor departure from the aerofoil decides the lift and drag coefficients on the aerofoil for different paces. This eventually chooses the (L/D) proportion. Examination has been completed in Familiar and Ploy to discover the best pressing factor inclination of the aerofoil.

The better the pressing factor variety, the better will be the streamlined presentation of the aerofoil. The pressing factor coefficient is the distinction between the nearby static pressing factor and the free stream pressure. The aerofoil calculation and the pressing factor are straightforwardly identified with one another. Hence drag force varies [13].

Test Design

Today, it is standard to specially craft the airfoils for each new plane on a PC. Airfoil configuration has turned up every day. In the early years, each new plane may get another airfoil. The equivalent is genuine today-however today we presently don't plan new airfoils in obliviousness of how they work [14].

Airfoil (aero plane surface as wing) is put in the test segment of air stream with 5000 mm/sec or 5 m/sec.

3 Dimensional flows considered during simulation in the solid works.

Simulation

At the point when a fluid streams around the airfoil, in a similar heading, at that point a pressing factor is created over the surface. The net amount of the multitude of powers brings about a net pressure, this is known as Drag, to think about the numerical plans, we utilize the idea of Drag coefficient.

$$C.D = D/v^2 A \tag{4}$$

The drag coefficient is also influenced by lot of other factors; hence it is represented as function of [15]:

$$C.D = f \text{ (form, Re, Fr, } \varepsilon)$$ (5)

Here
  Re = Reynolds number
  Fr = Froude number
  ε = Dissipation rate

Results and Discussion

This Different values of coefficient of drag can be originated at different velocities. Table 99.1 shows the distant values of Parameters and coefficient of drag.

Table 99.1: Distant values of parameters and coefficient of drag.

S.No.	Velocity (V)	Density of Air (ρ)	Area (A)	Drag Force (N)	Coefficient of Drag (Cd)
1	300	1.225	0.023325	28.6	0.022
2	250	1.225	0.023325	28.772	0.032
3	220	1.225	0.023325	19.855	0.029
4	200	1.225	0.023325	9.506	0.016
5	195	1.225	0.023325	8.499	0.015
6	180	1.225	0.023325	4.4275	0.0009
7	160	1.225	0.023325	3.981	0.0108
8	150	1.225	0.023325	4.17	0.0129
9	130	1.225	0.023325	4.298	0.0178
10	100	1.225	0.023325	6.41	0.0448

Formula Used

$$C_d = 2F_d / \rho A V^2$$ (6)

Where,
  C_d is coefficient of drag, F_d is drag force.
  ρ is density of fluid.
  A is curved surface area of half cylinder which is given by πrh.
  V is the velocity of air.
  Figure 99.1 shows the values of pressure acting on the airfoil during the simulation. These values are obtained by accessing all the flow trajectories on the airfoil.

Dynamic Viscosity and Specific Heat (Cp)

Figure 99.2 shows the variation in dynamic viscosity with temperature (Dynamic viscosity vs temperature plot) as we know viscosity depends on temperature and for air viscosity

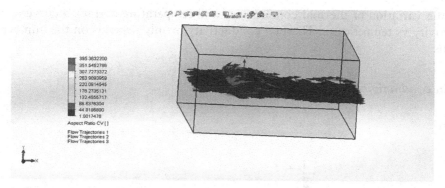

Figure 99.1 Shows the values of pressure acting on the airfoil during the simulation.

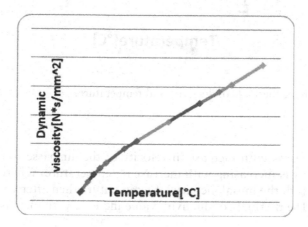

Figure 99.2 Variation in dynamic viscosity with temperature.

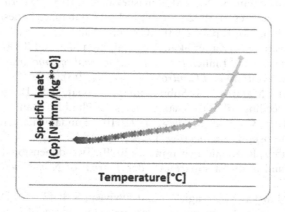

Figure 99.3 Variation specific heat and temperature.

increases with increasing temperature. Figure 99.3 shows the variation specific heat and temperature basically it's a specific heat vs temperature curve. We know that it's a property of substance and it varies with temperature according to curve. Figure 99.4 This figure shows the variation of thermal conductivity and temperature and it's a curve of thermal conductivity vs temperature. Thermal conductivity mainly depends on the number of free electrons.

Thermal conductivity

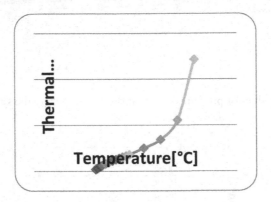

Figure 99.4 Variations of thermal conductivity and temperature.

Conclusion

The drag force increases with increase in velocity of the air passes on the aerofoil. Also, the dynamic viscosity is increasing with increase in temperature. Whereas the specific heat found to be stable with the initial increase of temperature then afterwards it also increases with temperature. The thermal conductivity value increase with increase in temperature.

References

1. Nandini, K. S., Subhashini, K. N., and Somashekar, V. (2018). Experimental investigations of aerodynamic performances of S9023 airfoil. In IOP Conference Series: Materials Science and Engineering (Vol. 376, No. 1, p. 012060). IOP Publishing.
2. Kakate, V. L., Chavan, D. S., Karandikar, P. B., and Mahulkar, N. (2014). Study of measurement and control aspects of wind tunnel. *International Journal of Innovative Research in Electrical, Electronics, Instrumentation and Control Engineering*, 2(3), 1–4.
3. Venkatakrishnan, L., Majumdar, S., Subramanian, G., Bhat, G. S., Dasgupta, R., and Arakeri, J. (eds.). (2021). Proceedings of 16th Asian Congress of Fluid Mechanics. Singapore: Springer.
4. Cattafesta, L., Bahr, C., and Mathew, J. (2010). Fundamentals of wind-tunnel design. *Encyclopedia of Aerospace Engineering*, 1–10.
5. Sahu, A. K., Chhabra, R. P., and Eswaran, V. (2009). Two-dimensional unsteady laminar flow of a power law fluid across a square cylinder. *Journal of Non-Newtonian Fluid Mechanics*, 160(2–3), 157–167.
6. Maurya, N. K., Maurya, M., Tyagi, A., and Dwivedi, S. P. (2018). Design & fabrication of low speed wind tunnel and flow analysis. *International Journal of Engineering & Technology*, 7(4.39), 381–387.
7. Narsipur, S., Pomeroy, B., and Selig, M. (2012). CFD analysis of multielement airfoils for wind turbines. In 30th AIAA Applied Aerodynamics Conference (p. 2781).
8. Patel, K. S., Patel, S. B., Patel, U. B., and Ahuja, A. P. (2014). CFD analysis of an aerofoil. *International Journal of Engineering Research*, 3(3), 154–158.

9. Khurana, M., Winarto, H., and Sinha, A. (2008). Airfoil geometry parameterization through shape optimizer and computational fluid dynamics. In 46th AIAA Aerospace Sciences Meeting and Exhibit (p. 295).

10. Timmer, W. A., and Bak, C. (2023). Aerodynamic characteristics of wind turbine blade airfoils. *In Advances in wind turbine blade design and materials*, 129–167. Woodhead Publishing.

11. Mallick, M., Kumar, A., Tamboli, N., Kulkarni, A., Sati, P., Devi, V., and Chandar, S. S. (2014). Study on drag coefficient for the flow past a cylinder. *International Journal of Civil Engineering Research*, 5(4), 301–306.

12. Prasad, K. S., Krishna, V., and Kumar, B. A. (2013). Aerofoil profile analysis and design optimisation. *Journal of Aerospace Engineering and Technology*, 3(2), 1–10.

13. Prasad, K. S., Krishna, V., and Kumar, B. A. (2013). Aerofoil profile analysis and design optimisation. *Journal of Aerospace Engineering and Technology*, 3(2), 1–10.

14. Mohaideen, M. M. (2012). Optimization of backward curved aerofoil radial fan impeller using finite element modelling. *Procedia Engineering*, 38, 1592–1598.

15. Matsson, J. E. (2013). An Introduction to SolidWorks Flow Simulation 2013. SDC Publications.

100 A study on smart irrigation system using IOT

Sheetal Solanki[1], Arko Bagchi[1,2], Prashant Johri[2], and Ankit Tyagi[3,a]

[1]Department of CSE, Faculty of Engineering & Technology, SGT University, Haryana, India

[2]Department of CSE, Galgotias University, UP, India

[3]Department of Mechanical Engineering, Faculty of Engineering & Technology, SGT University, Haryana, India

Abstract: At present, the major issue faced by the most of the countries in the world is scarcity of water. This scenario is exacerbated in India because agriculture is one of the key economies. In India, traditional agriculture systems are still in use. These systems aren't properly monitored, plants receive enough water even when the soil is damp leading to the over and under irrigation of plants resulting in water wastage. Hence, there is urgent need to inculcate advanced technology like Smart Irrigation System in agriculture. This paper proposes a smart Irrigation system based upon Internet of Things which analysis soil moisture and meteorological conditions to automate the irrigation process. This system is made up of sensors that monitor the condition of the crops and communicate data to an Arduino, which regulates the water motor and switch it ON/OFF. The Objective is to not only provide the correct amount of water to plants depending on their needs but also enhance water productivity, reduces manpower effort, fertilizer and water requirement. Farmers will benefit from this system since they will be able to monitor their fields from anywhere. Thus, Irrigation automation will result in increased yield productivity with less human oversight.

Keywords: Agriculture, Arduino, Bluetooth, IoT, smart irrigation, smart irrigation system.

Introduction

Agriculture makes a significant contribution to our economy and has a huge impact on it [1]. It is an important aspect of our life because we will not be able to survive without food grains. If the world's population grows, so does the market for food grains. Agriculture is something that yields the food grains. The most crucial phase in agriculture is irrigation. It is a system that allows us to supply the required water to the crops. Irrigation will improve agriculture and increase yield production if done correctly. Irrigation necessitates a large amount of fresh water [5]. Rainfall is the primary source of fresh water for agriculture. Water was once abundant, but increasing conditions such as global warming, climatic changes, and other factors have resulted in erratic weather patterns. Agriculture's growing demand consumes more water than rain, resulting in a scarcity of water for farming. Farmers in some rural areas are unable to irrigate their crop fields on time due to electricity problems [1]. The productivity of the soil suffers as a result of the lack of water, which reduces yield production and encourages poor crop health. Many times, due to a lack of

[a]tyagiankit10506@yahoo.com

care and control of the fields, crops are over-irrigated, causing them to deteriorate and cause water damage [2]. Traditional methods take longer and are less effective because they need more human effort and they are entirely manual. This method involves bringing water from the canal to the crop field at predetermined intervals, resulting in water waste, under and over irrigation of crops, and uneven irrigation of large areas, all of which contribute to low crop production [3]. As we all know, water is the most essential requirement for our survival, and the demand for water is increasing all the time, with a greater amount of water needed in agriculture [4]. To address the problem of water scarcity caused by the factors mentioned above, such as a lack of automation, a lack of expertise, and climatic conditions, there is an urgent need to implement a smart irrigation system in agriculture. IoT technology is used to power the smart irrigation system. This fully automates the irrigation process and delivers the required water to the crops based on conditions such as soil moisture, humidity, and weather. This method is efficient because it eliminates manual labour and allows farmers to stay informed about crop conditions such as the amount of fertiliser needed, the amount of water required by specific crops, and so on. Internet of Things (IoT) technology, allows us to track our field by allowing devices that are far apart to communicate and share data. It is made up of various sensors such as soil moisture sensors, temperature sensors, and humidity sensors, among others. These sensors detect the state of the crops and send data in the form of signals to Arduino, which processes the data and controls the water motor according to the needs of the crops [6]. This system will aid in the resolution of issues such as unequal irrigation of large areas, over-under irrigation, low yield production, and will allow farmers to track their crop fields from anywhere without being physically present there, improving water quality, yield production, reducing manual labour, fertilizer and water requirements, and automating the entire irrigation system [4].

Literature Review

The intelligent water saving irrigation system proposed by Zhang et al. [7] solves the issues of water scarcity and unequal soil moisture distribution across the crop field. The system is developed using agricultural internet of things and is based on the principle of water balance. In order to solve the watering issues and make the system more efficient Archana and Priya [8] proposed an automatic plant watering system which include sensors that are placed beneath the soil for sensing different factors such as moisture content of soil, water requirement of crops based on which signals are sent to a microcontroller which activates the pump or motor for watering the plants. This system's limitation is that it does not keep farmers up to date on field conditions and does not allow them to track them. Kumar et al. [9] suggested a smart irrigation system for the purpose of monitoring the moisture level of soil using moisture sensor circuit and for determining the nature of soil based on different conditions. The result of this system is better yield production, sowing crops according to future requirements, irrigating crops based on their needs, and precision farming. To ensure that irrigation is automated and effective Parameswaran and Sivaprasath [10] proposed an automated drip irrigation system which completely depends upon the humidity of the soil. For the purpose of sensing humidity of soil, a humidity sensor is placed beneath the soil. The value given by the humidity sensor in the form of signal is further transferred to a microcontroller which activates the functioning of the valve for irrigating the crops. Using a PC, the entire status of this irrigation method is updated on the local host which is monitored throughout by farmers. This system will help in better yield production [20] and effective water usage.

Existing Systems

Using Wireless Sensor Networks for Collection of Data [11]

Kaewmard and Saiyod says that the greatest issue in the twenty first century is feeding the globe, especially for smart businesses. Instead of tradition agriculture, the smart farm employed an agricultural automation system. Traditional farming practices used by the locals are quite sustainable, despite the fact that the whole cost is not cheap. The objective of this research is to lay out a long-term sustainable solution for agricultural automation. Agricultural automation uses a variety of techniques to collect data from vegetables. As a result, we created a portable measuring system that includes three types of sensor namely, a sensor for measuring the amount of moisture in the soil, sensor for checking the humidity in the air and the sensor that checks the temperature in the air. Furthermore, these sensors have been deployed as part of an irrigation system that uses a wireless sensor network to gather the data form the environment and manage irrigation system through an application in the phone [21]. To identify better and different ways to operate this system manually as well as automatically using the mobile device is the main goal of this project. In order to regulate the proper functioning of the irrigation system they created a wireless sensor network communication technology for collecting environmental data and transmitting control commands to switch on/off irrigation systems. The working of this is well is done appropriately to manage irrigation and channelize the flow of water towards roots of the pants. The attempt has been made to propose and automatic irrigation system which is valuable for the farm businesses and is more pleasant than conventional agriculture by utilizing a smart phone to monitor and manage the system in this article. As a result, the cost has been lowered in the long run.

Arduino and Raspberry pi Systems [12]

The Raspberry Pi 3, Arduino microcontrollers, Wi-Fi module, GSM shield, relay boards, and a few of sensors are used in this research to present a concept for an autonomous water providing system in farms. Depending on the level of moisture in the farmland and the intensity of sun, system determines right time for delivering the water to the plants and maintaining the proper tracking of the level of water in order to avoid the accumulation of water roots of the saplings. The analogous data from sensor is sent the Raspberry Pi 3 as a digital single through the Arduino Wi-Fi module. The system may change the admin's track and communicate with the system by sending SMS with a certain phrase. This technique may be used in both agricultural and tiny potted plants. This approach yields a very promising result in terms of maintaining and cherishing plants in a more scientific manner. The sensor readings were examined once the system was assembled. They put the technology through its paces in a variety of scenarios.

Using the Xbel Modules for Operating the System

Agrawal and Singhal offer a concept for a home automation system that includes Arduino boards (microcontroller boards), modules of Xbel, ready to use relay boards and Raspberry Pi [11]. The relay boards used saves energy and are also cost effective, and. The employment of these components results in a system that is Robust, scalable and a cost-effective system. Raspberry Pi uses the Python programming language to process the commands from the user. The ZigBee protocol is utilized to accept on/off commands from the raspberry pi utilizing Arduino micro controllers. The star topology of ZigBee technology acts like a backbone in order to perform communications between the end device and raspberry Pi. The end devices and Raspberry Pi serves as routers and coordinators respectively. As proof of concept, an energy—efficient and a low-cost system of drip irrigation is used. This idea may be used to irrigate plants in large farm areas along with gardens. This article covers all elements of system installation, including hardware and software.

ZigBee Communication [13]

The major goal is to create an automated water delivery system for home gardens as well as an irrigation system for farm fields. It's done using two different sensors, one for measuring the amount of moisture in the soil and the other for measuring the temperature. Both these sensors are attached to the plant's root system. The information detected by sensors is then sent to a base station. The collection of data from the field is the primary goal performed by the base station and then the data is uploaded over the internet, as well as to alert users to any unusual conditions such as low moisture or excessive temperature. They claim that their procedure has been authorized in a variety of regions with varying moisture content, particularly in red chilly weeds. Many people's interest is home gardening, and the application of this system is the same task as for irrigation systems in farm areas. After receiving the information from the sensors, it is transmitted through ZigBee module. The received data for the sensors of soil and temperature are uploaded to the web from the ZigBee receiver, and graphs for temperature and moisture are plotted using Visual Basic. The results of this system are that the irrigation system automation was deployed at a location that was acceptable and affordable for obtaining water sources for crop management. This technology allows cultivation in water-scarce places while also improving water retention.

Using Wireless Sensors and Data Monitoring

Prof. Jain, Shaikh, Sood and Kulkarni discuss the importance of agriculture and food production in this study [14]. They claim that agriculture in our nation is reliant on the monsoons, which are insufficient water sources. As a result, irrigation is employed in agriculture. Water is delivered to plants in an irrigation system based on the soil type. Greenhouse-based contemporary agriculture enterprises are becoming increasingly popular in India's agricultural sector. Plant humidity and temperature are accurately managed with this method. Due to varying atmospheric conditions, these variables might change from place to place in a big farmhouse, making it impossible to maintain consistency at all points across the farmhouse manually. For the first time, an Android phone-controlled irrigation system has been presented, which might provide the ability to maintain consistent environmental conditions. The major goal of this research is to create a smart wireless sensor network (WSN) for use in an agricultural setting. It's important to keep an eye on the agricultural environment for things like soil moisture, temperature, and humidity, among other things. Individuals manually collecting measures and verifying them at various intervals was a conventional technique of measuring these elements in an agricultural context. The purpose of this article is to look at a remote monitoring system that uses an RF module. These nodes communicate data wirelessly to a central server, which collects, stores, and analyses the data before displaying it as needed, as well as sending it to the client mobile.

GSM Based Systems

He discusses about the issue of food shortage due to an alarming increase in the rate of population growth. Water shortage is another big concern that farmers face. Karan's main goal is to develop an autonomous irrigation system that saves the farmer time, money, and energy. He was told that the benefits of an automated irrigation system include its ease of installation and configuration. It conserves energy and resources so that they can be used in the most efficient and effective manner possible. It saves time and reduces human mistake when altering available soil moisture levels, among other benefits. It employs an android phone to handle a GSM-based autonomous irrigation system for effective resource utilisation and crop planning [15].

Start process:

- initialize power is supplied to GSM
- check moisture level
- If the level less
- start irrigation
- stop the process.

The biggest advantage of this system is that it saves time and water. On the other hand there is a disadvantage too, this system is just limited to automation of irrigation systems and lacks in extra ordinary feature.

Using Arduino for Manually Switching ON/OFF the Water Pump

Agriculture is the most significant employment for the majority of Indian households, according to this article. It is critical to the agricultural country's growth. Agriculture accounts for around 16% of overall GDP and 10% of total exports in India. Agriculture's primary resource is water. Irrigation is one means of supplying water, although it can waste a lot of water in other circumstances. As a result, we have presented a project termed Arduino based autonomous irrigation system utilising IoT to save water and time in this area. Various sensors, such as temperature, humidity, and soil moisture sensors, are used in this suggested system to sense various soil characteristics. Furthermore, land is automatically watered by ON/OFF of the motor dependent on soil moisture value. The user's android application will display the detected parameters and motor status. The major goal of this project is to develop an autonomous irrigation system, which will save the farmer time, money, and energy. Traditional farmland irrigation methods need human involvement. Human intervention can be reduced using irrigation equipment that is automated. Agriculture networking technology is not only a requirement of contemporary agricultural growth, but it is also a key indicator of agricultural development's future level. With greater progress in the field of IoT projected in the next years, these systems might become more efficient, quicker, and less expensive. In the future, this system might be upgraded to an intelligent system that forecasts patterns in rainfall, best time for harvesting, user activities, and other aspects that would make this system self-contained or could operate without any human intervention [16].

Components

Arduino Board (Microcontroller)

Arduino is a microcontroller board which is the combination of hardware and software. A microcontroller is a single-chip computer that functions as a tiny computer. It's known as a system on a chip in current parlance. There are two parts of Arduino on is the software referred as an IDE at and other is a circuit which is programmable (this circuit is basically known as the microcontroller). It is an open—source system which is capable of taking inputs. There can be differ-ent types of inputs like—clicking a button, sensing light on a sensor, and these in-puts are converted into outputs as programmed in the IDE. These outputs can be turning an LED on or starting a motor or even post something online. A proper instruction set is provided to the microcontroller board. These instructions are the programs that tell the board what has to be done. The hardware part can have multiple CPUs, input and output ends which can be programmed and a memory chip. Applications of this includes remote controls, machinery used in offices, various devices in medical sector and other gadgets [17].

Bluetooth

Kardech was the person who coined the term Bluetooth in the year 1997. It is a wireless technology which is generally used when there is requirement of transmission of data from one device to another over shorter distance, around 30 feet. This distance keeps on reducing as the obstacles (for example a wall) between both the devices start increasing. It uses UHF (Ultra High Frequency) radio waves over in the ISM (Industrial, scientific and medical) bands. The frequency ranges from 2.402 GHz–4.8 GHz. It is a communication protocol standardised for low power consumption. When the device is connected via Bluetooth, there is no need for them to be placed in a line-of-sight positioning. The technology is based on existing wireless LAN approaches, but it stands out because of its tiny size and inexpensive cost. The present prototype circuits fit on a 0.9 cm square circuit board, but a significantly smaller version is being used nowadays.

Sensor for Measuring the Amount of Moisture in Soil

To determine the volumetric water content of the soil, Soil moisture sensors are used [18]. This sensor includes two probes that carry current through the soil before detecting the resistance of the soil to determine the moisture level. As a result, the increase of water in the soil makes it more electrically conductive which leads to less resistance, and if the water level is low or the soil is dry the conductivity will be low hence resulting in more resistance.

Sensor for Measuring Humidity and Temperature

With the help of a humidity sensor and a thermostat the ambient air is measured. This sensor is used to determine the relative humidity in the soil. It measures the electrical resistances between two electrodes to detect water.

Proposed System

Irrigation may be automated utilising sensors, microcontrollers, Bluetooth, and an Android application. We employ a low-cost soil moisture sensor as well as a humidity and temperature sensor. These sensors keep a close watch on the situation. The Arduino board is linked to the sensors. The sensor data is sent to the user through wireless transmission for properly managing the irrigation system. Appropriate configuration of the application installed in the mobile device has to be done for the assessment of received data and make a comparison with threshold values of moisture, temperature and humidity. The application can either be programmed to take the decision independently itself with user intervention or it can be done manually by the user. The motor is turned on if the threshold value is greater than value of moisture in soil, and it is turned off when the value of moisture in soil becomes more than threshold value. Arduino board used is linked with sensors. These components connect through Bluetooth wireless transmission, allowing the user to view the data via his smartphone, which includes an Android application that can receive sensor data from the Arduino over Bluetooth. Bluetooth is used to communicate between user and Arduino in a smart irrigation system. Bluetooth technology has a short range and the range depends on the type of application being used. The Arduino and mobile application will be used to establish and save the threshold values for humidity, temperature and soil moisture. Depending on the weather conditions the values of the sensors changes. Summer and winter soil moisture levels, as well as temperature and humidity levels, will change. After taking into account all of these environmental and climatic factors, the threshold value is determined. If the reading of soil moisture goes below the threshold, then the motor will automatically turn on, and vice versa. Once receiving the control from mobile device, the watering system is automated. The result is conveyed to Arduino over Bluetooth, and the switches of motor are activated as a result. To keep the real tack of the reservoir's water level, the ultrasonic sensor is utilised. The piezoelectric technique is used to power the ultrasonic sensor. It consists of two pins: an echo pin and a trigger pin. The eco pin serves as a reflector and the trigger pin acts as a transmitter. When the trigger pin is activated, it emits ultrasonic vibrations. When it starts working, it produces ultrasonic waves. These ultrasonic waves get reflected off the water. The time it takes to get an echo is computed, and the result is the water level. The water level is checked before turning the motor on. This checking of level of water is done in order to verify whether the required amount of water for irrigation is available or not. The motor will either not turn on or it supply a very little amount of water if the required amount of water is not available. An alert is transmitted to the user mobile device through the application. With the help of application, the user can turn on and off the motor according to the need [19].

Conclusion and Future Work

The goal of this article was to study the entire notion of a smart irrigation system utilising the Internet of Things. The research focuses on the purpose of irrigation, which is to provide the proper amount of water to the plant at the correct time (just when needed) depending on the soil's state. The presented autonomous watering system can ensure a scientific and systematic approach to plant care, resulting in increased yield. Such a system is simple to construct and inexpensive. The system will become more efficient and helpful as sensor technology advances. For example, sensor values can aid in supplying suitable amounts of water and decreasing water waste. All environmental characteristics will be sensed by this system, and data will be sent to the user through the cloud. This asset enables the farmer to enhance the cultivation in a way that is beneficial to the plants. It results in a larger crop output, a longer production period, higher quality, and fewer usage of pesticides. This technique also saves electricity that is used to provide water. This irrigation technique improves sustainability by allowing farming in water-scarce areas. This technology significantly cuts water use. It just requires minor upkeep.

References

1. Patel, J., Patel, E. K., and Pati, P. (2019). Sensor and cloud based smart irrigation system with arduino: a technical review. *International Journal of Engineering Applied Sciences and Technology,* 3, 25–29. ISSN no. 2455–2143.
2. Sen, D., Dey, M., Kumar, S., and Dr. Boopathi, C. S. (2020). Smart irrigation using IOT. *International Journal of Applied Science and Technology,* 3, 3080–3090.
3. Manimegalai, V., Judy, A. L., Gayathri, A., Ashadevi, S., and Mohanapriya, V. (2020). Smart irrigation system with monitoring and controlling using IOT. *International Journal of Engineering and Advanced Technology,* 2249–8958.
4. Ashwini, B. V. (2018). A study on smart irrigation system using IOT to Surveillance of crop–field. *International Journal of Engineering and Technology,* 370–373.
5. Leh, N. A. M., Kamaldin, M. S. A. M., and Muhammad, Z. (2019). Smart irrigation system using internet of things. In IEEE 9th (ICSET).
6. Singla, B., Mishra, S., Singh, A., and Yadav, S. (2019). A study on smart irrigation system using IOT. *International Journal of Advance Research, Ideas and Innovations in Technology,* 2454–132X.
7. Zhang, S., Wang, M., Shi, W., and Zheng, W. (2018). Construction of intelligent water saving irrigation control system based on water balance. IFAC Papers Online 51–17, pp. 466–471.
8. Archana and Priya (2016). Design and implementation of automatic plant watering system by "Archana and Priya". *International Journal of Advanced Engineering and Global Technology,* 4(1).
9. Ravi Kumar, G., Venu Gopal, T., Sridhar, V., and Nagendra, G. (2018). Smart irrigation system. *International Journal of Pure and Applied Mathematics,* 119(15), 1155–1168.
10. Parameswaran, G., and Sivaprasath, K. (2016). Arduino based smart drip irrigation system using IOT. *Presented at International Journal of Engineering Science and Computing (IJESC).*
11. Kaewmard, N., and Saiyod, S. (2014). Sensor data collection and irrigation control on vegetable crop using smart phone and wireless sensor networks for smart farm. In IEEE Conference on Wireless Sensors (ICWiSE) (pp. 106–112).
12. Dlodlo, N., and Kalezhi, J. (2015). The internet of things in agriculture for sustainable rural development. In IEEE (pp. 13–18).
13. Kumar, M. K., and Ravi, K. S. (2016). Automation of irrigation system based on Wi-Fi technology and IOT. *Indian Journal of Science and Technology,* DOI-10.17485/ijst/2016/v9i17/93048.
14. Aruna, M., Narayana, P. B., Kumar, S. N., Walid, M. A. A., Patra, J. P., and Kumar, B. S. (2023). Arithmetic Optimization Algorithm with Machine Learning based Smart Irrigation System in IoT Environment. *In 2023 Fifth International Conference on Electrical, Computer and Communication Technologies (ICECCT),* 01–05. IEEE.
15. Vallejo-Gómez, D., Osorio, M., and Hincapié, C. A. (2023). Smart Irrigation Systems in Agriculture: A Systematic Review. *Agronomy,* 13(2), 342.
16. Mahesh, R., and Ramya, G. (2023). Efficient utilization of water in smart irrigation system using bluetooth and IoT to increase the crop production. *In AIP Conference Proceedings,* 2822(1). AIP Publishing.
17. Singh, S. S. (2019). Smart irrigation system using IOT. *International Journal of Innovative Technology and Exploring Engineering (IJITEE),* 8(12S).
18. Karpagam, J., Merlin, I. I., Bavithra, P., and Kousalya, J. (2020). Smart irrigation system using IoT. *In 2020 6th International Conference on Advanced Computing and Communication Systems (ICACCS),* 1292–1295. IEEE.

19. Ragab, M. A., Badreldeen, M. M. M., Sedhom, A., and Mamdouh, W. M. (2022). IOT based smart irrigation system. *International Journal of Industry and Sustainable Development*, 3(1), 76–86.
20. Krishnan, R. S., Julie, E. G., Robinson, Y. H., Raja, S., Kumar, R., and Thong, P. H. (2020). Fuzzy logic based smart irrigation system using internet of things. *Journal of Cleaner Production*, 252, 119902.
21. Kashyap, P. K., Kumar, S., Jaiswal, A., Prasad, M., and Gandomi, A. H. (2021). Towards precision agriculture: IoT-enabled intelligent irrigation systems using deep learning neural network. *IEEE Sensors Journal*, 21(16), 17479–17491.

101 Bond graph modeling of soft contact interaction between a rigid sphere and a soft material

Dinesh Deshwal[1,a], Anil Kumar Narwal[2,b], and Monika Deshwal[1,c]

[1]Mechanical Engineering Department, Shree Guru Gobind Singh Tricentenary University Gurugram, Haryana, India

[2]Mechanical Engineering Department, Deenbandhu Chhotu Ram University of Science and Technology Murthal, Sonipat, Haryana, India

Abstract: The deformation effect of the soft material on the Robot's fingertips is a useful feature in general robotics applications. Besides, the mechanics of contact with the soft material plays a vital role in the object's safe handling stability during grasping of an object. Therefore, the induced forces and contact moments because of contact area change must be analyzed. This paper employs a multi-bond graph approach, in conjunction with the finite element method, to depict the dynamics of soft contact between a rigid sphere and a pliable material. The soft material is discretized into numerous brick-shaped nodal elements using the finite element method. For these nodal elements, the mass and stiffness values at each node are also calculated. A series of sphere balls with varying weights were individually placed on a soft material pad to observe the deformation effects. Using the model, the researchers were able to determine the contact area and force distribution at each contact node. Bond graph method is used to develop the mathematical model and the system equations are generated from it. The formulation of the MATLAB code is done from a bond graph by using the algorithms. The system equations generated are solved by using the ODE's available in MATLAB for solving differential equations.

Keywords: Bond graph, dynamics, robot grasping, soft contact.

Introduction

Contact mechanics was studied by Hertz in 1882, two linear elastic materials contact behavior was investigated. The model is only applicable to frictionless linear elastic surfaces that deform little during contact. In 1971, K. L. Jhonson et al. discussed the effect of surface energies on the elastic solids contact. The equations were derived for their influence on the amount of contact and adhesion force between the forces of adhesion between two lightly charged solid spherical surfaces [1].

Compatible materials have been tested under environmental pollution, clean and dry conditions. Cutkosky et al. determine which materials are best suited for the robot hand's contact areas and create accurate models of the friction behavior of the materials [2]. Vaz and Hirari have devised a more straightforward model for this type of joint, which effectively captures the majority of the skeleton's behavior. They implemented the model within a spherical joint system featuring soft cartilage. They used the Bond graph technique to

[a]deshudinesh@gmail.com, [b]aknarwal@dcrustm.org, [c]monikamech.nain@gmail.com

study the dynamics [3]. Narwal and others present the assessment of the dynamics of smooth contact lamination by taking the circular disk which rolls on the silicone rubber using the multi bond graph method. The disk motion is controlled by applying controlled force with the help of proportional and derivative regulator. The model was validated through simulations to assess the deformation, contact area, and force distribution on the soft material. It's important to note that all the research performed earlier was limited to two dimensions [4, 5].

The contact between the two objects can be categorized into two types: point contact and surface contact. Surface contact is considered more realistic since it enables the creation of force distribution within the contact area and induces moments due to the extended contact region [6]. There are two cases where the contact point is frictionless and the other is with friction. At a non-friction contact point, the end effector can only exert its normal strength at contact point. In the case of friction point contact, the end effecter can exert a force directed on the friction cone at the contact point, which contains the components of normal and tangential force, which are also referred to as hard finger contact. The contact conditions greatly influence the kinematics and dynamics of dexterous handling [7, 8].

Soft contacts have a vital application as the force is distributed over the contact surface between the interacting bodies due to the material's compliance effect. The contact area changes dynamically as the contact interface moves. Therefore, it is important to study the mechanics of the contact, especially from robotics and control point of view [9–11].

In the current study, the smooth contact between the rigid sphere and the soft material is extended to a three-dimensional scenario. Unlike the previous work that focused on plane stresses, this new research has broadened its scope to encompass the general three-dimensional case.

In order to demonstrate the soft contact interaction between the rigid sphere and the pliable material, a bond graph model is formulated. The experiment involves placing spheres with varying loads individually onto the soft material to observe the corresponding deformations. The model is implemented in MATLAB, incorporating the finite element method as the primary technique. The resulting system equations are then solved using MATLAB's ordinary differential equation (ODE) solver.

Methodology

FEM Formulation

In finite element modeling entire domain is discretized into small elements to get the displacements in terms of value at the discrete point. Here domain is discretized into hexahedral elements with each element having eight nodes. In this three-dimensional problem, each node is limited to acquiring displacement in only three directions along the x, y, and z axes. Thus, the three degrees of freedom is associated with each node. A three-dimensional array is used to store the nodal coordinates. The array representation of the system consists of the total number of nodes, with each node being defined by its three coordinates along the x, y, and z axes. For linear elastic materials, the Hooke's Law is stated for the stress-strain relation. There are two material properties young's modulus (E) and Poisson Ratio (v) are very important for isotropic materials. From, Hooke's Law relationship and the material law, $\sigma = D \in$ Where D is the symmetric (6×6) material matrix.

For an eight-node hexahedral element, we consider ξ, η, and ζ as natural coordinates. Where (ξ_i, η_i, ζ_i) represent the coordinates of node i of the element in the local system.

Displacement *(u, v, w)* in the element is interpolated from nodal displacement (u_i, v_i, w_i) using shape functions N_i.

$$\{u\} = [N]\{\vec{d}\}$$

The displacement of the 3D hexahedral element is determined by the Shape function denoted as N. For the later element, the Lagrange shape function is mentioned by eq. (1). Where i=1 to 8

$$N_i(\xi, \eta, \zeta) = \frac{1}{8}(1 + \xi_i\xi)(1 + \eta_i\eta)(1 + \zeta_i\zeta) \tag{1}$$

The hexahedral element comprises eight nodes, and in the natural coordinate system (ξ, η, and ζ), there exist eight shape functions. Across the entire domain of the element, the sum of all these shape functions equals one. Utilizing the displacement equation, we can express the three-dimensional strain-displacement relations as the given equation. (2).

$$\vec{\varepsilon}_{(6\times1)} = [B]_{(6\times24)}\vec{d}_{(24\times1)} \tag{2}$$

[J] is the Jacobian Matrix and [K] is the element stiffness matrix which is given by eq. (3)

$$[K] = \int_V BDB^T dV \tag{3}$$

Matrix [B] is expressed in natural coordinates and it is necessary to carry out the integration in natural coordinates too, using the relationship as mentioned in eq. (4)

$$dV = dxdydz = \det[J].d\xi d\eta d\zeta \tag{4}$$

In all integrals, the limit of integration is from –1 to 1. Consequently, the stiffness matrix (K) for a Hexahedral eight-node element is calculated as follows:

$$[K] = \int_{-1}^{+1}\int_{-1}^{+1}\int_{-1}^{+1} B^T DB \det[J].d\xi d\eta d\zeta \tag{5}$$

In the case of a hexahedral eight-node element, three degrees of freedom are associated per node, and there are eight nodes in the element. The stiffness matrix (K) for the Hexahedral eight-node element has a size of 24x24. To integrate the equation, a 6x6 Gauss Quadrature scheme is employed. (5).

Gaussian Quadrature has been proven to be the most efficient method of numerical integration of this type of problem. Consider the n-point approximation as taken in eq. (6)

$$I = \int_{-1}^{+1} f(r)dr = w_1 f(r_1) + w_2 f(r_2) + \cdots + w_n f(r_n) \tag{6}$$

The weights ($w_1, w_2, ..., w_n$) and sampling points or Gauss points ($r_1, r_2, ..., r_n$) play a crucial role in Gauss Quadrature. By using n Gauss points and their corresponding weights, this method yields an exact solution for polynomials f(r) with the highest possible degree.

Now, extending Gaussian Quadrature to three-dimensional integral forms.

$$[K] = \int_{-1}^{+1}\int_{-1}^{+1}\int_{-1}^{+1} B^T DB \det[J].d\xi d\eta d\zeta$$

Follows readily

$$[K] = \sum_{i=1}^{n}\sum_{j=1}^{n}\sum_{k=1}^{n} w_i w_j w_k B_{ijk}^T DB_{ijk}\left(\det[J]\right)_{ijk} \tag{7}$$

For the Hexahedral eight-node element, we use 6-point Gaussian quadrature in eq. (7).

After finding out the stiffness matrix of each element, we assembled the matrix for calculating the stiffness matrix (global stiffness matrix) for the whole domain or geometry [12].

Bond Graph Modelling

Using the finite element method, the global stiffness and mass matrices for the soft material are calculated. These computed matrices are denoted as the C-field and I-field when constructing the bond graph model. Subsequently, a bond graph model for the soft material is developed and validated under various loading conditions. The soft material patch, made of silicon rubber and with dimensions of 0.1m in length, 0.05m in thickness, and 0.05m in width, is taken into consideration.

The bond graph is specifically designed to represent the contact model between the soft material and a rigid sphere, as illustrated in Figure 101.1. The translational inertia of the sphere is incorporated into the model using the inertia element I: m. Additionally, the soft

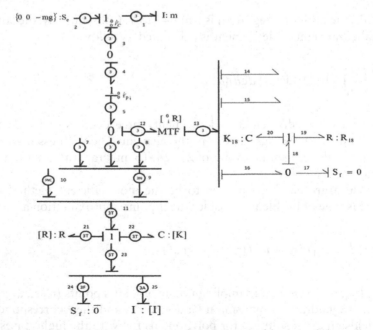

Figure 101.1 Bond graph model for soft contact between a rigid body and the soft material.

material is discretized into multiple hexahedral elements (brick-type elements with eight nodes). Within the bond graph, the inertia, compliance, and damping properties of the soft material are respectively represented by the I, C, and R fields.

Where, T = Total nodes in discretized soft material

N = Nodes in contact with the sphere

NC = the nodes which are not in contact with sphere

F= Nodes on bottom layer of the soft material

A = T – F, number of nodes having flow

Now, the interaction between the soft material and the sphere is considered. When the bond graph subsystems of the rigid sphere and soft material are combined, derivative causality emerges at the inertia of contact nodes. This signifies that the momentum states related to these nodes are influenced by other states within the system. To resolve this concern, the contact between the sphere and the soft material is represented as viscoelastic in the model, effectively resolving the derivative causality issue [14].

Result and Discussion

The models were tested by simulation. The simulation is performed for 2 seconds in MATLAB and the results are discussed here. The various parameters taken are tabulated in Table 101.1.

Table 101.1: Parameters used in simulation [4, 5].

PARAMETERS	VALUE
Length	0.1m
Width	0.05m
Thickness	0.05m
No. of elements in the x-direction	10
No. of elements in the y-direction	6
No. of elements in the z-direction	6
Modulus of Elasticity (Young's Modulus)	1.90 Gpa
Poisson's Ratio	0.48
Damping Coefficient	0.20Ns/m
Density	1100 kg/m3
Mass of Sphere	100 kg
Radius of sphere	0.03m

Case 1: The rigid sphere, weighing 100 kg, is positioned on the central node of the top layer with its base fixed. As a result, the soft material undergoes deformation due to the gravitational force and the mass of the sphere, both acting in a downward direction. The deformation of the soft material in the X-Z plane, caused by the force exerted by the sphere, is illustrated in Figure 101.2.

In Figure 101.3, the soft material's deformation is demonstrated when a 150 kg sphere is positioned on the central node of the top layer. The layers of the soft material are

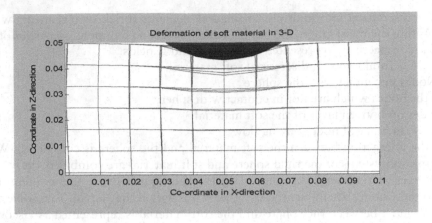

Figure 101.2 Deformation of soft material in X-Z plane.

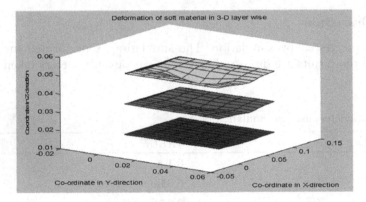

Figure 101.3 Deformation of top, middle, and bottom layer of the soft material.

distinguished by colors, with the top layer shown in green, the middle layer in red, and the bottom layer in blue. The illustration depicts the effect of the force exerted by the sphere on the soft material layers.

   Once the sphere is placed on the soft material, the center of mass coordinates of the sphere undergo changes due to the deformation of the soft material. These altered coordinates of the sphere's center of mass are illustrated in Figure 101.4.

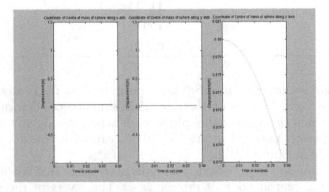

Figure 101.4 Coordinates of centre of mass of the rigid sphere.

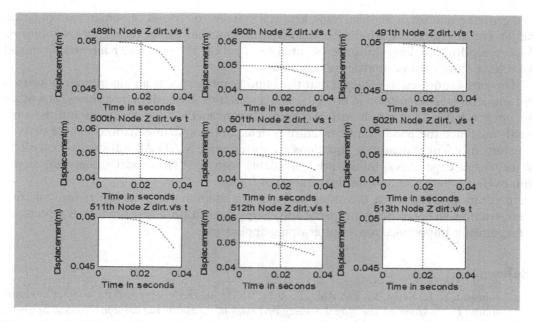

Figure 101.5 Effect of deformation on centre node and nearby nodes on the top layer.

The deformation of center node and its nearby nodes when the sphere is placed on the center node of the top layer of the soft material is shown in Figure 101.5.

Case 2: When different weights of the sphere are placed on the soft material different nodes are coming in contact with them. The contact area increases as the normal force on the soft material increases or vice-versa.

As the mass of the sphere is responsible for the deformation of the soft material, different-different masses of the sphere are placed and the corresponding deformation is noted, which are tabulated in Table 101.2.

Table 101.2: Showing the corresponding deformation and contacting nodes for different masses of sphere.

Sr. no.	Sphere mass (kg)	Maximum Deformation (mm)	No. of Nodes in Contact
1.	0	0	1
2.	5	4.7	9
3.	10	7.2	9
4.	25	14.2	21
5.	50	24.6	23

Conclusion and Future Scope

The main problem with holding the object with the Robot is choosing the clamping forces, avoiding or minimize slipping risk. Modeling contact interaction is the most critical skill in skillful manipulation, as grasping and manipulating is determined by contact behavior. The Robot next generation will interact directly with humans, due to which researchers are growing interest in implementing artificial systems to reproduce human manipulation skills. The interface between the Robot and its environment will be through the soft fingers. Hence, skillful handling skills are essential for a robot. The soft finger interaction model enables a more realistic analysis in robotics.

The simulation results indicate that the Bond graph serves as an algorithmic approach for modeling the interaction in flexible contact scenarios. Extensive efforts have been made to integrate the Bond graph with the finite element method for modeling and dynamic analysis. Once the Bond graph structure is defined, the system equations can be derived conveniently. The system state equations are solved using MATLAB software, which offers unrestricted capabilities, unlike other available software. The model successfully determines the contact area and the distribution of forces.

In future investigations, this work will explore additional aspects like tangential stiffness, friction, shape recovery property, and energy dissipation. These functional aspects are of significant importance for robot grasping applications.

References

1. Johnson, K. L., Kevin, K., and Roberts, A. D. (1971). Surface energy and the contact of elastic solids. *Proceedings of the Royal Society of London. A. Mathematical and Physical Sciences*, 324(1558), 301–313.
2. Cutkosky, M., Jourdain, J., and Wright, P. (1987). Skin materials for robotic fingers. In Proceedings IEEE International Conference on Robotics and Automation, (Vol. 4, pp. 1649–1654).
3. Vaz, A., and Hirai, S. (2004). A simplified model for a biomechanical joint with soft cartilage. In 2004 IEEE International Conference on Systems, Man and Cybernetics, IEEE Cat. No. 04CH37583, (Vol. 1, pp. 756–761).
4. Narwal, A. K., Vaz, A., and Gupta, K. D. (2014). Study of dynamics of soft contact rolling using multibond graph approach. *Mechanism and Machine Theory*, 75, 79–96.
5. Vaz, A., Narwal, A. K., and Gupta, K. D. (2014). Understanding soft contact interaction between a non circular rigid body and a soft material using multibond graph. In 11th international conference on Bond graph modeling (ICBGM), 6–10 July, (pp. 690–697).
6. Park, K. H., Kim, B. H., and Hirai, S. (2003). Development of a soft-fingertip and its modeling based on force distribution. In 2003 IEEE International Conference on Robotics and Automation (Cat. No. 03CH37422), (Vol. 3, pp. 3169–3174).
7. Ren, X. J., Smith, C. W., Evans, K. E., Dooling, P. J., Burgess, A., Wiechers, J., and Zahlan, N. (2006). Experimental and numerical investigations of the deformation of soft materials under tangential loading. *International Journal of Solids and Structures*, 43(7–8), 2364–2377.
8. Gilardi, G., and Sharf, I. (2002). Literature survey of contact dynamics modelling. *Mechanism and Machine Theory*, 37(10), 1213–1239.
9. Zhang, X., Xu, S., Jia, C., Wang, G., and Chu, M. (2020). Generalized modeling of soft-capture manipulator with novel soft-contact joints. *Energies*, 13(6), 1530.
10. Lippiello, V., Siciliano, B., and Villani, L. (2011). Online dextrous-hand grasping force optimization with dynamic torque constraints selection. In 2011 IEEE International Conference on Robotics and Automation, (pp. 2831–2836).
11. Bakhy, S. H., Hassan, S. S., Nacy, S. M., Dermitzakis, K., and Arieta, A. H. (2013). Contact mechanics for soft robotic fingers: modeling and experimentation. *Robotica*, 31(4), 599–609.
12. Chandrupatla, T. R., Belegundu, A. D., Ramesh, T., and Ray, C. (2002). Introduction to Finite Elements in Engineering (Vol. 10). Upper Saddle River, NJ: Prentice Hall.
13. Rao, S. S. (2017). *The Finite Element Method in Engineering*. Butterworth-Heinemann.
14. Karnopp, D. C., Margolis, D. L., and Rosenberg, R. C. (2012). System Dynamics: Modeling, Simulation, and Control of Mechatronic Systems. John Wiley & Sons.

102 Design and analysis of skirt supports for vertical cylindrical pressure vessels

Kapil Muni Singh[a] and Nilamber Kumar Singh[b]

National Institute of Technology, Patna, India

Abstract: This paper investigates the design and analysis of skirt supports for vertical cylindrical pressure vessels based on American Society of Civil Engineers (ASCE). The 3D models of skirt support and pressure vessel with ellipsoidal heads are created using CATIA V5 software. These models are imported in ANSYS workbench and static structural analysis is done using finite element method (FEM). Convergence test is performed to optimize the results with 5% allowable change fortwo iterations. Mesh independent test is done for different mesh qualities such as skewness, orthogonality and aspect ratio (≤5) with varying mesh element sizes, 35–50 mm for validation of results.

Keywords: Convergence test, pressure vessel, skirt support, structural analysis.

Introduction

Pressure vessel is defined as a vessel or tank that can carry, store and receive liquids or gases at a pressure substantially different from the ambient pressure. Compressed gas cylinders, vacuum chambers, storage tanks, submarines etc. are common examples of pressure vessels. Pressure vessel is mainly used in power generation engineering for production of fossils, petrochemical engineering for storing crude petroleum oil and chemical engineering for chemical reactor. Skirt support is one of the important supporting structures for vertical cylindrical pressure vessels in the form of rolled cylindrical or conical shells (called skirt). This support minimizes the local stresses and distributes the direct load uniformly which are acting throughout the circumference at the point of attachment with the pressure vessel. Several researchers [1–6] have worked on the support structures of the vertical cylindrical pressure vessel. Yahya et al. [1] designed the vertical pressure vessel using ASME codes. Mitesh et al. [2] calculated stresses on the skirt support based on combined load considerations, primarily internal pressure, dead weight and seismic load, and found to be safe. The thermo-mechanical analysis of skirt support was done by Sawant and Hujare [3] using finite element method. Dalvi et al. [4] performed the finite element analysis of weld joint for curved plates (overlap) specially for designing pressure vessel skirt support. Structural design of overloading pressure vessel support was done by Ganesh et al. [5].

[a]kapilmunisingh400@gmail.com, [b]nilambersingh@nitp.ac.in

Patil et al. [6] found that more floor space below the vessel can be achieved by making legs unsymmetrical distributed and inclined to some suitable angle.

In this paper, the author has designed the skirt support for vertical cylindrical pressure using ASME code and performed its static-structural analysis to determine von-Misses stresses, elastic strains and total deformation using ANSYS. Convergence test is performed to optimize the result of stress, strain and deformation with 5% allowable change for two iterations. Mesh-independent test is performed for 35-, 40-, 45- and 50-mm mesh element sizes with different mesh quality such as skewness, orthogonality and aspect ratio to validate the result.

Methodology

Design Data

Inner diameter (d_i) is 2134 mm, outer diameter (d_o) is 2188 mm, thickness of ellipsoidal head (t_h) is 27 mm, major axis of ellipse is 1067 mm and minor axis of ellipse is 533.5 mm.

Analytical Calculation

Wind load calculation (in lateral direction)

$$\text{Force acting on the cylindrical portion of pressure vessel, } F = q_z \times G_f \times C_f \times A_f \qquad (1)$$

$$q_z = \text{velocity pressure at height Z above ground} = 0.00256 \times K_z \times I \times V^2 \qquad (2)$$

Here, K_z = end connection coefficient, I = importance factor = 1.15, V = Basic wind speed = 40mph = 23.846 K_z, G_f = gust response factor for flexible vessel = 1.15, C_f = force coefficient, shape factor 0.7 – 0.9 = 0.9, A_f = projected area, sq meter

$$\text{where, } A_f = L_e D_e \qquad (3)$$

L_e = effective length of shell
D_e = effective vessel diameter
    = 6096 × 3491.2 $(D_e = 1.6D)$
    = 21282355.2 mm²
Hence, from equations 1, 2 and 3, force calculation is
F = 23.846 × 1.15 × 0.9 × = 24.68 × K_z
F = 24.68 × 212.823 × 0.8 = 18698.76 N
For factor of safety, force should be considered as, F = 20,000 N

Finite Element Analysis

Finite element analysis (FEA) has been carried out on cylindrical pressure vessel having geometric discontinuity at the junction of cylindrical portion, ellipsoidal end cap and skirt support with different geometry and thickness.

Step 1–Material selection:
In this work, structural steel material is decided and its properties are assigned in ANSYS workbench.

Step 2–Modeling:
The pressure vessel geometry such as ellipsoidal head, shell and skirt support are to be modeled in CATIA V5 and save as .igs file as shown in Figure 102.1.

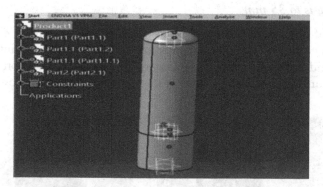

Figure 102.1 Modelling of skirt support.

Step 3–Meshing:
Tetrahedral elements are used for complex geometry with small distortion of mesh and computational cost is lower for assembling the global stiffness matrix.

Step 4–Boundary condition:
Bottom of skirt supports need to be fixed to the ground, internal pressure of 1.03142 MPa applied to the vessel, wind load of 20,000 N acting on the vessel (lateral direction) and acceleration due to gravity in downward direction (in –Z direction).

Step 5–Static structural analysis:
Static structural analysis is performed for vertical pressure vessel with skirt support in ANSYS workbench 2019 R2 to calculate total deformation, equivalent elastic strain and von-Misses stresses. Convergence test helps to optimize the solution and obtain the precise result by increasing element or number of iterations. The Newton Raphson method are used to converge the result in ANSYS. Convergence test is performed for elastic strain, stress and total deformation with 5% allowable change to optimize the result. Convergence test are broadly classified into two types:

1. h-type convergence: In h-type convergence, the number of elements is increased and element size is decreased whereas order of polynomial is kept constant.
2. p-type convergence: In p-type convergence, the number of element and element size are kept to be constant whereas order of polynomial is changing.

Mesh-independent test is performed on skirt support for pressure vessel assembly varying mesh element size (35, 40, 45 and 50 mm) for 3 different mesh quality such as skewness, orthogonality and aspect ratio to validate the result.

Results and Discussion

Static structural analysis is performed for skirt support structures for vertical cylindrical pressure vessels by considering the effect of wind load acting in lateral direction on shell of cylindrical pressure vessels. Results are obtained in the form of von-Misses stress, total deformation and elastic strain.

Mesh element size is to 35 mm

At the first, FEA of vertical pressure vessel with skirt supports are to be done for fixed skirt support at bottom with ground, internal pressure of 1.03142 MPa at internal diameter of vessel (cylinder and both end cap) in downward direction, acceleration due to gravity in downward direction and wind load acting in longitudinal direction as shown in Figure 102.2.

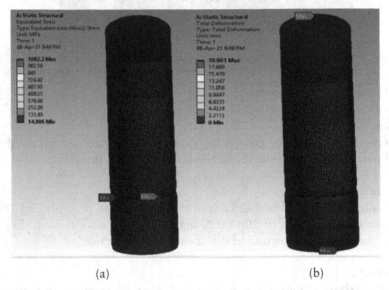

(a)   (b)

Figure 102.2 Result of static-structural test (a) von-misses stress (b) Total deformation.

Table 102.1: Variation in mesh element size with skewness, orthogonality and aspect ratio mesh quality.

S.No	Mesh element size(mm)	Von-misses stress (MPa)		Equivalent elastic strain(mm/mm)		Total deformation(mm)	
		Maximum	minimum	Maximum	Minimum	Maximum	Minimum
1	35 mm	1082.2	14.906	0.0054109	0.000078426	19.901	0
2	40 mm	1103.6	11.554	0.0055284	0.000061201	19.93	0
3	45 mm	1094.7	12.22	0.0054813	0.000066928	19.912	0
4	50 mm	931.8	7.687	0.0046591	0.000050738	19.88	0

In Table 102.1 variation of mesh element size 35, 40, 45 and 50 mm is shown for different mesh quality. It has been found that, result is independent of mesh quality and mesh element size.

From Figure 102.3 determined that by varying mesh element size stress vs strain graph is linear and overlapping for all different mesh elements.

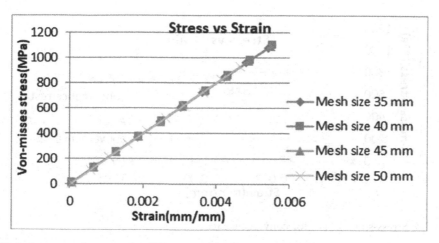

Figure 102.3 Stress-strain curve for different mesh element size.

Convergence Test

Convergence test (Figure 102.4) is performed to optimize result with help of two iterations for mesh element size 35 mm by considering the allowable change as 5%.

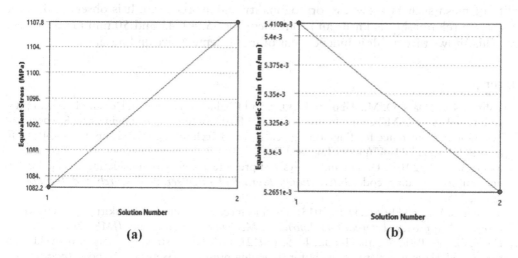

Figure 102.4 Convergence test (a) Equivalent stress (b) Equivalent elastic strain.

From Figure 102.4 after convergence, the equivalent stress is changed by 2.3396%, elastic strain is changed by –2.7328% and total deformation is changed by –0.22555 % in two iterations.

Figure 102.5 represents the relation between strain on X-axis and von-Misses stress on Y-axis. After convergence test, stress-strain curve is overlapping up to 0.12% strain then start increasing for higher stress and strain value.

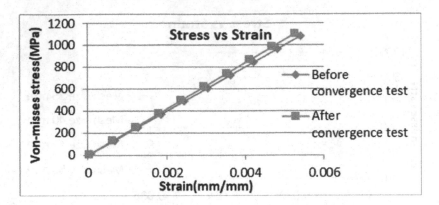

Figure 102.5 Stress-strain curve for convergence test.

Conclusions

The present investigation shows that results of von-Misses stress, elastic strain and total deformation are optimized in two iterations. When mesh element size is 35 mm, maximum von-Misses stress is 1082.2 MPa and optimized stress is 1107.8 MPa, maximum strain is 0.0054109 and optimized strain is 0.0052651, maximum total deformation is 19.901 and optimized total deformation is 19.856 by keeping the allowable change 5% while convergence test is performed. Results are not changing in mesh independence test by changing quality of mesh such as skewness, orthogonality and aspect ratio. It is observed that the stress-strain relationship is linear and coincident for 35, 40, 45 and 50 mm mesh element sizes. This shows that result is independent of mesh element size and mesh quality.

References

1. Yahya, N. A., Daas, O. M., Alboum, N. O. F. and Khalile, A. H. (2018). Design of vertical pressure vessel using ASME codes. In Proceedings of First Conference for Engineering Sciences and Technology Conference for Engineering Sciences and Technology, (Vol. 2, pp. 653–664), DOI: https://doi.org/10.21467/proceedings.4.33.
2. Mungla, M. J. (2016). Design and analysis of pressure vessel skirt considering seismic load as per uniform building code. *International Journal of Engineering Research and Technology*, 4(10), 1–5.
3. Sawant, S. M., and Hujare, D. P. (2013). Thermo-mechanical analysis for skirt of pressure vessel using FEA approach. *International Journal of Mechanical Engineering (IJME)*, 2(3), 13–20.
4. Dalvi, M. V., Patil, V., and Bindu, R. S. (2012). FEA based strength analysis of weld joint for curved plates (overlap) specially for designing pressure vessel skirt support. *International Journal of Recent Technology and Engineering (IJRTE)*, 1(3), 17–23.
5. Rakate, G. N., Desavale, R. G., and Jamadar, I. M. (2013). Structural design of overhanging pressure vessel support used in pyrolysis process using non-linear finite element analysis for structural stability. *International Journal of Engineering Research and Technology*, 2(6), 2036–2042.
6. Patil, V. G., Bajaj, P. S., and Patil, V. G. (2013). Determination of safety of inclined leg support for pressure vessel subjected to arbitrary wind load using FEA. *International Journal of Innovations in Engineering and Technology*, 2(1), 345–351.

103 Nonlinear static and dynamic characteristics of MEMS and NEMS beams with different boundary conditions

Uma Shankar[a], Shivdayal Kumar[b], Mainakh Das[c], and Anand Bhushan[d]

Department of Mechanical Engineering, National Institute of Technology Patna, India

Abstract: Micro-Electromechanical System and Nano-Electromechanical System beam resonators have numerous applications as ultrasensitive sensors and switches. In this paper, we have studied comparative analysis of two different boundary conditions—clamped and simply supported—on nonlinear static and dynamic characteristics of electrostatically actuated MEMS/NEMS beams. The nonlinear governing equations have been developed using Euler-Bernoulli theory including geometric nonlinearities and electrostatic forcing nonlinearities. Galerkin's discretization technique has been used to develop a reduced order model from the partial differential equation of motion of beams. The ROM results have been validated by comparing with the results of Finite Element simulation results. The investigation has been carried out by determining the effect of boundary conditions on static and dynamic characteristics of electrostatically actuated MEMS and NEMS beams with different aspect ratios. In static characteristics, we determine the static deflection of the beam for applied DC voltage, whereas the variation of fundamental resonant frequency for applied DC voltage is determined during the study of dynamic characteristics. Our investigation shows that there is a range of aspect ratios where the mathematical model with both boundary conditions provides nearly similar results of static and dynamic characteristics.

Keywords: Boundary conditions, electrostatic, MEMS/NEMS beam, reduced order model.

Introduction

Micro-Electromechanical System (MEMS) and Nano-Electromechanical System (NEMS) based devices such as actuators and sensors are the most exciting research area in today's world as they are very economical in terms of manufacturing cost and power consumption. These devices have a wide range of applications in automotive industries as pressure sensors, accelerometers [1]fabricated and tested for navigation application, using one mask process. The accelerometers consist of an out-of-plane (for z-axis and biomedical applications [2].

Several research works have been carried out on MEMS and NEMS devices to study their static and dynamic behavior. Younis et al. [3] have presented analytical as well as reduce order model (ROM) method based solutions to study the characteristics of doubly clamped MEMS beams. Chan et al. [4] have used finite element modeling software

[a]uma2.me18@nitp.ac.in, [b]shivdayal.me17@nitp.ac.in, [c]mainakhdas@gmail.com, [d]anand.bhushan@nitp.ac.in

ABAQUS, Intelli CAD and Quasi 2-D to study the static characteristic of the fixed-ends beam. Using Galerkin's method based on ROM Hung and Senturia [5] have investigated the dynamic behavior of clamped-clamped beams. Researchers have used various techniques to characterize static and dynamic behaviors of MEMS and NEMS resonators. Lei et al. [6] study the static and Eigenvalue characteristic of graphene nano ribbon in COMSOL by assuming that constant electrostatic force is acting on ribbon throughout the deflections. Lee et al. have applied the atomic finite element method [7] to study the sensitivity of the monolayer graphene resonator. Najar et al. [8] devised a pressure sensor using dynamic response of clamped-clamped MEMS beam. Khaniki et al. [9] have reviewed the size dependent effects on the static and dynamic behavior of NEMS/MEMS beam. Bhushan et al. [10] have used Galerkin based multi-modal ROM technique to study the effect of internal stress on the pull-in and natural frequency of carbon nanotube.

Many investigations have been carried out to determine static and dynamic characteristics of electrostatically actuated micro and nano beams having clamped-clamped (CC) and simply supported (SS) end conditions individually. To the best of the author's knowledge, a detailed comparative analysis has not been conducted between electrostatically actuated CC and SS beams. Here, we have presented a detailed investigation related to the comparison of static and dynamic characteristics of electrostatically actuated CC and SS beams.

Mathematical Models

Figure 103.1(a) and 103.1(b) shows schematic diagrams of electrostatically actuated rectangular cross-section beams under consideration. We have taken two boundary conditions: CC and SS. Both the beams have Young's modulus E, length $\hat{l}$, width $\hat{b}$ and thickness $\hat{h}$. It is in capacitive arrangement with a fixed electrode plate, where $\hat{g}$ is the initial distance of separation between the beam and the fixed electrode plate.

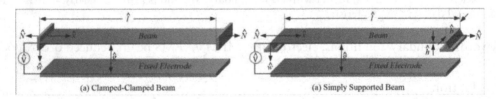

(a) Clamped-Clamped Beam   *(a) Simply Supported Beam*

Figure 103.1 Diagram of an electrostatically actuated (a) clamped-clamped and (b) simply supported beam.

Euler-Bernoulli beam theory is being used to device the models of CC and SS electrically actuated microbeams. The governing equation of motion such beams are partial differential equations [10]. By following the method of Bhushan et al. [10], we have developed Galerkin based reduced order models, using the first mode shape, to investigate the static and dynamic characteristics of these beams. The ROM for static analysis for SS beam is

$$-W - 0.5\alpha_1 W^3 + \frac{2\alpha_2 V^2}{(\pi)^4} \int_0^1 \frac{\phi_{ss} dx}{(1 - W\phi_{ss})^2} = 0 \tag{1}$$

and for CC beam we get

$$-W - 0.3047\alpha_1 W^3 + \frac{\alpha_2 V^2}{(1.5\pi)^4} \int\limits_0^1 \frac{\phi_{cc} dx}{(1 - W\phi_{cc})^2} = 0 \qquad (2)$$

The ROM to determine dynamic characteristics as a variation of first natural frequency ω_d for applied DC voltage V for the SS beam is

$$\omega_d^2 = (\pi)^4 + 146.11358\alpha_1 W^2 - 4\alpha_2 V^2 \int\limits_0^1 \frac{\phi_{ss}^2 dx}{(1 - W\phi_{ss})^3} \qquad (3)$$

and for CC beam.

$$\omega_d^2 = (1.5\pi)^4 + 450.7978\alpha_1 W^2 - 2\alpha_2 V^2 \int\limits_0^1 \frac{\phi_{cc}^2 dx}{(1 - W\phi_{cc})^3} \qquad (4)$$

where, the constants are $\alpha_1 = 6\left(\dfrac{g}{h}\right)^2$ and $\alpha_2 = \dfrac{6\varepsilon_0 l^4}{Eh^3 g^3}$.

The variables ϕ_{ss} and ϕ_{cc} are the first mode shapes of SS and CC beams respectively.

$$\phi_{ss} = \sin(\pi x),$$

$$\phi_{cc} = \cosh\left(\frac{3\pi}{2} x\right) - \cos\left(\frac{3\pi}{2} x\right) + \left[\frac{\cos\left(\dfrac{3\pi}{2}\right) - \cosh\left(\dfrac{3\pi}{2}\right)}{\sinh\left(\dfrac{3\pi}{2}\right) - \sin\left(\dfrac{3\pi}{2}\right)}\right]\left(\sinh\left(\frac{3\pi}{2} x\right) - \sin\left(\frac{3\pi}{2} x\right)\right)$$

Model Validations

Initially, validation of the ROM of the CC and SS beams has been done by comparing them with the results of finite element model (FEM) analysis. The properties of the validating SS and CC beams have been taken as length (l) 610 μm, width (b) 40 μm, thickness (h) 2.1 μm and gap (d) 2.3 μm. and Young's modulus (E) 149 GPa [11]dynamic loads, and the coupling between the mechanical and electrical forces. It accounts for linear and nonlinear elastic restoring forces and the nonlinear electric forces generated by the capacitors. A new technique is developed to represent the electric force in the equations of motion. The new approach allows the use of few linear-undamped mode shapes of a microbeam in its straight position as basis functions in a Galerkin procedure. The macromodel is validated by comparing its results with experimental results and finite-element solutions available in the literature. Our approach shows attractive features compared to finite-element softwares used in the literature. It is robust over the whole device operation range up to the instability limit of the device (i.e., pull-in. Then a finite element model is developed of the given dimensions in COMSOL.

Figure 103.2(a) and 103.2(b) show the comparison of static deflection of the mid-point of the beam with applied DC voltage of SS and CC beam respectively between ROM and COMSOL results. The comparison of static deflection of the mid-point of the SS and CC beam in COMSOL with the reduced order model shows that the ROM results are well in agreement with the COMSOL results.

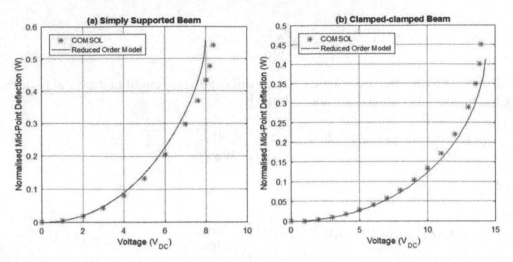

Figure 103.2 Comparison of mid-point deflection as function of DC voltage between our (a) COMSOL results with our reduced order model (ROM) results for SS beam (b) COMSOL result with our reduced order model result for CC beam.

Results and Discussion

The dimensions of SS and CC beam under investigation have been taken as length (l) 500 nm, width (b) 20 nm, thickness (h) 0.35 nm, initial gap (d) 200 nm and Young's Modulus (E) 1000 GPa [6]the relationship between resonance frequency and driving force was explored. The effects of built-in tension, adsorbates and graphene size on the performance of resonator including resonance frequency and tunability were also studied. It was shown that resonance frequency could be tuned by the electrostatically induced average tension due to driving force, and exponentially increased with increasing driving force. When the single-layer graphene resonator without any adsorbates had no or very small built-in tension, the tunability of resonator was greater. However, for a high-frequency-range resonator, the resonator with high built-in tension should be used. The simulation results suggested potential applications of graphene resonators tuned by a driving force, such as widely tunable or ultrahigh frequency nanoelectromechanical systems (NEMS).

After validating the model, the static and dynamic characteristics of the beams of dimensions mentioned earlier are investigated after solving the ROMs. The ROMs have been solved in the MATLAB environment and the results are shown in Figure 103.3(a) and 103.3(b).

Comparison of Static and Dynamic Characteristics between SS and CC Beams

- For comparing static characteristics of SS and CC beams, the voltage is compared for the same amount of deflection in both the beams.
- For comparing dynamic characteristics of SS and CC beams, the voltage is compared at the same first natural frequency at the deflected position of both the beams.

From the plot of Figure 103.3(a), the difference in voltage for normalized mid-point deflection of 5%, 25% and 50% of initial gap for SS and CC beam is given in Table 103.1.

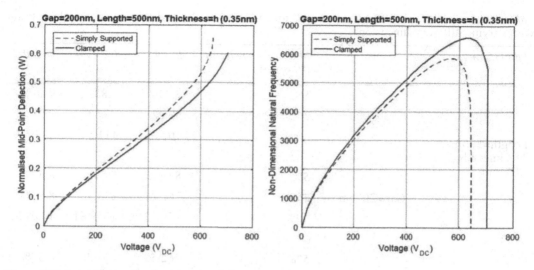

Figure 103.3 (a) Normalized mid-point deflection and (b) Variation of non-dimensional natural frequency of CC and SS beam as a function of applied DC voltage.

Table 103.1: Percentage difference in voltage for the same mid-point deflection of SS and CC beam.

	Mid-Point Deflection (% of gap)	SS (V)	CC (V)	% Diff
	5%	30.50	32.94	7.98%
Gap=200nm, Length=500nm, Thickness=h (0.35nm)	25%	278.41	303.35	8.96%
	50%	573.75	631.18	10.01%

It is observed that for noted deflections the voltage difference is 7.98%, 8.96% and 10.01% respectively.

From the Figure 103.3(b), the percentage difference in voltage of SS and CC beam for 10%, 50% and 90% of the maximum frequency of the SS beam is given in Table 103.2.

Table 103.2: Percentage difference in voltage for the same frequency of SS and CC beam.

	% of maximum of SS frequency	SS (V)	CC (V)	% Diff
	10%	18.49	17.56	5.28%
Gap=200nm, Length=500nm, Thickness=h (0.35nm)	50%	189.15	177.09	6.81%
	90%	446.87	414.55	7.80%

It is observed that for the noted frequency, the percentage difference in voltage of SS and CC beam are 5.28%, 6.81% and 7.80% respectively.

Further, we examine the percentage difference in voltage of SS and CC beam for same mid-point deflections and varying the different dimensions of the beam. Keeping all the dimensions same the gap, the length and the thickness are varied one by one and observations are noted in Table 103.3. The maximum percentage difference in the voltage is found to be within 10.42% irrespective of change in the individual dimensions of the system.

Table 103.3: Percentage difference in voltage for the same mid-point deflection of SS and CC beam with varying different dimension.

Dimensions		Mid-Point Deflection (% of gap)	SS (V)	CC (V)	% Diff
Length=500nm, Thickness=h (0.35nm)	Gap=100nm	5%	5.34	5.83	9.14%
		25%	49.22	53.63	8.97%
		50%	101.38	111.54	10.03%
	Gap=300nm	5%	82.82	89.91	8.55%
		25%	767.21	835.93	8.96%
		50%	1581.31	1739.61	10.01%
Gap=200nm, Thickness=h (0.35nm)	Length=100nm	5%	762.58	812.70	6.57%
		25%	6960.30	7583.86	8.96%
		50%	14343.78	15781.43	10.02%
	Length=250nm	5%	121.77	130.62	7.27%
		25%	1113.64	1213.41	8.96%
		50%	2295.00	2534.06	10.42%
Gap=200nm, Length=500nm	Thickness=5h	5%	68.61	75.42	9.93%
		25%	622.66	678.86	9.03%
		50%	1282.84	1411.64	10.04%
	Thickness=10h	5%	97.39	106.72	9.58%
		25%	881.13	962.49	9.23%
		50%	1815.12	1998.31	10.09%

To study the dynamic characteristics with a varying gap, length and thickness for SS and CC end conditions, the gap, length and thickness are varied one by one keeping all the other dimensions and properties of the beam the same. The result obtained after the simulation of the dynamic ROM in MATLAB is presented in Table 103.4. Varying the initial gap of 200nm to 100nm and 300nm, there is percentage difference in voltage of SS and CC beam to a maximum of 7.8%. Varying the length to 25% and 50% of the original length, we find that the variation in percentage difference in voltage is below 8%. With variation in thickness also the percentage difference in voltage is again below 8% for similar frequency up to the 90% of SS beam frequency.

Conclusion

In our investigation, we have analyzed the static and dynamic behavior of simply supported and clamped beams varying their length, width, thickness and gap. A Comparison

Table 103.4: Percentage difference in voltage for the same frequency of SS and CC beam with varying different dimension.

Dimensions	.	% of maximum of SS frequency	SS (V)	CC (V)	% Diff
Length=500nm, Thickness=h (0.35nm)	Gap=100nm	10%	3.63	3.41	6.42%
		50%	33.49	31.37	6.78%
		90%	79.12	73.40	7.80%
	Gap=300nm	10%	50.04	46.39	7.86%
		50%	521.76	488.64	6.78%
		90%	1233.09	1143.91	7.80%
Gap=200nm, Thickness=h (0.35nm)	Length=100nm	10%	462.19	439.01	5.28%
		50%	4728.79	4427.36	6.81%
		90%	11171.84	10363.63	7.80%
	Length=250nm	10%	72.50	69.96	3.62%
		50%	756.93	708.88	6.78%
		90%	1788.65	1659.60	7.78%
Gap=200nm, Length=500nm	Thickness=5h	10%	42.48	41.71	1.85%
		50%	422.94	396.26	6.73%
		90%	998.64	926.92	7.74%
	Thickness=10h	10%	58.47	58.36	0.18%
		50%	599.36	563.40	6.38%
		90%	1415.77	1314.88	7.67%

was made on the model based on ROM between SS and CC beams. We have found that at the lower voltage the deflection and natural frequency curve of both beams remains quantitatively similar. After that, the deviation in the difference between static and dynamic profiles between SS and CC beams has been observed to have a similar pattern. We can also conclude that the effect of boundary conditions (SS or CC) on static and dynamic characteristics is not much significant particularly for the lower magnitude of applied DC voltage. It is well noted that the first mode shape of the SS beam is a much simple function than the first mode shape of CC beam. So, for a quick approximate analysis, we can use SS mode shape for the investigation of electrostatically actuated CC beams.

References

1. Kim, H. C., Seok, S., Kim, I., Choi, S. D., and Chun, K. (2005). Inertial-grade out-of-plane and in-plane differential resonant silicon accelerometers (DRXLs). In Digest of Technical

Papers—International Conference on Solid-State Sensors, Actuators and Microsystems, TRANSDUCERS'05, (Vol. 1, pp. 172–175).

2. Burg, T. P., Mirza, A. R., and Milovic, N. (2006). Vacuum-packaged suspended microchannel resonant mass sensor for biomolecular detection. *Journal of Microelectromechanical Systems*, 15(6), 1466–1476.

3. Younis, M. I., and Nayfeh, A. H. (2003). A study of the nonlinear response of a resonant microbeam to an electric actuation. *Nonlinear Dynamics*, 31(1), 91–117.

4. Chan, E. K., Garikipati, K., and Dutton, R. W. (1999). Characterization of contact electromechanics through capacitance-voltage measurements and simulations. *Journal of Microelectromechanical Systems*, 8(2), 208–217.

5. Hung, E. S., and Senturia, S. D. (1999). Generating efficient dynamical models for microelectromechanical systems from a few finite-element simulation runs. *Journal of Microelectromechanical Systems*, 8(3), 280–289.

6. Lei, Y., Sun, J., and Gong, X. (2015). Mechanical and vibrational responses of gate-tunable graphene resonator. *Physica B: Condensed Matter*, 461, 61–69. doi: 10.1016/j.physb.2014.12.012.

7. Lee, H. L., Hsu, J. C., Lin, S. Y., and Chang, W. J. (2013). Sensitivity analysis of single-layer graphene resonators using atomic finite element method. *Journal of Applied Physics*, 114(12), 123506.

8. Najar, F., Ghommem, M., and Abdelkefi, A. (2020). A double-side electrically-actuated arch microbeam for pressure sensing applications. *International Journal of Mechanical Sciences*, 178(March), 105624.

9. Khaniki, H. B., Ghayesh, M. H., and Amabili, M. (2021). A review on the statics and dynamics of electrically actuated nano and micro structures. *International Journal of Non-Linear Mechanics*, 129(July 2020), 103658. [Online]. Available: https://doi.org/10.1016/j.ijnonlinmec.2020.103658.

10. Bhushan, A., Inamdar, M. M., and Pawaskar, D. N. (2011). Investigation of the internal stress effects on static and dynamic characteristics of an electrostatically actuated beam for MEMS and NEMS application. *Microsystem Technologies*, 17(12), 1779–1789. doi: 10.1007/s00542-011-1367-y.

11. Younis, M. I., Abdel-Rahman, E. M., and Nayfeh, A. H. (2003). A reduced-order model for electrically actuated microbeam-based MEMS. *Journal of Microelectromechanical Systems*, 12(5), 672–679. doi: 10.1088/0960-1317/14/7/009.

104 Finite element modelling of EDM based on single discharge for pure aluminium and Al/SiC MMC

Ravinder Kumar[1], Alok Kumar[2], and Abhishek Singh[2,a]

[1]Division of Research and Development, LPU, Punjab, India

[2]Department of Mechanical Engineering, NIT, Patna, India

Abstract: A 2-D axi-symmetric model has been investigated for a single spark machining during electric discharge machining (EDM) process using ANSYS 12.0 software. Finite element models of aluminium and aluminium matrix composite (i.e. Al/SiC) has been developed and comparison has been drawn between the two. Gaussian distribution heat source is assumed for the temperature distribution and material removal that have been studied using transient thermal analysis. The temperature distribution in Al/SiC MMC is investigated considering three cases: (i) SiC particle exactly under the spark (ii) SiC particle highly displaced (iii) SiC particle far from the spark. A parametric study of variation of parameters (i.e. electric current, pulse duration and spark radius) has also been done and its effect on material removal has been analysed. The model predicts that the position of SiC particle with respect to the single discharge affects the material removal rate of Al/SiC metal matrix composite.

Keywords: EDM, finite element method, material removal rate, single discharge, temperature distribution.

Introduction

Electric Discharge Machining is an unconventional machining process that uses electrical energy for producing electrical sparks between the electrodes. The electric spark has sufficient energy to melt and even vaporise workpiece material and thus forming a crater. This process is repeated again and again until desired amount of material is removed. EDM is mainly used for difficult to machine materials like tool and die steels, ceramics, metal matrix composites (MMCs) etc.

Modelling the response is one way of predicting the response values prior to the actual experimentation. EDM is generally considered to be based on electro-thermal mechanism for modelling the MRR [1–6]. In this case, electrical energy (i.e. high densities of electrical current) is converted to the thermal energy thus increasing the temperature of plasma channel causing the material to erode. Modelling of the MRR includes the calculation of the temperature distribution inside the workpiece by applying a specific modelled heat source. It is assumed that the heat transfer between the plasma and the electrodes is taking place by conduction and radiation [7] and volume of material is assumed to be removed

[a]abhishek.singh@nitp.ac.in

which achieves temperature greater than its melting point temperature [3]. Thus, modelling of single discharge is generally carried out to calculate MRR based on single discharge.

Numbers of researchers have focused in modelling the material removal rate (MRR) tool wear rate (TWR), plasma channel and crater size thermal distribution and thermal stress. Fan and Desilva [1,2] studied electro-thermal erosion process of EDM with Al/ SiC as cathode and observed that if Aluminium surrounding the reinforcement (SiC) is melted, it is flushed and removed along with reinforcing material. Weingartner et al. [3] observed that by applying a time dependant heat source and considering the latent heats of fusion and vaporization, a better approximation between the simulated and experimental results can be achieved. Salonitis et al. [4] developed a thermal based model for the determination of the MRR and the average surface roughness (SR) for die-sinking EDM. The developed model predicted that increasing the discharge current and sparks duration results in increased material removal rates as well as increased surface roughness. Joshi and Pandey [5] developed a thermo-physical model for EDM process based on single discharge using Gaussian distribution of heat flux to determine MRR and crater size. Parametric and experimental studies were conducted to evaluate the effect of EDM process variables on its performance. And it was found that model prediction is close to the experimental results. Yadav et al. [6] developed a finite element model for EDM process based on single discharge using Gaussian distribution of heat flux to determine temperature field as well as thermal stresses. The effects of process variables were reported on thermal stress and temperature distribution. The regions of large stresses and high temperature gradient zones were seen where the thermal stress exceeded the material yield strength.

In this paper, a 2-D axi-symmetric model has been investigated for a single spark electric discharge machining using ANSYS 12.0 software assuming a Gaussian distribution heat source. Thermal models of aluminium and aluminium matrix composite (i.e. Al/SiC) have been developed and temperature distribution as well as crater size have been studied using transient thermal analysis. A parametric study of variation of parameters (i.e. electric current and pulse duration) has also been done and its effect on temperature distribution has been analysed.

Thermo-Physical Model

In EDM process, heat transfer is mainly considered to take place through conduction, convection and radiation. However the transfer of heat through convection and radiation mode is very less and hence is neglected in this paper. So all the heat received/transferred through the workpiece is assumed to be by conduction only. The EDM plasma channel is supposed to be a uniform cylindrical column and the heat generated by the spark falling on the workpiece propagates symmetrically in all directions. This axisymmetric heating of the workpiece because of a single discharge is based on the following equation [6] below:

$$\rho c \frac{\partial T}{\partial t} = \frac{1}{r} \frac{\partial}{\partial r}\left(k \frac{\partial T}{\partial r} \right) + \frac{\partial}{\partial z}\left(k \frac{\partial T}{\partial z} \right) \tag{1}$$

Where ρ is density, c is specific heat, t is time, T is temperature, r and z are coordinate axes and k is thermal conductivity.

Heat Flux Due to a Single Spark

The total heat given, spark radius and the thermo-physical properties of workpiece are considered as the important factors that can affect the Material Removal Rate in single spark Electro Discharge Machining model. Number of researchers had modelled the EDM process by considering two heat source models i.e. point source and uniformly distributed heat source models. But in actual practice these models are unsophisticated and cannot predict accurately. The present work used a Gaussian distribution of heat flux to estimate the amount of heat from the plasma channel. One can represent the heat flux q(r) entering the workpiece by the equation given below:

$$q(r) = \frac{4.45FUI}{\pi R^2} exp\left\{-4.45\left(\frac{r}{R}\right)^2\right\} \tag{2}$$

Where, F is the fractional part of discharge power entering the workpiece, I is the current released, U is the voltage released and R is spark radius at the work surface.

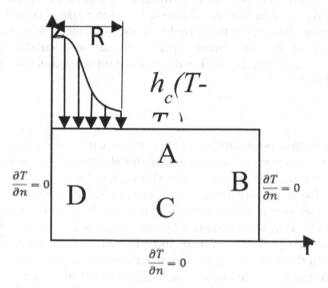

Figure 104.1 An axisymmetric model.

Results and Discussion

Aluminium as Cathode

In this section, the modelling of EDM process using ANSYS 12.0 software was done by taking aluminium as cathode (workpiece) and copper as anode (tool electrode). The input parameters involved in the simulation are mentioned in Table 104.1 and the temperature distribution in the workpiece is predicted by solving equation 1. The energy input to the workpiece after a single discharge at the end of pulse duration provides temperature isotherms which are depicted by colour and indicated values in Figures 104.1 and 104.2 helps identifying the various heating regions i.e. heat effected zone, molten region and boiling region. The maximum temperature generated by the heat input is at the centre of the discharge which is at the upper left corner of the model (Figure 104.2). The temperature decreases moving away from the centre of the discharge as the heat flux distribution is Gaussian [8]. The elements of the material which achieves the temperature greater than its melting point temperature are said to be killed/eroded. The determined number of elements killed in case of aluminium workpiece is 1868.

Table 104.1: Experimental condition in FEM simulation.

Discharge current I [A]	Discharge voltage U [V]	Pulse duration τ_{on} (μs)	Spark radius R (μm)	Coefficient of energy distribution, F
2	100	30	15	0.08

Figure 104.2 shows a close up image of temperature isotherms at the end of pulse duration subsequent to material removal. As the cooling rate in EDM is very high because of the dielectric fluid in contact, some material also gets solidifies on to the surface of the workpiece called recast layer. Top three isotherms in Figure 104.3 are recast layers and subsequent four layers are heat affected zones. The crater size is considered as the major factor in determining the surface topography of the workpiece. By calculating the crater volume, the volume of material removed per spark can be calculated. The crater radius and crater depths are calculated on different values of input parameters as shown in Tables 104.2, 104.3 and 104.4.

Parametric Study

A parametric study had been carried out for investigating the effect of current, pulse duration and spark radius on crater size and material removal. From Table 104.1 we can see that the peak temperature in the model shows near linear relationship with current. Moreover, MRR also starts ascending as the current goes up. It should also be noted that crater depth increases as the current ascends that finally increases the surface roughness.

The effect of pulse duration predicted by the model is the same as that by intuition. Table 104.3 shows that peak temperature and MRR both increases as the pulse duration ascended because of the increased heat energy with the pulse duration. This increased heat energy also affects the crater size in the same manner as that of current. But experimentally, the increase of pulse duration also results in increased spark radius the effect of which is to reduce the crater radius and MRR [9]. So the net result is a product of both discharge duration and spark radius. The effect of spark radius on peak temperature and MRR is tabulated in Table 104.4.

Table 104.2: Effect of current on peak temperature and material removal.

Current (A)	Peak temp. (K)	Crater radius (μm)	Crater depth (μm)	Ratio (Radius/ Depth)	Elements killed
1	4970	36	34	1.06	221
2	10718	50	48	1.04	467
4	20070	64	62	1.03	815
6	30075	72	72	1	999
8	40071	78	76	1.03	1194

Table 104.3: Effect of pulse duration on peak temperature and material removal.

Pulse duration (μs)	Peak temp. (K)	Crater radius (μm)	Crater depth (μm)	Ratio (Radius/ Depth)	Elements killed
10	7860	24	24	1	115
20	9190	40	40	1	314
30	9957	50	48	1.04	467
40	10578	56	56	1	621
50	11026	64	62	1.03	775

Table 104.4: Effect of spark radius on peak temperature and material removal.

Spark radius (μm)	Peak temp. (K)	Crater radius (μm)	Crater depth (μm)	Ratio (Radius/ Depth)	Elements killed
5	42285	72	72	1	4159
15	9957	50	50	1	1868
30	4143	36	32	1.12	895
60	1352	22	10	2.2	155

Al/SiC as Cathode

In this section, the modelling of EDM process using ANSYS 12.0 software is done by taking Al/SiC as cathode (workpiece) and copper as anode (tool electrode). All the conditions and input parameters are the same as that with previous case. The temperature distribution is predicted by equation 1. The SiC is reinforced into aluminium in the particulate form which makes the difference of properties between pure aluminium and Al/SiC MMC. The individual properties of aluminium and SiC are much dissimilar with respect to the electrical and thermal conductivities. The individual properties of both the materials are shown in Table 104.5. The SiC particles used in this study are 15 μm in radius and spherical in shape.

Table 104.5: Material properties.

Material	Density (g/cm³)	Melting point °C	Vickers hardness	Modulus of elasticity (GPa)	Electrical conductivity (V⁻¹ m⁻¹)	Thermal conductivity (W/mK)
SiC	3.21	2730	2500	430	1×10^3	132
Al6063	2.7	600	50	69.5	3.7×10^7	200

To account for the effect of Si-C particles doped into aluminium, three cases were considered:

Case I: SiC Particle Exactly Under the Spark

SiC particle exactly under the spark, the model shows it by the white line concentric to the edge as shown in Figure 104.2 (c). The location of the spark is at the centre of the SiC particle. As SiC has extremely low conductivity, the effect of this can be seen as the heat concentration in a localised region. This heat concentration leads to very high peak temperature (i.e. 45190 K) much higher in amount than in the case of pure aluminium. Most of the heat given by the single discharge is stuck within the SiC particle because of its lower thermal conductivity. Temperature of the aluminium beside the SiC boundary reaches to 931 K shown by the temperature isotherm (third from left). In this case the number of elements killed at the end of pulse duration are 580 (Table 104.6).

Case II: Particle Highly Displaced
The position of the SiC particle is shown by the white line adjacent to the edge of the model Figure 104.2 (d). The centre of the particle is 15 μm away from the spark (edge of the model). In this case the peak temperature drops to 35000 K (compared to case I) and number of elements killed are increased to 703. The spark falls on the adjoining area of aluminium and SiC as shown in Figure 104.2 (d). Here the depth of the temperature isotherm is higher than the radius as it is difficult for the heat to flow through the less conductive SiC particle.

Case III: SiC Particle far from the Spark
In this case, the SiC is considered to be far from the spark as shown in Figure 104.2 (e). The centre of the particle is 27 μm away from the spark (edge of the model). The distance of the SiC particle from the spark affect the number of elements killed as in this case it is 979 (more than the above two cases). The greater the conductivity of any material causes wider areas of boiling region, molten region and HAZ than the material of a low thermal conductivity. The presence of SiC particle created an obstruction to the heat flow in the direction where it is present as shown in Figure 104.6.

The comparison among all the four cases (i.e. pure aluminium and three cases of Al/SiC) in terms of number of elements killed is shown in Table 104.6. The number of elements killed in case of pure aluminium is greater than all the cases of Al/SiC MMC which is a direct indication of higher material removal rate.

Table 104.6: Comparison of number of elements killed in different cases.

Workpiece	Current (A)	Voltage (V)	Pulse duration (μs)	Spark radius (μm)	Elements killed
Aluminium	2	100	30	15	1868
Al/SiC (Case I)	2	100	30	15	580
Al/SiC (Case II)	2	100	30	15	703
Al/SiC (Case III)	2	100	30	15	979

The low MRR in case of Al/SiC MMC is due to the presence of lower conductivity of SiC particles. Among the three cases of Al/SiC MMC, case III indicates higher MRR. MMC as industrial point of view is desired to have high mechanical properties which can be achieved by addition of higher amount of reinforcement particles into the matrix. But the higher volume fraction of reinforcements affects its machinability. In case of EDM of MMC, at higher volume fraction of reinforcement, the particles will be closely packed into the matrix causing to fall more sparks on the low conductive SiC particles resulting in lower MRR. Therefore a balance should be maintained between the mechanical properties and the machinability to make it suitable for the particular application and economical for the machining and repair.

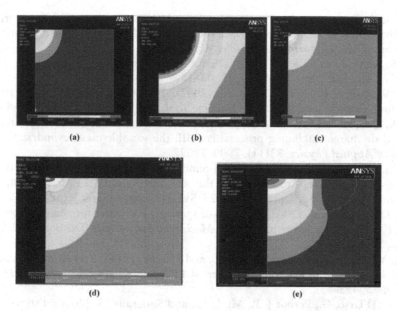

Figure 104.2 (a) Temperature isotherm just before material ejection; (b) Temperature isotherms subsequent to material removal; (c) Temperature isotherms when SiC is exactly under the spark; (d) Temperature isotherms when SiC is highly displaced; (e) Temperature isotherms when SiC is far away from the spark.

Conclusion

The following conclusions can be extracted from the finite element modelling of EDM based on single discharge taking aluminium and Al/SiC as workpiece materials.

* Here an ascending trend of MRR is found with increasing spark radius whereas increasing trend is found with the increase in current and pulse duration.
* Surface finish decreases with ascending current and pulse duration but enhances with increasing spark radius.
* The presence of reinforcement particles (SiC) in aluminum give much lower MRR under the action of single spark as compared to MRR of pure aluminum.
* The position of SiC particle with respect to the single discharge affects the temperature distribution and hence the material removal rate.

References

1. Mohanty, S., Routara, B. C., and Bhuayan, R. K. (2017). Experimental investigation of machining characteristics for Al-SiC12% composite in electro-discharge machining. *Materials Today: Proceedings*, 4(7), 8778–8787.
2. Fan, W. W. (1993). EDM characteristic of the sic particles reinforced aluminum matrix composite. Master's thesis, National Central University, Taiwan.
3. De Silva, A., and Rankine, J. (1995). Electrical discharge machining of metal matrix composites. In Processor International Symposium for Electro Machining XI, Switzerland, (pp. 75–84).
4. Weingartner, E., Kuster, F., and Wegener, K. (2012). Modeling and simulation of electrical discharge machining, In 1st CIRP Global Web Conference: Interdisciplinary Research in Production Engineering. *Procedia CIRP*, 2, 74–78.

5. Salonitis, K., Stournaras, A., Stavropoulos, P., and Chryssolouris, G. (2009). Thermal modeling of the material removal rate and surface roughness for die-sinking EDM. *International Journal of Advanced Manufacturing Technology*, 40, 316–323.
6. Joshi, S. N., and Pande, S. S. (2010). Thermo-physical modeling of die-sinking EDM process. *Journal of Manufacturing Processes,* 12, 45–56.
7. Yadav, V., Jain, V. K., and Dixit, P. M. (2002). Thermal stresses due to electrical discharge machining. *International Journal of Machine Tools and Manufacture*, 42, 877–888.
8. Eubank, P. T., Patel, M. R., Barrufet, M. A., and Bozkurt, B. (1993). Theoretical models of the electrical discharge machining process. Part III: the variable mass, cylindrical plasma model. *Journal of Applied Physics,* 73(11), 7900–7909.
9. Pradhan, M. K. (2010). Modelling and Simulation of thermal stress in electrical discharge machining process. In Proceeding of the 4th International Conference on Advances in Mechanical Engineering, September 23–25, SVNIT Surat, Gujarat India.
10. Panda, D., and Bhoi, R. (2005). Analysis of spark eroded crater formed under growing plasma channel in electro-discharge machining. *Machining Science and Technology: An International Journal*, 9(2), 239–261.
11. Zhang, Y., Liu, Y., Shen, Y., Li, Z., Ji, R., and Cai, B. (2014). A novel method of determining energy distribution and plasma diameter of EDM. *International journal of Heat and Mass Transfer*, 75, 425–432.
12. Bigot, S., D'Urso, G., Pernot J. P., Merla C., and Surleraux, A. (2016). Estimating the energy repartition in macro electrical discharge machining. *Precision Engineering*, 43, 479–485.

105 Application of nanomaterials in solar desalination systems for enhancing productivity: an article survey

M. Murugan[1,a], A. Saravanan[2,b], P.V. Elumalai[2,c], Pramod Kumar[1,d], P.S. Ranjit[2,e], and V.R. Lenin[3,f]

[1]Department of Mechanical Engineering, Aditya College of Engineering & Technology, Surampalem, Andhra Pradesh, India

[2]Department of Mechanical Engineering, Aditya Engineering College, Surampalem, Andhra Pradesh, India

[3]Department of Mechanical Engineering, Anil Neerukonda Institute of Technology & Sciences, Sangivalasa, Andhra Pradesh, India

Abstract: Solar desalination system is one of the technologies for converting brackish water to potable water which has been thoroughly researched in terms of yield, effectiveness, and financial side during the last several decades. Improving solar still by optimizing the solar evaporation process using nanomaterials has recently been offered as a possible technique for overcoming the bottleneck of conventional solar desalination systems. Because of their favourable thermo-physical and optical properties, nanomaterials have been generally exploited to augment the effectiveness of many energy systems in recent years. In particular, solar desalination method has benefited by the use of nanomaterial technology as an economical and dependable method of providing freshwater. This paper conducts an article survey about the use of nanomaterial knowledge in both passive and active solar desalination systems. The advances accomplished and the problems that remain are considered, and some major findings and recommendations for research opportunities are suggested. According to the present study, utilizing nanomaterials and increasing the volume percent of nanomaterials improves the daily productivity of solar desalination systems. Future study in this sector should take these considerations into account.

Keywords: Effectiveness, nanomaterial, productivity, potable water, solar desalination system.

Introduction

As water occupies more than 70% of the total Earth's surface, the majority of it is unhealthy for human utilization. Only 2.5 percent of the world's total freshwater supply comes from freshwater lakes, rivers, and subterranean aquifers. Freshwater is unfortunately scarce and unevenly distributed, in addition to being limited. Many regions of the world are facing demand for potable water as well as short of conventional energy sources and are unable to afford costly fossil fuels. More lately, there has been a growing realization that the utilization of fossil fuels adds to climate change and global warming, even if they were considerably cheaper.

[a]murukar@gmail.com, [b]saran_thermal@yahoo.co.in, [c]elumalimech89@gmail.com,
[d]pramodknitp@gmail.com, [e]ranjitsinghaec619@gmail.com, [f]murukar@rediffmail.com

Natural resources will be practically gone by 2050, thus everyone should focus on finding alternate sources of drinking water production method and other essential requirements. Solar desalination methods are an alternate approach for producing potable water [1,2]. Two common methods for distilling saltwater are electrodialysis and reverse osmosis. Salt ions are moved from one liquid to another via electrostatic membranes due to an electrical potential difference in electrodialysis [3]. Semipermeable membranes are used in reverse osmosis to remove molecules, ions, and bigger particles from saline water [4]. Electrodialysis and reverse osmosis processes have a number of drawbacks, includes high manufacture cost, repair, and maintenance expenses, as well as the requirement for electrical power usage. Furthermore, due to their enormous capacity, these technologies are unsuitable for regions with a dispersed population.

For these applications, solar distillation devices offer a possible alternative to get fresh water without any power source. Solar energy is utilized in solar distillation systems to heat and evaporate water while also separating it from contaminants and salt. After evaporation, the condensate is cooled to obtain the pure water. Low fabrication, service, and operating costs, as well as simplicity, portability, and the use of renewable energy sources, are all advantages of solar distillation systems. Traditional solar desalination systems were inefficient and large, taking up a huge space. To address the shortcomings and enhance the efficiency of these devices, researchers have applied a variety of passive and active strategies [5–8].

When compared to all other strategies, the preceding literature analysis [1–8] clearly showed that the production of freshwater using solar energy is the best. The prime objective of the present literature survey is to focus on application of nanomaterials for enhancing the yield of solar desalination systems. Even though numerous research works are being carried out to augment the yield of solar stills, use of nanomaterials in the production of freshwater in solar desalination systems have significant impact. Hence, in this survey we mainly focused on the influence of nanoparticles in solar desalination systems that will help the beginners to step up their research in this area.

Potential of Nanofluids

If nanoparticles are uniformly distributed and steadily floating in base fluids, their partial contribution improves the thermal characteristics of the base fluid. Nanofluid is a colloidal dispersion of nanoparticles ranging in size from 1to100 nm in a base fluid, which was first invented by Choi in 1995. Nanoparticles are made from a variety of materials, including ceramic carbides, oxides, and nitrides and also carbon nanotubes, materials and compound materials, depending on the purpose. Nanofluids have been generally used for different applications, and they are emerging as the next-generation heat transfer fluids because they offer effective new ways to augment heat transfer rate when compared with the pure liquid. Generally, nanofluids are prepared by one step method or two step method. Although nanoparticle synthesis techniques have previously been scaled up to commercial production levels, the two-step process is the most cost-effective way to create nanofluids on a wide scale.

Nanofluids in Single Slope Solar Stills

The thermo-physical properties of aqua nanofluids of Zinc Oxide (ZnO), Aluminum Oxide (Al_2O_3), Tin Oxide (SnO_2), and Iron Oxide (Fe_2O_3) of various concentrations have been investigated by Elango et al. [9] using single basin single slope solar stills (SSSS), and appropriate nanofluids have been identified for data acquisition in solar stills. The result revealed that the system with Al_2O_3 nanofluid produces 29.95% more freshwater than the system with ordinary water, whereas the stills with ZnO and SnO_2 nanofluids produce 12.67% and 18.63% more respectively. The photographic view of the experimental facility is depicted in Figure 105.1.

Figure 105.1 Photograph of experimental setup [9].

Kabeel et al. [10] examined the effect of adding an external condensation unit and nanofluids to an SSSS experimentally. When using Al_2O_3 and CuO nanofluids with a constant vacuum fan and an integrated external condensation unit, the revised solar still yields 73.80% and 84.20%, respectively. According to Rashidi et al. [11] fluid volume simulation model has been developed to study the vaporization and precipitation characteristics of SSSS using Al_2O_3-water nanofluid. The numerical findings revealed that increase in the solid volume percentage of nanoparticles boosts freshwater production. As the solid volume percentage grows from 0% to 5%, higher productivity by roughly 25%. Since the solid volume percentage increases from 0% to 5%, the mean increment in Nusselt number is by around 18%. The behaviour of copper oxide nanofluid in a SSSS with an exterior thermoelectric condensation duct is presented in this experimental and theoretical study by Nazari et al. [12]. The results disclosed that the production of freshwater, exergy, and energy efficiencies are increased by around 82.4%, 92.6% and 81.5%, respectively, when 0.08% volume portion of Cu_2O nanoparticles is added to water in the basin. Figure 105.2 shows the arrangement of thermoelectric condensation unit with SSSS.

For the first time in July 2018, the productivity of two nanofluid-based SSSS at the mountaintop Tochal approximately 4000 m above the sea level and in the capital of Tehran has been investigated for four days continuously by Parsa et al. [13]. They came to the conclusion that the standard energy and exergy performance of the proposed system filled with

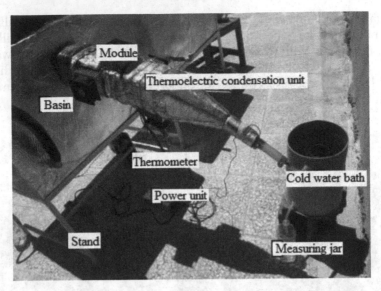

Figure 105.2 Attachment of thermoelectric condensation unit with SSSS [12].

nanofluid has improved by roughly 106% and 196%, respectively, when compared to the system without nanofluid.

Nanofluids in Double Slope Solar Stills

Koparde and Cummings [14] investigated the energy and exergy analysis, and effectiveness of a dynamic double slope solar still (DSSS) using nanofluids and basefluids. The schematic view of DSSS is portrayed in Figure 105.3. For this investigation, three distinct nanofluids (Al_2O_3, TiO_2, and CuO–water based) were explored. Anatase, brookite, and rutile are three distinct types of titanium oxide. In the nanocrystalline form, anatase is the most

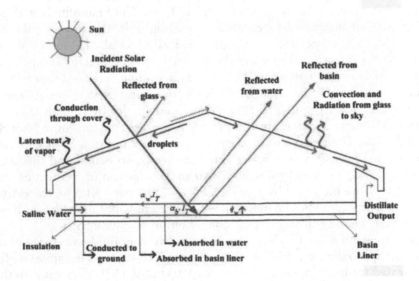

Figure 105.3 Schematic view of DSSS [15].

thermodynamically stable, although the rutile stage is added stable in the bulk appearance. Bulk form of anatase and brookite are metastable, and when heated, they convert irreversibly to rutile. Sahota and Tiwari [15] used the constructed model to analyze the basefluid (excluding nanoparticles) and the nanofluid with three different concentrations (0.04 percent, 0.08 percent, and 0.12 percent). The effect of varying concentration of Al_2O_3 nanoparticle on thermal conductivity, fluid yield, internal heat transfer coefficients (HTC), and fluid temperature was also investigated.

Chaichan and Kazem [16] have disseminated the nanoparticles in paraffin wax (a phase change substance) to boost the latter's heat conduction over the base material in their investigation. When nanoparticles are mixed with wax, the thermal and physical properties of the wax are altered. The results revealed that when nano-Al_2O_3 has been added to paraffin wax, the daily distillation output increased by 10.38%, while the distillate yield increased by 60.53% when correlated with the simple distiller output owing to continued distillation after sunset. The experimental performance evaluation of DSSS has been carried out by Jani and Modi [17] using hollow fins with square and circular cross-sectional shapes. The result revealed that the daily productivity of solar stills with circular and square hollow fins for the water depth of 10 mm are 1.4917 kg/m^2 and 0.9672 kg/m^2, respectively than the stills without fins. Photographic view of DSSS as shown in Figure 105.4.

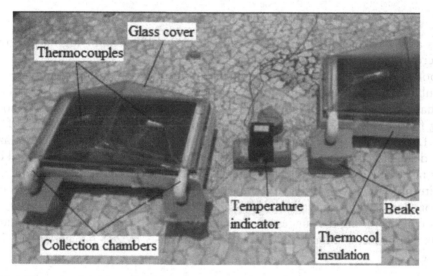

Figure 105.4 Photographic view of DSSS [17].

Nanofluids in Solar Stills Integrated with Solar Collectors

A study of the influence of saline water on the effectiveness of a DSSS with a tracking parabolic trough collector (TPTC) has been reported by Hassan et al. [18]. The data shown that adding wire mesh and sand to the redesigned system increases the daily freshwater production by about 3.1 percent and 13.7 percent in the winter and around 3.4 percent and 14.1 percent in the summer. Subhedar et al. [19] have described an experimental study that used nanofluid to determine the freshwater production of a standard SSSS plant combined with a PTC which is shown in Figure 105.5. In the integrated system, experimentation trails were executed with water and Al_2O_3 nanofluid with 0.05 percent and 0.1 percent volume

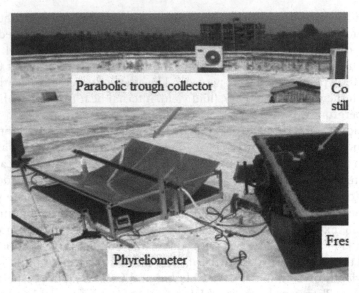

Figure 105.5 Photograph of solar still integrated with parabolic trough collector [19].

fractions as a working medium. The results suggested that employing an integrated still plant increases pure water yield significantly.

Dawood et al. [20] improved the productivity of standard SSSS by combining an evacuating tubular on the concentrating line of a PTC and a heat exchanger loop as an under basin phase change material. The result revealed that the solar still with nano-oil as a working fluid has produced the daily yield of freshwater at mass flow rates of 1.5, 1.0, and 0.5 L/min, as 6.21, 8.79, and 11.14 L/m²/day, respectively. Sadeghi and Nazari [21] compared the effectiveness of a customized SSSS with a thermoelectric-equipped duct and a focussing evacuated tube collector using an antimicrobial Ag@Fe3O4/deionized water composite nanofluid to that of a standard solar still in an observational investigation. When correlated with the typical solar still, the customized solar still incorporated with the solar collector using the suggested composite nanofluid at 0.08 percent volume concentration has improved the daily yield and energy efficiency by 218 percent and 117 percent, respectively.

Conclusion and Future Recommendation

This paper briefly explained about the application of nanomaterials in various solar stills for augmenting the production of freshwater. Based on previously evaluated scientific research results, the following essential points are highlighted.

- Various experimental and numerical assessments have been with several solar water distillation concepts.
- Changes to the solar still designs, as well as the use of nanofluids, resulted in a considerable improvement in performance.
- Connecting the solar still to various solar collectors (evacuated tubes, parabolic trough collector, and so on) in the presence of nanoparticles, the system operated effectively.

References

1. Delyannis, E. (2013). Historic background of desalination and renewable energies. *Solar Energy*, 75, 357–366.
2. Arunkumar, T., Ao, Y., Luo, Z., Zhang, L., Li, J., Denkenberger, D., and Wang, J. (2019). Energy efficient materials for solar water distillation—A review. *Renewable and Sustainable Energy Reviews*, 115, 109409.
3. Tedesco, M., Hamelers, H. V. M., and Biesheuvel, P. M. (2016). Nernst–planck transport theory for (reverse) electrodialysis: I. effect of co-ion transport through the membranes. *Journal of Membrane Science*, 510, 370–381.
4. Gokcek, M. (2018). Integration of hybrid power (wind-photovoltaic-diesel-battery) and sea-water reverse osmosis systems for small-scale desalination applications. *Desalination*, 435, 210–220.
5. Kabeel, A. E., Omara, J. M., Essa, F. A., and Abdullah, A. S. (2016). Solar still with condenser—a detailed review. *Renewable and Sustainable Energy Reviews*, 59, 839–857.
6. Rashidi, S., Rahbar, N., Valipour, M. S., and Esfahani, J. A. (2018). Enhancement of solar still by reticular porous media: experimental investigation with exergy and economic analysis. *Applied Thermal Engineering*, 130, 1341–1348.
7. Murugan, M., Saravanan, A., Murali, G., Kumar, P., and Reddy, V. S. N. (2021). Enhancing productivity of v-trough solar water heater incorporated flat plate wick-type solar water distillation system. *ASME Journal of Heat Transfer*, 143(3), 032001.
8. Murugan, M., Saravanan, A., Kumar, P., Reddy, V. S. N., and Arif, A. (2021). An overview about influence of wick materials on heat and mass transfer in solar desalination systems. In IOP Conference Series: Materials Science and Engineering, (pp. 1057), 012044.
9. Elango, T., Kannan, A., and Murugavel, K. K. (2015). Performance study on single basin single slope solar still with different water nanofluids. *Desalination*, 360, 45–51.
10. Kabeel, A. E., Omara, J. M., and Essa, F. A. (2017). Numerical investigation of modified solar still using nanofluids and external condenser. *Journal of the Taiwan Institute of Chemical Engineers*, 75, 77–86.
11. Rashidi, S., Akar, S., Bovand, M., and Ellahi, R. (2018). Volume of fluid model to simulate the nanofluid flow and entropy generation in a single slope solar still. *Renewable Energy*, 115, 400–410.
12. Nazari, S., Safarzadeh, H., and Bahiraei, M. (2019). Experimental and analytical investigations of productivity, energy and exergy efficiency of a single slope solar still enhanced with thermo-electric channel and nanofluid. *Renewable Energy*, 135, 729–744.
13. Parsa, S. M., Rahbar, A., Koleini, M. H., Javadi, Y. D., Afrand, M., Rostami, S., Amidpour, M. (2020). First approach on nanofluid-based solar still in high altitude for water desalination and solar water disinfection (SODIS). *Desalination*, 491, 114592.
14. Koparde, V. N., and Cummings, P. T. (2008). Phase transformations during sintering of titania nanoparticles. *ACS Nano*, 2, 1620–1624.
15. Sahota, L., and Tiwari, G. N. (2016). Effect of Al_2O_3 nanoparticles on the performance of passive double slope solar still. *Solar Energy*, 130, 260–272.
16. Chaichan, M. T., and Kazem, H. A. (2018). Single slope solar distillator productivity improvement using phase change material and Al2O3 nanoparticle. *Solar Energy*, 164, 370–381.
17. Jani, H. K., and Modi, K. V. (2019). Experimental performance evaluation of single basin dual slope solar still with circular and square cross-sectional hollow fins. *Solar Energy*, 179, 186–194.
18. Hassan. H., Ahmed, M, S., and Fathy, M. (2019). Experimental work on the effect of saline water medium on the performance of solar still with tracked parabolic trough collector (TPTC). *Renewable Energy*, 135, 136–147.
19. Subhedar, D. G., Chauhan, K. V., Patel, K., and Ramani, B. M. (2020). Performance improvement of a conventional single slope single basin passive solar still by integrating with nanofluid-based

parabolic trough collector: An experimental study. *Materials Today: Proceedings,* 26, 1478–1481.

20. Dawood, M. M. K., Nabil, T., Kabeel, A. E., Shehata, A. I., Abdalla, A. M., and Elnaghi, B. E. (2020). Experimental study of productivity progress for a solar still integrated with parabolic trough collectors with a phase change material in the receiver evacuated tubes and in the still. *Journal of Energy Storage,* 32, 102007.

21. Sadeghi, G., and Nazari, S. (2021). Retrofitting a thermoelectric-based solar still integrated with an evacuated tube collector utilizing an antibacterial-magnetic hybrid nanofluid. *Desalination,* 500, 114871.

106 Reliability of lead-free solder based on wettability

Niranjan kumar[1,a], Sushant Kumar[2,b], Ambrish Maurya[1], Md Irfanual Haque Siddiqui[3], and Aman Kumar[1]

[1]Department of Mechanical Engineering, National Institute of Technology Patna, India

[2]Department of Material Science Engineering, National Institute of Technology Srinagar, India

[3]Department of Mechanical Engineering, King Saud University, Riyadh, Saudi Arabia

Abstract: Lead-free solder materials are gaining attention by the manufacturers over the leaded solder. It is because of the hazardousness caused by the lead present in electronic wastes and various new legislations and tax benefits in different countries on the use of lead-free solder. The aim of the researchers is to develop the lead-free solder that have the properties close to the leaded solder. Wettability of solder during the joining of electronic component is the property which is needs to be maximized. The effort has been paid to provide the advancement in lead-free solder, explicitly focused toward the wettability property of the solder. A reliability context has been reviewed for the lead-free solder based on wettability and analyzed the effect of temperature, tin whisker, vibration and electromigration for reliability of lead-free solder.

Keywords: Electromigration, electronic component, lead-free solder, surface tension, wettability.

Introduction

The present world is headed towards the advancement in computerization and robotics, which increases the dependency on electronic components and gadgets. Solder plays an important role in electronic components as a joining, filler or conducting glue material. The term "solder" is derived from the Latin word "SOLIDARE," which itself is derived from the ancient French word "SOUDURE." "To combine or fasten together" is what it implies. Humpston and Jacobson [1] stated that solder is a eutectic alloy that has melting point lower than the two components to be joined. It melts and join when the temperature reaches to its eutectic point. Soldering is a metallurgical process of joining of two materials below 425°C. A solder must be electrically and thermally conductive along-with imparting appropriate mechanical strength for electronic assemblies.

Solder is mainly of two types, lead based solder and lead-free solder. Tin-lead (Sn-Pb) is a common example of the Lead based solder generally used in the electronic packaging industry because it is soft in nature and has a small wetting angle with a low eutectic temperature (183°C) on copper. Hongtao et al. [2] have reported that the commonly used soldering alloys are 60Sn-0Pb and 63Sn-37Pb, who's tensile and shear strength depends

[a]niranjank.phd20.me@nitp.ac.in, [b]sushantkumar2024@gmail.com

upon the concentration of Sn used in the solder. Pb is mixed with Sn due to its poor wetting characteristic. Lead is highly neurotoxic, and its high exposure may lead to anemia, kidney damage, weakness and brain damage, damaging the nervous system of developing babies in pregnant women [3]. Several Laws and Legislations were proposed to ban the use of lead in electronic component and industry.

USA banned lead-based solder since 1990, senate and House of Representatives passed some rule such as Lead-based paint Hazard Abatement Act of 1991, Lead Exposure Reduction Act of 1991 and Lead Exposure Act of 1992. Ervina et al. [4] reviews some points like the Environmental Protection Agency (EPA) gave notice to the USA, Japan and Korea for Pb content in the Environment under Toxic Substance Control Act containing regulatory implications. China used Restriction of Hazardous Substances Directive (RoHS), to exclude Lead in electric soldering. European union were restricted to use mercury, lead, hexavalent chromium, Cadmium and certain flame retardants. Japan prohibited the lead and shifting them towards the use of lead-free solders.

Several researchers developed various lead free solders to replace the leaded solder, example: Sn-Bi [7], Sn-Cu [4], Sn-Zn [4], Sn-Ag [4], Sn-Ag-Cu [4], Zn-Bi [11] Sn-Zn-Bi [5], Sn-Cu-Bi [6], Sn-Cu-Ni [7], Sn-Au [8], Sn-In [9], Sn-Sb [10], Sn-Pd [11], Sn-Ag-Bi[12] etc. Jayesh and Elias [6] states that the basic requirement of Sn based system is to improve the wettability and reduce the surface tension, enhance the development of intermetallic compounds between substrate and solder, improve ductility, improve mechanical properties and prevent the excess growth of tin whiskers, restrict the phase change of β-Sn to α-Sn because this leads to degradation of reliability and structural integrity of solder.

Wettability

Wettability is the ability of soldering material to maintain contact with base material via intermolecular interaction. Wettability is usually characterized by the contact angle. Nanjing [13] has used for evaluating the contact angle with the help of Young formula. Figure 106.1 is shows that the surface tension under equilibrium state.

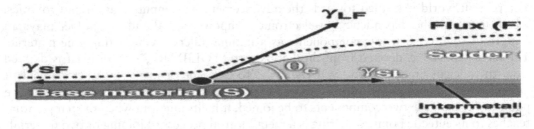

Figure 106.1 Schematic diagram of the surface tension under equilibrium state [16].

$$\cos\theta = \frac{\Upsilon_{SF} - \Upsilon_{SL}}{\Upsilon_{LF}}$$

Where θ = wetting angle, Υ_{SF} = solid-gas surface tension, Υ_{SL} = liquid-solid surface tension, and Υ_{LF} = gas-liquid surface tension.

In general, if degree (θ) less than 90° proper wetting takes place. If wetting angle lies between 0° $\leq \theta \leq$ 20° it means very good wetting quality, If wetting angle lies between 20° $\leq \theta \leq$ 40° which is acceptable [14].

For an ideal lead-free solder joint, surface tension should be less. Chen et al. [15] states that an increase in temperature leads to decrease in the contact angle; they also claim that among Ni, Au, organic solderability preservatives (OSP) and Sn, Sn has better surface finish. Ervina et al. [16] described that wetting angles are affected via (*printed circuit board*) PCB, type of solder alloy and flux used, along with the surface roughness, testing temperature and soldering environment. The 'wetting balance test' and 'Area of spread test' are the considerable tests for wettability.

Surface Tension

Surface tension is the ability of the liquid surface to minimize its surface area determining the shape of a liquid. Decrease in surface tension decreases the contact angle enhancing the wettability of liquid.

The Sn-Cu system behaved differently on Cu and Ni substrates in high vacuum, and Cu-substrate was shown to be more wettable up to 300°C. Considering the wetting behaviour of SAC 305 solder on Sn, Ni/ Au, and OSP finished substrates in high vacuum and the inert gaseous atmosphere, Goncalves et al. [17] discovered a sudden spread in solder at 225°C on the Sn substrate, slightly decreasing the contact angle with temperature; later, it was discovered to be completely independent of temperature. It was observed that the oxidation of the Tin surface was prevented using OSP film ultimately enhancing the solder wettability, it was seen that the angle of contact between melted solder on the OSP and Ni/Au-finished substrate was temperature dependent.

Reliability

The capacity of lead-free solder junctions to withstand functional deterioration of electronic devices is referred to as reliability. Temperature, vibration, tin whiskers, and electro migration, according to them, are the extrinsic causes that cause solder connection failure. Hu et al. [18] investigated the influence of Ag on the inter-face interaction of Sn Bi/Cu solder and discovered that Ag increases brittleness while decreasing reliability. Kanlayasiri and Sukpimai [19] observed the effect on reliability of low-silver SnAgCu solder while aging on adding *In*, and concluded that an increase in the content of In enhances the reliability by decreasing the thickness ratio of Cu_6Sn_5 and Cu_3Sn layers.

Effect of Temperature

The temperature changes the microstructure as well as the internal stress which leads to joint failures, to avoid this the surrounding temperature is being maintained via thermal protection to ensure its reliable operation under extreme conditions [20].

Aging

Aging process coarsens the microstructure and initiates the growth of IMC (intermetallic compound—a brittle phase forms at the interface of molten solder and base material or substrate) that initiates and expands the crack in solder joints. Increase in aging time, thickens the IMC reducing its strength and reliability finally resulting in failure of solder joints. Mustafa et al. [21] observed that the fatigue life of solder joints decreases with increase in aging time by increasing its grain size and, IMC plays a crucial role in coarsening the grain. Wang and Nishikawa [22] studied the effect of isothermal aging on reliability of SAC 305 solder joints and observed that an increase in aging time reduces its tensile and shear strength and found that Cu substrate with Ni, Ag or Au plating can be used to inhibit the interfacial growth of IMC.

Thermal Cycling

Thermal cycling occurs due to periodic change in ambient temperature resulting in generation of alternate stress-strain in the solder joint that visualizes large extent of micro-cracks, which can lead to thermal fatigue crack near interfacial regions in bulk solder joints. Quan Zhou et al. [23] found that continuous thermal cycling in solder joint results in higher degree of recrystallization and less recovery, whereas interrupted thermal cycling creates a strong and stable recovered microstructure. Sun et al. [24] observed that adding 0.1% Al nanoparticles on low-Ag (in SnAgCu solder) enhances the mechanical properties and the solder reliability by inhibiting the germination as well as the growth of IMC layer. It was found that adding elements like Nickel, Mn, and Bismuth to the low-Ag in SnAgCu alloy solder improves its thermal reliability.

Thermal Shock

Thermal shock testing is used to improve the thermal durability of solder junctions by subjecting them to a specified number of hits. With increased temperature shocks, the IMC layer at the solder connection contact thickens, reducing its dependability. Excess thickness in the IMC layer is prone to stress resulting in crack initiation, ultimately failing the solder joint Sharma et al. [25] studied the effect of thermal shock on the growth of IMC in Sn-Ag-Cu solder joint in range 65 ~ 50°C and found that with an increase in the period of thermal shock, the Cu_6Sn_5 IMC gets thinner due to pore densification, thickening of the long scallops and finally transforming it into a flat interface of lower surface energy. Tian et al. [26] observed that the tensile strength of SAC 305 (Sn3.0Ag0.5Cu) solder decreases with an increase in thermal shock cycle in QFP (Quad Flat Package), the thermal shock resistance can be enhanced by adding trace amount of Zn to Sn-Ag-Cu solder.

Effect of Vibration

The breakdown of the solder connection is produced by low-cycle fatigue, which is brought on by temperature load, whereas high-cycle fatigue is more difficult and brought on by vibrational stress. Zhao et al. [27] found that the fatigue crack growth rate of the Sn-Ag alloy increases due to vibration load, increasing the stress ratio and frequency, changing the fracture mode from trans-granular to inter-granular and changing the crack propagation from cyclic dependency to time dependency.

Vibration Shock

The BGA package's four corner solder connections were the first to fail as a result of higher relative stress there under vibration load. Large bending and deformation are seen when there is a vibration load, and this produces a lot of alternate stress and strain, which starts, grows, and widens fractures in solder joints. According to Song et al. [28], adding Cu can improve the toughness of Cu6Sn5 IMC, which in turn improves its vibrational life. Lead-free solder joints perform better mechanically and anti-vibrationally when their Cu content is increased by 1% or more.

Drop

Internal solder joints in portable devices fail owing to drop impact (impact from dropping), external impact, and up-and-down bending of the PCB during the drop process. [29] Xia et al. The product-level drop test is only used after the electronic product has been completed, making it unfavourable for the product development cycle and increasing the cost. There are two types of solder drop impact: product grade and board grade. The pulse-controlled drop test and drop test are used in the test standard to analyse the dependability of electronic components at the board level. Wu et al. [30] considered the Board level drop test for mobile employing hypermesh and discovered its failure criteria under the influence of the dropping. Luan et al. [31] monitored entire failure process of the solder joint under a dynamic monitoring system and the dynamic resistance method was found suitable in analyzing failure of different types of solder joint.

Tin Whiskers

Tin whiskers occur on diffusion of elements of low-melting point near room temperature and grow under slight pressure on coatings. Sun et al. [32] states its length varies from a few micrometers to 10mm with 0.2 ~ 5µm diameter in shapes like columnar, bifurcated, needle-like, tortuous, knurled, short columnar and bunched. At room temperature its needle-like while at high temperature its mound and entangled flock shaped. It has a strong current load capacity; its certain growth connects nearby solders resulting in short circuit causing irreversible damage to the electronic product. Zhang et al. [33] obtained that compressive stress is responsible for the growth of Sn whiskers due to stable grain boundary and recrystallization of tin in the coating. Britton [34] found that Ni intermediate can reduce the growth of whiskers and its growth at room temperature can be inhibited by heat or reflow treatment at 150°C. Barbara et al. [35] observed that annealed samples stored at room temperature were whisker free whereas un-annealed samples were full of whiskers and concluded that whisker growth can't be suppressed via Ni intermediate plating or heat treatment because of oxidative corrosion.

Electromigration

Electromigration can be defined as the disconnection due to the formation and growth of voids in thermally activated metal traces or solder joints caused by elevated temperature and electric current Ko et al. [36]. High temperature and current density gives rise to electromigration in the solder joints, acting as an important indicator of reliability of joints. According to researchers, employing graphene enhances the electronic packaging's ability to resist electromigration. Kang et al. [37] Multi-layered graphene (MLG) was applied on top of the Cu wire above the wafer, and by lowering the resistance, it was discovered that the breakdown of current density increased by 18%.

Conclusion

The demand of lead free solder drastically increased due to hazardousness caused by the lead present in electronic wastes and various new legislations and tax benefits in different countries. There are several challenges or tests to ensure reliable lead-free electronic products. Reliability of lead-free solder depends on solder alloy composition, manufacturing process, circuit board packaging, test conditions, etc. The oxidation of the Tin surface was prevented using OSP film ultimately enhancing the solder wettability. Addition of Indium in lead free solder alloy enhances the reliability. Addition of Bi and Ni in lead free solder, improve the thermal Reliability. Whisker growth can't be suppressed via Ni intermediate plating or heat treatment because of oxidative corrosion. Graphene improves the electro-migration property of electronic packaging.

References

1. Humpston, G., and Jacobson, D. M. (eds.), (2004). Principles of Soldering. ASM International.
2. Ma, H., and Suhling, J. C. (2009). A review of mechanical properties of lead-free solders for electronic packaging. *Journal of Materials Science,* 44, 1141–1158.
3. Noor, E. E. M., Sharif, N. M., Yew, C. K., Ariga, T., Ismail, A. B., and Hussain, Z. (2010). Wettability and strength of In-Bi-Sn lead-free solder alloy on copper substrate. *Journal of Alloys and Compounds,* 507(1), 290–296. doi: 10.1016/j.jallcom.2010.07.182.
4. Mhd Noor, E. E., Mhd Nasir, N. F., and Idris, S. R. A. (2016). A review: lead free solder and its wettability properties. *Soldering and Surface Mount Technology,* 28(3), 125–132.
5. Wu, J., Xue, S. B., Wang, J., et al. (2018). Effects of alpha-Al2O3 nanoparticles-doped on micro-structure and properties of Sn-0.3Ag-0.7Cu low-Ag solder [J]. *Journal of Materials Science: Materials in Electronics,* 29, 7372–7387.
6. Jayesh, S., and Elias, J. (2019). Experimental Investigation on the effect of Ag addition on ter-nary lead-free solder alloy–Sn–0.5Cu–3Bi. *Metals and Materials International,* 26, 107–114.
7. Bhavan, J. S., and Elias, J. (2019). Finite element modeling and random vibration analysis of BGA electronic package soldered using lead free solder alloy, Sn-1Cu-1Ni-1Ag. *International Journal for Simulation and Multidisciplinary Design Optimization,* 10, A11.
8. Ciulik, J., and Notis, M. R. (1993). The Au-Sn phase diagram. *Journal of Alloys and Compounds,* 191, 71–78.
9. Lin, S. K., et al. (2014). Effects of zinc on the interfacial reactions of tin–indium solder joints with copper. *Journal of materials science,* 49, 3805–3815. DOI 10.1007/s10853-014-8092-8.
10. El-Daly, A. A., et al. (2011). Microstructural evolution and tensile properties of Sn–5Sb sol-der alloy containing small amount of Ag and Cu. *Journal of Alloys and Compounds,* 509, 4574–4582.
11. Ho, C. E., et al. (2010). Strong effect of Pd concentration on the soldering reaction between Ni and Sn–Pd alloys. *Journal of Materials Research,* 25(11), 2078–2081.
12. Vianco, P. T., and Rejent, J. A. (1999). Properties of ternary Sn-Ag-Bi solder alloys: part II—wettability and mechanical properties analyses. *Journal of Electronic Materials,* 28(10), 1138–1143.
13. Nanjing Electron Society (1978). Soldering in Electronics [M]. Nanjing: Nanjing Agriculture Science Press. pp. 12–14.
14. Salleh, M. A. A. M., Al Bakri, A. M. M., Kamarudin, H., Bnhussain, M., and Somidin, F. (2011). Solderability of Sn-0.7 Cu/Si3N4 lead-free composite solder on Cu-substrate. *Physics Procedia,* 22, 299–304.
15. Chen, W., Xue, S., and Wang, H. (2010). Wetting properties and interfacial microstructures of Sn-Zn-xGa solders on Cu substrate. *Materials and Design,* 31(4), 2196–2200. doi: 10.1016/j.matdes.2009.10.053.
16. Ervina, E. M. N., and Tan, S. Y. (2013). Wettability of molten Sn-Zn-Bi solder on Cu substrate. *Applied Mechanics and Materials,* 315, 675–680.
17. Goncalves, C., Leitao, H., Lau, C. S., Teixeira, J. C., Ribas, L., Teixeira, S., Cerqueira, M. F., Macedo, F., and Soares, D. (2015). Wetting behavior of SAC305 solder on different substrates in high vacuum and inert atmosphere. *Journal of Materials Science: Materials in Electronics,* 5106–5112. doi: 10.1007/s10854-015-3037-9.

18. Hu, F. Q., Zhang, Q. K., Jiang, J. J., et al. (2018). Influences of Ag addition to Sn-58Bi solder on SnBi/Cu interfacial reaction. *Materials Letters,* 214, 142–145.

19. Kanlayasiri, K., and Sukpimai, K. (2016). Effects of indium on the intermetallic layer between low-Ag SAC0307-xIn lead-free solders and Cu substrate. *Journal of Alloys and Compounds,* 668, 169–175.

20. Choubey, A., Yu, H., Osterman, M., et al. (2008). Intermetallics characterization of lead-free solder joints under isothermal aging. *Journal of Electronic Materials,* 37(8), 1130–1138.

21. Mustafa, M., Suhling, J. C., and Lall, P. (2016). Experimental determination of fatigue behavior of lead-free solder joints in microelectronic packaging subjected to isothermal aging. *Microelectron Reliab,* 56, 136–147.

22. Wang, J. X., and Nishikawa, H. (2014). Impact strength of Sn-3.0Ag0.5Cu solder bumps during isothermal aging. *Microelectron Reliab,* 54(8), 1583–1591.

23. Zhou, Q., Zhou, B., and Lee, T. K., et al. (2016). Microstructural evolution of SAC305 solder joints in wafer level chip-scale packaging (WLCSP) with continuous and interrupted accelerated thermal cycling. *Journal of Electronic Materials,* 45(6), 3013–3024.

24. Sun, L., Chen, M. H., and Wei, C. C., et al. (2018). Effect of thermal cycles on interface and mechanical property of low-Ag Sn1.0Ag0.5Cu(nano-Al)/Cu solder joints. *Journal of Materials Science,* 29(12), 9757–9763.

25. Sharma, A., Kumar, S., Jung, D. H., et al. (2017). Effect of high temperature high humidity and thermal shock test on interfacial intermetallic compounds (IMCs) growth of low alpha solders. *Journal of Materials Science,* 28 (11), 8116–8129.

26. Tian, R. Y., Hang, C. J., Tian, Y., et al. (2018). Growth behavior of intermetallic compounds and early formation of cracks in Sn-3Ag-0.5Cu solder joints under extreme temperature thermal shock. *Materials Science and Engineering A,* 709, 125–133.

27. Zhao, J., Miyashita, Y., and Mutoh, Y. (2001). Fatigue crack growth behavior of 96.5Sn-3.5Ag lead-free solder. *International Journal of Fatigue,* 23(8), 723–731.

28. Song, J. M., Liu, Y. R., Lai, Y. S., et al. (2012). Influence of trace alloying elements on the ball impact test reliability of SnAgCu solder joints. *Microelectron Reliab,* 52(1), 180–189.

29. Xia, Y., Lu, C., and Xie, X. (2007). Effect of interfacial reactions on the reliability of lead-free assemblies after board level drop tests. *Journal of Electronic Materials,* 36(9), 1129–1136.

30. Wu, J., Song, G., and Yeh, C. (1998). Drop/impact simulation and test validation of telecommunication products. In Proceedings of the Inter Society Conference on Thermal Phenomena; Austin, TX, USA. (Vol. 847, pp. 330–336).

31. Luan, J. E., Tee, T. Y., Pek, E., et al. (2007). Dynamic responses and solder joint reliability under board level drop test. *Microelectron Reliab,* 47(2), 450–460.

32. Sun, M., Dong, M., Wang, D., et al. (2018). Growth behavior of tin whiskers on Sn Ag micro bump under compressive stress. *Scripta Materialia,* 147, 114–118.

33. Zhang, L., Yang, F., and Zhong, S. J. (2016). Whisker growth on SnAgCu-xPr solders in electronic packaging. *Journal of Materials Science,* 27(6), 5618–5621.

34. Britton, S. C. (1974). Spontaneous growth of whiskers on tin coatings: 20 years of observation. *Transactions of the Institute of Metal Finishing,* 52(1), 95–102.

35. Barbara, H., Balazs, I., and Shinohara, T., et al. (2012). Whisker growth on annealed and recrystallized tin platings. *Thin Solid Films,* 520(17), 5733–5740.

36. Ko, Y. H., Son, K., Kim, G., Park, Y. B., Yu, D. Y., Bang, J., and Kim, T. S. (2018). Effects of graphene oxide on the electromigration lifetime of lead-free solder joints. *Journal of Materials Science: Materials in Electronics,* 2334–2341. doi:10.1007/s10854-018-0506-y

37. Kang, C. G., Lim, S. K., Lee, S., Lee, S. K., Cho, C., Lee, Y. G., Hwang, H. J., Kim, Y., Choi, H. J., and Choe, S. H. (2013). Effects of multi-layer graphene capping on Cu interconnects. *Nanotechnology,* 24, 115707.

107 Reliability of lead free solder based on corrosion resistance properties

Niranjan Kumar[1,a], Prince Kumar[2,b], Ambrish Maurya[1], Vikas kumar[3], and Mitesh Pratap[1]

[1]Department of Mechanical Engineering, National Institute of Technology Patna, India

[2]Department of Material science Engineering, National Institute of Technology Srinagar, India

[3]Department of Mechanical Engineering, National Institute of Technology Kurukshetra, India

Abstract: Lead free Solder alloys have been widely examined to incorporate conventional Sn-Pb solder. Since scaling down of electronic items and application improvement, protection from consumption of solder alloys plays a significant ingredient in the dependability of electrical items over the long haul. In this article, we survey the new advancement on corrosion performance of solder compounds without lead by summing up the outcomes for agents of Sn-Bi, Sn-Cu, Sn-Zn, Sn-Ag and Sn-Ag-Cu potentiodynamic polarization have been utilized to examine the destructive properties. Utilizing solder in various arrangements of HCl and NaCl. The polarization bends demonstrated that the corrosive in NaCl was smaller serious compared to HCl arrangement dependent at consumption current and passivation get treats and furthermore showing destructive obstruction on expansion of some material like phosphorus, bismuth, Aluminum, iron, magnesium, etc at specifically condition. The corrosion instrument of Sn solder without lead after joining of compound components or particles is summed up. It is trusted that this outline will give some valuable data in explaining the corrosion component and advancement of lead free solder. Besides, remaining difficulties and future patterns in this exploration field are proposed.

Keywords: Corrosion, lead free solder, reliability.

Introduction

Electronic parts are welded together with solder to get mechanical reconciliation and electrical associations in the electrical fabrications. As we are aware, Sn-Pb solder are extensively utilized in electronic industry because of an ideal blend of acceptable consumption opposition, minimal expense, great wettability and mechanical properties and so forth despite the fact that, Pb has a place with substantial metal with toxic nature, which is hurtful health for human as well as environment [1]. Scientists all throughout the world much consideration has been given to the testing of solder without lead, which is examined as a share encouraging possibility to replace traditional Sn-Pb solder lately [1,2].

Many significant Sn solder were accounted for in applicable tests for example Sn-Zn [3], Sn-Cu [4], Sn-Ag [5] and Sn-Bi [6] etc. To increase combination properties of solder without lead further, are planned by ternary or multi-elements alloys. Sn-Ag-Cu [7] is soldered without lead, viewed as the most ideal alternative to supplanting ordinary Sn-Pb solder.

[a]niranjank.phd20.me@nitp.ac.in, [b]princekr8698@gmail.com

The primary trouble in the improvement of solder without lead need to accomplish the ideal blend of wettability, mechanical achievement, electrical properties, corrosion opposition, expense and so forth [8]. The corrosion issue of solder significant factor in long haul administration, particularly in high temperatures and high moisture conditions [9]. In spite of the fact that corrosion is happening in electronic gadgets, they are similar seen in steel structure spans, vehicles, pipelines and other natural in day to day life. Corrosion items containing nanograms or less may bring about complete gadget failure [5]. Subsequently, testing on corrosion resistance of solder without lead mostly centers on the improvement of microstructure and right now copying the help climate [3]. Fostering the corrosive characteristic of solder in neutral, alkaline and acidic arrangements has been investigated as of now research on consumption execution Sn solder are disconnected and hand-off possible outcomes are dispersed, even if a portion of the ends aren't right conventional. Therefore, a basic audit on rust and the characteristics of Sn-solder without lead are essential. The current article, the writers attempt to gather all the concerned information and data on rust's presentation, late examinations of Sn solder without lead compounds and afterward give some direction in future work. As of now, the examination is principally centered on Sn-Zn [3], Sn-Cu [4], Sn-Ag [5], Sn-Bi [6], Sn-Ag-Cu [7] and other multi-components of Sn-solder without lead compound [1–9].

Corrosion and its Methodology of Resistance

Essential of corrosion for solder without lead. The advancement of the consumption characteristic of Sn-solder combinations for fundamentally connected with the difference in microstructure [1]. For Sn-Zn lead without solder, galvanic consumption framed linking the Zn-rich stage and the Sn-rich stage. The corrosion of Sn-Bi solder in Zn-rich and is ascribed to the possible contrast between stage and bi-rich stage [2]. Though other solder alloys like Sn-Cu, Sn-Ag and Sn-Ag-Cu also having galvanic consumption shaped linking the Zn-rich stage and β intermetallic consolidation, and the intermetallic consolidation primarily comprise of Ag-Sn and Cu-Sn, which are practically indissoluble in corrosion state because of their synthetic steadiness. Erosion execution Sn solder without lead are dissected through different various experimental methods. Pollutants play a significant aspect in the corrosion interaction of solder amalgams, particularly Cl [2]. Subsequently, huge insightful endeavors have been suffer to consider the advancement of the consumption weakness of without lead solder amalgams in NaCl arrangement. Submersion corrosion experiments are performed to assess the consumption of solder NaCl arrangement [2]. Experiments of corrosion and its results on solder without lead.

This strategy can be at first applied to survey the corrosion execution of a solder when it is separated from real assistance conditions. Then at that point, a salt shower administration climate and high moisture and high temperature administration conditions are established to additionally reproduce the assistance climate. The analysis of corrosion in bad impact of electronic hardware is identified with the atmosphere for this the electrochemical response that happens under a dainty electrolyte layer. Thus, the corrosion opposition Solder is tested under a slight electrolyte layer, while the corrosion instrument should be dissected by electrochemical analyses. Open circuit potential, potentiodynamic polarization and electrochemical Impedance spectroscopy tests are continually coordinated in electrochemical examinations. Then, at that point, the consumption component is talked about as acquired electrical Parameters in electrochemical analyses [8, 9].

Sn-Ag-Cu

The ternary SnAgCu [7] solder is put in electronic bundling. As of now, the overall suggested synthetic organization of SnAgCu solder compounds [7]. The SAC305 solder has lower corrosion weakness contrasted with conventional Sn–Pb bind in 0.1 M NaCl and HCl arrangement, which is related with the development of a more minimized oxide film of SnO and SnO2 on the outside of solder [2]. Examine the solder compounds to improve the corrosion opposition. assessed the influence of augmentations of Fe & Bi about corrosion opposition of SAC105 solder compound through potentiodynamic polarization investigation, and called attention to that the fuse of 1 wt% Bi and 0.05 wt% Fe disintegrated the corrosion opposition of SAC105 in impartial arrangement where change in microstructure by galvanic corrosion [10]. The corrosion sensitivity improved by adding more Zn contents in solder by examining the 3.5wt% NaCl solution. Al and Ni are doped in NaCl by increasing the contents range 0 to 0.5 to improve corrosion opposition [11, 12].

Examine corrosion resistance of SnAgCu under real situations such as high–humidity as well as temperature. Resulting same during cooling in air or furnace under temperature about 70°C and high-humidity [13] in this examine found degradation of corrosion due to change in cooling rate and also change in intermetallic compound [14]. Showed that the treatment of maturing beneficially affected the improvement of imperviousness to rust of Sn-Ag-Cu bind composite, and the justification advancement is chiefly worried about the appropriation, size and change of Morphology of Cu_6Sn_5 and Ag_3Sn intermetallic compounds. Corrosion of solder in electronic segments constantly confronted air disintegration, because of which a slight fluid film joined to the outside of the solder in states of high temperature and high dampness [13, 14]. Potentiated was utilized to go through potentiodynamic polarization tests. Checking the rate was, with a potential likely scope of –2.0 to 2.0 VSCE for the both arrangements. Electrochemical portrayal was performed multiple times to guarantee reproducibility of the result. Energy dispersion scanning electron microscope spectroscopy was utilized to decide the microstructure and corrosion occurring of Sn-Ag-Cu solder after corrosion tests. Sn-Ag-Cu corrosion sensitivity of solder alloy solid-magnet fed with stirring. He tracked down that the affectability of Sn-Ag-Cu solder amalgam diminished with lasting magnet mixing and consumption affectability changes Solder was worried about the difference in microstructure and crystallographic surface [15, 16].

Sn-Cu

The binary Sn-2.8 wt% Cu solder is unique as possible replacements of supplanting Sn-Pb solder compounds because of the upside of minimal expense and great complete execution. Eutectic compound with a dissolving point of 227°C is framed when Cu has 0.69% [17]. Erosion of corrosion Cause of SnCu Solder is the arrangement of galvanic joints linking Sn-enriched lattice and Cu-Sn intermetallic compound, and the destructive change happens because of the development of the microstructure-sensitivity of solder [18]. It brought up that solder without lead exist Lower consumption affectability than Sn-Pb solder in 3.49 wt% Na-Cl arrangement. Nonetheless, Sn-Pb (73-37%) solder compounds present the most minimal of corrosion defenselessness and Sn–Ag–Cu solder combination has the least consumption resistance of corrosion among these four solder Sn–Pb, Sn–Cu, Sn–Ag and Sn–Ag–Cu on the understanding that of free smoke created by the consuming of poly-Vinyl chloride [19]. Likewise, it has been tracked down that the center increased rosin affectability of the four examined solders with an expansion in the convergence of smoke [19]. The cooling pace of hardening assumes a significant part in development of consumption properties of solder without lead in 0.49 M NaCl. The affectability of Sn-Cu solder alloy in solutions was evaluated with various cementing cooling methods. Variety of corrosion opposition SnCu solder composite with various cooling rate has a place with the two strains limited linking the Sn-rich phage and CuSn intermetallic accumulate and cathode/anode region proportion [20].

Adding alloying components or particles to solder without lead as Sn-Cu. The impact of inclusion of Fe or Bi on the consumption defenselessness of Sn-Cu solder amalgam in 3.49 wt% Na-Cl arrangement [20]. It was tracked down that the consumption opposition decayed in the wake of doping-Ingestion of Fe and Bi because of the arrangement of FeSn2 and configuration Bi-rich stage [21]. While S may contain improving the corrosion properties of Sn-Cu solder because of the arrangement of detached consumption compounds conduit of Sn_3O (OH) 2Cl2. Improvement of imperviousness to rust of Sn-Cu–Al solder [22]. It was to explore the impact of P on the corrosion sensitivity of SnCu solder in aggress-siev conditions. It has been tracked down that doping of P can reduce the consumption affectability of SnCu solder. In expansion, Sudden corrosion pace belonging to solder is less in basic arrangement than in acidic arrangement, show that P. amount of had no undeniable impact oath of corrosion affectability of SnCu solder compound in reproduced aquatic climate [23].

Sn-Ag

The Sn-3.5Ag solder is an expected promising candidate for replacing Sn-Pb to solder because of the predominant mechanical strength of Sn-Pb. For the Sn-Ag solder framework Sn-3.5Ag happens at a softening temperature of 221.1°C [5]. The affectability of the Sn-Ag solder amalgam is mostly connected with galvanic consumption between Sn-enriched stage and $A_{g3}Sn$ intermetallic compound. It was called attention to a well-known scale, size and circulation Sn-rich encourage in lattice or Ag3Sn particles Solder combination played a significant factor on the impact of corrosion resistance, mechanical and warm characteristic [24, 25].

The corrosion reliability of Sn–Ag–Zn solder evaluation of composites with various cooling strategies was done as indicated by electrochemical analysis with NaCl arrangement [25]. It was tracked down that the rust suspended the limit of water-cooled solder combination was not as much as heater cooled and air-cooled solder compounds, which have been relating to the kind, size and appropriation of accelerate test. It likewise showed that the obstruction property of the past the shaped with various cooling techniques was unique. Affectability of Sn-Ag solder amalgam with high cooling: the rate is smaller in comparison to solder combination with a smaller cooling rate [26].

The advancement of consumption opposition was related with the development of the morphology of Ag_3Sn. The microstructure of the weld composite can be altered by alloying expansion of alloying components or particles, the outcome is corrosion obstruction and change of mechanical legitimate connection that the incorporation of Ce further developed the corrosion obstruction in Sn-1.0Ag solder [24–27]. In normally circulated air through Na_2SO_4 with Chloride particle expansion arrangement, which is a Ce speed increase of aloof film arrangement. Moreover, it was tracked down that the doping of CeO_2 nanoparticles can influence corrosion sensitivity of Sn-Ag Solder Alloy in 3.49 wt% best doping of NaCl arrangement and CeO_2 nanoparticles about 12.1% decided by corrosion obstruction [27]. Corrosion resistance of Sn-Ag is improved by adding of this metal like In, Bi, Cu, etc to form lead without solder [28]. It was discovered that Sn–2.29Ag–9.1In and Sn–3.0Ag–10.4Bi had higher consumption affectability than solder Sn-Pb. notwithstanding corrosion opposition the Sn-Ag solder has particularly further developed Cu with addition, which is higher than that of Sn-Pb solder. The alloying expanded with the expansion of Ag content. They were tracked down that the volume part of the 3Sn-Ag intermetallic compound expanded with the increment of Ag content, bringing about less in galvanic corrosion. It found that Ag content had little impact on the corrosion affectability of the Sn-Ag solder amalgam [28, 29].

Sn-Zn

The Sn-Zn solder is viewed as an alluring contender for supplant Sn-Pb solder amalgam on account of a good understanding of destructive properties. The eutectic creation with Sn–Zn is at the temperature of 197°C, which is generally near conventional eutectic Sn-Pb [30]. However, Sn-Zn solder has a high consumption affectability medium in the climate because of the presence of Zn. Accordingly, scientists have given incredible consideration to the corrosive nature of Sn-Zn solder [31]. Consumption opposition of five Sn solder assessed in 0.1M NaCl arrangement They were tracked down that the Sn–Zn solder has the most elevated consumption presence of NaCl [32]. It was tracked down that the erosion affectability of the patch combination expanded with the increment of Zn content, which is the development of coarse Zn-rich haste is related with Sn-Zn [32].

The affectability of metastable pitting the solder amalgam is fundamentally connected with latent film legitimate Growth and improvement of tiny construction, and affectability with the expansion of Zn content [33]. What's more, it was tracked down that the consolidation of Ti and Mn clearly affected further developing the consumption opposition of Sn-Zn solder and solder related corrosion mechanisms along the expansion of various amalgam components were microscopic construction and arrangement dissected from the advancement of detached flume and the improved Ti content was 0.06%. Likewise, the impact of Cr content of range 0 to 0.6 wt% on corrosion sensitivity of Sn-Zn solder in 3.49 wt% Na-Cl arrangement was dissected by the potentiodynamic outcomes of polarization [34, 35]. The doping because of Cr the difference in anodic polarization can best consumption opposition of Sn-Zn solder amalgam. Furthermore, Ag and Ga were added to Sn-Zn to lessen the corrosion affectability of the solder amalgam, while Al doping had the opposite impact. The examination referenced above for the most part centered around Corrosion sensitivity of Sn-Based Solder in NaCl. It can exist in antacid climate electronic gear administration [36, 37].

Sn-Bi

The SnBi solder is additionally considered a possible advancement of Cesar for standard Sn-Pb solder. SnBi (48–58%) is eutectic structure and liquefying for the Sn-By arrangement of softening point is 137 °C. Contrasted with erosion of corrosion obstruction Sn–Bi solder amalgams Sn–Ag and Sn–Pb with solder amalgam in 0.149M NaCl arrangement solution [38]. The sharp difference in corrosion affectability is identified with dissolvable material, which decides the microstructure Changes in anode region proportion, eutectic portion and cathode. The helplessness to consumption is in the scope of three solder [39]. The coming arrangement: Sn–Pb > Sn–Ag > Sn–Bi. Incidentally, the rust has turned sour mechanical properties of Sn-Bi solder amalgam. The doping researched the impacts of corrosion susceptibility on Sn5-6 Cu particles of SnBi solder and exhibited the consumption pace of Sn-Bi solder with nanoparticles is smothered noticeably [38–40].

Conclusion

In this rundown, we propose a short survey of the corrosion behavior on lead free solder, the development instrument of consumption obstruction of solder amalgams and expansion of alloying components or particles. The outcome of investigations affirmed that advancement of corrosion of Sn solder amalgam basically has a place with It is feasible to change and work on the microstructure Corrosion opposition of solder amalgam and other alloys, element or particle. Behavior is quantitatively founded on staggered characterization. It is notable that solder exists as Solder joint and bundle substrate in viable condition, increasing the danger of galvanic corrosion.

References

1. Shuai, L., et al. (2020). Corrosion behavior of Sn-based lead-free solder alloys: a review. *Journal of Materials Science. Materials in Electronics,* 31(12), 9076–9090.
2. Nurwahida, M. Z., et al. (2018). Corrosion properties of SAC305 solder in different solution of HCl and NaCl. *IOP Conference Series: Materials Science and Engineering,* 318(1). IOP Publishing.
3. Hou, Z., et al. (2019). Intermetallic compounds formation and joints properties of electroplated Sn–Zn solder bumps with Cu substrates. *Journal of Materials Science: Materials in Electronics,* 30(22), 20276–20284.
4. Huang, H.-Z., et al. (2016). EffectsL of phosphorus addition on the corrosion resistance of Sn–0.7 Cu lead-free solder alloy. *Transactions of the Indian Institute of Metals,* 69(8), 1537–1543.
5. Li, J., Dai, J., and Johnson, C. M. (2018). Comparison of power cycling reliability of flexible PCB interconnect smaller/thinner and larger/thicker power devices with topside Sn-3.5 Ag solder joints. *Microelectronics Reliability,* 84, 55–65.
6. Zhang, S., et al. (2017). Effects of acrylic adhesives property and optimized bonding parameters on Sn58Bi solder joint morphology for flex-on-board assembly. *Microelectronics Reliability,* 78, 181–189.
7. Zhang, S., et al. (2019). Wettability and interfacial morphology of Sn-3.0 Ag–0.5 Cu solder on electroless nickel plated ZnS transparent ceramic. *Journal of Materials Science: Materials in Electronics,* 30(19), 17972–17985.
8. Satizabal, L. M., et al. (2019). Microstructural array and solute content affecting electrochemical behavior of SnAg and SnBi alloys compared with a traditional SnPb alloy. *Materials Chemistry and Physics,* 223, 410–425.
9. Chidambaram, V., et al. (2009). A corrosion investigation of solder candidates for high-temperature applications. *Journal of Management (JOM),* 61(6), 59–65.
10. Liyana, N. K., et al. (2019). Effect of Zn incorporation on the electrochemical corrosion properties of SAC105 solder alloys. *Journal of Materials Science: Materials in Electronics,* 30(8), 7415–7422.
11. Ramli, M. I. I., et al. (2020). The effect of Bi on the microstructure, electrical, wettability and mechanical properties of Sn-0.7 Cu-0.05 Ni alloys for high strength soldering. *Materials and Design,* 186, 108281.
12. Fayeka, M., Fazal, M. A., and Haseeb, A. S. M. A. (2016). Effect of aluminum addition on the electrochemical corrosion behavior of Sn–3Ag–0.5 Cu solder alloy in 3.5 wt% NaCl solution. *Journal of Materials Science: Materials in Electronics,* 27(11), 12193–12200.

13. Wang, M., et al. (2012). Effect of Ag3Sn intermetallic compounds on corrosion of Sn-3.0 Ag-0.5 Cu solder under high-temperature and high-humidity condition. *Corrosion Science*, 63, 20–28.

14. Xiong, M., and Zhang, L. (2019). Interface reaction and intermetallic compound growth behavior of Sn-Ag-Cu lead-free solder joints on different substrates in electronic packaging. *Journal of Materials Science,* 54(2), 1741–1768.

15. Shuai, L., and Yan, Y.-F. (2015). Intermetallic growth study at Sn–3.0 Ag–0.5 Cu/Cu solder joint interface during different thermal conditions. *Journal of Materials Science: Materials in Electronics*, 26(12), 9470–9477.

16. Liew, M. C., et al. (2012). Corrosion behavior of Sn-3.0 Ag-0.5 Cu lead-free solder in potassium hydroxide electrolyte. *Metallurgical and Materials Transactions A*, 43(10), 3742–3747.

17. Zhong, X., et al. (2017). Electrochemical migration of Sn and Sn solder alloys: a review. *RSC Advances*, 7(45), 28186–28206.

18. Xiao, Q., Nguyen, L., and Armstrong, W. D. (2005). Anomalously high tensile creep rates from thin cast Sn3. 9Ag0. 6Cu lead-free solder. *Journal of Electronic Materials*, 34(7), 1065–1075.

19. Li, Q., Liu, X., and Lu, S. (2018). Corrosion behavior assessment of tin-lead and lead free solders exposed to fire smoke generated by burning polyvinyl chloride. *Materials Chemistry and Physics*, 212, 298–307.

20. Osório, W. R., et al. (2014). Electrochemical behavior of a lead-free Sn–Cu solder alloy in NaCl solution. *Corrosion Science*, 80, 71–81.

21. Jaffery, H. A., et al. (2019). Electrochemical corrosion behavior of Sn-0.7 Cu solder alloy with the addition of bismuth and iron. *Journal of Alloys and Compounds*, 810, 151925.

22. Lin, K.-L., Chung, F. C., and Liu, T. P. (1998). The potentiodynamic polarization behavior of Pb-free XIn-9 (5Al-Zn)-YSn solders. *Materials Chemistry and Physics*, 53(1), 55–59.

23. Huang, H.-Z., et al. (2016). Effects of phosphorus addition on the corrosion resistance of Sn–0.7 Cu lead-free solder alloy. *Transactions of the Indian Institute of Metals*, 69(8), 1537–1543.

24. Ameer, M. A., Fekry, A. M., and Ghoniem, A. A. (2009). Electrochemical behavior of Sn-Ag alloys in alkaline solutions. *Corrosion*, 65(9), 587–594.

25. Sharma, A., Das, S., and Das, K. (2015). Electrochemical corrosion behavior of CeO2 nanoparticle reinforced Sn–Ag based lead free nanocomposite solders in 3.5 wt.% NaCl bath. *Surface and Coatings Technology*, 261, 235–243.

26. Gain, A. K., and Zhang, L. (2018). Thermal aging effects on microstructure, elastic property and damping characteristic of a eutectic Sn–3.5 Ag solder. *Journal of Materials Science: Materials in Electronics*, 29(17), 14519–14527.

27. Sharma, A., Das, S., and Das, K. (2015). Electrochemical corrosion behavior of CeO2 nanoparticle reinforced Sn–Ag based lead free nanocomposite solders in 3.5 wt.% NaCl bath. *Surface and Coatings Technology*, 261, 235–243.

28. Rosalbino, F., et al. (2008). Corrosion behaviour assessment of lead-free Sn–Ag–M (M= In, Bi, Cu) solder alloys. *Materials Chemistry and Physics*, 109(2-3), 386–391.

29. Zhang, L., and Liu, Z.-Q. (2020). Inhibition of intermetallic compounds growth at Sn–58Bi/Cu interface bearing CuZnAl memory particles (2–6 μm). *Journal of Materials Science: Materials in Electronics*, 31(3), 2466–2480.

30. Grobelny, M., and Sobczak, N. (2012). Effect of pH of sulfate solution on electrochemical behavior of Pb-Free solder candidates of SnZn and SnZnCu systems. *Journal of Materials Engineering and Performance*, 21(5), 614–619.

31. Liu, J.-C., et al. (2015). The role of Zn precipitates and Cl– anions in pitting corrosion of Sn–Zn solder alloys. *Corrosion Science*, 92, 263–271.

32. Grobelny, M., and Sobczak, N. (2012). Effect of pH of sulfate solution on electrochemical behavior of Pb-Free solder candidates of SnZn and SnZnCu systems. *Journal of Materials Engineering and Performance*, 21(5), 614–619.

33. Liu, J.-C., et al. (2016). Effects of intermetallic-forming element additions on microstructure and corrosion behavior of Sn–Zn solder alloys. *Corrosion Science*, 112, 150–159.

34. Liu, J.-C., and Zhang, G. (2018). Effect of trace Mn modification on the microstructure and corrosion behavior of Sn– 9Zn solder alloy. *Materials and Corrosion*, 69(6), 781–792.

35. Liu, J.-C., et al. (2015). Ti addition to enhance corrosion resistance of Sn–Zn solder alloy by tailoring microstructure. *Journal of Alloys and Compounds,* 644, 113–118.
36. Liu, M., et al. (2015). The electrochemical corrosion behavior of Pb–free Sn–8.5 Zn–XCr solders in 3.5 wt.% NaCl solution. *Materials Chemistry and Physics,* 168, 27–34.
37. Xue, P., et al. (2012). Study on properties of Sn–9Zn–Ga solder bearing Nd. *Journal of Materials Science: Materials in Electronics,* 23(6), 1272–1278.
38. Satizabal, L. M., et al. (2016). Immersion corrosion of Sn-Ag and Sn-Bi alloys as successors to Sn-Pb alloy with electronic and jewelry applications. *Corrosion,* 72(8), 1064–1080.
39. Wang, Z., et al. (2019). Influences of Ag and In alloying on Sn-Bi eutectic solder and SnBi/Cu solder joints. *Journal of Materials Science: Materials in Electronics,* 30(20), 18524–18538.
40. Li, X., et al. (2017). Effects of nanoscale Cu6Sn5 particles addition on microstructure and properties of SnBi solder alloys. *Materials Science and Engineering: A,* 684, 328–334.

108 Mechanical, dielectric and piezoelectric properties of PVDF based composites: a review

Atul Kumar[1,a], Amit Kumar[1], and S. K. Pradhan[2]

[1]Department of Mechanical Engineering, National Institute of Technology, Patna, India

[2]Department of Mechanical Engineering, BIT Mesra, Patna Campus, Patna, India

Abstract: The purpose of the paper is review the research progress in the field of PVDF based composite material. PVDF based composite has attended wide attention due to its properties like piezoelectric, thermal stability and good chemical resistance. The direct and reverse piezoelectric properties have industrial applications. The review article focuses on the effect of various fillers on the properties like dielectric, piezoelectric and mechanical properties.

Keywords: Composite materials, MWCNTs, piezoelectricity, PVDF, PZT.

Introduction

Technology is driving at its fastest pace in the 21st century. The technology has its presence in almost everything from the fastest computers to smart gadgets of our daily life. The consumption of energy has almost increased many folds, so the harvesting and generation of energy is among the top priority for human race.

In the era of Internet of Things (IoTs), piezoelectric material has its own space. In 1950s, Piezoelectricity was detected in certain materials only. Post 1970s, researchers and scientists have found many materials like composites, Biomaterials, Polymers which exhibits the piezoelectricity.

Composite material has significant role in the development of society since start of civilization. In new, Industrial revolution 4.0, the composite material will be the leader in the material especially in electrical and electronic industry. The role of polymer based composite in new industrial revolution is going to be decisive.

Polymer PVDF exhibits better piezoelectric properties as compared with others polymer as well as it also possesses better properties suitable for piezoelectricity.

The polymer composite doped with ceramics are in the center of the research. Polymer PVDF is piezoelectrically active with low value of dielectric constant whereas ceramics such lead zirconate titanate (PZT) possess excellent piezoelectric and ferroelectric properties but possess high hardness and low mechanical strength. The conglomeration of Polymer and ceramics would to be solution to improve the piezoelectric and mechanical properties.

---

[a]atulkumar@nitp.ac.in

Generally, the ceramics materials, metal oxides, hydrated salts and their mixture are added to improve its mechanical properties as well as electrical properties. The materials which are often used include lead zirconate titanate, iron oxide, multiwall carbon nanotubes, nano-clay, barium titanate, silver, graphene and etc.

PVDF based composite has found the application in wearable function devices, chemical sensor, biosensor and flexible actuators devices, ultrasonic sound waves generation (Reverse piezoelectric effect). Piezoelectric effect has application in wide range of field, from detection to production of sound and wide range of microscope. It also finds application in separator/binder in supercapacitor and batteries. Piezoelectricity has application field ranging from domestic to daily life use gadgets to modern day musical instruments for amplification purpose.

Composite Preparation

The preparation method of polymer based composite has also effect on properties like piezoelectric and dielectric on the composites as well as mechanical properties. The various fabrication techniques include compression molding, solvent casting, tape casting, spin coating, jet machining, dielectrophoresis, ultrasonic cutting, dice and fill, tape lamination, relic processing, co-extrusion, hot pressing, lost mold method.

Dice and fill methods are employed when the composites with various connectivity is to be obtained. In lost mold method, lithography Galvano-forming and plastic molding (LIGA) method employed for preparing the mold and preparation of the structure of composites.

Injection molding is used where the rate of production needs to be high and it involves low material wastage and provide flexibility it wide range of transducer design.

Tape casting method is extensively used where wide range of connectivity is need to be manufactured as well as with changing the powder type, composite with 2-0-2 connectivity can be obtained.

In microfabrication by co-extrusion technique, the thermoplastic ceramics is forced thorough the die with given reduction ratio. It is used for production of complex shapes.

Solid freedom fabrication is used when the internal structure of composite is complex and symmetrical. CAD models are prepared for analyzing the strategy and tool paths.

Hot Press Method is one of the mass production methods, generally employed for production of 2-2 connectivity composites. In this method, Polymer is dissolved in solvent and at certain viscosity other fillers are added stirred well for proper mixing, then the solvent is dried with help of non-solvent.

Solvent Casting Method is most commonly used for manufacturing thin and thick films. In this method, Solution of PVDF-PZT are well mixes with magnetic stirrers and solution is heated in furnace for certain time. It is mainly employed for producing 3-1 connectivity.

Literature Review

Guo et al. in the review of the PVDF fabrication and application, polymer PVDF is doped with inorganic materials, bio material and nano materials and fabricated with different techniques, improved the flexibility and mechanical properties and higher voltage coefficient [1]. Satyaranjan Bairagi et. al. studied about the inducing CNTs at 0.1% in the composite resulting in increase of performance of the nanogenerator, such that voltage improved to 23.3 V, current to 9.1 µA and power density to 52µW per sq. cm when compressed mechanically [2]. Twinkle et. al investigated in field of energy storage application, PVDF doped with Multi walled carbon nanotubes (MWCNTs), XRD result shows the increase in the interlayer spacing at peak angle of 24.16 degree between PVDF/MWCNTs. The covalent bonds between the atoms were confirmand by FTIR Spectroscopy. FESEM microscopy shows the embedment of CNTs in PVDF membrane pores and crosslinking of nanotubes [3]. Kaeopisan et. al. investigated the composite by doping of CNTs in PVDF indicates that different weight percentage of CNT in PVDF shifted the deflection angle to higher value with increase in the percentage of CNT. FTIR analysis indicates the value of peaks increased with increase weight percent. Thermal Stability is also improved and no chemical interaction between PVDF and CNT [4].

Chen et al. studied the composite with molecular dynamics simulation (MDS). It is used to speculate the mechanical properties of PVDF and CNT composites. Embedment of CNTs improved the Young's modulus and tensile strength. Fractures is seen in PVDF rich region and induction of CNTs has increased the stress concentration which increased the brittleness at low strain rates. The structure of PVDF which is biaxially oriented and stretched at different deformation rates shows variation in the piezoelectric properties when studied with Fourier transformation infrared spectroscopy (FTIR) and wide-angle X-ray diffraction (WAXRD). PVDF is generally found in α-phase, β-phase, γ-phase in nature. Among which β-phase possess most ferroelectric properties. The degree of crystallinity and β-phase content effects is crucial for ferroelectric properties of the PVDF. When CNTs doped at 24%, the value of Young's modulus increased by 25.93% and the ultimate strength increased by 16.93%.

The free volume increase is maximum when the CNTs are doped at 24% [9].

PZT has been prepared by sol-gel method and PVDF-PZT composite are prepared by solvent casting method. PVDF doped with different percentage of PZT prepared with solvent casting method and analyzed by XRD, FTIR and Raman exhibits enriched β-phase which shows good ferroelectric property. PZT concentration enhances the electrical properties and tensile modulus and degree of crystallinity increase the piezoelectric properties. FESEM micrograph of PVDF-PZT shows sample is homogenous without wrinkles, cracks and deformation. PVDF with 4% of PZT shows the prominent grains and grain boundaries whereas 8% and 1% PZT shows small grain and grain boundaries. PVDF-PZT with variation from 0–1% of PZT shows the dielectric constant value decrease with increase in the frequency. The lattice strain is also induced in composite when the value of PZT is 4%. The β-phase contribution is maximum when the 4% of PZT is present in composite. The value of dielectric constant increases more rapidly at higher temperature. At higher temperature, the dielectric constant is attaining maximum value when the PZT is 4% due to presence of prominent β phase and lattice strain.

The dielectric loss of composite decreases with increase in the frequency up to 10 kHz. The dielectric loss at 4% of PZT is less as compared to other composition.

For improvement in the electromechanical properties in pressure sensing applications, PVDF is doped with CNTs and reduced graphene oxide and its FTIR result shows that PVDF crystallizes in α-phase independent of filler and polymer type. Inclusion of fillers like reduced graphene oxide has increased the tensile strength, stiffness, yield strength. As compared with CNTs, percolation threshold is lower for reduced graphene oxide. Pressure sensibility is larger at 0.5 wt% of CNT [8].

Silakaew et al. investigated the tertiary composites of PVDF, BT and CNTs. It has improved the piezoelectric properties and mechanical properties. The morphological studies show nano-barium titanate are spherical and micro-Barium titanate are not spherical.

At room temperature, dielectric loss is the function of room temperature. The DC conduction path formation increase the loss tangent in the PVDF Matrix [5].

Caccaoti et al. also investigated the tertiary composites of PVDF-BaTiO$_3$-MWCNTs, when examined with XRD, PVDF and PVDF/CNT find co-presence of Alpha and Beta phase in both conditions. FTIR spectroscopy of solvent casted PVDF and PVDF/CNT films are compared which has presence of all three crystals of PVDF. The conductivity of PVDF/CNT is more as compared with PVDF/CNT/xBT due to presence of well dispersed CNTs. Morphological allocation shows that all the solvent casted films are uniform and pore free. Increase in the percentage of barium titanate, electrical conductivity variation is not linear. The permittivity of both sample are similar at constant frequency and increases with barium titanate content. The addition of dialectic fillers strongly influences the permittivity of composite [6].

González-Benito et al. investigated the tertiary composite with the addition of barium titanate, MWCNTs, or its mixture does not affect the polymorphism of PVDF. With induction of nano-fillers, there is change in surface to volume ration due to which there is significant increase in the capacitance of the composite. The dielectric properties of the materials are based on the frequency and it was modelled by a Debye equivalent circuit below the percolation threshold respect to the quantity of MWCNTs. The tertiary nanocomposites have effect on the piezoelectric behaviour when the high dielectric losses occurred above the percolation threshold. The differential scanning calorimetry (DSC) peaks shows same nature irrespective of filler nature and composition, DSC traces a cooling curve of 10 degree per minute [9].

Kim et al. also investigated the tertiary composite prepared with 3D printing. In 3D printed nanocomposites, the dielectric loss and dielectric constant decrease with increase in the frequency due to diploe mobility reduction. The dielectric loss increases with increase in the BT Content. The dielectric constant increases with increase as CNT content increases. The 3D printed films as compared to solvent casted film shows improved dielectric property because of reduction in dipole of polymer molecule and alignment of molecule in 3D printing. In 3D printed nanocomposites, the energy density of 0.455 J/cm at 39MV/m has been attained within dielectric loss of 0.11 at 1 kHz, at 1wt.%-CNT/60wt.%-BT [10].

Wang et.al studied about single-layer ternary composites, CNTs and barium titanate are uniformly distributed in the complete PVDF matrix, whereas Barium Titanate and CNT are arranged alternately in stacked layers. With an increase in the content of barium titanate from 60–70 wt%, there is significant improvement of permittivity from 4000 at 10 kHz to 7000 at 10 kHz. There is significant change in the dielectric constant, loss and breakdown strength when there is change in spatial distribution of fillers. The breakdown strength of trilayer composites are better compared to CNTs and Barium Titanate co-filled ternary composites. In tertiary composite, with increase in the barium titanate, the dielectric permittivity increases, it should be attributed to Maxwell–Wagner–Sillars polarizations at the enlarged BT/PVDF interfaces [14].

Kim et.al studied, higher amounts of Barium Titanate agglomerates as Barium Titanate contents increase, when examined with SEM. The size of agglomeration increases when the barium titanate content increases. Scanning electron microscope (SEM) image shows that no Micro-cracks are observed in any nanocomposites. With increase in the temperature from 25 to 150°C, there is increasing trend at 1 kHz. The films start deforming and melting at temperature above 150°C.

The value of capacitance remains changing and remain maximal at 1wt% of CNT, when it is added at 12 wt% BT/PVDF. The ultimate tensile strength for BT/PVDF nanocomposite without addition of CNTs. Although it is observed that 12 wt% Barium Titanate shows higher ultimate tensile strength than 40 wt.% BT around 50% strain [11, 15].

Sanchez et al. observed that higher amount of Barium Titanate, the higher number of brighter domain and high energy ball milling (HEBM) permits the barium titanate particles to disperse within the PVDF. The young's modulus increased linearly when the barium titanate increases linearly. There is only an increase in the stiffness, when CNTs increase at 0/40 wt% of barium titanate below the percolation threshold. With increase in degree of crystallinity, the strength of crystalline and the modulus [12].

Xiang et.al investigated the composites and the results shows with increase in the content of NiO, there is increase in the hardness and Young's modulus. At room temperature, dielectric constant is the function of room temperature. NiO redefines the grain boundaries so the microstructure is redefined which is crucial in mechanical and electrical properties of the composite [16].

Tuan et al. studied about the PVDF-barium titanate composite and concluded that microstructure has important role in the mechanical properties. Abnormal grains are produced as temperature and time is prolonged. Abnormal grains led to the formation of Microcracks due to which there is increase in the toughness. In case of abnormal grains, fracture originates from there. It is generally found in coarse grained specimen and its strength is low [17].

Zhang et al. investigated that the PVDF composites doped with iron oxide and barium titanate, XRD results at diffraction angle of 42 degree, splitting of peaks at 200 and 002 which shows the barium titanate has tetragonal phase structure. SEM result shows BT particles has diameter of 200nm and spherical shape. TEM result showed iron oxide particles were randomly distributed on the Barium Titanate surface. With increase in the volume percentage of BT-iron oxide, there is increase in dielectric permittivity [13].

Elnabawy et al. investigated the doping PVDF with TPU, the morphological studies show the PVDF and TPU ratio in 3:1 shows diameter 230nm and ratio of 1:1 shows 311 nm and ratio pf 1:3 shows the 212 nm. The FT-IR studies shows the β-phase reduction with increase in the concentration of TPU. The mechanical properties also improves when the ratio of PVDF/TPU is 1:3 and attains the value of 13.2 Mpa and strain of 85%. With addition of 25% of TPU increases the strength five times and elasticity of the sample, increases four times. The PVDF/TPU (1:1) attains the toughness value of 389.3 J/cubic meter and piezoelectric sensitivity at 1.5 mV/gm [7].

Research Gap

Based on literature survey and research work done by different investigators, PVDF composites doped with single fillers has been studied and polymer PVDF doped with more than two fillers need to study further. The tertiary composite constituting PVDF as matrix and PZT and CNTs as fillers or reinforcement in the application such as sensor, actuators, electromechanical devices etc., so we have to prepare the tertiary composite by solvent casting method with different weight %. The sample structural properties need to be characterized with the help of XRD. The microstructure of the sample are studied to correlate its electrical and mechanical properties. The electrical properties and mechanical properties such as dielectric and ferroelectric properties using atomic force microscopy (AFM), universal testing machine (UTM), and dynamic mechanical analysis (DMA) are to be studied.

Conclusion

The addition of barium titanate significantly affects the properties of PVDF. The tertiary composites with reduced graphene oxide and multi walled carbon nanotubes helps to improve the mechanical properties, which has general application in pressure sensing applications. The addition of lead zirconate titanate enhances the electrical properties. The electrical properties are affected by the temperature and frequency when the PVDF doped with PZT. The PVDF composites doped with MWCNTs and PZT also shows the improved piezoelectric properties and dielectric properties.

References

1. Shuaibing, S., Duan, X., Xie, M., Aw, K. C., and Xue, Q. (2020). Composites, fabrication and application of polyvinylidene fluoride for flexible electromechanical devices: a review. *Micromachines*, 11(12), 1076.
2. Bairagi, S., and Ali, W. (2020). Investigating the role of carbon nanotubes (CNTs) in piezoelectric performance of PVDF/KNN based flexible electrospun nanogenerator. *Soft Matter*, 10.1039.
3. Twinkle, Kaur, M., Gowsamy, J. K., Kumar, P., and Kumar, S. (2020). Synthesis and characterization of CNT/PVDF paper for electronic and energy storage applications. *Emergent Materials*, 3, 181–185.
4. Kaeopisan, A., and Hassakorn W. (2020). Step synthesis, electrical generator of PVDF/CNTsPiezoelectri. *IOSR Journal of Applied Physics*, 12(6), 35–41. ISSN: 2278-4861.
5. Silakaew, K., and Thongbai, P. (2019). Significantly improved dielectric properties of multiwall carbon nanotube-BaTiO/PVDF polymer composites by tuning the particle size of the ceramic filler. *RSC Advances*, 9(41), 23498–23507.
6. Cacciotti, I., Valentini, M., Raio, M., and Nanni, F. (2019). Design and development of advanced BaTiO3/MWCNTs/PVDF multi-layered systems for microwave applications. *Composite Structures*, 224, 111075. ISSN 0263-8223.
7. Elnabawy, E., Hassanain, A. H., Shehata, N., Popelka, A., Nair, R., Yousef, S., and Kandas, I. (2019). Piezoelastic PVDF/TPU nanofibrous composite membrane: fabrication and characterization. *Polymers (Basel)*, 11(10), 1634.
8. Jo, E. H., Kim, S. K., Chang, H., Lee, C., Park, S. R., Choi, J. H., and Jang, H. D. (2019). Synthesis of multiwall carbon nanotube/graphene composite by an aerosol process and its characterization for supercapacitors. *Aerosol and Air Quality Research*, 19, 449–454.
9. González-Benito, J., Olmos, D., Martínez-Tarifa, J. M., González-Gaitano, G., and Sánchez, F. A. (2019). PVDF/BaTiO3/carbon nanotubes ternary nanocomposites prepared by ball milling: piezo and dielectric responses. *Journal of Applied Polymer Science*, 136, 47788.
10. Kim, H., Johnson, J., Chavez, L. A., Rosales, C. A. G., Tseng, T. L. B., and Lin, Y. (2018). Enhanced dielectric properties of three phase dielectric MWCNTs/BaTiO3/PVDF nanocomposites for energy storage using fused deposition modeling 3D printing. *Ceramics International*, 44(8), 9037–9044. ISSN 0272-8842.
11. Chen, H. L., Su, C. H., Ju, S. P., Chen, H. Y., Lin, J. S., Hsieh, J. Y., and Lin, C. Y. (2017). Predicting mechanical properties of polyvinylidene fluoride/carbon nanotube composites by molecular simulation. *Materials Research Express*, 4(11), 115025.
12. Sanchez, F. A., and González-Benito, J. (2017). PVDFBaTiO3/carbon nanotubes ternary nanocomposites: effect of nanofillers and processing. *Polymer Composites*, 38, 227–235.
13. Zhang, C., Chi, Q., Dong, J., et al. (2016). Enhanced dielectric properties of poly(vinylidene fluoride) composites filled with nano iron oxide-deposited barium titanate hybrid particles. *Scientific Reports*, 6, 33508.

14. Chatterjee, J., Nash, N., Cottinet, P., and Wang, B. (2012). Synthesis and characterization of poly(vinylidene fluoride)/carbon nanotube composite piezoelectric powders. *Journal of Materials Research*, 27(18), 2352–2359.
15. Yu, S., Zheng, W., Yu, W., Zhang, Y., Jiang, Q., and Zhao, Z. (2009). Formation mechanism of β-Phase in PVDF/CNT composite prepared by the sonication method. *Macromolecules*, 42(22), 8870–8874.
16. Xiang, P. H., Dong, X. L., Feng, C.-D., Zhong, N., and Guo, J-K. (2004). Sintering behavior, mechanical and electrical properties of lead zirconate titanate/NiO composites from coated powders. *Ceramics International*, 30(5), 765–772. ISSN 0272-8842.
17. Tuan, W. H., and Lin, S. K. (1999) The microstructure–mechanical properties relationships of BaTiO3. *Ceramics International*, 25(1), 35–40. ISSN 0272-8842.

109 Synthesis, processing and wear characterization of ultra high temperature ceramics composite (UHTC)

Sandeep Kumar[a] and Abhishek Singh[b]

Department of Mechanical Engineering, NIT, Patna, India

Abstract: Ultra-high temperature ceramics composites (UHTC) are materials that are physically and chemically resilient at elevated temperatures and in reactive environment. Due to its high thermal-conductivity and high-melting points, the diboride, dicarbide, and dinitride family of materials is typically utilized to make ultra-high temperature ceramics. The development of UHTC and tribological characterization of the developed sample are the primary goals of the current effort in order to increase wear resistance. In the current study, zirconium diboride (ZrB_2) based UHTC is produced using three powder constituents B_4C (35%), ZrB_2 (55%), and Cr (10%) with particles size of 30 μm 15 μm, and 4 μm, respectively. They are produced by grinding zirconium balls for 0 to 48 hours at 500 revolutions per minute. Additionally, a mixture of powders and binder retained in a cylindrical compaction machine mold weighing 20 grams of milled powder mixed with 5 ml of polyvinyl alcohol. A compacted sample of composites was created using a cylindrical compaction machine with a 12-mm cylinder diameter that applied 8 tons of pressure and 150 seconds of dwell time. The sample was then sintered in a traditional furnace at 1600°C and 1700°C at 15–20 bars for 4 hours while being exposed to an inert gas (Ar). Additionally, dry compacted samples of composite are sintered using the sparks plasma sintering (SPS) method at 1900° while an inert gas (Ar) is present and the process is held for 1 to 2 minutes. The ball-on-disc wear testing machine was applied to examine the wear characteristics of an ultra-high temperature ceramics composite that had been developed. The results showed that the conventionally sintered composite at 1600°C and 1700°C had deeper penetration, ranging from 550–1300 microns, compared to the sparks plasma sintered composite at1900°C, which has a deeper penetration of 70–75 microns. To further investigate the bonding and porosity in the UHTC microstructure, scanning electron microscopy (SEM) is used.

Keywords: Conventional sintering, SEM, SPS, UHTC, wear characterization.

Introduction

ZrB2 and boron carbide (B_4C) are prospective UHTC materials due to their exceptional mechanical, thermal, and electric properties, high stability, resistance to corrosion and wear, low specific weight for cutting tools, heavy-duty wear applications, and neutron absorber in nuclear reactors [1–8]. Monolithic ZrB_2 ceramics' insufficient oxidation resistance and insufficient degradation and disintegration tolerance prevent its use in ultrahigh

[a]sandeep06mech@gmail.com, [b]abhishek.singh@nitp.ac.in

temperature applications. Despite an exceptional mix of qualities, monolithic ZrB_2 ceramics cannot be employed for ultrahigh temperature applications due to low densification brought on by a strong covalent bond and a restriction to brittle failure [9–11]. Therefore, the adding together of second phases is required to address these issues, and there are two main factors to take into account during the design of ZrB_2 based UHTC: Distribution in second phase (continuous or discontinuous), and secondary phase between ZrB_2 and secondary phase, which is the highest temperature at which the composite is to be used [12–15]. Because of its hardness and abrasion resistance, B_4C is excellent for heavy-duty wear applications. It has a low density (2.52 g/cm^2) and a maximum hardness of 35 to 45Gpa at higher temperatures (more than 1300 degrees). Due to this, despite its exceptional features; it is a legitimate option for high temperature wear. B4C has a limited application in engineering due to its poor densification and brittle breakdown due to a strong covalent bond [16–19]. Hence, composite of Zr_2B and B_4C have excellent properties to overcome above mention problems therefore many researchers investigated to resolve the problems like microwave sintering method is excellent than CS method used two phase ceramic materials containing ZrB_2 (4%wt) and B_4C (96%wt) and results the sintered particulate composite achieve greater 98% relative density at lower then 1700°C sintering temperature. When microwave sintering is increased at lower temperatures, between 1600°C and 1900°C, the grains size of ZrB_2 metal matrix increases, 2 μm to 15.9 μm. hardness and fracture toughness of a 96%ZrB_2-4%B_4C microwave sintered particulate composite sample are 17.5 Gpa and 3.8 Mpa.m$^{1/2}$, respectively [20]; To improve the densification and fracture toughness 4.52 to 7.98 Mpa m$^{1/2}$ and strength 424 to 450 Mpa, the ZrB_2-Mo composite was developed using a hot pressing sintering technique at 1950°C using Mo with different volume fractions of ZrB_2 and Mo like 95%ZrB_2-5%Mo and 90%ZrB_2-10%Mo [21]. To create B_4C-30%ZrB_2 ceramics composite, both the pulse electric current and traditional sintering techniques are applied. To determine the microstructure, phases and effect of temperature on the chemical composition of composite, SEM and XRD are used. A ceramics phase called ZrO_2 is added during traditional sintering, creating a totally dense composite with small grain size. The mechanical parameters of composite created using both methods vary from 30 to 32 Gpa in hardness, 2.4–2.9 MPam$^{1/2}$ in fracture toughness, and 630–730 Mpa in flexural strength [22]. By using pressure-free sintering to create ZrB_2 ceramics with a composition of ZrB_2-15%Mo-10%Fe-1%C, alloying elements Mo, C, and Fe are added as additives, resulting in a composite of densified ceramics with improved mechanical properties and microstructure, like strength of 12.5 MPam1/2 and fracture toughness of 1% and 12.5 MPa fracture toughness gradually improves with carbon content [23]. The fraction of ZrO_2-B_4C must be optimized to generate single phase ZrB2 Composite, which are produced by chemical reaction between ZrO_2 and B_4C between 1020°C and 1600°C. When Si is introduced to ZrB_2-B_4C composite, Si content builds up in SiC, and ZrB_2-SiC-B_4C develops in powder. Via chemical reaction and evenly dispersed over the entire ZrB_2 matrix. In the continuous furnace, ZrO_2+B_4C+Si is heated at increased temperatures between 900°C and 1700°C for 3–5 min. To identify the presence of ZrB_2 with SiC at temperatures above 1300°C, XRD methods and SEM examination are used. Consequently, powder particle agglomeration is observed between 1300° and 1700°C, and at this temperature, small-shaped particles are present [24]. The hot pressing process is used to create ZrB_2-B_4C composites for elevated range of temperature application with stability and maximum hardness because the characteristics of ZrB_2-B_4C modify in sintering at high temperature (1800°C), enhancing porosity and density. The SEM is used to estimate the surface features of the developed sample with volume fraction components element during sintering when the density of the develop sample is reduced and the porosity changes from a closed to an open structure. Additionally, heating activity during the sintering process regulates the ZrO_2 transient phase [25]. The coefficient of friction for monolithic boron carbide, B_4C-BN, and B_4C-SiC-Si-Cermets is reported to range from 0.03 to 0.9, depends on load applied, the chemical makeup of the surface, the characteristics of the contacting material, and the relative humidity. Additionally, it was discovered that B_4C oxidized during wear tests, forming B_2O_3 and then H_3BO_3 film, which decreased the coefficient and served as lubricants [26–29]. As a result, no researchers have worked on the creation of composites including ZrB_2, B_4C, and Cr, or on the wear characterization of ZrB_2-B_4C–Cr Composite. However, in the current study ZrB_2-B_4C-Cr Composites are created using spark plasma sintering (1900°C) and conventional sintering (1600°C and 1700°C), and wear characteristics of UHTC composites sample produced using Ball-on-disc wear tester machine were compared. The matrix-bonding and porosity of microstructure of UHTC are also observed using scanning electron microscopy.

Material and Methods

In the current study, ZrB_2-based UHTC is created using three components, B_4C (35%), ZrB_2 (55%), and Cr (10%), with particles sizes of 30µm, 15µm, and 4µm, respectively. To reduce the particle size, they are synthesized by ball mill with zirconium ball for 0–48 hours at 500 rpm. Three samples of the synthesized powder are produced, with the average particle size being 30µm before mechanical milling, 15 µm after 35 hours, and 6 µm after 48 hours. Additionally, mixture of powders and binder kept in compaction machine mold (cylindrical) weighing 20 grams of milled powder mixed with 5 ml of polyvinyl alcohol. A compacted sample of composites was created using a cylindrical compaction machine with a 12-mm cylinder diameter that applied 8 tons of pressure and 150 seconds of dwell time. The sample was then sintered in traditional furnace at 1600°C and 1700°C at 15–20 bar for 4hours while being exposed to inert gas (Ar). Additionally, compacted composite

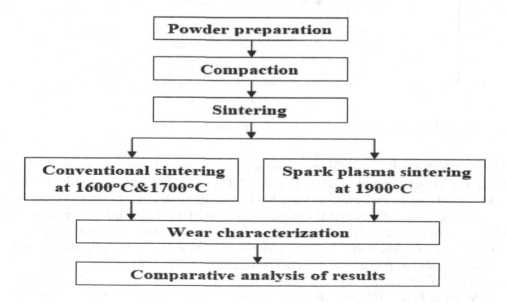

Figure 109.1 Work flow chart of development and characterization of ZrB_2-B_4C-Cr composites.

pallets are sintered using the sparks plasma sintering process at 1900°C when an inert gas (Ar) is present, with holding time 1–2 minutes. The ball-on-disc wear testing machine was used to examine the wear characteristics of the created ultra-high temperature ceramics composite (ZrB_2-B_4C-Cr). The stainless steel ball had a diameter of 6mm and rotated over the composite sample under conditions of 30N load and 50 ram speeds for various durations. To ascertain the microstructure and porosity of the composites sample, the JSM 6480LV, a JEOL microscope, is used for scanning electron microscopy. In Figure 109.1, the step-by-step chart is displayed.

Results and Discussion

Figure 109.2 illustrated the wear characterization of ZrB_2-B_4C-Cr composites made by conventional sintering (at 1600°C and 1700°C) and Spark plasma sintering (1900°C). It can be seen that the SPS-produced composite has significantly better wear resistance than that made by conventional sintering. Due to the liquid-phase sintering of Cr in the metal matrix and the development of ZrC and Cr_2BC that improve the hardness. The depth of penetration on CS composites at 1600°C and 1700°C is higher in the range of 550–1300 microns as compared to the depth of penetration on SPS-sintered composites at 1900°C, which is in the range of 70–75 microns. This was confirmed by scanning electron microscopy (SPS).

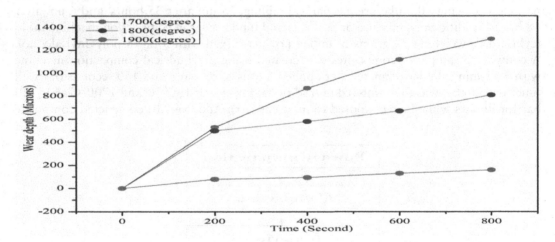

Figure 109.2 Wear V/S sliding time of ZrB_2-B_4C-Cr composites.

Using a JSM 6480LV, a JEOL microscope, the scanning electron microscopy of ZrB_2-B_4C-Cr composite sintered at 1600°C and 1700°C conventionally in a tube furnace in the availability of Ar environment for 2 hours and 30 minutes is shown in Figure 109.3. The outcome demonstrates that the particles are not precisely bound as shown in Figure 109.3(a) and Figure 109.3(b) due to partial sintering, but that the composite generated at 1600° C is more sinterible than the composite formed at 1700° C. Furthermore, both sintered samples had porosity in their microstructure because only partial sintering at temperatures of 1600°C and 1700°C was able to create a solid link between the particles.

As shown in Figure 109.3(c), the ZrB_2-B_4C-Cr composite was sintered at 1900°C using the SPS method for 1–2 minutes. The scanning electron microscopy image reveals that the

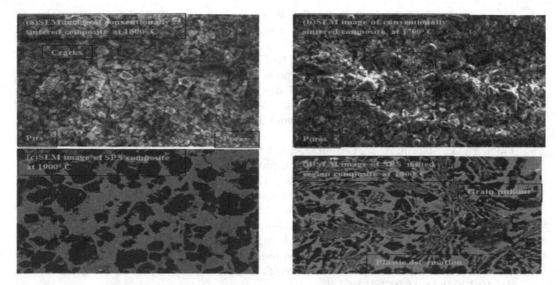

Figure 109.3 SEM Images of the sintered sample of ZrB$_2$-B$_4$C-Cr composites.

particles are strongly bound, with the excellent phase of ZrB$_2$ and the dark-phase of B$_4$C being free of pores.

Conclusion

In present research work 55%ZrB$_2$-35%B$_4$C-10%Cr based UHTC composite material is developed by CV and SPS methods. The wear properties and microstructure are studied and following are the results of present work:

1. 55%ZrB$_2$-35%B$_4$C-10%Cr composite is developed using CS and SPS method.
2. The wear characterization of 55%ZrB$_2$-35%B$_4$C-10%Cr composites are performed found that depth of penetration on the conventionally sintered composites at 1600°C and 1700°C are higher in range of 550–1300 micron.
3. The wear characterization of 55%ZrB$_2$-35%B$_4$C-10%Cr composites are performed found that the depth of penetration on sparks plasma sintered composite at 1900°C is in range of 70–75 micron.
4. The scanning of electron microscopy (SEM) is performed on developed composite sample to determine the grain pull out, porosity, micro crack and pit formation.
5. In the SEM analysis, pores, pits and micro cracks found are large in UHTC sample produced by conventional sintering technique at 1600°C then 1700°C.
6. In the SEM analysis of UHTC composite sample produced by spark plasma sintering is found that the strongly bonded particles with bright phase of ZrB$_2$, dark-phase of B$_4$C, Drark-phase of Cr and grain pullout without porosity.

References

1. Upadhya, K., Yang, J. M., and Hoffman, W. P. (1997). Materials for ultrahigh temperature structure applications. *American Ceramic Society Bulletin*, 58, 51–56.

2. Norasetthekul, S., Eubank, P. T., Bradley, W. L., Bozkurt, B., and Stucker, B. (1999). Use of zirconium diboride copper as an electrode in plasma applications. *Journal of Materials Science*, 34, 1261–1270.

3. Riedel, R. (2000). Handbook of Ceramic Hard Materials. (Vol. 2). Wiley-VCH Verlag GmbH.

4. An, Q., and Goddard, W. A. (2017). Nanotwins soften boron-rich boron carbide (B13C2). *Applied Physics Letters*, 110, 111902, 1–4. doi:10.1063/1.4978644.

5. Watts, J. L., Talbot, P. C., Alarco, J. A., and Mackinnon, I. D. R. (2017). Morphology control in high yield boron carbide. *Ceramics International*, 43, 2650–2657. doi:10.1016/j.ceramint.2016.11.076.

6. Pillai, H. G., Madam, A. K., Chandra, S., and Cheruvalath, V. M. (2017). Evolutionary algorithm based structure search and first-principles study of B12C3 polytypes. *Journal of Alloys and Compounds*, 695, 2023–2034. doi: 10.1016/j.jallcom. 2016.11.040.

7. Gunnewiek, R. F. K., Souto, P. M., and Kiminami, R. H. G. A. (2017). Synthesis of nanocrystalline boron carbide by direct microwave carbothermal reduction of boric acid. *Journal of Nanomaterials*, 2017, 1–8. Article ID 3983468. doi:10.1155/2017/3983468.

8. Ahmed, Y. M. Z., El-Sheikh, S. M., Ewais, E. M. M., Abd-Allah, A. A., and Sayed, S. A. (2017). Controlling the morphology and oxidation resistance of boron carbide synthesized via carbothermic reduction reaction. *Journal of Materials Engineering and Performance*, 26, 1444–1454. doi:10.1007/s11665-017-2548-3.

9. Suri, A. K., Subramanian, C., Sonberand, J. K., and Murthy T. S. R. C. (2010). Synthesis and consolidation of boron carbide: A review. *International Materials Review*, 55(1), 4–40. doi10.1179/095066009X 12506721665211.

10. Sonber, J. K., Murthy, T. S. R. C., Subramanian, C., Fotedar, R. K., Hubli, R. C., and Suri, A. K. (2013). Synthesis, densification and characterization of boron carbide. *Transactions of Indian Ceramic Society*, 72(2), 100–107. doi:10.1080/0371 750X.2013.817755.

11. Guo, S. Q. (2009). Densification of ZrB2–based composites and their mechanical and physical properties: a review. *Journal of the European Ceramic Society*, 29(6), 995–1011.

12. Wuchina, E. J., Opila, E., Opeka, M., Fahrenholtz, W. G., and Talmy, I. (2007). UHTCs: ultra-high temperature ceramic materials for extreme environment applications. *The Electrochemical Society Interface*. 16, 30–36.

13. Fahrenholtz, W. G., Wuchina, E. J., Lee, W. E., and Zhou, Y., editors (2014). Ultra-High Temperature Ceramics: Materials for Extreme Environment Applications. Hoboken, NJ: John Wiley & Sons, Inc. (pp. 112–143).

14. Guo, S. Q. (2009). Densification of ZrB2-Based composites and their mechanical and physical properties: a review. *Journal of the European Ceramic Society*, 29, 995–1011.

15. Guoa, S. Q., Nishimura, T., Mizuguchi, T., and Kagawa, Y. (2008). Mechanical properties of hot-pressed ZrB2–MoSi2–SiC composites. *Journal of the European Ceramic Society*, 28, 1891–1898.

16. Jianxin, D., and Sun, J. (2009). Microstructure and mechanical properties of hot-pressed B4C/TiC/Mo ceramic composites. *Ceramics International*, 35, 771–778.

17. Yue, X. Y., Zhao, S. M., Lü, P., Chang, Q., and Ru, H. Q. (2010). Synthesis and properties of hot pressed B4C–TiB2 ceramic composite. *Materials Science and Engineering A*, 527, 7215–7219.

18. Topcu, I., Gulsoy, H. O., Kadioglu, N., and Gulluoglu, A. N. (2009). Processing and mechanical properties of B4C reinforced Al matrix composites. *Journal of Alloys and Compounds*, 482, 516–521.

19. Wenbo, H., Jiaxing, G., Jihong, Z., and Jiliang, Y. (2013). Microstructure and properties of B4C-Zrb2 ceramic composites. *International Journal of Engineering and Innovative Technology (IJEIT)*, 3, 163–166.

20. Zhu, S., Fahrenholtz, W. G., Hilmas, G. E., Zhang, S. C., Yadlowsky, E. J., and Keitz, M. D. (2008). Microwave sintering of a ZrB2–B4C particulate ceramic composite. *Composites*, 39, 449–453.

21. Wang, H., Chen, D., Wang, C. A., Zhang, R., and Fang, D. (2009). Preparation and characterization of high-toughness ZrB2/Mo composites by hot-pressing process. *International Journal of Refractory Metals & Hard Materials*, 27, 1024–1026.

22. Huang, S. G., Vanmeensel, K., and Vleugels, J. (2014). Powder synthesis and densification of ultrafine B4C-ZrB2composite by pulsed electrical current sintering. *Journal of the European Ceramic Society*, 34, 1923–1933.

23. Mousavi, M. J., Zakeri, M., Rahimipour, M. R., and Amini, E. (2014). Mechanical propertiesofpressure-lesssintered ZrB2 with molybdenum, iron and carbon additives. *Materials Science and Engineering*, 13, 3–7.

24. Krishnaraon, R. V., Alam, M. Z., Das, D. K., and Prasad, V. V. B. (2014). Synthesis of ZrB2–SiC composite powder in air furnace. *Ceramics International*, 40, 15647–15653.

25. Asl, M. S., Kakroudi, M. G., and Nayebi, B. (2014). A fractographical approach to the sintering process in porous ZrB2–B4C binary composites. *Ceramics International*, 41(1), 379–387.

26. Ang, C., Seeber, A., Wang, K., and Cheng, Y. B. (2013). Modification of ZrB2 powders by a sol-gel ZrC precursor—A new approach for ultra-high temperature ceramic composites. *Journal of Asian Ceramic Societies*, 1, 77–85.

27. Sonber J. K., Limaye, P. K., Murthy, T. S. R. C., Sairam K., Nagaraj, A., Soni, N. L., Patel, R. J., and Chakravartty, J. K. (2015). Tribological properties of boron carbide in sliding against WC ball. *International Journal of Refractory Metals and Hard Materials*, 51, 110–117. doi:10.1016/j.ijrmhm.2015.03.010.

28. Sahani, P., and Chaira, D. (2017). Nonlubricated sliding wear behavior study of SiC–B4C–Si cermet against a diamond indenter. *Journal of Tribology*, 139, 051601, 1–13. doi:10.11151.4035344.

29. Larsson, P., Axen, N., and Hogmark, S. (1999). Tribofilm formation on boron carbide in sliding wear. *Wear*, 236, 73–80. doi:10.1016/S0043-1648(99)00266-5.

110 Design and process optimization of front lower control arm based on dynamic behavior

Santosh Billur[1,a], Gireesha R. Chalageri[1], Shreeshail M. L.[1], G. U. Raju[2] and B. B. Kotturshettar[2]

[1]Assistant Professor School of Mechanical Engineering, KLE Technological University, Hubli, Karnataka, India

[2]Professor, School of Mechanical Engineering, KLE Technological University, Hubli, Karnataka, India

Abstract: Front lower control arm (FLCA) is one part of the suspension sub assembly. FLCA of any automobile is the mediator link which takes the maximum force exerted by the tire and transmits it to the vehicle body. The objective of present work is to optimize manufacturing process for mass of the FLCA. Initially cast iron casting base model component of FLCA is selected. The dynamic parameters such as mode shape, frequency and static stiffness parameters of this base model are extracted. Altair HyperWorks is used for preprocessing and post processing and OptiStruct solver is used to run the analysis. Next design optimization of FLCA is carried out with respect to mass. Stamping process is selected in all the modified design components. Finally the natural frequency and stiffness values of design modified stamping components are determined. The optimized component compare to casting base model having lower mass and nearer values of natural frequency and stiffness is suggested for durability and fatigue test run.

Keywords: Control arm, natural frequency, optimization, stiffness.

Introduction

Lower suspension control arm of any automobile is the mediator link which takes the maximum force exerted by the tire and transmits it to the vehicle body [1]. Front lower control arm (FLCA) aids the vehicle wheels to respond to varying road conditions. Extreme road conditions like bumps, unscientific speed breaker, pits or other obstructions on road are encountered by the vehicle wheels and then that force is transmitted to the vehicle body through the lower control arm [2]. Vehicle dynamic effects influence the FLCA fatigue life [3]. In order to improve the durability and fatigue life of the lower arm, one has to perform shape optimization process. To absorb vibrations which are caused by uneven terrains and unwanted road obstacles, a perfect suspension system is needed. To have this perfect suspension system, steering stability and control arm's safe design are also the requirements [4]. Ansys Workbench rotor dynamic tool was used to evaluate modal and unbalanced mass response of high speed vertical spindle supported by hybrid bearing set. Natural frequency and critical speeds of new bearing sets are in good agreement with conventional bearing system [5]. The part design is important because faulty part designs not only reduce the system performance but also cause damage to the system [6]. Modal analysis

---

[a]santosh@kletech.ac.in

is used to predict the performance of the automobile parts. Both simulation results and numerical studies will have some degree of inaccuracy. The component's stiffness model is created and analyzed in order to predict the error [7]. Finite element analysis used to study the use of heat energy, produced from burnt fuel to recharge the battery [8].

The FLCA of a four-wheeled vehicle is taken in the present work for full safe design process, modal analysis, and static stiffness analysis. The simulation results and the key features of the safely designed FLCA are discussed.

Methodology

A FLCA connects the wheel hub and steering knuckle to the frame of the vehicle. Considering the difficulty level of solving analytical approach and also time and resource consuming experimental approach make both methods the least preferred ones. Finite element method (FEM) based numerical study is selected for the analysis of FLCA design as compared to the analytical and experimental approaches.

Initially analysis carried on FLCA made of cast iron using die casting process. Next different design change FLCA steel models made of stamping process are analyzed for modal and stiffness parameters. FLCA cast iron base model and finite element (FE) model are as shown in Figure 110.1.

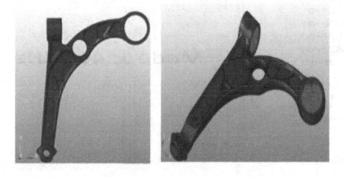

Figure 110.1 FLCA cast iron base model and FE model.

CAD modeling was done in Catia V5 and imported to Altair Hypermesh to perform geometry clean-up. The elements considered in the present analysis were 3-dimensional (3D) P-solid with four nodded tetra elements having three degree of freedom at each node. Global element size used for FE modeling or discretization process is 4 mm and near the rib structure, side walls and curvature area the mesh was refined. During FE modeling or meshing, some of the quality parameters including tetra collapse for 3D elements and minimum/maximum angle of elements, lowest element size are to be maintained for skin 2 dimensional (2D) elements.

Modal Analysis of Base Model

The natural frequencies and mode shapes of the component are determined using modal analysis. The typical behavior of the structure and the component's natural frequency are studied using modal analysis. The component's stiffness may be evaluated, and resonance could be prevented by knowing the natural frequency [9]. Free-free (without constraints) and forced-free boundary conditions are used for carrying out the modal analysis. For both boundary conditions, natural frequencies of base line cast iron component are shown Table 110.1. Further, Figures 110.2 and 110.3 shows first mode shapes.

Table 110.1: Modal frequencies of base line model for free-free and forced-free boundary condition.

Modal frequencies	Free-free boundary condition (Hz)	Forced-free boundary condition (Hz)
1	466.66	287.4
2	527.44	388.6
3	837.5	446.2
4	918.74	785

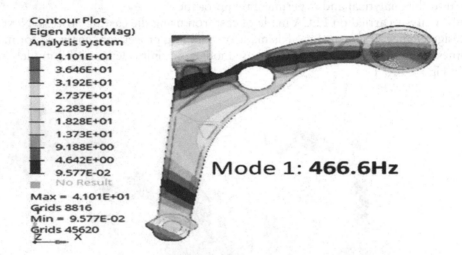

Figure 110.2 Mode 1 of Baseline Model Free-free boundary condition.

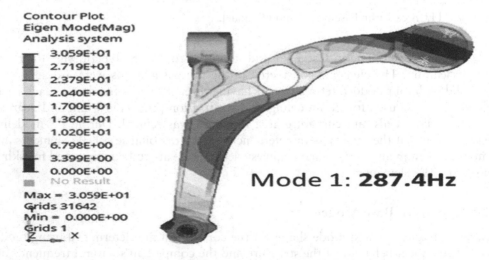

Figure 110.3 Mode 1 of Baseline Model Forced-free boundary condition.

Static Structural Analysis of Baseline Model

For the specified loads and boundary conditions, static structural analysis was done to determine the component's stiffness at that position. In the analysis, displacement in the directions are fixed at the bush location as shown in Figure 110.4. The force of 1000 N was applied along X and Y directions at the Ball joint location as shown in Figure 110.5. From the study, the displacement is determined in the X and Y directions. Further, the stiffness of the component was found using displacement values. The study employed the Optistruct solver, and the results are shown in Table 110.2 for the baseline model's displacement and stiffness parameters.

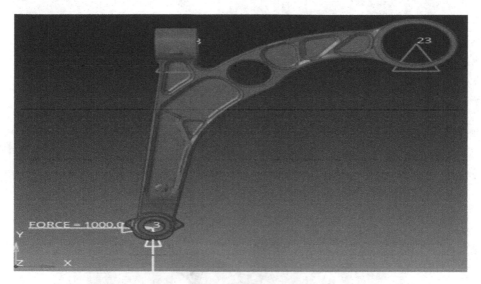

Figure 110.4 Loading and boundary conditions applied to Baseline FLCA.

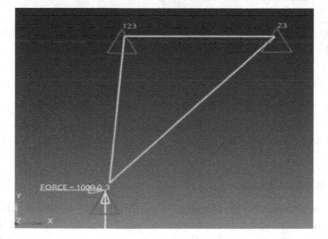

Figure 110.5 Loading and boundary conditions applied (line diagram).

The Figures 110.6 and 110.7 shows the displacement of the control arm in X and Y loading direction respectively.

Table 110.2: Displacement and stiffness values of base model.

Iterations	Direction	Displacement (mm)	Stiffness (kN/mm)
Baseline model	Longitudinal direction = X	0.25	4000
	Lateral direction=Y	0.006	166667

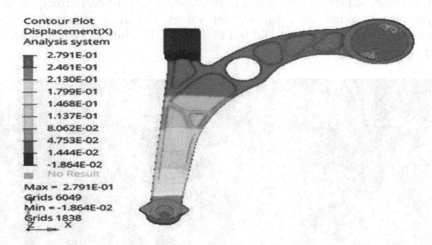

Figure 110.6 Displacement of the baseline model in X direction loading.

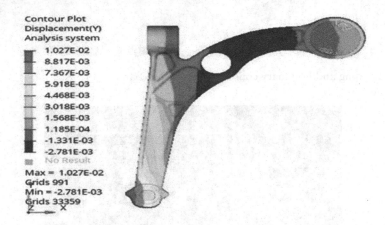

Figure 110.7 Displacement of the baseline model in Y direction loading.

Design Modification of FLCA

Design modification, known as "re-design," is the process of achieving the required specifications, that minimizes critical factors of the model. The modification is based on behavior and loading conditions of the model. Design changes for the present FLCA are given in Table 110.3 along with mass comparison. Also stamping process with steel as a core material is used for all the models. The detailed parts variation in design change models is given in Table 110.4.

Table 110.3: Design and process changes of FLCA.

Model	Description	Mass (kg)
Base model	Single piece casting.	3.52
Design 1	The baseline design is modified into two piece stamping model.	3.2
Design 2	Design 1 model is converted into 4 piece stamping model.	2.9
Design 3	Design 2 model is changed near the bolt connection and new model is generated.	2.9
Design 4	Single piece stamping.	2.4

Table 110.4: Total number of parts in different models.

Description	Models	No of components	Control Arm Upper	Control Arm Lower	Bushing Sleeve	Stud (Bolt connection)	Sleeve tool	Bush Sleeve attach Bracket
Stamping method	Design 1	4	Yes	Yes	Yes	Yes	No	No
	Design 2	6	Yes	Yes	Yes	Yes	Yes	Yes
	Design 3	6	Yes	Yes	Yes	Yes	Yes	Yes
	Design 4	2	Yes	No	Yes	No	No	No

Modal and Stiffness Analysis of Modified FLCA

For design 1 to design 4 models, modal analysis and stiffness analysis were carried out. Table 110.5 represents modal frequencies of the models 1 to 4 for free-free boundary condition and corresponding mode shapes are shown in Figures 110.8–110.11.

Table 110.5: Modal frequencies (in Hz) with free-free boundary condition.

Sl. No	Baseline model	Design 1	Design 2	Design 3	Design 4
1	466.66	689.53	796.06	815.75	193.92
2	527.44	916.86	910.4	926.79	218.6
3	837.5	971.89	1006.7	1004.66	478.78
4	918.74	1454.2	1438.6	1402.5	617.86

All the design modifications are giving higher modal frequencies except design 4 in comparison with base model. As design 4 is single piece stamping component the deviations observed is more.

Table 110.6 represents modal frequencies of the models for Forced-Free boundary condition and corresponding mode shapes are shown in Figures 110.12–110.15.

For modified designs, the static structural analysis was carried out using the same loading and boundary condition as that of baseline model. Table 110.7 shows the results for displacement and stiffness in the X and Y directions. The displacement values in Y direction are negligible when compare to X direction loading in all process models as strength against Y direction loading is more. Also, stiffness values are very close in design 1 stamping model with base model and remaining design model stiffness values are in good agreement.

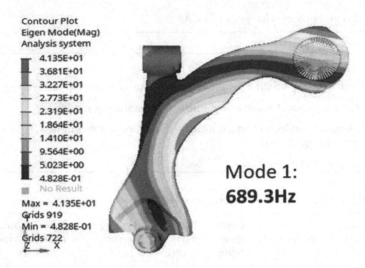

Figure 110.8 Mode 1 of design 1 model.

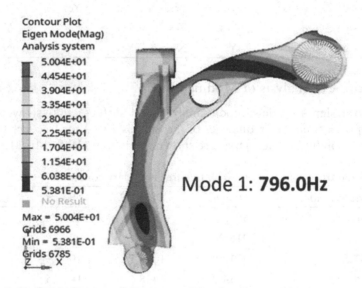

Figure 110.9 Mode 1 of design 2 model.

As the manufacturing process of FLCA changes from casting to stamping and with variation of design modelling aspects, the performance of vehicle also affects.

Conclusions

Front lower control arm (FLCA) with base casting model and design changed stamping models are analyzed and results are evaluated in HyperWorks based on FEM. Modal frequency of FLCA increased in stamping components except design 4 which is single piece stamping design. Design 3 gives higher modal frequency values in both free-free and forced-free boundary conditions. Stiffness and displacement values of stamping models are

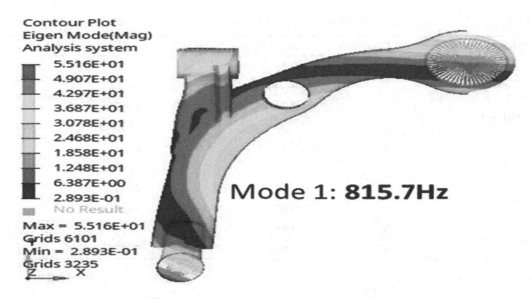

Figure 110.10 Mode 1 of design 3 model.

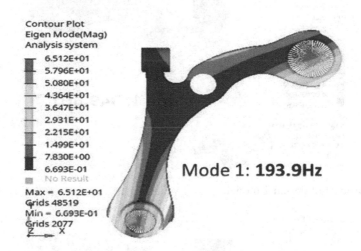

Figure 110.11 Mode 1 of design 4 model.

Table 110.6: Modal frequencies (in Hz) with forced-free boundary condition.

Sl. No	Baseline model	Design 1	Design 2	Design 3	Design 4
1	287.4	349.6	308.7	351.7	180.1
2	388.6	512.2	507.5	454.1	182.4
3	446.2	721.3	753.6	774.4	281
4	785	1025	1075.7	1097	317.8

Table 110.7: Displacement and stiffness values of FLCA models.

Iterations	Direction	Displacement (mm)	Stiffness (kN/mm)
Base line model	X	0.25	4000
	Y	0.006	166667
Design 1	X	0.24	4167
	Y	0.008	125000
Design 2	X	0.37	2703
	Y	0.008	125000
Design 3	X	0.34	2941
	Y	0.009	111111
Design 4	X	0.85	1176
	Y	0.047	21277

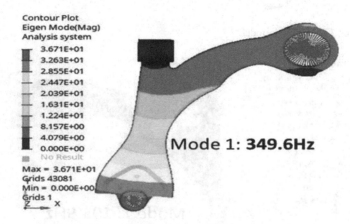

Figure 110.12 Mode 1 of design 1 model.

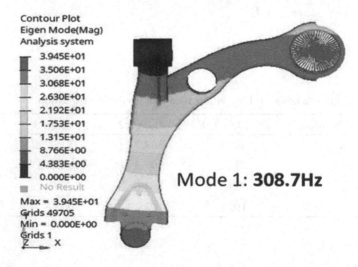

Figure 110.13 Mode 1 of design 2 model.

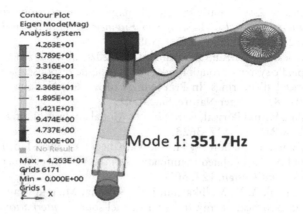

Figure 110.14 Mode 1 of design 3 model.

Figure 110.15 Mode 1 of design 4 model.

in good agreement with base line casting model except design 4. There is a reduction of 9–30% of mass when we move process from casting to stamping.

References

1. Taksande, S. P., and Vanalkar, A. V. (2015). Design, modeling and failure analysis of car front suspension lower arm. *IJSTE—International Journal of Science Technology & Engineering*, 2(01), 235–249. ISSN (online): 2349-784X.
2. Gunjan, P., and Sarda, A. (2018). Design and analysis of front lower control arm by using topology optimization. *International Journal of Advance Research and Innovative Ideas in Education (IJARIIE)*, 4(2), 1982–1986. ISSN (O): 2395-439.
3. Kang, B. J., Sin, H., and Kim, J. H. (2007). Optimal shape design of the front wheel lower control arm considering dynamic effects. *International Journal of Automotive Technology*, 8(3), 309–317.

4. Singh, J., and Saha, S. (2015). Static structural analysis of suspension arm using finite element method. *International Journal of Research in Engineering and Technology (IJRET)*, 04(07), 402–406. eISSN: 2319-1163, ISSN: 2321-7308.
5. Chalageri, G. R., Bekinal, S. I., and Doddamani, M. (2020). Evaluation of dynamic characteristics of a VMC spindle system through modal and harmonic response. part 1: spindle supported by angular contact ball bearings. In Proceedings of the 6th National Symposium on Rotor Dynamics (pp. 29–38). Springer Nature, Singapore.
6. Singh, D. P., Mishra, S., and Porwal, R. K. (2019). Modal analysis of ultrasonic horn using finite element method. 18(Part 7), 3617–3623.
7. Shi, S., Wu, H., Song, Y., Handroos, H., Li, M., Cheng, Y., and Mao, B. (2017). Static stiffness modelling of EAST articulated maintenance arm using matrix structural analysis method. *Fusion Engineering and Design*, 124, 507–511.
8. Shreeshail, M. L., Patil, A. Y., Mallikarjun, J. K., Krishna, M., and Mahantesh, M. M. (2015). Effective use of unused heat energy from burnt fuel source. *International Journal of Applied Engineering Research*, 10(48), 639–641.
9. Chalageri, G. R., Bekinal, S. I., and Doddamani, M. (2020). Dynamic characteristics of drilling spindle supported by radial permanent magnet bearings. *Materials Today: Proceedings*, 28, 2190–2196.

111 Modelling and simulation of concentrated solar powered Stirling engine based water pumping system

Qusai Alkhalaf[1], Arvind Singh Bisht[2], Amar Raj Singh Suri[1], and Shyam Singh Chandel[2]

[1]Department of Mechanical Engineering School of Mechanical, Civil and Electrical Engineering, Shoolini University, Solan, India

[2]Solar Thermal Research Group, Centre of Excellence in Energy Science and Technology, Shoolini University, Solan, India

Abstract: In recent years, there has been a significant surge in energy demand, leading to the rapid depletion of traditional energy sources. In response to this growing need, this paper aims to present the design and performance evaluation of a solar-based water pump system utilizing a concentrated solar power (CSP) Stirling engine, known as the Solar Dish Stirling Pump System (SDSPS). This system employs various reflector materials for modeling and analysis. In this study, we investigate twenty-one different pumping capacity scenarios, each characterized by unique criteria such as parabolic dish diameter, reflective materials, and direct normal irradiance (DNI). The primary objective of this research is to develop a highly reliable Stirling engine-based water pumping system, specifically tailored for remote areas devoid of grid electricity access. By examining the performance of the SDSPS under various conditions, the aim is to contribute to sustainable energy solutions that can serve the energy needs of off-grid communities, thereby addressing the challenges posed by the escalating demand for energy while reducing dependence on traditional energy resources.

Keywords: Concentrated solar power, solar thermal, Stirling engine, water pump.

Introduction

The renewable energy share is increasing rapidly due to increase in conventional fuel costs, energy demand and environmental concerns. Solar energy is an effective alternative for clean renewable power generation. The concentrated solar dish-based Stirling engine system is a promising technology for mechanical and electrical power applications. The primary goal of this research is to assess how the following factors impact the performance of a water pump: the dish diameter, direct normal irradiance, and the material used for the receiver in the system. Specifically, the aim is to determine how these variables influence both the pump's head (pressure) and discharge (flow rate).

In the current study, the performance of a solar dish-based Stirling engine water pumping system (SDSPS), utilizing the mechanical power generated by a solar Stirling engine, is examined under various operating situations. The effect of climatic conditions, diameter, rim angle, material of the dish and receiver are also investigated. The case study results of Solan city (Lat.30.86801° N, Long.77.13961° E) in the state of Himachal Pradesh, India, are presented.

Modeling, Simulation, and Validation

The SDSPS comprises a power conversion unit that is linked to a concentrated solar dish, along with a centrifugal water pump. The power conversion unit includes a receiver and Stirling engine. The conversion unit converts heat into mechanical power by the Stirling engine. The mechanical power from Stirling engine operates the water pump. The Solar dish concentrates the direct normal irradiance on focal point at the receiver. The temperature of receiver increases and receiver transfers that heat to the fluid. The fluid (helium or hydrogen) of Stirling engine, is heated and because of the temperature difference between two sides of engine the Stirling engine operates. The mechanical power from the Stirling engine operates the water pump. The system's performance is assessed using mathematical modeling. The model consists of two sub-models. The work carries on selecting and computing the optimal parameters namely dish diameter, receiver material, direct normal irradiation, water pump capacity and head and water pump discharge. Then the performance of the system is evaluated using direct normal irradiance satellite data of Solan city taken from NASA [1].

Optical Geometry of Dish

The objective of this sub-model aims to compute the optical parameters of dish like diameter, focal point, concentrator ration. The concentrator ratio (C) is the concentrator dish area to receiver area [2] as:

$$C = \frac{A_{Con}}{A_{rec}} = \frac{\frac{\pi}{4} D_{con}^2}{\frac{\pi}{4} D_{rec}^2} \tag{1}$$

$$D_{con}^2 = C \cdot D_{rec}^2$$
$$D_{con} = \sqrt{C} \cdot D_{rec}$$

Equations (2) and (3) are employed to calculate the focal length (f) and the distance between the focal point and the receiver (d_f), with these calculations depending on the rim angle (ϕ_{rim}).

$$\frac{f}{D_{con}} = \frac{1}{4 \cdot \tan\left(\phi_{rim}/2\right)} \tag{2}$$

$$d_f = \frac{D_{rec}}{D_{con}}\left(f - \frac{D_{con}^2}{16f}\right) \tag{3}$$

The distance between the receiver focal point and the concentrator surface (P) is computed in equation (4):

$$P = \frac{2f}{1 + \cos\phi_{rim}} \tag{4}$$

The height of the dish (h) depends on the diameter of dish (D_{con}) and focal length (f) as per the equation (5):

$$P = \frac{2f}{1 + \cos\phi_{rim}} \tag{5}$$

Thermodynamic Analysis Model

The heat loss in SDSPS occurs due to convection conduction, radiation losses. The modelling is established to compute the output power and efficiency of the system. Solar energy reaches the dish of SDSPS ($Qcon$) is given in equation (6).

$$Q_{con} = A_{Con} \cdot I = \frac{\pi}{4} D_{con}^2 \cdot I \tag{6}$$

The energy available for receiver ($Q_{in\text{-}reciver}$) depends on four main parameters, the concentrator ratio (C), direct solar radiation (I), material dish reflectivity (ρ), and thermal conductivity of fluid (K) and calculated by equation (7) [8]:

$$Q_{in\text{-}reciver} = C \cdot I \cdot \rho \cdot K \tag{7}$$

The useful energy (output of receiver $Q_{out\text{-}reciver}$) which is used to operate Stirling engine is defined in equation (8):

$$Q_{out\text{-}reciver} = Q_{in\text{-}reciver} - Q_{loss} \tag{8}$$

Where Q_{loss} is denote to the total heat losses in receiver.

Temperature of Cavity Receiver

In this model, it is integrated the empirical equations by Steinfeld and colleagues [3] to calculate the operational receiver temperature. This adaptation considers key operating parameters, including concentrator efficiency (η_{con}), degradation factor (F_D), and excess heat removal factor (F_{EX}), aiming to calculate receiver temperatures. The concentrator efficiency (F_{EX}) accounts for optical losses and industrialization imperfections, with an adopted FD value of 0.8, as per reference [3].

$$T_{cav} = \sqrt[4]{\frac{\alpha_{abs,r} \cdot C \cdot I}{\varepsilon_r \cdot \sigma}} \cdot \eta_{con} \cdot F_D \cdot F_{EX} \tag{9}$$

Parabolic Dish Efficiency

The concentrator efficiency ηcon depends on concentrator reflector materials and the parabolic defects resulted from manufacture process, is given by equation (10):

$$\eta con = \rho \cdot FS \cdot \Gamma \cdot \cos(Pi) \tag{10}$$

Where, ρ, Γ, and Fs are the factors under consideration include the mirror soiling factor, the intercept factor, and the reflectance of the reflective material. Additionally, Pi represents the solar radiation incidence angle, which is influenced by tracking control mechanisms that ensure the collector surface remains aligned with the sun's rays., the value of Pi is set at zero.

The Efficiency of the Receiver

The efficiency of the receiver is calculated by equation (11) that represented by Castellanos et al. [4]. The receiver efficiency increases by increasing transmittance, absorbance of receiver surface.

$$\eta_{the-reciver} = \tau_r \cdot \alpha_{abs,r} - \frac{h_{conv}\left(T_{cav} - T_a\right) + \varepsilon_r \cdot \sigma\left(T_{cav}^4 - T_a^4\right)}{\eta_{con} \cdot C \cdot I} \tag{11}$$

Stirling Engine Efficiency

The efficiency of engine is measured by the technology coefficient (Ωst) constant that describe the irreversibility due to friction and heat losses in the engine. The constant in this study is given as 0.55 as suggested by Castellanos et al. [4] and Stirling engine's efficiency is computed by using following equation (12).

$$\eta_{st} = \Omega_{st}\left(1 - \frac{T_a}{T_{cav}}\right) \tag{12}$$

Here, *Tcav* represents the receiver temperature, whereas *Ta* denotes the temperature of the working fluid in the compression space of the engine.

Centrifugal Pump

The pump is a device that moves fluids by mechanical action into hydraulic energy. The efficiency of water pump in this study is 0.76 [5]. Head is calculated by equation (13):

$$P_m = \frac{Q' \cdot H \cdot \rho' \cdot g}{100 \cdot \eta_{pump}} \tag{13}$$

Total Efficiency and Output of System

When the efficiency of the parabolic dish, Stirling engine, and receiver of SDSPS are computed, the total efficiency of SDSPS is calculated using equation (14):

$$\eta_o = \eta_{con} \cdot \eta_{th-recevier} \cdot \eta_{St} \cdot \eta_{pump} \tag{14}$$

$$\eta_o' = \eta_{con} \cdot \eta_{th-recevier} \cdot \eta_{St} \tag{15}$$

The net mechanical power is given by equation (16):

$$P_m = \eta_o' = A_c \cdot I = \eta_o \cdot \frac{\pi}{4} D_{con}^2 \tag{16}$$

Solution Method

The performance of various water pumping systems is influenced by several key factors, including dish diameter, concentrator ratio, rim angle, dish material reflectivity, and receiver material. These factors are considered in a mathematical simulation tailored for the SDSPS. The primary objective is to determine the most suitable design parameters for a water pumping system based on the availability of direct normal irradiance (DNI) at a specific location. The initial step involves selecting the appropriate reflectivity and absorber materials for both the dish and the receiver. For the purposes of this study, a water pump efficiency of 0.76 is assumed, and a Stirling constant of 0.55 is used.

In order to achieve a maximum concentrator ratio of 2862, a rim angle of 45° is selected. Additionally, ceramic is considered the most effective material for optimizing receiver efficiency [6based on the Eurodish system, is one of the several Country Reference Units of the EnviroDish project. The system has achieved a maximum thermal efficiency (solar to electricity]. Solar concentrator dishes can be constructed using materials such as aluminum and mirrors, which exhibit reflectivity values ranging from 0.8 to 0.95. To assess solar radiation data for Solan city in Himachal Pradesh, India, throughout the year 2019, information from NASA [1] was utilized.

Validation of Model

In the new SDSPS system, the water pumping system is operated using mechanical power. To validate our previously presented model [7], we compared its results with those of the Eurodish system, which also generates mechanical energy. The comparison revealed an error of approximately 5.46% between the mechanical power output of the Eurodish system and that of the SDSPS. Additionally, there was a margin of approximately 2.57% in the error observed for the receiver's cavity temperature between the Eurodish system and the SDSPS. Table 111.1 provides a comparative overview of the experimental data obtained from Eurodish and the SDSPS model.

Table 111.1: Comparison between SDSPS and experimental data of Eurodish.

	DNI (W/m²)	T_a (°C)	D_{con} (m)	ϕ_{rim} (°)	T_{cav} (°C)
Eurodish [7]	900	303	8.5	45	815
SDSPS	900	303	8.5	45	836
Error (%)					2.56

Results and Discussion

The size of the dish is considered as the main parameter of the SDSPS because the dish collects the solar energy and concentrate it on the receiver. It is critical for determining the capacity of water pump. Obviously, each water pump of different capacities, has different diameter of dish for running the pump. Selecting wrong size will affect the performance of the SDSPS system. Twenty one water pumping systems range from 0.1 hp to 2 hp for SDSPS are investigated and observed that it required 1.5m to 30m dish to operate the SDSPS system, when reflectivity of the dish material varies from 0.45 to 0.95 as shown in Figure 111.1. Increasing the reflectivity of material decreases the diameter of the dish. When the reflectivity increases more than 0.8, which include glass aluminum and thermoplastic, the range of dish diameter according to the capacity of water pump reduces from (1.5–18 m) to (1.5–5 m).

The relationship between diameter of dish and DNI are represented in Figure 111.2. When the DNI reduces from its average value the minimum required dish diameter changes from minimum 2 m to 18 m and when the DNI increases the diameter of dish reduce to 2 m to 0.6 m with respect to the capacity of water pump.

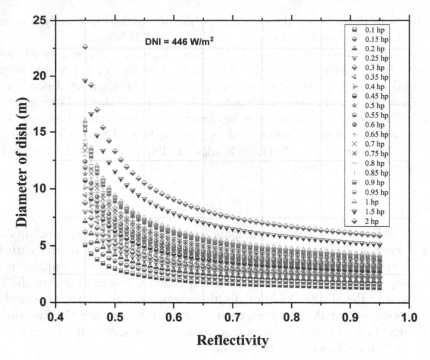

Figure 111.1 Correlation between material reflectivity and solar dish diameter for different water pumping configurations.

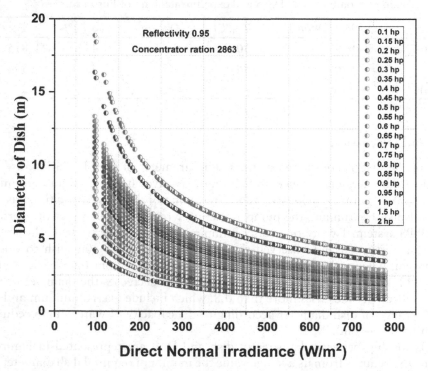

Figure 111.2 Correlation between DNI and solar dish diameter for different water pumping configurations.

As shown in Figure 111.2, dish diameter is inversely proportional to DNI. Minimum dish diameter is calculated with maximum solar normal irradiance. Maximum DNI of 778 W/m^2 gives minimum dish diameter 0.9 m to 4.05 m, and for average DNI of 446 W/m^2 gives minimum dish diameter 1.32 m to 5.98 m for different water pumping systems ranging from 0.1 hp to 2hp.

Conclusions

In this paper a novel mathematic modeling technique is adopted for modeling and simulation of a concentrated solar power-based water pumping system known as the Solar Dish Stirling Pump System (SDSPS). A concentrated parabolic solar dish with Stirling engine is used to convert heat energy into mechanical power and then to operate the water pump. Twenty-one different pump configurations are investigated to determine the minimum diameter of the dish with respect to material of the dish and direct normal irradiance (DNI). A mathematical model is established, simulated and validated solar radiation data of Solan, Himachal Pradesh, India. Results indicate that minimum dish diameter has an effect on overall efficiency of the system minimum dish diameter gives maximum efficiency of the system and reduces total cost of the system. When the DNI increases, it reduces the minimum diameter of the dish. It indicates that if a region has higher solar radiation as compared to the present study area, to run the same capacity pumping system that area required lesser minimum diameter of dish which assures its use for small scale water pumping system for individual households.

References

1. Orte, F., Lusi, A., Carmona, F., D'Elia, R., Faramiñán, A., and Wolfram, E. (2021). Comparison of NASA-POWER solar radiation data with ground-based measurements in the south of South America. *In 2021 XIX Workshop on Information Processing and Control (RPIC)*, 1–4. IEEE.
2. Hafez, A. Z., Soliman, A., El-Metwally, K. A., and Ismail, I. M. (2016). Solar parabolic dish stirling engine system design simulation, and thermal analysis. *Energy Conversion and Management*, 126, 60–75.
3. Steinfeld, A., and Schubnell, M. (2020). Optimum aperture size and operating temperature of a solar cavity-receiver. *Solar Energy*, 50(1), 19–25.
4. Castellanos, L. S. M., Noguera, A. L. G., Caballero, G. E. C., Souza, A. L. D., Cobas, V. R. M., Lora, E. E. S., and Venturini, O. J. (2019). Experimental analysis and numerical validation of the solar dish/stirling system connected to the electric grid. *Renewable Energy*, 135, 259–265.
5. Ding, H., Li, Z., Gong, X., and Li, M. (2019). The influence of blade outlet angle on the performance of centrifugal pump with high specific speed. *Vacuum*, 159, 239–246.
6. Granados, F. J. G., Pérez, M. A. S., and Ruiz-Hernández, V. (2008). Thermal model of the eurodish solar stirling engine. *Journal of Solar Energy Engineering, Transactions of the ASME*, 130(1), 0110141–48.
7. Reinalter, W., Ulmer, S., Heller, P., Rauch, T., Gineste, J. M., Fernere, A., and Nepveu, F. (2008). Detailed performance analysis of a 10kW dish/stirling system. *Journal of Solar Energy Engineering, Transactions of the ASME*, 130, 011013-1.

112 Investigations on improved trenched cooling effectiveness with modified film cooling holes

Krishna Anand V.G.[a]

Assistant professor, Department of Aeronautical Engineering, Malla Reddy College of Engineering and Technology, Hyderabad, India

Abstract: The film hole embedded in trenches is a promising method applied on gas turbine surfaces to safeguard them against hot thermal loads. The computational study is performed to analyze the cooling performance of modified film hole models embedded in trenches at three injection ratios of 0.6 to 1.4. The base trench model of this study shows close agreement with the literature trench data. The study results show that the rectangular-shaped hole model has produced improved effectiveness, due to effective cooling distribution with reduced jet separation from the bottom surface. The results show that the rectangular configuration has shown an improved area-averaged cooling effectiveness of 54.78%, 91.44%, and 112.12% for an M of 0.6, 1, and 1.4 respectively.

Keywords: Blowing ratio, film cooling effectiveness, film hole configurations, trench.

Introduction

Combustion turbines have wide uses in the areas of electricity production and high-speed propulsion. Advanced cooling methodologies are essential to safeguard the turbine components against the higher TIT. Bunker [1] showed that the thermal barrier coating (TBC) could be effectively used to develop trenches on the FC surfaces without complicated machining. Lu et al. [2] showed that the trench's presence had reduced the momentum of jet ejection from the FC holes. The trench has also assisted in wider film coolant distribution along the test surface's spanwise direction with better FC performance. Sundaram and Thole [3] performed FC investigations with cylindrical holes embedded in bump and trench modification. The study shows that the row slot model has improved the adiabatic effectiveness compared to the individual slot and bump models. Oguntade et al. [4] studied the FC performance of various slot outlets embedded with cylindrical film holes The results showed that the filleted trench outlet has shown superior cooling performance. Khalatov et al. [5] performed investigations on a slot embedded with a row of circular FC holes and showed that the slot models have shown better FC performance at a higher blowing ratio and increased external flow turbulence. Lu et al. [6] showed that a hole within the trench had produced a significant spread of coolant with complete wall surface coverage and higher cooling effectiveness. The method of trenching with TBC is cost-effective in comparison with conventional models.

---

[a]kakrishnaanand@gmail.com

Lee et al. [7] carried out computation investigations on a transverse trench adopted with circular cooling holes and showed that the model space between the slot and cooling hole showed better cooling performance. Zhang et al. [8] carried out computational studies on different slot configurations with circular cooling holes. The study shows that the narrow trench with 2D width and 0.5D depth has delivered higher FCE at a higher blowing ratio (BR) than the inclined trench configuration. Waye and Bogard [9] conducted studies to identify the useful trench configuration and found that the narrow slot model has produced an effective later spreading of coolant jets with lower jet separation Kross and Pfitzner [10] showed that tetrahedron upstream of the trench have favorable effects on cooling performance. Wang et al. [11] reported that the slot model with a circular FC hole produced effective cooling effectiveness on the turbine vane surface. Harrison and Bogard [12] showed that the narrow trench produced a better FC performance with lower coolant jet separation. Zuniga and Kapat [13] showed an increase in the P/D ratio of cylindrical holes embedded in the trench doesn't increase FC effectiveness. Rhee et al. [14] reported that the rectangular film hole shape had higher FC performance than the cylindrical film hole on flat surface film cooling. Abdala et al. [15] showed that the rectangular hole shape has shown higher FC performance than the circular cooling hole. Most of the studies [2–9] carried out a trench FC using circular or shaped FC holes. The studies reported on the flat surface model tested with other FC configurations viz.., rectangular, square, and semicircular have shown higher FCE compared to circular FC holes [14, 15]. However, no clear studies are reported to investigate the effect of other film hole configurations' influence on trench FC. Hence the present study was performed to examine the performance of slot embed with modified cooling hole models viz., circular, semicircular forward model, semicircular backward model, square, and rectangular model at three different injection ratios (0.6, 1.0, 1.4).

Computational Methodology

The trench configuration used in this study is adapted from Waye and Bogard [9]. The circular FC hole's diameter is maintained at 4.11 m, and the inclination angle is 30°. The length-to-diameter ratio and pitch-to-diameter ratio of the FC hole are maintained at 5.7 and 2.775. The FC hole configurations examined are circular (case 1), forward semicircular (case 2), backward semicircular (case 3), square (case 4), and rectangular (case 5). The hydraulic diameter of the cylindrical cooling hole (case 1) is 4.11 mm. The hydraulic diameter of other investigated cooling hole configurations was also maintained at 4.11 mm to compare with the circular film hole. The computational domain has a longitudinal, vertical, and lateral extent of 50D, 6D, and 2.775D respectively. The distance between the inlet of the computation domain to the coordinate origin is 11D. The hexahedral mesh cells were employed for the mainstream domain and are highly concentrated on the lower wall region to capture the thermal effects. The first mesh cell was placed at a distance to make the Y+<1. The unstructured mesh cells were used for film hole and plenum models. The mesh independence check is conducted with three different mesh cells viz., 0.5 million cells, 1.01 million cells, and 1.58 cells. The 1.01 million cells were chosen after GIS with a balance of accuracy and computational time. The velocity and temperature of the mainstream flow are 30.82m/sec and 300k, respectively. The velocity of the film coolant at the plenum inlet is varied to match the required injection ratio, and the thermal condition of the jet flow is set to 230.77k. The SIMPLE method is employed for the pressure velocity coupling. The 2nd order upwind scheme is applied for the discretization of flow, turbulence, and energy equation. The criteria on set for the convergence of the solution are the drop in residual to 10^{-5} for flow, and turbulence equations, and 10^{-9} for the energy equation.

Results and Discussions

Numerical results are validated against several trench FC studies available in the open literature. The comparison of experimental data for the spanwise averaged cooling effectiveness (M of 1.0) with various turbulence models viz., realizable k-ε, Renormalization group k-ε, Shear stress transport k-ω is shown in Figure 112.1. The results of LAFCE predicted with the realizable k-ε turbulence model were in close agreement with the experimental data. The spanwise averaged cooling effectiveness along the bottom wall surface (normalized with D) for different film hole configurations viz., cases 1–5 at M of 0.6 is shown in Figure 112.2. Case 5 has produced higher LAFCE along the bottom wall surface than other tested cases (1–4) mainly due to the effective span-wise and stream issuance of coolant jet along the bottom wall region.

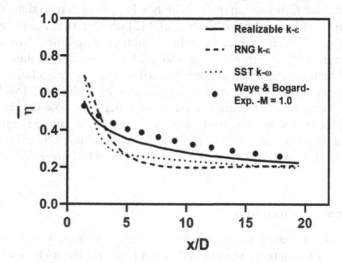

Figure 112.1 Comparison of experimental and computational lateral averaged effectiveness with various turbulence models at M of 1.0.

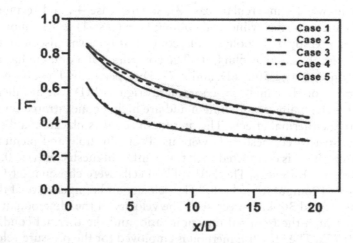

Figure 112.2 Lateral averaged effectiveness along the bottom surface for different cases at M of 0.6.

Case 2 has delivered lower LAFCE than case 5, and higher LAFCE than case 3. For a lower M of 0.6, case 1 has delivered lower LAFCE than other tested cases due to the ineffective distribution of film jet along the test surfaces. LAFCE of different film hole configurations for an M of 1.0 is shown in Figure 112.3. Case 5 continued to deliver a higher LAFCE along the tested wall region for both low and mid-M. Case 3 has shown higher cooling performance up to x/D ≤ 3 after which case 2 showed a higher LAFCE distribution up to x/D ≤ 20. In comparison with case 1, case 4 showed a lower LAFCE in the region (2 ≥ x/D ≤ 20) of the bottom wall surface. For a higher M of 1.4, case 5 has delivered a higher LAFCE as shown in Figure 112.4. This is due to the effective span-wise distribution of coolant jet with lower jet separation on the tested wall surface even at higher M. Compared with case 1 and case 4, the former has delivered higher LAFCE in the range of 2.5 ≥ x/D ≤ 14. Case 1 and case 4 have delivered lower LAFCE than other tested cases due to the higher coolant jet separation with the lower span-wise issuance of coolant jet on the tested wall surface.

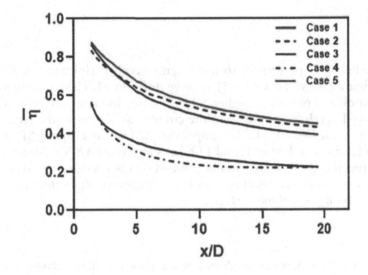

Figure 112.3 Lateral averaged effectiveness along the bottom surface for different cases at M of 1.0.

The region averaged cooling effectiveness (AAFCE) was computed for the test surface region of 0 ≥ x/D ≤ 20, for all film hole configurations at three different M (0.6, 1, 1.4) For low M, case 5 delivered higher AAFCE than other tested cases (1–4). Case 5 has delivered 54.78% and 51.56% higher AAFCE than case 1 and case 4. The higher AAFCE in case 5 is due to the effective issuance of the coolant jet along the test surfaces in streamwise and spanwise directions. For an M of 1.0, case 5 has delivered 91.44% and 110.86 % higher AAFCE than cases 1 and 4 respectively. This higher AAFCE of case 5 is due to the close attachment of the ejected film coolant on the bottom wall region. For an M of 1.4, case 5 has delivered 112.12% and 127.6 % higher AAFCE than case 1 and case 4 film hole configurations. The higher AAFCE of case 5 is due to the lower coolant jet liftoff and higher spanwise issuance of the coolant jet on the test wall surface. The studies [14, 15] also confirm that the rectangular film hole (case 5) produces higher cooling performance on the flat wall surface in comparison with other tested cooling hole configurations.

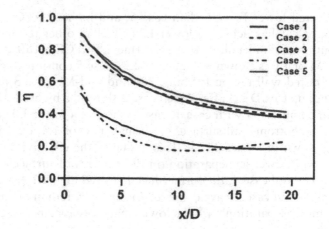

Figure 112.4 Lateral averaged effectiveness along the bottom wall surface for different cases at M of 1.4.

Conclusion

Numerical studies were performed to investigate trench performance with five different film hole configurations viz. cases (1–5) at three different M. The rectangular (case 5) film hole configuration has produced higher LAFCE along the test surface at all three tested M viz., 0.6, 1, and 1.4. In comparison to the circular cooling hole, the rectangular-shaped model (case 5) has delivered 54.78% improved AAFCE at a lower M of 0.6, 91.44% improved AAFCE for a mid-M of 1, and 112.12% improved AAFCE for a higher M of 1.4. The trenched rectangular film hole can be studied on the pressure and suction surfaces of gas turbine blades. Also, investigations can be performed to study the rotational effects of this configuration on gas turbine surfaces.

References

1. Bunker, R. S. (2005). A review of shaped hole turbine film hole technology. *Journal of Heat Transfer,* 127, 441–453.
2. Lu, Y., Nasir, H., and Ekkad, S. V. (2005). Film cooling from a row of holes embedded in transverse slots. In Proceedings of the ASME Turbo Expo 2005: Power for Land, Sea, and Air. Volume 3: Turbo Expo 2005, Parts A and B. Reno, Nevada, USA (pp. 585–592).
3. Sundaram, N., and Thole, K. A. (2008). Bump and trench modifications to film-cooling holes at the vane end-wall junction. *Journal of Turbomachinery,* 130, 041013-1–041013-9.
4. Oguntade, H. I., Andrews, G. E., Burns, A. D., Ingham, D. B., and Pourkashanian, M. (2013). Improved trench film cooling with shaped trench outlets. *Journal of Turbomachinery,* 135, 021009-1–021009-10.
5. Khalatov, A. A., Borisov, I. I., Dashevskiy, Y., Kovalenko, A. S., and Shevtsov, S. V. (2012). Flat plate film cooling from a single-row inclined hole embedded in a trench: effect of external turbulence and flow acceleration. *Thermophysics and Aeromechanics,* 20, 713–719.
6. Lu, Y., Dhungel, A., Ekkad, S. V., and Bunker, R. S. (2009). Effect of trench width and depth on film cooling from cylindrical holes embedded in trenches. *Journal of Turbomachinery,* 131, 011003-1–011003-13.
7. Lee, K. D., and Kim, K. Y. (2014). Film cooling performance of cylindrical holes embedded in a transverse trench. *Numerical Heat Transfer,* 65, 127–143.

8. Zhang, D. H., Wang, Q. W., Zeng, M., and Sun, L. (2008). Numerical research on the influence of different slot configurations on film cooling characteristics. *Progress in Computation Fluid Dynamics*, 8, 518–525.
9. Waye, S. K., and Bogard, D. G. (2007). High-resolution film cooling effectiveness measurements of axial holes embedded in a transverse trench with various trench configurations. *Journal of Turbomachinery*, 129, 294–302.
10. Kross, B., and Pfitzner, M. (2012). Numerical and experimental investigation of the film cooling effectiveness and temperature fields behind a novel trench configuration at high blowing ratio. In Proceedings of the ASME Turbo Expo 2012: Turbine Technical Conference and Exposition. Volume 4: Heat Transfer, Parts A and B. Copenhagen, Denmark (pp. 1197–1208).
11. Wang, C., Sun, X., Fan, F., and Zhang, J. (2020). Study on trench film cooling on turbine vane by large-eddy simulation. *Numerical Heat Transfer, Part A: Applications*, 78(7), 338–358. DOI: 10.1080/10407782.2020.1791523.
12. Harrison, K. L., and Bogard, D. G. (2007). CFD predictions of film cooling adiabatic effectiveness for cylindrical holes embedded in narrow and wide transverse trenches. In Proceedings of the ASME Turbo Expo 2007: Power for Land, Sea, and Air. Volume 4: Turbo Expo 2007, Parts A and B. Montreal, Canada (pp. 811–820).
13. Zuniga, H. A., and Kapat, J. S. (2009). Effect of the increasing pitch to diameter ratio on the film cooling effectiveness of shaped and cylindrical holes embedded in trenches. In Proceedings of the ASME Turbo Expo 2009: Power for Land, Sea, and Air. Volume 3: Heat Transfer, Parts A and B. Orlando, Florida, USA (pp. 863–872).
14. Rhee, D. H., Lee, Y. S., and Cho, H. H. (2002). Film cooling effectiveness and heat transfer of rectangular-shaped film cooling holes. In Proceedings of the ASME Turbo Expo 2002: Power for Land, Sea, and Air. Volume 3: Turbo Expo 2002, Parts A and B. Amsterdam, The Netherlands (pp. 21–32).
15. Abdala, A. M. M., and Elwekkel, F. N. M. (2016). Pressure distribution effects due to chevron fences on film cooling effectiveness and flow structures. *Applied Thermal Energy*, 110, 616–629.

Printed in the United States
by Baker & Taylor Publisher Services

Printed in the United States
by Baker & Taylor Publisher Services